The Collected Works of Eugene Paul Wigner

Part A · Volume V

Photo: 1958

Eugene Paul Wigner

The Collected Works of Eugene Paul Wigner

Part A

The Scientific Papers

Editor: Arthur Wightman

Annotated by

Nandor Balazs   Herman Feshbach   Brian Judd   Walter Kohn
George Mackey   Jagdish Mehra   Abner Shimony   Alvin Weinberg
Arthur Wightman

Part B

Historical, Philosophical, and Socio-Political Papers

Editor: Jagdish Mehra

Annotated by

Conrad Chester   Gérard Emch   Jagdish Mehra

The Collected Works of Eugene Paul Wigner

Part A

The Scientific Papers

Volume I
Eugene Paul Wigner: A Biographical Sketch
Applied Group Theory
Mathematical Papers

Volume II
Nuclear Physics

Volume III
Part I: Particles and Fields
Part II: Foundations of Quantum Mechanics

Volume IV
Part I: Physical Chemistry
Part II: Solid State Physics

Volume V
Nuclear Energy:
Part I: Eugene Wigner and Nuclear Energy
Part II: Memoir of the Uranium Project
Part III: Articles, Reports, and Memoranda on Nuclear Energy
Part IV: The Wigner Patents

Part B

Historical, Philosophical, and Socio-Political Papers

Volume VI
Philosophical Reflections and Syntheses

Volume VII
Historical and Biographical Reflections and Syntheses

Volume VIII
Socio-Political Reflections and Civil Defense

The Collected Works of

Eugene Paul Wigner

Part A

The Scientific Papers

Volume V

Nuclear Energy

Edited and Annotated by Alvin M. Weinberg
with the Assistance of Alfred M. Perry

Springer-Verlag

Berlin Heidelberg New York
London Paris Tokyo
Hong Kong Barcelona
Budapest

Arthur S. Wightman
Department of Physics, Princeton University
Joseph Henry Laboratories, Jadwin Hall
Princeton, NJ 08544, USA

Alvin M. Weinberg
Oak Ridge Institute for
Science and Education
Medical Sciences Division
Oak Ridge, TN 37831-0117, USA

Alfred M. Perry
Oak Ridge National Laboratory (Retired)
Oak Ridge, TN 37831, USA

ISBN-13:978-3-642-77427-0   e-ISBN-13:978-3-642-77425-6
DOI: 10.1007/978-3-642-77425-6

© Springer-Verlag Berlin Heidelberg 1992
Softcover reprint of the hardcover 1st edition 1992

Typesetting: Springer TEX in-house system

55/3140-543210 - Printed on acid-free paper

Editors' Preface

The papers have been divided, necessarily somewhat arbitrarily, into two parts

    Part A: The Scientific Papers
    Part B: Historical, Philosophical, and Socio-Political Papers.

Within each part, the papers have been divided by subject, and within each subject printed chronologically. With some exceptions, every scientific paper is reprinted in its original form. One class of exceptions consists of papers that are simply translations into Hungarian from German or English; they are omitted, but listed in the bibliographies. Scientific papers originally in Hungarian have been translated into English. The papers of Volume V3, Uranium Project Reports, have been reset and the figures redrawn. The originals were declassified reports, some in nearly illegible shape. Some reports and patents in Volumes V3 and V4 are listed by title only. In contrast to the scientific papers where the coverage is essentially complete, in Part B, a selection has been made. We believe it is representative of Wigner's far ranging concerns.

The preparation of Volume V, Nuclear Energy, required a more extensive effort than that for Volumes I–IV. Alvin Weinberg undertook the considerable labor of locating the declassified reports of the Metallurgical Laboratory in Chicago and the patents, of selecting those to be reproduced here and of interpreting them. The retyped manuscript was proofread with the assistance of Alfred M. Perry of Oak Ridge National Laboratory. The Editors wish to thank Drs. Weinberg and Perry for their indispensable effort.

Publication in its present degree of completeness has been made possible by a grant from the Lounsbery Foundation and by support from the American Nuclear Society. Weinberg and Perry join the undersigned in thanks for this generosity which has made it possible to do full justice to the contributions of Eugene Wigner as the founder of nuclear engineering.

Arthur S. Wightman
Jagdish Mehra

Preface

This collection of Eugene Wigner's contributions to nuclear energy consists of four parts:

I. A summary and appreciation of Wigner's contributions to nuclear engineering.
II. Wigner's memoirs, written between 1941 and 1948.
III. A selected list of 42 reports, articles and memoranda, mainly written between 1941 and 1945.
IV. Twelve of Wigner's 37 patents relating to nuclear energy.

I have chosen those reports and patents that I consider to be most important for reproduction in their entirety. The less important items are listed by title.

This collection, in its full completeness, has been made possible by a grant from the Lounsbery Foundation; additional support has been provided by the American Nuclear Society. I am most grateful to the Foundation and the ANS for their generosity, which makes it possible to do full justice to the major contributions of Eugene Wigner, the founder of nuclear engineering. I also wish to thank Dr. Alfred M. Perry for assisting me in the arduous task of checking for accuracy the transcription and reformatting of Wigner's original, often barely legible, Metallurgical Laboratory Reports.

Oak Ridge, Tennessee   *Alvin M. Weinberg*
August, 1992

Preface

Contents

Reactor Engineering at Chicago, 1942–1945, and at Clinton, 1946–1947

PART IV

The Wigner Patents

Breeders and Converters

Materials Testing Reactor

Miscellaneous Patents

PART I

Introduction:
Eugene Wigner and Nuclear Energy

Introduction:
Eugene Wigner and Nuclear Energy[1]

By Alvin M. Weinberg

I. The Pre-Chicago Days

When fission was discovered, Eugene Wigner was the classic case of the prepared mind. After all, Wigner had by this time made many of the most central contributions to theoretical nuclear physics. For example, it was he who made the first quantum mechanical calculation of the scattering of neutrons by protons (1); who first explained (but never published) that the large rise in neutron scattering cross-section of the proton around 20 keV was attributable to the contribution of the singlet state of the n-p system; who introduced the isobaric spin (2) and the super-multiplet theory. In a sense, then, Wigner must be regarded as a dominant figure, if not the founder, of theoretical nuclear reaction physics. His contribution here culminated in his visualization (independent of Bohr, who is usually credited with this discovery) of the compound nucleus. Wigner came very naturally to this view of nuclear reactions since, as a graduate chemical engineering student under Michael Polanyi, he was very familiar with Polanyi's transition state in chemical kinetics; he and Polanyi had exploited this concept in their famous paper on the quantum mechanical interpretation of the simplest chemical reaction, the formation of the hydrogen molecule (3). The compound nucleus for Wigner was merely the nuclear analogue of the chemical transition state. The Breit-Wigner resonance formula was worked out during Wigner's short stay at Wisconsin (4).

Wigner and Leo Szilard were close friends, having attended and taught at, along with John von Neumann, the University of Berlin. (It was here that Wigner first developed his ideas on the application of group theory to quantum mechanics, and Szilard introduced information theory into thermodynamics). Wigner was therefore familiar with Szilard's ideas about nuclear chain reactions, ideas that culminated in Szilard's secret British Admiralty Patent of 1934 (5). At the time of this patent, fission was unknown; however, the $^9Be(n,2n)2^4He$ reaction was believed, erroneously, to be exothermic, and so Szilard's ideas even then were more than idle speculation. When fission was discovered, Szilard and his friend Wigner immediately grasped its significance.

[1] Based in part on A. M. Weinberg: "Eugene Wigner and Nuclear Energy: A Reminiscence", in *Advances in Nuclear Science and Technology*, Vol. 19 (J. Lewins, M. Becker, eds.), Plenum Press, New York, 1987, pp. 1–39. Permission to reprint granted by the publisher.

In those earliest days, I suppose Szilard was the more deeply absorbed in the fission problem – after all, he had worked out the theory of a nuclear chain reaction even before fission had been discovered; and on March 20, 1939, he applied for a U.S. patent on what he called an "Apparatus for Nuclear Transmutation" – i.e. a neutron chain reactor based on the fission of uranium (6). Wigner was fully aware of Szilard's thinking; and with his powerful grasp of the mathematical and physical principles underlying the chain reaction, Wigner was, even then, able to make independent estimates of the critical conditions.

The famous letter from Einstein to Roosevelt in August of 1939 was translated into English by Wigner from Einstein's dictated German, and followed an earlier discussion by Wigner and Szilard with Einstein. By this time many others, including E. O. Lawrence and Harold Urey in the United States, had become seriously concerned with the prospect of military use of fission. But at the very beginning in the United States the political momentum was generated mostly by the refugee scientists – Szilard, Wigner, Teller and, possibly with a lesser sense of urgency, Fermi.

During this time Wigner was working out his version of the theory of the chain reaction. Szilard had espoused graphite as moderator, and had, along with Fermi, pointed out the advantage of a lattice over the homogeneous arrangement. Wigner estimated the gain in multiplication constant achieved by lumping while he was still in Princeton (Paper 1). By this time the four-factor formula (at first a three-factor formula because the significance of fast fission in ^{238}U had been overlooked) was known (Wigner, among others, had independently derived the formula); and the conditions for criticality had been worked out. During this period Louis Turner had stressed the likelihood that ^{239}Pu was fissionable, a point brought out in Bohr and Wheeler's famous paper on fission; and when this was verified by Seaborg at California, the attainment of a slow neutron chain reactor based on ordinary uranium as fuel acquired great military significance.

While still at Princeton, Wigner and Breit applied transport theory to estimate the critical size of a bomb (Paper 2). Similar calculations had already been performed by Peierls and Pryce in England, and by Fermi. Wigner and Breit's calculations confirmed the earlier estimates that an explosive device would require only a few kilograms of fissile material.

Wigner, having developed the theory for calculating both the resonance escape probability, and the thermal utilization (a word which he coined) in a lattice, embarked on both an experimental and theoretical program to establish optimum sizes for uranium lumps in graphite lattices. The theoretical work at Princeton was carried out with John A. Wheeler and Gilbert N. Plass, and the experimental work with R. R. Wilson, E. Creutz, T. Snyder and H. Jupnik. Using neutrons from the Princeton cyclotron, they measured both the resonance absorption and the thermal utilization in spheres of uranium. Some of this work was done in collaboration with Fermi's group at Columbia. Wigner used these experiments to derive semi-empirical values for the diffusion lengths in uranium, which he and Plass then used to map out the variation of multiplication constant with dimensions in graphite lattices. The division of the resonance absorption into a surface and a volume effect also resulted from these experiments. These experiments, together with Wigner's analysis, are summarized in Papers 5, 6, 7, 8 and 9.

Wigner by this time (1941) had worked out many of the most familiar results from neutron transport theory (Papers 3 and 4): for example, he showed that the distribution of monoenergetic neutrons around a point source diminished asymptotically as $(1/r)e^{-\kappa r}$, where κ was the solution of the transcendental equation

$$(\Sigma_s/\kappa)\tanh^{-1}(\kappa/\Sigma) = 1 \tag{1}$$

or, for Σ_a small,

$$\kappa^2 \approx 3\Sigma\Sigma_a(1 - (4/5)\Sigma_a/\Sigma) \tag{2}$$

where Σ, Σ_s, and Σ_a are respectively the total, scattering and absorption macroscopic cross-sections. Equation (2) was used by Wigner and his group in all subsequent calculations of thermal utilization and of resonance absorption.

II. Reactor Engineering at Chicago Metallurgical Laboratory, 1942–1945, and at Clinton Laboratories, 1946–1947

By early 1942 when Wigner arrived to lead the theoretical group at the Metallurgical Laboratory, his theory of the neutron chain reactor was a year or more old. He had fairly reasonable values for the cross-sections derived both from the Princeton work and from the work done by Fermi's group at Columbia. And he had begun a systematic investigation, with Plass, of the various lattice configurations. This is not to say that Wigner's was the only formulation of the criticality problem. In the United States, Fermi had independently worked out the theory and had, for a year at least, been conducting exponential experiments along with W. H. Zinn and H. L. Anderson on graphite uranium-oxide lattices; others in the United States, notably Carl Eckart at Chicago, and James Fisk and William Shockley at Bell Laboratories had also worked out many elements of the theory. Somewhat similar efforts had been going on in Germany, mainly under W. Heisenberg and C. F. von Weizsäcker; in France, before its fall, by Joliot-Curie, von Halban and Perrin; in England with R. Peierls and G. Placzek; and a little later in the Soviet Union (Y. Zeldovich and Y. Khariton) and in Japan. But Wigner's work was possibly the most systematic. Wigner himself had been trained as a chemical engineer, and he approached the problem in a very practical way; explore systematically the various configurations, and determine the one that gives the highest multiplication constant. Thus the whole of the theory of reactor lattices, as it has developed, had much of Wigner's imprint, though Fermi's always independent approach is also much in evidence.

Wigner found at Chicago an enthusiastic, but very young group of theoretical physicists, many of whom knew little of either nuclear physics or of engineering. The most notable exception, of course, was the brilliant John A. Wheeler, who had worked out the theory of fission with Bohr, and had helped in the work at Princeton, and who, along with Robert Christy, worked essentially independently. But the others – F. L. Friedman, H. Ibser, G. N. Plass, G. Young, K. Way, and I – pretty much took our cues from Wigner.

Wigner's style is best illustrated by the "recipe sheets" he would distribute to his group. Whether the matter at issue was computation of the multiplication constant, or heat transfer in pipes, or elastic deformation, he would prepare a one-page "recipe sheet" on which he summarized the relevant formulae including appropriate constants. (His heat transfer sheet has enabled me to hold my own in arguments about heat transfer for more than 40 years!)

Each member of the group was assigned an area to investigate. Gale Young, who held an engineering degree, was, along with Lee Ohlinger, Wigner's personal assistant for engineering and general physics; Francis Friedman, the brilliant young physicist who had worked with Gregory Breit, was a roving generalist; G. Plass, H. Ibser, K. Way, and I were put in charge of the multiplication constant. Later, when Frederick Seitz, M. L. Goldberger, and H. C. Schweinler arrived, they worked on questions of solid state; and S. Dancoff became our nuclear physics expert.

Wigner's mastery of the theoretical phases of the plutonium project was complete – at least equal to Fermi's and, in engineering, chemistry and material science, probably beyond. Thus, there is hardly an aspect of reactor science and engineering, as we now define the fields, that does not bear Wigner's stamp. In the following pages I shall review Wigner's contributions in each of the sub-fields to which he contributed.

When Wigner arrived in Chicago the general outline of the first plutonium-producing reactor had been settled: graphite was to be the moderator; natural uranium, either metal or oxide, the fuel. The coolant was still in dispute, with Szilard suggesting liquid bismuth (since the thermal capture cross-section of bismuth was so small), and the engineers suggesting helium under pressure. For several months the helium-cooled design was the main line of development; but with this decision, Wigner strongly disagreed. Wigner's views of the matter are contained in Paper 11. He recognized that the materials and handling problems of hot helium were formidable, even open-ended, and that a helium-cooled reactor could not be built quickly. Instead he argued (though not in this paper) that ordinary water at modest temperatures be used as coolant.

This proposal for a water-cooled pile met with serious objections since, at the time it was made, the highest k measured in an exponential experiment was still below unity, and water as well as the aluminium that lined the channels and clad the slugs would reduce k by about 3 percent. But Wigner, with his confident command of reactor theory and of the results of the exponential experiments, was convinced that a water cooled system could be made critical. He therefore proceeded to design, with the help of Gale Young, "A Plant with Water Cooling" (Paper 14). In this design the uranium rods were disposed vertically. This design was followed by a more detailed design "On a Plant with Water Cooling" (Paper 15), issued on July 18, 1942 and finally by Paper 16, "Preliminary Process Design of a Liquid Cooled Power Plant

Producing 5×10^5 kW," by M. G. J. Boissevain, M. C. Leverett, L. A. Ohlinger, A. M. Weinberg, E. P. Wigner, and G. Young, issued on January 9, 1943. This report outlined the basic design of what became the "W" reactors, which were eventually built by the du Pont Company at Hanford (code-named site "W").

In both of the earlier designs (Paper 14 and 15), the uranium was disposed vertically as long hollow cylinders. The interior of the uranium cylinders was coated with aluminium and the water coolant flowed through these interior channels. The graphite ran at high temperature, $\sim$1000K, and so the calculation of the neutron flux was complicated since neutrons in the graphite were hot, but cooled down when they reached the water channels. Significantly, Wigner realized that cladding the uranium would be a prime difficulty, a realization that turned out to be all too true.

During these months Wigner fought hard to convince the project management to switch from helium cooling to water cooling. In this he ultimately succeeded, and the du Pont Company, which was brought in by the U.S. Army to take over the construction and operation of the plutonium-producing reactors, chose Wigner's water-cooled design over the helium-cooled design.

In preparation for du Pont's taking over, the Metallurgical Laboratory prepared two "instructional" reports: "Possibility of a Chain Reaction" by Fermi, Wigner, and Weinberg; and "Survey of the Power Plant Problem" in which various authors reviewed different possibilities for cooling a Pu-producing reactor. Both reports were dated November 26, 1942, just one week before Fermi established the first chain reaction. We reproduce here Wigner's contribution to these reports, Paper 12, "General Considerations Concerning the Lattice Structure" by Wigner and Weinberg; and Paper 13, "Relative Advantages of Several Cooling Systems" by Wigner, and "Heavy Water Plant" by Wigner and Ohlinger.

The actual preliminary design of the W-piles was almost entirely dominated by Wigner. There was hardly a detail that he did not personally attend to, though he was always careful to involve each member of the design team in the discussions. Actually two versions of the reactor were developed: one cooled with water, the other with diphenyl, though Wigner realized that the radiation stability of diphenyl could be a problem. Looking at the matter from the perspective of almost 50 years, I realize that creating a workable design in little more than four months must be regarded as Mozartean! One had to decide dimensions of the lattice and of the reactor; the disposition of the control rods; whether the reactor be vertical or horizontal; the cladding and tube material (aluminium); whether the uranium fuel-rods were to be hollow rods cooled internally, or solid slugs cooled externally; the design of the shield; the method for loading and unloading; the pressure drops and heat transfer; the design of the slugs; how to manage the after-heat – to name some of the prominent design elements. Wigner would decide each of these matters after spirited discussions with each of his team, as well as with other members of the project, particularly E. C. Creutz, a Princeton physicist-turned-metallurgist for the duration of the project, and John Wheeler. That the actual Hanford reactors built by du Pont were very close to Wigner's preliminary design attests to his excellent engineering judgment, and overall grasp of every element of the design; he even reviewed personally every reactor blueprint prepared by du Pont! I have no doubt that his extraordinary attention to detail, combined with an uncanny ability to spot weaknesses – which derived from his ability to estimate almost any physical situation in his head – were key elements in the eventual success of Hanford.

The reactor proposed by Wigner's group was limited by the war-time scarcity to 200 tons of U metal. This was contained in 1695 process tubes disposed in a cylindrical block of carbon. As the reactor was laid out at Hanford, the reactor contained only 1500 tubes that formed a cylinder of 4.65 meters radius in a squared-off block of graphite; thus there was room for an additional 504 process tubes in the four unoccupied corners, making a total of 2004 process tubes. The du Pont designers, including John Wheeler, in order to gain reactivity in case the actual k turned out to be lower than had been calculated, filled in the corners. This added 0.27 percent to k over its value had the original 1500 tube design been adhered to, but it required an additional 67 tons of uranium. All of this was done, of course, before the ^{135}Xe poisoning had been discovered. The poisoning of the Hanford pile by ^{135}Xe has often been described – particularly the remarkable sleuthing by Wheeler and Fermi who, once the kinetics of the phenomenon had been established, identified ^{135}Xe as the culprit. (Wheeler had considered the possibility of fission product poisoning even before the Hanford reactors had been built; it was therefore natural for him to be the one who first identified ^{135}Xe.) At Hanford the loss in k due to xenon amounted to about one percent. Had the corners not been filled, the Hanford reactors could have been operated at about 35% of their original design power. And as the boron impurity in the graphite was burned out and ^{239}Pu produced, the power could have reached the nominal design within about 350 days. Wigner in his memoir points out that had du Pont adopted his original tube configuration with 60 cm, instead of 20 cm, long slugs the ^{135}Xe poisoning would have been less severe; nevertheless he agrees that du Pont's filling-in of the corners was very fortunate since it allowed the reactors to operate at almost full power from the start.

Breeders. Wigner was much intrigued by the breeder (Papers 18, 19, 20). At the time the total known uranium amounted to thousands rather than millions of tons; and Wigner recognized that with so little uranium, nuclear energy could not be very important unless the breeder were developed. When the values of η (neutrons emitted per neutron absorbed) for ^{233}U and ^{239}Pu as functions of energy were measured it became clear that in principle a thermal breeder based on ^{233}U and a fast breeder based on ^{239}Pu or ^{233}U were possible.

Wigner was not attracted to the fast breeder; he regarded its engineering problems as formidable. Nevertheless, he and Harry Soodak sketched out the design of a fast breeder (Patent 27P). The core of the reactor consisted of a small block of plutonium traversed by many cooling channels. The coolant was liquid sodium; thus, except for its small size, this first design for a fast breeder resembled today's breeders.

Wigner's real love was the thermal breeder, which he regarded as a more practical engineering device than was the fast breeder, even though the breeding ratio was lower than in a fast breeder. In 1944 he sketched out the design of a light-water-moderated converter with plate-type fuel and a thorium blanket for converting ^{239}Pu to ^{233}U (Patent 19P): this was prompted by the discovery of spontaneous fission in ^{240}Pu and the possibility that plutonium from Hanford would not be suitable for a nuclear bomb. From this design to a plate-type reactor, moderated by either light or heavy water was, of course, a small step.

But Wigner's real hope was for a homogeneous thermal breeder – say, a slurry of UO_2 and ThO_2 carried in heavy water. The idea for the slurry was given impetus by Harold Urey who, in 1943, left the Columbia project and came to Chicago with Karl Cohen and Irving Kaplan to work on heavy water reactors. Considerable work was done on the chemistry of UO_2 slurries, and Wigner followed this with interest. Wigner sketched out designs for a slurry reactor – one in particular involved pumping the slurry back and forth through the heat exchanger by means of fluctuating gas pressure. This ingenious idea, however, was never embodied in an actual design.

Wigner, the chemist, was aware that the breeder would require rapid chemical reprocessing; this was a primary reason for his preoccupation with homogenous reactors. And in due course he examined several uranium compounds as carriers of the fissile material. Uranium hexafluoride had been suggested by the chemists, and Wigner asked me to investigate breeding in the resonance region in a fluid fuel system in which ^{233}U hexafluoride was dissolved in a fluorocarbon. This was the first attempt to analyze a resonance reactor in some detail.

In later years Wigner has maintained his skepticism about large fast breeders cooled with sodium; he has always been uncomfortable with the thought of 3–5 tons of plutonium in a reactor – despite the extraordinary advances that have been made in fast reactor technology since he and Soodak designed a fast neutron, sodium-cooled breeder more than 45 years ago.

Water Reactors. After pressure to design Hanford eased, Wigner and his group turned to examination of other combinations of moderators, coolants, and fuel. The most thoroughly studied were the light-water cooled and moderated, and the heavy-water moderated systems.

Wigner suggested to Robert Christy, who spent 1942 at Chicago before he went to Los Alamos, and me that we investigate lattices moderated with water. These earliest calculations on water moderated systems were extended by R. Williamson and myself. It was from these calculations that we noticed that the multiplication constant in a water moderated, closely packed lattice might be quite high; and this, along with Arthur Snell's experiments on large blocks of natural uranium, led to the first experiments at Clinton Labs in 1944 on water lattices. This in turn led to the proposal to use pressurized water as coolant and moderator in a power reactor. Throughout this development Wigner maintained a very close overview, even though the actual experimental work was carried out in Clinton.

Heavy Water Reactors. (Paper 17) By the Fall of 1943, Wigner's group had pretty well finished its involvement with the Hanford Project. A group under H. C. Vernon joined the Metallurgical Laboratory to look into heavy-water moderated reactors as possible back-ups to the Hanford reactors, should they encounter difficulties. Vernon's group worked very closely with Wigner's people, and the work was carried out under Wigner's general direction. By August the group had investigated heavy-water moderated reactors with six different coolants: light water, heavy water, helium, homogeneous slurry, heterogeneous slurry, and circulating uranium hexafluoride. A so-called P-9 Committee (P-9 being the code name for heavy water) consisting of H. D. Smyth, Wigner, and Vernon reported their findings in August of 1943: they concluded that the light water, the heavy water, and the homogenous slurry "justified serious consideration."

The New Piles Committee. Many of the ideas for new designs of nuclear chain reactors were put forth in a series of meetings during the Spring of 1944. Attending these meetings were Fermi, Wigner, Szilard, Allison and James Franck, as well as a half-dozen lesser-known members of the Metallurgical Laboratory. At one of these meetings (April 26, 1944), Fermi and Szilard each outlined ideas for fast reactors that produced more plutonium than they consumed – though the word "breeder" had not yet been coined by Szilard. Wigner presented ideas for homogenous reactors, and for a reactor cooled by CO_2 at such high temperature that the bulk of the heat went into the endothermic dissociation of CO_2 into CO and O_2. Many other ideas were put forward there (including P. Morrison's observation that the residual uranium in granite could be burned in breeders with a positive energy balance).

The Materials Testing Reactor (MTR). Wigner came to Oak Ridge as research director of Clinton Laboratories in 1946. At that time the laboratory was designing a heavy-water moderated, light-water cooled, enriched-uranium research reactor. Wigner suggested two modifications of the design, which were incorporated in the MTR: first, he pointed out that since most of the moderating power was provided by the light-water coolant, the heavy water was rather superfluous. Moreover, with the heavy water squeezed out and the fuel stacked compactly, the fast flux would be increased. The resulting closely packed plate configuration is used in most light-water research reactors today.

Wigner's second contribution to the MTR was the curved plate fuel element. A curved plate is stiffer and its thermal expansion more predictable, than is a flat plate; this suggestion was incorporated into the design of MTR fuel, and has been used in many other pool-type reactors.

Though MTR was not approved for construction until after Wigner had left Clinton Laboratories, the precepts that he laid down during the early conceptual design gave engineers who completed the design a sound foundation. The MTR operated at Idaho very successfully for 20 years.

The MTR has profoundly affected subsequent developments of both power and research reactors based on light water (LWR). The first nuclear submarine,

Nautilus, was powered by a pressurized version of the MTR, and insofar as all subsequent LWR's derive from Nautilus, Wigner's MTR may be regarded as a sort of technological ancestor of most of the world's power reactors.

Gas-Cooled Reactors. Though Wigner regarded the original proposal for a helium-cooled production reactor to be infeasible, he recognized that high temperature gas cooling might be useful for power production. He therefore was very supportive of Professor Farrington Daniels who had been espousing a gas-cooled pebble bed pile. Daniels' ideas originated in his design for a nitrogen fixation system based on a bed of magnesium oxide pebbles operating at extremely high temperature and cooled by nitrogen. At high temperature the magnesium oxide is converted into magnesium nitride. When Daniels came to the Metallurgical Laboratory in 1944, he envisioned a power reactor as a bed of uranium oxide pebbles similarly cooled by gas.

Daniels' idea lay dormant until Wigner took over the Clinton Laboratories in 1946. By that time the Army had made a commitment to build a gas-cooled reactor along the general lines suggested by Daniels. A group drawn from various industries and headed by Rogers McCullough came to Clinton to engineer the Daniels' pile. The main line of research centered on a semi-homogenous system in which highly enriched uranium was dispersed in beryllium oxide, and was cooled by circulating helium at high pressure. In many ways, the Daniels pile resembled DRAGON, and later pebble bed reactors, though of course the idea of coated particles had not yet been invented.

Wigner, as research director at Clinton, assumed responsibility for the project. In this he was, unfortunately, not supported very well by the "old timers" (meaning those of us who had already been there for a couple of years); we could not understand how uranium at $100 per gram (1946 $) could ever be an economical fuel; and besides, we were too involved with the MTR. Perhaps this lack of support was an important reason why the Daniels pile was never built, although many similar reactors have since been operated successfully.

The Nautilus. Wigner was one of the first to propose a parallel plate reactor cooled and moderated by light water (although plates of uranium in light water had, unbeknownst to him, been investigated experimentally in Germany under Heisenberg's direction). The pressurized version of this scheme was embodied in the prototype for the Nautilus, which went critical in Idaho in 1952. The idea for Nautilus originated at Clinton Laboratories in 1946 while the then Captain Rickover was there learning about reactors. With the demise of the Daniels pile, many of the engineers involved with that project shifted to design of Nautilus: the most notable of these was Harold Etherington, who became the chief designer of the Nautilus prototype. Another very important contributor to Nautilus was Samuel Untermyer, who was Wigner's engineering assistant. The entire project soon moved to Argonne where W. H. Zinn was Director. Wigner by this time had returned to Princeton; however, he was retained by Zinn as a consultant on Nautilus, and he served as a technical conscience for Nautilus designers.

Savannah River Reactors. Wigner's last active participation in reactor design was in connection with the Savannah River heavy-water production reactors in 1952. The du Pont Company, which had responsibility for the Savannah River Project, called on Wigner as general consultant; to carry on this work, he took leave from Princeton and, as he did in the Hanford design, he served as an overall technical adviser on the Savannah River design.

Project Hope. (Paper 21) By 1953 Wigner, though he no longer participated in detailed reactor design, nevertheless was concerned about the delays in development of competitive nuclear power. Since he still regarded the breeder as essential, he decided to try to design a chemical plant that would reprocess spent fuel at a cost affordable in a full-fledged breeder. The design for "Project Hope", as it was called, was carried out by chemical engineers from ORNL headed first by Wigner during a six-month stay, and then by Robert Charpie. The summary of the report, published in 1953, states:

"A rather embarrassing situation exists at the present time in that the cost of recovering a gram of enriched fissionable material from a spent fuel element is very often as large as the initial cost of producing a new gram of fissionable material in the diffusion plant ... if we are ever going to establish a nuclear power economy based on heterogeneous reactors which use metallic fuel elements, it will be absolutely necessary to drive the costs of reprocessing down from the present high levels to such a point that the chemical costs represent only a small fraction of the total cost of power".

The Project Hope plant was designed to reprocess MTR fuel. Several new ideas, especially for coping with criticality, and for waste handling, were proposed. The estimated cost for reprocessing came to about $1 per gram of ^{235}U. Thus, at least on paper, the project had achieved its hoped-for goal. Unfortunately, chemical processing technology moved along rather different paths, so Project Hope had little direct influence. Nevertheless, it did illustrate Wigner's extraordinary versatility – he was equally at home in detailed chemical engineering design and in the quantum mechanics of nuclear systems.

III. Reactor Theory

The line between reactor physics and reactor engineering is not very distinct, and Wigner laid much of the foundation for both fields. Thus during that extraordinary three years at Chicago (1942 to 1945), and the year at Clinton (1946–47), Wigner invented many of the techniques that we now teach in textbooks of reactor design. (Papers 27, 28, 29, 30)

Microscopic (Lattice) Reactor Theory. Wigner's version of the theory of the multiplication constant in the uranium lattice had been worked out by the time he arrived at Chicago in 1942. What remained to be done was its systematic application to the design of reactors, and to the interpretation of Fermi's exponential experiments. (I should stress that Fermi was his own, somewhat separate and rather private, theoretical department. Because Fermi was less involved in the actual design at Hanford, Wigner's, rather than Fermi's, techniques became the basis for subsequent lattice design). In addition, many details still needed clarification.

Wigner initiated and supervised the examination of as many combinations of fuel and moderator as the group could manage. By the end of 1945 he had explored lattices moderated with H_2O, D_2O, CO_2, Be, BeO, and C, as well as homogenous mixtures of U, D_2O and H_2O. The constants such as diffusion length, resonance absorption, and cross-sections used in this exploration were based on experiments at Columbia, Princeton, and, later, on the exponential experiments in Chicago.

Spherical Harmonics Method. Wigner recognized that the diffusion theory approximation in all of these calculations was deficient. However, he was reluctant to invoke more complicated transport theory because the neutrons were not monoenergetic, and any transport correction to the diffusion theory would be obscured by the error caused by the assumption that the thermal neutrons were monoenergetic. He and Breit had used the spherical harmonics method in their original calculations of the size of a bomb, and he supervised the later work at Chicago on the application of the spherical harmonics expansion to calculate the thermal utilization.

Correction to the Sphericized Cell Calculation. (Paper 30) Wigner also worked out the correction to the spherical cell approximation in a cubic lattice. Here his previous work on cohesive energy in crystal lattices stood him in good stead, since the Wigner-Seitz method of calculation of cohesive energy also involved sphericizing a cubic cell.

The Fast Effect. Though Szilard seems to have been the first to point out the importance of fast fission in uranium-238, Wigner and his group were the first to systematize the calculation, particularly the calculation of the first collision escape probability in various geometries. (Wigner made the latter calculation by first solving the diffusion equation, where the kernel is of the form e^{-ax}/x, and then integrating with respect to a to get the transport kernel, e^{-ax}/x^2. This led to his making the only calculational error I have ever known him to make – when he used this technique to compute the fast effect in a hollow slug. M. H. L. Pryce several years later pointed out that Wigner had calculated the escape probability for a ring of fissions in a solid slug, not uniform fissions in a hollow slug – which perhaps proved that Eugene Wigner is human after all!)

Macroscopic Reactor Theory. Although Fermi's idea of the age of neutrons provides a general structure for Macroscopic Reactor Theory, Wigner contributed many of the most familiar techniques that are still in use today. Among these are:

Two-Group Theory. The group approximation was used in Chicago before it was developed anywhere else, though I cannot say exactly who first introduced the idea. Wigner's formulation of the resonance escape probability in effect lumped the fast neutrons into a single group whose spatial distribution was determined by one-group diffusion theory; thus Wigner's calculation of the resonance disadvantage factor was a forerunner of more formal two-group theory.

The standard and much-used formulae of control rod statics, perturbation theory, and reactor kinetics were derived by Wigner in papers 34, 36, and 37; and the theory of the pile oscillator was given in Paper 35.

Shielding. (Papers 22, 23) Wigner estimated the thickness of both gamma ray and neutron shields. He did not use elaborate transport theory methods, but this was sufficient since at the time the Hanford shield was being designed, Fermi's reactor was in operation, and experiments on actual shields could be performed.

Fission Products and After-Heat. (Paper 25) Wigner, together with K. Way, analyzed the data on fission products and they were able to deduce a semi-empirical formula for the decay of fission products. The Way-Wigner expression for the decay has been since modified, but the overall approach remains valid.

Temperature Effects: The Wigner-Wilkins Distribution. (Paper 31) From the beginning, Wigner and his group estimated the temperature coefficients of the criticality factor in graphite lattices. Some of these, like the diffusion length coefficient, could be estimated theoretically; others, like the resonance coefficient, had to be measured. In any event, the estimated temperature coefficients agreed well with the observed ones.

One of the best known of Wigner's contributions is the velocity distribution of neutrons that become thermal in an absorbing medium at temperature T (the so-called thermalisation problem). We knew that because of the absorption, neutrons would be hotter than the moderator, but the exact distribution of thermalized neutrons posed a formidable mathematical problem.

In 1945, Wigner, in collaboration with the young mathematician, J. E. Wilkins, Jr., solved the problem (Paper 31) in the special case of moderation by hydrogen.

IV. Solid State Physics

The "Wigner Disease". Wigner had contributed to the quantum mechanics of solids with his calculations, together with Seitz, of the cohesive energy of crystals. It was therefore natural for him, especially with his background in chemistry, to worry about the effect of neutron bombardment on the properties of graphite. He, together with F. L. Friedman and F. Seitz, estimated the number of displacements that a moderator atom would suffer in the course of operation of the reactor. These effects quickly became known as the "Wigner disease", a name that Professor Wigner has always disliked. Though the calculations were hardly conclusive, Wigner's anticipation of the effect led to a major experimental investigation of radiation damage at the Metallurgical Laboratory. The Wigner disease in its innumerable manifestations is a continuing concern of both reactor engineers and material scientists.

V. Nuclear Physics at Chicago and Clinton

Though there was little time at Chicago to work on nuclear physics, Wigner was challenged by a remark of Fermi that, despite its great success, the Breit-Wigner resonance theory, and the compound nucleus, had no rigorous basis in quantum mechanics. Thus even before the war ended, Wigner was working out a rigorous formulation of the resonance theory – and this led eventually to the R-matrix theory. This was worked out in a series of papers with L. Eisenbud, most of which were written while Wigner was Research Director at Clinton. (During his year at Clinton, Wigner observed "holy Tuesday" – a day he devoted to his own research, and on which he was not accessible). The R-matrix theory has since developed into a major tool for analysis of the many resonances now routinely measured, but only suspected at the time the theory was developed. It has also led to analysis of the statistical distribution of level spacing and in particular, to the Wigner distribution of level spacing as well as to renewed interest in random matrices by mathematicians.

VI. Wigner and the Founding of Oak Ridge National Laboratory

Even while the war was still on, Wigner thought about ways of continuing the research and development on nuclear power. In this he was attracted to the possibility of expanding the Clinton Laboratories into a large, permanent nuclear research center. During the war, Clinton Laboratories had been primarily a chemical pilot plant for developing the methods used at Hanford to extract plutonium. Though there were many able physicists and biologists there, the chemists and chemical engineers dominated. Wigner envisioned the laboratory developing into a large, permanent center with primary emphasis on reactor development and with all the ancillary disciplines represented – physics, biology, materials, as well as chemistry and engineering. To carry out his plan, Wigner accepted the research directorship of Clinton Laboratories, and spent 1946–47 in that job. He brought with him several members of his wartime theoretical group from Chicago – Gale Young, Katherine Way, and myself.

At Clinton two main projects were being pursued – the high-flux reactor (which culminated in the Materials Testing Reactor, MTR); and the Daniels Power Pile. In addition, Captain Rickover and his energetic navy aides came to learn nuclear energy, and to develop what became the Nautilus.

During Wigner's one-year tenure as Research Director, most of the organizational structure of the Oak Ridge National Laboratory was established: in particular, a powerful biological research laboratory under Alexander Hollaender, and a metallurgical laboratory, as well as the already-established physics, chemistry, and technical divisions.

But perhaps the most important contribution to nuclear energy at the time was the Clinton Laboratories Training School, which later became the Oak Ridge School of Reactor Technology. Wigner understood the importance of spreading the knowledge gained during the war to the industrial and academic people who were to carry on the enterprise. He therefore persuaded Frederick Seitz, later to become President of the U.S. National Academy of Sciences, to organize a school of reactor technology. The student body was drawn from industry, the navy, and academia. Among the students were many who later became prominent figures in nuclear energy – people such as S. Siegel, J. Simpson, H. G. MacPherson – and, of course, H. G. Rickover. The faculty were drawn from the staff at the Laboratory, and included several of Wigner's Chicago associates.

The Oak Ridge School of Reactor Technology greatly influenced the way in which reactor theory was taught in this country. One output of the school was the famous text *The Elements of Nuclear Reactor Theory*, by S. Glasstone and M. Edlund which was based on lectures given by Edlund in 1950.

Wigner and I wrote a somewhat more detailed account of reactor physics, *The Physical Theory of Neutron Chain Reactors*, which appeared in 1958. Copies of the book were presented to all the delegates at the 1958 Geneva Conference on the Peaceful Uses of Atomic Energy. The book summarized what we knew about the physics of chain reactors at that time.

Wigner wrote most of the less standard parts of the text. In particular, he wrote the first six chapters, which beautifully summarized nuclear physics for nuclear engineers, and remain even today a compact account of much of the nuclear physics relevant to nuclear reactors. Wigner used the book as a means of clarifying several loose theoretical ends that had been left untied during the wartime effort. Two such loose ends are notable: first, the realization that the transport integral equation admitted of singular solutions which had a continuous spectrum, and second, that the surface resonance absorption in lumps of surface S could be better represented by a term proportional to $S^{1/2}$, as proposed by Gurevich and Pomeranchuk in the Soviet Union, rather than S, which was the expression used in the design of the Hanford lattice. Although my name appears first in the title (since We comes before Wi), this was a courtesy that Professor Wigner extended to me, the sort of generous gesture that he has unfailingly extended in all of his dealings with collaborators who were younger, and less gifted, than he. Indeed, this trait of Wigner's explains why so much not only of reactor theory, but of theoretical physics from 1930 to 1965, though it may not bear Wigner's name, actually has its origin in a suggestion made or question asked by Professor Wigner.

VII. General Energy Policy

(Papers 38, 39, 40, 41, 42) Eugene Wigner was one of the early contributors to the general debate on the role of nuclear energy. In these early papers he displayed an openness and honesty that is refreshing in these days of bitter polarization over nuclear power. As he put it "If nuclear power is not economic, it makes little difference whether it is only slightly uneconomic; it will not be used as long as other sources of energy are cheaper" (Paper 39). In the very long run, nuclear power based on breeders, whose cost is hardly dependent on the cost of uranium ore, will place a limit on the price of energy – and this limit should not be very much higher, and might even be lower, than the cost of today's energy. Thus, Wigner argued that unless the breeder were developed, nuclear power would not amount to very much – but that with the breeder, nuclear power in the long run might eventually become the dominant energy source.

In these early papers Wigner conceded that breeders might develop simply through improvements in non-breeders (e.g. the Shippingport light-water breeder); or they might spring up as a wholly new technology, e.g. the Liquid Metal Fast Breeder (LMFBR). He was unwilling to choose between these two possible paths; and, although the main line of breeder development has certainly favored the latter path, the very high capital cost of the LMFBR suggests that incrementally improved converters may yet have their day.

Wigner argued that although development of the breeder was central to nuclear energy, non-breeders should be built, mostly as a means of gaining experience in the handling of radioactive wastes. Though he did not regard waste disposal as being a particularly difficult technical problem, he realized that we needed experience in handling radwastes; on this point he was right on target.

VIII. The Wigner Memoir of the Uranium Project

While at Chicago, 1942–1945, Wigner always carried a large grey-covered notebook in which he recorded the happenings of the day. This original diary is not available. However, Wigner also wrote personal accounts of the project, beginning in 1941 and continuing to 1948, based largely on his diary. It is these memoirs that are reproduced here.

The memoirs reveal Wigner as sorely troubled by what he considered to be unnecessary delays that plagued the project. The main focus of his frustration was the du Pont Company which was responsible for construction and operation of Wigner's water-cooled reactor. du Pont, rather than incorporating Wigner and his people into an integrated engineering group to reduce the Hanford design to practice, kept the Wigner team at arm's length. In Wigner's view, this delayed the project by months, if not longer. He regarded any delay as unconscionable; in his view every day lost meant that the Nazis were that much closer to having the bomb.

At the time, Wigner attributed du Pont's attitude toward him to the Company's concern about patent rights after the war. Wigner later realized he was mistaken, and in his 1948 memoir he offers à "silent apology" to du Pont for misconstruing its motives. Actually, du Pont had the very highest regard for Wigner. This is attested by Wigner's being retained as a senior technical adviser by du Pont during the building of the Savannah River tritium-producing reactors in 1951.

A reader of the Wigner memoir in 1992, a halb-century after the events depicted there took place, may be puzzled by the intensity of emotion Wigner displays, especially in respect to the du Pont Company, and particularly to Crawford H. Greenewalt, who was in charge of liaison between Chicago and du Pont. Indeed, as editor of these papers, I had to decide whether to publish those parts of the memoir that dwelled on the animosity that Wigner felt at the time toward the du Pont Company. I have decided to leave the manuscript as Wigner wrote it for two reasons: first, because unless Wigner's frustrations with du Pont were described, the memoirs would not explain, for example, why Wigner offered to resign in the midst of the Hanford project; and second, because Prof. David Hounshell of Carnegie Mellon University, in a paper "du Pont and the Management of Large-Scale Research and Development" (in *Big Science*: P. Galison and Bruce Hevey, Editors, Standford University Press, 1992) quotes liverally from Dr. Greenewalt's diaries of that period. These diaries are to be published after Dr. Greenewalt's death; it seemed to me that it was proper for Wigner's memoirs to be available in their entirety so that future historians of technology could make their own judgment as to the merits of the Wigner and the du Pont views.

As one who lived through those stirring days, I can only say that both Wigner and Greenewalt, in very different ways, are remarkable people. All we really know is that the Hanford project worked, and, in my view, the contributions of both Wigner and of the du Pont Company were essential to Hanford's success.

IX. Eugene Wigner and Nuclear Energy – an Evaluation

One must understand that the uranium problem mobilized the best scientific brains of the time. Szilard, the brilliant visionary, and Fermi, the all-powerful experimental and theoretical genius, who actually achieved the first chain reaction, are unique and in a way unmatched. On the other hand, one must also recognize not that the chain reaction was so difficult to achieve, but that it was so easy. Lesser people than those who gathered at Chicago and Los Alamos could have succeeded – but could they have succeeded in only two years, the time from the first exponential experiment that showed k to exceed unity to the Hanford reactors; and another year to Alamogordo?

I believe not; that though the underlying ideas are straightforward, only scientists and engineers of genius could push the project through in this incredibly short time. And of all the able people assembled at the time, Wigner was unique in possessing, in himself, a complete command of nuclear physics, immense mathematical power, an aptitude and liking for detailed engineering, a powerful grasp of chemistry – and perhaps most of all, an unmatched zeal, and sense of responsibility.

Perhaps the last was the most important. Wigner knew from first-hand contact with Nazi Germany, that the project had to succeed. No matter the administrative obstacles or the criticism he might encounter, Wigner committed himself absolutely and completely. He is, as John Wheeler said, a person who cares. I saw this in innumerable ways, but perhaps most strikingly when Wigner, the same person who had published the monumental paper "On Unitary Representations of the Inhomogeneous Lorentz Group", would pore over the detailed blueprints of the Hanford reactor – and who was sufficiently confident of his judgment to insist on changes when he spotted a deficiency.

The standards Wigner has set for reactor engineering are of the highest – standards of physical intuition, standards of knowledge, standards of diligence – and perhaps most of all, standards of commitment and responsibility. No wonder that in the beginning Wigner dominated reactor engineering and reactor theory. Many of the major thrusts in the business go back to him, and many ideas of his are often rediscovered.

I write these words as the nuclear enterprise has fallen upon bad days. Though we tend to blame our travail on others – the Nuclear Regulatory Commission, the activists, the media – if we are honest, we must confess that some of our troubles are of our own doing – that some of our engineering decisions have been less than they should or could have been.

I hope this account of Eugene Wigner's contribution to nuclear energy will serve to remind the present generation of nuclear people that one of the shoulders upon which they stand is that of Eugene Wigner. He has demonstrated how nuclear engineering and nuclear science at the highest level can be done. How can nuclear technologists rescue nuclear energy from its present impasse? No policy more strongly recommends itself than scrupulous adherence to the standards of technical excellence set by Eugene Wigner.

References

(1) E. P. Wigner: Physical Review *43*, 252 (1933).
(2) E. P. Wigner: Physical Review *51*, 106 (1937).
(3) M. Polanyi and E. P. Wigner: Zeit. Physik. Chemie *139*, 439 (1928); H. Pelzer and E. P. Wigner: Zeit. Physik. Chemie B *15*, 445 (1932).
(4) G. Breit and E. P. Wigner: Physical Review *49*, 519 (1936).
(5) Personal Communication from E. P. Wigner.
(6) L. Szilard: Collected Papers.
(7) O. Halpern and M. H. Johnson: Physical Review *51*, 992 (1937); *52*, 52 (1937).

PART II

Memoir of the Uranium Project

Memoir of the Uranium Project

Eugene P. Wigner

April 16, 1941

I came into contact with the ideas on the development of nuclear power at a rather early stage. In the following, I wish to record that part of the history of this development which I know from first hand information and record it as I remember it. This history will be centered to a large extent around the person of Szilard who became interested in this question and convinced of its importance very early. It is quite possible, of course, that others had similar or even further-reaching ideas at the same time, or even earlier. Naturally, I have no way of telling. The following pages are therefore not intended to give the history of a development – they are intended only to conserve my recollections and in the best case, to serve as one source for a history of nuclear power.

Ideas Before the Discovery of Nuclear Fission

I first heard the possibility of nuclear power seriously discussed in the spring of 1934 when I saw Szilard during a visit to London. The efficiency of collisions between neutrons and nuclei was realized by Szilard simultaneously with and independently from Fermi. He visualized the possibility of chain reactions involving neutrons even at this time. His interest was centered around the reaction

$$Be^9 + n \rightarrow Be^8 + 2n \,, \tag{1}$$

which he considered to be a potential neutron source. His belief in the usefulness of Be in this connection was based on the erroneous (too high) value of Be^9 mass which was accepted at that time. He made some experiments, however, which raised suspicion in his mind as to the correctness of this mass. When the energy release from (1) was later found to be much smaller than it was believed in 1934, Szilard's hopes temporarily faded.

Szilard showed me copies of some of the patent applications which he made in 1934 and 1935. (I may remark that he presented his patents to the British Admiralty, according to a copy of a letter from this authority to Szilard.) These patent applications contain rather detailed plans and calculations and many of these calculations were very useful when the plans later reached a more acute stage. One of his applications contains the following passage:

"(a) Pure neutron chains, in which the links of the chain are formed by neutrons of the mass number 1 alone. Such chains are only possible in the presence of a metastable element. A metastable element is an element the mass of which (packing fraction) is sufficiently high to allow its disintegration into parts under liberation of energy. Elements like uranium and thorium are such metastable elements; these two elements reveal their metastable nature by emitting alpha particles. Other elements may be metastable without revealing their nature in this way"

These were certainly almost prophetic words in 1934 or 1935. However, Be^9 was considered to be metastable at that time and was the element which was probably in Szilard's mind in the first place.

I myself followed Szilard's considerations with great interest. I did not actually believe that any particular one of his mechanisms would be successful but felt that the production of nuclear energy must be possible one way or another. Its impossibility would only be the consequence of some deep law of nature (similar to the energy principle) and I did not believe in the existence of such a law. Hence, I believed that the utilization of nuclear energy was only a question of time.

I expressed this view at the end of 1935 to a representative of the General Electric Company who consulted me for another purpose, and of course to several friends in conversations. Later, in the spring of 1936 I gave a short popular lecture to the Γ. A. Society in Madison, Wisconsin, and set the time of the realization of nuclear energy production to five years from that date. I had no basis for that number.

Nuclear Fission

The news of Hahn, Meitner and Strassmann's discovery of nuclear fission was brought over by Niels Bohr at the very beginning of 1939. At that time I was laid up in the hospital with jaundice for about six weeks and Szilard was in town most of the time. He came down to the Infirmary every day and we discussed the uranium fission regularly.

Since I was interested in chemical reactions some time before, it was not difficult for us to realize the mechanism of the process and we developed in our daily conversations all the essential points of the theory which was published later by Bohr himself, in collaboration with Wheeler. In fact, I continue to believe that there were several points which we saw better than they are presented by Bohr and Wheeler's paper. These concern mainly the comparison of the stabilities of ordinary and neutron-excited uranium nuclei. This topic served later as the subject of several discussions between Fermi and myself who objected to certain passages in the Bohr-Wheeler paper.

While I could claim half authorship in these considerations, carried out by Szilard and myself, I wish to emphasize two points.

First, that neither Szilard nor myself realized the fact that the slow neutron fission is due to the 235 isotope alone. This so extremely important realization is due to Bohr alone. In fact, we did not quite believe in the soundness of his arguments until his prediction (which did appear to us reasonable but by no means certain) was verified experimentally.

Second, that the prediction that neutrons would be emitted during the fission process is entirely due to Szilard. I remember quite well when he mentioned it first. This prediction was extremely important because it naturally revived Szilard's pre-fission plans. Bohr and Wheeler were extremely skeptical towards this idea. Even when the neutron emission was found experimentally they attributed it to a process which succeeds rather than precedes the various β activities. Wheeler interpreted the neutron emission in this way in his lecture at the Princeton meeting of the American Physical Society, and indeed, a delayed neutron emission has been found by Tuve. This delayed neutron emission is, however, very much weaker and much less important than the instantaneous emission.

There was a Conference on Theoretical Physics in Washington some time in the early spring which I could not attend because of my jaundice. I heard, however, that Fermi gave a report at this meeting on the theory of fission and its practical consequences. Although I did not hear his lecture I have no doubt that he also developed the essential points of the Bohr-Wheeler theory of fission independently and foresaw many of its consequences just as Szilard did, although his attitude was much more cautious than Szilard's.

Attempts to Prevent Publication

Szilard was the one who made the most serious efforts to enhance research on fission on one hand and to prevent leakage of this information to the outside on the other hand. It became apparent by April 1939 that there are two distinct processes which invite further research. The slow neutron process (i.e., a chain reaction in which the fission is induced by slow neutrons) which is suited, probably, only for power production, and the fast neutron process which may cause heavy and violent explosions.

Unfortunately, Szilard's efforts were all doomed to fail. Bohr and Wheeler said that (1) a chain reaction is impossible, (2) even if the reaction would go at low temperature, the increase of temperature would automatically stop it, (3) that Fermi spoke about the question in Washington in public and secrecy is impossible henceforth. In spite of this, Szilard, Teller, Weisskopf and I persuaded Bohr to agree to a delay of the publications if Joliot's group in France and the British also agreed to such a delay and to the temporary substitution of circulated manuscripts for publications. Consequently letters and telegrams were dispatched to Joliot, Halban, Blackett and Dirac with this suggestion. The British agreed but Joliot declined. It is hard for me at this point to refrain from making a comment. This attitude of the French caused the collapse of Szilard's plans.

Szilard and Fermi were at this time both working on the fission question and published various letters in the *Physical Review* when the above efforts were frustrated. Joliot, Halban and Kowarski also published some relevant papers.

I myself did not want to get into competition with Szilard and partly for this and partly for other reasons did not desire to assume the role of more than an interested spectator.

Approaching the Government

However I felt that the United States Government should be advised of the developments and the possible implications of a chain reaction. Szilard originally opposed this plan. He saw, however, that our floundering on the publication question was caused by the lack of any authority behind us which made it possible for us to speak in the name of *all* American colleagues. In fact Joliot's fears that he would be the personal loser if he withheld his publications received some apparent justification through some sensational reports which appeared in American newspapers. (Allegedly the articles were instigated by Dunning.) All this convinced Szilard that it was desirable to contact the Government.

We talked to Pegram about the question. He phoned to Edison (Under Secretary of Navy) in Washington. As a consequence Fermi was asked to a conference with some representatives of the Navy, including Gunn. This conference did not yield any commendable results, except that Beams received $2,000 for isotope separation.

In spite of this, with Teller's consent, Szilard and I visited Einstein in July 1939 and persuaded him to write a letter to Roosevelt. Einstein of course saw the importance of the matter at once. His letter reached Roosevelt through Sachs in the Fall and led to the meeting of a committee on the 21st of October under L. J. Briggs, about which I shall report later.

The summer brought some of the most important ideas of a practical nature. First, Fermi and Szilard decided to concentrate on the slow neutron chain's exploration. Second, they abandoned hydrogen as slowing down substance and proposed to use carbon, i.e., graphite, instead. This became necessary because of the relatively high neutron absorption of hydrogen. The use of carbon was first proposed by Szilard but Fermi may have thought of it independently. Third, instead of a uniform mixture, a lattice-like arrangement of the U and C was proposed. It is my impression that Fermi deserves somewhat more credit for this idea than Szilard. It is not necessary to describe here the advantages of these innovations as they were described in a paper by Szilard which he sent to *Phys. Rev.* about January 1940.

Meanwhile, as was mentioned above, Einstein's letter to Roosevelt bore a conference. Briggs was Chairman, Colonel Adamson and Commander Hoover were there to represent the Army and Navy, also Meggers to represent the Bureau of Standards. Szilard, Teller and I, "the Hungarian conspiracy" as Tuve put it, furthermore Roberts and Abelson, to represent Tuve's views. Szilard proposed the allocation of funds for experiments, the acquisition of U-supplies, the setting up of a permanent committee to serve as intermediary between physicists and the Government. Teller and I supported Szilard; Roberts and Abelson opposed him. The Government representatives remained rather aloof.

Adamson remarked that wars are not won by armaments but by morale. I proposed thereupon to disband the army.

Naturally we became rather nervous when it became apparent that nothing decisive was happening to further experiments on U. I constantly urged Szilard to write up his ideas for publication because much more rapid action on the part of the authorities and much more interest on the part of colleagues could be expected if the former were confronted with an immediate need for action, the latter by a definite priority claim. Meanwhile an article by the French physicists, reported by Turner, definitely convinced Szilard of the possibility of a chain reaction and he wrote a paper, asserting such a possibility, on the basis of the ideas discussed above. The manuscript was sent to the *Physical Review* about January 1940 with the suggestion that it should not be published.

All these events helped to break the ice and the Government-physicist co-operation began to materialize, even though rather sluggishly. Breit and Urey deserve great merits in hastening matters and so probably do many others. It was through Breit's efforts that a Reference Committee was established to referee papers, a "delay" in the publication of which may be advisable. This committee probably did some useful work.

There is one more very important development which I would like to report on. Turner drew attention in a proposed Letter to the editor of the *Physical Review* in about May 1940 to the fate of the U_{239}. This nucleus is obtained by neutron capture from U_{238} and this neutron capture forms the most important hindrance to the slow neutron reaction. Turner pointed out that the U_{239} goes over by β-emission into element 93 and this element may be susceptible to fission itself. If it is, this would improve the slow neutron reaction. If it is not, it will absorb an additional neutron until finally a slow neutron sensitive nucleus is produced.

Szilard points out that this element would be produced as a by-product in the projected slow neutron machines. It could be separated from the bulk of U by chemical means. One could obtain in this way a pure element which is slow neutron sensitive and could be used probably instead of U_{235} which must be obtained by the more complicated process of isotope separation.

I wish to repeat here the reason which I gave at the outset for my friends' and my playing such a large role in this aide-memoir. I quite realize that this is in no proportion with our actual merits. But, as I said, this is not a history, but only a record of those events with which I came into direct contact.

Addition on March 12, 1944

On rereading my recollections which I put down on April 16, 1941, there are only a few things that strike me and that I would like to correct.

The first is that those pages do not reproduce the mood in which we were. It was evident, all that time, that Germany was out for a big adventure to subjugate all the other nations and races of the world. It was evident that her rule would be cruel and ruthless and if she were successful, the next century would be spent in whipping the rest of the world into subordination. It was equally evident that most of the rest of the world was not aware of Germany's intentions although most of Germany's population, and much of the surrounding countries', was aware of them in an intuitive and somewhat vague way. It is against this background that those pages should be read.

Neither are our fears in connection with the new possibility very clearly described. I guess that both emotions were so vivid in my mind that I considered them to be permanent and therefore did not feel that it is necessary to describe them. We did not know what the U would do. But we felt that it would be, in some way, a powerful weapon, full of potentialities and that it surely would be disastrous negligence not to explore them fully. Energy concentrations more than 10,000 times higher than in any explosive cannot be harmless and it will be possible, probably in more than one way, to bring them to action. It is because this appeared self-evident that I did not fully recognize the importance of Turner's remark which, together with Szilard's remarks, showed one way to bring those energies to bear.

The fears and impressions described above, while originally not shared by many, slowly penetrated into wider and wider circles. After the fall of France, in the summer of 1940, they were so general that a letter requesting, I believe, the National Research Council to support war research was signed in Ann Arbor by practically everyone present. I believe that letter did not refer specifically to uranium fission but this subject was in the mind of everyone who signed the letter.

There is, finally, a last point that is omitted in those pages. It refers to the role which Mr. B. Liebowitz played. He participated, mainly financially, in the non-profit organization which Szilard founded and which financed some of the early work. He was entirely unselfish in this and never even claimed credit for his foresightness.

March 1944

The pages written on April 16, 1941 mostly refer to the time before April 1940. It is more difficult for me to write about the subsequent time because I began to participate in the uranium work in the winter of 1941 and may therefore not be quite unbiased in my narrative of the events.

The Reference Committee

In May 1940 (about a year after Szilard and I visited Einstein) I heard about Breit having urged the National Research Council to support research on uranium. Condon agreed with the suggestion and a committee was set up consisting of Breit, Condon and Beams. The most lasting effect of their work was the establishing of a Reference Committee under Breit's chairmanship. This committee was to review papers before publication and thus prevent important information from being available to potential enemies of the country. The membership of the committee was, as I remember it, Breit, Condon, Fermi,

Urey, Beams, Pegram and myself. This committee continued to function for about a year.

Papers which were received for publication were sent by the Editor of the Journal to Breit who circulated them among the members of the committee. If the majority thought that the paper should not be published, the writer was asked to withdraw it. However, he was under no legal obligation to do so. The work of the committee was much hampered by misunderstandings and by incomplete collaboration of the editors. It happened more often than once that a paper, while being reviewed by us, was also made available to others. At any rate, until our recommendation came in to withhold the paper from publication, the author was under no obligation whatever to refrain from discussing his work with others. Even after we made a recommendation, he was under no legal obligation. Trying to control publications under such conditions was an uphill work and probably caused more irritation than the good it achieved was worth. Altogether, perhaps 8 or 10 papers were withheld from publication. We would have run into more trouble were it not for the fact that most physicists started on some war work by that time and very few papers were written on nuclear physics.

One definitely good thing emerged from the action of the Reference Committee: Its members got acquainted with each other and with the literature on the subject. Both proved to be very fruitful later. Very few papers were received after Pearl Harbor and the Committee became dormant about that time.

Advisory Committee on Nuclear Physics Research

A second event of importance was the Conference, called by L. J. Briggs for the 13th of June 1940 at the Bureau of Standards. The members of the Committee which assembled were Urey, Breit, Pegram, Tuve, Fermi, Szilard, Teller and I. Teller did not come. There was another Committee (Urey, Beams, Gunn, Fermi, Kistiakowsky) appointed at the same time to study isotope separation. Both were under Urey's leadership and Urey also sent out invitations. Most members of the first committee met on the train from New York to Washington and we discussed the function of the Reference Committee and, in particular, an article on X-ray induced fission which originated in the Westinghouse Laboratories.

When opening the meeting of the Committee next morning, Briggs welcomed us but said that the Committee as constituted at the time would be dissolved and substituted by one all members of which will be "Americans of old standing."

This was, unquestionably, not a very hearty welcome, but I more or less expected it. I inquired, however, as to the purpose of the meeting under the conditions. Briggs' answer was that it is for general orientation. We had an interesting if somewhat desultory discussion and left Washington the same evening. I remember that Urey spoke a few words to me on the train. When I said that I often feel as if I were swimming in syrup, he said that those are the exact words he wanted to use.

A few days later, I wrote to Urey from Princeton saying that I did not wish to participate in the work of his committee any more. I gave as reason that my principal function was to direct the Government's attention to the field and this function had been performed. I also hoped that if I resign from participating in the work, the cooperation of the "other foreigners", Szilard and Fermi, will be more highly valued and that my resignation will therefore strengthen their position.

I discussed this point a few weeks later with Breit. He said that he would like to centralize the work on a military reservation and he believed that if Fermi and Szilard agree to move to such a reservation, they will be well supported and fully informed of all progress and developments. The plan appeared reasonable to me but Fermi refused to work on a reservation. Knowing what I know now, it appears to me doubtful that full confidence would have been bestowed on Fermi and Szilard even if they had consented. On the contrary, their efficiency and influence might have been even more seriously impaired than under the arrangement which did go into effect.

Under this arrangement, they received limited clearance which did extend, however, to the whole field of nuclear chain reactions. This situation is essentially in effect even now, except that Fermi is cleared both for fast and slow, Szilard only for slow chain reactions.

Work at Columbia University

Szilard informed me, as much as secrecy restrictions permitted, about the progress of the work at Columbia. It was not too fast. All dealings and communications with the outside had to go through Prof. Pegram, and he was too busy with other affairs to attend to them promptly. Nevertheless, Szilard struck up a number of important business acquaintances – I always wondered how he did it – although, I believe his evaluation of their value was very optimistic.

Szilard was much perturbed by the slowness of the development. He wanted to get the materials together, if necessary obtain them on loan, and see if one can establish a chain reaction. He was inclined to lay the blame for the delays on the organization which made it necessary that every decision concerning the acquisition of materials be relayed to different authorities and he complained that it is impossible to come to an agreement with manufacturers if the one who conducts the negotiations has no power whatever to act. Nobody besides himself had any time to negotiate with manufacturers.

Fermi adopted a much more philosophical attitude. He knew that no progress is as fast as it could be and considered the delays and the lack of efficiency as quite natural. He said that matters will not be handled perfectly in Germany either and that one only wastes one's energy and runs into hostilities if one tries to improve on a situation of this nature. As far as the situation in Germany is concerned, he must have been right – otherwise the Germans would have been ready with atomic bombs a long time ago.

It will be remembered that Fermi and Szilard were promised some sort of help at the first meeting at the Bureau, in October 1939. This was delayed several times but they finally received some graphite about March 1940. Fermi started on measurements of fundamental constants, such as diffusion length in C, etc., and he made good progress with his work in the following months. The order of magnitude of most fundamental constants was established during this period.

Fermi gave a few lectures in Princeton during the summer term of 1940 (on Cosmic Rays) and I got somewhat better acquainted with him during this time. He is a man of enormous energy, extraordinarily sound and sober judgment and great honesty. On one occasion he asked me whether I could fully recommend Szilard as a collaborator. My reply was that I consider Szilard to be one of the most honest and unselfish people and though his behavior is often irritating, his motives are always noble and that most people soon get used to his unusual manners. Fermi was glad to hear this.

Soon afterwards I heard that they decided to cooperate on the uranium research, on an absolutely equal basis. I was not very sure how long it would be possible to maintain this situation but was glad that two such exceptionally able men were working together in this field.

The work in Columbia remained no secret from other physicists. I remember that, during the summer, Uhlenbeck mentioned it to me and told about the many counters ticking in Fermi's office. I am sure that he had this work in mind, possibly in addition to other work, when we wrote a letter to the National Research Council from Ann Arbor (Summer 1940).

First Princeton Project

Unfortunately, much jockeying for positions took place all this time. Pegram may have had some intentions of keeping a substantial part of the work in Columbia and this, if true, was not an unnatural reaction on his part. The predominant role played by Columbia was much resented, however, at other places. In particular in Princeton, Turner and Wheeler got quite annoyed and expressed their feelings in letters to Briggs. This again was much resented by Breit who had Wheeler's earlier reluctance to support any secrecy policy still vividly in his mind. I did not want to take sides and Breit advised me not to participate in any work that might be started in Princeton. He probably felt that the number of "foreigners" who participated in the work should not be increased unnecessarily. This caused a somewhat anomalous situation since I was a member of the Reference Committee all the time and knew more about the work in general than anybody else in Princeton.

On the whole, I felt that it was neither fair nor wise to restrict the work to Columbia. I expressed this view rather strongly in a letter to Breit, in July 1940. On the other hand, I was a little surprised by the attitude of Wheeler who was, the day before, dead set against any Government support and therefore secret work, and who wanted a leading role in it the next day. After a great deal of shilly-shallying which lasted from about May to September, a Project was approved for Princeton in September 1940. It was carried out during the school year 1940/41. Wheeler was to lead it and Snyder was to be his most important collaborator. However, the whole thing was on a very small scale and Wheeler soon lost interest. Very little came of this work which concerned a measurement of the width of the resonance level in U^{238}.

I did not participate in this work, partly because of Breit's advice and the events recorded under Advisory Committee on Nuclear Research, and partly because I felt that it was quite fully staffed and thought that Wheeler was friendly but not very intimate and he might have resented my giving too much advice.

Work During the Summer Term of 1941

The reader will realize that the condition of the U work was very confused all this time. The authorities had no full confidence in Fermi and Szilard – I myself had to dispel some suspicious they had – and, in spite of this, their work on the chain reaction received practically all the support that was available for this field. I understand that Allison had work in progress on the same field but the authorities did not expect as much from this work as they did from the work at Columbia. There was the Reference Committee to supervise publications but no full confidence was bestowed on all the members of this either. The physicists clamored for a more consistent and more vigorous policy on the part of the authorities which were criticised severely. Changes were promised constantly but nothing happened. The impression of swimming in syrup remained. The situation may have been better in the field of isotope separation, but I had no information on that point.

Breit was in Washington by this time, first at the Navy Yard and later with Briggs at the Bureau of Standards. He acted as a sort of aid to Briggs in matters pertaining to U and, in addition to correlating the work at various places, conducted theoretical research himself also. He had, I believe, full realization for the difficulties under which the work labored and did his best to improve the situation. He had very little power and, perhaps, too much patience with red tape. I am not at all sure, however, that he could have achieved more if he had been more vigorous because I learned since that the high authorities are immensely efficient in blocking any changes.

Breit felt that the work on chain reactions should be extended. For one, he asked Teller (who was at Columbia for that year) and me to visit him in Washington and we discussed theoretical questions relating to the chain reaction. It was during this time that Teller conceived the idea of the fissionability of rare earths. Breit was interested in accurate solutions of the transport equation and I began to be interested in resonance absorption. Much of my acquaintance with the subject was acquired during this time (Dec. 1940–Jan. 1941).

Szilard visited me quite frequently in Princeton during the school year 1940–1941. He told me, even before the June meeting in Washington about his confidence in the possibility to establish a slow chain reaction in the newly proposed U-C lattice and I strongly urged him to write down his ideas. On one occasion, he asked me about the diffusion in a strongly absorbing medium and

I worked out (about January 1941) the theory which is the subject of Report A-20.[1]

All this awakened my interest in the problem of chain reaction and I became involved in it, although I had not publicly revised my decision not to work on it. I also felt that all forces must be united if we are to overcome the advantage which the Germans have in this field through their realization that the war is for their hegemony over the world which they cannot keep if they fall behind in the field of nuclear power.

Little sense remained under these conditions in my not participating in experimental work in Princeton. This impression was shared also by Breit as well as Smyth. Smyth, in particular, considered that Princeton could handle a larger share in U work.

Breit called a meeting at the Carnegie Institution in which the program for Princeton should be discussed by himself, Fermi, Szilard, Wheeler, Teller and me. The usual secrecy restrictions between Columbia and Princeton were to be lifted for this day but, as Breit repeatedly emphasized, not permanently.

The main topic of the meeting was Fermi's review of the experiments at Columbia. In particular, he developed the theory of the pure graphite pile (σ pile) and the measurement of the diffusion length. As always, he spoke well and clearly and even a beginner could easily follow his talk.

The end of the meeting had a discordant note which forebode no good for the relation between Breit and Wheeler. They broke out in open quarrel because Wheeler did not take the secrecy restrictions as seriously as Breit thought it to be necessary.

Nevertheless, the meeting was followed by one at Columbia in which various propositions for Princeton were considered. As I remember it, Breit proposed that a lattice of U and C slabs be built, Wheeler that the fission spectrum be investigated by the H recoil method and I that measurements on the resonance absorption be started. Fermi was very much against Breit's proposition and the measurement of the fission spectrum appeared both difficult and not very urgent. Thus we decided on the measurement of the resonance absorption by measuring the depression in the density of resonance neutrons in the neighborhood of a uranium lump embedded in graphite but that it appears to be very difficult to carry out such an experiment. Similarly Szilard mentioned that he contemplated using a two-dimensional lattice, i.e., a rod lattice which is particularly suited for experimental purposes because the lattice constant can be varied so easily. Unfortunately, he had no opportunity to carry out this experiment.

Once the subject was decided upon, the work in Princeton got under way very soon. Fermi and Anderson came down to help Creutz and Wilson with the first experiments and these were finished in a few days. Fermi and Anderson were very helpful in every way and helped us over the first difficulties which would have been very time-consuming otherwise.

---

[1] Reference to Volume V, I

The results of the experiments at Princeton are incorporated in reports and it is not necessary to go into details concerning them. At the beginning, Creutz and Wilson showed considerable tendency to return to experiments (on proton-proton scattering) which they had planned before. However, after a couple of months, they became interested in the subject and it was a real pleasure to work with them. The work covered not only the question of resonance absorption but also that of the diffusion length in U and U_3O_8 which soon became very important. At the time when I am writing this, Anderson and Hughes are engaged in remeasuring the diffusion length in U.

We had weekly meetings in Smyth's office, which were, on the whole, very pleasant and enlightening. Smyth, Wheeler, Creutz, Wilson and I participated in them. The other project, that of Wheeler and Snyder, went on simultaneously but we did not discuss it in these meetings as they were considered to be more or less Snyder's and Wheeler's private affairs. However, Wheeler did not take a strong interest in either work and Snyder's and his problem languished somewhat. Snyder's work was finished by the fall of 1941 and he joined our group at that time.

Toward the fall, Wilson developed a very nice idea concerning isotope separation. Although it soon turned out that the idea had been conceived before by others, among them by Breit, he was generally encouraged to work on his idea. He left, therefore, our work toward the end of the summer 1941. We missed his help greatly and he himself became by that time so interested in the work on chain reactions that he would have preferred to continue working with us. However, it was realized by all that his idea on isotope separation should be tested experimentally. It developed into a large project which was continued in Princeton for about a year. After that time it was considered by the authorities in Washington that it will not be able to compete with other methods of separation and was closed down.

Status of the Work by the Fall of 1941

The work on uranium was completely reorganized early in the winter of 1941 and it seems worthwhile to describe its status at various places around this time.

The main work was going on in Columbia. There Fermi and his collaborators built an exponential pile towards the end of the summer 1941 and obtained a multiplication constant which they first reported as .89 and later on decreased to about .87. This pile contained graphite of reasonably high purity but an oxide which later on proved to be very badly contaminated with neutron absorbing materials. Furthermore, the uranium oxide was contained in iron cans which also had a considerable neutron absorption. In addition to this, there was naturally little reason to believe that the geometry of this experiment was even approximately optimal, i.e., that the size of the cell and the ratio of uranium to graphite were chosen so as to give a high multiplication constant. In spite of this, this was the first real experiment to measure a multiplication constant by a method which everyone recognized to be adequate. Fermi was very much pleased with his achievement as was everyone in spite of the very low value for the multiplication constant which the experiment gave. This latter point, though, acted as a considerable damper on the optimism concerning the possibility of a chain reaction. Fermi went so far, in fact, as to say that he did not believe that a chain reaction was possible with unseparated uranium because he figured that improvement in the purity of materials and the geometry would hardly yield more than a few percent increase of the multiplication constant.

Szilard, also working at Columbia, became interested around this time in what is now called the fast effect. The fast effect, is the increase in the multiplication constant obtained by the emission of neutrons by U^{238} which is induced to fission by the fission neutrons before they are slowed down. Szilard measured both the cross section of such fission neutrons to induce fast fission and also their inelastic cross section, i.e., the probability for their being slowed down below the fast fission threshold by an inelastic collision with uranium. He concluded on the basis of these measurements that one may obtain an increase of as much as 6–8% in the multiplication constant by using large and metallic lumps of uranium. Szilard was also somewhat discouraged by the low multiplication constant which Fermi's experiment gave but was far from giving up hope.

It may be interesting to note that Szilard carried out these experiments, which later on proved to be very important, more or less because he felt thwarted in his endeavors to procure materials of higher purity. He said that since he could not do the things which he considered to be really important, he would play around for a while in nuclear physics.

I knew very little at that time about the work that was being carried in Chicago and what I can say about it is necessarily very sketchy. It was only a short time ago that I found out that V. Wilson wrote a memorandum to Mr. Compton during the summer of 1940, that is, only about six months after Szilard sent his article to the Physical Review. In this memorandum Wilson attempted to calculate the multiplication constant if one uses beryllium instead of hydrogen as a moderator. On the basis of nuclear constants which we now know to be rather far from their actual values, he arrived at the conclusion that a chain reaction with unseparated uranium and beryllium as a moderator may just barely be possible. V. Wilson apparently was not familiar with the idea of the lattice as developed in Columbia but assumed a homogeneous mixture of beryllium and uranium.

As far as I know it was mainly the beryllium line that was pursued in Chicago with Mr. Allison being the leader of the work. This work was not dissimilar to that carried out by Fermi except in its emphasis on beryllium. They remeasured the diffusion length in graphite and were planning to make exponential experiments when the reorganization mentioned above took place. Also a table of nuclear cross section for thermal neutrons was prepared which proved to be very useful in later work.

The work in Princeton was concerned with three problems: the diffusion length of thermal neutrons in uranium oxide and that in metal was measured, which permitted a more exact calculation of the "thermal utilization" than was possible before. In addition, the experiments on the resonance absorption began to bear fruitful results and showed that the resonance absorption consists of two parts: a mass absorption and a surface absorption. The magnitude of both was determined with what we thought good accuracy for the oxide and lower accuracy for the metal. However, the experiments on the metal were not completely successful until the spring of 1942. The third line consisted in an investigation of a possible resonance effect in the fission process. This line started from an accidental observation of R. Wilson and Creutz who found a very considerable fission activity in samples which were protected from thermal neutrons by cadmium. However a more accurate calculation of the effect that should be expected under such conditions did not indicate any departure from l/v law. I was particularly interested in these experiments since if successful they would have proved a point in which Szilard's and my ideas were different from those expressed in the theory of Bohr and Wheeler. As a matter of fact, we now know that resonance effects of the kind suspected by Creutz and Wilson do exist and that the reason they could not prove their existence is purely accidental inasmuch as a negative and a positive effect accidentally just compensate each other.

The results obtained on the first two subjects were used by me to calculate the optimal geometry and multiplication constant for oxide lattices and to obtain estimates for the same quantity in metal lattices. I arrived at the conclusion – around the middle of November – that the multiplication constant in Fermi's Columbia pile could be increased by about 5% by going over to a lattice with a considerably smaller lattice constant. I expected a further increase of another 5% if one could replace the oxide by metal. This last increase was not really based on the measurements of Creutz and Wilson, but on a theory of the resonance absorption which I developed around this time. On the basis of these calculations and because it became evident that a very considerable improvement in the multiplication constant can be achieved by using materials of a higher purity, I became convinced that a chain reaction is possible in a graphite-uranium mixture and estimated the multiplication constant obtainable with an oxide-graphite lattice as 1.02, with a metal-graphite lattice as 1.07.

The fast effect discovered a short time before by Szilard was omitted in these calculations and this proved to be a serious error, although we now know that the fast effect is less than half as great as Szilard calculated. I believed at that time that it is even smaller than that. The lattice which I proposed for the oxide proved later on to have a multiplication constant only about 2% higher than that of Fermi's Columbia lattice. The reason that it was so much smaller (about 3%) than I expected was mainly due to the fact that the fast effect was much smaller in it than in Fermi's lattice. However I did not quite believe in the fast effect and argued furthermore that if it existed we would recover the loss incurred by the transition to the smaller lattice when we replace the oxide by metal.

This is where we stood when the work was taken over by Dr. Compton on December 7, 1941.

Period of Scientific Research
Under Dr. Compton's Leadership

September 9, 1944

The whole uranium work was completely reorganized early in December 1941. Lawrence took over the work on isotope separation by electromagnetic methods, Urey the isotope separation by chemical methods such as diffusion, centrifuging, etc, and Compton was assigned the methods of nuclear physics. Compton's work was to extend originally not only to the establishing of a nuclear chain reaction but also to the final use of the product. It will be seen that the second task was transferred around June 1942 to Oppenheimer's group.

Compton turned up in Princeton on Pearl Harbor day. I was in the laboratory by accident that Sunday and I got the message about the Japanese attack there. About an hour later I heard that Compton was in the laboratory discussing with Smyth the uranium program. It was unavoidable that Compton's taking over the work became inseparably connected in my mind with the awakening of my friends to the realization that we have to fight a war. The Japanese attack put an end to an equivocal situation and I hoped that Compton would bring clarity and vigorous leadership to our project.

Later in the evening Smyth asked Creutz, Robert Wilson, Wheeler and me to his office to participate in the conference with Compton. I was asked to summarize the results obtained so far in Princeton and gave an outline also of the compilation of relevant constants which I undertook at Smyth's request. This compilation, together with calculations mentioned before, strongly indicated the feasibility of a chain reaction. Compton listened very attentively and it was evident from his remarks that he followed the discussion closely. He appeared at once as a man of considerable charm and quick thorough comprehension. He raised confidence in all of us and when he left we were feeling that perhaps after so many disappointments the work would be vigorously pursued under competent and able leadership. In the course of the following weeks Compton came to Princeton repeatedly and on some of the occasions Szilard also participated in the discussions. I remember that on one occasion Szilard and I drove him to Trenton where he had to catch a train. He conducted a lively discussion with Szilard during the trip and I remember that both Szilard and I were full of hopes and in fine spirit when we returned from Trenton.

The informal discussions which Compton had at the places where work was going on were soon followed by some larger conferences in Chicago and at Columbia. One of these conferences on January 18, 1942 at Columbia was attended by 26 physicists including Lawrence, Pegram, Fermi, Breit, Szilard and myself. Both general questions of policy and technical questions were discussed in a somewhat popular fashion. Allison discussed the measurement of the diffusion length and of the cross section of beryllium, Fermi measurements on the cross section of boron and the intermediate experiment; Creutz gave an account of the measurements which were carried out in Princeton. The question of secrecy was emphasized as at practically all such meetings. Lawrence made a strong plea not to lose time with a too scientific approach but try to do something practical even if we did not understand all phases of it and even if it did appear that we are yet unable to choose the best system. He emphasized in this connection the need for early decisions between alternatives. The question naturally came up whether it was wise to continue work on three lines such as chain reaction and two methods of isotope separation. Lawrence gave the impression that he would rather put all the efforts on one direction and have one horse come in first rather than three horses come in second. This was appreciated by every participant in the discussion. We were strongly under the influence of the events at Pearl Harbor and it appeared to us more clear than it ever did before or since that success or failure in the uranium work would mean victory or defeat. We expressed agreement with his point of view and expressed our willingness to abandon our own line of work and switch to backing the best horse. We also asked him whether there was any disagreement anywhere with his point. Lawrence answered that people profess to agree with his point but that there are few in responsible positions who follow it. I remember that I was quite perturbed by this remark since I could not interpret it in any other way but as an attack against the new leadership of the project and particularly against Compton. Even though the younger participants in the conference heartily agreed with Lawrence, I must admit that we soon conveniently forgot his remarks.

Among the questions of policy which were discussed the most important one was a tentative time schedule.

It was expected that by July 1, 1942 we would know the characteristics of a lattice that is chain reacting.

By January 1943 we would have a chain reacting pile.

After another year, by January 1944, we would be ready for extracting 49, that is, by that time it was expected that a large pile will have operated sufficiently long to have accumulated plutonium worth while for extraction.

By January 1945 the whole project should be over in the sense that the device for ultimate application would be ready.

This time schedule was repeated on numerous other occasions and I remember that several of us were even somewhat dissatisfied with it, thinking that it was too slow. Creutz in particular wrote a memorandum about an accelerated time schedule and I agreed with the possibility of it. Early in June 1942 it was expected (CS-134)[2] that a few bombs would be ready early in 1944.

Looking back now at the time schedule we know that the first two dates were successfully met. In fact, the laboratory did beat both dates by one or two months. The third date is based on the estimate that a production pile can be designed and constructed in less than a year. This still appears completely reasonable both to myself and all members of the laboratory with whom I have close contacts. It is supported also by the following statements: Fermi is willing

---

[2] Reference to Volume V, I

to go on record any time that the Clinton pile could have been completed in six months and Compton made a statement (in the Policy Meeting February 16, 1944) that it would have been possible to convert the Clinton pile to water cooling within four months. These two statements, put together, leave an upper limit of ten months for the construction of a water cooled pile (supporting our estimate) and I know for sure that it is practically impossible to design a water cooled plant with a power output below 50,000 kW. A more likely output is about 150,000 kW even if one does not actually strive for high power. Hence the plant could have operated for at least two months by January 1944. Unfortunately we are already more than 8 months behind schedule with respect to the third date.

Comments regarding the slowness of purchases in the past were made at the same meeting (January 18, 1942) by Dana Mitchell and Compton asked him and Szilard to put the matter in the form of a report to him. It was understood at that time that Murphree, the vice-president of the Standard Oil Company of New Jersey, was in charge of procuring the materials and it was desired to have the arrangements with Murphree clarified. Szilard also asked whether it would not be desirable to concentrate the work on the chain reaction in one place and whether it would not be advisable to choose the site for a plant with 100 watt capacity. A free for all discussion ensued on these points. A common feature of all suggestions was a desire to complete the project ahead of schedule.

I am describing the conference of January 18 because I have Breit's minutes of it available. There were, however, several conferences of a similar nature, all very enthusiastically attended, and most of the discussion turning around the same points. Some of the meetings were followed by a conference of a smaller group in which questions of policy were discussed more in detail. I remember rather clearly the conference which followed the above meeting. Compton again brought up the matter of procurement of materials and reiterated his desire that Szilard do something about it. On the whole, Compton leaned heavily on Szilard in all matters during the first few months. This was quite natural because he was most foresighted among all of us and least lost in details of technical problems. This smaller conference which followed the general meeting and which was attended by about ten people was followed by an additional meeting in Compton's hotel room. I do not know what he discussed there with his closest associates, but I presume it had to do with the best assignment of responsibilities.

The question of concentrating the work at one location soon came strongly into the foreground. Szilard pushed for it most vigorously. I am afraid he expected to assume a sort of supervisory capacity over everything. New York, Chicago, Berkeley, and even Princeton were discussed. Pegram was quite willing to put two floors of Pupin Laboratories at Compton's disposal over which Compton would have had full authority. This plan was strengthened by Doan's finding a site across the river in New Jersey where a pilot plant could have been built. The discussion was rather agitated at times. Finally Compton caught a heavy cold and decided during his illness that the best place to work would be in Chicago. Szilard was very strongly opposed to this as he felt that Compton should give a good example in making a sacrifice and leave his accustomed surroundings. In addition, both he and I were afraid that Compton would not be able to devote full attention to the project if he were at the University where he would be called upon so frequently for advice and help.

Not everybody was very enthusiastic for the concentration. I was rather swept away by Szilard's enthusiasm but secretly I agreed pretty much with Fermi who thought that the atmosphere for quiet and thorough work would be more favorable if the work were not centralized to such a large extent. He would have preferred to stay in New York and I felt myself it would be unwise to abandon the work in Princeton prematurely. Szilard, who was the main proponent of centralization, disliked the idea of centralization in Chicago. Thus nobody was in favor of the decision taken, excepting probably Compton's personal surrounding, that is, Hilberry and Doan.

However, the disappointment about the location of the central laboratory soon died away. We felt that we had a leader whom we had to follow and if his decisions were not always to our liking we must abide by them anyway. Compton took a vigorous leadership and the thoroughness of his understanding of many of the intricate technical points connected with the chain reaction gave a new life to the whole project. This resurgence of enthusiasm was accelerated by a set of meetings which he had organized in Chicago. Problems connected with the chain reaction pile were discussed at this meeting such as engineering, cooling, control, etc. Szilard and Fermi were the principal contributors and I remember vividly the pleasure which I derived from the discussions. Szilard spoke about Bi cooling, intermittent water cooling and problems connected with both systems. Fermi proposed a He cooling system in which, in some mysterious way (without pumps), the pile heat would give the power to circulate the gas. Van Vleck was also present and he and I undertook to investigate the possibility of water cooling to which I took a liking in the early days. Stability of the pile, the effect of Szilard's differential cooling (hot graphite, cold U) and the control of the chain reaction were other subjects discussed. Few discussions ever gave me more pleasure than did these and all participants greatly increased their familiarity with the problems to be solved.

The work itself continued in the meantime essentially at three locations: at Chicago where Allison's group was considerably strengthened, at Columbia where Fermi continued with his (exponential) experiments, at Princeton where we investigated the resonance absorption. It was planned that all these investigations be completed as speedily as possible so that the whole group then should be able to move to Chicago. In addition, Spedding was brought into the group to head a newly formed chemistry section.

The whole spirit of the enterprise was adventurous in these days. Every person who participated in the work was a colorful personality and it was a pleasure even to see them all together. There was a somewhat older friend of Compton's, a chemist, McCoy, who had been in retirement for some time and joined us, as I understand it, without salary. He knew more about the chemistry of uranium than anybody we could lay our hands upon and explained the relation of the different oxides with gusto. Wheeler and Breit both came to Chicago among the first ones and promptly got into heavy disagreements. Allison was phlegmatic as ever but always had good advice ready. Doan amused us always when he explained the excellence of the Chicago Laboratory where the doors had doorknobs so that they could be opened from both sides. V. Wilson and H. Jones were pleasant and helpful, whatever one asked them. On the whole, one would hesitate little to call the early months of 1942 the golden time of the project.

The projected scale of operations was not yet a very large one. We considered the enterprise as an experimental one which had a considerable chance to shorten the war and bring about victory. However, we thought that our work would not be permitted to interfere with the production of the ordinary weapons of warfare and we tried to make only demands which were compatible with this idea. I can quote as an example that we were perturbed when we found that about 4,000 tons of steel would be needed eventually for construction and shielding of a pile. We realized that this is enough steel for a destroyer and there was considerable doubt whether the Navy would give up a destroyer to make a uranium pile possible. We did not agree with this attitude of the authorities but we did not much want to fight it. The requirements were discussed quite frequently in meetings and as late as June 11th it was estimated that the site would have an area of 1,000 to 1,500 acres and employ 800 people. Of this number, 160 were to be executives or professional men, the rest workers.

I paid about one visit a month to Chicago which was spent mostly by straightening out personal difficulties. In Princeton the work went on with exemplary enthusiasm and I believe very successfully. The importance of it was realized to such an extent that Doan and Jones were sent to help us for a couple of weeks. However, by the middle of April I began to feel that I could be more helpful if I were to move to Chicago and since the work at Princeton appeared to come to a conclusion we moved up toward the end of the month. My wife and I expected to stay for three or four months.

Unfortunately the procurement of materials still lagged badly behind. We understood from Compton that Bush assigned the responsibility for the procurement of materials to Murphree. It did not appear, however, that Murphree considered this as an important responsibility and the procurement of materials made very little progress. What little progress was made was made by infraction of the directive. Thus, for instance, Creutz virtually forced the project to give an exploratory order to the Westinghouse plant at Bloomfield, New Jersey, where they had some experience in the manufacturing of metal by a combined photochemical-electrolytic method. This source gave the first pure metal. Similarly attempts were made, mostly by Hilberry, to procure graphite and these met with considerable success. We were not aware of any successful effort, or in fact of any effort, made by the Standard Oil Company. Creutz was much perturbed by this situation and decided to do something about it. He called up Bush for an appointment and we saw him after a meeting with Briggs on March 13, 1942. The conversation with Briggs was mildly reassuring and he said that procurement of graphite appeared to be satisfactory and the purity of the material good. He was somewhat less assuring on uranium and seemed to be less familiar with the situation in that respect. On the whole we had the impression that not much was being done toward this very urgent goal. In particular, he told us that no change in policy is expected. We therefore went to see Bush and tried to explain to him that procurement of the materials was not a job that can be handled as a side-line and that a thorough understanding of the problems and a full scale effort is necessary. We explained to him the unusual requirements concerning purity and told him that we did not see how Murphree could be familiar with these. Bush reacted very vigorously to our suggestions, unfortunately in a negative manner. He told us that he had arranged things in a certain way and that Murphree was and would be responsible for the procurement. He asked whether we had explained to Murphree the purity requirements and when we said no, he told us to attend to this. This was, of course, contrary to the organization which he himself had set up because we had no authority to report to Murphree directly.

In spite of Bush's negative reaction, both Creutz and I were quite pleased with the meeting. We found a man with a vigorous mind, giving clear statements instead of mild reassurance to which we became so accustomed. It must be admitted, however, that our visit had no results and the procurement of the materials remained in an unsatisfactory stage until it was finally handed over to the Laboratory early in June. Soon after that both graphite and uranium began to flow in in substantial amounts.

Compton's project was given a definite organization as early as February 25, 1942. There was a planning board consisting of Compton, Allison, Breit, Doan, Fermi, Hilberry, Spedding, Szilard, Thiele, Wheeler and myself. The work was divided into three main sections: the Chicago Laboratory was placed under the directorship of Doan; Szilard and Thiele were assigned the task of starting an engineering development group; and Hilberry ("not a yes man") was made responsible for the extra laboratory activities. The scientific personnel of the laboratory was organized into seven groups under the group leaders Allison, Fermi, Whitaker, Manley, V. Wilson, Spedding and me. Allison and Fermi, the former in Chicago and often represented in this work by Mitchell, the latter first at Columbia, were doing exponential experiments. Manley was in charge of the D-D source; Whitaker had a physics group which soon took over the exponential experiments which were conducted under the West Stands. He replaced Mitchell who took over the cyclotron group. V. Wilson was responsible for the building of instruments such as counters which were badly needed. Spedding had to organize a chemistry group but maintained at the same time his group at Ames. In my absence Wheeler was in charge of the theoretical group in Chicago. Stearns was personnel director, the procurement was in the hands of Byers. There was general dissatisfaction with the procurement. H. Jones had the supervision of the shop and he worked in close collaboration with Wilson.

The Chicago Laboratory occupied the third and fourth floors of Eckhart, the cyclotron building and the West Stands. In addition, part of the basement and the first floor of Eckhart were occupied by Laboratories and the D-D source. The office space was sufficient but the laboratories were quite crowded and the shops were entirely inadequate. The shop was located in the Ryerson annex. A good deal of space was available in the West Stands but the facilities needed in a laboratory were insufficient. They were improved as time went on. The chemists occupied half of the fourth floor of Jones and about three rooms on the first floor of Kent. These were taken over when the chemistry groups were organized early in the Spring 1942. There was a lot of planning done to improve the laboratories but little of it was carried out. It was not expected in these days that the project would stay in Chicago for any length of time and all arrangements were made on a temporary basis.

The laboratories outside of Chicago over which Hilberry had general supervision were those at Columbia, Princeton and Berkeley; Seaborg was in charge at the last place.

A proposal to start technological work was made in the very early days by Szilard who clearly foresaw the technical difficulties of putting together a large pile and of manufacturing the materials in suitable shapes and form. He made a request as early as February 23, 1942 to bring in Foote to help with the technological problems. It is interesting to compare his foresightedness in this matter with the disregard for these problems which prevailed in most other quarters. On June 22nd I had to write a memorandum enumerating technological problems to convince Fermi of the need of setting up a group for this purpose. I mention this memorandum because it contains the first written reference to the "crumbling" of graphite which was later often called the Wigner disease. The problems which Szilard had in mind included fabrication of the uranium in suitable shapes, coating of the uranium to prevent corrosion, manufacturing of pumps and shielding material, developing methods for manufacturing beryllium, and similar objects. Unfortunately, although he clearly saw the need for technological research at a very early date, he did not succeed in organizing much work in this direction.

The research made quite rapid progress during the first few months of 1942. Several exponential piles were put up and tested for their multiplication constant. On the whole the calculations of the multiplication constant which were carried out by Plass and me were quite well confirmed except that we underestimated the importance of the fast effect and thus somewhat underestimated the optimal size of the uranium lumps. The first lattice, the multiplication constant of which was quite close to 1 (.995), was measured as early as May 1942 and one with a multiplication constant slightly larger than 1 was obtained in July. Both these experiments employed the lattice which we considered optimal. Very important for this success was the tremendous improvement in the quality of the oxide. Compton and Hilberry deserve most of the credit for this and the Mallinckrodt Company which undertook the purification. The rest of the work of the Chicago Laboratory was less obviously connected with the ultimate goal. It was concerned with very important but more general scientific problems. The cross sections of various elements were measured including those of boron, bismuth, lead, beryllium. The boron was measured by Manley, the others by Fermi's and Allison's groups. The diffusion length in graphite was remeasured by both Allison and Fermi and that in water was measured by Manley. The number and periods of the delayed neutrons were measured by

Snell in Mitchell's group and the amount of γ radiation from fission products by Borst. A preliminary measurement of the number of neutrons emitted per slow neutron absorbed in uranium was made by Fermi. In addition, considerable work was expended to get better equipment and machinery. Most of the chemistry work on the products was carried out in these days by Seaborg's group in California and the most important properties of Np and Pu were ascertained before the end of March.

The theoretical group was built up rather slowly. My original plan was that I remain in charge of the calculations of the multiplication constant and that Wheeler familiarize himself with the problems of heat transfer. He agreed to that at first but actually did not give any attention to the problem. This annoyed me somewhat because I hoped to rely on him in this respect. However, I should have understood that a man of his capabilities would always work on whatever interested him in the minute. He did much useful and important work on other subjects. The group in Chicago consisted on March 31, 1942 of Miss Gish, Messrs. Christy, Weinberg, Friedman, Ibser, Wheeler, Young, and Mrs. Monk. Christy and Gish helped Fermi in evaluating the exponential experiments. Friedman made numerical tables for Breit on the distribution of thermal neutrons in graphite originating from a fast neutron source. He also was active in organizing a library. Ibser made calculations for Wheeler on the effect of the spontaneous fission in the pile. Mrs. Monk made comparisons between the multiplication constants in cylindrical and spherical arrangements. This was also a suggestion of Wheeler, who was a very strong promoter of the cylindrical geometry which became so important later. Young did some work on the cooling of the pile. Weinberg attempted an interpretation of experiments conducted at Allison's request on the (n,2n) reaction in beryllium. Plass and I in Princeton were engaged in calculating multiplication constants and optimal sizes for uranium oxide and metal lattices with graphite as moderator. In addition, I took an active interest in the experiments of Creutz, Snyder and Jupnik on resonance absorption.

Breit did not belong to the theoretical group but was in charge of the dissemination of information and the maintaining of secrecy. Unfortunately, he did not get along very well with other members of the Laboratory, particularly with Wheeler. As a result he left the project in May 1942. Even before, his work was practically divorced from that of the rest of the laboratory and concentrated around the problems of the use of the final product. Little headway was made and could be made on this line in those days.

Period of Engineering Under Mr. Compton's Leadership

In about June 1942 everybody became convinced that a chain reaction in a graphite moderated pile was feasible with ordinary uranium. As a result emphasis began to shift from problems of nuclear physics to problems of engineering and arrangement. The work on nuclear physics continued to go on but the principal (and often emotional) interest was shifted to questions of cooling and general engineering.

Simultaneously with the Laboratory obtaining the conviction that the chain reaction was possible, the authorities in Washington who were in charge of the uranium work also shifted from skepticism concerning the possibilities of our work to the belief that the problems of a chain reacting pile, as far as physics is concerned, were all solved. Early in 1942 we even heard that Compton's assignment in connection with the chain reaction in reality was to prove that the chain reaction was *not* feasible. From June to about September the conviction gained more and more ground that the chain reaction was not only possible but that the job of the physicists was essentially done and the leadership should be handed over to the engineers. There were early indications in the same direction (fall of 1941) which should have foreshadowed to us this attitude, but we did not attach much significance to those indications and the engineers were made rulers so suddenly that the whole project was very badly shaken and, in spite of repeated statements, it took a long time before we actually believed in the reality of the change in leadership. Had we believed what unfortunately became only too true, probably many of us would have left the project. However, we could not fathom that people could change their attitude so completely in such a short time and looked for explanations in other directions. This was probably very beneficial for the project because I hate to imagine what would have happened if the physicists had left alone what were called engineering problems and left them to the professional engineers.

Hand in hand with this development went a general expansion of the expenditures contemplated. Even in June 1942 amounts of steel of the order of a few hundred tons were considered almost unattainable and the total scale of expenditures moved in the order of a dozen million dollars. By February 1943 nobody cared how much steel would be necessary and the scale of expenditures has been multiplied by more than ten. From the point of view of the physicists the period under review was the most agitated one and the conflicts and arguments often assumed a heated nature which I had not seen before. The physicists were in one way at a great disadvantage as compared with the engineers. Their whole training and experience was in objective reasonable arguments which they were educated to carry on only on an absolutely fair basis. I cannot imagine a physicist using an argument which he knows can easily be contradicted in the hope that his opponent is unfamiliar with the counter-argument. Engineers, on the other hand, are much closer to business life where you are trained to take advantage of the weakness of your opponent and where it is anything but unusual to use an argument which one knows not to stand up under very close scrutiny. The "lowest possible price" which a business man asks for his wares is not necessarily the lowest for which he is willing to sell.

I have often wondered about the reasons which made the authorities have so little confidence in the abilities of the physicists – assuming that that was the reason for placing the engineers in command. At first sight it seems remarkable: The physicists just achieved something which the authorities themselves considered to be impossible. Furthermore, many of the people who were in charge in Washington knew scientists much better than they knew engineers. I do not think that I know the correct explanation for their behavior but I believe that they had no confidence in us who were younger than they were for the same reason for which older people in prominent positions are often inclined to distrust younger people. They notice that there are subjects about which they know more than those to whom they "handed over the job" and this makes them distrustful although they should realize that it is impossible for anybody to know all that another person knows and that the material commonly known changes with changing interest of the generations. Thus the fact that a 20-year old man nowadays does not know much about the Hg spectrum does not indicate that his knowledge is superficial although it would have indicated that 20 years ago. It is difficult to remember this every time one should.

In addition even people in prominent positions come into contact with a set of younger people who represent the average and soon realize that these are on the whole not as able as they are themselves. They become convinced that the younger generation has no thoroughness and cannot be relied upon. They see, on the other hand, that the big industrial organizations can achieve a great deal and feel that their responsibilities would be smaller if they gave the command to people who are members of a big organization that can take the responsibility itself. What this point of view fails to take into account is that there is no organization as smoothly working as one human mind and that the work on the chain reaction was in these days and is probably now far from the stage where its different aspects can be separated and handed over to specialists, the work of whom can then be put together in a routine fashion. The latter situation prevails certainly in the automobile industry and the work of the big organizations bears fruit in that field. However, the work on chain reactions was and probably is now in the same stage in which the automobile industry was 40 years ago and in those days it was individual knowledge and initiative which carried weight. I offer the above only as a tentative explanantion for the behavior of the authorities and must admit that I am not sure that it is the real reason for the distrust of prominent people in the younger generation. A certain amount of overestimation of themselves may also play a role in it. The fact itself is, however, undeniable.

Materials

It was mentioned in the preceding section that although the procurement of materials was officially Mr. Murphree's responsibility, practically the only progress achieved in this field was by the efforts of the Laboratory which were made more or less in contravention of the directives. The situation was officially discussed with the physicists, as far as I know, on only one occasion when a meeting with Murphree was arranged to discuss needs of material. Five to six tons of oxide per month were mentioned in this meeting as a goal. The attitude of the meeting surprised me so that I wrote on the blackboard, during an intermission, in tremendous letters "The Germans have 2500 Tons". When Compton and Murphree saw this they came up to me and Murphree gave me more reassuring figures. He mentioned very large amounts which are in Congo, in Canada, etc., with the result that I became quite satisfied that the situation was reasonably good. Only a few weeks later did I learn that the figures to which Murphree referred concerned material which was yet under the ground, i.e., unmined material. The incident is quite characteristic of the disadvantage at which the scientists were in such discussions.

On May 7, it was expected that 40 tons of graphite would be available "very soon", 250 tons by Sept. 1, these quantities from Nat. Carbon Co. St. Mary's would furnish 7 tons "the same week and 100 tons by Sept. 1. Nat. Carbon would furnish 350 more tons by Nov. 1 and 350 more by Feb. 1st, 1943.

The U procurement stood as follows: 15 tons of relatively pure oxide were expected by May 15; 30–35 additional tons from then on every month. 8 tons of metal were scheduled to have arrived from Westinghouse by Sept. 1. These dates are taken from notes on a meeting and are illustrative of the situation, particularly the difficulties of metal procurement. One W plant uses more than 200 tons of metal. Nor did anybody consider the figures to be final, they changed almost from week to week.

When the Laboratory took over in June, it was expected that 80 tons of graphite would be delivered per week from September on and two tons of oxide per day from August on. It was furthermore hoped that 600 pounds of metal would be fabricated daily from an unspecified date on. These figures are already considerably higher than the figures envisaged up to that time and were to a large extent due to the efforts of Hilberry, Doan and probably some others.

Oxide and graphite came in at slow but reasonable rates. There remained, however, considerable difficulty in obtaining metal because no good method of manufacturing was known at that time and considerable research and development had to be done before substantial amounts of high purity metal could be manufactured. There were two manufacturers who made metal, the Westinghouse Company which was brought into the picture by Creutz, and the Metal Hydrides Company, contacted by Szilard in the early days. However, the Metal Hydrides metal was of a poor quality while the Westinghouse metal was very expensive. There was therefore a clear need for an improvement in the process and when the attention of the Laboratory was directed toward this goal a suitable process soon was found. Both Szilard and Spedding hit on the system of reducing the fluoride by calcium or magnesium though they favored different methods. Szilard proposed as early as July 25th to get the Brush Beryllium Company to try out his system and it proved to be quite successful. Even more successful, however, was Spedding who used the same process in a bomb and satisfactory metal began to flow in from Ames in October. More than a ton of metal was produced there by the end of that month although only 600 lbs. of it were received at Chicago. In addition to that, Chicago held three tons of the Westinghouse metal and an even higher amount of the Metal Hydrides metal. However the Ames production soon surpassed the production at Westinghouse and Metal Hydrides and was so much cheaper that it is the only process now in use. Other manufacturers also make metal now by the Ames process, which they copied rather faithfully. I understand that even now the metal made at

Ames is considerably less expensive than that of "commercial producers". This is a case indeed where university people not only could handle a job but beat all companies in a field which is certainly less remote from the ordinary activities of these companies than the building of a chain reacting pile.

The contrast of the success of the materials provision, after the Laboratory took over, with its earlier stagnation did not go by unnoticed in the Laboratory. There were many who were inclined to blame Mr. Murphree for the delays. I never agreed with this point of view and believe that there are few left now who still maintain it. It was the system which was to be blamed which did not realize that human beings are not machines and are inefficient if they are used as machines. Nobody in the Standard Oil Development Company was or could be really interested in the procurement of the materials for us and all they tried to do was to carry out instructions. There was nobody, however, who could have given definite instructions. The whole affair is a sad commentary only for those (and Mr. Murphree does not belong to that class) for whom "visionary" is a word of scorn and eagerness and enthusiasm are reasons for contempt. It is more sad than this that the authorities learned exactly nothing from this experience and continued to prefer machines to men whenever the occasion arose.

Story of the Engineering Council
and the Fight of the Various Cooling Systems

Some time in May an Engineering Division was created within the Laboratory with T. V. Moore in charge of it. Mr. Moore was a pleasant and jovial fellow with a good sense of humor and a very likeable personality. He used to work for the Humble Oil Company and I understand that he came to us on Mr. Murphree's recommendation. His principle interest before he came to the project was to investigate how much more oil one can extract by forced drillings from oil field within a short length of time than one could extract during the same time by regular operations. He created an engineering council, the members of which were himself, Allison, Compton, Doan, Fermi, Hilberry, Leverett, Szilard and Wheeler. Leverett was Moore's assistant and they had worked together on previous occasions. As the occasion arose other people were invited to the meetings, such as Spedding and V. Wilson. The topics of discussion were systems of cooling, methods of controlling the chain reaction, acquisition of a site, and similar subjects. Among the cooling systems, helium and bismuth cooling received particular attention and a method of operation proposed by Zinn in which uranium rods were pushed in and pulled out of the pile with their temperature increasing while in the pile and decreased again by some sort of cooling while they were outside the pile. However, the discussions soon settled down to the problems of helium cooling which was the only one that could be discussed without further experiments being made. A good deal of the time was devoted to the question whether the cooling system, that is, in the case of helium cooling, the helium pumps, should be operated by power purchased from the outside or whether the energy of the pile should be used at least for providing the power for its own cooling.

In connection with the He cooling there was a certain divergence of opinion between Moore and Szilard, the former favoring a series type of cooling, the latter a parallel system. In every type of gas cooling the amount of gas that has to be pumped through the pile assumes quite formidable volumes if the heat to be removed is to become appreciable. Moore's idea was to pump the helium through long channels extending all through the pile and let it pass on its way through all the lumps of uranium of one row. Szilard's idea was to subdivide the helium stream into many branches and let every branch flow through only one, two, or three lumps. Although there was considerable discussion on the point, everybody finally agreed that the parallel cooling is theoretically the more efficient one. However, the claim was made and never definitely refuted that the construction of the pile is simpler in case of series cooling. On the other hand, the series cooling required using the helium under very large pressure (about 10 atmospheres) so that a very large pressure tank became necessary for the pile. It is conceivable therefore that the parallel cooling, which is more efficient theoretically, may have been more practical also because one may have been able to get away with a smaller pressure. There were intermediate schemes proposed, as the one by Creutz and me in the early spring, or another one proposed by me during the summer, but these received very little attention. On the whole it was not easy to advise Moore as he went along on his own ideas and paid little attention to what other people thought. I vividly remember two conversations which I had with him. On one occasion in Wheeler's office I was presented with some calculations which looked very optimistic to me. I asked Moore whether the fact had been taken into account that the power output at the center of the pile is very much larger than on the outside. He appeared quite surprised to hear this and the figures which he quoted a few weeks later were much less optimistic. It is pretty evident that it must be awfully difficult to design a cooling system for a pile, the properties of which are not known to one as a matter of course. On the second occasion in Moore's own office I presented him the system which was considered before by Creutz and me and which represented intermediate half parallel, half series cooling, which I still think has considerable advantages[3]. No point at which Moore's system had any advantage over ours came to the fore while numerous points to the contrary became evident, but this apparently did not ruffle his belief in his own method. It is probable that he considered that by having the discussion with me (which was more or less forced on him) he had done his duty and that the purpose of the discussion was not to arrive at any conclusion but to pacify a young, quite talented person. All the above discussions were conducted on a perfectly pleasant and amiable basis and there was, as far as I am aware, no personal antagonism present.

---

[3] The Brookhaven reactor which is being built now will operate on this principle. (Note on April 27, 1948)

The anomaly of the Engineering Council was that the theoretical section had only Wheeler as its representative on it. Wheeler was the one among us who was least interested in heat transfer and engineering and, although he tried his best to help Moore in his design, it is my impression that the cooperation between them was rather superficial. I have a description of one of Moore's systems by Wheeler which makes it evident, by the errors contained in it, that he had not acquired a real familiarity with the ideas. The fact that I was omitted from the Engineering Council bothered me considerably. I had worked on the problems which came up for discussion as much as anyone else and since the possibility of the chain reaction was a well-established fact at this time as far as we were concerned, I did not see what else the theoretical group should do unless it was to cooperate on the engineering and design. This same point of view was expressed to me also by Fermi a few months later. Szilard was so much annoyed by this system that he did not attend the first meetings of the Council. Not only was I not present at the meetings of this Council, but the minutes of their meetings and therefore the content of their deliberations was not made accessible to me in the first few months, nor could I see any engineering reports. This made my situation rather anomalous because obviously we could not do useful work if we did not know in what direction the plans were moving. When the few months for which I originally contemplated coming to Chicago were up, I made a trip to Princeton and seriously considered returning there. When this transpired, Hilberry arranged within a few minutes that from then on I should receive the engineering reports and I was invited to the Engineering Council from July 21st on. He also agreed that an engineer should be added to my group. Unfortunately, because of various obstacles, in spite of repeated urging this became possible only by the middle of October when Mr. Ohlinger joined us.

My entering the engineering Council did not abolish the difficulties. A crude review of the possibilities of He cooling convinced both Szilard and me (in about April 1942) that this system had serious drawbacks and would lead at best to a very clumsy solution of the problems. For this reason, we looked for other cooling methods. Szilard favored cooling with liquid Bi which he proposed in his paper to the Physical Review as early as January 1941. My impression was that although the future may belong to Bi cooling, we could not wait until all the problems and difficulties inherent in such a radically new system could be solved. Saving a couple of months appeared to us in those days a matter of utmost significance. For this reason I first favored the system of "intermittent water cooling" in which the chain reaction is going on in the pile cooled by water. I did not remember it at the time when I proposed this, but soon learned that it was discussed in the early meetings in Chicago (January 1942) quite thoroughly by Szilard. Later, however, when I became convinced around the middle of May that the multiplication constant in a metal-graphite lattice may reach 1.10, I decided that ordinary water cooling holds the greatest promise for a rapid construction and early conclusion of our project.

Gale Young and I wrote a report on water cooling in the middle of June 1942, which gave a favorable outlook to this possibility. The laboratory was small enough for our ideas to go around in a short time and almost simultaneously with our writing Burton was able to report on some technical aspects of the water cooling system. The laboratory seemed to be with us right in the early stages and gave us its unstinted support ever since. Soon after this report came out, Dr. Compton asked me to prepare, with the help of my group, within two weeks a more detailed report on water cooling. Practically the whole theoretical group participated in the work on this report (197)[4] which came out early in July. Furthermore, about a week after Young's and my report on water cooling came out, I wrote a report (June 22, 1942) warning against the difficulties caused by the heavy radioactivity which the He would acquire in a pile. This was one of the points which went by unnoticed by the engineering group.

As a result of these activities, there was a definite difference of opinion between the engineers and our group by the time I was permitted to join the Engineering Council. They favored He cooling, we-water cooling. This difference of opinion never became quite resolved and caused more than one clash between Moore and myself.

About one week after our second report on water cooling came out, Compton arranged a meeting to discuss the advantages and drawbacks of the two systems. Moore, Szilard, Young, Creutz, and I were present at the beginning of the meeting. Szilard enumerated the possibilities: He, Bi, water, water-steam and fluorocarbons can be used as coolants. He gave a summary of the advantages and drawbacks of all but said that, for the present, one can compare only the He and water systems because the others contain too many unknown elements. Even as far as these are concerned, one can make a comparison only between Moore's plans for He cooling and those of the theoretical group for water cooling, not between the systems themselves. What he had in mind was doubtless that the parallel He cooling scheme should be further explored. After this, I gave a summary on the advantages and drawbacks of the two systems and Moore and I agreed on most points in abstracto. However, Moore made several objections as to details. He declared about several arrangements that they "cannot be done" – a statement which I considered then and consider now to be somewhat unfair in most cases. He expected that every connection which we wanted to install would leak. When his plans for the He cooled pile came out, they contained connections which were much more difficult to keep tight than ours were.

Compton came in for the second part of the meeting. I felt that he took, on the whole, our part. He helped me out at several occasions when I could not express myself clearly and became reasonably outspoken when Moore categorically declared that the first pile must be a He cooled one. Compton seemed to be inclined toward my point of view that work on both types should be continued and if both are promising, both should be built and that the first pile should be the one which would become ready first. Szilard and Young did not speak much although there was no doubt as to where their sympathies

---

[4] Reference to Volume V, I

were. I remember that I was very much pleased when Young gave the method of discharging the U in our system without omitting a single step. This meeting cheered us, of course, no end, and went far in abolishing the effects of Compton's July 1st letter which will be discussed later. As a practical result, it emphasized the importance of better engineering help for our group and this was promised to us soon after the meeting. I must admit that what I hoped most from the engineer was to have a man who could authoritatively contradict the other engineer when he said that "cannot be done". Mr. Ohlinger, who joined our group in October, certainly surpassed this expectation and his help on various problems in connection with the water cooled pile was of decisive value. In the meantime, however, we always felt unjustly bested when we were told that something "cannot be done".

One of the difficulties of this and many subsequent meetings was that the time was usually insufficient to give a detailed description of plans. This was particularly true if somebody as busy as Compton was present but was true often also in his absence. As a matter of fact, it was my impression during the above meeting that Compton knew fully that we had considered points in more detail than I presented them and that he realized that we did not try to make unjustifiable claims. This realization was less vivid in Moore's mind although we could give him more details than we gave Compton.

It was hoped that the type of cooling would be decided by July 20. As I said, the time schedules were very rigorous in these days and the wasting of a few weeks was considered a major calamity. Nevertheless, our ideas were taken sufficiently seriously so that the decision was postponed a couple of weeks, particularly because Moore was not yet ready with any report. Col. Nichols was expected in Chicago early in August and Compton told us that opportunity would be given to us, as well as to Moore, to present our thoughts to him on August 15. The reasonableness of this solution was doubtful since it was not obvious that Nichols would be able to understand the problems in the course of an afternoon sufficiently well to make a sound judgment. in such an involved, highly technical problem. Still, on the correctness of that decision hinged the success of our project and possibly much, much more.

In fact, we never presented our ideas to Nichols. I waited in vain in my office to be called in until finally Compton entered my room and told me that the decision had fallen for the He cooled pile. A brief summary of our plans as well as of our official reaction to this decision is given in my monthly report of August 15 (CP-235)[5].

As I pointed out there, we quite understood that the He cooled plant had to be chosen if one took it for granted that only one type of plant should be built. It was more certain at that time that, given enough time, the He cooled plant could be constructed than it was that a water cooled plant could. It did appear to us, on the other hand, that the water cooled system constituted the better bet in the sense that the ease of its construction and operation together with the possibility of a much faster construction more than compensated for

---

[5] Reference to Volume V, I

the small difference in certainty that it was at all possible. In the light of information acquired since, this estimate does not stand up any more, and it does appear that the water cooled system should be preferred to the helium cooled system in all respects if one is after the single purpose of producing 49. This, however, was and is now the only purpose for which the project works.

The decision mentioned above did not surprise us greatly. We realized that Compton was under heavy pressure from many sides and that it would be difficult for a man stronger than he was to stand up under those pressures. We were unquestionably very much disappointed as we felt that the decision was a wrong one but seeing the vacillation of the administration between complete submission to the engineers and a very small amount of self assertion toward them, we thought that this result had to be expected. Neither did we completely give up the idea of a water cooled pile, as just this vacillation taught us that things may turn again for the better and our work may yet become useful.

There is one more remark which I made to Compton when he told me of his and Col. Nichols' decision for which I am sorry now but which it would be unfair to leave unmentioned. When I told him that I more or less expected that this would be the decision I also told him that it would be unwise in my opinion to give all the responsibility for our project's future to Moore. I pointed out that Moore is a very pleasant and amiable person but his experience and past did not quite qualify him to be the chief engineer on a problem of such complexity as is now in front of him. Compton seemed to pay attention to my remark and in a later memorandum referred with remarkable bluntness to the question of the competence of the engineers.

For me personally this defeat was coupled in one way with a victory. I learned that I can rely on the loyalty of my group and on the benevolence of the whole Laboratory. During the two weeks which were given us to complete our report we did calculations and sketches to cover more than 30 pages of text and 16 figures. The report has six authors (Christy, Weinberg, Young, Plass, Wigner and Williamson) so that its mere elaborateness constitutes a considerable achievement.

About a month later (September 16, 1942) Moore submitted the main lines of his plans on a helium cooled plant in a meeting. He said that the plans are entirely preliminary and do not represent a final design in any sense of the word. "However, they are sufficiently final to permit a decision whether or not such a plant should be built." He may not quite have meant all this. The detailed report (277) came out September 25, 1942. The plant was to have a production of 100 gm/day (100,000 kW). It was to contain 55 tons of metal, 760 tons of graphite (the idea that the plant provide its own power was abandoned), and the power requirement for driving the pumps, etc., was 12,000 kW. The pile itself was to be in a pressure shell weighing 250 tons through which 39 m^3 He were to be pumped per second at 7 atmospheres. The velocity of the He in the center columns was to be about 120 m/sec, the maximum surface temperature of the metal about 525 °C. A committee consisting of Allison, Cooper, Fermi, Szilard and myself was appointed to study the ideas of this report and make a recommendation on that basis. None of us was entirely convinced that the plans constituted a workable arrangement and each of us submitted a long list of questions, objections, and points to be checked. I was assisted in the work on this report by Christy, Friedman, and Ohlinger and our remarks comprise the equivalent of 15 double-spaced pages. In addition to the separate reports which every member of the committee submitted, the committee as a whole gave a summary of the most significant points and should have also made a recommendation that the plant should be built. However, as I remember it, we avoided making a recommendation that any action should be taken. Whether the lack of success of Moore's report was due to inherent difficulties of the system, or whether it was due to his lack of experience and the shortness of time (four months) which was available for him to familiarize himself with the subject, I cannot tell. The problems were indeed quite formidable and when the du Pont Company made a start at the same problem it could not do a very much better job. However, the number of improvements which offered themselves was quite large and there is little question in my mind that many of them had been offered before by Moore's collaborators, Leverett and Wheeler. My report contained, among others, the following statements:

"The question of multiplication constant has been fully discussed in yesterday's meeting ... It does appear that for the chosen lattice constant the uranium units are too small so that there is a question whether the pile would operate even at ordinary temperatures ... "

"There is a multitude of stuffing boxes at dump rods and other places from the reactor to the outside. When discussing the water cooled plant we were led to believe that there is considerable difficulty in the safe operation of a great number of stuffing boxes."

Similar remarks were offered by the other members also but I do not have those available.

However, the ordering of the steel shell and of the compressors was authorized by Compton on October 6. Also, extensive experimentation with pumps went on under the North Stands – the house in which we lived, although more than a block away, often was sent into strong vibrations by them. The question of He versus water cooling thus appeared to have been settled and only our faith in our ideas and the realization of the many weaknesses of the He unit prevented us from dropping the whole matter. The order for the steel shell and the compressors had to be canceled some time in December.

The above report on the helium plant of Moore and Leverett was submitted on November 22nd and it is best now to return to the time when this plan was first discussed. The great difficulties of the helium cooled plant became evident even in a short discussion and gave new impetus to the work on water cooling. The matter came to a head when we were requested, in a meeting of the Technical Council on October 1, to make recommendations to Whitaker concerning the cooling of a 10,000 kW unit for X. Four plans were submitted on October 5. The time for submitting them was very short and it was impossible for me therefore to have extensive consultations with my group and I had to send my recommendation to Fermi under my own name. Of the four schemes,

Fermi has proposed two: one in which tubes crossed the graphite between the uranium lumps and water was circulated through these tubes. The heat floated from the metal to the graphite and from the graphite to the water. Because the heat had to cross such a long path the temperatures would have increased very much if the power output had been substantial. Fermi proposed therefore to decrease the power output to 3,000 kW. In his second scheme the pile would not contain any cooling arrangement but would have to be cooled from the outside. This would have permitted really rapid construction but the power output would have been reduced to 300 kW. Cooper suggested using the metal in the form of shot and cool it with a helium stream. It was assumed that the discharge of the metal would be facilitated very much by its being in the form of shot. The scheme which I proposed was the only one which permitted operation at 10,000 kW. As a matter of fact, it would have been very easy to double or triple this power output. The metal was in the form of rods which were surrounded by water annuli which in their turn were bounded on the outside by aluminium tubes. The aluminium tubes would go through the graphite structure. The flow of the water was under gravity.

There was a heated discussion between the proponents of the four schemes. Fermi doubted the practicability of my scheme although during the meeting he did not produce any arguments against it. Later on he told me that he thought the air could not be replaced in the pile by helium (which is desirable in order to increase the multiplication constant) because the concrete in which the whole system was encased would not be air tight. This objection can be easily met and I doubt that it was the only thing that was in Fermi's mind. He did not have the custom of reading memoranda carefully and the time was not sufficient in such meetings to discuss schemes fully. He had the impression therefore that many of the features of the proposition presented by me would be impracticable. Moore did not present many arguments but was very much against the water cooled scheme.

Looking back we know now that the scheme as I proposed it had one serious weakness and that was the difficulty of protecting the uranium against water. It is also evident, however, that the scheme as proposed by me would have worked perfectly satisfactorily for the purpose and at the power for which it was proposed. The principal weakness which was recognized at the time was the uncertainty in the multiplication constant but the proposition provided for this also by suggesting that if k turned out to be smaller than expected, the plant should be operated intermittently with diphenyl. The reason I proposed to build such a plant was not so much that I thought it is the only suitable one for the purpose given – the production of very small amounts of 49 for the chemists – but because I thought that it would prove the value and efficiency of the water cooling.

The discussion did not get anywhere in the meeting because it was too agitated. It was lucky in a way that Compton and General Groves entered. This interrupted the meeting and the discussion was postponed by two days. On October 7th the outside cooling scheme which would have yielded only 300 kW was adopted without discussion. I did not make any objections to this because I did not think that the recommendations would be put into effect anyway.

On October 7th I also received a memorandum from Leverett which commented on the water cooled plant proposed by me. It praised this very highly and in effect recommended that it should be tried out. Leverett thought that it would be possible to build a plant as described in my memorandum in seven to eight months if sufficiently high priorities were given. Leverett's memorandum and the attitude which he took toward us around this time cheered me very considerably. He in fact said that he would rather work with us directly than through an intermediary because it would enable him to work with those who knew the subject.

However, this was not to come true. Compton realized that Moore's predilection for the helium cooled plant might not be justified. He suggested therefore that all three types of plants be investigated and named Wheeler as the sponsor for the helium cooled plant, Szilard as sponsor for the bismuth cooled plant, and me as the sponsor for the water cooled plant. He said that all three of us should work in collaboration with Moore. Unfortunately Moore's interpretation of Compton's directive was very different from mine and Moore sent a directive to me in which he said (October 8th) that "As you know, we are starting at once on the design of a water cooled plant for a maximum capacity of 100,000 kW (?) or as close thereto as it appears practicable to design such a plant. I would appreciate it if you will have your group furnish the necessary information on the permissible geometry and configuration of the uranium and graphite pile, the range of permissible volumes and locations of the water ducts relative to the uranium and the amount of alloy other than uranium, for example, aluminium, magnesium and copper that can be tolerated..." "I would prefer that the coordination of the efforts of the theoretical and engineering groups be handled through Wheeler." The first sentence and the last one were in direct contradiction to Compton's directive. Compton did not specify that we should work under Moore but in collaboration with Moore and it certainly was not specified that we should take directives from him. Also, according to Compton, Wheeler should not have coordinated the work on water cooling but should have devoted his attention to the helium cooled plant. Moore began to realize the advantages of the water cooled system around this time and wanted to assume direction of it. The second sentence which I quoted will sound ridiculous to everyone who knows something about the problem. It is, of course, impossible to furnish information as requested by Moore not only in the short time that was available but even after ten years of intensive study. For these reasons Moore's letter annoyed me considerably and I never answered it. I gave all information that I had to Leverett although he was by that time scared back into line by Moore. However, my relations with Moore became quite strained and we avoided each other as much as possible.

The work on water cooling might have bogged down entirely had it not been for a fortunate accident. Just before the middle of October Fermi made an experiment with a metal graphite lattice and found a multiplication constant which was only $2\frac{1}{2} - 3$ % better than that of the oxide. He estimated thereupon that the maximum multiplication constant obtainable with a metal graphite lattice would be 1.06. Fermi telephoned to me this result on a Sunday morning, the same day on which a meeting between Moore, Leverett and my group was arranged to discuss the method of procedure for designing the water cooled pile. If the multiplication constant had been as low as Fermi indicated at the time it would have been impossible to use the scheme I proposed. I must admit that Fermi's communication came to me as a relief. I knew that if it were quoted to the engineers they would abandon the water cooling again. On the other hand, I did not abandon hope that Fermi's result was preliminary and that the multiplication constant might be raised considerably above 1.06. The prospect offered itself to us that we may be able to go ahead designing a plant undisturbed by the directives of the engineers.

These expectations were fulfilled to a considerable extent. When Fermi's result was quoted in the meeting the meeting disbanded without much discussion. On the other hand, I suggested to Fermi a few days later to increase the amount of metal in his pile so as to improve the multiplication constant. When he did this the multiplication constant in fact started to creep up and we know now that it could have reached a value of at least 1.08 with materials that we had at hand then. With the present higher purity materials it can be augmented to as much as 1.10. Unfortunately, matters did not go quite as smoothly as I hoped and we received numerous other directives to submit to the engineers and it would not be surprising to hear that the engineers were told also that they should not be quite so (argumentative – Editor's addition). There was a plan, furthermore (October 22, 1942), to bring W. K. Lewis as chief engineer and it would have been perfectly satisfactory for us to work under him. On the whole our position became much better in October, particularly when Mr. Ohlinger was added to our group. A directive was issued by Compton to Young and Ohlinger to design a water cooled pile which uses 200 tons of metal and can produce 500 grams of product per day. Boissevain and Leverett were expected to help with this work as much as they could do so without interfering with their work on helium cooling. Three months were given to complete a report on the assignment and the report was submitted on January 11, 1943. It serves now as the basis of the plants which are being built at W.

There were a few more meetings into the details of which I do not want to go because it must be evident what the general trend of discussions was. There was one meeting in which Compton announced that a reviewing committee would come to Chicago on November 26th and that it would include W. K. Lewis and Murphree. Several high du Pont officials also functioned on the committee. It is very worthwhile to read the minutes of the meeting (CS-356) in which Compton told us about the committee. He defended the arrangements made under which the committee consisted only of engineers and did not include any physicists. On the other hand, he suggested that a report be written about the desirability of good coordination and close collaboration between engineers and physicists. I wrote such a memorandum and have it enclosed here although I do not know that anybody except for myself ever read it. The main result of the meeting in which Compton announced the coming of that committee was the so-called

"Feasibility Report" which was eventually submitted to the du Pont Company. It consists of articles written by most members of the Laboratory who have played an important role in the developments of the past and is an important and significant document.

I cannot say much about the meeting of that committee because I was present only during the deliberations in the first morning. This meeting was, as many of the deliberations of important committees which came to review our work, quite formal. An opportunity was given to several of us including Moore to present our ideas within a limited time and I spoke for about five minutes. The discussion was quite perfunctory. It was amusing that Murphree, who is an engineer, pointed out that I was over-cautious in my appraisal of the dangers of the first type of water cooling and he entirely discounted the dangers of "boiler explosion" which was raised as a grave danger by our engineers. I do not want to go into the details of the second meeting of this committee because they properly belong in the chapter of collaboration with du Pont which I hope to write later.

When going through the events of this half-year, one may be surprised that any work could be done under the conditions which prevailed and with all the excitement around. It now seems to me, however, that the conditions might have been as suitable for work as they ever are. Once the multiplication constant was in doubt, the engineers did not really put their hearts into trying to direct our work. As a result we were left more or less to ourselves and could think over most questions reasonably well. Consequently we had developed for most problems solutions in which we had real confidence. The experimental people helped us in every possible way. In fact, it would have been impossible to arrive at solutions for many of the problems without the help of Burton, Howe, and, in particular, of Creutz. The reason that we were left alone to the extent to which we were is essentially due to the fact that we had a monopoly on the knowledge of the multiplication factor which we always trusted (and rightly so) to come up to our expectations. Few other people did. Had this not been the case we would have been interfered with a great deal more and I do not know whether it would have been possible to get any satisfactory result. The lack of confidence of the engineers which was present to some degree to the very end may be illustrated by the statement which Leverett made when he finished our report "I now know that I have no reason to regret that I placed my principal interest into the helium system." It must not be forgotten that the steel shells and the pumps for the Moore-Leverett He plant were on order all this time.

General Organization

In addition to the Engineering Council a Planning Board was organized of which Allison, Compton, Doan, Fermi, Hilberry, Moore, Spedding, Szilard, Wheeler and I were members. The purpose of this Planning Board was for Compton to inform us of the general organization and discuss scientific questions. We were told (June 6, 1942) that the leadership of the project had been reorganized. An Executive Committee consisting of Conant, Briggs, Urey, Lawrence, and Compton was created which was responsible to another committee of which Bush, Conant, Briggs, and General Styer were members. In the laboratory the chemistry division came under Allison's leadership. This became necessary because there was friction within the division, both Spedding and Seaborg desiring to exert leadership. Fermi was placed in charge of the work in physics, Moore in charge of engineering, Szilard and I were given the responsibility of surveying the "whole picture." This was an indefinite assignment which was not taken seriously either by those who gave it nor by us. Evidently I could not survey the whole picture if I was not informed of the plans made in the Engineering Council.

The responsibility of the laboratory was not entirely clearly defined. The chain reaction definitely belonged to our sphere of interest but the isotope separation remained Murphree's responsibility much longer. Toward the end of June it was decided that a contractor should be chosen who should be familiarized with the problems so that the building of the plant could go on without additional delays. It was planned originally that the contractor be chosen by Murphree and Compton. Compton consulted members of the Laboratory on these questions and there were two points of view presented: One was to get in touch with one of the larger engineering firms of the country and Stone and Webster's name was most often heard in this connection. The other point of view was that it would be better to choose a smaller contracting firm so that its whole organization would be devoted to the building of our pile. I was more in favor of the second method because I felt that it is important that the contractor shall be on the job heart and soul and feared that this would not be the case if our job were one of the many responsibilities of a contractor. While these consultations were going on we suddenly learned that Stone and Webster had been appointed as our contractor by the Army. Compton was taken by surprise not much less than we were but did not make any vigorous objection to this procedure as far as we know. It turned out later that the matter was of greater significance than we thought at the time, when the main apprehension this incident caused was the fear that Compton would not stand up for the rights and the interests of the Laboratory as it would be desirable. Stone and Webster sent a group of engineers to Chicago under Mr. Steinbach's leadership he became a member of the Engineering Council about simultaneously with me. Unfortunately the engineering ability represented in Mr. Steinbach's group did not go beyond routine matters and the laboratory derived very little inspiration from the contact with them. However, they built a few buildings and the Argonne Laboratory in a perfectly satisfactory manner.

Simultaneously with the above events a vigorous discussion went on within the Laboratory as to how the cooperation with the engineers should be organized. The physicists all favored the method of close and immediate collaboration and stronger representation of the physicists in the Engineering Council. The engineers, particularly Mr. Moore, on the other hand, favored a system of funneling the information through one of the physicists, Mr. Wheeler, to the engineers, who would then work on their ideas. Compton attempted to settle the question in a letter written on July 1, 1942 which outlined the plans for different piles to be built and then made some "further remarks". In these "further remarks" he handed over all authority to the engineers and the letter culminated in the statement that this arrangement is so suitable because under it "the research men will not be interfering with construction by requests for non-essential changes in initial plans." This was a slap in our faces and every physicist resented it. Even Wheeler's reaction, which he laid down in a memorandum, was quite negative. I replied essentially that it was good to know that I had no responsibility for anything. I also went to see Compton and told him that I did not believe in the success of the Laboratory if his arrangement was adhered to and that I believed that he had done a good deal toward destroying the so far excellent morale of his group. Compton was very pleasant as always and assured me of the importance of my work and also that he did not want to exclude the physicists from anything. The impression prevailed in the Laboratory that Compton must have been acting under some outside pressure and we hoped that the decisions made would be revoked soon. Nevertheless when I left a few days later for Princeton I was not at all sure whether I should not arrange for my return there. In Compton's absence I spoke a few words to Hilberry, who reassured me and, as mentioned before, arranged immediately that I receive the engineering reports. I asked at the same time for the inclusion of an engineer in my group and he agreed even to this. When I came back I was in fact introduced to an engineer who was interested in a position with the Laboratory. Also, on July 2nd there was a meeting of the Engineering Council about which I heard by accident, in which Compton protested against the idea favored by the engineers that the decision had been made for a helium cooled plant. All this served to alleviate considerably the effects of the letter of July 1st without ever formally withdrawing it.

Perhaps most characteristic of the general situation was a discussion in a meeting on July 9th a few sentences of which I will quote from Wheeler's minutes:

> E. Fermi: At the present stage of experiments we should not limit tests to helium alone.
> Wigner: Experiments are proceeding on water.
> Friction.

This represents the situation rather clearly. The engineers resented our work on water cooling but the general personnel of the laboratory wholeheartedly supported it. Fermi was on the whole in favor of helium cooling but he realized that the setup which had been established for the work on this type of cooling might not lead to satisfactory results. I hoped that our work on water cooling would finally be recognized and might come in very handy if the great difficulties with the helium cooling would become apparent. In spite of the letter of July 1st, nobody left the laboratory although I, for instance, certainly had officially no work to do that appeared to me useful. I hoped, however, that what I was in fact doing would assume significance later (which it did).

Simultaneously with the above events plans went on for the Argonne Laboratory. It was first contemplated that the first chain reacting unit would be erected either there or at the final location which was considered to be X at that time[6]. These plans had to be revised numerous times and when it finally became evident that Argonne would not be ready in 1942, it was decided to erect the pile in the West Stands. This was done and the chain reaction was established on December 2. This will be described more closely later in connection with the participation of the duPont Company. There was considerable confusion about who would be in charge of this laboratory and first Stearns and then Whitaker were put in charge (September 9, 1942), until finally Fermi was made director at a later date. Considerable complaint was voiced by the Laboratory on account of all these delays and it was felt that the Argonne Laboratory should have been ready at an earlier date.

The relation between the Laboratory and the physicists was not the same in these days as it used to be. The distance between the workers in the field and the director grew bigger and bigger as it perhaps had to in an expanding organization. Naturally the physicists did not like this and felt forsaken by their leader who surrounded himself with his own personal advisers and listened to the engineers much more than to the physicists if he went outside his personal group for advice. One of the most noticeable signs of this development was that the meetings of the Group Leaders were slowly abolished. In the early days of the project all relevant questions were discussed in this group and we all had the impression that our voice was heard and that we knew about the most important events. The meeting of the Group Leaders was first replaced by a Planning Board which was referred to at the beginnning of this section. However, this Planning Board held very few meetings and of course had no authority whatever. It was in its turn superseded by a Technical Council composed pretty much of the same people and also possessing the same authority. In order to re-establish the early situation, a sort of revolution took place which started during a visit which I made to Princeton. When I came back from this visit – I went there with a question in my mind whether to remain in Chicago – Wheeler came to my office and told me that the Group Leaders got together in the evening during my absence and discussed the general situation of the Laboratory. He said that he hoped that I would participate in the deliberations of the Group Leaders in the future. He mentioned that both Szilard and Creutz were very much interested in the matter. Stearns was chairman. Among the concrete propositions of this group, the most important one was that made by Creutz on September 28 to replace the present Technical Council by a Steering

---

[6] Oak Ridge, Tenn. (Note on April 27, 1948)

Committee which would be more closely informed of events and would have a larger influence on them than the Technical Council which was in operation at that time. This request was referred to Compton who seemed to agree with it and appointed Allison as the head of the Steering Committee. Unfortunately, he again did this without consulting anybody and, although everybody liked his choice, the method of procedure did not please everyone. This held in particular for Szilard, who felt that this was not quite the right way to begin a more parliamentary regime. I am afraid that on the other hand Compton resented Szilard's interference and that this might have been a partial reason for the events to be related below.

The Group Leaders met, on the whole, about three times and I participated only in the last two meetings. Wheeler was first enthusiastic for the matter but abandoned it, I believe, from the second meeting on. The meetings were held in the evenings so as not to interfere with the work during office hours. One official meeting was held after the three unofficial ones, but the more active participation of the younger people in the decisions was soon halted again.

Many of us were very much disappointed by these events. We began to feel that administrators have a mind that is very different from ours, that they know little tricks of how to handle people, and how to keep their power, much better than we liked it. And that they devoted a great part of their attention to these tricks rather than to making the work of the Laboratory more effective and to improve conditions in general. Of course, we continued to like Compton because we all realized that he tries to be fair and is entirely unselfish. We felt, however, that he might be too much under the influence of the higher-ups and that he, like other administrators, had learned to evaluate "what is possible under the circumstances" so well that he may temporarily forget to be interested in the question "what is necessary."

A new reorganization of the leadership of the uranium research took place on September 29th when Bush became personally director of the main committee in charge of the uranium work. Conant was his substitute and the committee also contained General Styer of the Army and Admiral Purnell of the Navy. They were assisted by an advisory committee consisting of Conant, Murphree, Briggs, Urey, Lawrence, and Compton. In the same meeting in which Compton told us about these changes – the significance of which we could of course not very well appreciate – he told us that a site called X had been chosen for the pilot plant and that probably a big concern such as du Pont would carry out the construction and operation. This was a distinct progress from the arrangement with Stone and Webster but foreshadowed all the difficulties under which we have had to labor ever since. I would like to quote that I mentioned in the meeting that we "hand over in this way an immensely valuable store of knowledge to a single private concern." The executive of the main committee was General Groves, who appeared at a meeting of the Planning Board on October 5th and by his strong plea for speed and action gained the sympathy of all of us.

Unfortunately, the relation between engineers and physicists had deteriorated by that time a great deal, and the engineers had the ear of the administration to a much larger extent than we had it. This led to an explosion early in November when Szilard walked into Fermi's office where Fermi and I were discussing some technical problem and told us that Compton had just asked him to leave the Laboratory within 48 hours. He would have remained on the pay-roll but would have had to move to New York. Fermi and I were quite taken aback but I remember that we did not say anything at first. There was a meeting in the afternoon and I saw Compton after it. I told him that I believed Szilard's leaving the laboratory would sound the death knell to the morale of all physicists and that I did not believe he knew what he did. I also told him that I knew that he personally liked Szilard and asked him whose influence prevailed when he made this move. My recollection is that he attributed it to Moore but told me that he did not believe anybody aside from myself would be much upset about Szilard's leaving because he had not heard anything so far. Our conversation was interrupted by a long distance telephone call to Compton and he was surprised to see me in his foreroom when he finished the call. I told him that I had not said all that I wanted to tell him but that I must ask him to dismiss me simultaneously with Szilard as I could not conceive of staying if Szilard were forced to leave. Compton was taken aback and suggested an arrangement under which Szilard would not be as active in taking decisions as he had been before but would remain useful. I remember that he likened Szilard to an eagle whose wings should not be clipped even if he is not given complete freedom. It became rather late by this time and we decided to have a few minutes more talk about it next morning.

In the meantime several other members of the Laboratory saw Compton. Zinn insisted on seeing him the same night and told him in rather strong words his disapproval of the decision. Creutz talked to him next morning and reported what most everybody felt in the Laboratory. Even Fermi was about as strong as he ever was and told me that he would see to it when the occasion arose that Moore's influence disappeared in the Laboratory. The only one who seemed to understand Compton's motives was Allison and he told Szilard so directly. Szilard always praised Allison for his directness in this connection. When I saw Compton next morning his mood was entirely changed. He proposed a new arrangement which was in agreement with Szilard's desires that Szilard should not participate in the meetings of the "important committees" any more but remain a general adviser of the laboratory. He also told me that I misunderstood him the day before when I thought that it was at Moore's request that he wanted to send Szilard away. We had a very friendly conversation in which we talked as much man to man as we ever did. I expressed again my concern about the dominating influence of the engineers and he asked me whether I thought it would be worth while to call a meeting of all physicists in which he would explain the reasons for the position of the engineers. The meeting took place the same afternoon but Szilard did not participate in it any more.

At the meeting Compton read a passage of the Bible in which Gideon sent back most of his warriors and beat the enemy with a small fraction of all his people. He then pointed out that not everybody who started on an enterprise can expect to go on with it to the end and that we could not expect cooperation from the engineers unless we gave the authority to them. When he finished his speech he invited comments to it.

There was no lack of comments. The only one who seemed to agree with Compton was the first speaker, Wheeler. About 20 to 30 people gave good and strong arguments against it. Teller pointed out that physicists have done engineering jobs of considerable magnitude, such as the cyclotrons, big telescopes, and Van de Graaff generators. Somebody pointed out how much greater the success of the Wisconsin van de Graaff generator was than of that at Round Hill and attributed this directly to the fact that the engineers tried to take over that job completely. It was pointed out that no matter how able the man is who funnels information from one group to another, a good part of the information will be lost on its way. It also was emphasized that we do not see that, under the present arrangement, the engineers do succeed in familiarizing themselves with the problems. I do not remember all the speakers but they included Fermi, V. Wilson, and Anderson. All of them spoke clearly, forcefully and to the point. Every speaker after Wheeler favored the idea of putting the engineers and physicists into the same room, mixing them as thoroughly as possible so that each group would learn the point of view of the other and a good understanding of problems and methods could be arrived at universally. It was pointed out that the chain reaction is not yet in the stage where it would be amenable to the methods of routine engineering and that what we have seen so far does not give cause to changing this view. The great successes and the smooth functioning of the Radar Laboratory were mentioned. This Laboratory was far from being dominated by engineers. I did not want to speak at this meeting and spoke only in reply to a direct question from Compton as to whether I agreed with the point of view that physicists and engineers should be mixed as thoroughly as possible. I replied that I indeed did and when he asked me what I thought of the question as a whole I answered that I had heard only one argument in favor of handing the authority to the engineers and that that argument was perfectly symmetric. Compton seemed to waver under the influence of our arguments and finally he himself drew a picture on the blackboard in which he showed that instead of putting the physicists into a room and the engineers into another, one could alternate in the same room desks of engineers with desks of physicists. I believe that all the physicists left the meeting with the impression that we succeeded in convincing Compton of the necessity of a close collaboration between the two groups. Unfortunately we were badly mistaken. Matters continued to drag on without any clear decision and the physicists operated as if they were to collaborate with some engineers in the future while the engineers continued to assume the job for themselves. This situation was checked towards the end of the year when the du Pont Company took over authority to build the chain reacting units for production.

The above summary of events hardly mentions the second assignment to Compton's laboratory, the problems of the final device. In fact, practically no work was done on this question after Breit left us. The attention and interest of all of us was focused toward establishing the chain reaction and we felt that it had little sense to worry about the use of a material which we did not know whether we could make. Just about the time when it became evident that the manufacturing of the explosive is possible (June 1942) Oppenheimer appeared one day with an idea of a defensive measure against atomic bombs. Although his idea did not stand up under scrutiny, and I doubt that he himself seriously thought it would (he wanted to fill the air around the object to be protected with neutrons which would predetonate the bomb), the discussions with him did serve to bring out one point. This point was that the work on the final device should be reorganized and handled by a group which has access not only to the results of the work on chain reaction but also to the work on isotope separation. This necessitated the creation of a new group which was placed under Oppenheimer's leadership and set up at Y [7]. It grew in the course of time and now contains more physicists than Compton's group.

At first there was a loose kind of collaboration between Oppenheimer's group and Compton's laboratory which, however, became constantly more tenuous until finally no member of the Laboratory was allowed to know about the group on the final device. A certain relief in this matter came towards the end of September when Oppenheimer's group was authorized to discuss their problems with Messrs. Allison, Fermi, Manley, Hilberry and me. This arrangement did not last long and the authorization was soon withdrawn, at least as far as I was concerned. I did not learn much from Oppenheimer's group while it lasted because they were just beginning their investigations.

Oppenheimer's group will not be referred to very much in the next few sections. It was a vigorous group and its personnel policy was always rather to employ one person too many than one too few. The Laboratory lost many of its best members to Oppenheimer's group in the year that followed and even more when the plans for W [8] approached completion. Among the first ones whom we lost was Manley, who was made director of the experimental research in Oppenheimer's organization. I understand, though, that he was soon superseded by others. Teller also left Chicago and Christy too not very much later. We regretted to see these people go, but every member of the Laboratory realized that Oppenheimer's group had a very difficult and critical function to fulfil and was entitled to all possible help.

The organization of the Chemistry Division underwent a couple of changes in the period under review. The friction between Seaborg and Spedding necessitated the appointment of a division director above both of them. Allison assumed this position in June and complained bitterly most of the time about the difficulties to keep peace in his division. He was replaced in October by Franck, who did a very good job in organizing the Chemistry Division. He established four sections in it under the leadership of Seaborg, Coryell, Burton, and Boyd. Spedding withdrew to Ames to a large extent and his group there had a subcontract. They did an excellent job as did also the whole Chemistry Division in spite of all the friction for which Franck thought that Seaborg was mainly responsible.

---

[7] Los Alamos, N. M. (Note on April 27, 1948)
[8] Richland, Wash. (Note on April 27, 1948)

Collaboration with the du Pont Company in 1943 on Matters of the W Piles

The most important activity of the Laboratory in 1943 was aimed at the design and eventual construction of piles to produce 49 in quantities of the order of 1 kg per day. This task was achieved eventually in collaboration with the du Pont Company. When the du Pont Company first came into the picture, late in 1942, there was some opposition in the minds of most members of the Laboratory against handing over to the Company the full responsibility for the chain reaction work. We thought that it was not only desirable but also necessary for the timely completion of our job that the Laboratory retain a strong voice in the taking of decisions. When it became evident that such would not be the case it was hoped that the Company would see other ways to use the knowledge and experience of those who have been, in the Laboratory, most active in the field. These expectations were doomed to be thwarted and all hopes for an effective collaboration were abandoned toward the end of 1943. The following pages try to tell the story of this development.

The first contact between the du Pont Company and the Laboratory took place in the first days of November 1942 when a Committee of du Pont engineers visited the Laboratory. They were sent by the Army, partly to help in a rational evaluation of the Laboratory's chances for success and the comparison of the outlooks of the different Projects, partly to interest the Company in the work of the Laboratory. The second purpose was successfully achieved and it could be announced on November 16 that the du Pont Company is ready to take over the design and construction of any chain reacting unit that may be built.

It may be interesting to quote the Company's evaluation of the Laboratory's chances for success. They estimated that the chances are about 1 % and they favored the He cooled system as the most promising. Their second choice would have been a heavy water moderated unit, the third a Bi cooled unit and the water cooled unit was the last. An almost perfect score, with the wrong sign!

Many a member of the Laboratory felt that it was an admission of weakness on our part to call in the du Pont Company to take over, and many of us felt that it was an unjustified admission of weakness. This was felt particularly by those who sensed that all authority would soon be assumed by the Company, and that this may secure a privileged position in the field of chain reaction which would be undesirable for several reasons. Many of us felt that the chain reaction and liberation of nuclear energy should not become the property of a single Company and I voiced this feeling in the meeting in which the announcement of the du Pont Company's taking over was made. Security reasons were usually brought in to answer such points and, in addition, that no other Company was ready to assume the task.

The future organization was further clarified at the next meeting when the proposed method of collaboration with the du Pont Company was presented. It was pointed out that the responsibility for the design (meaning the authority to make decisions) will rest with the du Pont Company which will make also the arrangements concerning the liaison with the Laboratory. It was mentioned that the Company favored the funneling of the information from the Laboratory to the Company through a Chief Engineer. We gained the impression that the Company will have all rights and the functions of the Laboratory will be reduced, at best, to those of a calculator. The "funneling" through the Chief Engineer also worried us considerably since we saw little reason for the introduction of a bottleneck into the path of communications.

Among other members of the Council, I voiced these and some other worries and Dr. Compton asked me to write down my objections to the proposed arrangement. This was done the next day (November 20) and when I presented my memorandum to Dr. Compton, I did not gain the impression that he personally disagreed with it. He read parts of the memo in my presence and said, at one occasion: "This seems so evident that I do not understand they do not see it." I myself still subscribe to all views expressed in that memorandum, in particular to the view that the total handing over of the job to a Company which is new in the field would cause a delay of at least six months. And more, if the information is "funneled" to it. Also, that no job should be handed over to anybody who sees only a 1% chance for accomplishing it. His chances *will be* 1%.

The meeting of the Technical Council at which Compton asked me to write the above memo was held on the 19th of November and its main purpose was the discussion of the Feasibility Report to be ready by November 26. Under Dr. Compton's leadership, we surveyed all solved and unsolved problems of the Project and attempted to obtain an estimate for our chances of solving them. Multiplying the estimates agreed upon, we obtained a figure of about .5 which would have shown that we have an about 50% chance for being successful. However, this estimate appeared too low to most participants in the discussion and was jacked up some. The results of the discussion were laid down in the following week in the Feasibility Report to which most members of the Laboratory in leading positions contributed. It was an important and impressive document which was presented to a du Pont Committee on November 26. This was the most important du Pont Committee to visit us and included also two outside members, Messrs. Murphree and W. K. Lewis. The committee left for California from Chicago but returned in time to witness the first establishment of the chain reaction on December 2. During its presence in Chicago the Committee had occasion to get acquainted with our work and to listen to the various propositions concerning the future big pile. They devoted a full morning to survey Moore's plans on the helium cooled unit and gave an afternoon to Wheeler, Young, and Szilard to present the advantages of the helium cooled unit, the water cooled unit, and the bismuth cooled unit, respectively. I was not present at the morning session but the afternoon session gave me the impression that the members of the Committee were highly competent and that they made a serious effort in the short time available to obtain a real understanding of our problems. However, I also felt that even the wisest person could obtain only a partial picture in the short time they had. Naturally, the presentation of Wheeler, Young, and Szilard could not be very complete but they did the best they could in the half hour at their disposal. I am afraid that Young was a little peeved on account of the brevity of the time and the presumption of the committee that it could within an afternoon arrive at a more satisfactory evaluation of our problems than we could ourselves during several months of work. As a consequence he leaned over backward and was unusually cautious in all his statements. Szilard did not have many of the details of the bismuth cooled plant available and he frankly admitted this. Under such conditions it is not perhaps very surprising that the opinion of the committee still gave first choice to the helium cooled plant and continued to consider the water cooling as the least attractive choice.

I was also very much impressed by the difficulty of the Committee's task. It looked to me as though their evaluation of the possibilities had to depend to such a large extent on what they heard that it was impossible for them to properly evaluate the varying amounts of optimism or pessimism with which the different propositions were presented. Neither was I convinced that they realized that we knew our job too. They checked our heat transfer calculations – which any child could make – right then and there. This amused me a good deal. Of course, it was the only calculation they could check.

As mentioned before, the return trip of the second du Pont committee co-incided with the first successful operation of the chain reacting unit. I was not in at the deliberations of the committee on their return trip but had, with several other members of the laboratory, luncheon together with members of the committee and a few other visitors who were around. Oppenheimer, Greenewalt and I sat at a common table and the discussion naturally shifted toward the possibilities of chain reaction and of the future of our work. Greenewalt declared that he and his Company were still far from believing that a chain reaction could be established and placed the probability for it a good deal below 50%. It annoyed me that he presumed to give an authoritative opinion on a question on which he could have only very perfunctory knowledge and to do this in the presence of people who have devoted many months' thinking to this question. I said that of course I did not know what the probability for establishing the chain reaction was but that I did know that it would be established in the very near future. Greenewalt was quite shocked as he was evidently not used to open contradiction and Oppenheimer, who was considerate and diplomatic, smoothed over the situation by saying that what I wanted to express was only that the probability cannot be defined for events which may happen only once.

Fermi's pile stood assembled in the West Stands by this time. It was not easy to gain admission to the West Stands for the afternoon's proposed demonstration because it was feared that the interest in the event might be too great and the whole place might be overcrowded if all members of the Laboratory were admitted to witness the occasion. Somehow I did not apply for admission in time and was on the verge of giving up the idea of going there when I accidentally met Compton and he offered to take me over. Greenewalt was walking with us. I had a paper bag in my hand which contained a bottle of Chianti – I had bought it in Princeton about a year before in anticipation of this occasion. By this time Greenewalt seemed more convinced that the chain reaction would be established and he discussed with Compton the probability whether the product of the chain reaction could be made into an explosive. Compton noted the difference between the thinking of the physicists and the people mostly concerned with engineering. We were worried lest technical difficulties would interfere with successful operation of a chain reacting plant (which is a very different proposition from the establishment of such a reaction) while the engineers were worried lest the number of neutrons per fission of 49 would be too small. Compton felt that the fact that everyone was sure of that part of the problem which he understood best was a good omen for the future of our work.

Fermi manipulated the control rods of the pile in the West Stands and there was an automatic recorder to register the pile activity. In addition to the control rod which was moved in and out there was another "safety" rod which could have been released to enter the pile on short notice if something untoward had happened. In addition, I understand, a bottle of cadmium sulphate solution was handy to be used if this control had also failed. Nothing of this sort happened and the pile behaved exactly as it was supposed to behave and after about half an hour's experimentation Fermi pulled out the moveable control rod sufficiently to make the pile critical. The neutron density started to increase and increased with a positive curvature which was the sign for me to hand over the bottle of Chianti to Fermi together with my congratulations.

Soon after that we left in a rather cheerful mood. The occasion marked the successful conclusion of the efforts of the Laboratory to prove the possibility of the chain reaction and although there was probably nobody in the Laboratory who expected anything different for the last six months, it was most satisfactory to have this ultimate confirmation. There were celebrations in commemoration of this date six months, one year, and two years after the event, with speeches and a fine dinner furnished by Compton.

The next two months were spent by the Laboratory essentially waiting for the du Pont Company's decision as to the type of pile they decided to build. In the meantime, a good deal of effort was spent to get the Laboratory acquainted with the members of the du Pont Company and the members of the du Pont Company with the members of the Laboratory. The organization of the Company was explained to us by Greenewalt himself and Compton surveyed Greenewalt's past accomplishments. He told us that it was Greenewalt who put the nylon plant into operation and that this was an extraordinary accomplishment. Greenewalt was very nice and modest about this, saying that he only continued Carothers' work and put it into practice. He gave a brief description of this discovery and the time it took to have the first plant started. He looked forward with considerable optimism to the completion of our work and adopted the attitude that "it has got to be done." This meeting went far in allaying our fears that a second-rate personnel would be placed in charge of the work, as the way in which Greenewalt spoke, and also what Compton told about him, made it amply clear that he is a person of unusual abilities. In addition to that, he had great charm which few of us could escape. I myself was greatly impressed by him and excused his somewhat unfortunate remark at the December 2nd dinner by saying that he probably did not realize how much time we devoted to the subject of chain reaction.

It might appear that our attitude toward the du Pont Company was not too friendly even at the start. It is undeniable that we feared right from the beginning that the setting of the tremendous organization of such a big company into motion would require considerable time and that this delay and the delay inherent in the transfer of any assignment would be quite considerable. In addition, the transfer was to be made to a group of people who were up to that time ignorant of the problem and who naturally could not be expected to show the same enthusiasm for it which we had. We must admit that we fought on one front against the transfer of the whole assignment to the du Pont Company. On the other hand, however, we worked hard on the second front to give the Company as much information and knowledge as possible and to enable it to go ahead if it should get the assignment. We considered the chances for this to be extremely large. We fought on the first front, although most of us considered the fight to be hopeless, if for no other reason than for the record's sake. But this fight was not permitted to interfere with our help to the du Pont Company in technical matters. I do not believe that we could have strived more wholeheartedly toward informing and advising the company in all technical matters even if we had considered them to be a godsend which would relieve all our worries and carry us to the goal on fast wings. This attitude was much facilitated by Greenewalt's personality: he was quick-witted and grasped easily even rather involved technical details. He soon acquired a sufficient knowledge of most technical problems to be able to assess the importance of the advantages and drawbacks of the different schemes and all of us always enjoyed his occasional visits to our offices. Unquestionably there were moments of irritation also, such as when, in later days, after almost every Policy Meeting Greenewalt summarized the thinking of their Company and expounded details of design and propositions of the Company which he had gone through many times months ago. Most of his propositions were such that we could have shown right then and there why they involved undesirable features for which they would have to be abandoned. Of course, in due course of time he and his collaborators found out about these reasons also and most of the early propositions which he made did not live longer than a couple of weeks. However, we were never given a chance to comment on his propositions, and ideas which we thought to be outright faulty were often admired pretty much as strokes of genius. I presume if the Company had to take over it was good for them to sharpen their wits and to learn by their own mistakes. We, however, felt as a relay runner would feel whose baton is taken over by his successor, looked at from each side, turned around in his hands, rather than carried on as swiftly as possible. Also, it is only too human to be irritated if somebody else is admired for what one knows to be a mistake and a mistake at that which one is quite familiar with. I even believe that most other people would have been more irritated than I was as my indifference to such matters is often a subject of jokes in the group with which I am working.

The du Pont Company started with a survey of the helium cooled plant and invited for this purpose Messrs. Cooper, Moore, Wheeler and Whitaker to Wilmington. Their conference lasted three days and they apparently surveyed the problems rather thoroughly. The purpose of this conference was to improve on Moore's design and to raise the power level from 100,000 to 250,000 kW. It is my impression that their conference was quite successful, although none of the ideas conceived in it remained accepted for more than a couple of weeks. I remember that Dr. Wheeler, with whom the Company personnel got well acquainted during this visit, praised the method of operation of the Company very highly when he returned to Chicago. He returned early in 1943 to Wilmington and worked for the Company since that time. Both he and Moore made propositions to increase the power output and the proposition of one of them was to replace the uranium cartridges of the earlier design by a set of washers places on top of each other so as to form a tube. The washers were to have radial grooves which would have provided communication between the inside and the outside of the tube. The helium would have entered the inside of the tube at the bottom, streamed through the grooves to the outside and left the pile on the top through an annular space which was left open between the tube and the graphite.

I liked this proposition a good deal and do not know why it was abandoned finally. It constituted a form of parallel cooling which was favored by Szilard right from the start. Anyway, the design which Greenewalt presented to us a few weeks later (February 4th) did not contain this arrangement and although it embodied some of the suggestions which the committee reviewing the Moore-Leverett design made, it did not constitute a radical departure from the latter.

In particular, the problem of charging and discharging the pile did not look very easy and also remained somewhat vague. The survey of the blueprints presented by Greenewalt depressed all of us and when I met Compton next morning, I told him what I so often said before, that it may be best if the fate of the nation is not staked on one single type of chain reacting plant but if several types are tried out. Compton averred disappointment with what we saw the day before and admitted that he had hoped for more. We parted in a somewhat gloomy spirit but both of us realized that we had to leave the Company alone to complete its survey. In spite of that, I brought up again the suggestion which I also had made several times before that we find another company to survey water cooling in the same way as the du Pont Company was surveying helium cooling.

Somewhat before this (January 11, 1943) we (Boissevain, Leverett, Ohlinger, Weinberg, Young and I) submitted our long report on water cooling. Soon thereafter rumors began to come through that the du Pont Company considered these plans very seriously. Hilberry estimated that they were about half and half divided on the matter. Greenewalt reviewed the advantages and disadvantages of the water and helium cooled systems on January 22, 1943, and we were all pleased to listen to his comparison. Of course, we all were familiar with all points that he brought up. We admired, however, the precision of his expression, the insight which he and his collaborators had acquired in the short time that they could devote to the matter. Even if we admit that they knew the essential features of the water cooled plant from earlier meetings and conversations, it was a real achievement to have not only read but also fairly digested our report which numbered 50 single spaced pages and contained over 30 figures. At this occasion, we thoroughly enjoyed his exposition without any sidetaste in our mouths. Greenewalt concluded that the matter is about 50-50: neither system has a definite edge over the other and both have difficulties but appear feasible in spite of them.

Our own feeling was very much the same. Naturally, we hoped that, eventually, our plans would turn out to be the better ones. But we never even considered recommending that the He cooling should be abandoned. On the contrary, we urged that both systems be tried out. Only in that case, we felt, would the success of the Project be assured and we did not want to gamble. We were strengthened in our belief of the advisability of this procedure by the fact that it was, evidently, very difficult to make a choice between the two plans. For this reason, and because the choice between the two systems was dragged out considerably, after consultation with Dr. Compton I renewed on February 9 in writing the proposition that the du Pont Company be asked to build a He cooled plant (they appeared to veer toward the system at that time) and another contractor be found to build a water cooled unit. Not much time had elapsed between the 22nd of January and this date, but we felt that there was no point in spending time in a comparison of the two cooling systems which are, evidently, so very difficult to compare. Rather, both should be *built*. We also hoped that a healthy competition would arise between the two contractors and both might be more willing to listen to advice than, we realized, the du Pont Company would be if it were left alone in the field. The time between the 11th of January and the 8th of February may appear to be short now but time counted much more in those days that it does now. It was one third of the time which we had been given to make the plans.

Soon after my memorandum was submitted to Dr. Compton on February 8, we heard through the "grapevine" that the water cooled system was adopted by the Company. The fact that we heard this through the "grapevine" rather than officially was another little source of irritation. We worked hard on the plans and nobody in an official position took the trouble of informing us that the plans were accepted. The reason may have been though that the Laboratory was reorganized around this time and that I was dropped from the group which attended the Policy Meetings. This again may have been the result of a partial misunderstanding and bothered me less than it might have. Both Szilard and I hoped that the Laboratory would soon be united with the Company (nobody hoped very seriously that any attention would be paid to my memo) or that if this did not happen, those who were most intimately connected with immediate plans would be taken over by the Company. I thought that the group participating in the preparation of the "water cooled report" certainly would be among those who would participate in the further detailing of the plans and I did not expect to stay long in Chicago anyway. As far as the decision to build a water cooled unit itself was concerned, it pleased us, of course, that the du Pont Company's appraisal of the different cooling systems coincided with our own when they had an opportunity to review them in a most thorough fashion. On the other hand, we continued to feel that the abandonment of the helium cooled system may have been premature. It is pardonable perhaps that we felt that the choice made was better than if they had been chosen the other way around.

Unfortunately, our hopes concerning the collaboration with the Company were doomed to be thwarted. The members of the Company showed a considerable reluctance to discuss details of their plans with us and although Greenewalt often came to my office for half an hour or so, and although he was always friendly and cordial on these occasions, we soon sensed that the Company wanted to carry out the job alone, with as little help from the Laboratory as possible. This belief was further strengthened by a conversation which I had with Dr. Compton. He called me to his office and asked me whether we could let Christy go to Wilmington to collaborate there with Wheeler whom they took over early in the year. I answered that we could not afford not to let Christy go, as the future of our work is now in the hands of the Company and we should strengthen them as much as possible. I said on the same occasion that we should send even Fermi to Wilmington and the implication was, of course, that the writers of the water cooled report should also go there. Compton looked at me and said, "They do not want Fermi." I understood then what the situation was and that there was very little chance for a really close collaboration between the Company and ourselves.

Naturally, all this bothered me considerably. I looked for the fault in myself and thought that perhaps a more diplomatic person who has not given his opinions in the past so explicitly would be a more acceptable partner for the Company. I also feared that the efficiency of the members of the group which worked with me would be diminished by the fact that I was their leader. After some hesitation I went to see Compton on about the 26th of February and asked to be relieved of my responsibilities.

Of course, the whole matter was a bitter disappointment to me. The reasons which I gave Compton for my resignation were, first that I was tired, second that the leader of the Theoretical Group must have the confidence of the Company with which the Laboratory works, and third that he may participate in the meetings in which the policy of the Laboratory is promulgated. (The Policy Meetings never did more than listen to the changes which the policy of the Laboratory underwent since the last meeting.) I said that the lack of cooperation between the Company and the Theoretical Group which made the plans for the now adopted water cooling was appalling, that we never were asked for advice or explanation on any point, and that all contacts were made by Greenewalt personally who, although he is obviously extraordinarily bright, naturally cannot concern himself with too many technical details. As a result, I said, his visits to us became more and more friendly social calls. I said that I did not know whether I was to be blamed for this situation but I did not want to be responsible for the fact that a vital and extremely urgent task would be carried out by people who first have to be taught the fundamentals of the art when there are people eager to do the job and who have considerable experience in it.

Compton assured me of his confidence (which I never really doubted) but did not promise to readmit me to the Policy Meetings. He was very cordial, however, and explained to me that nobody is as useful as he could be if he had the confidence of everyone else. He said that this applied to himself as well as to anyone else. He agreed emphatically with the statement that I was tired and asked me to go on a vacation for the full month of March. He hoped that I would find during the vacation that my place remains with the Laboratory and that I would return refreshed to my duties. I suggested that he offer my position to Van Vleck and offered to write to him. This he accepted but with the proviso that it should not bar my remaining. Christy was to substitute for me during my absence.

Soon after I started on my vacation matters did show some improvement. Weinberg and Young were invited to Wilmington for a week and the question of lattice constants was discussed with them. Weinberg and Young convinced the Company's representatives that the lattice, etc. dimensions given in our report constituted as good a choice as one could make. On their return, Weinberg and Young related their impressions which were both favorable and unfavorable.

Into the latter category belongs the fact that they were not permitted to see the engineers who did the actual designing, but conferred only with members of Greenewalt's division. Weinberg and Young had the impression that most of the work was done by rather second rate people who, in addition, knew only a very small segment of the work. Only a few top people, such as Greenewalt, were appraised of the whole picture. Weinberg always could find out matters in a day which other people would not hear about for a year and he said that the purpose of this arrangement was to make nobody in the lower ranks too independent or too indispensable. This was the unfavorable side. On the favorable side they found that the work progresses, John Wheeler conducts an excellent "progressive school for Nuclear Physics" for the benefit of the du Pont big shots, chiefly Greenewalt, that the number of people engaged in the work is ample and that they accepted our lattice dimensions. I did not like the brevity of the visit.

About a week later Hilberry called me up and told me that the Company invited me also for a visit. This pleased me greatly, and I hoped that, together with their invitation of Weinberg and Young, it would herald a new era in our relationship. I asked permission to take Mr. Ohlinger along for the visit as he began to feel somewhat left out of matters. We left Chicago on March 16.

The visit lasted $3\frac{1}{2}$ days and proved to be quite interesting. Greenewalt was absent most of the time, first on a visit to the Aluminum Company of America. He came back very happy that the canning of the slugs is "in the bag." During his absence, we had discussions with Messrs. Babcock, Wheeler and Worthington which were very pleasant. When Greenewalt returned, we had a little informal discussion on more or less general human matters which I enjoyed a good deal. As I later found out, this was not true of Ohlinger who resented Greenewalt's reference to the Chicago Laboratory as a screwball organization. Neither was it true of Greenewalt on whom Ohlinger tried out some of his magician's tricks (he is an excellent amateur magician). Greenewalt resented being fooled. Poor Ohlinger did it in self defense, everybody tried to kid him.

The problems that were discussed before Mr. Greenewalt returned were not of great importance as far as the pile is concerned. Discussion of the important points was relegated until Greenewalt's return when a meeting would be called in which the problem of the "wet shield" was to be thrashed out. The first day I saw Wheeler most of the time and he showed me the outline of his book and the chapters which were ready. Evidently, he did not participate in the planning a great deal although he was the only person in Wilmington who knew what a chain reaction is. Most of his time was taken up by book writing, the answering of specific questions and of conducting the above mentioned Nuclear Seminar for the Company bigshots.

The much heralded Conference took place next day and we were actually permitted to see the Engineers in charge. This was a departure from the accepted Company Policy and was made, probably, in deference to the importance of my person. I felt very proud. John Wheeler, who was a regular employee of the Company could not, of course, attend the meeting with the Engineers. The "wet shield" (about which more later) was not discussed but Greenewalt gave a talk on the dangers that the graphite layers in the pile may shift with respect to each other during operation. Nobody believed in this danger very much but nobody voiced his lack of belief in it, except Ohlinger. This was a grave mistake which Greenewalt never quite forgave him. More than a year later he referred to Ohlinger's lack of tact at that meeting. I kept still, fortunately, and a few weeks later an employee of the Company who was present at the meeting commended me on my discretion and also told me that he heard nothing about graphite shifting since that meeting.

I made my mistake the next day. Greenewalt was absent again and I thought that I might speak to the Engineer in charge without witnesses. Somebody called him down and I told him that I was worried that so little knowledge of the actual process was present in his group which did the actual designing. He was a very fine fellow and, of course, freely admitted that he himself had only very vague notions about the matter. He told me that he understood my worries and that it would make the start of the real designing very difficult. He was confident, however, that the work would proceed expeditiously once it got started. He proved to be essentially right and I instinctively had confidence in him. Unfortunately, when we got this far, Greenewalt appeared unexpectedly and I won't forget the glance that he threw at me when he saw me with an Engineer. The conversation was terminated in a matter of minutes and I fear that somebody got a long nose for having arranged it.

The rest of our visit was taken up by reading and evaluating the Report of the Shielding Committee which Friedman brought down to Wilmington. This Shielding Committee was organized during my furlough and made a new calculation on the thickness of the shields that were needed.

Socially, the visit was very pleasant and I had a couple of nice luncheons with the men in Greenewalt's section and one or two with Greenewalt himself. The last day I received a telegram to be back in Chicago in a couple of days to attend a Conference with Urey on P-9. This was the end of my "vacation". It would be wrong to believe that we were blind toward the accomplishments of the du Pont Company. As a matter of fact, I believe that any able group going over a report written in the course of three months could have suggested many useful improvements and the Company personnel were certainly both experienced and able. Among the technical improvements which they suggested and which were finally adopted was the elimination of the boiling disease, the method of dumping the uranium slugs into water rather than pushing them into coffins and, most important of all, the dry shield. More important than these technical improvements was the general attitude of the Company which made much greater requirements than the Laboratory would have dreamed of making. The first thing was that they more or less reprimanded the Army for not having had a more vigorous program for producing uranium metal and they pointed out very rightly that the work must not be permitted to be hampered by lack of the basic materials. They also bluntly declared that the location X, where the Army wanted to have the units built, was entirely unsatisfactory and they received a really excellent location known at present as W. The water at W was much more abundant and much colder than it would have been at X, and its quality was so much superior that it made many problems very much easier. As a matter of fact, we also tried to protest against the location X where the water was barely sufficient, warm, and of low purity.

However, our suggestions toward a better location were not even transmitted to anyone who could have transmitted them to the Army. One might even think that the Company went too far in its requirements and had the whole work laid out, if anything, on too large a scale. Making everything too perfect also consumes time.

While we heartily approved of many of the suggestions and improvements which the du Pont Company made we disagreed with others. Furthermore, we could not avoid feeling that it would be better to have the pile built, at least one of them, rather than to spend the time with improvements. We looked at the whole problem as a very simple one. A big pile of graphite had to be erected, but certainly Fermi's group had shown how to pile graphite. Holes had to be drilled into the graphite, but certainly Creutz had shown how to machine graphite. Aluminium tubes had to be put into these holes, but Creutz had had manufactured aluminium tubes of the type needed[9] (much to the surprise of our engineers who declared that this job "cannot be done"). The whole graphite block had to be surrounded by a thick shield but we did not feel that this was a major difficulty and we knew that we had given a solution for the transit of the aluminium tubes through the shield. Uranium slugs had to be made and

---

[9] He left Chicago for an about 10 day's visit with the Wolverine Co. in Detroit. When he came back, he had two long pieces of the "impossible" tubes and carried them around with a great deal of clatter in the corridors of Eckhart Hall. (Note on April 27, 1948)

jacketed before being put into the tubes and this was a real problem, but as far as we could see, the only real problem. It was sufficient, furthermore, to be ready with the solution of this problem by the time the pile was to be finished. Creutz's and his group's attempts toward a solution of this problem had very promising results.

As I said, the whole problem looked to us to be extremely simple and we were convinced that it could be completed in seven to eight months. We realized, on the other hand, that it could not be completed during this time if the so-called question-answer game was continued. This consisted in the du Pont Company thinking up some improvement on our design. They usually then had some misgivings as to certain points concerning this improvement. They then sent Mr. Greenewalt to tell the Laboratory about the improvement or communicated it to us in some other way and asked for a written reply concerning that point. According to us, such a procedure could not lead to rapid progress and even Fermi emphatically agreed with us. The point that they asked could very rarely be answered in a clear cut fashion and the answer, if it could be given, involved questions as to their proposition concerning other parts of the pile. A lengthy exchange of views resulted in the best case. It appeared to us that the whole formal exchange of letters and communications was an unnecessary game. As a matter of fact, most of the arrangements they proposed in the early days, and most of the blueprints which they presented, had to be eliminated for reasons very different from the points concerning which they had doubts and asked questions. For these reasons it was our continued desire to have a close cooperation between those parts of the Laboratory which were to be intimately connected with the "W work" as it came to be called, and I offered to Greenewalt repeatedly, first by implication and later in so many words, to move the group which made the plans for W to Wilmington. The answers which I received were always evasive and the suggestion was never followed.

My chance to make the suggestion so that it would be on record came after the P-9 meeting at the end of my vacation. In the evening after the meeting Greenewalt called up our home and offered to come over to talk to me. He was concerned about the P-9 work and whether and how his Company should participate. I said that Szilard would be with us earlier in the evening, but when he said that he would shut up like a clam in Szilard's presence, I cancelled the appointment with Szilard. He came and we had what I believe was an open and frank conversation concerning the difficulties of cooperation. The whole conversation was reproduced in an aide memoir which I sent to him on March 29th, and in which I naturally tried to lean over backwards and bring out his points at least as strongly as I brought my own. Greenewalt started out by saying that he felt such a strong animosity among "my people" toward his group that he would not permit collaboration on the P-9 pile even if they had sufficient personnel for this – which they did not. I told him that we really had no animosity, either against him or against his people, and that he must know that our personal relations were friendly and cordial. We did have objections, though, against the method of procedure which was adopted for designing the

W piles. It did not seem reasonable to us that a "Progressive School of Nuclear Physics" had to be established in Wilmington when there were people available who were already familiar with the job and eager to participate in it. I also said that we considered it much safer to have one unit erected in a hurry and to have the others built on the basis of the experience gained with the first pile. I said that we appreciated their suggestions for improvement very much but the investigation of the possibility of many of these improvements would waste valuable months. Greenewalt disagreed with me and said that when they built something, they want to be and will be 100% sure that it works without major changes. He also said that the job is so big that the construction of the pile itself would be only a very minor part in it. He quoted figures concerning the cost of the pile and the cost of the whole enterprise; the latter one was so high that he said that one simply could not spend this much money in the short length of time in which we thought the enterprise should be completed. This was, of course, tantamount to admitting that the enterprise is put up on too large a scale. In spite of the disagreement a clarification of the points of view of both of us resulted which I was glad of. We also had some technical discussion concerning the dry shield which was perhaps the most serious bone of contention between Laboratory and Company. I finally told Greenewalt that I hoped he was convinced that "my group" had no animosity toward the du Pont Company and that he would send some responsible people to work with us on the P-9 problems. He said that he would look into the question.

I told Compton about the conversation soon and also showed him the aide memoir in which I tried to reproduce it. I asked him whether he thought it would be worth while to send it on to Greenewalt for his records and he read it and telephoned to me that it might be a good idea.

The result of this letter was certainly unexpected. I received a note from Compton the 6th of April saying that he had a telephone call from Greenewalt asking that my aide memoir be withdrawn and destroyed. Compton identified himself with this request and said that he removed his copy of my aide memoir from his files and asked me to do the same. The reason given was that such descriptions of conversations make informal discussions difficult.

Of course I considered Greenewalt's procedure to be insulting. He could have asked me to withdraw my letter, but to ask my superior to tell me so is not a procedure that can be considered amicable. I did not understand either how Compton became a party to such a procedure but we all knew that he was overburdened with work and did not often appraise properly the effects of his actions on the emotions of those with whom he had to deal. Szilard advised me to write a letter that I did not know that the conversation with Greenewalt was off the record, but that I wanted to go on the record with the following statements, after which I should reproduce those parts of my aide memoir which referred to my own remarks. I did not follow this advice because, first, Fermi had a copy of my aide memoir which was not affected and second, because Compton's note was so phrased, perhaps intentionally, that it was consistent with it to take my copy of my aide memoir home and keep it there. This is what I did and, to clear my conscience, informed Compton that I greatly regretted it, but I considered my action correct and that I would not feel right had I not appraised my superiors of the content of such an important conversation. Fermi and he were my superiors at that time.

All these events should have convinced us that we should not expect too close an understanding with the leading members of the du Pont organization. There was too much difference between them and us, both as far as general outlook on life was concerned and also in our attitudes toward the work. A scientist's main interest is in the esteem of his friends and collaborators, in the abstract knowledge of the work well done and in the satisfaction of having understood something. Of course, we are not free from vanity, but it is the esteem of our friends and colleagues which we value most and which means the same thing for us as money and power mean for so many other people. After all, money and power do not contribute more to the comfort of the individual than those things we strive for and our desires are therefore not more unreasonable – only less common – than the usually accepted desires. If we wield any influence and power it is rarely based on formal position, but on the recognition of all concerned that this is the better way of collaboration. If it is necessary for us to point to our authority embodied in the organization chart to someone, we may just as well terminate the relation with that collaborator – he would not do us much good.

High standing members of the du Pont organization are much more normal beings. They are proud of their salaries, look down to some extent on their subordinates and treat them only too often in such a way that they "know their places". I should have known that from the way Greenewalt shut up Cooper a couple of times in meetings in front of everybody. On the whole, Cooper took it gracefully and I am not sure that he resented it as much as I would have. In spite of their attitude toward their inferiors, it is admirable how much loyalty and sincere devotion to the Company they command on the part of their employees. I am afraid in this respect they are much ahead of us. I presume that the du Pont officials must have felt quite virtuous when they treated us courteously and observed the rules of politeness rigorously. But they always did.

There is also a difference in the kind of work we allot to people. It is always my intention to distribute work in such a way that the man who does it not only accomplishes something but benefits from the work by increasing his knowledge and experience. This consideration does not enter the work assignments of industrial organizations to the same extent and the man is usually given an assignment which, by his past experience and education, he is merely able to carry out. As a result, our people consider their work part of their life and education, stay overtime, and take great pride in their accomplishments. Of course, it does happen that some people misuse their privileges, but I am referring to my colleagues who are academic workers not only formally but also in spirit. The work done by the workers in industrial laboratories is a duty. I know of very few cases when people stayed beyond office hours and in many cases they think that staying beyond office hours shows incompetency. This makes our whole life rather different from theirs. The life of a scientist is integrated into one unit. The employee of the du Pont Company really has two lives – one during the office hours and the other outside, during which he tries, and usually successfully tries, to forget completely the office work with its problems and difficulties. I do not mean to say that our method is the more efficient one. One often feels that the same person does more in an industrial job than in an academic one. However, the same man becomes a different person in an academic career than he would in industrial life since the stature of the academic people usually increases with their work. The stature of industrial people increases less often and as a result there are more important results obtained per unit person by academic people than by those in the industrial laboratories. Evidently all the above contains sweeping and therefore unjustified generalizations. It is my conviction, however, that its essence is sound and the difference between our integrated life and their divided life cannot escape even a casual observer.

The difference between members of the du Pont organization and ourselves was very great also as far as our attitude toward the work at hand was concerned.

Very few of us expected ever to participate in the design of another industrial installation after we had completed this job. This gave to our endeavors the character of uniqueness and we were willing to go far in exerting ourselves and in using uncustomary methods to do well the single job of designing a large scale installation that we were expected to design in our lives. For the Company personnel the designing of large scale installations was a routine matter and they wanted to carry it out in routine fashion. The directive of the first du Pont committee which visited us said in part that the purpose of the review shall be to ascertain whether du Pont methods are applicable to the problem at hand. They went at the job as they would have gone at any other production job, without undue haste, and with methods which would have enabled them to construct not three but a hundred plants. The plants had to be such they could be operated essentially by unskilled labor. This proved to be possible to a large extent but it took, of course, much more time than the building of plants would have taken for the operation of which people well acquainted with the problems would be provided.

The desire to have the plant designed by people who knew little about the basic process, for operation by people who were unskilled in the art, forced the Company to take risks which we never would have taken. Chief of these risks was to plan for three plants to be erected on exactly the same pattern, completed so nearly at the same time that the experience gathered with the operation of the first could in no way be utilized in the planning of the others. Our attempts to provide for all possible contingencies went, however, also in detail much further than theirs and, of course, we wanted the first pile to be built quickly, operated at rather high power, and the design of the others adjusted, if necessary, following the experience obtained with the first pile. Such a procedure would have been incompatible with the Company's methods: all details of their plans were embodied in so many blueprints that a request for even the smallest change upset them completely. A point in question is the He circulating system. This, as it now stands, is an extremely costly installation which, however, according to our judgment would be unable to perform the function for which it was built: the location of leaks in the aluminium tubes. When this was pointed out to the Company representatives and alternatives proposed, they admitted that their helium circulating system cannot be expected to perform its function of leak detection and that it could be much simpler and be able to locate leaks. However, they pointed out that it would be more costly to design and have blueprints made for a new helium system than to have this unnecessarily costly one installed.

Similar examples can be found by the dozen – the most amusing one of which dates from the time of the completion of the last unit. For this, as for every other unit, a deärator was planned (which, by the way, was so designed that it could have deärated water a couple of order of magnitudes more thoroughly than there was any conceivable purpose to deärate it). At any rate, by the time the third pile neared completion, it was known that the deärator would not be used at all. However, it was on order and was to be installed. The equipment was manufactured in Minneapolis and sent on to W in due course of time. Unfortunately the shipment was misdirected and "lost" for some time. It was finally found near Detroit and sent on its way toward W. It arrived a couple of months too late – and the operation of the plant was held up this time pending the arrival of the deärator – which was never to be used. The whole construction, as the design, had to go systematically, with no last minute changes. Their attitude toward the building of piles was the same as it would have been toward any other article that has to be mass produced. We considered the erection of piles as a matter of art in which not even we were well versed. We wanted them to be built and operated under constant expert supervision and were willing to participate in this since we felt that if we succeeded in obtaining a sizeable production at an early date, we have done as much for the war effort as we could expect to do.

These incidents go far to show the other point in which our attitude was very different from the Company's. This is the apprecation of speed. We felt, in the critical months during 1944 that the war was far from won, that the enemy probably realized that it could not win the war with weapons it already had and was concentrating on the production of new weapons. In this we proved to be right. We felt, furthermore, that the nuclear weapon was the most likely one on which it might concentrate. In this, of course, we were wrong. We felt that we would be in a very poor position if we had no answer to their new weapon when it was pulled out, and it is often asserted nowadays that we would have been indeed in a very poor position had the enemy been ready only one month earlier with its new weapon. Our attitude was, therefore, even during 1944, that our job is an extremely urgent one and that a month's delay might cause a catastrophe for the nation. It is now, in our security, difficult indeed to recall the danger that has gone by and to appreciate the extremely grave responsibility that we believed we carried.

This view of urgency was not shared by the Company. Their representatives never expected that the weapon they forged would be used in the war and the stereotyped answer to our query – what if we are too late – was "That would be a serious loss of prestige for the Company". This attitude of course, exasperated us and gave us a sense of frustration, particularly because it was tolerated by the higher-ups. We felt that we may be sailing toward a rock and that the Captain gave the helm, in spite of our repeated warnings, to a part of the crew which did not realize the danger.

If I were to summarize the difference in the attitudes of the Laboratory and the Company I would say: we were out on a romantic mission, to avert danger from the nation. They were out on business, to fabricate a weapon. This also explains that our attitudes are pretty much reversed now. We feel that now that the danger is averted one might almost as well stop work on the new weapon. Many of us are, at any rate, opposed to its rise. The Company people see no reason to change their attitude and probably see no possible reason for not using the weapon. A product should be put on the market when it is ready.

Collaboration with the du Pont Company in 1943 on Matters of the W Piles II.

I met Van Vleck in New York when returning from Wilmington to Chicago and I asked him to take over my job. I explained to him that the work that I had done in Chicago may have been all right from a technical point of view but from that point of view the work was practically finished. The work that remained was not purely technical and required a great deal more diplomacy that I possessed. I told him that I considered his personality and makeup to be very well suited for the position which I was supposed to fill and that it would be a relief to me if he took it over. He told me, however, that he had a laboratory of his own called Radio Research Laboratory which he did not want to give up for the sake of this position. I talked to him a long time but could not persuade him to replace me. As a result I had to tell Compton on March 31st that for the time being we had nobody to take over my assignment and that I did not know what to do. He said that he wanted to re-admit me to the Policy Meetings and hoped that I would reassume my duties. I agreed to that and after some delay received a letter, dated March 10th, asking me to attend the Policy Meetings of the Laboratory Council. I thanked him in a note, which as I come to think of it now, may not have been particularly tactful.

There was another occasion, soon after this one when my replacement was seriously considered. I learned about it in a roundabout way but very reliable. This time, the initiative was not mine and the negotiations were carried out without my being aware of them at the time. Uhlenbeck was approached to join the Project and the idea was that he would replace me after some time. Of course, Uhlenbeck was not told of this part of the plan when he was approached as even leadership of our Project was aware that he would not knowingly become a party to such a game. However, the plan fell through for reasons which are not yet known to me and it may have been just as well. Uhlenbeck, I am convinced, would have strengthened my point of view rather than weakened it.

I understand that the reason Smyth was asked to join the Project was a similar one. But even he supported me, rather than curbed me, once he became acquainted with the situation. More about that in the next section.

These intrigues, as well as my being dropped from the Policy Council in January, annoyed Szilard no end. He said, not entirely without justification, that the Project should have been grateful to me and the Theoretical Group for having provided them with a workable design. He said that without our work, the Project wold have been in a hole and the present opinion of the du Pont Company concerning the feasibility of a helium cooled unit fully supports this view. According to Szilard it was under these conditions particularly ungraceful to have tried to demote me. Szilard's view was not altogether shared by me, though, because I believed and still believe that my being dropped from the Policy Council was the result of misunderstanding.

From April to September, the main duty of the Theoretical Group was on the P-9[10] piles. We did not devote much attention to the W problems for many months, which may have been just as well because the blueprints which came through those days for being reviewed by the Laboratory did not constitute workable arrangements. This would have irritated us much more had W been our main concern. Even this way we shook our heads most vigorously after each blueprint session, the early ones of which were in the form of lantern slides with popular explanatory lectures. We wanted to have as little to do with these blueprints as possible because we did not believe that the Company paid any attention to what we said and because we were too much interested in the work on P-9 piles. This situation continued until about the first of September when blueprints began to come through which required serious review.

Meantime, of course, the irritation spread all over the Laboratory and we became a bunch of restless and dissatisfied people. The restlessness was, in fact, less prevalent in the Theoretical Section than in many other parts of the Laboratory because we "wrote off" W in those days and staked our hopes on the P-9 pile which we were trying to design. For most of the rest of the Laboratory, however, W remained the main objective and the inadequacies in the organization of the W work made them nervous. It was in order to keep this nervousness within bounds and to ban the feeling of frustration which was spreading over the Laboratory that Compton and Allison called a meeting for the evening of the 7th of July (1943).

Allison is not a great orator. However, he speaks simply and directly and makes a point in saying whatever he has to say in simple and unadorned sentences. He spoke after Compton and only demonstrated the fact that whatever can be said can be said also in simple words.

I sincerely like Compton in spite of the many disagreements of opinion which we had. He is a truly good-hearted person, who, I am convinced, often stood up for me and others without letting us know about it. In fact, I suspect that he often concealed favors which he did to others. He is also a person of extraordinarily quick and good judgment in technical matters even though

---

[10] P-9 ("product 9") was the code-word for heavy water (AMW)

he is not closely connected with them any more. However, not even his most ardent admirer can say that he is very good at judging the effect of his words and actions on his collaborators. This is what became manifest again at the occasion of his speech on July 7th. The speech was nicely phrased and very well delivered. It emphasized the importance of our work and painted an attractive picture of the usefulness of the training we were acquiring. It tended to reassure us in every respect. However, it entirely omitted any mention of the crucial question to which all the Laboratory was eager for an answer: Why this method of collaboration? How can one claim that this will lead fastest to the goal? It was the weakness of Compton's understanding of our mental makeup that he thought that by avoiding these issues he would erase them from the minds of his listeners.

The result of his speech was very different from what he expected. The next day Irving Lowen came to my office and told me that several of our colleagues tried to convince him before that there is something wrong with our "organization". He said that he always scoffed at such speech since he knew from past experience how easily unjustified dissatisfaction spreads in projects such as ours. He said, however, that Compton has convinced them that there *is* something wrong. In all his speech there was no mention of the urgency of our task. Lowen also said that he wanted to do something – he did not say what.

Lowen was working in the Theoretical Group for a couple of months when this happened. He was an exceptionally unselfish, straightforward and widely read person for whom I had much respect even though he was new on the Project and, naturally, did not have the same experience and knowledge which the old hands, such as Young or Weinberg, had. I never discussed organizational matters with him before this occasion and was a little surprised to hear that he had discussed it with others and that Compton's speech had such a profound effect on him. As for myself, I have heard so many similar speeches of Compton that I almost forgot this particular one by the next morning. At any rate, as I see it now, the reason Compton avoids all controversial issues is that he does not trust his audience to understand his arguments on them and sometimes also because he secretly agrees with his audience but is forced, in his actions, to a course in which he does not believe. He is, then, loath to defend this course because he does not believe in it and equally loath to attack it because he has to follow it.

Lowen was, however, full of fight and he left no stone unturned to get the ear of the highest authorities for his story. We must not forget that the question of our progress was of vital importance in the eyes of everyone who knew about the Project. Lowen succeeded in seeing some of the most influential people in Washington and to tell them his story. He even saw the President of the United States around the 31st of July.

The Project Leadership learned about this visit to the President practically as soon as it was made. It appears almost that they have had their "observers" in the President's office. There was tremendous excitement among our leaders and Compton even came to see me to find out what I knew about the matter. We had a somewhat painful discussion in which, I am afraid, I was not willing to take full responsibility for Lowen's actions even though I made it clear that I approved of Lowen's motives. Compton told me that there would have been no need to go over the heads of the Project Leaders, since, e.g., I could see Conant whenever I wanted and since a P-9 investigating committee would be in Chicago in September to listen to all grievances that we might want to bring up. Of course, it was strange that it took Conant $1\frac{1}{2}$ years to declare that we can see him. Both he and Bush evaded requests for interviews as much as possible both before and after this occurrence and if they did not know about the dissatisfaction in the Laboratory they must have buried their heads in sand.

The two points which Lowen brought out in his interviews were, first, that the work does not progress as fast as it could and that this may jeopardize the safety of the Nation. He said that the reason for the sluggishness of our progress was that the Army does not trust the scientists but has blind confidence in the industrial concerns with which it works – a confidence misplaced indeed since those industrial concerns knew, even a few months ago, as much about uranium as the average man on the street knows now, and they knew nothing about a fission chain reaction. Lowen's second point was that it is very dangerous from the standpoint of the continued safety of the nation to give all responsibility to one single industrial concern and to create this way a virtual monopoly. He also stressed the salutary effects of a healthy competition on the sincerity of the efforts toward speed.

These were, I presume, also the points which Lowen stressed when Conant saw him at the President's request. Conant countered Lowen's arguments in a rather strange fashion. He asked Lowen whether he knew that, in addition to the du Pont Co., several other industrial concerns are participating in the work. Lowen, of course, did not know this since, as a matter of fact, the du Pont Co. was at that time the only industrial concern which had any connection with the work on nuclear chain reaction. Conant was referring, however, to the work on isotope separation in which, indeed, several companies participated. Lowen's ignorance concerning this point was then used to discredit Lowen. Conant pointed out to the other people with whom Lowen had interviews that Lowen is a young man who, no doubt, means well but who, of course, is informed only of a very small segment of the whole enterprise. What Conant failed to emphasize was that Lowen's information was correct concerning the work on chain reaction though incorrect as far as other Uranium Projects are concerned. This way he apparently successfully counteracted Lowen's efforts.

In spite of this Lowen's interviews had a very good influence both on the Project and its Leadership. The Project was much heartened by the possibility of going above the head of the Project Leadership. The Project Leadership, on the other hand, obtained some food for thought by the difficulty with which it could refute Lowen's arguments in an open and aboveboard fashion. Lowen never became convinced that this was done and his opinion of the basic truthfulness of some of the people he came into contact with is not too high.

Lowen's visit to the White House was followed by a very strange incident. On August 4th, in the afternoon, during a discussion with Young and Weinberg, a document was delivered to me. It was signed by Compton and bore the date July 23rd. It essentially contained the accusation that I sabotage the collaboration with the du Pont Co. and that, as a result of this, the progress of the W plant is suffering. His letter was so full of statements concerning the du Pont Co.'s and our past actions, the inaccuracy of which could easily be ascertained from the records, that I was flabbergasted. It said, in essence that we were responsible for the failure of effective collaboration.

Evidently, Compton did not want to accuse me of anything that could be refuted as easily as most of what was said in his memo. I should have realized this even if I doubted, temporarily, his intentions. As it was I thought that he wanted me to resign in a rage and that it could then be represented to the P-9 investigating committee that I ran away from the investigation.

I was enraged enough to show Compton's letter to both Young and Weinberg the following day – the letter asked me to do this – and to write a reply (dated August 5th and delivered August 6th) in which I enumerated all the inaccuracies of his memorandum. My reply also mentioned the discrepancy between the date given on his memo and the date of delivery – the implication being that the date of the memo preceded Lowen's interview with the President, the delivery succeeded it. My letter ended with the sentence, " In view of the above, I cannot assume the responsibility for any lack of collaboration with du Pont in matters of the W pile".

Szilard shared my view concerning the purpose of Compton's letter. Young went even further and wrote a note to me which teemed with expressions like "bold lie" in reference to Compton's letter. All of us were baffled as we could not quite believe that Compton misjudged the situation to such an extent, nor that his memory was so poor concerning some salient points, as appeared from his letter. Of course, I filed away Young's note and never showed it to anybody, not even to Szilard.

Smyth saw the two memoranda about a week later. He took, of course, my part as far as the merit of the matter was concerned. There really was no question as to who – du Pont or we – refused the collaboration and who offered it. Smyth told me, however, that in spite of the explicit request in Compton's memo, I should not have shown it to anybody but should have seen Compton and explained to him that he had his facts wrong. I daresay he was right. However, Compton's memo would have upset the equanimity of a more even tempered man than I was. In addition, the top of the memo said "no. 1 of 6 copies" so that I knew that several other people had copies of it. That my confidence in Compton's integrity was in no way shaken in spite of all these incidents is, perhaps, best demonstrated by the fact that I always knew that he had forwarded the copies of my reply to his memo to everyone who received his. In fact, he seems to have shown it to Greenewalt who was most annoyed by the whole incident.

It is not easy to give a picture of the mental frame of the Laboratory in those days. It was a mixture of frustration, rebellion and bitterness which is hard to describe. I took notes on one day of all the impressions that reached me and will reproduce that sheet here.

"July 8, 1943 – one day's crop:

Chilton said that, six months ago, he expected that the plant would be running by now.

Vernon said that he once asked 'the Army', how long they can resist after Germans have the bomb. 'They' answered, 3 to 6 months.

J. R. O. said (rumor): What is wrong with Y is that there is too little water and too much general.

Lowen came to speak to me: 'This time not about physics. I cannot comprehend the complacency which has gripped this project. There were two long speeches yesterday by Compton and Allison. Both agreed that whichever side is ready first will win the war. In spite of that, neither of them emphasized speed. Yes, Germany is doing something but we do not know what, is what they said, and they seemed to be quite satisfied. Look how things can be done speedily: You remember the promises made by the M.I.T engineers, how they expected to have 20 mev particles. Herb put a little machine together, with few men and little expense, and had ten times more results than M.I.T.'

I should add that I also saw Dancoff and asked him 'en passant' what he thought about Compton's and Allison's speeches. He said that they were mildly interesting. I did not comment."

And July 8th differed very little from other days. Almost every day, there were two or three people in my office, complaining and "not understanding". This stream of complaints with which I was forced to agree created a very depressed atmosphere.

Of the July 8th impressions, Mr. Chilton's admission is the most significant. He is a du Pont official in a very prominent position, of an unusual integrity who, as one sees, put sincere endeavour and truthfulness high above the Company interests. The situation with the schedule for W was the following: On February 8, 1943 Mr. Greenewalt declared that the helium cooled plant would be ready by July 1, 1944. This date came as a shock to the Laboratory and it was attempted to mitigate its effect by the statement that the du Pont Co. always gives schedules which it hopes to beat by several months.

On July 14th, Mr. Greenewalt said that the "the first pile is scheduled to operate on May 1st (1944) and followed by the second and third piles at one month intervals." This was not considered to be a substantial improvement of the schedule since it referred to a water cooled pile which was supposed to permit a faster construction. It was still hoped that this schedule would be beaten by several months. The probable schedule which was submitted to the P-9 Investigating Committee expected the first pile to go into operation some time in April – and the Committee based its recommendations on this completion date of the graphite pile. Or were they less naive than that?

Soon after the P-9 Committee left Chicago an accelerated schedule was made public (Dec 20, 1943). It was rumored that the General himself pressed for the faster schedule. This accelerated schedule was announced by Greenewalt himself who said that the du Pont Engineering Department had committed itself in a letter to the General to have the first unit ready the 1st of June, 1944.

This was announced so officially that I almost paid up a bet to Hilberry. This bet I made in the middle of the summer of 1943 in a moment of desperation, as I often made such bets, partly to show the seriousness of my worries and partly to get some, though insignificant, insurance. I fully expected to lose the bet. Hilberry, on the other hand, may have entered the bet to reassure me and to show me that he, at any rate, had confidence that the progress would not be *that* slow. At any rate, I did not pay up the bet and collected it later when the first pile was not ready by September 1, 1944.

The accelerated schedule[11] was in effect for a few weeks only, although this fact was not advertised as the accelerated schedule itself was. The events around the accelerated schedule clearly showed that the Leadership did not really expect us to believe in the statements that were circulated. Otherwise that schedule could not have been called accelerated. While the accelerated schedule was still in effect as far as we were concerned, Mr. Conant called Mr. Fermi and me into Dr. Compton's office for a Conference one day in the middle of January, 1944. He asked us to prevail upon Mr. Lowen, who again became active in Washington, not to interfere with Project matters any more. I said that we might be able to convince Lowen that nothing could be done right then if we received permission to tell him about the accelerated schedule. To a further inquiry we both said that we had, at that time, no suggestions to make, supposing that the accelerated schedule would be adhered to. Mr. Conant took cognizance of our attitude and did not attach any remarks. He knew, however, at that time already that the accelerated schedule was revoked.

These were the kinds of matters which caused us to be dissatisfied in 1943. As far as the difference of opinion between du Pont Co. and ourselves concerning technical matters was concerned, I would have been ready until a few weeks ago to admit that if we take their time schedule for granted, all the changes in our plans which they proposed and did not abandon after our first objections, were reasonable. Of course, all of them, but one, involved uncertainties and required research which could not have been carried out within our (or Mr. Chilton's) schedule. The "once through" water system was the only exception, i.e., the proposition not to recirculate the cooling water but to return it to the river after it left the pile. This was hailed by us, as soon as we saw the chemical analysis of the W water, as a great improvement and our enthusiasm was only slightly dampened later when we realized that the water treatment which the Company deemed necessary required enormous installations at a price much higher than that of the pile itself. Much of this installation was obviously purposeless.

The second technical point brought up by the du Pont engineers was the "boiling disease", the claim that our cooling system could, under certain conditions, become unstable and that if the water would start boiling in a tube, this boiling would persist even if the exit temperature of the cooling water from all other tubes were many degrees below the boiling point. In this case, also, we recognized at once that the du Pont engineers had a real point al-

---

[11] Wigner's original manuscript inserts the phrase "which advanced the complete date from April 1944 to June 1, 1944", which appears to be inconsistent.

though the point did not appear to us as significant as it appeared to them. Thus, e.g. at the present level of operation the "disease" could not possibly appear. In spite of this, we missed few occasions to point to the importance of the Company's contribution in having recognized this condition and to praise the solution which they proposed to it. Greenewalt told me that he himself suggested the solution (insertion of orifices causing pressure drops).

The situation was different as far as the other suggestions of the Company were concerned. The main bone of contention was the dry shield which they proposed, against our wet shield. It was apparent to us, and was also accepted by the Company, that the shield must contain some hydrogeneous material to slow down the fast neutrons, which tend to escape from the pile, so that they could be absorbed easily. We proposed that the hydrogeneous material be water which was to be circulated in the shield and thus provide, at the same time, for its cooling. They were afraid that the water may leak from the shield toward the pile and spoil it. For this reason, they proposed to use a solid hydrogeneous material, a sort of abietic acid which was to fill the interspaces of sintered iron blocks. We did not like this mainly because we feared that the abietic acid, or whatever it was to be, would decompose under the influence of radiation and spoil the pile. The abietic acid was soon replaced by masonite and in order to protect it from radiation, the graphite reflector was thickened a good deal and a "thermal shield" inserted between the pile and the masonite layers. With these changes, and because Burton's group found very little gas evolution from the masonite under the influence of radiation, we finally accepted the dry shield. The dry shield did cost a good deal of graphite (which was so critical a material at that time that the plan for a power pile at the Argonne had to be abandoned because of lack of it), it also cost a lot of money and, what we had resented most, a lot of time because experimental results had to be awaited before it could be decided upon. In spite of this, until a few weeks ago, I would have said without hesitation that the change to the dry shield was wise because the water connections for the wet shield would have crowded up the faces of the piles even more than they are crowded up now. A few weeks ago we heard, however, that some masonite which received the same irradiation as the thermal shield, completely disintegrated. This now again tends to indicate that some unnecessary risks were taken with the dry shield even in its finally adopted form. The history of the dry shield is quite characteristic of the controversies in technical matters between the du Pont Co. and ourselves. Although, particularly at first, we did not like the idea of the dry shield at all, many of the suggestions which finally made it possible (the thickened graphite reflector and the thermal shield, called by us "shield of shields") originated with us.

Other changes on our plans which were suggested by the Company and discussed more than the usual ones were the elimination of the water manifolding at the exit end (this suggestion was discarded) and elimination of the coffins in which to push the uranium after irradiation. This last suggestion was finally adopted. Not much discussed but adopted against our better judgment were, in addition to some mentioned before, the reduction of the length of the individual pieces of uranium rods from about 60 to about 20 cm and the replacement of a much simpler shield at the sides and top of the pile by the composite shield which was used at the faces. This last change, in our opinion, made the construction more expensive and difficult and, at the same time, made the pile and shield less stable from a mechanical point of view. However, we became aware of these changes at a time when the directive was definitely to play the question – answer game, i.e. to tell the Company only whether we have any very good reason to believe that their plans would make the pile inoperable.

It was realized, on the part of the du Pont Co., that they were severely criticized because they did not make efficient use of the help which the Laboratory could offer to them. Even very conservative and calm individuals such as Z. Jeffries who, in addition, certainly understands the point of view of large companies, concurred in this criticism. In order to take the edge off this criticism, it was decided to call another member of the Laboratory to Wilmington. The Theoretical Section's demonstrated lack of veneration of the Company excluded it from consideration for this purpose and H. L. Anderson was selected to go to Wilmington. Anderson was one of the most able members of Fermi's group. He was on the Project right from the start and his personality should have been excellently suited for the job at hand: the survey of blueprints and advice in design and constructional matters. In fact, Anderson seized the job wholeheartedly, familiarized himself with the design in a remarkably short time and soon was in a position to make suggestions for its improvement. He even repeated some of our calculations on the multiplication constant and pointed out at least one substantial abbrevation in these which has been in use ever since.

Unfortunately, he was too sincere and outspoken to be really acceptable to the Company. He hinted at the desirability of direct contact between himself and the Engineers, at the desirability of asking Fermi for advice on certain question ("Yes, and I could then withdraw from the work" was Greenewalt's reply) and pointed too bluntly to the unnecessarily complicated nature of the helium circulating system. As a result of this and probably other matters better known to the Company than to me, after a few months Greenewalt told him that he felt that the design of W was in such a shape that he could not further claim the need for Anderson's presence in Wilmington. Anderson was, of course, to some extent upset as nobody likes to be fired. An aftermath of this incident occurred when the du Pont Co. wanted to to recruit physicists from our Laboratory to supervise the operation of the W units in their earlier stages. When this happened (March 10, 1944), the Company asked for the transfer of almost every person in leading position in the Argonne Laboratory who were not scheduled to join the Y Laboratory. They did not, however, ask for Anderson. This was clearly an anomalous situation since among all members of the Argonne Laboratory, Anderson knew most about the pile and he was probably the only one who would have been glad to go. Yes, but you must not offend du Pont.

A second attempt at reconciliation with the Laboratory and more specifically with our group was made in October while Anderson was still in Wilmington. We had just given up the work on P-9 in order to devote ourselves more fully to the review of the W blueprints and it was felt by Roger Williams of du Pont, who was in Wilmington formally in charge of the work, that the antagonism between ourselves and the Company served no useful purpose. He asked me to visit the Company for a week during which I should learn how the Company operates, how it is organized. He hoped – and I hoped so too – that a better knowledge of the Company's methods of procedure would provide also a better understanding and appreciation of these methods. Greenewalt told me that he was, personally, not in favor of the visit but it took place anyway between the 25th and 28th of October.

It was, indeed, a very instructive visit which I shall always keep in mind. Not only was I treated most courteously and considerately but I was introduced to the highest officials of the Company, including Mr. Carpenter and Mr. L. du Pont. It was my impression that all discussions were conducted in a very friendly and pleasant atmosphere and a serious effort was made to win me over to the Company's point of view.

This last part was not entirely successful. Although I realized more clearly than before that the pile building itself is only a small part of the whole enterprise as far as erecting buildings and routine engineering was concerned, it did not appear necessary to me that this part be handled exactly in the same way as the other parts were handled. Granted that Chicago could not make a worthwhile contribution toward expediting the building of roads, fences, houses for the personnel, even the river pumps, it still could have, and should have, been given a share in the design of the pile building. It is only as far as this building is concerned that we asked for a greater share of the responsibility. Against this, Roger Williams stressed the desirability of adopting a uniform procedure for the design and construction of all parts of the plant – a point of view which appeared debatable to me. However, we did not go too much into the details of these questions as we both realized that it is impossible to consider a change in the procedure which was not only adopted but already in operation at that time. The important thing was, of course, to arrive at an understanding for the future roles of the two organizations. In addition, I suspect that Roger Williams hoped to win me over to his point of view more by impressing me than by arguments. And the show he gave was impressive.

Although my visit to Wilmington had no very concrete results, it did serve to alleviate the tension between the Company and the Theoretical Section. This had favorable effects on the main assignment of the Theoretical Section as fas as W was concerned: on the review of the blueprints for W which were prepared by the Company. With respect to these, the procedure set up by the Army stipulated that they could not be executed before the Laboratory gave its approval. This approval was given by Allison who was the Director of the Laboratory but, at least in the case of the pile building which was, as far as I know the only building seriously reviewed by the Laboratory, Allison's approval was always based on our evaluation of the plans embodied in the blueprint. Our evaluation was communicated to him in a memo, usually written by me.

As mentioned before, the review of blueprints really started only about the middle of August. A set of blueprints would arrive in Chicago and be deposited in an office which was first under the supervision of Mr. Diver, later of Mr. Thompson. Mr. Diver, a most pleasant and helpful person, sent notices to certain people letting them know of the arrival of the prints. By the time we came down, he had already subjected them to a preliminary scrutiny. Usually Ohlinger, Young, Weinberg and I went down together and Mr. Diver had a first round to explain the prints. Quite often, Ohlinger had to supplement his explanation since blueprints are not easy to read if one does not know exactly what the purpose of the installations is which it shows. Ohlinger had an exceptionally quick grasp to see what was meant, even if the print was not quite correctly drawn, and he drew little diagrams which made everything clear to the rest of us. We debated then some and reviewed the possible dangers which may arise in connection with the installation shown, such as overheating of parts, escape of radiation through cracks, stresses caused by thermal expansion, etc. Having discussed the matter some in the blueprint room, we would retire to our offices and make, as far as this was possible, more detailed calculations on the questionable points. G. Young usually took over calculations concerning thermal expansion, thermal stresses, temperature increases, etc. Mr. Weinberg checked the effects on the multiplication constant, the level of operation and similar items. Mr. Friedman and later Miss Way checked the questions concerning the escape of radiation and Mr. Ohlinger surveyed questions of arrangement and similar matters. I tried to make sample calculations concerning every point in order to keep the "feel" of matters and also in order to provide an independent check.

We communicated our results, if they were unfavorable, quite often informally to Mr. Diver because we felt that there was a better chance of having something changed if it was not evident, from the record, that Chicago suggested the changes. Otherwise, we wrote a memo to Allison who usually transmitted the information contained in it. The time which was available for reviewing a set of print was, on the average, not more than 4 days – a very short time indeed if one considers that, in some cases, reasonably elaborate calculations had to be made before approval. If we could not complete these, we gave a preliminary opinion and later a final one. We may have omitted doing this latter in some cases but I hope that we have not missed noticing any major point that is likely to cause difficulties.

The task of reviewing the blueprints was neither easy, nor pleasant. The installations shown in the prints were interconnected, not only physically but also functionally. For this reason, it was difficult and in some cases impossible to appraise the adequacy of one installation without having a picture of all the installations before us. This, however, was not the case since we had to review the prints as they came in.

On the whole, we did not like the prints too much. It was our impression, in the majority of the cases, that the Company designers have learned from Rube Goldberg a good deal. For security reasons, the draftsmen did not know the purpose of the installations which they designed and this was bound to hamper their style.

We often felt that we could do better than the prints did but were not encouraged to present alternative designs. For one, the addition of a draftsman to our group was not approved. However, we did not feel unhappy only on account of this, as we recognized the limited scope and time of our review of the blueprints. This latter circumstance we had to point out repeatedly.

The last two chapters were written very much from the point of view of the 4th floor of Eckhart where the Theoretical Section was housed. Thus the above narrative hardly does justice to the contributions of other parts of the Laboratory many of which were of fundamental nature but about which I naturally know less than about our own work.

First, the Experimental Physics Section entered the W work rather reluctantly. The reasons were the same as ours: as long as only answers to definite but ill-defined questions were requested, the Section did not feel that its contributions would be really worthwhile. In addition, the Experimental Section was probably not as soon convinced as we were that the decision to build water cooled piles was final. When it became evident that the water cooled piles would be built and that the success of the Project might depend on the solution of certain problems which could be solved only by them, W work obtained top priority at least on their schedules. Of course, most work was done as a duty and without the interest and enthusiasm that usually accompanies the work of a scientist. As a result, it was not always pushed so wholeheartedly to the final conclusion as it would have been otherwise. Neither was it attempted to make a secret of this condition: I once heard myself when Fermi told Mr. Miles who represented the Company in Chicago, "What do you expect? The Company treats us awfully badly."

The most urgent single contribution of the Experimental Group was the verification of our calculations concerning the multiplication constant. A considerable error in those calculations would have rendered the whole plan of the W piles questionable. The work was first undertaken with some hesitation, but once it was started it was carried out by P. Morrison's group with great care and accuracy and was essentially completed in June, 1943. It corroborated our calculations so accurately that one is inclined to believe that the degree of agreement between our calculations and Morrison's experiments was partly accidental.

Another point which had to be checked experimentally was the thickness of the shield. Fermi himself undertook this work and determined the shield thickness that appeared necessary. This work was done rather in a hurry and Fermi advised to use a considerable safety factor over the shield thickness that appeared to be absolutely necessary. The shield is more than a foot thicker than the thickness estimated to be necessary by our calculations.

Other points of importance included the radioactivity of the cooling water that leaves the pile, the efficiency of the control rods and many others. Fermi estimated that in June, 51% of the effort of his section was devoted to work on W problems and while this might have been somewhat exaggerated, the actual figure could not have been very much lower.

The experimental physicists felt just as strongly as we did that the du Pont Co. was dawdling with the job. They were less upset emotionally about this than we were, first because both Fermi and Zinn were convinced that the Germans had no secret weapon close to readiness and their evaluation of the urgency of our work was therefore different from ours. Second, the plans were not theirs so that they felt less responsibility for their prompt and efficient execution. Fermi, who often gave me advice in a very pleasant and fatherly way once said: "Do not worry, it is not your trouble. Let it blow up into their faces. I hope it will."

The Chemists got along with the du Pont Co. better than we did. First, they were more familiar with the methods and ideas of the chemical industry than we were and second, the Company could contribute more toward a solution of their problems than it could toward the solution of ours. The whole arrangement of the chemical plant was the idea of an engineer (Cooper) and the design was detailed in Wilmington by du Pont engineers, under Mr. Generaux's leadership. There was some disagreement because the Company wanted the chemical process to be "frozen" by June 1, 1943 while the Laboratory wanted to carry along research on alternate processes after that date. However, the discussions never became quite as vigorous as they were with the physicists.

Collaboration with the du Pont Company in 1944 in Matters of the W Piles

While our blueprint review continued in 1944 it did not have the importance and significance which it had earlier. Instead, problems of a technical-engineering nature and those connected with the radiation-caused changes of the mechanical and thermal properties of the materials inside the pile assumed greatest significance. The most important technical-engineering problems were those of corrosion and canning, the most dangerous radiation-caused changes were expected to occur in the graphite.

The dangers of corrosion were, I believe, very much exaggerated. Everyday experience indicates that aluminium is very resistant against corrosion by water and this appears to be true even if the water is far from being free of dissolved salts and also if the temperature of the water is above the ordinary boiling point. Recent experiments indicate that this everyday experience is found even in the laboratory but for some reasons which are not altogether clear any more, it was feared for a long time that the water will dissolve the aluminium tubes as if they were made of sugar. In the early days of the project, there was some reason to dread the unusual conditions prevailing in a pile and, particularly, the fact that hydrogen peroxide is formed in the water under the influence of radiation caused many a headache. It soon turned out, however, that the peroxide rather hinders than enhances the corrosion so that is was, on one occasion, even proposed to add some of this substance to the cooling water to minimize corrosion. The question of corrosion was fundamentally settled by the end of 1942. There was only one temporary scare after this date when Morrison found that hydrogen is evolved from the water in his exponential experiments. It was ascertained, however, that this effect occurs only on fresh surfaces of aluminium.

Film formation replaced the corrosion scare in 1943. During the corrosion experiments which were carried out by the du Pont Company in W, it was found that a thin film deposits from the water on the tubes and the slugs inside the tubes. This film increases in thickness in the course of time and stops up the tube to such an extent that it requires an increased pressure to pump the required amount of water through the tubes. How this effect went by unnoticed in the innumerable tests which the Technical Division carried out in Chicago is not very clear. It caused a moderate scare when it was detected. A very simple calculation showed that a film which is thick enough to increase the pressure drop appreciably could spoil the heat transfer very, very badly. This, however, does not seem to happen – a circumstance which was never completely cleared up.

The additions to the water and the method of operation which avoid film formation without increasing the corrosion unduly were cleared up in Burton's section, particularly with the help of Mr. Shapiro and the Company agreed to follow their advice after some parleying.

The most serious difficulty encountered in the whole W work was the canning of the slugs. The uranium does get badly corroded by water and it is necessary, therefore, to surround it with a water resistant material, i.e., either to coat it or to put it into a can. Both coat and can must adhere to the uranium so closely that the heat transfer between them remains very easy. Specimens of coats were investigated both by Creutz's group and also at other sites, particularly by H. B. Wahlin in Wisconsin. The general agreement was, however, that it always will be difficult to trust a coat to be entirely free from flaws and pinholes. As a result, they were abandoned one after another and it was decided to can the uranium slugs. Aluminium was selected as the canning material and an aluminium can, almost one mm thick, certainly looked more trustworthy than the thin and often not very well adhering coatings that were proposed before.

Many, many types of cans were proposed, some as early as the summer of 1943. Szilard wanted to solder the can on the slug, with some alloy which contracted after cooling. Creutz proposed to replace that alloy by zinc which adheres well both to the uranium and the aluminium. Another proposition of Creutz's was to put all the slugs into one common tube which would have to be closed only at the two ends. Ohlinger proposed to draw the aluminium tube over the uranium in a drawbench and it was proved that the mechanical pressure which is created in the drawing process produces a sufficiently close thermal contact between the slug and the can. The du Pont Company's plant at Cleveland made a proposition that was very similar to Creutz's except that the zinc was replaced by the alloy of aluminium and silicon.

It soon became apparent that the canning process had to overcome two difficulties. If the can, in spite of its sound appearance, should have a pinhole water would reach the uranium and attack it. If this happens, the volume of the products formed by the chemical attack is very much larger than that of the metal which is being replaced and the slug begins to swell. This swelling can reach such proportions that the slug gets stuck in the tube and cannot be removed. This must be avoided under all conditions. The second problem of the canning is the closing of the can, into which the slug has been placed, in a reliable and watertight fashion.

There were very strong objections on the part of the Company against Ohlinger's proposition of the unbonded slugs. These were first based on the mistaken idea that the heat transfer between slug and can is too poor. When they had to admit that this was not the case, they pointed to the very fast swelling – which the unbonded slugs undergo if a pinhole develops. It was for this reason that the unbonded slugs were eliminated.

There was no valid objection to the zinc bonded slugs and they were relatively easily manufactured. Even a few months before W started operation, it was the almost unanimous opinion of the Laboratory that the zinc bonded slugs are more reliable than the aluminium-silicon bonded ones. Mr. Jeffries concurred with this opinion. However, the Company did not like the zinc bonded cans and they were eliminated from further consideration. This may have been decided because, with the design adopted by the Company, they would have had an unduly large loss in the multiplication constant.

The problem of canning was constantly "essentially solved" during 1943. The Technical Division of the Laboratory worked on a "main line", prescribed by the du Pont Company, with an enormous manpower (more than one hundred people) in greatly expanded facilities. The work was singularly unsuccessful. The solution always was within reach but escaped every time with a tantalizing consistency. The situation was aggravated by an exaggerated optimism by those in charge who started production before they had good evidence that they were on the right track. When this mistake was finally recognized, the Director of the Technical Division wrote, in the hope that his remark would go unnoticed, that, "Last month, the canning problem progressed from the production to the experimental stage." This remark hit the nail on the head in more than one way. Brute force methods were trusted to solve the canning problem in reference to a combination of experimental and theoretical work and the group at B [12] was always hammering away at some method proposed by the company. I did not participate very actively in this work which was supposed to be, after all, definitely outside my sphere of interests. The incident which I remember is when Allison, on one occasion, called the physicists together and announced the "main line" which the company had just decreed, the "double can". The proposal was to protect the slug not by one, but by two cans. Fermi, Young, Weinberg, Cooper and I participated in that meeting. It was obvious, at least to me, and probably to others also, that the "double can," as proposed might

---

[12] B was the code-name for the site near the university of Chicago where the development of canning was conducted. The abbrevation "B" was chosen because the site originally housed a brewery (AMW).

not be better at all than a single can. I voiced this opinion then and there. The double canning was the "main line" for several months and remained the "main line" even at the end of 1943 but it was finally abandoned for the reason, good or bad, given by me many months earlier. This incident illustrates that it is best to think a little before starting many people working on a subject and also that one should not follow any method blindly, even one that is proposed by one's boss. Most of the members of the Technical Division, including its Director, Mr. Cooper, were du Pont employees on leave.

The attitude of Greenewalt to this difficulty was expressed to me in October 1943 when I visited the company. He said that the situation was "crystal clear" that the canning was research and, therefore, the Laboratory's responsibilty. The situation with respect to canning was typical in many other fields. In effect, the Company had the authority, the Laboratory the responsibility if anything went wrong.

Greenewalt's statement concerning the responsibility for the canning was contradicted by the initial monthly report of X [13] and Chicago to the Army. This report had du Pont's approval. It classified both corrosion and canning as engineering: "930 – Engineering Studies – Phenomena basic to corrosion and other studies essential to continued engineering development of processes." The method of jacketing the slugs must be chosen about Sept. 1, (1943).

To continue the story of the canning – Szilard and Young proposed the thick end cap early in 1944 and Greenewalt had the emphasis of the Laboratory's Technical Division shifted from double canning to this method. In spite of these developments the canning situation looked almost desperate in the summer of 1944 and at that time, Cooper gave a frank presentation of the progress of the canning problem. He said that, "Current gross production is 740 per day of which 250 are non-classifiable." Non-classifiable cans were such that they could not be used at W. The pile needed 50,000 slugs and the first one was expected to be loaded a few weeks hence. This was indeed a very tight hole. As a result, Creutz was taken out of the bag and charged with finding methods by which the swelling of slugs could be detected before it would become dangerous. Creutz and his group carried out this assignment, as all other assignments that they obtained, in a short time and they could report in a couple of months on at least two possible detecting methods. The one involved a collar around the slugs which was pushed down by the water stream if there were no swells and got stuck if there were any. The second involved a periodic pushing of the slugs back and forth and measuring the force necessary therefor. An increase of this force indicated that a swell was touching the tube and the slugs have to be discharged.

Another measure that the Laboratory took was to revive Creutz's old suggestion of placing all the slugs of a tube into a single jacket. Ohlinger was charged with the development of this method but the progress on it was not as rapid as was hoped for at first.

---

[13] Site X = Clinton Laboratories (A.M.W.)

The real solution of the canning problem was obtained, however, by the manufacturing outfit of the du Pont Company at W. The difficulty of the aluminium-silicon bonding method was that the melting point of the aluminium silicon eutectic was only about 20 °C below the melting point of the aluminium. As a result, unless the operation of canning was performed at a very accurately determined temperature and unless it was completed in a very short time, the liquid eutectic penetrated and dissolved the aluminium can. This manifested itself in "pock marks" which were a sure sign that the can is faulty (although earlier these were not necessarily "unclassifiable"). The canning process required almost split second accuracy and this is what was ultimately virtually achieved by the Manufacturing Department of the Company. The last very pessimistic report was given by Cooper June 21st, when among 1600 slugs, only 18 were in class A, i.e., usable everywhere in the pile. On July 5th, Cooper reported that the canning situation was much better and by the time the first pile was to be loaded, a sufficient number of classifiable slugs accumulated. This was a very great success for the Company.

The time schedule, with respect to the starting date of the pile, changed in 1944 as follows: early in the year, the accelerated schedule was revoked and July 1st "re-established" as starting date. On March 22nd, in response to Fermi's query as to whether the starting date had been postponed, Dr. Compton answered "July 15th ± two weeks still holds." On April 19th, returning from W, Fermi commented, "First impression is size. One is also impressed by the excellence of graphite machining – there is a question whether the degree of excellence of the work in general may not have slowed down the completion appreciably." On May 17th, Dr. Compton said that the first pile would be under way late in August or early in September. On June 7th, Hilberry declared that labor caused the delays. The pile actually became critical in September and was under power a couple of weeks later. In addition to its success in solving the canning problem, the company should be commended also for its skill in postponing the starting date by six months without ever having it apparently shifted back. It was a masterpiece of diplomacy and fooled everyone who wanted to be fooled.

While thus the problems of a technical-engineering nature were brought to perhaps tardy but essentially successful solution, the same may not be true of the other set of problems concerning the Laboratory's work on W which played an important role in 1944: the elimination of the radiation-caused changes in the materials.

The radiation-caused changes were early recognized to be most likely to be serious in the graphite. The main reason for this is that the uranium is periodically replaced while the graphite has to stay in the pile all the time. The aluminium tubes also remain in the pile but aluminium, as a metal, is much more likely to "heal" than is the rigid graphite. The investigation of the "graphite disease," as I liked to call it, was carried out by Burton and Franck, under the latter's initiative, beginning the first days of 1943. In spite of the low radiation densities which were available in the early stages of the investigation, Burton and his collaborators soon found a variety of changes in the graphite, the most important of which appeared to be a decrease of the heat conductivity. Burton and Franck called the changes "Wigner effect" because it was I who first pointed out that changes of this sort are to be expected as a result of continued neutron irradiation. In fact, on the basis of some preliminary calculations, I estimated that the operation of the pile may run into difficulties at about 50,000,000 kilowatt days. This fear was mentioned also in the report which we issued on water cooling in January 1943.

Franck's and Burton's work was much pooh-poohed in its early stages. It takes a good deal of objectivity to believe in an effect for which there is no experimental evidence, even if there is a good reason to expect it theoretically. However, Franck, Burton and their collaborators soon overcame the skepticism at the Laboratory and the reports on their work were listened to with great interest by everyone, excepting a few diehards.

The du Pont Company's representative in Chicago, Mr. Miles, first belonged to the skeptics. He even went over to see Burton and told him to stop his boon-doggling. He had some other work in mind for Burton. However, the findings of the Laboratory did not fail to influence Miles and he soon began to follow the work on the effect of radiation, not only on graphite but also on other substances, with great interest.

Since neither Burton, Franck, nor any of their collaborators were very familiar with involved theoretical calculations, I asked Seitz to give some time to Burton's group for the evaluation of their results. It became evident toward the end of 1943 that the changes in the graphite *might* seriously affect the life-time of the contemplated piles and Seitz, with Burton's and Ohlinger's help, soon spent a good deal of effort to design equipment for a test hole in the cyclotron where irradiations of graphite could be carried out. The idea was that since there was no radiation available which would have been even nearly as strong as that of the W piles would become, hence, the graphite of the pile would undergo, within a few days, a stronger irradiation than any sample could be subjected to before W's operation. Even a sample of the W graphite, by the time it was investigated, told only in what condition the graphite was a few days before. This was not considered satisfactory and ways and means were proposed to obtain, from the W pile, samples of graphite which were irradiated more strongly than the rest of the pile graphite. In order to achieve this, it was attempted to create, inside the pile, regions of particularly high radiation density. The graphite samples to be investigated were to be placed into these regions. A number of minor difficulties had to be overcome before such a scheme could be done. Obviously the samples had to be removable from the pile. They also had to be cooled while in the pile. Seitz and Ohlinger solved all the problems and submitted a blueprint to Allison for presentation to the Company. The blueprint proposed a mechanism to be put into one of the holes built in the pile for experimental purposes. Needless to mention, the Company rejected this blueprint and brought out one of its own. The Company's blueprint looked satisfactory from the point of view of its effects on the pile but it did not fully achieve the purpose which was in Seitz's and Burton's minds. This was realized at a rather late stage and, since the effect of radiation on the graphite assumed an increasingly serious aspect, other arrangements were introduced as last minute substitutes. These arrangements, called "papooses", were similar to arrangements which were also proposed by Burton and Seitz but were not identical with them. As a result, their value is quite doubtful and I do not believe that anybody can safely interpret the results derived from investigation of the du Pont papooses.

Seitz, who has a very conciliatory character and who has a demonstrated ability to work with great commercial concerns, went to Miles to inquire as to the causes of all these difficulties. Miles was a jovial, openhearted fellow who was devoted to his Company and also to Greenewalt personally. He said that he realized than he was in a somewhat queer position but he had directives from the Company on how to handle these jobs and he wanted to follow these directives. At the same time, Miles asked Seitz to write a memorandum to him, stating in his own way how the operation of the pile could be affected adversely by the changes in the properties of the graphite. This Seitz did. He also incorporated into his memorandum the remark, made by Szilard some time before, that the energy accumulated in the graphite, as a result of the dislocations of atoms, may be released by temperature fluctuations and cause an overheating of the pile.

Seitz also asked Miles to be permitted to go to W, together with Burton, in order to confer on the spot with the people who were going to position the samples into the pile. This request was refused at that time under rather ridiculous circumstances. When the trip was authorized very much later, all the samples had already been placed. Agreements which Burton and Seitz believed to have been arrived at during their trip to W, between themselves on the one side, and a very obliging crew on the side of the Company, were repudiated in a letter less than a week later.

I do not see any benefit that the Company could have expected to derive from this very uncooperative and uncordial attitude. For this reason, I am surely at a loss to explain it. The only possible motivation that I can see is that the Company officials in higher position, particularly Greenewalt, whose hand was apparent in all disputes, did not want to admit that the Company was still dependent on the Laboratory in many important matters. That their conscience is not clear is evident from two incidents.

On June 20, 1944, in a meeting, Franck and Burton, in response to a direct question by Compton, asserted that, "It is improbable that graphite as moderator can go on for two years." Mr. Graves, who was Mr. Greenewalt's aid, took strong exception to this statement which "might seriously embarrass the project." He wrote a memorandum saying that it was regrettable that such statements appeared in the records.

Evidently, Franck's statement was more conservative than my statements of a couple of years before, some of which were in the records when the Company took over, others contained in a document which they chose as a basis of their engineering efforts. They had, therefore, no reason to become excited and I am surprised that Allison chose to transmit Graves' reprimand to Franck. After Franck made his statement in response to a direct question, posed by Compton, all matters of this nature are to be concealed from the readers of records, and from Compton who should have been reprimanded for having asked the question. Franck could not very well refuse to answer, nor could he be expected to say anything but what he believed.

Under ordinary conditions, graphite is a good lubricant but in our project, it certainly caused a lot of friction. The result of this friction is still unforseeable. There is a good chance that the change of the properties of the graphite will not greatly affect the operation of the pile. [14] One can certainly hope this. On the other hand, if it should, the lack of cooperativeness on the part of the Company may have serious effects. The irradiation of the samples that can be investigated cannot be gauged very well but it is certainly much lower than it could be. However, one may rest assured that should this trouble arise, Burton will be blamed for it.

There is only one more point concerning the relationship of our Laboratory to the du Pont Company that I would like to relate here. This is the controversy of the famous "know how" stamp. It was the custom of the du Pont Company, from time immemorial, to put on every blueprint that it makes, a stamp saying essentially that the "know how" which is embodied in that blueprint is the property of the du Pont Company and that reproduction of it must bear a similar stamp and that no use of this "know how" is permitted without written authorization by the Company. I understand that this stamp is omitted only when it is very obvious that the ideas underlying the blueprint do not originate with the Company. Evidently the Company did not think that this is the case with regard to the blueprints of the W pile and all the blueprints which we reviewed bore the "know how" stamp. This annoyed us, of course, no end. Since most of the blueprints were only an elaboration of our ideas, it was simply foolish to say that they were the property of the du Pont Company. Nevertheless, by not objecting to this "know how" stamp we feared that we tacitly yielded the ownership of these ideas to the Company. This might have been unfortunate from the point of view of the post war world.

Although Fermi was not personally concerned in this matter, because he never surveyed blueprints nor was he one of the originators of the W plans, he very nobly took up our cause and broached the question in the meeting of January 3, 1944. Compton replied that he knew about the situation and that it was under investigation and clarification could be expected in the near future. However, nothing happened on this score and, learning from past experience, I wrote a letter to Mr. Williams who, it may be remembered, was extremely cordial during my visit to Wilmington and who promised to take a personal interest in any question affecting the relationship of his Company to the Laboratory. I believed that this letter might serve two purposes. First of all, it would answer the question as to what extent the Company was willing to implement its promises concerning fair treatment of the Laboratory personnel. Moreover, a lawyer advised me that our fears were justified concerning the yielding of

---

[14] Actually the original Hanford reactors operated for many years even though the graphite deteriorated (A.M.W.)

the ownership of the ideas for W to the Company. This lawyer assured me, however, that if we communicate to the Company that we do not agree with the statement of the "know how" stamp, there will be no presumption of our having given up any rights concerning these ideas. I, of course, phrased my letter to Mr. Williams very politely and very carefully so as to avoid offense. I add, I showed it to Mr. Allison who fully approved of it.

The answer to this letter came after four weeks and was rather unexpected. It just reiterated the statement of the "know how" stamp and after that it asked me quite rudely to channel all communications to him through either Mr. Allison or Mr. Compton. My impulse was, of course, to answer the letter in the same tone and Smyth, to whom I showed William's letter, dictated to me an answer which was even sharper than mine was. On further consideration, however, I dropped this matter because I felt that the dual purpose of my letter was achieved: William's reply made it evident that if there was, for sometime, a tendency in the Company to conciliate the Laboratory and to treat it fairly, this tendency certainly had been abandoned; secondly, the lawyer advised me that no matter what the Company replied, my letter abolished any surmise of my abandoning any rights. Of course, I do not know whether Mr. Williams, whom I still consider to be a rather fair minded person, wrote his reply of his own volition or whether he was gently persuaded to do so by others.

I am afraid this chapter of the history of our Project is rather full of gripes and of causes for our dissatisfaction with our relations with the du Pont Company. A couple of chapters ago, I described the general reasons that I believe to underlie the antagonism between ourselves and the Company. It may be, however, that those general causes would not have manifested themselves as strongly had it not been for one of the personalities involved. In fact, many a member of the Laboratory, even employees of the du Pont Company in prominent positions, came to the conclusion toward the end of the Project that the cause for the deterioration of the relations may be looked for in one single person. They did not always say who that person was but everybody knew that Greenewalt was meant.

Greenewalt was indeed an unusual personality. He was bright and quick and had a faculty of picking up a good deal of information in meetings and conversations. In addition, he had an extraordinary charm and although I knew him quite well after a few month's contact, I still remained pleased whenever he visited my office and had a little chat with me. I knew from some meetings that he could be rude indeed to people who were under him but he was never that way with any of us and I must admit that I often was under the spell of his personality. Coupled with these extraordinary qualities, however, Mr. Greenewalt possessed an exceptionally selfish attitude when the interests of the du Pont Company could be furthered. He did not hesitate to take any step, or to override anybody, not even Mr. Williams who was supposed to be above him, if he thought that he could ferret out anything for his organization. In addition, he was as fully imbued with pride in the achievements and perfection of his Company as anybody could be. He was probably the only person who did not disagree when General Groves said in a meeting early in 1943 that,

"The du Pont Company probably noted some inaccuracies in your calculation of the multiplication constant." Of course, he recognized the childishness of such statements which, by the way, greatly irritated us, but he considered them as a just tribute to his Company's past achievements in other fields and the role which it would play in the field of Nuclear Physics in the future.

Secrecy

In the early days of our work on uranium, the interest in maintaining secrecy was stronger among those who wanted to do the work than it was with the authorities. Szilard was the first one who emphasized the importance of keeping our results from our prospective enemies and most of those who had a glimpse of the future agreed with him. It was described before how his efforts broke down essentially on the resistance of the French. In spite of his original failure, Szilard did not give up his efforts to bring about a reasonable arrangement to prevent divulging too much information. The situation with respect to secrecy was rather satisfactory when Breit's Reference Committee was organized and prevented publications which were too obviously connected with the possibility of a chain reaction. Later, when the project was organized, the importance of the Reference Committee diminished because there were very few physicists in the country who were not engaged in the war work and the number of articles on nuclear physics which were submitted to the journals suffered a very sharp drop. Breit's functions then became those of the Secrecy Chief of the Project. He had to see that classified material was properly handled, that people did not talk too much to each other about the secret work, and he also was in charge of the dissemination of the information in the cases when the results of the work of one person were likely to help the investigations of another.

Breit discharged his duties as Secrecy Chief very conscientiously and he was most accurate in keeping his records and in accounting for the reports which he received. Unfortunately, he was not always as tactful as would have been desirable and there were quite a few flare-ups between Breit and other members of the Laboratory. This was not really surprising considering his somewhat rigid outlook on life. He expected others to be as conscientious as he himself was in following instructions, even if this interfered to some extent with the efficiency of the work. Not everyone was willing to go as far as he went in observing those regulations and restrictions which were generally recognized as purposeless.

The end of our interest in actively participating in the maintenance of secrecy came when Conant declared on a certain date that from then on, every written communication between employees of the Project should be either unclassified or classified "secret." Secret letters were quite cumbersome to handle, particularly since in the early days of the Project nobody had efficient secretarial help and nobody knew which of the written regulations should be taken seriously and which should be disregarded. Mailing a secret letter always took me half a day, during which I naturally cursed the inventor of the system.

Conant's decree about the two classes of communications resulted not only in minor inconveniences. One of its unintended results was to hamper our contacts with the manufacturers of uranium. Before the above rigid regulation was introduced, we could and did inform the prospective manufacturers of uranium, particularly the Westinghouse Corp. at Bloomfield, which impurities were to be eliminated most strictly and what the relative harmfulness of the different impurities was. This suddenly became impossible when we received a secret document written by Fermi which contained a compilation of the "Danger Co-efficients." Before this document came through it was a few minutes work to calculate these quantities from the data available in the literature. There was no objection to informing a uranium manufacturer of the result of the calculation. After Fermi's document marked "Secret" was received, the same information could not be imparted any more to anyone without definite authorization and we were certainly stymied in our efforts of counseling the manufacturers of uranium. Of course, matters were not as bad as they might have been had we blindly followed the rules laid down. It is evident, however, that the situation in which everyone had to exert his own judgment as to which regulation to take seriously and which to disregard was not a salutary one and naturally led to friction within the Project. Breit himself was, of course, as well aware of the abnormality of the situation as the rest of us and I remember that we had a good laugh together when, in the early days, the meetings were held on Sundays because there would be no spies around the Laboratory on these days. In May 1942, just before Breit left the project, he commented to me that he did not believe that the secrecy was still directed against the Germans. He thought that, by that time, its purpose was to maintain the power of the higher authorities. I did not quite understand at that time what he meant but understand it very well now.

There were two circumstances in connection with secrecy which remained throughout the life of the project. The first was the absence of any sign of activity of the enemy's espionage. As early as June 6, 1942 Compton commented on this fact with some worry. This situation prevailed all through the life of the project and our activities were, as far as I know, entirely unhampered by any actions of enemy agents. As I see it now, the reason for this must have been that the Germans knew very well that the project was dealing with uranium research but did not believe that uranium would become an important weapon during the war. As a result, they may not have been much interested in how far we got and may have been happy in the thought that some of the energies of the country were wasted in this work.

The second point in connection with secrecy which was permanent on the Project was our fundamental belief that it was handled with little intelligence. Hundreds of anecdotes can be told on the Security Department, some of which will be repeated here and all of which take it for granted that the Security Department did not know what it wanted to keep secret and that it often believed that information which had been divulged could be made secret again by decree.

Of course, the Security Department is likely to contradict this and boast of its excellent record. One must realize, however, that it is much easier to keep something secret from the general population and even from the Congress and the Army Ordnance than it is from the enemy. Most people were pretty busy during the war and did not have much time to gather news and collect information in which they were fundamentally not much interested. It is a different thing to keep a secret from the enemy if he has a few able observers on the job and I doubt very much that our objectives could have been concealed from the Germans with the methods employed.

On the other hand, we on the Project may have overestimated the intelligence of the Germans and the German spies. This was perhaps most strongly manifested in a time schedule which I prepared on June 4th, 1942, at Compton's request. I maintained in that time schedule that it was conceivable that the Germans might have atomic bombs within one year of that date and I sincerely believed this. As a matter of fact, we now believe to know that the Germans were not much interested in the whole uranium work, even after Stalingrad when they must have realized that they could win the war only with unusual weapons.

On the other hand the methods of our security people were such that they could have fooled only an enemy so unintelligent that he would not have been worth fooling. One of the first signs of this was that they hoped that it could be concealed that we were working with uranium. Of course, every single physicist in the country knew this as a matter of course and most of them did not even know that it was supposed to be a secret. In 1942 I received a Christmas Card addressed to Nuclear Research Laboratory. It was addressed in this way for the simple reason that the writer of the card just did not know any better address. Since it was believed that the fact that we were working on uranium could be kept secret, much less emphasis was placed on keeping secret the nature of the other materials which were necessary for the chain reaction. Blocks of graphite were lying in the court in front of the West Stands for anyone to pick up, although there might have been some hope of concealing that we used graphite as a moderator.

A lot of fuss was made not to attract attention to our project but there was, of course, no better way to attract attention than to create vast reservations at X and W with no declared purpose. It was attempted, of course, to conceal the connection of our Laboratory with X and W but with the constant traveling and moving of personnel between these two places, this indeed was a futile attempt. Every moving firm knew about it and anyone could easily find out about it by obtaining the new address of any people who moved away.

Among the more amusing incidents in connection with security were those connected with pseudonyms. It was realized that the names of some of the members of the Project actually gave away the secret of the purpose of the work. It was, however, only in 1944 that it was decided to do something about this. Among others, Fermi, Szilard, Whitaker and I were given pseudonyms. Of course, this had very little sense because by that time our real names were known by everyone, not only where we were permanently located, but also at the other sites. Whitaker once told me the story that he was given a talk by some Security person whether the use of pseudonyms should be handled in the same way as last year. When the man answered in the affirmative, Whitaker said that that was fine and of course did nothing.

I remember that once I went down to X under the name of Winston. Of course, this was as unsuitable a name for me as any that could have been selected because after a few minutes talk everyone could realize from my accent that my name could not possibly be Winston. I was hoping for some complication when it turned out at the gate of the reservation that my badge was obsolete and I was taken to the Sergeant. He asked me my name and we had a little talk, in the course of which I hoped he would arrest me for impersonating somebody else. However, he did not seem to notice any discrepancy and after some telephoning, wrote out a new pass for me, also in the name of Winston. When I arrived at the Laboratory, the man at the desk in front of the guards recognized me and wanted to give me a pass right away. Knowing, however, that I should be Mr. Winston there, I telephoned to the Security Officer saying that this was the person whom he helped out at the gate of the reservation a short time ago and that he should help me to get into the Laboratory. He told me to wait a minute and came out with a pass in my right name. The man at the desk smiled a little, showed him the pass that was made out in my pseudonym and asked him whether this was the one that I might have used otherwise. From there on I used my right name and the only man whom we actually fooled was the Sergeant at the gate of the reservation. He still believes that he passed Mr. Winston into the reservation.

Only once was there a little more complication with my pseudonym: when I was challenged by a guard at W. I had a temporary pass without a picture on it and he asked me whether I had any other identification. Of course, I had to tell him that I did not because all other identifications that I possessed were in my real name. He thereupon refused to let me go through and was adamant even when Greenewalt took the guard to the side during his persuasive attempts and I suspect that he told the secret of my pseudonym to the guard. When this did not help either, I was taken to the guard house and a lot of telephoning ensued, after which I was passed through the gate by the Sergeant. The poor guard still doesn't know what my real name is but this is about all the effect my carrying a pseudonym had. Next time I had to pass through that gate, Fermi with a big "Farmer" badge was behind me and when the guard challenged me again, he shouted, "I vouch for him." This time we were more lucky and I got through at once.

Not all the incidents were as amusing as these. It annoyed me a great deal that I was accused of numerous technical mistakes before people who had not been cleared and I could not correct these accusations because I did not want to violate security regulations. The du Pont people complained to Urey about the very impractical shield that I wanted to have installed on the W piles and I could not explain the situation to Urey because he was not cleared at that time. Another time a person without clearance told me that he learned from the du Ponts how lucky it was that they built enough multiplication constant into the pile. Again I could not explain to him that had the du Ponts followed our plans, the excess multiplication constant would have been much greater than it was. The only case when I could counteract a similar rumor was when Howe came back from W saying that it was lucky that the du Ponts built more control into the pile than we wanted. This was, of course, the exact opposite of the truth.

It will be clear from the above that although we did not think very highly of the regulations to maintain secrecy, we observed them to a much larger extent than the other parties connected with the Project. We heard the story that a few weeks after the du Ponts were brought into the picture, there was a dance in Wilmington during which somebody entirely unconnected with the Project whispered into the ear of one of Greenewalt's people, just to tease him, "element 94."

The du Ponts were, in general, much less conscientious than we were but the carelessness of the Army itself and in particular of the Army Intelligence Division was much beyond the carelessness of any other group. The head of the Army Intelligence Division, which was in charge of maintaining secrecy, was a Colonel Lansdale. Towards the termination of the Project, he gave us a security talk, pointing to the importance of keeping absolutely to ourselves everything we ever heard on the Project and not divulging even the smallest detail of our work, not to mention such important things as that we were working on atomic power with uranium. All the academic personnel had to go to this talk and most of us regarded it as one of the unavoidable hardships of working for the Army. However, two of the people were quite annoyed and decided to do something about these talks. When the meeting was over, they left the Project building and walked around the corner of the street to the back door of the room in which the meeting was held. This was, of course, an unguarded area but they could very well hear the proceedings of the next meeting in which Col. Lansdale spoke to the non-academic personnel pretty much in the same vein as he spoke to us. These two people then wrote a memorandum relating their experience. Everyone had a good laugh except, of course, Col. Lansdale, who flew into a rage when he read his copy of the memorandum.

Another story that I heard about Col. Lansdale, but which I do not know from first hand experience, was a security talk which he gave to Mrs. Mayer. At the time of the talk, Mrs. Mayer could be cleared only for a small part of the Project information. Col. Lansdale explained to her that she was not cleared and that she would not know much about what was going on but that he expected her to keep secret those details which she was going to learn. He tried to explain to her that a detail harmless in itself could give the clue to more complete information which should be kept secret. As an example, he quoted that it was not a secret that Fermi, Szilard and some others were in Chicago; that neither was it a secret that they were nuclear physicists; nor would it matter much if it became known that the Chicago Laboratory is in close contact with the du Pont Company, although one tried to keep this information in as narrow a circle as possible. In addition, he stated that it was perfectly well known that the du Pont Company had constructed a big plant somewhere in Washington. He pointed out that while none of these facts were secret, their connection should be kept secret because a great deal of inference could be arrived at from placing them all together. This was indeed an excellent way to illustrate that it is advantageous to keep even non-secret information restricted to as small a circle as possible. The only drawback of Col. Lansdale's explanation was that the facts which he quoted, and particularly the obvious inference which could be derived from them, should have been kept secret from Mrs. Mayer also. She was very much amused when she left Col. Lansdale's office after having found out so much that she should not have known.

What made the Intelligence Division very unpopular in the later years was that it so openly revealed itself as being more concerned with keeping people in line than keeping matters secret. They intimated that they had a book in which the misdeeds of everyone were recorded and that this book might come in very handy after the war. When Szilard sent to Oak Ridge for circulation a petition in which the President was asked to forbid the use of the atomic bomb, it was attempted to suppress this petition and to accuse Szilard of a security violation because the members of the Project were not supposed to know that the product of the Laboratory would be used for the atomic bomb. Of course, everyone knew this and the Security Division knew as well as anybody that everyone knew it. They considered it, however, a possible means to suppress the petition which, of course, they did not like. When I was questioned in connection with this, I asserted of course that everybody knew the objective of the Project. I was told thereupon that that was probably our fault again because we handled the whole secrecy question in too light a manner. It was purely by accident that I was reminded by Weinberg that the General gave a speech to all group leaders about three weeks before in which he said that "he would like to drop one of these things on the heads of our little yellow friends." He also asked that the substance of his speech, which was a general pep talk, be communicated to all the academic personnel. When I later brought out this fact, the criticism of Szilard's petition from this particular point of view was silenced. However, had it not been for the accidental fact that Weinberg remembered a phrase in the General's speech, a great deal of fuss would have been made over Szilard's security violation.

Another practice which annoyed us a great deal was one which showed the very strong political bias of the Security department. This was not unexpected, of course, considering that the General missed very few occasions, if he saw a small group of people, to give a vigorous speech against Roosevelt. In spite of this, it was most disturbing to hear them ask questions of this sort: "Do you think he voted for Roosevelt" or "Was he a premature anti-fascist?"

All of the above was very irritating at times and I am sure that I often wished the whole Security Department were taken away to some very hot place. In spite of this, I do not know whether they did not provide more amusement than annoyance. The seriousness of our work and our struggles were often relieved by the battles on the sideline with the Security Department and we enjoyed very much telling stories of the above kind to each other. In addition, many a tense situation was relieved by the personality of the Chicago head of the Security Department, Capt. Chapman. He was a somewhat corpulent, very jovial fellow with a good sense of humor. Although he naturally always observed his instructions, we all realized that he did not attribute a superhuman importance to himself and that he realized that the various rules which he wanted us to follow could not possibly be always uppermost in our minds if we wanted to do our work in addition to obeying the rules.

While the secrecy regulations did not make the Army exactly popular in the Laboratory, one can aver that their behavior as far as discrimination against Jews or colored people was concerned was blameless and most creditable. A most liberal atmosphere was introduced into the Laboratory by Dr. Compton, whose views in this respect are well known and which he always vigorously maintained. No change in this respect came when the Army took over and the peace and collaboration within the Laboratory was excellent, at least as far as the scientific personnel was concerned. There were minor squabbles in the mail room where colored and white girls worked together but these were soon squelched by the unanimous attitude of the scientific personnel who evidently were not willing to tolerate any injustices. As a result, we felt completely free to talk over, not only with Dr. Compton, but also with representatives of the Army, such problems as arose. For instance, when a move of the theoretical group to X was contemplated, we did not know what to do about Wilkins, who was a colored boy. Evidently, the general conditions in the South would have made it undesirable for him to come along as he would have had to live in a different part of the town than the rest of us. As a result, it was decided that he should be transferred to the Argonne in case we should move to Oak Ridge and all arrangements were made for this with the full consent of the Army. Since we remained in Chicago, it was not necessary to make use of these arrangements.

There is another incident in connection with Security that I cannot resist relating. Miss Way was out one evening and she left her bicycle in front of the Laboratory. When she returned to pick up her bicycle at a rather late hour, she noticed a ladder leaning against the Laboratory and reaching up to the third floor. She went to see the guard immediately and told him about the ladder. The guard, as is customary in such cases, did not want to believe her and subjected her to a sort of cross-examination as to the kind of ladder, where she believed she saw it, etc. When she became insistent that he do something about it, he said he couldn't leave the gate. After some delay, he called another guard. They then went to search for a flashlight and left Miss Way alone at the gate. After about ten minutes, they found a flashlight and one of the guards remained at the gate, the other one going with the flashlight and Miss Way to investigate the ladder. When the guard saw it, he said, "Oh, I guess it's O.K., this is the kind of a ladder painters are using."

May 3, 1948

An Intermezzo: Work on Heavy Water Reactors

I. Preliminary Negotiations

Until recently, I would have been inclined to consider the work which we did on heavy water reactors, during about six months in 1943, as a classic example of frustration and futility. I now realize, however, that the work actually did serve two good purposes: It helped us to bridge over the time during which the du Pont Company officials studied in Wheeler's Progressive School of Nuclear Education and it kept us from being even more irritated and impatient than we were. To divert our attention temporarily from the main problems may not have been quite unintentional on the part of the Project Leaders, and if it was not, it demonstrates real skill and imagination on their part. Second, a survey of the possible uses of heavy water had to be made some time and we learned a great deal from this study. The actual purpose of the study, as we saw it – to design a reactor which could be ready before the W reactor would be completed – was never accomplished. However, we have explored many possibilities and obtained an excellent introduction to the later, more extended and more general, studies on nuclear reactors. In addition, one of our two most favored propositions was, though with several substantial changes, executed later on at Chalk River. For me personally, it was a second grade course in my school on human nature.

The project started, as far as I am concerned, with a telegraphic invitation from Compton which I received toward the end of my three and one half day's visit in Wilmington. I was to join him and Urey in a conference on the possible uses of P-9. Urey was reponsible not only for the discovery of heavy water but also for developing some of the most efficient processes to obtain it on a large scale.

The meeting took place a few days after I returned to Chicago. Mr. H. C. Vernon who was to play an important role in the P-9 Project was present also. He was lent, by the du Pont Company, to Urey's project and had completed his assignment at Columbia when the P-9 Project was to be undertaken. Vernon was an agile and industrious engineer, without being actually brilliant and had a reputation commensurate with his abilities both with du Pont Company and also in Urey's Project. He had an obvious loyalty to his parent Company and many commendable words of praise for his previous superiors, particularly Mr. Cooper.

I cannot recollect the list of participants at the initial heavy water conference, nor am I quite sure as to its exact timing. The circumstance that sticks in my mind is that Vernon came to see me in the morning and asked me to give him a practical definition of the multiplication constant. I likened it to the rudder of the boat, or to the steering wheel of a car, which do not determine the position of the vehicle but only the change in direction of its motion. Similarly, the multiplication constant does not determine the actual level of operation of a reactor but only the rate of increase or decrease of that level. Strange as it may sound, the circumstance that the multiplication constant determines only the rate of change of the power level, but does not determine the power level directly, was for most outsiders one of the most difficult things to comprehend and to believe. Vernon, when he asked this question, actually demonstrated that he was not wholly unfamiliar with our work.

I heard a few days later that, when he left my office, Vernon called on Weinberg and Young and asked them the same question and was given a similar explanation. He must have felt very secure of his grounds after these two explanations because he declared at the meeting, which was held during the afternoon, that he was ready to take over the direction of the heavy water work if he were asked to do so. When Urey asked him whether this applied also to the aspects which concerned nuclear physics, he answered with a curt yes. My admiration for the self confidence of engineers was much increased by this but my admiration for their judgment was not.

I do not remember what else the meeting accomplished. We were probably informed of the production schedule for heavy water and a few other pertinent matters.

My conversation with Compton, upon my return to Chicago from Wilmington, and my re-admission to the Policy Committee was recorded before. Naturally, I discussed the heavy water work also with my group, in particular with Young and Weinberg. We soon made the resolve to push this work so vigorously that the heavy water unit would be ready for operation before the W units were. We based this hope on a somewhat optimistic estimate of the effects of the high multiplication constant which moderation by heavy water would afford, on a correct appraisal of the delays and procrastination with which the work on the W units would proceed, but on wholly unrealistic hopes for freedom from red tape and administrative fetters with which our work would be permitted to proceed. It was with respect to these matters that Young admonished me when we ended our conversation: Be tough, boss!

It was largely under the influence of this advice that I wired Dr. Compton, with whom I had an appointment to visit Columbia University and the engineering group there, asking him for an interview before we went up to Columbia. He saw me at his hotel in New York and I told him that, much as I was inclined to be conciliatory and yielding, I felt that I had to resist these tendencies in the present instance. Hence my willingness to undertake the P-9 work was subject to three conditions: First, that Urey be asked to join the Project in a leading capacity, and if he did not, I should be in charge, at least in the early phases until it would be felt that the engineers had acquired reasonable familiarity with the properties of chain reacting units. As a background to this, it should be mentioned that Urey exhibited a great deal of interest in the use of heavy water in chain reacting units. He felt that he, as the discoverer of this material, had a natural claim for furthering its applications. Urey's work at Columbia appeared at that time to approach completion and it was therefore natural to think of him in connection with the direction of the P-9 work in Chicago. The difficulty with this request was not unknown to me: Urey's outspokenness and directness did not ingratiate him with the Army leaders. My second request was that a group of experimental physicists, preferably Creutz's group, be assigned to collaborate on the P-9 project. It was clear to me that purely theoretical work could not be adequate to solve the many technical problems which we were sure to encounter. I also had very vividly in mind how much the W plans owed to the help which we received from Creutz and many others. The third request was that the du Pont Company be given the choice either to participate in the P-9 work wholeheartedly – which would have included their sending representatives to Chicago to whose judgment they would eventually acquiesce – or keeping their hands off.

I was most agreeably surprised when Dr. Compton not only agreed to all three of my conditions but actually started to implement them at once. Unfortunately, it soon became apparent that the work at Columbia would continue to need Urey's attention for some time to come. In spite of this I felt that some measure of success was achieved when Urey was invited to head up the heavy water work and thus at least an infinitesimal part of the honor was bestowed upon him which he so well deserved. Actually, our request for Urey may have strengthened somewhat his position with the Army. When it transpired that Urey would not be available, Compton was quite ready to agree that I be put in charge. He said that nobody was willing to state unequivocally that Vernon, the leader of the engineering group which was to join us from Columbia, was entitled by his past accomplishments to the leading role in the enterprise. Compton's agreement to this point was all the more gratifying because, in an earlier conversation, I had not raised very vigorous objections against subordinating the theoretical group to Vernon. As to the second point, Compton stated that he wanted our work to succeed and would give us, therefore, all help that we needed. He was not sure whether it would be Creutz's group – he had not consulted either Creutz or Cooper to whose division Creutz was assigned at that time. However, some help, and substantial help, would be forthcoming. Concerning the third point, Compton assured me that he would see to it that the du Pont Company's interest in the work be clarified and that we should not be hampered in our work by outside interference.

Compton stood by these decisions, particularly the first, also at Columbia, and Urey certainly did not seem to object to them. Vernon agreed to join the P-9 Project at Chicago together with his group and accepted the position as second in command. I do not know how wholeheartedly he did this. Actually, when the official meeting broke up, I accidentally overheard him drawing Compton into a corner and asking him whether his proposed relation to me was expected to be permanent.

We spent the rest of the day by my giving an introductory talk on chain reacting units to Urey and to his group. Again, I do not remember all who were present but the group did include Vernon, Huffman, Hiskey, Karl Cohen and Irving Kaplan. We discussed the calculation of the multiplication constant, the control of the pile and similar subjects. It pleased me very much that the comments which were made by the group were all clear and to the point. After the meeting, Vernon, Huffman and Hiskey invited me to dinner at the Faculty Club and I became much impressed by the variety of their interests, by the facility with which they spoke and their friendship to each other. It seemed to me that collaboration was bound to be easy with such imaginative and interesting people. Actually, I now believe that I must have somewhat overestimated them. Vernon is agile and industrious but is far from being brilliant. Huffman, I am sorry to say, is not particularly able. Hiskey was an able experimenter and even though I did not become as well acquainted with him as I did with Vernon and Huffman, I am inclined to believe that he was the most imaginative of the three.

II. Implementation of the Agreement

One should think that, all our requests having been granted, the work on the P-9 reactors should have proceeded rapidly and with a minimum of friction. Unfortunately, this was not the case. A large part of the blame for this attaches to me: while I was always able and willing to fight for major issues until final victory or, more often, defeat, I was both emotionally and intellectually unsuited for fighting the war of attrition which followed. Never was I able to refuse a request on my time and, as a result, had to hasten from one appointment to another. Also as soon as I had achieved a minor victory in the war of attrition, I began to be sorry for the man whom I had bested, gave up part of what I had achieved and spent a lot of time conciliating and further convincing him that matters were best as they had been decided. This may sound as a sympathetic trait; actually it was irresponsible. The war of attrition consists of minor victories and their ruthless exploitation. Furthermore, all the reputation one can earn by weakness is a reputation for softness, weakness and lack of consistency. This is fully justified. As Shaw remarked, a very considerate way of driving a cab in London would inevitably result in a traffic snarl and a Saint's way of doing business would only result of confusion. At least a Saint would soon lose his business but I was not enough of a Saint to lose my influence. I was only unable to use it as it should have been used.

The question is: what created the war of attrition? This can be best explained by relating how the three agreements which were made with Compton were implemented.

Urey was unable to come to Chicago. Instead, Smyth was asked to join us to coordinate the heavy water work with the rest of the Laboratory and maintain liaison with the higher-ups (the Policy Committee to which I had been readmitted was used by this time only to relay information to the Laboratory and obtain reports from it. It took no decisions).

Smyth is an agreeable person, perfectly honest and I have a very high regard for his administrative abilities. He can analyse a situation in cool, unbiased fashion and even though he may err occasionally, I have yet to witness him representing an obviously foolish or shortsighted point of view. He is bound to be an asset to any enterprise with which he may become associated and was a valuable asset to our Project also. Reasonable and sufficiently clear arguments never failed to convince him and we could harmonize our points of view almost in every instance. Naturally, Smyth felt that he had to represent the point of view of the "higher-ups" but he was by no means dogmatic and did not fail to see that the higher-ups often committed serious blunders. He soon came around to my point of view in all the larger issues involved and my complaint against him could only be that it took a constant effort to keep him convinced – which was natural, if somewhat time consuming. And, of course, that he was not really willing to go to bat for us but ceded this prerogative to me whenever it was convenient. He also was out of town too much for being really effective in pressing a point through.

Clearly Smyth had neither the enthusiasm, nor the technical skill which Urey would have brought with himself. On the other hand, he did not think that he had them and this made up for 90% of his failure to have them. If it is taken for granted that there was a need for an overall coordinator, and there probably was, we could hardly have hoped for a better man than Smyth turned out to be.

The agreement was that, at least in the early stages of the work, I be the technical director. Stricly speaking, this part of the agreement was never implemented and Vernon joined the Technical Division under Mr. Cooper rather than our group. This did not matter too much in the early parts of our work because my greater familiarity with the problems secured the leadership for me anyway. My acquiescence in Vernon's administrative independence probably would have proved to be a fatal mistake later on – but our Project never got beyond the exploratory stages.

During the exploratory stage of an investigation, it is not the number of collaborators that counts, it is their intelligence, their interest in the problem, the ease with which they can talk over matters, appreciate the good in each other's thoughts and discover flaws in each other's reasoning. The theoretical physics group almost ideally satisfied these conditions: we knew and liked each other, we were interested in the problem and in the criticism which we could offer to each other's propositions.

Vernon's collaborators would not have added much to this, even though some of them did not lack engineering experience and judgment. The worst result of my failure to press for the title of director was that it further encouraged Vernon's separatist tendencies. In addition, I missed the standing which the administrative head of the P-9 Project would have had – and would have needed. There are many questions which can be decided only by experiment and which can be decided by an experiment very easily. Such questions were answered for us during the work for W by Creutz, by Burton and many others who were always eager to help. By the time the P-9 project got under way, the Laboratory was so well organized that even though Creutz and many others were most anxious to help us, they were unable to do so without directive from above. Creutz, in particular, was in the Technical Division, most unhappy there and felt thwarted. He wanted to work on our problems. Even though I spent hours on end asking for his transfer, we never got anywhere.

These negotiations were particularly painful for me. Creutz's superiors, with whom I had to negotiate, had all the time in the world on their hands – they were administrators pure and simple. By repeating the same sentences in a sufficiently boring fashion, they could always outsit me. I wanted to find out how large the chain reacting unit would become in a certain arrangement instead of listening to tedious and repetitive arguments and my patience soon wore thin. Consciously or unconsciously, they took advantage of this circumstance. Had I had a dignified title of a Project Director or even of the Technical Director of the P-9 work, I could have pressed home my point much better.

This relates also to the fate of the second part of the understanding with Compton. However, most unhappiness was caused by my inability of having the third part of the agreement kept. Indeed, if I look back at the way in which agreements were adhered to on the Project, I cannot understand how our negotiators in international affairs can be naive enough not to know that promises always can be circumvented or reinterpreted. If it is by these means that we are bested in our international negotiations, it would only be necessary to replace our negotiators with some of the early administrators of the Uranium Project.

We now know that the du Pont Company has not created for itself a monopoly in the field of nuclear reactors – in fact it is not in this business at all. There is no reason to believe that they ever intended to do so and I owe a silent apology to them for having constantly suspected them of this intention. All that can be said as my excuse is that they certainly acted as if they had this intention and that, as a result, most of the Laboratory personnel thought that they did.

The method of collaboration in the W plans constantly loomed before us as something to be avoided at all costs. We were told that all our conclusions and choices would be carefully rechecked by the du Pont engineers and this appeared to us a monstrous and irritating nonsense. Actually, this rechecking was more a pretense than not and served mostly as an explanation of the use of time which the training of the du Pont engineers took. However, we did not know this at the time. We were absolutely resolved to prevent a repetition of the question-answer game and of the other characteristics of the W arrangement. It is for this reason that my third and strongest request to Compton had been: the du Pont Company, if it wants to build the P-9 reactor, should attach responsible engineers to our project so that these engineers become familiar with the thinking in Chicago, become convinced of the correctness of the decisions taken and can transmit this conviction to the parent company. This arrangement would have made it unnecessary to reopen, before detailed design, every question to which we would propose an answer. If the du Pont Company was not willing to send engineers to Chicago and give them the responsibility of approving the major decisions which were to be taken there, we wanted them to stay out of the job entirely. In this case, we were to be permitted to look for another contractor to detail and execute the plans which we were to formulate.

I must admit that, among the above two alternatives, we would have preferred the second one. As a matter of fact, after my memorable conference with Greenewalt, in the middle of March, this appeared the only practical alternative. However, the du Pont Company successfully thwarted these plans. If we assume that the reason was not that they wanted to remain alone in the field

– as all the Laboratory including ourselves had assumed at that time – the only motive that I can think for their actions is that they were actually afraid that the P-9 reactor might get ready before theirs and that their prestige might suffer as a result.

To the end, the du Pont Company refused to send any engineers to Chicago to collaborate with us. Instead, Mr. T. H. Chilton came up for occasional visits. These occasions were enjoyable enough both because of Mr. Chilton's engaging personality, his clean and obviously straightforward character and also because of his superior technical knowledge from which we benefited on many occasions. However, his visits were short and far between and were clearly not intended to serve the purpose of committing the Company to the plans which we were to draw up. As a further irritation, although nobody asked them to do so, the du Pont Company drew up a time schedule for the P-9 work which purported to show that the P-9 reactor could not be completed before the W reactors. By the time this happened we had already wasted so much time that I was quite willing to believe that their P-9 schedule might be realistic. However, on the basis of the same experience, the time schedule for W appeared most doubtful to me.

All this time I was pressing Smyth hard to see to it that the third part of the understanding with Compton be implemented. However, action was constantly postponed, under all possible pretenses. Finally, on July 17, I got Smyth and Vernon to sign with me a request to the Army, asking for permission to consult with the Dow Chemical Company and with the Monsanto Company on problems of P-9 reactors.

No reply to this request was ever received. On repeated urgings, we were told to wait until a reviewing committee, "which had already been appointed", would survey the plans and advise on further action. It is probably unnecessary to say that none of us believed that this promise was made in good faith, and the reader will see whether it was.

Some of the above are bitter words but they cannot be otherwise if they are to reflect our exasperation. The question naturally comes up why we were treated this way, why we "were given the run around" as Vernon put it. As very often, the superficial disagreements had a more fundamental reason in this case also.

The objectives of the P-9 work were viewed very differently by Compton than they were viewed by us. In fact, as was said before, it now appears to me quite possible, even though not very likely, that the heavy water project was only invented to keep us busy and out of mischief until we would be needed again for W.

The objectives of the P-9 work were summarized in an early memorandum by Drs. Compton and Smyth. The first objective that was stated was to guarantee the success of the Plutonium Project by providing an alternate method for the manufacture of plutonium in case the W plants were to develop unexpected difficulties. In the absence of this contingency, the work was to provide, for future use, more efficient methods for utilizing uranium, with a higher and more easily accessible production of radioactive materials and with the possible production of useful power. Thus the purpose of the work, in Compton's mind, was very different from its purpose as it appeared to us; Compton did not want the heavy water reactors to compete with the W reactors, except in the case of unexpected technical difficulties. We considered the organisational difficulty, into which the W planning ran, to be sufficiently serious to warrant the promotion of competing plans. Compton was perhaps willing to admit to himself that the organization of the W work was not brilliantly conceived. However, he felt that that mistake had been made and could not be corrected any more. We were not willing to admit this and wanted to fight the windmills.

The reader may be inclined to say at this point, that we should have considered the above statement of the objectives of our work on the part of Compton and Smyth as a cancellation of the earlier understanding. Hence, we should not have expected that that earlier understanding would be implemented. He may be right. However, all such statements and directives were so loose and so vague that we have not come to this conclusion. In addition, I believe that a more direct language could have been used if the intention was to inform us of a changed attitude – an attitude which, for all I now know, may have been the correct one.

III. The Work: Some Progress

It was surprising that any work could be done under the conditions – but some could, even if I had very little part in it.

Young and Weinberg carried the ball most of the time and investigated the different systems of piles, one after the other. Many of these systems were not our ideas: the homogeneous reactor was originally suggested by the du Pont engineers, the UF_6 pile even earlier by Harrison S. Brown and supported later also by H. L. Anderson. However, the comparison and detailed evaluation of these – and of our own – suggestions was due to the theoretical group and I believe that they did a very good job.

Vernon's group – about ten of them moved to Chicago soon after the meeting at Columbia University – was initially badly handicapped by the lack of knowledge and experience of their leaders in the field under investigation. In addition, the leaders had some idea that their men have to be kept busy. Since there was, in the early weeks, very little routine work to be done, such work had to be invented. There were endless and most hectic meetings on all the time, and it was attempted to thrash out a working program for all of Vernon's group in a short time. This was, of course, very difficult because nobody had as yet any very definite idea concerning the direction the investigations should take.

Unfortunately, I was pulled in for many of these purposeless meetings and I was too weak and too unsure of myself to put a stop to them by declaring that "the Vernons" should read reports or take a vacation for ten days and come back when the directions were ascertained in which routine work could and should be done. Such a request would have been resented, perhaps, because it might have been deleterious to the morale of Vernon's collaborators. However, it would have been better to insist on it rather than to live from hand to mouth in the matter of directives.

There was some misunderstanding between Vernon's group and "my people" which had its origin in the different attitudes of the two groups toward the notions of priority and intellectual property. Vernon's collaborators tried, in their monthly reports, to summarize the status of the thinking on P-9 reactors and they paid little attention to the question of whose ideas and work they described. They felt that priorities are a matter of no importance and that the project leaders, for whom the reports were intended, were little interested in where the idea, suggestion, or result originated. The physicists, Gale Young in particular, felt that this led to a misrepresentation of the actual accomplishments of the two groups – between which a split was developing. He did not need to be concerned: the reports of Vernon's group, although well written and nicely phrased, were so full of errors and misstatements that it would have been easy, any time, to demonstrate the relative competency of the two groups. There was, for instance, a detailed proposition about thermal syphoning, of which much was made in one report, although any reasonably experienced person, such as myself, could at once tell that it was the result of some miscalculation. "New results" were given in such cases in the report of the following month. Such obvious miscalculations and covering up attempts on the part of Vernon's collaborators went far toward destroying my early illusions about Vernon's and Huffman's leadership, technical reliability and judgment.

Actually, the tasks were probably somewhat too difficult for a group of inexperienced engineers. They were never taught to look at the problems from a distance, to compare their results by shortcuts, to obtain them by more than one method. They just completed their college courses – many of them hurried college courses – and have not yet learned much more than to substitute numbers into formulas. It was not always so easy to know which formula to use. Their leaders, who would have checked them, and whose main function in my opinion should have been to check them, spent their time in assigning problems, keeping everyone busy, scheduling work and writing reports. In addition, they had no flair for looking through things.

The rivalry between the two groups broke out into the open when the report for the reviewing committee was to be written. In the middle of July, Chapin and Huffman submitted Summary Reports for this purpose, describing about ten or twelve types of heavy water units in some detail. About 6 or 8 pages were devoted to each type. Unfortunately, these summary reports were quite full of errors, inconsistencies and unprovable statements. I wanted, instead, to have a relatively brief summary to give the general results of our investigations, each statement of which to be pertinent and supportable. For each type, the summary should give the best estimate for the amount of heavy water needed for one unit, the production rate of one unit, the production rate per unit amount of heavy water, what the most important unsolved problems of the system were and about how soon a unit could be put into operation. As a matter of fact, I had repeatedly written such summaries as our work progressed and had a pretty fair idea of how these summaries should be rewritten and rephrased when they were brought up-to-date for the reviewing committee. A more extended appendix then should contain all the reports which deal with particular types in detail. The appendix was to contain, in particular, Ohlinger's report, based on calculations of Weinberg and Young, on the reactor type which was later selected for Chalk River, Young's calculations on a very attractive, if slightly problematic system to which du Pont's thinking contributed some, and other reports which its writer was willing to defend.

This is what was finally decided upon, after a few somewhat lengthy meetings toward the end of July. The first of these, on July 22nd, did nothing but to tear one of the Summary Reports of Chapin and Huffman to pieces. This was, of course, a somewhat cruel thing to do but could not be avoided. The main body of the report for the reviewing committee was finally written by me, with many improvements and suggestions of others incorporated. It was submitted to the reviewing committee over the signatures of Smyth, Vernon and myself. The reports of Ohlinger and Young were attached in the Appendix. I cannot quite recollect what happened with the Chapin-Huffman Summary Reports, whether they were appended as a second Appendix to be used with caution or whether they were left out entirely.

Recollecting the scope and results of this Report, I feel that a good deal had been accomplished, in spite of the many difficulties and the hectic atmosphere. Ohlinger's, Weinberg's and Young's work was mentioned before. It would be unfair not to mention that Vernon's collaborators have also contributed a number of good ideas, mostly concerning arrangements and equipment. Vernon himself, even though perhaps not brilliant, has a vivid imagination, and Chapin, I am sure, is a very capable and conscientious engineer. It is not surprising, on the other hand, that they were not able to digest the vast amount of already accumulated information during the very short period of six months and that, as a result, their judgment was not yet reliable. After all, as Fermi said, they had to begin by learning that "the neutron is a tiny particle". However, judgment is what is most needed in the early parts of an investigation as we undertook it.

IV. Visit of the Reviewing Committee

By the time the reviewing committee arrived, on August 11, the Laboratory was so much looking forward to the event that it is a wonder indeed that the visit did not end in a great disappointment. We had a great deal of material to present, both technical and policy matters. In addition, we hoped that the Committee would listen to our grievances and recommend the dilution of the syrup in which we were struggling until it became at least of the consistency of heavy oil. We had been promised that the Committee would advise the higher ups whether we should be permitted to consult another company, in addition to the du Pont Company – a permission from which we expected the birth of competition between the companies and of a serious effort for speed and efficiency. We all expected that the visit would be a turning point not only for the heavy water work but for the whole Plutonium Project. The

Committee consisted of four regular members: Dr. Richard Tolman, chairman, Dr. W. K. Lewis, Dr. E. V. Murphree and Dr. E. B. Wilson, Jr. Dr. Tolman had been connected with the Manhattan Project for some time although we did not know exactly in what capacity. He was well known for the courageous stand which he took on several questions during the first world war. Between the two wars, he was a greatly respected member of the staff of California Institute of Technology. Dr. W. K. Lewis was an equally highly respected chemical engineer at Massachusetts Institute of Technology, one of the most able consultants I ever met, either on or outside the Project. Dr. Murphree's role was touched upon before; he was Vice President of the Standard Oil Development Company. Dr. Wilson, the youngest member of the Committee, was the head of a section of the NDRC; in peacetime he was a professor of physical chemistry at Harvard University. In addition to the regular members, Dr. W. Thompson, of the Standard Oil Development Company, attended all the meetings, he was a "ringer" for Murphree. Mrs. O'Leary, General Groves' secretary, took shorthand notices of all the proceedings. We do not know, and did not care, whether she showed her notes to General Groves when she got back to Washington.

An open session occupied the first morning. As I remember it, in addition to the Committee, at least 15 people were present, including Compton, Smyth, Colonel Peterson to represent the Army, Vernon, some of Vernon's group, myself and a few members of my group. Chilton and Greenewalt were present some of the time. A review of the objectives and status of the P-9 reactor work started the proceedings: Vernon and I spoke. The Committee followed our explanation with interest and obvious understanding. There was rather little difference in Vernon's and my exposition, except that Vernon illustrated his address with a few wooden models.

Very soon, however, the discussion shifted to questions of policy, in particular to the crucial issue: whether du Pont or another contractor should detail and execute the plans.

If the higher ups have considered our work only as a time filler, they certainly have not admitted that to the Committee. I reproduce here the notes which I put down into my notebook soon after the Committee left.

"*A. H.*[15] said: Considering du Pont's preoccupation with W pile it is an open question whether they or a new company could build the P-9 pile more quickly.

"*Smyth* said that this is so but a new company would bring in new points of view. *I* tried to say that, in addition to this, the insurance would be increased if a new company were brought in. The W plans depend not only on physics but also on personalities and systems of procedure. In addition to physics, these could be changed. Furthermore the du Ponts should be kept available if something goes wrong on the W pile. Apart from the insurance angle, the morale of the project would be improved by bringing in a new company.

"*Vernon* said that he discounts the point concerning the desirability of the du Ponts being available in case trouble develops with the W pile. He emphasized that the du Ponts not only have learned from us but will have taught themselves a great deal by the time they become available.

---

[15] Means Dr. A. H. Compton

"In the afternoon next day, Chilton emphasized as very important the point which Vernon discounted. On his visit on the 19th, J. A. Wheeler brought up the same point."

Considering the intense interest I took in the proceedings of the Committee, I do not believe that these notes are far from being verbatim. Vernon's favoring the du Pont Company, a fact which he has not mentioned before, struck me like a small scale betrayal and I resented it. This is not expressed in the notes but is evident also from the last paragraph which shows that it was not only I who considered his discounting the preoccupation of the du Pont Company with the W plans as nonsensical.

It is not sure that my words were well chosen during the meeting. However, I believe that the Committee understood me: it seemed to me unreasonable to force on us, without any good technical or administrative reason, cooperation with a company, between which and ourselves a deep seated antagonism had already developed. It would have made cooperation on W – which had to be accomplished – even more tense and difficult and we would have lost all pleasure in the P-9 work. If two partners of an enterprise are in constant quarrel already, they should not start a new joint venture. In addition, other technical and administrative reasons spoke for a new contractor. The P-9 work would become most important if trouble were to develop with the W plans. Such trouble would, however, further strain the already heavily overloaded capacity of the du Pont Company's engineering section and force it to neglect the P-9 work just under those conditions under which it would be most important. Lastly, what I said about changing not only the moderator but also the method of planning in order to obtain full insurance, still appears to me as relevant as it did at the time.

I had, probably, another motive for wanting to bring in a new company: we felt oppressed by the du Ponts and hoped that we could breathe a freer atmosphere of we would not have all our dealings with them. The officials of the du Pont Company, particularly Mr. Greenewalt, probably felt just the other way around: they thought that they could work better with us if we were made to understand that we could not escape their control. This, together with their possible fear that another Company would beat their W with its P-9, was probably responsible for their unrelenting opposition to the creation of a competitive situation. In addition, they may have felt that it was not compatible with the standing of the du Pont Company to be just "one contractor" on the Plutonium work. These are, of course, guesses, but they are the best guesses I am able to make.

To continue with the meeting, the method of operation had to be explored next for both procedures. The supporters of the du Pont Company had to admit, though rather reluctantly, that they had, at the time, no personnel to put on the P-9 job. We had to admit that we did not know any Company in which we would have absolute confidence that it can do the P-9 job. I claimed, however, that this was not necessary and that all we expected from the contractor was bookkeeping, procurement and some standard engineering and construction jobs which could even be subcontracted. As a result, I said, even Sears Roebuck could act as the contractor.

It was after this statement that Tolman asked us what those radical views were, about which he had heard so much. So far, he said, all the discussion was most reasonable. I said that the view about Sears Roebuck's suitability as a contractor was the most radical view I was holding.

Soon after this, the open session came to a conclusion and the Committee proceeded to listen to some of us individually. I believe that I was the first one they wanted to talk to privately. They even asked Colonel Peterson, who apparently wished to remain, to leave them alone.

It is not very clearly in my mind what I said. Of course, the antagonism between the du Pont Company and ourselves must have come up and I explained that that antagonism was stronger on the part of the Company than on our part because, I said, they came to see that they were wrong on many of the controversial issues and because almost everybody dislikes the person who proves to be right against oneself. I must have quoted a number of examples in the W plannning in which our judgment prevailed and explained that any group of people with the overwhelming self-confidence which characterised the du Ponts must have resented that. I tried to be objective in my statement, as I tried to adopt an objective attitude in the work; to what extent I succeeded, I do not know. Tolman took notes of my statement, which he read back when I was through; but his notes gave such a warped version of my views that Wilson spoke up and said that even though I said all that Tolman took down, the sense of my statement was very different from what it would appear from his notes. Mrs. O'Leary's notes, which were rather accurate, were thereupon accepted as the proper reproduction of my statement. Much of this is surely trivial; it is all that remained in my mind from the interview with the Committee.

Naturally, I do not know anything about the other statements which the Committee received. Vernon, Smyth and many others were listened to separately and Greenewalt also demanded to be heard. Next day, we heard that Compton reversed his stand and came out for du Pont as the contractor. Smyth was annoyed because Compton did this without even consulting him and took a rather strong stand.

My general impression of the Committee and its method of procedure was most favorable. I remember that I went to see Weinberg and Young on the third day of the Committee's stay and told them that now we were given an opportunity to speak before an impartial group of men who listened to us carefully, tried to understand us and were intelligent and experienced not only to appreciate the technical problems but also the organizational ones. Whatever their decision would be, we must abide by it and make the best of it.

The Committee devoted the last day of its stay, August 19, to the writing of its report. We were informed of its recommendation in a most straightforward fashion, right the following day, by a copy of their report. I quote from this the part which refers to the P-9 work, omitting only technical details which should remain secret.

"12. *Heavy Water Piles for Production.* The Committee recommends immediate studies with the purpose of selecting and designing the most appropriate type of heavy water piles for the production of amounts of 49 of the order of the rated capacity of one or more Hanford piles. . . .

"The studies on such piles should be carried out by Dr. Wigner's group in close cooperation with a contractor who should be brought in as early as possible to handle engineering phases and detailed design, and who would presumably be the contractor to do the ultimate construction if that were later authorized. It is regarded as important that this contractor should set up headquarters for the work in close proximity to the location of Dr. Wigner's group. This might be in Chicago. Serious consideration, however, should be given to the possibility of relocating and concentrating design and engineering research work on production heavy water piles at some other center, preferably under the auspices of some university.

"In spite of the circumstances that the work of the du Pont Company on the graphite pile has been satisfactory, and in the opinion of the Committee would not have been surpassed by any other organization, the Committee regards it as preferable that some other contractor if possible should be selected for the proposed heavy water work, since this work is planned as insurance against low performance of the Hanford graphite piles and work by du Pont on heavy water piles would divert their effort from the graphite piles, perhaps particularly in the early stages of operation when they may be performing badly.

"The Committee realizes that it will be difficult to find an appropriate contractor other than du Pont, and that this would have the serious disadvantage that time and energy would be lost in educating the new contractor. As possibilities for a new organization, the Committee suggests Kellog, Union Carbide and Carbon, Badger, Foster-Wheeler, Dow and Koppers. The Committee feels that the construction of heavy water piles, if later authorized, should be undertaken in Hanford in order to make use of the separation process installed there. It is realized that this would then make it necessary to arrange proper cooperation between du Pont and the contractor for the heavy water piles."

Clearly, our point of view was accepted by the Committee. We were cautiously jubilant, very cautiously, knowing how often the higher ups had found reasons before for the reinterpretation of promises.

This caution soon proved to be fully justified. Colonel Nichols appeared on the scene with a dispatch which we never experienced if some promise had to be implemented. We had a conference of about four hours which would have been very amusing to a disinterested observer. Its amusing nature, however, was probably lost on all who were actually present: Smyth, Vernon, Colonel Peterson in addition to Colonel Nichols and myself.

Colonel Nichols started out by complimenting the Committee for the excellent work it had done and appeared pleased by the recommendation it made. The only point, he said, which would be difficult to implement is the one in which a new contractor is recommended for the P-9 work. It seemed best to him to entrust this work to the du Pont Company, which we knew already and which has proved itself so eminently able to shoulder its obligations and to cooperate with the Metallurgical Laboratory.

The discussion became, from this point on, a dialogue between Colonel Nichols and myself with almost no variation in our statements. I said that if one returns to the recommendation of the Committee, the first company to approach, perhaps was Kellog or Foster-Wheeler. It would be nice if we could get permission to do that, at an early date, because this permission was to be dependent on the recommendation of the Reviewing Committee from which we just had heard and because we had to wait for this recommendation several months. Colonel Nichols thereupon returned to the discussion of the possibility of entrusting the heavy water work to the duPont Company, which we knew already, etc. I thereupon returned to the recommendation of the Committee and suggested that permission be granted at an early date, etc. At 1 o'clock we adjourned for luncheon. On the way there, Smyth chided me gently for being so stubborn.

Actually, the procedure of the Army was a grave breach of faith. The Reviewing Committee was so often referred to as the ultimate forum for the decision on which we would have to wait but whose decision would be inviolable that we should have been outraged when it was set aside so promptly and lightly. Actually, we were not even greatly surprised.

V. The End

Even though we were not surprised when the report of the Reviewing Committee was discarded when it came out in our favor, this action did kill our enthusiasm. Our confidence in the promises and even the good faith in which promises were made was more than shaken. In fact, if I try to recollect our mood, it does appear to me that we developed a good deal of contempt for the higher ups. Procedures as related before, such as their reaction to Lowen's visit to President Roosevelt, or Compton's subsequent letter to me, were not calculated to restore our confidence.

In spite of this, there was no crisis in our relations, partly because even Colonel Nichols did not say in so many words that the recommendations of the Reviewing Committee will be disregarded, but particularly because we felt that even should we prevail in a fight, so much time had already been lost that the decision was sure to be made on the battlefield or in the German laboratories. This last possibility still haunted us and we felt particularly bitter that we had to compete with the German efforts with such heavy ballast as the Army tied to our feet. This bitterness manifested itself often in, I am afraid, contemptuous jibes such as the one which I made when Captain Parsons and Oppenheimer visited me and I said: "My evaluation of our situation is that the Army fights a delaying action. It fights it very successfully". I now regret, of course, having said this and Parsons, who first wanted to give me advice how to proceed in such a "delicate situation" gave up in a rather obvious fashion when he saw how clumsily I acted during an important consultation. The fact is that I was tired of trying to make those succeed who did not appear to care much one way or the other, and we were too disillusioned to believe that any fight might succeed in time to make much difference. The invasion of Europe was getting underway: Sicily was captured while the Reviewing Committee was in Chicago and the Russians made good progress in the Ukraine. We felt that if the Germans were not ready with an atomic weapon within a few months, the war would be won for us on the battlefields, and that if they had progressed enough to be ready within a few months, we could not beat them anyway. These considerations were false – we now know that the German rockets could have turned the tide of the war had they been completed a couple of months earlier and that we would have been, in that case, sorely in need of the atomic weapon. However, we were tired of fighting the windmills and tired of trying to make our higher ups strive with real vigor for success.

The higher ups covered up their failure to implement the recommendation of the Reviewing Committee by a series of reorganizations. A P-9 steering committee was organized with Smyth, Vernon, Hiskey and myself as members. When this Committee was appointed (Sept. 27) we were assured that there was no indication that the emphasis on the P-9 work would be changed. However, hardly was this Committee appointed, when we were summoned to a conference in Dr. Compton's office in which we were informed, in the presence of Army representatives, that the P-9 work would be given two objectives. The first objective, which would be pursued under Zinn's and Fermi's direction, would be the construction of a lower power experimental unit, to be erected in the Argonne. Some of the attention of Vernon's group would also be diverted toward this objective. The second objective would be the old one of designing a high power reactor but the priority of this objective was to be lowered considerably. It would be, furthermore, the designing rather than the construction of such a reactor. Young and Ohlinger came with me to Compton's office when the decision was communicated and Ohlinger whispered into my ears, quite audibly, that we were again sold down the river.

The experimental unit was indeed erected, essentially by Zinn, in a few months and has served ever since, and is serving even now, as an excellent neutron source, to the work with which we owe a great number of important scientific results. In spite of a very auspicious beginning, Zinn's collaboration with Vernon was of short duration and practically the whole work was done by Zinn and his group.

The work on the second objective, the design of the high powered reactor, was soon reorganized to be carried out by two Committees. The Steering Committee was to make the administrative and policy decisions. It consisted of Allison, Hiskey, Vernon, Zinn, Hogness to represent the Chemistry Division, and myself. The Information Committee, with a much larger membership (15) and including Creutz and Szilard, was to be a working group, mainly concerned with technical aspects. It is doubtful that anybody took either of the two committees seriously. I did not. They were a waste of time.

We made, at Creutz's instigation, another vigorous attempt to get permission to approach other concerns, as recommended by the Reviewing Committee. Of course, nothing came of this. The du Ponts, on the other hand, decided to tempt us with some honey and invited Vernon and me to visit one of their plants where heavy water was to be produced. It was a nice visit and I actually believed it when I was told that the plant was "so much ahead of schedule that it is no fun any more." Actually, it turned out to be a flop, because of an error in the design and never produced more than one third of the design figure.

Even though I appreciated the attitude of the du Pont Company to stretch out a friendly hand and to try to impress us with their skill and accomplishments, I could not be convinced that the future of the heavy water work was with them, in fact I was not sure that it was anywhere. When my attitude became evident, another method was attempted: Vernon received authorization to increase his group so much that I would be automatically in the shadow. Even this did not perturb me greatly.

Meantime, around the middle of December (1943), blueprints began to come in for the W units which required serious review. Young, Ohlinger, Weinberg and I, together with Friedman and Miss Way, carried the burden of this work. We soon found that we could not attend to it properly if we were also to attend the endless meetings of the P-9 committees. I therefore requested, at Young's advice, that we be freed of one of the two responsibilities.

General Groves happened to pass through Chicago soon after we made this request. He saw me, together with Colonel Peterson, in his car and was most gracious and reasonable. At this occasion, and at a subsequent meeting in the Army office at which Dr. Compton was present also, it was decided that the second objective of the P-9 work be abandoned "temporarily". I suggested that Vernon be authorized to continue the work alone but this suggestion was not accepted. None of us believed that it was a good idea.

This decision was, of course, a heavy blow to Vernon who was made to play, during much of the P-9 work, the role of a straw bogey. He was discarded unceremoniously as soon as his role was over. Even though I had not seen eye to eye with him much of the time, I was sorry for him because he did deserve better. I helped him to place his people and to keep his face as much as I could. He commented: You are a strange kind of a man, Eugene.

Glossary of Names Mentioned
in Memoir of the Uranium Project

Abelson, Philip H.
  Physicist, University of California

Adamson, Col. Keith F.
  Army ordnance specialist

Allison, Samuel K.
  Professor of Physics, University of Chicago, and first Director of Chicago Metallurgical Laboratory

Anderson, Herbert L.
  Physicist at Columbia, later at Chicago

Babcock, Dale
  du Pont engineer

Beams, Jesse
  Profesor of Physics, University of Virginia

Boissevain, M. G. J.
  Belgian engineer, attached to Chicago Project as part of an agreement under which uranium from Belgian Congo was made available to U.S.

Borst, Lyle
  Physicist at Chicago, later at Clinton Laboratories

Boyd, George
  Chemist at Chicago, later at Clinton Laboratories

Breit, Gregory
  physicist, University of Wisconsin

Briggs, Lyman J.
  Director, National Bureau of Standards

Brown, Harrison
    Chemist at Chicago, later at Clinton Laboratories
Burton, Milton
    Chemist at Chicago, later at Clinton Laboratories
Bush, Vannevar
    Head of Office of Scientific Research and Development
Byers, Horace
    Administrator at Chicago

Carpenter, Walter S., Jr.
    President, du Pont Company
Chilton, Thomas
    du Pont engineer
Christy, Robert
    Physicist at Chicago, later at Los Alamos
Cohen, Karl
    Physicist at Columbia, later at Chicago
Compton, Arthur H.
    Professor of Physics, University of Chicago, and Head of Metallurgical Project
Conant, James
    Director, National Defense Research Committee
Condon, Edward
    Physicist, member of Uranium Section of NDRC
Cooper, Charles
    Senior du Pont engineer at Chicago
Coryell, Charles
    Chemist at Chicago, later at Clinton Laboratories
Creutz, Edward C.
    Physicist at Princeton, later at Chicago

Dancoff, Sidney
    Physicist with Wigner's group
Dirac, P. A. M.
    British physicist
Doan, Richard
    Physicist at Chicago and first Research Director of Clinton Laboratories
Dunning, John
    Physicist at Columbia University

Fermi, Enrico
    Head of Physics Division, Chicago Metallurgical Laboratory

Foote, Frank
    Metallurgist at Chicago
Franck, James
    Physical chemist at Chicago
Friedman, Francis
    Physicist with Wigner's group

Generaux, R.
    du Pont engineer
Greenewalt, Crawford H.
    du Pont engineer, responsible for liaison with Metallurgical Laboratory
Graves, Alvin
    Physicist at Chicago, later at Los Alamos
Groves, General Leslie R.
    Head of Manhattan Project
Gunn, Ross
    Technical Adviser, Naval Research Laboratory

Hahn, Otto
    German chemist, co-discoverer of fission
Halban, Hans von
    German physicist at Chalk River
Herb, Ray
    Physicist, University of Wisconsin and builder of van de Graaff accelerators
Hilberry, Norman
    Physicist at Chicago, later Director, Argonne National Laboratory
Hogness, Thorfin
    Professor of Chemistry, University of Chicago
Hoover, Gilbert C.
    Commander, U.S. Navy
Howe, John
    Metallurgist at Chicago
Huffman, John
    Engineer at Columbia, later at Chicago
Hughes, Donald
    Physicist at Chicago

Ibser, H. I.
    Physicist with Wigner's group

Jeffries, Zay
    Metallurgist, General Electric Company
Joliot-Curie, Frédéric
    French physicist

Jones, Haydn
Physicist at Chicago, later at
Clinton Laboratories
Jupnik, H.
Physicist at Princeton

Kaplan, Irving
Physicist at Columbia, later at
Chicago
Kistiakowsky, George
Professor of Chemistry, Harvard
University, later at Los Alamos
Kowarski, Lew
French physical chemist, later at
Chalk River

Lansdale, John, Jr.
Lieutenant-Colonel, U.S. Army,
in charge of security for
Manhattan Project
Lawrence, Ernest O.
Professor of Physics, University of
California and Head of Electromag-
netic Separation Project
Leverett, Miles
Engineer at Chicago
Lewis, W. K.
Professor of Chemical Engineering,
Massachusetts Institute of Technol-
ogy
Lowen, Irving
Physicist with Wigner's group

Manley, John
Physicist at Chicago, later at Los
Alamos
Meitner, Lise
German chemist
Mitchell, Dana
Physicist at Chicago
Monk, Ardis T.
Mathematican with Wigner's group
Moore, T. V.
Engineer at Chicago
Morrison, Philip
Physicist at Chicago
Murphree, Eger V.
Chemical engineer with Stan-
dard Oil Development Company
and head of Planning Board on
Uranium of OSRD

Nichols, Kenneth D.
Colonel, U.S. Army

Ohlinger, Lee
Engineer with Wigner's group
Oppenheimer, J. R.
Professor of Physics, University
of California and Director of Los
Alamos

Parsons, William S.
Captain, U.S. Navy
Pegram, George
Chairman, physics department,
Columbia University
Peterson, Arthur V.
Colonel, U.S. Army
Plass, Gilbert
Physicist with Wigner's group
Purnell, William R.
Admiral, U.S. Navy

Roberts, Richard B.
Physicist, Carnegie Institution

Sachs, Alexander
Economist with Lehmann Corpora-
tion
Seaborg, Glenn
Chemist at Chicago, co-discoverer
of plutonium
Seitz, Frederick
Physicist with Wigner's group
Shapiro, Edward
Chemist at Chicago, later at Clin-
ton Laboratories
Smyth, Henry D.
Chairman, physics department at
Princeton
Snell, Arthur
Physicist at Chicago, later at Clin-
ton Laboratories
Snyder, Thoma A.
Physicist, Princeton University
Spedding, Frank
Professor of Chemistry, Iowa State
University
Stearns, Joyce
Physicist at Chicago
Strassmann, Fritz
German chemist, co-discoverer of
fission
Styer, Wilhelm D.
General, U.S. Army
Szilard, Leo
Physicist at Columbia and Chicago

Teller, Edward
Physicist at Chicago, later at Los Alamos
Thiele, Ernst
Chemical engineer, Standard Oil of Indiana Company
Thompson, W. I.
Engineer, Standard Oil Development Company
Tolman, Richard C.
Professor of Physics, California Institute of Technology
Turner, Louis
Professor of Physics, Princeton University
Tuve, Merle
Physicist, Carnegie Institution

Uhlenbeck, George
Professor of Physics, University of Michigan
Urey, Harold C.
Professor of Chemistry, Columbia University and Head of Gaseous Diffusion Project

Van Vleck, John H.
Professor of Physics, Harvard University
Vernon, H. C.
du Pont engineer

Way, Katherine
Physicist with Wigner's group

Weinberg, A. M.
Physicist with Wigner's group
Weisskopf, Victor
Physicist at Los Alamos
Wheeler, John A.
Physicist at Chicago, advisor to du Pont
Whitaker, Martin D.
Physicist at Chicago, first Director of Clinton Laboratories
Wilkins, J. Ernest
Mathematician with Wigner's group
Williams, Roger
Head of du Pont TNX (plutonium project) Division
Williamson, Robert
Physicist with Wigner's group
Wilson, E. B.
Professor of Physical Chemistry, Harvard University
Wilson, Robert R.
Physicist, Princeton University
Wilson, Volney
Physicist at Chicago
Worthington, Hood
du Pont Engineer

Young, Gale
Physicist with Wigner's group

Zinn, Walter H.
Physicist, Fermi's chief assistant and first director of Argonne National Laboratory

Glossary of Code Words and Abbreviations Used at Chicago

Because of war-time secrecy, code words were used to designate elements, sites, and materials that disclosed the nature of the work at Chicago. These code words are the following:

Heavy Elements

Designated by the last digit of the atomic number followed by the last digit of the atomic weight: thus ^{235}U: 25; ^{239}U: 28; ^{239}Pu: 49; etc.

Sites

Site B: A site in Chicago for development of uranium and aluminium metallurgy
W: Hanford, Washington
X: Oak Ridge, Tennessee
Y: Los Alamos, New Mexico

Materials

P-9 ("Product #9"): Heavy Water
Tube-alloy: Uranium

Organizations

NDRC: National Defense Research Committee
OSRD: Office of Scientific Research and Development

PART III

Articles, Reports, and Memoranda
on Nuclear Energy

The following reports bearing Wigner's name are compiled primarily from the files of the Technical Information Center of the U.S. Department of Energy and from the laboratory records at Oak Ridge National Laboratory. The reports were originally numbered according to the following system:

A: Reports written while Wigner was still in Princeton and before the Metallurgical Laboratory was organized.

C-: Chicago Metallurgical Laboratory reports. The suffixes F, P, E, C, H denote: F – fast neutron (bomb-related); P – physics; E – engineering; C – chemistry; H – health physics. Memoranda received various designations, e.g., MUC, N, etc.

Mon-: Clinton Laboratories reports during Monsanto Company's tenure, 1946–1947.

MDDC: Manhattan District declassified reports. These were re-issues of wartime reports.

AECD: Declassified Atomic Energy Commission reports.

The most significant papers are reproduced in their entirety. Preliminary versions of the papers, numerical calculations of specific chain-reacting systems, or brief notes, are quoted by title and are designated with the letter T. Copies of these reports are on file at ORNL.

Many of the papers are co-authored. However, I should explain that, almost without exception, whenever a paper carried Wigner's name, he either wrote it, or checked it in detail, being in every sense the senior author.

The Pre-Chicago Days

Annotation by Alvin M. Weinberg

1. Wigner, E. P., "For discussion of Homogeneous and Lattice Arrangements", C-1 (No date, probably 1941).

In this paper Wigner suggests that the lumps of uranium in a heterogeneous lattice ought to be smaller than had originally been proposed by Szilard. Note that the moderator is referred to as the "slower", the word "moderator" not yet having been invented (by Wheeler).

2. Wigner, E. P. and Breit, G., "Approximation for Radius of Sphere Sufficient for Chain Reaction in Light Isotope", CF-3, January 9,1942.

Though this report is dated January 9, 1942, it was written at the latest in 1941 and certainly before Wigner arrived in Chicago. Wigner and Breit had realized almost from the beginning that a bomb would be feasible since the critical mass of ^{235}U was so small. Curiously, Breit and Wigner assumed that the neutron flux vanished at the geometric surface of a critical sphere rather than at the extrapolated end-point. Their estimates of the size of a bare sphere were therefore too high. Notable in the Wigner-Breit calculation was the expansion of the neutron flux in spherical harmonics. This is the first appearance of this widely-used technique in the nuclear project literature.

3. Wigner, E. P. "Diffusion of Slow Neutrons in Absorbing Materials", A-20, probably 1941.
4. Wigner, E. P., "Solution of Boltzmann Equation for Monoenergetic Neutrons in an Infinite Homogeneous Medium", (CP-1120) (A-1608), November 30, 1943.

Paper 3 & 4 enabled Wigner to estimate the error involved in using simple diffusion theory rather than transport theory for calculations of the optimal size of a multiplying lattice. Paper 4, written in Chicago deduces the non-asymptotic behavior of neutrons near a source.

5. E. Creutz, H. Jupnik, T. Snyder, and E. P. Wigner, "Review of the Measurements of the Resonance Absorption of Neutrons by Uranium in Bulk", *J. Appl. Phys.* 26, 257 (1955).

6. E. P. Wigner, E. Creutz, H. Jupnik, and T. Snyder, "Resonance Absorption of Neutrons by Spheres", *J. Appl. Phys.* 26, 260 (1955). (Original version by R. Wilson, E. Creutz, and E. P. Wigner appeared as Report A-183, May 29, 1942).
7. E. Creutz, H. Jupnik, T. Snyder, and E. P. Wigner, "Effect of Geometry on Resonance Absorption of Neutrons by Uranium", *J. Appl. Phys.* 26, 271 (1955).
8. E. Creutz, H. Jupnik, and E. P. Wigner, "Effect of Temperature on Total Resonance Absorption of Neutrons by Spheres of Uranium Oxide", *J. Appl. Phys.* 26, 276 (1955).
9. E. Creutz, R. R. Wilson, and E. P. Wigner, "Absorption of Thermal Neutrons in Uranium", A-40, September 26, 1941.

The papers 5–9 record results of experiments on absorption of resonance and thermal neutrons in spheres of uranium performed under Wigner's guidance at Princeton in 1941. These papers appeared originally as project reports but all except Paper 9 were subsequently published in the open literature. These experiments provide the data on resonance absorption that Wigner used in all calculations of multiplying lattices; Paper 9 gives a value for the diffusion length of thermal neutrons in uranium.

Wigner has told me that at the time Szilard and Fermi pointed out the advantages of the heterogeneous over the homogeneous arrangement of uranium, the resonance absorption was believed to occur almost entirely at the surface of the lump. Szilard therefore visualized the uranium lumps as being very large so as to minimize the surface. Wigner's calculation, Paper 5, as well as the experiments, showed that the volume resonance absorption was very important. Wigner therefore realized that the optimum size of the uranium lumps was much smaller than had been estimated at the time the advantage of lumping was first recognized. Wigner suspected this at the time he wrote Paper 1.

10. E. P. Wigner, F. L. Friedman, A. T. Monk, G. N. Plass, A. M. Weinberg, "Density of Neutrons in Carbon Block with and without Absorbing Material", C-186, July 16, 1942.

This paper, though written after Wigner came to Chicago, is largely based on calculations performed at Princeton in support of the experiments reported in papers 5–9. This is the earliest use by Wigner of Fermi's age theory for the slowing down of fast neutrons.

T. Halban, H. V. and Wigner, E. P., "Possibility of a Resonance Absorption in Carbon", A-134, March 30, 1942.
T. Wigner, E. P. and Smyth, H. D., "Radioactive Poisons", A-73 (No date, probably 1941).

The possibility of using fission products as agents of radiological warfare was often discussed at the Chicago Metallurgical Laboratory during the war. This paper contains the earliest estimate of the feasibility of using fission products for this purpose.

1.

For Discussion of the Homogeneous and Lattice Arrangements for Power Plants

E. P. Wigner

No date, probably 1941

1. It is recognized at present that, for a power plant using natural U, the lattice arrangement[1] has great advantages compared with a homogeneous mixture of the U and the slowing down material. In spite of this, it may be worth while to investigate the possibility of a power plant with a homogeneous mixture, particularly since such an investigation also affords, on the basis of some more recent experimental and theoretical results, a quick appraisal of the possibilities of a power plant which uses the lattice arrangement. The consideration of the power plant with a homogeneous mixture of U and slowing down material is so simple that one may have considerable confidence in it. It is possible also to discuss the uncertainties inherent in the consideration of the lattice arrangement.

If a neutron is slowed down in a non-absorbing material, its energy will decrease discontinuously. The probability that, during the slowing down process, its energy once fall into the energy region E, $E + dE$ is[2]

$$dE/\xi E \qquad (1)$$

where, for a simple substance with atomic weight M the ξ of (1) becomes

$$\xi_M = 1 - \frac{(M-1)^2}{2M} \ln \frac{M+1}{M-1} \qquad (1\,a)$$

and in general

$$\xi = \frac{\sum N_M \xi_M \sigma_M}{\sum N_M \sigma_M} . \qquad (1\,b)$$

In (1b), N_M is the relative number of atoms with mass number M, and σ_M is the scattering cross section for these atoms. The value of ξ_M is very close to $2/(M+1)$: it is 1 for H (hydrogen); .725 for D (deuterium), .158 for C; etc.

Let us now consider a mixture of slowing down material and U. The relative number of U atoms be N_U. The absorption cross section of U for neutrons of

---

[1] L. Szilard Report A-55. This report already contains many of the results to be derived here.

[2] G. Placzek

energy E be $\sigma(E)$. If a neutron has the energy E, the probability that it shall be absorbed rather than slowed down further is

$$\frac{N_U \sigma(E)}{N_U \sigma(E) + \sum N_M \sigma_M} < \frac{N_U \sigma(E)}{\sum N_M \sigma_M}. \tag{2}$$

The probability that the energy of the neutron once assume a value between E and $E + dE$ during the slowing down process is given by (1). The probability that it be absorbed while its energy is in that region is the product of (1) and (2). The probability that it shall *not* be absorbed in that energy region is

$$1 - \frac{dE}{\xi E} \frac{N_U \sigma(E)}{N_U \sigma(E) + \sum N_M \sigma_M} > 1 - \frac{dE}{E} \frac{N_U \sigma(E)}{\sum N_M \xi_M \sigma_M} \tag{2a}$$

where (1 b) also has been used. The probability p_1 that the neutron be slowed down to thermal regions without being captured by the U is

$$p_1 = e^{-\beta N_U} \tag{3}$$

where

$$\beta = \int \frac{\sigma(E)}{E} dE / \sum N_M \xi_M \sigma_M. \tag{3a}$$

The integral in (3 a) must be extended from the thermal region to the initial energy of the neutrons. In the following, p_1 will be assumed to be equal to $e^{-\beta N_U}$ which will make the chances for a chain reaction in a homogeneous mixture appear smaller than they actually are.

The probability p_2 that the thermal neutron be absorbed by U rather than slowing down material is

$$p_2 = \frac{N_U a_U}{a_s + N_U a_U} = 1 - \frac{a_s}{a_s + N_U a_U} \tag{4}$$

where a_U and $a_s = \sum N_M a_M$ are the capture cross sections of U and the slowing down material for thermal neutrons. If for any N_U the product $p_1 p_2$ is larger than $1/\eta$, a divergent chain reaction is possible in the corresponding homogeneous mixture:

$$p_1 p_2 = \left(1 - \frac{a_s}{a_s + N_U a_U}\right) e^{-\beta N_U} > \frac{1}{\eta} \tag{5}$$

η is the number of fission neutrons produced per thermal neutron absorbed by U. Naturally one will choose N_U so that $p_1 p_2$ be as large as possible. This gives for $Y = a_s + N_U a_U$ the equation

$$Y^2 - a_s Y - a_s a_U / \beta = 0. \tag{6}$$

It is sufficiently accurate to write $Y = \sqrt{a_s a_U / \beta} + 1/2 a_s$ or

$$N_U = \sqrt{a_s / \beta a_U} - 1/2 \, a_s / a_U \tag{6a}$$

and in most cases one can drop the second term of (6 a). By choosing the somewhat inaccurate N_U of (6 a) the chances for divergent chain reaction will appear only very little smaller than they really are. Inserting (6 a) into (5) gives for the condition of the divergent chain reaction that the multiplication factor

$$\eta p_1 p_2 = \eta \frac{1 - z/2}{1 + z/2} \exp(-z + z^2/2) > 1 \tag{7}$$

where

$$z^2 = \frac{\beta a_s}{a_U} = \frac{a_s}{a_U} \frac{\int \sigma(E)\,dE/E}{\sum N_M \xi_M \sigma_M}. \tag{7 a}$$

The second factor of (7) corresponds to the loss of neutrons caused by radiative capture by U, the first factor corresponds to the loss of neutrons caused by the absorption of thermal neutrons by the slowing down material. Of course, (6 a) and (7) have a physical significance only if $N_U > 0$ i.e. $z < 2$.

2. The following data are available for the calculation of z^2:

$$a_s = 1.2 \times 10^{-27}\,\text{cm}^2 \qquad\qquad\qquad \text{for D}$$
$$a_s = 4.5 \times 10^{-27}\,\text{cm}^2 \qquad\qquad\qquad \text{for C} \tag{I}$$
$$a_s = (2 \times 1.2 + 1 \times 4.5) \times 10^{-27} = 6.9 \times 10^{-27}\,\text{cm}^2 \text{ for CD}_2.$$

The first result is due to E. O. Lawrence (Report A-44), and Halban and Kowarski (Report B-8); the second to Anderson and Fermi (Report A-21, amended value); also Halban and Kowarski (Report B-8). The

$$\int \sigma(E)\,dE/E = 240 \times 10^{-24}\,\text{cm}^2 \tag{II}$$

is also due to Anderson and Fermi (Report A-2). The

$$a_U = 6 \times 10^{-24}\,\text{cm}^2 \tag{III}$$

has been found by Fermi and Weil (Report A-1) and by Creutz, Wilson and Wigner (Report A-40). For σ_M one has to use a scattering cross section which is valid between the original energy of the fission neutrons ($\sim$ 2 Mev) and thermal energies. This is, for D, an average between 2.2 and $3.3 \times 10^{-24}\,\text{cm}^2$ (Aoki, Proc. Phys. Math. Soc. of Japan, 21, 75, 1939; Zinn, Seely, Cohen, Phys. Rev. 56, 260, 1939 and Hanstein, Phys. Rev. 59, 494, 1941) and was taken to be $2.8 \times 10^{-24}\,\text{cm}^2$. For C, it is an average between about 2.7 and $4.8 \times 10^{-24}\,\text{cm}^2$, (Good and Scharff-Goldhaber, Phys. Rev. 59, 917, 1941; Goloborodko-Leipunski, Phys. Rev. 56, 891, 1939 and Hanstein, l.c.) and was taken as $3.8 \times 10^{-24}\,\text{cm}^2$. This gave, with the ξ_M quoted above

$$\sum N_M \xi_M \sigma_M = .725 \times 2.8 \times 10^{-24} = 2.0 \times 10^{-24}\,\text{cm}^2 \quad \text{for D}$$

$$\sum N_M \xi_M \sigma_M = .158 \times 3.8 \times 10^{-24} = .60 \times 10^{-24}\,\text{cm}^2 \quad \text{for C} \tag{IV}$$

$$\sum N_M \xi_M \sigma_M = (2 \times 2 + .60) \times 10^{-24} = 4.6 \times 10^{-24}\,\text{cm}^2 \text{ for CD}_2.$$

From these, we obtain for

$$z = .147 \text{ for D} \qquad .545 \text{ for C} \qquad .245 \text{ for CD}_2 . \qquad (8)$$

For $p_1 p_2$ this gives with (7)

$$p_1 p_2 = .76 \text{ for D} \qquad .38 \text{ for C} \qquad .64 \text{ for CD}_2 . \qquad (9)$$

This has to be compared with $1/\eta$ for which the most recent value is, according to recent measurements of Fermi and Anderson, $1/\eta = .72$ showing that a chain reaction is probably possible in a homogeneous mixture (1:840) of U and D but impossible in a homogeneous mixture of U and C or even U and heavy paraffin. The most uncertain of the above data is, no doubt (II) but there are uncertainties involved also in the others.

3. The preceding considerations permit to draw some reasonably well founded conclusions for the prospects of a chain reaction in a lattice arrangement. In such an arrangement, for the same ratio N_U of U and slowing down molecules, the loss caused by the absorption of the latter is larger, the loss caused by the radiative capture is smaller. In both the thermal and the non-thermal region, the inner part of the U sphere is protected from the neutrons by the outer portions. This has a favorable effect as far as the radiative capture absorption is concerned but an unfavorable as far as the absorption of thermal neutrons is considered.

The reduction of the radiative capture can be estimated from recent measurements by Creutz and Wilson. They show that the absorption caused by radiative capture in a solid sphere is 13 times smaller than the absorption in very thin samples, i.e. in dilute material. This factor 13 seems to be, even to somewhat disconcerting extent, independent of the size and density of the sphere. We can take it into account by decreasing the value of the integral (II) by 13.

On the other hand, the absorption of thermal neutrons is smaller in a solid U sphere than in the same amount of U if it is spread uniformly over the whole material. The corresponding factor can be obtained by comparing the ratio of the absorption in the slowing down material and in the U in case of uniformly spread U with the same ratio in case of lumped U. In the former case, $1/p_2 - 1$ is given by (4) to be $a_s/N_U a_U$; in the latter, a more elaborate calculation is necessary which will be reported separately. The results for $1/p_2 - 1$ are summarized in the following Table I for $U_3 O_8$ and in Table II for the metal; r_0 is the radius of the U sphere, $r_0 + r_1$ the radius of a sphere the volume of which is equal to the volume of a cell. The $U_3 O_8$ is assumed to have a density 6, the metal a density 18.

As is to be expected, the "disadvantage" of the lattice arrangement, from the point of view of thermal neutron utilization, increases with increasing size of the spheres. This, together with the relative independence of the "advantage" of the lattice arrangement from the point of view of avoidance of radiative capture, seems to indicate that a lattice with relatively small U spheres offers most prospects. However, this is not a safe conclusion, since the reason for the independence of the resonance absorption on the size of the sphere is not understood at present and may be spurious.

Table I. Values for $(1 - p_2)/p_2$ for oxide, graphite-moderated[3]

	Uniformly spread oxide	Solid spheres	Ratio
$r_0 = 11$ cm $r_1 = 6$ cm	.0132	.029	2.17
$r_0 = 11$ cm $r_1 = 12$ cm	.04	.111	2.77
$r_0 = 3.65$ cm $r_1 = 3$ cm	.025	.031	1.26
$r_0 = 3.65$ cm $r_1 = 6$ cm	.087	.112	1.29
$r_0 = 3.65$ cm $r_1 = 9$ cm	.201	.267	1.33

Table II. Values for $(1 - p_2)/p_2$ for the metal, graphite-moderated[4]

	Uniformly spread U	Solid spheres	Ratio
$r_0 = 5.92$ cm $r_1 = 9$ cm	.0208	.07	3.35
$r_0 = 5.92$ cm $r_1 = 15$ cm	.0527	.23	3.85
$r_0 = 3.94$ cm $r_1 = 9$ cm	.0480	.11	2.30
$r_0 = 3.94$ cm $r_1 = 15$ cm	.153	.38	2.50

One can correct for the increased value of p_2 in the lattice arrangement as compared with the homogeneous arrangement by increasing a_s by a factor 2 (a 3.65 cm oxide sphere is probably too small, a 5.9 metal sphere certainly too large). On the whole, z will be decreased by a factor of about $\sqrt{2/13} = .39$ so that the z for the solid spheres will become for D, C, and heavy paraffin

$$z = .051 \text{ for D} \qquad .21 \text{ for C} \qquad .095 \text{ for } CD_2. \qquad (8\,a)$$

[3, 4] "graphite-moderated" added by A.M.W. and A.M.P.

The corresponding $p_1 p_2$ become

$$p_1 p_2 = .90 \text{ for D} \qquad .67 \text{ for C} \qquad .83 \text{ for CD}_2 . \tag{9a}$$

The number of U atoms per unit (D, C, CD$_2$) of slowing down substance is readily obtained from (6a) which can be rewritten to

$$N_U = \frac{1}{\beta}(z - z^2/2) \tag{10}$$

where we have to put this time

$$\beta = 1/13 \int \frac{\sigma(E)}{E} \, dE / \sum N_M \xi_M \sigma_M .$$

This gives

$$N_U = .0054 \text{ for D}, \quad .00617 \text{ for C}, \quad .0224 \text{ for CD}_2 , \tag{10a}$$

and for the optimal ratio of weight of slower to weight of U

$$\frac{w_s}{w_U} = 1.56 \text{ for D}, \quad 8.17 \text{ for C}, \quad 2.62 \text{ for CD}_2 . \tag{11}$$

4. Evidently, the preceding figures mean, in the case of deuterium or heavy paraffin, only that the chain reaction could be easily established with these materials if they were available in sufficient amounts. In the case of C, one can attempt an evaluation of the possibility of a chain reaction in the following way.

In the intermediate scale experiment of Fermi and Anderson, the ratio of volumes of carbon and of U_3O_8 was 7:1. The densities were 1.6 and 3.2. Of the oxide, 238/281 is U, the rest oxygen, which we shall count as C. Hence the ratio of weights is $(7 \times 1.6 + 3.2 \times 43/281) : 3.2 \times 238/281 = 4.3 : 1$ instead of a considerably higher optimal ratio, as given in (11). Similarly $N_U = .0117$ instead of .0062 and $p_1 = .698$, $p_2 = .89$ and $p_1 p_2 = .62$ showing that an 8 percent improvement in $p_1 p_2$ and hence also in the multiplication factor could be achieved by decreasing the amount of oxide in the arrangement. This estimate should be substantiated by showing that the reducing of the amount of oxide does not necessarily involve a decrease of the above mentioned "advantage factor" of the lattice arrangement (ratio of resonance absorption of uniformly spread U to the absorption of U concentrated in a lattice) below, or an increase of the "disadvantage factor" (ratio of absorption of thermal neutrons by C if U is concentrated to absorption by C if U is uniformly spread) above, the value realized in the experiment of Anderson and Fermi.

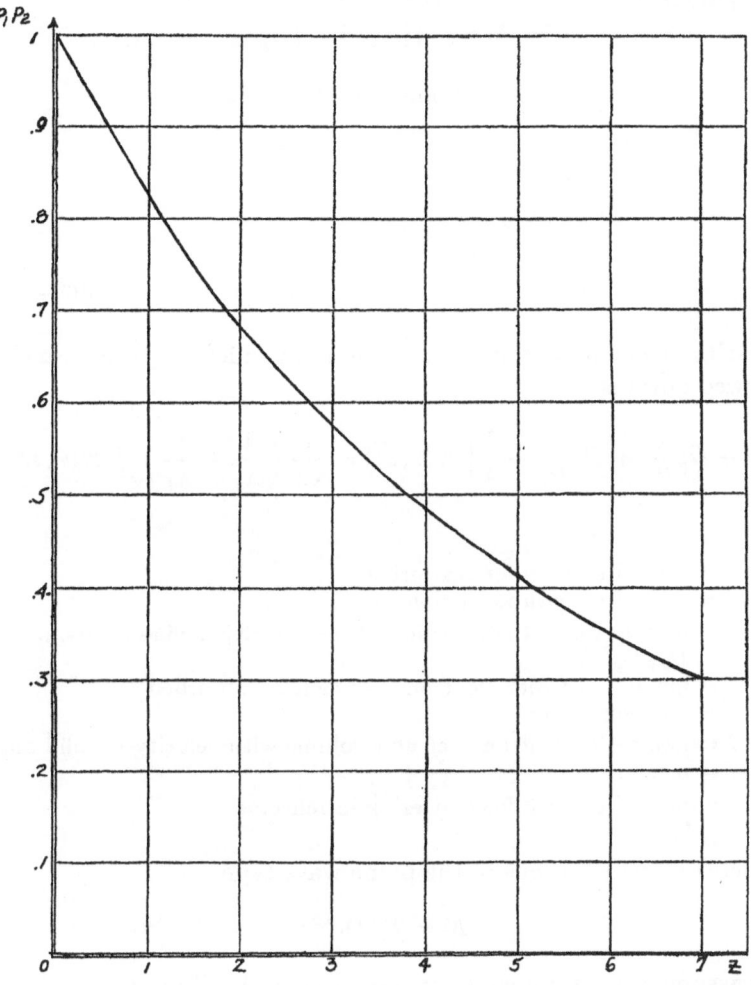

2.

Approximation for Radius of Sphere Sufficient for Chain Reaction in Light Isotope

E. P. Wigner and G. Breit

January 9, 1942

1. The Boltzmann equation inside a medium in which neutrons travel without being slowed down is

$$\left[\Omega_x \frac{\partial}{\partial x} + \Omega_y \frac{\partial}{\partial y} + \Omega_z \frac{\partial}{\partial z} + \frac{1}{\lambda} \right] N(\vec{r}, \vec{\Omega}) = \frac{1}{4\pi} \left(\frac{1}{\lambda_s} + \frac{\eta}{\lambda_f} \right) \int N(\vec{r}, \vec{\Omega}) \, d\vec{\Omega} \quad (1)$$

where

λ_s	= mean free path for scattering
λ_f	= mean free path for fission
$\vec{\Omega}$	= unit vector in the direction of the velocity having components $\Omega_x, \Omega_y, \Omega_z$
η	= number of fission neutrons per neutron absorbed
$\vec{r}$	= position vector
$N(\vec{r}, \vec{\Omega}) \, d\vec{\Omega}$	= number of neutrons per unit volume with velocity in solid angle $d\vec{\Omega}$
λ	= mean free path; $\frac{1}{\lambda} = \frac{1}{\lambda_s} + \frac{1}{\lambda_f}$
λ_r	= mean free path for scattering in reflector.

For neutron distributions of the plane wave type

$$N = f(\vec{\Omega}) e^{i\beta z} \quad (2)$$

one may assume axial symmetry for the angular distribution

$$f(\vec{\Omega}) = f(\mu), \quad \mu = \Omega_z \quad (3)$$

The function f may be expanded in Legrendre functions of argument μ

$$f = f_0 + f_1 P_1(\mu) + \ldots + f_L P_L(\mu) + \ldots \quad (4)$$

Substituting (4) into (1) one has

$$f = \left(\frac{1}{\lambda_s} + \frac{\eta}{\lambda_f} \right) f_0 \bigg/ \left(i\beta\mu + \frac{1}{\lambda} \right) \quad (5)$$

which gives the angular dependence of the velocity distribution for assumed β. The value of β follows by integrating the above expression with respect to μ. This eliminates f_0 and one obtains

$$\frac{\beta\lambda}{1+(\eta-1)\lambda/\lambda_f} = \tan^{-1}\beta\lambda \tag{6}$$

with the understanding that $\tan^{-1}\beta\lambda < \pi/2$. The above procedure is practically a repetition of Wigner's considerations in Report A-20. For $\eta > 1$ equation (6) has real roots for β and in this respect the present case differs from the case of neutron absorption.

A sphere of U^{235} if it is not surrounded by a reflector of neutrons can be assumed to have zero neutron density on the surface. One may set

$$\beta = \frac{\pi}{R} \tag{7}$$

where R is the radius of the sphere. The transcendental equation determining R is thus

$$\frac{\tan^{-1}(\pi\lambda/R)}{(\pi\lambda/R)} = \frac{1}{1+(\eta-1)\lambda/\lambda_f}. \tag{8}$$

An approximation to this equation is

$$\frac{1}{3}\left(\frac{\pi\lambda}{R}\right)^2 = (\eta-1)\frac{\lambda}{\lambda_f} \tag{9}$$

which can be solved for R:

$$R = \frac{\pi}{\sqrt{3}}\sqrt{\frac{\lambda\lambda_f}{\eta-1}} = 1.81\sqrt{\frac{\lambda\lambda_f}{\eta-1}}. \tag{9'}$$

Equation (8) gives somewhat smaller R for the same η than (9). In turn (9) gives smaller values of R than one obtains by substituting $\log\eta$ for $\eta-1$ as has been recommended by some authors. The relation of (8) to (9) is shown in Fig. 1 which shows also the values of $\pi\lambda/R$ and of λ/R that correspond to an assumed value of $(\eta-1)\lambda/\lambda_f$.

Formula (6) can also be understood in terms of the approach used by Peierls and Pryce. The neutron density $\rho(\vec{r})$ satisfies the equation

$$\rho(\vec{r}) = \frac{1}{4\pi}\left(\frac{1}{\lambda_s}+\frac{\eta}{\lambda_f}\right)\int \frac{e^{-|\vec{r}-\vec{r}'|/\lambda}}{|\vec{r}-\vec{r}'|^2}\rho(\vec{r}')\,dr' \tag{10}$$

which is simpler than the Boltzmann equation because the angular distribution of velocities does not enter. For an infinite medium the integration on the right side of (10) can be performed on the assumption that the neutron density is a plane wave

$$\rho(\vec{r}) = e^{i\beta z}. \tag{11}$$

and this leads to

$$1 = \frac{1}{2i\beta}\left(\frac{1}{\lambda_s}+\frac{\eta}{\lambda_f}\right)\ln\frac{1+i\beta\lambda}{1-i\beta\lambda} \tag{11'}$$

which is just another way of writing (6). The "stationary states" of type (11) can be superposed to give a spherical wave of type

$$\frac{\sin\beta r}{\beta r}$$

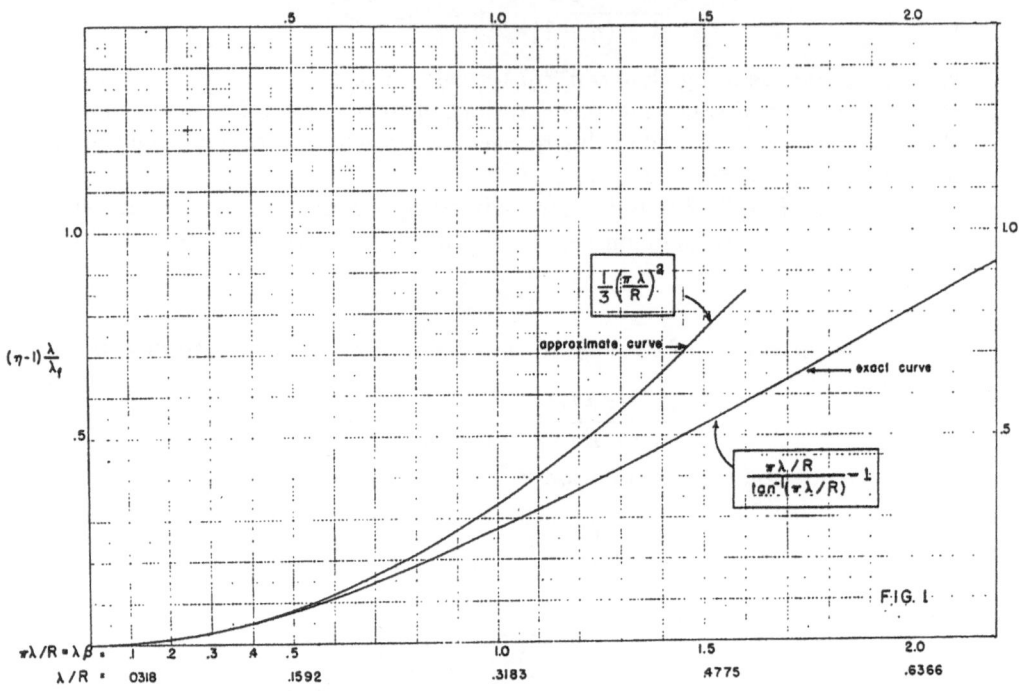

Fig. 1

which can be made to vanish at $r = R$ by making

$$\beta R = \pi.$$

Solutions obtained in this way are exact for an infinite medium and give a convenient approximation for a finite sphere. At the boundary of the sphere there is an inaccuracy due to the fact that the neutron sinks and sources outside the sphere assumed for the infinite medium are not actually present. This error cannot be serious because at the boundary of the sphere the neutron density vanishes.

The two methods used are equivalent. The advantage of (10) is that the superposition of plane waves to form a spherical wave is simpler. A solution of (10) can be used to construct a solution of (1) by setting

$$N(\vec{r}, \vec{\Omega}) = \text{const.} \int_0^\infty e^{-R/\lambda} \rho(\vec{r} - R\vec{\Omega}) \, dR. \tag{12}$$

This can be verified by substituting (12) into (1) and making use of (10).

Formula (8) has been mentioned in report A-64. For $\lambda/\lambda_f = .1$ and $\eta = e = 2.718$ the approximation (9′) is fair. It gives $\pi\lambda/R = .72$, $R = 4.4\lambda = .44\lambda_f$ while formula (8) gives $\pi\lambda/R = .77$, $R = 4.1\lambda = .41\lambda_f$. This difference is not very serious and it appears justifiable to neglect it. For $\lambda/\lambda_f = .2$, $\eta = 3$ the approximation (9') gives $\pi\lambda/R = 1.095$, $R = 2.88\lambda = .58\lambda_f$ while formula (8) gives $\pi\lambda/R = 1.259$, $R = 2.50\lambda = .50\lambda_f$. The error due to the approximation corresponds to a factor .64 in the mass in this case.

Numerical Results

The fission cross section is, at 300 kV, 1.4×10^{-26}. This must be due to the lighter isotope since recent measurements indicate that the heavier isotope is not induced to fission by neutrons of less than 1000 keV energy. Since the abundance of the lighter isotope is about 1/140, the above cross section corresponds to

$$\sigma_f = 2 \times 10^{-24} \text{ cm}^2 \tag{13}$$

for pure U^{235}. It will be assumed that this cross section holds for U^{235} for all fission neutrons – which is, of course, not justified though reasonable. The total cross section of U^{235} is not known and, instead, the total cross section of natural U, i.e.

$$\sigma = 8 \times 10^{-24} \text{ cm}^2 \tag{14}$$

will be used. This gives $\lambda = 2.75$ for a density 18. With

$$\eta = 2.8 \tag{15}$$

one obtains $\beta\lambda = 1.36$; $R = 6.35$ cm, a volume of 1080 cm^3 and a weight of 19 kg. If $\sigma = 10 \times 10^{-24}$ cm^2 is assumed $\beta\lambda = 1.18$; $R = 5.85$ cm and the weight decreases to 15 kg. For $\sigma_f = 2.5 \times 10^{-24}$ cm^2 and $\sigma = 8 \times 10^{-24}$ cm^2, $R = 5.5$ and the weight becomes 13 kg.

2. These weights can be considerably decreased by imbedding the U sphere into a reflecting material, such as Carbon or Bismuth. The advantage of a heavy material has been discussed in Report A-46. The advantage of a lighter material may lie in the lower energy of the reflected neutrons which causes a larger fission cross-section. However, it is improbable that the resulting gain is appreciable, since, in order to reach the region in which the $1/v$ law gives a cross-section exceeding (13), one would have to go down to about 20 keV. It will be assumed, therefore, that the cross sections of the neutrons have not changed by the "reflection" from the surrounding material which permits the use of (1) and the succeeding equations.

Even for radii as obtained in the preceding section, the absorption of the reflecting material can be altogether neglected so that the density of the neutrons in this material will be proportional to $1/r$. Hence the neutron loss per cm^2 U surface will be $v\lambda N/3R$ where N is the density of neutrons at the U surface, $v\lambda_r/3$ is the diffusion constant of the reflecting material (λ_r the mean free path of neutrons therein, v an average velocity of the neutrons) and R the radius of the U sphere.

Following the calculation presented in Report A-20 and leading to equation (8b) of that report one arrives at the following expression for the flux of neutrons satisfying equation (1)

$$-\frac{v(\eta - 1)}{\lambda_f \beta^2}\operatorname{grad} N .$$

This must be equal to the neutron loss calculated above so that we have, at the surface of the sphere

$$\frac{v\lambda_r N}{3R} = -\frac{v(\eta - 1)}{\lambda_f \beta^2}\operatorname{grad} N .$$

(grad N must be taken inside the sphere, the derivative of N is discontinuous at the surface). Since, inside the sphere, the density N of neutrons is proportional to $(\sin \beta r)/r$, this gives

$$R\frac{\operatorname{grad} N}{N} = 1 - \beta R \operatorname{ctg}\beta R = \frac{\lambda_r \lambda_f \beta^2}{3(\eta - 1)} = \frac{\lambda_r \lambda_f}{\lambda^2}\frac{(\beta\lambda)^2}{3(\eta - 1)} . \tag{17}$$

Equation (17) takes the place of (7) if the U sphere is imbedded in scattering material. ($-\beta R \operatorname{ctg}\beta R$ is plotted against βR in Fig. 2.)[5]

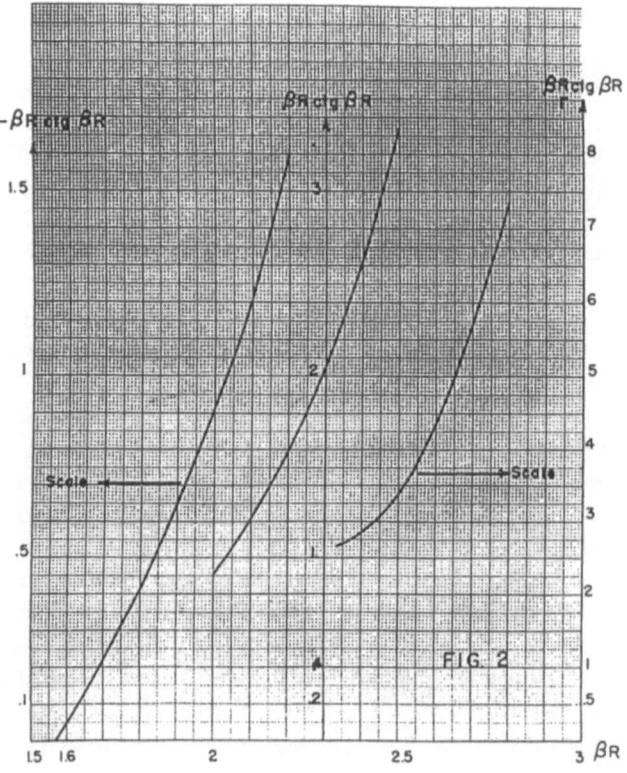

Fig. 2

---

[5] Parenthesis added by A.M.W. and A.M.P.

Numerical Results

In case of C, the scattering cross section is, for the average fission neutron energy of about 2 MeV, about 1.5×10^{-24} (cf. e.g. Good and Scharff-Goldhaber, Phys. Rev. 59, 917, 1941, Kikuchi, Aoki and Wakatuki, Proc. Phys. Math. Soc. Japan, 21, 410, 1939). With a density 1.6 this gives

$$\lambda_r = 8 \text{ cm}. \tag{18 a}$$

This is evidence that the boundary of the U and C should not be treated in the way we do ($\lambda_r > R$). Applying (17) in spite of this gives 4.0 for the right side for $\lambda = 2.75$ cm, $\beta\lambda = 1.36$. From this it follows that $\beta R = 2.45$ rather than $\beta R = \pi$. As a consequence, $R = 4.95$, the volume of the U sphere becomes 510 cm^3, its weight 9 kg. This, altough probably inaccurate, should give at least an approximate value for the reduction of the size of the U sphere which one can attain by surrounding it with C.

For Bi we take a scattering cross section 7×10^{-24} cm^2 which gives with a density 9.8

$$\lambda_r = 5 \text{ cm} \tag{18 b}$$

and 2.47 for the right side of (17). From this $\beta R = 2.15$, $R = 4.35$ and a weight of 6.2 kg follows.

Evidently, natural U is as efficient a reflector for the neutrons originating in the U^{235} sphere as one is likely to find. With the $\lambda = 2.75$, as calculated above, one can trust the present treatment of the boundary more than in the preceding cases. Evidently, the reflection on the surface is much more complete than in case of C or Bi and the amounts necessary of the light isotope are much smaller. The following table summarizes the results for different values of σ_f and σ_{total} of the light isotope. The fission and absorption of the natural U is neglected and its scattering cross section is assumed to be 8×10^{-24} cm^2. All densities are 18. The Table gives the radii and weights of the spheres which are just unstable. The first and third columns give the radii (R) and weights (W) for an isolated sphere, as calculated at the end of the first section, the second and fourth for a sphere embedded in natural U with $\sigma = 8 \times 10^{-24}$. For the fifth column it was assumed that σ for the reflector is 10×10^{-24} cm^2.

	σ_{total}	8×10^{-24}		10×10^{-24}		
σ fission	R	6.35 cm	3.59 cm	5.85 cm	3.50 cm	3.24 cm
2×10^{-24}	W	19 kg	3.6 kg	15 kg	3.2 kg	2.6 kg
2.5×10^{-24}	R	5.5 cm	3.18 cm	5.08 cm	3.11 cm	2.88 cm
	W	13 kg	2.4 kg	10 kg	2.3 kg	1.8 kg

It is hardly necessary to mention that the above figures are estimates. There is an inaccuracy in the theory, inasmuch as the treatment of the boundary layer is inaccurate and involves the assumption that the dimensions of the system, in particular the radius R of the sphere, are large compared with the mean free path. This is hardly true in the above cases. The error caused by this inaccuracy may well amount to 25 percent in the radius. Another and equally serious uncertainty of the final results can be seen from the above Table where the calculations are carried out for four sets of constants.

The above figures give the critical radii and critical masses of the U^{235} spheres which are just on the limit of instability. In order to have an explosive arrangement, according to Fermi, (Report A-46), about 1.4 times more U^{235} is needed.

3.

Diffusion of Slow Neutrons in Absorbing Materials

E. P. Wigner

No date, probably 1941

1. In the following, a simplified treatment of the problem of diffusion of slow neutrons will be given. The reason that this problem can be solved so much more easily than the well known transport problems of the kinetic theory of gases lies in the fact that the collisions involving *two* molecules of the diffusing substance can be neglected in the neutrons diffusion case. This makes the fundamental Boltzmann equation linear.

The number of neutrons per unit volume present at time t at x, y, z, the velocities of which are in the range between v_x and $v_x + dv_x$, etc. will be denoted by $f(x, y, z, v_x, v_y, v_z, t)\, dv_x\, dv_y\, dv_z$. Boltzmann's equation then reads in the absence of external forces

$$\left(\frac{\partial}{\partial t} + v_x \frac{\partial}{\partial x} + v_y \frac{\partial}{\partial y} + v_z \frac{\partial}{\partial z} + v\sigma(v)\right) f(\mathbf{r}, \mathbf{v}, t) = B. \tag{1}$$

Herein σ is the total (scattering + absorption) cross section of the material contained in unit volume, $\mathbf{r}$ and $\mathbf{v}$ stand for x, y, z and v_x, v_y, v_z, respectively, B is the number of neutrons which acquire, by a collision, the velocity v_x, v_y, v_z. It is a linear expression in f. If the neutron scattering is elastic (slow neutrons) and the probability of an angular deviation θ is $\sigma_s + \sigma_p \cos\theta$ per unit solid angle, we have

$$B = v\sigma_s f(\mathbf{x}, \mathbf{v}, t) + (\sigma_p/v)(v_x \overline{f(\mathbf{r}, \mathbf{v}, t)v_x} + v_y \overline{f(\mathbf{r}, \mathbf{v}, t)v_y} + v_z \overline{f(\mathbf{r}, \mathbf{v}, t)v_z}). \tag{1a}$$

The bar means averaging of the expression below it over all possible directions of the velocity so that a barred quantity depends only on the position, the time and the absolute value of the velocity. Expression (1a) for B is the only one which we shall use explicitly although, of course, it applies only under very special conditions. In particular, the absolute value of the velocity can be treated as a parameter if (1a) applies since no physical process occurs that changes its value. The total scattering cross section is σ_s and $\sigma_p = 0$ holds in case of spherically symmetric scattering. If there are neutrons entering the system from the outside, a further term $P(\mathbf{r}, \mathbf{v}, t)$ must be added to the right side of (1), which then becomes a linear inhomogeneous equation.

2. The purpose of the transformation of (1) into a diffusion equation is to eliminate the variables v_x, v_y, v_z and obtain an equation for $\overline{f(\mathbf{r}, \mathbf{v}, t)}$ alone.

It should be evident that, strictly speaking, such an elimination is impossible since the further fate of the system will depend, in general, not only on the density of neutrons at $t = 0$ but also on the distribution of their velocities over different directions. In fact, all treatments which achieve an elimination of the $v_x v_y v_z$ from (1) use, at some point of the treatment, some arbitrary assumption. On the other hand, if one asks for stationary states $(\partial f/\partial t = 0)$ satisfying (1), the above problem may have a definite sense since the condition of stationarity may determine the dependence of f both on $\mathbf{r}$ and $\mathbf{v}$. We shall see that this is actually the case so that it is impossible to speak of an equation for $\bar{f}$ alone.

A situation very similar to that outlined above is encountered in the problem of the elimination of the light particle field in nuclear theories (or the elimination of the electromagnetic field in the theory of Coulomb interaction). It is clearly impossible to give a rigorous justification of the time dependent Schrödinger equation out of which the light particles are eliminated. Such an equation would permit one to calculate the fate of the heavy particles if only their initial state is known. However, the fate of the heavy particles depends also on the initial state of light particles and this is not specified by the initial heavy particle wave function. On the other hand, if one looks for a stationary state, the possible wave functions are determined by the condition of stationarity and it may be possible to find equations which refer only to the properties of the heavy particles. This could be done, in principle, by solving the original equations, involving both light and heavy particles, and giving on the basis of the explicit solutions expressions for the distribution etc., of the heavy particles. It will be true in general, then, that the state of the heavy particles cannot be described by a single wave function but only by a statistical matrix, but this need not bother us.

3. We now return to our problem of solving equations (1), (1 a). Since the problem is linear, it suffices to find a basic set of solutions. Since it shows translational symmetry it seems sufficient (this step is not quite rigorous) to consider solutions which depend exponentially on the space coordinates, i.e., correspond to plane waves. It follows furthermore from the rotational symmetry of the problem that the wave vector can be assumed to have the direction of the x axis. This causes the vanishing of $\partial f/\partial y$ and $\partial f/\partial z$ while $\partial f/\partial t$ vanishes because of the stationarity condition. For $\partial f/\partial x$ we write κf where the constant κ is the length of the wave vector. Hence (1), (1 a) becomes

$$(v_x \kappa + v\sigma)f = v\sigma_s \bar{f} + (\sigma_p/v)v_x \overline{v_x f}. \tag{2}$$

As we pointed out before, the absolute value of v enters the equations only as a parameter and will be assigned a definite value. The average values $\overline{v_y f}$ and $\overline{v_z f}$ vanish because our solution is supposed to have rotational symmetry about the x-axis. Both $\bar{f}$ and $\overline{v_x f}$ are, of course, constant multiples of $e^{\kappa x}$. From (2) we have

$$f(r,v) = \frac{v\sigma_s \bar{f} + v_x(\sigma_p/v)\overline{v_x f}}{v\sigma + v_x \kappa}. \tag{3}$$

The right side determines the dependence of f on $\mathbf{v}$. This is the step necessary for the transformation of (1) into a diffusion equation which is essentially achieved herewith. We have only to make (3) self-consistent by making sure that the averages of f and $v_x f$ over the velocity directions are equal to the $\bar{f}$ and $\overline{v_x f}$ of the right side of (3). For this purpose we write $S = v_x/v$ so that the average becomes $\frac{1}{2}\int_{-1}^{1}\ldots dS$ and we obtain the equations

$$\bar{f} = \frac{1}{2}\int_{-1}^{1}\frac{\sigma_s\bar{f} + S(\sigma_p/v)\overline{v_x f}}{\sigma + \kappa S}\,dS \tag{4}$$

$$\overline{v_x f} = \frac{1}{2}\int_{-1}^{1}\frac{S\sigma_s\bar{f} + S^2(\sigma_p/v)\overline{v_x f}}{\sigma + \kappa S}\,dS. \tag{4a}$$

Carrying out the integrations gives

$$v\bar{f} = (\sigma_p/\kappa)\overline{v_x f} + \frac{1}{2}[v(\sigma_s/\kappa)\bar{f} - (\sigma\sigma_p/\kappa^2)\overline{v_x f}]\ln\frac{\sigma + \kappa}{\sigma - \kappa} \tag{5}$$

$$\overline{v_x f} = \frac{1}{2}\left(\frac{v\sigma_s}{\kappa}\bar{f} - \frac{\sigma_p\sigma}{\kappa^2}\overline{v_x f}\right)\left(2 - \frac{\sigma}{\kappa}\ln\frac{\sigma + \kappa}{\sigma - \kappa}\right). \tag{5a}$$

From these

$$\overline{v_x f} = -v(\sigma_a/\kappa)\bar{f} \tag{6}$$

and

$$\ln\frac{\sigma + \kappa}{\sigma - \kappa} = 2\frac{\kappa(\kappa^2 + \sigma_p\sigma_a)}{\kappa^2\sigma_s + \sigma\sigma_p\sigma_a} \tag{7}$$

follow where $\sigma_a = \sigma - \sigma_s$ is the total absorbing cross section contained in unit volume of the material. Our solution thus becomes, if we set $\bar{f} = e^{\kappa x}$

$$f(\mathbf{r}, v) = \frac{v\sigma_s\kappa - v_x\sigma_p\sigma_a}{(v\sigma + v_x\kappa)\kappa}e^{\kappa x} \tag{8}$$

while κ is determined by the implicit equation (7). More generally, we have for any $|\boldsymbol{\kappa}| = \kappa$ the solution

$$f(\mathbf{r}, \mathbf{v}) = \frac{\sigma_s v\kappa^2 - \sigma_p\sigma_a(\mathbf{v}\cdot\boldsymbol{\kappa})}{(v\sigma + (\mathbf{v}\cdot\boldsymbol{\kappa}))\kappa^2}e^{(\boldsymbol{\kappa}\cdot\mathbf{r})} \tag{8a}$$

for which $\bar{f} = e^{\boldsymbol{\kappa}\cdot\mathbf{r}}$. The flow of neutrons is

$$\overline{\mathbf{v}f} = -(v\sigma_a/\kappa^2)\operatorname{grad}\bar{f}. \tag{8b}$$

4. What has been essentially demonstrated by the foregoing is that if κ is a solution of (7) and $\bar{f}$ a solution of the "diffusion equation"

$$\Delta\bar{f}(\mathbf{r}) = \kappa^2\bar{f}(\mathbf{r}) \tag{9}$$

then there is a stationary solution $f(\mathbf{r}, \mathbf{v})$ of the Boltzmann equation (1), (1a) the average of which is $\overline{f(\mathbf{r}, \mathbf{v})} = \bar{f}(\mathbf{r})$. This follows from the fact that the solutions of (9) can be written as superpositions of plane waves $e^{\pm(\boldsymbol{\kappa}\cdot\mathbf{r})}$ and that (8 a) gives us the $f(\mathbf{r}, \mathbf{v})$ for which $\bar{f}(\mathbf{r}, \mathbf{v}) = e^{(\kappa \cdot \mathbf{r})}$. It is, in particular, easy to find the spherically symmetric solution of (1) by superposing waves (8 a) with equal amplitudes and all possible directions. The density of neutrons in this solution is proportional to $Sh\kappa r/r$.

Figure 1 gives the solution of (7) in case of spherically symmetric scattering, $\sigma_p = 0$. For practical purposes

$$\kappa = \sqrt{3\sigma_a \sigma}(1 - \frac{2}{5}\sigma_a/\sigma) \qquad (9\,\text{a})$$

will be sufficiently accurate. It is given in the figure as a broken line.

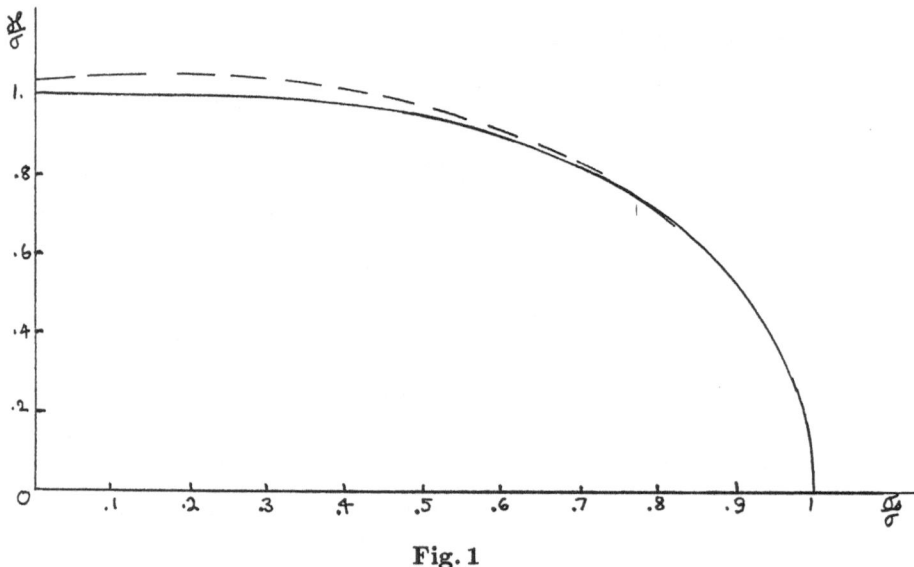

Fig. 1

It should be remembered that it was necessary, in order to derive (9), to assume that the Boltzmann equation (1), (1 a) holds throughout the whole space. If this is not the case (9) will not hold even in that part of the sphere where (1), (1 a) does. In particular, if the whole space contains two different media, (9) will not hold in either medium in the neighborhood of their boundary. It is clear, however, that the deviations will be restricted in both media to a domain the thickness of which is of the order of the mean free path $1/\sigma$.

5. We now want to solve (1) with a production term P on the right. We shall consider as the place of production the point where the neutron suffers its first collision. The scattering being spherically symmetric, the scattering center can be considered to produce neutrons with a spherically symmetric velocity distribution $P(x, y, z)$. Using this method one has to bear in mind that the total neutron density consists, in addition to those to be described by our f, of these which have not yet suffered a collision since their birth.

We can decompose $P(x, y, z)$ into a Fourier integral[6] and it is sufficient, therefore, to solve (1) with a production term $e^{i\nu x}$ on the right side. The corresponding f will depend on x also as $e^{i\nu x}$ and we have

$$(v_x i\nu + v\sigma)f = v\sigma_s \bar{f} + v e^{i\nu x} \tag{10}$$

whence with $\bar{f} = b(\nu)e^{i\nu x}$

$$f = \frac{\sigma_s b(\nu) + 1/v}{\sigma + i\nu v_x/v} e^{i\nu x} . \tag{10a}$$

We have to satisfy the equation $\bar{f} = b(\nu)e^{i\nu x}$ which gaves

$$b(\nu) = \frac{\sigma_s b(\nu) + 1/v}{2i\nu} \ln \frac{\sigma + i\nu}{\sigma - i\nu}$$

or with $\operatorname{tg} \beta = \nu/\sigma$

$$b(\nu) = \beta/(\nu - \sigma_s \beta)v . \tag{10b}$$

One can add to (10a), of course, an arbitrary solution of (9).

These formulas provide, apart from their original purpose, an approximate treatment of the above mentioned problem in which there are two different media present[7]. One can consider both media infinite and treat them separately. However, in the region which is actually filled with the second medium, one assumes, when calculating the density in the first medium, a negative production which just reduces the number of neutrons in this region to zero. Conversely, in the region which is actually occupied by the first medium one assumes at every point a positive production equal to the number of neutrons which, coming from the second medium, suffer their first collision at this point. Naturally, the total positive production assumed for the calculation of the density in the first medium must be equal to the total negative production assumed in the calculation of the density in the second medium. Conversely, the total negative production in the first is equal to the consumption in the second medium. This method is most easily applicable if the properties (σ and σ_s) of the two media are similar.

---

[6] The following calculation was presented recently by G. Placzek and is reproduced here only for easier reference's sake. The publication of Placzek's paper has been delayed.

[7] Halpern, Lueneburg and Clark, Phys. Rev. *53*, 173 (1938), gave a rigorous solution of the closely related problem in which only the space below a plane is filled with matter, the space above it is empty.

4.

Solution of Boltzmann's Equation for Monoenergetic Neutrons in an Infinite Homogeneous Medium

E. P. Wigner

November 30, 1943

Abstract. The Boltzmann's equation is solved in the case of monoenergetic neutrons created by a plane or point source in an infinite medium which has spherically symmetric scattering. The customary solution of the diffusion equation appears to be multiplied by a constant factor which is smaller than 1. In addition to this term the total neutron density contains another term which is important in the neighborhood of the source. It goes with $1/r^2$ in the neighborhood of a point source.

1. The following represents a solution of Boltzmann's equation for neutrons which do not change their energy during their diffusion in a material. The solution can be easily obtained on the basis of Placzek's work, which is also reproduced in A-21. The contents of the first part of the following note were also published by Bothe in the Zs. f. Physik.

We first consider a plane source of neutrons at $x = 0$. Evidently, the angular distribution of neutrons will be independent of y and z and have at every point axial symmetry with respect to the x axis. The number of neutrons per unit volume for which the velocity component in the x direction v_x lies between μv and $(\mu + d\mu)v$ will denoted by

$$f(x, \mu)\, d\mu. \tag{1}$$

The density of neutrons at x is obtained by integration over the direction cosine μ

$$n = \int_{-1}^{1} f(x, \mu)\, d\mu = 2\bar{f} \tag{1a}$$

where $\bar{f}$ is the average value of f over all directions of the velocity.

The Boltzmann equation for spherically symmetric scattering is

$$\left(-\frac{v_x}{v}\frac{\partial}{\partial x} - \frac{v_y}{v}\frac{\partial}{\partial y} - \frac{v_z}{v}\frac{\partial}{\partial z} - \sigma\right) f + \sigma_s \bar{f} + P/v = 0 \tag{2}$$

where σ, σ_s and $\sigma_a = \sigma - \sigma_s$ are total, scattering and absorption cross sections per unit volume of the medium in which the diffusion occurs; P is the production of neutrons per unit volume and unit velocity range. For an f of the form (1) one can write instead of (2)

$$\left(-\mu\frac{\partial}{\partial x} - \sigma\right)f + \sigma_s\bar{f} + \frac{1}{2v}\delta(x) = 0. \tag{2 a}$$

The production is assumed in (2 a) to be one neutron per unit area of the $x = 0$ plane, with equal probability for every direction: there are $\frac{1}{2}d\mu$ neutrons produced per unit area with direction cosines between μ and $\mu + d\mu$.

In order to solve (2 a) we write

$$\delta(x) = \frac{1}{2\pi}\int\limits_{-\infty}^{\infty} e^{i\nu x}\,d\nu\,;\;\; f(x,\mu) = \frac{1}{2\pi}\int\limits_{-\infty}^{\infty} f_\nu(x,\mu)\,d\nu\,. \tag{3}$$

Introducing this into (2 a) we obtain

$$\left(-\mu\frac{\partial}{\partial x} - \sigma\right)f_\nu + \sigma_s\bar{f} + \frac{1}{2v}e^{i\nu x} = 0 \tag{4}$$

$$\bar{f}_\nu = \frac{1}{2}\int\limits_{-1}^{1} f_\nu(x,\mu)\,d\mu\,.$$

We try the assumption

$$f_\nu = a(\mu)e^{i\nu x} \tag{5}$$

Herein, μ is the cosine of the angle between the velocity of the neutron and the x direction. This gives us for a the equation

$$-i\mu\nu a(\mu) - \sigma a(\mu) + \sigma_s\bar{a} + 1/2v = 0 \tag{5 a}$$

whence we obtain

$$a(\mu) = \frac{1/2v + \sigma_s\bar{a}}{\sigma + i\mu\nu} \tag{5 b}$$

where

$$\bar{a} = \frac{1}{2}\int\limits_{-1}^{1} a(\mu)\,d\mu = \frac{1}{4v}\int\limits_{-1}^{1}\frac{1 + 2v\sigma_s\bar{a}}{\sigma + i\mu\nu}\,d\mu = \frac{1 + 2v\sigma_s\bar{a}}{2v\nu}\,\text{arctg}\,\frac{\nu}{\sigma}\,. \tag{5 c}$$

In order that the above solution be self-consistent, we must have, therefore,

$$2v\bar{a} = \frac{\text{arctg}\,\nu/\sigma}{\nu - \sigma_s\,\text{arctg}\,\nu/\sigma} = \frac{\ln\frac{\sigma+i\nu}{\sigma-i\nu}}{2i\nu - \sigma_s\ln\frac{\sigma+i\nu}{\sigma-i\nu}}\,. \tag{6}$$

From this and (5) we obtain

$$2v\bar{f}_\nu = 2v\bar{a}e^{i\nu x} = \frac{e^{i\nu x}}{2i\nu - \sigma_s\ln\frac{\sigma+i\nu}{\sigma-i\nu}}\ln\frac{\sigma+i\nu}{\sigma-i\nu} \tag{6 a}$$

and (5 c), (6) and (5) give

$$f_\nu = \frac{1}{2v}\frac{e^{i\nu x}}{\sigma + i\mu\nu}\frac{2i\nu}{2i\nu - \sigma_s\ln\frac{\sigma+i\nu}{\sigma-i\nu}}\,. \tag{6 b}$$

This finally gives for f in the case of a plane source because of (3)

$$f(x,\mu) = \frac{1}{4\pi v} \int\limits_{-\infty}^{\infty} dv \frac{e^{ivx}}{\sigma + i\mu v} \frac{1}{1 + \frac{i\sigma_s}{2v} \ln \frac{\sigma + iv}{\sigma - iv}} \qquad (6\,c)$$

and for the density of neutrons we have from (6 a)

$$n = 2\bar{f} = \frac{1}{2\pi v} \int\limits_{-\infty}^{\infty} dv \frac{e^{ivx}}{2iv - \sigma_s \ln \frac{\sigma + iv}{\sigma - iv}} \ln \frac{\sigma + iv}{\sigma - iv}. \qquad (6\,d)$$

Of course, one can write $2i\mathrm{arctg}v/\sigma$ for $\ln \frac{\sigma + iv}{\sigma - iv}$. The analysis so far closely follows Placzek's work.

This is Bothe's result. In order to bring it into a more suitable form which also shows the asymptotic behavior, the derivation of which was the subject of A-21, we push the path of integration toward the upper part of the positive imaginary axis. Evidently if the imaginary part of v is infinite, the integrand vanishes for positive x. However, the integrand has two singularities in the upper half plane, one where the denominator under e^{ivx} vanishes and the other at $v = i\sigma$. As a result of the deformation the path of integration will become as indicated in Fig. 1. The first singularity (at $i\kappa$) is a pole and corresponds to the value of v, which is i times the macroscopic absorption coefficient given in A-21. The other one is an essential singularity. The contribution to the first part can be easily calculated and expressed in a closed form. It is for $x > 0$

$$n_1 = I \frac{\kappa}{2\sigma_a v} e^{-\kappa x}; \quad I = \frac{2\sigma_a}{\sigma_s} \frac{\sigma^2 - \kappa^2}{\kappa^2 - \sigma_a \sigma}. \qquad (7)$$

In this κ is the value of v/i for which the denominator vanishes. It is, therefore, given by the equation

$$2\kappa = \sigma \ln(\sigma + \kappa)/(\sigma - \kappa). \qquad (7\,a)$$

According to A-21, it is closely approximated by the expression

$$\kappa = \sqrt{3\sigma\sigma_a}(1 - 2\sigma_a/5\sigma); \quad \sigma_a = \sigma - \sigma_s. \qquad (7\,b)$$

The contribution of the second part cannot be expressed in a closed form because the singularity is essential. However, it can be expressed as a real integral if one introduces a variable along the imaginary axis. It is with $v = i\sigma(1 + \eta)$

$$n_2 = \frac{1}{2\pi v} \int i\sigma \, d\eta \frac{e^{-\sigma(1+\eta)x}}{-2\sigma(1 + \eta) - \sigma_s \ln \frac{-\eta}{2+\eta}} \ln \frac{-\eta}{2 + \eta} \qquad (8)$$

where the path of integration corresponds to the loop around $i\sigma$ in Fig. 1. It goes, for η, from $\eta = \infty$ around $\eta = 0$ back to $\eta = \infty$. On the first part of the path, $\ln \frac{-\eta}{2+\eta}$ must be replaced by

$$\ln \frac{\eta}{2 + \eta} - i\pi = -\ln(1 + 2/\eta) - i\pi,$$

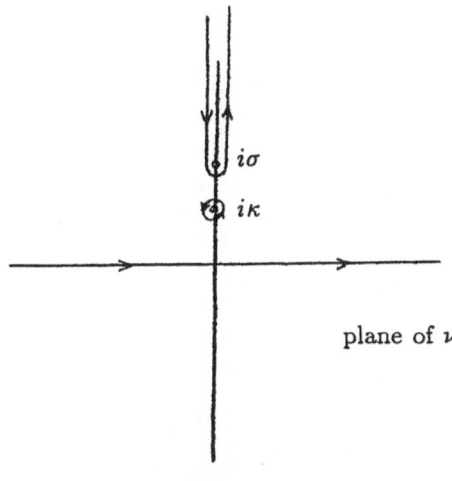

plane of ν

Fig. 1

on the second part by

$$\ln\frac{\eta}{2+\eta} + i\pi = -\ln(1+2/\eta) + i\pi.$$

We have, altogether

$$n_2 = \frac{i\sigma}{2\pi v}\int\limits_0^\infty e^{-\sigma(1+\eta)x}d\eta$$

$$\times\left[\frac{-\ln(1+2/\eta)-i\pi}{2\sigma(1+\eta)-\sigma_s\ln(1+2/\eta)-\sigma_s i\pi} - \frac{-\ln(1+2/\eta)+i\pi}{2\sigma(1+\eta-\sigma_s\ln(1+2/\eta)+\sigma_s i\pi}\right]$$

$$= \frac{2\sigma^2}{v}\int\limits_0^\infty \frac{(1+\eta)e^{-(1+\eta)\sigma x}\,d\eta}{[2\sigma(1+\eta)-\sigma_s\ln(1+2/\eta)]^2 + \pi^2\sigma_s^2}. \qquad (8\,a)$$

The first part of $n = n_1 + n_2$. i.e., n_1 is the only one that is important at distances from the source which are larger than a mean free path. Its functional dependence on x is the same as given by the diffusion equation. However, its absolute value is everywhere smaller by a factor

$$I = \frac{2\sigma_a}{\sigma_s}\frac{\sigma^2-\kappa^2}{\kappa^2-\sigma_a\sigma}. \qquad (8\,b)$$

The reason for this "initial absorption" is that the density in the neighborhood of the source is greater (by n_2) than given by the diffusion equation so that the absorption in the neighborhood of the source is also greater. As a result, fewer neutrons get away from the source than the diffusion equation would indicate. The initial absorption is plotted in Fig. 2 as a function of $1 - \sigma_s/\sigma = \sigma_a/\sigma$.

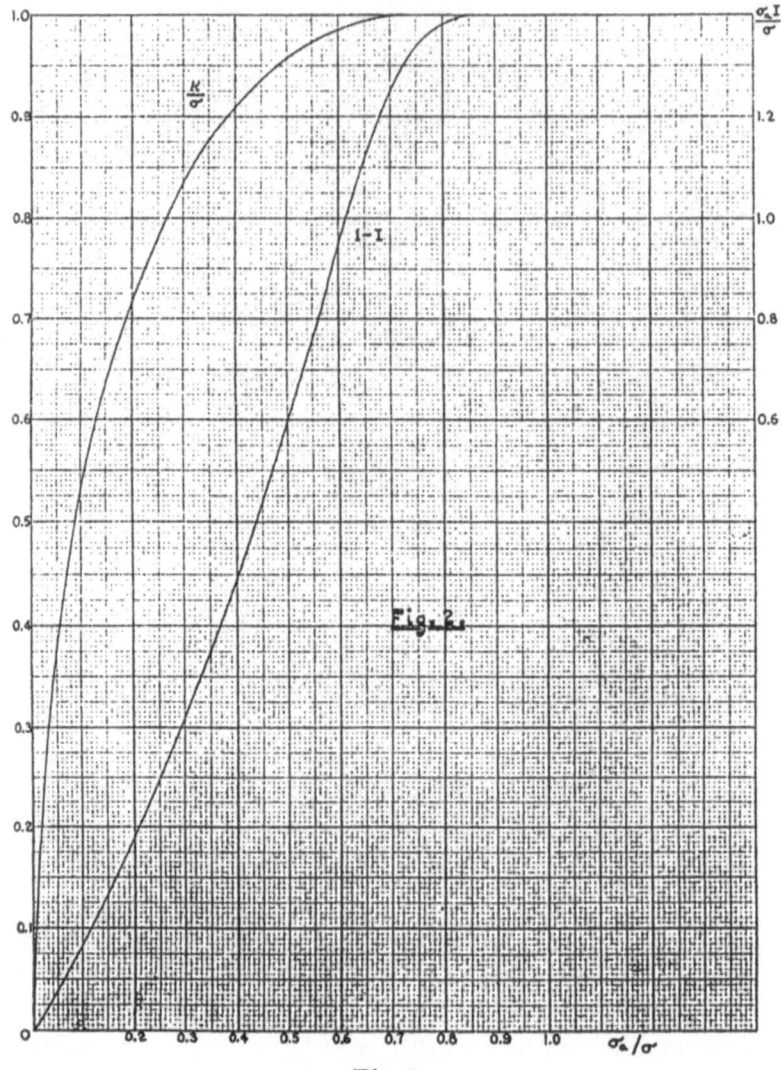

Fig. 2

In order to calculate the second part of n, i.e., n_2, it is advisable first to go over from the density distribution around a plane source n to the density distribution n around a point source. This can be done by the well-known equation

$$N(r) = -\frac{1}{2\pi r}\frac{dn(r)}{dr}.$$ (9)

If this transformation is applied to n_1, etc., it only gives the customary solution of the diffusion equation multiplied by I, i.e.,

$$N(r) = N_1(r) + N_2(r); \quad N_1(r) = \frac{I}{4\pi}\frac{\kappa^2}{\sigma_a vr}e^{-\kappa r}.$$ (10)

The second part becomes

$$N_2(r) = \frac{\sigma^3}{\pi r v} \int_0^\infty \frac{(1+\eta)^2 \, d\eta \, e^{-(1+\eta)\sigma r}}{[2\sigma(1+\eta) - \sigma_s \ln(1+2/\eta)]^2 + \pi^2 \sigma_s^2} \,. \tag{10 a}$$

This expression, as well as (8 a), shows that the second part of the density (n_2 and N_2) is a sum of exponentials, all of them with positive coefficients. However, the relaxation lengths of all exponentials are equal to or shorter than the mean free path, $1/\sigma$. It does not seem to be possible to express n_2 or N_2 in a closed form. The coefficients of the different exponentials

$$\frac{(1+\eta)^2}{4\pi r v} e^{-\sigma(1+\eta)r}$$

(normalized so that the integral over all space be σ) are given in Fig. 3 as function of η for various $\sigma_s/\sigma = 1 - \sigma_a/\sigma$.

2. It appears reasonable to approximate N_2 by a single exponential which has the right singularity at $r = 0$

$$N_2 \approx \frac{1}{4\pi r^2 v} e^{-\sigma' r} \,. \tag{11}$$

Such an approximation cannot be very accurate since N_2 contains not one but an infinite range of exponentials. For σ' one can choose the value of $(1+\eta)\sigma$ for which the normalized exponential contributes most to (10 a), i.e. the point where the curves of Fig. 3 have their maxima. The corresponding $1+\eta$ is given by the transcendental equation $2\sigma(1+\eta) = \sigma_s \ln(1+2/\eta)$ which gives for σ'

$$2\sigma' = \sigma_s \ln \frac{\sigma' + \sigma}{\sigma' - \sigma} \,. \tag{11 a}$$

The formal similarity between this equation and that determining κ (7 a) is remarkable. The resulting σ'/σ is given in Fig. 4 as function of σ_a/σ (curve a).

An alternative procedure is to choose σ' in such a way that the integral of (11) over all space have the right value. Since the production of neutrons is 1 per unit time, the absorption of neutrons also is 1

$$\sigma_a v \int (N_1 + N_2) 4\pi r^2 \, dr = 1 \,.$$

From this and (11) $I + \sigma_a/\sigma' = 1$ and

$$\sigma' = \frac{\sigma_a}{1 - I} \tag{11 b}$$

follows. The σ'/σ corresponding to (11 b) is the curve marked (b) in Fig. 4. For small σ_a/σ the two curves in Fig. 4 are close to each other indicating that (11) is a reasonably good approximation in this case. For large σ_a/σ, the σ' of (11 b) becomes smaller than σ which shows that the approximation is poor in this case.

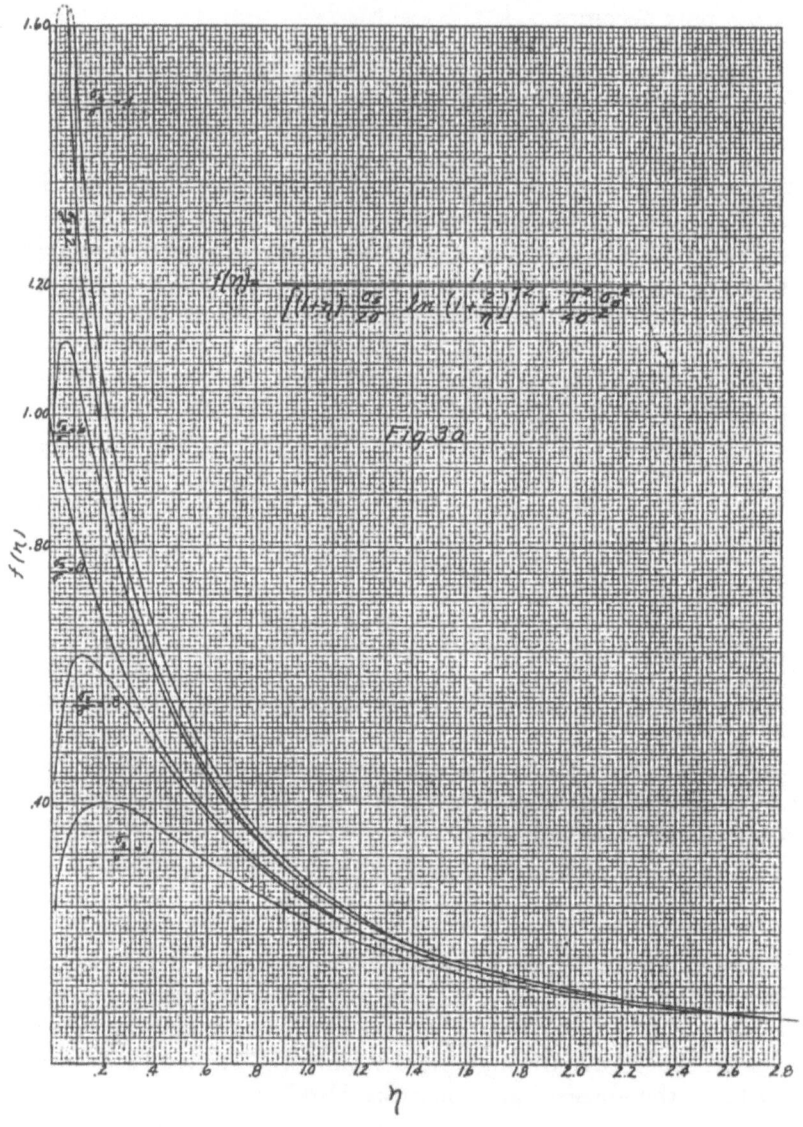

Fig. 3a

The above formulae give a representation of the whole neutron density due to a point source. The first part of the density N, i.e. N_1, is dominant at large distances from the source (except if σ_a/σ is nearly one). The second part, N_2, dominates in the neighborhood of the source. As σ_a/σ approaches 1, the most probable exponential in N_2 becomes the one with $\sigma' \approx \sigma \approx \sigma_a$. At the same time, the exponent κ of N_2 also approaches $\sigma \approx \sigma_a$. In this limiting case N_1 and N_2 have the same exponential. However, N_1 contains the factor I/r, N_2 the factor $1/r^2$. Since I goes, at the same time, to zero, the total N becomes in this limiting case $(4\pi r^2 v)^{-1} e^{-\sigma_a r}$ – as evident from physical considerations.

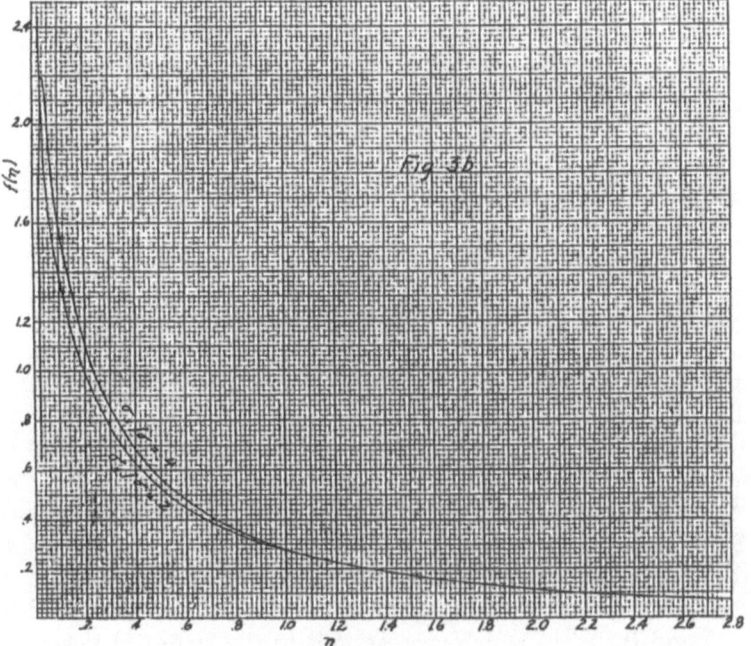

Fig. 3b

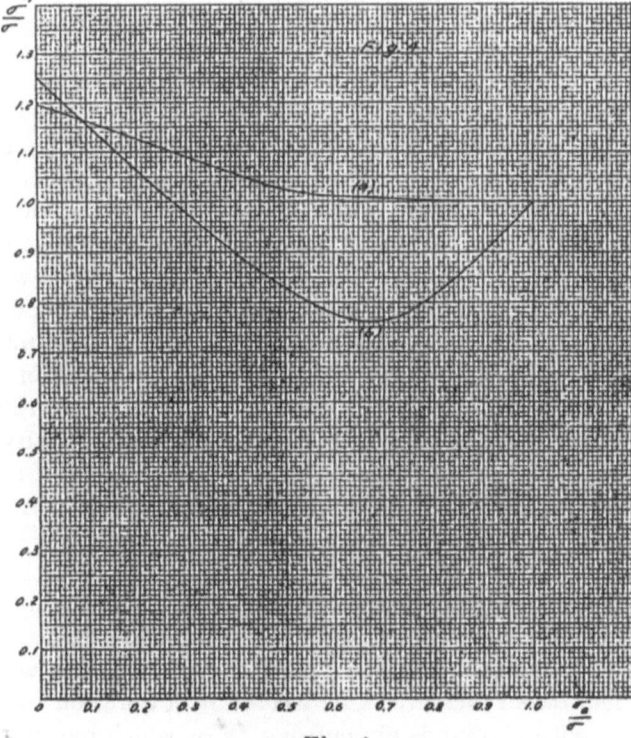

Fig. 4

A third way of writing (10 a)

$$N_2(r) = \int\limits_{\sigma}^{\infty} \frac{e^{-\xi r}}{4\pi r v} \frac{\xi^2 \, d\xi}{\left(\xi - \frac{\sigma_s}{2} \ln \frac{\xi+\sigma}{\xi-\sigma}\right)^2 + \frac{\pi^2 \sigma_s^2}{4}} \tag{10 b}$$

$[\xi = (1 + \eta)\sigma]$ can be obtained from (10 b) by partial integration. This gives

$$N_2(r) = \int\limits_{\sigma}^{\infty} \frac{e^{-\xi r}}{4\pi r^2 v} \frac{d}{d\xi} \frac{\xi^2}{\left(\xi - \frac{\sigma_s}{2} \ln \frac{\xi+\sigma}{\xi-\sigma}\right)^2 + \frac{\pi^2 \sigma_s^2}{4}} \, d\xi \,. \tag{12}$$

This integral, incidentally, converges for all r, including $r = 0$ and thus shows the behavior of $N_2(r)$ for small r. It gives $N_2(r)$ as a sum of functions $e^{-\xi r}/4\pi r^2 v$ with $\xi \geq \sigma$. The coefficient is infinite (but integrable) for $\xi = \sigma$, drops hence and goes to zero as ξ^{-3} for large ξ. It stays positive for all ξ if $\sigma_s/\sigma > 8/\pi^2$, otherwise it becomes negative and approaches 0 for large ξ from below. Since the second factor has a relatively sharp maximum in the neighborhood of $\xi = \sigma$, it seems reasonable to replace (12) by a single $e^{-\sigma' r}/4\pi r^2 v$ curve, the coefficient of which is equal to the integral of the second factor (i.e. equal to 1) and the σ' of which corresponds to the center of mass of the same factor, i.e. is equal to

$$\sigma' = \int\limits_{\sigma}^{\infty} \xi \frac{d}{d\xi} \frac{\xi^2}{\left(\xi - \frac{\sigma_s}{2} \ln \frac{\xi+\sigma}{\xi-\sigma}\right)^2 + \frac{\pi^2 \sigma_s^2}{4}} \, d\xi \,. \tag{11 c}$$

One obtains from this a σ' rather similar to that of (11 a).

It may be, in many cases, just as simple to use the accurate formula (10 b) than any approximate expression. The functions $f_\xi = e^{-\xi r}/4\pi r v$ obey a simple differential equation $\Delta f_\xi - \xi^2 f_\xi$ which makes it often possible to obtain the combined effect of a distribution of sources. If the source density is ρ, the equation $\Delta f_\xi - \xi^2 f_\xi + \rho/v = 0$ gives f_ξ. This integrated over ξ with the weight factor appearing in (10 b) gives the "second part" of the neutron density created by sources of the density ρ. Similarly, the "first part" of the neutron density is in most cases most easily obtained by solving the equation $\Delta f - \kappa^2 f + (I\kappa^2/\sigma_a v)\rho = 0$. On the whole, one can say that the assumption of a single energy value makes it almost as easy to obtain the neutron density in an infinite homogeneous medium as it is to obtain the electric potential if the charges are given. It is much more difficult, however, to take into account the variation of energy as is also to consider a problem with two media.

3. The approximate expressions given above for the neutron density tempt one to try the following procedure: Take the path before the first collision into account rigorously and take the place where the first collision occurs to be the source of the neutrons which are to be treated by a diffusion equation. Such a procedure deals with two kinds of neutrons, those which have not yet suffered

any collision and those which have already suffered a collision. The sources of the latter are more extended than the original sources and it appears more justifiable to treat them by means of the diffusion equation. The procedure can be further generalized by treating not only the first but the first two collisions rigorously etc. Evidently the above method will give more accurate results in the neighborhood of the source than the straight application of the diffusion theory. It will be seen, however, that no matter how many collisions one takes into account rigorously, the asymptotic behavior remains the same as in the straight diffusion theory, i.e. does not contain the factor I of (10). Of course, one can obtain a very good expression for the density at any point by taking into account sufficiently many collisions rigorously, but no matter how many collisions one treats this way, there is always a distance where the result becomes inaccurate. Let us consider the case of a point source of unit strength. The density before the first collision is, evidently

$$N_{b_1} = \frac{1}{4\pi r^2 v} e^{-\sigma r} = \int\limits_{\sigma}^{\infty} \frac{e^{-\xi r}}{4\pi r v} \, d\xi \, . \tag{13}$$

The density after the first collision N_{a_1} will be assumed to obey the equation

$$\frac{\sigma_a}{\kappa^2} \Delta N_{a_1} - \sigma_a N_{a_1} + \frac{\sigma_s}{\sigma} N_{b_1} = 0 \, . \tag{14}$$

One may be tempted to introduce a factor I into the last term of (14) (which would give the correct asymptotic behavior to N_{a_1}) but this is not justifiable since, evidently, the integral of $\sigma_a N_{a_1}$ must be equal to the integral of the production $(\sigma_s/\sigma) N_{b_1}$. One can solve (14) most easily by using for N_{b_1} the last expression in (13) and writing for N_{a_1}

$$N_{a_1} = \int a(\xi) \frac{e^{-\xi r}}{4\pi r v} \, d\xi + a \frac{e^{-\kappa r}}{4\pi r v} \, . \tag{14a}$$

One obtains in this way

$$N_{a_1} = -\frac{\sigma_s}{\sigma} \int\limits_{\sigma}^{\infty} \frac{\kappa^2}{\sigma_a} \frac{1}{\xi^2 - \kappa^2} \frac{e^{-\xi r}}{4\pi r v} \, d\xi + \frac{\kappa^2}{\sigma_a \sigma} \frac{e^{-\kappa r}}{4\pi r v} \, . \tag{14b}$$

The coefficient of (14a) is determined so that the Laplacian of N_{a_1} shall contain no δ-function at $r = 0$. Hence, N_{a+1} has only a logarithmic singularity at $r = 0$.

The solution of the problem in the present approximation is the sum of (13) and (14b). Had we used the diffusion equation right from the start, the solution would have been only the last term of (14b). One sees that the present procedure gives a much more nearly correct result at small distances. In particular, the density is $\sim 1/4\pi r^2 v$ near the origin, as it is in the accurate solution. However, at large distances the behavior of the sum of (13) and (14b) is the same as that of the solution of the diffusion equation. Comparison with (10) shows that it is too large by the factor $1/I$. Neither will this behavior change if one takes further collisions rigorously into account before going over to the diffusion equation: the neutrons which suffered their first collision at a point P will, even if another collision is taken into account rigorously, give the same density at large distances as in the foregoing treatment in which they were treated, from P on, by the diffusion equation. This holds for all points P and thus for the whole distribution. The reason for this surprising behavior is that the diffusion equation assumes a bias in the velocity distribution of the neutrons in the direction of decreasing density right from the start. In reality, this bias develops only after a distance of about one mean free path.

It may be worthwhile to remark that the calculation of the first two sections could be carried out also in case of a not spherically symmetric scattering. The exponent in N_1, i.e. κ, has already been calculated in A-21 for the case that the differential cross section contains, in addition to a constant term, a term proportional to the cosine of the scattering angle.

We had occasion to derive and use the above results for the calculation of the multiplication constant of a water cooled pile. In this case, some of the neutrons are made thermal in the water and a certain loss by initial absorption was to be expected. σ_a/σ is in this case about .01, and thus much greater than it is in graphite. However, Fig. 2 shows that the initial absorption remains quite small and amounts to only .8% causing a loss in the total multiplication constant of less than .02%.

5.

Review of the Measurements of the Resonance Absorption of Neutrons by Uranium in Bulk

E. Creutz,* H. Jupnik,† T. Snyder,‡ and E. P. Wigner

Journal of Applied Physics 26 (3), 257–259 (1955)

Received December 11, 1954

Nuclear Reactor Issue

The first four articles in this issue describe some of the early experimental investigations leading to the nuclear reactor. Although these papers have just recently been declassified, they have already become "classics" in the science of reactor technology. The editors are indebted to Professors Wigner and Creutz for their cooperation in presenting these papers and for the review of other work which is provided by the first paper.

THE work reported in the following three articles dates back to the end of 1941, or the early part of 1942. The editors of this Journal suggested that we preface our reports by a short description not only of the work which preceded ours in the field of our investigation but also by a brief review of the work done subsequently. This task proved more difficult than we anticipated because little of the work on resonance absorption of material in bulk was published in regular scientific journals and because the references even to the published literature were difficult to collect. It is quite likely, therefore, that we missed some important papers and we wish to offer our apologies to those concerned. We hope that our omissions will be brought to our attention.

At the time our work was undertaken, the opinions of our colleagues were quite divided concerning the possibility of establishing a self-sustaining fission chain reaction with natural uranium. Most of our responsible physicists were inclined to question this possibility. However, Fermi and Szilard had made their historic decision to explore first the possibility of using thermal neutrons as chain carriers. This decision implied that the fission neutrons would have to be slowed down somehow to thermal energy; in the course of this "moderating process" they would have to pass through the energy region in which U^{238} was known to have strong absorption lines. That the neutron absorption at these resonance lines would not result in fission and fission neutrons and that it would therefore interrupt the chain, was surmised already by Meitner, Hahn, and Strassmann,[1] and it followed from N. Bohr's theory[2] that only U^{235} can undergo fission when absorbing low-energy neutrons. Bohr also noted that even if the chain reaction could be established while the uranium in the system was cold, the fission heat would increase the temperature of the uranium and thereby increase its resonance absorption. He suspected that this might bring the chain reaction to an early standstill. The point that not only one but several lines participate in the resonance absorption was particularly emphasized by G. Placzek. On the other hand, both Fermi and Szilard recognized that the resonance absorption could be diminished, to a certain degree, by "lumping" the uranium, i.e., concentrating it into finite chunks, rather

* Now at Carnegie Institute of Technology, Pittsburgh, Pennsylvania.
† Now at American Optical Company Research Laboratory, Southbridge, Massachusetts.
‡ Now at Knolls Atomic Power Laboratory, Schenectady, New York.

[1] Meitner, Hahn and Strassmann, Z. Physik, 106, 249 (1937) and 109, 538 (1938). The processes called 1 and 2 in this article were later shown by these workers to be the fission process. See also Halban, Kowarski, and Savitch, Compt. rend. 208, 1396 (1939), also Haxley, Shoupp, Stephens, and Wells, Phys. Rev. 57, 1088A (1940), 58, 199A (1940).
[2] See N. Bohr, Phys. Rev. 55, 548L (1939).

than distributing it uniformly over the reactor. They also recognized that the resonance absorption can be decreased, to any desired degree, by increasing the amount of the moderator, so that the neutrons would pass through the resonance region very fast. The amount of moderator is limited, however, because it itself acts as chain interrupter by absorbing the neutrons after it has slowed them down to thermal energy, thus decreasing the number of neutrons which may cause thermal fission in uranium.

Therefore, the problems which faced the would-be designers of chain reactors early in 1941 were (1) the choice of the proper moderator to uranium ratio, and (2) the size and shape of the uranium lumps which would most likely lead to a self-sustaining chain reaction, i.e., give the highest "multiplication factor." In order to solve these problems, one had to understand the behavior of the fast, of the resonance, and of the thermal neutrons. We were concerned with the second problem which itself consisted of two parts. The first was the measurement of the characteristics of the resonance lines of isolated uranium atoms, the second, the composite effect of this absorption on the neutron spectrum and total resulting absorption. One can liken the first task to the measurement of atomic constants, such as molecular diameter, the second one, to the task of kinetic gas theory which obtains the viscosity and other properties of the gas from the properties of the molecules. The first task was largely accomplished by Anderson[3] and was fully available to us when we did our work. Anderson's and Fermi's work on the absorption of uranium, and on neutron absorption in general, also acquainted us with a number of technics which will be mentioned in the third and fourth of the reports of this series. Finally, Fermi, Anderson, and Zinn carried out, in collaboration with us in Princeton,[3a] one measurement of the resonance absorption. This will be discussed in the third article of this series.

The foregoing facts fairly summarize the information which was available to us when we undertook the investigations to be reported below. Earlier or simultaneous work on resonance absorption includes, first of all, Bethe's calculation[4] of the absorption of uniformly distributed material in a hydrogeneous moderator. Bethe's results were adapted to the particular case of resonance absorption by uniformly distributed (i.e., not lumped) uranium and generalized to other moderators in a classified report by J. A. Wheeler (A-88). We did not learn about this report until some time in 1943. In addition, both J. Fisk and W. Shockley, and also Eckart, treated the resonance absorption of uranium lumps embedded in a moderator. We still know very little about Fisk's and Shockley's work. Eckart's work

became available later in 1942. It was only part of a much larger undertaking and attributed all the resonance absorption to the surface of the lump and endeavored to express it in terms of an energy interval in which the surface is black. The absorption outside this interval was considered negligible.

None of the foreign work on uranium fission was available to us and much of it is unavailable now. At present, we know the German work best.[5] The lumping of the uranium was suggested in Germany by Harteck in 1939, carbon as moderator by Heisenberg in the same year. This was practically simultaneous with the American work. The resonance absorption of uniformly distributed uranium was treated theoretically by Flügge and by Flügge and Sauerwein, as early as 1939. The absorption by U^{238}, both in the thermal and in the resonance region, was studied experimentally by Volz, Haxel, and Sauerwein toward the turn of the year 1942— just about one year after our studies. The importance of the higher absorption lines was recognized and experimentally demonstrated but the magnitude of the resonance absorption, as of all absorption by U^{238}, grossly underestimated (by a factor of about 10). We are unable to account for this error unless it was caused by some error in calibration.[6] It re-emphasizes the value of the careful measurements of Fermi and his collaborators. The effect of temperature on resonance absorption was anticipated and experimentally demonstrated. (The results are, in this case, considerably in excess of ours.) An improved theory of the resonance absorption was given later by V. Weizsacker. His work remains unreported but we surmise that it must have applied to lumped uranium.

The early French and British work closely paralleled the United States and German work.[7] The heterogeneous arrangement, as well as the use of moderators other than ordinary hydrogen, was proposed as early as 1939 by Halban, Kowarski, and Joliot.[8] It may be of interest to note that the arrangement for a chain reaction with hydrogeneous material and U_3O_8 proposed by these authors is not far from the arrangement which comes closest to being chain-reacting. Measurements were carried out to obtain the resonance escape probability of uniformly distributed uranium in hydrogeneous moderators and again, the values obtained are close to those now considered correct.[9] The effect

[3] H. L. Anderson, Phys. Rev. **80**, 499 (1950). See also reference 1.
[3a] Smyth, Fermi, and Anderson, Report A-12, June, 1941.
[4] H. A. Bethe, Revs. Modern Phys. **9**, 69 (1937). See also W. H. Furry, Phys. Rev. **51**, 592 (1937); O. Haxel and H. Volz, Z. Physik **120**, 493 (1942).

[5] See W. Heisenberg, Nature **160**, 211 (1947), also *Nuclear Physics* (Methuen and Company, Ltd., London, 1953), and the so-called Fiat report *Kernphysik und Kosmische Strahlung* (Dieterich'sche Verlagsbuchhandlung, Wiesbaden, 1948). Part 1, Articles by Hahn, Seelmann-Eggebert, Gotte, Flammersfeld, Droste, Flügge, Sauerwein, and Jentschke, pp. 171–224.
[6] Dr. W. K. Jentschke kindly informs us that the error was indeed one of calibration. However, Sauerwein's estimate [Z. Naturforsch. **2a**, 73 (1947)] of the resonance integral of the higher absorption lines of U^{238} is entirely reasonable.
[7] Cf. L. Kowarski, J. phys. radium **7**, 253 (1946).
[8] Cf. Halban, Kowarski, and Joliot, Compt. rend. **229**, 909 (1949).
[9] Halban, Joliot, Kowarski, and Perrin, J. phys. radium **10**, 428 (1939); Halban, Kowarski, and Savitch, Compt. rend. **208**, 1396 (1939).

of self-reversal was noted.[10] However, as far as we could ascertain, no calculations or measurements on the resonance absorption of lumped uranium were carried out at this early date.

When writing the present article we were informed of measurements of the resonance absorption in heterogeneous media. The reports on these measurements by Broda, Fennig, Graham, Guéron, Halban, Kowarski, and Seligman are still classified. Their dates extend from 1942 to 1944. Several of the French-British references were received through the courtesy of Dr. Cockcroft, Dr. Halban, Dr. Kowarski, and Dr. H. D. Young and we wish to express our appreciation for their help.

The Russian work is, since the war, much more closely guarded than the American. We found the article of Zeldovic and Chariton[11] very interesting and informative both from the technical point of view and also in its indication of the state of progress in Russia around 1940.

Our own results are described in the articles which follow. The experimental work was since repeated and extended by A. C. G. Mitchell and his collaborators. C. O. Muelhouse and S. Untermyer obtained values for the resonance absorption by a method fundamentally different from ours in 1949, Arfken, Caldwell, Cuykendall, Risser, and Stephenson by a more similar method in 1950.

The measurements of Anderson[3] and Fermi on the resonance lines of U^{238} were repeated also, in addition to Germany, by Hellstrand and Persson in Sweden, and by Levin and Hughes[12] at Brookhaven. Their results differ from those of reference 3 quite drastically. In particular, the lowest resonance line, the only one measured by Anderson, appears to be several times narrower than given in reference 3. On the other hand, the absorption at the center of the line is several times greater so that the total absorption of the line remains close to the figure which we used. It is not clear whether the cancellation of errors in Anderson's work is accidental. The measurements of reference 12 have not yet been reported in detail.

Among the theoretical work which is subsequent to ours, we wish to mention the calculations of S. M. Dancoff and M. Ginsburg, contained in report CP-1589. This report gives, on the basis of Anderson's measurements,[3] a quantitative explanation of the results of our third and fourth paper. Our second paper, written before the results of the third and fourth became available, gives only the functional dependence of the resonance absorption on size and shape of the uranium lumps and, as will be seen, only a crude numerical estimate of the absolute magnitude of the resonance absorption of lumps. No calculations have yet been carried out on the basis of the data of reference 12; they are planned, however, when the detailed results become available.

---

[10] Halban, Joliot, and Kowarski, Nature 143, 680 (1939) (last sentence of article).

[11] J. Zeldovic and J. Chariton, J. Exptl. Theoret. Phys. (U.S.S.R.) 10, 29 (1940).

[12] Hellstrand and Persson, Arkiv. Fysik 6, 57 (1953); J. S. Levin and D. J. Hughes, Phys. Rev. 95, 645A (1954).

6.

Resonance Absorption of Neutrons by Spheres

E. P. Wigner, E. Creutz,* H. Jupnik,† and T. Snyder,‡

Journal of Applied Physics 26 (3), 260–270 (1955)

Received September 15, 1954

By concentrating uranium into lumps, one can decrease its resonance absorption as compared with the resonance absorption of the same amount of uranium spread out uniformly in a moderator. It is shown here that this decrease is caused principally, not by the diminished over-all density of the neutrons at the lump, but by their changed energy distribution. This shows deep minima at the resonance lines, i.e., where the uranium absorbs most strongly. The considerations here presented show that the decrease in absorption is largest for the low energy resonances of U^{238} which are responsible for the bulk of the absorption in the case of uniform distribution. As a result, the resonance absorption of uranium lumps extends over a very large energy region—up to about 5000 ev. It can be decomposed, approximately, into two parts: the mass absorption, proportional to the number of U^{238} atoms present; and a surface absorption, proportional to the surface of the lump. The total decrease of the resonance absorption produced by lumping (the "advantage factor" of lumping) can be estimated to reach, under favorable conditions, a value as high as 20.

INTRODUCTION AND SUMMARY

THE work incorporated in the present article was done during the summer and fall of 1941. Its purpose was to provide a preliminary estimate of the magnitude of the resonance absorption of uranium lumps embedded in a moderating medium and, even more important, to enable us to carry out and interpret the experiments which are described in the following two articles. The most important result of the following considerations was that the resonance absorption of lumps embedded in a moderator can be divided into two parts: the mass absorption and the surface absorption. The former is proportional to the total number of U^{238} atoms present in the lump, the proportionality constant depending principally on the total scattering cross section (not the slowing-down power) associated with a U atom. It is, therefore, larger for U_3O_8 than for U metal. This fact showed the importance of eliminating any container of extraneous material from the uranium sample, the absorption of which sample was to be considered representative of the absorption of the whole lump: The scattering by the container material invariably increased the absorption of the sample contained therein and often led to erratic results. The importance of the second part of the resonance absorption, that of the surface absorption, turned out to be much smaller than was estimated before and is not much more than a correction to the mass absorption as long as the diameter of the lumps is not less than a couple of centimeters. Our considerations implicitly assume that the lumps have at least this dimension. The surface absorption also depends on the total scattering cross section of the materials admixed to the uranium, but this dependence is weaker than that of the mass absorption and decreases with increasing scattering cross section of the admixtures.

The work is reported here with hardly any alterations from the original manuscript,[1] except for the omission of less significant material, adoption of a more customary notation, correction of some minor numerical errors, and the redrawing of the figures to save space. It should be noted that metallic uranium was not available at the time the present work and most of the experiments described in following articles were done. Hence, our attention and interest were focused principally on the properties of U_3O_8. This is mentioned explicitly because the present treatment of the absorption in the lowest resonance level of U^{238} is very crude. The absorption of this level plays only a subordinate role in the oxide but is rather important in the metal. A more accurate consideration of the absorption of this level was given in CP-1589 by S. M. Dancoff and M. Ginsburg. This report also contains, implicitly, much of the work reported at the end of this article.

GENERAL DESCRIPTION OF THE ABSORPTION LINES

The so-called resonance absorption differs from ordinary absorption mainly because, over the most important range of energy, the absorption coefficient is a rapidly fluctuating function of energy. It is caused by rather narrow absorption lines, with an average spacing of about 20 ev. The first of these lies at 5.8 ev, the second, probably, at 30 ev, but nothing is known about the location of the higher ones.[2] Anderson and Fermi[3] measured the relevant constants of the absorption. They concluded from their measurements that the most important line lies around 10 ev. The self-absorp-

* Now at Carnegie Institute of Technology, Pittsburgh, Pennsylvania.
† Now at Optical Company Research Laboratory, Southbridge, Massachusetts.
‡ Now at Knolls Atomic Power Laboratory, Schenectady, New York.

[1] The material originally appeared in Document C-4. It was declassified by authority of the U. S. Atomic Energy Commission on December 29, 1953.
[2] Baker, Bacher, and McWilliams, Report A-23. The experimental information now available can be found in Supplement 2 to the Neutron Cross Section Compilation (AECU-2040) issued by the U. S. Atomic Energy Commission on June 15, 1953. See also J. S. Levin and D. J. Hughes, Phys. Rev. 95, 645A (1954).
[3] Report A-2. Also H. L. Anderson, Phys. Rev. 80, 499 (1950).

tion of the resonance activity is 5000×10^{-24} cm^2, and they find for the integral over the absorption cross section $\sigma_a(E)$

$$\int \frac{\sigma_a(E)}{E} dE = 239 \times 10^{-24} \text{ cm}^2. \qquad (1)$$

It will be attempted to show at least a rough picture of the absorption on the basis of the foregoing data.

It will be assumed that the absorption cross section can be represented in the neighborhood of a line at E_i by the formula

$$\sigma_a(E) = \frac{0.65 \times 10^{-18}}{E^{\frac{1}{2}}} \frac{a\Gamma}{(E-E_i)^2 + \frac{1}{4}(\Gamma + aE^{\frac{1}{2}})^2} \text{cm}^2. \qquad (2)$$

Although it is probably far from being true, we shall assume that the γ-ray width Γ and the constant a in the neutron width $aE^{\frac{1}{2}}$ are the same for all lines and use the foregoing data of Fermi and Anderson to find their values.[4]

This will be done in two ways, to be distinguished as assumptions a and b. The first one assumes that the lines are at 5.8, 3×5.8, 5×5.8, etc. ev and that the above data give the sum of the effects of all lines. It follows that the integral (1) is a sum of the integrals due to the various lines

$$\int \frac{\sigma_a(E)}{E} dE = \sum_i \frac{2 \times 0.65 \times 10^{-18} \pi a\Gamma}{E_i^{\frac{3}{2}}(\Gamma + aE_i^{\frac{1}{2}})}. \qquad (3)$$

When integrating (2), it was assumed that the line is sharp, i.e., $\Gamma + aE_i^{\frac{1}{2}}$ is much smaller than the distance of the lines—a condition which is satisfied for the relevant lines. Since, furthermore, only the low lines play an important role in (3) and since for these $aE_i^{\frac{1}{2}} \ll \Gamma$, we can write

$$1.3 \times 10^{-18} \pi a \left(\frac{1}{5.8^{\frac{3}{2}}} + \frac{1}{(3 \times 5.8)^{\frac{3}{2}}} + \frac{1}{(5 \times 5.8)^{\frac{3}{2}}} \cdots \right)$$
$$= 239 \times 10^{-24}$$

from which

$$a = 4.8 \times 10^{-4} (\text{ev})^{\frac{1}{2}} \qquad (3a)$$

follows (the number of all equations and figures based on assumptions a will be followed by an a). If we make the assumption that the energy spectrum of the neutrons used for the measurement of the self-absorption followed the dE/E law, we can interpret this quantity in a similar way. We obtain

$$\Gamma = 3.8 \times 10^{-2} \text{ ev} \qquad (4a)$$

for the γ-ray width.

Assumption b postulates that both the integral (1) and also the self-absorption are due to a single line at 5.8 ev. It gives, for the foregoing quantities,

$$a = 8.2 \times 10^{-4} (\text{ev})^{\frac{1}{2}} \qquad (3b)$$

$$\Gamma = 9 \times 10^{-2} \text{ ev}. \qquad (4b)$$

Furthermore, we shall assume unber b that the higher lines are located at 30, 50, 70, etc., ev and that their a and Γ are given by (3b) and (4b). It follows from this assumption that the integral of $\sigma_a(E)/E$ over the whole energy spectrum is larger than 239×10^{-24} cm^2 since the first line already contributes this amount. The total integral is, under assumption b, 310×10^{-24} cm^2.

The purpose of carrying out the following consideration with two sets of constants is to obtain some information concerning the sensitivity of the results with respect to variations of these constants. It may be argued that assumptions a are more reasonable[5] than assumptions b. It is sure, however, that neither set is correct. Both lead to values of a and Γ of the same order of magnitude as was found by Fermi and Anderson for other elements,[1] except I, viz., $10^4a = 1.80$, 8.5, 8.0, 20.5, and $10^2\Gamma = 12.5$, 9.8, 9, 11 for Rh, Ag (22-sec activity) In, Au, respectively. For the ratio of the reduced neutron width γ^2 and the level spacing D, assumptions a and b both give 0.09×10^{-13}. Both are below the theoretical values but agree, as far as order of magnitude is concerned, with the values of γ^2/D for other heavy elements.[6] The connection between γ and a is $a = 4.4 \times 10^9 \gamma^2$.

For the thermal absorption, assumptions a and b give 2.2×10^{-24} cm^2 and 9×10^{-24} cm^2, respectively, if one neglects the effect of all but the lowest line. The experimental value[2] is 3×10^{-24} cm^2 indicating that the actual value of a may be between the values given by assumptions a and b.

One can obtain, by integrating (2), an expression for the average absorption coefficient. If we denote the average spacing of the lines by D, this becomes D^{-1} times the integral of (2), i.e.,

$$\langle \sigma_a(E) \rangle_{Av} = \frac{1.3 \times 10^{-18} \pi a\Gamma}{E^{\frac{1}{2}}(\Gamma + aE^{\frac{1}{2}})D}. \qquad (5)$$

With

$$D = 11.6 \text{ ev} \qquad (6a)$$

$$D = 20 \text{ ev} \qquad (6b)$$

this gives the two curves of Fig. 1. For $E = 200$ kev, assumptions a and b give 0.57×10^{-25} and 0.74×10^{-25}, respectively. Both values are considerably lower than the one given by J. Marshall and Szilard, viz., 5×10^{-25}.

---

[4] The consequences of possible fluctuations of the quantities a and Γ have been explored by Dancoff and Ginsburg report CP-1589. In an earlier report (CP-1092) they have analyzed the data of reference 3 much more carefully than they are analyzed here. In the notation here adopted, they conclude that for the 5.8-ev line $a = 9 \times 10^{-4}$ ev$^{\frac{1}{2}}$, $\Gamma = 0.065$ ev.

[5] The results of the measurements reported in the following two papers later led us to believe that assumptions b may be the ones closer to reality.
[6] See T. Teichmann and E. P. Wigner, Phys. Rev. 87, 123 (1952).

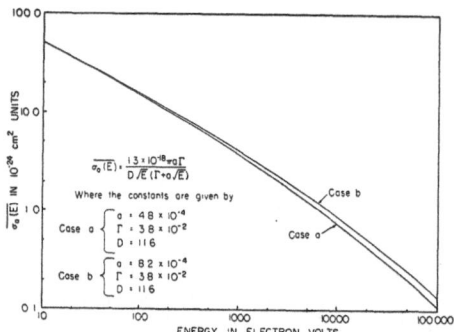

FIG. 1. The average absorption coefficient of uranium for neutrons as a function of energy under assumptions a and b described in the text.

However, this value for the absorption cross section may refer to a lower energy.[7]

We shall denote the ordinary scattering cross section of the material associated with one atom U, by σ_{sn}. Thus, for the metal

$$\sigma_{sn} = 12 \times 10^{-24} \text{ cm}^2 \qquad (7)$$

is the scattering cross section of the U nucleus.[8] For the oxide U_3O_8, this must be augmented by 8/3 times the scattering cross section of O, giving

$$\sigma_{sn} = 22 \times 10^{-24} \text{ cm}^2. \qquad (7a)$$

According to the usual theory of resonance processes, an anomalous scattering cross section is associated with the absorption. It is given, in the neighborhood of the line E_i, by

$$\sigma_{sa}(E) = \frac{0.65 \times 10^{-18} a^2}{(E - E_i)^2 + \frac{1}{4}(\Gamma + aE^{\frac{1}{2}})^2} = \frac{aE^{\frac{1}{2}}}{\Gamma} \sigma_a(E). \qquad (8)$$

The total scattering cross section is $\sigma_s = \sigma_{sn} + \sigma_{sa}$.

It is probably useful to remark that the total natural line width $\Gamma + aE^{\frac{1}{2}}$ is essentially γ-ray width below 6500 ev and that the neutron width, $aE^{\frac{1}{2}}$, can be neglected compared with Γ if one is well below this energy. Above 6500 ev, Γ becomes smaller than $aE^{\frac{1}{2}}$.

The magnitude of the anomalous scattering is smaller than the absorption below 6500 ev, larger above. Below this limit its value at the center of a line is about $4 \times 10^{-22} \text{ cm}^2$; i.e., very much larger than the ordinary scattering cross section. Above 6500 ev, its value at the center of the line decreases, it becomes equal to the ordinary scattering cross section of the metal at 150,000 ev.

THE DOPPLER EFFECT

The foregoing considerations neglect the temperature motion of the U nuclei which causes a broadening of the

[7] This possibility was suggested by Marshall and Szilard. The discrepancy is not fully cleared up even now.
[8] The value now accepted is $9.5 \times 10^{-24} \text{ cm}^2$. See reference 2.

lines the significance of which, in connection with our problem, was first pointed out by J. A. Wheeler.

Let us consider a neutron of energy $239/238E$, colliding with a U atom which has velocity components $v_{||}$ and $v_{\perp}$ parallel and perpendicular to the velocity of the neutron. The relative energy E' of neutron and U is then

$$E' = \frac{1}{2} \frac{238}{239} M \left\{ \left[\left(\frac{2 \times 239E}{238M} \right)^{\frac{1}{2}} + v_{||} \right]^2 + v_{\perp}^2 \right\}$$

$$\approx E + (2EM)^{\frac{1}{2}} v_{||}, \qquad (9)$$

where M is the mass of the neutron, 238 M that of the U nucleus. From (9), the probability that the relative energy be between E' and $E' + dE'$ becomes

$$\frac{dE'}{(2EM)^{\frac{1}{2}}} \left(\frac{238M}{2\pi kT} \right)^{\frac{1}{2}} \exp\left[-\frac{238M}{2kT} \frac{(E'-E)^2}{2EM} \right]. \qquad (10)$$

In order to obtain the cross section for neutrons of energy $239E/238$, (10) must be multiplied with the cross section for the relative energy E' and integrated over E'. For the former, we use (2), assuming that E and E' are in the neighborhood of E_i

$$\sigma_a(E) = \frac{0.65 \times 10^{-18} a\Gamma}{E} \left(\frac{238}{4\pi kT} \right)^{\frac{1}{2}}$$

$$\times \int \exp\left[\frac{-238}{4kTE}(E'-E)^2 \right]$$

$$\times \frac{dE'}{(E'-E_i)^2 + \frac{1}{4}(\Gamma + aE_i^{\frac{1}{2}})}. \qquad (11)$$

This integral has been tabulated recently.[9]

THE PRACTICAL WIDTH

It is useful for the following to define a practical width of the absorption lines. This width, 2Δ, is the distance of the two points, at the wings of the line, for which the sum of absorption and anomalous scattering cross sections $\sigma_a + \sigma_{sa}$ becomes equal to the ordinary scattering cross section σ_{sn}. Of course, for very high levels, $\sigma_a + \sigma_{sa}$ does not become equal to the ordinary scattering cross section even in the middle of the line and we shall say, in this case, that the practical width is zero. In fact, one has no real line absorption at these levels.

The practical width was calculated on the basis of the approximate evaluation (12) of the integral (11) and is given in Figs. 2(a) and 2(b). The calculation will not be given in detail here; its approximate nature is evident from the kinks in the figures.

The origin of these kinks is as follows: The absorption due to the natural line width decreases, far from the

[9] Report BNL 257, issued by the U. S. Atomic Energy Commission.

center of the line, as $(E-E_i)^{-2}$; that due to the Doppler, broadening decreases exponentially. It is possible, therefore, to approximate the integral (11) by

$$\sigma_a(E) = \frac{0.65 \times 10^{-18} a\Gamma}{E_i} \left\{ \frac{E_i^{\frac{1}{2}}}{(E-E_i)^2 + \frac{1}{4}(\Gamma + aE_i^{\frac{1}{2}})^2} + \frac{(238\pi/kT)^{\frac{1}{2}}}{\Gamma + aE_i^{\frac{1}{2}}} \exp \frac{-238(E-E_i)^2}{4kTE_i} \right\}. \quad (12)$$

The first term predominates far from the line, where the absorption is already relatively low, the second one is larger close to E_i where the absorption is still relatively high. If one equates $\sigma_a(E)$ to σ_{sn}, the first term of (12) will be important at the low E_i where σ_{sn} is relatively small. This is the region to the left of the kinks and will be called the natural width region. For the lines at high E_i, the normal scattering cross section σ_{sn} is a relatively large cross section and the second term of (12) is the more important one in the region where $\sigma_a(E)$ is equal to σ_{sn}. The E_i for which this condition prevails are said to lie in the Doppler region which is to the right of the kinks in the figures. Different approximations are useful in different parts of the energy spectrum and in the following the different regions will be treated separately.

The natural width region extends for the metal to 70 and 200 ev under assumptions a and b, respectively, at 300°K. At 900°K, the corresponding limits are 32 and 90 ev. For the oxide, the natural width region extends at 300°K to 50 and 140 ev, at 900°K to 25 and 55 ev for assumptions a and b, respectively.

One sees that the Doppler region is very important under assumption a, but becomes narrower if the constants b are adopted. The temperature dependence of the resonance absorption is due to the Doppler region. It will be seen from the experimental data to be pre-

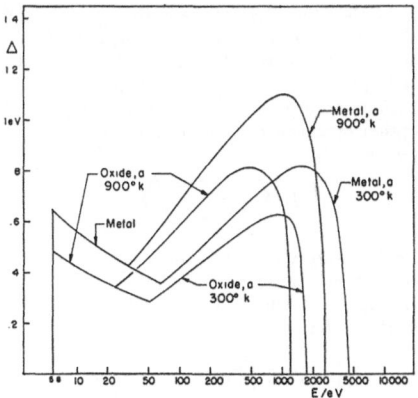

FIG. 2(a). One-half of the practical width of the uranium absorption lines plotted as a function of neutron energy for assumption a given in the text.

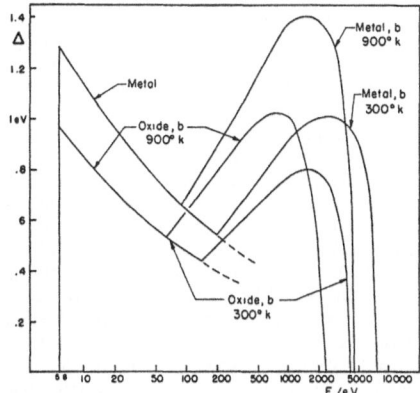

FIG. 2(b). One-half of the practical width of the uranium absorption lines plotted as a function of neutron energy for assumption b given in the text.

sented later that the temperature coefficient of the resonance absorption is actually very small. This favors the constants b.

The energy range beyond the upper limit of the Doppler region will be called continuous region.

GENERAL FEATURES OF LINE ABSORPTION

Over the most important part of the spectrum, the absorption is concentrated in lines. The absorption of a block is, in these regions, much smaller than that of the same amount of material spread out thinly over the slowing down material. It is also much smaller than it would be if the absorption were not concentrated in lines but spread out uniformly over the whole energy range. The reason for this is that the number of neutrons with energies within an absorption line is strongly decreased by the absorption and much fewer absorbable neutrons remain than would be present in a material with continuous absorption.

Let us denote the number of neutrons per cm³ and unit energy range by Q/Ev where $v=(2E/M)^{\frac{1}{2}}$ is the velocity corresponding to energy E. The Q is the number of collisions which some material with 1 cm² cross section would suffer per second from the neutrons of unit logarithmic energy range (one hundred times the number of collisions with neutrons with energies between E and 1.01 E). Let us further denote the average of a quantity over an energy range containing several absorption lines by a bar.[10] The barred quantities, such as $\bar{Q}$, will vary only slowly with energy (and position). The consideration of such average quantities is particularly useful in the energy region in which the widths of the lines are small compared with the average energy loss of a neutron per collision. Even in the metal, this

---

[10] For typographical reasons, some of the bars had to be changed to $\langle \ \rangle_{Av}$. This symbol and the bar are equivalent.

region extends from about 120 ev to the upper limit of line absorption, i.e., the upper limit of the Doppler region. The considerations of the present section apply only to that part of the spectrum where the lines are narrow in the above sense.

The line absorption consists of two parts: the mass absorption and the surface absorption.

We first consider the *mass absorption*. If the U block had no absorption, the number of neutrons which are, within a cm³, thrown out of the energy range between E and $E+dE$ by collisions would be $N\sigma_s Q dE/E$, where N is the number of U atoms per cm³. Under the same conditions, this would be also the number of neutrons entering the same energy range from above—in fact $N\langle\sigma_s Q\rangle_{Av} dE/E$ is the correct expression for the number of neutrons entering this energy range in spite of the absorption because these neutrons originate over a considerable energy range which contains several lines. On the other hand, the number of neutrons leaving the same energy range is determined by the local density and cross sections so that we have

$$N\langle\sigma_s Q\rangle_{Av} dE/E = N(\sigma_a+\sigma_s)Q dE/E. \qquad (13)$$

This gives

$$Q=(\sigma_a+\sigma_s)^{-1}\langle\sigma_s Q\rangle_{Av} \qquad (14)$$

showing that Q has sharp minima where $\sigma_a+\sigma_s$ has maxima, i.e., in the absorption lines. If averaged over E, (14) gives $\bar{Q}=\langle(\sigma_a+\sigma_s)\rangle_{Av}^{-1}\langle\sigma_s Q\rangle_{Av}$. Since $\sigma_s+\sigma_{sa}$ is larger than σ_{sn} only within the relatively small practical width, the average of the reciprocal value of $\sigma_a+\sigma_s$ $=\sigma_a+\sigma_{sa}+\sigma_{sn}$ will be practically equal to the reciprocal of σ_{sn}. It will be smaller by an amount of the order $1-2\Delta/D$ where 2Δ is the "practical width" and D is the average spacing of the λ lines, but the accuracy of the present calculations hardly justifies carrying on such factors. We shall write, therefore, σ_{sn}^{-1} for $\langle(\sigma_a+\sigma_s)\rangle_{Av}^{-1}$ and we have $\sigma_{sn}\bar{Q}=\langle\sigma_s Q\rangle_{Av}$. The absorption in the energy range dE becomes,

$$N\sigma_a Q dE/E=\frac{N\sigma_a}{\sigma_a+\sigma_s}\sigma_{sn}\bar{Q}\frac{dE}{E}, \qquad (15)$$

and the average mass absorption $\bar{A}_v$ per unit energy interval

$$\bar{A}_v=N\frac{\bar{Q}}{E}\left\langle\frac{\sigma_a\sigma_{sn}}{\sigma_a+\sigma_s}\right\rangle_{Av}. \qquad (16)$$

The *surface absorption* is caused by the fact that the energy spectrum of the neutrons which strike the surface from the outside is uniform, $Q=\bar{Q}$. It does not show the deficiency of neutrons within the line which the Q calculated from (14) manifests. The number of neutrons of unit energy interval which strike unit area of the surface from the outside is $\frac{1}{4}\bar{Q}/E$. It shows a surplus of

$$\frac{\bar{Q}}{4E}-\frac{\bar{Q}}{4E}\frac{\sigma_{sn}}{\sigma_a+\sigma_s}=\frac{\bar{Q}}{4E}\frac{\sigma_a+\sigma_{sa}}{\sigma_a+\sigma_s} \qquad (17)$$

over the number of neutrons of the same velocity and direction in the inside. Of the neutrons (17), the fraction $\sigma_a/(\sigma_a+\sigma_s)$ will be absorbed, the rest suffers a scattering collision and is, from then on, one of the regular neutrons of the inside. Hence, the surface absorption per unit surface and unit energy range $\bar{A}_s$ is

$$\bar{A}_s=\frac{\bar{Q}}{4E}\left\langle\frac{(\sigma_a+\sigma_{sa})\sigma_a}{(\sigma_a+\sigma_s)^2}\right\rangle_{Av}. \qquad (18)$$

Even from the comparison of (18) and (16) it is evident that, while the mass absorption is proportional for given volume to N, and therefore to the density of the block, the surface absorption is independent of this quantity. The absorption of an average U atom in a lump of volume V and surface S becomes per unit logarithmic energy range

$$\frac{E\bar{A}}{N}=\frac{E}{N}\left(\bar{A}_v+\frac{S}{V}\bar{A}_s\right)$$

$$=\bar{Q}\left\langle\frac{\sigma_a\sigma_{sn}}{\sigma_a+\sigma_s}\right\rangle_{Av}+\frac{S\bar{Q}}{NV}\left\langle\frac{(\sigma_a+\sigma_{sa})\sigma_a}{4(\sigma_a+\sigma_s)^2}\right\rangle_{Av}. \qquad (19)$$

Naturally, this formula does not hold for the bottom part of the energy region where the number of resonance levels per unit logarithmic energy range is small. One may remember that $\bar{Q}$ is essentially inversely proportional to the slowing down power of the moderator.

The two averages in (19) have simple interpretations. If the nucleus with absorption cross section σ_a were not shielded by the other nuclei, its absorption would be, per unit logarithmic energy range, $\bar{Q}\sigma_a$. Hence, if we replace σ_a by

$$\sigma_{eff}=\left\langle\frac{\sigma_a\sigma_{sn}}{\sigma_a+\sigma_s}\right\rangle_{Av}, \qquad (19a)$$

we can neglect the effect of the mutual shielding of the nuclei and obtain their mass absorption. Similarly, nuclei with total cross section S would have an absorption $\bar{Q}S$ per unit logarithmic energy range. Hence, the surface absorption of the lump is equal to the absorption of unshielded atoms with the total surface of Sf where

$$f=\left\langle\frac{(\sigma_a+\sigma_{sa})\sigma_a}{4(\sigma_a+\sigma_s)^2}\right\rangle_{Av}. \qquad (19b)$$

The total absorption of the lumps can be described as the absorption of unshielded nuclei with a total surface of $(NV)\sigma_{eff}+Sf$ where NV is the total number of nuclei and S the surface of the lump

$$A(\text{lump})=\bar{Q}\int[(NV)\sigma_{eff}+Sf]dE/E. \qquad (19c)$$

This formula neglects possible variations of $\bar{Q}$ over energy and position within the lump. If the lumps are of considerable size these may have to be corrected. In such a case, $\bar{Q}$ has to be under the integral sign and $NV\bar{Q}$ has to be replaced by integration of $N\bar{Q}$ over the volume of the lump. This was not necessary for the interpretation of the experiments to be described in the two articles which follow, but has been necessary for the calculation of the reproduction factor of some of the chain reacting units.[11]

We now go over to the detailed consideration of the absorption in the different energy regions.

THE DOPPLER REGION

The preceding considerations can be applied directly in the Doppler region. For the calculation of the average in (16) we need the integral

$$\int \frac{\sigma_a}{\sigma_a+\sigma_s}dE=\int \frac{\sigma_a dE}{\sigma_a+\sigma_{sa}+\sigma_{sn}} \qquad (20)$$

over a line. Evidently, the integrand is zero far away from the center of the line and nearly 1 near to the center. For $\sigma_{sa}\ll\sigma_a$ it is $\frac{1}{2}$ at the edge of the practical width. In order to calculate (20), we need an accurate value of σ_a and $\sigma_a+\sigma_{sa}=\sigma_a(1+aE^{\frac{1}{2}}/\Gamma)$ only in the neighborhood of the practical width. Thus we can substitute for σ_a the second term of (12) except, perhaps, at the lower end of the Doppler region (which will turn out to be a relatively unimportant region). Thus,

$$\int \frac{\sigma_a dE}{\sigma_a+\sigma_s}=\frac{\Gamma}{\Gamma+aE^{\frac{1}{2}}}\int \frac{dE}{1+\gamma \exp[238(E-E_i)^2/4kTE_i]}$$

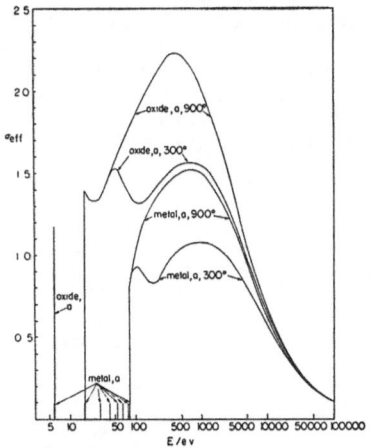

FIG. 3(a). The effective absorption cross section plotted as a function of neutron energy under assumption a given in the text.

[11] Cf. S. Glasstone and M. C. Edlund, *The Elements of Nuclear Reactor Theory* (D. Van Nostrand Company, Inc., New York, 1952), Secs. 9.36–9.41, pp. 261–263.

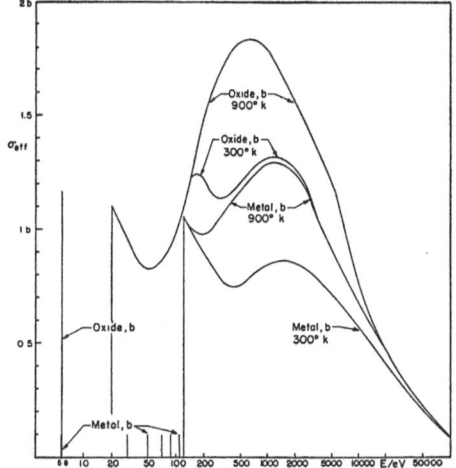

FIG. 3(b). The effective absorption cross section plotted as a function of neutron energy under assumption b given in the text.

where

$$\gamma=\frac{\sigma_{sn}E_i(kT)^{\frac{1}{2}}}{1.8\times10^{-17}a}. \qquad (21)$$

It is not to be confused with the reduced width. Furthermore, this expression remains valid even in the continuous region where the practical width becomes meaningless. In this case, the 1 in the denominator can be neglected and one obtains the ratio of the integral of (2) to σ_{sn}. If one substitutes a new variable for the argument of the exponential, the integral in (21) assumes the form which occurs in the electron theory of metals and can be evaluated in the same way.

For practically all of the Doppler region, $\gamma\ll1$ which corresponds to strong degeneracy. In this case, the exponential in the denominator of (21) changes very rapidly in the neighborhood of the edge of the practical width. Hence, the integrand is practically 1 within the practical line, 0 outside, and the integral (20) becomes 2Δ. The mass absorption is, therefore,

$$\bar{A}_v=(N\bar{Q}/E)\sigma_{eff}=2N\bar{Q}\sigma_{sn}\Delta/[ED(1+aE^{\frac{1}{2}}/\Gamma)], \qquad (22)$$

where D is the spacing of the lines, $1/D$ the number of lines per unit energy interval. The σ_{eff} is plotted in Figs. 3(a) and 3(b).

Near the upper end of the Doppler region σ_a does not change rapidly at the distance Δ from the center of the line. In this case, one has to evaluate the integral (21) graphically or use the tabulated values therefor. For $\gamma=0.5, 1, 1.33,$ and 2, the value of the integral in (21) is $(4kTE/238)^{\frac{1}{2}}(1.58, 1.07, 0.9, 0.67)$.

For the calculation of the surface absorption we need

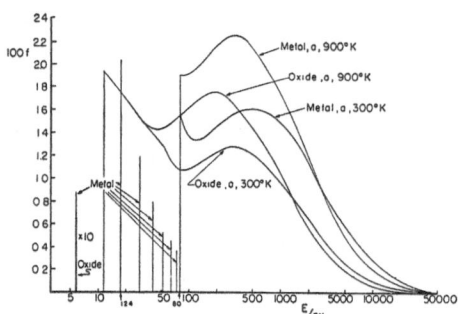

FIG. 4(a). 100×function f plotted as a function of neutron energy under assumption a given in the text.

the integral

$$\int \frac{\sigma_a^2(1+aE^{\frac{1}{2}}/\Gamma)dE}{\sigma_a(1+aE^{\frac{1}{2}}/\Gamma)+\sigma_{sn})^2}$$

$$=\frac{\Gamma}{\Gamma+aE^{\frac{1}{2}}}\int \frac{dE}{[1+\gamma \exp(238(E-E_i)^2/4kTE)]^2}. \quad (23)$$

Again, in the lower part of the Doppler region the exponential increases very rapidly at $|E-E_i|=\Delta$. The integrand is nearly 1 for $|E-E_i|<\Delta$, it is very small for $E-E_i>\Delta$. Hence, we have for the surface absorption

$$\bar{A}_s=(\bar{Q}/E)f=\bar{Q}\Delta/[2ED(1+aE^{\frac{1}{2}}/\Gamma)]; \quad (24)$$

f is plotted in Figs. 4(a) and 4(b). The approximation (24) is much poorer than (22). Actually, the value of the integral is $(4kTE/238)^{\frac{1}{2}}(1.35, 0.80, 0.40, 0.16)$ for $\gamma=0.25, 0.50, 1$, and 2.

In the lower part of the Doppler region the absorptions per unit logarithmic energy interval, $E\bar{A}_v$ and $E\bar{A}_s$, actually increase with increasing energy because they are proportional to Δ which is roughly proportional to the square root of the energy. The mass absorption is, furthermore, proportional to σ_{sn}; it is for the metal 12/22 times smaller than for the oxide. In the upper part of the Doppler region the absorption becomes increasingly independent of the temperature and σ_{eff} and becomes, in the continuous region, equal to the ordinary absorption.

The ratio of the second, or surface term of (19), to its first or mass term is

$$S\bar{A}_s/V\bar{A}_v=S/4NV\sigma_{sn}. \quad (25)$$

This is 8.3 $S/\rho V$ for the metal and 6.15 $S/\rho V$ for the oxide where ρ is the density of the material in the block.

It may be mentioned here that the actual mass absorption is smaller than (22) because $\bar{Q}$ is smaller in the sphere than is its average value in the cell.[11] For a similar reason, the ratio of surface and volume absorptions is somewhat larger in the lower Doppler region than (25).

NATURAL WIDTH REGION

If we continue to assume that the average energy loss is large compared with the line width, (16) and (18) will give the mass and surface absorptions. An easy integration over the natural line shape gives for these

$$\bar{A}_v=(N\bar{Q}/E)\sigma_{eff}=\pi N\bar{Q}\sigma_{sn}\Delta/ED(1+aE^{\frac{1}{2}}/\Gamma) \quad (26)$$

$$\bar{A}_s=(\bar{Q}/E)f=(\pi/8)\bar{Q}\Delta/DE(1+aE^{\frac{1}{2}}/\Gamma). \quad (27)$$

At the dividing point of the natural width and Doppler width region, (26) should become equal to (22). This is not the case, however, because the same width is attributed in (22) and (26) to different causes: the Doppler width in (22) and the natural width in (26). This causes a discrepancy by a factor $\pi/2$ at the point where the two terms of (12) are equal and where (12) is least accurate. The curves corresponding to (22) and (26) had to be joined as smoothly as possible in the neighborhood of this point and this could not be done without any arbitrariness. It causes the characteristic wiggles in the curves for σ_{eff}. It also resulted in a slightly higher σ_{eff} than either (22) or (26) alone would have given. This is in the right direction to compensate the inaccuracy of (12). Also, the natural width region is slightly extended as far as mass absorption is concerned. Hence, the Δ for (26) and (27) was needed somewhat beyond the point where the values calculated from the first and second terms of (12) become equal. These are given by the broken lines in Figs. 2(a) and 2(b) and were used to calculate σ_{eff} for (26) somewhat beyond the kinks in Figs. 2(a) and 2(b).

The ratio of surface and volume absorptions becomes, according to (26) and (27), half as great as in the Doppler region. The reason is again the difference in the shape of the lines of the two regions. The absorption as given by (26) and (27), is independent of the temperature and inversely proportional only to the square root of σ_{sn}. This gives to the metal a smaller advantage over the oxide than it has in the Doppler region.

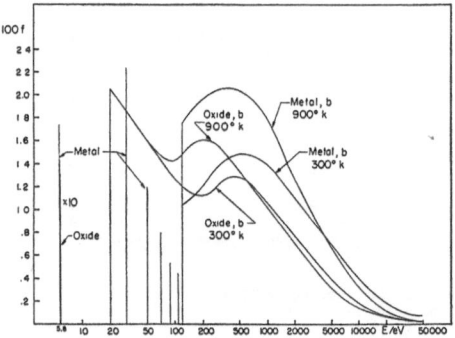

FIG. 4(b). 100×function f plotted as a function of neutron energy under assumption b given in the text.

According to (26), the absorption per unit logarithmic energy interval $E\bar{A}_v$ increases in the natural width region with decreasing energy. In spite of this, if one adopts the constants a, even the low energy portion of the natural width region fails to become as important as the upper part of the Doppler region is. On the other hand, if one adopts the constants b, the lowest energy region is about as important for the oxide as the top of the Doppler region. For the metal, the bottom of the Doppler region is always relatively unimportant which makes the inadequacy of our approximation at the dividing line between Doppler and natural width regions less objectionable.

As mentioned at the beginning of this section, (26) and (27) hold only if the average energy loss is large compared with the practical width of the line. This is true for scattering collisions on oxygen, down to the lowest level, but it is not true in the metal below 80 ev under assumptions a and below 120 ev under assumptions b. This limits the region of validity of (26) and (27), particularly in the case of the metal, to a very narrow energy band. The reason is that, in the metal, the neutron cannot jump over the levels below 80 or 120 ev, respectively, and the absorption of these levels becomes equal to the number of neutrons which would be slowed down below this level if no absorption were present. This number is Fermi's "slowing down density" $QN\sigma_{sn}(2/239)$ so that the volume absorption of every level below 80 or 120 ev becomes

$$A_v = 2\bar{Q}N\sigma_{sn}/239. \qquad (28)$$

The average absorption as given by (28) is lower than (26) would indicate, the ratio being just the ratio of the average energy loss $2E/239$ to $\pi/2$ times the practical width 2Δ of the level. The reason for this diminution of the absorption is that the neutrons cannot reach at all the lower part of the line since all have been caught in the upper part. Thus, the low energy side of the line has practically no absorption. It follows from (28), which is valid only for the metal, that the absorption per logarithmic energy interval becomes again smaller below 80 or 120 ev. One has the paradoxical situation that the volume absorption of a block is smallest where the absorption coefficient is largest.

The absorption of the lines below 120 ev in the metal is given in Figs. 3(a) and 3(b) and Figs. 4(a) and 4(b) directly, i.e., it is not distributed continuously over an energy range. This is reasonable because the number of levels per unit logarithmic energy range is not large enough for defining the average of the absorption per unit $d \ln E$. Hence, if one defines $\int \sigma_{eff} dE/E$ so that the total mass absorption becomes equal to $\bar{Q}N\int \sigma_{eff}dE/E$, the ordinates at the levels (which represent $2\sigma_{sn}/239$) must be added to the integral under the curves of Figs. 3(a) and 3(b),

For the oxide, (26) remains valid except that one has to use, below 80 and 120 ev, for σ_{sn} only the cross section of the oxygen, i.e., 10×10^{-24} cm². Only the lowest level is too broad to be jumped after a collision with an oxygen nucleus and its absorption is

$$A_v = \bar{Q}\times10\times10^{-24}N\times2/17. \qquad (29)$$

This also is given as a separate ordinate in Figs. 3(a) and 3(b) and Figs. 4(a) and 4(b).

Although the mass absorption in the lowest energy region is smaller than (26) would indicate, the surface absorption is larger than (27) since (see (17)) of the $\frac{1}{4}\bar{Q}dE/E$ neutrons entering through unit surface, practically none would be present, were it not for the surface. Hence, particularly for the lowest levels, the surface absorption is

$$A_s = -\frac{1}{4}\frac{\bar{Q}}{E}\int \frac{\sigma_a(E)}{\sigma_a(E)+\sigma_{sn}}dE = (\pi/4)\bar{Q}\Delta/E \qquad (30)$$

or twice larger than (27). The absorption of the metal for the first few resonances is almost entirely a surface absorption, compared with which the volume absorption is negligible.

In the oxide (27) remains valid except for the lowest resonance. However, below 80 and 120 ev, when calculating Δ, one has to use for σ_{sn} only the cross section of oxygen, i.e., $10b$. For the lowest level, (30) was used.

The characteristics of the neutron absorption vary very strongly within the natural width region. At the lowest end, the mass absorption of the metal should be negligible as compared with that of the oxide (and even the oxide's absorption may be less than 1/30th of the absorption of uniformly distributed material). On the other hand, at the high energy end, the absorption of the oxide is greater than that of the metal only by a factor $\sqrt{22/12}$, while in the Doppler region this ratio becomes $22/12$. Similarly, the ratio of surface absorption to mass absorption is extremely large at the low energy end for the metal and even for the oxide larger than it is in the Doppler region. At the high energy end of the natural width region, the ratio of surface and volume absorptions is only half as great as in the Doppler region. Finally, the whole importance of the natural width region depends strongly on the assumptions one makes for the constants; the whole region is of lesser significance if the constants a are valid, it is an important region if the constants b hold.

THE CONTINUOUS REGION

We have already dealt with the absorption in the lower part of the continuous region, i.e., beyond the point where Δ drops to zero, in the section dealing with the Doppler region. The problem reduces, mathematically, to the evaluation of the integrals (21) and (23) for large γ. For this we found the approximate formulas

$$\int_{-\infty}^{\infty} \frac{dx}{1+\gamma \exp x^2} \approx \frac{\pi^{\frac{1}{2}}}{\gamma}\left(1-\frac{1}{8^{\frac{1}{2}}\gamma}\right), \qquad (31)$$

$$\int_{-\infty}^{\infty} \frac{dx}{(1+\gamma \exp x^2)^2} \approx \frac{\pi^{\frac{1}{2}}}{\gamma^2}\left(1-\frac{2}{3\gamma}\right), \qquad (32)$$

quite useful and sufficiently accurate. Neither (31) nor (32) have any good mathematical foundation, except for $\gamma = \infty$, but they give roughly correct results for $\gamma > 1$. In the very high energy region, the mass absorption is as large as that of uniformly spread out material and the advantage factor is 1. Neither the temperature, nor the substitution of metal for oxide influence the absorption beyond about 15 000 ev. Whether our equations describe the cross section up to the highest neutron energies of about 2 Mev is, of course, not known. It seems rather unlikely that no changes should occur in the density of levels or in the constants a or Γ at these high energies. All levels which play any role at low energies have angular momentum $\frac{1}{2}$, and levels with higher angular momenta may play a role at high energies. Szilard and Marshall's value[7] of 0.4×10^{-24} cm² for the absorption cross section at 200 000 ev may be already an indication in this direction. However, the absorption is presumably quite small at high energies and the inaccuracy of our picture in the high-energy region surely causes less errors than its inaccuracy at low energies.

The surface absorption per unit logarithmic energy range $E\bar{A}_s$ is, for very high E, inversely proportional to E^2 while the mass absorption was inversely proportional to E. This is only another expression for the absence of surface absorption at very high energies. Furthermore, for the same energy, the surface absorption is lower for higher temperatures and also lower for the oxide than for the metal. The reason for the latter is that an incoming neutron in the oxide has a chance to hit an oxygen nucleus first, in which case it did not contribute to the surface absorption. The reason for the decrease of the surface absorption with increasing temperature is that at higher temperatures the lines are more completely washed out and the neutron density Q is a more uniform function of the energy, i.e., is more nearly equal to $\bar{Q}$.

This completes the discussion of the mass and surface absorptions in the different energy regions for given $\bar{Q}$. The remaining section of the original report, dealing with the variation of $\bar{Q}$ within a cell, will be omitted. Instead, we insert a short comparison of the experimental data of the succeeding two papers with the preceding calculations. These data were, or course, not available when the present report was written and it would have been surprising to find a close agreement between the two. After all, the constants given in (3), (4), and (6) constituted only crude estimates.

COMPARISON OF THE CALCULATIONS WITH THE EXPERIMENTAL RESULTS

Even a crude survey of the Figs. 3a and 3b and Figs. 4a and 4b leads one to the following qualitative conclusions:

1. That the total resonance absorption of a sphere can be represented in a good approximation by a mass

TABLE I. Mass absorption.

	Metal		Oxide	
	300°K	900°K	300°K	900°K
Experimental	7.3 barns	8.4 barns	10.9 b	12.5 b
Constants a	3.9 b	5.1 b	7.7 b	9.1 b
Constants b	4.85 b	8.15 b	8.7 b	10.2 b

absorption and a surface absorption. The former is, for a definite material, proportional to the total mass of the sphere, (hence, for a given radius, proportional to the density) the second is proportional to the surface. The ratio of the surface and mass absorptions depends mainly on the importance of the absorption in the continuous region (above 6000 ev) and is the smaller this is. It should be of the order of $5S/M$.

2. The resonance absorption of the metal should be smaller than that of the oxide by a factor of about 1.4. The greater the continuous absorption is, the smaller is this ratio.

3. Increasing the temperature increases the absorption proportionally with about the cube root of the absolute temperature. The dependence on the temperature is weaker if the high energy absorption is larger.

The foregoing qualitative statements, taken from the original report, can now be supplemented by Table I, a more detailed comparison with the experimental results taken from the following two articles.

The measurement of the absorption, leading to the constants of Table I, will be discussed in the ensuing two articles. The calculated values are the areas under the curves of Figs. (3a) and (3b) augmented by the lengths of the lines representing the absorption of the low levels. The experimental results are usually expressed in the form

$$\int \sigma_a(\text{total})_{eff} dE/E = A(1 + \mu S/M) \qquad (33)$$

and Table I gives the first part of it

$$\int \sigma_a(\text{mass})_{eff} dE/E = A. \qquad (33a)$$

The A, as determined experimentally, is given in the first row. (It should be remembered that Figs. 3(a) and 3(b) refer to the mass absorption only.)

The agreement between calculation and experiment is surely not good, but it could be hardly expected to be much better. Naturally, one could obtain higher values for the mass absorption, by either increasing both a and Γ, or by decreasing D. It appears more likely, however, that the actual situation differs from that assumed in the body of the report in another way. It is possible that the absorption decreases less rapidly with increasing energy than previously assumed or that part of the absorption which we have attributed to surface absorption is measured as mass absorption. Discussion of the

TABLE II. Surface absorption.

TABLE II. Surface absorption.

	Metal(%)		Oxide(%)	
	300°K	900°K	300°K	900°K
Experimental	5.8	5.5	4.6	4.2
Constants a	18.2	19.1	11.2	11.8
Constants a 1st line	8.8	8.8	6.5	6.5
Constants b	27.7	29.3	19.6	21.0
Constants b 1st line	17.5	17.5	13.1	13.1

second possibility will follow. The oft quoted measurements of Szilard and Marshall appear to support the first possibility. Increased absorption in the continuous region would add the same amount to all calculated numbers of Table I and thus decrease both the temperature coefficient of the absorption and the difference between oxide and metal.

While the preceding calculations give at least the right order of magnitude for the mass absorption, they are entirely incorrect as far as the surface absorption is concerned. Table II gives the integrals over f, i.e., $\int f dE/E$, augmented again by the absorption in the lines. In order to convert the experimental values for the same quantity, we note that the surface absorption of one uranium atom, in units of 10^{-24} cm^2, will be denoted by $A\mu S/M$ in the two articles which follow. For the whole lump, this is $(NV/M) \times A\mu S \times 10^{-24}$ cm^2 and $(NV/M)A\mu \times 10^{-24} = \int f dE/E$ per unit surface. Since NV/M is the reciprocal of the mass associated with one uranium atom, we have, in the case of the metal

$$\int f dE/E = 2.5 \times 10^{-3} A\mu, \qquad (33b)$$

and for the oxide

$$\int f dE/E = 2.15 \times 10^{-3} A\mu. \qquad (33c)$$

In addition to the calculated total, Table II also gives the contribution of the 5.8 ev line alone, and one sees that the contribution of this line alone already exceeds the whole experimental value by a factor of the order 3.

When reviewing the calculation presented above, we have indeed found a point where the approximations are particularly poor. This is after (17) where it is implicitly assumed that all the neutrons entering the lump will suffer some kind of collision therein. This assumption may be permissible in very large lumps but is not permissible in lumps of the size used in our experiments or the usual chain-reacting units. It is not difficult, however, to correct for this circumstance, i.e., take into account the fact that some of the neutrons entering the lump will cross it without suffering a collision in the lump. We shall carry out the calculation for a spherical lump.

The argument leading to (17) shows that the excess number of neutrons, crossing unit area of the surface in

unit time with a direction cosine between μ and $\mu + d\mu$ is

$$\frac{\bar{Q}}{2E} \frac{\sigma_a + \sigma_{sa}}{\sigma_a + \sigma_s} \mu d\mu. \qquad (34)$$

The distance which a particle entering with the direction cosine μ travels inside the sphere is $2r\mu$. Hence, the probability that it suffers any collision in the sphere is $1 - \exp[-2\mu r N(\sigma_a + \sigma_s)]$ and the probability that this collision leads to absorption is $\sigma_a/(\sigma_a + \sigma_s)$. This gives for the surface absorption, instead of (18)

$$A_s = \int dE \frac{\bar{Q}}{2E} \int_0^1 \mu d\mu [1 - e^{-2\mu r N(\sigma_a + \sigma_s)}] \frac{\sigma_a(\sigma_a + \sigma_{sa})}{(\sigma_a + \sigma_s)^2}. \qquad (35)$$

It differs from (18) by the square bracket.

The integration over μ can easily be carried out in (34) and yields ($\Sigma = (\sigma_a + \sigma_s)N$)

$$\tfrac{1}{2} + e^{-2\Sigma r} \left(\frac{1}{2\Sigma r} + \frac{1}{(2\Sigma r)^2} \right) - \frac{1}{(2\Sigma r)^2} \approx \frac{\Sigma r}{3 + 4\Sigma r}. \qquad (35a)$$

The right side is an approximation to the correct value; it is introduced to enable one to carry out the integration over E. This gives, when extended over a single line in the natural width region and divided by D

$$\bar{A}_s = \frac{\bar{Q}}{E} f = \frac{\pi \bar{Q} \Delta}{DE} \left(1 + \frac{aE^{\frac{1}{2}}}{\Gamma} \right) \frac{2}{1 + (1 + \tfrac{3}{4} \sigma_{sn} N r)^{\frac{1}{2}}}. \qquad (36)$$

The scattering cross section σ_{sn} is neglected in (35) when compared with the absorption at the center of the line. This is justified except at the high energy end of the Doppler region where the surface absorption is very small anyway. Expression (36) differs from (27) in the last factor which results from the square bracket in (35). It not only decreases the surface absorption, it also renders (36) dependent on Nr which will result in part of (36) appearing as mass absorption and only part of it as surface absorption.

Even though (36) was derived only for the natural width region, its correction factor can be used also in the Doppler region. We conclude that the area under the curve f should be multiplied with the factor

$$c_2 = \frac{2}{1 + (1 + \tfrac{3}{4} \sigma_{sn} N r)^{\frac{1}{2}}}. \qquad (36a)$$

The possibility of a collision-free traversal of the lump can be taken into account in the same fashion for the calculation of the surface line absorption (i.e., on the expression (30)). This must be replaced by

$$A_s = \int dE \frac{\bar{Q}}{2E} \int_0^1 \mu d\mu [1 - e^{-2\mu r N(\sigma_a + \sigma_s)}] \frac{\sigma_a}{\sigma_a + \sigma_s}$$

$$\approx \int dE \frac{\bar{Q}}{2E} \frac{\sigma_a N r}{3 + 4(\sigma_a + \sigma_s)N r} \qquad (37)$$

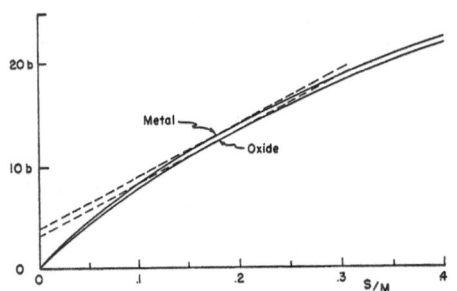

FIG. 5. The surface absorption of spheres plotted as a function of their surface to mass ratio.

where the approximate formula (35a) was used again for the integral of the square bracket. The integration can be carried out again, and neglecting σ_{sn} against the cross section at the middle of the line, gives instead of (30)

$$A_s = \frac{\pi}{4}(\bar{Q}\Delta/E)(1+\tfrac{3}{4}\sigma_{sn}N\tau)^{-\frac{1}{2}}. \qquad (38)$$

The correction of (30) is the last factor

$$c_1 = (1+\tfrac{3}{4}\sigma_{sn}N\tau)^{\frac{1}{2}} \qquad (38a)$$

with which the contributions of the lines of Figs. 4(a) and 4(b) must be multiplied.

As a result of the correction factors (36a) and (38a), the surface absorption is not strictly proportional to the surface. Instead, assumption b gives, in the case of the metal at 300°K,

$$\int f dE/E = 2.5 \times 10^{-3} A\mu = 17.5c_1 + 5.2c_1 + 5c_2. \qquad (38b)$$

The first term on the right side is due to the lowest line, the second to all other resonance lines up to 120 ev, the last term to the integral of the curve of Fig. 4(b) above 120 ev. Similarly, in the case of the oxide one has

$$\int f dE/E = 2.15 \times 10^{-3} A\mu = 13.1c_1 + 6.5c_2. \qquad (38b')$$

The surface absorption of spheres, obtained from (38b) and (38b'), is plotted in Fig. 5 against S/M. The ordinate on the left side is $A\mu$. One sees that the surface absorption is not proportional to S/M but can be well approximated by a straight line for S/M values around 0.2 where most measurements were carried out. The equation of the line for the metal is $4+50\ S/M$, that for the oxide $3+50\ S/M$. Adding to these expressions the mass absorption from Table I, one obtains the total absorptions given in Table III. The apparent mass absorptions now agree almost exactly which is, unquestionably, fortuitous in view of the arbitrariness in the choice of the constants. (One might remember also that the values of σ_{sn} of (7) and (7a) are too high. Reducing them to the now accepted values would decrease the calculated values.) The calculated value of the apparent surface absorption, however, remains too high by a very considerable factor. This absorption is due, principally, to the lowest line (at 5.8 ev) and we come, reluctantly, to the conclusion that the constants

TABLE III.

	Metal 300°K	Oxide 300°K
Experimental	7.3 $+23.5\ S/M$	10.9+21 S/M
Recalculated b	8.85+50 S/M	11.7+50 S/M

given for this line are inaccurate, in particular that Γ must be too high.[12] As was mentioned before, such a change is indicated also by the relatively low thermal absorption cross section of U^{238}.

A variety of assumptions could be made to explain the low temperature coefficient of the resonance absorption. The calculated temperature coefficient is due to the absorption in the Doppler region. The extent of the Doppler region could be decreased, most simply, by increasing the a and Γ of the lines. If this interpretation is correct, it re-emphasizes the great variability in the constants of the various lines. In view of the experimental data which have been accumulated in the last few years[6] on the constants of resonance lines, such a variability should be even less surprising now than it was when our reports were written.

It seems to us that the preceding calculations show that the measurement of the resonance absorption of lumps may in many cases yield the average values of the constants of resonance levels more directly and, perhaps, even more easily than a detailed measurement of several levels.

[12] This conclusion is borne out, however, by the measurements of Levin and Hughes, reference 2.

7.

Effect of Geometry on Resonance Absorption of Neutrons by Uranium*

E. Creutz,† H. Jupnik,‡ T. Snyder,§ and E. P. Wigner

Journal of Applied Physics *26* (3), 271–275 (1955)

Received September 15, 1954

The total resonance absorption of neutrons by spheres of uranium and uranium oxide imbedded in graphite and bombarded with essentially a dE/E spectrum is found to be given by $1/\alpha = c[1+(\kappa/\rho R)]$ where α is the advantage factor, that is, the ratio of absorption of uranium when distributed uniformly throughout graphite to the absorption when it is lumped in a sphere. R is the radius of the sphere and ρ is its density. c and ρ were measured in various parts of a large graphite block exposed to cyclotron neutrons from the reaction Li (p,n). Values of c and ρ, respectively, are 0.051 and 5.1 g/cm² for U_3O_8. In a newer notation, $A = 10.9$ barns and $\mu = 1.95$ g/cm².

INTRODUCTION

WHEN uranium with its natural mixture of isotopes is exposed to neutrons in a chain-reacting pile there is a competition by several processes for the absorption of the neutrons. One of the chief competitors to the absorption by U^{235} to cause fission is the absorption by U^{238} to produce the compound nucleus U^{239}. This system may re-emit the neutron or emit a gamma ray. In the second case, the remaining beta active nucleus U^{239} decays with a 24-min half-life.[1]

The capture of neutrons above thermal energies is described rather well by the Breit-Wigner formulas,[2] and an analysis of the case, assuming that one resonance level predominates, is given by Anderson.[3] A more accurate analysis was given subsequently by S. Dancoff and M. Ginsburg.

Since the neutrons with energies above thermal are predominantly absorbed in narrow resonances, there should be a significant difference in the specific absorption of neutrons near the surface and in the volume of a chunk of uranium, so that the total loss of neutrons by this process might be reduced by choosing the proper size of chunk. Thus, the "self-reversal" effect of the strong resonance absorption would produce a spectrum of neutrons in the chunk which would be deficient in the energies of the resonances. This would suggest making the chunk, for example, a sphere, with a small surface-to-volume ratio. But clearly a limitation on the optimum size of a sphere would be reached because (1) in too large a sphere fission-producing neutrons would be depleted before the center portions were reached, and (2) the uranium of the sphere itself would produce some energy losses by scattering of the neutrons and thus destroy the self-reversed character of the neutron spectrum.

EARLY WORK

In June, 1941, preliminary work on the absorption of resonance neutrons in spheres of uranium oxide was carried out by Fermi and Anderson together with R. R. Wilson and others of the Princeton group. This work verified the fact that the total absorption could be divided roughly into a surface effect and a volume effect with much higher specific absorption taking place at the surface.

The spheres for the first work described below were made of U_3O_8 powder tamped into thin Cu shells which in turn were covered with cadmium to remove thermal neutrons. They were imbedded in a large graphite block 81 by 81 by 107 cm³ which was bombarded with neutrons from the reaction $Li^7(p,n)Be^7$ produced by 8-Mev protons in the Princeton cyclotron. The resonance absorption was measured by sampling the material in the sphere, removing fission products and UX by the method described by Anderson,[3] and counting the induced 24-min beta activity. This method suffered from certain inaccuracies. A second improved method was therefore used later. Both methods and the results obtained with them are described below.

FIRST METHOD

Spheres of diameter $2\frac{1}{4}$, 3, 4, 5, 6, 7, and 8 in., consisting of thin copper shells packed with U_3O_8, were prepared. Into a cylindrical cavity extending to the center of each shell, a cadmium cone could be placed. This cone, which was intended to sample the surface and the volume in the correct ratio, was filled with U_3O_8 to furnish the activity for measuring after bombardment. The bombarded material was thoroughly mixed and a sample of it was painted thinly on paper, using a small amount of dilute collodion as a binder, or else the powder was pressed onto Scotch tape. This latter method gave a very uniform and rather thin sample. If a still thinner sample was desired, a second piece of Scotch tape was stuck to the coated one and pulled away when a very thin layer of U_3O_8 particles would adhere to it. Each bombardment was monitored with a thin U_3O_8 sample in a standard place, so all runs could be compared to a standard irradiation. The

* This work was supported by the NDRC in 1941. It was declassified by authority of the U. S. Atomic Energy Commission on May 27, 1952.
† Now at Carnegie Institute of Technology, Pittsburgh, Pennsylvania.
‡ Now at American Optical Company Research Laboratory, Southbridge, Massachusetts.
§ Now at General Electric Company, Schenectady, New York.
[1] Meitner, Hahn, and Strassman, Z. Physik **106**, 249 (1937).
[2] G. Breit and E. Wigner, Phys. Rev. **49**, 519 (1936).
[3] Herbert L. Anderson, Phys. Rev. **80**, 499 (1950).

amount of resonance absorption in the sphere was thus measured by the 24.2-min U^{239} activity counted with an argon-alcohol-filled thin glass-wall Geiger counter. To compare the absorption in different spheres, the relative weights of the samples prepared from them had to be determined. For comparing the effective weights of very thin samples it had been found satisfactory to bombard them all on a motor-driven rotating wheel in a cavity inside the graphite block, so that they would, on the average, all be in the same position and receive the same irradiation. Their weights were then assumed proportional to their induced activities. This effective weight method (in samples of various thicknesses) corrects for self-absorption of the beta rays, but also is affected by self-absorption of the neutrons in the sample. This is not what was wanted. Since, for samples prepared from the material in the spheres, it was necessary to know a quantity actually proportional to the activity per milligram, some method of calibrating them must be found which was independent of neutron self-absorption in the samples.

The method used was to irradiate them inside a vessel having thick walls of U_3O_8, so only those neutrons having the property of not being easily absorbed (hence, not suffering self-absorption) would induce the calibrating activity. The samples were thin enough so the beta absorption was negligible.

To compare the activity per milligram produced in the spheres with that which would have been induced had the U_3O_8 been uniformly distributed throughout the graphite, a very thin sample was prepared and bombarded. This sample was calibrated for thickness in the thick-walled vessel as were the sphere samples.

The ratio of the activity per milligram induced in a thin sample to the average that is induced in a given sphere gives a measure of the advantage of thus lumping the uranium, and hence may be called the advantage factor, α, for the sphere. This factor will vary with the size of the sphere since the specific absorption of the surface layer is considerably greater than that of the volume. Measured α's are listed below. The advantage is seen to increase as larger spheres are used as expected since the surface becomes smaller relative to the volume. These data were taken with the spheres 20 cm from the front of the graphite block which was surrounded with borax and paraffin so that few neutrons could enter except from the front, which was shielded with $1\frac{1}{2}$ in. of B_2O_3 to pass only neutrons of energy $\gtrsim 1000$ ev.

Radius of sphere (cm)	Advantage
2.86	8.92
5.35	10.7
7.55	12.7

SECOND METHOD

Because of the difficulty of obtaining calibrations of the effective thickness of the samples made from the spheres by the first method, and also because of the small activity obtainable in the thin samples, a thick sample method was used. In this method the counter samples were all made infinitely thick for the beta rays, so no calibration was necessary. This was accomplished by making dishes by sealing 0.002-in. Al foil between two brass rings $2\frac{1}{2}$ in. in diameter and $\frac{1}{8}$ in. deep, which could be filled with the bombarded U_3O_8. This dish was supported in a Bakelite cradle which fitted tightly inside the lead counter shield so that the geometry could be accurately reproduced. That this was possible was proven by the fact that the slopes of the growth curves of UX activity taken with various U_3O_8 samples from which the UX had been previously removed chemically were the same to within about 1 percent. About 25 grams of U_3O_8 were required to fill the dish. The spherical shells used in this method were spun in two hemispheres from cadmium sheet. Each was filled with U_3O_8 and covered across the equatorial plane with a sheet of 0.002-in. Al, held in place by Scotch tape. The two halves thus covered were held together with more tape. After bombardment of such a sphere inside a graphite block one of the hemispheres was opened and its contents were sampled. This eliminated the difficulty with the earlier cadmium cone method, namely, the difficulty of insuring that the base of the cone was flush with the sphere surface. The cadmium cone also slowed neutrons into resonances inside the sphere to an uncertain extent. The sampling of the hemisphere was done by placing its contents in a large bottle, which was thoroughly shaken to mix. The mixing method had been previously tested with black and white powders, which gave a uniform gray after a few seconds. A portion of the U_3O_8 so shaken was then dissolved in HNO_3, and ether was added. The ether solution of $UO_2(NO_3)_2$ was made neutral with Na_2CO_3, dried with $CaCl_2$, and decanted into a separatory funnel after which about 1 cc of water to 100 cc of ether was added and the contents vigorously shaken. The separation was then made to remove fission products and UX which are selectively soluble in the water. The ether was then blown off, and the residue calcined, ground, and put in the Al foil-bottomed dish for counting. Up to ten times, the UX background was observed for the initial 24-min counting rate one hour after bombardment.

The magnitude of the 24-min activity relative to the various monitors was computed from the counter data by a least squares method. The decay period was assumed known as 24.2 min and the slope of the growth curve of UX was frequently measured and known to about 1 percent; thus, the net strength of the U^{239} activity at a certain time could be calculated as the difference between the total observed activity at that time and the extrapolated UX activity.

MONITORS

At first, the neutron beam was monitored with two thin PbI_2 samples placed on either side of the cyclotron in cadmium envelopes. It was supposed that if the ratio of the activities in these two monitors remained the same from run-to-run, the neutron beam probably was reasonably consistent in its spectrum since it must have originated from the same part of the cyclotron. Since the frequency and other conditions of the cyclotron were not entirely constant, some variation in this ratio was found. Therefore, the later experiments were monitored inside the graphite block. A small correction to the relative irradiations had to be made in comparing the 25-min iodine activities with the 24.2-min uranium activities, but this amounts to only about 4 percent for a one-hour bombardment. A calculation showed that this correction did not change appreciably even if the beam was very unsteady.

In most of the experiments, a thin iodine monitor was taped to the outside of the sphere to measure the neutron density at that point. To obtain the specific volume activity, a surfaceless or "infinite" sphere was made by surrounding a small sphere of U_3O_8 by a larger one. The activity induced in the central sphere was only measured then. In most cases a $2\frac{1}{4}$ in. sphere imbedded in a 6 in. sphere was used for this purpose. A monitor was placed in the center of the sphere and also on the surface.

"NO-SPHERE" EXPERIMENTS

To obtain the activity of a thin sample of uranium, a piece of filter paper was soaked in $UO_2(NO_3)_2$ solution. This was then dried, weighed, wrapped in Cd, and bombarded inside the graphite block. It was then possible to obtain the advantage factors for spheres by purifying and calcining the material from this thin sample, and measuring the activity in the same way as was done with the sphere samples with which it was compared. This type of experiment was called a "no-sphere" experiment. In order to obtain the absolute reciprocal advantage for a given configuration it was only necessary to divide the value of the absorption relative to one kind of monitor for that configuration, by the no-sphere absorption relative to its monitor of the same kind. Figures 1 and 2 are typical and show the

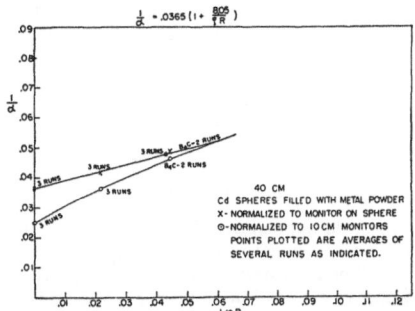

FIG. 2. Reciprocal advantage factor for spheres of metal powder plotted as function of reciprocal product of sphere radius and sphere density.

measured reciprocal-advantage factor is plotted as a function of $1/\rho R$ where ρ is the density of the material in the sphere and R is the radius of this sphere. In each figure two curves are shown, the upper one represents the data obtained by using as monitor the detector fastened to the surface of this sphere, the lower curve shows the data calculated by using as monitor a detector near the front of the graphite block (called 10-cm monitors). Because of depression of the true neutron density in the vicinity of the sphere, the more meaningful values are those obtained with the surface monitor, that is, those shown on the upper curve in each case. The theoretical form

$$\frac{1}{\alpha} = c\left(1 + \frac{\kappa}{\rho R}\right)$$

was assumed and straight lines were drawn.

METAL POWDER

Rough values for absorption by spheres of metal powder were obtained. Since U metal powder is highly pyroforic, these spheres were filled in a covered washtub containing a CO_2 atmosphere maintained by several large chunks of dry ice. After bombardment, the powder was dumped into a flask with small chunks of dry ice. The flask was kept in the washtub, with its neck protruding through a hole in a cover to the tub, and shaken to mix. No explosions were encountered. The metal powder was divided into ten portions, each of which was added slowly to a separate beaker of dilute HNO_3. Considerable H_2 and N_2O_4 were evolved. In one case when the powder was added too fast, it ignited. The fire was easily put out with a CO_2 extinguisher. In subsequent experiments the H_2 was blown off by a stream of air as it was evolved, and no more trouble was experienced.

RESULTS

The total resonance absorption by a sphere of U or U_3O_8 has been found to consist of a term proportional

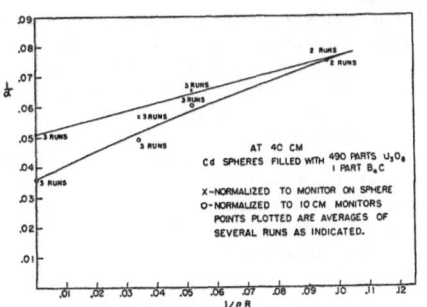

FIG. 1. Reciprocal advantage factor for spheres of U_3O_8 containing B_4C plotted as function of reciprocal product of sphere radius and sphere density.

to the total mass and another term proportional to the surface of the sphere. Thus, the reciprocal of the advantage factor may be expressed as

$$1/\alpha \propto M + dS$$

where d is a constant, M the mass of the sphere, and S is the surface. This may be written

$$1/\alpha = a + (b/R)$$

since to a good approximation the specific volume absorption is independent of the radius, R. The effective "thickness" of the surface is inversely proportional to the number of atoms of uranium per cm³, so finally:

$$\frac{1}{\alpha} = c\left(1 + \frac{\kappa}{R\rho}\right)$$

where ρ is the density of the sphere and constants c and κ are to be determined. Some qualitative knowledge of the variation of c and κ with the neutron spectrum was obtained by making measurements at 20, 30, 40, and 45 cm from the front of the block since the average neutron energy decreases with the increasing depth.

Since this work was carried out, it has become customary to express these and related results in terms of the resonance integral. Here,

$$\int \sigma_c(E)dE/E,$$

assumes a dE/E neutron spectrum. For an extremely dilute mixture of uranium in a noncapturing, slowing-down medium this integral is accepted as having the value of 240 barns. For lumped arrangements of uranium and its oxide, these experiments show that this integral may be expressed as follows:[4]

$$\left(\int_{\text{lump}} \sigma_c(E)\frac{dE}{E}\right)_{\text{effective}} = A\left[1 + \mu\frac{S}{M}\right],$$

where S is the surface of the lump in square centimeters and M is its mass in grams. A and μ are then the experimentally determined quantities. Comparing this expression with ours for the reciprocal advantage factor,

$$\frac{1}{a}c\left(1 + \frac{\kappa}{R\rho}\right) = \left(\int_{\text{lump}} \sigma(E)\frac{dE}{E}\right)_{\text{effective}} \Big/ 240 \text{ barns},$$

it is seen that $\mu = \kappa/3$ and $A = 240c$ barns.

TABLE I. The constant A measured for various materials.

Material	Average σ_s per U atom $= \bar{\sigma}_s$	$(\bar{\sigma}_s)^{\frac{1}{2}}$	A (measured)
U	9.5	3.08	7.4
U_3O_8	20.7	4.55	12.2
5 U_3O_8+2C (By weight)	66	8.10	22.6

[4] Phys. Today 5, 32 (May, 1952).

To study the effect of slowing-down-material in the sphere, such as the oxygen of U_3O_8, spheres were filled with a mixture of U_3O_8 and graphite of such proportions that the total scattering of the graphite in the mixture was twice as much as that of the uranium and the oxygen, namely, 5 parts U_3O_8 to 2 parts of graphite by weight. As was expected, the activity per milligram of U_3O_8 was larger under these conditions. As an extreme dilution, an 8-in. cube was filled with a mixture of 10 parts of graphite to 1 part of U_3O_8. The U_3O_8 was removed after bombardment by panning, as in separating gold from sand. The advantage of this homogeneous mixture was only about 3 showing it to be much less effective for depressing the resonance absorption than are spheres of pure U_3O_8.

According to theory the volume absorption should vary as the square root of the average scattering cross section, per uranium atom, of the material in the sphere. Values of this cross section,[5] its square root, and meas-

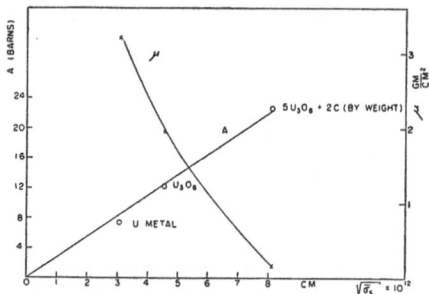

FIG. 3. Effective volume absorption A (barns) and effective mass absorption coefficient, μ, of spheres of uranium to which are mixed different amounts of scattering material, plotted as a function of the square root of the average scattering cross section, per uranium atom, of the material in the sphere.

ured values of A, are given in Table I. These values are plotted in Fig. 3 and are seen to fall near a straight line. In the same figure are plotted values of μ for various materials.

A 2-in. cube was filled with U_3O_8 of different densities, including the standard one of 3.76, to determine the effect of density, as well as the relative advantages of a cube and a sphere. The cube of standard density showed 12 percent more resonance absorption than would a sphere of the same density and mass.

To estimate the effect of the $1/v$ neutrons, spheres filled with 1/491 parts by weight of B_4C were used. This is enough to absorb about 80 percent of the $1/v$ neutrons. Its effect was to reduce the 24-min activity by less than 10 percent in general. However, the data are just good enough to justify applying a correction to the A values obtained with spheres not containing

[5] At the time our measurements were undertaken, the accepted value for the scattering cross section of uranium was $12b$.

boron. This correction, due to the $1/v$ tail above the cadmium cutoff, is given by

$$\int_{0.45}^{\infty} \sigma_{\text{thermal}} \left(\frac{E_{\text{thermal}}}{E} \right)^{\frac{1}{2}} \frac{dE}{E}$$

$$= 2.85(0.025)^{\frac{1}{2}} [-2/E^{\frac{1}{2}}]_{0.45}{}^{\infty} = 1.35 \text{ barns.}$$

Applying this correction to the constant A improves the consistency of the data. This correction is applied only to the volume effect since the penetrating power of these neutrons is too great to cause surface absorption. The correction to A changes μ slightly since μ is calculated in terms of A. The corrected μ is listed with A and the uncorrected κ, c, and μ in Table II.

Because the c's depend on the "no-sphere" experiments, their absolute values are less accurate than the ratios of those for various spheres at a given distance in the block.

TABLE II. Constants for U_3O_8 spheres.

Material	κ	c	$\mu = \kappa/3$	$A = [240c$ -1.35 barns]	μ corrected g/cm^2
U[a]	7.1	0.036	2.7	7.3	3.2
U_3O_8	5.1	0.051	1.7	10.9	1.95
5 U_3O_8+2C (By weight)	0.43	0.078	0.14	17.4	0.155

[a] Values for U metal less accurate than for U_3O_8 due to greater experimental difficulties.

ACKNOWLEDGMENT

Much credit for finishing these experiments is due Mrs. E. Creutz for her work in compiling and computing data, to T. Lanahan for his consistent efforts to keep the cyclotron in operation and help in numerous other ways, and to E. Frank, who constructed and maintained the Geiger counters. Similar experiments were carried out later by A. C. G. Mitchell and his collaborators at the University of Indiana. The accuracy of their experiments was greater than that of ours, but the general results were similar.

8.

Effect of Temperature on Total Resonance Absorption of Neutrons by Spheres of Uranium Oxide*

E. Creutz,† H. Jupnik,‡ and E. P. Wigner

Journal of Applied Physics *26* (3), 276–279 (1955)

Received September 15, 1954

In the expression $1/\alpha = c[1 + (\kappa/\rho R)]$ for the advantage (α) of decreased resonance absorption of uranium in a dE/E neutron spectrum brought about by lumping UO_2 into spheres, the temperature dependence of c is measured and κ is measured also. ρ is the density of UO_2 in g/cm³; R the radius of the sphere in cm. In the range 20° to 1000°C, $\Delta c/\Delta T$ averages 1.8×10^{-6} per degree centigrade and is positive as expected. This temperature dependence decreases by about a factor of three from the lowest to the highest measured temperatures. κ is 5.06 g/cm³ leading to a value of μ in the expression [Phys. Today *5*, 32 (May, 1952)]

of 1.69 g/cm².

$$\left(\int \sigma(E) \frac{dE}{E} \right)_{\text{eff}} = A \left[1 + \mu \frac{S}{M} \right]$$

$$\frac{\Delta A}{\Delta T} \text{ is } 4 \times 10^{-4} \frac{\text{barns}}{°C}.$$

INTRODUCTION

THE reaction rate of a moderated uranium pile depends on temperature. For example, in a reactor in which the neutron spectrum is approximately Maxwellian, the average velocity will increase with temperature, thus decreasing the absorption cross sections for the low-energy neutrons which vary as $1/v$. Resonance levels will be broadened by the Doppler effect, and if lumping of the uranium has been made use of to decrease the total resonance absorption as proposed by Szilard,[1] and later found experimentally to be effective,[2] increased temperature will decrease the advantage thus gained.

In order to design a method for controlling the reaction rate of the nuclear chain reaction a knowledge of the variation with temperature of the number of neutrons absorbed at the resonances by uranium is, therefore, important.

METHOD

For studying this effect a furnace was constructed, eliminating as much as possible elements whose neutron absorptions are large or unknown. This furnace was immersed in a graphite block 100 cm square by 125 cm long, and bombarded with neutrons from the reaction $Li^7(p,n)Be^7$ using 8-Mev protons from the Princeton cyclotron. Various arrangements of uranium oxide were surrounded by material containing B or Cd to absorb thermal neutrons, and the 24-min activity induced by resonance neutrons was measured.

The furnace itself consisted of a block of graphite 16 in. square by 12 in. long surrounded on all sides by a 4 in. layer of lampblack. Twelve standard type Globar (silicon carbide) heating elements were arranged symmetrically in a circle of radius 18 cm about a central block 4×4×10 in. which could be removed and loaded with the uranium sample. A sketch of the furnace is shown in Fig. 1. As may be seen, the center of the sphere was separated from the front face (towards cyclotron) of the graphite block by 35 cm of graphite and 10 cm of lampblack (density ∼0.1). The graphite outside of the lampblack layer stayed below about 30°C during all experiments. The temperature of the furnace was measured by two thermocouples, one placed 2½ in. from the center of the sphere, and one placed about 1 in. from one of the Globars. The furnace was heated by passing current through the Globars until the thermocouple near one of them registered a temperature about 50°C above the desired temperature, at which time the temperature near the sphere was about 50°C less than that desired. The power was then turned off for about 15 min while the two temperatures equalized, after which bombardment was started. After an hour, the sphere was usually a few degrees hotter than the Globars. During the actual bombardment all the furnace graphite and the sphere were at the same temperature within about 25°C.

After bombardment, the hot central-sphere block was removed and the uranium oxide from one hemisphere was taken out and broken up in a mortar, dissolved in HNO_3, and the ether purification made, as described by Anderson,[3] after which a thick-dish sample was prepared for counting, as in previous work.[2] The cyclotron beam was monitored from run-to-run with (a), two PbI_2 samples placed at 10 cm in the block (at room temperature), (b), with a PbI_2 sample placed on the cyclotron outside the block, and (c), with a thin

* This material originally appeared in Document C-110 in 1942 reporting research supported by the NDRC. It was declassified by authority of the U. S. Atomic Energy Commission on May 27, 1952.

† Now at Carnegie Institute of Technology, Pittsburgh, Pennsylvania.

‡ Now at American Optical Company Research Laboratory, Southbridge, Massachusetts.

[1] Private communication.

[2] Creutz, Jupnik, Snyder, and Wigner, J. Appl. Phys. (to be published).

[3] Herbert L. Anderson, Phys. Rev. **80**, 499 (1950).

layer of KI melted onto a steel or platinum dish behind and at 10 centimeters from the center of the sphere and at the temperature of the sphere. The KI layer contained less than 7 mg/cm² of iodine in each case and was prepared in the same way each time, by evaporating 1.5 cm³ of an alcohol solution, so the correction for self-absorption should be negligible and not vary appreciably from run-to-run. After bombardment, the KI was converted to PbI₂ which was washed and pressed onto Scotch tape for counting.

All monitors were surrounded by cadmium or CdSO₄. When half a dozen or so of these monitor samples had accumulated they were placed in a tilting rotating peanut-roaster type cylindrical mixer and bombarded with neutrons including thermals, inside the large graphite block, after which their relative activities were assumed to give their relative weights. Since they were only about 2 to 3 mg/cm² thick, this method should quite accurately give the relative numbers of atoms present.

Since KI melts at 723°C and has an appreciable vapor pressure below that temperature, a substance was sought for use as a monitor which was physically and chemically stable at higher temperatures and became radioactive after resonance-neutron capture with a period not far different from 24.2 min. The material chosen was Ga_2O_3, since it is stable and not appreciably volatile at 1200°C, and $Ga(n,\gamma)$ gives a period of 21 min. There is also a period of 14 hours, which introduces a correction, small for short bombardments.

In an attempt to determine how the actual number of resonance neutrons inside the furnace relative to those outside varied with temperature, caused by the different temperature-dependence of absorption by the carbon and other materials, two types of check experiments were made. At each temperature several bombardments were made of the cyclotron monitor, the 10-cm monitor, and the hot-dish monitor, but with no uranium present, leaving a hole in the graphite where its uranium oxide sphere was otherwise placed. These experiments were not very accurate, but they show that for a given bombardment of the furnace the number of neutrons absorbed by a hot iodine or Ga sample inside the hot furnace is the same within 8 percent as when the furnace is cold.

The interpretation is complicated by the fact that the iodine levels may change their effectiveness by Doppler broadening as the temperature is raised. Therefore, a second type of check was made in which the dish monitor inside the hot furnace was separately water-cooled. In this case, the sensitivity of the monitor could not depend on the temperature of the graphite. Although there appears to be a slight increase in the resonance-neutron density within the graphite at higher temperatures, these experiments indicate no effect on the cold iodine in the hot furnace greater than 6 percent, which is within their accuracy. Part of this effect was probably caused by the leakage of thermal

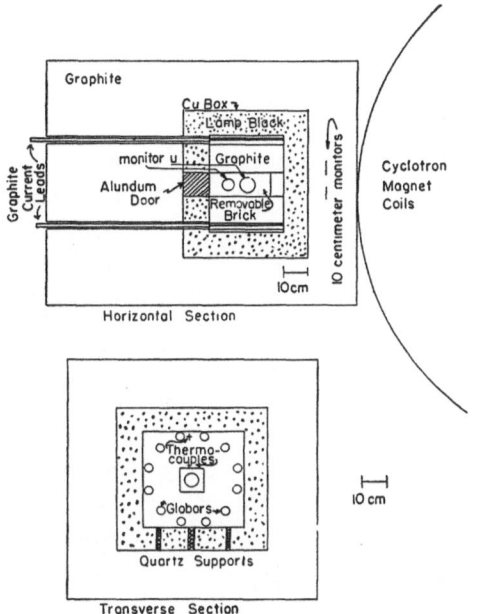

Fig. 1. Furnace.

neutrons through the cadmium shield of the monitor as discussed later. Because of the uncertainties, the values of the absorption relative to the 10-cm monitors are believed to be more reliable than those relative to the hot monitors.

The uranium oxide was held in most cases in steel shells 6.8 by 10⁻² cm thick, surrounded by shields to absorb thermal neutrons. No difference was found if copper shells were used in place of the steel. The thermal neutron shields were of two types: one a mixture of CdSO₄ and Al₂O₃ such that there was 0.49 g/cm² of Cd around the sphere, and the other B₄C and Al₂O₃, sufficient to place 0.13 g/cm² of B around the sphere. Above 700°C, in the CO atmosphere generated by diffusion of air into the furnace, the CdSO₄ decomposed, and CdS and Cd boiled off. Therefore, a chemical analysis of the shell was made after use, showing that some of the experiments were done with less Cd around the uranium than were others. To partially remedy this trouble, CdSiO₃ shells were prepared, but they also partly decomposed to Cd which evaporated.

Three types of spheres were used: an approximately 2 in. sphere of uranium oxide of density 4.12, a 3¼ in. sphere of the same density; and an "infinite" sphere, i.e., a 1 in. sphere imbedded in a 3¼ in. sphere, from which only the activity of the material in the 1 in. sphere was measured, giving a sphere with little or no surface effect for absorption.

CREUTZ, JUPNIK, AND WIGNER

TABLE I. Heat experiments—typical data.

2 in. spheres, 10-cm monitors.
Density of all materials in spheres was 4.12 g/cc

Material	Outer shell	Sphere, cm	$1/\rho R$	Temp., °C	No. sphere/adv.
UO_2	$CdSO_4$	Fe; $R=2.45$	0.0990	room temp.	weighted av: 0.03065±0.0003
UO_2	$CdSO_4$	Fe; $R=2.45$	0.0990	300	weighted av: 0.0302±0.0003
UO_2	$CdSO_4$	Fe; $R=2.45$	0.0990	700	weighted av: 0.0350±0.0003
UO_2	$CdSO_4$ (0.344 g/cm²) of Cd	Fe; $R=2.45$	0.0990	1000	0.0358±0.0006
UO_2	B_4C	Fe; $R=2.55$	0.0953	room temp.	weighted av: 0.0197±0.0003
UO_2	B_4C	Fe; $R=2.55$	0.0953	300	0.0212±0.0002
UO_2	B_4C	Fe; $R=2.55$	0.0953	700	weighted av: 0.0227±0.0002
UO_2	B_4C	Fe; $R=2.55$	0.0953	1000	0.0260±0.0007

3¼ in. spheres, 10-cm monitors.
All spheres were filled with UO_2; density = 4.12 g/cc

Material	Outer shell	Sphere, cm	$1/\rho R$	Temp., °C	No. sphere/adv.
UO_2	$CdSO_4$	Fe; $R=4.11$	0.0591	room temp.	weighted av: 0.0252
UO_2	$CdSO_4$	Fe; $R=4.11$	0.0591	300	weighted av: 0.0269±0.0003
UO_2	$CdSO_4$	Fe; $R=4.11$	0.0591	700	weighted av: 0.0276±0.0002
UO_2	B_4C	Fe; $R=4.11$	0.0591	room temp.	weighted av: 0.0164±0.0001
UO_2	B_4C	Fe; $R=4.11$	0.0591	300	0.01731±0.0003
UO_2	B_4C	Fe; $R=4.11$	0.0591	700	weighted av: 0.0191±0.0006

∞ spheres, 10-cm monitors.
All spheres were filled with UO_2; density = 4.12 g/cc

Material	Outer shell	Sphere, cm	$1/\rho R$	Temp., °C	No. sphere/adv.
UO_2	$CdSO_4$	Fe; $R=1.27$ inside $R=4.11$	0	room temp.	weighted av: 0.0190±0.0001
UO_2	$CdSO_4$	Fe; $R=1.27$ inside $R=4.11$	0	300	weighted av: 0.0203±0.0007
UO_2	$CdSO_4$	Fe; $R=1.27$ inside $R=4.11$	0	700	weighted av: 0.0211±0.0006
UO_2	$CdSO_4$ (0.108 g/cm²) of Cd	Fe; $R=1.27$ inside	0	1000	0.0224±0.0003
UO_2	B_4C	Fe; $R=1.27$ inside $R=4.11$	0	room temp.	weighted av: 0.0142±0.0001
UO_2	B_4C	Fe; $R=1.27$ inside $R=4.11$	0	300	0.0145±0.0002
UO_2	B_4C	Fe; $R=1.27$ inside $R=4.11$	0	700	weighted av: 0.0147±0.0003
UO_2	B_4C	Fe; $R=1.27$ inside $R=4.11$	0	1000	0.0148±0.0004
U_3O_8	None	none (thin sample)	∞	20	weighted av: 0.333

Besides using these concentrated arrangements of uranium, measurements were made to obtain the activity that would be induced in uranium oxide if it were thinly dispersed throughout the graphite. These were made by painting a thin layer of UO_2 and rubber cement on the inside of a steel shell, which was then filled with a graphite sphere. Such an experiment was called a "no-sphere" experiment. The self-absorption of the layer was rather small, and could be corrected since its thickness was only about 5 mg/cm² of uranium. The shells were always held together by Scotch tape which gave enough support even at 700°C, due to the formation of a rather rigid ash.

RESULTS

The sphere experiments are listed in Table I with activity given relative to the 10-cm monitors. To obtain the reciprocal advantage factor for a given configuration and temperature, the quantity "no-sphere"/advantage is to be divided by the "no-sphere" (thin sample) value of 0.333 given at the end of Table I.

At room temperature both UO_2 and U_3O_8 spheres were used. The resonance absorption in the spheres of the higher oxide was greater, as expected, due to the increased scattering of the neutrons by the extra oxygen, thus dropping more neutrons into uranium resonance levels inside the sphere. At the higher temperatures it was known that U_3O_8 would have been reduced to UO_2 to some uncertain extent, so only UO_2 was used. On opening the red-hot sphere in air after bombardment, the UO_2 was partly oxidized to U_3O_8. A chemical analysis of some of these used under various conditions shows how far this proceeded. These analyses indicate that the original "UO_2" was partly U_3O_8. Analysis before heating gave 71.5 percent UO_2, 28.5 percent U_3O_8. (After heating to 700°C and opening in air, 40 percent UO_2, 60 percent U_3O_8.)

To reduce errors because of changes in the composition of the shields or the furnace during the experiments, data were taken first at room temperature (20°C), then at 700°C, 300°C, room temperature again, 700°C again, and 1000°C. The return to room temperature and 700°C gave checks on the first results. New $CdSO_4$ shells prepared gave a check on the values obtained with those that had been used several times and had lost some Cd.

A 2-in. copper shell filled with UO_2, not surrounded with Cd, showed about 15 percent less absorption at 700°C than at 20°C. This is due to the smaller effectiveness of the "thermal" neutrons at the higher temperature since the cross section decreases as $1/v$.

From Table I we obtain the average values given under $[\int \sigma(dE/E)]_{eff}$ uncorrected of Table II. The corrected values were obtained by subtracting the epicadmium absorption of $1.35b$.[2]

Table III gives the calculation of μ. One sees that the values of μ obtained at the two values of $1/\rho R$ differ somewhat, but hardly in excess of the accuracy of the measurements. They give for μ the value 2.27 at 20°C

TABLE II.

	$[\int (dE/E)\sigma]_{eff}$ uncorrected				$[\int (dE/E)\sigma]_{eff}$ corrected			
$1/\rho R$	20	300	700	1000°C	$1/\rho R$	20	300	700	1000°C
0	13.5	14.6	15.2	16.1	0	12.15	13.25	13.85	14.75
0.059	18.1	19.4	19.9	⋯	0.059	16.75	18.04	18.65	⋯
0.099	21.8	21.9	25.2	27.3	0.099	20.55	20.65	23.85	25.95

TABLE III.

	Total absorption Mass absorption					μ		
$1/\rho R$	20	300	700	1000°C	20	300	700	1000°C
0.059	1.38	1.36	1.35		2.27	2.17	2.04	
0.099	1.69	1.56	1.72	1.76	2.27	1.85	2.43	2.54
Av					2.27	2.01	2.23	2.54

TABLE IV.

$[\int \sigma(dE/E)]_{eff}$ Cd shields	$[\int \sigma(dE/E)]_{eff}$ B₄C shields
$= 10 \ \ +23 \ \ S/M$ at 20°C	$= 10.3+18.7 \ S/M$ at 300°C
$= 11.5+21.5 \ S/M$ at 700°C	$= 10.4+19.8 \ S/M$ at 700°C
$= 11.3+21.7 \ S/M$ at 600°C	$= 10.5+19.8 \ S/M$ at 1000°C

TABLE V. Values of $[\int \sigma(dE/E)]_{eff}$.

	20°C	700°C
U	$7.3+23.5 \ S/M$	$8.4+22 \ \ S/M$
U_3O_8	$10.9+21 \ \ S/M$	$12.5+19.5 \ S/M$

which is almost exactly equal to the average of the μ values[3] of 1.95 and 3.2, for U_3O_8 and U, taken with weights 3:1. While this is a satisfactory agreement with the data of the preceding article, the absolute value of the absorption, i.e., the mass absorption coefficient, is considerably in excess of the similar average of the mass absorption coefficients of U_3O_8 and U. We attribute this to difficulties in monitoring (i.e., the 0.33 "no-sphere" value) and write for UO_2 the results in Table IV. The figure at 600°C is simply interpolated. It is satisfactory that the surface absorption coefficient remains practically independent of temperature.

We have assumed in the first article here presented that the temperature dependence of the absorption of U and U_3O_8 is the same as that of UO_2. In view of the inaccuracy of the measurements it does not seem worthwhile to attempt a more refined interpretation. This then gives the figures of Table V.

ACKNOWLEDGMENT

The authors wish to thank E. O. Wollan and R. J. Stephenson for constructing the furnace and for their many suggestions and aid in getting the experiment going. Also, thanks are due to Heyroth and Bovee of the Globar Division of the Carborundum Company for their advice on the design of the furnace, to E. Frank for help in taking the data, to T. Lanahan for his work on the cyclotron, and to Mrs. E. Creutz for a very great amount of data-compiling and computation. The experiments could not have been carried out in the relatively short time of three weeks without the excellent cooperation of all those named, as well as others called on from time to time.

9.

Absorption of Thermal Neutrons in Uranium

E. C. Creutz, R. R. Wilson, and E. P. Wigner

September 26, 1941

Abstract. A knowledge of the absorption processes for neutrons in uranium is important for planning a chain reaction experiment. The absorption of thermal neutrons in uranium and uranium oxide has been studied. Neutrons from the cyclotron were slowed down by passage through a graphite block. A uranium or uranium oxide sphere was placed at various positions in the block. The neutron intensity at different points in the sphere and in the graphite was measured by observing the activity induced in detectors of uranium oxide or manganese. It was found that both the fission activity in the uranium oxide and the activity induced in manganese were affected by non-thermal neutrons. An experimental correction for such effects was made by making measurements with the detectors surrounded by cadmium. After such corrections the results from three methods of procedure with the uranium oxide detectors and from the manganese detectors were consistent to within a few per cent. If f, the density of thermal neutrons in the absorbing material, is given by $\Delta f = \kappa^2 f$, the experimental data are shown to give a value $\kappa = 0.41$ cm^{-1} for a U metal sphere of density 8.6 in contrast to 0.35 cm^{-1} from constants given by Fermi and his collaborators. For a U$_3$O$_8$ sphere of density 5.3 the data give $\kappa = .24$ cm^{-1} in agreement with Fermi's results. Possible sources of error have been discussed.

I. Introduction

Neutron absorption in uranium is believed to be principally of two general types: an absorption by U-238, chiefly without fission, in the thermal region and in a resonance region; and absorption by U-235 which leads to fission. A detailed knowledge of these absorption processes is of interest in calculating the optimum size and spacing of the spheres to be used in the experiments on the chain reaction. The object of the present series of experiments is the study of the absorption processes in uranium and in particular is the accumulation of data from which can be calculated absorption coefficients for thermal and resonance neutrons.

The absorption of resonance neutrons was dealt with in a report submitted on June 1, 1941. In that report, Appendix A represented results of measurements carried out at Princeton jointly by the Columbia and Princeton groups on "The Capture of Resonance Neutrons by a Uranium Sphere Imbedded in Graphite". Appendix A (written by Fermi and Anderson) arrived at a figure of 4800 cm^3 for the volume of a black body equivalent with respect to resonance absorption to a sphere of 8.5 cm radius containing 9170 gm of U$_3$O$_8$. Appendix

B investigated what proportion of the resonance absorption takes place in various portions of the U_3O_8 sphere. It was found that the absorption could be divided into a surface effect and a volume effect. This result made it possible to derive a formula giving the absorbing power for resonance neutrons of U_3O_8 spheres of radii greater than or less than 8.5 cm.

Further studies of the absorption of *resonance* neutrons have been carried out at Princeton, with the important difference that the absorbing spheres were composed of compressed U_3O_8 and of uranium metal. Results of these studies will be described in a subsequent report. The present report deals with the absorption of *thermal* neutrons in both uranium metal and uranium oxide. A sphere of the substance to be investigated was placed in the interior of a graphite block. The surface of the block was exposed to an intense source of neutrons of about 4 Mev. The neutrons were slowed to thermal energies in the graphite and penetrated into the absorbing sphere. The relative number of thermal neutrons was measured at various points in the graphite and in the sphere. The measured falling off of the number of thermal neutrons toward the center of the sphere allows the determination of a certain absorption coefficient.

The absorption coefficient is a macroscopic property characteristic of the material of the sphere. It depends not only upon the absorption cross sections of the constituent atoms and upon their number per cubic centimeter but also upon their scattering cross sections. From the calculated absorption coefficient for thermal neutrons and the measurements of resonance absorption already reported for U_3O_8 and in progress for uranium metal and compressed U_3O_8, it is the intention to deduce the optimum dimensions for the typical cell in the proposed lattice.

The following measurements of density of thermal neutrons have been carried out in some cases by observing the radioactivity of the products resulting from fission by thermal neutrons; in some cases by measuring the activity induced in a detector made of manganese, which is sensitive to thermal neutrons; and in a few cases by separating out chemically the radioactive iodine resulting from fission produced by thermal neutrons. The three methods of measurement, when compared, were found to give results in satisfactory agreement with one another. However, certain internal discrepancies which appeared in the first experiments were only eliminated when it became apparent that two unexpected effects were present and would have to be taken into account. One of these effects is apparently fission produced in the uranium detectors by neutrons of energy between 1 ev and 100 ev. It was corrected for in later experiments by covering the uranium detectors with cadmium and subtracting the activity so obtained from the total. The other important effect revealed in the first experiments was activity produced in the manganese detectors by non-thermal neutrons. Cadmium shielding also made it possible to correct for this effect in later measurements. Neither effect was expected on the basis of what was previously known. The apparent production of fission by epi-thermal neutrons is of considerable interest. Further experiments are in progress to investigate this point in more detail.

The cyclotron was used as the source of neutrons. In the previous experiments[8] beryllium was bombarded at the target chamber with about one microampere of 8 Mev protons. This corresponded to about 3500 curies radon beryllium equivalent. Lately it has been found that the neutron intensity at the graphite block is increased by a factor of about ten if the beryllium is bombarded internally on a probe; this increase occurs even though the probe is about 30 cm farther away from the block than is the usual target.

The neutrons were slowed down, as in previous experiments, by a graphite block (61 × 61 × 91 cm), the front face (61 × 61 cm) of which was 60 cm from the internal beryllium target. The uranium sphere was placed in a spherical cavity which could be located at various distances from the front face of the graphite block. Experiments were made on U_3O_8 spheres of 16.4 and 15.2 cm diameter and of density 4.0 and 5.3 g/cm^3 respectively, and on a pure uranium powder[9] sphere of 11.4 cm diameter and of density 8.60 g/cm^3. In each sphere was a cylindircal cavity about one cm in diameter which extended somewhat past the center as shown on Fig. 1. As the same procedure was used for all spheres, we shall speak only of the uranium sphere except in the statement of the final results.

II. Absorption of Fission-Producing Neutrons

We used four different methods for measuring the distribution of fission-producing neutrons. These methods will be described in this section. The first will be called the Wide Sample Physical Method, the second the Narrow Sample Physical Method, the third the Manganese Method, and the fourth the Chemical Method.

Wide Sample Physical Method. The cavity in the uranium sphere was filled to the surface with U_3O_8 which had just been purified from UX and which was contained in a cylindrical bag of cellophane that fitted snugly into the cavity. Purification of the U_3O_8 from UX was carried out by the standard ether extraction method recently described by Anderson.[10]

The average number of neutrons present just outside the sphere was measured by a thin purified U_3O_8 detector, the center of which was placed 13 cm from the center of the sphere as shown in Fig. 1. In some cases, as a check on the method, two detectors were placed together. To correlate experiments made with the sphere at different positions in the graphite block, a purified U_3O_8 detector was placed each time at the position (M) indicated in Fig. 1. This detector will be referred to as the monitor.

With the sphere placed in the graphite so that the U_3O_8 cylinder was horizontal and parallel to the front face of the block, a neutron exposure of

[8] Preliminary report submitted to NDRC by H. D. Smyth on June 1, 1941.
[9] We were informed by Dr. Szilard that the U metal contained 50–200 parts of boron per 1,000,000 parts of metal.
[10] H. L. Anderson, Resonance Capture of Neutrons by Uranium.

several hours was made. The cellophane tube containing the U_3O_8 was then withdrawn and cut into four or five measured segments. The U_3O_8 from one of these sections was removed and finely ground using an agate mortar and pestle. Several drops of a dilute solution of collodion in amyl acetate were mixed with the ground-up U_3O_8 and about 30 mg of the mixture were then painted thinly and uniformly over an area of 4.5 × 6.5 cm on a piece of paper 6.0 × 6.5 cm which was subsequently covered with Scotch Cellulose Tape. It was in just this way that the monitor and outside neutron detector had been prepared from the purified but unexposed U_3O_8. The U_3O_8 from the other sections was treated in the same way and the activities of the samples prepared from the sections of the cylinder together with the monitor and detector were observed continuously one after another for a period of from five to seventeen hours. The activity of a sample was measured by enclosing it in a lucite tube which could be slipped over a thin-walled silvered glass Geiger-Muller counter. About 20,000 counts were taken for each measurement. The consistency of the Geiger-Muller counter was checked at frequent intervals using a standard uranium sample and, as a further check, an unexposed sample of purified U_3O_8 was also followed.

In some of the earlier experiments, the recording equipment, a hard tube scale of thirty-two and a Cenco mechanical counter, missed counts at counting rates above 5000 counts per minute. This was ascertained by observing the decay of a sample of iodine which had been exposed to neutrons. The decay curve is plotted in Fig. 2. It is seen that, at low counting rates, the activity decays with a 25-minute half-life as it should, but that, at higher counting rates, the curve deviates from the straight line and the observed curve is a measure of the counts lost, and was applied as a correction to the data. Usually the correction amounted to but a few per cent. In the later experiments the resolving time was decreased so that no counts were lost below 20,000 counts per minute.

In Fig. 3 the measured activities of a typical run are plotted on semi-log paper as a function of the time after the end of the exposure. Presumably the measured activities consist of three parts, that from U-239 of half-life 24 minutes, that from the fission products, and that from UX newly generated by the uranium. The first two activities will die out, the 24-minute activity rapidly, the fission activity more gradually, but the UX activity will grow, giving the upward turn to the curves. To obtain the amount of UX activity to be subtracted, the samples were measured several days later when the UX had grown to such an extent that the fission products were negligible. The UX should grow as $A(1 - e^{\lambda t})$ where A is a constant proportional to the mass of uranium in the sample, λ is the decay constant of UX, and t is measured from t_0 which is approximately the time of UX separation – actually a few hours earlier as the separation is not complete. For about a day after the separation the growth is very nearly linear. The activity of an unexposed sample was followed, and when these measurements were plotted as a function of time, the straight line obtained could be accurately extrapolated to obtain t_0. After a day or so the growth is no longer linear but can be closely approximated by $At(1 - 0.59 \times 10^{-5}t)$ where t is measured in hours. It was found expedient to plot the observed activity plus the product of it by $0.59 \times 10^{-5}t$ as this should reduce the data to a straight line continuation of that which would be obtained at earlier times. Such growth curves are shown in Fig. 4. It is seen that the curves are essentially straight lines after a couple of days, and that all of these lines can be extrapolated back to very nearly the same zero time given by the unexposed example. In some cases, the activities of the monitor and detector were measured before the neutron exposure. The slopes of the straight line part of the build-up curves for all samples were measured. These are proportional to the masses of U_3O_8 in the samples, and enable one to calculate how much to subtract from the earlier measurements in allowing for the UX activity.

In Fig. 5, the activities measured within the first twenty hours after exposure are plotted after the subtraction of the UX contribution. If the neutron energy distribution at each sample had been the same, one would expect these curves to be parallel, i.e., the ratios of the activities of different samples to be independent of time. Actually the neutron energy distribution was not the same and therefore the relative proportions of the 24-minute U-239 activity and of fission products of various life-times will not be the same. The effect of the 24-minute U-239 can be ignored after about four hours when it has fallen to less than one tenth of one per cent of its first measured value. After this time, if the only activity were from fission products created by thermal neutrons, the ratios of the activities of various samples should remain constant in time and should give the relative intensities of thermal neutrons at the positions of the various samples. For samples taken from the sphere these ratios did remain constant to within a few per cent over a range of about 15 hours as can be seen from Fig. 5. However, the ratio of the detector and monitor intensities to those of the sphere samples increased in the same time range by as much as ten per cent, although the ratio of the detector activity to the monitor activity remained constant. This result indicates that we are dealing with an effect that does not arise entirely from thermal neutrons and, as might be expected, that the neutron energy distribution at the monitor and detector are about the same but differ from that prevalent throughout the sphere. Since this effect does not enter into the question of the relative activities within the sphere, its further consideration will be postponed until after the final reduction of data has been explained and the results of this group of experiments presented.

In Fig. 5 we have the activity of various samples corrected for UX and the 24-minute U-239. In Fig. 4 we have the UX activity build-up curves of the various samples. The slopes of these build-up curves are proportional to the amount of uranium in the samples so that dividing the activities at a given time (Fig. 5) by the slopes of the corresponding build-up curves gives the relative activity per mg of the various samples. Figure 6 exhibits the result of three separate runs made with the center of the sphere located 20 cm from the front face of the graphite block. The results are presented in the form of histograms where the relative fission produced (taken four hours after the neutron exposure) per mg is plotted as a function of the distance from the center of the sphere. The relative strengths have been adjusted to unity at the center of the sphere; and the monitor and detector activities are indicated by the horizontal lines. It is seen that the agreement is to within three per cent for all points within the sphere, and that the agreement with the results of the chemical method, which is also shown and which will be explained presently, is about the same except for the detector activity which is about seven per cent lower.

As an additional check on the method, in a few of the experiments, two outside detectors were placed together. The relative fission activity per mg of the two samples was the same to within one per cent. The average results of runs made with the uranium sphere at 10, 20, 30, 40, and 60 cm from the front face of the graphite block are shown in Fig. 7 a. In Fig. 7 b are shown the results of runs made with two different uranium oxide spheres. Thus, Figs. 7 a and 7 b present the data from this first group of experiments. Before reporting other methods and results, we will consider further the curves in Fig. 5.

It was mentioned before that the decay curves of the samples inside the sphere were not parallel to those of the detector and monitor samples (Fig. 5). Furthermore, it is noted in Fig. 7 a that the absorption of neutrons appears appreciably less when the sphere is closer to the front face of the graphite block. This effect might be attributed to the 2.3-day activity of element 93, which is the daughter substance of U-239. An estimate indicates the number of 93 disintegrations per minute comparable to those of the fission products, after four hours, but they will contribute considerably less to the measured activity because the maximum energy of the electrons of 93 is .47 Mev while the maximum energy of those of the fission products is about 1.2 Mev. Therefore a much greater fraction of the 93 electrons will be absorbed in the counter wall. To see how much of the activity was due to element 93 or fission products resulting from fast neutron collisions, we enclosed in cadmium the uranium oxide cylinder and the detector but not the monitor. The usual exposure was made and the samples were prepared, measured, and the data worked up as before. The curves for the samples enclosed in cadmium were flatter than that for the unenclosed monitor, but still showed the slow change of half-life that is characteristic of fission products instead of a simple composition of 24-minute and 2.3-day component curves. This would indicate fast neutron fission and its nature will be discussed later. The relative activities per mg four hours after the end of the exposure, due to whatever cause, were reduced to the same monitor activity in the experiments with and without cadmium. Then a corrected absorption curve was obtained by subtracting the cadmium-covered relative activities from the non-cadmium-covered values. The relative values so obtained with the sphere at 20 cm from the front face are shown in Fig. 8. When the above experiment was repeated, but with the sphere 60 cm from the front face of the graphite block, no measurable activities except those due to UX were observed in the samples covered with cadmium. Furthermore, the difference curve obtained with the sphere at 20 cm is in reasonable agreement with the absorption curve obtained with the sphere at 60 cm if the width of the samples is taken into account. This was to be expected since both activities are caused by thermal neutrons produced outside the sphere. The result was checked also by the Mn method (see below) and forms the basis of the theoretical interpretation.

Narrow Sample Physical Method. A modification of the general method described above was to prepare all the samples in a form suitable for measurement before the exposure. These samples were prepared as before except they were one cm wide and 5.5 cm long. They were then wrapped around a thin-walled brass tube so that they could be slipped snugly into the cavity in the sphere. The brass tube was filled with uranium to the surface of the sphere, the rest of the space being filled with a carbon plug. The advantages of this arrangement are that the positions of the samples are more definite and that more points can be taken. Indeed, many samples were placed in the graphite at intervals of a few centimeters on a horizontal line parallel to the front face and extending from the center of the sphere to the edge of the block.

In this modified procedure, as in the earlier method, the samples were measured after the neutron exposure. However, the weight of the U_3O_8 in the samples could now be obtained by igniting and then weighing them rather than by the more tedious device of following the build-up of UX. This method was not possible with the larger sized samples used before because the counter wall did not have uniform thickness over its length. This is illustrated in Fig. 9 where the counts per minute of a narrow source of uranium in equilibrium with the UX is plotted as a function of the position of the source along the counter. It is seen that only over a short distance near the middle of the tube is the thickness uniform. The samples were placed in this region. By counting small weighed uranium samples in equilibrium with UX, it was possible to calculate the amount of UX activities to subtract from activities of the exposed samples. To determine t_0, the growth of an unexposed sample of the purified uranium was also followed. Results of two exposures using this method are shown in Fig. 10.

Manganese Method. Thermal neutron distributions in and around the sphere were also obtained by substituting MnO_2 in place of the uranium oxide in the narrow samples used in the method just described. The decay curves of these samples after exposure all displayed a pure 2.59-hour half-life as is shown in Fig. 11 for a typical experiment. The relative neutron density was then obtained by dividing the intensity of a sample by its weight. The weight was obtained by piling all the samples at one place in the graphite block, exposing then to neutrons, and then measuring the activity induced. To obviate absorption by the samples themselves, several other exposures were made with the samples piled in different orders. An average of the resultant intensities was taken as the relative weight.

When the manganese samples were exposed covered with Cd, considerable activity was still noticed. Hence, in some cases, a second exposure was made with the samples in the same positions except that all but the monitor sample were covered with cadmium. Thus the non-thermal component could be subtracted. The non-thermal part will be discussed later. Figs. 12 a, b, c, d, e show thermal neutron distributions so obtained in the graphite block in various directions with and without the uranium spheres. Figure 8 shows the agreement with the Physical Method in those cases when the non-thermal components have been subtracted.

Chemical Method. In the chemical method of measuring the fission production, a cylinder of purified U_3O_8 was prepared for exposure in the same way previously described. The outside detector sample of several grams was contained in a thin cellophane envelope. It was found necessary to make a neutron exposure of about six hours in order to get measurable activities. After the exposure, the cellophane cylinder was taken from the sphere and cut into measured segments as before. The U_3O_8 from each segment was then weighed. The U_3O_8 which constituted the outside detector was mixed and then divided into two parts which were also weighed. To each sample was added an equal weight of iodine to serve as a carrier, and then a chemical separation of iodine was made. This separation yielded silver iodide which was finely ground up, mixed with alcohol, and painted on a weighed piece of aluminium over an area of 2×6 cm. A second weighing yielded the weight of the silver iodide which indicated the efficiency of the separation. The chemical separation of all the samples was made as nearly simultaneously as possible. The order of the samples was the same for every operation. First we took part of the detector sample, then the samples from the sphere, and finally the other fraction of the detector sample. Thus any error introduced because of not making the separations exactly simultaneously would be indicated by different activities per mg in the two detector fractions.

After the samples were prepared for measurement in the usual way, their activities were followed for several hours. These were plotted on semi-log paper, and the smooth curves so obtained were closely parallel. The activities of the samples at a given time were read from the curves. The relative fission activity per mg of a U_3O_8 sample was then found by dividing the activity obtained from the curve by the corresponding weight of the U_3O_8 and by the weight of the silver iodide of the measured sample. The results of such a run made with the uranium sphere at 20 cm from the front face of the graphite block have been shown in Fig. 6, where it was seen that they agreed to within seven per cent with the results obtained by the physical method. The two parts of the outside detector agreed with each other to within five per cent.

Because the internal consistency of the results of the chemical method was not so good as the results given by the physical methods, and because the labor and time of bombardment involved were much greater, it was decided not to perform more extensive experiments using the chemical method.

III. Evaluation of the Macroscopic Absorption Coefficients

The thermal neutrons obey in an absorbing material, apart from the immediate neighborhood of the surface, the equation

$$\Delta f = \kappa^2 f \tag{1}$$

where f is the density of neutrons and κ is given, with a sufficient accuracy, by the equation

$$\kappa = \sqrt{3\sigma\sigma_a} \left(1 - \frac{2\sigma_a}{5\sigma}\right). \tag{1a}$$

Herein σ and σ_a are total and absorbing cross sections contained in unit volume. The above described experiments will be used to determine κ for our U metal of density 8.6 to be

$$\kappa = .41 \text{ cm}^{-1}. \tag{2}$$

This is sufficient for the calculation of the neutron density within a sphere in the "final arrangement" where the disposition of material can be considered to be spherically symmetric about the center of the sphere. Thus the density of thermal neutrons in the final arrangement will be, in the inside of the sphere, given by

$$f = \text{const.}\frac{1}{\kappa r}Sh\kappa r \approx \text{const.}\left(1 + \frac{\kappa^2 r^2}{6} + \frac{\kappa^4 r^4}{120}\right). \tag{3}$$

For our experimental arrangement, (3) is not accurately obeyed because the distribution of the neutrons may deviate from spherical symmetry. This effect will be discussed and shown not to be large enough substantially to influence our results. It will be shown, furthermore, that the validity of (1) (and hence of (3) in the final arrangement) should extend practically to the surface, i.e., that the surface effect is very small. It will be noted finally that the energy distribution of thermal neutrons is not very sharp but extends over a considerable range. This necessitates some modification of (1) and (3) which will be considered in some detail in a later report.

Our result (2) is a good deal higher than that which follows from the constants given by Fermi and collaborators, i.e.,

$$\sigma = 18N \times 10^{-24} \qquad \sigma_a = 6N \times^{-24} \tag{4}$$

where $N = 2.2 \times 10^{22}$ is the number of atoms per cm^3. The κ derived from (4) is 0.35 cm^{-1}.

Discussion of Experiments with U Metal Sphere. The density of thermal neutrons was obtained from three essentially independent experiments: fission activity inside the sphere when sphere was at 60 cm from source; same activity for 20 cm distance of sphere from source; activity of Mn detectors inside of sphere for 60 cm distance. The "source" means, in this connection, the front of the block, i.e., the place where the fast neutrons enter.

A separate experiment showed that there is no observable non-thermal fission activity induced in the sphere at the 60 cm position. In this case, the experimental material can be used without any correction. In the two other cases, the activity induced by non-thermal neutrons had to be subtracted from the whole activity.

In all cases, first of all, the histograms were replaced by smooth curves. Then the activity of the Cd covered samples (giving the non-thermal neutrons' activity) was reduced to the same monitor intensity which was observed in the experiment with the uncovered sample. The reduced activity of the Cd covered sphere was subtracted from the activity of the uncovered sphere and the distances r determined at which the activity is 1.1, 1.2, 1.3 etc. times greater than the activity at the center. This gave, for the above described three experiments, the three sets of radii given in the second, third and fourth columns of Table I.

Table I

Ratio to activity at center	Distance from center in				Derived value of κ in cm^{-1} from activity		
	60 cm exp	20 cm exp	Mn exp		60 cm exp	20 cm exp	Mn exp
1	0	0	0	0	–	–	–
1.1	1.6 cm	1.7 cm	1.65 cm	0.765	0.48	0.455	0.465
1.2	2.25	2.35	2.3	1.065	0.47	0.455	0.465
1.3	2.85	2.9	2.95	1.29	0.45	0.445	0.440
1.4	3.3	3.45	3.55	1.475	0.445	0.430	0.415
1.5	3.65	3.95	3.95	1.62	0.445	0.410	0.410
1.6	4.3	4.4	4.3	1.76	0.41	0.400	0.410
1.7	4.65	4.75	4.55	1.87	0.40	0.395	0.410
1.8	4.95		4.8	1.975	0.40		0.410
1.9			5.0	2.05			0.410
2.0			5.2	2.17			0.415
2.1			5.4	2.25			0.415

The values of κr for which $Sh\kappa r/\kappa r$ assumes the values of the first column are given in the fifth column and the ratio of this κr to the r of the second, third, and fourth columns is given in the sixth, seventh, and eighth columns. These ratios should be all the same, i.e., κ. It is seen that in all three experiments the values of κ decrease considerably from the center. However, the experimental accuracy is least in near the center and the uncertainty of the magnitude of the activity at $r = 0$ may partly be responsible for this discrepancy. At higher r, the values rather consistently tend to .41. The agreement within the last values of one column does not mean, of course, much more than that they were all taken from a smooth curve, but the agreement between the three curves must be considered to be significant.

Experiments with U_3O_8 Sphere. The procedure of evaluation was the same as in the experiments with the metal sphere. Table II summarizes some of the results, viz. Exp. 10 B and 12 E. In the first one, the fission activity was measured directly at 60 cm from the front of the block; in the second, the Mn activity was measured at the same position and corrected by subtracting the fission produced by non-thermal neutrons. It is seen that the Mn experiment agrees with the $Sh\kappa r/\kappa r$ curve very satisfactorily. In fact, in view of the limited accuracy of the measurements, the agreement must be, to a large extent, accidental. The κ in the fission experiment increases with increasing r, a behavior which must be expected on theoretical grounds but which is surprising in view of the opposite behavior of the more accurate data used in Table I. On the whole, the data indicate a

$$\kappa = .24 \text{ cm}^{-1} \tag{5}$$

for this U_3O_8 sphere of density 5.3.

Table II

Ratio to activity at center	Distance from center in Fig. 10	Distance from center in Fig. 12 E		κ derived from Exp. 10 B	κ derived from Exp. 12 E
1	0	0	0	–	–
1.1	3.9 cm	3.3 cm	0.765	0.195	0.23
1.2	5.0	4.6	1.065	0.215	0.23
1.3	6.0	5.7	1.29	0.215	0.225
1.4	6.3	6.5	1.475	0.235	0.23
1.5	6.8	7	1.62	0.24	0.23
1.6	7.2	7.4	1.76	0.245	0.24
1.7	7.4–7.5		1.87	0.25	

Discussion of Errors. The chief difficulty in the interpretation of the experimental data is persistent increase of κ with increasing r. This behavior is manifested in the very accurate experiments summarized in Table I. It is, of course, possible that this behavior is due to the inhomogeneity of the material, but it is impossible to discuss this possibility on theoretical grounds.

One might think, next, that the brass tube which contained the U_3O_8 for the analysis may be responsible for the effect. This brass tube had a diameter of 1.25 cm and a wall thickness of .08 cm. The absorbing cross section of Cu is 1.6×10^{-24} (F. Rasetti, Phys. Rev. 58, 869) that of Zn is about 1×10^{-24} (Coltman, ibid, 59, 917). Since the total concentration of Cu and Zn atoms is about four times higher in brass than that of U atoms in our metal, this gives a σ_a for brass which is just about equal to the σ_a of U. The total σ is, on the other hand, higher in brass than in our U metal by about 40 per cent. This corresponds to an increase in κ by 20 per cent. However, the oxide powder contained in the tube decreases the average κ of the center to an even greater extent than the brass increases it, and the cover of the tube has too little effect to explain the variation of κ by much more than 1 per cent.

The solution (3) holds only if the distribution of neutrons is spherically symmetric within the sphere. It is evident from the figures reproduced in the experimental part that this condition is not realized in the experiments. One has to consider, therefore, other solutions of (1), corresponding to higher spherical harmonics. It follows from the symmetry of the problem that among the spherical harmonics with $L = 1$, only the $m = 0$ can enter corresponding to an angular distribution $\sim \cos \theta$. This, however, vanishes in the direction $\theta = 90°$ in which the measurements were made. The effect of the next spherical harmonics $L = 2$) can be estimated as follows.

The solution in question is

$$C15(3 \cos^2 \theta - 1) \left(\frac{Sh\kappa r}{\kappa r} - \frac{3Ch\kappa r}{\kappa^2 r^2} + \frac{3Sh\kappa r}{\kappa^3 r^3} \right) \approx C(3 \cos^2 \theta - 1) \left(\kappa^2 r^2 + \frac{\kappa^4 r^4}{14} \right) \tag{6}$$

where C is a constant. Since κr is about 2 on the boundary (6) gives $C \times 2(4 + 16/14) = 10.2\, C$ for $\theta = 0$; $-5.1\, C$ for $\theta = 90°$; $10.2\, C$ for $\theta = 180°$. The experimental values are, in one particular case (Fig. 12 A, 12 B) 2.0, 1.75, 1.6. The combination $f(0) - 2f(90°) + f(180°)$ gives 0 for the solutions with $L = 0$ and $L = 1$ and can be attributed to the $L = 2$ solution. It gives $0.1 = (10.2 - 2 \times 5.1 + 10.2)C = 10.2\, C$, i.e., $C = 0.01$. This will give the correct order of magnitude of C, but the experimental values are too inaccurate to exclude even a three times higher value for C.

Adding (6) to (3) gives

$$1 + \frac{\kappa^2 r^2}{6} + \frac{\kappa^4 r^4}{120} - C \left(\kappa^2 r^2 + \frac{\kappa^4 r^4}{14} \right) \tag{7}$$

for the dependence upon distance of the activity at $\theta = 90°$. If one calculates for this function the product of the constant terms and of the coefficient of r^4 and divides the result with the square of the coefficient of r^2, one obtains a quantity which is independent of κ, viz.

$$.3(1 + 3.43\, C). \tag{8}$$

The experimental value of this quantity, as derived from the figures underlying Table I, is around 0 and certainly not larger than .04. This would involve a $C \approx -1/3$ – which seems impossible.

There are two further effects which should be discussed in detail, viz. the surface effect and the effect that not all neutrons have the same velocity. Both discussions involve a fair amount of theoretical work and will be given, therefore, in another report. It should be mentioned, however, that it does not seem that they can explain the trend exhibited by κ in Table I and the only suggestion we can offer for explaining this trend, therefore, is a possible inhomogeneity of the metal. The absence of the trend for the oxide sphere (Table II) appears to give some support to this view.

Discussion of Results. The value of κ given by (1) is incompatible with a total cross section 18×10^{-24} cm^2. The minimum value of σ which is compatible with our κ is just this κ itself, which gives with $N = 2.2 \times 10^{22}$ the value 18.5×10^{-24} cm^2 as the smallest possible value for the total cross section. However, this would involve $\sigma_a = \sigma$, $\sigma_{sc} = 0$, i.e., that U is black toward thermal neutrons. It is more reasonable, therefore, to use a somewhat higher σ, such as corresponding to a total cross section of $20 \times^{-24}$ cm^2. This gives

$$\sigma = 20N10^{-24}; \qquad \sigma_a = 8.2N10^{-24}. \tag{9}$$

From these figures, the κ for the oxide of density 5.3 can be derived if one assumes 4×10^{-24} cm^2 as the scattering cross section of oxygen and further assumes that oxygen has no appreciable absorption ($\sigma_a < 10^{-25}N$ cm^{-1}) for thermal neutrons. This gives

$$\sigma = 0.35 \text{ cm}^{-1} \quad \text{and} \quad \sigma_a = 0.095 \text{ cm}^{-1} \tag{10}$$

for the oxide of density 5.3. These figures do not take possible contaminations of the metal into account. This gives a $\kappa = 0.28$ cm^{-1} for the oxide, a value considerably larger than given by (5). This indicates either an even higher σ (and lower σ_a) than adopted in (9) or a considerable amount of absorption by contaminations in the metal. It is not necessary to discuss this since only the immediately measured κ enter into the consideration of the "final" arrangement. The absorption coefficient of the oxide is in substantial agreement with the value which can be derived from the cross sections given by the Columbia group.

Our thanks are due to Professors H. D. Smyth and J. A. Wheeler for their invaluable help in connection with the formulation of our results and the preparation of this manuscript.

Summary of Figures

Fig. no.	Exp. no.	Type	Sphere material	Sphere density	Sphere radius	Sphere position (Z)	Bomb. time	Method	Remarks
1									Experimental arrangement
2									Correction for finite resolving time of counter
3	4	Fission	U Metal	5.7	8.6	20 cm	2–1/3 H	Wide Sample	Total activity curves
4	4	Fission	U Metal	5.7	8.6	20 cm	2–1/3 H	Wide Sample	Growth of UX
5	4	Fission	U Metal	5.7	8.6	20 cm	2–1/3 H	Wide Sample	Activity with UX subtracted
6	3 4 9	Fission	U Metal	5.7	8.6	20 cm	2–1/3 H 2–1/3 H 6–1/4 H	Wide Sample	Comparison of physical and chemical methods
7 A	14} 17}	Fission	U Metal	5.7	8.6	10 cm	2 H	Wide sample	Summary of wide sample results, with U metal sphere, uncorrected for non-thermal fission
7 A	3} 4} 9} 18}	Fission	U Metal	5.7	8.6	20 cm	2–1/3 H 2–1/3 H 6–1/4 H 2 H	Wide sample	
7 A	10} 15} 22}	Fission	U Metal	5.7	8.6	30 cm	? 2 H 2–2/3 H	Wide sample	

Summary of Figures

Fig. no.	Exp. no.	Type	Sphere material	Sphere density	Sphere radius	Sphere position (Z)	Bomb. time	Method	Remarks
7 A	6)	Fission	U Metal	5.7	8.6	40 cm	2 H	Wide sample	
7 A	5) 16)	Fission	U Metal	5.7	8.6	60 cm	2-1/2 H 4 H	Wide sample	
7 B	8) 12) 13)	Fission	U_3O_8	8.2	4.0	20 cm	1 H 2 H 4 H	Wide Sample	Results with two U_3O_8 spheres
7 B	38	Fission	U_3O_8	7.6	5.3	60 cm	4-1/6 H	Narrow Sample	
8	3) 4) 9) 18	Fission	U Metal	5.7	8.6	20 cm	2-1/3 H 2-1/3 H 6-1/4 H 2 H	Wide Sample	Thermal fission and thermal manganese activities in U metal sphere
8	5) 16) 22	Fission	U Metal	5.7	8.6	60 cm	2-1/2 H 4 H 2-2/3 H	Wide Sample	
8	49	MnO_2	U Metal	5.7	8.6	60 cm	1-1/2 H 1 H	MnO_2	
9	35								Counter sensitivity

Summary of Figures

Fig. no.	Exp. no.	Type	Sphere material	Sphere density	Sphere radius	Sphere position (Z)	Bomb. time	Method	Remarks
10 A	50	Fission	U Metal	5.7	8.6	40 cm	4-1/12 H	Narrow Sample	Narrow sample results
10 B	38	Fission	U_3O_8	7.6	5.3	60 cm	4-1/6 H	Narrow Sample	
11	43	MnO_2						MnO_2	MnO_2 decay curves
12 A	43	MnO_2	No Sphere	5.7	8.6	20 cm	1-3/4 H	Narrow Sample	Mn neutron distribution in block along Z (thermal + resonance)
	28		U Sphere				?	Narrow Sample	
12 B	29	MnO_2	U Metal	5.7	8.6	20 cm	2/3 H	Narrow Sample	Mn neutron distribution across block with sphere (thermal + resonance)
12 C	33	MnO_2	U Metal	5.7	8.6	40 cm	1 H	Narrow Sample	Mn neutron distribution across block with sphere (thermal + resonance)
12 D	35	MnO_2	U Metal	5.7	8.6	60 cm	1 H	Narrow Sample	Mn neutron distribution, corrected for non-thermal part
	49						1-1/2 H		
12 E	37	MnO_2	U_3O_8	7.6	5.3	60 cm	1-1/2 H	Narrow Sample	

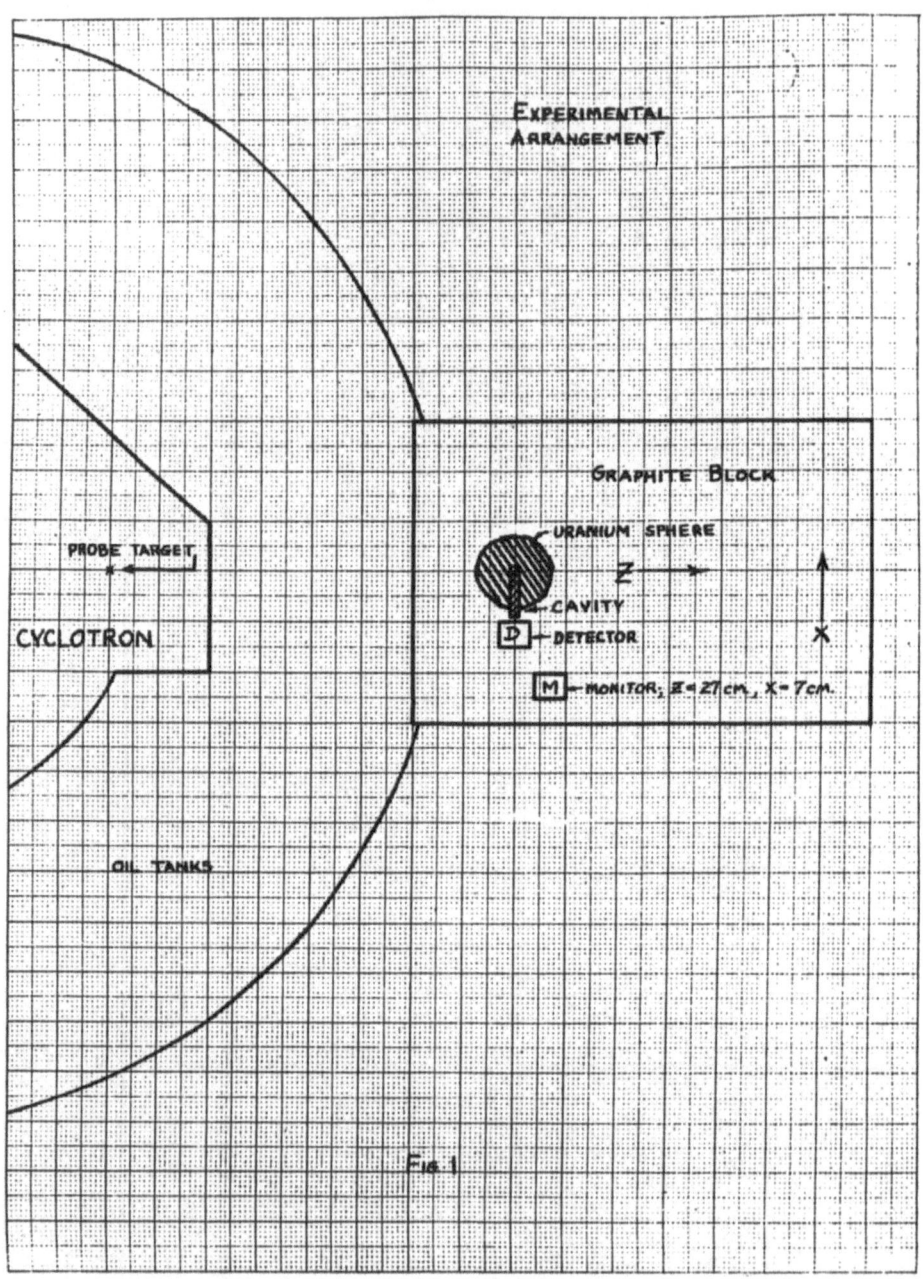

Fig. 1

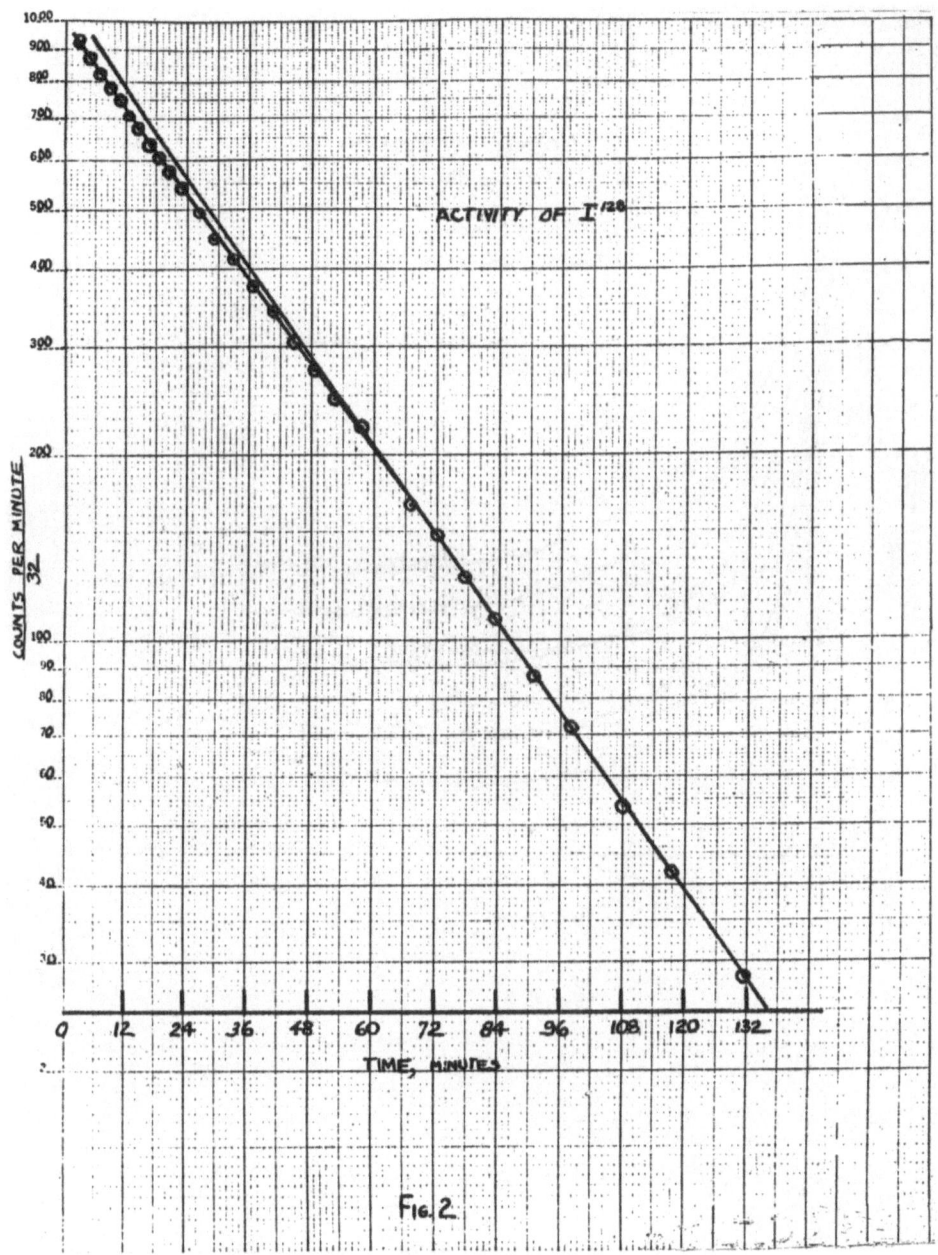

Fig. 2

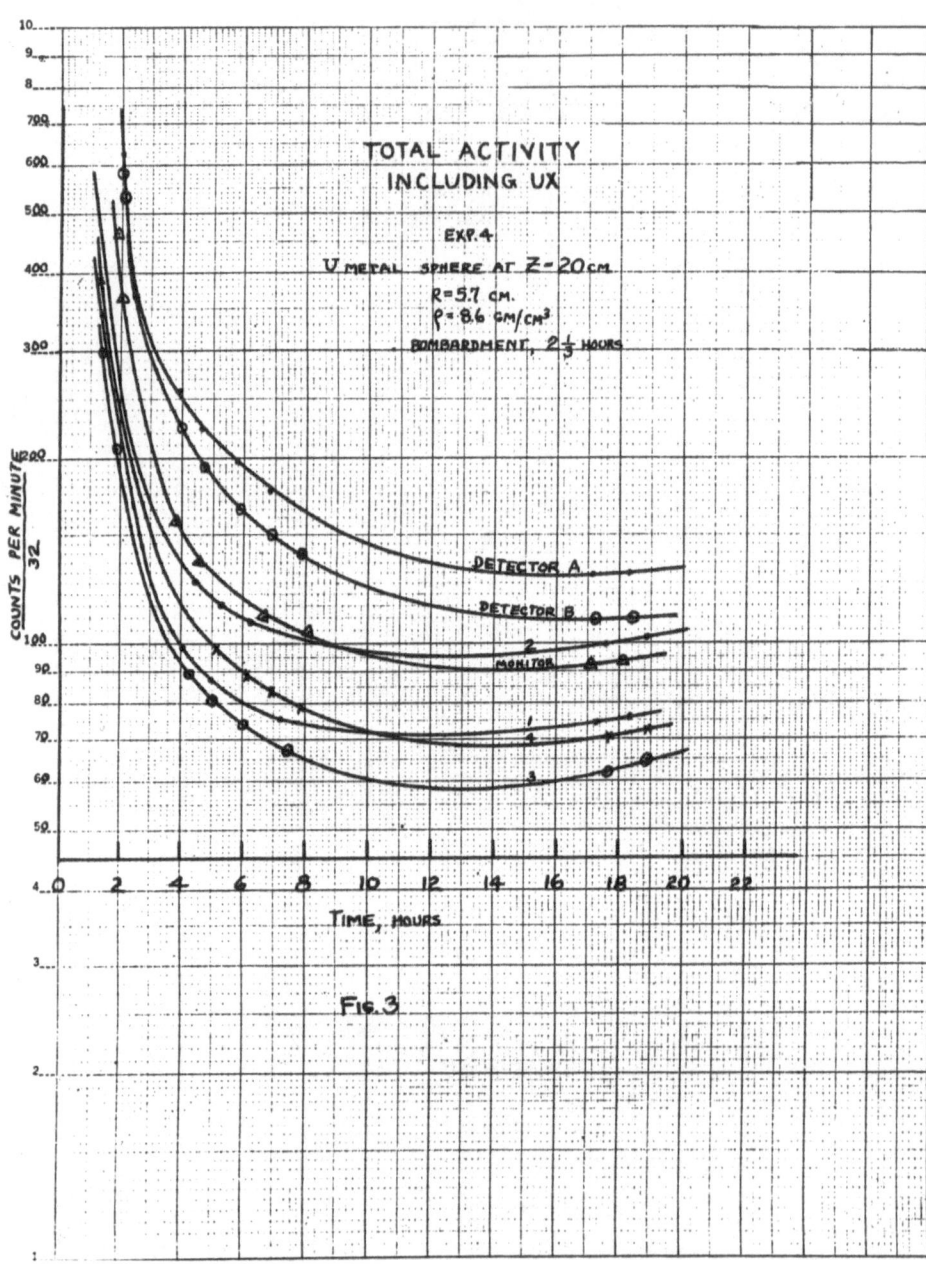

Fig. 3

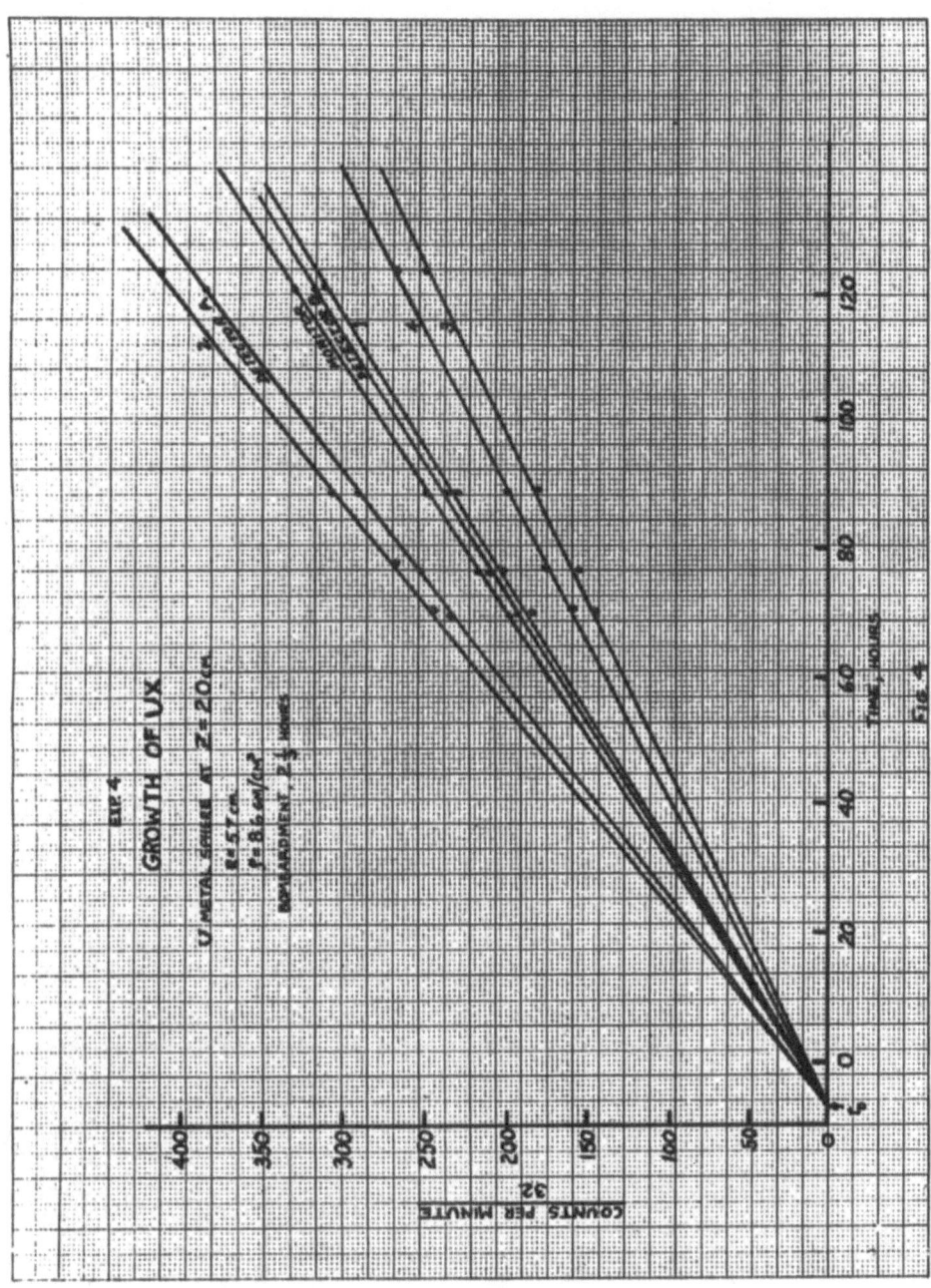

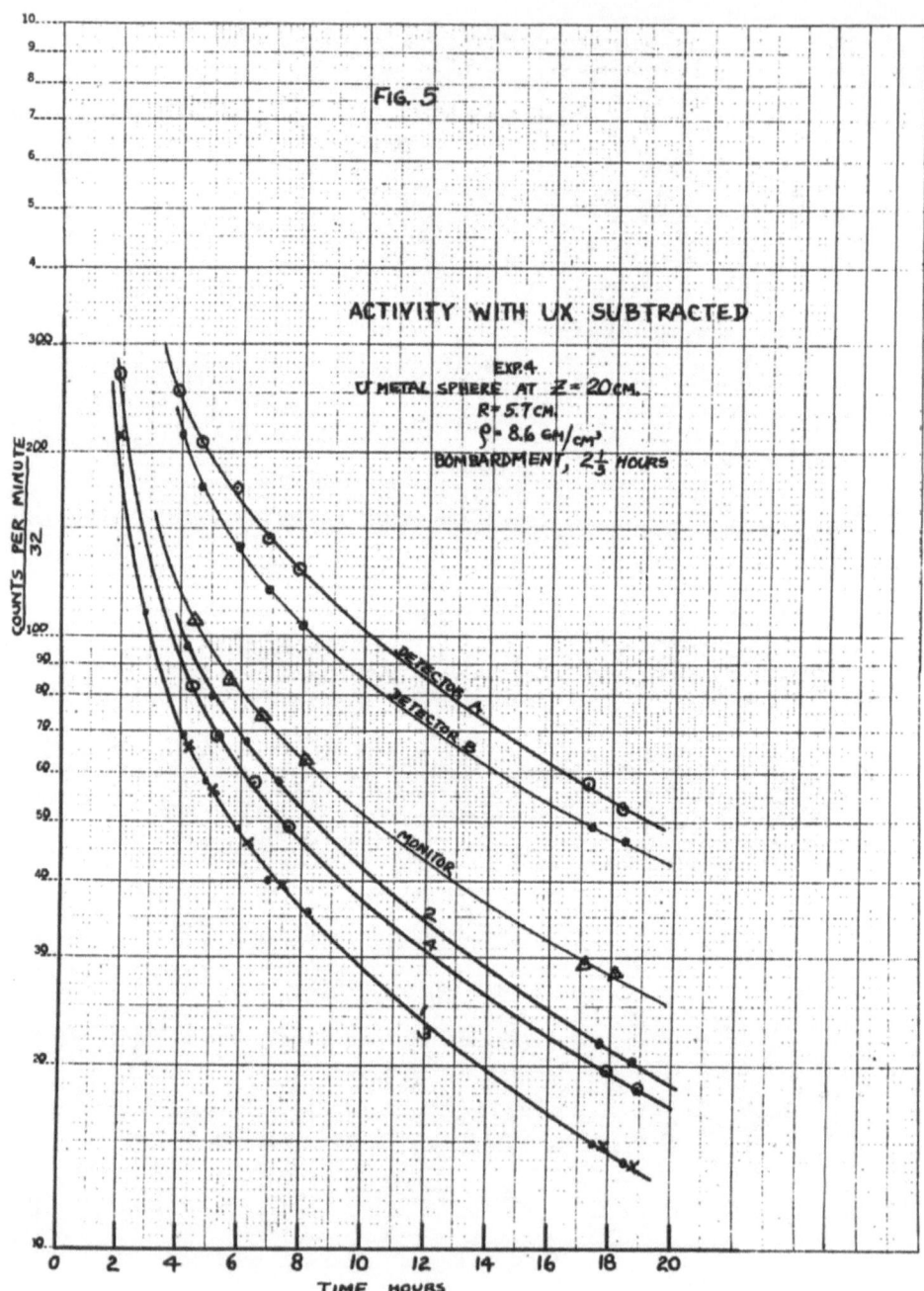

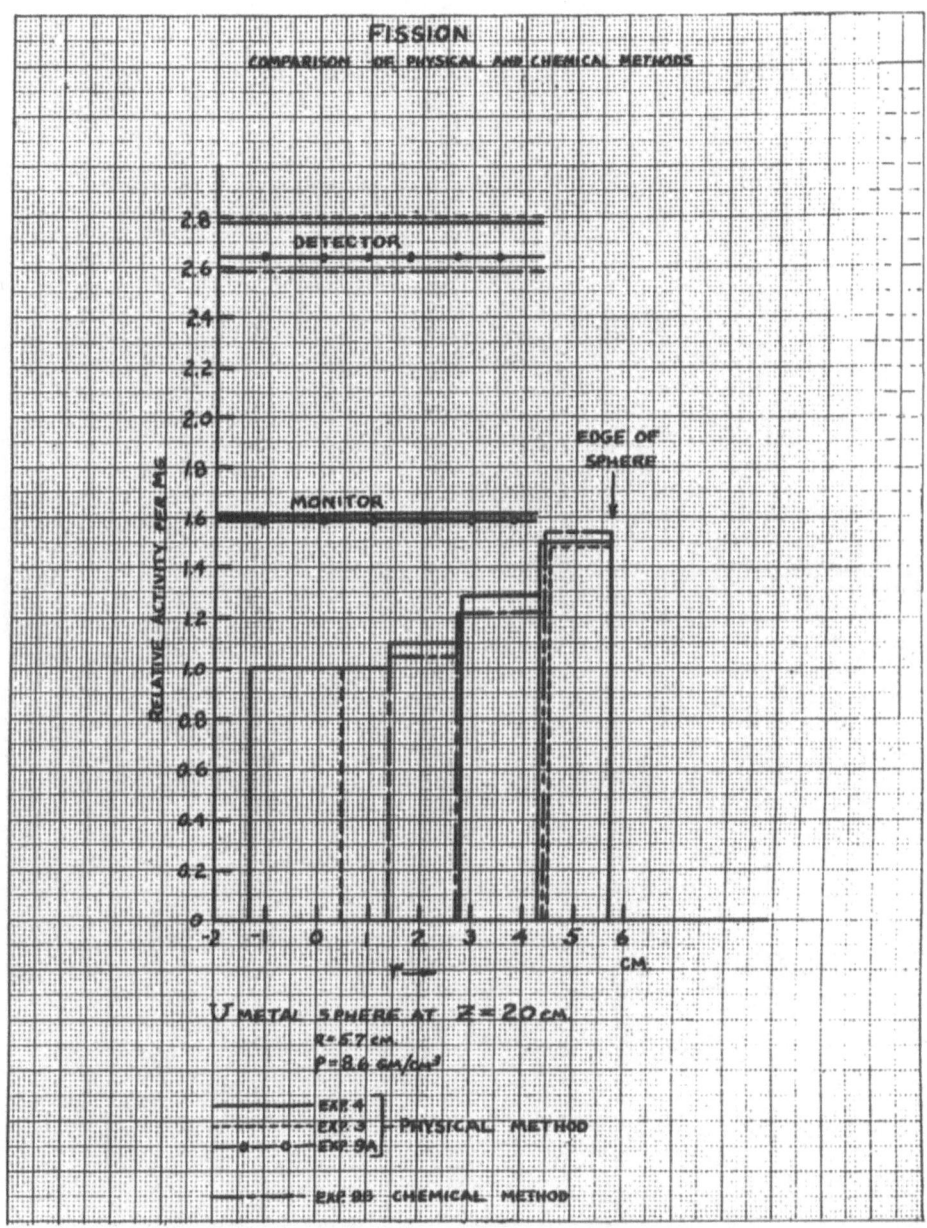

Exp. 9A consisted in measuring, by the Physical Method, the activity of an extra detector and monitor exposed simultaneously with the samples of experiment 9B.

Figure 6

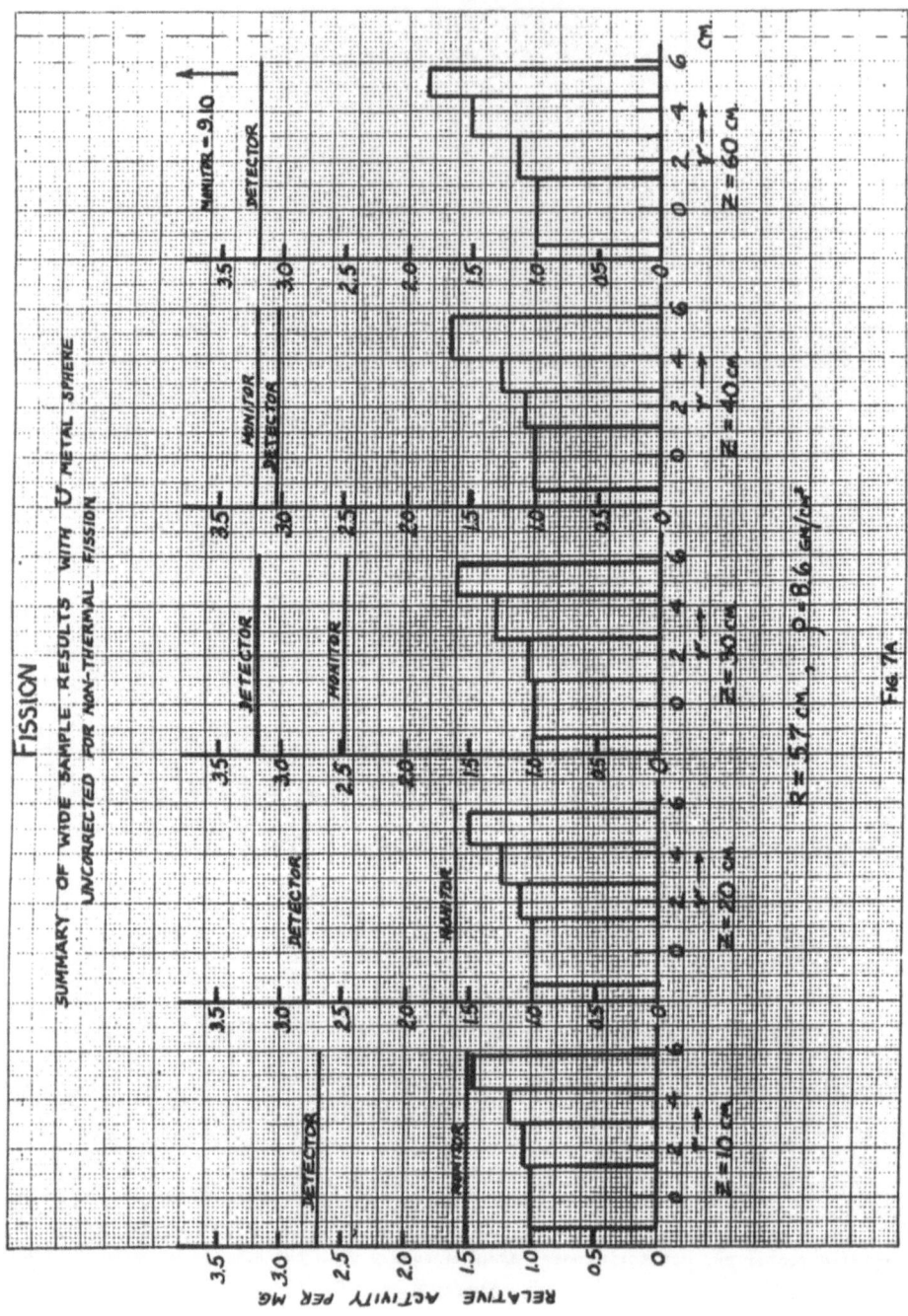

FISSION

SUMMARY OF WIDE SAMPLE RESULTS WITH U METAL SPHERE
UNCORRECTED FOR NON-THERMAL FISSION

RELATIVE ACTIVITY PER MG

$R = 5.7$ cm., $\rho = 8.6$ gr/cm^3

Fig. 7A

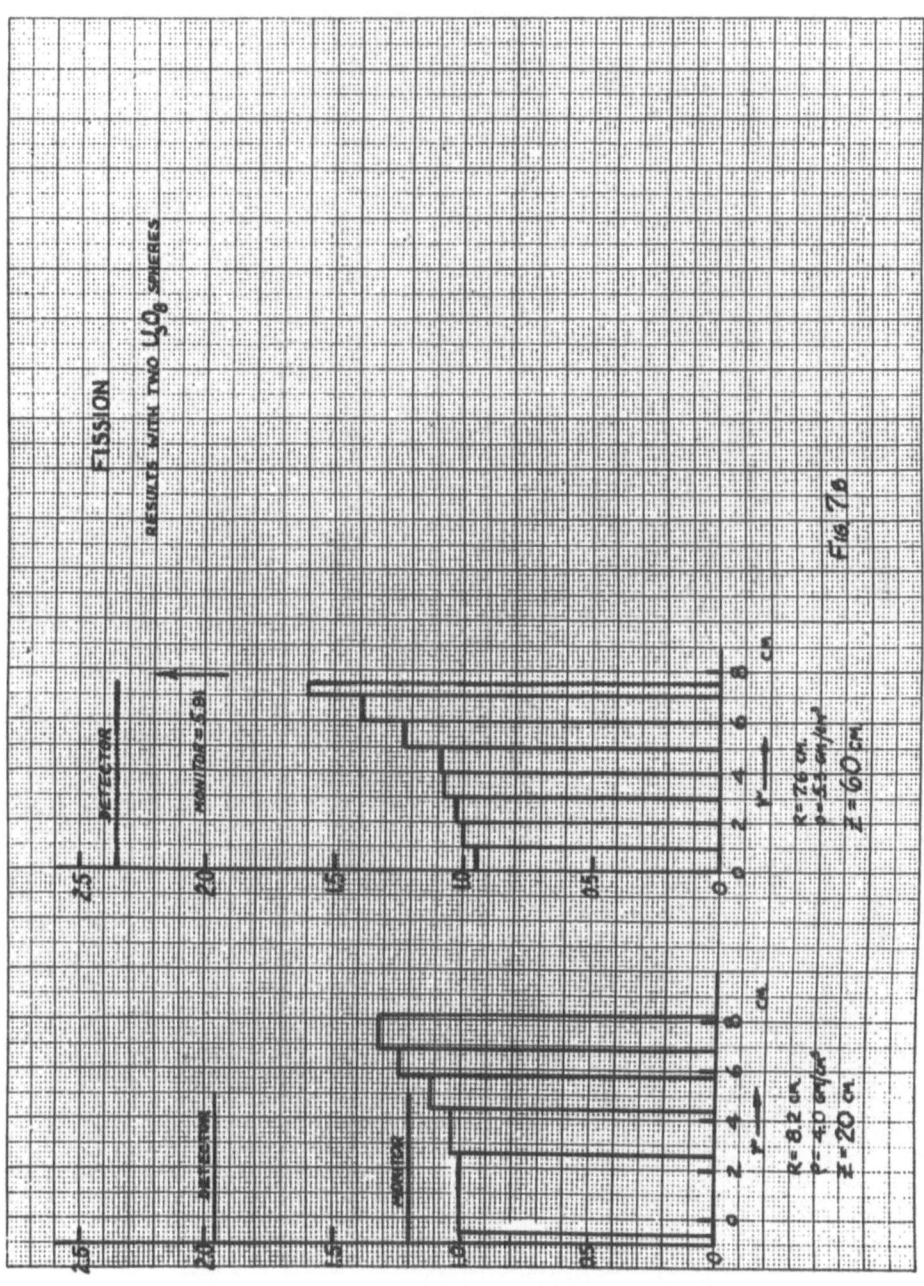

FIG. 7b

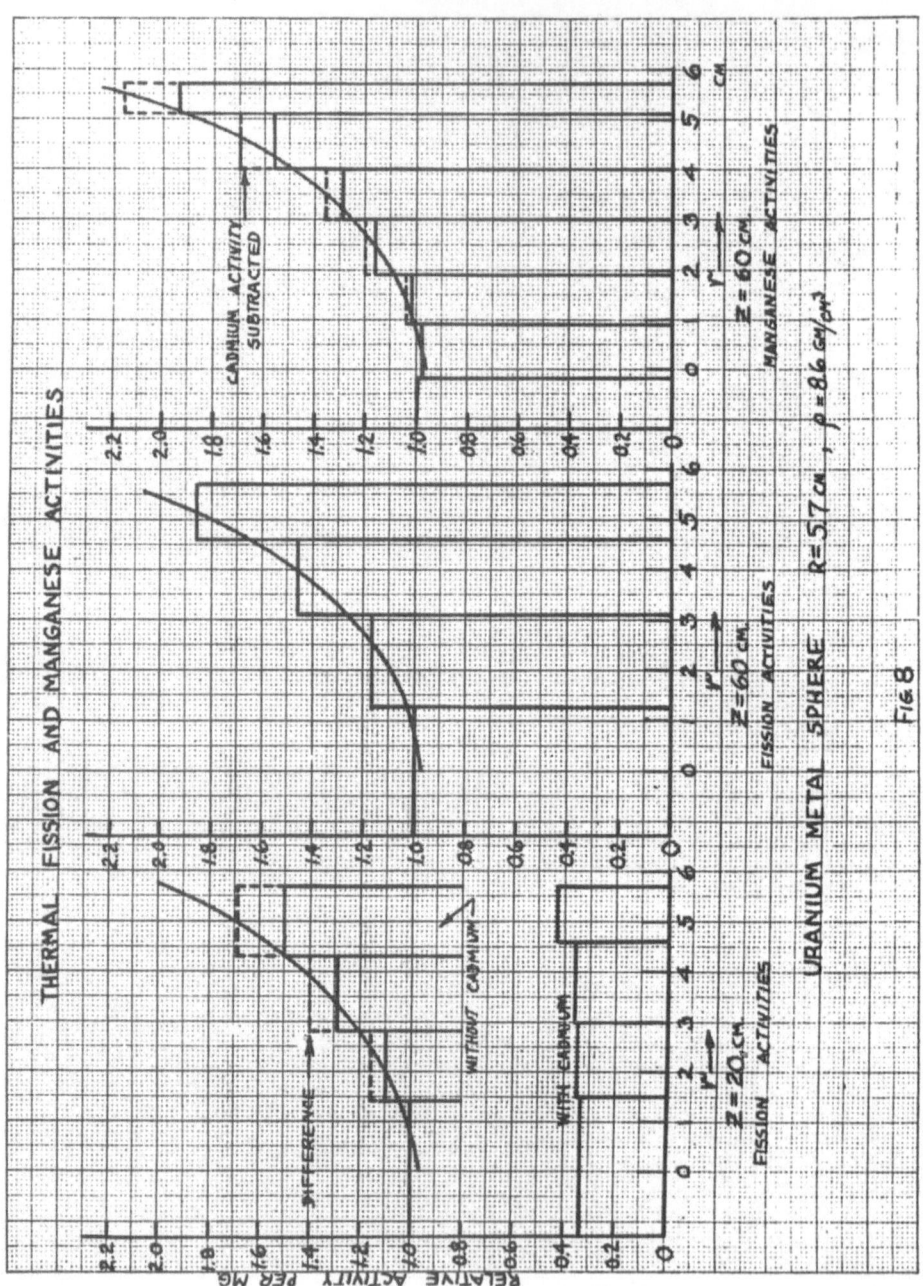

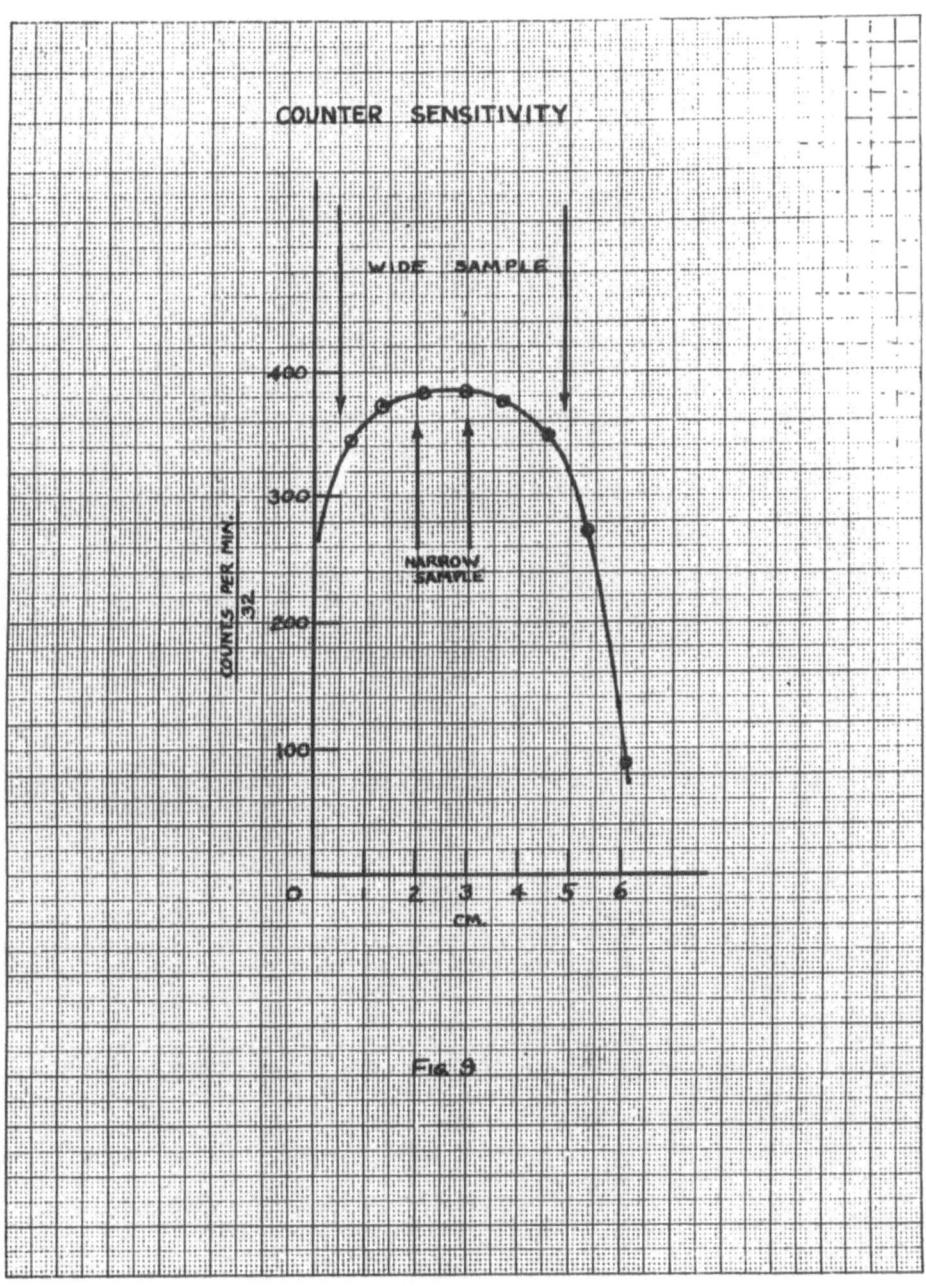

Fig. 9

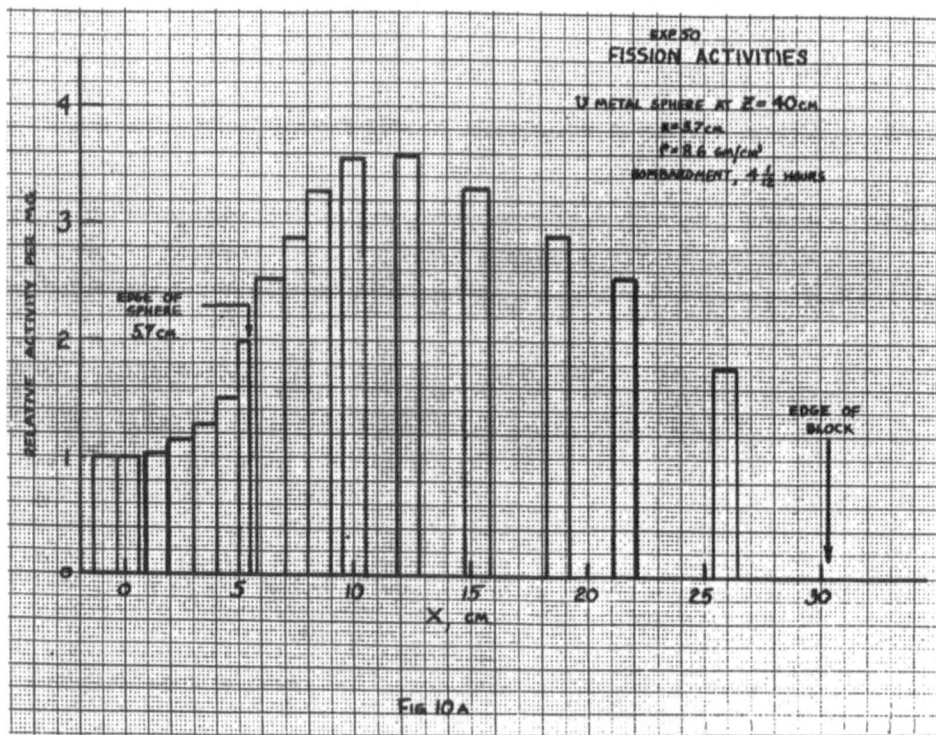

Fig. 10 a

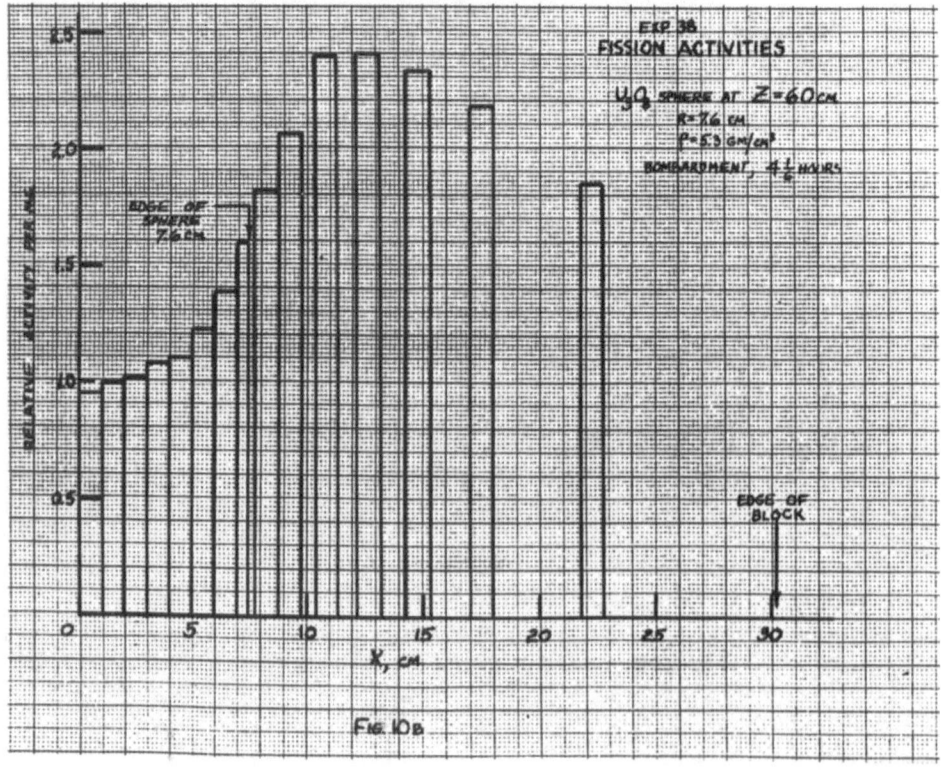

Fig. 10 b

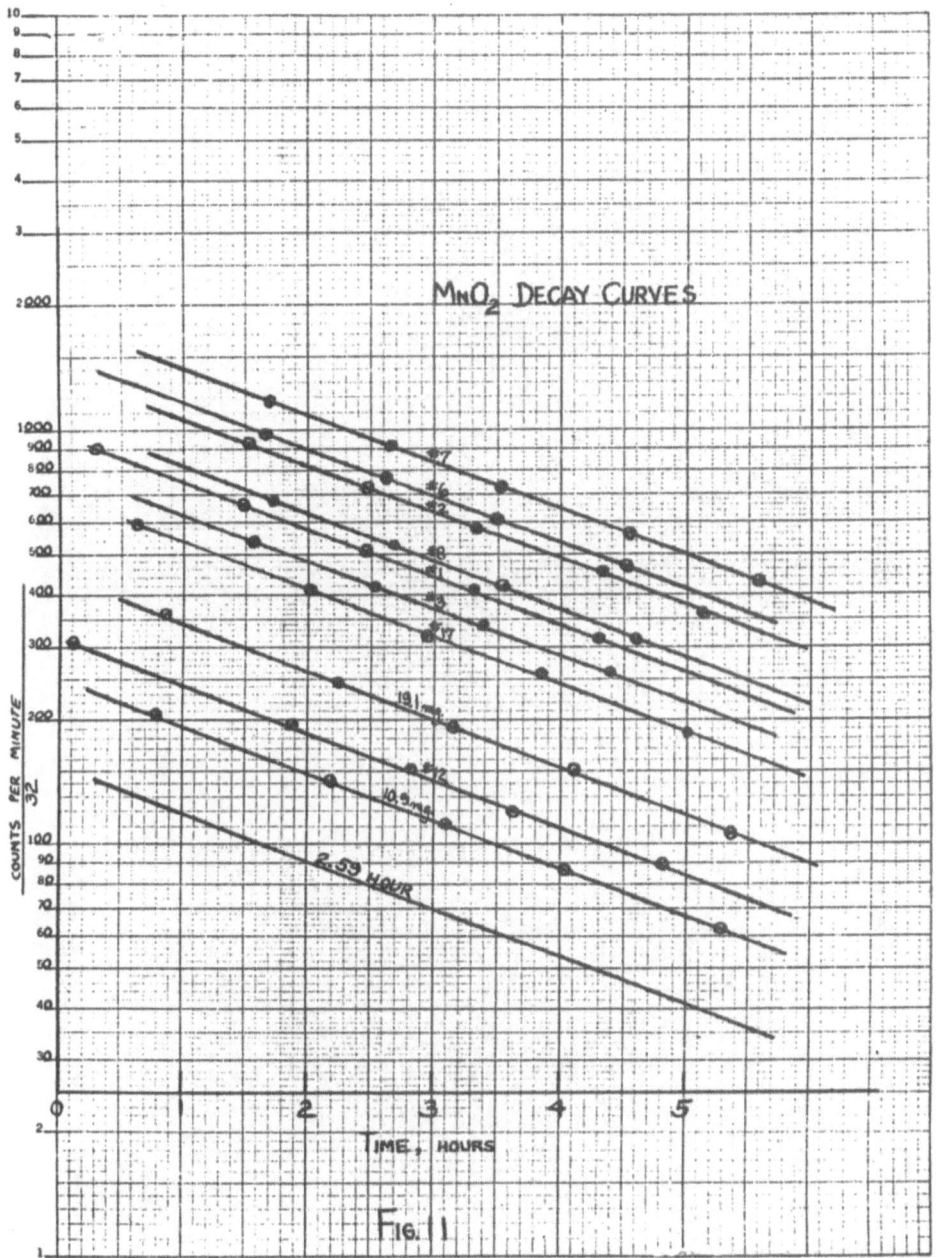

Fig. 1

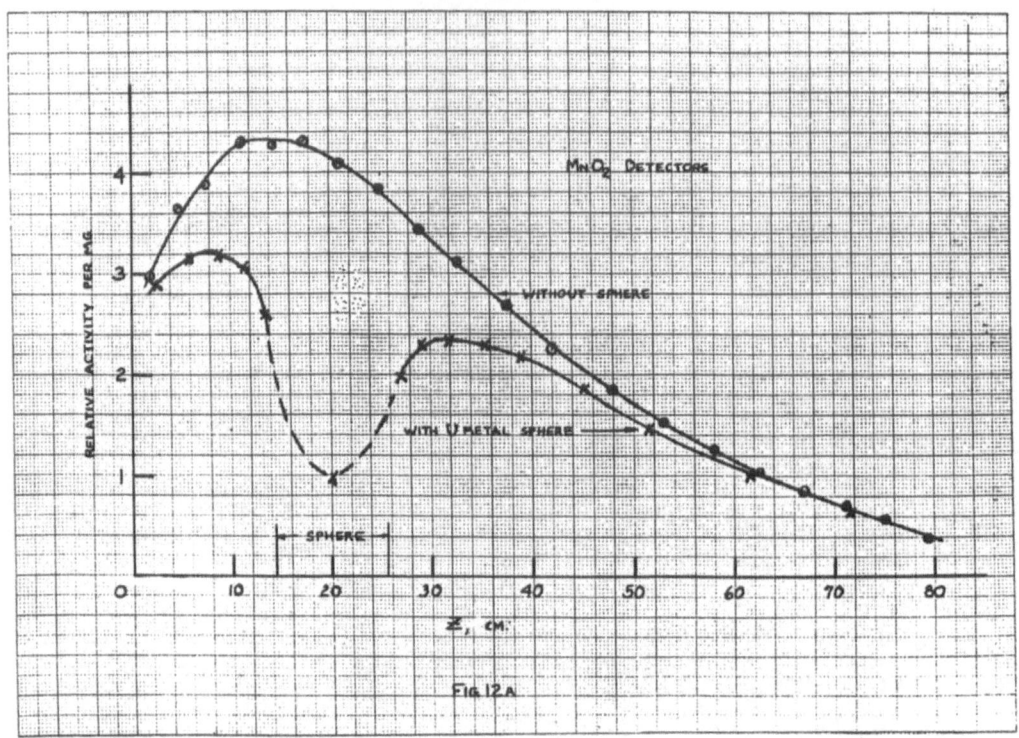

Fig. 12A

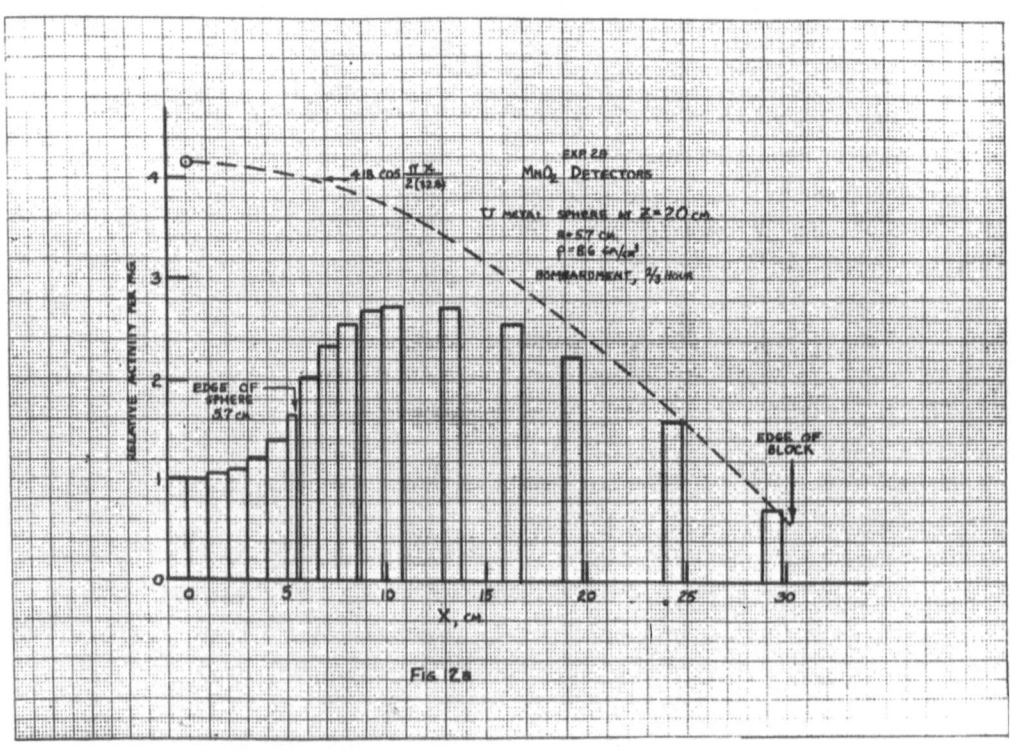

Fig. 12B

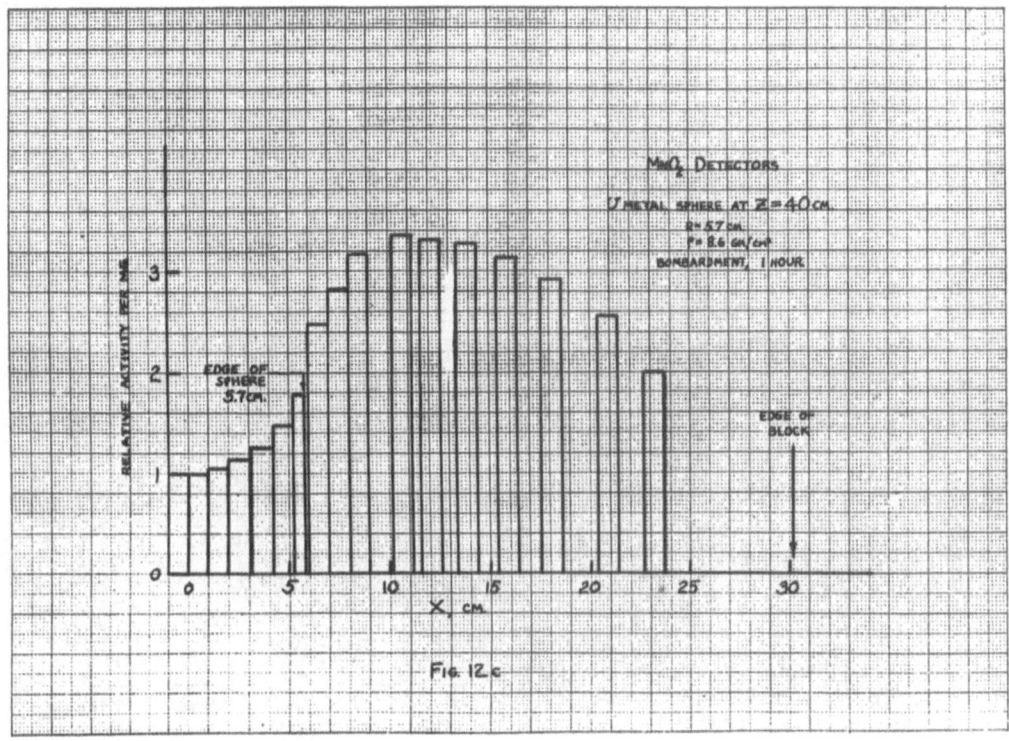

Fig. 12 c

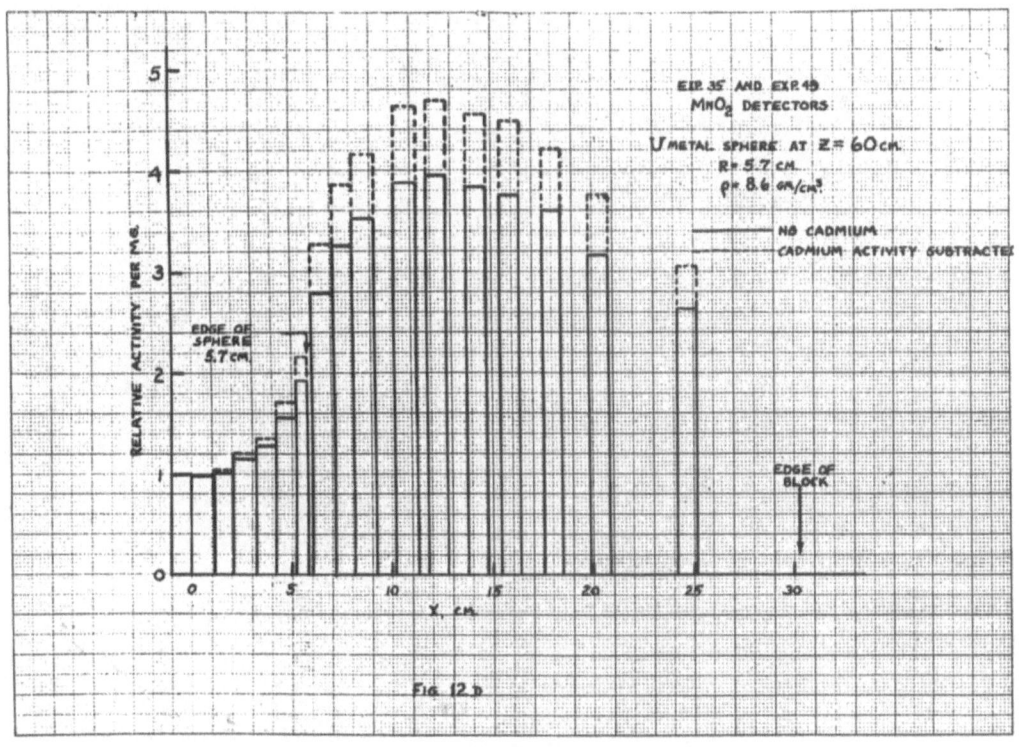

Fig. 12 D

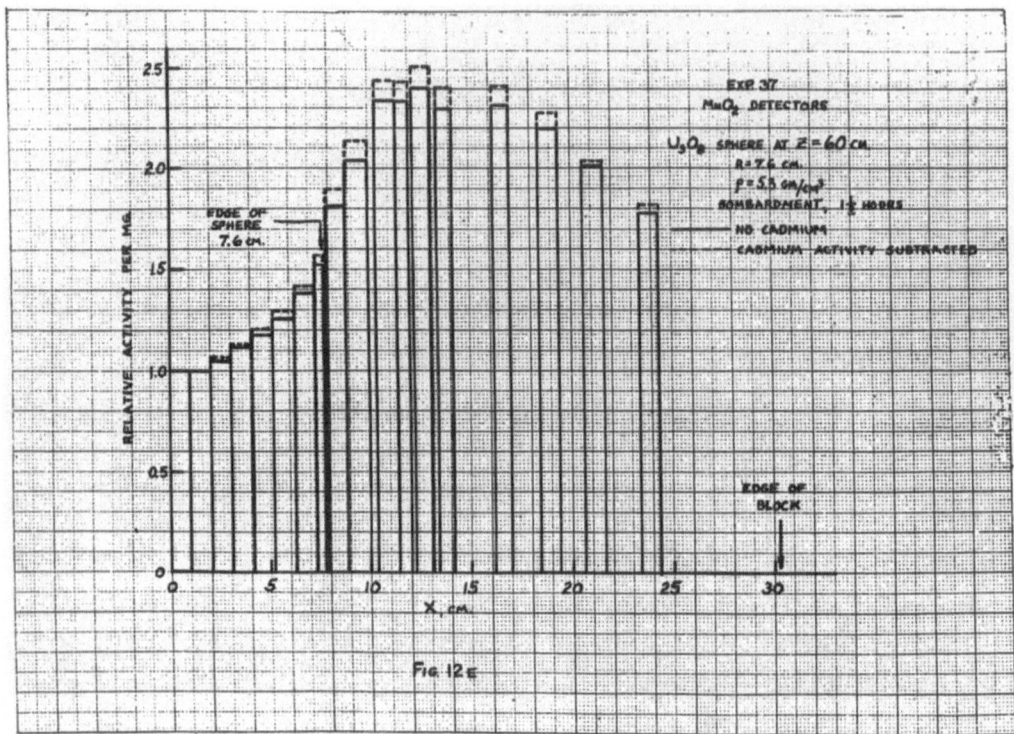

FIG. 12E

10.

Density of Neutrons in Carbon Block
with and Without Absorbing Material

E. P. Wigner, F. L. Friedman, A. T. Monk, G. N. Plass, and A. M. Weinberg

July 16, 1942

The following report is the result of a rewriting of an older note of the first writer which was used on connection with the interpretation of the experiments carried out in Princeton. Most of the work reported here had been done in Princeton but was not officially reported due to the urgency of other work.

The report consists of sections with the following contents:

1. Equation of slowing down with absorption (uses several results of Fermi and of Placzek).
2. Separation of this equation and approximate solution of the absorption equation.
3. Exact solution of the absorption equation with numerical results of Mrs. Monk.
4. Solution of the diffusion equation for the Princeton block.
5. Calculation of the energy distribution of neutrons in the Princeton block, partly on the basis of Mr. Friedman's and Mr. Weinberg's calculations.
6. Density of thermal and epicadmium neutrons, mainly on the basis of Mr. Plass's calculations.
7. Comparison with experimental results obtained at Princeton.

1. Elementary considerations, based on the assumption of spherically symmetric scattering, give for the number $C(r, E)\, dE$ of scattering collisions which the neutrons of the energy range $E, E + dE$ make in unit time and unit volume

$$\frac{1}{3}\frac{\sigma_s}{N^2\sigma^3}\Delta C + \frac{2}{M+1}\left(C + \frac{\sigma_s}{\sigma}E\frac{\partial C}{\partial E}\right) - \frac{\sigma_a}{\sigma}C + P' = 0. \tag{1}$$

Herein, σ_s, σ_a, σ are scattering, absorbing and total cross sections, N the number of atoms per cm³, M the mass of the scattering nucleus, P' the production of neutrons per cm³/sec and unit energy range. If we substitute

$$E = E_0 e^{-\tau}, \quad Q = CE, \quad \sigma_s = \sigma \tag{2}$$

Q will be the number of scattering collisions which those neutrons make in a cm³ and sec for which the logarithm τ of the energy is in an interval of unit length. The subtitution (2) gives

$$\Delta Q - \frac{6}{M+1} N^2 \sigma_s^2 \frac{\partial Q}{\partial \tau} - 3N^2 \sigma_a \sigma_s Q + 3E_0 e^{-\tau} N^2 \sigma_s^2 P' = 0. \tag{3}$$

If we compare this with Fermi's well-known equation

$$\Delta Q = \frac{\partial Q}{\partial \tau'} \qquad \tau' = F \int_E^{E_0} \frac{dE}{N^2 \sigma_s^2 E} \tag{3a}$$

we see that the coefficient of the second term of (3) should be changed to $1/F$. The inaccuracy of this coefficient in (3) results from the inaccuracy of the differential equation (1) which, in particular, neglects the persistence of directions in collisions. Similarly, if we set $M = \infty$, i.e. $F = \infty$, $P' = 0$, equation (3) should go over into the diffusion equation with absorption and a combination of s and p scattering. This means that $N\sqrt{3\sigma_a \sigma_s}$ should be replaced by

$$\kappa = N\sqrt{3\sigma\sigma_a} \left(1 - \frac{2}{5}\frac{\sigma_a}{\sigma}\right)\sqrt{1 - \overline{\cos\Theta}} \tag{4b}$$

$$\overline{\cos\Theta} = 2/3M$$

(Cf. Report A-20). We shall replace in the following, the second factor of κ by 1 because, as we shall see, $\sigma_a \ll \sigma$ holds over the energy region in which we are interested. The resulting equation

$$\Delta Q - \frac{1}{F} N^2 \sigma_s^2 \frac{\partial Q}{\partial \tau} - \kappa^2 Q + 3E_0 e^{-\tau} N^2 \sigma_s^2 P' = 0 \tag{4}$$

$$\frac{1}{F} = 3\xi\left(1 - \frac{2}{3M}\right) \qquad \xi = 1 - \frac{(M-1)^2}{2M}\ln\frac{M+1}{M-1} \tag{4a}$$

is by no means accurate, as probably no differential equation can describe the slowing down process accurately. Nervertheless, we shall adopt it for the following calculations.

2. It is remarkable that (4) permits a "separation of variables". i.e., a decomposition of Q into a product

$$Q = Q_0(r, \tau) S(\tau) \tag{5}$$

where Q_0 is the distribution if the absorption coefficient is zero, and S is the probability of survival which corrects for the finite absorption. Q shall satisfy the equation

$$\Delta Q_0 - \frac{N^2}{F}\sigma_s^2 \frac{\partial Q_0}{\partial \tau} + 3N^2 \sigma_s^2 P = 0 \tag{6}$$

where

$$P = E_0 e^{-\tau} P'$$

is the production of neutrons per cm^3 and sec and unit τ interval. One obtains for $S(\tau)$ if one assumes that $S(\tau) = 1$ over the range in which $P(r, \tau)$ is different from zero

$$N^2 \frac{\sigma_s^2}{F}\frac{\partial S}{\partial \tau} + \kappa^2 S = 0. \tag{7}$$

Hence with (4 a) and (4 b)

$$\ln S = -\int \frac{\sigma_a \sigma}{\xi \sigma_s^2} \left(1 - \frac{2}{5}\frac{\sigma_a}{\sigma}\right)^2 d\tau \approx -\frac{1}{\xi}\int \frac{\sigma_a}{\sigma_s} d\tau. \tag{7a}$$

3. Let us assume now that σ_s is independent of the energy in the region in which σ_a is appreciable and that σ_a is inversely proportional to the velocity

$$\sigma_a = \sigma_s e^{\frac{1}{2}(\tau - \tau_a)} \tag{8}$$

where τ_a is the value of τ for which absorption and scattering become equal. These assumptions correspond to a scatterer and a $1/v$ absorber e.g. carbon with an admixture of boron. We now have from (7 a)

$$\ln S = -\frac{2}{\xi}e^{\frac{1}{2}(\tau - \tau_a)}. \tag{9}$$

The broken line in Fig. 1 gives (9) for $M = 12$, $\xi = .158$.

One can calculate the probability of survival more accurately than has been done above by assuming a distribution which is homogeneous in space. Then, the slowing down is accurately described in a material with mass M in which the scattering is spherically symmetric in the center of mass coordinate system, by the integral equation

$$\left(1 + \sqrt{\frac{E_a}{E}}\right) C(E) = \frac{\eta^2}{\eta^2 - 1} \int_E^{\eta^2 E} \frac{C(E')}{E'} dE'. \tag{10}$$

Herein, E_a is the energy for which absorbing and scattering cross sections are equal, and

$$\eta = \frac{M+1}{M-1}. \tag{10a}$$

Introducing the transformation (2) but writing S for Q of (2), this gives

$$\left(1 + e^{\frac{1}{2}(\tau - \tau_a)}\right) S(\tau) = \frac{\eta^2}{\eta^2 - 1} \int_{\tau - 2\ln \eta}^{\tau} S(\tau') e^{(\tau' - \tau)} d\tau'. \tag{11}$$

The solution of this can be expanded into a series

$$S(\tau) = \sum_{\mu=0}^{\infty} b_\mu e^{\frac{1}{2}\mu(\tau - \tau_a)} \tag{12}$$

where $b_0 = 1$ and

$$b_\mu = \frac{(\mu+2)(\eta^2 - 1)}{\mu(\eta^2 - 1) - 2 + 2\eta^{-\mu}} b_{\mu-1}. \tag{12a}$$

Mrs. A. T. Monk calculated the series (12) as function of $\tau - \tau_a$ and her results are represented in Fig. 1 as a full line. One sees that (9), derived from (4), agrees quite closely with the true value of the probability of survival which gives some confidence to (4).

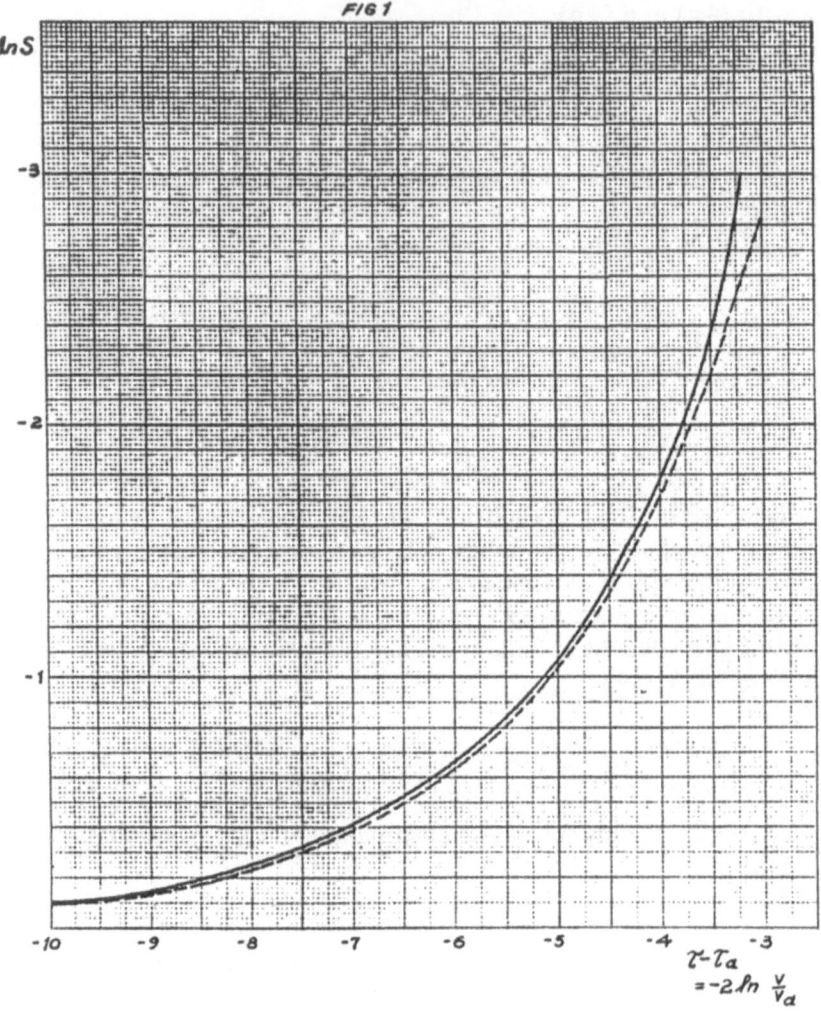

FIG 1

We can apply (9), as a somewhat different example, to air. Let us assume that $\sigma_a = \frac{4}{5} \times 1.3 \times 10^{-24}$ cm^2 for $E = .04$ eV, 1.3 being the absorption cross section of nitrogen. We assume, furthermore, $\sigma_s = \frac{1}{5} \times 4.5 + \frac{4}{5} \times 11 = 9.7 \times 10^{-24}$ cm^2 and $\xi = .130$. Then $1.3 \times \sqrt{.04}/\sqrt{E} = \sigma_a$ and $\sqrt{E_a} = .26/9.7 = .027$ and the number of neutrons which arrive at thermal energies is, according to (9)

$$S = \exp\left(-\frac{2}{.130}\frac{.027}{\sqrt{.04}}\right) = .125$$

i.e. 88% of all neutrons are absorbed in the nitrogen before being slowed down to thermal energies.

4. Let us assume that we have a carbon column the basic sides of which have the height and width a, and the length b. In the actual block used in the

experiments of Creutz and Wilson, the dimensions were $80 \times 80 \times 110$ cm^3. This is best represented by taking a larger than 80 cm by $2\lambda/\sqrt{3}$, i.e. about $a = 83$ cm, similarly $b = 113$ cm. The neutrons enter the quadratic face and we shall assume that their energy is E_0.

Equation (6) can be solved essentially by Fermi's method. One can substitute

$$Q_0 = \sum g_{kk'l}(\tau) \sin\frac{\pi k x}{a} \sin\frac{\pi k' y}{a} \sin\frac{\pi l z}{b} \tag{13}$$

and obtain

$$-\pi^2 \left(\frac{k^2 + k'^2}{a^2} + \frac{l^2}{b^2}\right) g_{kk'l} - \frac{N^2}{F}\sigma_s^2 \frac{\partial g_{kk'l}}{\partial \tau} + 3N^2\sigma_s^2 P_{kk'l} = 0 \tag{13a}$$

in which

$$P = \sum_{kk'l} P_{kk'l} \sin\frac{\pi k x}{a} \sin\frac{\pi k' y}{a} \sin\frac{\pi l z}{b} . \tag{13b}$$

One now substitutes Fermi's "age" τ'

$$\tau' = F \int_0^\tau \frac{d\tau}{N^2\sigma_s^2} = 2.23 \int_0^\tau \lambda^2 \, d\tau$$

which gives

$$\frac{\partial g_{kk'l}}{\partial \tau'} = (3N^2\sigma_s^2 P_{kk'l})\left[-\pi^2 \left(\frac{k^2 + k'^2}{a^2} + \frac{l^2}{b^2}\right)g_{kk'l}\right] . \tag{14}$$

Let us assume that the production of neutrons corresponds to an incoming beam of neutrons of energy E_0 with a single sin-sin distribution.

$$P(x, y, z, \tau) = \delta(\tau)\sin\frac{\pi k x}{a} \sin\frac{\pi k' y}{a} e^{-z/\lambda}/\lambda, \tag{15}$$

with λ of the order of the mean free path $1/N\sigma_s$. Then

$$P_{kk'l} = \frac{2\pi l \lambda}{b^2 + \pi^2 l^2 \lambda^2}\delta(\tau) = \frac{2\pi l \lambda}{b^2 + \pi^2 l^2 N^2 \sigma_s^2}\delta(\tau') . \tag{15a}$$

This gives for $\tau' > 0$

$$g_{kk'l} = \frac{6\pi l \lambda F}{b^2 + \pi^2 l^2 \lambda^2} \exp\left[-\pi^2 \tau' \left(\frac{k^2 + k'^2}{a^2} + \frac{l^2}{b^2}\right)\right] \tag{16}$$

$$Q_0 = \sum_{l=0}^\infty \frac{6\pi l \lambda F}{b^2 + \pi^2 l^2 \lambda^2} \sin\frac{\pi k x}{a} \sin\frac{\pi k' y}{a} \sin\frac{\pi l z}{b} \exp\left[-\pi^2 \tau' \left(\frac{k^2 + k'^2}{a^2} + \frac{l^2}{b^2}\right)\right] \tag{16a}$$

while, of course, $Q_0 = 0$ for $\tau' < 0$.

The summation over l can be carried out in (16a) by means of Poisson's summation formula. The $\pi^2 l^2 \lambda^2$ will play a role in (16a) only for l for which it is of the order of b^2. However, if $\pi^2 l^2 \lambda^2 \approx b^2$, the exponent $\pi^2 \tau' l^2/b^2$ will be of the order of τ'/λ^2 and the corresponding terms extremely small, except for very small τ'. Hence, we can write

$$Q_0 = -\frac{3\lambda F}{b} \sin \frac{\pi k x}{a} \sin \frac{\pi k' y}{a} \sum_{l=-\infty}^{\infty} \frac{\partial}{\partial z} \exp\left[\frac{i\pi l z}{b} - \pi^2 \tau' \left(\frac{k^2 + k'^2}{a^2} + \frac{l^2}{b^2}\right)\right]$$

$$= -\frac{3\lambda F}{b} \sin \frac{\pi k x}{a} \sin \frac{\pi k' y}{a} \frac{\partial}{\partial z}$$

$$\times \sum_n \int_{-\infty}^{\infty} dl \exp\left[2\pi i l n + \frac{i\pi l z}{b} - \pi^2 \tau' \left(\frac{k^2 + k'^2}{a^2} + \frac{l^2}{b^2}\right)\right]$$

$$= -\frac{3\lambda F}{b} \sin \frac{\pi k x}{a} \sin \frac{\pi k' y}{a} \sum_n \frac{b}{\sqrt{\pi\tau'}} \frac{\partial}{\partial z} \exp\left[-\pi^2 \tau' \frac{k^2 + k'^2}{a^2} - \frac{(2nb+z)^2}{4\tau'}\right].$$

Thus

$$Q_0 = \frac{3\lambda F}{\sqrt{\pi}\tau'^{3/2}} \sin \frac{\pi k x}{a} \sin \frac{\pi k' y}{a}$$

$$\times \sum_n \left(\frac{z}{2} + nb\right) \exp\left[-\frac{\pi^2 \tau'(k^2 + k'^2)}{a^2} - \frac{(2nb+z)^2}{4\tau'}\right]. \tag{17}$$

Apart from $z = b$, only the first term ($n = 0$) plays any role in (17). Around $z = b$, the $n = -1$ term also becomes significant and corresponds to the well known reflection with opposite sign across this boundary, similar to the one occurring in the theory of the exponential pile.

Equation (17) shows that the total intensity is proportional to λ, i.e. the penetration of the primary neutrons into the pile. This is natural, since the neutrons which are deeper inside the pile have a greater chance not to escape through the front surface. In fact, (17) can be considered as the difference of two Gauss error distributions with centers at $z = \lambda$ and $z = -\lambda$.

Evidently, if the incident neutron beam does not have a simple sin-sin distribution in the xy plane, the neutron density in the inside will be also a superposition of functions of the form (17) with different k and k'.

It is remarkable, that the distribution of neutrons of a certain energy (i.e. definite τ') is independent of the transverse distribution, i.e. of k and k'. However, the ratio of neutrons with different energies at the same point depends on k and k' and the low energy side in the spectrum is, at every point, strongest for the lowest values of k and k'. This goes so far that the $k = k' = 1$ is the only appreciable component for lower energies, practically independently of the distribution of the different components in the incoming beam. For this reason we can restrict ourselves henceforth to $k = k' = 1$ in (17).

5. The neutron density in (17) is not directly given as a function of energy but rather, as a function of Fermi's age τ'. It is necessary, therefore, to establish a connection between both and this must be done by integrating (3 a). This is done in Fig. 4, on the basis of the data on cross sections collected in Fig. 3, for graphite of density 1.64. In the high energy region above 1 MEV, the formula

$$\tau' = \frac{1}{3\xi} \int \frac{dE}{E N^2 \sigma_s^2 (1 - \overline{\cos \Theta})} = 2.11 \int \frac{\lambda^2 d\tau}{(1 - \overline{\cos \Theta})} \tag{18}$$

was used and the average cosine of the angle between incident and scattered neutron estimated, on the basis of Friedman's report, to be .15. The ordinate in Fig. 4 is the age difference between neutrons of energy E (abscissa $\ln E$) and Indium resonance neutrons of energy 1.35 eV (Report C-25 by Bacher, Baker, and McDaniel). The two upper curves of Fig. 4 refer to two different assumptions for the cross section. On the basis of available experimental evidence, there is no reason to favor either curve.

Figure 2 gives the density of In resonance neutrons, according to experiments by Creutz and Wilson. The zero of the abscissa corresponds to the front of the block. However, the zero of z must be taken about $\lambda/\sqrt{3}$ to the left of this, i.e. at about the point where the curve (which is extrapolated in this region) intersects with the abscissa.

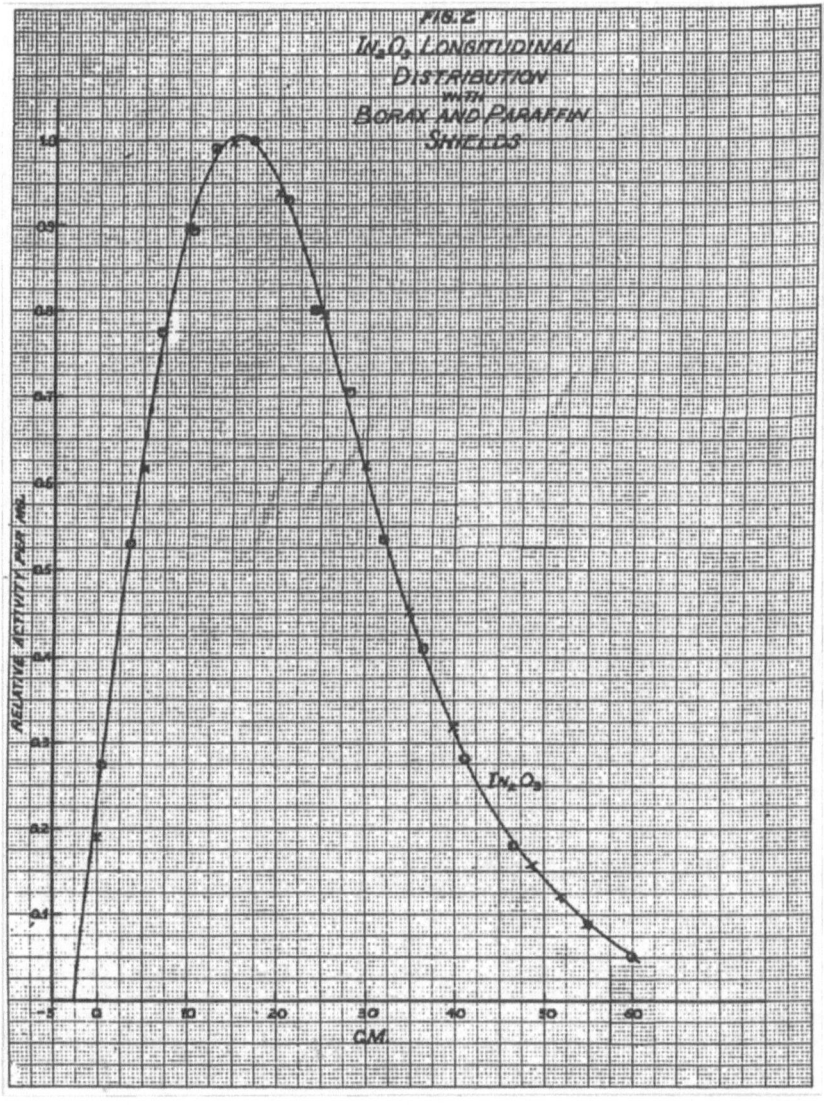

One can soon convince oneself that it is impossible to represent Fig. 2 by a single $cz \exp(-z^2/4\tau')$ curve. This would have to be possible if the incident neutrons were monoenergetic (the $n \neq 0$ terms of (17) are entirely negligible up to 60 cm). Mr. Weinberg gave a good analytic representation of Fig. 2 by

$$c(\exp(-z^2/4 \times 109) + .7\exp(-z^2/4 \times 246)). \tag{19}$$

The circles are measured values in Fig. 2, the crosses calculated by (19). This representation of the curve of Fig. 2 is, of course, to some extent arbitrary but one easily convinces oneself that one cannot change the constants of (19) greatly without introducing a considerable discrepancy. This is quite remarkable because Fig. 4 shows that (19) indicates incident neutron energies of 2.5 and 1200 keV. The former of these is very low indeed as compared with an estimated primary neutron energy of 6000 keV. Some of the neutrons may not have come directly from the Be target. However, neutrons below 200 eV hardly could have penetrated the 6.5 gr/cm^2 B$_2$O$_3$ shield which was in front of the block and although the energies are evidently averages over very wide bands, the lower value appears very remarkable. One recalls, in this connection, Fermi's decomposition of somewhat similar data relating to RaBe neutrons in which the age differences to In resonance neutrons were 130, 342, and 815 cm^2. The first two of these correspond to 11.5 and 200 keV primary neutron energy, respectively. Similar data, also by Fermi and Anderson, on the RaBe photo neutrons, indicate primary neutron energies of 5.3 and 750 keV.

The interpretation that some of the neutrons measured by the In resonance method have considerably higher energies does not appear very probable since Creutz found that 3 gr/cm^2 of B$_4$C decrease the activity of In resonance detectors to less than 1%. Particularly because of the similarity of the results obtained by Creutz and Wilson and by Fermi and Anderson, one is tempted to attribute the anomaly to the graphite.

6. Assuming that the reason for the presence of two terms in (19) is the polyenergetic nature of the incident neutrons we can give, by (17), an expression for the density of neutrons in the block as function of their energy. For this purpose the constants in two expressions of the form (17) have to be adjusted in such a way as to give (19) for In resonance neutrons. We can drop all terms in (17) apart from the $n = 0$ term and also assume $k = k' = 1$ in the low energy region. We obtain

$$Q_0 = cz \sin\frac{\pi x}{83} \sin\frac{\pi y}{83} \left\{ \frac{1550}{(\tau''+109)^{3/2}} \exp\left(-\frac{2\pi^2(\tau''+109)}{83^2} - \frac{z^2}{4(\tau''+109)}\right) \right.$$
$$\left. + \frac{5500}{(\tau''+246)^{3/2}} \exp\left(-\frac{2\pi^2(\tau''+246)}{83^2} - \frac{z^2}{4(\tau''+246)}\right) \right\} \tag{19 a}$$

where τ'' represents the age difference with respect to In resonance neutrons. The connection between τ'' and the energy is given in Fig. 4.

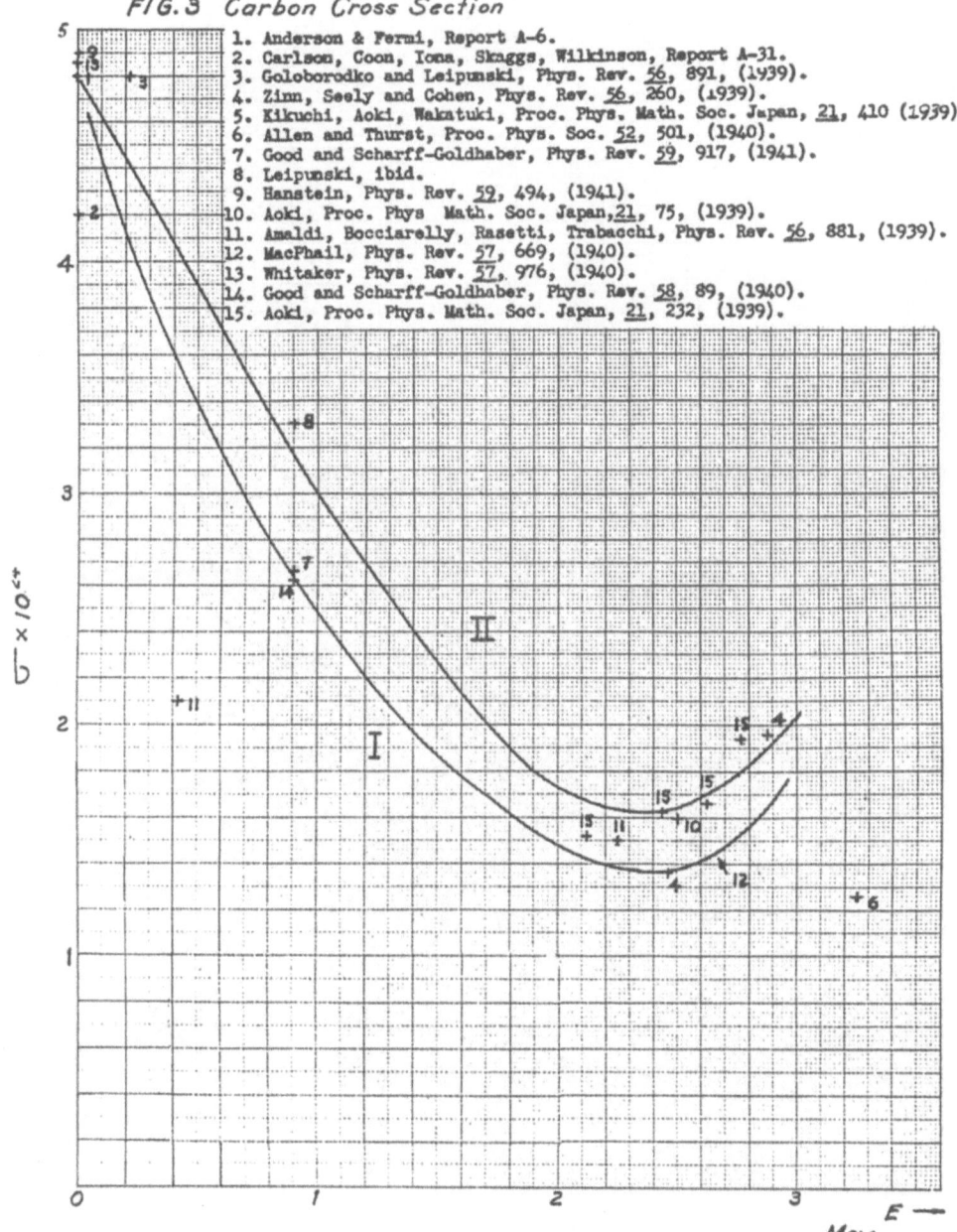

FIG.3 Carbon Cross Section

1. Anderson & Fermi, Report A-6.
2. Carlson, Coon, Iona, Skaggs, Wilkinson, Report A-31.
3. Goloborodko and Leipunski, Phys. Rev. 56, 891, (1939).
4. Zinn, Seely and Cohen, Phys. Rev. 56, 260, (1939).
5. Kikuchi, Aoki, Wakatuki, Proc. Phys. Math. Soc. Japan, 21, 410 (1939).
6. Allen and Thurst, Proc. Phys. Soc. 52, 501, (1940).
7. Good and Scharff-Goldhaber, Phys. Rev. 59, 917, (1941).
8. Leipunski, ibid.
9. Hanstein, Phys. Rev. 59, 494, (1941).
10. Aoki, Proc. Phys Math. Soc. Japan, 21, 75, (1939).
11. Amaldi, Bocciarelly, Rasetti, Trabacchi, Phys. Rev. 56, 881, (1939).
12. MacPhail, Phys. Rev. 57, 669, (1940).
13. Whitaker, Phys. Rev. 57, 976, (1940).
14. Good and Scharff-Goldhaber, Phys. Rev. 58, 89, (1940).
15. Aoki, Proc. Phys. Math. Soc. Japan, 21, 232, (1939).

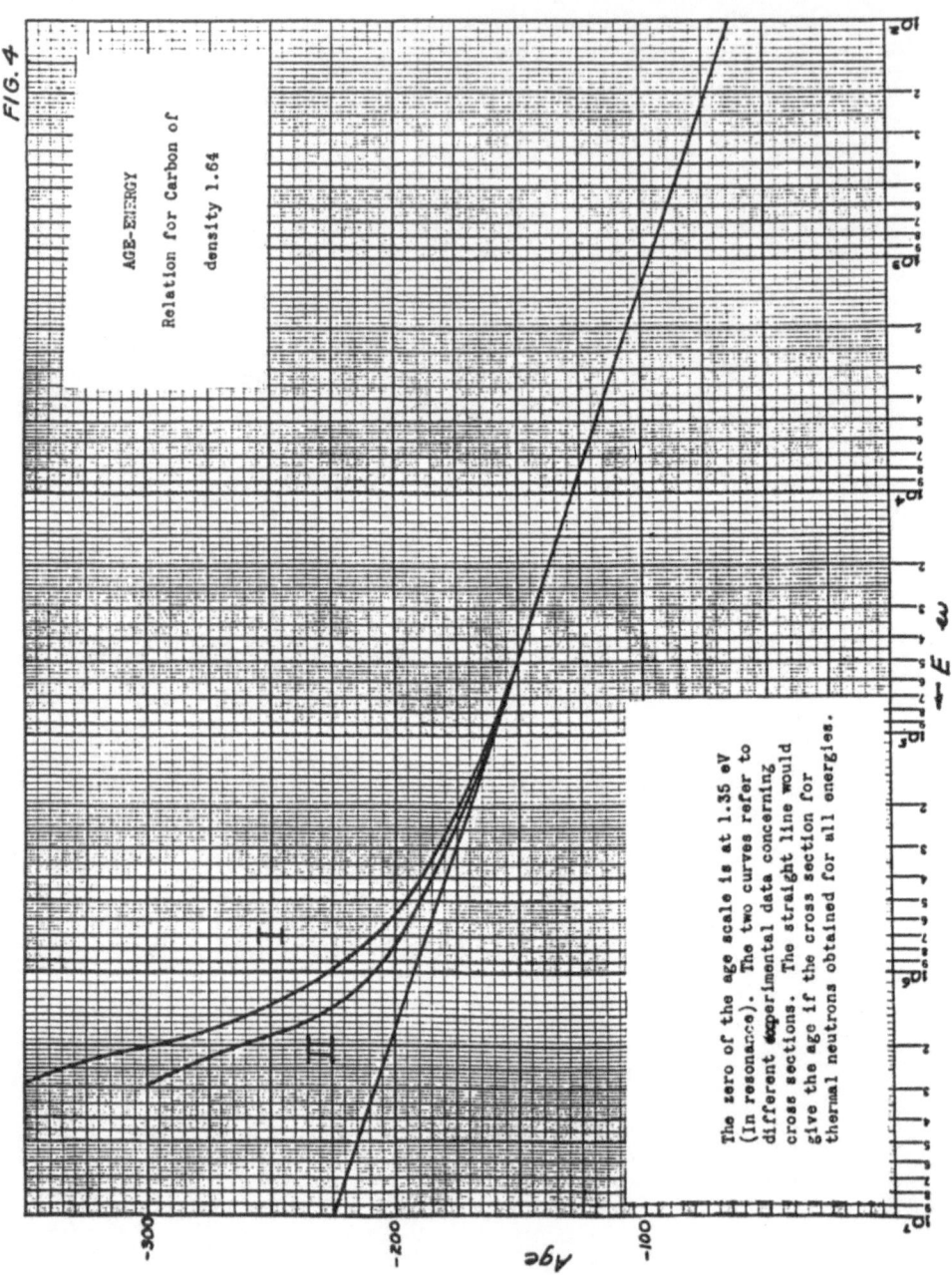

FIG. 4

AGE-ENERGY

Relation for Carbon of

density 1.64

The zero of the age scale is at 1.35 eV
(in resonance). The two curves refer to
different experimental data concerning
cross sections. The straight line would
give the age if the cross section for
thermal neutrons obtained for all energies.

This shows that, since the scattering cross section below about 100 keV is constant, there is a linear relationship between $\ln E$ and τ or τ'' in this region:

$$\tau'' = -14.5 \ln(E/1.35) = 14.5(\tau - \tau_{1.35}).$$ (20)

This can be substituted into (19 a) to give the density of collisions per logarithmic energy interval as a function of τ:

$$Q_0 = cz \sin \frac{\pi x}{83} \sin \frac{\pi y}{83} \left\{ \frac{28}{(t+7.5)^{3/2}} \exp\left[-.0415(t+7.5) - \frac{z^2}{58(t+7.5)}\right] \right.$$
$$\left. + \frac{99.5}{(t+17)^{3/2}} \exp\left[-.0415(t+17) - \frac{z^2}{58(t+17)}\right] \right\}.$$ (20 a)

where

$$t = -\ln(E/1.35) = \tau - \tau_{1.35}.$$ (20 b)

One can use (20 a) to determine the energy spectrum at different distances from the block. Inserting 37, 42 and 47 into (20 a), one obtains the spectrum at 35, 40 45 cm from the front of the block.

Since there are no neutrons with energies greater than those corresponding to $t < -7.5$ in the group of neutrons represented by the first term of (20 a), this term should be taken to be zero when $t < -7.5$. A similar remark applies to the second term when $t < -17$. The results are given in Fig. 5. The experiments on the resonance absorption should be carried out where the neutron distribution is most nearly "natural" in the relevant region (from 5 eV to 15000 eV), i.e. where Q_0 is, between $t = 1.3$ and $t = 9.3$, most nearly independent of t.

7. The original purpose of the experiments in Princeton was to study the resonance absorption with neutrons of relatively well defined energies. It was proposed to obtain these neutrons by adding different amounts of boron to the carbon in the block in order to absorb the neutrons below a certain energy. The neutrons above a somewhat higher energy limit were to be excluded by carrying out the experiments at such values of z at which the number of high energy neutrons is already negligible. Although it was not possible to carry out this plan fully, it remains of some interest to see to what extent it could be carried out.

According to (5), the number of collisions of neutrons at a given point is the product of the survival probability S and of Q_0. The logarithm of this number is the difference between $\ln Q_0$ and $\ln S$ i.e. the ordinates of Figs. 5 and 1 for corresponding abscissae.

Hence, if one places the two figures on each other so that one can read both simultaneously, the difference between the ordinates gives directly the logarithm of the density of neutrons as function of their energy. The position of the abscissae must be adjusted so that the abscissa 0 of Fig. 1 should fall over that energy for which absorption and scattering coefficients are equal. Since, according to Baker, Bacher, and McDaniel, and Manley, Haworth, and Luebke, (Reports C-19 and C-22), the absorption cross section of a B atom can be

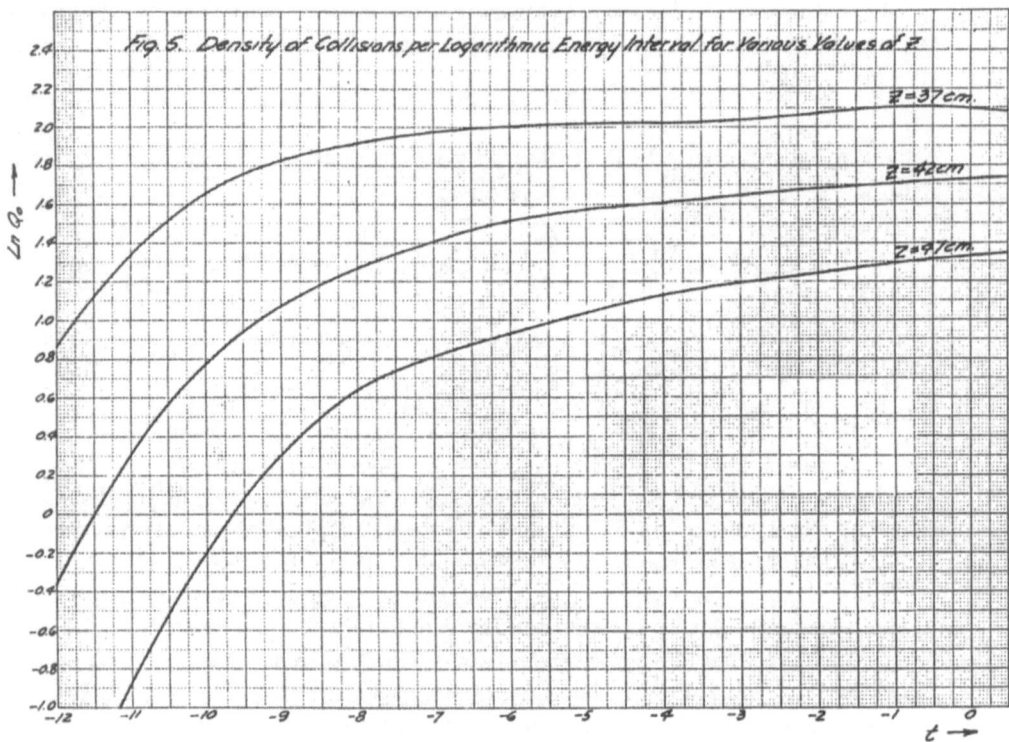

Fig. 5. Density of Collisions per Logarithmic Energy Interval for Various Values of z

represented as $(116/\sqrt{E})10^{-24}$ cm^2 where E is measured in eV, we have, if N_B is the number of B atoms per cm^3.

$$N_B\sigma_a = \frac{115N_B}{\sqrt{E}} \times 10^{-24} = 10^{-22}N_Be^{\frac{1}{2}t}. \qquad (21)$$

Hence, $N_B\sigma_a = N\sigma_s = 1/2.55$ for

$$t_a = 2\ln\frac{10^{22}}{2.55N_B} \qquad (21\,\text{a})$$

and the abscissa $\tau - \tau_a = 0$ of Fig. 1 must be placed to coincide with the abscissa (21 a) of Fig. 5.

A few trials will convince one that the above filtering technique permits one to obtain neutron bands with a half width of about 1 in the τ scale. It would be very difficult to achieve much more although the use of a monoenergetic primary beam of lower energy would improve the situation considerably. [Cf. M. Goldhaber and R. D. O'Neal, Phys. Rev. 60, 834, 1941 and a later (unpublished) article by R. D. O'Neal.] Because of the natural limitation on N_B the highest energy for which such a band could be obtained is about 50,000 eV. For an energy of this magnitude, so much B$_4$C must be added to the carbon that the scattering by the B is not negligible any more. However, in the above estimates, the (unknown) scattering by the B was disregarded.

8. The estimates of the preceding section 7 tacitly assume an absorption of such magnitude that practically no neutrons can come down to the thermal region. The present section deals with the thermal and so-called epicadmium neutrons.

Naturally, both (9) and (20 a) cease to be valid as soon as the neutron energy approaches the average energy of the atoms. This is true even if one disregards the effects of chemical binding since the neutron energy naturally approaches the thermal energy, only asymptotically. Fermi lately adduced strong evidence that the approach to thermal energy may be even slower in a graphite crystal. This means that the effective age of thermal neutrons is even higher than the preceding calculation indicates. However, this age difference plays only a subordinate role in what follows and the crystal effect will be disregarded in the following.

One can calculate the density of thermal neutrons on the basis of the following consideration.

The path of the thermal neutrons will be the same as if their τ continued to increase (their energy decrease) even beyond the value τ_{thermal}. Hence we can use (4) and (20 a) to calculate the number of collisions they suffer at any point. The only change in their behavior, as compared with the assumptions under which (4) and (20 a) were derived, is that they go through their path more rapidly than they would if they continued to lose energy even beyond the thermal energy point. This, however, does not change the number of collisions which they suffer before they leave the system through one of the boundaries of the block. However, the probability of survival S must be calculated with an absorbing cross section σ_{a0} which is independent of τ and is the absorbing cross section of the block for neutrons. This gives, according to (7)

$$S_0 = \exp\left(-\frac{\kappa^2 F}{N^2 \sigma_s^2}(\tau - \tau_{\text{thermal}})\right) = \exp(-.0072(\tau - \tau_{\text{thermal}})) \tag{22}$$

if we neglect the absorption above the thermal energy. We have set in (22) $\kappa = .0222$, $N\sigma_s = 1/2.55$, $F = 2.23$. If we refer, furthermore, the τ to the indium resonance energy, i.e. introduce the t of (20 b) and use $E = .04$ eV for the thermal energy, this becomes

$$S_0 = 1.025 \exp(-.0072t). \tag{22 a}$$

According to the preceding consideration, we obtain the number of collisions by thermal neutrons per cm^3 and sec by multiplying (20 a) with (22 a) and integrating the product from the t corresponding to thermal neutrons to infinity.

$$Q_t = c \sin\frac{\pi x}{83} \sin\frac{\pi y}{83} Q_t^* \tag{23}$$

$$Q_t^* = \int_{3.5}^{\infty} \frac{28.8z}{(t+7.5)^{3/2}} \exp\left[-.0487(t+7.5) - \frac{z^2}{58(t+7.5)}\right] dt$$

$$+ \int_{3.5}^{\infty} \frac{102z}{(t+17)^{3/2}} \exp\left[-.0487(t+17) - \frac{z^2}{58(t+17)}\right] dt. \tag{23 a}$$

The t corresponding to thermal energy is $\ln 1.35/.04 = 3.52$.

The coefficient of t from (20 a) is $-.0415$, from (22 a) $-.0072$. The first corresponds to the escape of thermal neutrons through the side of the block, the second to the absorption in the block. One sees that the former effect is almost six times greater than the second. In a block with a base of about 200×200 cm, the two effects would be almost equal. In the Princeton block, however, the ratio of absorption by carbon to other losses had about the same value which it has in the piles.

Equation (23 a) contains integrals of the form

$$I = \sqrt{t_0} \int_{t_0}^{\infty} \frac{\exp(-\alpha^2 t/t_0 - \beta^2 t_0/t)}{2t^{3/2}} \, dt \qquad (24)$$

which, according to Binet, Plass and Weinberg, can be evaluated as follows. One first substitutes $\sqrt{t_0/t} = x$ and obtains

$$I = \int_0^1 \exp(-\alpha^2/x^2 - \beta^2 x^2) \, dx \qquad (24\,\mathrm{a})$$

which gives (C.R. *12*, 958 (1841))

$$I = \frac{\sqrt{\pi}}{4\beta} \left[e^{-2\alpha\beta}(1 + J(\beta - \alpha)) - e^{2\alpha\beta}(1 - J(\beta + \alpha)) \right] \qquad (24\,\mathrm{b})$$

where α and β are assumed to be positive and J is the probability integral (an odd function)

$$J(q) = \frac{2}{\sqrt{\pi}} \int_0^q e^{-x^2} \, dx . \qquad (24\,\mathrm{c})$$

The integrals of (23 a) were evaluated in this way and plotted in Fig. 6 against z to give the number of collisions of thermal neutrons per cm^3 and sec. Curves I and II represent the first and second terms of (23 a) respectively, their sum giving Q_t^*. If (23 a) is multiplied by σ_a/σ_s it gives the number of thermal neutrons absorbed per cm^3 and sec.

9. The epicadmium activation of a $1/v$ detector can be calculated from (20 a) very easily if one assumes that the Cd cutoff is sharp, i.e. that the usual Cd shield of .045 cm thickness is impermeable to neutrons below this energy and completely transparent to neutrons of higher energies. In this case, we can multiply the Q_o of (20 a) with the sensitivity of the $1/v$ detector for which we take $\exp \frac{1}{2}(\tau - \tau_{\text{thermal}}) = .172 \exp \frac{1}{2} t$ and integrate over t from a low value (corresponding to a high energy) to τ_{Cd} corresponding to the Cd cutoff. The above assumption for the sensitivity corresponds to unit sensitivity in the thermal region so that the ratio of the integral to be obtained to (23) is equal to the ratio of activations by epicadmium to that by thermal neutrons. We have for the epicadmium activity

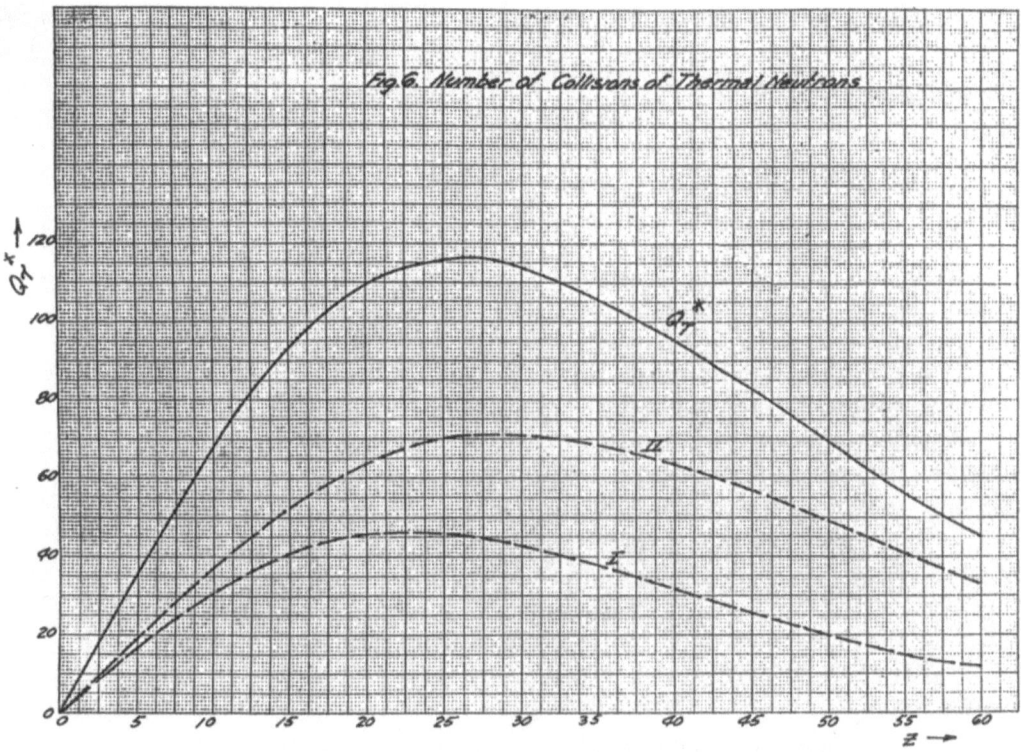

Fig. 6. Number of Collisions of Thermal Neutrons

$$A_{\text{Cd}} = c \sin \frac{\pi x}{83} \sin \frac{\pi y}{83} A_{\text{Cd}}^* \tag{25}$$

$$A_{\text{Cd}}^* = \int^{1.1} \frac{4.8z}{(t+7.5)^{3/2}} \exp\left[.458t - .312 - \frac{z^2}{58(t+7.5)}\right] dt$$

$$+ \int^{1.1} \frac{17.1z}{(t+17)^{3/2}} \exp\left[.458t - .705 - \frac{z^2}{58(t+17)}\right] dt. \tag{25 a}$$

The upper limit in the integrals (25 a) corresponds to the t of the Cd limit .45 eV. The lower limit corresponds to some suitable energy for which the exponential is already very small (because of the low sensitivity of the detector).

One can evaluate the integrals of (25 a) in the following way. One first substitutes $\zeta = 1.1 - t$, which gives

$$A_{\text{Cd}}^* = \int_0 \frac{5.8z}{(8.6 - \zeta)^{3/2}} \exp\left[-.458\zeta - \frac{z^2}{58(8.6 - \zeta)}\right] d\zeta$$

$$+ \int_0 \frac{14z}{(18.1 - \zeta)^{3/2}} \exp\left[-.458\zeta - \frac{z^2}{58(18.1 - \zeta)}\right] d\zeta. \tag{25 b}$$

This can be written in the equivalent form

$$A_{Cd}^* = 0.230ze^{-0.00200z^2} \int_0^\zeta \frac{1}{(1 - \frac{\zeta}{8.6})^{3/2}} \exp\left[-.458\zeta - \frac{z^2\zeta}{499(8.6 - \zeta)}\right] d\zeta$$
$$+ 0.182ze^{-0.000953z^2} \int_0^\zeta \frac{1}{(1 - \frac{\zeta}{18.1})^{3/2}} \exp\left[-.458\zeta - \frac{z^2\zeta}{1050(18.1 - \zeta)}\right] d\zeta .$$

In the first integral, the last term in the exponential may be written $\frac{z^2\zeta}{499 \times 8.6}$ neglecting the ζ in the denominator. This is a valid approximation for all z of interest in the block ($z < 60$), since for smaller values of ζ, the ζ in the denominator makes a negligible contribution to the total value of the exponential and for large values of ζ, the exponential becomes very small.

A similar argument will justify omitting the corresponding ζ in the second integral. The equation can now be written as

$$A_{Cd}^* = 0.230\, ze^{-0.00200z^2} \int_0^\zeta \frac{1}{(1 - \frac{\zeta}{8.6})^{3/2}} e^{-\beta\zeta} d\zeta$$
$$+ 0.182\, ze^{-0.000953z^2} \int_0^\zeta \frac{1}{(1 - \frac{\zeta}{18.1})^{3/2}} e^{-\gamma\zeta} d\zeta , \qquad (26)$$

where

$$\beta = 0.458 + 0.000232z^2 , \quad \gamma = 0.458 + 0.0000526z^2 . \qquad (27)$$

The above integrations can be done approximately by integrating by parts once and substituting in the remaining integrals $e^{\frac{5}{2}\frac{\zeta}{8.6}}$ and $e^{\frac{5}{2}\frac{\zeta}{18.1}}$ for $(1 - \frac{\zeta}{8.6})^{-\frac{5}{2}}$ and $(1 - \frac{\zeta}{18.1})^{-\frac{5}{2}}$ respectively. One thus obtains as a convenient expression for A_{Cd}^*,

$$A_{Cd}^* = \frac{0.230ze^{-0.002z^2}}{\beta} \frac{(17.2\beta - 2)}{(17.2\beta - 5)}$$
$$+ \frac{0.182ze^{-0.000953z^2}}{\gamma} \frac{(36.2\gamma - 2)}{(36.2\gamma - 5)} , \qquad (28)$$

where β and γ are given by (27). In Fig. 7 the curve marked A_{Cd}^* is a plot of equation (28) showing the epicadmium activity as a function of z. The curves I and II refer to the first and second terms of (28) respectively.

Instead of the sharp cutoff at 0.45 eV, assumed in the previous work, the transmission of cadmium is actually as shown in Fig. 8 (Baker and Bacher, Phys. Rev. *59*, 332 (1941)). Due to instrumental difficulties their curve is very uncertain for energies greater than 1 eV. For this reason the curve of Fig. 8 has been changed somewhat from Baker and Bacher's in the region above 1 eV in order to bring it more nearly in agreement with theory. At no place does the curve of Fig. 8 deviate from an experimental point by more than the probable error of this point.

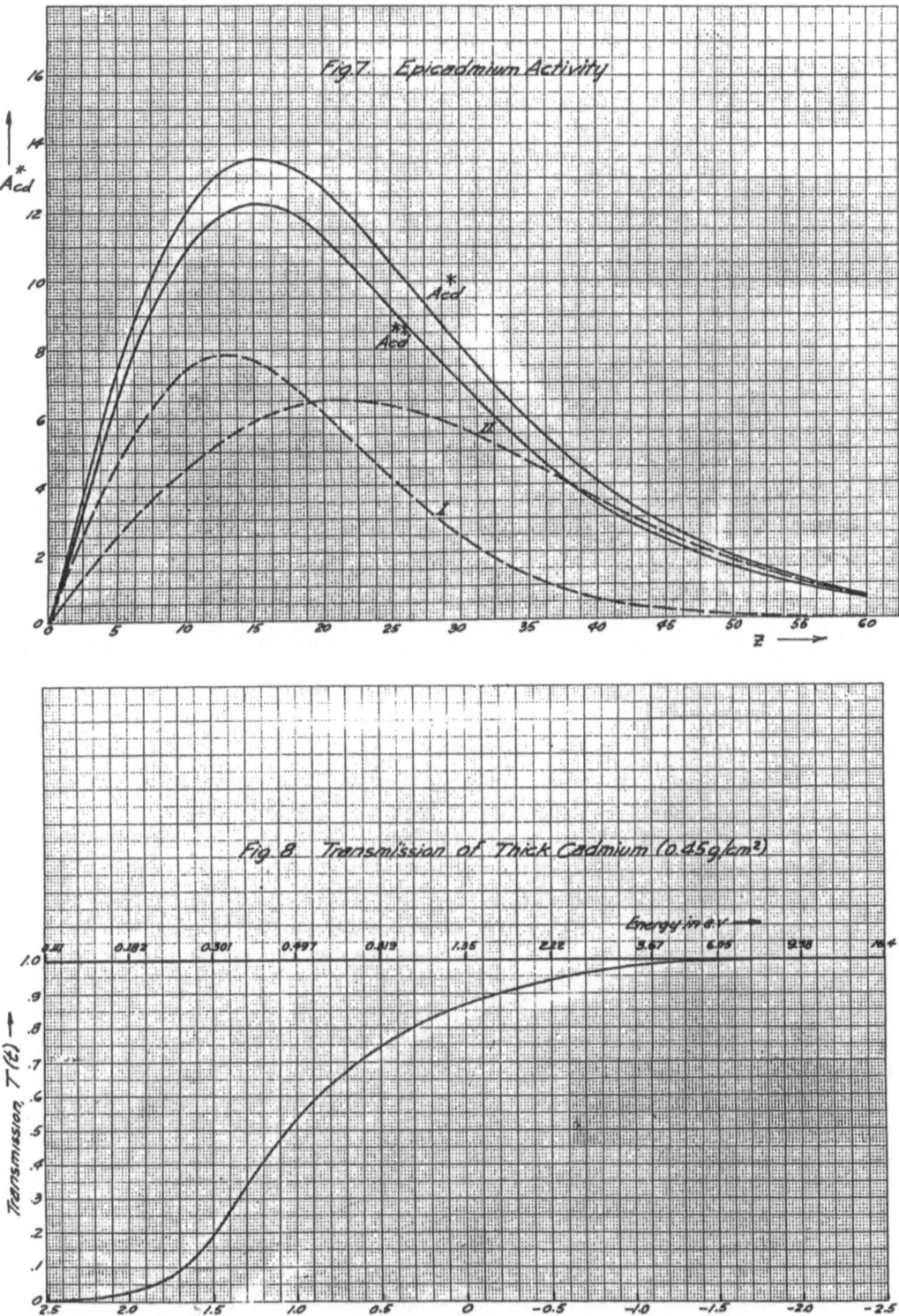

The correction that this will introduce in equation (25 a) can be calculated in the following manner. Let the transmission curve as given in Fig. 8 be represented by $T(t)$ and let the integral of one of the integrals of (25 a) be $I(t)$. Then the correction to this integral can be written as

$$C = \int_{-2.5}^{2.5} T(t)I(t)\, dt - \int_{-2.5}^{1.1} I(t)\, dt\,. \tag{29}$$

(The interval -2.5 to 2.5 is chosen to be sufficiently large so that a further increase of it would not affect the results.)

Expanding $I(t)$ in a Taylor's series around the point 1.1 and rearranging the terms gives

$$C = I(1.1)\left\{\int_{-2.5}^{1.1}[T(t) - 1]\, dt + \int_{1.1}^{2.5} T(t)\, dt\right\}$$
$$+ I'(1.1)\left\{\int_{-2.5}^{1.1}[t - 1.1][T(t) - 1]\, dt + \int_{1.1}^{2.5}[t - 1.1]T(t)\, dt\right\}\,. \tag{30}$$

After evaluating the necessary integrals by numerical integration and substituting for $I(1.1)$ and $I'(1.1)$, it is found that the correction to be applied to both terms of (25 a) is

$$C = -0.0649\, ze^{-0.002z^2}[1 + 0.000278z^2]$$
$$- 0.0558\, ze^{-0.000953z^2}[1 + 0.0000493z^2]\,. \tag{31}$$

When this correction is made to A_{Cd}^*, the curve marked A_{Cd}^{**} is obtained. Thus the correction is seen to reduce considerably the amount of the activity, but does not shift the position of the maximum appreciably.

The transmission curve for the cadmium used in these experiments would actually rise more slowly as the higher energies are approached than the one shown on Fig. 8. This is because Baker and Bacher made their measurements with the neutrons incident perpendicularly on the cadmium sheet; in the experiments they were incident from all angles, thus causing a greater absorption at a given energy. Thus a further correction of an unknown amount should be made which would act in the same direction as the previous correction.

10. The calculations of the last two sections were made in order to interpret the experiments on the $1/v$ law of fission. These experiments were carried out toward the end of 1941 in Princeton by Creutz, Wilson and Kamm and were reported this year under C-109. The investigation was motivated by the surprisingly small ratio of thermal activity to epicadmium activity which appeared to indicate a deviation from the $1/v$ law for energies as low as .5 eV. The suspicion of such a deviation (at least to any considerable extent) was dispelled by the experiments and their interpretation to be gone through below. Unfortunately, due to the constant pressure of other work, the experiments are less accurate than they would have been made had more time been available.

Let us first compare the distribution of thermal neutrons, i.e. our Fig. 6 with the difference of curves A and B of Fig. 2, C-109. The points on curve

A scatter considerably and, of course, it is not altogether justified to consider all neutrons as thermal which are below the Cd limit. Furthermore, as Fig. 4, C-109 shows, the curve A should be increased by about 18% as pointed out there.

The question arises next as to the point (value of z) to which the measurements refer. The abscissae in Fig. 4 refer to the front of the fission chamber and the sensitive region extends only about 1 cm to the back. However, the neutron density in the whole detecting apparatus – which was about 8 cm long and had a diameter of 5 cm – is different from what it would be if its place were filled with graphite. From the point of view of neutron distribution, the whole chamber is a hole with very little scattering material in it. The neutron density in such a hole can be roughly estimated by assuming that the neutron distribution outside the hole is the same as if the hole were not present. Then, the density of the neutrons inside is approximately the average of the neutron densities on the boundaries of the hole, each boundary being given a weight proportional to the solid angle under which it appears from the point at which we want to obtain the neutron density. We estimated in this way that the density measured in the fission chamber corresponds to the density prevailing in the bulk of the graphite about 2.5 cm farther back than the front of the chamber was. Hence, the z of Fig. 4, C-109 must be increased by this value when comparing it with the calculations. A further correction of 2 cm must be applied because of the $\lambda/\sqrt{3}$ correction mentioned before and we compared, therefore, in Table I the experimental data at a given z with calculated data at a z which was larger by 5 cm. The z of the first line in the table is the experimental z.

Table I

z(exp)	5	15	25	35
exp	109	159	161	128
calc	68	109	104	95
ratio	1.6	1.45	1.5	1.25

The second line of Table I gives the experimental values, calculated as the difference of curves A and B of the figure of Creutz, Wilson, and Kamm. The third line gives the readings of our Fig. 6 for a z which is larger by 5 cm than the z of the first line. The last line gives a ratio of the two preceding lines and should be a constant.

One sees that the agreement is not perfect but is, probably, as good as one can expect on the basis of the data. We are inclined to attribute the fluctuations of the last line to experimental errors since Fermi and Anderson, in a careful set of experiments (cf. C-92), found good agreement in a similar case.

We now go over to the comparison of epicadmium activities. Table II is constructed as Table I was, except that curve C, Fig. 2, C-109 is used instead of the difference between curves A and B and that is compared with curve A_{Cd}^{**} of our Fig. 7.

Table II

z	5	15	25	35
exp	13	14	10	4.5
calc	10.9	11.3	7	3.4
ratio	1.2	1.25	1.4	1.3

The ratios obtained here should be equal to the ratios obtained in Table I. One sees that this is true probably within experimental accuracy. If anything, it seems that the epicadmium activity's ratio to thermal activity is smaller (1:13.3 at 20 cm) than one would expect (1:11.6). Thus the deviation from the l/v law which one would suspect on the basis of this comparison is a more rapid drop of the cross section with increasing energy than this law indicates (contrary to the original surmise). However, it is doubtful whether one can infer a real discrepancy because both calculation and experiments are rather inaccurate. For the latter, it is well to point to the relative uncertainty in the "thermal energy" and to the correction for transparency of the Cd absorber (Fig. 8) which we applied as if the neutrons had perpendicular incidence. This latter fact introduces an error in the same direction in which the discrepancy is.

Reactor Engineering at Chicago, 1942–1945, and at Clinton, 1946–1947

Annotation by Alvin M. Weinberg

Most of these reports were written while Wigner was director of the theoretical section at the Chicago Metallurgical Laboratory, 1942–45. As Wigner notes in his diary, he regarded his job at Chicago to be primarily one of reactor engineering; the reactor theory he developed was almost always directed at answering questions that arose in the design of the plutonium producing reactor or of breeders.

Plutonium-Producing Reactors

11. E. Fermi, S. K. Allison, C. Cooper, E. P. Wigner, "Report of the Committee for the Examination of the Moore-Leverett Design of a He-Cooled Plant", CE-324, October 22, 1942.

Although helium cooling was originally the main line, early in 1942, Arthur Compton, Director of the Metallurgical Project began to share Wigner's doubts as to the feasibility of building and operating a helium-cooled reactor in a short time. Compton appointed a committee consisting of Fermi, Allison, Cooper, Wigner, and Szilard to re-examine the whole question of helium cooling. Paper 11 is excerpted from the report of the committee. (Szilard dropped out of the committee before the report was written). Wigner's written report hardly reflected his doubts about the helium reactor. He was convinced, probably by spring of 1942, that operating a reactor at very high temperatures and cooling it with intensely radioactive helium posed insuperable difficulties.

The Moore and Leverett mentioned in this report were T. V. Moore, an engineer in charge of the helium-cooled design, and his young associate Miles Leverett. Mr. Moore left the project shortly after this report appeared. Dr. Leverett remained at Chicago, and subsequently became one of the country's leading nuclear engineers.

12. A. M. Weinberg and E. P. Wigner, "General Considerations Concerning the Lattice Structure", from "Possibility of a Chain Reaction" by Fermi, Weinberg, and Wigner, CP-383, November 16, 1942.
13. E. P. Wigner, "Relative Advantages of Several Cooling Systems", and E. P. Wigner and L. Ohlinger, "Heavy Water Plant", from "Survey of the Power

Plant Problem" by T. V. Moore, E. P. Wigner, M. C. Leverett, L. Szilard, L. Ohlinger, M. D. Whitaker, G. Young, A. M. Weinberg, J. Ashkin, R. F. Christy, B. T. Feld, CE-374, November 26, 1942.

Papers 12 and 13 record Wigner's contributions to the "instruction manual" prepared by the Metallurgical Laboratory for the du Pont Company which had just assumed reponsibility for the Plutonium Project. Even at this early time Wigner had analyzed various combinations of coolant and moderator; of particular note is his analysis of the engineering problems associated with a heavy-water moderated reactor.

14. Gale Young and E. P. Wigner, "A Plant with Water Cooling", CE-140, No date, probably Spring, 1942.
15. A. M. Weinberg, G. Young, R. Christy, G. Plass, E. Wigner, and R. Williamson, "On a Plant with Water Cooling", C-197, No date, probably Summer, 1942.
16. M. G. J. Boissevain, M. C. Leverett, L. A. Ohlinger, A. M. Weinberg, E. P. Wigner, and G. J. Young, "Preliminary Process Design of Liquid Cooled Power Plant Producing 5×10^5 KW", CE-407, January 9, 1943.

Papers 14 and 15 are forerunners of the ultimate Hanford design report, Paper 16, CE-407. In the two earliest versions, the uranium rods are hollow, and are cooled internally. Though this possibility is considered in Paper 16, the final design used solid uranium slugs cooled externally by a thin film of water or diphenyl. At the time these reports were written, it was generally believed that successful operation for as little as 100 days would supply enough plutonium to end the war. As the Hanford reactors were actually built by du Pont, they continued to operate for many years. The original reactors were replaced in the late 50's not because they were no longer operable but because they were unstable against loss of coolant, that is they had a strong positive void coefficient. The possibility of complete loss of cooling, and a subsequent catastrophic power surge was hardly considered during the design of the Hanford reactors. On the other hand, the du Pont engineers did worry about an obstruction in a single tube causing the water in that tube to flash into steam, and the uncooled fuel rod to melt. To avoid such local burn-out, du Pont orificed each inlet, raised the inlet pressure, and took most of the pressure drop across the orifice; a small, local increase in flow resistance down stream would not cause the entire column of water to flash into steam.

T. E. P. Wigner "Radioactivity of Graphite Remaining in Pile." C-151, No date, probably Spring, 1942.
T. E. P. Wigner "Radioactivity of the Cooling Water", CP-499, March 1, 1943.
T. A. Cahn, E. P. Wigner and M. Ginsburg, "Heat Transmission Coefficient and Temperature Fields Around Various Combinations of Pipes", CP-1693, May 1943.
T. E. P. Wigner et al. "Planning Experiments on Liquid Cooling", CC-141, June 23, 1942.

T. E. P. Wigner and M. C. Leverett, "Report on Design and Construction of 10^4–10^5 KW Liquid Cooled Plant", CP-G Memo 12, probably 1942.

17. H. D. Smyth, E. P. Wigner, and H. C. Vernon "Memorandum from the P-9 Committee", CS-583, August 10, 1943.

This paper records the results of a study of heavy water ("P-9" in war-time code) reactors as plutonium producers. As Wigner points out in his diary, he strongly urged that a second line based on heavy water be pursued even though du Pont was already constructing the original water-cooled graphite reactors at Hanford. A special group, nominally headed by H. D. Smyth, was organized to investigate heavy water reactors. H. D. Vernon, a du Pont engineer was in charge of that part of the group that had been associated with Harold Urey at the SAM laboratories in New York. The group had just moved to Chicago and included such able people as Karl Cohen and Irving Kaplan; but because chain reactions were new to them, their contribution tended to be overshadowed by the contributions of Wigner's group. Paper 17 is excerpted from a fuller report; the excerpted portion bears Wigner's stamp, and was almost certainly written by him.

Breeders

18. E. P. Wigner, "Breeders and Converters", CF-2860, April 7, 1945 (Minutes written by L. A. Ohlinger).

In this paper Wigner analyzes breeders operating with thermal neutrons, resonance neutrons, and fast neutrons. This characterization of breeders according to the average energy of the neutrons that induce fission is still valid.

19. E. P. Wigner, A. M. Weinberg, and G. Young, "Preliminary Calculations on a Breeder with Circulating Uranium", MUC-EPW-134, May 17, 1945.

This paper, written by Wigner, is the original analysis of circulating fuel thermal breeders based on the Thorium-U^{233} cycle. Five experimental homogeneous, circulating fuel reactors, four at Oak Ridge, and one in the Netherlands, were built during the next twenty years. The heavy water homogeneous reactor experiments (at Oak Ridge and in the Netherlands) encountered serious problems of fuel stability and corrosion, and they were dropped, even though the Homogeneous Reactor Test at Oak Ridge achieved its goal of 5 megawatts. Two high-temperature circulating-fuel reactors in which the uranium is dissolved as a mixture of molten fluorides were operated at Oak Ridge. The most successful of the experiments, the Molten Salt Reactor Experiment (MSRE), operated on U^{233} for long periods at about 7 megawatts. Though molten fluoride salts are not mentioned explicitly in Paper 19, Wigner had given thought to their use. Since no insuperable problem was encountered in the operation of MSRE, several groups, particularly in Japan, France, India, and USSR, remain interested in molten salt reactors.

20. Eugene P. Wigner, "New Ideas for Nuclear Reactors", *Nuclear Science and Engineering*, 6, 420, (1959).

This paper is the first Annual Distinguished Lecture presented at the meeting of the American Nuclear Society, Gatlinburg, Tennessee on June 15, 1959. The lecture gave Wigner an opportunity, 15 years after the Hanford project, to review the status of nuclear reactor technology, especially in the context of general energy policy. Wigner brought to the review several original ideas – for example, in his analysis of direct conversion of fission heat to electricity by means of thermionic converters.

Note also that the Wigner-Soodak fast breeder is described in U.S. Patent 27P.

21. E. P. Wigner, et al., "Project Hope: A Chemical Reprocessing Plant for a Nuclear Power Economy, ORNL 1635, Dec. 19, 1953 (Summary).

Wigner chaired the group that designed a chemical plant for a breeder economy. Because the aim was to reprocess for only one dollar per gram of fissile material, the project was dubbed "HOPE".

The MTR and the Daniels Power Pile

Though Wigner wrote no reports specifically on reactor engineering at Clinton, he closely supervised the design of the Materials Testing Reactor. His ideas are embodied in patents 28-P, 29P-T and 30P-T.

T. E. P. Wigner, "Present Status of Power Pile", CL-EPW-2428, March 5, 1947.

The "Power Pile" referred to in this memorandum is the helium-cooled, BeO-moderated, high-temperature reactor of Professor Farrington Daniels.

Reactor Shielding

22. E. P. Wigner, "Protection Against Radiation", C-137, No date, probably early 1942.

In this paper Wigner estimates the amount of shielding needed around a high-powered reactor. Wigner attributes the calculations of the γ-ray shielding to his wife, Mary Wheeler Wigner, though almost certainly Wigner guided his wife's calculations.

23. E. P. Wigner and G. Young, "Penetration of Fission Neutrons Through Water", Mon P-283, April 9, 1947.

The 30 megawatt Materials Testing Reactor was at the bottom of a pool of water. Paper 23 gives an estimate of how much water was required to shield against neutrons leaking from the top of the MTR.

24. E. P. Wigner, "On the Available Information Concerning the Interaction of Neutrinos with Matter", CP-720, June 2, 1943.

The Hanford reactor would produce 3.5×10^{19} neutrinos per second. Wigner worried about whether the neutrinos could cause physiological effects in the workers at Hanford. In this paper Wigner dispels this concern. Neutrinos were first detected directly at Hanford by Reines and his associates some fifteen years after Wigner had made this calculation.

Fission Products

25. K. Way, and E. P. Wigner, "The Rate of Decay of Fission Products", *Phys. Rev.* 73, 1318 (1948).

T. Wigner, E. P. and Way, K., "Summary and Correlation of Data on the Rate of Decay of Fission Products", CC-R-3032, June 13, 1945.
T. Way, K. and Wigner, E. P., "Radiation from Fission Products", MDDC-48 (No date, probably 1947).
T. Wigner, E. P., "Neutron-Produced Radioactive Isotopes with $T_{1/2} \geq 1$ day", CP-730, June 11, 1943.

Wigner Effect

T. Wigner, E. P., "Graphite Disease in the W Pile", N-1585, September 14, 1944.

Wigner early in the project had estimated the rate at which the graphite moderator would deteriorate under intense fast neutron bombardment, (indeed, as he says in paper 26 "It would be unscientific to claim a useful life longer than about 100 days.")

26. Seitz, F. and Wigner, E. P., "Effects of Radiation on Solids", *Scientific American*, 195, 76 (1956).

This is a semi-popular review of the field of radiation damage some 13 years after Wigner had first called attention to the phenomenon in connection with high-powered chain reactors. Frederick Seitz, one of Wigner's colleagues at Chicago, gave detailed accounts of the Wigner effect in two papers, "The Influence of Operation on Tube Alloy", CP-1598, April 21, 1944; and "The Motion of Massive Particles through Solid Matter", CP-1661, April 27, 1944.

11.

Report of the Committee for the Examination of the Moore-Leverett Design of a He-Cooled Plant

Fermi, Anderson, Cooper, and Wigner

October, 1942

On September 26, 1942, the members of the Committee received copies of Report No. CE-277 by Messrs. Moore and Leverett. The work of examining in detail the proposed structure of the plant was partitioned as follows between the various members of the Committee:

Steel and graphite structures – Fermi
Uranium structures – Allison
Helium purity – Allison
Helium circulation and heat transfer – Cooper
Neutron efficiency – Wigner
Radiation Shielding – Wigner
Controls – Wigner
Relative advantages of parallel and series flow – Szilard

After a preliminary investigation of the projected plant by the members of the Committee and some discussions with Messrs. Moore, Leverett and Wheeler on particular points, the Committee met on October 22, and October 29, 1942. Mr. Szilard did not participate in this last meeting and expressed his views in a letter that is attached to the present report.

The following main conclusions were reached:

General Dimensions and Steel Shell

Partly due to new evidence as to the properties of metal that became known after Moore and Leverett's plans were completed, the steel shell has been redesigned in order to accomodate a pile of dimensions considerably larger than had been planned originally. Since the new design is not available to the Committee, no specific recommendations can be made here as to its adoption. From the point of view of increasing the probability that the plant may work efficiently as far as neutron reproduction is concerned, it is desirable to have the steel shell capable of accomodating a very large pile. On the other hand, a limitation of the dimensions permits us to economize on critical materials and on time. It seems to the members of the Committee that a reasonably safe compromise could be the following:

A cylinder of 28 ft. diameter and 26 ft. height with uranium throughout the cylinder except for the outer thickness of one foot – or a somewhat different shape equivalent from the point of view of the chain reaction.

The fact that the dimensions are considerably larger than according to the original plan makes it very desirable to design the lattice so as to achieve a reduction in the amount of metal. The problems connected were discussed at a conference on October 21 in which Steinbach, Leverett, Moore, Wigner, Wheeler and Fermi took part. Wheeler summarized the results of that discussion in Memorandum No. 15.

It is the opinion of the Committee that the amount of uranium in each lump should be increased if the side of the cell of 11″ is left unchanged. A possibility that was considered favorably at the conference on October 21 was to use a non-cubic cell having one side of 11″ and the other two sides somewhat smaller, perhaps 9″. Such an arrangement would permit leaving the final decision as to these dimensions to some future time in the hope that additional information may be available at the time when the decision must be taken.

Circulation of the Helium

Mr. Cooper checked in detail the calculations on the pressure drops in the helium circulation system and on the heat transfer and reported to the Committee that he was in essential agreement with the conclusions of Moore and Leverett.

It is the opinion of the Committee that centrifugal compressors would be more dependable than reciprocating compressors and it is the understanding of the Committee that Mr. Moore has revised his original plans in this sense.

It is the opinion of the Committee that use of a parallel flow of the helium inside the pile would reduce very considerably the pressure drop and, consequently, the power requirement of the compressors. Such a change, however, would require presumably a complete new design of the internal structure of the pile and possibly also of the dumping mechanism of the uranium. Definite conclusions in this respect could we reached only if such a new design were actually completed.

Corrosion Problems

Mr. Allison stressed the necessity of using helium of very high purity to avoid corrosion of uranium by impurities of the circulating helium. He referred in particular to recent experimental results that indicate that considerable amounts of gases are emanated by the graphite on heating after being previously evacuated.

It is important that fully adequate provision should be made therefore, not only to purify initially the commercial helium but also to eliminate the impurities that will accumulate in the helium when the pile begins to operate at high temperature.

It might be desirable to provide facilities for heating artificially the pile to the temperature of operation. This would make it more easy to outgas the graphite and could be used also to adjust the final amount of uranium to be put inside the pile to operation at the desired temperature.

Radiation Protection

Mr. Wigner reported to the Committee on the problems of the radiation protection. The Committee agrees with his opinion that the shielding against radiation is essentially adequate. The circulating helium, however, should be prevented from depositing active materials on the parts of the system that should be accessible after shut down of the operation and in particular on the walls of the upper section of the pressure tank above the chamber, so as to prevent helium from the pile diffusing into the upper section.

In a similar manner it might be advisable to introduce purified helium at a pressure somewhat higher than that of the circulating helium in some of the stuffing boxes to prevent gas that may carry active materials from leaking to the outside.

Considerable attention should be given to the problem of disposing of the helium after shut down of the operation, since it appears possible that amounts of activity up to the order of thousands of curies may be collected in the gas. If it is intended to discharge the helium in the atmosphere, some suitable shielded stack at some distance from the plant and connected to it by a shielded pipe should be provided.

It is important in this respect to give considerable thought to the problem of the emergency measures to be taken in case of a serious leak of the helium. In such an event, it would be desirable to discharge in a short time most of the helium. Some provision should be taken, however, in order to prevent the activated uranium from melting or, in case that it should melt, to avoid its breaking through the steel shell.

It is apparent that may of these difficulties could be eliminated or minimized by coating the uranium with a protective layer in order to prevent or reduce the escape of fission products in the circulating gas. The final decision on the proper procedure for safeguarding against most radiation dangers will depend on the success in developing such a coating.

Controls

The controlling devices are only outlined in the report and consequently no definite recommendation can be made concerning them. Mr. Wigner pointed out the advantages of operating the controlling rods on the neutron density rather than on the temperature in order to get a more prompt response to changes in the intensity of operation.

Graphite

The general type of arrangement proposed for the graphite appears in the main lines satisfactory. Some doubts have been expressed, however, as to the adequacy of the provisions taken against the danger of displacements due to thermal expansions that may jam some of the uranium bearing cartridges inside the structure or interfere with the operation of the control rods. Perhaps such danger could be reduced by either strengthening the dowels that pin together the graphite structure or by leaving them somewhat loose so that major displacements of the graphite blocks could be avoided without too much danger of shearing the dowels.

Some concern has been felt due to the possibility of displacements and possible ruptures inside the structure produced by the somewhat indefinite pressure that may be found at various places inside the pile on account of accidental differences in the leakage of the helium through the narrow channels among the chamfered edges of the bricks.

Miscellaneous

A number of further suggestions and recommendations on these and on other points are found in the memoranda presented to the Committee by Messrs. Allison, Cooper and Wigner. Such memoranda are attached to the present report.

Conclusions

The Committee is of the opinion that a plant of the general type proposed by Moore and Leverett has a fair probability of operating in a satisfactory manner. It appears, however, that a number of points mentioned in the present report and in the attached memoranda should be worked out in some detail in order to eliminate as much as possible beforehand the serious difficulties that may arise in the actual operation.

October 22, 1942

To: E. Fermi

From: E. P. Wigner

The following constitutes the remarks to the helium cooled plant of Moore and Leverett. Aside from myself, Messrs. Friedman, Christy and Ohlinger have read the report quite carefully and many, if not most, of the remarks are due to them. When writing down the following lines I thought it better to include three unnecessary remarks than to omit one that might be helpful.

General Remark

Evidently at this stage of the game it is not reasonable to take up questions of series and parallel cooling, as such fundamental questions would involve a very considerable modification of the plans. However, I think it might be useful to consider a cylindrical arrangement instead of the spherical arrangement used in the report. The obvious disadvantage of the cylindrical arrangement is that k is somewhat smaller. On the other hand it appears that the cooling is made somewhat easier. Instead of the uranium units being separated by graphite spacers, the whole column would be made up of uranium-containing cartridges. This would avoid the unnecessary pressure losses at the entrance and exit of the cartridges and would increase the useful helium surface considerably. A saving in pumping power would result. Furthermore, the natural spacing is larger in a cylindrical arrangement than in the spherical arrangement, and yesterday's discussion leads me to believe that this would be of advantage. There is however a question whether these advantages would compensate for the loss of k.

The Multiplication Constant

The question of the multiplication constant has been fully discussed in yesterday's meeting. It is unnecessary therefore to go into this in detail. It does appear that for the chosen lattice constant, the uranium units are too small so that there is a question whether the pile would operate even at ordinary temperatures.

Controls

The second subject which I was supposed to investigate is the question of controls. Unfortunately the description of the controls is somewhat scanty so that it was difficult to do this. There are statements only with regard to the number of control and safety rods, and the general statement that the controls are operated on the temperature of the outgoing gases.

This last feature appears to me to be undesirable, although lacking further details, I cannot exactly say why. As the power output of the pile increases, the temperature of the materials will increase only rather slowly. The temperature of the outgoing helium will lag even further behind the operating intensity, and the action of the controls will be considerably delayed. If the safety rods operate on the inner temperature, it may well happen that, under certain conditions, the safety rods will operate before the controls had a chance to restore the operating intensity to the desired value. A delayed action of the controls is also likely to cause hunting. Just off hand I would argue that the controls should be operated on the neutron density rather than anything else, because this would insure the quickest response. There are only very general statements concerning the material of the rods, their cooling and their erosion. There is a certain danger that the helium will enter the cavities left for the rods and stream upwards in these cavities. The resulting power loss would be, of course, very small.

However, the control rods might be brought both into vertical and horizontal oscillations which is evidently undesirable. The question of the erosion of the control rods should be further discussed.

There is one more remark in this connection which I would like to make. The calculated distribution of heat production applies only if the control rods are entirely out. If the control rods are in at least to some extent, the energy production towards the center is much smaller than calculated, and the distribution of the helium flow will not be ideal anymore.

Radiation

Mr. Christy investigated the question of the radioactive shields outside the pile and his remarks follow:

Radioactivity of the Helium in the Helium Plant. A check on the effects of the radioactivity of the helium revealed no particular disagreement with the previous estimates.

Activities were estimated on the basis of long time operation where saturation exists. The escape of fission products by recoil was estimated to be about 10^{-4}. This could be stopped by an exceedingly thin coating (.1mm.). I am informed by Mr. Mulliken that recent work on the diffusion of fission products in U metal at 600 °C (hotter than in the He plant) fails to disclose any mechanism for escape other than recoil. Of order 1/10 of the recoil activity will be transmitted through a permanent gas (1/20 activity in permanent gases and 1/20 resulting from their decay). Another 1/20 of the recoil activity will be volatile at the helium temperature of 400 °C (largely Iodine). We may then expect 10^{-4} of the total activity to deposit on walls as soon as it can. 10^{-5} of the activity will deposit as soon as it can after the helium is cooled, and 10^{-5} will remain as a noble gas.

The coolers, being the first location of large surface after the He leaves the pile, may be expected to gather most of the condensable activity (10^{-4}). The filters coming after will not greatly reduce the activity since 10^{-5} is non-condensable. However they may be expected to considerably reduce the activity deposited in the compressor cylinders. The decay of Xe leads to a deposition of 220 cu. ft/46,500 cu. ft. $\times10^{-5} = 5 \times 10^{-8}$ in the cylinders. Fission products escaping previous deposition might increase this considerably, but even as it stands it would be impossible to be near a compressor for more than a minute a day without washing out the cylinders. Thus oiling would have to be automatic and repair would be contingent on waiting or washing away these depositions quite effectively. There is evidence for long periods in the decay products of Xe so that after 100 days of operation, waiting might only reduce the activity by 10.

The coolers, collecting most of the fission products, would require considerable shielding. A concrete shell of 4 or 5 feet in thickness around each cooler should suffice. Similarly all pipes should be protected by the equivalent of about 4 feet of concrete on top.

There appears to be no danger from the irradiation of oil in the compressors by quite a large margin (at least 1000).

The storage tanks should have provision for shielding if purification of the He from Xe is not almost 100% efficient.

There is a question concerning the storage of the He at end of operation. The tanks have a volume of 16,000 cu. ft., the system of 46,500 cu. ft.

Above the pile, the He radiation is small compared to that directly from the pile and doesn't increase the internal radiation shield requirements. Leakage of active Xe in He through plugs or otherwise above the shield would be dangerous. If that region is to be occupied at any time, it should be kept free of active He at all times by a suitable pressure differential or otherwise. On the other hand, the object of such a chamber is somewhat obscure. Reloading of a 10^5 kW plant is almost equivalent to admission that the project can't succeed in any reasonable time. It is a very inefficient use of U and should be resorted to only in the event that we can't find a better use for it. The only valid reason for such a chamber seems to lie in aiding the dumping of the U if the cartridges stick.

Anyway, the reloading is rather difficult without further precautions. Many of the fission products may be deposited in the carbon and these scattered around the inner shield should be considered.

The radioactivity of the asbestos may not be negligible in this connection. (Ca, Mg)?

Essential Conclusions. One cannot rely on large decreases in activity deposited by Xe decay by simply waiting since these products seem to include long periods (10 days, 1 yr.,...). Provision must always be made for removal of such decay products insofar as possible by washing.

In addition to the internal radiation shield, active Helium must be prevented from leaking into the upper chamber if it is ever to be entered. If leaks persist the gas there must be rapidly replaced by clean Helium.

A thin coating of the U is recommended if, as it now appears, diffusion is small compared to recoil, a thick coating to reduce diffusion must be weighed against the loss of k entailed.

Radioactivity of the Reactor

The following constitutes a summary on Mr. Friedman's calculation of the protection against radiation of the reactor.

I. Gamma Rays. Under the assumption that there are five 10 Mev gamma rays per fission we have about 10^{11} gamma rays per second passing out of the pile per cm^2 of the surface. In order to cut this radiation down to .1R/day 9 feet of water would suffice. There is a minimum of 10 feet of water all around the pile in addition to the concrete, and 1' of carbon. On the other hand there are two 3' wide helium tubes traversing the water shield vertically. This however will not alter the fact that the general radiation protection is sufficient. All this is under the assumption that the water shield has a minimum width of 10'.

On the other hand, if the statement on the bottom of page 16 concerning the inner diameter of the concrete column were correct, the radiation shielding would be insufficient.

The general conclusion of the first paragraph must be restricted however, on account of the different conduits from the reactor to the outside. The conduits for the ionization chambers have 6" diameters and are 10' long. One easily calculates that the radiation at the exit of these conduits is about 3000 times greater than at other parts of the external surface of the water. The increase in radiation density at the end of the 1" thermocouple pipes is 100 times greater than the average density. The former may amount to 300R/day, the latter to 10 R/day. It is not entirely clear from the report how the readings of these instruments will be taken and how the ionization chamber will be serviced. Evidently the space which forms the continuation of these conduits is dangerous. This danger could be avoided by bending, or in some way plugging, the pipe in water.

For purposes of experimental biology, some big cavities within the pile are contemplated. The possible dangers involved in these arrangements cannot be calculated without more detailed information.

II. Neutrons. One easily calculates that about 10^{11} neutrons per second escape from the pile per cm^2 on the average. Evidently the slow ones among these are very soon stopped. The slowing down and capture of the fast neutrons can be calculated as in C-137, or by using an absorption coefficient of .1 cm^{-1}. Both methods give the result that even the fast neutron intensity is reduced to about 1 neutron per cm^2 and sec. The reason that the outgoing neutron intensity is so small is that the water shield is dimensioned sufficiently thick to protect even against gamma rays.

When the neutrons are absorbed by the water, gamma rays are produced. If we denote the number of incident neutrons[1] per cm^2 and sec by N_0, the number of neutrons absorbed, or the number of gamma rays produced per cm^3 and second at the distance x is .1 $N_o e^{-.1x}$. If all the gamma rays were to travel straight on with an absorption coefficient of .05, the number of gamma rays passing at x_1 would be

$$\int_0^{x_1} .1 N_0 e^{-.1x} e^{-.05(x_1-x)} \, dx = 2N_0(e^{-.05x_1} - e^{-.1x_1});$$

with $N_0 = 10^{11}$ this gives a radiation of 10 R/day. However, this will be reduced by the concrete wall to a fraction of one R/day.

III. Internal Radiation Shield. The internal radiation shield's dimensions would permit, as far as gamma rays are concerned, entering of the pile even during operation. A short wait after the operation of the pile has ceased, the

---

[1] Original paper read "gamma rays" for "incident neutrons". (A.M.W. and A.M.P.)

gamma ray intensity is so low that one could enter the upper chamber even if the internal shield had only 1/2 its actual thickness. At that time of course, no neutrons would come out of the pile any more.

As we understand it, the purpose of the internal radiation shield is twofold. First, if a cartridge should stick to the wall and fail to drop out, it could be rammed out from above. The second purpose is the reloading of the pile. There is a certain danger involved in the first operation because the personnel carrying out the ramming would be just above the opening, more or less directly exposed to the radiation of an uranium cartridge. This difficulty could be avoided by plugging the hole and ramming out the plug together with the cartridge.

As pointed out in Christy's report, it is very important to prevent the leakage of the contaminated helium into the upper chamber. It is not clear that this is taken care of in the present arrangement. If the internal shield remains as thick as contemplated at present, there certainly is no danger of having a substantial radioactivity of the shell of the upper chamber. Even if this is reduced, the danger from this source appears to be negligible, if ordinary iron without particularly dangerous impurities is used.

IV. Essential Conclusions. The radiation shielding of the reactor seems to be generally sufficient.

The space for biological experiments should be more closely designed to permit calculation of the radiation dangers inherent in their construction.

If stuck uranium cartridges are to be rammed out by using the upper chamber, plugs should be provided which can fall into the cooling channels and protect the personnel from the radiation of the uranium cartridge.

The conduits for the ionization chamber and thermocouple leads should be angled or otherwise better shielded.

The possibility of radioactive contamination of the stuffing boxes should be investigated.

Questions Concerning Engineering

Responding to the request to look for troubles in general in the system, we are submitting the following questions:

It is pointed out in the report that the pressure drop of 11.4 p.s.i between the bottom and the top of the pile is not sufficient to lift up the whole pile. However, the pressure drop is not uniform within the pile. It is much smaller at the bottom of the pile than at the top, both because the velocity is higher at the top, on account of the increased temperature, and also because the channels at the top are more narrow. As a result it is not excluded that the pressure drop in the top of the pile may be higher than the weight of the top of the pile and it may be lifted up. More accurate calculations are desirable on this point. The danger mentioned here is further increased by the fact that the pressure in the bulk of the pile, that is between the bricks, is given by the accidental leakages between the bricks. If the carbon is stacked by accident more loosely at the bottom than it is at the top, the high pressure of the bottom may extend rather high, and the top of the pile may be lifted.

We were afraid also that some of the bricks would fall into the gas intake nozzles when the uranium is withdrawn.

There is a multitude of stuffing boxes at the dump rods and other places from the reactor to the outside. When discussing the water-cooled plant we were led to believe that there is a considerable difficulty in the safe operation of a great number of stuffing boxes. Mr. Ohlinger has devised a method to avoid the stuffing boxes at the exit of the instruments, but this still leaves the dump rods.

The operating personnel of seven men appears to be rather low. Further experiments would be necessary on the danger of corrosion of the compressors and heat exchangers by the acid wash. Pumps are needed for the original evacuation of the system and also for the original compression of helium. I do not know how the heat exchangers stand up under evacuation.

Alternative Arrangements

There are two alternative arrangements mentioned in the report and these are not as carefully considered as the rest of the report. They are, in a sense, unessential, but it seems to me to be worthwhile to remark on them nevertheless.

The first alternative is the use of carbide instead of Uranium. This appears to be difficult in the present arrangement because the volume of the carbide is considerably greater than the space available for it.

The second point is the coating of the Uranium by Zn or Al. The former is evidently impossible at the temperatures at which the helium plant operates, but the present experience indicates that even the Al will rapidly react with the uranium even at the average surface temperatures.

12.

General Considerations Concerning the Lattice Structure

A. M. Weinberg and E. P. Wigner

November 16, 1942

Every cycle in a chain reaction has three parts. Let us trace the course of the cycle by following the life of a neutron. Each neutron is created, together with 1 or 2 brothers, by the fission process, i.e. the disintegration of a U nucleus into two fragments with the simultaneous emission of 2 to 3 neutrons. This is the first part of the cycle. The neutron, when created, has a very large energy, about 2 Mev (a velocity of 12,000 miles/sec) and it must be slowed down to thermal velocities (1 mile/sec). This is the second part of the cycle; it occurs in the damper by collisions with relatively light atoms (carbon in the usual arrangement). The slowing down of the neutrons is necessary because slow neutrons are much more easily absorbed by U than are fast neutrons. The absorption of the neutrons by the U is the third part of the cycle; it causes the U atom by which it is absorbed to undergo fission and thus emit, in its turn, new neutrons.

If matters were as simple as described above, the chain reaction would be easily established. Unfortunately, there are three effects which complicate matters somewhat and may cause a premature death of the neutron, without its having induced a fission. The first effect is the so-called "resonance absorption" which may occur in the second phase of the cycle. While the neutrons are being slowed down and pass through intermediate energies , (so called "resonance energies", i.e. velocities of 15 to 1,000 miles/sec.), they may be absorbed by the U^{238} without causing fission. This parasitic neutron absorption makes it more difficult to establish a chain reaction and must be made, therefore, as small as possible. Its magnitude is decreased by using a large amount of damper. However, the amount of damper cannot be increased indefinitely for reasons which will be given below. It was a very important discovery, therefore, when Szilard and Fermi realized, that, for a given damper to U ratio, the resonance absorption can be decreased by lumping the U rather than distributing it uniformly over the damper. One can decrease the resonance absorption in this way almost by a factor of 25, so that only about 14% of all the neutrons are absorbed at resonance in the usual lattice arrangement.

The last two kinds of premature neutron absorption may occur in the third part of the cycle. First, the damper as well as the U absorbs slow neutrons to some extent. Evidently, the neutrons absorbed in the damper will not cause fission; this reduces the chances for establishing a chain reaction. As a consequence, those substances can be considered to be good dampers which slow down the neutrons efficiently without absorbing them at thermal energies. Evidently, the thermal loss will be increased by using a larger amount of damper and this is what makes the use of a very large damper to uranium ratio impossible. Furthermore, for a given damper to U ratio, the thermal loss will be diminished by distributing the U more uniformly over the damper. This explains why we use a lattice of many smaller spheres rather than lumping all the U into one single sphere. One can say rather generally that those geometric arrangements which diminish the absorption in the damper increase the resonance absorption and vice versa so that one has to strike a compromise to keep both losses in limits.

The last cause of neutron loss is the absorption of the thermal neutrons by the U^{238} rather than the U^{235}. Only the latter absorption leads to fission and the creation of daughter neutrons. The relative amount of absorption by the U^{238} cannot be diminished except by increasing the percentage of U^{235} in the U or by adding 49, i.e., by "enriching" the U.

The so-called lattice calculations are designed to obtain the geometries (carbon to U ratio and size of lumps) in which the sum of the resonance and thermal losses is least. It turns out, for instance, that a carbon to uranium mass ratio of 6 with uranium spheres of $2''$ diameter gives a particularly small premature loss of neutrons. The calculations are based on numerous experimentally established nuclear constants. They have been, on the whole, well confirmed by direct observations.

Thermal Stability

A chain reacting system will be thermally stable if it becomes less chain reacting as the temperature increases. On the other hand, if the system becomes more chain reacting as the temperature increases, it is thermally unstable. These possibilities have been described already above, by Fermi.

Evidently, the thermal stability depends on whether or not the above detailed losses increase with increasing temperature; we shall give a brief analysis of the relevant factors. It must be emphasized, however, even at the outset, that it is not possible, at present, to predict theoretically whether a given system will be thermally stable or unstable. As was pointed out by Fermi, an empirical answer to this question must also await further work.

The resonance absorption *increases* with increasing temperature and thus tends to make the system thermally stable. However, while the direction of the effect can be predicted, very little is known about its magnitude.

The thermal loss *decreases* with increasing temperature and thus tends to make the system thermally unstable. This tendency is more pronounced if the size of the lumps is large, i.e. if the lattice spacing is large. It may be possible to give reasonably accurate estimates concerning this effect but lacking information about the behavior of the resonance loss, we cannot settle the major question of overall stability at the present time.

In enumerating the causes of neutron losses, we have, so far, omitted one cause which occurs only in finite lattices, namely, the escape of neutrons through the sides of the pile. This loss increases with increasing temperature and makes it, in fact, probable that the usual lattices are thermally stable. It should be mentioned even here, however, that an eventual thermal instability can be and will be easily compensated by suitable controls.

13.

Survey of the Power Plant Problem

E. P. Wigner and L. Ohlinger

November 26, 1942

Relative Advantages of Several Cooling Systems

E. P. Wigner

Abstract. So far, three agents have been considered for conveying the heat out of a uranium-graphite pile: helium, a liquid (water) and molten bismuth. This gives three different plans for a uranium-graphite pile. Furthermore, the graphite can be substituted by heavy water, the uranium by some enriched material so that one has at least five possible setups.

None of these is better in every respect than all others and neither of them can be guaranteed to give a smooth operation, uninterrupted by unforeseeable troubles. For this and other reasons it appears desirable to erect several types of plants in order to ensure a safe and sufficient supply of 49 in time.

All the plans will be described in more detail in the ensuing sections. In the first three, He, water or Bi streams through the system and carries with itself the heat generated. The heavy water plant can be cooled in numerous ways and so can enriched plants. The latter two have been considered, so far, in less detail because the critical materials (heavy water and enriched U) are not yet available.

The attached table is to summarize the merits and disadvantages of the five different schemes. Evidently, the statements contained in this table as well as some of the statements made below, do not represent absolute truths but hold only under the conditions which are usually considered reasonable. They are designed to serve for an orientation of the reader.

The main advantage of the He cooled plant is that, if properly built, it certainly will operate according to expectation for some time. Its main disadvantage is the vastness of the external equipment required, which hardly permits a fast construction. The difficulty of exchanging the U within the pile may limit the useful life and the yield of the pile even if no unexpected trouble arises.

The main advantage of the water cooled plant is that the external equipment is exceedingly simple and the construction of the pile, therefore, fastest. However, water corrosion may seriously limit the operating time of the pile and

Table I

Type of plant	Possible trouble and complications	To what extent can troubles be forseen and eliminated	Usefulness for power production	Speed of construction	Speed of production of 49. Amount produced during life of pile
He cooling	High temperatures Strong radiation from fission products	Almost completely	Useful	About as fast as Bi cooled plant	Relatively low rate of production. Not very high utilization of U
Water cooling	Corrosion of pipes	Probably not completely	Very small	Fastest	Higher rate of production. Not very good economy in long run.
Bi cooling	Probably little trouble although large sizes and high temperatures necessary	To a large extent	Best adapted for power production	Less fast than water cooled plant	Very high rate of production. Low utilization of U
Heavy water	Corrosion trouble Cooling difficult	To a large extent	Useful, particularly for small units	Fast if 4–5 tons of heavy water available	For 4–5 tons of heavy water, production not high. Utilization of U best among all plans
Enriched plant	Depends on system	Probably to a large extent	useful, particularly for small units	Fast, if 10–20 kg 25 or 49 available	At present open to question

the extent of this will remain, to a certain extent, unpredictable. The U within the pile can be exchanged more readily than in the He cooled pile. It is, on the other hand, not suitable for power production.

The Bi cooling has great advantages as compared with both systems above. It can readily be used for power production and unless the liquid Bi erodes some of the materials in which it flows, no major trouble can be expected. This plan has not been worked out in the same detail as those above and this still may require some time. It is already clear, however, that the rate of production of a Bi cooled plant per unit amount of U is greater than that of either of the above plants.

The great disadvantage of the heavy water plant is that it requires heavy water. It appears that about 3,000 l (800 gallons) would be needed for the smallest unit, the power output of which would be very low. For a reasonable power output (100,000 kW) one can estimate that twice that amount will be required. In all other respects, the heavy water-U plant is superior to the graphite-U plants. There is a considerable leeway in the multiplication constant, part of which can be expended to further an efficient cooling and to minimize the results of corrosion. Another part can be expended to give a longer life to the pile and to permit the production of a larger amount of 49 per unit amount of U than one can obtain with the graphite-U piles. It is important to note, furthermore, that the heavy water plant can produce 23 and other by-products more efficiently than any of the above plants. It must be emphasized, on the other hand, that the details of the heavy water plant have not yet been worked out and one must expect to encounter some serious difficulties of detail.

The economy of the enriched plants for 49 production is at present an open question. Anyway, no such plant can be built unless much larger amounts of 25 or 49 are available than we expect to have in the near future. One way in which enriched plants can surely be expected to be important is as concentrated sources of high power.

Heavy Water Plant

L. Ohlinger and E. P. Wigner

Abstract. Due to the lack of sufficient heavy water for large scale experimentation, the possibilities of heavy water plants have not been well explored. Nevertheless, it is certain that heavy water is superior to graphite as a slowing down agent and would permit the use of a much smaller chain-reacting unit for the same power output than would be required by graphite as the slowing down agent. As a very preliminary estimate, it would appear that an output of 10^5 kW or more might be anticipated from less than 8 tons of metal, and about 8 tons of heavy water.

The major problem to be expected with heavy water plants is that of heat dissipation from systems as small as can be built with the amounts of heavy water which will be available in the near future. In spite of this, the early design and construction of such a plant is considered prudent, not so much because of its own advantages, but because unforeseen difficulties may delay the actual conpletion and satisfactory operation of the graphite plants.

As mentioned above, the nuclear properties of heavy hydrogen are not sufficiently well known to permit the detailed design of a heavy water plant at the present time. The best available information indicates that a uranium-heavy water lattice will have a multiplication constant of about 1.23 and a volume of 3,000 to 4,000 liters (about 1,000 gallons). This is the order of magnitude of heavy water which might be available by the end of next year. One sees that the multiplication constant is much higher than in a U-graphite lattice and, as a consequence, the size of the chain reacting unit much smaller. The amount of U metal required is about $1\frac{1}{2}$ tons. The problem is to convey the heat out of as small a unit as this and the following lines give a brief discussion to this (unsolved) problem.

It is evident that the heat extraction encounters serious difficulties. First of all, in the smallest possible pile, discussed above, the U is in lumps of about 4 cm diameter. If heat is constantly generated at a reasonable high rate in such lumps, their inner temperature will soon rise far above the 660 °C which we consider to be the upper limit. This shows that the surface to volume ratio of the lumps must be considerably increased and this in itself decreases the multiplication constant and increases the amounts required. One can gain something by Zinn's device of constantly removing part of the metal and returning it into the pile after having cooled it down outside the pile. This requires much larger amounts of metal than the $1\frac{1}{2}$ tons given above, because, at any given time, only a fraction of the total amount of metal is in the pile. This is no serious objection in the present case because, even so, the requirements of metal remain reasonable. A modification of this scheme has been suggested lately by Brown who wants to circulate the liquid UF_6 through the pile. It is evident that all these arrangements will be useful only after considerable engineering development.

All the other schemes arrive at a better heat transfer from the U at the expense of the multiplication constant and require, hence, larger than the minimum amount of heavy water. We enumerate some such schemes.

Du Pont proposed to use the pile as a large boiler which carries a condenser on top of it. We do not believe that this scheme is, in this form, a promising one, because of the huge volume of steam formed.

The heavy water in the pile might be drawn off, circulated through a cooling and degassing system, and returned to the pile continuously. This recirculation of the heavy water itself would obviously require a larger quantity of heavy water in the system.

Evidently, all the cooling media used for a graphite pile can also be considered for a heavy water pile. However, Bi may have its particular difficulties because the chemical decomposition of deuterium compounds at the temperature of hot molten bismuth. Helium appears to be a promising possibility, particularly because it decreases the multiplication constant so little. We have considered cooling by ordinary water which can be introduced in rather large quantities without increasing the necessary amount of heavy water too much (7,000 to 8,000 liters of D_2O with 7 to 8 tons of U will give a running pile even in the presence of about 300 liters of ordinary water). However, neither of these schemes is worked out sufficiently to permit final conclusions to be drawn.

It is evident, furthermore, that all the above schemes (excepting, perhaps, the He cooling) will require considerable engineering development. The difficulties which the various cooling media present when used in a graphite pile are in evidence also when they cool a heavy water pile. To these, the chemical decomposition of the heavy water must be added, which must be counteracted in some way.

On the other hand, the advantages of a heavy water pile are also manifold. We would consider as the principal of these that it dispenses with the use of graphite. Hence, it is free of the difficulties which any chemical or physical change, induced by the powerful nuclear reaction, may introduce into this system. As the second advantage, we would name the high multiplication constant which permits a large fraction of the neutrons to be used for creating "by-products" such as 23 or radioactive materials. The abundant production of H^3 is also to be mentioned. Lastly, the small size of the unit has also advantages.

For all these reasons, the design of a heavy water plant is strongly recommended. A request for enough heavy water for experimental purposes has already been made and we hope to receive it as soon as it becomes available.

14.

A Plant with Water Cooling

G. Young and E. P. Wigner

No date, probably Spring 1942

The present system is an alternative to the commonly considered He cooled plants which has some advantages, although perhaps, the main reason that its difficulties are not apparent, is that the plan is relatively new. The general nature of the estimates to be described below seem to indicate that it is worth a more detailed consideration.

The general ideas underlying are as follows:

One uses metal in the center 1/8 of the plant, and, preferably, carbide outside. Both metal and carbide are arranged in the form of cylinders which are cooled by a liquid circulating through a pipe at the axis of the cylinder. The calculations below are carried out for water as a cooling liquid, but it is evident that other liquids, such as Dowtherm or carbon fluoride also can be considered. There are several advantages in such an arrangement, such as the low power requirements, possibility of both differential heating and of gap structure, simplicity of arrangement, and absence of radioactivity in the cooling medium.

The use of the uranium in the form of cylinders rather than spheres has been proposed independently by Szilard and by Wheeler. The water cooling is one of the subjects which have been allotted for consideration to Van Vleck and one of the writers in an early meeting. The use of aluminium tubes was proposed by Creutz. The differential heating was proposed by Szilard. The present plan is rather a compilation of generally known ideas than anything else.

The main difficulties of the plan that we can see at present are the difficulty of establishing a heat contact between the aluminium tubes and the uranium, and the possibility of a corrosion of the aluminium under the joint action of water and radiation. The present plan does not provide for a cooling of the graphite, but we believe that this can be arranged in numerous ways and will not present serious trouble.

The form of the pile is a cylinder with a diameter of about 8 m and a height of 5.5 m. Its volume is 280 m^3 and it requires about 65 tons of uranium, 1/8 of which is to be in the form of metal. The total amount of graphite is 440 tons. The uranium is arranged in the form of cylinders about 5.5 m long and with a diameter 3.6 cm, where it consists of the carbide. Its diameter is 3 cm where it consists of metal. The holes in the graphite have everywhere at least 3 cm diameter leaving a gap around the metal. The cooling pipe has a diameter of 1.5 cm towards the center of the plant, but can be diminished to 1 cm towards the sides of the cylinder. The cylinders are arranged in a triangular pattern similar to that in an early plan of Wheeler. The distance between cylinders is 20 cm, and there are 26 cylinders per m^2. There are altogether 1300 pipes. The cylinders occupy a total volume of 5 m^3. The aluminium tubes are assumed to have a wall thickness of 1 mm which should be sufficient for supporting the weight of a cylinder which is about 52.5 kg. The water is supposed to circulate at 16 m/sec through the center tubes. The velocity will be lower in the peripheral tubes. The following are the calculations for the change of k and the cooling.

The number of mols of water present at any time in the pile is 70,000. The number of mols of graphite is 36.5×10^6. The ratio of the two is 1.9×10^{-3}, giving a total decrease in k of 2 percent. The aluminium gives a decrease of .4 percent. There is a further decrease in k due to the use of cylinders and to the inner surface which we estimate to be 2 percent. Against this there is an increase of k due to the differential heating gap structure and the specific action of water which we estimate to be 1.5 percent. Thus the total decrease in k is about 2.9 percent. The value of k for the conventional arrangement was estimated to be 1.09 in the metal and 1.055 in the carbide. The above decrease of k due to the presence of water can be considerably reduced by use of Dowtherm or carbon fluoride as a cooling liquid. The dimensions of the pile as given above are ample on the basis of these estimates, considering that the structure has great compactness.

The water enters the central tube at 10 atmospheres pressure and assumes a velocity of 15.8 m/sec. The temperature increase of the water in the central pipe is assumed to be 60 °C. The heat transfer coefficient under these conditions is 1.4 cal/cm^2 sec °C, giving a surface temperature drop of 41 °C, in the central pipe. The central pipe carries 1.3×10^5 cal/sec. The total power output is 7.5×10^7 cal/sec, or 310,000 kW. The heat drop in the central cylinder is 400 °C in the uranium at the worst place, so that the temperature at the outer surface is more than 500 °C. At this point the water is under considerable pressure so that its boiling point is sufficiently increased (to about 150 °C) to avoid boiling with a margin of safety. The power requirement for the water circulation is about 1800 kW to overcome friction and about 280 kW for other losses. This, however, does not include heat exchangers. On the whole we have the impression that the power output is too large, and that it would be better to content oneself with about 200,000 kW. This would permit one to use somewhat narrower pipes and also somewhat lower pressures. On the other hand we believe that there are some corrections in the cooling calculation which are favorable because the above calculations were done as if the temperature of the uranium were constant throughout the pile. In actuality it is highest at the center so that the power output is more nearly uniform than we have assumed, (see the graphs of Monk and Wheeler). Hence, the cooling problem is not quite as serious in the central pipe as represented in the calculation.

15.

On a Plant with Water Cooling

Weinberg, Young, Christy, Plass, Wigner, and Williamson

No date, probably Summer 1942

Abstract. An investigation of the feasibility of a 10^5 kW plant with water cooling has been carried out, and rough, preliminary designs for such a plant are included. The effect of the cooling fluid on k is shown to be less than 2.4 percent (including cylindrical arrangement and absorption of aluminium tubes carrying water) even for a 2×10^5 kW output, and the possibility of water as a cooling agent cannot be ruled out on the basis of its adverse effect on k. The estimated amount of metal and carbide required for a cylindrical plant is 9 and 63 tons, respectively. One advantage of a water plant lies in the greater simplicity, compactness, and availability of the external equipment required; a disadvantage is the possible chemical action of water on metal under intense radiation and the difficulty of construction of the lattice cell. It is concluded that each of the various types of proposed plants must be subjected to an exhaustive study, and as detailed designs as possible for each of them drawn up before any objective choice between them can made.

Two weeks ago today Dr. Compton requested us to make as detailed plans for a 100,000 kW plant as we can do within two weeks, and the present note is a summary of our work. We were hoping, at the time we received this request, that we shall have the help of a construction engineer in some phases of our work. It is very likely that it will be evident from several of the constructional details which we present that we had no such help.

We specialized from the beginning on a water cooled plant, and the present note is, substantially, an elaboration of the plans engendered in Report C-140. In that report we mentioned that possibly the main reason for our seeing many advantages in water cooling as compared with helium cooling, is the fact that we know so much less about water cooling. This situation has materially changed since, and we now believe that we see the difficulties of water cooling perhaps to a greater extent than we see those of the helium cooling. The main difficulties are still the establishing of a good heat contact between the uranium cylinders and their lining, and corrosion of the uranium, if we do not use any lining. These difficulties will be discussed more in detail in the next section.

The present report contains a discussion of not one, but at least three, possibilities. On the basis of the present information and with the present knowledge of the purpose of the plant, we were unable to decide among three alternatives. The main alternative is whether or not it should be possible to exchange the uranium rods without dismantling the whole plant. Such an exchange would be, of course, desirable for informative purposes, and also because the pile could be kept in operation in this way for an almost indefinite length of time, and the carbon contained in it would not be wasted. Furthermore, a possible usefulness of the radioactive fission products also requires such a continuous operation. If the plant is to be used only for a few months, and then dismantled, and the resulting materials worked up at that time, all the problems would be simplified very considerably. We do not include sketches for such a plant because they are only too easily seen from the sketches of the other two alternatives included. The whole arrangement would very closely resemble then, that of the usual steam boiler, in which the water is inside of the tubes which are surrounded by a hot material. The main difference which, however, hardly effects the construction, is that most of the heat is produced not in the material surrounding the pipes, but in the pipes themselves. There is, of course, the added difference that the material of the pipes is essentially uranium, rather than steel, and the difficulties caused by this difference are common with those of the two other alternatives.

Both other alternatives have removable uranium pipes that differ by the arrangement at the bottom of the pile. In the first case the pipes stick at the bottom through a stuffing box into a water tank, and one relies essentially on these stuffing boxes for preventing the escape of water vapor into the pile. In the second alternative, the lower connection of the pipes is more elaborate and is made and released by the personnel operating the pile. In this case it is necessary to put a shield below the pile in order to protect the operating personnel below the pile. This shield causes the main difference between these two alternatives. There is a shield above the pile in both cases.

The sketches and the general description of the erection of the pile were done by Mr. Young. Most of the calculations concerning the neutron reproduction factor were done by Mr. Weinberg, with the help of Messrs. Plass and Williamson. We drew also on the help of many other members of the group, and we had interesting discussions with Messrs. Creutz, Szilard, as well as with several members of Dr. Burton's group.

Heat Contact and Corrosion

As emphasized in Report C-140, perhaps the greatest difficulties of water cooling are the chemical problems. Considerable information has been collected since on this question which we want to summarize in the following:

Experiments by Mr. Howe proved that boiling water attacks uranium to a considerable extent. This makes it undesirable, if not impossible, to use water within unlined uranium pipes.

Two paths are open from here on. Either one substitutes some other liquid, or introduces some lining within the uranium. In connection with the first alternative, liquids without oxygen content suggest themselves, such as, biphenyl. The main difficulty in the use of these liquids is the polymerization of these liquids, which would make a constant operation of a distilling plant necessary. For this reason this possibility has not been explored so far, but we can return to it if necessary. Use of some of these liquids may offer advantages also from the point of view of heat transfer and reproduction factor over the use of water.

There are only a few metals which can be considered for thick lining of the uranium pipes. These are, beryllium, magnesium, aluminium, tin, lead and bismuth. Among these aluminium has been rather extensively tested for its corrosion properties if in contact with hot water and radiation. These tests give favorable results, and although they evidently should be further continued, it seems likely that aluminium will have the necessary resistance. Unfortunately Creutz did not succeed so far to mould the aluminium lining into the uranium without considerable alloying between the two materials. The test on the resistivity of aluminium to corrosion was made in pure aluminium. However, the possibilities are far from being exhausted in this connection, and at least three possibilities of procedure remain open.

Anderson has sprayed tin on freshly sandblasted uranium and obtained a surface that adhered reasonably well. If it should prove possible to make a well adhering tin layer, it is very likely that 2 half-pipes of uranium could be soldered together in the ordinary fashion, and then welded together on the surface for added strength. This procedure has been suggested and is being tried out by Creutz.

The beryllium-aluminium alloys offer another path that has to be tried out. Practically nothing is known of the behavior of these alloys in contact with uranium.

On the whole, we feel much better on the question of heat contact now than we did a few weeks ago. It seems that several materials probably stand up under the conditions to be expected, and it also seems that it is not as difficult to get good metallic contact between uranium and other metals as we feared. Furthermore, Creutz has shown that uranium is an agreeable metal to work with, that pieces of uranium can be welded together, and that the metal casts well. However, we can expect to use uranium metal only in the center of the pile. UO_2 or UC_2 will have to be used at the outer portions. Unfortunately we hardly know yet anything about the resistivity of these materials, and the possibility of lining them with other metals. In the following considerations it will be assumed, nevertheless, that we can establish very good heat contact between the lining and the pipes, whether these be metallic or not. This is to some degree justified by the apparently perfect continuity of the transition between some pieces of aluminium and beryllium welded to uranium and the uranium itself.

Comparison with Previous Plan

The most important change as compared with the plan envisaged in C-140, is the lower power output. This makes it possible to use narrower water pipes than we expected to use when making that report, and thus, have a smaller decrease in k. The inner diameter of the pipes will be assumed as only 1 cm as contrasted with the 1.5 cm used before. This decreases the drop in k, probably to less than 2 percent. However, exact calculations were made only on the older arrangement which gives a somewhat even lower decrease of k than estimated there, i.e., of the amount of 2.4 percent, rather than the 2.9 percent estimated.

The other changes are of less substantial nature. Most apparent of them is perhaps the proposed use of iron shields (instead of the lead shields). This makes it necessary to increase the thickness of the shield by a factor $2\frac{1}{2}$. However, the new shields are more effective in slowing down neutrons than the old ones were, and we estimate that they give almost the same protection against both neutrons and gamma rays, if they contain 16 volume percent water, as the old lead and water shields together. Furthermore, they are probably less expensive and can be used also as supports. Another addition is several storage tanks for the helium which surrounds the pile. These storage tanks serve for the purpose of taking care of the expansion of the helium caused by the increased temperature of the plant, and also to make it possible to flush the helium over some drying agent in order to take out the last traces of water. At the same time some of the radioactive materials will be eliminated.

All the rest of the changes are of an inconsequential nature, and can be readily seen from the sketches and their descriptions.

Design and Construction

The aim of the present section is to give a sketch of what appears to be a possible design for a 100,000 kW water cooled plant. Not all of the details have been worked out, nor are we confident that we have made the best choices from among various alternatives. It is hoped, however, that we have made possible ones, and that the outline to be presented is not misleading in any serious way.

The pile proper is in the form of a cylinder about $5\frac{1}{2}$ meters high and 8 to 9 meters in diameter. The uranium is placed in the form of vertical tubes with cooling water flowing down their centers. These tubes hang from the bottom of a water tank above the pile, pass down through vertical holes in the graphite structure, and discharge into another water tank beneath the pile. Upon removing the lid of the upper water tank the uranium tubes may be lifted out of the pile.

The structure is to be placed underground, resting on a concrete floor and enclosed laterally within a concrete wall which in turn is surrounded by earth. The pile is shielded on top and made gas tight so far as possible, so that it can be approached safely from above. The gas within the structure is allowed to expand into an underground atmospheric pressure well.

The control rods work vertically through the top of the pile, extending upward into a room above the pile. Over this is a ceiling about level with the earth's surface, and this is covered with earth.

We pass to a more detailed description of the plant by outlining the approximate procedure which would be gone through in building it. It is to be understood that much of what follows is suggestive only, but for the sake of definiteness we shall state dimensions and show drawings even when these have not been precisely worked out.

(1) Dig a hole in the earth 15 meters deep and 15 meters in diameter, within a suitable earth restraining wall, if such is needed.

(2) Lay a concrete floor of suitable thickness to cover the bottom of the hole.

(3) Concentric within the hole erect a concrete or reinforced concrete wall of 9 meters inside diameter and 6 meters in height, of suitable thickness and taper and with proper footings and stiffening columns. Through the wall near the floor pass two pipes, one leading to the water pumps and one to the gas expansion chamber.

(4) Install the bottom water tank. This is in the form of a shallow tray about 1/2 meter deep resting on the floor and extending to the wall. A hole in one side of it joins to the water pump pipe. The water in this tank is not under pressure.

(5) Place supports resting on the tank bottom and extending to the height of the tank sides, and on these lay the floor for the pile which is also the tank lid. This is pierced with openings through which the ends of the cooling tubes will later descend, and fastened around each opening on the underside of the floor is a spring collar which grips around the tubes as they come through. This is to give somewhat of a seal around the rod to prevent water splashing up into the pile, and yet permit the rod to be pulled up freely from above.

(6) The graphite is then stacked in the space above the floor up to the height of and out to the surrounding concrete wall. In it are to be left cracks to allow for its expansion when it becomes hot; some 1300 holes of about 5 cm diameter running vertically through it and kept in vertical alignment with the holes in the floor; and 7 vertical holes of about 12 cm in diameter for the central control rods, one in the center and 6 on a circle about the center of some 4 meters in diameter. The pipe leading to the gas expansion chamber is just above floor level, and enough path should be left for the pile to be able to easily expand its gases out through this opening.

The vertical holes must be aligned with some care so that the uranium and control rods will be able to slide freely up and down. This will presumably be checked with dummy rods hung from an overhead frame as the stacking of the graphite progresses. The expansion cracks are to be differentially patterned about the center in accordance with the expected distribution of temperature in the operating plant.

(7) As the graphite stacking progresses, or perhaps after it is finished, earth is filled in outside the concrete wall to the sides of the hole and up to a little below the top of the wall. The time at which this is done is immaterial, except that its pressure on the wall is to be taken into account.

(8) Install the beams over the pile which rest on the wall and are to support the shield and upper water tank.

(9) Install the shield sections, being careful to align the holes therein with those in the graphite.

(10) Install the control rod guide tubes and seal the shield sections together so that they represent a continuous gas-tight disc except for the control and uranium rod openings.

(11) Sealed to the shield and extending outward and downward from it place a thin metal gas cover which extends a ways down the outside of the concrete wall, is fastened to it, and overlaid with earth or possibly another thickness of mortar.

(12) Install the upper water tank sides and water mains and connections.

(13) Install roof beams over the entire hole.

(14) Install pile derrick.

(15) Install control rods and their mechanism.

(16) Add uranium rods until pile will operate. The left over holes around the edge of the structure may be plugged, or fitted with water pipes to cool the graphite there.

(17) Put on water tank lids and run preliminary tests on plant.

(18) Close over the roof and cover with earth.

Figures 1 and 2 show the pile structure as a whole. Earth is filled in around it ultimately, to the height of the pile proper or perhaps a little above this. The roof passes above the tops of the control rods. A more detailed sketch of the base construction is given in Fig. 3.

Figure 5 illustrates how the uranium rods appear in the pile, and Fig. 4 shows the spacing of the openings left for these in the graphite and shield. The holes in the pile floor above the bottom water tank are similarly spaced but the openings are smaller. Fig. 11 gives more details about the top of the uranium rod; Fig. 7 of the bottom; and Fig. 6 of the cross-sections in the intermediate regions.

The radiation and neutron shield is of iron shot and water and about 60 cm in thickness. The holes left in it to accomodate the uranium rods are partially plugged by an iron sheath around the water tube, as shown in Fig. 11.

For the control rod disposition we picture the following arrangement: the 6 outer rods are the emergency controls, and the center one is the continuous adjustable control. The outer are simply solid iron, with perhaps a salting of boron to increase the neutron absorption. They must be adequate to stop the chain reaction and keep it stopped regardless of the position of the central rod, the temperature of the pile, and of whether the cooling water is in position or not. These solid rods are themselves radiation plugs in the holes they require in the shield, and since they are either completely in or completely out of the pile it can easily be arranged to have them effect a good gas seal in either of these extreme positions, as illustrated in Fig. 15. Only when the rod is moving will there be gas leakage, and it is hoped that this can be sufficiently reduced by having the rod slide through stuffing boxes which also serve to guide its motion.

Unlike the emergency rods, the central rod may be at any position in the pile. It has to be cooled, which is most simply done by water circulation, and by dissolving boron in the water it can be arranged to have most of the heat produced directly in the solution itself without having to get it transferred there from the tube walls. This leads us to the type of control rod pictured in Fig. 16. The plug at the top gives a radiation seal when the rod is entirely down. A lead tube projecting upward around it above the water tank can be employed to make it safe to be nearby when a rod is raised.

Like the emergency rods the central one is to work through stuffing boxes. An auxiliary device can be used to effect a tight seal by clamping around the rod whenever it is at rest, and to electrically release from the control circuit whenever the rod is to be moved. This device is merely indicated on Fig. 15.

Alternatives would include such things as operating the control rods from the side instead of the top, or sealing the central tube in the pile and varying the height of the boron solution column in it by controlling a faucet.

The pressure tank arrangement shown in Fig. 13 breaks the whole shield and tank into 24 sections of about equal size, and permits pressure reduction in the outer sections (except for the one used as a water main to the inner tanks) where the cooling requirements are less.

Construction at top of pile is shown in Fig. 12.

A more positive seal against water vapor at the bottom of the pile is shown in Fig. 3. A still stronger one could be obtained by increasing the height of the bottom water tank and putting a shield above it, so that a man could go below to work, and having the rods clamped tightly with nuts. The connection to the upper water tank can of course be screwed down tightly in any case if desired.

Better conditioning of the pile gas can be obtained by having another gas pipe near the top, in addition to the one pictured previously near the bottom, and maintaining circulation of the gas over drying materials to absorb water vapor and substances to absorb as much of the harmful gases as possible.

Referring again to Fig. 8, if the bottom sealing fixture be threaded, one can get a positive screw seal at the bottom by turning the rod from the top, without need for men to go beneath the pile. Turning of the rod may be effected by prongs sticking downward from the removable plug into slots left in the uranium, or attached to the uranium and projecting upward into slots in the plug.

Problems Remaining for Water Cooled Design

(1) Most important is the construction of the uranium rod itself, which apparently must be protected from the action of water. This may be done (a) by a thin protective coating sprayed or plated on, or (b) by a thicker lining tube of low absorbing material such as Al or Be. To get adequate heat transfer it appears that this tube must be welded or alloyed or soldered to the uranium. See also (7).

(2) Estimate pile dimensions as accurately as possible, and the error therein, before building begins.

(3) Run hydraulic flow and heat transfer tests on a sample rod to get a closer estimate on water pressure and pumping power required to give necessary cooling.

(4) Design external heat exchanger, etc., taking into account the available environment temperatures. This may possibly change estimate for water temperature rise in the pile.

(5) Decide rod arrangement to seal against water vapor at bottom. It is still possible that a gravity held gasket, with pile gas circulated past a drying agent, should be entirely feasible.

(6) More careful calculation on relative heats produced in iron and boron (or other) in the central control rod.

(7) Construction of carbide rod sections, if this is needed. This is related to (1), and perhaps more troublesome.

(8) More details on gas expansion, storage, and drying chambers.

(9) Perhaps temperature of concrete should be estimated, and cooling pipes allowed for, if needed.

Water Cooling of 10^5 kW Plant

The pile is in the form of a cylinder, so that the activity in general and the heat production in particular vary about as

$$J_0 \left(\frac{2.4r}{R} \right) \cos \frac{\pi z}{H} .$$

Here R is the radius of the cylinder and H is its height.

The uranium is in the form of tubes placed parallel to the axis of the cylinder. These tubes are cooled by water flowing through them; while the heat produced in the carbon is handled primarily by radiation to the outer surfaces of the tubes. This maintains the carbon at higher temperature than the uranium, and thus achieves differential heating. The annular gap between uranium and carbon also allows for the tubes to be easily changed; helps to keep the fission products from diffusing into the graphite; and has a beneficial effect on k by making it easier for thermal neutrons to move to the uranium. On the other hand, such gaps slightly increase the leakage of neutrons from the pile.

The heat load of the central pipe is 2.3 times the average load for all pipes, and the heat load per unit length at the center of a pipe is 1.6 times the average for the entire pipe.

The present plant is to have heat production of 10^5 kW in the metal, which gives to some extent a measure of the chemical output. In addition, some 6000 kW is produced in the graphite. The desired output has been reduced since report C-140 was written, and we take advantage of this to reduce the amount of water in the pile and hence also the loss in k caused by its presence. Otherwise the dimensions are the same as before.

The heat load per tube decreases from the center outwards, and corresponding continuous gradation of pressure or water cross-section could be made. Here we have kept the cross-section constant and provided 2 pressure values by separating the water tank into 2 divisions.

Some typical values and estimates are as follows:

Water stream diameter	= 1 cm
Water velocity in central tubes	= 10 meters per sec
Water velocity in outer tubes	= 7 meters per sec
Pressure in central tanks	= 100 lbs per in^2
Pressure in outer tanks	= 50 lbs per in^2
Pumping power for pile	= 500 kW
Temperature rise of water in center tube	= 60 °C
Maximum film drop	= 30°C
Maximum metal temperature above that of incoming water, excluding drop at Al-U boundary	= 325 °C
Maximum carbon temperature	= 750 °C

These values are to be regarded as about the best obtainable, and safety factor margins allowed therefrom.

The Effect of Water on k

1.) In determining the feasibility of a plant with water cooling, the effect of the water on the multiplication factor k is, after all, the critical consideration – for if the increased absorption of thermal neutrons due to the presence of the water reduces k below unity, a water plant is automatically ruled out. Fortunately this is not the case; preliminary calculations (C-149) give 2.9 % as an upper limit to the decrease in k caused by the water. If k with metal but without water is 1.10, the effective k in a water cooled plant would then be about 1.07. This preliminary estimate was based on rather rough calculations in which, among other things, the effect of differential heating was ignored. The present calculation was undertaken to get a more accurate estimate of the effect of water on k, in particular to estimate the effect of the differential heating, the results will still overestimate the *deleterious* effect of the water for 3 reasons:

1.) The effect of gaps will be neglected.

2.) The resonance neutron "reservoir" effect caused by the particular variation of the mean free path in water will be neglected.

3.) The production of hot neutrons (1000°K) in the graphite and the production of cold neutrons (400°K) in the water will be assumed to be uniform throughout each of the media. Actually *a majority of the cold water* neutrons on entering the hot carbon, will come into thermal equilibrium with the carbon close to the surface; consequently they may be absorbed by the metal with less probability of capture in the carbon than neutrons which are made hot near the periphery of the lattice cell.

2.) The method of calculation used here is a generalization to three regions (water, metal, graphite) of the scheme used by R. F. Christy and A. P. Monk to compute thermal utilizations in two-phase cylindrical systems (C-104, C-171). Because of the greatly increased complexity of the 3-region problem the computation of thermal utilization and resonance absorption was made only for a single geometric configuration: namely, a water cylinder (.75 cm radius) surrounded by a metal shell (thickness 0.75 cm), both encased in a graphite cylinder of outer radius 11.1 cm. The carbon is supposed to be at 1000°K, the water at 400°K; hot neutrons (1000°K) are assumed to be produced only in the carbon, cold neutrons (400°K), only in the water. The distributions of 1000° and 400° neutrons are computed separately, and from the results of this calculation the overall thermal utilization, p_2, defined as absorption of neutrons in metal/overall production of thermal neutrons per cell is calculated. Finally the resonance disadvantage factor and the resonance absorption, $\frac{1}{p_1}$, are determined, and the critical reproduction factor $\eta' = \frac{1}{p_1}\frac{1}{p_2}$ is found. A corresponding calculation for a 2-region system consisting of a 1.25 cm radius metal cylinder in an 11.1 cm radius graphite block is presented for comparison purposes.

We have, in our problem, three regions. The inner region, containing water, will be region 0, the second one, which contains the metal, will be region 1, the third, containing the graphite, region 2. The calculation neglected the gap and was carried out only for metal of density 18. The constants used in the 2 and 3 region calculations were the same, except that in the thermal region, the substances of the 2-region comparison problem were at the lower temperature (400°K). In this case, region 0 is metal, region 1, carbon.

As mentioned above, we shall distinguish only two kinds of thermal neutrons: the "cold" or 400°K neutrons which are produced only in the water, and the "hot" or 1000°K neutrons which are produced in the hot graphite. The usual diffusion equations hold for both kinds of neutrons. However, the cold neutrons, upon reaching the graphite, will be warmed up and this will appear as an additional absorption of the cold neutrons. The neutrons "absorbed" in this way will appear, on the other hand, as hot neutrons and there is an additional production of hot neutrons on this score. This production will be concentrated in the neighborhood of the U but will be assumed, in the calculation, to be uniform over the whole graphite. This is the pessimistic assumption 3 mentioned above.

Similarly, hot neutrons, created in the graphite, will be further slowed down if they reach the water. This process again will be taken into account as an absorption of hot and production of cold neutrons.

The constants used for the thermal region were as follows: The carbon absorption cross section was assumed to be 1060 times smaller than the U cross section. This is a reasonable average between the ratios given by Fermi for different types of graphite. The H absorption cross section is 18.5 times smaller than that of U. Specifically, we assumed at 400°K for water, U and C, respectively $\sigma_a = .3 \times 10^{-24} \times 2$, 5.6×10^{-24}, $.0053 \times 10^{-24}$ cm^2; at 1000°K (the absorption cross sections are inversely proportional to the square root of the temperature) we assumed the values $.19 \times 10^{-24} \times 2$, 3.56×10^{-24}.

$.00336 \times 10^{-24}$ cm^2. In ordinary problems, only the ratio of the absorption cross sections of the different substances present plays any role. In the present problem, however, the ratio of these cross sections to the apparent absorption cross section, corresponding to the removal of hot neutrons by cooling and of cold neutrons by heating, is also relevant. The determination of these figures is at present not possible without ambiguity and our values are, therefore, only guesses. We may mention, however, that Dr. Fermi, whom we asked for advice at this point, made substantially the same guesses which we made. In order to decrease the temperature difference between a neutron and carbon to $1/e$ of its original value, one needs $1/.158$ collisions. Since the scattering cross section of Carbon is 4.8×10^{-24}, we took for the apparent absorption coefficient of cold neutrons in hot graphite the value $4.8 \times 10^{-24} \times .158 \times .7$ cm^2, the last factor .7 being designed to take into account the fact that even after $1/.158$ collisions there remains a certain energy difference between neutron and graphite. This gave $.53 \times 10^{-24}$ for the apparent absorption cross section of C for cold neutrons which, together with the real absorption gives $\sigma_{a2} = .5353 \times 10^{-24}$ cm^2 for cold neutrons.

The estimate for the "cooling" of hot neutrons in water is even more ambiguous. We took 30×10^{-24} cm^2 for the scattering cross section, because of the somewhat elevated temperature. However, because of the chemical effect, we assumed that only $\frac{1}{2}$ of all collisions are effective, the rest not leading to an energy exchange. The factor .158 is 1 in case of hydrogen but we kept the factor .7 for safety also in this case. We were thus led to an apparent absorption coefficient of hot neutrons per H$_2$O molecule of $30 \times \frac{1}{2} \times 1 \times .7 \times 2 \times 10^{-24}$ cm^2. Adding to this the real absorption coefficient, we obtained $\sigma_{a0} = 21.38 \times 10^{-24}$ cm^2 for hot neutrons. In addition to the ratio of the σ_a, the usual calculation of thermal utilization needs the κ. These were obtained on the basis of the values at ordinary temperatures, by the usual formula, in which the temperature dependence of the absorption cross section was taken into account. We used for cold neutrons

$$\kappa_0 = .325 \qquad \kappa_1 = .71 \qquad \kappa_2 = .22 \quad \text{cm}^{-1}$$
$$N_0 \sigma_{a_0} = .020 \qquad N_1 \sigma_{a_1} = .255 \qquad N_2 \sigma_{a_2} = .0434 \text{ cm}^{-1}.$$

The second line gives the absorption cross section per cm^3 in the three regions. N_0, N_1, N_2 denote the number of molecules or atoms per cm^3. For hot neutrons, we used

$$\kappa_0 = 1.44 \qquad \kappa_1 = .600 \qquad \kappa_2 = .0180 \text{ cm}^{-1}$$
$$N_0 \sigma_{a_0} = .715 \qquad N_1 \sigma_{a_1} = .162 \qquad N_2 \sigma_{a_2} = 2.76 \times 10^{-4} \text{ cm}^{-1}.$$

When calculating κ_2 for cold and κ_2 for hot neutrons, the apparent absorptions had to be taken into account along with the real absorption. In the comparison problem, with two regions, we assumed $\sigma_{a0} = 5.6 \times 10^{-24}$. Hence

$$N_0 \sigma_{a_0} = .255 \qquad N_1 \sigma_{a_1} = 4.34 \times 10^{-4} \text{ cm}^{-1}$$
$$\kappa_0 = .71 \qquad \kappa_1 = .022 \text{ cm}^{-1}.$$

The constants for the calculation of the resonance absorption were the usual ones. Creutz's formula for the effective absorption in spheres

$$\left(\int \frac{\sigma \, dE}{E} \right)_{\text{eff}} = 240 \times 10^{-24} \times .0365 \left(1 + \frac{8}{r\rho} \right)$$

shows that the ratio of surface and mass absorptions is 8/3 gr cm^{-2}. Hence, in our case of pipes, we have[2]

$$\left(\int \frac{\sigma \, dE}{E} \right)_{\text{eff}} = 240 \times 10^{-24} \times .0385 \left(1 + \frac{16}{3(r_1 - r_0)\rho} \right)$$

where $\rho = 18$ was the density of the metal. The slowing down powers of H_2O and C were taken as $19 \times 1 \times 2 + 4 \times .12 = 38.48$ and $4.8 \times .158 = .76$. For the calculation of the improvement factor, the density of the neutrons in the three regions is needed. These were obtained by solving a three region problem, similar to that occurring at the calculation of the thermal utilization. The absorption cross section of U was assumed, (on the basis of Creutz's measurement) to be the same as in the case of the oxide i.e.

$$\sigma_{a_2} = \frac{240 \times 10^{-24} \times .05 \times 1.2}{\ln 1500} = 1.97 \times 10^{-24} \text{ cm}^2$$

This, together with the scattering cross section of 13×10^{-24} cm^2 gave

$$\kappa_1 = .405, \qquad N_1 \sigma_{a_1} = .0895$$

Since we assumed the same absorption cross section for oxide and metal, the width of the resonance absorption band must be made smaller in the latter case than it was in the former. The ratio of the absorptions was assumed to be $.0365/.050 = .73$ which gives $\ln E_{\text{max}}/E_{\text{min}} = .73 \ln 1500 = 5.35$. This is a slightly pessimistic assumption because the surface absorption plays a greater role in the oxide than in the metal, or $E_{\text{max}}/E_{\text{min}} = 210$. In the carbon, $1/.158$ collisions are necessary to change the energy of a neutron by e, and

$$\frac{\sigma_2}{\sigma_{a_2}} = \frac{\ln 210}{.158} = 34$$

collisions are necessary to carry a neutron over the resonance region. Of course, σ_{a_2} is an apparent absorption, just as in the thermal region. In water, we need

$$\frac{\sigma_0}{\sigma_{a_0}} = \frac{\ln 210}{38.48} \times 42 = 5.8$$

collisions to carry a neutron over the resonance region. This gave with $N_2\sigma_2 = .392$ and $N_0\sigma_0 = 1.4$

$$\begin{array}{lll} N_0\sigma_{a_0} = .241 & N_1\sigma_{a_1} = .0895 & N_2\sigma_{a_2} = .0114 \\ \kappa_0 = .316 & \kappa_1 = .405 & \kappa_2 = .108 \ . \end{array}$$

The latter were calculated by the usual formulae.

---

[2] Note discrepancy between .0365 and .0385. (A.M.W. and A.M.P.)

3.) The neutron distributions in the three regions are determined as solutions of diffusion equations subject to the conditions that the neutron and neutron current densities be continuous across the interfaces and the neutron current vanish at the periphery of the cell. (cf. C-104). The distributions so determined are:

$$vn_0 = AI_0(\kappa_0 r) + \frac{q_0}{N_0 \sigma_{a_0}}$$

$$vn_1 = BI_0(\kappa_1 r) + E\,K_0(\kappa_1 r)$$

$$vn_2 = DC_{01}(\kappa_1 r_1,\,\kappa_2 r) + \frac{q_2}{N_2 \sigma_{a_2}} \tag{1}$$

where subscripts 0, 1, 2 refer to the water, metal, and graphite, respectively, (except where they are used as indices of Bessel functions) and the symbols have the following definitions:

$$A = \frac{1}{S}\left\{\frac{q_2}{N_2 \sigma_{a_2}} - \frac{q_0 T}{N_0 \sigma_{a_0}}\right\}$$

$$B = AP + \frac{q_0}{N_0 \sigma_{a_0}} G$$

$$E = AQ + \frac{q_0}{N_0 \sigma_{a_0}} H$$

$$Q = -\frac{\kappa_1 r_0}{N_1 \sigma_{a_1} r_0}\{N_0 \sigma_{a_0} \kappa_1 I_1(\kappa_0 r_0) I_0(\kappa_1 r_0) - N_1 \sigma_{a_1} \kappa_0 I_0(\kappa_0 r_0) I_1(\kappa_1 r_0)\}$$

$$P = \frac{1}{I_0(\kappa_1 r_0)}\{I_0 - K_0 Q\}$$

$$G = \frac{1}{I_0(\kappa_1 r_0)}\{1 - \kappa_1 r_0 I_1(\kappa_1 r_0) K_0(\kappa_1 r_0)\}$$

$$H = \kappa_1 r_0 I_1(\kappa_1 r_0)$$

$$S = PI_0(\kappa_1 r_1) + QK_0(\kappa_1 r_1)$$
$$- \frac{N_1 \sigma_{a_1} \kappa_2}{N_2 \sigma_{a_2} \kappa_1} \frac{PI_1(\kappa_1 r_1) - QK_1(\kappa_1 r_1)}{C_{11}(\kappa_2 r_1,\,\kappa_2 r_2)} C_{01}(\kappa_2 r_1,\,\kappa_2 r_2)$$

$$T = GI_0(\kappa_1 r_1) + HK_0(\kappa_1 r_1)$$
$$- \frac{N_1 \sigma_{a_1} \kappa_2}{N_2 \sigma_{a_2} \kappa_1} \frac{GI_1(\kappa_1 r_1) - HK_1(\kappa_1 r_1)}{C_{11}(\kappa_2 r_1,\,\kappa_2 r_2)} C_{01}(\kappa_2 r_1,\,\kappa_2 r_2)$$

$$C_{01}(x, y) = I_0(x) K_1(y) + K_0(x) I_1(y)$$
$$C_{11}(x, y) = I_1(x) K_1(y) - K_1(x) I_1(y)$$

$\frac{1}{\kappa}$ Diffusion length
σ_a Capture cross-section
N Number of atoms per c.c.
v Neutron velocity
q Rate of production of neutrons (per c.c. per second)

The 400°, 1000° and resonance neutron distributions computed from (1) are given in Figs. (1), (2), and (3). The values of the constant used are indicated on the figures. It should be noted that the production of 400° neutrons is zero in the carbon, the production of 1000° neutrons is zero in the water. In Table 1 the total thermal neutron absorption in each of the three regions per unit production are given:

Table 1. Total neutron absorption in three regions

	Abs. in H_2O	Abs. in metal	Abs. in graphite
400°	.019609	.32044	.6600
1000°	.4738	.4416	.0846

In order to better estimate the effect of the water, a comparison calculation for a 2 region system consisting of a 1.25 cm radius metal cylinder concentric with an 11.1 cm radius graphite cylinder has been carried out. The distributions at thermal and resonance energies are shown in the figures[3]. The thermal utilization factor, $\frac{1}{p_2}$, for such an arrangement is 1.185, the resonance absorption factor, $\frac{1}{p_1}$, is 1.100 and so $\eta' = \frac{1}{p_1} \times \frac{1}{p_2} = 1.304$.

To compute the thermal utilization from the thermal neutron densities in the three regions, we must first compute the total slowing down powers, P_0 and P_2 of the water and the graphite. Taking the effective hydrogen scattering cross section to be 15×10^{-24} (which underestimates the scattering power of the water), we have

$$P_0 = 15 \times 2 \times .5625\pi \times 1 \times \frac{.602}{18} = .564\pi$$

$$P_2 = 4.8 \times .158 \times 120.96\pi \times \frac{1.64 \times .6}{12} = 7.55\pi \,;$$

these numbers may be taken to be proportional to the production of neutrons due to the slowing down. The number of cold neutrons produced in water per c.c. per second by cooling of hot neutrons (normalized to unit density of production) is obtained from Table 1 by multiplying the average hot absorption, .4738, by the relative cooling cross-section, $\frac{.38}{21.38}$. Thus of every .4738 hot neutrons absorbed in water, .4653 are "absorbed" by cooling, .0085 are absorbed by capture. In the same way one finds that of every .6600 cold neutrons absorbed in graphite, .654 are "absorbed" by heating, the rest by true capture. The total production of cold neutrons in water, P_0^*, and the total production of hot neutrons in carbon, P_2^*, are therefore related by

$$P_0^* = .564 + .465 P_2^*$$

$$P_2^* = 7.55 + .654 P_0^*$$

from which we have

$$\frac{P_0^*}{P_0 + P_2} = .721 \qquad \frac{P_2^*}{P_0 + P_2} = 1.403\,.$$

---

[3] These figures are not available in original. (A.M.W. & A.M.P.)

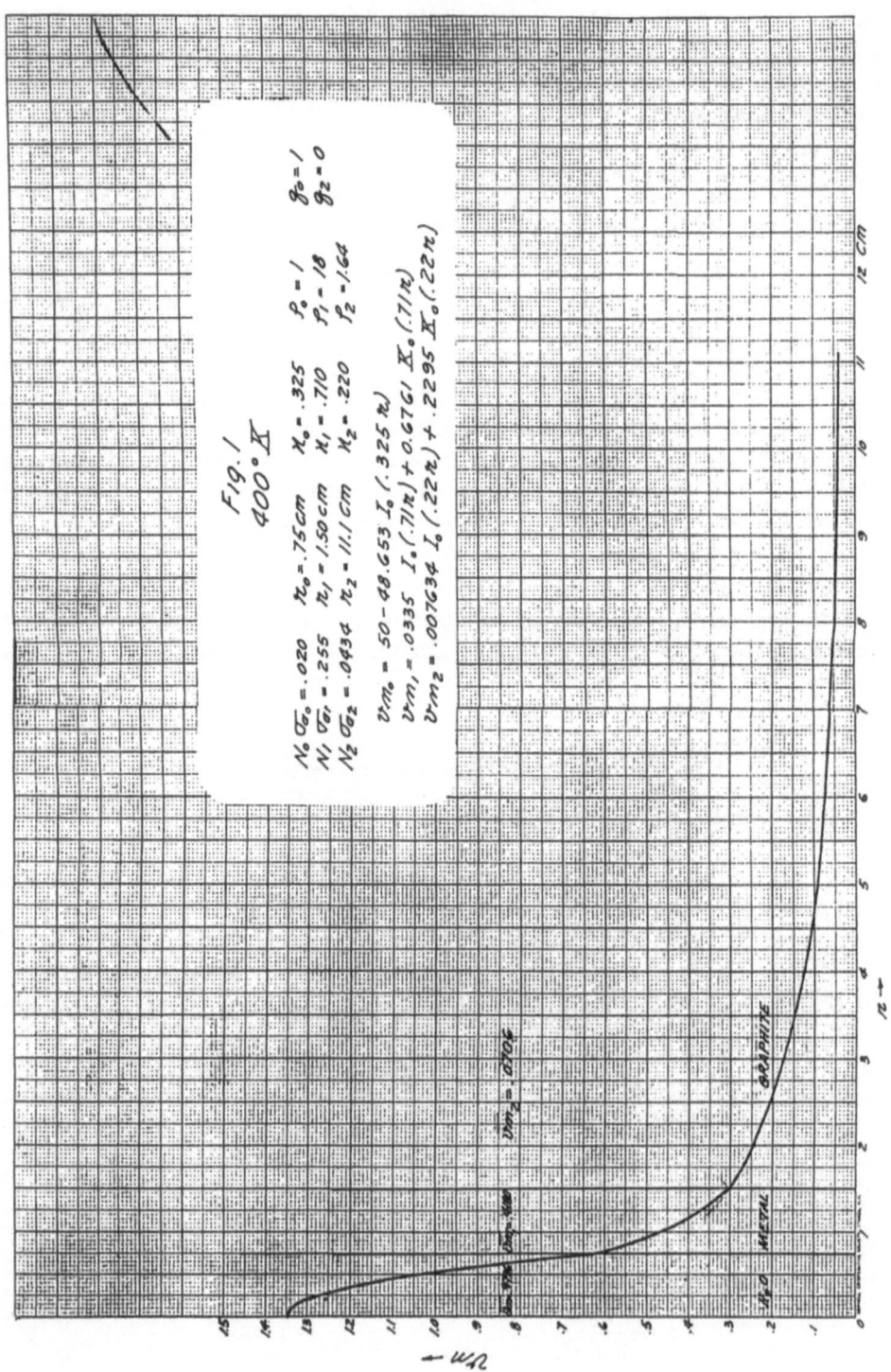

Fig. 1
400° K

$N_0 \sigma_0 = .020$ $\pi_0 = .75\,cm$ $x_0 = .325$ $\rho_0 = 1$ $g_0 = 1$
$N_1 \sigma_1 = .255$ $\pi_1 = 1.50\,cm$ $x_1 = .710$ $\rho_1 = .18$ $g_2 = 0$
$N_2 \sigma_{G_2} = .0434$ $\pi_2 = 11.1\,cm$ $x_2 = .220$ $\rho_2 = 1.64$

$\nu m_0 = 50 - 48.653\, I_0\,(.325\,\pi)$
$\nu m_1 = .0335\, I_0\,(.71\pi) + 0.6761\, K_0\,(.71\pi)$
$\nu m_2 = .007634\, I_0\,(.22\pi) + .2295\, K_0\,(.22\pi)$

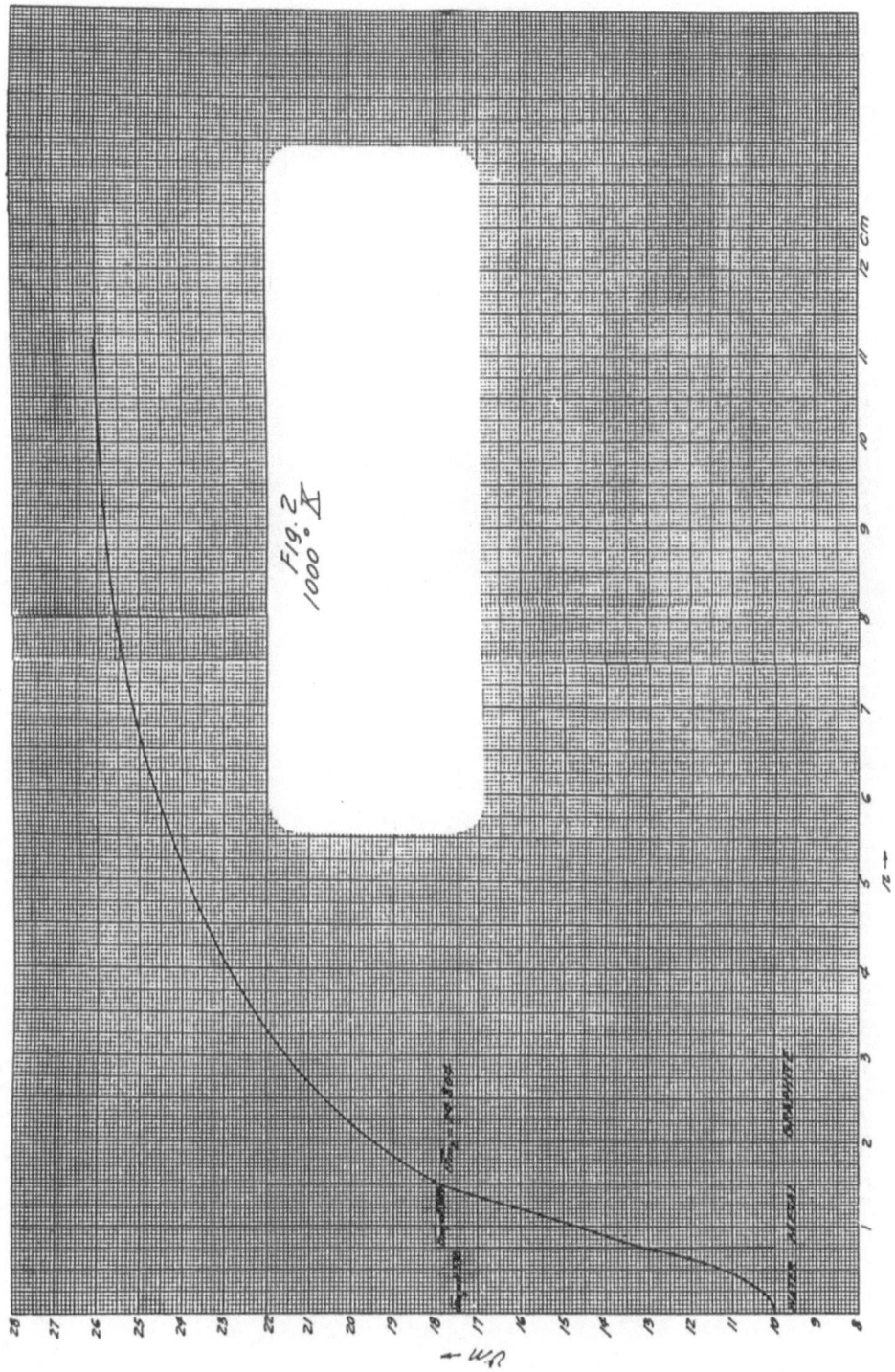

Fig. 2

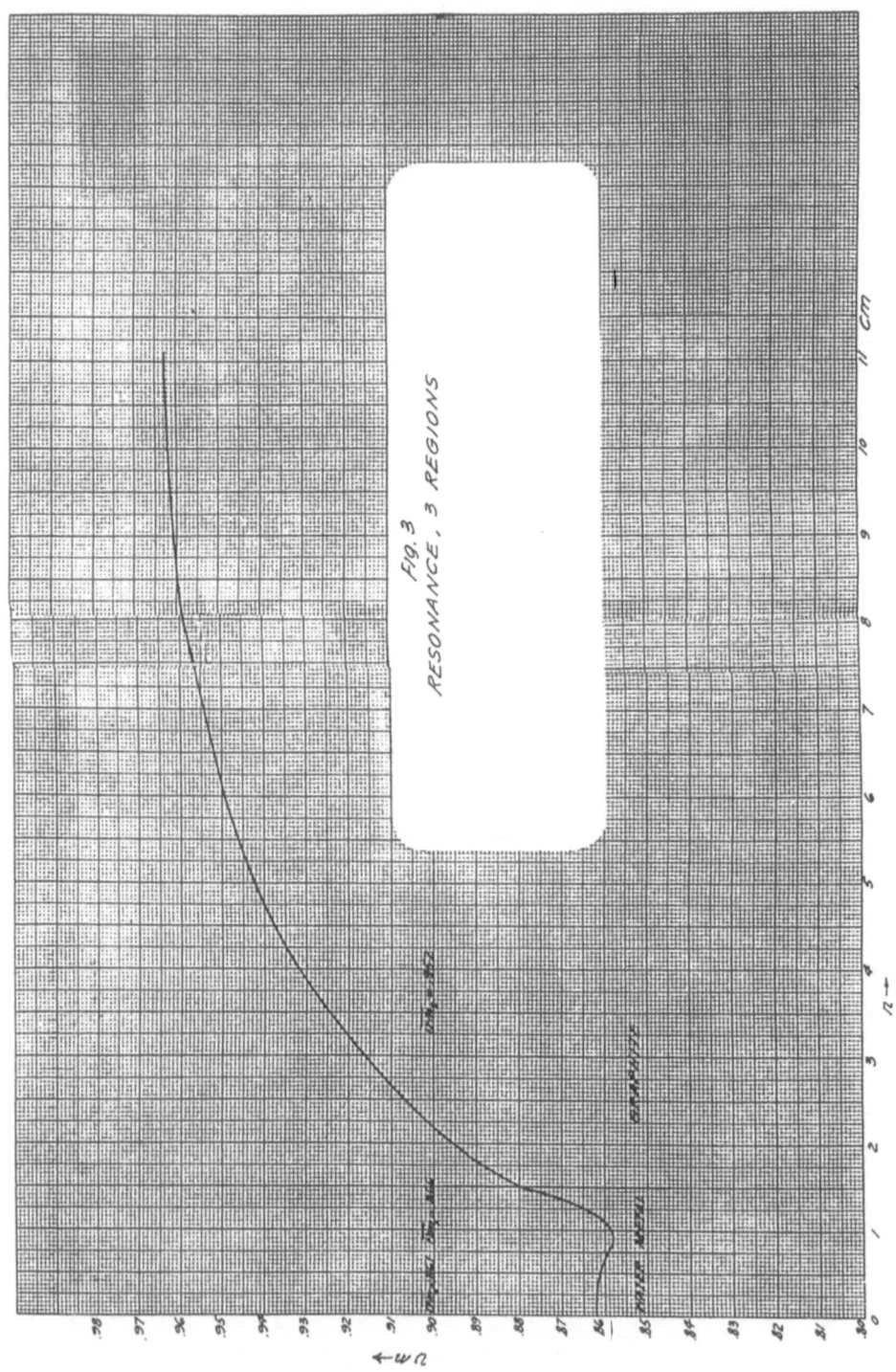

Fig. 3

RESONANCE, 3 REGIONS

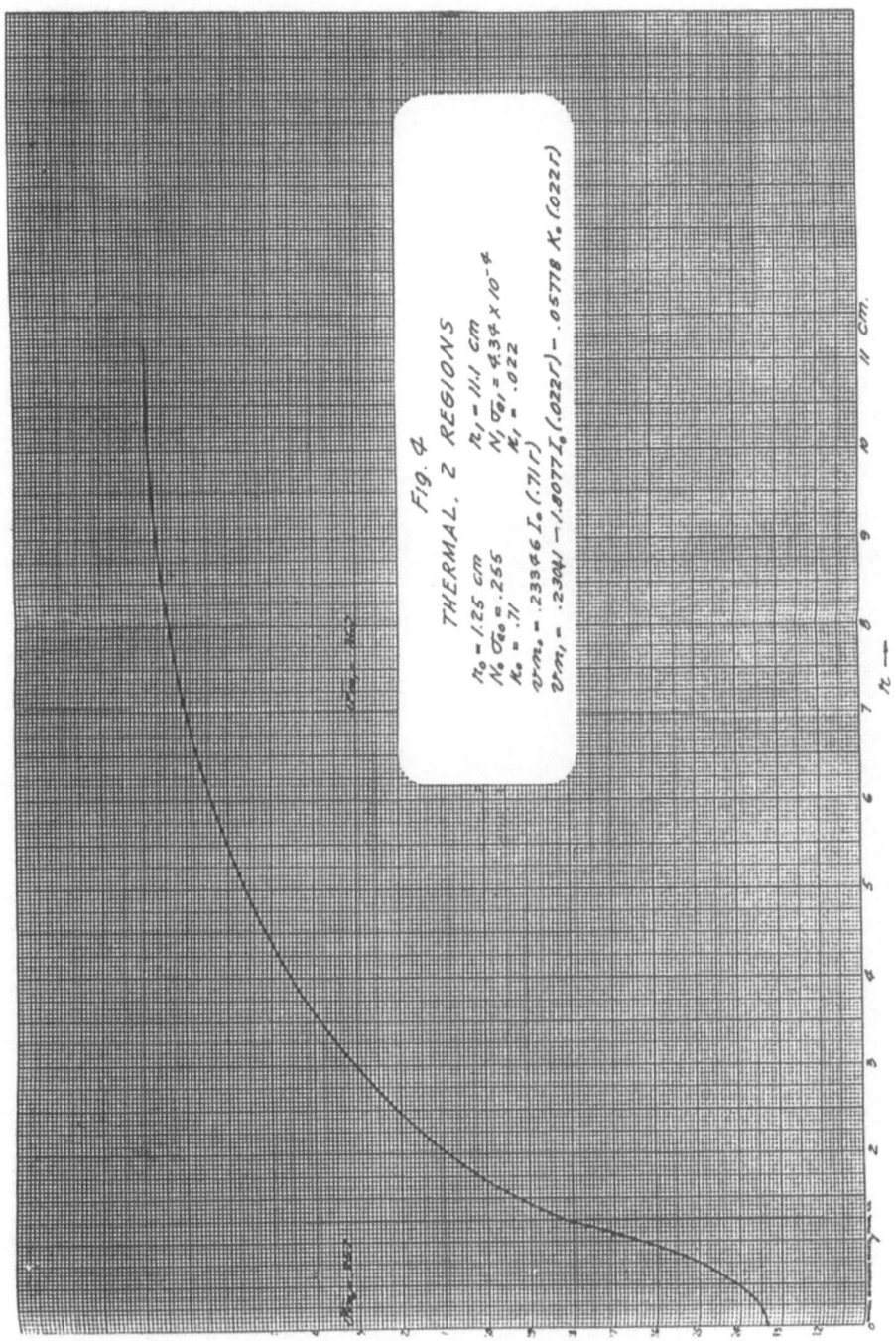

Fig. 4

THERMAL. 2 REGIONS

$r_0 = 1.25$ cm $r_1 = 11.1$ cm
$N_0 \sigma_{t0} = .255$ $N_1 \sigma_{a1} = 4.34 \times 10^{-4}$
$K_0 = .71$ $K_1 = .022$
$v/n_0 = .23346 \, I_o(.711\,r)$
$v/n_1 = .2304\, I_o(.022\,r) - 1.80772 \, I_o(.022\,r) - .0578\, K_o(.022\,r)$

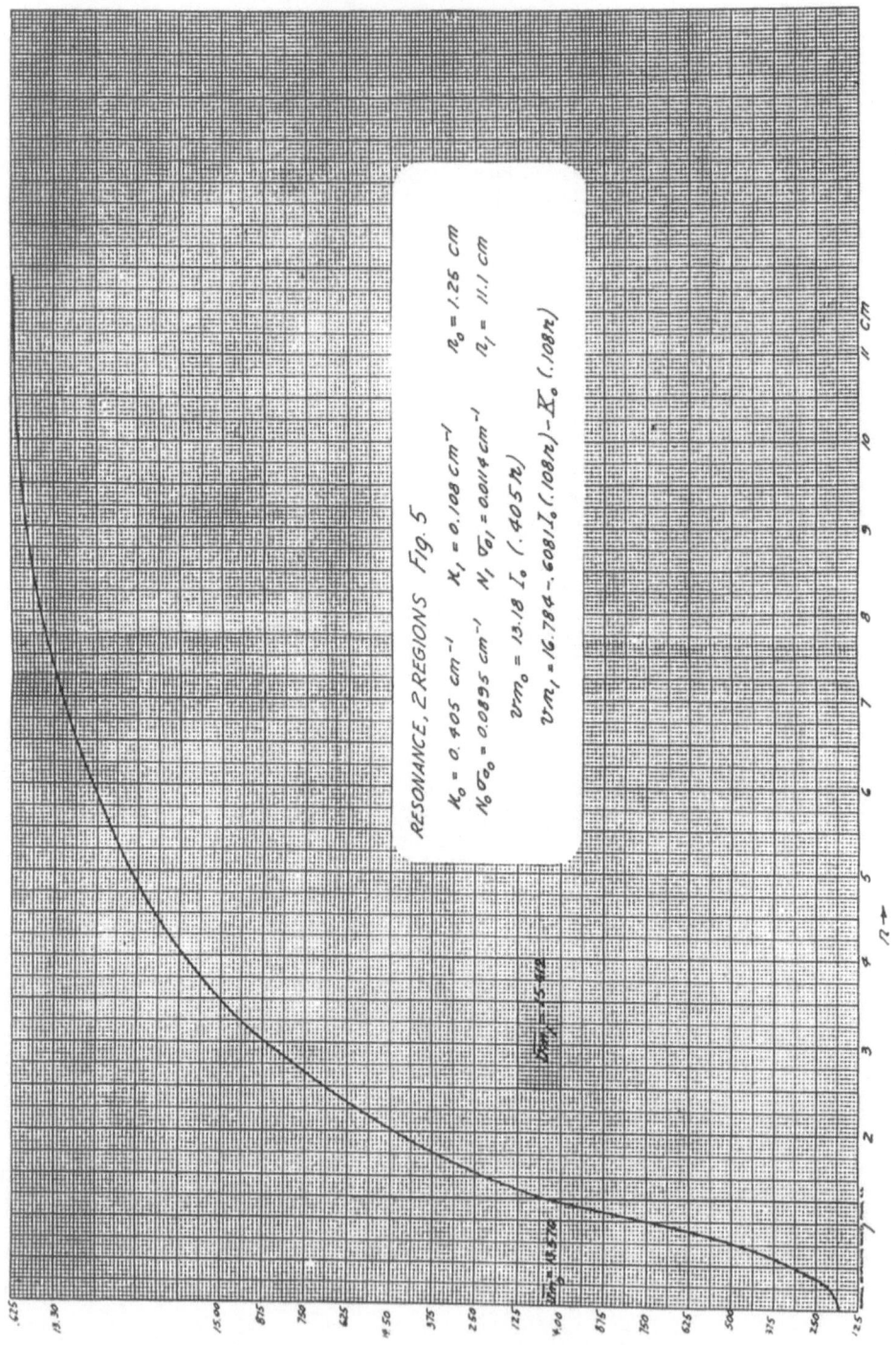

RESONANCE, 2 REGIONS   Fig. 5

$K_o = 0.405$ cm^{-1} $K_1 = 0.108$ cm^{-1} $r_o = 1.26$ cm

$N_o \sigma_{a_o} = 0.0895$ cm^{-1} $N_1 \sigma_{a_1} = 0.0114$ cm^{-1} $r_1 = 11.1$ cm

$v m_o = 13.18 \, I_o \, (.405 r_1)$

$v m_1 = 16.784 - .608 \, I_o (.108 r_1) - K_o (.108 r_1)$

The total capture of cold neutrons per c.c. in the metal is found by multiplying the thermal capture cross-section by the cold neutron density, and similarly for the hot neutrons. In this computation the densities must be reduced to the normalization, $P_0 = 0.564\pi$, $P_2 = 7.55\pi$. The total thermal capture divided by the total neutron production is thus found to be .850; hence $\frac{1}{p_2} = 1.176$. This may be compared with 1.185 for the two region comparison system.

The ordinary two region formula for the resonance absorption factor must be modified to include the fact that there are two metal surfaces exposed to good moderating materials, and that the disadvantage factor is rather more complicated than before. It may be shown that the proper generalization of $\frac{1}{p_1}$ for the three region problem is

$$\frac{1}{p_1} = \exp \frac{N_1 V_1 \bar{n}_1}{N_0 V_0 \bar{n}_0 S_0 + N_2 V_2 \bar{n}_2 S_2} \left(\int \frac{\sigma \, dE}{E} \right)_{\text{eff}}$$

where V_i is the volume of the ith region, S_i is the slowing down power per c.c. of the ith medium, and $\int \frac{\sigma \, dE}{E}$ is given above.

Using the average neutron densities of Fig. 19 for the $\bar{n}$, we find $\frac{1}{p_1} = 1.116$; this may be compared with 1.100 in the two region comparison system.

4.) Using the preceding values for $\frac{1}{p_1}$ and $\frac{1}{p_2}$ we compute for the critical value of the multiplication constant

$$\eta' = \frac{1}{p_1} \times \frac{1}{p_2} = 1.312 \quad \text{in water system}$$

$$\eta' = \frac{1}{p_1} \times \frac{1}{p_2} = 1.304 \quad \text{in no-water system}.$$

The value of k is inversely proportional to η'; hence

$$\frac{\text{multiplication factor in water system}}{\text{multiplication factor in no-water system}} = \frac{1.304}{1.312} = .9939.$$

The value of k is reduced 0.61 percent by the presence of the water.

The geometric proportions used in this calculation were chosen somewhat arbitrarily; presumably a better k could be found (for both the 3 and 2 region system) by choosing geometry closer to the optimum. To estimate the order of magnitude of the correction introduced by using optimum geometry, we note that, in general, the optimum geometry is reached when $\frac{1}{p_1} \approx \frac{1}{p_2}$; i.e., when $R = \frac{1/p_1 - 1}{1/p_2 - 1} \approx 1$. The value of R in the present 3 region case is $\frac{.116}{.176} = .659$; in the 2-region case it is $\frac{.100}{.185} = .540$; while in Christy's and Monk's oxide calculations $R = .9$ at optimum. In going from $R = .54$ to $R = .9$, Christy and Monk find that η' is improved from 1.299 to 1.286, an improvement of 1.0 percent; in going from $R = .659$ to $R = .9$, η' is improved by 0.31 percent. If we improve the η's calculated here in the same ratio (a procedure which is admittedly crude), we find

$$\eta'_{optimum} = 1.308 \quad \text{in water system}$$

$$\eta'_{optimum} = 1.291 \quad \text{in no-water system.}$$

The ratio of the optimum k's is .987; in other words, if optimum geometry is used, the value of k is reduced 1.3 percent by the presence of water.

The value of $R = .659$ corresponds in Christy's & Monk's case to a volume of oxide cylinder which is slightly (12%) low compared to optimum. In arriving at the maximum value of k in the water system one should therefore increase the volume of the metal rods by about the same amount, preferably by reducing the bore of the water tubes to, say, 0.5 cm radius.

Increasing the relative volume of metal tends to increase the resonance absorption effect; this is desirable since it means a greater production of 49 by resonance capture.

The conclusion that may be drawn from these computations is that the presence of water cooling tubes will reduce k by 1.3 percent if only the effect of differential heating is considered. The added absorption of the aluminium tubes is the most important residual factor which reduces k and so should be added to get an overall estimate of the water effect; if the aluminium reduces k by 0.4 percent, an upper limit to the adverse effect on k of water cooling is therefore 1.7 percent.

Size of Plant

One of the merits of a water cooled plant is its compactness, i.e., that there is no need for wide cooling channels with a resulting decrease of the total density. The water pipes, as contrasted with the gas channels, reduce rather than increase the migration length of the neutrons. This results in a partial compensation of the decrease in k, as far as the size of the pile is concerned, even at low temperatures.

Furthermore, the operating temperature is relatively low so that the size of the plant is not so much bigger at operating temperature than at normal temperature as it is in a gas cooled plant.

One can define a characteristic distance $1/\kappa$ for every lattice. This is the migration length, divided by $\sqrt{k-1}$

$$\frac{1}{\kappa} = \frac{l}{\sqrt{k-1}}$$

The overall neutron density obeys the differential equation

$$\Delta n = \kappa^2 n$$

As a consequence, $1/\kappa$ is proportional to the linear dimension of a going plant. In order to calculate κ, we have to calculate l first. The migration length consists of two parts

$$l = \sqrt{\tau + \mathcal{L}^2}$$

where τ is the "age" of thermal neutrons, $\mathcal{L}$ their diffusion length. We assume a $\tau = 380$ cm^2 for a pure graphite pile. A simple calculation then gives

$$\tau = (380 - 14.5 \ f) \ \text{cm}^2$$

where f is 1000 times the mol ratio of water and carbon. In our case, $f = 1.9$ so that the τ are, for our water system and for the usual plant, respectively

$$\tau = 353 \ \text{cm}^2 \qquad \tau = 380 \ \text{cm}^2 \ .$$

The calculation of the diffusion lengths is somewhat more complicated, because of the presence of two kinds of thermal neutrons (hot and cold). It is given in Table I. Columns 2, 3 and 4 of this refer to the water cooled plant, columns 5 and 6 to the gas cooled plant. The second line gives the mol ratios of the substances present in the pile. The third line gives the volume ratios. These were somewhat changed from those assumed in the preceding section, so as to bring k nearer to its maximum. The fourth line gives the average densities of "cold neutrons", the fifth the average densities of "hot neutrons". These were taken from the calculations and graphs of the preceding section. The next two lines give the absorption cross section per cm^3 for cold and hot neutrons and includes, of course, only the real absorptions. The absorption of "hot" neutrons is somewhat larger in the gas pile than in the water pile because their temperature was assumed to be only 600 °C, as contrasted with the hot temperature in the water pile of about 700 °C. The next two lines (8 and 9) give the contributions of the various media to the total absorption. The probability that a neutron be absorbed as cold neutron in water, e.g., is equal to the $(N\sigma_a)_{\text{cold}}$ of water, multiplied with the probability that the neutron shall be cold and be in the water. This last number was calculated using the volume fractions (line 3) and relative densities of the neutrons (lines 4 and 5) in the corresponding media. In order to obtain $\mathcal{L}^2$, one must divide the diffusion constant by the velocity

Table I

	Water cooled pile			Gas cooled pile	
	Water	U	C	U	C
2) mol fractions	.0019	.0074	.9907	.0065	.9935
3) volume fractions	.0045	.0130	.9825	.0113	.9887
4) n_{cold}	.978	.417	.068		
5) n_{hot}	1.29	1.77	2.77	1	1.35
6) $(N\sigma_a)$ cold	.020	.255	4.3×10^{-4}		
7) $(N\sigma_a)$ hot	.0127	.162	2.7×10^{-4}	.171	2.9×10^{-4}
8) absorption cold/cm^3	3.2×10^{-5}	46×10^{-5}	$.80 \times 10^{-5}$		
9) absorption hot/cm^3	2.7×10^{-5}	123×10^{-5}	21×10^{-5}	135×10^{-5}	22×10^{-5}

and the sum of the numbers in lines 8 and 9. We obtained for the diffusion constant over the velocity 2.69 and 2.72 cm, respectively, for water cooled and gas cooled systems. Hence, we obtain, respectively,

$$\mathcal{L}^2 = 457 \text{ cm}^2 \quad \text{and} \quad \mathcal{L}^2 = 576 \text{ cm}^2 .$$

For the migration lengths l, this gives

$$l = 28.5 \text{ cm} \quad \text{and} \quad l = 30.9 \text{ cm}$$

for water-cooled and gas-cooled plants, respectively.

The preceding calculation applies to the lattice, underlying the calculation of the preceding section, in which the water tubes had 1.5 cm diameter. We estimate that the reduction of this diameter to 1 cm will increase the migration length to 29.8 cm. On the other hand, the channels in the gas cooled plant, the volume of which we estimate to amount to 10% of the total volume, will increase the migration length in that case by 10%. We find, thus

$$l = 29.8 \text{ cm} \quad \text{and} \quad l = 34 \text{ cm}$$

for the water cooled plant here proposed and a gas cooled plant.

The decrease in k was (including .7% for the cylindrical geometry of Wheeler, Monk, Ibser, C-86) 2.4% for the water cooled plant underlying the calculations of the preceding section. We estimate again that the reduction of the water volume will diminish this to 1.6% loss in k. We assume hence

$$k = 1.084 \quad \text{and} \quad k = 1.10$$

in the part of the pile which contains metal and

$$k = 1.019 \quad \text{and} \quad k = 1.035$$

in the part containing oxide or carbide. It is in fact, rather likely that the loss in k is somewhat smaller in the oxide or carbide lattice than in the metal lattice but we disregarded this fact. We now obtain

$$\kappa_0 = 9.7 \times 10^{-3} \text{ cm}^{-1} = \frac{1}{103 \text{ cm}} \quad \text{and} \quad \kappa_0 = 9.3 \times 10^{-3} \text{ cm}^{-1} = \frac{1}{108 \text{ cm}}$$

for the seeds of water cooled and gas cooled plants, respectively. For the outside of the plant which contains oxide or carbide, we obtain

$$\kappa_1 = 4.6 \times 10^{-3} \text{ cm}^{-1} = \frac{1}{216 \text{ cm}} \quad \text{and} \quad \kappa_1 = 5.5 \times 10^{-3} \text{cm}^{-1} = \frac{1}{182 \text{ cm}} .$$

The migration lengths were assumed to be equal for metal- and oxide-containing lattices. We now go over to calculating the size of a pile, the center one eighth of which contains the uranium as metal, the outside as oxide or carbide.

If the shape of the pile were a sphere, the above estimates would permit us to calculate its radius R_s at once. Outside $r = R_s/2$, the neutron density would be proportional to

$$n = \frac{\sin(4.6 \times 10^{-3}(R_s - r))}{r \sin(2.3 \times 10^{-3} R_s)} \quad r > R_s/2$$

in the seed it would be

$$n = \frac{\sin(9.7 \times 10^{-3}r)}{r\sin(4.85 \times 10^{-3}R_s)} \quad r < R_s/2.$$

The constants in these equations are so chosen that n is continuous at the boundary $r = \frac{1}{2}R_s$ of seed and reflector. The condition that the derivatives be also continuous gives

$$-4.6 \times^{-3} \mathrm{ctg}2.3 \times 10^{-3}R_s = 9.7 \times 10^{-3}\mathrm{ctg}4.85 \times 10^{-3}R_s .$$

The solution of this is $R_s = 395$ cm, or a total volume of 260 m^3, one eighth of which contains the uranium in metallic form, the rest as an oxide or carbide.

If one uses a cylindrical rather than a spherical plant, the height H and radius R of these are connected with the radius R_s of the sphere by the equation

$$\frac{\pi^2}{R_s^2} = \frac{\pi^2}{H^2} + \frac{2.40^2}{R^2} .$$

This equation applies, strictly speaking, only for a uniform pile (pile without seed) but it is unlikely that one commits a relevant error by applying it to our system. Substituting $H = 600$ cm, one finds $R = 405$ cm. This corresponds to a volume of 306 m^3, one eighth of which, i.e. 38 m^3 contains metal. It requires therefore .50 m^3 or 9 metric tons of metal, about 63 tons of oxide or carbide and about 480 tons of graphite.

These figures should be compared with the amounts needed, under similar assumptions for the multiplication factor and migration length, for a He-cooled system. We assumed that the multiplication constant is not affected in this case by the higher temperature of the U – an assumption which favors the He-cooled system.

A calculation similar to that given above gives 400 cm for the radius of a sphere-shaped He-cooled plant, again with a seed of one eighth containing the uranium as metal. A cylindrical plant of 6 m height would have a radius of 412 cm, a volume of 320 m^3 and a seed of 40 m^3. It would contain .41 m^3 or 7.35 tons of metal and again about 7 times more oxide or carbide. It needs about 450 tons of graphite.

If the pile contains only metal, the above calculation does not give the correct ratio of materials needed. It is, however, very easy to calculate the sizes and amounts of materials needed in this case and the results are given below.

The figures for a pile containing only metal mean only that a plant of the size given above would be chain reacting. The calculations for heat output etc., were made for a plant of which only the center one eighth is metal and the size estimates etc. for this are given above. For the plant without oxide, new calculations on the power output would have to be made.

It is probably unnecessary to emphasize that the fact that we made a somewhat detailed investigation of water cooling does not mean that we believe that the water cooling is better than the other schemes of cooling. On the contrary, we believe that, if the time permits, similar investigations should be

	Water cooled	Gas cooled
Sphere shape		
radius	324 cm	339 cm
volume	142 m^3	163 m^3
U needed	33 tons	30 tons
4.5 m high cylinder		
radius	3.57 m	3.94 m
volume	180 m^3	220 m^3
U needed	42 tons	39 tons

made of all other cooling systems, so that an objective choice for the best one can be made. We are eager to give priority to any cooling system that is better than the present one.

Drawings

Figure 1 is a side view sketch of the installations sunk in the earth; Figure 2 shows aligned top and side views of pile.

Figure 3 indicates more detail of the foundation construction for the case in which no shield is provided below the pile.

Figure 4 shows the pattern of vertical holes left in the graphite·for the uranium rods.

Figures 5 and 6 illustrate the uranium rods. Figure 7 shows the gravity seal against water vapor at the bottom of the pile; and Figures 8, 9, 10 suggest alternative schemes. In 8 and 9 the rod is tightened by turning it from the top; in 10 it is tightened from below. In this case a bottom shield becomes necessary. Figure 11 shows the arrangement at the top of the tube, including the spiral to allow for vertical expansion of the rod.

Figure 12 shows the constrcution at the top of the pile, supplemented by Figures 13 and 14.

Figure 15 shows mounting of the control rods, and Figure 16 the construction of the central control rod.

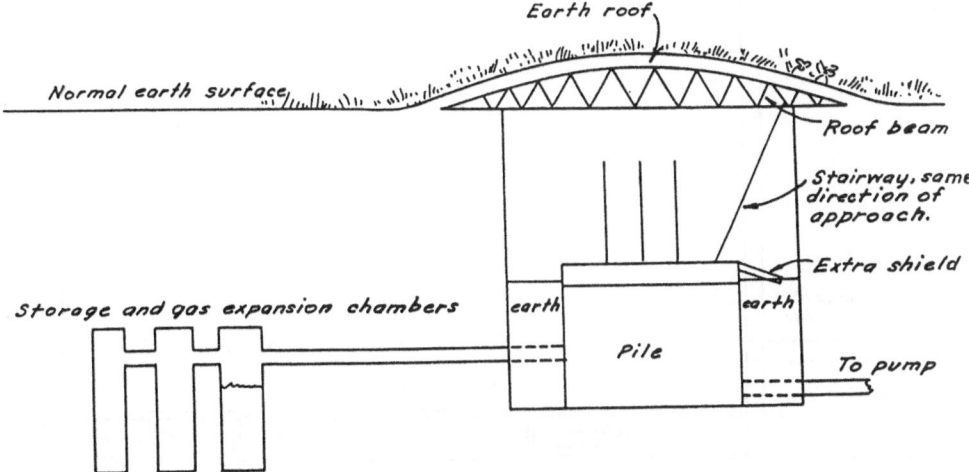

Earth roof

Normal earth surface

Roof beam

Stairway, same
direction of
approach.

Extra shield

Storage and gas expansion chambers   earth   earth

Pile

To pump

Fig. 1

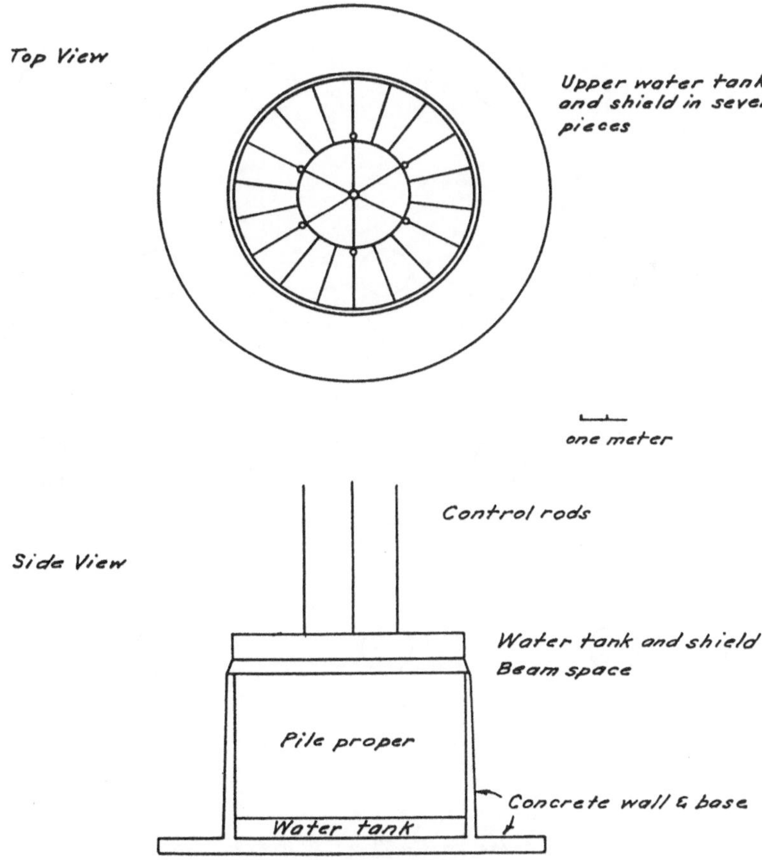

Top View

Upper water tank
and shield in several
pieces

one meter

Control rods

Side View

Water tank and shield
Beam space

Pile proper

Concrete wall & base

Water tank

Fig. 2

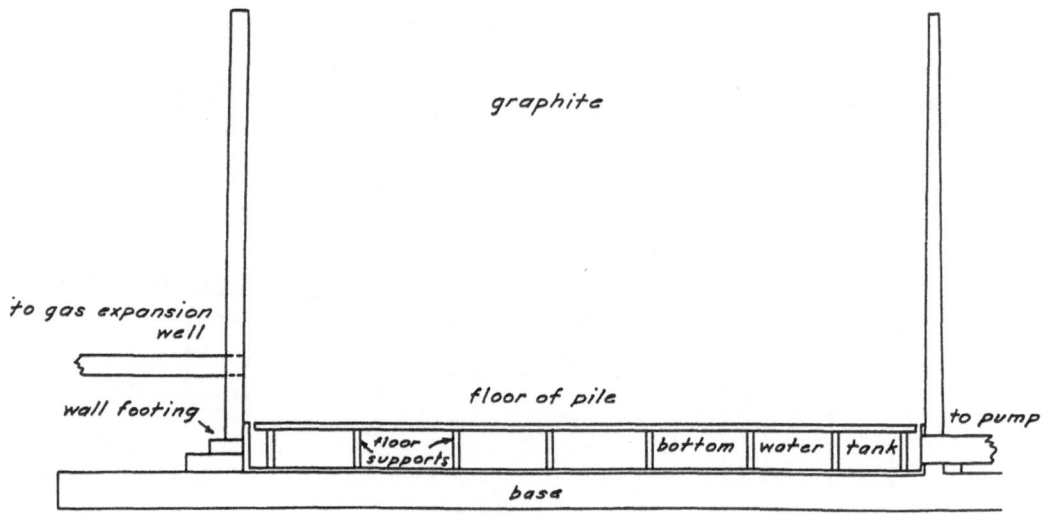

Fig. 3. Pile sides and bottom

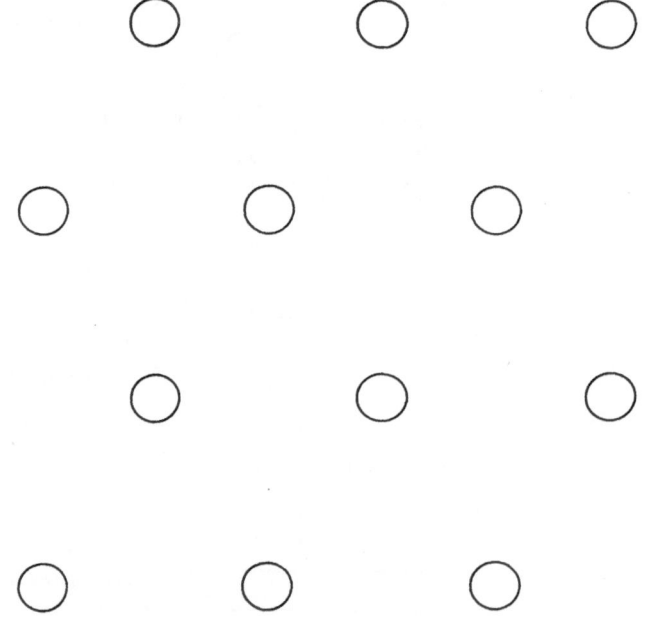

template for holes in Carbon: $\frac{1}{4}$ actual size

Fig. 4

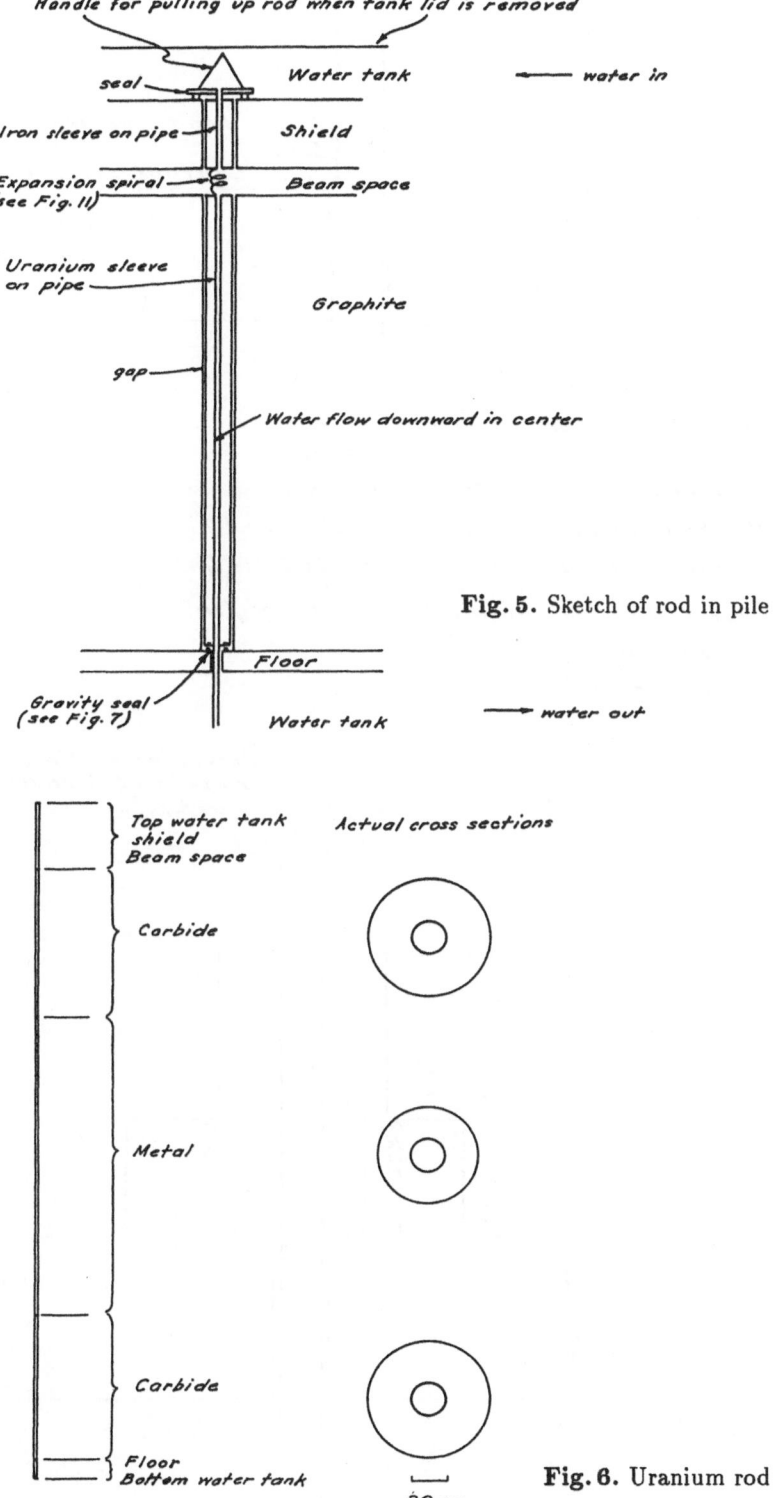

Fig. 5. Sketch of rod in pile

Fig. 6. Uranium rod

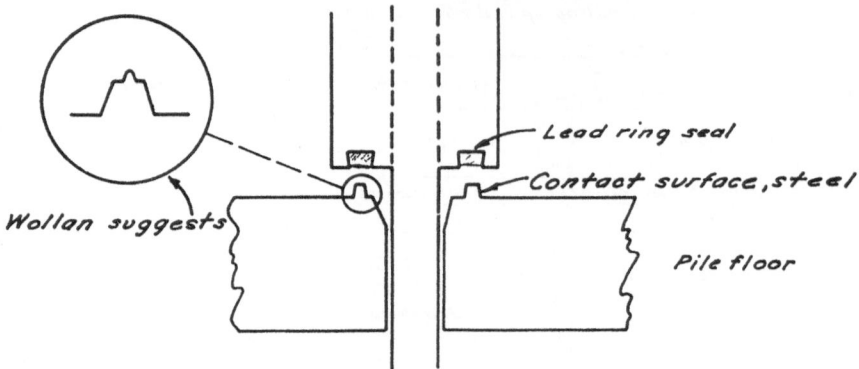

Wollan suggests

Lead ring seal

Contact surface, steel

Pile floor

Diameter of contact ridge = 1 inch
Width   of   "   "   = $\frac{1}{8}$ inch
Area of contact surface   = 0.4 in²
Weight of rod assembly   = 120 lbs.
Pressure at contact surface = 300 lbs./in²
Gas pressure to be held by seal = 15 lbs./in²

$$\frac{contact\ pressure}{gas\ pressure} = 20$$

According to Marks (p.1049 f.)
a ratio of 5 to 10 should be
sufficient.

Fig. 7

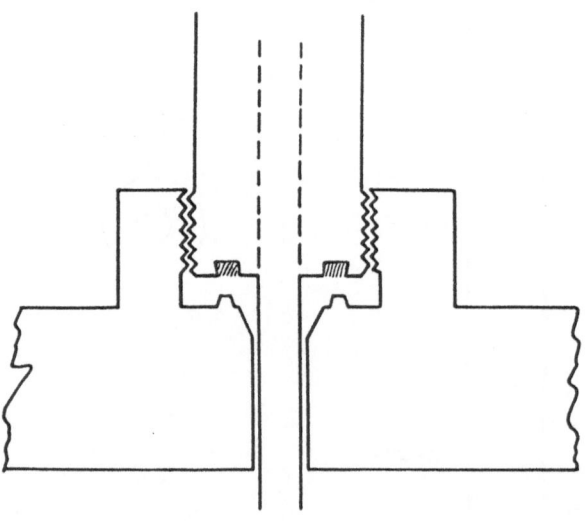

Fig. 8

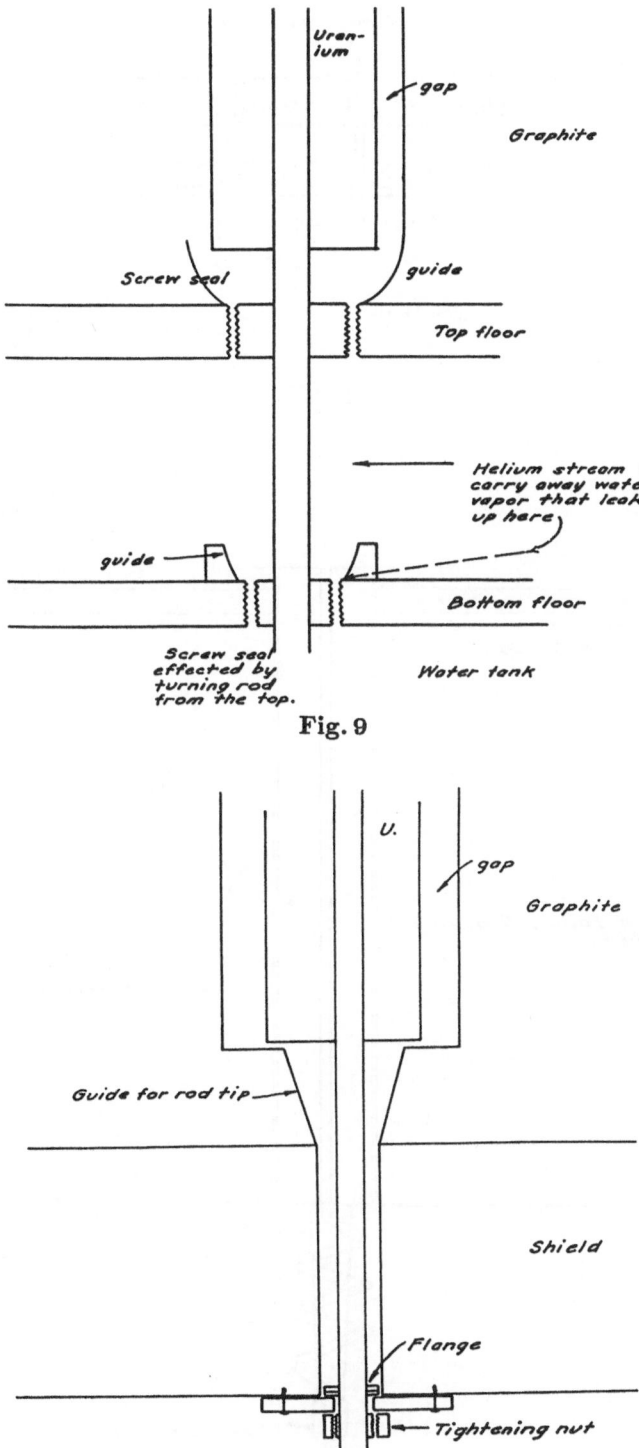

Fig. 9

Fig. 10

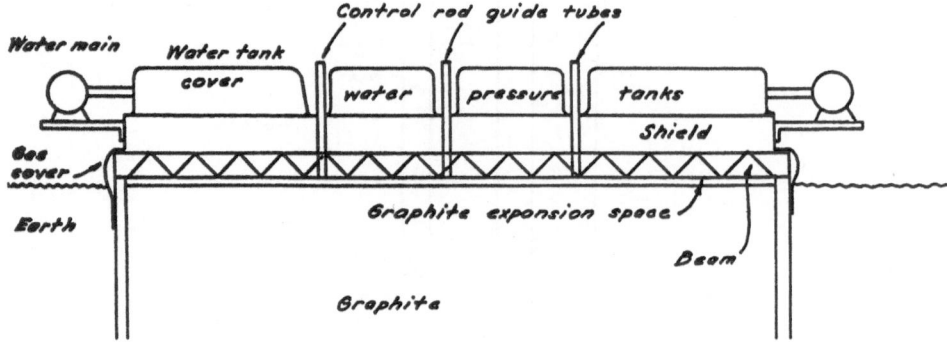

Control rod guide tubes

Water main   Water tank
cover   water   pressure   tanks

Gas cover   Shield

Earth

Graphite expansion space

Beam

Graphite

Fig. 11

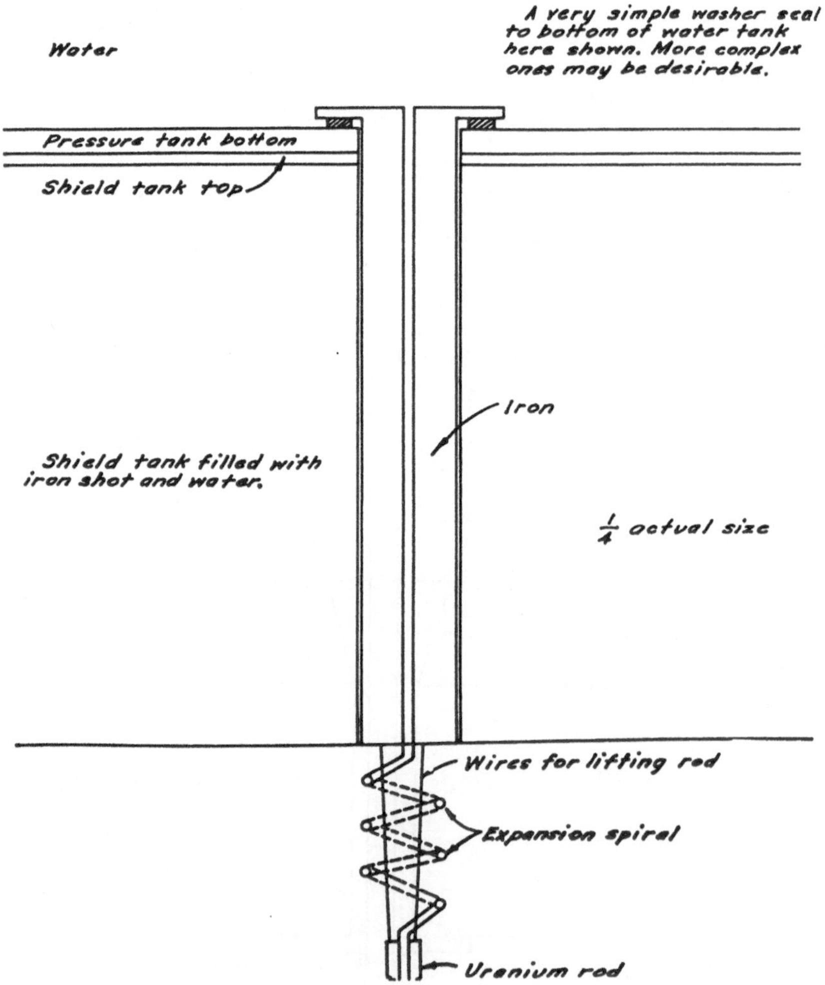

Water

A very simple washer seal
to bottom of water tank
here shown. More complex
ones may be desirable.

Pressure tank bottom
Shield tank top

Iron

Shield tank filled with
iron shot and water.

$\frac{1}{4}$ actual size

Wires for lifting rod

Expansion spiral

Uranium rod

Fig. 12. Top of pile

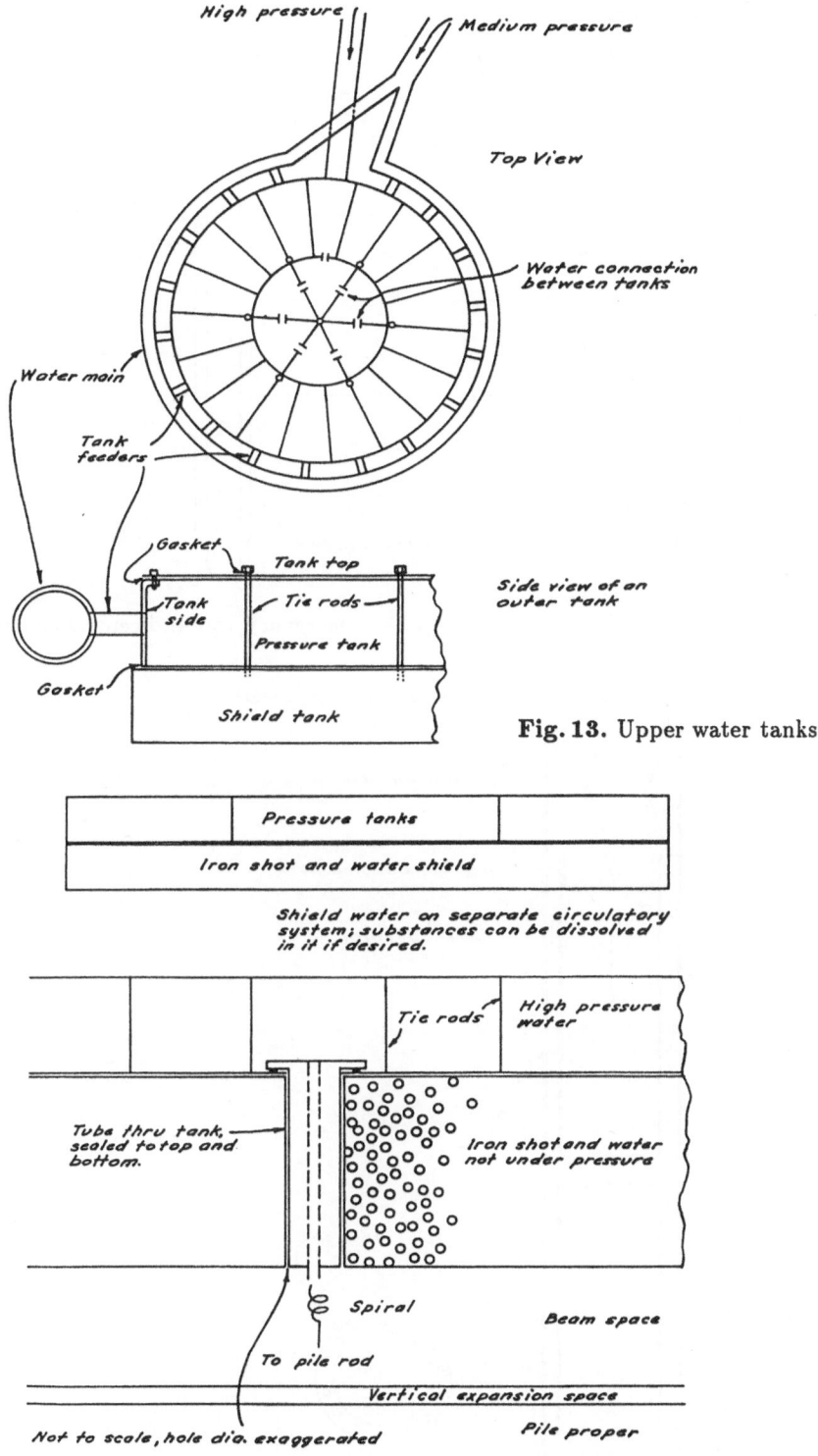

Fig. 13. Upper water tanks

Fig. 14. Shield and tank

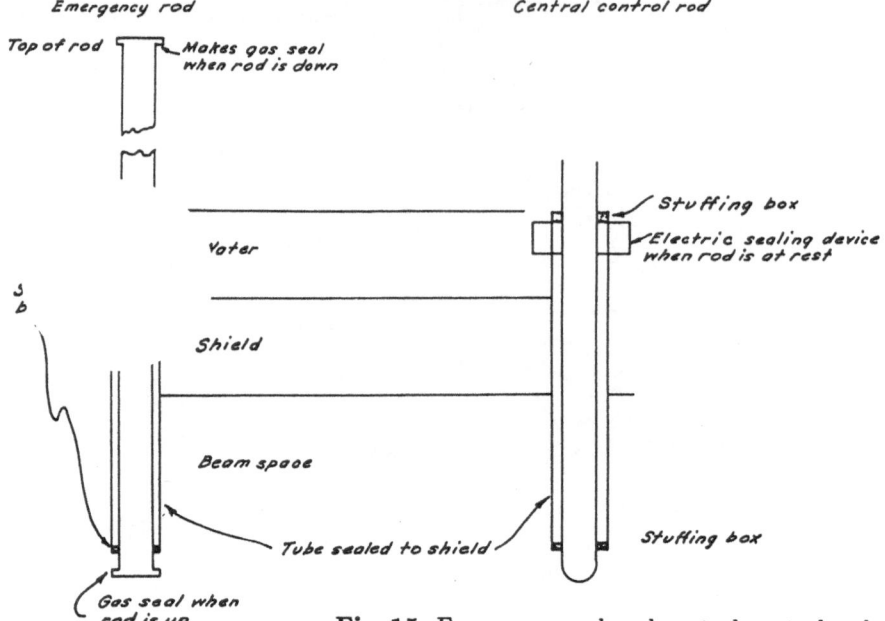

Fig. 15. Emergency rod and central control rod

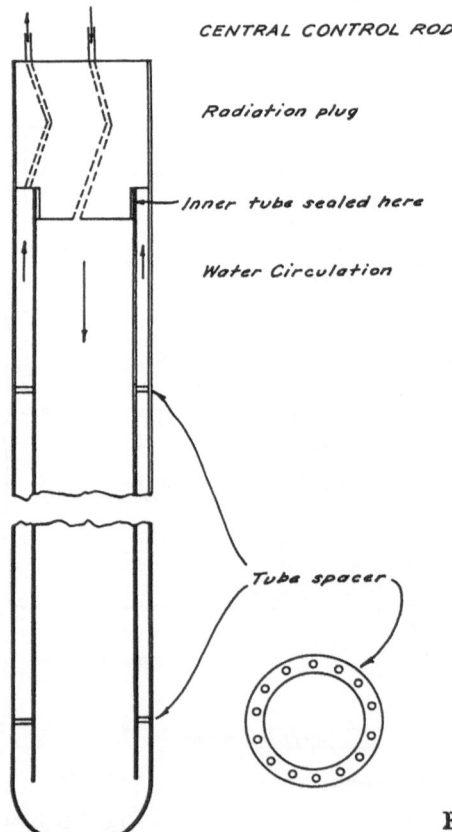

Fig. 16. Central control rod

16.

Preliminary Process Design
of Liquid Cooled Power Plant Producing 5×10^5 kW

M. G. J. Boissevain, M. C. Leverett, L. A. Ohlinger, A. M. Weinberg,
E. P. Wigner, and G. J. Young

January 9, 1943

Abstract. A water-cooled plant is described, with a rating of 500,000 kW from 200 tons of metal, and a potential output somewhat higher. The design allows for ready exchange of the metal without releasing pile gas or fission products, and provides a thickness of material at corrosion points which should be easily ample for a lifetime of 100 days according to the results of corrosion experiments.

Among the principal requirements are 1200 tons of graphite, 2500 tons of iron, 8000 kW power supply with 2000 kW standby, a site area of 140,000 ft^2, 25 gallons per minute of distilled water and 4200 gpm of treated water for makeup.

With minor changes the plant can be cooled with diphenyl.

Thanks are hereby extended to Messrs. Creutz, Burton, Howe, Kratz, Christy, Friedman, Sacher, Plass, and others for their kind and able assistance in obtaining the information from which this report is compiled.

Table of Contents

I. Introduction

This report presents a general description of the physics and engineering problems involved in the design of a liquid cooled plant for the generation of element 49 at a rate of about 500 grams/day. Although this report is not intended as a final and complete basis for the detailing and construction of such a plant without further development, it does indicate, not only the many problems, but the best probable solutions based on current information, the probable amount of material and utilities involved, and a background for mapping the future course of development of this phase of the project.

This report is not intended to render earlier reports obsolete but rather to bring up to date the development of the liquid cooled power plant and present a practical design for further development. It originated from a request by the project leader, to Messrs. Young and Leverett, to develop plans for a liquid cooled plant which uses 200 tons of metal and generates power at a rate of 500,000 kW.

II. Physical Problems and Data

We shall present a specific design for a liquid cooled plant, and discuss a number of the problems involved.

General Description. We consider a graphite-uranium metal pile with metal in the form of long parallel rods spaced regularly throughout the graphite. Long aluminium tubes are fitted fairly closely into holes in the graphite; and centered in these tubes by means of longitudinal ribs are the somewhat smaller uranium rods which have thin aluminium tubes drawn tightly over them. The cooling fluid flows along the rods in the annular spaces between the rods and the outer tubes, touching only aluminium.

Heat is produced throughout the uranium and flows outward to the surface of the rod, through the aluminium coating, and into the liquid stream. A smaller amount of heat is produced throughout the graphite and flows inward to the surface of the hole in the graphite, across a gap filled with helium to the outer aluminium tube, through the tube wall, and into the cooling stream.

The pile proper is in the from of a cylinder with the uranium rods parallel to its axis. This active cylinder is surrounded on all sides by a layer of graphite without uranium, to reflect escaping neutrons back into the pile.

Principal Dimensions

Axial length of active cylinder = 7 meters = 23 feet
Radius of active cylinder = 4.94 meters = 16.2 feet
Thickness of reflector = 50 cm = 1.64 feet
Total weight of metal = 200 metric tons
Weight of graphite in pile = 850 metric tons
Weight of graphite in reflector = 315 metric tons
Radius of metal rods = 1.7 cm
Thickness of inner aluminium tube = 0.5 mm
Thickness of outer aluminium tube = 1.5 mm
Thickness of liquid layer = 2.2 mm with water, 4 mm with diphenyl
Number of rods in pile = 1695
Weight of aluminium in pile proper = 8.7 metric tons
Rod spacing in square array = 21.3 cm = 8-3/8 inches

Distribution of Activity in Pile. The activity is most intense at the center and decreases toward the edges, the decrease being as a cosine function along the axis and as a J_0 function along the radius. If there were no reflector the activity would have to reach zero at the edges of the active cylinder. With a reflector the activity is still finite at the surface of the cylinder, and extrapolates to zero at some distance outside the surface. With 50 cm of graphite this distance is about 35 cm.

Without a reflector the ratio of maximum activity to the average activity throughout the cylinder is 3.64; and the total activity of a rod at the axis is 2.32 times the average activity per rod. Use of a reflector decreases these ratios, in our case to about 2.9 and 2.0 respectively.

There is a small further flattening of the activity distribution due to the rise in metal temperature when the pile operates. Control rods operating in the center of the pile produce a similar effect, of which more will be said later.

Heat Flow in Rods. To a first approximation the rate of heat production per unit volume is constant over the cross-section of a rod. On this assumption, together with that of a constant value for the thermal conductivity k, the power output per unit length of rod is $P = 4\pi kT$, where T is the temperature drop from center to edge. This is independent of the size of the rod, which makes the ouput per unit amount of metal inversely proportional to the square of the rod radius, for a given temperature drop.

Actually the rate of heat production increases somewhat toward the surface of the rod, because of decreased neutron penetration into the interior, and this increases the power output for a given temperature drop. For a rod of 1.7 cm radius this effect increases the output by a factor of 1.1.

The average temperature rise in the rod interior is $\frac{1}{2}$ that of the rod center, both being measured from the temperature at the surface. If the thermal conductivity varies linearly with the temperature, it should be taken at the mean temperature of the rod for use in the above formula.

For numerical values we have the following:

Power rating of plant = 500,000 kW
Power produced in metal = 94% = 470,000 kW
Average power in metal = 2.35 watts/gram
Maximum power in metal at center of pile = 6.8 watts/gram
Density of metal = 18.6 gm/cm^3
Thermal conductivity of metal = 0.05 cal/sec/cm^2/°C/cm
Maximum temperature rise in metal at center of pile = 400 °C
Average temperature rise in metal at center of pile = 200 °C
Average temperature rise in metal throughout pile = 69 °C

There is also a temperature drop where the heat passes from the uranium rod to its tightly fitting aluminium coating.

Weight of uranium rod per unit length = 169 grams/cm
Maximum heat from rod per unit length = 1150 watts/cm = 275 cal/sec/cm
Maximum heat from rod per unit area = 26 cal/sec/cm^2
Transfer coefficient from uranium to aluminium $\sim$ .6 cal/sec/cm^2/°C
Maximum U-Al drop $\sim$ 43 °C
Average U-Al drop $\sim$ 15 °C

Heat Flow in Graphite. Around each rod there is essentially a graphite shell in which the heat production rate is about uniform over the cross-section.

Outer radius of graphite = 12 cm
Inner radius of graphite = 2.13 cm
Maximum heat from graphite per unit length = 17.6 cal/sec/cm
Conductivity of graphite = 0.17 cal/sec/cm^2/°C/cm
Maximum temperature drop in graphite at center of pile = 21 °C
Average temperature drop in graphite at center of pile = 18 °C
Average drop in graphite throughout pile = 6 °C

When this heat reaches the surface of the hole in the graphite it passes over to the aluminium tube, partly by conduction through the helium which fills the gap and partly by direct conduction where the aluminium tube is in contact with the graphite. We base a computation on the helium conductance alone, which should give an upper limit to the temperature drop.

Conductivity of helium = 0.00036 cal/sec/cm^2/°C/cm [4]
Thickness of gap = 5 mils = .0127 cm
Maximum power from graphite per unit area = 1.3 cal/sec/cm^2
Maximum C-Al drop = 46 °C
Average C-Al drop = 16 °C

Heat Pickup by Cooling Liquid. When the heat has passed through the aluminium coating on the uranium rods, it has to be transferred to the flowing liquid and carried away. This involves a temperature drop from the aluminium wall to the interior of the liquid stream (this drop being expended across the non-turbulent film of liquid adjacent to the wall), also a rise in temperature of the liquid itself as it moves through the pile. The wall temperature at any point is made up of the rise in liquid temperature to that point plus the film drop at that point. The liquid enters the pile at considerable pressure, and the pressure decreases as the liquid moves along the tube. Thus the temperature at which the liquid will boil decreases as the liquid moves through the pile. Local boiling in the film may hamper the flow of heat into the liquid, and so we keep the wall temperature everywhere lower than the boiling temperature of the liquid. By keeping the liquid under some excess pressure as it leaves the pile the boiling temperature is raised all along the tube, and the liquid can be worked through a greater temperature rise.

Calculations for water as the cooling fluid run as follows:

Pressure drop along the rod = 115 lbs/in^2
Total flow rate in tube neglecting ribs = 1790 cm^3/sec
Velocity of flow = 7 meters/sec = 23 feet/sec
Allowance for reduction of stream cross-section by ribs = 10%
Flow rate with ribs = 1600 cm^3/sec
Average output per rod = 295 kW = 70,500 cal/sec

---

[4] /cm added by A.M.W. & A.M.P.

Output of central rod = 141,000 cal/sec
Temperature rise of water along the central rod = 88 °C
Total volume of water in pile = 2.75 cubic meters
Maximum heat from rod per unit length = 275 cal/sec/cm^2
Maximum heat into water per unit area = 25 cal/sec/cm^2
Film transfer coefficient = 1.2 cal/sec/cm^2/°C
Maximum film drop = 21 °C
Input temperature of water to pile = 35 °C
Maximum metal temperature = 540 °C
Average metal temperature throughout pile = 165 °C
Maximum graphite temperature ~ 150 °C
Average graphite temperature ~ 90 °C

The only construction change in the pile proper in going to diphenyl is to increase the thickness of the annular cooling space to 4 mm. The temperature rise is about the same as with water, but the film drop is now about 5 times as large. The highest temperature in the metal will be 100–150° higher than with water cooling, so that if this be a limiting factor the plant output is smaller with diphenyl.

Returning to water cooling, it is seen that the emergent temperature of the central stream is 123° for the assumed input temperature, so that the emergent pressure must be at least 20 pounds/in^2 to avoid boiling in the tube. A typical calculation for the central water stream is as follows:

Location	Pressure	Boiling temp.	Water bulk temperature	Film drop	Wall temp.	Safety margin for boiling
Input end	150 psi	185 °C	35 °C	4 °C	39 °C	146 °C
Center of tube	92	167	79	21	100	67
3/4 along tube	54	150	107	16	123	27
Output end	35	138	123	3	126	12

If the temperature of the water entering the central tube can be lowered below 35 °C, the margins of safety will be increased by the same amount.

Note that the water is closest to its boiling point in the exit headers, where overheating can be most easily detected.

Multiplication Factor. For our structure we have:

Square of migration length = 640 cm^2
Required k without reflector = 1.028
Required k with reflector = 1.024
Base k = maximum obtainable with graphite-metal lattice = 1.082
k for our cylinders in graphite = 1.074

$$\frac{\text{Weight of aluminium in pile}}{\text{Weight of uranium}} = 4.35\%$$

k loss due to aluminium $= 1.3\%$

$$\frac{\text{Weight of water in pile}}{\text{Weight of uranium}} = 1.37\%$$

k loss due to water $= 2.3\%$
k for our pile $= 1.0334$
k loss due to heating of the metal $= 0.3\%$
Safety margin in $k = 0.95\%$

Note that if the k loss due to the water were calculated simply on the basis of a danger coefficient equal to 1, as has been suggested by some experimental results, then it would be only 1.4% instead of the 2.3% which has been allowed above. This difference arises mainly from the fact that, in our arrangement, the liquid completely separates the uranium and the graphite, and so interposes a diffusion barrier to slow neutrons returning from the graphite to the metal. This blocking effect forces the density of the thermal neutrons in the graphite to be higher, and so increases the graphite absorption. The loss due to a 4 mm cooling layer of diphenyl is the same as that due to 2.2 mm of water, so that the plants have the same k.

In our k estimates it is assumed that the air has been removed from the pile so that the loss (of some 0.6%) due to absorption of neutrons by nitrogen does not enter. Of the various gases that might be used to replace the air, helium is preferable because of its high thermal conductivity and freedom from chemical reactions.

Control Rods. We have designed the pile so as to have about a percent larger k than is actually needed to make the chain reaction go, so that if our estimates are correct, the control rods will have to absorb 1% in k to keep the plant under control. As the reaction proceeds the plant is enriched by 49 and poisoned by fission products. If the poisoning dominates it will gradually use up the 1% margin; if enrichment dominates the margin will increase. Current opinion suggests that the enrichment will outweigh the poisoning, but this is not certain. The extent to which this goes in either direction depends upon how long the rods are run before being removed. The control rod capacity should be sufficient to handle the 1% margin + errors in calculation which may have made k too small + such enrichment change as may occur.

The control rods will operate in the central region of the pile, and will thus flatten the activity distribution. This decreases the ratios of maximum to average activity per rod and per unit length of rod, just as does the reflector around the outside of the pile. This flattening by the control rods increases the attainable pile output. The amount of the increase depends upon the k margin which the rods are using up, and upon the geometrical arrangement of the rods. If the (nearly) 1% margin which we have allowed is used by controls in the best possible manner, the output of the pile is multiplied by about 4/3 without increasing the maximum metal temperature or maximum power per rod. This means that at 500,000 kW the plant is operating comfortably below peak capacity and at substantially lower temperatures, or if additional capacity in the external system is provided a somewhat higher output can be obtained.

Other Cooling Cross-Sections. Other arrangements are possible, and some of these have been given consideration. The system discussed above has been called *external cooling* since the liquid flows on the outside of the metal rods. In this same category *hollow rods* might be used with the center left empty. Hollowing decreases the temperature drop within the metal since it makes the wall thinner, but for the same reason it decreases the fast fission and hence causes a loss in k. Such a procedure, possible if the above estimates on k are realized, if adopted in the central rods, would further materially increase the power output – or permit a lower maximum operating temperature in the metal. However, in the following, we do not contemplate to make use of this possibility in this design.

By making the rods hollow and putting the aluminium lining and cooling liquid on the inside instead of the outside we have an *internal cooling* system. In this arrangement the k loss due to the liquid is smaller, since it does not interpose a diffusion barrier to neutrons moving from the graphite to the metal. It does, however, involve the loss in fast effect due to the hollowing; and to get a good power output with the heat flowing inward and hence "bottle-necking" toward the center the amount of hollowing out must be appreciable. In this arrangement the outer surface of the rod runs hot, and it is more difficult to keep the fission products confined in the metal. The graphite also runs hot in the internal cooling designs which have been proposed (C-140, C-197). This improves k substantially by a differential heating effect, which is only partly compensated by the increases in the migration length.

A modification of internal cooling is to center a solid metal rod within a larger hollow one and flow the cooling fluid between them. This so-called *annular cooling* is capable of rather large power outputs, but the k situation has not been sufficiently investigated as yet. It has the high temperature graphite and engineering disadvantages of internal cooling, as discussed elsewhere in this report.

A modification of external cooling is to use hollow rods with the liquid flowing inside as well as outside. This *double cooling* has high power possibilities, but large k losses.

Another suggestion, which might be called *bundle cooling*, is to put together several small internally cooled rods in a bundle to represent a single rod in the graphite. It is not known what k losses are involved here.

The power outputs and thermal stresses in these various arrangements have been discussed in CP-249 and CP-319.

Shielding. It is not intended to discuss shield thicknesses in detail here. These depend considerably upon how long and how close it is intended that people shall remain near the pile. Since the shielding is outside the pile proper its dimensions and composition can be varied readily, except perhaps for the portions at the ends of the pile cylinder where the rods slide in and out. For these end shields we favor the use of heavy metals to keep the thickness down; the lateral shielding may be of other material.

To give rough values only, the shield may be something like 50 cm of iron + 100 cm of water. The amount of iron in the two end shields is then of the order of 700 tons.

Heat is dissipated in the shield and cooling must be provided. The heat generated in one of the end shields will be of the order of 100 kW.

Corrosion and Erosion

General Remarks. The problem of corrosion and erosion is not specific with the liquid cooled plants. A similar problem arises in one form or another in all cooling schemes which have been seriously suggested so far. What is specific in the case of liquid cooling is the absolute necessity of protecting against direct corrosion of the U. However, we do not consider this to be a serious handicap of liquid cooling, since the need for a covering of the U with some sort of layer is compensated by the advantages which such a covering offers. In the ideal case, it confines the fission products to the U proper and eliminates the need for radiation shields for all pipings, etc., containing the cooling agent outside the pile. Also, in case of leakage of the cooling agent to the outside, there is no danger for the operating personnel. These two points would justify the application of a coating of some sort on the U even if there were no absolute need for it, as evidenced by the fact that the application of coatings is very seriously considered, also for a He cooled unit where there is no absolute need for it (at least not in order to avoid corrosion). This is being considered in spite of the fact that the application of coatings is more difficult if they have to resist the high temperatures at which these piles operate.

The work on linings and coatings in Chicago has been carried by Messrs. Creutz, Simmons, Kratz, Gurinsky, Plott and Lanahan. Without their untiring efforts we would not be in a position now to point to three possible solutions of the problem. The work on testing has been carried out, with equal perserverance, by the groups of Burton and Howe. They have checked the present section of this report.

While the desirability of coating the U is not specific with the liquid cooling, it is, perhaps the most serious problem for such a plant. The statement that it can be solved in three different ways means only that with these linings we have no reason to doubt a satisfactory performance. Considering all the unpredictable conditions which will prevail in a high powered nuclear pile, it would be unscientific to claim a definite life time for any liner. Similarly, we believe it would be optimistic to claim any very long life time with great assurance for any pile.

Both for water and for diphenyl cooling several possibilities have been investigated. Most of these were found naturally to be unsatisfactory and will not be referred to in this report. A survey of all the attempts made can be found most easily in the monthly reports of our project.

The tests made include a radiation test and a chemical test. They are briefly described and their results summarized below. They are being continued with a view of both selecting the most promising method among the three and improving the same and also in order to find an even better new method.

Naturally, in a system containing about 1700 tubes, it is wise to guard against the consequences of failure in a few tubes. For this reason, all arrangements discussed provide for the possibility of quickly removing any rods the sheath on which may have become injured. Similarly, if a tube may become injured, the U would be removed from that tube and the water flow in it shut off. This would, of course, reduce the effective multiplication constant of the pile, but as long as only a few tubes are out of commission, the pile would continue to function satisfactorily.

Water Cooling. In case of water cooling, two arrangements are considered to be satisfactory at present. Both involve Al as the material of both tubes and metal covering sheath. In the first, the cooling agent is pure distilled water, in the second it is distilled water with an inhibitor. A mixture of potassium or ammonium phosphate and silicate is expected to give the best results.

The choice of Al was made for the following reasons: It is desirable to use a not too thin liner, particularly for the tubes which cannot be readily replaced but must be stopped up if injured. Among all the elements, only Be, C, O, F, Na, Mg, Al, Si, P, S, Ca, Sn, Pb, Bi have sufficiently small neutron cross sections to be permissible in such amounts. Some of the above are evidently impracticable. Among the others, Be and Al stood up best under the various tests. Fabrication difficulties and other reasons prevent the use of Be, which left Al as a rather unique choice.

The advantage of distilled water is the low conductivity which makes small electrolytic potential differences harmless. However, the inhibitors have a very similar effect inasmuch as they prevent the Al from going into solution. At any rate, at every contact point of two different metals, both could be thickened so that a small amount of corrosion would not lead to serious trouble. The distilled water would be led from the pile to a degasser, to free it from O_2 and H_2 and over scrap iron (or similar material) which reduces its H_2O_2 content. According to Mr. Shapiro's experiments, a 10 sec contact time with scrap iron reduces the H_2O_2 content of the hot irradiated water to 10^{-6} normality. This amount of H_2O_2 is practically harmless on the external equipment. It corrodes in 100 days 0.5 mm iron. The behavior of the Al will be described below. The only drawback of this kind of operation is the minute amounts of colloidal iron oxide which enter the water from the scrap and remain even after a filtration through a layer of sand. No other filtering has been tried to date.

The iron scrap is probably less efficient in reducing the H_2O_2 content of inhibited water. However, inhibited water, even if it contains some H_2O_2, hardly attacks the iron equipment so that its elimination is not necessary. It could be eliminated, probably, by another reducing agent but no experiments on this score have been tried so far. The disadvantage of the inhibitor is that it becomes, if it contains Na, slightly radioactive in time. If the inhibitor were

Na_2SiO_3 and present in 60 parts per million parts of water, one among every 6×10^6 neutrons will be absorbed by the Na. Thus 5×10^{12} radioactive Na atoms are formed per sec. This gives about $5000/4\pi$ R/sec at a distance of 1 cm. At an average distance of 10 m, it would give about 50 R/day without an absorber. Potassium, which is to be used instead of Na, behaves much more favorably; only the radioactivity of K^{42} is to be feared and this is not known to emit any gamma-rays. Furthermore, its parent, K^{41} is present only to the extent of 7% in K and although its cross section is 5 times that of Na, this gives it an edge of a factor of 3. Similarly, Si, P and O have no gamma-rays so that the gamma-activity of the cooling water can be expected to remain inappreciable.

The most important question remains, of course, the corrosion of Al by the water. The following tests were carried out on this question:

A. Direct Action of Radiation. About 5×10^5 watts energy are dissipated in the water in form of "ionization". This is equivalent to 3×10^{18} MeV/sec for the whole water or 3×10^{11} MeV/sec per cm^2 surface. Since the pile is expected to run 100 days, it has to stand 3×10^{18} MeV/cm^2 energy in form of ionization. This has been tested, under similar conditions, by 6 MeV deuterons of which a cm^2 should stand about $\frac{1}{2} \times 10^{18}$ or about 100 millicoulombs. Actually, a considerably greater amount of bombardment should be harmless.

Al, in contact with distilled water, showed in the steady state, a loss of weight of .08 mgr/cm^2 at a bombardment with 3.5 millicoulombs. For the whole time of operation (100 days at 5×10^5 KW), this corresponds to an average loss of .01 mm thickness of Al (Howe and Hirt). The results with inhibited water (Burton, Hirt and Howe, Report CC-298) are even more encouraging but perhaps somewhat less convincing because $K_2Cr_2O_7$ has been used as an inhibitor. This substance is reduced to the chromic salt by the radiation.

B. Effect of H_2O_2. In the same inhibited water with an Al-Fe couple both Al and Fe gained during a day less than 10^{-4} gr when a hot 10^{-4} normal H_2O_2 solution was circulated past them for a day (same report). The gain in weight is, no doubt, due to oxidation but is so small that it does not give cause to serious worry. It seems safe to assume that the concentration of H_2O_2 in water can be kept at this level even if scrap iron cannot be used for reducing the H_2O_2. Without an inhibitor, the effect of H_2O_2 on Al was negligible (Burton, Howe and Hirt, Report CC-258. There is an error in this report when it quotes the results of irradiations).

C. Erosion Due to Dissolved H_2 *and* O_2. One easily calculates that, even in the central rod, the formed H_2 is 1/3, the formed O_2 is 1/6, of the amount necessary to saturate the water at 100 °C and 1 atm pressure. Actually, the pressure is higher than this. Nevertheless, tests would be desirable concerning the erosion by such a liquid and the pressure would have to be increased if erosion appears to occur.

The above is, of course, only a brief summary of the tests that have been carried out and which are being continued. They give, at the present time, no indication that a failure of the linings must be feared. If such indications were to develop, it would be desirable – but also in all probability possible – to increase the thickness of the Al sheath on the central rods to 1 mm.

D. Diphenyl Cooling. The advantages and disadvantages of diphenyl from the point of view of k, tolerances, heat content are described in another section. It is worth repeating also here that, diphenyl not being a conductor, no electrolytic action is to be feared if it is adopted as the coolant.

U itself does not seem to stand up in diphenyl under irradiation. This is not quite certain (Cf. reports CC-342 and CC-389) but even if it were not corroded it would be worth while to apply a liner on it in order to prevent the fission products from entering the coolant.

Al seems to be a satisfactory liner and to stand up under irradiation not only in contact with water but also in contact with diphenyl. A neater solution of the lining problem is provided for, however, by Creutz's discovery of the easy formation of a nice dip-coat of Zn on U. This dip coat is so thin (3 mils) that its neutron absorption is not greater than that of the Al sheath. Such a thin sheath is not impermeable to β-rays from the U which have to be considered together with the Compton recoil electrons.

According to Garrison and McClinton (Report CC-389) a 10 day's operation in a 10^5 kW plant (or a 2 day's operation in a 5×10^5 kW plant) gives a loss of 4×10^{-5} gr/cm^2. A 100 day operation would give an average loss of 2×10^{-3} gr/cm^2 or 3×10^{-4} cm (a tenth of a mil). This is not excessive in a 3 mil coating of zinc.

In addition to the corrosion problem, diphenyl offers the problem of chemical breakdown under radiation. The corresponding problem in H_2O is less serious because water is more easily replaced and the products of disintegration (H_2 and O_2) more easily eliminated. It is important, therefore, to investigate how diphenyl itself is affected by radiation.

One easily calculates that during the whole operation of the pile about 160 watt hours energy is dissipated in the form of ionization in a cm^3 of diphenyl which is constantly in the pile. The above report describes diphenyl which has been exposed to 60 watt hours/cm^3 radiation. The product is brownish, its melting point 58 °C instead of 60 °C and its average molecular weight 320 instead of 150. Apart from the discoloration, it shows to the naked eye no obvious sign of the harsh treatment it has received.

Actually, the diphenyl will spend only a small fraction of the operating time in the pile. The above test, is therefore, far too severe and its result, as well as the result of shorter irradiation tests, perfectly satisfactory (Burton, Garrison, McClinton, same report).

It should be emphasized that the testing of diphenyl is by no means as complete as that of the water and it would be probably unwise to draw too far-reaching conclusions from them at the present time (Franck). Even the water tests, the writers feel, should be further refined, repeated and completed.

So far as the experiments have gone, the corrosion picture is favorable. Additional testing, some of it in semi-plant scale, is needed to provide answers to such problems as the effect of radiation on the efficiency of inhibitors, the tendency toward pitting, the effect of local zones of high turbulence on corrosion and erosion at the comparatively high velocities (25' per second) found in the pile, and large scale elimination of H_2O_2 by scrap iron and MnO_2.

III. Engineering Problems

The actual pile design to be adopted for further detailed study and, it is hoped, ultimate construction, is contingent upon a choice between a number of alternate design details and the solution of several problems. Although the experimental answers to some of the factors affecting a decision have not been completely determined, there is reasonable assurance that the decisions and answers given in the following discussion will justify the choice of design made in this report. This Section III will include some of the designs considered in deciding upon the design proposed in Section II. (Throughout the following report, the word "rods" will refer to the uranium rods, whether solid or hollow, and the word "tubes" to the Al liners in graphite for the "external" type cooling, unless otherwise noted.) However, except in Sections A-9, where the relative merits of water and diphenyl are discussed, water has been mentioned throughout this report as the cooling medium because it is common, easy to obtain, easy to handle, and its properties are well known.

A. In Pile Proper

Design Alternatives and Decisions

1. "Internal" vs "External" Type Cooling. Until further considerations make the more complex cooling water cross-sections (such as in Figs. 2 and 5) necessary, the pile designs will be confined to those employing the simpler water cross-sections shown in Figs. 1, 3, or 4.

The "internal" type cooling, where the cooling liquid flows through a lined tube of uranium inserted loosely in a hole in the graphite, presents some design and construction problems distinctly different from those encountered in the "external" type cooling. Although these problems are not insurmountable, they appear much more difficult of solution and more complex as far as construction and operation of the pile is concerned. The greater expansion introduced by the hotter graphite, the more difficult uranium handling, and the construction problems are a few of the obstacles that have made "internal" cooling less attractive than "external " cooling. Therefore, it was decided to employ the "externally" cooled rods for design in this report.

(However, since "internal" cooling does appear a little more favorable than "external" from the standpoint of "k" (see Section II, Other Cooling Cross-section), it has not been completely disregarded, and one possible design has been included herewith. Figure 8 shows a *vertical* pile with "internal" cooling. Figures 6 and 7 show a few possible details of this pile. Other arrangements are possible for all details. A horizontal pile might also be adopted but, for this type of cooling, the vertical pile appears to be the preferable design. All other drawings are for piles with "external" type rod cooling.)

2. Pile Above Ground vs. Pile Below Ground. There are two main reasons for placing the pile below ground, – camouflage and shielding. Since no attempt or request has been made to sink the helium-cooled pile, the chemical-extraction plant, or other units, structures, or contigent equipment below ground, there is no reason to believe that the water-cooled pile will be unique and have to be buried for camouflaging. For shielding, it is obviously economical to take advantage of the shielding value of the earth to replace a portion of the artificial shield required around a pile above ground. However, the extra excavation, particularly in a terrain where deep excavations may present some difficulty, and the increased difficulty of "charging" and "dumping" the rods from a pile below ground make it appear much preferable to have the pile above ground, with the possible exception of one alternate, Fig. 13. It is self evident that, if excavating is not too difficult, this particular design is readily and preferably adapted to an underground installation.

3. Square vs. Hexagonal Rod Spacing in Pile. Since there is no preference in lattice geometry from the standpoint of physics, the square lattice spacing of rods will be adopted for its easier adaption to construction of the tank shields, piping, valve arrangements, etc.

4. Removal of the Rods by Pushing vs. by Pulling. It is generally easier to pull an object through a tube than to push it through, if the object can be easily and successfully gripped. However, gripping a small rod in a small tube may present quite a problem. Also, if each rod is made up of several independent segments, the process of "fishing" through a long, small diameter tube, gripping a segment, and withdrawing it, repeated about ten times for each of hundreds of rods presents an unpleasant picture. Therefore, pulling the rods will be abandoned as a possibility for those pile designs employing segmental uranium rods. An experiment to determine any unforseen difficulties in pushing segmental rods having Al ribs on them through an Al tube is now being organized by Mr. Simon. Pulling is probably preferable for the long one piece rods. The method of removing rods is indicated on each of the pile sketches.

5. Tank Type vs. Header and Manifold Type Distributing and Collecting Systems for Water. Tanks on the water inlet side of the pile for distributing water to the tubes through the pile present the problem of designing large flat tanks which will withstand pressure, permit the insertion and/or withdrawal of the rods through the tank (under reduced pressure), seal against leakage of water into the graphite even while removing rods, and permit access and means of plugging one or more tubes against water flow at full or reduced pressure. This type distribution tank is most readily applicable to the vertical pile shown in Fig. 20.

Tanks on the water outlet side of the pile for collecting water from the tubes would be of two types, – one, a simple collecting pan where the rods are inserted and withdrawn from the same end (opposite the pan), and the other, a tank through which segmental rods ejected from the pile could fall to a "coffin", such as in Fig. 13 for a vertical pile, and Fig. 9 for a horizontal pile. The first type has no particular advantage over a manifolded piping system. In either case, means must be provided for plugging the outlet end of one or more of the pile tubes discharging into the tank.

A piping system having main and branch headers, with separate valved lines to each tube through the pile, lends itself readily to either inlet or outlet of water through the pile with complete throttling and shut off control of the water to any tube in the pile. It permits easier and more accurate detection of isolated tube or sheath failures, if any. It permits better temperature reading and control throughout the pile tubes. Examples of this type are shown in Figs. 8, 9, 13, 18, and 23 with details in Figs. 12, 14, 15 and 26.

6. Tubes in Pile Horizontal vs. Tube Vertical. A pile is referred to as vertical or horizontal according to the position of the rods and tubes. (The axis of the cylindrical pile its always parallel to the direction of the tubes and rods.) Each has its own advantages and disadvantages, peculiar to the position of the rods and tubes and irrespective of details of design. The major advantages of the horizontal pile over the vertical pile are:

(1) The horizontal pile will undoubtedly be easier to construct and operate than the vertical pile.

(2) In the event of the water failure in any tube due to obstructions or other unforeseen causes, the local overheating and possible ultimate melting of part of the rod will be self relieving in the horizontal pile and self-aggravating in the vertical pile. In a horizontal pile, the molten metal would run to the lower side of the tube, opening a passage above the rod for cooling water. Even if it melted through the liner it would tend to seal the wound itself and prevent pile impregnation with water. In a vertical pile, melting of any portion of a rod might completely seal off the annular water space and prevent all cooling water flow in that tube.

(3) More clearance between the rod and the liner ribs can be allowed for the horizontal pile than for the vertical pile. The horizontal pile really requires only two ribs although the third (a smaller rib) is desirable from a standpoint of wear. The vertical pile requires three ribs as spacers which gives less play and more likelihood of binding. This is probably the greatest of the advantages listed herein.

(4) In a horizontal pile, the weight of the rods and liner will hold the liner in better thermal contact with the graphite at least on the lower side of the hole. In a vertical pile, actual contact, if present, is unpredictable. Examples of horizontal piles are shown in Figs. 9, 10, 18, 23.

The major advantages of a vertical pile over a horizontal pile are:

(1) The greater ease of adjusting for differential thermal expansion.
(2) The benefit of gravity flow at reduced or zero pressure.

(3) Wider end openings (outside of the pile) for the tubes are possible, permitting easier entry of the rods in charging. Examples of vertical piles are shown in Figs. 8, 13, and 19.

7. Evacuation vs. Displacement for Replacing Air in Pile with Helium. Replacing the air in the graphite and in the pile with helium for its favorable effect on k can be performed either by simple gravitational displacement, with or without the aid of an intermediate displacement by CO_2, or by first evacuating the air and then letting in helium. Obviously the former is preferable since it requires only a gas tight shell around the pile, whereas the latter requires a shell capable of withstanding an atmosphere of external pressure.

Experiments are now being conducted by Mr. Kratz which seem to indicate the simple displacement will be quite practical, the actual efficiency begin around 98% of the air in the graphite replaced. If necessary, the air could be first replaced by CO_2 (which is also favorable from the physical point of view) and the CO_2 then by He. In this way one can expect to eliminate the harmful N_2 almost completely.

8. Handling U from Pile to Chemical Extraction Plant, in Solid Form vs. in Solution. Dissolving the rod or rods at the pile site introduces the problem of disposing of large volumes of liberated gases containing xenon and possibly other radioactive elements, and pumping a solution containing other fission products over some distance. Since the Chemical Extraction Plant will already have equipment designed to safely handle the xenon and other fission products, it appears much preferable to send the coated rod over to the extraction plant as a solid rod where it can be handled in a safer manner. The only problem involved by such an arrangement is one of suitable carriers which will protect workmen from the radiations from the rod from the time it emerges from the pile till it is safely in the vats at the extraction plant, and which will keep the rod adequately cooled in transit. These will be discussed later in this report.

9. Diphenyl vs. Water as the Cooling Medium. The increased annular space for the flow of cooling liquid along the rods, the ease in the elimination of corrosion, and the reduction in the external heat exchange surfaces due to the higher initial temperature for the diphenyl may make diphenyl appear more attractive than water as a cooling medium. However, the probable polymerization of the diphenyl under radiation plus the higher metal temperatures induced by its use leave sufficient doubt about its immediate practicability to eliminate it in favor of water.[5] The comments in Section III-B following on diphenyl vs. water for the cooling system outside of the pile are also applicable here.

(The following alternates are more applicable to the "external" type pile.)

10. Permanent vs. Replaceable Al Liners in Graphite. Since the experiments of Messrs. Burton and Howe on the corrosion of aluminium in the presence of H_2O_2 and O_2 while under radiation have not been completed, estimates of the anticipated life of the aluminium[6] liners are uncertain and so it might be

---

[5] (Cf. the section on Corrosion + Erosion).
[6] Original reads "graphite" (A.M.W. & A.M.P.)

desirable to be able to replace the liners in routine maintenance. However, since the present indications are that the Al corrosion is unimportant, it was decided that this should not be a major factor in the choice of pile design and that it would be kept in mind only as a desirable feature but not a necessity.

11. Individual vs. Group Removal (Including Total Dumping) of Rods. At this time, there appears no necessity for quick simultaneous "dumping" of all the rods.

From an operating standpoint, it is most desirable to have individual control over each rod and tube. This permits removing "sample" rods and rods with coating failures ad lib, and permits plugging any tube where a liner fails. In a vertical pile with segmental rods, individual control might prove difficult and group discharging might offer a better solution but since most other designs, including the one chosen herein, will permit individual control, this requirement has been adopted.

12. Long, One-piece Rods vs. Segmental Rods. Long, one-piece rods require inordinately long "coffins" for handling the rods from the pile to the chemical extraction plant since the "coffin" must hold the long rod plus the extensions through the shields. Likewise, a long rod does not lend itself readily to dissolving at the pile site for transporting in liquid form, and presents the same problem in handling and dissolving at the extraction plant.

Long rods are awkward to handle and appear prone to warp and bind more than short ones, increasing the problem of charging and discharging the rods in the pile. On the other hand, long rods permit a more nearly streamlined flow and eliminate the possibility of corrosion at the ends of segmental rods. It must be remembered that there is no substantial water flow above the end caps of the segmental rods so that a small steam bubble may form at these surfaces. However, the advantages of the segmental rod arrangement were considered to outweigh this disadvantage and so the segmental arrangement was adopted. Examples of piles employing long rods are shown in Figs. 8, 13, 18, and 19. Examples of piles employing short segmental rods are shown in Figs. 9, 13, and 23.

13. Space Ribs on Inside of the Tube vs. Outside of Rod Sheath. For charging rods into the pile from a cartridge, it is preferable to have the rods smooth, so the ribs must be on the tube.

Since the tube is relatively thin and subject to wear as rods are inserted and withdrawn, it is desirable to have the ribs on the tube to take the wear and save the tube itself.

Since it has been found that, with the use of helium in the pile, it will not be necessary to have a tightly expanded thermal contact between the tube and the graphite, there is no longer any objection to having the ribs in the tube. Since the tube with ribs is extruded or drawn as a unit independent of other parts, it is obviously simpler to fabricate than in the case of ribs on the rod sheath which must not only be made with ribs but drawn snugly onto the uranium rod. Therefore the ribs will be inside the tubes.

Design Problems and Solutions

14. Chemical and Electrochemical Corrosion of Pile Materials Under Radiation. Since uranium is readily oxidized, it must be protected from the water by a sheath or coating. Also, for the "external" cooling, a lining tube must be placed in the graphite. With water as the coolant, these two (the sheath and tube) must be of the same material or of two materials close together in the electrochemical series to suppress electrochemical corrosion. Of the few suitable materials, which have low danger coefficients and are not attacked appreciably by hot water, etc., the most promising are, in order, beryllium and aluminium. While the former is more desirable also for its hardness and resistance to wear, the limited supply of beryllium available plus the lack of commercial development make aluminium with its more extensive commercial production and development practically the only alternative, until such time as beryllium or beryllium-aluminium alloys can compete or surpass aluminium or unless further corrosion tests prohibit the use of aluminium and force the development of Be or Be-Al.

A second danger point for electrochemical corrosion is at the junction of the aluminium and the steel which must comprise the water cooling system outside of the pile. Tests on inhibitors and corrosion from H_2O_2, O_2, H_2, etc. in hot water while under radiation seem to indicate no great danger from corrosion at these junctions, if proper precautions are taken.

However, in all this question of corrosion and life of the parts, it must be pointed out that the important consideration is the immediate production of 49 and not the construction later of a plant for long time power production.

15. Shielding Pile. There are three types of shielding required for the pile:

a. Side shields: These can be the simplest type such as a water tank around the pile. Examples are shown in Figs. 8, 9, 10, 13, and 24.

b. End shields: these must be as thin as possible, and permit the insertion or withdrawal of rods without permitting the escape of the helium at any time. They must, however, be adequate to permit the proximity of workmen during pile operation, absorbing fast and slow neutrons and primary and secondary gamma-ray radiation. Since water or concrete alone requires so thick a shield, a shield composed of iron shot and water in tanks will be used. Examples are shown in Figs. 8, 9, 12, 13, 15, 18, 19, and 23.

c. Shielding during removal of rods: a shield must be provided to protect against the lateral gamma-ray leakage from the rods during withdrawal through the unshielded space between the pile end shield and the "coffin" or "coffin-apron". This must be easy to "open" at any point for access to piping. Examples are shown in Figs. 12, 20, and 26.

d. Cooling shields: all shields must be adequately cooled to eliminate the differential thermal expansion problems.

16. "Coffins" for Handling Rods

a. "Charging" the Pile. The coffin or chamber used to hold a new rod to be inserted into the pile need not have a cooling system of its own. It merely needs adequate material to shield against gamma-rays and a seal to prevent leakage of water from the pile during the insertion of the rod.

b. "Dumping". If the new rod is used to push out the active rod in the pile, no other coffin is needed at the entry end. If the active rod is to be removed without replacement, a separate flexible or ram-rod plunger in a coffin must be provided, similar to the charging coffin.

The coffin for receiving the rod "dumped" or discharged from the pile may be one of several types depending on the pile design. In the order of preference, they are: (1) a short "cartridge" holding several rod segments and mounted on a small car (Fig. 33), (2) a covered pot which may be tipped and which is mounted on a railroad car (Figs. 9 and 10), and (3) a long cylinder in a cradle platform (Fig. 18).

c. Handling "Coffins". At the pile site, cartridge type coffins would be handled by a small elevator tower on a crane traversing the face of the pile (Figs. 30 and 31). Long cylinder coffins would be handled by overhead crane and hoist arrangements (Fig. 18). Pots would be handled by railroad car directly (Figs. 9 and 10).

The cartridge and long coffins would, in turn, be handled on railroad cars to and from the extraction plant, where facilities would be provided for dumping the contents of the coffins into storage and/or solution vats. The pot type coffin cars would have their own dumping mechanism for tipping the pots at the extraction plant.

d. Cooling Rods After Removal from Pile. There are two ways for performing this operation, (1) continuous cooling of the rod by circulating water in the coffin and (2) cooling by allowing the water in the coffin to boil. The former requires a pump and drive, reservoir, heat exchanger, and piping mounted on the coffin car while the latter requires no more than a reservoir to replace the water as it boils away. Unless the ultimate correlation of the problems involved in this plant, the extraction plant, and other plants indicate the need for the more complex scheme, the second method mentioned above will be adopted.

17. Thermal Stresses and Differential Expansion. There are several different expansion problems involved here. The aluminium liners in the graphite will expand out of the graphite in varying amounts depending on their location in the pile. The graphite in the pile and the various adjacent and/or contiguous steel sections will expand differently. The shield unless carefully controlled for temperature will expand differently than the pile. Each design attempts to allow for these factors as well as the normal misalignment to be expected in actual construction.

18. Control Rods. The design of control rods suitable for a chain-reacting pile is a problem of major proportions in itself. For this reason the physical and electrical problems, and to some extent the mechanical ones, of such rods has been made the responsibility of V. C. Wilson. However, we may set down some of the engineering problems involved, using Wilson's present ideas regarding the other features of control rods.

Probably 9 rods will be required, 4 of them safety rods, to be used for stopping the reaction completely in emergencies, one of them, centrally located, the operating control, to be automatically moved by changes in the neutron density, and 4 of them "shimming rods" to be manually adjustable and to serve to control the "effective size" of the pile.

The safety rods must be so designed that they can be driven into the pile very quickly. Wilson is of the opinion that two of the safety rods should be operated by some force which will give faster response than simple free fall into the pile, and the Stone and Webster group is at present working out a fast-acting pneumatically driven safety rod. Gravity actuation is desirable for the other two rods because of its certainty and independence of failures of other parts of the plant.

The safety rods should act in response to the following events:

a. 5% too high neutron density
b. Too high coolant outlet temperature
c. Failure of plant power supply
d. Too great radioactivity in the coolant issuing from the pile
e. Too low a back-pressure on the pile

Audible alarms should be connected to advise the operators of any of the above reactions.

Coolant exit temperatures probably should be measured in each of the several manifolds, so that the averaging effect of the total stream from the pile will not so easily mask the effects of too high temperature locally.

The control rod should act in response to changes in the neutron density of ±1% approximately.

We made no detailed study of the control rod problem. However, the following now appear to be the principal elements of it:

(1) Size of Control Rods. The rod should probably be about 4″ to 6″ in diameter, depending on the fraction of the neutrons in the pile which it is intended that the rod shall absorb.

(2) Cooling of Control Rod. This can best be accomplished with water, possibly containing a dissolved boron compound e.g. sodium borate. This procedure will have the advantage that the rod itself can be made of any structurally convenient material, since the bulk of the neutron absorption will occur within the liquid itself. Troublesome heat transfer and pressure drop problems are thus somewhat alleviated also. Connections to the cooling rod can be made by flexible metal hose.

(3) Actuation of the Control Rod. will depend somewhat on whether the rod is horizontal or vertical. The vertical case is a little easier from the engineering standpoint since no supporting horizontal track for the rod need be put into the pile. In the vertical case the rod would either be suspended by a cable passing over a motor driven drum, counter-weighted and fitted with a stuffing box, or it would be hung from a vertical rack and fitted with a counter-weight,

pinion gear drive and a stuffing box. In the horizontal case the rack-and-pinion drive would undoubtedly be preferred, and the rod would have to travel on a graphite or water-cooled aluminium track in the pile. In either case the cooling solution probably would be introduced through a central tube running through the control rod, and would return through the annulus to an external cooler. The stuffing box would not, in this case, be very troublesome since it would need to hold only a few pounds of helium pressure, and will be accessible for servicing at all times. The jacket into which the control rod slides must pass through the radiation shielding and connect on to the gas shell inside. Care must be taken that there is no radiation leakage of consequence around the control rod or the shimming and safety rods. Very probably this can be done by an arrangement of offsets and small solid angles.

From the standpoint of preserving the symmetry of the heat distribution in the pile and the consequent ease of control and flow adjustment it undoubtedly is desirable to insert the rods, both shimming and control, parallel to the axis of the pile. In the horizontal case this will involve some complications at the face of the pile, as well as some internal mechanism, but this probably is not serious.

19. Control of Helium Through the Pile. In order to detect water leakage into the graphite and to hasten its removal from the graphite it is desirable to circulate the helium in the pile continuously. This will require an external helium system consisting of a reservoir such as a balloon gas holder floating on the line, a pump and drive with spare, a stack, driers, and possibly exchangers.

20. Instrumentation, Including Detectors for Tube or Rod-sheath Failures. Except for the special controls for the safety and control rods (see Section B-18), the intrumentation for the pile proper will be only nominal. The cooling water temperature will be measured at the inlet header and several places in the outlet headers in the conventional manner. Water pressures will be measured at several points on the inlet and outlet headers and branches. Somewhere in the system, a flowmeter will measure the flow of cooling water. If feasible, several thermocouples in the graphite would be desirable for measuring graphite temperatures. Provisions should be made for several neutron and gamma-ray counters or intensity indicators and openings should be provided for inserting ionization chambers. These may be located in the graphite reflector around the pile or elsewhere.

If the Al sheath on a rod fails, the uranium will be eaten away by the water. This corrosion will contaminate the water with the radioactive metal and with fission products. Detectors should be provided to check the pile at continuous, regular intervals for indications of sheath failures. These detectors should notify the operator in the control room either by audible or visual alarms or both, so he can "dump" the faulty rod and replace it if desired.

If the Al liner in the graphite corrodes thru or breaks, water will leak into the graphite. Water vapor detectors in the helium recirculating system would advise the operators by alarms, as above, of such a failure. Means should be provided for identifying the tube at fault, such as by control rod actuation or other methods. The rod could then be dumped, and, if the liner is not readily replacable, the tube could be plugged and abandoned.

21. Commercial Production of Pile Parts. The main pile parts for which the commercial facilities have been incompletely developed are the uranium bars or tubes, the aluminium ribbed tubes for lining the graphite, the assembling of the rods and sheaths and end caps, and the graphite blocks.

Mr. Creutz has been ably exploiting the facilities at Wolverine Tube Company, Summerill Tubing Company, and others for extruding and drawing uranium bars and tubes, and for drawing an aluminium sheath on the rods.

There has been some little difficulty in getting the aluminium ribbed tube but it now appears that Wolverine and maybe others will attempt to extrude, draw, or roll the tubes with integral ribs. As an alternative, it is possible to weld separate ribs inside the tube. Sciaky Brothers have expressed their assurance that this process is entirely feasible, although it is still preferable to have extruded or drawn ribs.

The National Carbon Company have succeded in making graphite blocks $8'' \times 8'' \times 50''$ of the desired purity. These will be fastened together in long columns by tube and socket arrangement.

B. In Liquid Cooling System Outside of Pile

Selection of Type of Cooling System. The function of the external system is to cool the water circulated through the pile and partially to remove from it the hydrogen and oxygen formed in the water under the strong radiation in the pile. Although several possible ways of doing this probably can be made to work the one to be outlined below is preferred. A number of cooling schemes have been suggested. The ones considered here are listed below.

1. Once through passage of raw water, discarding the hot water into the stream from which it was taken cold. Numerous objections to this procedure justifiably can be raised, among them the facts that the quantity of water required would be a considerable fraction of the river from which the water would have to be taken, that scale formation in the pile would undoubtedly occur, that a coating failure would result in contamination of the stream and thereby constitute a health hazard for down-stream communities.

2. Cooling of the water after its passage through the pile by pumping it over a cooling tower and thus evaporating a portion of it. The disadvantages of this arrangement are:

(1) The contamination resulting from the water exposure in the cooling tower requiring redistillation.
(2) The radioactive contamination of the plant area to the leeward of the cooling tower, in the event of a serious rod coating failure.
   Also, considerable make-up water of high quality would be required.

3. Cooling of the water after its passage through the pile by countercurrent exchange against water cooled, in turn, by passage over a cooling tower. The

principal objections to this scheme are the large quantities of heat exchanger surface required and the possible corrosion problem involved in circulating the pile-water.

4. Cooling the water after its passage through the pile by flash vaporization of a part of it at a low pressure. The dissolved gas content of the water resulting from this cooling would be very low, but the heat transfer surface required would be several times that necessary for proposal *3*.

5. Cooling the water after its passage through the pile by countercurrent exchange against cooling tower water as in *3*, with the addition of a distillation column for removing the dissolved gases from the hot water stripping with steam. This scheme is to be preferred to *4* if a low oxygen content in the water in the heat exchangers is necessary to prevent corrosion of the exchangers, since it requires no more heat exchange surface than *3*. Some additional equipment is, however, required, and *3* is to be be preferred if as seems likely, use of scheme *4* would not reduce corrosion troubles.

With a few exceptions the engineering problems in the external system of a water cooled plant have straightforward answers in conventional engineering practice. These we need only mention.

The question of purity of the circulating water is not at all critical. For example, the boron tolerance for a loss in k of 0.0001 is roughly 10 parts per million. This is easily attained by distillation, and if it were not for the possibility of oil contamination, power boiler condensate would be entirely satisfactory. However, it probably will prove desirable to have a water evaporation unit available for concentrating radioactive wastes anyhow, and this same unit may well be used for producing distilled water for the pile circulating system.

The power required for pumping is relatively small as compared to the proposed helium plant, and aside from the unusual importance of adequate and reliable standby power, capable of being brought into service in fractional minutes, no new problems are involved.

Relatively large quantities of make-up water (estimated at 4225 gpm) will be required. This should be softened and filtered before introduction into the cooling tower circulating system, since steel tubes will be used in the heat exchangers and these foul and scale badly if the water is not properly treated.

Waste disposal presents no unusual problems except for radioactive wastes. These cannot be discharged into streams, and probably must be disposed of by pumping into deep wells.

Some problems unique in the water cooled plant may now be mentioned Probably the most serious problem is corrosion by the pile circulating water. This problem has been discussed in Section II.

Corrosion of External Equipment. The results of corrosion tests on 1020 steel, in approximately 1×10^{-4} normal hydrogen peroxide (representative of the circulating solution) show an average penetration rate of 0.016" per month or about 0.19" per year. The heat exchanger tubes which probably would be used are only 0.083" in wall thickness. Obviously this ordinary steel cannot be used without corrosion protection of some sort.

Several possible solutions of this corrosion problem have been suggested, and are listed below:

1. Use of inhibitors
2. Cathodic protection
3. Catalytic decomposition of H_2O_2
4. Decrease of H_2O_2 content by removal of oxygen
5. Use of corrosion resistant alloys
6. Prevention of H_2O_2 formation
7. Reduction of H_2O_2 with scrap iron

Some inhibitors for iron corrosion already have been investigated. Potassium dichromate has been found to be effective at $6 \times 10^{-4}N$ in $1 \times 10^{-4}N$ H_2O_2, but is reduced by the H_2O_2 so that it must be added continuously. The resulting chromic hydroxide is only sparingly soluble, and must either be removed continuously, probably a troublesome operation, or it must be kept in solution by the addition of an acid. In the latter case the solution probably will turn out to be so acid as to endanger the aluminium tubes in the pile. Sodium phosphate also is effective, and although not quite as good as the dichromate is probably the most practical inhibitor yet investigated. As mentioned previously, it now seems likely that a mixture of NaH_2PO_4 and sodium silicate at pH~8 will give adequate protection to both aluminium and steel. Sodium sulfite was found slightly to accelerate corrosion, although it presumably reduces the free oxygen content of the water.

Cathodic protection of the iron does not now seem attractive, because its use would very considerably complicate the mechanical problems in the heat exchangers. It may be suitable for protection of other types of equipment, however.

It seems possible that the H_2O_2 could be decomposed catalytically while the water is hot, thus removing it and its corrosive influence before it reaches the heat exchanger, pumps, etc. A large number of catalysts for the reaction

$$H_2O_2 \rightarrow H_2 + O_2$$

are recorded in the literature, among them the finely divided noble metals, iron hydroxide, activated charcoal and others. Unfortunately no reports of the decomposition rates attained with these catalysts have been found for the rather dilute solutions contemplated here, so that the possible effectiveness of this approach is in doubt. MnO_2 is considered promising.

On the assumption that the reactions

$$H_2O_2 \rightleftharpoons O_2 + H_2$$
$$H_2O_2 \rightleftharpoons \frac{1}{2}O_2 + H_2O$$

can be forced to the right by removal of the products O_2 and H_2, it has been proposed to control the H_2O_2 concentration and corrosion in most of the external sytem by very complete removal of the dissolved oxygen and hydrogen, by stripping with steam or by vacuum flash cooling. However, the H_2O_2 concentrations reached in deuteron bombardments of water are already much larger than would correspond to a thermodynamic equilibrium between the H_2O_2 and any possible concentration of dissolved oxygen and hydrogen. It is therefore doubtful whether simple reduction of the O_2 and H_2 content of the water would materially reduce the H_2O_2 content and thus affect corrosion.

Tests have so far established the stability of aluminium under conditions somewhat similar to those likely to be found in the external system. In addition it is likely that 18-8 stainless steel will be suitable, since this is the material conventionally used for handling H_2O_2. However, a glance at the table showing the material requirements of the process makes it clear that relatively large quantities of any of these alloys would be required.

The suggestion to prevent H_2O_2 formation is attractive, but as yet untried. The use of scrap iron for reduction of H_2O_2 has been discussed in Section II.

Consideration of the facts and possibilities outlined above leads to the conclusion that the sodium phosphate and sodium silicate inhibitor or the scrap iron reduction of H_2O_2 appear to offer the best present probabilities for solving the corrosion problem. However, no definite decision can be reached until more actually is known regarding the stability of phosphate silicate inhibited systems, and of the characteristics of the other methods proposed.

For the purpose of this report, we shall assume that the present favorable prospects for eliminating corrosion are in fact justified, and the external system for the water plant to be described is based on this assumption.

Radiation Shielding. One of the attractive features of the water cooled plant is the fact that in its simplest conception the fluid does not become dangerously radioactive. As a practical manner, however, it would be extremely foolish not to provide for the installation of substantial radiation protection. Some additional outlay is thus needed, but it can only be regarded as cheap insurance that minor coating failure, which is very probable, will not shut down the plant. In addition to coating failure and consequent appearance of fission products in the circulating water it can be foreseen that casual impurities in the water as well as intentionally introduced inhibitors, will possibly develop radioactivity, although for water which has been distilled the radioactivity of any casual impurity would be extremely small.

It would appear that earth, concrete and water are the most economical shielding materials where, as is here the case, bulk is no object. The application of these materials is described along with the rest of our preferred plant, in a later section of this report. The selection of shield thicknesses is difficult because there is no good approximation to the upper limit of the gamma ray energies to be expected. We can only be very conservative in this respect.

Dissolved Gas Removal. It has been pointed out previously that it is desirable, in order to avoid erosion by liberated gas, to keep the dissolved gas content of the pile to a minimum. For very complete removal of the dissolved gas something approaching a counter-current stripping column would be necessary. However, in the present case there is reason to think that no extraordinarily complete separation of the dissolved H_2 and O_2 from the water is desirable. Hence, the equivalent of a single "plate" in a stripping column, i.e., a single flash vaporization, will be adequate. Details appear later in this report.

Diphenyl vs. Water. From the engineering standpoint, the question of diphenyl vs. water in the external system seems better answered by choosing diphenyl. However, the use of a diphenyl creates a few new problems.

First, Burton's tests have established that radiation causes polymerization of the diphenyl, probably with the liberation of O_2. The extent of the polymerization is not definitely known, but an approximate doubling of the average molecular weight and some increase in viscosity was observed over the equivalent of a week or ten days operation at 500,000 kW. Evidently some removal of the higher polymers must be effected, and the purified diphenyl returned to the circulating system. A distillation unit of reasonable size probably could effect this separation. The removal of dissolved hydrogen from the circulating hydrocarbon probably is less important than the removal of gases from the distilled water circulation system, since erosion will probably be much less severe. If gas removal is necessary it probably can be accomplished by spraying the hydrocarbon mixed with superheated steam in a chamber similar to the "flash chamber" of the water plant.

Second, if diphenyl itself is used some freezing up in the external system may be expected unless it is guarded against. This can be done quite easily, since there is an established technique for handling such materials. The substitution of toluol or xylol would make this problem disappear.

Finally, it is likely that a diphenyl plant will require about 10% to 15% more pumping power, as well as more process steam, than a water plant. These are not considerations of the first importance.

Because the diphenyl circulates at a higher temperature level than the water, only about one half the heat exchanger surface is required in the external system. This represents a considerable fraction of the steel requirement for the plant, and is an advantage of medium importance.

IV. Description of Proposed Plant

In this section of our report we described a complete 500,000 kW water-cooled power plant. This is not to be construed as meaning that the plant described is our ultimate preference, of all liquid-cooled plants. However, in the light of present knowledge and after due consideration of this and the varied designs and details previously described, we do regard it as the soundest design, the most dependable, and the most likely to operate at the desired power output, with the possibility of realizing even more than 5×10^5 kW. Continued research and re-examination of our present opinions in the light of new facts, might cause changes either in the design or in our choice. However, the urgency of the situation forbids our following the obviously safe scientific course of awaiting these changes and renders the immediate presentation of this preliminary design desirable. The description to follow is for a plant cooled with water. The attractive features of hydrocarbon cooling, however, have not been set aside. The fact is that the diphenyl and water plants are in many essential respects quite similar, and in such a qualitative description as this are practically interchangeable. The corrosion data for water were better known than the stability of diphenyl at the time we undertook this study. Hence we describe a water-cooled plant.

A. Drawings and Description of Pile Proper

The pile will be of the horizontal type, – a large drum, in effect, lying on its side and supported in a cradle formed by filling in the lower corner spaces between the cylinder and the enclosing parallelopiped with second class, "dead" graphite or other suitable material. (See Fig. 24).

The active cylindrical pile will be formed of first class, high purity graphite formed into blocks $8\,3/8'' \times 8\,3/8'' \times 50''$ with all edges chamfered $1/4''$. Each block will have a central hole throughout its full length with a male projection at one end and a female counterbase or socket at the other. These projections and sockets will be machined to a press fit (Mr. Plott has demonstrated the practicability of this), and the blocks assembled to form long columns extending the full length of the cylindrical pile. The columns will be laid on their sides in tiers and rows to form the cylinder. Around the active cylinder will be laid a 20 inch layer of second class graphite as a reflector. The construction of the reflector will be similar to that of the active pile.

The rods and tubes (about 1695 in number) will be disposed in a square lattice arrangement with their axes all horizontal and parallel to the axis of the graphite cylinder. The square lattice spacing will be about $8\,3/8''$ center to center horizontally and vertically, with the rods and tubes grouped symmetrically about the center of the pile and located within the geometrical boundaries of an enclosing cylinder.

The holes in the graphite will be lined with ribbed aluminium tubes of about $1\,11/16''$ O.D. and 17 ga. wall thickness, inserted freely in the holes in the graphite with about $0.010''$ clearance on the diameter. The ribs will be ground off at the ends and the ends flared over to form a gasket and seal the junction between the tube and the piping manifold against water leakage. The backing-up for the flared tube end will be an extension nipple welded gas-tight onto the tube sheet (to be described later). This gasketed joint will then also seal against escape of the helium gas in the pile. See detail in Fig. 26.

The uranium will be in the form of rods of about $1\,5/16''$ O.D. in lengths of about two feet. Each segment will be covered by a 25 ga. aluminium tube or sheath drawn tightly onto the uranium. The ends will be covered by about a 17 ga. aluminium cap screwed on and welded to the sheathing tube. The end caps will have plug tips and sockets for lateral alignment in the event of local or individual warping. See detail in Fig. 26.

The pile will be surrounded by a gas tight shell. The sides, top, and bottom of this shell will be a continuous, welded plate completely encircling the graphite with provisions for some expansion in the graphite. The ends of the shell will be large continuous welded plates, having holes opposite the holes in the graphite, and special nipples welded gas-tight to the plate in the field so as to be directly centered with the holes in the graphite. (These end plates with nipples are referred to hereafter as the tube sheets). The nipples have a projecting section outside of the tube sheet and a section projecting back through the shield will have a large sloppy fit thread as shown. The tube sheet will be hung from overhead structural framing. The function of this gas-tight shell is to retain the helium which is used in the pile to replace the air in and around the graphite. See Figs. 23, 24, and 26 for details of this shell.

The pile will be completely surrounded by shields. The shielding medium for the sides and top will be water retained in a concrete tank in which the pile is immersed. A minimum water shield thickness of 12 feet will be obtained at sides and top of the pile. An automatic float-controlled valve will maintain the level in the tank and an alarm will be provided to warn the operator of the level drops below a safe minimum. The gas tight shell around the graphite will also serve to keep the shielding water out of the pile. See Figs. 23, 24, and 25 for this shield.

At the ends of the pile, the shields will be sectional, interlocking steel tanks containing water and graded iron or steel shot. The tanks will each be divided by a partition, parallel to the tube sheet, to separate the water in the tank and prevent water circulation within the tank from carrying any dissolved radioactive materials from the pile side of the shield to the outside face of the shield. The water in the pile side of each divided tank will be circulated and cooled to obtain temperatures that will keep the tank expansion the same as the pile expansion and thus maintain alignment between the shield and the graphite. This will be automatically controlled through thermocouples located at the graphite end surfaces, and a temperature controller which will vary the flow of water as needed. See Figs. 23, 26, and 29 for this shield.

In addition to the shielding tanks, a separate shield will be required at the water inlet (or "cool") end of the pile, where the rods are ejected, to protect against gamma-ray leakage while the rods are being pushed through the tube extension between the shield and the coffin. This shield must also allow access to the valves and piping. It will be a large honeycomb steel structure with removable sections of steel or lead (12″ thick if of steel, 8″ thick if of lead). See Figs. 23, 26, and 28 for this shield. No extra shield like this will be required at the "hot" end, where the new rods are inserted.

The foundation slab for the pile and all the shields will be one integral slab of such size as to prevent misalignment of the pile and end shields from differential settling. The top surface of the slab, immediately under the graphite pile proper, will be cooled by the water in the shielding tank which will be circulated under the pile by propeller mixers in the tank shield. See Figs. 23, 24, and 27.

Where the aluminium tubes in the graphite connect to the tube sheets, steel extension pieces will prolong the tube openings to the gate valves beyond which will be pipe nipples with outer ends levelled to fit the coffin socket. These steel extension pieces will have an annular space and holes for distributing the water uniformly into the tube. The connection to the piping manifold will be through a flexible piece of pipe of seamless steel connected to the extension piece and, by union, to an angle valve leading to the header; a main distributing pipe will feed or empty these headers. The piping may be screwed or welded. Where the end pipe goes through the auxiliary shield, oversize holes will be provided to allow for misalignment. See details in Fig. 26.

Parallel to either end face of the pile will be tracks for spotting standard flat cars especially rigged to hold several coffin cars. These flat cars are for hauling the rods to the Chemical Plant and for bringing back empty coffins. Between these tracks and the pile will be other rails on which cranes will travel back and forth across the end faces of the pile. On each crane will be mounted a tower structure enclosing and supporting a hoist platform and hoisting mechanism. The hoisting platforms will hold the independent coffin cars on each of which are mounted a coffin and a water reservoir. The coffin cars are moved toward or away from the pile on rails on the hoisting platform by means of a geared ratchet lever arrangement. Controls for the hoist and crane will be on the hoisting platform. Provision will be made at flat car level for running the coffin cars from the flat cars onto the hoisting platform and back. See details in Figs. 30 and 31.

The coffins for charging and dumping rods are lead and steel cartridge clips holding all the segments comprising one complete rod. They are mounted on little coffin cars so as to be easily moved from hoist platform to railroad flat car for transporting to the Chemical Plant. They will have plunger extensions for pushing the rod segments into the tube at the charging end and for centering the segmental rod at the dumping end when pushing out the old rod with the new one. A flexible connection will feed water into the dumping end coffin from the reservoir on the coffin car. A safety valve will permit the escape of steam formed in that coffin. A large apron on the coffin car will provide further shielding and will act as the valve for opening or closing the mouth of the coffin. See details in Figs. 32 and 33. About 12 coffin cars might be required for normal operation, 6 for charging and 6 for dumping.

In addition, one special plunger coffin will be required for pushing out a rod when it is intended to leave a tube empty temporarily or permanently. This will have a linked flexible plunger capable of reaching through the full length of a tube and will be pneumatically or hydraulically operated. See Fig. 34.

B. Drawing and Description of Liquid Cooling System Outside of Pile

General Description of System. In the external system the water goes through the following sequence of events: After leaving the pile at 2 °F above the boiling point at the existing atmospheric pressure and at $\sim$ 20 psi gage the water passes through a *throttling valve* which reduces the pressure to near atmospheric. Part of the water vaporizes, most of the H_2 and O_2 being carried into the water vapor. The vapor-liquid mixture passes into the *flash tank* where it is sprayed from a header pipe. The formation of the steam is completed in this tank, the remaining liquid water at its boiling point passing downward through a gravel *filter* in the bottom of the tank, and the steam-hydrogen-oxygen mixture passing off through an opening in the top of the tank to the *flash tank condenser*, where the water vapor mostly condenses and runs back into the flash tank. The remaining gases, hydrogen, oxygen and some water vapor, are immediately diluted with air and pass to the waste stack. The hot degassed water from the filter next passes through the *heat exchanger banks*, where its temperature is reduced to 95 °F (35 °C) by counter-current exchange against cooling tower water. The cold pile-water then passes to the pile-water pumps, and returns to the pile. The cooling tower water enters the heat exchangers at 85 °F (29.4 °C) and leaves at 130 °F (54.4 °C), passing over the cooling towers, and running into a pit beneath them from which it is picked up and recirculated through the heat exchangers again by the *cooling water pumps*. Air is blown through the cooling towers by motor driven fans, and it is the evaporation of the cooling water into this air which cools the cooling water from 130 °F to 85 °F.

The principal elements of this system are described more fully below, and a flow diagram is given in Fig. 36.

Throttling Valve. The throttling valves will be pressure reducing valves set for ~20 psi gage on the upstream side. Because of the high velocities in the throttling valves erosion must be guarded against. These valves, of which one for each flash tank is contemplated, will be 24 inches in diameter. Stainless steel is recommended. The Davis Regulator Company is a possible supplier.

Flash Tank. The flash tanks will be 22 feet in diameter and 40 feet high with conical tops and bottoms. Two such tanks will be provided, one to serve as a standby and for use when the other tank is being backwashed. In Fig. 36 is shown a diagrammatic cross section of a flash tank. Water from the spray header falls into the tank through a 5-foot spray space, and passes through the filter, located 30 feet below the surface of the water in the tank. The time of residence in the flash tank is about 3.3 min.

The gravel filter located in the bottom of the flash tanks will be made of 8–10 mesh gravel. The function of the filter is to catch suspended solids such as iron rust and aluminium hydroxide and to remove them from the circulating system. Periodic backwashing with water will be used to remove such material from gravel, and the backwashings will be put into a storage tank for settling. The water can then be drawn off and discarded or re-used. The gravel will be contained between screens to prevent its escape and supported on a false bottom.

The flash tanks will be set on five foot high foundations, so that the water level in them will be 40 feet above grade. This will permit gravity flow through the filter and heat exchangers. The water level in the flash tanks will be controlled by a float operated pump, which will take water from the distilled water storage tank and introduce it into the system so as to keep the water level in the flash tank between predetermined minima and maxima.

It should be observed that the conversion of a small fraction of the hot water stream to vapor is an essential part of the proper operation of the flash tanks. In the operation as planned 0.21% of the water is converted to steam. This gives a vapor-liquid volume ratio of 3.45/1, and provides for fairly good stripping out of the H_2 and O_2 in a single stage. This is desirable because, in order to avoid gas-bubble erosion in the pile, the dissolved gas content must be kept below saturation at all points in the pile, as explained elsewhere. This flashing does not, unfortunately, affect the H_2O_2 concentration in the water. Tables 1 and 2 summarize the design of the flash chamber.

Flash Tank Condensers. The flash tank condensers will be mounted directly on top of the flash tanks, and the condensed water will drip back into the flash tanks. Table 3 summarizes the design of the flash tank condensers. It will be noted that they are somewhat over-surfaced. This is done to allow some margin for variation of the exit water temperature from the 2 °F differential specified.

This condenser operates in an atmosphere rich in oxygen; hence corrosion probably will be too severe for steel tubes. Brass tubes and tube sheets are recommended.

Waste Disposal Stack. The stack for waste disposal will serve mainly to dilute the H_2O_2 mixture from the flash tank condenser to below the explosive limit and to discharge it, along with any other gaseous wastes from the process, to the atmosphere at a safely high elevation. In its base an induced draft fan will be located which should handle about 1400 cfm of air. Quite possibly this stack can be eliminated by use of others in other units of the project if these are located nearby.

Heat Exchangers. The heat exchangers will be of the shell-and-tube type, with the pile-water inside the tubes and the cooling tower water outside. Steel tubes are contemplated, with corrosion inhibitors in both the water streams. Scale formation should not be troublesome since the make-up cooling water will be softened before introduction into the system.

The heat exchangers are the principal heavy equipment. The requirements are summarized in Table 4.

Pile-Water Pumps. The pumps which will circulate the water from the heat exchangers back into the pile must have a combined normal capacity of 28,700 gpm. (1.435×10^7 pounds per hour). The present proposal is to install 8 pumps, any 6 of which can carry the normal load. As shown in Fig. 37 the pump inlets and outlets are manifolded, and two by-pass lines between these manifolds are provided. In these by-pass lines motor-operated valves will be installed, which will be automatically controlled by the average temperature of the water leaving the pile. In this way approximate constancy of outlet temperature will be obtained. ±0.5 °F regulation is desired.

Pumps suitable for this service should be capable of parallel operation, and should deliver water against a head of about 350 feet. (150 psi). Because of the corrosive nature of the circulating water and the high velocities in the pumps it is recommended that they be made of stainless steel, unless corrosion

Table 1. Oxygen and hydrogen contents of circulating water

	lbs. water/hr	(Mol Fractions)×(10^6)	
		O$_2$	H$_2$
Liquid water entering pile	1.435×10^7		
Liquid water leaving pile	1.435×10^7	.10	.20
Water vapor leaving flash chamber	2.96×10^4	1.53	3.06
Liquid water leaving condenser	2.95×10^4	692	1380
Water vapor leaving condenser	100	~ 6	~ 12
Assumed efficiency of removal of dissolved gas (realized % of theoretical removal)		97%	

Table 2. Flash chamber

Diameter	22 ft
Height	40 ft
Height of vapor space	5 ft
Depth of gravel filter	2 ft
Superficial water velocity in filter	0.17 ft/sec
Pressure drop through gravel	1.4 psi
Permeablility of gravel	1000 darcys
Volume water contained	13,300 ft^3
Material	Steel

Table 3. Flash chamber condenser

Water vapor condensed	29,500 lb/hr
Heat transferred	28,600,000 Btu/hr
Temp. condensate	212 °F
Temp. cooling water in	85 °F
Temp. cooling water out	130 °F
Heat transfer coefficient	~ 150 Btu/hr ft^2 °F
Heat transfer area	~ 1900 ft^2
Tube diameter	3/4 in
Wall thickness tubes	14 BWG
Tube length	4 ft
No. tubes	~ 2520
Shell diameter	60 in
Tube material	brass
Tube sheet material	brass
Shell and heads	iron

Table 4. Heat exchangers

Heat transfer rate	1.679×10^6 Btu/hr
Temperature dist. water in	212 °F
Temp. dist. water out	95 °F
Temp. cooling water in	85 °F
Temp. cooling water out	130 °F
Distilled water flow rate	28,700 gpm
Cooling water flow rate	74,600 gpm
Log mean temperature difference	34.2 °F
Heat transfer area, outside of tubes including spares	329,000 ft²
Overall heat transfer coefficient	186 Btu/hr ft² deg. °F
Tube fluid	Distilled water
Shell fluid	Cooling tower water
Tube length	20 ft
Tube O. D.	0.75″
Tube I. D.	0.584′
No. tube passes per exchanger	2
No. shell passes per exchanger	2
Shell I. D.	6 ft
No. tubes per exchanger	3500
No. exchangers, total	24
No. exchanger banks, in parallel including 1 spare bank	6
No. exchangers per bank, in series	4
No. banks in normal operation	5
No. stand-by banks	1
Pressure drop on shell side, total for battery	∼ 30 psi
Pressure drop on tube side, total for battery	13 psi
Material	Steel and iron

inhibitors of extraordinary potency can be found. Table 5 summarizes the pump requirements. A possible type of pump is the Byron-Jackson Company, Model DVS, 10 × 14 × 20 size.

Cooling Towers and Cooling Water Pumps. The cooling towers will be of conventional design. In the layout (Fig. 37) they are shown relatively close to the rest of the plant. It is possible, however, that they may be located at some distance, or that a central supply of cooling water for all units of the project might be established.

The pumps for the cooling water will be of conventional design and probably made of steel. The pump house and control stations are separated from those serving the pile proper because of the greater safety resulting from this arrangement.

Tables 6 and 7 summarize the requirements for the pumps and cooling towers. A possible type of pump is the Byron-Jackson Company, Model DVS, in an appropriate size.

Table 5. Distilled water pumps

Pumping rate	28,700 gpm
Minimum no. units desired	6
No. spare units desired	2
Discharge pressure	150 psi gage
Suction pressure	0 psi gage
Suction temperature	95 °F
Power required	3300 kW
Pump material	Stainless steel

Table 6. Cooling tower water pumps

Pumping rate	75,800 gpm
Minimum no. units desired	7
No. spare units desired	1
Discharge pressure	50 psi gage
Suction pressure	0 psi gage
Suction temperature	85 °F
Power required	3000 kW
Pump material	Iron or steel

Table 7. Cooling tower and cooling water

Temperature water in	130 °F
Temperature water out	85 °F
Loading density	2 gpm/ft²
Hot water rate	75,800 gpm
Tower area required	37,900 ft²
Tower area required	37,900 ft²
Water evaporation rate	3,300 gpm
Maximum probable water draw-off rate	165 gpm
Windage water loss rate	760 gpm
Treated water make-up rate	4,225 gpm
Power required for cooling tower fans	1,000 kW [7]
Tower material	Wood

---

[7] Original reads "gpm". (A.M.W. & A.M.P.)

Water Still. It may be necessary to discard the water from the pile circuit because of contamination with various impurities or if the inhibitor concentration should rise too high. This latter may occur even with sodium phosphate since, although there is no chemical reaction between O_2, H_2 or H_2O_2 and sodium phosphate, it is reported to lose its effectiveness as time goes on, and more must be added. Disposal of such wastes will be difficult since there is almost certain to be radioactivity of dangerous intensity associated with the solids dissolved in the water. The water still will serve to concentrate these wastes into a volume small enough to permit injecting them into deep wells. The still also can be used to make fresh distilled water to replace evaporation and leakage losses from the system, although these latter probably will be negligibly small.

The requirements of the still are summarized in Table 8 and 9.

The still capacity is chosen so that it will be capable of evaporating the entire water content of the circulating system in about one week.

Piping in External System. All piping will be so laid out that operation can be continued at 50% or more of the normal rate in case of failure of any element in the plant except the main power supply. Power supply failure would make it necessary to stop the chain reaction. Isolating valves are provided in the headers so that entire areas with pumps and banks of exchangers can be cut out.

An emergency connection is provided from the distilled water storage tanks directly to the distilled water pumps suction header to insure that water can be supplied to pumps at all times. A second emergency connection is provided from the plant water system standpipe and a third from the cooling water pumps, so that exhaustion of the distilled water reserves will not prevent cooling the pile.

Radiation Protection in External System. The plant is so laid out that all units can be provided with radiation shielding as indicated. The line going from the pile to the flash tanks will be buried and brought up alongside the flash tanks themselves and inside the concrete shields around the flash tanks. Sufficient room near the flash tanks will be left so that these shields can be erected if they are later found necessary. The flash tank condensers will be shielded by the same shield which goes around the flash tanks. These shields will be approximately 60 ft high, thus making it possible to submerge both flash tank and condenser in water to a depth of several feet.

The heat exchangers will be arranged in 6 groups of 4 each. Each group will be placed in a concrete enclosure, as shown in Figs. 36 and 37. This enclosure can be flooded with water, submerging the exchangers to a depth of several feet if necessary.

At present it is extremely difficult to make a justifiable estimate of the amount of radiation protection required on the external system. Since coating failures in the pile might occur, certainly no part of the plant should be so designed that a minor failure makes operation impossible.

Controls and Instrumentation. In addition to the pile control and safety rods described elsewhere, there must be automatic temperature and pressure con-

Table 8. Still

Capacity	10,000 lb/hr
Steam consumption	12,100 lb/hr
Steam pressure	40 psi gage
Steam superheat	0 °F
Temperature water entering	70 °F
Temperature vapor leaving	212 °F
Number of tubes required	630
Outside tube diameter	0.75 in
Wall thickness	14 BWG
Length of tubes	4 ft
Diameter outside shell	4 ft
Height outside shell	8 ft
Tube pitch	$1\frac{1}{16}$ in
Material	Steel and iron

Table 9. Still condenser

Two passes on tube side, floating head Cooling water in tubes Vapor in shell Number of tubes per pass	~ 310
Total no. tubes	~ 620
Outside tube diameter	0.75 in
Wall thickness	18 BWG
Length of tubes	4 ft
Diameter shell	3 ft
Temperature cooling water in	85 °F
Temperature cooling water out	100 °F
Rate of flow cooling water	1290 gpm
Vapor to be condensed	10,000 lb/hr
Temperature vapor	212 °F
Materials	Iron and Steel

(Temperatures given above assume 760 mm Hg barometric pressure)

trols on the water as it leaves the pile, as well as temperature, pressure and flow-rate measurements elsewhere in the external system. The outlet pressure control will be an automatic back pressure regulator or throttling valve located just before the flash tanks. Pressure regulations to ±3 psi will be satisfactory. The temperature control will consist of a temperature sensitive element located in the outflow line just ahead of the throttling valve, which will actuate synchronized motor-driven by-pass valves between the pump inlet and outlet manifolds. Temperature regulation to ±0.5 °F is desirable. This will require special equipment and some compromise on this point may be necessary. In addition to these two controls the safety rods will be actuated by a temperature rise of the outlet water to 220 °F, or if the pressure of the outlet water falls below 15 psi gage.

The control of the outlet temperature of the pile probably is most easily effected by making the control operate on the difference between the outlet temperature and the boiling point of water. This can be accomplished by a thermopile with one set of junctions in the discharge line and the other in water boiling at the prevailing atmospheric pressure. The resulting potential can be made to operate the valve control motors through suitable amplifiers and relays.

Orifices for flow rate measurements will be located in the cold pile water return lines and in the cold pile water lines from each of the 6 heat exchanger banks. The temperatures of both streams in and out of the 6 heat exchanger banks will be measured but not automatically recorded as will also the pressures in each of the four main pump manifolds, at the bottom of the flash tank below the filter and above the filter, and at other points to be specified.

A recording voltmeter for each supply circuit to the pumps and indicating ammeters for all motors over 50 HP will be provided. Radiation meters of suitable types will be placed at critical locations in the plant, such as near the discharge lines from the pile, in the flash tank condenser vent, etc.

All the instruments having to do with the operation of the plant except those directly connected with the cooling tower pumps and fans will be located in a central intrument and control house, as shown in Fig. 37. Operating headquarters for the plant will be in this building. The instruments for the cooling tower pumps and fans will be located near these units for convenience in control.

The principal valves in the plant will be motor operated, and controllable by push buttons in the central control room, as well as at the site of the valve.

Operating Crew. The operating crew will consist of one chief operator, one instrument man, two operators and two oilers per shift, when the pile is in operation. The pile unloading crew will probably consist of one unloading foreman, 2 pusher operators, 2 elevator coffin operators, three transfer car operators, with several helpers.

Power Requirements. Power requirements for operating the plant at a heat rate of 500,000 kW are estimated at 8000 kW. Most of this is used in the motors which drive the various circulating pumps. It is recommended that at least 2000 kW of electrical power generating capacity be installed as standby, capable of being put on the line in less than 30 seconds. Any of the pumps can serve as standby for circulating the water at the 1/6 rate required immediately after shutting off the chain reaction.

C. Materials Required

Estimated Principal Materials for Pile Proper

General

U-Metal	200	m Tons
Graphite	850	"
Second class graphite	315	"
Third class graphite	150	"

Steel

Plates	60	Tons
Rolled shapes	50	"
Forgings	120	"
Reinforcing	90	"
Pipe, fittings, valves	60	"

Concrete

Walls and foundations	1500	Cu. Yds.

Shields

Plates	250	Tons
Pipe	85	"
Steel or iron shot	120	"
Shapes	120	"
Lead	240	"

Cranes & Hoists

Steel shapes	25	Tons
Reinforcing steel	20	"
Concrete	360	Cu. Yds.

Coffins (12 total)

Steel	10	Tons
Lead	80	"

Material Requirements for Liquid-Cooling System Exclusive of Pile

Steel
Tubing (3/4 in. OD)

Heat exchangers (14 BWG)	450	
Still (14 BWG)	0.7	
Still condenser	0.5	
Total tubing		451.2 Tons

Piping (A.S.A schedule 30)

Distilled water system	263	
Cooling tower water system	237	
Total piping		500

Castings

Heat exchangers (heads and nipples)	136	
Valves	122	
Motor casings (16 at 700 HP)	14	
Cooling tower water pumps (8)	20	
Total castings		292

Rolled plate

Tanks (4 at 22 ft dia.)	85	
Heat exchanger shells	139	
Heat exchanger tube sheets	21	
Flash condenser shells	0.8	
Still shell	1	
Still tube sheets	0.5	
Still condenser shell	0.5	
Still condenser tube sheets	0.3	
Total rolled plate		248.1

Reinforcing steel

Foundations and shielding	50	
Total reinforcing steel		50

Structural steel

Cranes	50	
Total structural steel		50
Total steel		1591.3

Material Requirements (cont.)

Stainless Steel

Castings

Distilled water pumps	20 Ton	
Pressure reducing valves	4 Ton	
Total stainless steel castings		24 Ton

Brass

Tubing: Flash condenser	7 Ton	
Rolled plate: Flash condenser tube sheets	4 Ton	
Total brass		11 Ton

Copper

Wire

Motor windings (16 at 700 HP)	34 Ton	
Miscellaneous leads, switches and small motors	4 Ton	
Total copper wire		38 Ton

Concrete

Shielding	1890 Cu. Yds.	
Foundations	740 Cu. Yds.	
Total concrete		2630 Cu. Yds.

D. Utilities Required

Electrical Power Required

Distilled water pumps	3300 kW
Cooling tower water pumps	3000 kW
Cooling tower fans	800 kW
Miscellaneous equipment & lighting	900 kW
Total power required	8000 kW

Stand-By Steam-Generated Electric Power Required

Total power required	2000 kW

Water Required

Distilled water make-up	25 gpm
Treated water make-up	4225 gpm

V. Conclusions and Recommendations

Feasibility. It is the consensus of the writers that building a water-cooled power plant capable of developing 5×10^5 kW from 200 tons of metal appears entirely practical. (There is evidence that even more than 5×10^5 kW can be derived from this amount of metal with the plant proposed.)

The water-cooled plant runs at much lower temperatures than the helium cooled plant, and is safer in operation in that the fission products are kept away from the pile atmosphere and the coolant.

Principal Problems. The original three major problems of the multiplication factor k, the corrosion of the tubes and sheaths, and the commercial production of parts have all reached preliminary answers which indicate the practicability of this type of plant.

For water cooling, the estimated pile lifetime based on available corrosion tests is considerably more than 100 days even without exchanging all central rods. At present diphenyl is more uncertain and does not promise quite as high a power output as water.

Relative Construction Time. It is the consensus that a plant like this could be built in from two to three months less time than the lower powered helium cooled plants now under consideration, would involve less vital material and would require only a fraction of the total external power for the same power output.

Future Research and Development. Experiments now under way and design investigation should continue at the most rapid pace, with a view to starting the design and building of such a plant at once.

General. The advantages to be derived from the construction of a successful liquid cooled plant can be evaluated only in terms of the whole picture. It has already been pointed out that such a plant saves time – in construction, and more important, in chemical production. It must also be recognized there are unforeseen difficulties in every plant; the risk in "putting all eggs in one basket" is extremely great. Therefore, in view of the importance of the total task of the project and the irreparable loss if only one line of attack is tried and found wanting, it is a majority recommendation that a liquid plant be considered for building without delay.

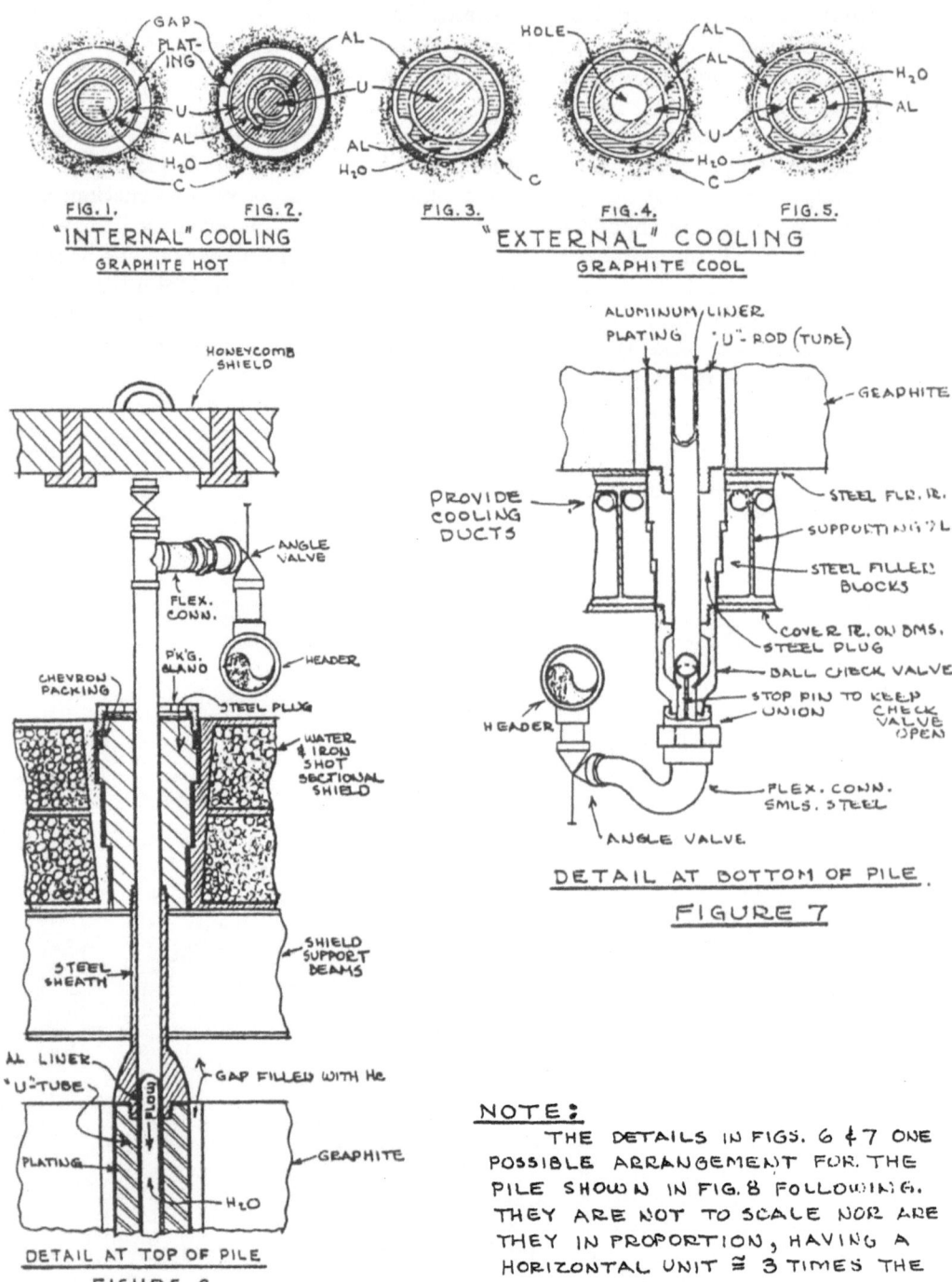

FIG.1.   FIG.2.
"INTERNAL" COOLING
GRAPHITE HOT

FIG.3.   FIG.4.   FIG.5.
"EXTERNAL" COOLING
GRAPHITE COOL

DETAIL AT TOP OF PILE
FIGURE 6

DETAIL AT BOTTOM OF PILE
FIGURE 7

NOTE:
THE DETAILS IN FIGS. 6 & 7 ONE POSSIBLE ARRANGEMENT FOR THE PILE SHOWN IN FIG. 8 FOLLOWING. THEY ARE NOT TO SCALE NOR ARE THEY IN PROPORTION, HAVING A HORIZONTAL UNIT ≅ 3 TIMES THE VERTICAL UNIT OF LENGTH.

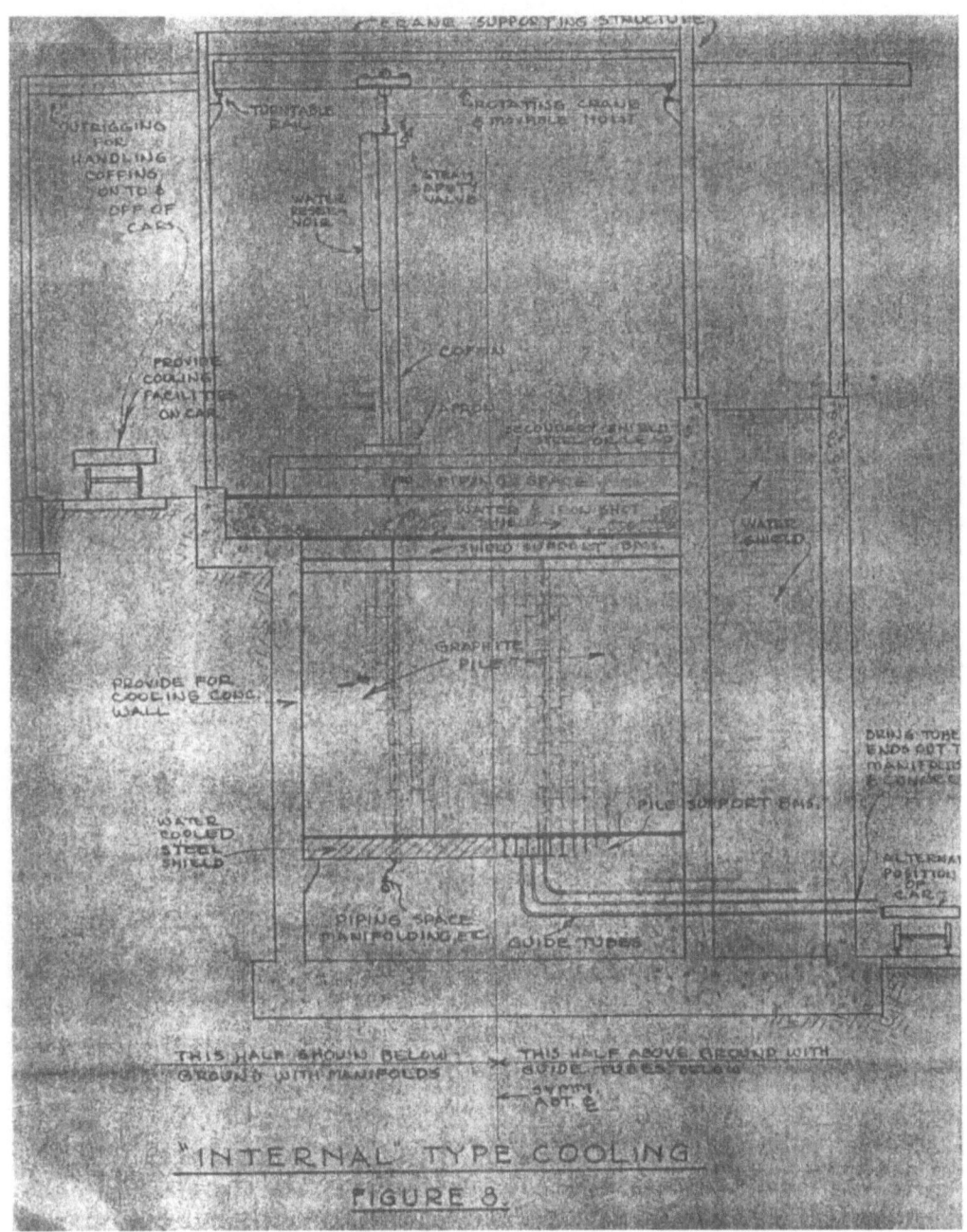

"INTERNAL" TYPE COOLING

FIGURE 8.

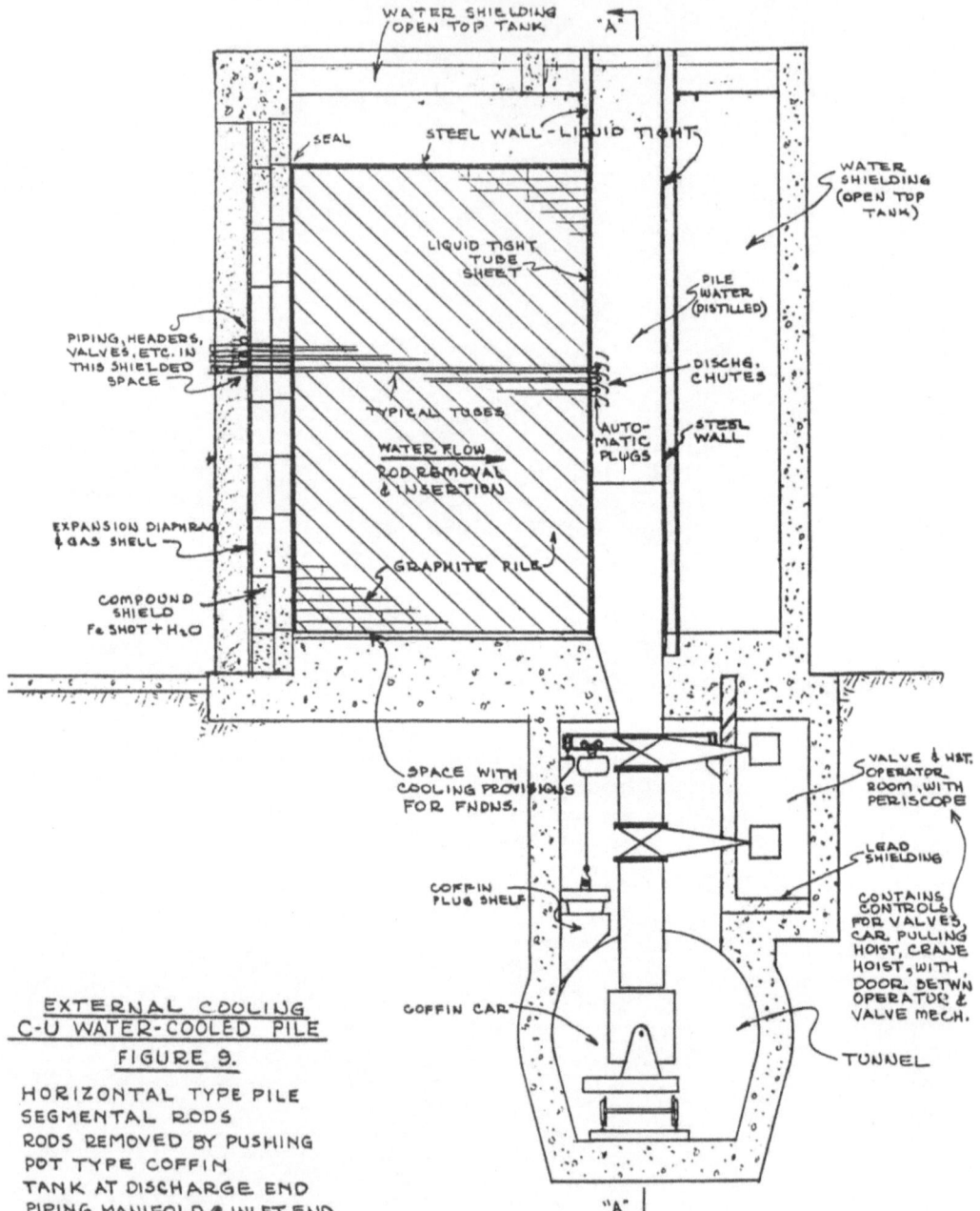

EXTERNAL COOLING
C-U WATER-COOLED PILE
FIGURE 9.

HORIZONTAL TYPE PILE
SEGMENTAL RODS
RODS REMOVED BY PUSHING
POT TYPE COFFIN
TANK AT DISCHARGE END
PIPING MANIFOLD @ INLET END

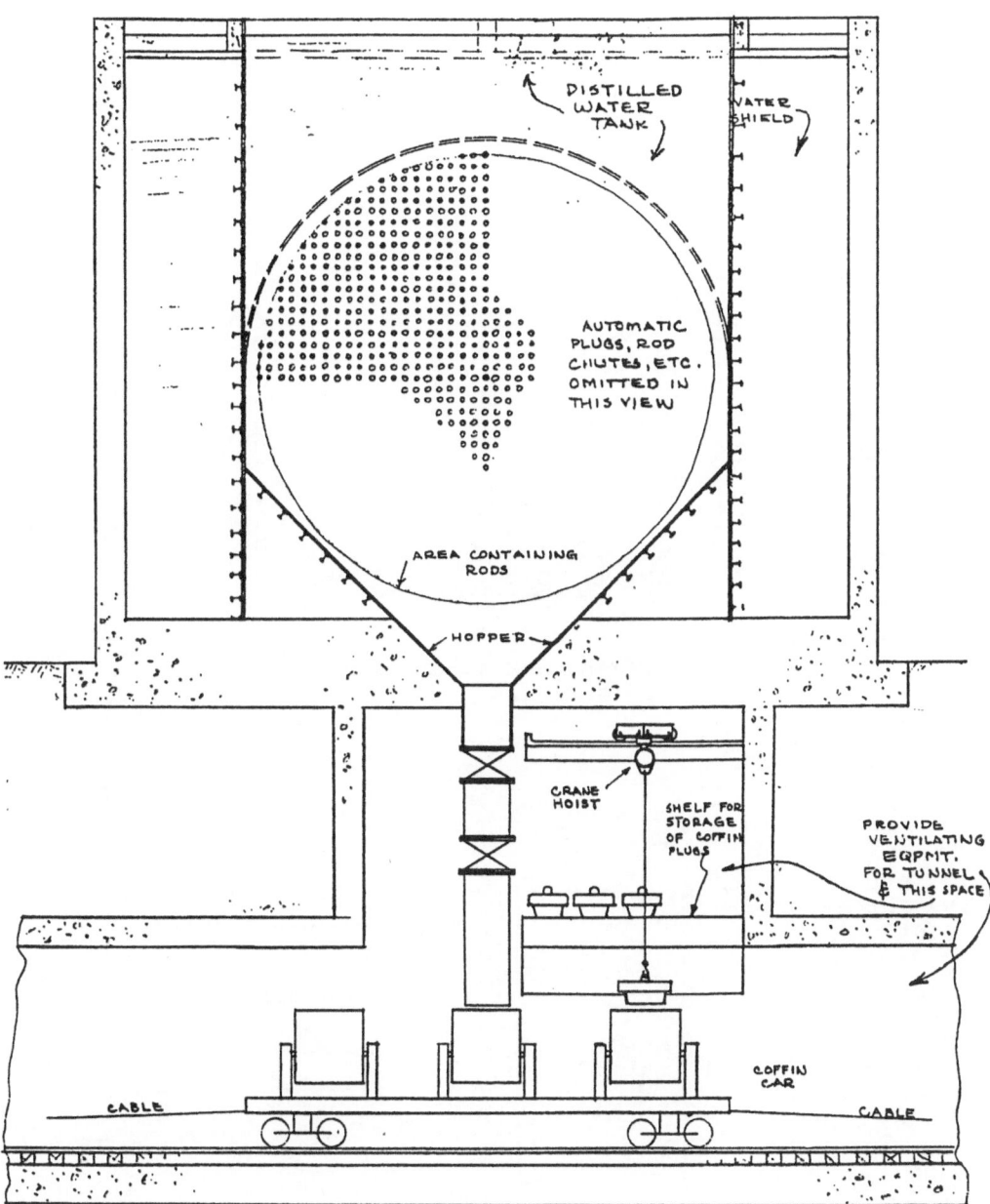

CROSS SECT. "A-A" FOR PILE SHOWN IN FIG. 9.

FIGURE 10.

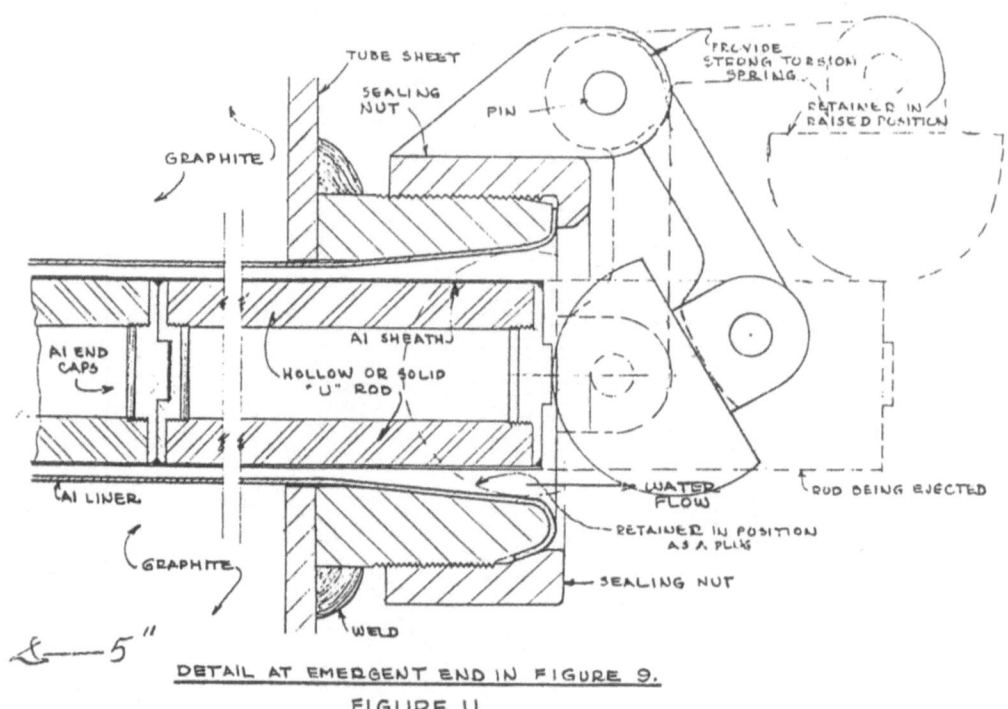

DETAIL AT EMERGENT END IN FIGURE 9.

FIGURE 11

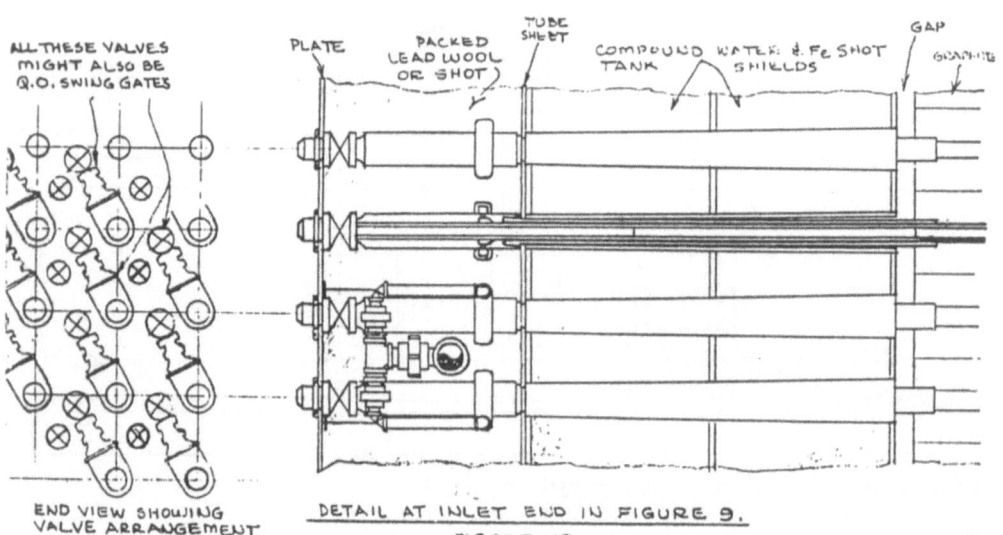

END VIEW SHOWING
VALVE ARRANGEMENT

DETAIL AT INLET END IN FIGURE 9.

FIGURE 12.

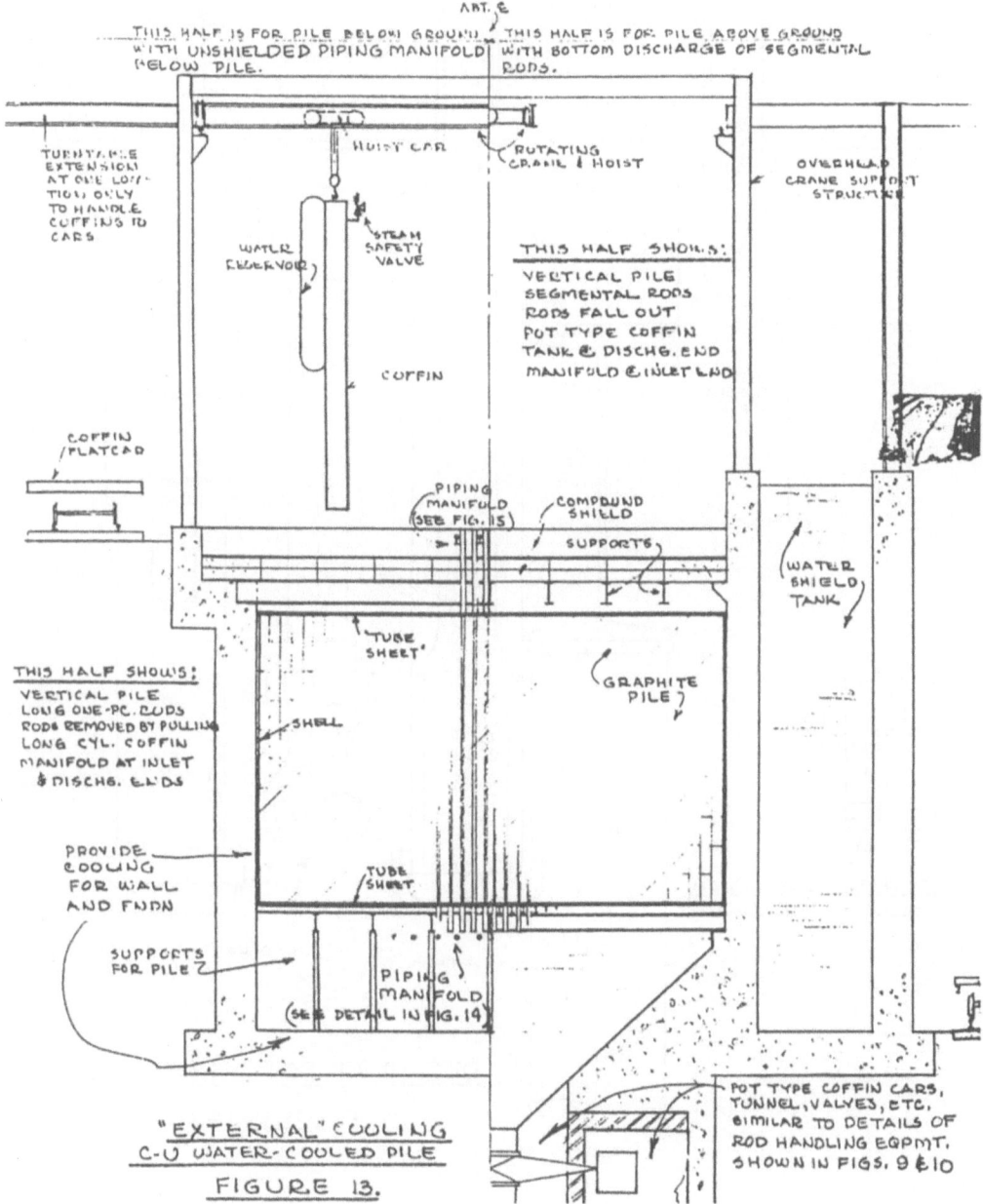

THIS HALF IS FOR PILE BELOW GROUND / THIS HALF IS FOR PILE ABOVE GROUND
WITH UNSHIELDED PIPING MANIFOLD / WITH BOTTOM DISCHARGE OF SEGMENTAL
BELOW PILE. / RODS.

SYMM.
ABT. ℄

TURNTABLE
EXTENSION
AT ONE LOCA-
TION ONLY
TO HANDLE
COFFINS TO
CARS

HOIST CAR

ROTATING
CRANE & HOIST

OVERHEAD
CRANE SUPPORT
STRUCTURE

STEAM
SAFETY
VALVE

WATER
RESERVOIR

THIS HALF SHOWS:

VERTICAL PILE
SEGMENTAL RODS
RODS FALL OUT
POT TYPE COFFIN
TANK @ DISCHG. END
MANIFOLD @ INLET END

COFFIN

COFFIN
FLATCAR

PIPING
MANIFOLD
(SEE FIG. 15)

COMPOUND
SHIELD

SUPPORTS

WATER
SHIELD
TANK

THIS HALF SHOWS:

VERTICAL PILE
LONG ONE-PC. RODS
RODS REMOVED BY PULLING
LONG CYL. COFFIN
MANIFOLD AT INLET
& DISCHG. ENDS

TUBE
SHEET

SHELL

GRAPHITE
PILE

PROVIDE
COOLING
FOR WALL
AND FNDN

TUBE
SHEET

SUPPORTS
FOR PILE

PIPING
MANIFOLD
(SEE DETAIL IN FIG. 14)

POT TYPE COFFIN CARS,
TUNNEL, VALVES, ETC.
SIMILAR TO DETAILS OF
ROD HANDLING EQPMT.
SHOWN IN FIGS. 9 &10

"EXTERNAL" COOLING
C-U WATER-COOLED PILE

FIGURE 13.

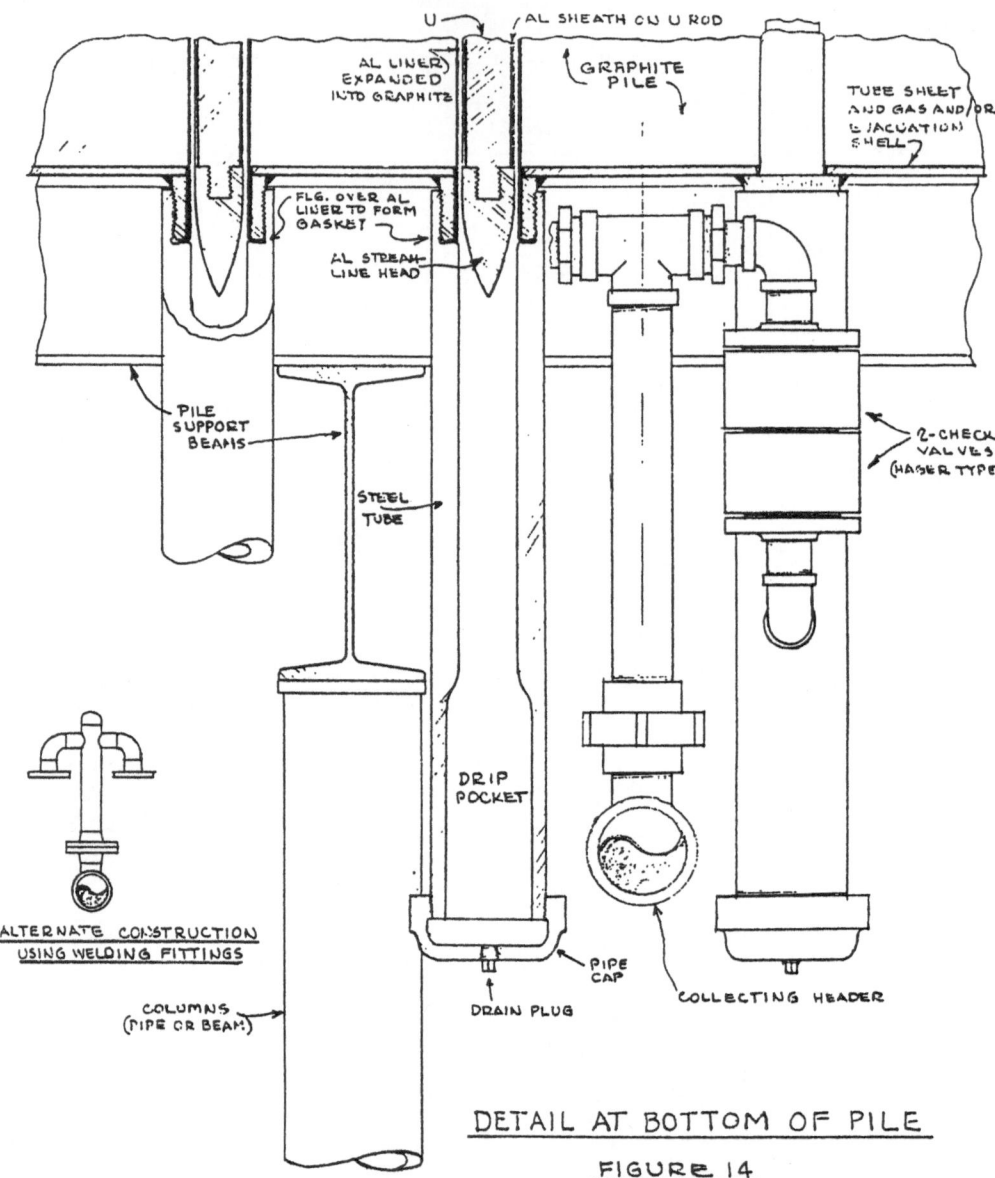

DETAIL AT BOTTOM OF PILE

FIGURE 14

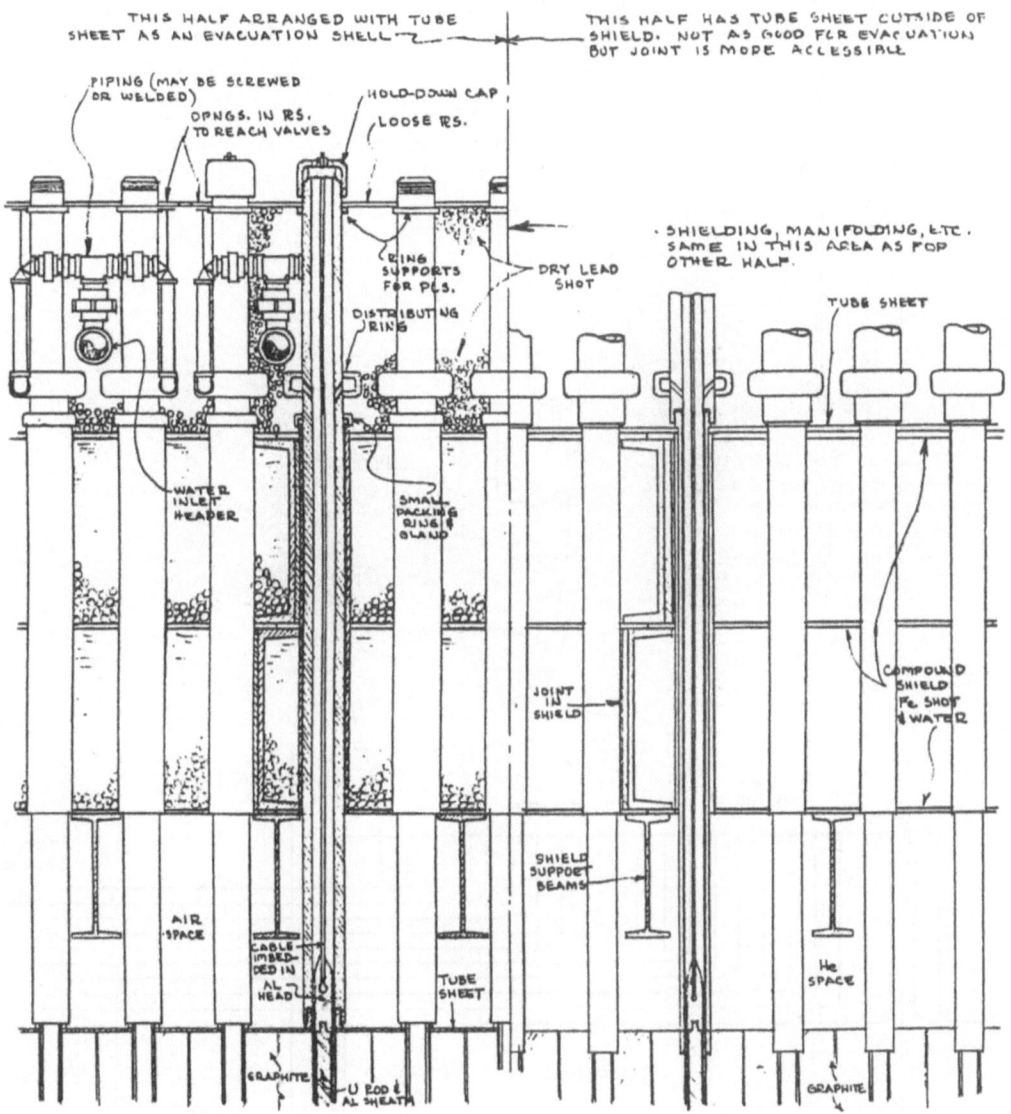

DETAIL AT TOP OF PILE IN FIGURE 13.

FIGURE 15

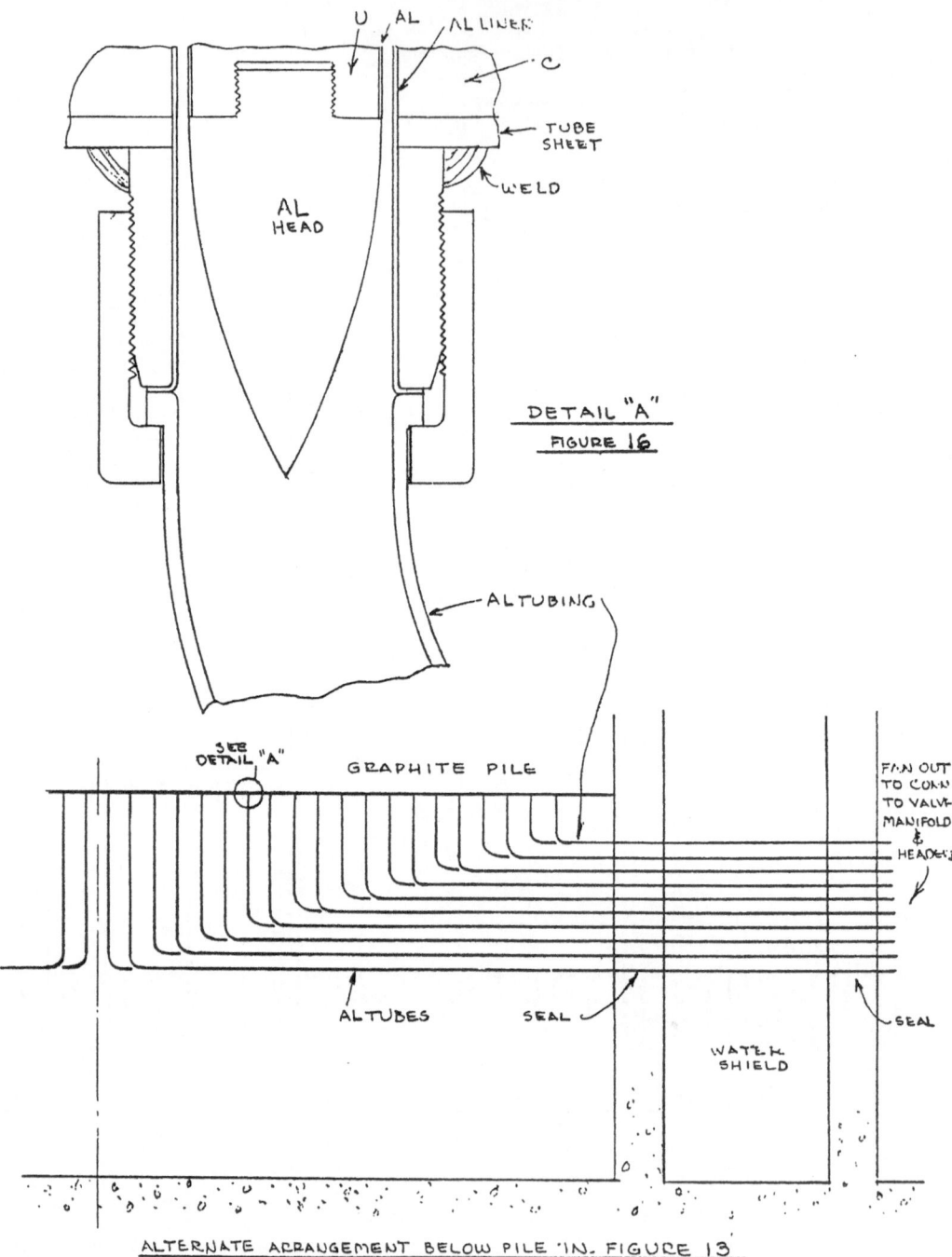

U AL AL LINER
C
TUBE SHEET
WELD
AL HEAD

DETAIL "A"
FIGURE 16

AL TUBING

SEE DETAIL "A"
GRAPHITE PILE
FAN OUT TO CONN. TO VALVE MANIFOLD & HEADERS

AL TUBES   SEAL   SEAL
WATER SHIELD

ALTERNATE ARRANGEMENT BELOW PILE IN. FIGURE 13
FIGURE 17

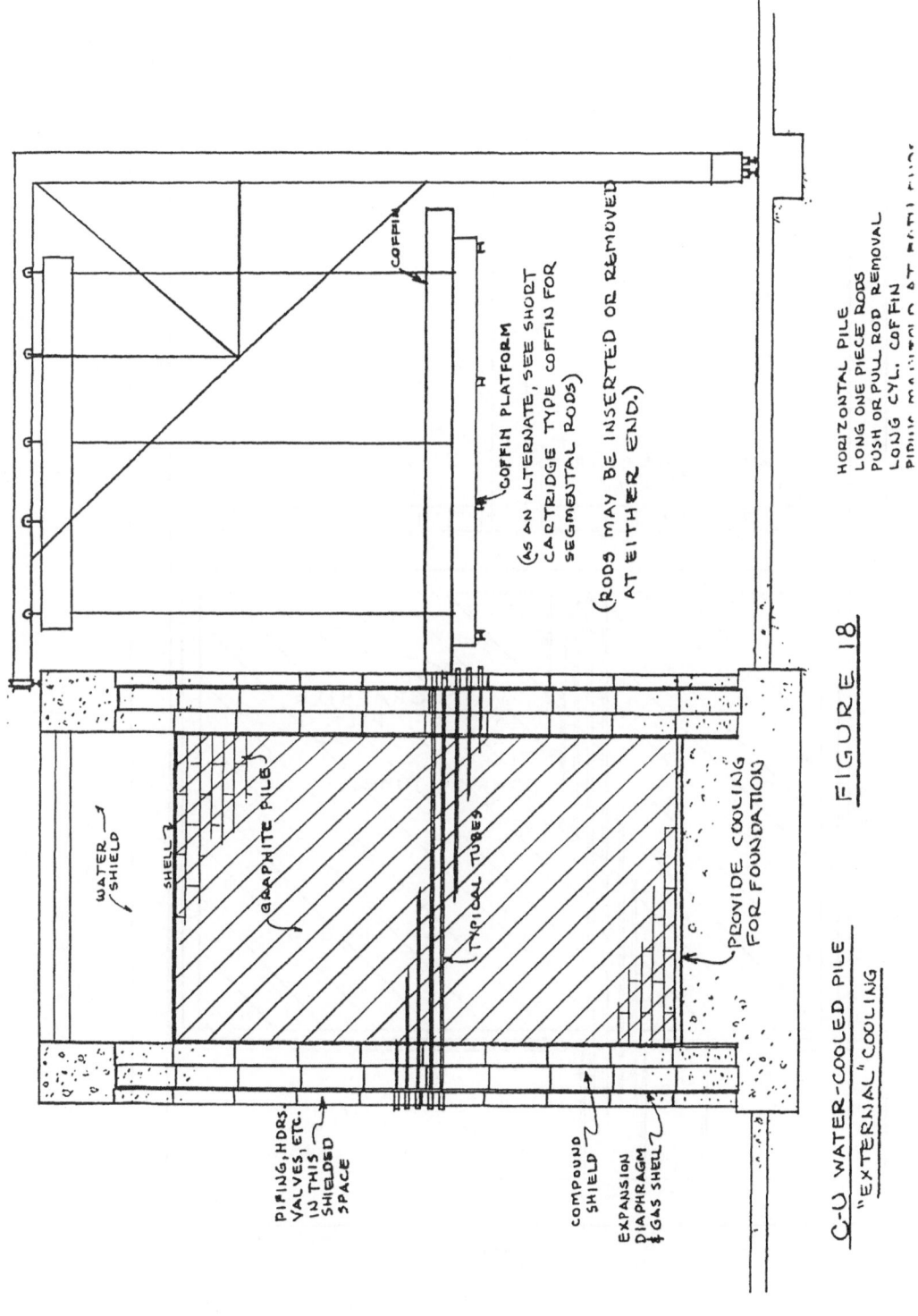

COFFIN

COFFIN PLATFORM

(AS AN ALTERNATE, SEE SHORT CARTRIDGE TYPE COFFIN FOR SEGMENTAL RODS)

(RODS MAY BE INSERTED OR REMOVED AT EITHER END.)

HORIZONTAL PILE
LONG ONE PIECE RODS
PUSH OR PULL ROD REMOVAL
LONG CYL. COFFIN

WATER SHIELD

SHELL

GRAPHITE PILE

TYPICAL TUBES

PROVIDE COOLING FOR FOUNDATION

PIPING, HDRS. VALVES, ETC. IN THIS SHIELDED SPACE

COMPOUND SHIELD

EXPANSION DIAPHRAGM & GAS SHELL

C-U WATER-COOLED PILE
"EXTERNAL" COOLING

FIGURE 18

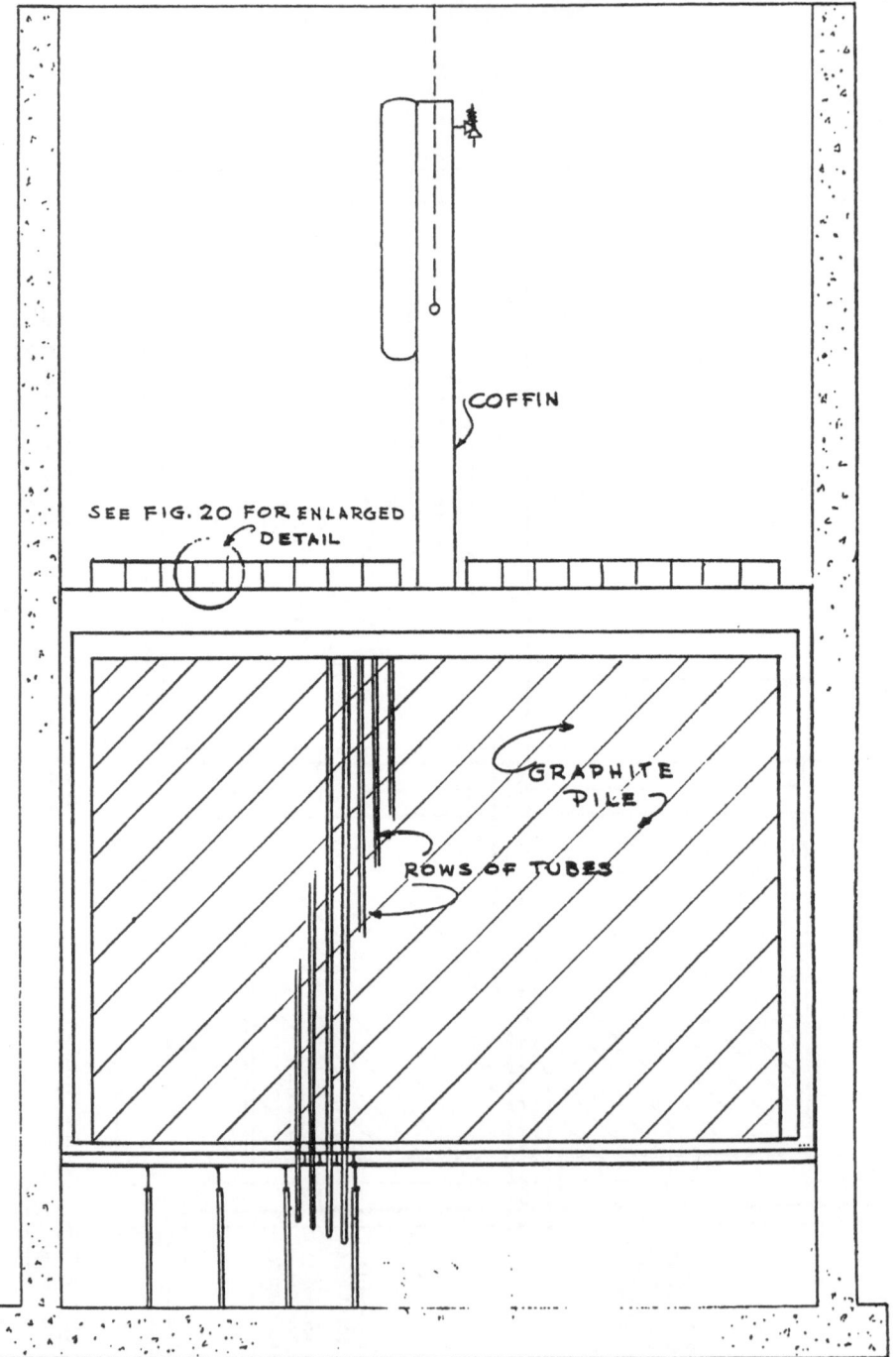

FIG. 19

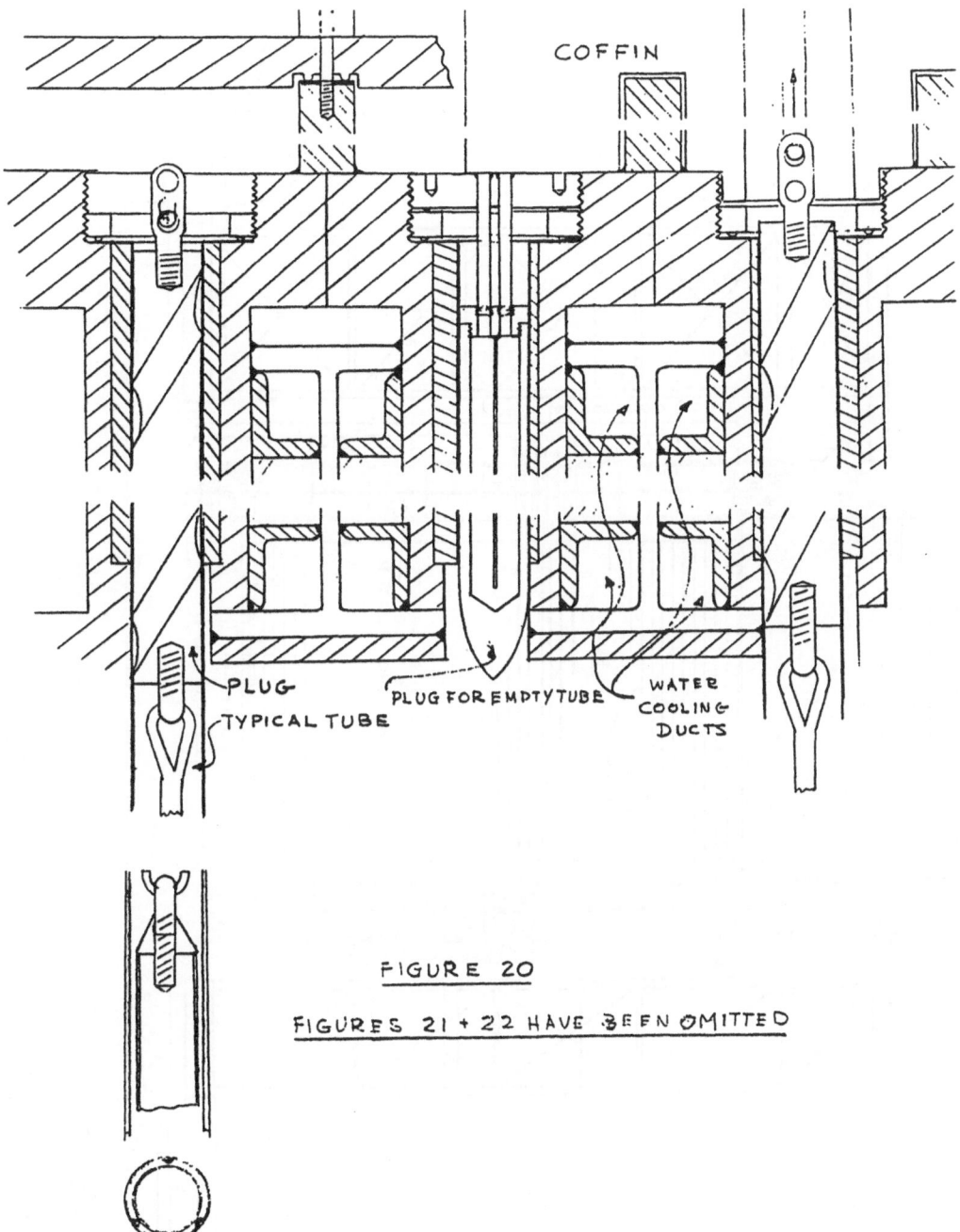

COFFIN

PLUG

TYPICAL TUBE

PLUG FOR EMPTY TUBE

WATER COOLING DUCTS

FIGURE 20

FIGURES 21 + 22 HAVE BEEN OMITTED

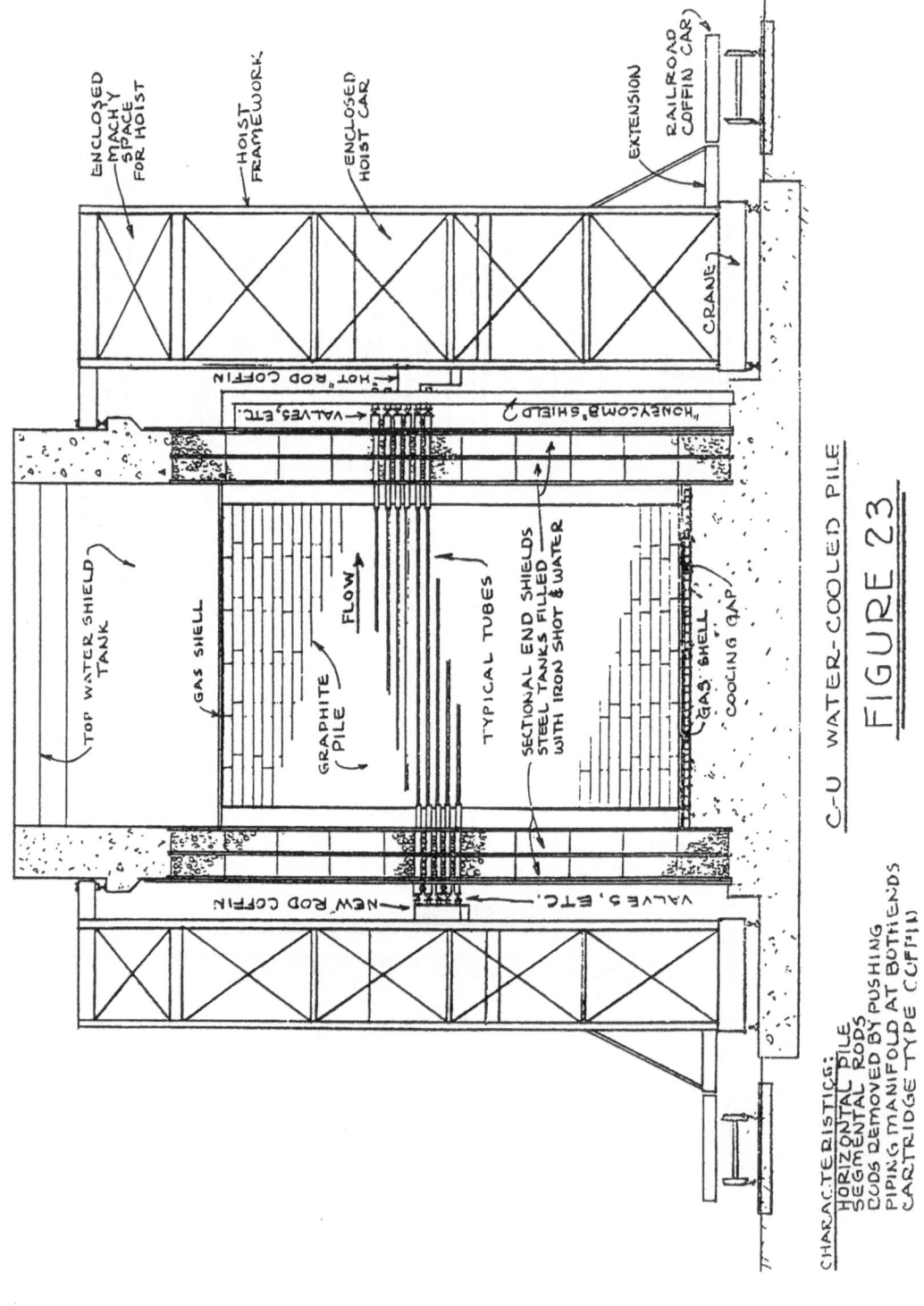

ENCLOSED MACHY SPACE FOR HOIST

HOIST FRAMEWORK

ENCLOSED HOIST CAR

EXTENSION

RAILROAD COFFIN CAR

CRANE

"HOT" ROD COFFIN

VALVES, ETC.

"HONEYCOMB" SHIELD

TOP WATER SHIELD TANK

GAS SHELL

FLOW

GRAPHITE PILE

TYPICAL TUBES

SECTIONAL END SHIELDS STEEL TANKS FILLED WITH IRON SHOT & WATER

GAS SHELL

COOLING GAP

NEW ROD COFFIN

VALVES, ETC.

C-U WATER-COOLED PILE

FIGURE 23

CHARACTERISTICS:
HORIZONTAL PILE
SEGMENTAL RODS
RODS REMOVED BY PUSHING
PIPING MANIFOLD AT BOTH ENDS
CARTRIDGE TYPE COFFIN

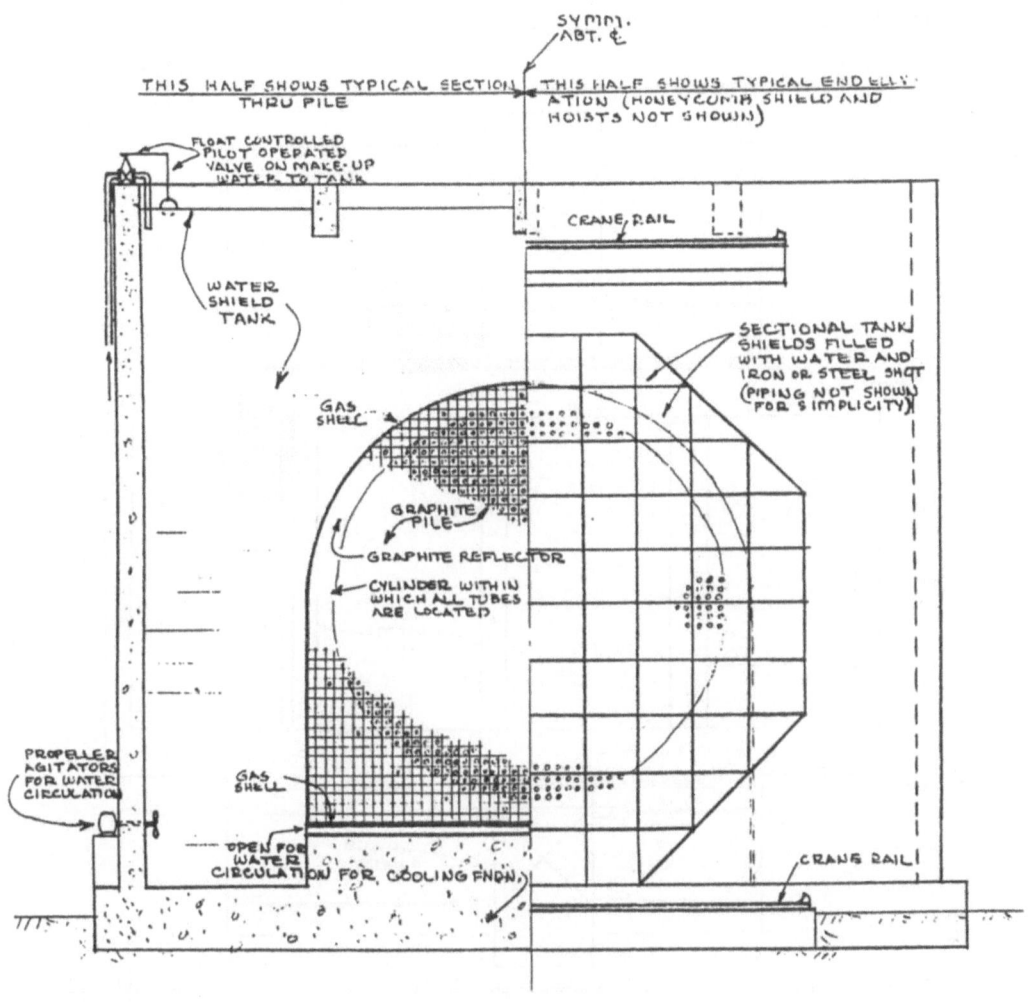

FIGURE 24

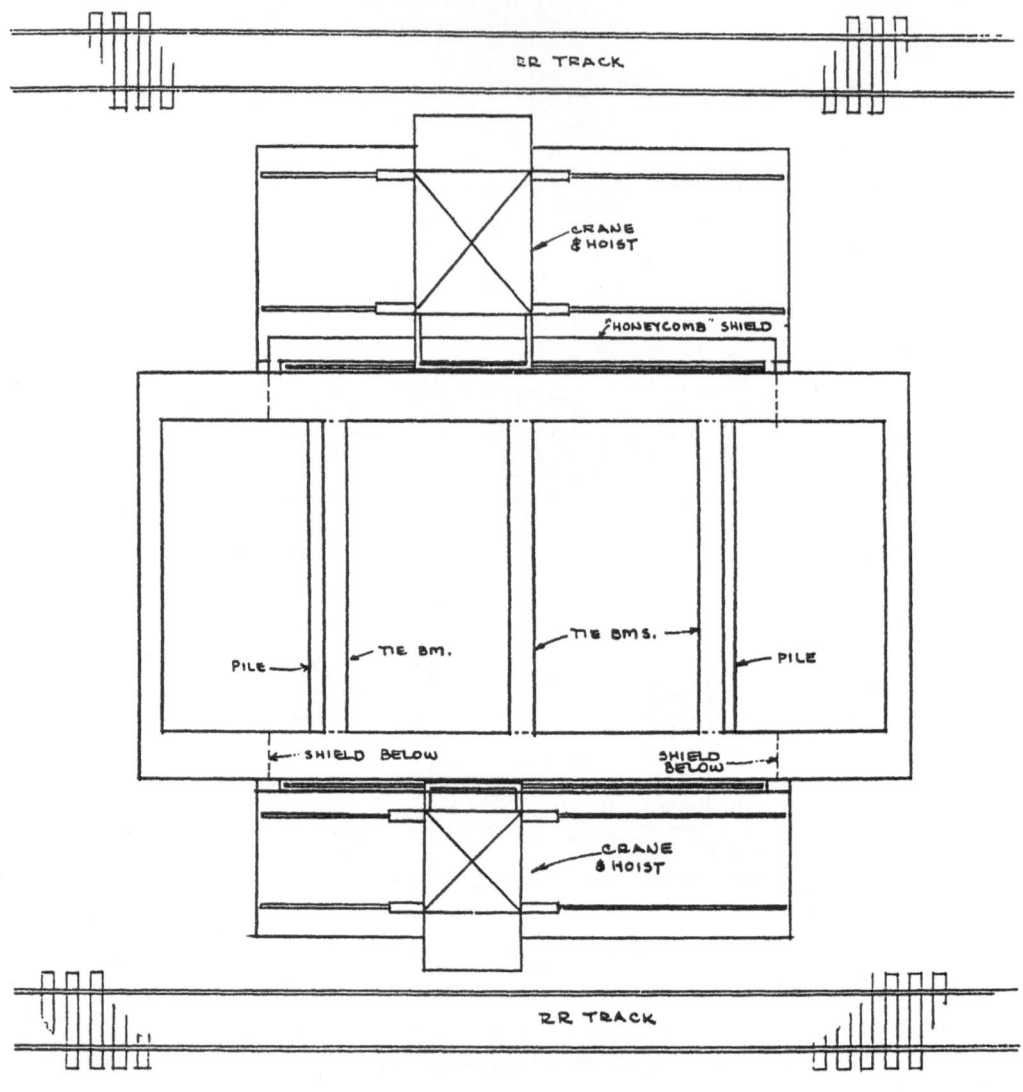

RR TRACK

CRANE & HOIST

"HONEYCOMB" SHIELD

TIE BM.

TIE BMS.

PILE

PILE

SHIELD BELOW

SHIELD BELOW

CRANE & HOIST

RR TRACK

PLAN VIEW

FIGURE 25

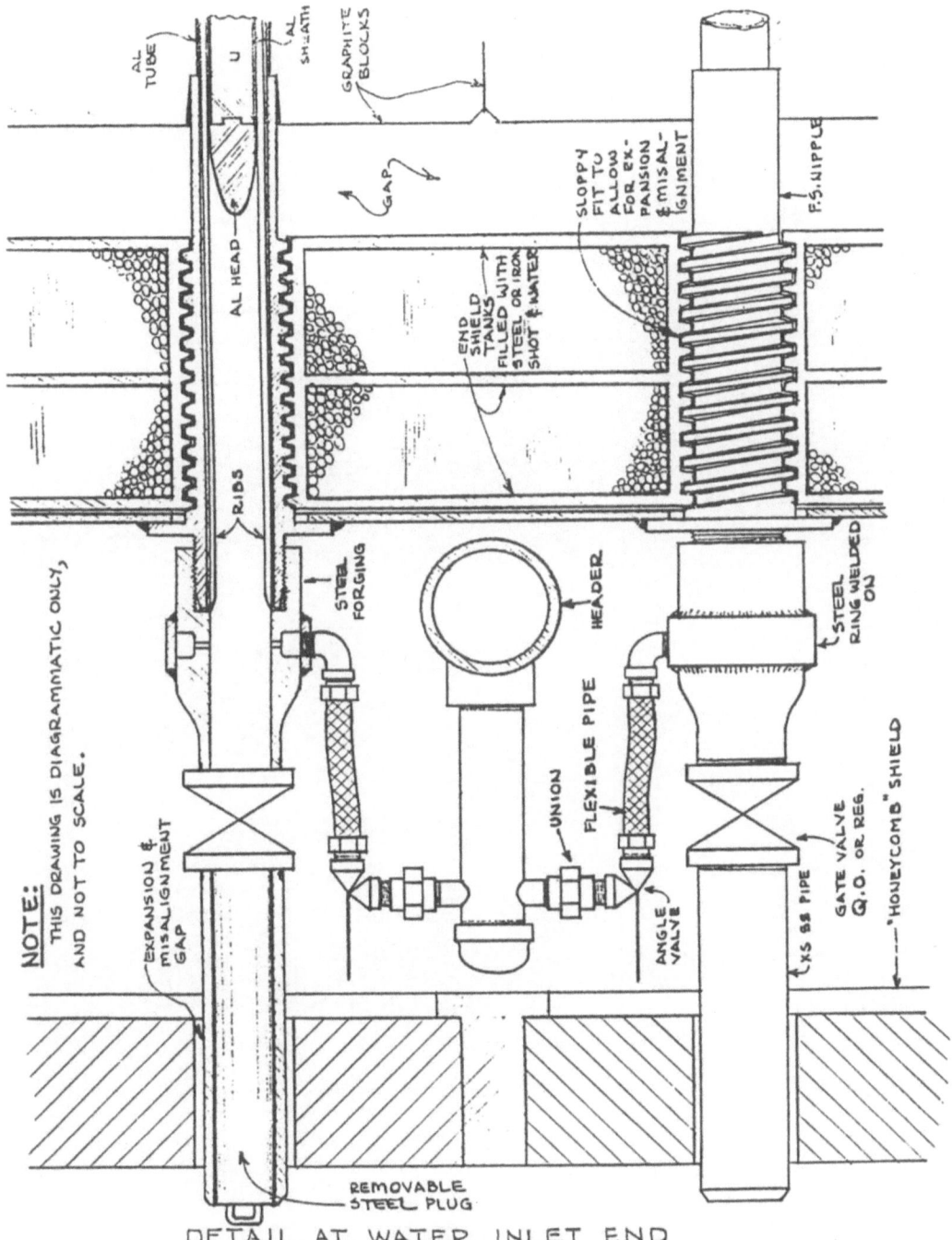

DETAIL AT WATER INLET END
(WATER OUTLET END SIMILAR EXCEPT "HONEYCOMB" SHIELD IS OMITTED AND
U-ROD END DETAIL IS DIFFERENT, WITH STEEL SPACER SEGMENTS TO VALVE)

FIGURE 26

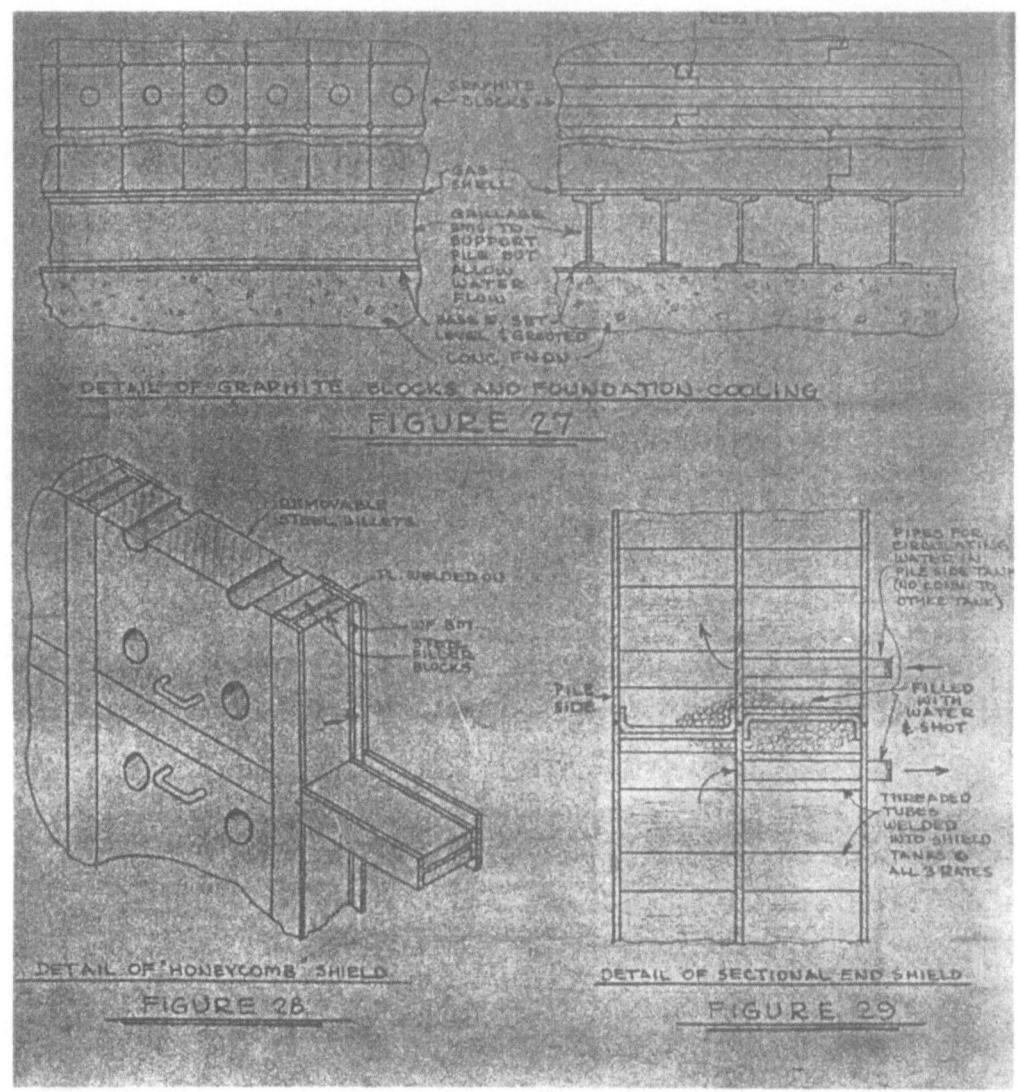

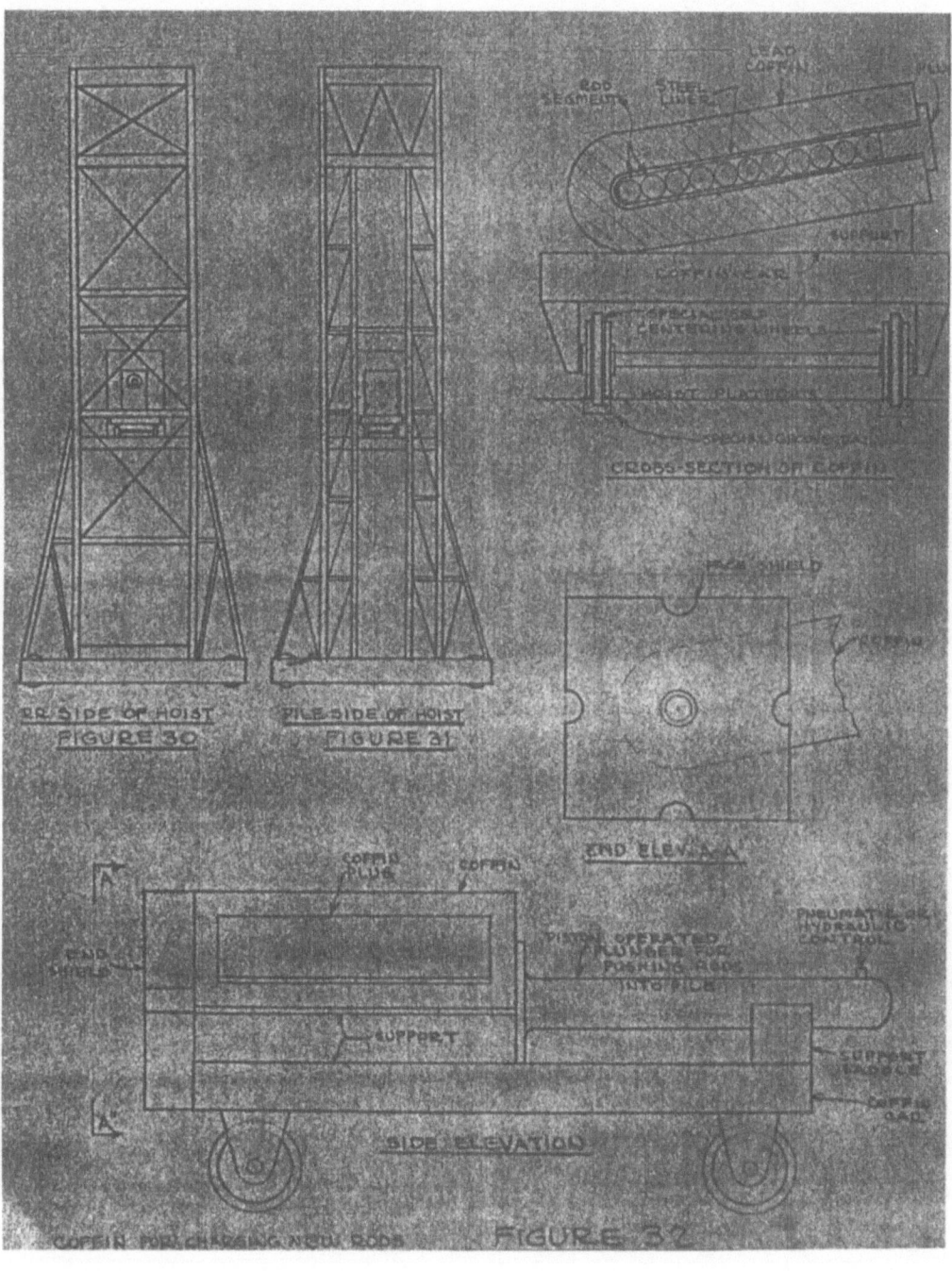

RR SIDE OF HOIST
FIGURE 30

PILE SIDE OF HOIST
FIGURE 31

CROSS-SECTION OF COFFIN

END ELEV. A-A

SIDE ELEVATION

COFFIN FOR CHARGING NEW RODS   FIGURE 32

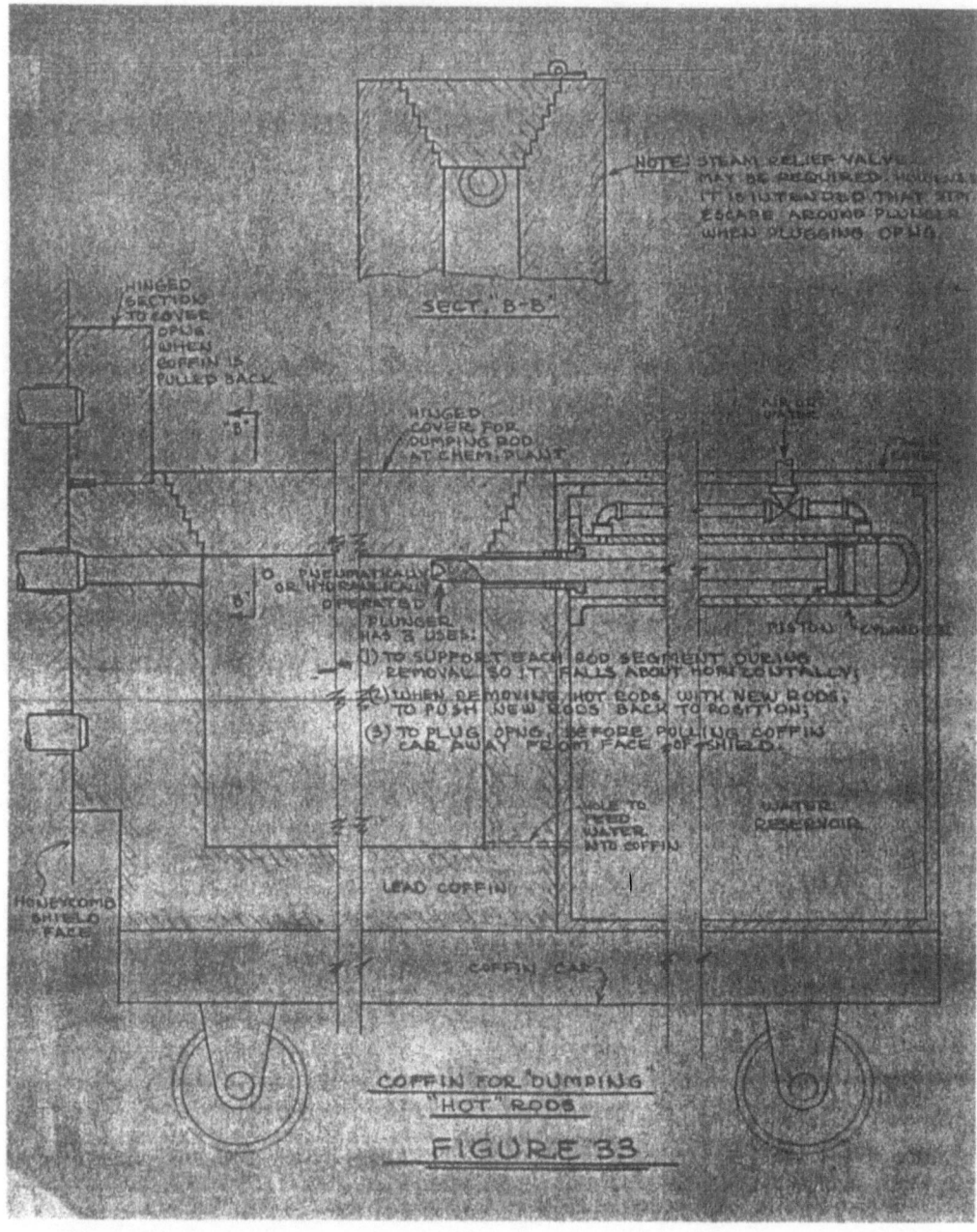

COFFIN FOR "DUMPING"
"HOT" RODS

FIGURE 33

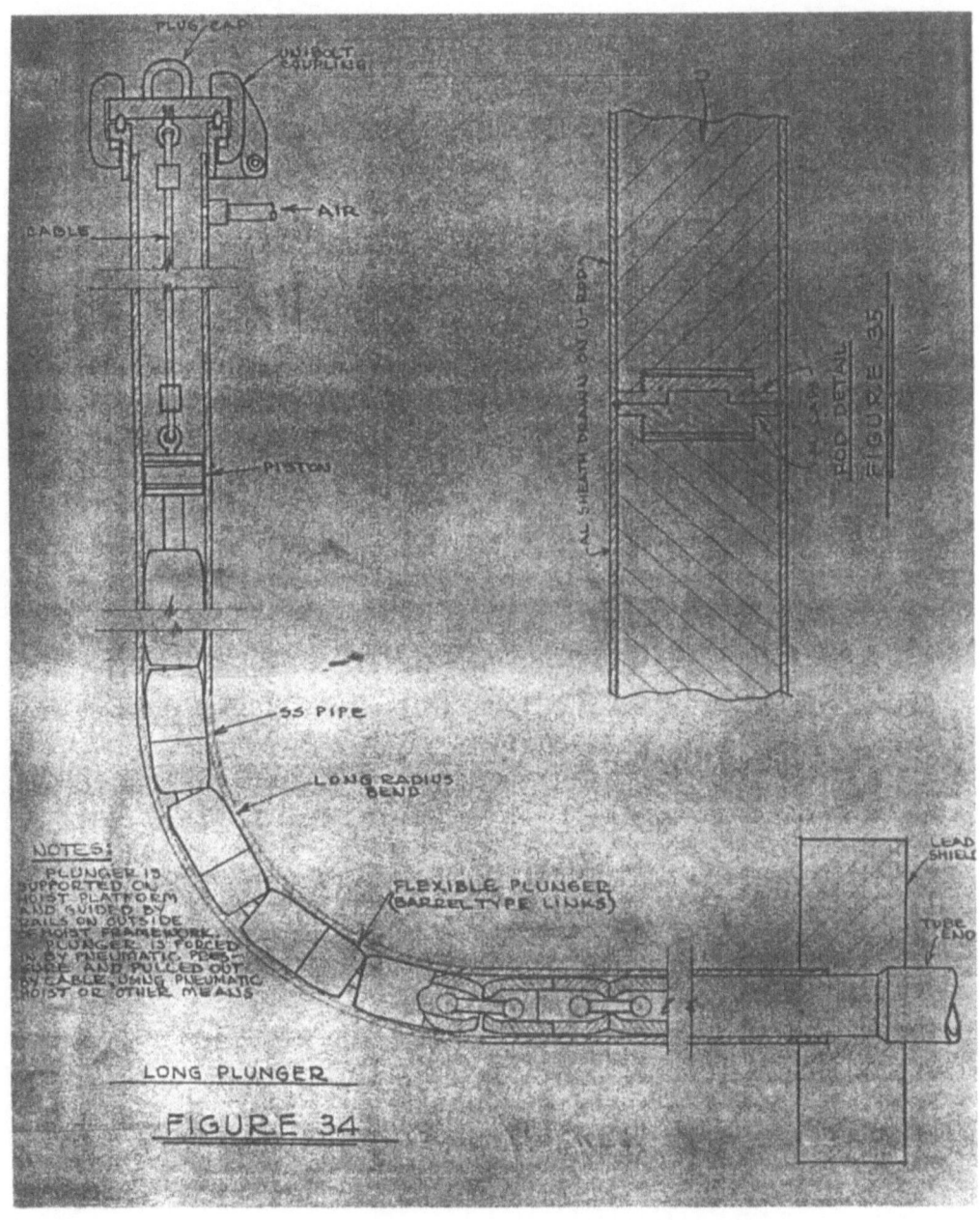

PLUG CAP
UNIBOLT COUPLING
AIR
CABLE
PISTON
SS PIPE
LONG RADIUS BEND
FLEXIBLE PLUNGER (BARREL TYPE LINKS)

NOTES:
PLUNGER IS SUPPORTED ON HOIST PLATFORM AND GUIDED BY RAILS ON INSIDE OF HOIST FRAMEWORK. PLUNGER IS FORCED IN BY PNEUMATIC PRESSURE AND PULLED OUT BY CABLE USING PNEUMATIC HOIST OR OTHER MEANS

LONG PLUNGER

FIGURE 34

AL SHEATH DRAWN ON (U ROD)
ROD DETAIL
FIGURE 35

LEAD SHIELD
TUBE END

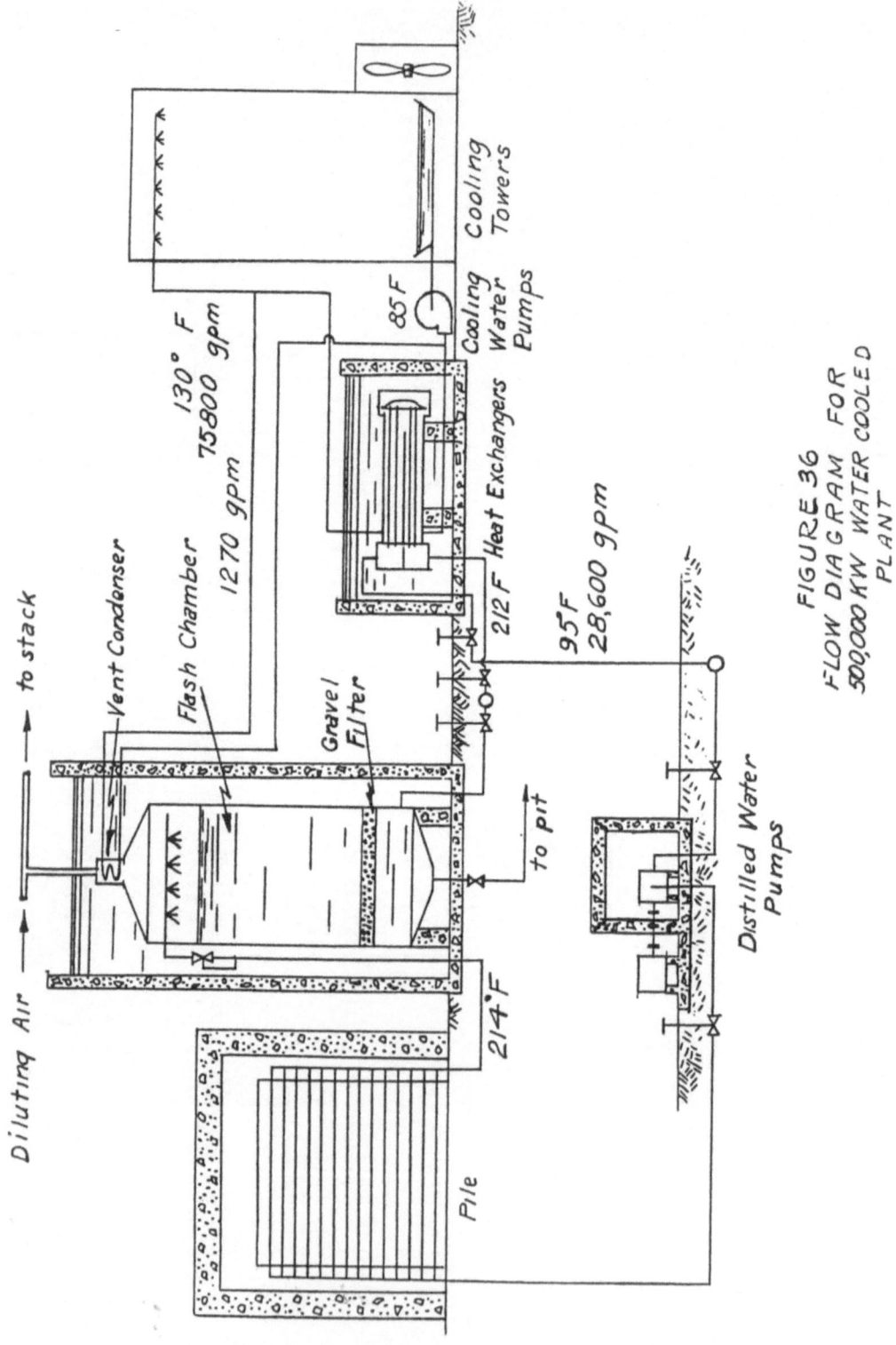

FIGURE 36
FLOW DIAGRAM FOR
500,000 KW WATER COOLED
PLANT

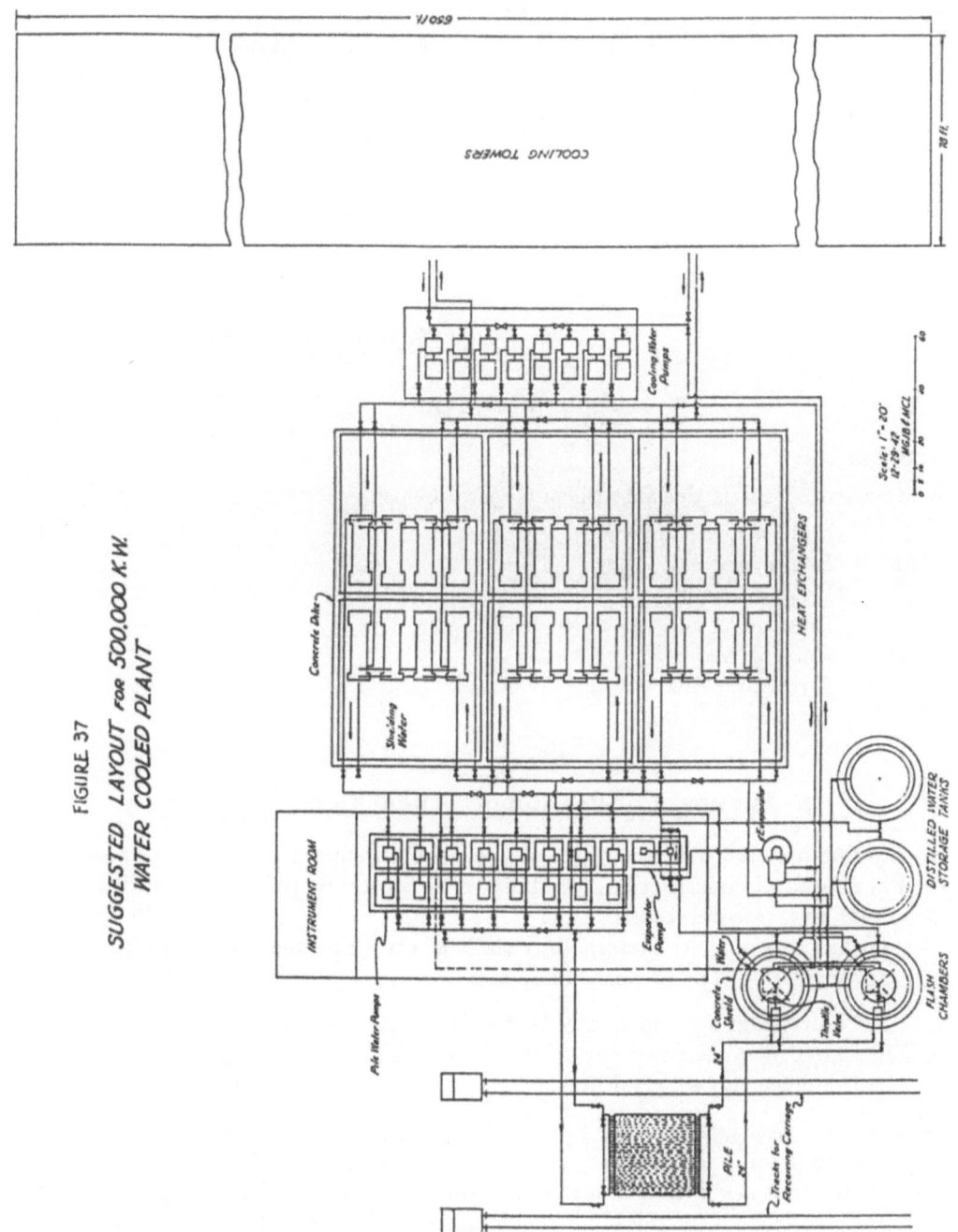

FIGURE 37

SUGGESTED LAYOUT for 500,000 K.W.
WATER COOLED PLANT

17.

Memorandum from P-9 Committee

H. D. Smyth, E. P. Wigner, and H. C. Vernon

August 10, 1943; revised November 3, 1943

Table of Contents

A. Summary

B. Review of Various Possible Piles: General Considerations

A. Summary of P-9 Report About to be Issued

In choosing between different types of piles, the principal factors to be considered are speed of construction, reliability as a prototype production unit and usefulness as a research tool.

On the basis of these factors, only three of the eight piles discussed justify serious consideration at present. They are:

1. The Light Water Cooled Heavy Water Moderated Heterogeneous Pile.
2. The Heavy Water Cooled Heavy Water Moderated Heterogeneous Pile.
3. The Homogeneous Slurry Pile.

Of those, the first two are very similar. At first sight, the heavy water cooling seems to have some advantages, such as greater simplicity of pile design and greater inherent stability. Further study shows these advantages largely illusory and outweighed by shielding problems. Therefore, the light water cooled is considered the better of the two. The homogeneous pile also appears to have advantages but they have not been adequately studied. The immediate objections to this pile are the amount of P-9 needed and lack of information. Study on it should continue but not delay planning the light water cooled heterogeneous pile.

B. Review of Various Possible Piles

We have investigated the basic properties of several types of plants from the point of finding out to what degree they achieve the above purposes. The systems which we have considered follow:

1. Light water cooling
2. Heavy water cooling
3. Helium cooling
4. Homogeneous pile
5. Heterogeneous Slurry
6. Circulated Hex
7. Bismuth cooling

Before going over to the comparison of the different types of pile, a few words may be said about the basis on which this comparison is made.

Every chain reacting system has a "critical size". For instance, a heterogeneous U-D_2O system cannot be made chain reacting (without a reflector) with less than about 5 tons of D_2O and 2 tons of U, a homogeneous U-D_2O system with less than about 30 tons of D_2O and about 7 tons of U. For a unit which develops power, the requirements increase for three reasons:

1. The higher operating temperature increases the critical size. The above critical sizes are given for room temperature.
2. Any extraneous materials, such as aluminium pipes in the pile, also increases the critical size.
3. In many systems, some of the D_2O and/or U is constantly outside the pile and is needed in addition to the critical amount.

It is evident from the foregoing, that the critical size increases with increasing power output. With 5 tons of D_2O, 2 tons of U virtually no power can be obtained ("Fermi pile"). As more D_2O and U become available, the power output can be increased because cooling equipment can be introduced into the pile and one can increase its temperature without stopping the chain reaction. It is, therefore, evident that the power output will increase with increasing excess of D_2O and U over the minimum requirements. How much it increases depends on the nature of the cooling system and cooling equipment used.

It would seem then, that in contradiction to the Fermi pile which develops virtually no power, a cooled unit has no natural size. However, this is not so. As the D_2O and U available increase, the power output and 49 production increase very rapidly until a "natural size" is reached. Once this natural size is reached, a further addition of D_2O and U will not make it possible to increase the power output much more than in proportion to the D_2O and U added. Thus, e.g. in the light water cooling system, an 8 ton D_2O, 3 ton U system will still only give nominal production (perhaps 1 gram per day), an 11 ton D_2O, 6 ton U system can give 40 grams per day. This is the natural size and a 22 ton D_2O, 15 ton U system will not give much in excess of 100 grams per day. It appears questionable whether 2 units of natural size are not preferable to one oversize unit. If the D_2O and U were available in arbitrary quantities and if the building of chain reacting piles were not an experimental enterprise, one may be inclined to prefer one big plant. As matters stand now, a small natural size is a distinct advantage.

The requirements to be given below for the different cooling systems all refer to the natural size characteristics for that cooling system. The listed D_2O requirement is all the D_2O that is needed for the operation at the production level given. As far as the U is concerned, one must remember that the U is consumed by the operation and that the pile operation is followed by a chemical one. However, the consumption of U does not depend on the cooling system. It is proportional to the 49 production and amounts to about 2000 kg of U for 1 kg 49 produced. Similarly, the chemical operation is also independent of the cooling system used (except for the 6th system). For these reasons, the U requirement includes only the amount of U which is tied up in the pile and subsidiary equipment. The data for other piles (cf. W pile) are also always given in the same form.

A last word about the accuracy of the figures given. No chain reaction with heavy water has been established at the present time and all figures which we gave for the water cooled graphite pile (W pile) proved to be accurate within a very small experimental error. This was, no doubt, partly accidental.

No great trouble would result if our estimates should prove to be somewhat in error, as the size of the pile can be regulated, within limits, by regulating the level of D_2O in the pile.

I. Light Water Cooling. With this type of plant, we believe that about 10 tons of D_2O and 6 tons of uranium would be necessary to produce 40 grams per day of 49. Given full authority and the highest priorities, the plant could be ready for operation, in our opinion, 3/4's of a year after it is authorized.

Pile Design. The pertinent features and several of the possible designs of a light water cooled heavy water pile are described in the two attached reports (MUC-AMW #5, G. Young and Weinberg; Report 805 by L. Ohlinger). Common to all designs is a tank of about $2\frac{1}{4}$ meters diameter and about equal height which contains the heavy water. This tank is traversed by about 400 aluminium tubes. The uranium is contained in these tubes in an Al sheathing; an annular space is left free between the uranium and the inner surface of the tubes. The heat created in the uranium is conducted away by light water which flows in these annular rings. A schematic diagram of the general arrangement is given in Figs. 1 and 2 of CE-805 report. About 10 tons of heavy water are in the pile and it is expected to have a production of 30 to 40 gms. of 49 per day. Naturally, the pile is surrounded by a thick shield. The velocity of the light water is about 10 meters per second.

Shielding. Only the pile proper has to be shielded. The effluent water will have no serious radioactivity. The pile shield itself can be a mixed iron-water shield or a dry shield of about 120 cm thickness at the critical places.

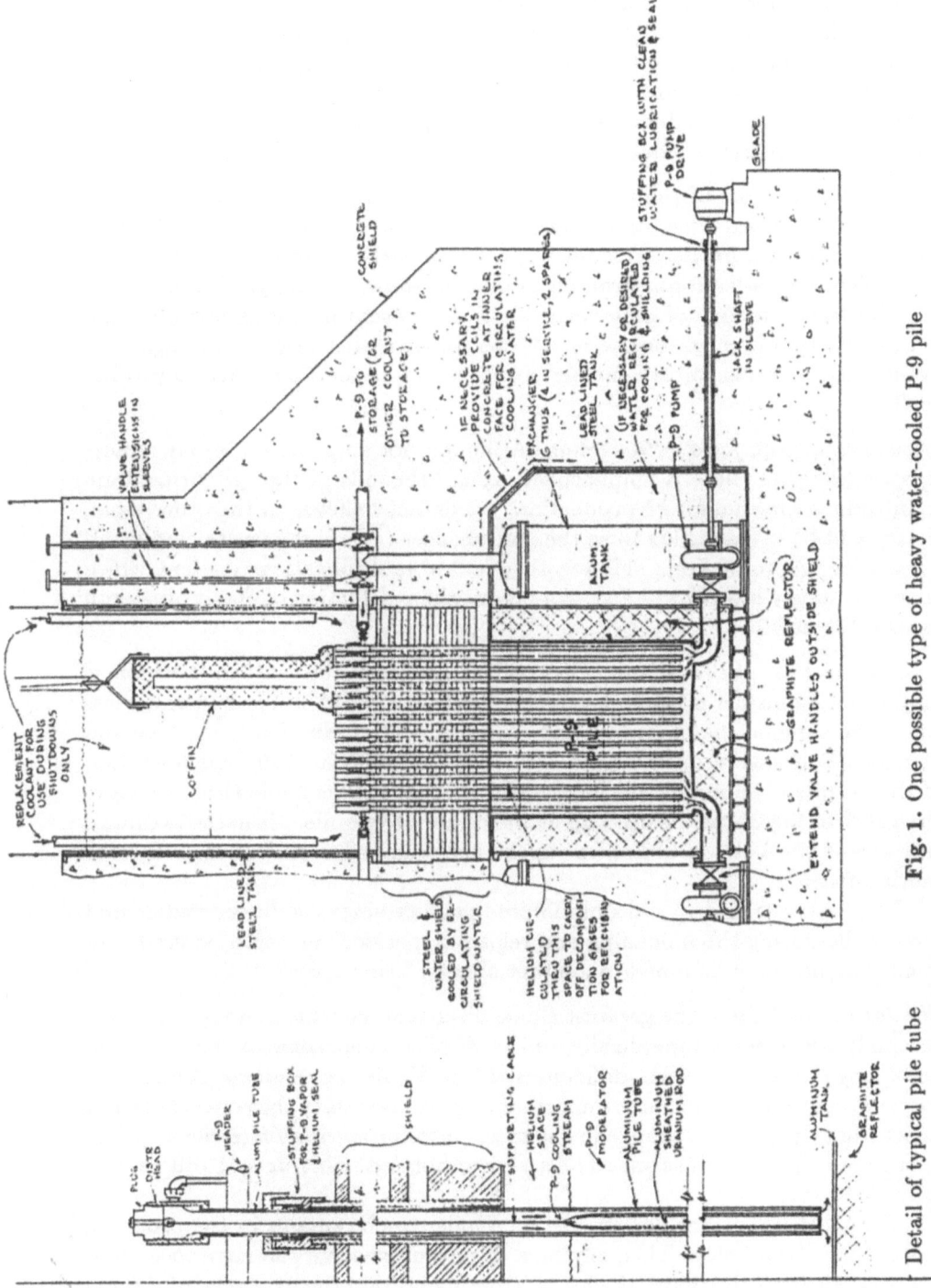

Fig. 1. One possible type of heavy water-cooled P-9 pile

Detail of typical pile tube

Removal of Product. This is also described in CE-805 and is, in our opinion, as simple as in the case of any of the other systems. The uranium rods which hang in the aluminium tubes are removed to coffins providing sufficient shielding for further transportation. The water circulation through the tubes must be maintained at a reasonable level during this operation, but this does not run into any insurmountable difficulties.

Time Schedule. One of the most important advantages of the present system is that in its broad lines its design is ready. This coupled with a relatively small requirement for heavy water promises an early completion of the plant, if sufficient authorization is granted to take all steps necessary thereto.

The large amount of experience that we possess in handling water and our knowledge of its reaction with Al constitute another great advantage. The availability of water makes further experiments easier than with any other coolant.

Easy Adaptability for 23 Production. Although not emphasized in our report, it is evident that there is ample space around the side of the cylindrical tank to introduce thorium for 23 production. We do not believe, on the other hand, that it will be practicable to use the end surfaces (where the water enters and leaves) for 23 production. This together with a multiplication constant which is not extremely high ($k = 1.10$), will permit the production of about one-tenth as much 23 as 49.

Sacrifice in Rate of Production. The production rate of the contemplated pile is only about 4 gms per day for each ton of heavy water. This is considerably less than with some of the other systems. However, the production is high per ton of metal. Four weeks operation in this pile correspond to 100 days operation of the W pile. The metal will have to be exchanged every month if not more than 10% of the 25 can be used up at the center of the pile. Since the exchange takes some time, it would not be worth very much effort to increase the rate much further.

This latter argument has only limited validity since the heavy water need not be idle during the unloading and reloading period, but could be used constantly by having it pumped to another already loaded pile.

Similarity to W Pile. The greatest single disadvantage which our choice has is its similarity, at least superficially, to the W pile. Thus, some of the expected, and many of the unforeseen, difficulties of the W pile are retained. Some, however, are eliminated. The most important single uncertainty that our pile avoids arises from the possible change in the mechanical properties of graphite during operation. This in itself seems to be a very important point since it will be very difficult to remedy.

The most important danger to the W pile which remains in the proposed pile is that of corrosion. This may be a disadvantage from the insurance point of view, but from the point of view of experimentation the existence of a small pile in which a day's run corresponds to a week's run of the big pile may be helpful. It seems to us, furthermore, that very much less corrosion can be expected in the proposed pile than may be feared in the W pile. The thickness of the outer Al tubes will be almost double that of the W tubes. The water temperature will be under 50 °C wherever it is in contact with these tubes. At one point the temperature of the sheathing of the U will reach 70 °C. However, the brevity of the operation, the absence of short pieces ("dogs") and of caps alleviates the situation even with respect to the sheathing. The flow of the water is streamlined throughout the critical section.

Instability. Another difficulty which this pile shares with the W pile is the instability with respect to the removal of the light water. In fact, the danger from this source is greater than in the W pile because of the higher power density. If the flow of the cooling water should fail, even for a short period, the light water in the pile would soon begin to boil decreasing the water density in the annulus. The multiplication constant would increase, and the heat production would also increase rapidly. This could lead to the melting of the uranium and to a minor explosion. Naturally, all precautions must be taken to avoid such a contingency. It is clear that a very rapid control will have to be installed which would have to act in not much more than 1/2 second. Such a rapid control is, however, perfectly possible, e.g. by the immersion of several aluminium coated cadmium plates which would stop the reaction practically at once. All water mains could be made double with a check valve on each so that the breakage of any main would remain harmless. It is also to be remembered that only the failure of the cooling in a very substantial fraction of all tubes would lead to the above sequence of events.

Power. Under the contemplated system of operation, no production of useful power could be obtained.

Difficulties. The light water cooling system has been investigated with a good deal of care and we do not know any difficulty not mentioned above. This indeed is a favorable point as compared with most of the other systems.

II. Heavy Water Cooling. In this type of plant we believe that the requirements of D_2O and uranium for the smallest practical unit would be about the same as the light water cooled type. The production would be perhaps somewhat but not substantially lower. The time schedule appears only slightly less favorable than with the light water type.

Pile Design. The heavy water pile can be given a much greater variety of designs than the light water cooled pile. The attached sketch shows only one possible design, the one which we at the present time believe could be put into operation most quickly. The pile proper is very similar to the pile in the case of light water cooling, except that its dimensions can be somewhat smaller and that one of the headers (probably the exit header) is eliminated by having the tubes open directly into the pile tank. This is possible because the coolant and moderator are the same.

The reason for introducing aluminium tubes to conduct the heavy water is that one can obtain a reasonable heat transfer only if the velocity of the heavy water in the neighborhood of the uranium is quite high. This evidently can be achieved in may other ways besides the one embodied in the sketch. However, any new design would bring up a variety of problems which we have not yet considered and which we believe are not amenable to solution within the short time considered.

The great difference between this type of pile and the light water pile is that the heavy water must be brought outside the pile and put through a heat exchanger before returning it to the pile. As a consequence, there is heavy water not only in the pile but also in the external system. It is estimated that the amount of water in the heat exchangers etc., is not much greater than the amount by which the D_2O in the pile is decreased by shifting from light to heavy water cooling.

Naturally, all the external equipment must be leak-proof to prevent loss of heavy water. We believe that this will require even greater care in the construction than is required to keep the light water and heavy water separate in the light water type.

Shielding. The shielding of the pile proper is the same as in the light water type. However, the effluent heavy water will have serious radioactivity. This is particularly so if the type given in the sketch is chosen, i.e., the pile itself is used as one header. In this case, the heat exchangers and the pumps will have to be shielded against γ-rays by about 6 cms of lead. It will be difficult to achieve this and have the external equipment accessible for repair and adjustment during operation. This and the reason to be mentioned in the next paragraph are the chief causes for our preference of light water cooling.

Removal of Product. When the pile is shut down, the uranium continues to need cooling. Since one does not want to tie-up the heavy water for this period, – which is somewhere between a week and a couple of months –, one has to replace the heavy water by another coolant in the pile. The procedure which involves the least delay is to replace the heavy water first by an immiscible liquid and the immiscible liquid then by ordinary water. When this is done, the removal of the rod can be carried out in the same way as in the light water type. However, the preceding operation is not simple and requires careful consideration.

Time Schedule. The simplicity introduced by eliminating the bottom header is more than compensated by the complications in the external equipment. When the added problems in removal are considered, the time schedule for this type appears less favorable than for the preceding type.

Production of 49 and 23. 49 production should be about 70% (30 gr/day) of that for the light water cooled system. Since a higher multiplication constant ($k \sim 1.15$) obtains and since an extra surface, the bottom, is available, the production of 23 relative to that of 49 will be higher, perhaps as high as 1/6 of the 49 production. This makes the production rate of 23 equal for light and heavy water cooling (4–5 gr/day).

Similarity to W Pile and Stability. The remarks made in the previous section concerning the light water type apply also to the present type. Corrosion of the tubes in the pile is, however, of less importance.

The present pile is stable against removal of the coolant and the multiplication factor would certainly not increase if the pumps should fail. As a result, boiling in the cooling tube would have less serious consequences than in the preceding type, and the control would not have to be quite as rapid. The U would not melt for $2\frac{1}{2}$ seconds after the cooling stops. On the other hand, failure of the pumps would lead to a stoppage of the coolant circulation, since in contradistinction to the light water type, no standpipes and reservoirs can be provided in this case. Also, duplicating the mains, etc., would lead to an increase in heavy water requirements.

Power Production and Difficulties. Power production would be as impractical in this as under the preceding system. Although the whole system was not investigated with the same care as the preceding one, we believe that there are no major difficulties except for those mentioned above.

III. Helium Cooling

Pile Design. The layout of a pile of this type would still be very similar to the ones described above, particularly to the second one. The amount of D_2O for the smallest plant is not much greater than for the light water system. The 49 production per ton of D_2O is, however, somewhat lower. In order to take out even this much heat, it is necessary to use the He under 10 atm pressure; the cross section for the flow of the coolant is made larger and the operating temperature is raised. Consequently, a pressure tank is needed around the pile and thermal insulation must be provided between the He and the D_2O.

A few points of comparison with the helium cooled graphite pile are worth making. One very important difficulty encountered with this system in the graphite pile, i.e., the procurement of pumps, seems successfully solved. It appears also that the comparison between helium and water is more favorable for helium in the P-9 pile than in the graphite pile as far as heat transfer is concerned. The reason for this is essentially the shorter distance which the helium has to travel within the pile. As a consequence, the heat capacity is of less importance and the pressure drops are much smaller. A great number of details have been worked out in connection with this type, a summary of which is given in the appendix.

Shielding. The problems of shielding are very much the same as in the helium cooled graphite pile and they will not be discussed in detail. The problems are much more serious than in case of any of the water cooled types because of the increased radioactivity and volume of the coolant.

The accessibility for servicing of the pumps and heat exchangers would be seriously impaired for this reason.

Removal of the Product, Time Schedule, and Difficulties. Removal of the product appears difficult. We wish to remind the reader in this connection of the difficulties encountered with the same problem in the case of the He cooled graphite pile. For this reason, we are unable to give a time schedule. Also,

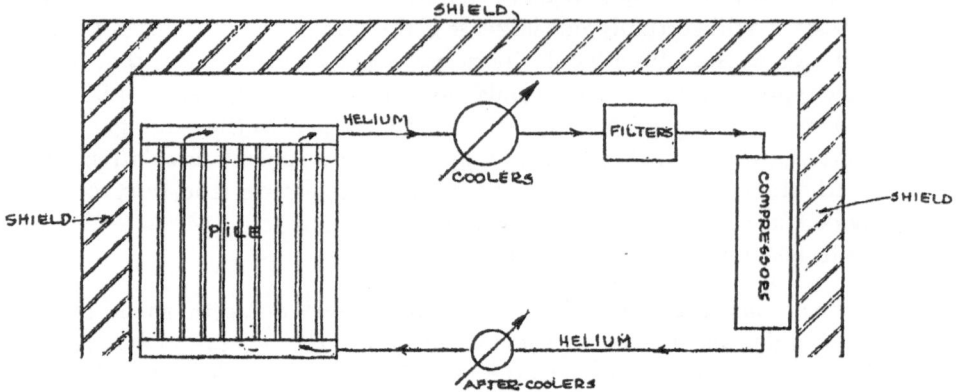

Fig. 2. Flow diagram for helium-cooled P-9 plant

we do not believe that sufficient studies have been made to reveal all of the difficulties which would be encountered.

23 Production. Although the multiplication constant would be probably higher than with any other type ($k \sim 1.17$), the pressure tanks would make it rather difficult to arrange for 23 production.

Other. The pile is quite different from the W pile. No instability similar to that encountered in the light water pile is to be feared. No power production was envisaged in any of the tentative plans, but the system is fundamentally adaptable to such production.

IV. Homogeneous Pile. With this type of plant, we believe that at least 55 tons of D_2O and 12 tons of uranium would be necessary to produce 600 gms per day of 49. Given full authority and the highest priorities, the plant could be ready for operation in about $1\frac{1}{2}$ to 2 years after its authorization. It would produce more 49 per ton of D_2O than any other plant.

Pile Design. In this pile and in the following two no coolant is used. The substance in which most of the heat is generated (slurry in the first two cases, hex in the last case) is brought out of the pile, cooled, and then returned to the pile. As a consequence, no substantial heat transfer from one substance to another is necessary within the pile. Among the three types which have this feature in common, we have a preference for the homogeneous type. The homogeneous pile consists of a slurry of an oxide of uranium in heavy water. The tank has no internal structure. On the basis of Creutz's measurements on the resonance absorption of mixtures, and Fermi's and Anderson's measurement of the cross section of D_2O for thermal neutrons, it appears certain that a homogeneous pile can be made to be chain reacting and will require as a minimum 30 tons of D_2O.

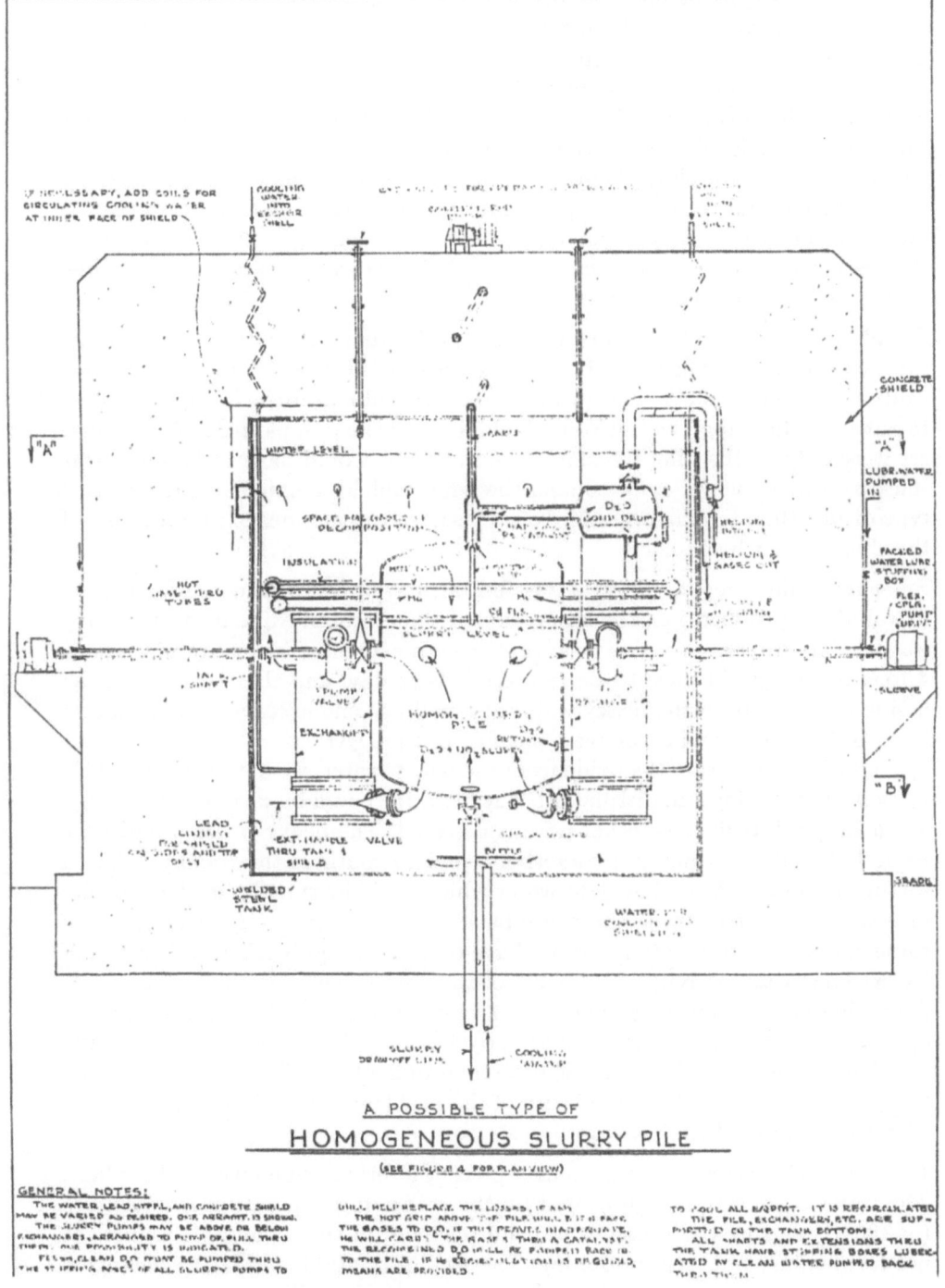

Fig. 3. A possible type of homogeneous slurry pile

When this type of pile was first considered, heat transfer by boiling the slurry was believed to be the most promising. Circulating the slurry through heat exchangers now appears more advantageous.

A sketch is appended to show the essential features of such a pile. The purpose of the pressure tank is to permit a higher operating temperature for the pile (although this increases its size) in order to facilitate the heat transfer outside the pile. The slurry is pumped through heat exchangers of a somewhat special design and reenters the pile again at the bottom. Great care must be exercised, of course, to prevent any leakage between slurry and light water. This is not an easy problem, because the heat exchangers are not accessible. The same holds for the pumps.

Shielding. The reason for the arrangement as shown in the sketch is the extreme radioactivity of the slurry. This consists not only of β and γ-rays but also of neutrons. In order to avoid great wastage of critical materials, it appeared best to put everything together into one shield. Even if the pumps and heat exchangers were outside the pile, the shield around them would have to be almost as thick as around the pile. Of course, the whole shield is much simpler for this type of pile than for the preceding ones because the number of perforations of the shield is much smaller.

Removal of the Product. Several possibilities have been considered for the removal of the product, i.e., the separation of the uranium oxide from the heavy water. The seriousness of this problem becomes evident if we realize that about 4 tons of slurry will have to be reworked every day, and that removal of the fission products from the heavy water also will become necessary from time to time, in order to avoid excessive poisoning of the D_2O.

In order to separate the oxide from the heavy water, we have considered filtration, centrifuging and distillation. The filtration could be carried out through a cooled graphite filter or other porous material. The heavy water would have to be followed immediately by another liquid to continue the cooling. This in its turn could be followed by light water which could be run also in the opposite direction in order to bring the oxide back into the slurry stage. During all of these operations, the material would have to be thoroughly shielded because of its extreme γ-ray activity and escape of radioactive gases must be prevented. A considerable holdup and some loss of heavy water must be expected in this operation and the amount of this has not yet been estimated with any certainty.

Centrifuging and distilling also have their difficulties. The above description was given to illustrate the complexity of the problem to be solved in connection with the removal of the product.

Corrosion. Corrosion and erosion may present a serious difficulty for this plant. The latter is, of course, aggravated by the rather high operating temperature ($\sim 130°C$) and by the presence of solid particles in the liquid. The danger is not from failure but from poisoning. The wall of the tank and heat exchangers can be made strong and thick enough to secure a long life. However, even at an erosion rate of .0004 mm/day in the heat exchangers, sufficient iron enters the slurry to give $\sim 1\%$ decrease in the multiplication constant in one week.

Time Schedule. The time schedule given in the introduction is necesssarily tentative since it will take some time before even the majority of the difficulties of this type of pile are discovered. We feel confident, however, that all the difficulties can be solved.

23 Production. One of the drawbacks of the present scheme is that it is not easily adaptable to 23 production. This is both on account of the relatively low multiplication constant (around 1.09), the loss due to the tank itself and because of the lack of space around the pile.

Comparison with W Pile. One of the favorable features of the present pile is its very great stability. An increase in the rate of production will cause an increase in the temperature of the slurry and a decrease in its density. This automatically decreases the effective multiplication constant. It may be possible to run this pile without regular control rods, although safety rods for stopping the pile must be provided.

The pile is about as different from the W pile as one can easily imagine. This makes it very advantageous from the insurance point of view, but it has little to offer from the point of view of experimentation. It must be admitted that it is clumsy and inaccessible. On the other hand, it appears to be relatively easily adaptable to power production. Also, it produces about as much 49 per 25 consumed as can be expected.

V. Heterogeneous Slurry

Pile Design. The general arrangement is again similar to a tube-shell heat exchanger. The slurry is inside suitable tubes which are immersed in heavy water. While the slurry is pumped through these tubes it is heated up by the chain reaction. After leaving the pile, the slurry enters a heat exchanger where it is cooled before it re-enters the pile.

The purpose of having a highly concentrated slurry separated from pure D_2O in this pile is to obtain a substantially higher multiplication constant than the homogeneous slurry pile gives. As a result, the smallest practicable pile becomes a good deal smaller than in the preceding system. No estimates concerning this point can be given because they depend on the tubing material. The problem of obtaining a suitable tubing is critical for this pile because of the high erosion by the slurry.

Use of a solution of a uranium salt instead of the slurry has been considered, but no satisfactory compound could be found. On the other hand, considerable progress has been made at Columbia in obtaining slurries which contain as much as 3 grams of uranium per cm^3. The properties of this slurry are satisfactory on the whole as long as there is no radiation present. There is a certain danger, however, that a further oxidation takes place under the influence of radiation which might result in caking. Furthermore, there is considerable danger of the slurry eroding the pipes. Aluminium has proved unsatisfactory as a material for the piping.

Because of the high erosion rate, only relatively thick pipes show any promise. The only metals permissible in substantial quantities inside the pile

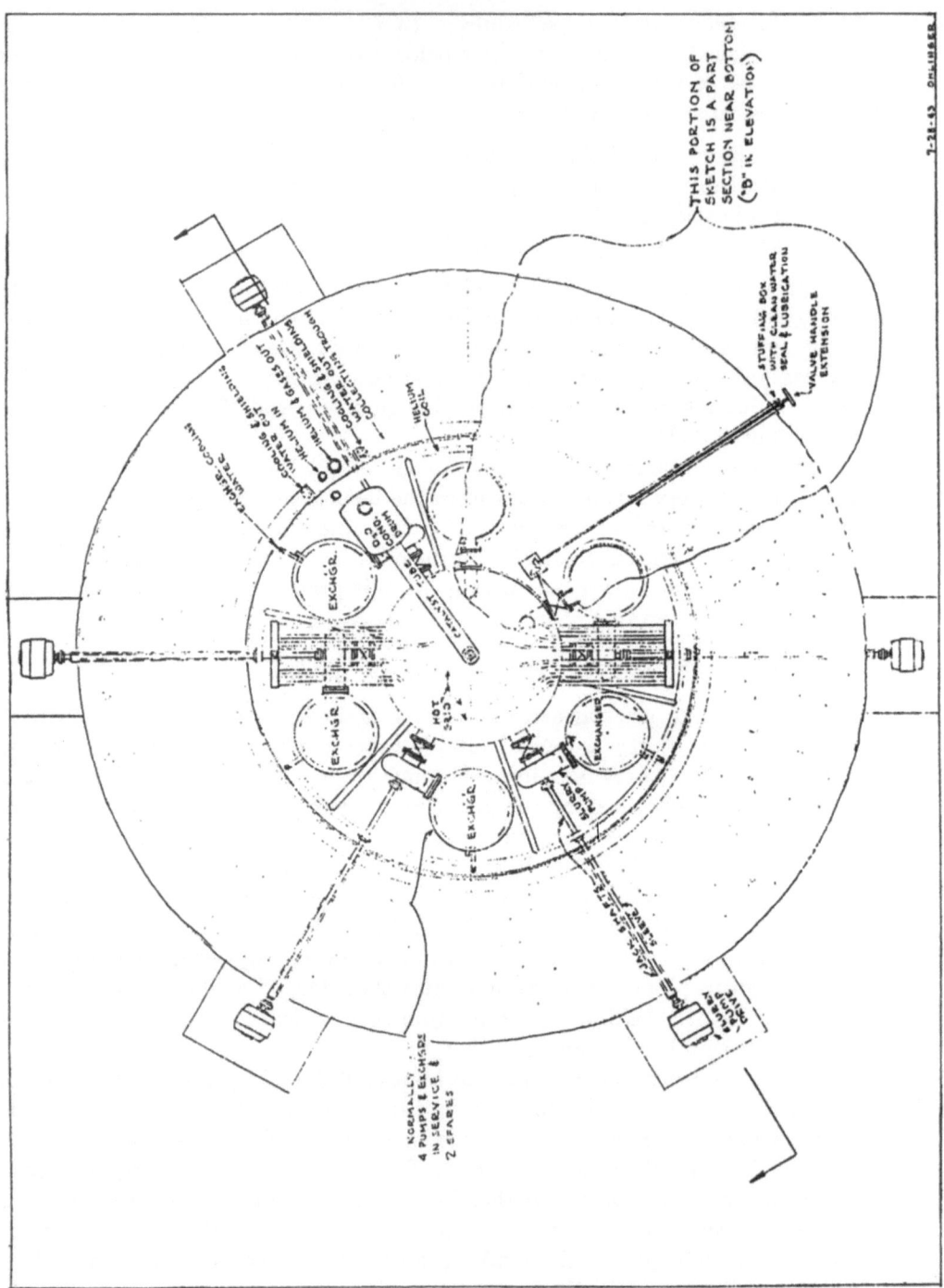

Fig. 4. Plan section "A-A" of homogeneous slurry pile

from the nuclear point of view are Be, Pb, Bi. Among these, Be offers the only real hope for a satisfactory tubing. However, considerable development work would be necessary to produce Be tubing. No tubing is inside the pile in the preceding system.

Shielding. The shielding problem is very much the same as in the homogeneous system. If we were to place the heat exchangers outside the shield, they would produce at about 10 m distance 10^8 times the permissible intensity of γ-radiation. As a result, the pumps and heat exchangers are inaccessible.

Removal of Product. The separation of the oxide from the heavy water presents the same problem as in the homogeneous slurry system. However, since the D_2O tied up with a given amount of U is about 10 times smaller more time is available for the filtration. One can allow the radioactivity to decay to some extent.

Comparison with Heavy Water Cooling and Homogeneous Pile. The heterogeneous slurry system occupies in almost every respect an intermediate position between the heavy water cooling and the homogeneous systems. The multiplication constant is higher than in the homogeneous pile (~ 1.09) but lower than for heavy water cooling ($k \sim 1.15$). Its value depends essentially on the tubing: for a tubing with as great an absorption as Al has, the multiplication constant would be hardly higher, the plant hardly smaller, than for the homogeneous pile.

If the production rate is limited by the heat capacity and velocity of the D_2O and slurry in the pile, it will not be in a better position than the heavy water cooling. The reason for this is that the specific heat of the slurry is hardly greater than that of water and that the permissible temperatures will be similar. If the limiting factors are heat transfer and temperature in the U and if the heat capacities play no role at all, it may have as high a production per ton D_2O as the homogeneous pile. Actually, the situation may be expected to be intermediate between the above extremes.

However, from the point of view of engineering and shielding problems, we believe that it presents as many, if not more, problems than the homogeneous pile.

Time Schedule and 23 Production. It is impossible to give a time schedule at the present time because no suitable piping material is known. From the point of view of 23 production, the situation is very similar to the light water cooled pile.

Power Production and Difficulties. Although the preceding remarks give a rather dark picture of the present status of the slurry pile, one should not disregard the possibilities that all these difficulties can be overcome at some future time. In that case, it might well be possible to adapt the pile for power production.

On the other hand, it is not believed that the preceding gives a full account of the difficulties which one would encounter if one tried to construct a heterogeneous slurry pile now. No layout has been made for this system and many more difficulties might crop up if one is attempted.

VI. Circulated Hex. Almost all the remarks, including those concerning shielding, made for the heterogeneous slurry pile apply also for a circulated hex pile. However, the absence of slurry particles alleviates the erosion problem. The problem of corrosion, on the other hand, is especially important because a leak between hex and water would be disastrous. Although several metals stand up very well in pure hex, nothing is known about the corrosion effect of hex under the influence of fission fragments, and Cannon found that free fluorine pits aluminium in the presence of β-radiation. The presence of free fluorine can hardly be avoided under pile conditions.

The most frequently quoted advantages of the hex system are the ease of removal of hex from the pile and the simplicity of the chemical separation, which follows the operation in the pile. The latter does not seem important at the present time because the problems of chemical separation appear to be solved for uranium in its usual form (metal or oxide). The same problem may be easier in case of hex, but it has not been worked out in detail.

VII. Bismuth Cooling

Pile Design. The general layout of the pile proper is very similar to the He cooling system. Bismuth is used instead of He and is circulated through a heat exchanger as well as through the pile.

The possibility of bismuth cooling (or cooling with a Bi-Pb alloy) depends on a great number of metallurgical, chemical and erosion problems, which limits the material of the piping. As a consequence, it is hardly possible at present to give a reasonable evaluation of this system. One advantage of bismuth, as compared with the heavy water, is that it will decrease the multiplication constant to a smaller extent than heavy water would if it is used inside the uranium. As a result, it is possible to use the uranium in the shape of tubes and cool both outside and inside (double cooling). For this reason and because one can obtain very high heat transfer coefficients with Bi, the power output per unit material can be substantially increased beyond that possible with light water or heavy water cooling.

Shielding. It appears very probable that only the pile proper has to be shielded. Bismuth becomes radioactive under the influence of neutrons but the resulting Po is known only to give α-rays, against which no substantial shielding is needed. It may be remarked here that Po is an extremely valuable by-product from the point of view of this project. There is a report in the German literature of another radioactivity induced in bismuth by the neutrons, but it is not known whether it emits γ-rays. Even if it did, it is unlikely that the shielding of the external system would cause very serious difficulties if the U is coated. The radioactivity is less serious than in the He cooled pile even if the U is coated.

Removal of the Product. The problem of the removal of the product is more difficult than in the case of the heavy water system. No satisfactory solution is known at present.

Time Schedule and 23 Production. No time schedule can be given at present. From the point of view of 23 production the system may be classed along with the light water system (about 10% of the 49 produced).

18.

Breeders and Converters:
Minutes of Lecture on April 7, 1945

E. P. Wigner

(Minutes written by L. A. Ohlinger)

The purpose of this discussion is to present a brief description of breeder piles and their details as we think of them today, not from a point of view of pure nuclear physics, but from a practical point of view.

Breeder piles may be divided into two classes; (1) those in which the active material is circulated and (2) those in which the active material does not circulate. In the latter, the problem of heat removal is complicated by the necessity of transferring the heat from the active material to another medium or coolant which circulates, while the former has the disadvantage that part of the active material is always outside of the pile proper. If the active material is very much diluted by the coolant and/or moderator, it may be more worthwhile to circulate the active material because of the large volume and large heat capacity of the dilutant. If the active material is highly concentrated, circulation is disadvantageous because a large amount of active material will be uselessly held up outside the pile.

As Mr. Weinberg pointed out, in the case of 23, η at thermal energy is high and does not increase much at energies above thermal. In the case of 49, the opposite is true. For example:

23 – η at thermal energy = ~ 2.4 – η at high energy = 2.4

49 – η at thermal energy = 2.05   – η at high energy = 2.52 – 2.9 .

(The value of η at thermal energy for 23 changes from week to week. The present Chicago value according to Mr. Dancoff is 2.52 while the value given by Y is 2.2. Accordingly, a reasonable average value is probably the figure 2.4 given above.) According to the above, 23 breeders may operate at thermal energy or above, but 49 breeders are only practical at energies above thermal.

Since thermal piles are practical for 23 breeders, the neutrons released by fission can be slowed down to thermal energy and therefore the concentration of fissionable material in the moderator may be low. Accordingly, this type pile lends itself more readily to circulating the active material. On the other hand, 49 breeders utilize the neutrons at resonance levels or above and therefore the concentration of fissionable material should be high, which lends itself more readily to piles in which the active material does not circulate. It would appear to be better terminology to replace the old title of "homogeneous piles" with "circulating systems" (which will, in general, refer to 23 converters and breeders) and to replace the old title of "heterogeneous piles" with "stationary systems" (which will, in general, refer to 49 breeders and converters).

In the case of 23 breeders, except for the value of η, all fundamental problems appear capable of solution. This does not mean that a breeder can be built tomorrow, but that the problems are not too difficult and that after reaching a certain point in our studies, the balance of the unknowns can be solved in a reasonable time. A large part of this is due to the chemists at X who have done an excellent job of preliminary investigation. Fundamentally, the 23 thermal breeder would probably look something like the diagrammatic sketch below.

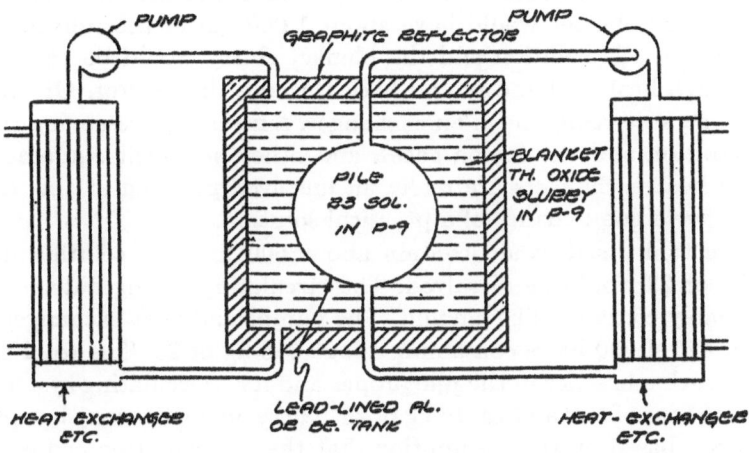

The chain reaction would be sustained in a solution of 23 in P-9. According to the chemists at X, the most promising form of soluble salt of 23 would be sodium uranyl carbonate, $Na_2UO_2(CO_3)_2$. This is best obtained, as I understand it, by precipitating a uranyl salt with sodium carbonate to form sodium uranyl carbonate which is filtered off and dissolved in a solution of sodium carbonate in P-9. The excess of sodium carbonate always present in the solution prevents the decomposition of the complex uranyl salt. This complex uranyl salt is very soluble. Even the presence of any amount of hydrogen preoxide in the solution does not cause a precipitation of uranium peroxide but, in fact, the hydrogen peroxide gives a more soluble complex. In a previous lecture, it was pointed out that the addition of colloidal Pd would reduce the gas evolution to 1/30th. Furthermore, the above 23 solution is sufficiently alkaline so that the colloidal Pd is stable therein. (It might be noted at this point that the Chicago chemists say that the high gas evolution that has been reported is only a transient phenomenon which later stabilizes and reduces to about 350 liters per second at 100,000 kw in an acid solution.) The same type of compound described above but formed with Pu instead of U was irradiated in a solution to which Pd had been added. Unfortunately, everything precipitated out because the Pu is reduced by the hydrogen peroxide to an unstable compound despite the excess sodium carbonate. It must be admitted, however, that a reasonable excess of sodium carbonate must be used in the solution of 23 also and this is not good because the sodium carbonate has a relatively high absorption cross section and, in fact, gives about 4% loss in fertility.

The pile proper would be a cylindrical tank about 90 cm in diameter and about 90 cm in height made of Al possibly with a lead coating inside. The Al would be about 2 mm thick. Surrounding this would be a blanket not less than 50 cm thick formed by a circulating slurry of thorium oxide in P-9 contained in a cylindrical tank 190 cm in diameter and in height. Surrounding the blanket would be a graphite reflector about 20 cm in thickness. The concentration of 23 in the pile solution would be about 8 gms of U per liter of solution while the slurry in the blanket would have about 1,000 gm of thorium per liter of slurry. Both the pile solution and the blanket slurry would be circulated by pumps through heat exchangers and back through the reactor. Of course, by heat exchangers we really mean an exchanger, a means for removing the gases of decomposition and the fission gases, and other pertinent equipment. The pumps and heat exchangers are really an unsolved problem at this time but appear to have a good chance of a practical solution.

The pile tank itself would contain about 500 liters of solution in which about 4 kg of 23 would be dissolved. The circulating piping outside the pile would probably contain a like amount. The pumps and heat exchangers would hold about 625–1,000 liters containing about 5–8 kg of 23. The total material involved in such a breeder in the pile proper and its recirculating system would be about two tons of P-9 and 13–16 kg of 23. These quantities may be optimistic since they are based on the assumption that the gas evolution can essentially be eliminated. Another 13–18 tons of P-9 will be involved in the Th slurry in the enclosing blanket and its recirculating system.

The loss in efficiency due to the various factors previously listed by Mr. Weinberg will probably be as follows:

(1) delayed neutrons – $\sim$ 3%,
(2) absorption by the moderator and anion – 1 to 5%,
(3) absorption by the fission and corrosion products (assuming a 2% loss in the chemical processing cycle) – 5 to 7%,
(4) leakage (mainly escape at the ends) – $\sim$ 1/2%,
(5) absorption by the pile tank – $\sim$ 1/2%,
(6) absorption by the moderator and anion of the Th slurry blanket – $\sim$ 1/2%,
(7) escape from the blanket – $\sim$ 1/2%,
(8) absorption by Pd (this depends upon the cross section of Pd about which not much is known. Assume a cross section to be about like that for Th which would be 1/2% if the Th were purified every month) – $\sim$ 1/2%,
(9) absorption by 24 is assumed to be negligible because it gives 25 which is fissionable,
(10) absorption by the controls – say 1/2% (?). Controlling a breeder pile of this type is difficult because k varies with the density and one cannot use a control rod such as is used for ordinary piles because of the neutron

absorption. Mr. Nordheim has suggested as a possible control a pocket of P-9 in the blanket immediately adjacent to the pile proper with a Th control rod which can be inserted into or withdrawn from this pocket at will. This is indicated diagrammatically at the right.

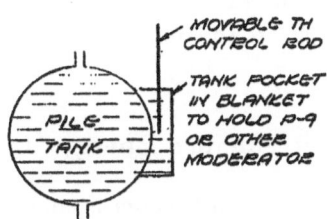

The total of all the above losses would be 12 to 18%. The larger figure is probably the safer one to use in preliminary studies.

Accordingly, the value of 2.4 for η should be reduced to ~ 2.22. Therefore, one atom of 23 destroyed gives 1.22 new atoms of 23. The greatest disadvantage of this system is that in burning up 1 kg of 23 and producing thereby 10^6 kw days of energy, one has only gained .2 of a kg of 23. This same power output at W yields 1 full kg of product. Another disadvantage of this sytem is the low temperature involved in the cooling cycle which makes it hard to produce power. Mr. Urey has suggested using higher temperatures. Al as corrosion protection might be okay at higher temperatures in distilled water but there are other problems involved.

It may be seen from the above that such an arrangement wastes power and is uneconomical because of the Th lost and the mechanical power required for recirculation. A better type of breeder would be the stationary high energy pile utilizing 49. However, our knowledge of this type is more limited and the most important problems are not solved. Most high energy piles have the following common principle – that Pu is dissolved in some solid material, kept in the piles instead of being recirculated and a coolant which may or may not also serve as a moderator recirculated to carry away the heat.

The first large difference in the high energy piles is the value of η at resonance energies vs high energies. To compare the two different types of units:

High Energy Units

The reactor for this type breeder will be small in physical size with a high concentration of Pu. One major difficulty is the manufacture of the Pu pile parts. Nothing is known about the corrosion of material containing a high concentration of Pu. The arrangement suggested by Mr. Soodak for such a breeder would be a block of Pu or a Pu-U alloy surrounded by a blanket of pure U. Cooling would be accomplished by forcing liquid Bi through holes bored through a block to form cooling streams or channels. The total volume of the cooling channels would be about half the volume of the block. There are several obvious problems in such an arrangement. The temperature drops would be high so that the thermal stresses would be very severe. It is known that the coolant, liquid Bi, dissolves U and probably will dissolve Pu too. It is not known whether Pu alloys with U. The removal of the Pu and of the U for processing would be difficult. In summation, the high energy unit offers many serious problems and its main advantage seems to be that we know the value of η to be about 2.9.

Resonance Energy Units

The main difficulty in this type of breeder is that η is quite uncertain although it probably lies between 2.45 and 2.75. The general arrangement of such a pile is to dissolve Pu in some other metal (such as Be) which is formed into shapes of the proper size. The Pu alloy is then placed in a large pile tank through which a coolant flows between and around the formed shapes. The Pu may be in any one of several forms with any of several solvents. Originally, it was thought to put the Pu in plates because of the heat transfer characteristics. However, the manufacture of such plates using reprocessed Pu appears too difficult and close packed rods or tubes are suggested as a more suitable arrangement. Liquid Na would probably be used as the coolant and would flow between and through such rods and/or tubes. A third form in which the Pu could be placed in the pile would be in small spheres as suggested by Mr. Daniels. As solvents, Al, Be, and BeO (Daniels) would all serve satisfactorily. Be is especially good as a solvent because the liquid Na does not seem to attack it according to the metallurgists. As coolants, Na, Bi, $(CF_2)_n$, He or steam might be used. Any of these coolants can be used at elevated temperatures to obtain considerable energy.

All in all, the larger resonance energy piles appear more reasonable than the terribly small high energy piles with their dubious thermal stresses.

The resonance and higher energy piles have no loss in efficiency from delayed neutrons and have a smaller absorption by the moderator because the resonance absorption is small and there are few thermal neutrons, but the absorption by the fission and corrosion products may be serious. If C equals the loss based on the chemical separation cycle (about 2% according to the chemists) and if f equals the ratio between the resonance absorption by the fission products and the resonance absorption of Pu, then the loss $= 2\sqrt{Cf}$. While f is unknown, if it is assumed to be approximately unity[8], then the loss is $2\sqrt{0.02 \times 1}$ or about 28%. This is rather large so it appears that the chemical losses must be minimized.

---

[8] Original reads "the same as for Pu". (A.M.W. & A.M.P.)

19.

Preliminary Calculations on a Breeder with Circulating Uranium

E. P. Wigner, A. M. Weinberg, and G. Young

May 17, 1945

The following constitutes our present ideas on the breeder which, no need to say, are very preliminary and which we hope will undergo considerable changes before they will crystallize. Under ordinary conditions we would not report on work which is still as much in flux as the present one is. However, we had to shift our attention, at least temporarily, to another subject and we want to record for this reason the results obtained so far.

A. General Considerations

The breeder is an arrangement in which 23 undergoes fission and the neutrons produced thereby are absorbed by thorium. When this thorium decays it produces again 23 via Pa. Since the η of 23 is, according to Anderson and May, 2.37 the net increase in the number of neutrons per 23 destroyed is 1.37. If all these neutrons were absorbed in thorium, we would obtain 1.37 23 atoms per 23 destroyed. It is well to remember, however, that if only as many as 15% of the 2.37 neutrons emitted are lost by escape from the system or parasitic absorption by other materials or otherwise, the efficiency of the breeder goes down to 1 and no increase in the amount of 23 results.

Another point that is important is the power to product ratio of such a breeder. Let us assume that the total losses in neutrons amount to only $7\frac{1}{2}$%. In this case 1.19 atoms of 23 are produced per 23 atom destroyed. Thus the net gain in 23 is .19 atoms per 23 atoms destroyed or per .89 atoms 23 undergoing fission. As a result, the power output per net production is about five times greater than in the conventional piles.

During wartime it might possibly be more important to operate the systems to be described as converters of 49 to 23. In this case the above considerations do not hold necessarily and the efficiency does not have the paramount importance which it will be given in the following pages. If we disregard this fact, the main importance of the breeders is that they eventually permit the utilization of all the natural thorium. If one wants to make use of the full amount of energy one will have to construct a system in which the power generated by the breeder is utilized, i.e., in which the breeder acts not only as a producer of 23 but also as a power source. This will demand high temperature operation of the breeder.

If only the energy of the 23 produced by the breeder is utilized, the energy of at least $1/1.19 = 84\%$ of the Th is wasted if we adopt the numbers of the preceding example.

An enumeration of the losses follows:

1. Delayed neutrons which may be released outside the pile.
2. Absorption by materials present in the chain reacting part, such as the moderator and anion in case of solution, Bi and tubing in case of a second system to be described below.
3. Absorption by fission and corrosion products in pile.
4. Absorption by tank.
5. Escape in wrong direction.
6. Absorption by 24 formed.
7. Absorption by Pa.
8. Absorption by fission and corrosion products in reflector.
9. Absorption by moderator and coolant in reflector.
10. Escape from reflectors, particularly of neutrons formed by fission in reflector.
11. Losses in chemical separations.

It may be well to introduce, at this point, the notion of the efficiency of the breeder. This is the net gain in the number of 23 atoms per 23 atoms destroyed. In the ideal case, it is 37%, in the example discussed above only 19%.

It will be seen in the next section that the arrangements which we think are most promising consist of essentially two parts: a chain reacting unit containing the 23 (the "pile") and a reflector containing all or most of the thorium. If all the thorium is contained in the reflector only $1/2.37 = 42\%$ of the neutrons will be absorbed in the pile and 58% will be absorbed in the reflector. Hence a relative neutron loss l inside the pile entails the loss of the fraction $.42l$ of all the neutrons. A relative neutron loss l in the reflector gives a loss of the fraction $.58l$ of all the neutrons. The loss in "efficiency" caused by some impurity which absorbs in the pile .01 times as many neutrons as the 23 is 1%, the loss in efficiency caused by an impurity which absorbs in the reflector 1% of all the escape neutrons is 1.37%. A loss of 1% of all the neutrons entails a loss of 2.37% in the efficiency. All losses must be subtracted from the original 37% efficiency.

The present report assumes that the active material in the pile is 23 because only this is known to have an η considerably in excess of 2. It is worth noting, however, that 49 is an equally or even more suitable material for the same purpose if 40 is slow neutron fissionable. According to Miss K. Way's calculations, this should be the case. Assuming that the properties of 40 are the same as those of 49 – which is perhaps a somewhat optimistic assumption – the number of fissionable atoms formed per 49 consumed is $\eta+(\alpha-1)/\alpha = 2.05+.33 = 2.38$, i.e., just the same as for 23. In spite of this, under the above assumption, 49 would be a somewhat more suitable material for the breeder than 23 because of its larger total cross section which would reduce the amounts needed and the concentrations necessary by the ratio of the cross sections, i.e. $630/1050 = .6$. Furthermore, a loss of p percent of all the neutrons would reduce the efficiency in the case of Pu only to $\eta(1-p) - 1 - 1/\alpha = .38 - 2.05p$ while the same loss would give, in the case of 23, an efficiency of $\eta(1-p) - 2 = .37 - 2.37p$. This calculation assumes that 24 is not slow neutron fissionable. The possibility of using 49 is presented also because there may be a certain experimental error in the η of 23, as determined by Anderson and May.

B. Brief Description of the Arrangements Considered

The general arrangement which at present appears most attractive consists of a relatively small pile proper which contains heavy water as moderator and all the 23 that is added to the system. The heavy water and the 23 are circulated to the outside of the system, cooled, stripped of the fission and decomposition gases and freed of the fission products outside. The chain reacting part is surrounded by the reflector. This also consists essentially of heavy water and contains most or all of the thorium. To give an idea we include a sketch at this point, although the detailed calculations will follow later on. The chain reacting unit has a volume of about 400 liters and contains of the order of 3 kg of 23. The thickness of the reflector is about 50 cm and contains about equal weights of thorium and heavy water. The pile is to be operated at 100,000 to 200,000 kW.

The first possibility (Fig. 1) envisages the uranium dissolved in the heavy water in the form of a salt, such as the nitrate, sulphate or fluoride. This solution is pumped into the pile through the neck which is on the bottom side and leaves the pile through the neck on the top side. The circulation rate is about 500 liters per second. The solution enters a heat exchanger and a degasser after it leaves the pile, whence it re-enters the pile again. The pile is surrounded by the thorium which is in the form of a slurry containing about one gram of ThO_2 per cm^3 of heavy water. If the wall of the tank separating the uranium solution from the thorium slurry is of a low absorbing material such as Be, Pb, Bi, or perhaps Cb, the neutron absorption of the tank may be sufficiently low. If, however, the tank has to be made out of a more absorbing material it may be necessary to back this up with a higher concentration of thorium, possibly in the form of small spheres ("beebees"), in order to reduce the thermal neutron intensity in that region.

The second arrangement shown (Fig. 2) contains the uranium dissolved in Bi, about 1 gm to 25 cm^3. This bismuth is circulated in suitable tubes and is cooled and freed of fission gases outside the pile. The shape of the pile is so chosen that the neutron escape at the ends shall be about 1% of the total neutron escape because this part of the neutrons will be lost. It may be possible to achieve the same purpose by another shape of the pile more similar to that in Fig. 1. This would involve, however, a more complicated inner structure of the pile. The tubes containing the bismuth are surrounded by heavy water which also has to be re-circulated, degassed and cooled outside the pile.

The interest in UF_6 as a pile material has been raised on this project by H. Brown and we are indebted to him for a discussion on both this and other problems. UF_6 could be used instead of metallic U in the pile of Fig. 2. The Bi would be replaced in this case by a fluorocarbon.

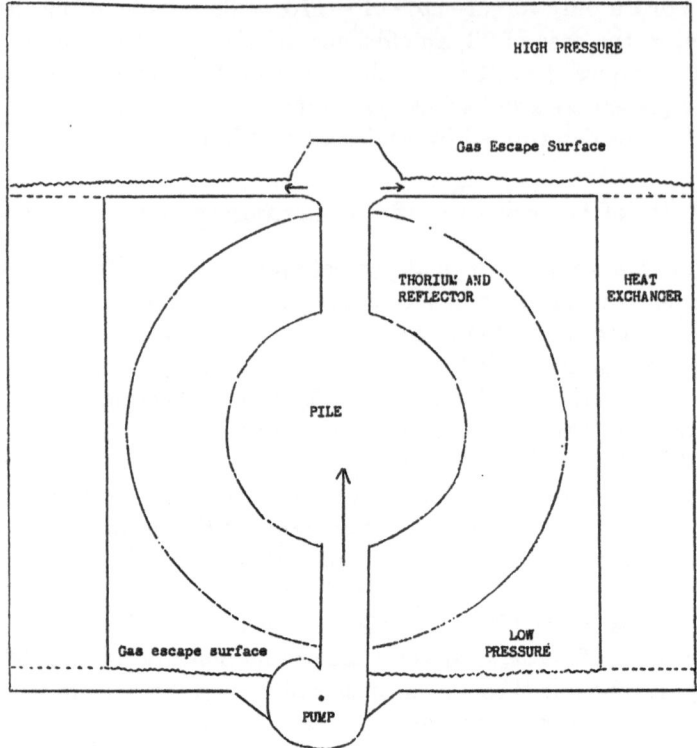

Fig. 1. The degassing method is only illustrative

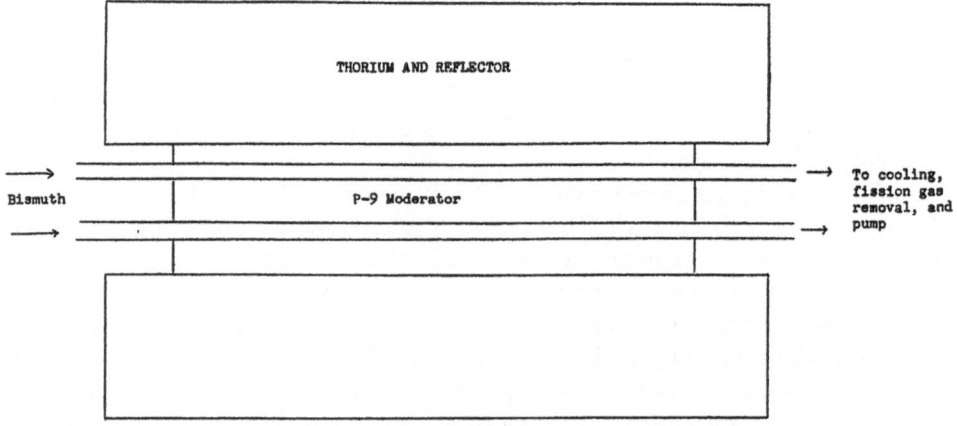

Fig. 2

The first arrangement is much simpler than the second one and has a very much greater production rate, but the power which it furnishes cannot be well utilized because the temperature at which the pile can be operated is quite restricted. The greatest unsolved problem is that of the degassing of the heavy water without increasing the holdup outside the pile unduly. The second system is much more complicated and involves many serious problems among which we believe that a suitable choice of the pipes and the separation of the fission products from the Bi solution are most serious. Its production rate is much lower than that of the first pile but it furnishes the energy at a sufficiently high temperature for purposes of power production. If UF_6 is used in a fluorocarbon, instead of U in Bi, the complications are much decreased, but so would be the temperature of operation. The problems of chemical separation would be greatly alleviated but one would have to cope with the effects of radiation on the fluorocarbons. According to arguments presented by Burton and Franck, one may expect a somewhat more favorable situation in case of unsaturated ringlike structures, if these are sufficiently stable chemically, than in case of saturated fluorocarbons.

C. Reasons for the Arrangements Discussed

The piles as described above assume pumping of the uranium solution by some suitable pump. Instead of this, the solution can be pushed alternately into and out of the pile and cooled and degassed while it is outside (pulsating pile).

The most definite reason for using the 23 in solution is the need for eliminating the Xe^{135}. If we assume that altogether 5 kg of 23 are in the system and that it operates at 100,000 kW, 100 gm of 23 undergoes fission every day, the amount of Xe produced per day is 3.3 gm. If this were permitted to remain in the system as long as one day its cross section would be five times greater than that of all 23 present. In order to reduce this cross section to 1% of the cross section of the 23, one has to eliminate the Xe formed every 5 minutes. This is possible only if the Xe can be flushed out of the system at frequent intervals and eliminates the possibility of using the 23 in a solid form. If the Xe is flushed out at very frequent intervals one gains an additional substantial elimination of other fission products which might otherwise poison the pile.

The above conclusion that the 23 must be used in a non-solid form is not absolute. The maximum loss of efficiency if all the Xe^{135} is left in the pile is 5.8% and although the permissible loss must be well below 37% a loss of 5.8% could be tolerated if all the other losses can be kept at a minimum. However, circulating the 23 has many other advantages, particularly from the point of view of heat transfer and it is not evident that using it in a solid form will facilitate the elimination of the other losses. We feel, therefore, that the conclusion to use the 23 in a non-solid form is a valid one. The only other possibility to eliminate the Xe^{135} loss which we realized is to operate in such a way that practically all neutrons be absorbed at higher energies (by resonance absorption) before they reach such energies at which the Xe^{135} cross section is large.

The Xe^{135} could be removed from the 23 inside the pile by letting a gas bubble through the liquid which contains the 23. This might be a practical possibility if the 23 is dissolved in Bi. We felt, however, that such an arrangement would introduce unnecessary complications and since it has great advantages to remove the 23 from the pile periodically for cooling and for separating the other fission products, we decided against it. This fixed one of the most important properties of the piles which were described above, that of circulating the 23 to the outside of the pile. Such a circulation of course increases the amount of 23 necessary for the pile but greatly facilitates the energy transfer from the 23 to the outside.

It will be seen that considerable difficulty is caused by the need of inserting a tank between pile proper and reflector because the tank is likely to absorb neutrons. It would be advantageous, therefore to place the thorium, or at least most of it, to the inside of the pile. This would result in the use of a synthetic "natural uranium" which would give a pile with a very small multiplication factor, an arrangement which would greatly help in decreasing the neutron losses. There is no question that there is great merit in this point of view. If we reduce the multiplication constant (which is nearly 2.37 in the pile as described above) to 1.05, only 5% of all the neutrons escape from the pile proper and only these have a chance of being absorbed by the tank. However, the diameter of the pile will be increased in this way to about 3 m and its volume to 14 m^3. The amount of 23 required for the pile proper would be of the order of 70 kg under these conditions. In order to keep the amount of 23 needed for the pile at a relatively low level, we have decided therefore to separate the chain reacting part from the reflector where the production of the 23 takes place. It might be advisable to add *some* thorium to the pile proper but this should be kept at such a level that the multiplication constant does not decrease below about 1.5. If this is done the absorption by the thorium within the pile proper will be about 60% of the absorption of the 23 and its amount in the pile will be about 60 times the amount of 23. Such an arrangement has the advantage that the number of neutrons which have to leave the pile is reduced by a factor of almost 3 and the losses in the tank and outside will be diminished by the same factor. If the pile proper contains the 23 in a solution in heavy water we have decided against this possibility because the thorium could be present only in the form of a slurry which would aggravate the pumping problems and the problems of chemical separation.

The next question is the question of the moderator. In this respect we believe that the choice of heavy water is rather unique. If we want the number of neutrons absorbed in the moderator to be only 1% of the neutrons absorbed by the 23, the volume g of the different moderators associated with 1 gm of 23 is given in Table I.

Among these C is undesirable because the radius of the pile would amount to about 60 cm, assuming a multiplication constant of 2.37 and a migration area corresponding to the age alone (i.e. assuming zero diffusion length for the thermals). Such a pile would contain far too much 23. Water is undesirable because the size of the pile would become too small so that one would run into

Table I

	$\sigma_a/$ molecule	Density	Volume associated with 1 gm to give loss of 1% (g)	Heat capacity of this volume in joules/°C (C_s)	Age
C	.0045	1.6	44.5 cm^3	65	320 cm^2
D_2O	.004	1.1	120 cm^3	500	115 cm^2
$D_2O + \frac{1}{2}\%H_2O$	.007	1.1	70 cm^3	290	110 cm^2
H_2O	.6	1	.8 cm^3	3.3	33 cm^2
Be	.01	1.8	13.3 cm^3	42.5	200 cm^2
Bi	.017	9.8	33 cm^3	48	–
$(CF_2)_n$	.023 n	1.8	32 cm^3	~ 70	

heat extraction difficulties. If the 23 is in the pile as short a time as 1/10 sec and if one wants a power output of 100,000 kW for 5 kg 23, the temperature increase in the water would amount to $20,000 \times 1/10/3.3 = 600\,°C$. Be in itself is not much more suitable than graphite and Bi cannot be seriously considered as a moderator. This leaves D_2O as a rather unique chioce. On the other hand, it is possible for the pile to contain substantial amounts of Bi and Be, in addition to the D_2O which has to do the bulk of the moderation.

Heavy water as a moderator gives considerable leeway for the concentration in which 23 can be used. The importance of this is the following: If we use a high concentration of 23 only a relatively small fraction of all the neutrons will reach thermal energies and many of them will be absorbed at resonance above thermal energies. There is not much reason to believe that the η is absolutely constant even in the very small energy region which extends from ordinary thermal energies to about 1 volt and if it should turn out that η increases with increasing energy it may constitute a considerable advantage to go to high concentrations and high resonance absorption. On the other hand, if it should turn out that η decreases with increasing energy of the neutrons, it would be best to keep the concentration of 23 in heavy water as low as possible. This question will be taken up again in section D.

It is well known that if the uranium is dissolved in water the fission causes a very large amount of disintegration. We estimate from data of Burton and Allen ($\Gamma = 25$) that such a pile operating at 100,000 kW would give 220 liters NTP hydrogen oxygen mixture per second. While the gases have the desirable effect that they carry with themselves the Xe and some other fission products when they leave the liquid, the disengagement of such amounts of gas causes considerable trouble even if it is done at relatively high pressures. There is considerable advantage therefore to have the 23 separated from the heavy water which can be done most simply by dissolving it in Bi and keeping the Bi separated from the heavy water by tubes of a non-absorbing material or thorium. This leads us to the two systems which we have considered most seriously and which will be described more closely in the next section.

There is another remark that ought to be made in this connection. At the concentrations of 23 which reduce the loss due to absorption by the moderator to the order of 1% a considerable number of neutrons will be absorbed before they reach thermal energies. If both the moderator and the 23 obeyed the $1/v$ law this would not change the ratio of the neutrons absorbed by the moderator. There is considerable indication, however, that the absorption of the 23 at resonance energies is considerably greater than would correspond to the tail of the $1/v$ absorption extrapolated from thermal energies. Mr. Dancoff estimates on the basis of experiments carried out in the Argonne Laboratory that the $\int \sigma_a \, dE/E$ is at least twice greater than it would be if 23 followed the $1/v$ law. As a result, the fraction of neutrons which are absorbed by the moderator will be about half as great for the neutrons absorbed at resonance than for the neutrons absorbed at thermal energies. Another effect which this resonance absorption has comes in in the calculation of the Laplacian which is increased by this effect. Since the ratio of the actual $\int \sigma_a \, dE/E$ to the value which is obtained by extrapolating the $1/v$ law is not known the calculations will be carried out under three different assumptions, viz., that that ratio is $\alpha = 1, 2, 4$.

D. Critical Size

The critical sizes for pure fissionable material have been calculated by Christy and Wheeler (CP-499), by Weinberg, and more recently these calculations have been extended by Nordheim.

It turns out that the two group theory and the three group theory both give results which are substantially below those given by Fermi's theory. The results of the three group theory are just about midway between those of the two group theory and the Fermi theory. Plass calculated the corrections to the Fermi theory and in the case of heavy water they turned out to be quite negligible, as was already indicated by Weinberg's work.

Our calculation includes the effect of the resonance absorption for which the $\int \sigma_a \, dE/E$ was estimated in the following way. The value of σ_a was taken to be $(98\alpha)/\sqrt{E}$ which gives for $\alpha = 1$ the correct value for thermal energy. However, three values of α were used, viz., $\alpha = 1, 2, 4$, as explained before. The integration was extended from $2kT$ to ∞ giving 900α. Evidently these assumptions are quite arbitrary but should give the right order of magnitude. The average energy of the neutrons in the pile will no doubt be considerably higher than .05 ev, but the slowing down just above the thermal region will be considerably slower than the formula for the resonance absorption implies. These two effects give corrections of opposite sign. The following table gives the probability for escaping resonance for $g = 100, 200, 300, 400$ cm^3 heavy water per gram 23.

The Laplacians L were than calculated by Mr. Wilkins on the basis of the formula developed by Messrs. Friedman and Weinberg

$$k_s \frac{\kappa_t^2}{\kappa_t^2 - L} + k_f = e^{-L\tau}$$

Table II. Number of neutrons reaching thermal energy

g	$\alpha = 1$	$\alpha = 2$	$\alpha = 4$
100	.875	.769	.592
200	.936	.875	.769
300	.957	.916	.840
400	.968	.936	.875

and his calculation will be given in more detail elsewhere. In this, $k_s + k_f = k$ is the multiplication constant; k_s is the number of neutrons produced per slow neutron absorbed, multiplied with the fraction p of neutrons which are thermal when absorbed; k_f is the number of neutrons produced per fast neutron absorbed, multiplied with the probability $1 - p$ of resonance absorption. In our case $k_s = p f \eta_s \approx p \eta$; $k_f = (1-p)\eta_f$. The results with Fermi's theory are given in Table III.

Table III

	$-10^3 L$ in cm^{-2}			R in cm			Volume in liters			23 in kg		
g	$\alpha = 1$	2	4	1	2	4	1	2	4	1	2	4
100	5.63	5.83	6.19	41.8	41.1	39.9	162	152	136	1.62	1.52	1.36
200	4.53	4.66	4.95	46.6	45.9	44.6	241	228	216	1.20	1.14	1.08
300	3.79	3.90	4.12	50.9	50.3	48.9	332	317	291	1.11	1.06	.97
400	3.30	3.39	3.57	54.8	54.1	52.6	430	411	372	1.07	1.03	.93

It was assumed in the calculation that the age τ of fission neutrons in heavy water is 115 cm^2, the diffusion mean free path 2.4 cm. This gave 50, 100, 150 and 200 cm^2 for the square of the thermal diffusion length $1/\kappa_t^2$ for the four different g. The radii R of a bare critical sphere are also given in Table III. According to calculations of Mrs. Monk and Mr. Friedman, which also will be reported elsewhere, it seems reasonable to subtract 8 cm from the radius in case of a ThO$_2$ slurry reflector. This gives the amounts of 23 necessary for the pile which are given in Table III in kg.

The advantage of using high concentration lies in the smaller losses of neutrons to the moderator, etc. On the other hand, a low concentration of 23 would give increased bulk to the system and thus facilitates a higher production rate. If α should be large there is an increased advantage in using high concentrations because the neutron losses to the moderator are further reduced by the resonance absorption. It is seen from Table III that the holdup of 23 within the pile is quite small. There is of course a holdup outside the pile.

If it should be necessary to use the cylindrical arrangement given in Fig. 2, the height of the cylinder will have to be chosen in such a way that the escape toward the two ends of the cylinder be about 1%. This gives the following equation for the height H of the cylinder in terms of its radius R

$$\frac{\pi^2}{H^2} = \frac{1}{100}\left(\frac{2.405}{R}\right)^2.$$

Calculating the volume from this equation it appears that it is 4.3 times greater than the volume of the sphere considered in Table III. This would necessitate a four times greater holdup in 23 which can be reduced, however, by reducing the diameter of the tubes containing the 23 solution toward the two ends of the pile.

The lattice spacing is not a critical quantity in the second arrangement. Of course, the total amounts of heavy water, beryllium, and solvent must be kept in such limits as can be obtained from Table I. In addition, one will not want the neutron density in the moderator to exceed substantially the neutron density in the solution. If $\kappa r_0 < 1$, the neutron density at the tube wall will be less than 10% higher than the average neutron density in the tube. Since $\kappa = .2$ for both the Bi and the fluorocarbon solution of Table I, this only gives $r_0 < 5$ cm. The ratio of the neutron density difference between surface of cell and surface of tube, divided by the density at the tube is

$$\frac{N\sigma_{a0}r_0^2}{2D}\left(\frac{r_s^2}{r_s^2 - r_0^2}\ln\frac{r_s}{r_0} - \frac{1}{2}\right).$$

In this $N\sigma_{a0} = .048$ cm^{-1} for the solutions of Table I, $D = .8$ cm is one third of the diffusion mean free path, r_s is the cell radius and r_0 is the tube radius. The above ratio will be under the conditions of Table I, smaller than .1 if $r_0 < 3.5$ cm.

E. Holdup Outside the Pile

The holdup outside the pile depends to a large extent on the diameter of the tubes in which the solution passes through the heat exchanger, the temperature difference between coolant and the temperature of the 23 solution, holdup in the pumps and, particularly in case of the arrangement of Fig. 1, on the time necessary for degassing. No calculations are available on these points at present but one can estimate that the solution will spend about 2 sec outside the pile before it re-enters it. Under this assumption the power output depends solely on the holdup and the temperature difference by which the solution is cooled down. It is given by

$$P = \frac{C_s T}{t}\frac{\text{kW}}{\text{kg holdup}}.$$

In this C_s is the heat capacity of the solvent per gram of 23 dissolved in it (cf. Table I), t is the time of holdup outside the pile, and T the temperature difference between the solution as it leaves and re-enters the pile. In case of pile of Fig. 1, $t = 2$ sec, $T = 50°C$, $g = 100$, this gives $P = 10,000$ kW/kg; for $g = 200$, $P = 20,000$ kW/kg.

One may consider flashing the liquid in order to achieve some cooling by evaporation as well as elimination of the fission gases. Messrs. Newel and Ginns proposed in some other connection to let the solution flow down on the inside of a cooled tube in a thin sheet and one could maintain sufficiently low pressure inside the tubes to permit evaporation and release of the bulk of the decomposition gases. These would carry the Xe and some other fission products with themselves and also a considerable amount of heat in the form of heat of vaporization. The advantage of this arrangement is that it unites the elimination of the decomposition gases with some cooling and also that the pressure drop in such a system is quite small. Considerations have not progressed sufficiently to know how large the saving in the holdup would be if this arrangement is used.

A possibility which might be applied to the arrangement of Fig. 2 is that of pulsating it. This avoids the holdup in the pump and also the need of pumping the radioactive solution. Figures for the theoretical efficiency of this cooling are given in Table IV in kW per gm holdup outside the pile. It was assumed that the temperature difference between coolant and 23 solution of the pile is 50° and that the temperature difference between the coolant and the tube of the heat exchanger is made negligible by vigorous circulation of the coolant and finning of the heat exchanger tubes. The pressure difference between the active pile and the end of the tubes alternates between .6 and −.6 atmospheres in such a way as to give a velocity of ≈ 700 cm/sec in the tubes. The holdup in the degasser is not taken into account in the figures in Table IV.

Table IV

Radius of tubes	Length of tubes	Length of cycle	kW/gr $g = 100$	kW/gr $g = 200$
1 cm	100 cm	.45 sec	37.5 (26)	75 (52)
1 cm	200 cm	.90 sec	31 (25)	62 (50)
.5 cm	50 cm	.23	82 (44)	164 (88)
.5 cm	100 cm	.45	65.5 (44)	131 (88)
.5 cm	200 cm	.90	45 (37)	90 (74)

The figures given for the power assume that the pressure adjustment at the end of the tubes can be carried out instantaneously. The bracketed figures assume an adjustment time of .1 sec. It was further assumed that the tubes can be entirely emptied in the "in" cycle. Assuming that the meniscus has to stay at least 10 cm away from the pile tank further reduces the power output by 20% to 5% for tube lengths of 50 cm to 200 cm. The advantage of longer tubes is, obviously, that they permit a greater total production, although with a somewhat decreasing efficiency. The advantage of wide tubes is the same and a greater mechanical simplicity.

F. Estimate of the Losses

It will be seen that, on the whole, the losses from the different sources enumerated under A become smaller as the production rate of the pile is diminished. It appears quite natural that the losses in a pile with a high power output are greater than if all attention can be given to a greater efficiency. We proceed with the discussion of the eleven separate points given under A.

(1) The number of delayed neutrons is about .8% in 25. Although its amount is not known for 23, Anderson found indications that it is greater for this element. As pointed out under section A, a loss of 1% of neutrons decreases the efficiency by 2.37%. Since all the delayed periods, except perhaps the first one, are long compared with the time of uninterrupted sojourn of the 23 within the pile, the number of delayed neutrons emitted outside the pile will be equal to the holdup outside the pile divided by the total amount of 23 present. From this point of view it is indeed advantageous to have a large pile and a very small holdup outside.

The flash cooling system may form an exception under this rule because it may be possible to arrange it in such a way that most of the neutrons still would be utilized. Similarly it may be possible to achieve this at least partially in the pulsating system. In spite of this there will be a loss of efficiency from delayed neutrons in every system in which the uranium is circulated at the outside. This constitutes a rather serious drawback of these systems since the loss in efficiency from this source is not very much smaller than that caused by the Xe^{135} would be.

(2) The absorption by the moderator has been discussed in detail before. It will amount to about 1% for the pile of Fig. 1 and one will be quite lucky if it is as little as 2% in the second case. However, in the first case the absorption by the anion also has to be taken into consideration. If the uranium is used in the form of a nitrate the nitrogen of the nitrate alone will give a loss of $\frac{1}{2}$%. In addition to this it appears likely that further nitric acid has to be added to the system in order to keep the uranium in solution in spite of the presence of the hydrogen peroxide. The situation would be very much better if N^{15} were available for use both in the nitrate and the nitric acid.

The reason that we gave the nitrate precedence over other salts is that we expect it to reduce the corrosion of the walls and also because it does not give products under radiation which may gum the system. The sulfate which has negligible absorption may precipitate sulphur although it is not clear that this cannot be avoided by the addition of peroxide. The parasitic absorption of the fluoride, the use of which was propsed by Allen, is also negligible but there is a strong opinion (Hiskey) that it will be difficult to avoid precipitation of the peroxide at such concentrations of the hydrofluoric acid which are tolerable from the point of view of corrosion. If the uranium is used in the form of a solution in Bi or fluorocarbons it may be necessary to eliminate the products of nuclear reactions of these substances along with the fission products.

(3) The estimate of the poison by the fission products is probably the most difficult at present. According to Miss Way only the Sm is a serious poison

among the stable fission products. Inasmuch as the yield of this is probably 1.6% its elimination is very important and may demand an average interval of one day between purifications. In order that this be permissible, it is necessary that the purification be carried out with a loss of about .1% of the 23.

The above applies to the stable fission products. There may be, however, numerous radioactive fission products (in addition to the Xe^{135}) which have large cross sections. Their effect has been estimated by Miss Way on the basis of a statistics of the cross sections and she finds that those with an odd number of neutrons are most dangerous. All elements may give substantial contributions, the lifetime of which is an hour or more. Only those with a lifetime of more than a day have been surveyed to date and Miss Way estimates that they may cause a loss in efficiency of $3\frac{1}{2}$% if the purification is carried out once a day. If this is done it is unlikely that the corrosion products will contribute much to the poisoning.

Probably the most efficient way of eliminating the fission products is by the solvent extraction method. We are much indebted to Mr. Tepe for discussing this system with us. The solvent extraction method would permit a continuous operation and cause a holdup of only one hour in the column. As a result it would appear that such a purification increases the amount of 23 required by only 4%. However, it is likely that it will be impossible to use the solution of uranium in heavy water directly in the column because the deuterium will exchange with the hydrogen of the solvent. For this reason it will be necessary to separate first the uranium salt from the heavy water which may be a more time consuming operation than the operation of the column itself and consequently increase the holdup more than the fission product separation itself does.

If the uranium is used in a solution in Bi the problem of chemical separation is even graver. Although many methods were discussed no really promising one turned up so far. One must remember that after the chemical separation has been carried out the uranium must presumably be reduced again to the metallic form. It is unlikely that this operation can be carried out with as small a loss as would be necessary for the successful operation of the breeder. This is the most serious drawback of the Bi system that has turned up yet. On the other hand, if the uranium is dissolved as a hexafluoride in fluorocarbons the problem of separation may be much smaller. As pointed out by Brown and Anderson, distillation may lead to the goal and the extraction column also could be used because reconversion to the fluoride can probably be handled with relatively great efficiency.

The Xe^{135} occupies a particular position. Let us first consider the case in which the uranium is dissolved in heavy water. In this case there is a considerable amount of water decomposition and the hydrogen and oxygen generated must be eliminated from the system every second. Even so, the equilibrium pressure of H_2 and O_2 will amount to a dozen atmospheres. During the degassing the hydrogen and oxygen will leave the water either in form of bubbles or through the surface. In both cases the fraction of Xe eliminated will be just about as great as the fraction of hydrogen and oxygen eliminated because the solubility of Xe is only insignificantly higher than the solubility of hydrogen and oxygen and because the diffusion of Xe to the surfaces will not be much slower than the diffusion of the oxygen. As a result, the amount of Xe will be reduced by a factor of the order of $\frac{1}{2}$ every second and the amount present will be that generated during a few seconds only. Since it is necessary to eliminate the Xe only about every couple of minutes, it is evident that the need of elimination of the hydrogen and oxygen will cause in this system an about 100 times better elimination of the Xe than is necessary. Even if it should prove feasible to reduce the gas evolution to 1/10 or even 1/30 by introducing some suitable catalyzer into the system which recombines most of the hydrogen and oxygen in the water the elimination of the Xe would be quite satisfactory. For a quantitative assessment of the problem let us assume that the mean life of the 23 in the pile is $\bar{t}$. This is the amount of 23 present, divided by the consumption rate and is for 100,000 kW power rate of the order of 6 kg/100 gr/day=60 days. Assuming a fission yield of 5% for the Xe^{135}, the production rate of Xe^{135} will be $.05\ N/\bar{t}$ where N is the number of 23 atoms. The total number of Xe^{135} atoms in the system will be $.05\ Nt_e/\bar{t}$ where t_e is the average time after which the Xe is eliminated. The poisoning due to this amount is $.05\ Nt_e\sigma_{Xe}/\bar{t}N\sigma_u$ and this will be less than 1% if

$$t_e < .2\frac{\sigma_u}{\sigma_{Xe}}\bar{t} = 4 \times 10^{-5}\,\bar{t}.$$

For $\bar{t} = 60$ days this gives $t_e \approx 3$ min.

The situation is somewhat different if the uranium is dissolved in bismuth or a fluorocarbon. In the former case at least it will be necessary to sweep out the Xe by bubbling some gas through the bismuth periodically. This has to be done, as pointed out before, every five minutes at least and will cause some additional holdup, the magnitude of which will depend on the size of the bubbles sweeping through the liquid and the area over which the solution is spread out during this operation. It does not appear, however, that the increase in holdup necessitated by this operation will go beyond a few percent.

(4) The absorption by the tank is principally determined by the material of the tank. If this absorption is low it will be quite possible to have the slurry of thorium which is outside the tank in immediate contact with the tank. Mr. Friedman and Mrs. Monk have calculated this effect. In the region of interest, their results can be approximated by the formula

$$\frac{\sigma_{at}l_t}{\sigma_{as}\sqrt{\tau}}$$

for the fraction of neutrons entering the reflector which are absorbed by the tank. In this, σ_{at} is the absorption cross section of the tank per cm^3, l_t the thickness of the tank σ_{as} the absorption cross section of the slurry per cm^3, $\tau = 115\ cm^2$ is the "age" in the reflectors. We assume a slurry which contains 1 gr Th per cm^3 which gives $\sigma_{as} = .016$.

The loss of efficiency is obtained by multiplying the above expression by 1.37. It is only 1% for 1.5 mm thick Al tank ($\sigma_{at} = .014$) and correspondingly less for a tank out of lead or beryllium or for some backing of the aluminium made out of beryllium.

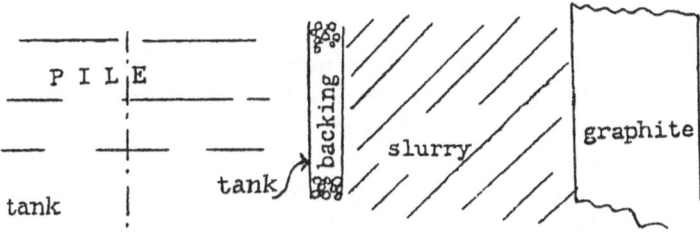

Fig. 3

If the tank has to be made out of a strongly absorbing material, it will be necessary to back it up with a layer of thorium in a more highly concentrated form (cf. Fig. 3). Beebees of metallic thorium which are cooled by heavy water circulated through them may be a suitable form. This arrangement has two effects: first the thorium reduces the density of the thermal neutrons in the neighborhood of the tank to such a level that the ratio of the thermal neutrons which are absorbed by the tank to the number of thermal neutrons which leave the pile toward the reflector is given by the formula

$$\sqrt{\frac{3\sigma_b}{\sigma_{ab}}}\,\sigma_{at}l_t\,.$$

In this σ_b is the scattering cross section in the backing per cm^3 of the backing, σ_{ab} is the absorption cross section of it, again per cm^3. One calculates that if the backing consists of 80% metallic Th and 20 volume percent of heavy water ($\sigma_{ab} = .13$, $\sigma_b = .33$) the absorption of thermal neutrons by a $\frac{1}{2}$ mm steel tank ($\sigma_{ab} = .22$) would give a loss in efficiency of about 1.5% if one assumes that $\frac{1}{2}$ of all neutrons escape as thermal. A reinforcing of the tank by a few mm aluminium would hardly add anything to this. Behind the backing there must be of course further heavy water containing thorium to slow down and absorb the neutrons which leave the tank at higher energies. It is easy to see, along the same lines as used in the argument by Friedman and Monk, that the fraction

$$\frac{\kappa_b\sigma_{at}l_t/\sqrt{\tau}}{\kappa_b\sigma_{as}Ch\kappa_bl_b + \kappa_s\sigma_{ab}Sh\kappa_bl_b}$$

of these neutrons will be absorbed by the tank. In this formula, as before, s and b refer to slurry and backing, σ_a, κ and l are absorption cross section per cm^3, reciprocal diffusion length and thickness of the corresponding region. For the above backing, containing 80 volume percent Th, 20 D_2O, $\kappa_b \sim .35$, and the above slurry, containing 1 gr/cm^3 Th $\kappa_s = .14$. As a result, if the backing has a thickness of about 5 or 6 cm the absorption of these neutrons is also quite small ($< \frac{1}{2}\%$). Evidently the introduction of the backing will somewhat reduce the efficiency of the reflector. However, the 8 cm thickness which was used in the previous calculations already takes this effect into consideration. Another point that might be mentioned is that the resonance absorption given in Table

II also will serve to reduce the losses due to the tank by the factors given in that Table.

(5) The shape of the tank as given in Fig. 1 or 2 is already designed to reduce the escape from the tank in the directions in which there is no reflector and it is sufficient to repeat that the extremely elongated shape given in Fig. 2 may not be necessary if the tubes toward the two ends of the tank are made thinner than at the center, thus reducing the multiplication constant in that region. This will, of course, cause a slight increase in the diameter of the tank but will still very substantially reduce the amount of uranium in the pile.

(6) The effect of the absorption by the 24 can be considered under the assumption that the 24 itself is not fissionable. In this case every atom 24 will eventually be transformed into 25 so that in equilibrium there appears to be a loss of 12% in efficiency due to the absorption by the 24 because in equilibrium, .12 as many neutrons are absorbed by 24 than by the 23. Of course if the cross section of the 24 is very small this equilibrium will be reached only after a very long time and until this happens the loss in efficiency is correspondingly smaller. However, even when equilibrium has been reached the loss in efficiency will be very much smaller than the 12% quoted because the absorption leads to a fissionable element 25 and the loss in efficiency is caused only by the fact that the 25 has a smaller η than the 23. When the absorption of 25 has come to an equilibrium one can consider the total effect in the following way. A 25 is formed by the absorption of two neutrons from the 24 and gives η_{25} new neutrons. If we denote, as in Section A, the fraction of neutrons which survive all losses by $1 - p$, the loss due to the absorption of 24 will be given at equilibrium by $2 \times .12 - .12\eta_{25}(1 - p) = .25p - .01$. One sees that this absorption will not cause a considerable decrease in efficiency. Strictly speaking, one should further consider the formation of 26 from the 25, but this evidently will not cause any major effect.

(7) The absorption by the Pa is a more serious effect. This element is, according to Miss Way, probably not fissionable and will therefore cause a loss in two ways. First of all, by the neutrons which it absorbs and second by forming an element of a relatively short half-life which decays into 24 instead of the fissionable 23. For this reason it will be necessary to keep the absorption by the Pa at a minimum. This can be done by extracting the Pa at sufficiently frequent intervals so as to reduce the absorption of neutrons by this element to about $\frac{1}{2}$% of the absorption by the thorium. This will be relatively easy if the absorption cross section of Pa is of the same order as the absorption of the thorium and if all the thorium can be used in the form of a slurry, since for a power output of 100,000 kW only about 120 gms of Pa are formed per day and since the mean life of Pa is 40 days, there will be, in steady state, 4.8 kg Pa in the reflector. Since the total amount of Th in the reflector is more than a ton, it will be sufficient to extract the Pa only n times a month if its cross section is n times that of the thorium. However, if it is necessary to use part of the thorium in form of a backing of the tank, a considerable fraction (about 25%) of all the neutrons will be absorbed by the first few cm of this backing. As a result, the Pa accumulates in this part of the thorium much more rapidly than in the bulk

of it and the purification would have to be undertaken more frequently. This is especially disconcerting because the removal and replacement of this backing is a somewhat awkward problem.

(8) Relatively frequent purification of the thorium seems also indicated by the accumulation of fission and corrosion products. The frequency of the need of this purification can be estimated in the following way: The absorption cross section of thorium is just about 100 times smaller than the absorption cross section of the 23. Therefore, if we divide the amount of thorium in the reflector by 100 times the amount of 23 in the pile, we obtain the number by which the cross section of the thorium in the reflector is larger than the cross section of 23 in the pile. This number is at least 4 to 5. On the other hand, the amount of fission in the reflector will be less than 1% of the amount of fission in the pile if all the thorium is distributed uniformly in the reflector. The relative poisoning of the reflector by the fission products will be about 400 times smaller than the poisoning of the pile and a correspondingly less frequent purification would be quite sufficient for the elimination of the fission products. It is seen from this that the need for purifying the thorium of the reflector will arise mainly from the absorption by the Pa, the absorption by the corrosion products, if any, and to reduce the holdup of 23 within the reflector. This last factor will demand a purification of at least once a month because after this time there will be about 3 kg of 23 in the reflector. Of course a more frequent purification may be demanded either because of the absorption of the Pa or the absorption by the corrosion products but we shall see that this does not seem to be the case.

The above discussion assumes that the thorium is uniformly distributed within the reflector but the situation is not very different if a backing of the tank is needed.

(9) As mentioned before, the absorption cross section of thorium is just about 100 times smaller than the absorption cross section of 23. Therefore, if we want to reduce the losses due to the moderator in the reflector to 1% of all neutrons absorbed in the reflector – this will give a loss of 1.37% in efficiency – the volumes associated with 1 gm of thorium are 100 times smaller than the figures given in the fourth column of Table I. It appears from this that only D_2O can serve as a carrier of the thorium in the reflector. Using the thorium in form of a slurry has the further advantage that a frequent mixing of the whole amount of thorium can be achieved and it will not be necessary to remove parts of the thorium at more frequent intervals.

(10) There are three causes for losses of neutrons from the reflector: The first one is the escape of neutrons toward the ends of the pile in case the arrangement of Fig. 2 is used. This, however, will be a very small quantity if the loss from the pile in the same directions is also very small. The escape of neutrons perpendicular to the main surface of the pile is governed by the tail of the slowing down density of neutrons. If one could adopt for this tail the Gaussian expression it would be sufficient to use such a thickness l_s that $l_s^2/4\tau = 4$ which would give $l_s = 4\sqrt{\tau} = 45$ cm. Actually it may be safer to surround the reflector by a second reflector which may be out of graphite.

In this case one can reduce the thickness of the reflector which contains the thorium to about 38 cm.

This figure is arrived at in the following way. The number of neutrons absorbed in the reflector divided by the number of neutrons entering the graphite reflector is approximately

$$\frac{4}{\sqrt{3}} \sqrt{\frac{\sigma_{ag}}{\sigma_g}}.$$

In this, σ_{ag}/σ_g is the ratio of absorption and scattering cross section in C. This formula assumes that all the neutrons entering the graphite are thermal which is a slightly pessimistic assumption. The fraction of neutrons entering the graphite reflector is given by

$$\frac{3\sqrt{\tau}\sigma'_s}{2\sqrt{\pi}} e^{-l_s^2/4\tau} - \frac{3l_s\sigma'_s}{2\sqrt{\pi}} \int_{l_s/2\sqrt{\tau}}^{\infty} e^{-y^2}\, dy + \frac{e^{-l_s^2/4\tau}}{4\sqrt{\pi}\tau\sigma_{as}}.$$

$\sigma'_s \approx .4$ cm^{-1} is an average fast neutron cross section of the slurry per cm^3. The first two terms refer to the fast neutrons entering the graphite, the last term to the slow neutrons entering it. The length, 38 cm, is so chosen as to reduce the total number of neutrons escaping from the system to less than .5% of all neutrons entering the thorium reflector. The above calculation should be somewhat improved to take into account the neutrons generated by fission in thorium. However, the number of these neutrons is quite small, less than 1% of all the neutrons, and the probability of their escape remains small even if they are found somewhat to the inside of the thorium reflector.

(11) The losses in chemical separation should really be considered in conjunction with the poisoning problem. Evidently the smaller losses the chemical separation involves, the more frequently it is worth while to carry out such a separation. The figures given under (3) were not calculated but estimated on the assumption that the loss of 23 during a chemical cycle amounts to .1%.

The total loss in efficiency therefore can be estimated as follows:

		System I (no power)	System II (some power)
1.	Delayed neutrons	2	2
2.	Parasitic absorption in pile	1.5	2.5
3.	and 11. Fission and corrosion products, chemical separation	3.5	5
4.	Tank	1	1
5.	Parasitic escape	.5	1
6.	24 absorption	1.5	1.5
7.	and 8. 13 absorption	1.5	1.5
9.	Parasitic absorption in reflector	1	1
10.	Parasitic escape	1	1.5
		13.5%	17%

It is unnecessary to remark that these figures give, in the best case, an orientation and that the chances are that they are optimistic. The losses which were enumerated are probably all present and it is unlikely that they are greatly overestimated in the systems considered. However, there may be other losses which we failed to recognize. In spite of this, it appears, at present, that the efficiency losses can well be held below the 37% figure which is the margin if Anderson and May's η for 23 proves accurate.

20.

New Ideas for Nuclear Reactors*

E. P. Wigner

Nuclear Science and Engineering 6, 420–432 (1959)

Received September 10, 1959

Early nuclear power reactor concepts were generated mainly by a few theoretical physicists during the relatively few quiet hours available in 1942–1945. Reactors were thought of in terms of structures "that a plumber could put together." A typical illustration of early thinking is presented. In spite of these primitive beginnings, the basic technical pattern of power reactor development was understood at an early date.

Estimates of world energy resources are summarized. It is pointed out that the great effort to exploit nuclear energy can be justified only if it is directed toward a full utilization of uranium and thorium. Without breeding, nuclear fuels will only supply energy for a few decades in the future energy-hungry world.

Recent findings on nuclear reactor stability are discussed and the value of computing machines in theoretical studies is noted. A commentary on current concepts in reactors is presented; the flux trap reactor, boiling reactors, gas-cooling, and breeders using beryllium or U^{238} for possible improvement of neutron economy.

Direct conversion of nuclear heat into electricity is briefly reviewed. Thermocouple batteries and thermionic converters are described and some remarks are made concerning their future development.

Many of you know that your speaker tonight is an old-timer in the field of nuclear energy. Now, an old-timer is a man who was in a field when the field was still undiscovered by most so that he was, comparatively speaking, isolated and lonesome in that field. He feels a bit baffled and uneasy now when he looks around and realizes his success in having made his field popular, and realizes even more keenly his failure to keep familiar with the events in his field. When the first newcomer arrived, and wrote a report, he said, "Now, this Mr. Read, he is evidently a very bright person, but if you look at the report that my friend Y wrote three years ago, you will find that he anticipated Mr. Read in almost everything." When the next report, by Mr. Yellow, came out, he said, "Now this Mr. Yellow, he impresses me very much, but if you read my report of last year, you will realize that I must have known all that he says." At the news of Mr. Greene's ideas he says, "Now this Mr. Greene, I do not know how bright he is, but he does have a good point there. I do not know why we did not think of it." A year or

* First Annual Distinguished Lecturer's address, presented to the Annual Meeting of the American Nuclear Society, Gatlinburg, Tennessee, June 15, 1959.

so later, his comment on Mr. Blue's report is, "Well, this Mr. Blue may not be so bright but he surely noticed something that nobody appreciated before." But finally, when the old-timer looked at Violet's report, he admits, "I just can't read or understand all this work that goes on here—it is just too much for me."

This is the state that I am in now but I will try, nevertheless, to review the very little that I do understand of the progress of the last few years, which seems to me to be truly remarkable, striking, useful, or interesting. I do not claim familiarity with all the amazing developments that took place in this country or abroad and do not even claim that all that is known to me will be presented. Much of what you will hear will not be more than the rambling of an old-timer, but let me start on it anyway.

The most striking difference between the work of the old days, and the present work, is that the old work was carried out on a shoestring. In the days of the Manhattan Project, there was only one group of people seriously concerned with the future of nuclear power as we now understand this term. This group consisted, at its height, of about 15 to 25 people. Virtually no experimental facilities were at their disposal to help them in thinking about projects of the future, though, occasionally, they could stimulate some experiment which did not interfere with the urgent and necessary work, to have their accepted dream, the Hanford reactors, realized. The group consisted mainly of theoretical physicists and was anxious all the time to find some engineering talent, that is, a couple of men with some experience in actually building something. This, they could not quite attain. It is small wonder, therefore, that the ideas of this group went into directions which did not require much more than everyday experience. As Dr. Znn put it, they thought of reactors which they believed that a plumber could put together. I was a member of this group, shared its dreams and was party to its errors. Among other members of the group, surely you are familiar with our President for the next year, Dr. Weinberg, and with Gale Young. You probably also know Szilard, who contributed more ideas to the early thinking on reactors than anybody else.

One of our principal difficulties was that we did not know how to cope with instabilities. It was almost axiomatic for us that instabilities cannot be adequately discussed, and those of you who followed the thermonuclear work realize that, *in abstracto*, we were not altogether wrong. However, many of our fears of dangerous instability eventually proved to have been exaggerated. We included under instabilities, quite properly, what some people call superstabilities, that is, the case when the restoring force is so great that it leads to oscillations, with increasing amplitude. In particular, we never dared to imagine something to boil in a reactor.

A second difficulty which we had, and which is still with us today, is that we did not know enough about materials and their compatibilities. In this regard, we were inclined to adopt the opposite attitude: we felt that materials suitable to our purposes will eventually be found. Finally, as to available machinery, such as pumps, valves, and similar equipment, we were quite intimidated by the very limited contacts which we had with engineers who understood the subject. Even they committed errors occasionally. Our diffidence in this regard was quite natural since it is difficult to imagine a theoretical physicist designing a usable pump.

In spite of all these handicaps, a good deal of useful thinking was done. Thus, the concept of the breeder was developed to the point that we could analyze the two types of breeders which are the principal types even today. Similarly, we were familiar with the majority, but only with the majority, of the concepts of burners which occupy the members

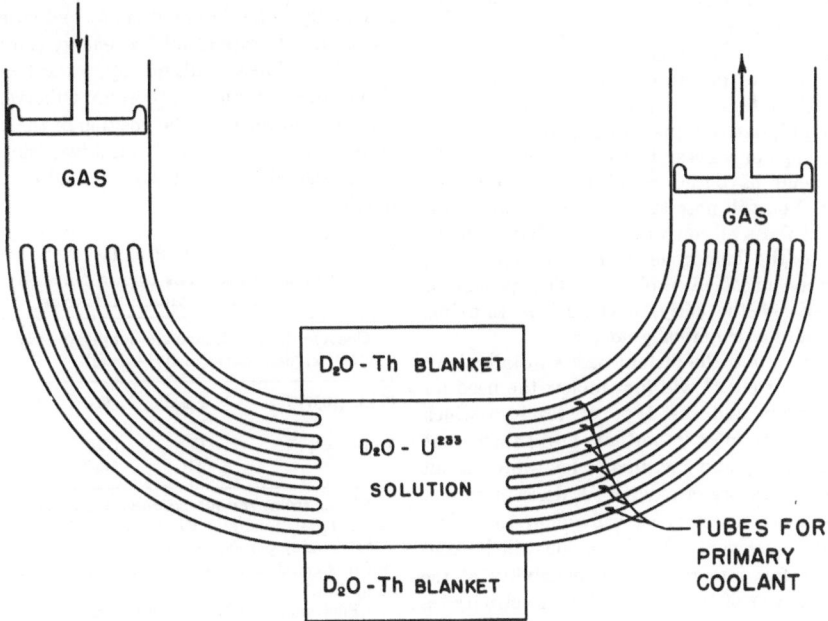

FIG. 1. Early design of a thermal breeder. See text for a more detailed description

of this group today. The situation to which I referred before, as being presented by Mr. Violet, or even by Mr. Blue, was not reached until about 5 years ago, but then it came with a "bang" and my principal subject tonight will be this "bang."

Figure 1 may illustrate the general character of our thinking. It shows one of the early designs for a thermal breeder. Admittedly, it is a somewhat weird design, much more weird than the majority of our ideas, but it appears to me characteristic of the strengths, and perhaps even more, of the weaknesses of the early thinking.

Evidently, Fig. 1 is not a very realistic picture because you have no trouble finding the reactor. The U shaped tube of the Figure has a diameter of about 100 cm and contains a solution of U^{233} (in the form of a salt) in heavy water. This solution is chain-reacting in the bottom part of the U-tube which does not contain cooling tubes. The total volume of the solution is about three times the volume of the chain reacting portion. At any rate, the solution does not fill the tube up to the pistons but has a free surface in both legs of the U-tube. The space between the free surface and the piston is filled with a gas (presumably helium) on both sides so that the piston does not come into contact with the solution. The purpose of the pistons is to move the solution alternately from left to right and back so that the liquid, which becomes hot as a result of the chain reaction, comes into contact with the cooling tubes. The upper part of the drawing is schematic but it does show the purpose of the arrangement: to move the highly radioactive liquid without having it pass through pumps or valves. The reactor of Fig. 1, as Dr. Zinn would have said, could be put together by a plumber. You will note that the control system is left out and those of you who are familiar with the problems of the homogeneous reactor will notice some further unsolved problems. The picture is characteristic of the early, overcautious thinking, greatly in need of empirical guidance.

Let me leave now the distant past and talk about the more recent one. I will first review the need for nuclear reactors and the type of reactors which appear to me the most important. I will then come to technical points and say, first, a few words about progress in the theory of reactors. A description of a few of the more recent types will follow then, first burners, then breeders. The discussion will be concluded by a few remarks on the possibility of the direct conversion of fission heat into electricity—a possibility of as yet not fully assured usefullness.

SIGNIFICANCE OF THE BREEDER

There is, in addition to the technical points which will be discussed later, a conceptual point the decisiveness of which emerged and was appreciated only in recent years. I am referring to the role of the energy resources in the national and international economy. The value of coal produced in 1956 in the United States was 2.65 billion dollars, the value of the crude oil and gas about 7.25 billion dollars. The total national income in the same year was 350 billion dollars, so that the value of coal and oil produced was not quite 3 % of the total gross income. Clearly, a reduction of these 3 % by a factor of ten would not greatly increase the national well-being but an increase by a similar factor is something that we have to guard against strenuously. The energy resources are shown in Table I. This shows that if the per capita energy consumption in the whole world were to reach but not to exceed the present per capita energy consumption in the United States— this is the basis for the numbers in the Table—then economic fossil fuels should last 950/12 = 80 years.

The numbers given for nuclear energy assume *complete* utilization of the fertile material, that is full breeding. Without this, the nuclear energy supply is much less than the energy stored in fossil fuels. This is a point which we should keep in mind constantly if we do not want to fool ourselves. The great effort to exploit nuclear energy can be justified only if it is directed, ultimately, toward a full utilization of the uranium and thorium. Otherwise, we may soon find ourselves in the position of the department store which started a city-wide advertising campaign for its shoe department when it had only two pairs in stock.

TABLE I[a]

ENERGY RESOURCES IN THE NATIONAL AND INTERNATIONAL ECONOMY

Consumption[b] (in 10^{16} kg cal/year)	Economic supply[c,d] (in 10^{16} kg cal)		Total supply[c] (in 10^{16} kg cal)	
U. S. (1957) 0.8	Fossil	950	Fossil	4900
World[b] 12	U	6300	U	43,000
	Th	1350	Th	8600

[a] Data from various sources, including AEC press release, "Fossil Fuel and Uranium and Thorium Resources."

[b] Assuming per capita energy consumption for the whole world is equal to the present U. S. per capita energy consumption.

[c] Fuel values of U and Th assume full breeding.

[d] Uranium ore reserves, U. S. 260 × 10^{16} kg cal.

It seems to me that we, technical people, should take this point very seriously. It is possible, of course, that our words will fall on deaf ears, but this does not absolve us from speaking. The data of Table I are actually optimistic in the sense that they assume that the only further increase in energy consumption will be an equalization of the per capita energy consumption throughout the world at the present U.S. level. This implies, in particular, that the consumption in the U.S. will not rise any further. Nor is the very probable increase in total population taken into account. It is very likely, therefore, that the figures for the energy consumption will undergo an upward revision.[1] As a result, the economic supply of fossil fuels will not last 80 years; shortages will appear sooner (1). Furthermore, the economic supply of uranium would provide, without breeding, only a relatively insignificant addition to the energy of fossil fuels. It may be even difficult to justify the extensive research and development effort that occupies some of our best engineers and scientists if it is directed only toward supplying us with energy for one or two decades. The same applies to the expenditures of money. Naturally, it is not possible to foresee the way which will lead to the practical breeder. It is quite possible that successive improvements on the neutron economy of one of the burners will lead to the breeder, rather than the direct approach of the breeder program. Be that as it may, we must not lose sight of the ultimate goal of utilizing essentially all uranium or thorium.

Let me just add a word of caution in regard to the accuracy of Table I. It seems that I prepare a table similar to Table I every two years and the numbers therein fluctuate by factors of the order of 2. In view of a lack of accurate data, this seems to be unavoidable. However, the inaccuracy of the data does not affect the conclusion concerning the insignificance of the "burners" from an over-all economic point of view and we should keep this point in mind throughout the rest of our discussion. The point was brought

[1] From a private industrial source I have recently received the results of a comprehensive survey and projection of energy requirements in the future, and the contribution nuclear energy may make in meeting them. These estimates included allowance for the probable increase in population and in energy requirement per person. They also allowed for the distribution between energy sources, including nuclear, from which the projected requirements would be satisfied. Although this study showed a somewhat lower percentage of nuclear contribution than most published surveys, the need for more rapid breeder development was indicated even more strongly than in the present paper.

out first by Gale Young, back in '43 or '44, and we have done far too little so far to hammer it in.

REACTOR THEORY

The first technical problem of a reactor is its criticality. It was possible to think, at one time, that the calculation of the nuclear behavior of a reactor had been reduced to standard operations. Right now, we do not know whether this is true. It is certain, however, that the standard operations of the present time are very different from the standard operations of five or ten years ago. The last five years or so brought an extension of the reactor theory to nonlinear cases, that is, to power densities high enough to affect the reactor's properties in periods comparable with the natural period of the reactor, that is comparable with the lifetimes of the delayed neutron emitters. If this situation prevails—and it was present in some reactors which were in operation more than ten years ago—the dynamic behavior of the reactor does not obey a linear equation any more. There are effects which are proportional to the square of the neutron density because they depend not only on the number of neutrons but also on the temperature increase created by these neutrons. A characteristic of such a behavior is the deviation from the exponential or sine-like time dependence of the neutron densities.

Figure 2 illustrates the situation. It shows the relative power as function of time, as calculated by Ergen and Weinberg, for a highly idealized reactor (2). You note that the oscillations have sine-like character only for very small amplitudes. At higher amplitudes, the maxima become high and narrow, the minima flat and long. Whether the reactor will remain intact under such oscillations depends on the maximum heat production which the fuel elements can stand and on the total amount of heat produced in the course of one excursion. Incidentally, the power-time curves of this diagram show a surprising similarity to the power *versus* time curves of the TREAT reactor. A short period of high-power density is followed by a longer period of low-power density—just the opposite of a desirable behavior, because the maximum power is very much higher than the average power. Figure 3 gives the relative power as function of the temperature. It is a so-called Poincaré diagram. We see the circles, characteristic of a sine-like time dependence, at low power, the strongly distorted curves at higher power.

In my opinion, the more detailed and more courageous investigation of the instabilities which are the objective of these inquiries into nonlinear reaction kinetics constitute the most important progress in reactor theory. Ergen and Weinberg, Lipkin, Nohel, and Brooks, Welton, and many others, have contributed to this field which is not only important but has also great mathematical beauty (3, 4, 5). The other striking change in reactor theory relates to the replacement of analytic and long-hand calculations by machine computations.

For a long time, many of us old-timers were inclined to scoff at the use of the computing machine. It has become clear, however, that the machines can serve not only to verify and polish up the approximate calculations which are based on direct

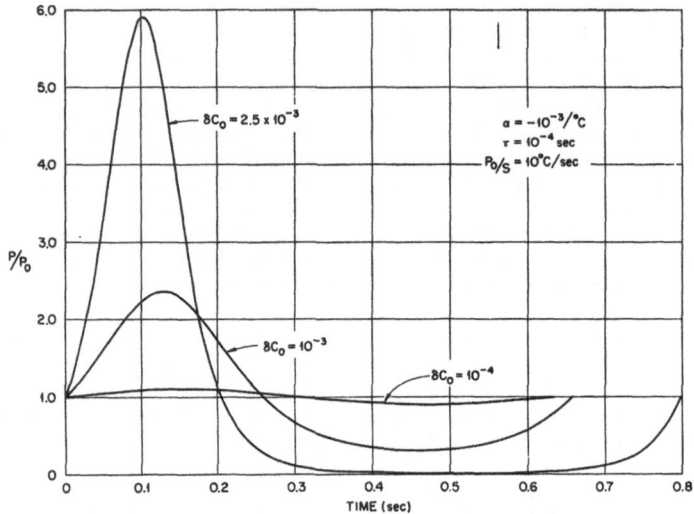

FIG. 2. Power *vs* time for a reactor with constant power extraction and various initial reactivity jumps. The effect of the delayed neutrons is disregarded. From W. K. Ergen and A. M. Weinberg, Physica, **20**, 413 (1954).

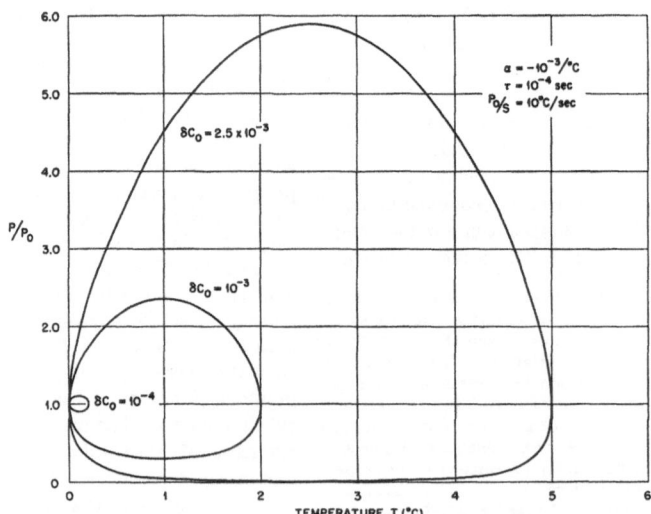

FIG. 3. Power *vs* temperature for the same systems as Fig. 2. From W. K. Ergen and A. M. Weinberg, Physica **20**, 413 (1954).

physical insight but, if used with understanding, can point to the existence of new effects or to the inadequacy of accepted approximations. Several examples could be quoted in which machine calculations have helped to correct or to improve the physical picture previously formed and thus sharpened our physical insight. I shall mention only a recent one: Tralli, Agresta, and Seibyl, when calculating the criticality of uranium-heavy water lattices, noticed that, at least in the case of large lattice constants, the production of resonance neutrons is not uniform over the cell. The reason for this is, unquestionably, that the fast neutrons, which are produced at the center of the cell, have not been distributed uniformly over the whole cell in the course of the moderation process. The possibility of such a situation has been considered repeatedly but has been shown not to prevail under the usual conditions (6). However, the low absorption cross section of heavy water makes it possible to use large amounts of it, i.e., a large lattice cell. The long distance from the cell center to the boundary, together with the relatively low age in heavy water, renders the spreading of the neutrons over the cell (during the moderation process) less complete than it is, for instance, in the usual graphite lattices. Hence, some further spreading takes place in the resonance region which causes a decrease of the flux with decreasing energy near the uranium. This is what the calculations show. The effect is somewhat obscured by the usual decrease of the flux with decreasing energy close to the uranium which is due to the resonance absorption itself. This is present in small cells as well as in large ones. However, whereas the former effect could also have been foreseen and taken into account, it actually was not appreciated until the machine calculation called attention to it. Whether or not the calculations of Tralli and collaborators offer the full explanation of the discrepancy between observed and calculated multiplication constants of heavy water lattices, a more thorough experimental and theoretical investigation of this discrepancy is most desirable.

TYPES OF REACTORS

Reactors have become highly specialized instruments, according to their functions. We have research reactors, burners, and breeders, and to these have been added, more recently, batteries and thermonuclear machines.

Even the special class of research reactors is undergoing a further specialization. We have shield-testing reactors, medical reactors, the Materials Test Reactor and its progeny and, of course, a great variety of training reactors. The high-flux reactors begin to separate into those with large high-energy flux and large thermal flux. It is in regard to these latter that the most striking idea is being actively pursued at present: that of the flux trap reactors.

With a little exaggeration, one could say that the idea of the flux trap reactor goes back to the ancient Romans. However, when it was first proposed, the need for it was not urgent enough and it was not seriously considered as a practical proposition until it was rediscovered, in 1956, by R. Stephenson, then at New York University. The underlying idea is to permit fast neutrons to diffuse into a medium in which the mean free path of thermal neutrons is so much smaller than that of fast neutrons that they remain "trapped" once they become thermal. If, furthermore, the absorption of the medium which traps the neutrons is very small, their density will grow until even this low absorption equals the diffusion of the fast neutrons into the trap. It follows that the trapping medium must have two properties: the diffusion coefficient, that is the mean free path, must decrease with decreasing energy, and the absorption cross section for thermal neutrons must be very small. Ordinary water satisfies the first criterion best, heavy water the second. The analyses at Oak Ridge and at the Argonne (7, 8, 9) show that, of the two, ordinary water forms the better trap but it is not quite clear that a suitable combination of the two is not better than either. Figure 4 was prepared by the Internuclear Company for Dr. Stephenson. It shows the fast, intermediate, and the thermal flux in a reactor which has a light water kernel, forming the trap, an active region, and a reflector. It shows that the density of thermal neutrons in the trap is several times greater than in the reactor. Hopefully, it will reach 5×10^{15} per cm^2 sec.

The burner in which the primary heat is used to boil water was invented at the Argonne National Laboratory and is new, by now, only for old-timers. It is commercially sponsored by the General Electric Company and is one of the most successful types of burners. The success of the organic moderated reactor, sponsored particularly by Atomics International, is more recent and as yet less complete. Nevertheless, it strikes me as a well-founded proposition, eliminating not only the instability, that is the tendency of "chugging," but also the high pressure of the water-cooled reactors. Its great drawback is, of course, the decomposition of the organic coolant

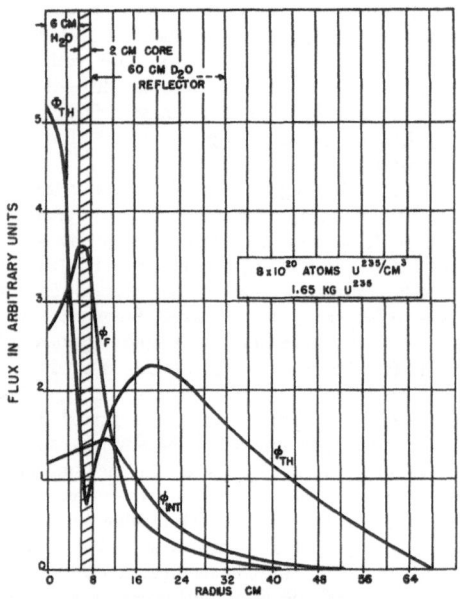

FIG. 4. Fast, intermediate, and thermal fluxes as function of position for flux-trap reactor. Note that the thermal flux in the water at the center is almost five times higher than the average thermal flux in the active portion. Calculations by Internuclear Co.

due to radiation. It would be, in my opinion, worth-while to learn to live with the decomposition or, rather, polymerization. The remarks which could be made about Aerojet's proposal of a liquid sulphur cooled reactor are quite similar.

The reappraisal of gases as heat transfer agents may become the most important development as far as burners are concerned. It is ironic that the pioneering work on gas-cooling was undertaken by the British, to whom the most suitable gas, helium, is not available. They have shown that we under-estimated the possibilities of gas cooling, that the heat transfer properties of gases, if under sufficient pressure, are higher than was realized before. It is true that the heat capacity of a unit gas volume is even under rather high pressure lower than the heat capacity of the same volume of liquid. There are, however, two compensating factors. The first of these is that the temperature range of gases is much higher than that of liquids, excepting liquid metals. Second, the lower heat capacity of gases is a direct consequence of their lower density. The lower density, however, decreases also the parasitic capture

by the gases so that one can afford a much larger volume of the gas-coolant in the reactor than of a similar liquid coolant.

The problems of gas cooling, which are many, appear to stimulate ingenious ideas for solving these problems. I am most familiar with the Oak Ridge-Kaiser and with the General Atomic designs. I understand that there is a similarity between this last and the British and also the Japanese designs, but certainly the Oak Ridge-Kaiser design, and the clever ideas incorporated in it, are very different from these.

Gas-cooling's principal advantage derives, in my opinion, from the chemical inertness of some of the gases. Helium, of course, is the prime example. The large temperature range under which gases can be used is a direct consequence of this inertness. The great disadvantage of gases is that they have to be used under high pressure. As one of the engineers admitted to me, although the high pressure does not present appreciable problems, the need to contain that pressure does. The mass of the coolant which has to pass through the reactor is given by the heat capacity of the coolant and the power output. The pumping power, for a given mass flow, is propor-tional to the square of the velocity, and hence in-versely proportional to the square of the density. Hence, the need for high pressures. The General Atomic design, for instance, stipulates 20 atmos pressure and even at this pressure, and a temperature rise of 400°C, the capacity of the gas to store heat is, per unit volume, more than fifty times smaller than the heat content of water at a temperature rise of 50°C. The effect of this low heat capacity is partially compensated by the larger volumes of coolant which one can use, but there is a limit for this also if one wants to keep the size of the reactor moderate and if one wants to prevent the escape of the fission products by a barrier which absorbs neutrons. The power for the auxiliary equipment, principally pumps, amounts to about 15% of the total power produced in the General Atomic design. Nevertheless, the total thermal efficiency is expected to reach about 35%. The figures for the Oak Ridge-Kaiser design, which does not use high enrichment, are, as a result, somewhat less favorable.

The gas-cooled reactors which have been proposed so far are all burners. This and the very severe competition from abroad are the principal drawbacks of this system so far as our national effort is con-cerned. As was mentioned before, in the long run, burners do not have the same economic promise as breeders. One will wonder, therefore, what advantages gas cooling can offer for the breeder program.[2] Since all absolutely tight barriers are also neutron absorbers, and since large gas volumes also entail large contact surfaces, one will have to compromise to a certain extent and reconcile ourselves to somewhat "dirty" operation, that is to the use of a radioactive coolant. Historically, the desire to avoid the radioactive contamination of the coolant was one of the prime reasons for the opposition of the old-timers to gas-cooling, at least as far as the first production reactors were concerned. Because of its high temperature possibilities—an advantage which one would not want to forego—the gas coolant which is much to be preferred to all others is helium; and it must be of very high purity. It may be possible to use graphite as moderator and overload the reactor with fuel to minimize the parasitic absorption by the moderator. Offhand, this appears preferable to using heavy water as moderator because the heavy water would have to be thermally insulated from the coolant. The fuel may be reasonably finely subdivided to provide a large surface. If it is, its heat conductivity need not be too high and the oxide or the carbide of U^{233} may be suitable. Since the fuel cannot be purified often enough, the parasitic absorption by Xe and Sm would have to be taken into the bargain. Nevertheless, the system appears at least worth investigating more closely and, before such an investigation is undertaken, appears to be rather simple and compact. All this may change, of course, at closer inspection. However, the short review given by the staff of the Oak Ridge National Laboratories does not point to any insurmountable obstacles (10).

We are now right in the midst of the discussion of breeders, and I should recall again the two old standbys, the fast breeder on the U^{238}-Pu cycle and the slow breeder on the Th-U^{233} cycle. This latter one has been beset lately by a number of engineering difficulties and it seems evident that it cannot be put together by a plumber. Nevertheless, the thermal breeder remains promising and there is little doubt in my mind that it can be made to work. This does not necessarily mean that it will be successful also economically. Whether or not it will be, may depend on the enthusiasm, skill, and ingenuity of those who are working on it at present and on the extent to which the chief advantage of the thermal

breeder, its high power level per unit mass of fissionable material inventory, can be realized.

There are two new ideas in the breeder program which are particularly worthy of attention. The first of these concerns the use of beryllium as moderator, as was first proposed, I believe, by Krasin et al. in Russia (11). It has been estimated that the (n,2n) reaction in Be may increase the total number of neutrons produced by fission[3] by a factor as high as 1.12. A more realistic estimate is, perhaps, 1.07. The proposition has two drawbacks; one economic and the other nuclear. The economic problem concerns the availability of beryllium. This seems questionable. The nuclear problem centers about the facts that the (n,α) reaction, which leads to a neutron loss, can be induced by neutrons above 0.695-Mev energy but the (n,2n) reaction, which leads to the increase in the neutron number, only by neutrons with 1.85 Mev or more. However, the high-energy neutrons can induce also fast fission in U^{238} and the question is, therefore, not really, "Shall we make use of the (n,2n) reaction in Be?" Instead, the proper question to be posed is, "Does the (n,2n) reaction in Be or the fast fission reaction in U^{238} yield more neutrons?"

The latter alternative was embraced by a proposal of the Nuclear Development Associates who want to breed plutonium in a thermal, or fast-thermal cycle. A crude sketch of the reactor is given in Figure 5. The reactor is heavy water moderated and has 12 composite fuel elements. The total radius of the core is 1 m; it is surrounded by a blanket of about 60 cm thickness, containing depleted uranium and heavy water. The blanket is expected to utilize the leakage neutrons and convert them into plutonium. Figure 6 shows the structure of the composite fuel elements. They have a diameter of about 37 cm, filled with niobium-jacketed uranium-plutonium rods of about 1 cm radius. The outermost layer of rods is only slightly enriched, the next layer somewhat more, the rest of the rods contain 10% fissionable material. The purpose of this arrangement is to have a uniform power density throughout the fuel elements but this could not be fully accomplished. The coolant is sodium and flows along the rods. There

---

[2] This question must have been in the minds of the staff of the Oak Ridge National Laboratory when they wrote their report ORNL 2510 (10).

[3] I am much indebted to W. Häfele for the remark that C. B. Mills and N. M. Smith (AECD-3973) gave already in 1951 a similarly high estimate for the contribution of the (n,2n) reaction to the breeding ratio. Dr. Häfele also made a very careful evaluation of this contribution which will appear as ORNL-2779. He further pointed to the adverse effect of the Li^6, formed in the (n,α) process, on the breeding ratio.

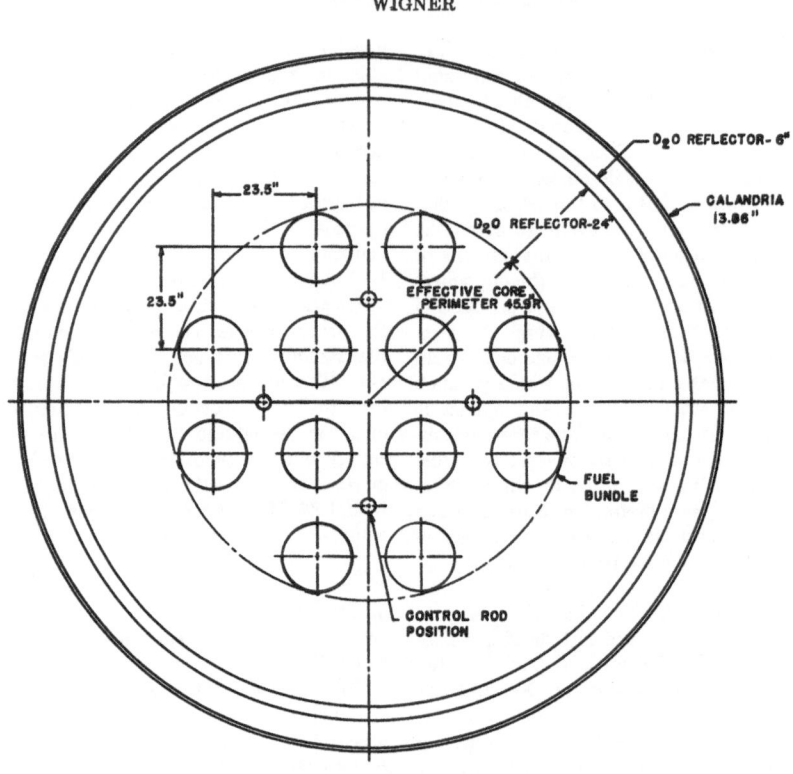

NDA 3-475

Fig. 5. Fast-thermal reactor. See text for a more detailed description

must be, of course, some insulation between the hot sodium and the cool moderator. The arrangement of the fuel element was chosen to make the fast effect as large as possible.

Table II shows the neutron economy. The value of η is 2.12. This is an average over the fast region, where η is high, over the thermal region, where it is low, and over the resonance region, where η is very low. The importance of the resonance region is decreased by the same self-shielding effect which reduces the resonance absorption of U^{238} in ordinary reactors. Nevertheless, the η given appears very high. It is further augmented by the fast effect which contributes 10 % to give a total theoretical breeding ratio of about 1.35. The parasitic losses are given to the right; they amount to about 17 %. The chemical losses are not estimated.

You will note that the breeding ratio of the mixed fast-thermal breeder is lower than that of the fast breeder, which may reach the value 2. The purpose of the thermal component of the system is to reduce the inventory of fissionable material. In this, the present design is not successful: it contains almost a ton of plutonium, for a total heat rating of about 200,000 kw.

DIRECT CONVERSION OF HEAT INTO ELECTRICITY

There is probably no reactor physicist who has not tried to invent a usable method for converting the energy liberated in fission directly and efficiently into electrical energy. There is certainly no reactor physicist who has succeeded.[4] However there are now several promising suggestions for the conversion of fission *heat* into electrical energy without using conventional engines. The proposals in this direction

[4] An analysis of the most direct method of converting the kinetic energy of the fission fragments into electrical energy (by letting the fragments run against a potential gradient) has been undertaken recently by A. Schock of the Fairchild Engine Division (ASTIA AD 216812).

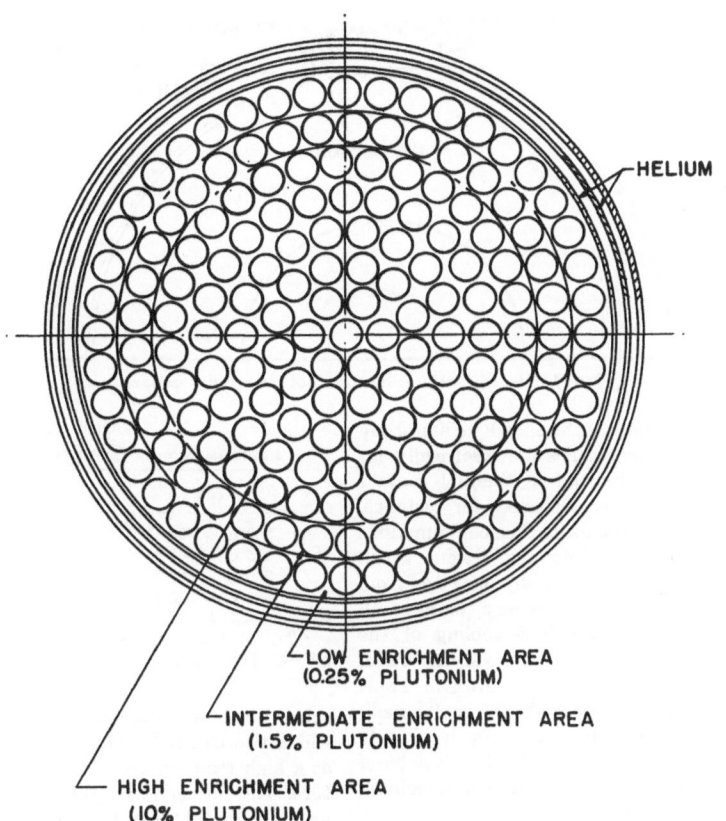

HELIUM

LOW ENRICHMENT AREA
(0.25% PLUTONIUM)

INTERMEDIATE ENRICHMENT AREA
(1.5% PLUTONIUM)

HIGH ENRICHMENT AREA
(10% PLUTONIUM)

FIG. 6. Structure of the composite fuel element of the reactor shown in Fig. 5

can be divided into two classes: the energy requirement is either so small that it would be too expensive to install a boiler-turbine or boiler-reciprocating engine system. Or, second, the energy requirements are so high that it is worth going to considerable trouble to simplify the system. The first type of systems are batteries with an energy output of 50 or 100 w. The second system has not been entirely invented.

Both systems are based on the phenomenon of the thermocouple; both are based on a new and more efficient couple than has been available until recently. Hence, one can say that the real discovery was made not in reactor engineering but in thermocouples. The hot junction of the couple is in thermal contact with a fuel element, the cold junction cooled as well as possible. But whereas 15 years ago the electromotive force was a few millivolts for a temperature difference of a hundred degrees, we now have

TABLE II

NEUTRON ECONOMY IN BREEDER OF FIGS. 5 AND 6

Neutron multiplication	Neutron losses	
$\bar{\eta} = 2.117$	Structural materials	0.081
$\epsilon = 1.104$	Moderator & coolant	0.033
$\bar{\eta}\epsilon - 1 = 1.348$	Xe, Sm	0.055
	Total parasitic capture	0.169
	Leakage	0.03
	Chemical losses	?

Breeding ratio $= 1.348 - 0.169 - x = 1.18 - x$.

thermocouples with much higher electromotive force *and* lower heat conductivity.

Figure 7 shows a particular battery. The core consists of some fissionable material; metallic beryllium is proposed as reflector. Although only two thermoelements are shown, actually about 180 are

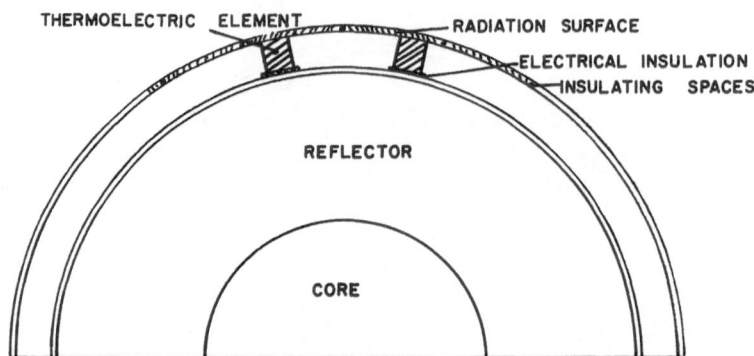

FIG. 7. Nuclear battery. See the text for a fuller explanation

disposed over the surface of the sphere as uniformly as possible. The thermocouples are electrically insulated from the reflector but in thermal contact therewith. This can be accomplished, for instance, by a thin sheet of beryllium oxide. The cold junctions are in contact with an aluminum sphere which also consists of 180 parts, electrically insulated from each other. It is possible, in this way, to have all 180 thermocouples in series. The cooling of the sphere is by radiation and hence not very effective. The performance of the system is shown in Table III. It gives 50 w at an emf of about 7 v. The efficiency is not very high, but efficiency is of only secondary importance in a battery. The weight of the system is about 40 lb, and it can remain in operation many years (12).

The thermionic converters form the beginnings of a much more ambitious program. Actually, the program is aimed at revolutionizing all conversion of heat into electricity, not only that of fission heat. The beginnings of the program again go back far in time, and it is not my intention to review it now. Two different approaches to the problem have been proposed. These two approaches have two features in common. First, one member of each thermoelectric junction is vacuum so that the junctions are not between two metals but between a metal and vacuum. Since the density of electrons in real vacuum is strictly limited by the space charge which they produce, the vacuum is replaced in both embodiments of the idea by a very low-density gas of positive cesium ions. Hence, the second point in common between the two systems is that the junctions are not between a metal and vacuum, as stated before, but between a metal and a low-pressure gas of cesium and cesium ions. Hence, both types of cells consist, essentially, of a diode, or a vacuum

TABLE III
PERFORMANCE OF BATTERY SHOWN IN FIG. 7[a]

T_{cold}	Efficiency (%)	Power (w)
260°C = 500°F	5	55
315°C = 600°F	3.9	60
370°C = 700°F	2.8	60
430°C = 800°F	1.8	55

[a] T_{hot} = 650°C = 1200°F; thermoelectric power 0.15 mV/°C. Total weight 40 lb.

tube, filled with low-density cesium vapor into which two electrodes protrude. One of these is kept at a high temperature, the other at a low temperature. The former forms the positive, the latter the negative pole of the cell.

The difference between the two embodiments concerns the relation of the mean free path of the electrons to the distance of the electrodes in the cell. In the modification which has been described in most detail by Hernquist, Kanefsky, and Norman[5] at RCA (13) [but was considered even before by Hatsopoulos, Kaye, Nottingham (14, 15, 18, 19) and by V. Wilson (16, 17), the vapor pressure of cesium is so low that the mean free path of the electrons exceeds the distance between the hot and cold electrodes in the tube. Electromotive forces up to 2.5 have been obtained from a single cell and thermal efficiencies up to 10 %. The electric power per cm² of hot surface was of the order of 10 to 20 w—a very respectable figure. The second approach has been described in most detail by Grover and his collaborators at Los Alamos (20, 21). In their system, the mean free path of the electrons is smaller than the

[5] The original suggestion seems to be due to W. Schlichter (Dissertation, Göttingen, 1915).

distance between the hot and the cold electrode in the vacuum tube. The vacuum tube actually was not a vacuum tube but had cesium vapor in it, up to a pressure 10^{-3} atmos or 1 mm Hg. In the system of Grover, the conditions at both hot and cold electrodes are close to thermal equilibrium and this system can be treated, at least approximately, by thermodynamic theory. This actually has been done a long time ago. No similar treatment is possible for the arrangement investigated by Hernquist and his collaborators because the conditions, in the neighborhood of the cold electrode at least, are far from equilibrium.

In first approximation, Grover's arrangement is a Carnot cell. We put in heat at the hot junction, the system rejects heat at the cold junction. The phenomenon of heat absorption (at the hot junction) and of heat liberation (at the cold junction) are called Peltier effect and are both proportional to the current passing through the junction. If the system formed a perfect Carnot engine, the electric energy of the system would be given by the Peltier heat absorbed by the hot junction, multiplied by the ratio of the temperature difference between hot and cold junctions to the absolute temperature of the hot junction. Unfortunately, the situation is not quite so favorable as this, principally because there is an unavoidable heat flow through the circuit from hot junction to cold junction, due to the heat conductivity of the material forming the circuit. There are also other complications. All in all, however, the thermoelectric power is proportional to the magnitude of the Peltier effect, that is proportional to the energy needed to transfer an electron from one side of the junction to the other. This energy is particularly large if one member of the junction is a metal, the other just empty space: it is of the order of a few volts so that the thermoelectric voltage will also be of the order of a few volts if the temperature difference between the two junctions is comparable with the absolute temperature of the hot junction. It is an order of magnitude greater than the thermoelectric potential of the best junction both sides of which are metals.

Figure 8 shows the experimental arrangement used by Grover, Roehling, Salmi and Pidd (20) to measure the characteristics of their cell. The purpose of the upper portion of the apparatus is only to heat the tantalum plate to a high temperature. Opposite this plate is a copper electrode which is oil-cooled. It is, of course, electrically insulated from the hot junction but has a lead to the outside. The cesium forms a small puddle at the bottom but its vapor

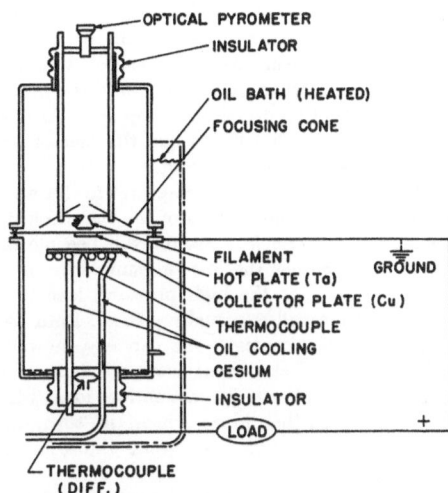

FIG. 8. Thermionic cell. The electrons emitted from the hot tantalum plate produce a potential difference of about 2.5 v. The upper portion of the figure shows the arrangement used in the particular experiment for heating the tantalum plate; the cell itself occupies the lower portion of the figure. The text gives a more detailed description of the cell.

fills the chamber. The hot junction is the tantalum surface, or, if one likes, the tantalum-cesium vapor interface. The cold junction is the copper-cesium vapor interface. The rejected heat is taken up by the oil cooling and is measured by the thermocouple at the bottom. The electric energy is obtained by connecting the hot junction with the lead from the cold junction through a resistance and measuring the current. The maximum electrical energy per square cm obtained was of the same order as in the arrangement investigated by Hernquist at RCA, that is about 10 w per cm^2. For helium at a pressure of 20 atmos as coolant, the *heat rating* is of the order of 25 w per cm^2 so that a thermoelectric cell, with an *electrical power* rating of 10 w per cm^2 would compare quite favorably with it.

If the thermoelectric cell should ultimately prove to be successful, the future fuel element may well run at 2000°C or more and have a coat of a refractory material such as tantalum. It would be surrounded by another tube, perhaps of aluminum, which is cooled vigorously. The interspace between the fuel element and the outer tube would be evacuated except for some cesium vapor. An insulated lead would go across the outer tube to the fuel element and this would form the positive pole, the outer tube itself would form the negative pole. Apparently, the introduction of the coolant into the reactor cannot be avoided, nor the coating of the fuel element. On the contrary, the reactor would be more complicated than at present; only the outside equipment would be drastically simplified. At least, it would not be so difficult to find the reactor on a drawing of the whole power plant.

It would evidently be premature to hazard a guess whether the combination of the thermoelectric principle with the nuclear reactor will become economical and whether the low pressure, high-mean-free-path system, or the high pressure, low mean-free-path system will eventually prevail. Both have strengths and weaknesses. One can say, however, that the thermoelectric heat conversion is an intriguing and intellectually attractive idea which will stimulate further ideas which, in turn, may render them attractive also from the economic point of view.

I would not like to close without thanking all those who made it possible for me, an old-timer, to retain some knowledge of the new developments in the field of reactors. It is difficult to believe that only 15 years ago I had the impression of knowing the field completely, and that 20 years ago, when I knew *nothing* of the field, I *did* know it completely. Whatever familiarity with reactors I have succeeded in maintaining is due to my numerous friends who instructed me patiently, and to all those who sent me reports of their new results. To all of them, I am greatly indebted.

REFERENCES

1. E. P. WIGNER, *Acta Phys. Austriaca*, **11**, 410 (1958).
2. W. K. ERGEN, A. M. WEINBERG, *Physica*, **20**, 413 (1954).
3. H. BROOKS, T. WELTON articles in forthcoming publication on reactor theory of the American Mathematical Society.
4. W. K. ERGEN, H. J. LIPKIN, AND J. A. NOHEL, *J. Math. Phys.* **36**, 36 (1957).
5. J. A. NOHEL, "Stability of Perturbed Periodic Motions" (Mathematics Research Center, University of Wisconsin, Technical Report 84, March, 1959).
6. See, e.g., A. M. WEINBERG AND E. P. WIGNER, "The Physical Theory of Neutron Chain Reactors" (University of Chicago Press, Chicago, Illinois, 1958). Chap. 18, particularly Eq. (18.15).
7. C. N. KELBER AND B. I. SPINRAD, *Nuclear Sci. and Eng.* **2**, 219 (1957).
8. LINK, KELBER, ARMSTRONG, DICKSON, HEINEMAN, RHODE, AND WARE, *Proceedings of the International Conference on the Peaceful Uses of Atomic Energy*, Geneva, 1958.
9. LANE, CHEVERTON, CLAIBORNER, COLE, GAMBILL, GILL, HILVETY, MCWHERTER, AND VROOM, "High Flux Isotope Reactor," ORNL 59-2-65 (1959).
10. ORNL-2510. "The ORNL Gas-Cooled Reactor: Advanced Concepts," by the Staff of ORNL.
11. A. K. KRASIN, B. G. DUBOVSKY, M. N. LANTSOV, Y. Y. GLAZKOV, R. K. GONCHAROV, A. V. KAMEYEV, L. A. GERASEVA, V. V. VAVILOV, E. O. INYUTIN, A. P. SENCHENKOV, *Proceedings of the International Conference on the Peaceful Uses of Atomic Energy*, Geneva, 1958.
12. R. L. MELA AND R. C. ROSS, NDA-TO-450.
13. K. G. HERNQUIST, M. KANEFSKY AND F. H. NORMAN, *RCA Rev.* **19**, 244 (1958).
14. G. N. HATSOPOULOS AND J. KAYE, *J. Appl. Phys.* **29**, 1124 (1958).
15. G. N. HATSOPOULOUS, Doctoral Dissertation, Massachusetts Institute of Technology, 1956.
16. V. C. WILSON, *Bull. Am. Phys. Soc.* **3**, 266 (1958).
17. V. C. WILSON, *J. Appl. Phys.* **30**, 475 (1959).
18. W. B. NOTTINGHAM, *J. Appl. Phys.* **30**, 413 (1959).
19. W. B. NOTTINGHAM, G. N. HATSOPOULOS, AND J. KAYE, *J. Appl. Phys.* **30**, 440 (1959).
20. G. M. GROVER, D. J. ROEHLING, E. W. SALMI, R. W. PIDD, *J. Appl. Phys.* **29**, 1611 (1958).
21. H. W. LEWIS AND J. R. REITZ, *J. Appl. Phys.* **29** (1958). This article includes references to earlier work.

21.

Project Hope: A Chemical Reprocessing Plant for a Nuclear Power Economy

R. A. Charpie, J. Halperin, R. J. Klotzbach, J. R. McWherter, F. Nelson,
E. L. Nicholson, C. H. Odom, R. W. Stoughton, E. P. Wigner, Chairman,
and H. R. Zeitlin

January 21, 1954

I. Summary of Results

A rather embarrassing situation exists at the present time in that the cost of recovering a gram of enriched fissionable material from a spent fuel element is very often as large as the initial cost of producing a new gram of fissionable material in the diffusion plant. It is evident that if we are ever going to establish a nuclear power economy based on heterogeneous reactors which use metallic fuel elements, it will be absolutely necessary to drive the costs of reprocessing down from the present high levels to such a point that the chemical costs represent only a small fraction of the total cost of power.

We feel fairly certain that a very substantial fraction of the cost of reprocessing at the present time is due to a combination of factors which is an outgrowth or holdover from our weapons' economy and which will not be present in a power economy. In fact, if one examines in detail the design of all existing chemical plants he finds that a very substantial fraction of the capital and operating expenses can be charged to need for SF accountability, preoccupation with criticality considerations, and uneconomic though well-engineered waste storage systems.

Since we felt at the beginning of the summer that there were enough unknown factors in chemical plant design that it should be possible to markedly reduce the cost of reprocessing by adopting a rational point of view, we set ourselves the problem of designing a completely rationalized reprocessing plant. This work has come to be known as Project Hope.

In order to have a definite chemical system to study, we have chosen a reactor which is probably not very realistic as a power producer, but one which has the advantage that we know both the design details of the fuel element and the cost and design data for the chemical processing plant in which this fuel element is presently handled. We have adopted an MTR type fuel element which is composed of enriched uranium aluminium alloy, but we have assumed that the U^{235} is replaced by U^{233}. Since any thermal reactor power economy will surely depend on U^{233} and not on U^{235} and since it is clear that physical operations with U^{233} are closely allied to the problems encountered in handling Pu, this assumption probably makes the study more realistic.

Before one can make an economic study it is necessary to choose a fuel processing rate. It is obvious from already existing studies that if one assumes a very large plant for study the unit cost will be low. In order to avoid this pitfall, we have chosen a plant which processes 5.5 kg equivalent of U^{233} per day. This is intermediate in size between the minimum and maximum rates for which the existing Arco plant was designed. For reference, these rates are about 1 and 8 kg per day.

This throughput is a convenient choice because it represents the reprocessing requirements of a power system which operates with a heat generation of a million kilowatts producing 200,000 kilowatts of electricity. The reactor required to produce this amount of power is more or less an average one and does not suffer from the criticism of being so large that the economic considerations involved in it will obviously lead to a lower cost. As an inital goal at which to shoot we assume that we are willing to pay 7 mills per kWh for power. Further, we arbitrarily assign about 0.9 mills per kWh for reprocessing. This turns out to be $1 per gram allocated to reprocessing for the assumptions of the study.

The chemical process we have chosen is solvent extraction using 1.5% TBP in Amsco. We chose this sytem over the Hexone system presently used at Arco because it is more flexible and does not require the precise concentration adjustment typical of Hexone. Laboratory studies and pilot experience show that adequate decontamination can be obtained with TBP in one extraction cycle.

The novel features of the Hope plant are continuous dissolving; relaxed, yet safe, criticality criteria; sampling only for process control; a realistic approach to waste storage; and a new operating-maintenance philosophy.

In the proposed design the processing plant is a concrete-lined pit which has a capacity of about 100,000 gallons. It is located outdoors and the top of the pit is formed of rolling concrete shielding blocks. The plant is operated dry but when maintenance is necessary the cells can be flooded with water. We have developed tools for the simple maintenance jobs which we know will be necessary and, in general, are very enthusiastic about the simplicity of design which is possible because of the maintenance philosophy.

Figures 1 and 2 summarize the economic results of this study for the most conservative assumptions which were examined; i.e., two cycles of extraction. For this case the processing cost is 96.1c per gram of U^{233} fed to the reactor; the total capital cost of the plant has been estimated to be $2,853,000. A physical plant has been designed to the point where we can make valid cost estimates for it. An artist's drawing of this plant is shown in Figure 3. It is clear from even the most cursory examination that this plant is simple in concept and small in size.

We believe that for fuel elements other than the MTR it will be no easier or harder to reach the same goal. It should be pointed out that $1 per gram is still high since no account has been taken of blanket processing, but we feel certain that routine process improvements will remedy this situation.

In summary, we can say that we are deeply convinced that it *is* possible to reprocess enriched fuel elements for about $1 per gram by being realistic in the plant design. We do not think that our choice of the MTR type fuel element abrogates the general applicability of this study to other fuel systems because it seems to be a general rule that the inert material is the controlling factor in the chemical process. Although by trading aluminium for zirconium or stainless steel one introduces new problems, at the same time you simplify some of the problems peculiar to the Al system. We hope that this study has broad general application to all of the chemical reprocessing problems of AEC.

22.

Protection Against Radiations

E. P. Wigner

No date, probably early 1942

Protection Against γ-Rays

The thickness of the lead shields necessary for protection against γ-radiation was calculated by M. W. Wigner a few months ago and the following section is a summary of her calculations. The same calculation was carried out by Dr. A. H. Compton.

Let px be the production of γ-rays per cm^3 and sec; b the absorption coefficient of γ-rays for the shield at perpendicular incidence. Then the number of γ-rays emerging through shield per cm^2 and sec is

$$N = \int_0^\infty px \, dx \, \frac{1}{2} \int_0^{\pi/2} e^{-\frac{ax+b}{\cos\theta}} \sin\theta \, d\theta \tag{1}$$

where a is the absorption coefficient of the pile material.

Hence

$$\begin{aligned} N &= \frac{p}{2} \int_0^{\pi/2} e^{-\frac{b}{\cos\theta}} \sin\theta \, d\theta \, \frac{\cos^2\theta}{a^2} \\ &= \frac{p}{2a^2} \int_b^\infty e^{-y} \frac{b^2}{y^2} \frac{b}{y^2} \, dy \approx \frac{p}{2a^2} \frac{e^{-b}}{b+3/2} \, . \end{aligned} \tag{2}$$

The last part of (2) is a rough approximation which is correct, however, both for $b = \infty$ and $b = 0$.

Let us denote the edge of the cube shaped pile by L. The total number of fissions in a 10^6 kW plant is 3.5×10^{19}/sec. The average number of fissions per cm^3 and sec at the distance x from the side of the pile becomes

$$\frac{1}{L^2} 3.5 \times 10^{19} \frac{\pi}{2L} \sin \frac{\pi x}{L} \approx \frac{1.8 \times 10^{20}}{L^4} x \tag{3}$$

for small x. The number of hard γ-rays is perhaps four times greater than the number of fissions but since the above calculation refers to an infinite area we set

$$p = \frac{1.8 \times 10^{20}}{L^4} \approx 1.4 \times 10^9 \gamma \, \text{cm}^{-4} \text{sec}^{-1} \, . \tag{3a}$$

[9] We can take the mass absorption coefficient of the pile for $L = 700$ cm to be equal to the mass absorption coefficient of air. This is for γ-rays of 5 MeV energy (cf. Heitler Quantum Theory of Radiation, p. 215) $.35 \times 10^{-4}/1.3 \times 10^{-3} = 2.7 \times 10^{-2}$. This gives for a density 1.6

$$a = 4.3 \times 10^{-2} \text{ cm}^{-1}. \tag{4}$$

For a lead shield of 1 cm thickness we can take[10]

$$b = .48 \text{ cm}^{-1}. \tag{5}$$

For γ-rays with 10 MeV energy we have

$$a = 3.1 \times 10^{-2}, \quad b = .61 \text{ per cm}.$$

The total absorption increases with both increasing and decreasing energy of the γ-ray.

Let us put, first, $b = 0$, i.e., no shield. N then becomes 2.5×10^{11}. This would be a dangerous dose in .2 sec. However, with a shield of 10 cm we have $b = 4.8$ and $N = 6 \times 10^8$, for a 20 cm shield $b = 9.6$ and $N \approx 2.5 \times 10^6$. The former can be tolerated for about 2 min, the latter for several hours.

The above estimates for the danger periods must be, of course, multiplied by 10 for a plant which yields only 10^5 kW. They are believed to be somewhat on the pessimistic side since it is unlikely that a fission gives two γ-rays of just about 5 MeV energy and since the absorption is larger for both harder and softer γ-rays. Furthermore, an additional shield is necessary because of the neutrons which will also somewhat decrease the γ-ray intensity. Nevertheless it is evidently inadvisable to build houses, etc., in the immediate vicinity of the pile.

The situation is much better once the pile has ceased to operate. According to estimates of Wheeler (Report CC-111), the radiation intensity decreases to about one twentieth of its original value two hours after the operation has been stopped. Even after one minute, the intensity is only one tenth of the original intensity.

Protection Against Neutrons

The number of neutrons leaving the pile per second is $k - 1$ times the number of neutrons absorbed within the pile. For a $k = 1.04$ this is $.04 \times 6 \times 10^{19}$ for a 10^6 kW pile. The number of neutrons leaving the pile per cm^2 and sec thus becomes, if the pile's edge is 7 m long

---

[9] Corresponds to $L = 600$ cm. (A.M.W. & A.M.P.)

[10] L. B. Borst – (Report C 40) gives .7 as the half thickness for lead. This would correspond to $b = .39$ cm^{-1}. No radiation of this hardness exists according to Heitler, the hardest radiation having about $b = .45$. If Borst's b is correct, the thickness of the shields should be increased by 20 percent.

$$N = \frac{.04 \times 6 \times 10^{19}}{6 \times 700^2} = 8 \times 10^{11}$$

i.e. about six times greater than we assumed for the number of γ-rays. Most of these neutrons can be very easily stopped by water but those with energies above 200 keV have a rather long mean free path. It is these which are somewhat difficult to eliminate.

We can calculate the number of fast neutrons which leave the pile per cm^2 and sec by the diffusion equation

$$N = \frac{\lambda^2}{3} \frac{dn}{dx} \tag{6}$$

where n is the number of collisions which these neutrons make per cm^3 and sec. The number of neutrons created is about twice greater than the number of fissions. In the energy range in question they make about 6 collisions. Their mean free path in carbon is about 7 cm. We have

$$N = 17 \times 6 \times \frac{7 \times 10^{19}}{L^2} \frac{\pi}{2L} \frac{\pi}{L} \cos \frac{\pi x}{L} = \frac{3.5 \times 10^{22}}{L^4} \approx 1.4 \times 10^{11} .$$

Since neutrons are just about as dangerous as γ-rays, if not more so, their number must be reduced to about 10^6 as that of the γ-rays was. This means that the number of neutrons must be reduced at least by a factor 1.4×10^5.

There are several ways to estimate the water layer necessary for such a reduction. The quickest safe way that one can easily think of is to assume that the collisions do not deviate the neutrons from their original direction. Under this assumption the neutrons evidently have a greater penetrating power than they have in reality.

The flux of neutrons at the distance $x = \xi\lambda$ from the source, within an energy interval one at the energy E becomes under the above assumption, if the original energy was E_0

$$\delta(E - E_0)e^{-\xi} + \sum_{n=1}^{\infty} \frac{(\ln E_0/E)^{n-1}}{(n-1)!} \frac{\xi^n}{n!} \frac{e^{-\xi}}{E_0} . \tag{7}$$

The n-th term in the sum refers to the neutrons which have suffered n collisions. One can write for (7)

$$e^{-\xi}\delta(E - E_0) + \frac{e^{-\xi}}{iE_0} \sqrt{\frac{\xi}{\ln(E_0/E)}} J_1\left(2i\sqrt{\xi \ln(E_0/E)}\right) . \tag{7a}$$

Expression (7a) should be integrated with respect to E over the dangerous region. Let us assume that all the high energy neutrons have the energy $E_0 = 2$ MeV and that the dangerous region extends to $E = E_0/10$. One easily sees then that the first term of (7a) can be neglected and that the maximum of the sum lies, for large ξ, at the lower energy limit. Hence the integral of (7a) is certainly less than E_0 times the value of (7a) for $E_0/E = 10$. For this E_0/E and large ξ, one can use the formula $J_1(i\eta) = ie^{\eta}(2\pi\eta)^{-\frac{1}{2}}$ and write for (7a)

$$\frac{e^{-\xi+2\sqrt{\xi \ln(E_0/E)}}}{\sqrt{2\pi}E_0 \ln(E_0/E)} . \tag{7b}$$

For $\xi = 20$ this is small enough to reduce the neutron intensity sufficiently.

The mean free path for neutrons in the dangerous zone is 7 cm. Thus a water shield of 140 cm is certainly more than sufficient to reduce the number of neutrons in the dangerous zone by more than 10^5. Once the neutron's energy is below 200 keV its mean free path is quite short and it will be stopped within a few inches. While the above figure of 140 cm is evidently too high, it is evident, on the other hand, that a thickness of 70 cm is insufficient since for $\xi = 10$ even the first term of (7) gives a too large contribution.

One arrives at a similar conclusion by using the experimental absorption coefficient for RaBe neutrons of about .1 cm^{-1}. (Anderson's report for the week of April 13 and personal communications). Since the theory of the slowing down in water at very large distances from the source is not easily accessible, one cannot interpret this result quite without ambiguity. It is most likely that .1 represents the probability per cm that the neutrons be thrown out of the dangerous region. This is to be compared with the probability for the same process, as estimated above, of .07. It would indicate that a somewhat thinner shield of about 90 cm would be sufficient. (This will be reduced later by about 20 cm). However, in view of the various uncertainties we continue to use the above, higher figure.

Once the plant is stopped, the neutron density decreases also very rapidly. According to Snell, Nedzel and Ibser (Report C-81) only about 1 percent of the neutrons is delayed and the longest period, responsible for about one third of all delayed neutrons, has a half-life of 57 seconds. On the basis of their figures one calculates that the ratio of the intensity of this longest period to the intensity during operation is

$$4 \times 10^{-4}e^{-.012\ t} .$$

For 5 minutes after the operation ceased $t = 300$. This gives 10^{-5} so that we can conclude, on the basis of the above figures, that no harm is to be feared from the neutrons even inside the water shield after five minutes after the plant has been stopped.

One sees in Heitler's book that the absorption coefficient of water for γ-rays of about 5 MeV (which are hardest to absorb) is .03. A shield of 110 cm gives a $b = 3$ for (2), permitting an about 6.5 cm decrease of the thickness of the Pb to be used. On the other hand, the lead shield also guards against neutrons. According to Friedman's report (C-99), the total cross section of Pb is about 6×10^{-24} cm^2 for fast neutrons, of which 3.6 is due to inelastic scattering. This gives .19 for the macroscopic absorption coefficient of lead for fast neutrons and the presence of 13.5 cm lead should permit one to decrease the thickness of the water layer to 105 cm.

The total amounts of material in the shields are quite formidable. For one single side of a cube shaped pile one needs $700^2 \times 13.5 \times 11.3 = 7.5 \times 10^7$ gr $= 75$ tons of lead and an almost equal weight of water. It is very desirable, therefore, to have natural (earthen) shields as much as possible, i.e., to build the pile under ground or into a hill. It may be remarked that it is somewhat better to have the lead on the inside and the water on the outside because lead has a substantial effect on fast neutrons but relatively much less on slow neutrons. Also, once the pile has ceased to operate for about five minutes, one can go inside the water shield, not, however, inside the lead shield. If one plans to do this to a considerable extent, it may be good to use for safety, a somewhat thicker lead shield (about 18 cm) than the one calculated above, if the plant is to operate at 10^6 kW. This will make it even safer to decrease the thickness of the water shield of 100 cm which should be sufficient even for a 10^6 kW plant.

23.

Penetration of Fission Neutrons Through Water

E. P. Wigner and G. Young

April 9, 1947

1. In the high flux experimental pile that is being planned, water will be, at least in some directions, practically the only neutron shield. The calculation of the effectiveness of such a neutron shield is greatly influenced by the variable cross section of the most important constituent of water, viz., hydrogen. This cross section is low at high energies, so that high energy neutrons can penetrate water relatively easily. However, once a neutron has suffered a collision, its energy will decrease and it will suffer a second slowing-down collision after a much shorter distance of travel. It will then again lose energy and the succeeding collisions will come at decreasing distances. As a result, one gains the impression that the penetration of neutrons is determined by the high energy end of the primary neutron spectrum and that the total penetration may not be much greater than the penetration before the first collision.

The above expectations are quite well borne out in the following calculations. They show, therefore, that the situation is quite different from that considered in CP-137, in which a calculation with an energy independent cross section was carried out. As a consequence of the energy independence of the cross section, the total penetration was much greater than the penetration before the first collision. Hence the picture obtained in the present note is very different from the picture of CP-137.

Since the high energy neutrons are responsible for most of the penetration, it will be all right to replace the neutron spectrum by a distribution which is accurate only for high energies. Such a distribution was given by G. N. Plass (CP-1261) and more recently by D. Hughes (C-3574). According to it, the high energy end of the fission spectrum can be represented by

$$F(E)\,dE = \beta e^{-\beta E}\,dE; \quad \beta = .625\ \mathrm{MeV}^{-1}. \tag{1}$$

For similar reasons, an accurate value for the cross section is necessary only for high energies. In this region the cross section of hydrogen can be represented by

$$\sigma(E) = \alpha'/E, \quad \alpha' = 9.7 \times 10^{-24}\ \mathrm{cm}^2\ \mathrm{MeV}. \tag{2}$$

It is remarkable that as long as the scattering is spherically symmetric, this is within 6% the maximum possible scattering cross section. According to measurements of Salant[1] and of Sherr[1], the above formula holds between 14 and about 25 MeV's, which comprises the region of interest.

It is uncertain, of course, whether the fission spectrum still has the form (1) in this region, but theoretical indications would rather postulate a more strong than exponential decrease for very large energies. Similarly, the spherical symmetry of the scattering is not very well established for high energies, although all present indications[3] speak in favor of it. If there is relevant forward scattering the formula to be obtained may be too optimistic. One may expect, however, that such an effect will be compensated by our neglect of the scattering by the oxygen. These are points which are well worth keeping in mind because the value of a formula is not greater than the accuracy of the constants which enter into it.

Let us first calculate the straight penetration to a distance x without any energy loss. This is clearly given by

$$\beta \int_0^\infty e^{-\beta E - \alpha x/E}\, dE = -\pi \sqrt{\alpha \beta x}\, H_1^{(1)}(2i\sqrt{\alpha \beta x}) \tag{3}$$

$$\sim \sqrt[4]{\pi^2 \alpha \beta x}\, \exp(-2\sqrt{\alpha \beta x}).$$

In this $\alpha = N\alpha'$ is the total scattering cross section of the hydrogen atoms contained in one cm³ of water for neutrons of 1 MeV energy. N is the number of hydrogen atoms per cm³ and hence $\alpha = .65$ cm⁻¹ MeV. The $H_1^{(1)}$ is the Hankel function as defined in Jahnke-Emde, Section VIII. The second line is the asymptotic value of (3) for large x. It can be derived easily, without the knowledge of the Hankel functions, by expanding the exponent $-\beta E - \alpha x/E$ of the integrand of (3) around its extremal value $\sqrt{\alpha x/\beta}$ [11]. Evidently (3) underestimates the penetration because it assumes that one single collision completely stops the neutron.

The next model which we shall consider is straight penetration with energy loss. While (3) gives an under estimate of the total penetration, the present model will give an over estimate. It will be assumed that a collision with hydrogen will only change the energy spectrum without affecting the direction of motion, the collisions with oxygen atoms will be neglected altogether. At first sight, this seems to be a rather crude assumption and the resulting over estimate of penetration might be expected to be quite large. It will turn out, however, that this is not the case, the reason being that a substantial change in direction entails a substantial loss of energy, which effectively stops the neutron by the mechanism described in the first paragraph. Hence, if we denote the flux of neutrons in unit energy range at E, at the distance x from the source, by $F(x, E)$ the differential equation which F must satisfy is

$$-\frac{\partial F(x, E)}{\partial x} - \frac{\alpha}{E} F(x, E) + \int_E^\infty \frac{\alpha}{E'} \frac{F(x, E')}{E'}\, dE' = 0. \tag{4}$$

A solution of (4) of sufficient generality is

---

[11] Original reads "$\sqrt{\alpha x/E}$". (A.M.W. & A.M.P.)

$$F(x, E) = -\int_E^\infty e^{-\alpha x/\varepsilon} f'(\varepsilon)\, d\varepsilon \tag{5}$$

in which $f'(\varepsilon)$ is an arbitrary function of ε. The solution of (5) was obtained from the remark that (4) is invariant with respect to the substitution $x \to ax$ $E \to aE$. Because of this invariance, (4) has solutions which are functions of x/E, multiplied with a power of x. One easily obtains a differential equation for the function of x/E and solving that differential equation one obtains (5) with an f' which is a power by ε. So far the procedure would be possible even if the cross section were proportional with an arbitrary power of E. If the cross section is proportional with the -1 power of E, the solution in the form (5) does not depend on the power of x which one has assumed. One is led, therefore, to a form (5) with an arbitrary f'. One can then verify a posteriori that (5) is actually a solution of (4), independently of the argument of this paragraph. For a physically significant solution, f' must be, of course, negative.

For $x = 0$, (5) assumes the form

$$F(0, E) = -\int_E^\infty f'(\varepsilon)\, d\varepsilon = f(E) \tag{6}$$

in which the primitive function $f(E)$ vanishes for $E = \infty$. In order to have at $x = 0$ the distribution corresponding to the fission spectrum, one has to choose

$$f(E) = \beta e^{-\beta E} \tag{7}$$

and one has hence

$$F(x, E) = \int_E^\infty \beta^2 e^{-\beta\varepsilon - \alpha x/\varepsilon}\, d\varepsilon \,. \tag{8}$$

If one had used the real fission spectrum in (7) the following calculations could not be carried out in analytical form. Since the integrand of (8) has a rather sharp maximum at $\sqrt{\alpha x/\beta}$, one sees that, for large x, the distribution is practically constant for low E. It begins to drop around $E = \sqrt{\alpha x/\beta}$ and drops to zero exponentially at high E. This is, of course, a rather different behavior from the one encountered in the model in which one single collision completely stopped the neutron. In this model, the energy distribution of the neutrons at x was given by the integrand of (3). This was, for large x, small at low E, rose to a maximum at $E = \sqrt{\alpha x/\beta}$ and dropped exponentially hence. In spite of this difference in the energy distribution the total flux will turn out to be quite similar in the two models. We now obtain for the total flux at x by a partial integration

$$\begin{aligned}
F = \int_0^\infty F(x, E)\, dE &= \int_0^\infty \beta^2 E e^{-\beta E - \alpha x/E}\, dE \\
&= -i\pi\alpha\beta x\, H_2^{(1)}(2i\sqrt{\alpha\beta x}) \\
&\sim \sqrt[4]{\pi^2 \alpha^3 \beta^3 x^3}\, \exp(-2\sqrt{\alpha\beta x})\,.
\end{aligned} \tag{9}$$

The last part of (9) can be obtained by differentiation from (3) divided by β.

Comparing (3) with (9) one sees that the latter which constitutes an over-estimate of the penetration, is larger than the underestimate (3) only by the factor $\sqrt{\alpha\beta x}$. For this reason, it does not seem worth while to inprove on the approximation which we have made.

There might be some question as to the validity of (8) at low energies at which the actual cross section is lower than given by (2). One easily convinces one, however, that if x is large enough so that one is well beyond the point at which (2) holds for $E = \sqrt{\alpha x/\beta}$, the actual F is not more than the F of (8) multiplied by the ratio of the actual cross section and the cross section given by (2). This ratio hardly exceeds 2 down 1 to MeV. The neutron flux below 1 MeV is clearly negligible.

2. There is an additional difference between the shielding calculation of a large pile, as carried out e.g. in CP-137, and the calculation of a small pile, such as the high flux experimental pile. In the former case, the shield is roughly flat. As a result, only those neutrons have a good chance at penetrating it which start off in the right direction, i.e. perpendicular to the shield. If the pile is small enough as compared with the shield and particularly if it is smaller than the mean free path of the most penetrating neutrons, the chance of a neutron to penetrate the shield will hardly depend on its initial direction. As against this, the intensity at the surface of the shield will be decreased in the ratio of the outer surface of the shield to the pile surface.

For calculating the effectiveness of a 400 cm thick water shield, let us assume that the pile is a point source of 10^{18} neutrons. This is clearly a pessimistic assumption. The geometrical factor of the last paragraph then reduces this to $10^{18}/4\pi(400)^2 = 5 \times 10^{12}$ neutrons/cm^2. Since $\sqrt{\alpha\beta x} = 12.75$, we have for this

$$\sqrt[4]{\pi^2 12.75^3} \times e^{-25.5} = 6.75 \times 10^{-10}.\,^{[12]}$$

As a result, $5 \times 10^{12} \times 6.75 \times 10^{-10} = 3400$ neutrons/cm^2 sec at the top of 400 cm water is clearly an overestimate. The value at the top of 500 cm of water is 210 neutrons/cm^2 sec.

The above calculations exhibit what is essentially a special case of a more general result given earlier by Greuling (MonP-172, pp. 23–38; MonP-130, pp. 7–9).

Footnotes

1. E. O. Salant, R. B. Roberts, and P. Wand, Phys. Rev. *55*, 984 (1939); E. O. Salant, and N. F. Ramsey, Phys. Rev. *57*, 1075 (1940).
2. R. Sherr, Phys. Rev. *68*, 240 (1945).
3. R. R. Wilson, Phys. Rev. *71*, 384 (1947); R. R. Wilson, E. J. Lofgren, J. R. Richardson, B. T. Wright, and R. S. Shankland, Phys. Rev. 71, 560 (1947).

[12] Appears to be an error; should read "$= 1.01 \times 10^{-10}$". (A.M.W. & A.M.P.)

24.

On the Available Information
Concerning the Interaction of Neutrinos with Matter

E. P. Wigner

June 2, 1943

Abstract The information available concerning the interaction of neutrinos with matter is summarized. The reverse process of the β activity gives an exceedingly small cross section. The cross section for interaction of neutrinos with electrons has been determined by Nahmias, and is less than 10^{-31} cm^2. The same experiments give an upper limit for the interaction of neutrinos with heavy particles. However, this upper limit is very high because the velocity imparted to a heavy particle is quite low and does not lead to ionization. It is estimated that the cross section might be 10^{-26} without having been detected by Nahmias. The possibility of detecting the neutrinos from the pile even at larger distances is discussed.

1. The only definite information that we have concerning the interaction of neutrinos with matter is the cross section for the inverse of the β disintegration. This is the process by which an original stable nucleus is converted into a nucleus which has the same mass but a charge by one smaller or by one greater than the original nucleus. In the former case, it captures an electron or may emit a positron. In the latter case it emits an electron. Both processes require energy, and this energy is furnished by the neutrino. In both cases the product nucleus would be radioactive in most cases and emit an electron in the former and a positron in the latter case. Both J. A. Wheeler and E. Fermi have estimated the cross section for this process and have found that it is barely detectable under conditions approximating that around the W pile.

There are two other simple ways of interaction of neutrinos with matter about which we have very little information. These are simple elastic collisions between the neutrino and heavy particles in the first case, electrons in the second case. All that the present information can yield in this connection are maximum cross sections. Whether the actual cross section is lower than this maximum by just a little bit, or by a very great amount is impossible to tell at this time. It seems worthwhile, therefore, to see under what conditions the neutrinos could be detected by these processes. If a neutrino collides with a nucleus of mass M, it transfers to it a maximum amount of energy

$$\frac{(2h\nu)^2}{2Mc^2}.$$

The velocity given to the nucleus becomes $(2h\nu)/(Mc)$ and the energy of an electron with this velocity is

$$\frac{m}{2}\left(\frac{2h\nu}{Mc}\right)^2.$$

Whether this heavy particle will ionize or not depends essentially on whether this lost energy is larger or smaller than the ionization energy of an electron. If it is smaller, the motion of the nucleus which obtains the recoil will be essentially adiabatic and no appreciable ionization will occur. By inserting values of a few mc^2 for $h\nu$ in the last formula, one sees that one can expect ionization only in the case of hydrogen. Since the effective energy for ionization is inversely proportional to the square of the mass number of the nucleus, practically no ionization can be expected even in the case of helium.

Since practically all efforts to detect a neutrino have been made by attempting to measure the ionization in air, it is evident that the negative results mean rather little concerning the type of interaction under question. The most far reaching information has been obtained by Mr. M. E. Nahmias, *Proc. Camb. Phil. Soc.*, 31, 99, 1935. His most effective source of γ rays was radium with its daughter products, among which RaC has the highest energy of 3.15 MeV. One can calculate from this that the highest recoil energy which the neutrinos can give to nitrogen is 1550 eV, which corresponds to an energy of .06 eV of the electrons of the struck nitrogen atom if considered from the coordinate system in which the nucleus is at rest. Practically no ionization can be expected under these conditions. In hydrogen the corresponding energy would be 11.6 eV, which could readily cause ionization.

Let us assume first that the extensive shielding used by Nahmias did not influence the energy of the neutrinos. In that case, it follows from his experiments that the effective cross section of the neutrinos for ionization in nitrogen is less than 6×10^{-30} cm^2, but hardly anything follows concerning the cross section of the interaction considered here which would not lead to ionization.

Nahmias makes no statement as to whether the air in his ionization chamber was dried. Assuming, arbitrarily, a partial pressure of 1.4 mm of water in his air of 76 mm pressure, his result indicates that the cross section of hydrogen for ionization is less than about 3×10^{-28} cm^2. The cross section for elastic neutrino-H scattering could easily be twenty times larger since an ionization can be expected only from the neutrinos of RaC, from these only if they have nearly maximum energy, and even in this case only in case of a scattering by nearly 180°. If his air was dried, the neutrino-H cross section could be as high as 10^{-25} cm^2 without having been detected. This is indeed a very unlikely figure. Even this figure would be increased if the extensive shielding used by Nahmias had softened the neutrinos, which it could have done if the cross setion had been in the neighborhood of the high amounts mentioned above.

To repeat what was said above, there is no indication whatever that the cross section for the scattering of neutrinos by heavy particles is actually as high as the above figures are. On the other hand, there is at present no obvious experimental evidence which would show that the cross section is below the above figures. Fermi has pointed out to the writer that if the neutrinos created in the sun could freely escape, they would have a density comparable to the density of the neutrinos from the pile in the neighborhood of the pile. Of course, if the cross section is anywhere near the above cross sections, the neutrinos would be unable to escape from the interior of the sun.

Turning to the other process, the scattering cross section of neutrinos by electrons, for which Nahmias' investigation was undertaken, there is no reason to modify his and Bethe's conclusions (ibid, p. 108). The cross section in this case is less than 10^{-31} cm^2.

2. The first question which may be of interest in this connection is the possibility of detecting the neutrinos at a considerable distance from a running pile. The following lines will give only a very crude estimate for this possibility which does not take into account possible practical difficulties in operating counters, etc. The only definite effect which would interfere with the detection of neutrinos is the cosmic radiation. According to Dr. N. Hilberry, who kindly assisted the writer in this work, the number of mesotrons at an elevation of 4750 meters (15,600 feet) is .04/cm^2 sec. The soft radiation which gives about 2.5 times more counts, can be excluded by a suitable lead shield. On the other hand, the total number of neutrinos from a 250,000 KW plant is about 3.5×10^{19}/sec. At a distance of 5000 meters this gives about 10^7 neutrinos/cm^2 sec. One has, therefore, $6 \times 10^{26} \sigma$ counts/cm^3 sec in hydrogen at one atmosphere. Assuming a depth of 10 cm for the counter, the pile activity will be equal to the mesotron activity if $5 \times 10^{26} \times 10\sigma = .04$, i.e., if σ is as large as 8×10^{-30} cm^2. It has to be taken into account, however, that σ is the cross section for the ionization of hydrogen which is close to the scattering cross section only in the case of relatively hard neutrinos.

3. It may be of interest to consider briefly the possible physiological effect of neutrinos. Assuming again 3.5×10^{19} neutrinos from the pile, the number of neutrinos/cm^2 sec at a distance of 10 meters from the center of the pile will be 3×10^{12}, This is under the assumption that the shield does not appreciably soften the neutrinos, which is a very likely assumption. In 1 cm^3 of the body, these neutrinos will create $2 \times 10^{35} \sigma$ free hydrogen atoms and radicals per second. Assuming an average absorption coefficient of .04 cm^{-1} for γ rays and 10^5 ions/γ ray, one has 4000 ions/sec in a cm^3 of the body if one γ ray is incident per cm^2. If one assumes that a hydrogen atom and a radical are just as harmful as an ion pair, the neutrino radiation will have the same physiological effect as $5 \times 10^{31} \sigma\gamma$ rays/cm^2 sec, or as $10^{23} \sigma r$ units/sec. The σ here is, of course, the cross section for elastic scattering of a neutrino by hydrogen. To this, the effect of the scattering by other nuclei should be added which will be of the same order of magnitude. If σ is smaller than 2×10^{-29} cm^2, no physiological effect must be feared at 10 m distance from the pile in an 8 hour day.

25.

The Rate of Decay of Fission Products

K. Way and E. P. Wigner

Physical Review 73 (11), 1318–1330 (1948)

Received January 28, 1948

By considering the fission products as a sort of statistical assembly, calculations have been made of the β-disintegrations per second and of the total energy emitted per second at any time after fission has taken place (cf. Fig. 6). The results are in good agreement with experiment. The theoretical work is based on the assumption that the mass of a nucleus of mass number A and charge Z is given by $a(Z_0(A)-Z)^2+b$. Empirical values for a and b are used. Use is also made of an approximate empirical relationship between half-life and disintegration energy. A further basic hypothesis which is important for the results at very short times after fission has taken place is that, in the most probable way of splitting, the chain lengths of the light and heavy fragments are equal and that there is not much deviation from this most probable mode of fission. (See L. E. Glendenin, C. D. Coryell,
R. R. Edwards, and M. H. Feldman, CL-LEG-1. A tentative explanation has been given recently by R. D. Present, Phys. Rev. 72, 7 (1947).) The average number of β-disintegrations per fission is found to be 6; the average energy of all radiations (β, γ, and neutrino) of the fission products is 21.5±3 Mev. Apparently, about half of this energy escapes in the form of neutrinos and a quarter is emitted in the form of β and in the form of γ rays.

A few remarks are made concerning the possible origin of delayed neutrons. It is also pointed out that the spread of the kinetic energy of a given pair of fission fragments cannot be easily explained on the basis of differences of chain length which result in differences in excitation energy of the fragments. It is possible that fluctuations in the production of fission neutrons are at least partly responsible for the kinetic energy spread.

I. INTRODUCTION

THERE are two ways to describe the total β- and γ-radiation emitted by the fission products. The first, more accurate method describes it as the sum of radiations emitted by the different fission products. The second, less accurate method considers the fission products as a sort of statistical assembly and tries to arrive at once at the total radiation from all the fission products together. While this second method is unquestionably less complete than the first one, it is so much simpler that it is much preferable in most practical calculations. Its validity is limited, of course, to times during which the radiation is emitted by many nuclei. At very long times after irradiation, when most of the radiation is caused by a few surviving species of nuclei, the first method is definitely preferable.

A complete description of the radiations would give the energy distribution of the β-rays for all times after irradiation and a similar distribution function for the γ-rays. No experimental data are available which would permit one to obtain this complete information, and it will not be attempted here to obtain it theoretically. Instead, we shall restrict ourselves to the number $\beta(t)$ of β-rays, emitted during unit time, t seconds after a fission has occurred, to the total energy B(t) of the β-rays which are emitted during the same time element and to the similar quantities $\gamma(t)$ and $\Gamma(t)$ for γ-rays.

If $N(E, t)dE$ be the number of fission products existing at time t after fission has taken place which have disintegration energies between E and $E+dE$ and if $\lambda(E)$ be the decay constant of nuclei with disintegration energy E, then $\beta(t)$ is given by

$$\beta(t) = \int_0^{E_m} N(E, t)\lambda(E)dE, \qquad (1)$$

where E_m is the largest disintegration energy found. Actually, there is no unique relation between disintegration constant and energy, as the function $\lambda(E)$ would imply. However, it will be easy to take care of this point in the course of the calculation.

The total energy emitted at time t following a fission is equal to the sum of the energy of the β-rays, γ-rays, and neutrinos. If one assumes that the energy of the neutrinos is twice greater than the energy of the β-rays (cf. Fig. 1), then this total energy is equal to 3B$(t)+\Gamma(t)$ and is given by

$$3B(t)+\Gamma(t) = \int_0^{E_m} EN(E, t)\lambda(E)dE. \qquad (2)$$

The functions and quantities which are necessary for the evaluation of the integrals are discussed in the next three sections.

II. RADIOACTIVE NUCLEI FORMED IMMEDIATELY AFTER FISSION

It has been assumed already in the earliest theoretical papers on nuclear physics that the energy difference between the isobars of an odd element can be written in the form $a(Z_0-Z)^2$ where a is a constant (depending on the mass number of the isobars). Z_0 (not necessarily an integer) is another constant which we shall call the charge number of the "most stable isobar" and Z is the charge number of the isobar in question. The value of a has been determined from empirical data[1] (Weizsäcker, 1935; Bethe and Bacher, 1936; Bohr and Wheeler, 1939; Feenberg 1947), and also from theoretical considerations[2] (Wigner, 1937). We attempted to obtain the constant a directly from the data on the disintegration energy of fission chains. We found, however, that the relative energies of isobars are, in most cases, rather irregular functions of Z, showing considerable fluctuations around any smooth function like $a(Z_0-Z)^2$. Thus the a in this formula cannot be determined in any precise fashion. On the whole, we found that $a=0.65$ Mev for the heavy fragment and 1 Mev for the light fragment represent the data as well as we could represent them. These values agree almost exactly with the early values (Weizsäcker, Bethe, and Bacher[1]) given to this constant $(78/A+0.58/A^{\frac{1}{3}})$ and are somewhat larger than the values obtained theoretically[2] $(55/A+0.59/A^{\frac{1}{3}})$. The energy of the radioactive disintegration[3] is given by the difference of the above

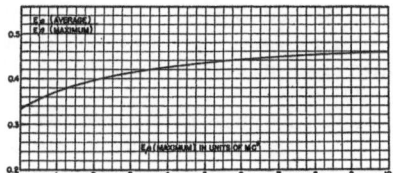

FIG. 1. The ratio of average β-energy to maximum β-energy on the basis of Fermi's formula for the energy distribution in an allowed transition. Effect of Coulomb field is neglected.

expression for two successive values of Z and is, therefore,

$$a(Z_0-Z)^2-a(Z_0-Z-1)^2=2a(Z_0-Z)-a.$$

This decreases as Z increases in the course of successive disintegrations.

The above expression is supposed to hold for isobars with odd mass number A. If A is even, the isobars with odd Z (containing an odd number of protons and also an odd number of neutrons) have, on the whole, higher energy contents than the isobars with even Z (containing an even number of protons and an even number of neutrons). This can be expressed mathematically by using, in case of even n, the expression $a(Z_0-Z)^2$ for the energy of nuclei with even Z and another expression $a(Z_0-Z)^2+b$ for odd Z.

The value given by Bethe and Bacher and by Weizsäcker for the constant b is $40/A+0.6/A^{\frac{1}{3}}$ which is about 0.34 Mev for the light and 0.26 Mev for the heavy fractions. The theoretical values[2] are much higher, viz. $92/A$, which gives $b=0.96$ Mev for the light and $b=0.66$ Mev for the heavy fragment. The actual values which will be determined in the third section are even greater than these, viz. 2.8 and 1.8 Mev, in agreement with Bohr and Wheeler.[1]

It follows from the above that the energies of successive disintegrations of fission chains with even A will not diminish steadily but that, particularly toward the end of the series, a high energy disintegration will be followed by a low energy disintegration, this again by a high energy disintegration, etc. The energy of transi-

[1] G. Gamow, Int. Conf. on Physics, London, 1934, Vol. I, *Nuclear Physics*, 60–66, Physical Society. W. H. Heisenberg, Rapport du VII me Congress Solvay, Paris, 1934. G. C. Wick, Nuovo Cimento 11, 227 (1937). C. F. v. Weizsäcker, Zeits. f. Physik 96, 431 (1935). H. A. Bethe and R. F. Bacher, Rev. Mod. Phys. 8, 82 (1936), §26, 28, 29, 30, Eq. (185), (185a) ff. N. Bohr and J. A. Wheeler, Phys. Rev. 56, 426 (1939). E. Feenberg, Rev. Mod. Phys. 19, 239 (1947).

[2] E. Wigner, Phys. Rev. 51, 947 (1937). Also Bicentennial Symposium, University of Pennsylvania Press, 1940. W. H. Barkas, Phys. Rev. 55, 69 (1939), also E. Feenberg, reference 1.

[3] By the energy of disintegration we mean the sum of the energies of all radiations (β, γ, and neutrino) emitted in cascade in a transition from the normal state of the parent to the normal state of the daughter. The energy of transition of a β-disintegration is, on the other hand, the high

energy limit of the β-rays emitted in that transition. In the case of a simple β-spectrum, with a succeeding γ-emission, the energy of transition differs by the energy of the γ-rays from the energy of disintegration.

tion from an odd Z to an even $Z+1$ will be represented by $2a(Z_0-Z)-a+b$, the energy of transition from an even Z to an odd $Z+1$ by $2a(Z_0-Z)-a-b$.

The radioactive nuclei to be considered are not all immediately created by fission but are mostly daughters of others. In the series with odd mass numbers, the energy of disintegration decreases. Since the lifetime increases with decreasing energy of disintegration, successive members of the radioactive series will have longer and longer half-lives. (See Fig. 2.) This conclusion is, in general, corroborated by the experimental material,[4] but it is not without exception. If we assume its validity, it is justifiable, for our purpose, to treat all the members of a radioactive series with odd A as immediately created by fission, since the ancestors of any nucleus will have much shorter lives than the nucleus itself, so that the time which elapses before a certain nucleus is formed can be neglected as compared with the lifetime of that nucleus.

If A is even, the situation is only slightly different. In this case, the energies of successive members of a radioactive series will not be steadily diminishing, but, particularly toward the end of the series, a high energy disintegration will be followed by a low energy disintegration, this again by a high energy disintegration, etc. This results in an alternation in the half-lives, such that a long-lived element will be followed by a short-lived element, this again by a long-lived element, etc. As mentioned before, this will be particularly true toward the end of a series while towards the beginning of the series the energies will be successively diminishing and the lifetimes increasing. In the present case (even A) one is not justified any more to consider all nuclei as being immediately formed by the fission. However, if a short-lived nucleus is the daughter of a long-lived nucleus, the lifetime of the short-lived nucleus may be neglected and the life history of the short-lived nucleus will be approximately the same as the life history of its parent. Hence the even-even nuclei (even number of protons and even number of neutrons) may be considered to be immediately formed by fission in the same sense as all odd-mass nuclei were.

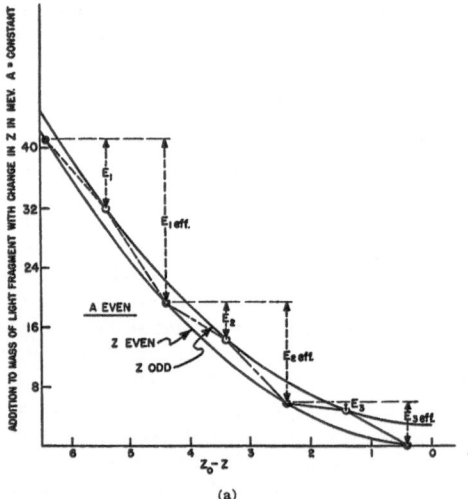

(a)

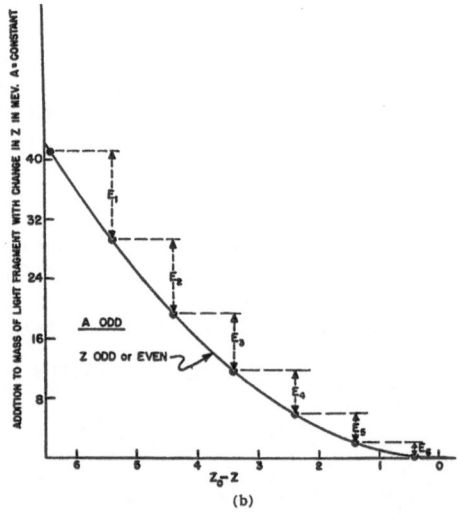

(b)

FIG. 2. All fission products with odd masses can be considered to be formed immediately upon fission. Of the fission products with even masses, only those with even charge are considered to have been formed immediately upon fission. The disintegrations of the daughters of these nuclei (which have odd charges) will be considered to occur simultaneously with the disintegrations of the parents. In Fig. 2a ● indicates nuclei "formed immediately in fission." ○ are odd-odd daughters whose half-lives are considered to be those of their even-even parents. In Fig. 2b ● are nuclei "formed immediately in fission."

[4] "Nuclei formed in fission: decay characteristics, fission yields and chain relationships," issued by the Plutonium Project, Rev. Mod. Phys. 18, 513 (1946).

TABLE I. Number of β-disintegrations per fission.

	130	131	132	133	134	135	136	137	138	139	140	141	142	143	144	145	146	147	148	149
Mass and yield (percent) of heavy nucleus	1.8	2.7	3.9	4.6	5.3	5.8	6.2	6.5	6.6	6.6	6.5	6.2	5.8	5.3	4.7	4.1	3.4	2.6	1.8	1.4
Smallest stable Z for this mass	54	54	54	55	54	56	54	56	56	57	58	59	58	60	60	60	60	62	60	62
Most stable Z for this mass ($Z_0(A_H)$)	53.5	54	54.5	55	55.5	56	56	56.5	57	57.5	58	58.5	59	59.5	60	60.5	61	61.5	61.5	62
Masses of corresponding light nuclei	104	103	102	101	100	99	98	97	96	95	94	93	92	91	90	89	88	87	86	85
Smallest stable Z for this mass	44	45	44	44	42	44	42	42	40	42	40	41	40	40	40	39	38	37	36	37
Most stable Z for this mass ($Z_0(A_L)$)	45.5	45	45	44.5	43.5	43.5	43	42.5	42	41.5	41	41	40.5	40	39.5	39	38.5	38	37.5	37
Number of β-particles emitted	6 7	7 6	6 6	7 5	4 6	8 6	4 4	6 4	4 6	7 5	6 7	8 7	6 6	8 8	8 7	7 6	6 5	7 6	4 5	7 6
$92-Z_{H0}-Z_{L0}$	7 6*	7 7	7** 7	7 6	7 7	7 7	7 6	7 6	7 6	7 6	7 7	7 7	7 7	7 7	7 7	7 7	7 7	7 7	7 6	7 6

* 6 stands for 6.5.
** 7 stands for 7.5.

Among the odd-odd nuclei (odd number of protons and neutrons) only those will be considered to be formed immediately upon fission which actually are, i.e., the first member of a chain if that first member is odd-odd. On the other hand, all odd-odd nuclei which are daughters of even-even nuclei will be assumed to disintegrate immediately after the disintegration of their parent. These parents will have a relatively long lifetime, corresponding to a relatively low real disintegration energy $E=2a(Z_0-Z)-a-b$ but, for that lifetime, will have an abnormally high effective disintegration energy, namely, the sum of their own and their daughters' disintegration energy:

$$[2a(Z_0-Z)-a-b]+[2a(Z_0-Z+1)-a+b] = 2E-2a+2b.$$

III. THE DISTRIBUTION FUNCTION $N(E,t)$: GLENDENIN'S RULE

One would think, offhand, that the most probable way of splitting of a nucleus into two fragments is the one in which the kinetic energy of the fragments can be highest. Therefore, if the splitting occurs into the two masses, A_L and A_H (L and H refer to light and heavy fragment), one would think that the most probable Z for the light fragment would be such that the radioactive energy of the two fragments,

$$E_r = a_L(Z_0(A_L)-Z)^2 + a_H(Z_0(A_H)-92+Z)^2, \quad (3)$$

is a minimum. In (3), $Z_0(A)$ is the "most stable nucleus" of mass A in the sense of the preceding section. If $a_L=1$ Mev, $a_H=0.65$ Mev, it would then follow from (3) that the most probable Z of the light fragment is given by

$$[Z_0(A_L)-Z]/[Z_0(A_H)-92+Z] = a_H/a_L = 0.65. \quad (4)$$

Since the numerator of the left side of (4) is, crudely speaking, the number of successive disintegrations in the light chain, and the denominator the number of disintegrations in the heavy chain, it would follow that the light chains are about half as long as the heavy chains.

It is important to note that the above picture (which was used in the past by the present writers[5]) is not confirmed experimentally. It has been shown[6] that the light and heavy chains are approximately *equally long* and we shall assume hence that, in the most probable type of disintegration, this is the case.

Since, as demonstrated in Table I, $Z_0(A_L)+Z_0(A_H)$ is, on the average, about 99.0, we can assume that the most probable Z differs by 3.5 from $Z_0(A_L)$ and that, hence, the most probable charge of the heavy fragment is also 3.5 units smaller than the most stable charge $Z_0(A_H)$ for the mass of the heavy fragment.

Knowing the most probable charges, the question arises as to the deviations from these most probable charges. The gas sweeping experi-

[5] E. P. Wigner and K. Way, Phys. Rev. **70**, 115 (1947) and CC-3032.
[6] L. E. Glendenin, C. D. Coryell, R. R. Edwards, and M. H. Feldman, CL-LEG-1. A tentative explanation has been given recently by R. D. Present, Phys. Rev. **72**, 7 (1947).

ments of R. M. Adams, A. Turkevich, and co-workers,[7] and the independent yield determinations of M. H. Feldman, L. E. Glendenin, R. R. Edwards, E. J. Hoagland, N. Sugarman, B. Finkle, and D. W. Engelkemeier[8] provide at least an approximate answer. The probability that the charge of the light fragment be Z is given by the expression

$$P(Z)dZ = (0.7\pi)^{\frac{1}{2}}$$
$$\times \exp[-0.7(Z_0(A_L) - 3.5 - Z)^2]dZ. \quad (5)$$

Since Z can assume only integer values, (5) is clearly only an approximate expression. However, it can be expected to give, in a general sort of way, the probability that the charge of the light fragment be Z, that of the heavy fragment $92 - Z$. According to (5), the probability for obtaining the charge $Z_0(A_L) - 3.5$ is highest, the probability for obtaining either of the charges $Z_0(A_L) - 4.5$ and $Z_0(A_L) - 2.5$ is only half as high.

We can substitute $Z_p = Z_0(A_L) - 3.5$ and, with the aid of $Z_0(A_L) + Z_0(A_H) = 99.0$, rewrite (3) to

$$E_r = a_L(Z_p + 3.5 - Z)^2 + a_H(Z_p - 3.5 - Z)^2$$
$$= (a_L + a_H)(Z_p - Z)^2$$
$$+ 7.0(a_L - a_H)(Z_p - Z) + 12(a_L + a_H) \quad (6)$$
$$= 1.65(Z_p - Z)^2 + 2.5(Z_p - Z) + 20.$$

It follows that the minimum value of E_r is 19.5 Mev. However, the average value of the energy which finally emerges in the form of radiation (γ, β, and neutrino) is

$$E_r = \int [1.65(Z_p - Z)^2 + 2.5(Z_p - Z) + 20]$$
$$\times (0.7/\pi)^{\frac{1}{2}} \exp(-0.7(Z_p - Z)^2)dZ = 21.5 \text{ Mev.}$$

Actually, as has been pointed out by A. Turkevich,[9] 1 Mev has to be subtracted from this. The reason is that the final nucleus is not the most stable nucleus but has, in most cases, a smaller charge. In the case of odd mass, the Z of the final nucleus can differ from Z_0 by as much as $\pm\frac{1}{2}$. Hence, the excess energy is, in this case, the average of aZ^2 for Z between $-\frac{1}{2}$ and $\frac{1}{2}$, i.e., $a/12 = 0.08a$. The condition, in case of even mass

[7] R. M. Adams and H. Finston, cf. Vol. 9B of the forthcoming Plutonium Project Records.
[8] M. H. Feldman, L. E. Glendenin, and R. R. Edwards, PPR, 9B, 7.36, E. J. Hoagland and N. Sugarman, PPR, 9B, 7.7, L. E. Glendenin, PPR, 9B, 7.42.2, B. Finkle, D. W. Engelkemeier, and N. Sugarman, PPR, 9B, 7.42.1.
[9] Personal communication.

number, that the even-even nucleus with charge Z be stable is a $(Z_0 - Z)^2 < a(Z_0 - Z - 1)^2 + b$ which gives $Z_0 - Z < 1.9$. Hence, the excess energy in the even-even end product is the average of a Z^2 for Z between 0 and 1.9 which is $1.2a$. Since half of the final fission products has odd mass, half even mass, the total excess energy in the stable end products is $\frac{1}{2}(a_L + a_H)1.2a = 1$ Mev. This correction is, however, compensated by one, due to the fact that one-quarter of all the fission fragments is an odd-odd nucleus with an excess energy of b. This gives an increase in the total radiation of $\frac{1}{4}(b_L + b_H) = \frac{1}{4}(2.8 + 1.8) = 1.15$ Mev which restores the validity of the above figure. The same result can, of course, be obtained also by integrating (14a) over the time t. We estimate that the above figure for the total amount of radiation may be accurate to 15 to 20 percent.

One can calculate, in a way similar to the above, that the root mean square deviation of E_r from its average is 2.7 Mev. This compares with a measured root mean square deviation[10] of the kinetic energy of the fission fragments from their average of 10 Mev. This shows that the variation of E_r from its average forms a not negligible contribution toward the variation of the kinetic energy of the fission fragments from its average, without being able to explain it by itself. Moreover, the observed energy distribution seems quite symmetrical while one would expect the energy variation from this variation in initial Z to be quite unsymmetrical. There would be quite a sharp falling off on the high energy side and a gradual tailing off on the low energy side. One of us (K. W.) has made some preliminary calculations which indicate that the kinetic energy spread may be accounted for by fluctuations in the number of fission neutrons.

The light fragment of charge Z will have, if it has an odd mass, in its first disintegration, a disintegration energy $E = 2a_L(Z_0(A_L) - Z) - a_L = 2a_L(Z_p + 3.0 - Z)$. As a result, the probability that the first disintegration energy of the light fragment be E is

$$P_{o1}(E)dE = \frac{1}{2}(0.7/\pi)^{\frac{1}{2}}(2a_L)^{-1}$$
$$\times \exp[-0.7(\frac{1}{2}E/a_L - 3)^2]dE. \quad (8a)$$

[10] W. Jentschke, Zeits. f. Physik 120, 165 (1943) A. Flammersfeld, P. Jensen, and W. Gentner, Zeits. f. Physik 120, 450 (1943). M. Deutsch and N. Ramsey, LA-510.

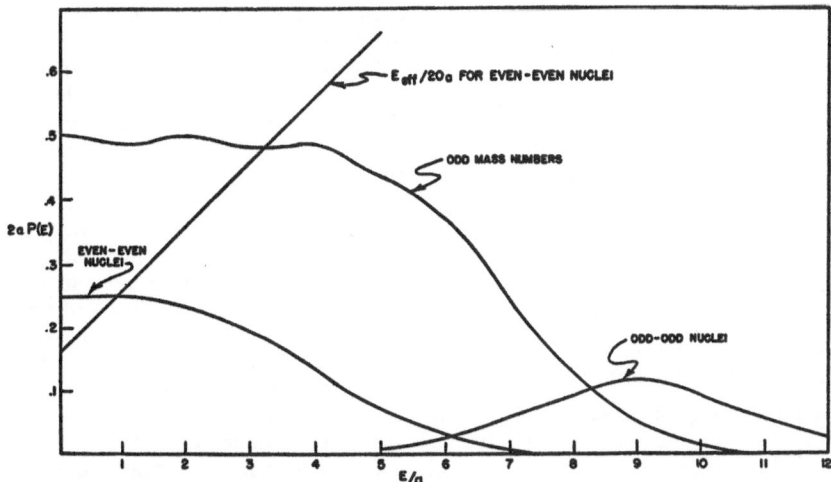

FIG. 3. Distribution of the disintegration energy of the nuclei which are created immediately after fission. This includes all nuclei which are really directly created by the fission, all nuclei with odd mass numbers, and all even-even nuclei. In the last case, the effective energy of disintegration is higher than E and is given, in units of $20a$, by the straight line. The curves apply both to the light and the heavy fragment.

The factor $\frac{1}{2}$ has been introduced because only half of the radioactive chains have odd masses. The energy of the second disintegration is

$$E = 2a_L(Z(A_L) - Z + 1) - a_L = 2a_L(Z_p - Z + 2.0)$$

so that the probability that the second disintegration yield the energy E becomes

$$P_{o2}(E)dE = \tfrac{1}{2}(0.7/\pi)^{\frac{1}{2}}(2a_L)^{-1}$$
$$\times \exp[-0.7(\tfrac{1}{2}E/a_L - 2)^2]dE, \quad (8b)$$

and so on. By adding the expressions $P_{o1}(E) + P_{o2}(E) + \cdots = P_o(E)$ one obtains the total probability for a disintegration energy E of the light fragment. The same expression, which is shown in Fig. 3, holds for the heavy fragment also, except that a_H must be substituted for a_L.

Needless to say, the wiggles at low E/a are not real but come from our taking too seriously our model of one definite a_L (or a_H) and a definite function (5) for the distribution of the Z. The constancy of $P(E)$ for E well below $2 \times 3a$ follows from the fact that the energies of successive disintegrations form an equidistant set and that the absolute position of the lines of this set is arbitrary. The absolute value of $P_o(E)$ for low E is simply the reciprocal distance $1/2a_L$ of the lines in this spectrum.

The above picture holds only for odd mass numbers. Since all even-even nuclei are also "immediately created upon fission," we can proceed with them in a similar way, except that the probability is only one-fourth that the nucleus with a given Z be even-even so that the $P_o(E)$ must be multiplied by the further factor $\frac{1}{2}$ in order to obtain $P_{ee}(E)$. In addition, it must be shifted to the left by b because the disintegration energy of an even-even nucleus with given $Z_0 - Z$ is smaller by b than the disintegration energy of an odd mass nucleus with the same Z. $P_{ee}(E)$, is also shown in Fig. 3. One has to remember that every disintegrating even-even nucleus gives rise to two β-rays, to its own and to its daughter's, and that the energy emitted during these disintegrations is not E but $2E + 2b - 2a$.

Finally, there is a chance of one-fourth that the nucleus formed immediately upon fission be odd-odd. If we continue to assume that its Z distribution is given by (5), its energy distribution will be

$$P_{oo}(E)dE = \tfrac{1}{4}(0.7/\pi)^{\frac{1}{2}}(2a_L)^{-1}$$
$$\times \exp[-0.7(\tfrac{1}{2}(E - b_L)/a_L - 3)^2]dE. \quad (9)$$

This function is also shown in Fig. 3.

The distribution function $N(E, t)$ at time $t=0$ can then be written as the sum $P_o + P_{ee} + P_{oo}$ and at any later time

$$N(E, t) = [P_o(E) + P_{ee}(E) + P_{oo}(E)]e^{-\lambda(E)t}.$$

The average number of β-rays emitted by the light chain is the integral of $P_o(E)dE$ (which is $\frac{1}{2} \times 3.5 = 1.75$) plus the integral of $P_{oo}(E)dE$ (which is $\frac{1}{4}$) plus twice the integral of $P_{ee}(E)dE$. This last quantity is $\frac{1}{4} \times 3.5 = 0.875$ minus b_L times the ordinate of $P_{ee}(E)$ at $E=0$ which is $0.25/2a_L$. Hence the average number of β-rays in the light chain, which is according to Table I equal to $\frac{1}{2} \times 6.1 = 3.05$, must be equal to

$$3.05 = 1.75 + 0.25 + 2(0.875 - 0.25b_L/2a_L). \quad (10)$$

It follows that

$$b_L = 2.8a_L. \quad (11)$$

The same holds for the heavy fragment. Hence $b_L = 2.8$ Mev, $b_H = 1.8$ Mev. This method of determining b is only apparently different from that of Bohr and Wheeler.[1] The average number of odd radioactive nuclei in the light (or heavy) fission chain is 1.75, the average number of even-even and of odd-odd radioactive nuclei in the same chain are 0.52 and 0.78. The same numbers hold for the heavy chain.

Figure 4 shows the energy distribution of the disintegrations of both chains. This should give the energy spectrum of disintegrations in a steadily running chain reacting unit. Of course, the energy spectrum of disintegration is not an immediately observable quantity because a disintegration contains, in general, β-rays, γ-rays, and neutrinos. Nevertheless, this curve may have some interest. For this reason, it also contains the odd-odd nuclei which are daughters and, according to our picture, are not "immediately formed upon fission." For even-even nuclei, it shows the effective energy of disintegration also, i.e., the sum of the energies of even-even parent and odd-odd daughter.

Before leaving this subject we wish to attract attention to one more circumstance.[11] The mass formulae clearly permit one to calculate the binding energy of a neutron, which is an important quantity from the point of view of the so-called "delayed" neutron emission. This is the emission of a neutron by a nucleus, after the nucleus is formed by a β-disintegration. Let us start with a nucleus of mass A and charge Z. Its mass is greater than that of a nucleus of mass A

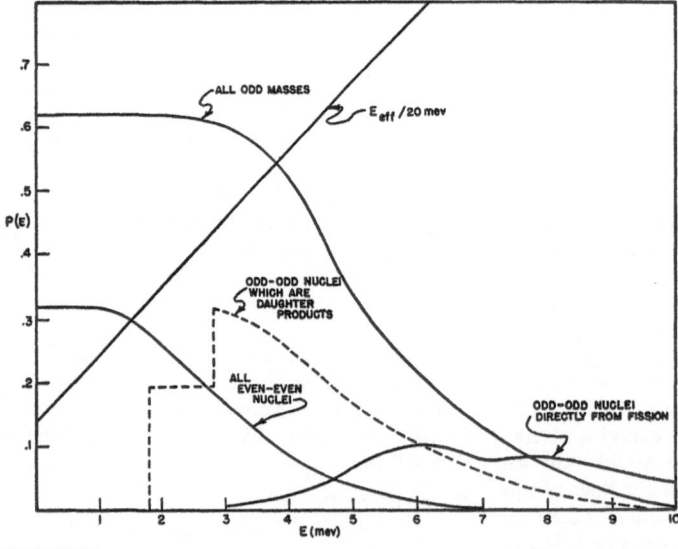

FIG. 4. Distribution of the disintegration energy of all fission products nuclei (both light and heavy fragments): P_o for nuclei with odd masses: P_{ee} for nuclei with even number of protons and of neutrons; P_{oo} for those nuclei with odd neutron and proton number which are fission fragments themselves and not daughters of others; broken line for the other odd-odd nuclei.

[11] We are very much indebted to C. Coryell and to his collaborators for discussions on this subject.

RATE OF DECAY OF FISSION PRODUCTS   1325

and charge $Z_0(A)$ by the amount $a(Z_0(A)-Z)^2$. The nucleus of mass $A-1$ and charge Z has a mass greater than that of a nucleus of mass $A-1$ and charge $Z_0(A-1)$ by $a(Z_0(A-1)-Z)^2$. $Z_0(A-1)$ is approximately equal to $Z_0(A)-0.45$. The mass difference of these nuclei (A, Z and $A-1$, Z) which is due to the fact that their charges do not have the "most stable" values is then $0.2a-0.9a(Z_0-Z)$. This quantity must then be added to the average energy B required to tear a neutron away from a nucleus of mass A to obtain the correct neutron binding energy. One must also add $+\frac{1}{2}b$ or $-\frac{1}{2}b$, depending on whether the number of neutrons was even or odd in the neutron emitting nucleus. The energy of disintegration which forms the neutron emitting nucleus is $2a(Z_0-Z)+a$ if A is odd. If A is even, $-b$ or $+b$ must be added to this, depending on the odd or even character of Z. Assuming the unlikely condition that all the disintegration energy remains in the nucleus A, Z, then neutron emission will be possible if

$$B+0.2a-0.9a(Z_0-Z)+\tfrac{1}{2}b < 2a(Z_0-Z) \\ +a+b \cdots (A \text{ even}, Z \text{ even}),$$
$$B+0.2a-0.9a(Z_0-Z)-\tfrac{1}{2}b < 2a(Z_0-Z) \\ +a-b \cdots (A \text{ even}, Z \text{ odd}),$$
$$B+0.2a-0.9a(Z_0-Z)-\tfrac{1}{2}b < 2a(Z_0-Z) \\ +a \cdots (A \text{ odd}, Z \text{ even}),$$
$$B+0.2a-0.9a(Z_0-Z)+\tfrac{1}{2}b < 2a(Z_0-Z) \\ +a \cdots (A \text{ odd}, Z \text{ odd}).$$

It follows that a delayed neutron is most likely to be emitted from a nucleus with even A, which results from a disintegration of a nucleus with an odd Z. As a condition of a neutron emission we obtain, in this case,

$$B < 2.9a(Z_0-Z)+0.8a+\tfrac{1}{2}b.$$

Again, this equation is most likely to be fulfilled for the light fragment. Even in this case and even under the most optimistic assumption which we used (that all the disintegration energy which produced the A,Z nucleus remain in it) we obtain $Z_0-Z > 2$. This shows that if a delayed neutron is emitted by a nucleus near the end of a radioactive chain, the fluctuations around the smooth mass formulae must be quite large to make this possible.[12]

[12] Similar deviations of the observed masses from smooth formulae occur in light nuclei also, e.g., at the ends of

IV. RELATION OF ENERGY TO HALF-LIFE. $\lambda(E)$

The next step is the establishment of a rule which connects the half-life $(\ln 2)/\lambda$ with the maximum energy of the β-ray. This is given, according to current theories,[13] by

$$\lambda = M^2/30[(E_\beta/mc^2)^5+5(E_\beta/mc^2)^4 \\ +10(E_\beta/mc^2)^3]. \quad (12)$$

In this, M is the matrix element for the nuclear transition, mc^2 the rest energy of the electron, and E_β the maximum β-ray energy of the transition in question. Equation (12) is not immediately useful for two reasons: first, because it contains the unknown quantity M, and second, because it refers to a β-transition to a definite level of the daughter nucleus, instead of referring to transitions to all levels of the daughter nucleus. For this reason, the energy of a transition E_β occurs in it instead of the disintegration energy. In spite of this, Eq. (12) indicates in a general way that for large E_β, the λ is proportional to the fifth power of E_β. For small E_β, the λ does not drop off as rapidly as the fifth-power law would indicate, so that one would be inclined, when giving a global expression for λ as a function of E_β, to use a lower power law and to write, e.g., $\lambda \sim E_\beta^4$. On the other hand, the total probability of the disintegration is the sum of the probabilities of the transitions to all possible levels of the daughter nucleus, and one can expect that there are more final levels available if the energy of disintegration is greater. Also, among the more numerous levels of the daughter nucleus in the case of high E, there is more likelihood to find one with a particularly large M. As a result, one may expect that λ is proportional to E^3 for small E and to an even higher power than E^5 in case of large E. Furthermore, one cannot expect a unique connection between E and λ, but a nucleus with a definite E will have a disintegration constant which may assume any value between a highest and a lowest limit. These expectations are, on the whole, corroborated by the experimental material shown in Fig. 5. This shows that there is a rather definite lower limit

closed shells. Very strong evidence for similar deviations in heavier nuclei will be given by Maria G. Mayer in a forthcoming paper in the present journal. Her article will also contain references to earlier literature.
[13] E. Fermi, Zeits. f. Physik 88, 161 (1938); also E. J. Konopinaski, Rev. Mod. Phys. 15, 209 (1943).

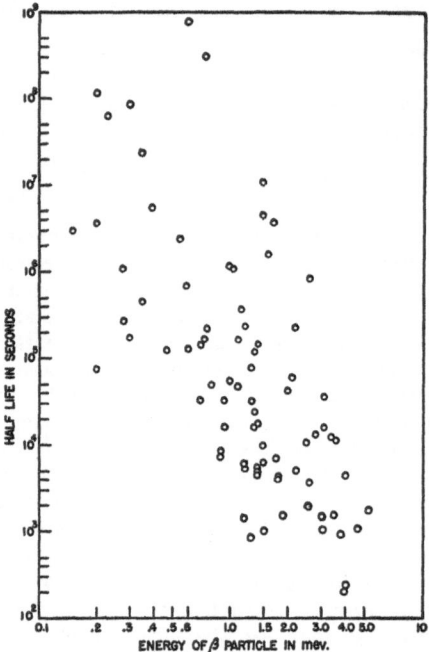

FIG. 5. Half-lives and disintegration energies of fission products.

for the half-life of a nucleus with a definite disintegration energy, corresponding to an upper limit of λ as function of E. (Incidentally, this upper limit of λ lies, as demanded by theory, considerably below the first Sargent curve, so that one obtains an E consistently too low if one tries to estimate it from λ and uses the first Sargent curve.) There is a considerable region of energy, from 0.5 to 1.5 Mev, in which the lower limit of the lifetime is proportional to E^5, and the lower limit behaves as expected outside this region.[14] The upper limit of the half-life, i.e., the lower limit of λ, is less pronounced, corresponding to the fact that there is no theoretical limit as to how forbidden a disintegration may be. In spite of this, one may recognize such an upper limit, running very roughly parallel to the lower. As a result, we can say that, very approximately,

$$\ln\lambda = -C + 5\ln E, \qquad (13)$$

where C has, with equal probabilities, values

[14] The single exception is Pd^{112}. The data was taken from reference 4.

between $C_u = 18.5$ and $C_l = 9$ if λ is measured in sec.$^{-1}$ and E in Mev. Although we realize its limitations, we have adopted (13) for the calculations. We must remember, however, at least the most obvious limitations of Eq. (13): it gives a too low λ for a very low E so that, for very large t when the lifetimes are determined by very low E, the activity will drop more rapidly than our formulae will indicate.

V. NUMBER OF DISINTEGRATIONS AND RADIO-ACTIVE ENERGY LIBERATED AS FUNCTION OF TIME AFTER FISSION

If (13) were valid for all nuclei with one single C, the number of β-rays emitted in unit time, at time t after fission, would be given by

$$\beta(t) = \int [P_0(E) + P_{00}(E) + 2P_{ee}(E)]e^{-C E^5}$$
$$\times \exp(-e^{-C}E^5 t)dE. \quad (14)$$

Similarly, the total energy (consisting of β-rays, neutrinos, and γ-rays) liberated in the same time interval would become

$$3B(t) + \Gamma(t) = \int [EP_0(E) + EP_{00}(E)$$
$$+ E_{eff}P_{ee}(E)]e^{-C E^5}\exp(-e^{-C}E^5 t)dE, \quad (14a)$$

where E_{eff} is the function given in Fig. 4. It is the sum of the disintegration energies of an even-even nucleus and of its odd-odd daughter.

Since, however, as explained in the preceding section, C varies from nucleus to nucleus, we have to average the above expressions over C, i.e., integrate them over C from $C_l = 9$ to $C_u = 18.5$ and divide the result by $C_u - C_l = 9.5$. This gives

$$\beta(t) = (9.5t)^{-1}\int [P_0(E) + P_{00}(E) + 2P_{ee}(E)]$$
$$\times [\exp(-x_u E^5 t) - \exp(-x_l E^5 t)]dE \quad (15)$$
$$x_u = \exp(-C_u) = 10^{-8.03};$$
$$x_l = \exp(-C_l) = 10^{-3.90}.$$

A similar expression can be obtained for the total disintegration energy liberated in unit time.

$$3B(t) + \Gamma(t)$$
$$= (9.5t)^{-1}\int [EP_0(E) + EP_{00}(E) + E_{eff}P_{ee}(E)]$$
$$\times [\exp(-x_u E^5 t) - \exp(-x_l E^5 t)]dE. \quad (15a)$$

In order to obtain $\gamma(t)$ or $B(t)$ and $\Gamma(t)$ separately, it would be necessary to know the fraction of energy that is emitted, on the average, in the form of γ-radiation during a disintegration with a definite disintegration energy. However, there is at present no theoretical foundation for determining this fraction and the quantities (15) and (15a) are the only ones for which a theoretical expression can be obtained with ease.

In general, i.e., for an arbitrary t, the integrals (15) and (15a) have to be evaluated numerically. Both square brackets in (15) as well as in (15a) tend to zero as E increases, but while at low t the first bracket goes to zero more rapidly, at high t the second one becomes very small first. As a result, one can evaluate both integrals more easily if either t is very small (smaller than 1 sec.), or if t is very large (more than one day).

In the former case, i.e., if t is very small, the exponentials in the second bracket can be expanded and the second bracket becomes

$$(x_l-x_u)E^5t-\tfrac{1}{2}(x_l{}^2-x_u{}^2)E^{10}t^2$$
$$\approx x_lE^5t-\tfrac{1}{2}x_l{}^2E^{10}t^2, \quad (16)$$

since x_l is so much larger than x_u. Hence, for very small t one has with

$$c_0=9.5^{-1}e^{-9}\int[P_0(E)+P_{00}(E)+2P_{ee}(E)]E^5dE, \quad (17)$$

$$c_1=9.5^{-1}e^{-18}\int[P_0(E)+P_{00}(E)+2P_{ee}(E)]E^{10}dE,$$

$$\beta(t)=c_0-\tfrac{1}{2}c_1t\sim0.38-2.6t \text{ per sec.} \quad (17a)$$

and with

$$C_0=9.5^{-1}e^{-9}\int[EP_0(E)+EP_{00}(E)+E_{eff}P_{ee}(E)]E^5dE,$$

$$C_1=9.5^{-1}e^{-18}\int[EP_0(E)+EP_{00}(E)+E_{eff}P_{ee}(E)]E^{10}dE, \quad (18)$$

$$3\mathrm{B}(t)+\Gamma(t)=C_0-\tfrac{1}{2}C_1t$$
$$\approx(3.8-0.61t)\text{ Mev/sec.} \quad (18a)$$

Unfortunately, the evaluation of these integrals, c_1 and C_1 in particular, is strongly dependent on the high energy tail of the P. Hence, the value of the constants in (18a) depends greatly on the accuracy of (5) which gives the probability of deviation of the original Z from $Z_0-3.5$. Hence, the relatively good agreement between (18a) and the measured values should rather be taken as an indication that (5) is not too inaccurate for values of $Z_0-3.5-Z$ which are several units great. One must remember that the experimental foundation[6] of (5) comes largely from the region in which $Z_0-3.5-Z$ is negative.

The situation is much more favorable for large t. In this case, the second bracket of (15) and (15a) drops so rapidly that the values of the P for $E=0$ can be used in the first bracket. As was pointed out in Section III, the curves of Figs. 3 and 4 are most accurate in the region of low E where they are essentially constant. We have $P_0=\tfrac{1}{4}(a_L{}^{-1}+a_H{}^{-1})\sim0.63$ (Mev)$^{-1}$, $P_{00}\sim0$, and $P_{ee}=\tfrac{1}{2}P_0\sim0.32(Mev)^{-1}$ in this region. As a result, the first bracket of (15) becomes 1.27 and that of (16a), $0.63E+0.32(2E-2a+2b)$. Substituting these expressions for the first brackets, the integrals can be carried out if one introduces a new variable y for x_uE^5t and x_lE^5t. One has, after a day,

$$\beta(t)=1.27\times9.5^{-1}t^{-6/5}(x_u{}^{-1/5}-x_l{}^{-1/5})1/5$$
$$\times\int_0^\infty y^{-4/5}e^{-y}dy\approx5.2\times10^{-6}d^{-1.2} \quad (19)$$

and

$$3\mathrm{B}(t)+\Gamma(t)=(3.9d^{-1.2}+11.7d^{-1.4})$$
$$\times10^{-6} \text{ Mev/sec.,} \quad (19a)$$

where d is the time in days. For very large t the first term of (19a) predominates. The origin of this term is the disintegration of an odd-odd nucleus, the formation of which has been long delayed because its even-even parent's disintegration energy was very small.

The variation of the number of β-particles with time as $t^{-1.2}$ depends essentially only on (13), i.e., on the proportionality of the disintegration constant with the fifth power of the disintegration energy. The $t^{-1.2}$ law follows at once from this and the circumstance that the number of radioactive nuclei per unit disintegration energy range is, for small disintegration energies, independent of energy. This last circumstance follows

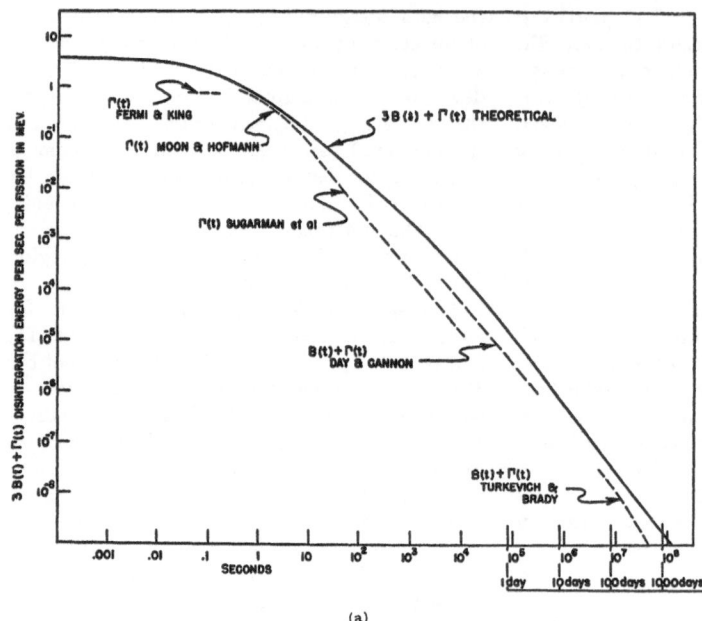

(a)

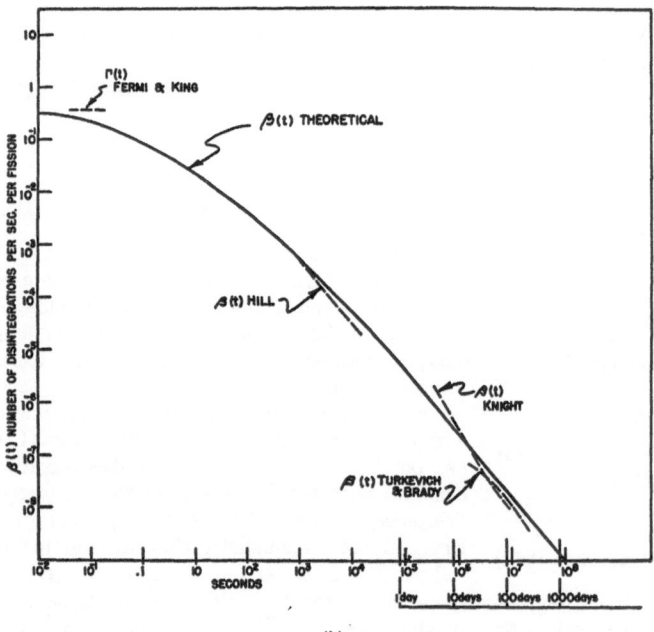

(b)

FIG. 6. Theoretical and experimental results for the radioactivity t seconds after fission. $\beta(t)$ = number of β-particles emitted per second, [$\gamma(t)$ same for γ-rays] $B(t)$ = energy of β-particles in Mev per second, $\Gamma(t)$ = energy of γ-rays in same units. $3B(t) + \Gamma(t)$ is total energy (including neutrino energy) liberated.

TABLE II. Summary of experimental results on rate of decay of fission products.

Quantity	Function of time t after fission	When valid	Reference
$\beta(t)$: Number of β-rays emitted per sec. per fission	$2.54\ t^{-1.23}$	10 min.–4 hrs.	D. L. Hill and L. Lanzl, Metallurgical Laboratory, CP-1827, June 25, 1944.
	$ct^{-1.7}$–$ct^{-1.2}$	1–100 day	J. D. Knight, Clinton Laboratories, CC-3479, May 1, 1946. PPR, 9B, 6.14.
	ct^{-1}–$ct^{-1.6}$	16–240 days	E. L. Brady and A. Turkevich, Metallurgical Laboratory, CL-697, III, D6, March, 1945.
$B(t)+\Gamma(t)$: Total γ-energy and average β-energy emitted per sec. per fission in Mev/sec.	$5.1\ t^{-1.23}$	1–100 hrs.	R. A. Day and C. V. Cannon, Clinton Laboratories, CP-2176, November 9, 1944.
	$5.6\ t^{-1.26}$ $98.0\ t^{-1.41}$	20 m–3 days 50–100 days	L. Borst, Metallurgical Laboratory, CL-697, VIII, C4.
	$29.4\ t^{-1.35}$	16–340 days	E. L. Brady and A. Turkevich. See above.
$\Gamma(t)$: γ-energy emitted per sec. per fission in Mev/sec.	$\sim$ const.	40–140 millisec.	L. D. P. King and E. Fermi, Los Alamos Laboratories, quoted in LA-253A, December 7, 1945.
	See Fig. 6.	0.1–10 sec.	J. A. Hofmann and P. B. Moon, Los Alamos Laboratories, LA-253A, April 7, 1945. I. Halpern, J. A. Hofmann, P. B. Moon, R. Perry, Los Alamos Laboratories, LA-253A, December 7, 1945.
	$0.90\ t^{-1.20}$	10 sec.–1 day	S. Katcoff, B. Finkle, N. Elliott, J. Knight, N. Sugarman, Metallurgical Laboratory, CC-1128, December 11, 1943.
	$4.2\ t^{-1.28}$ $49.0\ t^{-1.41}$	20 m–3 days 50–100 days	L. Borst. See above.
	$ct^{-1.28}$–$ct^{-2.0}$	16–240 days	E. L. Brady and A. Turkevich. See above.

from the assumption that the disintegration energies of successive members of a chain form an equidistant set. The same holds of the possibility of representing the energy liberation as the sum of two terms, proportional to $t^{-1.4}$ and $t^{-1.2}$, respectively. The present paper owes its origin to the recognition of this fact and the experimental confirmation of the $t^{-1.2}$ law.

VI. COMPARISON WITH EXPERIMENT

Figure 6 shows the theoretical curves for $\beta(t)$ and for $3B(t)+\Gamma(t)$ and various experimental results. The experimental work is also listed in Table II where analytical representations are given where possible.

The agreement between the theoretical $\beta(t)$ and the experimental results is seen to be fairly good. No experimental values are available for very short times. The fact that the theoretical curve lies above the experimental ones for the longer times is perhaps due to the approximation made in choosing the relation between λ and E.

The total energy released per second per fission, $3B(t)+\Gamma(t)$, includes the energy carried

by neutrinos. The total absorbable energy is only $B(t)+\Gamma(t)$. For times of the order of days after fission, calculations from known β- and γ-energies and fission yields of the different fission products show that $B(t)$ and $\Gamma(t)$ are approximately equal. At these times one would thus expect $3B(t)+\Gamma(t)$ to be twice $B(t)+\Gamma(t)$. Figure 6 shows that it is actually somewhat larger than twice the values of R. A. Day and C. V. Cannon which were obtained by calorimetric measurements. (See Table II for references.[15]) They are also more than twice the values calculated from the lifetimes, decay energies, and yields by A. Turkevich and E. L. Brady. Perhaps this is caused again by the relation between λ and E.

No experimental values of $B(t)+\Gamma(t)$ for very short times after fission are available, but several measurements have been made of $\Gamma(t)$ alone. Figure 6 and Table II give the results. At times of the order of 0.1 sec. after fission the value of $3B(t)+\Gamma(t)$ is about two and one-half

[15] See also articles by J. D. Knight; R. A. Day and C. V. Cannon; L. B. Borst; S. Katcoff, B. Finkle, N. Elliot and J. Knight in the forthcoming Plutonium Project Records, Vol. 9B.

times the experimental value of $\Gamma(t)$, which is certainly very plausible. At these short times, the calculated values depend very sensitively on the values of the disintegration energies and lifetimes of the primary fission products. The agreement with experiment at these times thus lends some support to the assumptions which governed the choice of the initial energies and lifetimes which were (1) that the parabolic mass formula holds for nuclei quite far removed from the region of stability and (2) that the chance of finding a given charge on the primary fission product is given by Eq. (5).

The authors are very much indebted to Mrs. N. Dismuke and Mrs. G. Haines for help in calculating values for the theoretical curves, and to Mrs. A. T. Monk who gave much appreciated assistance with an earlier report on which the present one is based. They are greatly indebted also to members of the Chemistry Divisions of both Argonne and Clinton Laboratories for many helpful discussions and clarification of experimental data. This article is based on work performed under Contract No. W-35-058-eng-71 for the Manhattan Project and the Atomic Energy Commission at Clinton Laboratories.

26.

The Effects of Radiation on Solids

F. Seitz and E. P. Wigner

Scientific American *195* (2), 76–84 (1956)

The orderly atomic arrangement characteristic of metals and other crystals determines many of their properties. Energetic radiation disturbs the order and thus can drastically alter the properties.

In the anxious days when atomic scientists were building the first chain-reacting pile in the "Metallurgical Laboratory" at the University of Chicago, no problem was more worrisome than the question concerning how the pile would be affected by its own radiation after it became active. On most of the other problems—the critical requirements for the chain reaction, controls, shielding, cooling—the physicists felt fairly confident of their calculations. But the radiation question was full of uncertainties. It was known that exposure even to weak natural radioactivity could change the structure and properties of materials. What would happen to the uranium rods in the reactor under the disruptive forces of intense neutron radiation, nuclear fissions and so on? More serious still, what would happen to the graphite moderator? Graphite was a part of the actual structure of the pile; unlike the uranium, it was not to be removed or replaced from time to time; and it was known to be subject to damage by radiation.

The group concerned with the future health of the new atomic "child" was so uncertain and pessimistic about the reactor's ability to survive radiation and other "diseases" that it reported: "It would be unscientific to claim a useful life longer than about 100 days." More

than 50 times that period has now passed and nearly all the original reactors are still alive and operating. What we did not realize at the time was that graphite, as well as metal, has some ability to recover from radiation damage —to heal its wounds, so to speak. Nevertheless, the effect of radiation on solids remains an important and absorbing study. It is still a major practical problem in the construction of reactors; besides this, it has become a valuable tool for fundamental research into the properties of solids. Research on radiation damage is now being carried on not only in the national laboratories of the Atomic Energy Commission but also at a number of universities and industrial laboratories. The AEC recently announced eight such research contracts totaling well over $250,000 a year. The program of study of radiation effects on solids has steadily grown both in magnitude and in scope.

Let us try to describe some of the facts we have learned about radiation damage. Metals and nonmetals react differently; we shall consider first the effects on a nonmetal—the graphite (crystalline carbon) commonly used as the moderator in a reactor. The neutrons released by uranium fission in a reactor have a kinetic energy of about one million electron volts. When a fast neutron strikes the nucleus of a carbon atom in the moderator, it transfers a substantial fraction of its kinetic energy to the atom, and the latter recoils from the impact. Since the carbon atom's recoil energy is much greater than the binding energy holding it in the crystal lattice (which is less than 10 electron volts), the atom is thrown out of its normal position. This results in two defects in the lattice: the dislodged atom occupies an interstitial space in the lattice (like a marcher out of his row in a parade), and it leaves behind a vacant site in the regular order.

The foregoing describes the direct effect of collisions between fast neutrons and atoms in the lattice. These collisions in themselves account for only a small part of the damage actually produced. A fast neutron dislodges about 60 carbon atoms, at most, before it is slowed to a harmless speed. It is the recoiling carbon atoms that produce most of the damage in the lattice. They have bulk as well as speed. The first carbon atom hit by a million-volt neutron, for example, recoils with an energy of about 150,000 electron volts. In effect it acts like a strong and husky man who decides to get out of a very crowded subway rather suddenly. It throws the other atoms to right and left until it reaches the end of its range, that is, until its energy is exhausted.

Now it develops that in the atomic world this series of events takes a turn which is the opposite of what one might expect if he thinks in terms of mechanical collisions. The charging atom creates more havoc near the end of its rush than it does at the beginning. The reason is that we are dealing here with interatomic forces rather than what we usually think of as physical contact. As the fast-moving atom begins its dash through the crowd of surrounding atoms, its encounter with each one is too fleeting to permit much transfer of its mo-

mentum. It therefore dislodges only an occasional atom from its lattice position. But as the traveling atom slows down, the interatomic forces have more time to act, and it displaces more and more atoms. Finally, when it drops to a certain low velocity, it transfers its remaining energy to a local cluster of atoms. As a result the tiny local region suddenly heats up, sometimes to a temperature as high as 10,000 degrees centigrade. This phenomenon, called a "thermal spike" or "displacement spike," lasts only about one hundredth of a billionth of a second, but it may damage or deform the crystal.

Its effects are fairly complicated and not yet well understood. It appears that the minute "spike" region melts. Evidence of this melting has been found in radiation experiments on a carefully prepared alloy of copper and zinc. The atoms were arranged in a regular lattice in which each copper atom was surrounded by eight zinc atoms and *vice versa*. Bombardment of this crystal with neutrons was done at very low temperatures, near the temperature of liquid helium, in order to "freeze in" any changes in the crystal. Analyses afterward showed that the atoms had become mixed in a disordered way, and that most of the disordering must have taken place in regions of thermal spikes.

Besides melting, the heated regions expand. Such swelling causes deformations of the crystal, some of which presumably remain after the hot regions cool, so that the material around them is permanently distorted.

In a crystal damaged by radiation it is very difficult to distinguish how much of the damage is due to these spikes and how much to simple displacement of atoms. We can assume that spikes are a more important source of damage in metals than in graphite, because in the heavier elements recoiling atoms produce spikes at a higher energy level and therefore have a larger fraction of their

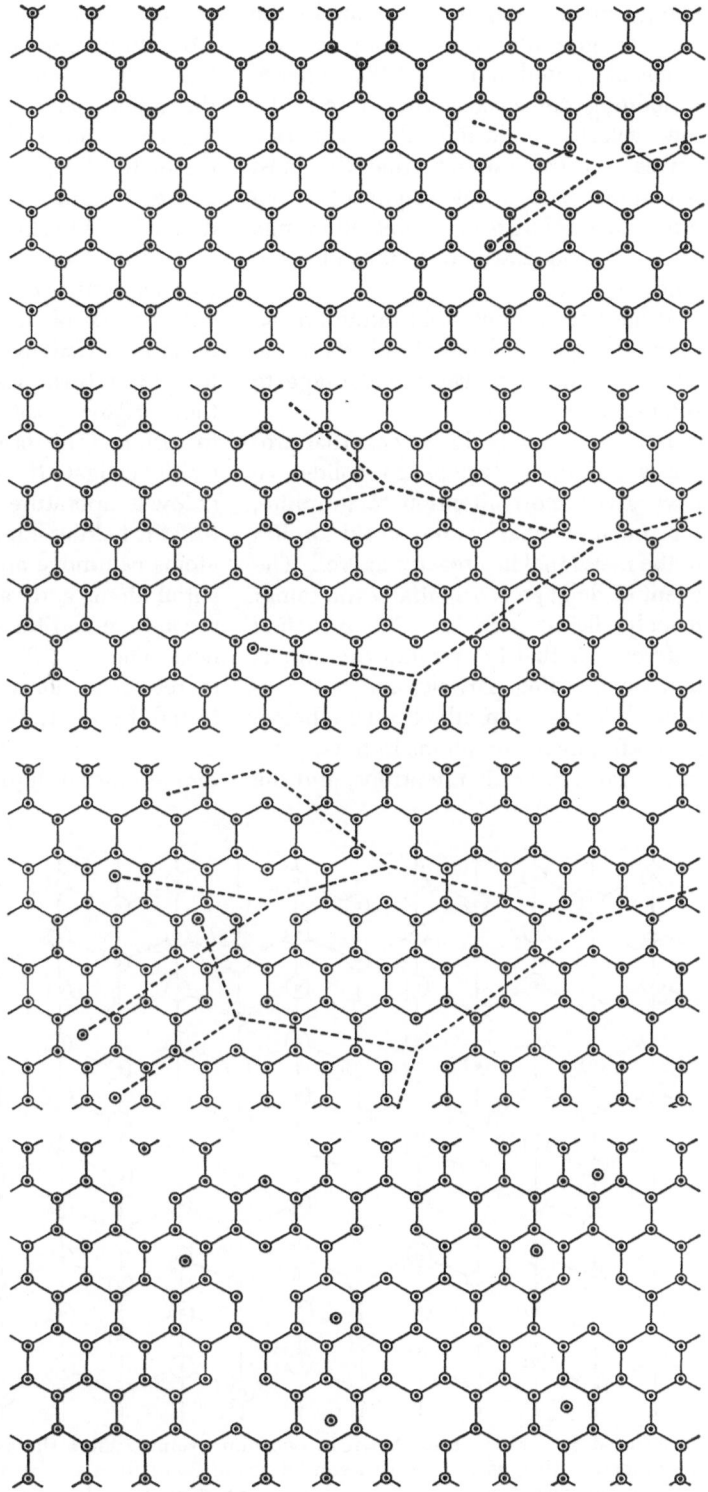

LATTICE DEFECTS are produced when a neutron strikes a graphite crystal. The hexagonal crystal structure is represented diagrammatically in two dimensions. At top the neutron (*colored dot*) has struck and dislodged a single atom. The next two drawings show how the process builds up, with both neutron and recoiling atoms acting to dislodge further atoms (*neutron path is in color; atom paths are in black*). At bottom is the final result: a lattice with a number of vacant sites and "interstitial" atoms. (Color not reproduced here.)

energy left for producing them. In the case of graphite, the moving carbon atoms have used up most of their energy dislodging atoms before they drop to the low velocity at which they generate spikes. We can estimate that the most damaging part of the flight of recoil atoms in graphite is in the velocity range from 100,000 down to 10,000 electron volts.

It has become clear that radiation can produce a great variety of defects in the lattice, resulting in varying damage to the material.

In this account of the process that produces radiation damage in solids we have given most attention to graphite, but much of what we have said applies to the metal fuel in a reactor as well. The agent of damage is essentially the same: namely, flying particles. The principal difference is that in uranium the important bombarding particles are not neutrons but fission products. The heavy fission fragments hit atoms in the crystal lattice far harder than neutrons, and the

atoms receive, on the average, about 1,000 times more energy. The damage is therefore much greater. In addition, the fission conversion of part of the uranium into other elements also weakens the metal. Fortunately metals are tough and can stand a lot, particularly if they do not have to stand it too long!

Let us consider now the recuperating powers of materials damaged by radiation. Usually dislodged atoms attempt to return to something resembling their original positions in the lattice and to restore their original properties. We can investigate the recovery process best at low temperatures. If the damaged material is held at a temperature where the atoms can move around a bit, the interstitial atoms and vacancies will begin to recombine and the lattice distortion will heal. The crystal's properties then tend to return to normal. This is well illustrated by a study of the recovery of copper after it was irradiated near the temperature of liquid helium. The prop-

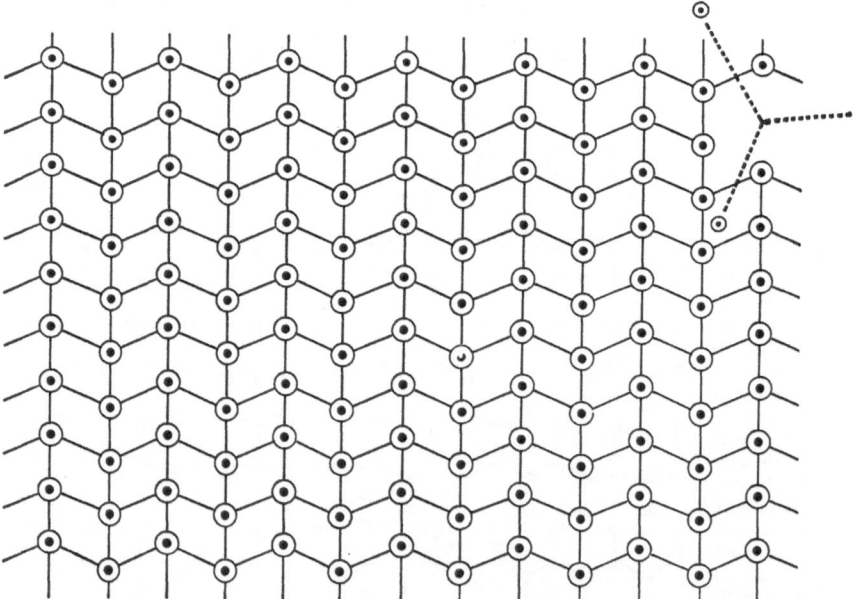

URANIUM LATTICE, also shown in two dimensions, suffers the same kind of damage as the carbon lattice, with an added complication. Sometimes the neutron is absorbed by the uranium nucleus, causing fission. The fission products may do more damage than neutrons.

erty measured was its conduction of electricity. Copper, which is a nearly ideal metal, recovers very rapidly if it is irradiated near room temperature; to "freeze in" all the damage and prevent recovery during irradiation it must be kept not far above absolute zero. Now when the temperature of the irradiated specimen is raised to about 35 degrees Kelvin, its electrical conductivity increases sharply. It is not yet known whether this abrupt and irreversible change, common to copper and many other metals, is a result of the reunion of vacancies and interstitial atoms which are very close to one another or whether it is due to healing of some of the distortion produced by thermal spikes. This is one of the critical questions being investigated at several laboratories.

It is interesting to note that each increase in temperature permits a little more of the damage to heal. This shows that there is a spectrum of different types of defects, some of which are more resistant to correction than others. We know that small traces of impurity atoms can have a significant influence on the rate of recovery. Some of the defects produced are so stable that one must heat the metals to temperatures nearly halfway to the melting point in order to remove them.

On the whole, pure metals are the most resistant of all materials to radiation damage and recover most easily, presumably because the atoms in metals are most mobile. But reactions like those in metals have been found in valence compounds such as diamond, silicon and germanium, and in simple salts and oxides such as sodium chloride and beryllium oxide. On the other hand, organic materials, particularly polymers such as plastics, are exceedingly sensitive to radiation and suffer permanent and irreparable changes. In these cases the damage is associated with the breaking of chemical bonds which are diffi-

cult to rejoin in the original way. Most polymers lose their ductility when given even moderate exposures. In brief, they behave in a way almost opposite to that of the metals.

From the practical standpoint, what are the types of damage—and the possible benefits—produced by radiation? As we have seen, the microscopic result of irradiation is the formation of lattice defects. How do these defects alter the properties of the material? There are four important kinds of predictable changes.

First, we know that properties such as conduction of electricity and heat depend on a regular and undistorted lattice. We are not surprised to find, then, that the conductivity of materials for both electricity and heat falls sharply with increasing irradiation [see charts at the left]. Losses of conductivity up to 30-fold have been measured. Fortunately reactors do not rely too heavily on the heat conductivity of the moderator, and that of metals is less severely affected by radiation. Hence the decreases in conductivity do not cause real concern from the point of view of reactor operation. We must quickly add, however, that these changes do affect the instruments stuck into the reactor and must be taken into account in problems concerned with instrumentation.

The second type of radiation damage is represented by a loss of ductility. The lattice defects have the effect of blocking the glide planes of the crystals. Thus the materials behave as if work-hardened, and in fact may become brittle. This damage affects the handling of uranium fuel elements and is a major cause for concern. The changes in ductility can be spectacular. The effect was demonstrated in a U. S. atomic energy display at Geneva. Every few seconds a light ball was thrown alternately at two copper cylinders, which looked

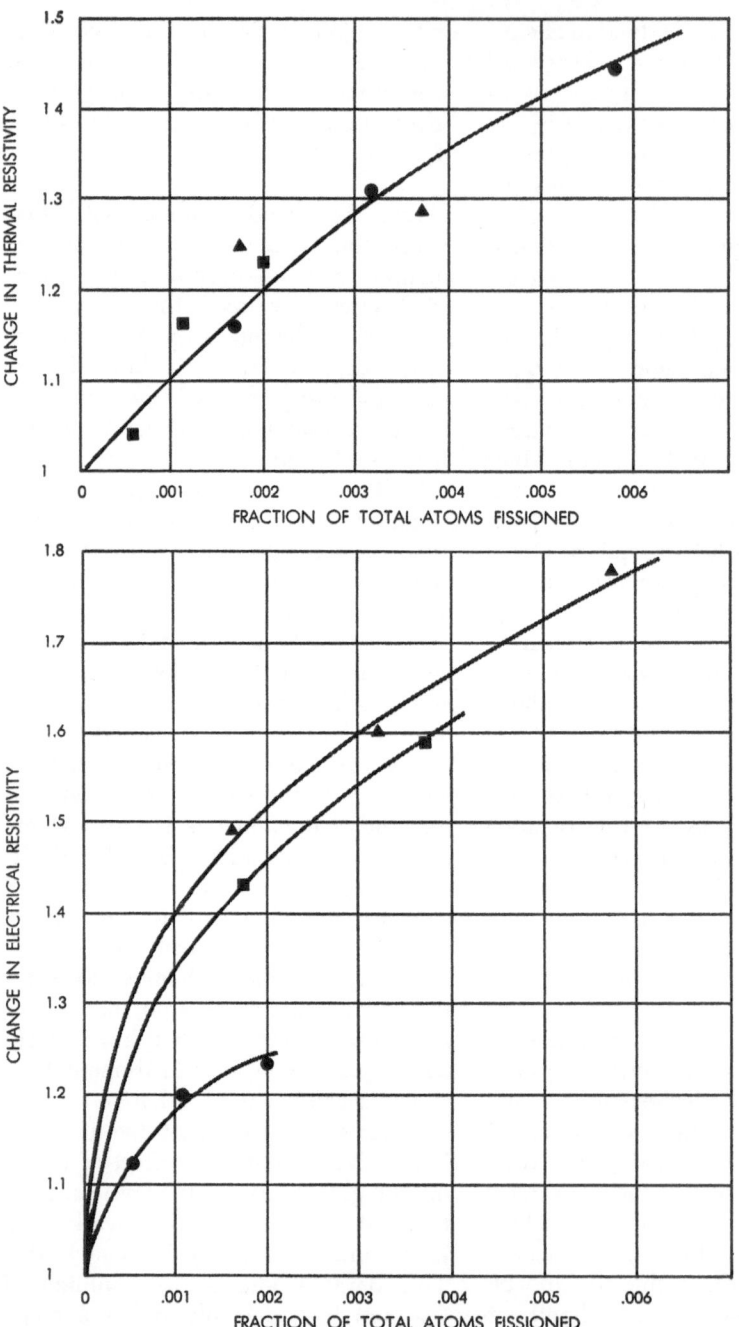

CHANGE IN PROPERTIES of uranium-aluminium alloys caused by fission of some of the uranium atoms was measured by D. S. Billington of Oak Ridge National Laboratory. The upper graph shows the increase in resistance to the flow of heat; the lower graph, the increase in resistance to electricity. Squares, circles and triangles represent observations on alloys containing respectively 5.7, 15 and 17.2 per cent uranium by weight. The vertical scales give the ratio of final to initial value. Thus the upper graph shows that when two thousands of the atoms have split, the thermal resistivity is 1.2 times its normal value.

identical but differed in the fact that one had been exposed to the neutrons of the Oak Ridge reactor. The normal cylinder, when hit by the light ball, gave no sound. But the irradiated one sang like a tuning fork. We understand that no amount of normal cold-working could endow copper with as much rigidity as this irradiated specimen possessed.

These first two types of effects—on conductivity and ductility—are the most striking but not necessarily the most harmful changes caused by irradiation. From the point of view of reactor operation there are two others which have caused more anxiety.

One is a swelling of the material. The displacement of atoms to irregular positions in the lattice expands the crystals. Hence the volume of a block of material increases as the dosage of radiation increases. When the Materials Testing Reactor of the AEC in Idaho went into operation with its new beryllium oxide moderator, the moderator expanded about 1 per cent the first day. Fortunately this expansion did not proceed linearly with time: after 10 days it was much less than 10 per cent. Nevertheless, it can be very disconcerting to have to use as structural elements materials which change their dimensions after they are installed.

The other disquieting effect of radiation is an unstable energy situation. The interstitial atoms represent a considerable amount of stored energy. When they move back into vacancies in the lattice, they release this energy. The amount of sword-of-Damocles energy stored in this way can reach values up to hundreds of calories per mole (one gram multiplied by the molecular weight of the material). Obviously a sudden release of it could lead to unpleasant complications. On the other hand, this property also has

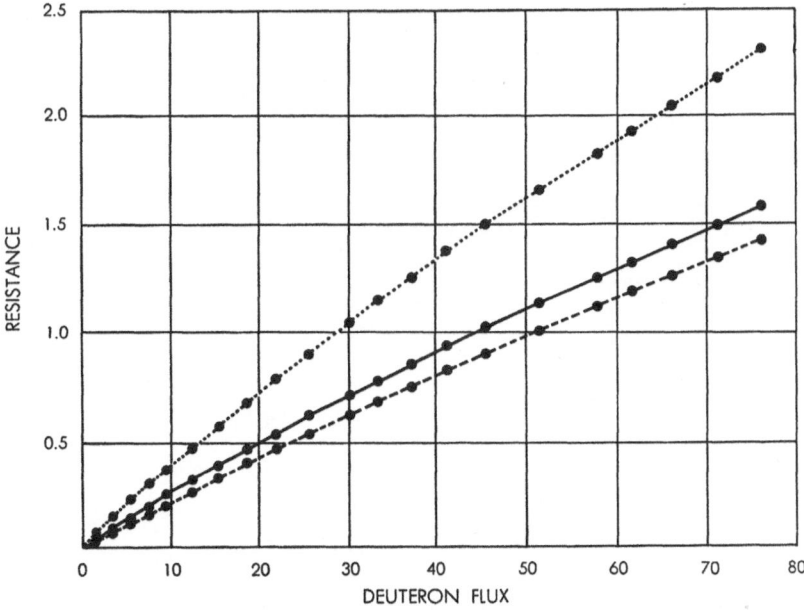

DEUTERON BOMBARDMENT of copper (*broken curve*), silver (*solid curve*) and gold (*dotted curve*) increases the electrical resistance of these metals. The horizontal scale gives the number of particles per second in each 10^{-15} square centimeters of cross section of the beam. The vertical scale shows the increase in resistance, measured in ten millionths of an ohm, of a one-centimeter cube. Measurements were made at 10 degree above absolute zero.

constructive possibilities: some have suggested using irradiated graphite as a kind of storage battery.

We call the various effects mentioned "damage" because they change critical properties of materials that have been placed in reactors to perform definite functions based in part on these properties. The changes in properties are regarded as harmful not because they would not be useful under certain circumstances, but because they impair the behavior for which the material was selected. To minimize the effects of these changes in a reactor, it has been suggested that materials might be deliberately irradiated before they go into the reactor. This stratagem might yield materials with desired properties and stability against further irradiation.

Indeed, we can expect that irradiated materials will be put to more and more uses as understanding of their properties and potentialities grows. Graphite storage batteries and the superhardening of copper are only a beginning of the list of possibilities. We have scarcely scratched the surface of knowledge of the radiation-induced properties of materials.

Speaking as individuals who have been interested in radiation effects on solids since the conception of the first large reactors, we find it gratifying that a phenomenon which originated as a pure nuisance promises to provide us with useful information about the solid state in general and about many of the materials we use every day.

Reactor Theory

Annotation by Alvin M. Weinberg

27. E. P. Wigner, "Theoretical Physics in the Metallurgical Laboratory of Chicago", *J. Appl. Phys.* 17, 857 (1946).

This is the first "unveiling" of the wartime work on reactors by Wigner and his group.

28. E. P. Wigner, "Mathematical Problems of Nuclear Reactor Theory", *Proceedings of Symposia in Applied Mathematics*, Vol. XI, Nuclear Reactor Theory, pp. 89–104, 1961. (American Mathematical Society).

Wigner had co-authored the *Physical Theory of Neutron Chain Reactors* in 1958. In paper 28, which was given in April of 1959, he reports progress on several mathematical problems that he had identified while writing the book.

Microscopic Reactor Theory

29. R. F. Christy, A. M. Weinberg, and E. P. Wigner, "Resonance Escape Probability in Lattices", CP-2602, August 19, 1944.

Though this report appeared in 1944, the methods described therein had been developed by Wigner by late 1941.

30. Wigner, E. P. and Plass, G. N., "On the Utilization of Thermal Neutrons – I", CP-103, (No date, probably written in 1941).

In Paper 30 Wigner derived the simplest expression for the thermal utilization in a spherical cell, and also showed that the error caused by sphericizing a cubic cell was very small.

Papers 29 and 30 served as the basis for the earliest calculations of the multiplication constant in an infinite heterogeneous reactor. By early 1942, the significance of fast fission was recognized, and all later calculations included estimates of ε, the fast effect. This gave the four-factor formula, $k = \eta \varepsilon p f$, η being the number of neutrons released per neutron absorbed in U. The quantity actually calculated in the following reports was $\eta' = (1/\varepsilon p f)$, $k = \eta/\eta'$.

T. Plass, G. N. and Wigner, E. P., "Values of the Thermal Utilization, Resonance Absorption, and Fast Neutron Effect for Oxide and Metal Spheres and Cylinders", CP-372, December 14, 1942.

T. Sacher, G., Weinberg, A. M., and Wigner, E. P., "η', Thermal Utilization and Resonance Absorption in Hex", CP-442, January 29, 1943.

T. Stephenson, J., and Wigner, E. P., "On the Multiplication Constant of Homogeneous Mixtures of U with Various Moderators", CP-668, May 15, 1943.

T. Wigner, E. P., Weinberg, A. M., and Stephenson, J., "Recalculation of the Critical Size and Multiplication Constant of Homogeneous UO_2-D_2O Mixtures", CP-1361, February 11, 1944.

Temperature Coefficients

T. Coon, J. H., Manly, J. H., Wigner, E. P., and Wheeler, J. A., "Temperature Coefficients of k: Notes on the Diffusion Length for Thermal Neutrons", C-54 (No date, probably early 1942).

31. Wigner, E. P., and Wilkins, J. E., Jr., "Effect of the Temperature of the Moderator on the Velocity Distribution of Neutrons with Numerical Calculations for H as Moderator", CP-G-2156, September 14, 1944.

Paper 31 gives the original derivation of the Wigner-Wilkins energy distribution of thermal neutrons in absorbing media.

T. Wigner, E. P., "The Magnitude of the η effect", AECD-3856.

This report estimates the contribution to the temperature coefficient of k contributed by the variations with temperature in η of natural uranium in the Savannah River Reactors.

Macroscopic Reactor Theory

32. Szilard L., A. M. Weinberg, E. P. Wigner, and R. F. Christy, "Approximate Boundary Conditions for Diffusion Equations at Interface Between Two Media", CP-189, July 10, 1942.

33. F. L. Friedman and E. P. Wigner, "On the Boundary Condition Between Two Multiplying Media", CP-1662, April 19, 1944.

Papers 32 and 33, separated by two years, deal with the general applicability of the one-group reactor equation, $\Delta \nu + B^2 \nu = 0$, even when there are several energy groups. Though $\nu = n \sigma_a M^2$ satisfies the one-group equation, this result was not used at Chicago in actual estimates of the critical size.

34. E. P. Wigner, "On Variations of the Power Output in a Running Pile", CP-351, November 11, 1942.

In Paper 34, Wigner derives the usual kinetic equations, in which the delayed neutrons dominate the variations of power following a small reacitivity changes. Similar results had been obtained by Fermi, by Wheeler and Ibser, and a little later, by J. Schwinger (who spent one month at Chicago in 1943 working with Wigner's group).

35. E. P. Wigner, "Use of the Pile Oscillator for the Measurement of Pile Constants", CP-G-3066, June 23, 1945.

In 1944 Wigner proposed deriving properties of a reactor by observing the response of the neutron flux to periodic modulation of the multiplication constant. He suggested that I work out the details of a simplified analysis of the scheme; this appeared as report CP-2907 by A. Cahn, A. Monk, and A. Weinberg. Paper 35 gives a much more sophisticated analysis of the problem. Wigner applied for a patent on the idea in 1947; the patent (31P) was issued on February 12, 1957. The method was used by H. Pomerance to demonstrate that hafnium-free zirconium has so low a neutron absorption cross-section, that it would make an admirable coating for fuel rods in water-moderated and cooled reactors.

36. E. P. Wigner, "Effect of Small Perturbations on Pile Period", CP-3048, June 13, 1945.

Wigner was the first to point out, in Paper 36, that the multigroup reactor equations were not self-adjoint. Thus the effect of a poison on the multiplication constant was proportional in general to the *product* of the neutron flux and its adjoint, not to the square of the flux as had been assumed before Paper 35 appeared. Of course in a large uniform reactor, the adjoint is everywhere proportional to the flux, so the simple recipe was usually valid; however in small enriched reactors, the adjoint is no longer proportional to the flux, and the adjoint has to be computed separately.

37. E. P. Wigner, A. M. Weinberg, and R. R. Williamson, "Efficiency of Control Rods Which Absorb Only Thermal Neutrons", CP-1461, February 24, 1944.

Paper 37 summarizes Wigner's method for estimating the number and disposition of control and safety rods at Hanford. Though the report appeared in February of 1944, the actual work was carried out during 1943, since the du Pont design was frozen during that year. The original du Pont design called for too few safety rods. As finally built, each reactor contained 29 rods, the number Wigner specified on the basis of these calculations.

27.

Theoretical Physics in the Metallurgical Laboratory of Chicago*

E. P. Wigner

Journal of Applied Physics *17* (11), 857–863 (1946)

November, 1946

DR. DARROW has suggested that a short introduction may enhance the usefulness of the more specialized papers to be presented to the Society by members of the old Theoretical Physics group of the Metallurgical Laboratory (Plutonium Project). Because the presentation of these papers will have to be very brief and because some of our collaborators could not present abstracts in time for the meetings, I was most happy to follow our Secretary's invitation to summarize our work in general terms.

There were, during the period extending from about the middle of 1942 until about the middle of 1945, that is for about three years, approximately twenty theoretical physicists assembled on the fourth floor of Eckart Hall who worked on those problems of chain reacting units which could be solved or at least attacked with the methods of theoretical physics. The membership of the group varied somewhat but I am happy to say, and do say it with a great deal of pride, that we formed a happy family and that I hardly remember a disagreement on non-technical points.

Most of our work was on very urgent problems and only a small fraction of it will bear publication in scientific periodicals. Our group had to do a great deal of engineering work, so much so that the calculation of liquid flow rates became to be considered to be part of our responsibilities. We calculated more than one I beam dimension, among similar matters, although we felt somewhat out of place when we did that. In addition, we had to maintain numerous contacts with the experimental groups and to take a lively interest in matters ranging from the fabrication and corrosion of aluminum tubes to the radioactivity induced in oxygen by neutron absorption. All this work was necessary and I do not hesitate to say in retrospect that our policy of assuming some of the functions which are usually reserved for engineers proved most useful. In the first place the Plutonium Project was not well provided with engineers in its early days—it had only two or three design engineers at the time when our plan for the W unit (which was later erected at Hanford) was virtually ready**; in the second place it was important, as in any new field such as that of chain reacting units was in 1942, that there be a few people at least who are sufficiently familiar with the whole picture to know about every difficulty and how it may be overcome. This second point will be very important in connection with Dr. Daniels' pile which is to be erected at Clinton. I fear that the present arrangement, which gives those who are

** It may be interesting to remark that the date on the W report is January 9, 1943 and that it was issued just 42 days after the first nuclear chain reaction was experimentally established by E. Fermi and his collaborators (December 2, 1942). Of course, the relevant dimensions and sizes, etc., of the W pile had to be fixed much prior to that date, although, of course, they had to be verified experimentally later.

* An address presented to the American Physical Society at the Chicago Meeting, June 22, 1946.

mainly interested in this pile even less responsibility than we had with respect to the W pile, will not help in meeting the time schedule which we read about in the newspapers. That time schedule was quite optimistic in the first place and I do not believe that the pile will be in operation before 1948.

As I said, knowing all details of the W plan was our main function and it took up most of our time. Practically all senior members of the group participated in it in one form or another although most of the burden was carried by Messrs. Friedman, Ohlinger, Weinberg, and Young and Miss Way. In addition, in the second year, Mr. J. A. Wheeler was transferred to Wilmington to give direct help to the DuPont Company. While work on the W plans was our most important function in the first two years, at the same time it was the one which is least suitable for being reported in public. In later years, work on the power production occupied most of our interest. However, this subject, to which Mr. G. Young has contributed so decisively, is still restricted from public discussion so that my report will deal, essentially, only with incidental studies which we have made. I shall deal with these under four headings:

1. Elementary theory of nuclear chain reactions.
2. More detailed theory of chain reactions.
3. Effect of radiation on matter.
4. Studies in theoretical physics.

I shall give an outline of our work in these fields in the above order.

1. ELEMENTARY THEORY OF NUCLEAR CHAIN REACTIONS

The great surprise about nuclear chain reactions was the ease with which they could be established. Szilard's paper of January 1940 already describes a workable arrangement. Our own early work in this field was not based on Szilard's paper but on Fermi's work, the concepts of which are less intricate than Szilard's. Ideas similar to Fermi's were developed also by others, notably by v. Halban; moreover, the whole work was duplicated, apparently without any major deviation, also by the German nuclear physicists.

In a chain reaction of the kind considered by us, uranium nuclei undergo fission and liberate

neutrons. These neutrons are first fast but are soon slowed down by the moderator which is carbon (graphite) in our case. After being slowed down, the neutrons still diffuse around for a period of time before being absorbed. Most of them are absorbed by uranium, which then undergoes fission and emits the neutrons of the next generation. The ratio of the number of neutrons in one generation to the number of neutrons of the preceding generation is called multiplication constant and was usually denoted by k. Fermi's theory divides the problem of multiplication constants and critical sizes into two parts. The first problem is the calculation of the multiplication constant in an infinite medium, k_∞, which is usually referred to briefly as the multiplication constant. It depends only on the geometry and the materials of the chain reacting system and gives the ratio of the numbers of neutrons in successive neutron generations under the assumption that the same materials, arranged in the same geometry, extend all over infinite space.

The second problem is the calculation of a critical length which does not depend on the inner structure of the chain reacting unit but only on its size and shape. This critical length, or itsr eciprocal κ, permits one to calculate, from k_∞, the second kind of multiplication constant, k_{eff}. This gives the ratio of the numbers of neutrons in successive generations in a finite pile. This second kind of multiplication constant is, of course, the relevant one from a practical point of view. It depends not only on the materials and their arrangement into a lattice which determine k_∞, but depends also on the actual extension of the lattice, i.e., the size and shape of the pile. In a steadily running pile k_{eff} is always 1 and it exceeds 1 only when the power of the pile is increased, e.g., during startup, and then only very little. The k_{eff} is always smaller than k_∞ because in an actual, finite pile some of the neutrons of every generation diffuse out of the pile and do not contribute to the next generation. No such "leakage" exists in an infinite pile.

It would seem that only the effective multiplication constant has real significance but it turns out that the calculation of k_∞ is an almost necessary preliminary for the calculation of k_{eff}.

I will only sketch the calculation of k_∞ which

is already given, in principle, in the Smyth report. In order to calculate the number of neutrons of the next generation produced by one neutron of the present generation, one may start at the birth of one neutron. This occurs in the uranium lumps and the neutron has, originally, considerable velocity. As a result, it will be able to induce fission not only in the U^{235} nuclei but, what is more important because of their larger numbers, also in the U^{238} atoms. Competing with this process are the process of inelastic scattering by U atoms by which the original neutrons may be slowed to a velocity below the fission threshold[1] of U^{238}, and the process of escape of the neutrons from the U lump into the moderator.

The importance of fast fission was recognized by Szilard and his collaborators. The rest of the factors making up k_∞ were all recognized before, and are contained also in Fermi's considerations.

Let us assume that the original neutron generates $\epsilon = 1$ further neutrons by fast fission. As Fermi has mentioned in his address to this meeting $\epsilon - 1 \approx .03$ so that we arrive with $\epsilon \approx 1.03$ neutrons just below the fission threshold. Most of these neutrons diffuse out into the moderator and are slowed down to thermal energies. Some of them occasionally enter the uranium and are absorbed there by one of the numerous resonance levels of the U^{238}. These absorptions do not lead to fission and constitute an actual loss of neutrons. The importance of this process has been already recognized by N. Bohr in 1940, and others. Only when the neutron has lost sufficient energy to be below the lowest resonance level of U^{238}—which is, according to data in the literature[2] at about 5 ev—is it safe from this fate. The probability that a neutron will escape resonance capture is usually denoted by p. It is a number smaller than 1. As a result of the resonance absorption, we arrive with ϵp neutrons below the energy of 5 ev instead of the ϵ neutrons which we had just below the fast fission threshold.

It would lead too far to describe the actual calculation of p. Among all the processes which contribute to the chain reaction, the resonance absorption is the only one which was not really understood when we started our work. S. M. Dancoff and I were the ones who were most interested in the physical principles which determine the resonance absorption of macroscopic bodies, but ideas similar to ours were developed also by others. The actual calculation of p was described by R. F. Christy, A. M. Weinberg, and myself, although many others, including H. L. Anderson, contributed to it. The material constants necessary for the calculation were measured by Creutz, Jupnik, Snyder, and R. R. Wilson in Princeton, and later by Mitchell's group at the University of Indiana.

We now have ϵp neutrons with an energy below the resonance levels of uranium. According to theory, they will be slowed down to thermal energies by the moderator. After that, they will be absorbed, some of them by the moderator and the impurities present in the pile, some of them by the uranium. Fermi denotes this last fraction by f so that, altogether, $\epsilon p f$ thermal neutrons are absorbed by the uranium, giving

$$k_\infty = \epsilon p f \eta \qquad (1)$$

secondary neutrons, η being the number of fission (fast) neutrons produced in the uranium per thermal neutron absorbed. The principles for the calculation of the "thermal utilization" f were established independently by Fermi, Placzek, and our group. The formulae which we used were derived by Christy, Mrs. Monk, Plass, and myself in a way which is similar to the calculation of wave functions in metals by the cellular method.[3]

On the whole, the calculation of the multiplication constant for an infinite lattice is quite "straightforward" and one of the great surprises of the Plutonium Project was how easy it was. Mr. G. N. Plass and myself happened to be the ones who attempted to calculate the "optimal lattice" (i.e., the lattice with the highest k_∞) early in 1942. Although the physical constants were not known at that time too accurately, the dimensions we obtained (later incorporated in the first chain reacting unit) are now believed to give a k_∞ just $\frac{1}{2}$ percent short of the k_∞ of the real optimal lattice. We are quite convinced that

---

[1] Haxby, Shoupp, Stephens, and Wells, Phys. Rev. **57**, 1088A (1940); **58**, 199A (1940).
[2] H. L. Anderson, Phys. Rev. **57**, 566 (1940).

[3] Cf. e.g., F. Seitz, *The Modern Theory of Solids* (McGraw-Hill Book Company, Inc., New York, 1940), Chap. IX.

any reasonably competent people would have arrived at the same results. In later years, computations of k_∞ were much facilitated by diagrams prepared for this purpose by Mrs. Monk and Mrs. Uchiyamada, under Professor Wheeler's direction. The calculation of k_∞ was extended to all sorts of lattices, containing heavy and ordinary water, etc. Most of this work was done by A. M. Weinberg and his collaborators, Mrs. Monk, Mr. Plass, Mrs. Uchiyamada, Mr. Stephenson and others. Qualitatively, the results were quite similar for all systems considered.

In spite of this, the properties which make a lattice optimal are not very simple. One may note that it is good if the high energy neutrons remain in the uranium to give a high ϵ. On the other hand, it is best if the lower energy (resonance) neutrons keep out of the uranium as much as possible so that p may remain reasonably close to 1. Again, the thermal neutrons should return to the uranium to give a high f—as close to 1 as possible. These conflicting requirements determine the geometry of the optimal lattice, i.e., give the ratio of the amounts of moderator and uranium as well as the lattice constant. However, even relatively large deviations from the optimal dimensions do not decrease k_∞ to a very great extent.

The foregoing describes the calculation of k_∞. Although Fermi has given a method for calculating k_{eff} from k_∞, I will not give his method here but will turn to the more advanced theory which permits a direct calculation if k_{eff}.

2. MORE DETAILED THEORY OF CHAIN REACTIONS

The more detailed theory of chain reactions should provide more accurate methods both for the calculation of k_∞ and also for the calculation of k_{eff}. However, as far as k_∞ is concerned, only few improvements were made. None of these improvements occurred in the calculation of ϵ and p, only one occurred in the calculation of f.

The behavior of "thermal" neutrons in a moderator-uranium lattice is far from simple. Evidently, it would take infinitely many collisions to establish real thermal equilibrium between the neutrons and the moderator, and in a well-designed lattice the neutrons will be absorbed by the uranium after a relatively small number of collisions. As a result, the energy spectrum of the neutrons will remain quite complex and their average energy will stay considerably above $\frac{3}{2}kT$. This average energy will be different even at different points of the lattice. The actual energy distribution will be influenced by the absorbing power of the material as well as by its moderating power. The latter is influenced in turn by the atomic weight of the moderator, by Fermi's chemical binding effect and by the crystalline nature of the moderator which gives a considerable anisotropy to the scattered (refracted) neutrons.

The only serious attempt to take these factors (excepting the crystal effect) into account is due to E. Teller and his collaborators, mainly N. Metropolis and P. Morrison. A more rigorous but much more formal attempt later by E. J. Wilkins and myself did not contribute much to the qualitative picture. Wick reported on Thursday about some work which he did on this problem. Teller's work gave, at least, an approximate measure for the difference in the effective temperature of the neutrons and the moderator. In spite of this we are far from having an adequate knowledge of the energy spectrum of the neutrons in a chain reacting unit.

Moreover, the problem of calculating f remains far from being simple even if the energy spectrum of the neutrons is known. It is, in fact, quite complicated even if one assumes that all the neutrons have the same energy. The reason for this is that the ordinary diffusion theory proves to be quite inadequate. G. Placzek carried out the most accurate calculations for the diffusion of monoenergetic neutrons. Some of his results were obtained also by German and Italian theorists[4] and published. Our work along this line was not pushed with much vigor because we were, perhaps, too well aware of the inadequacy of the model which uses monoenergetic neutrons. Actually, there is evidence that the errors in our primitive diffusion equations are quite substantial and in the direction indicated by Placzek's work.

There is no relevant difference between the fast effect in a finite and infinite lattice. However, the probability p that a neutron with an energy

---

[4] G. C. Wick, Zeits. f. Physik 121, 702 (1943).

just below fission threshold should become a thermal neutron is smaller in a finite lattice than in an infinite one because in addition to being captured by the uranium, some neutrons will be lost from a finite lattice by "leaking" (diffusing) out of it. This leakage was calculated by Fermi and his co-workers[5] even before fission was discovered. For a finite lattice, their work gives

$$p_{eff} = p \exp(-\tau \kappa^2), \tag{2}$$

where τ is one-sixth the mean square distance, in an infinite lattice, between the point where the neutron originated and the point where it becomes thermal. The quantity κ^2 is the ratio $-\Delta n/n$ where n is the average of the neutron density over a lattice cell and will be discussed presently.

According to (2) the effective p is smaller than it would be for the constant n of an infinite lattice, i.e., for $\kappa = 0$. The leakage depends on the "age" τ, which on its turn increases with increasing mean free path of the neutrons in the moderator and with the number of collisions which are necessary to slow down the neutrons to thermal energies. The quantity τ, and hence the leakage, is smallest in a water moderated pile and much greater in a graphite moderated pile.

Just as the fraction of neutrons which are slowed down to thermal energies in the pile is, because of the leakage, smaller in a finite than in an infinite pile, so is the fraction of thermal neutrons absorbed by the uranium decreased by the escape of some of the thermal neutrons from the pile. The equation analogous to (2) is

$$f_{eff} = f(1 + L_p^2 \kappa^2)^{-1}. \tag{3}$$

The significance of κ in (3) is the same as in (2), that of L_p^2 similar to that of τ in (2): L_p^2 is one-sixth of the mean square distance in an infinite lattice between the point where the neutron becomes thermal to the point to which it has diffused when it gets absorbed. L_p is also called the diffusion length of thermal neutrons in the lattice because the n decreases with an exponential relaxation distance L_p

$$n \sim \exp(-x/L_p) \tag{4}$$

in a region in which no thermal neutrons are

produced. G. N. Plass showed, by means of a calculation which is similar to Bardeen's work[6] on metallic wave functions, that

$$L_p^2 = L_m^2(1 - f) \tag{5}$$

is a very good approximation for L_p if L_m is the diffusion length in the pure moderator, without uranium lumps.

The condition that a lattice can maintain a chain reaction at a steady rate is that $k_{eff} = 1$, i.e., that

$$k_{eff} = \epsilon p_{eff} f_{eff} \eta = 1. \tag{6}$$

Using the expressions (2) and (3), this becomes

$$\epsilon p f \eta \exp(-\tau \kappa^2)(1 + L_p^2 \kappa^2)^{-1} = 1$$

or, by (1)

$$k_\infty = (1 + L_p^2 \kappa^2) \exp(\tau \kappa^2), \tag{7}$$

an equation essentially identical to one already obtained by Fermi.

This last equation can be considered to be an equation for κ which, in its turn, will be seen to depend only on the size and shape of the pile. Hence (7) gives us the size of a pile if its shape and internal structure, in particular its infinite multiplication constant k_∞, are given.

The connection between the quantity κ and the size and shape of the pile is established by the classical equation

$$\Delta n + \kappa^2 n = 0 \tag{8}$$

in which the average neutron density n is subject to the boundary condition that it vanish at the outer boundaries of the pile. It is well known that (8) allows a solution only for definite, discreet values of κ^2 which depend on the size and shape of the region on the boundary of which n has to vanish, i.e., on the size and shape of the pile. Only for the smallest of these κ^2 is n positive throughout and this smallest κ^2 is the one which occurs in (7). Equation (8), so to say, gives an effective dimension κ^{-1} to every size and shape and (6) shows how this effective dimension affects the effective multiplication constant. If the κ of the pile, as defined by (8), is larger than the solution of (7), the pile is under critical, its effective multiplication constant smaller than 1. If the solution of (7) is larger than the κ satisfying (8), the pile is above critical.

[5] E. Fermi and F. Rasetti, Ricerca Scient. 9, 472 (1938); G. Placzek, Phys. Rev. 69, 423 (1946).

[6] J. Bardeen, Phys. Rev. 49, 653 (1936). Cf. also reference 3.

The quantity n to which (8) applies is the average neutron density, the average to be taken over a cell. Evidently, an equation, applying to such an average as (8) does, can be accurate only if this average does not change too rapidly from cell to cell. The relation of the n of (8) to the actual neutron density is similar to the relation of the macroscopic density of bodies to their rapidly fluctuating density as given by their atomistic structure. The theory of Eq. (8) is therefore called the macroscopic pile theory while the quantities of Eqs. (1) to (7) are concepts of the microscopic pile theory. Actually, (8) is only the simplest equation of macroscopic pile theory, which applies if the spatial variation of the neutron density is independent of energy. This is an important particular case but does not hold in general. For instance, most control rods absorb only low energy, thermal neutrons. The surface of a control rod is, therefore, a boundary where the density of thermal neutrons vanishes. However, the density of fast neutrons does not vanish at the surface of the control rod and the densities of fast and of slow neutrons are not proportional any more. Problems of this nature call for more complicated equations than (8). The most important results toward the solutions of these problems are due to Messrs. F. L. Friedman, A. M. Weinberg, and J. A. Wheeler.

Even the simple Eq. (8) raises a number of interesting problems. If the shape of the pile is at all complicated—which is almost invariably the case if the chain reacting material is liquid—the solution of (8) could be obtained only by perturbation methods. Some of these show a remarkable similarity to the Rayleigh-Schrodinger method with which we are familiar from its application to quantum-mechanical problems. We owe many interesting results on (8) to Messrs. F. Murray, L. W. Nordheim, and H. Soodak.

A good part of the work in this connection is too special to be taken up in detail. Another part will be dealt with by the speakers following my address so that I may close the subject of the calculation of multiplication constants and critical sizes. I would not like to do this before emphasizing that, in my opinion, a good deal of work remains to be done in this field. In particular, the behavior of "thermal" neutrons in the pile and the transition from fast to thermal energies requires further clarification both from the experimental and from the theoretical side. But there remain interesting details to be worked out in almost any part of the theory. There are, also, some problems which have already commanded considerable attention but which I have not even touched. Chief among these is the change in the neutron densities with time if Eqs. (6), (7) are not exactly fulfilled and the pile is either below or above critical. Messrs. R. F. Christy, L. W. Nordheim, and J. E. Wilkins were particularly active in this field.

3. EFFECT OF RADIATION ON MATTER

The radiation densities, both γ and neutron, are higher in a plutonium producing pile than can be maintained outside the pile for extended periods. The effect of these radiations on the structure of materials was one of our early concerns from the theoretical point of view. The experimental work was carried out in the Chemistry Division. Dr. M. Burton reported at the Atlantic City meeting of the American Chemical Society about his, his collaborators' and Dr. J. Franck's work on the subject. On the theoretical side, M. Goldberger, R. S. Mulliken, and F. Seitz shared my interest in the subject which still has some aspects about which we cannot talk freely.

Clearly, the collision of neutrons with the atoms of any substance placed into the pile will cause displacements of these atoms. If the substance is a chemical compound, the displacement will result in chemical changes which were, of course, investigated already before chain reacting units came into being and are summarized, e.g., in the booklet of Lind.[7] All these changes are much more intense in the pile owing to the more intense radiation. But substantial effects can be expected in elementary substances also. The matter has great scientific interest because pile irradiation should permit the artificial formation of displacements in definite numbers and a study of the effect of these on thermal and electrical conductivity, tensile strength, ductility, etc. as demanded by theory. One may expect that

[7] Cf. e.g. S. C. Lind, *Chemical Effects of α Particles and Electrons* (Chemical Catalogue Company, New York, 1928).

studies of the solid state, particularly of the structure sensitive properties, will be greatly stimulated by the additional experimental facility given by the pile.

Before a final interpretation of the experimental results can be made, our knowledge of the ranges of low energy ions will have to be extended. It is on this subject that most of Messrs. Goldberger's and Seitz's work was concentrated and you will hear from them presently. A good deal of the rest of our work was speculation which will either be confirmed or refuted by future experiments.

4. THEORETICAL PHYSICS

As I emphasized before, real theoretical physics was always somewhat of a stepchild. This situation hardly could be remedied in view of our many pressing obligations. We tried to free Mr. S. M. Dancoff as much as possible from the pressure of urgent work and he carried out several investigations which are of considerable general interest. There is only one other function which we did not permit to be pushed into the background and this was the problem of keeping tables of nuclear constants up to date. Mrs. Uchiyamada was in charge of these tables but she received a good deal of help from the Project Information Department, in particular from Mr. Goldsmith who made a really brilliant contribution. We were happy to learn that these tables may be at least partially published soon.

The theoretical work of the group fell into two categories; help with the evaluation and planning of experimental work, and real theoretical work. Into the first category falls the work of Messrs. Cahn, Schweinler, Weinberg, and others on the so-called pile oscillator. This is an instrument which permits an absorber of known or unknown neutron absorbing characteristics to be put into periodic motion in the pile. The oscillation of the neutron absorber causes intensity waves to spread all over the pile. These waves are similar to the temperature waves in the earth, generated by the daily and yearly heat fluctuations of the heat input on the earth's surface. The amplitude and wave-length of the waves permits one to evaluate the characteristics

of the neutron absorbing oscillators and of properties of the pile.

The work on neutron diffraction received considerable attention on the part of Goldberger and Seitz. They interpreted and extended Weinstock's results[8] considerably and took into account phenomena not previously considered. Their work is being continued by Mr. M. Moshinsky in Princeton. There is little need for my going into details on this question since Seitz and Goldberger will tell you about them presently.

As a last example, I would like to mention Dancoff's work on short range α-particles. This work actually started because of some acute problem which was practically forgotten by the time Dancoff took over. He noticed that the intensity of short range α's is often anomalous in the light of Gamow's theory[9] which stipulates that the α-particle is emitted by an excited residual nucleus. Dancoff investigated several other mechanisms among which the excitation of the residual nucleus by the α-particle after it has already penetrated the potential barrier seems to be the most important. These theoretical investigations have now received added interest in view of Chang's[10] experimental results. Chang has discovered Dancoff's mechanism independently. Dancoff will tell you about this phase of his work in the afternoon.

I would not like to close my review of the work of our group without expressing my sincere thanks to all members of the group for their most unselfish and loyal cooperation. I have to extend my apologies to those whose work I may have slighted. As I said before, we had a very concrete objective in mind during the course of the work and the most important problems solved were not always the ones which now appear most worth remembering. We have, however, encountered a good many interesting problems, several of which will bear a great deal of further study. A glimpse at these problems, together with the knowledge of the importance of our aims, contributed a great deal to making the relations between our group cordial.

[8] R. Weinstock, Phys. Rev. 65, 1 (1944).
[9] Cf. G. Gamow, *Structure of Atomic Nuclei* (Clarendon Press, Oxford 1937), p. 104 ff.
[10] W. Y. Chang, Phys. Rev. 69, 60 (1946).

28.

Mathematical Problems of Nuclear Reactor Theory

E. P. Wigner

Proceedings of Symposia in Applied Mathematics, vol. XI:
Nuclear Reactor Theory, pp. 89–104 (1961)

The unanswered mathematical questions of reactor theory can be divided roughly into two classes. The first class concerns the mathematical theory of the basic transport equations and of the approximations thereto. Thus, the multiplication and criticality equations are characteristic value equations but their operators do not belong to the class for which the characteristic value theory is well established: they are not normal. Much progress was made recently concerning the character and properties of the highest characteristic value and the corresponding characteristic vector but the properties of the lower characteristic values and vectors are not known. In particular, the extent of a continuous spectrum and the completeness of the whole set of characteristic vectors are not established in general. A simple example is given in which the transport operator has a continuous spectrum which has not been recognized to date. The need for a generalization of the characteristic value problem is pointed out and such a generalization is proposed.

The second class of problems is concerned with methods for obtaining or at least discussing the solutions of the reactor equations. The coding of these equations for calculating machines belongs to this class. There are, however, more subtle methods among which the transformation of the reactor equations into a variational principle has proved, so far, the most effective. Again, this transformation has been carried out so far only for the most simple problems and it is not known whether all problems of reactor theory can be reformulated as variational problems. Some remarks are made on a reformulation of diffusion theory which gives, at least in the cases considered so far, much more accurate results than the conventional theory.

Introduction. It took less than 24 hours to erect the first chain reacting unit and this was done without elaborate planning; nowadays it takes more than 24 months to build a reactor, after years of planning. The dedication of those who built the first unit may have much to do with the early accomplishment and over-cautiousness with the present schedules. However, this is not the full story. Present reactors are expected to perform definite functions for extended periods in addition to maintaining a chain reaction. The demonstration of the chain reaction was the only function of the early experiment. In the reactors of the present day, the chain reaction is only one requirement. As a result, the planning of today's reactors involves the solution of a host of problems of chemistry, metallurgy, mechanical engineering, heat transfer, radiation shielding, mechanical and nuclear stability and, last but not least, nuclear physics. The experiment of December 2, 1942, was, from today's point of view, simply a critical experiment. However, when we speak about reactor theory, we still mean only that comparatively narrow set of problems the solution of which immediately preceded and made

E. P. WIGNER

possible the establishment of the chain reaction and which concerns itself with the calculation of the neutron densities or neutron fluxes as functions of the position, time, and energy of the neutron.   Furthermore, the calculation is based on an assumed knowledge of the cross sections, that is the basic properties of neutrons and reactor materials.   My remarks will be entirely restricted to these problems which, as was mentioned before, were first successfully discussed more than sixteen years ago.   There is another common element between the situation in which we are now and the situation of which the reactor theorist complained sixteen years ago : The fundamental constants, that is the cross sections, were not known well enough sixteen years ago ; they are not known well enough now.

The assumptions of transport theory.   As you well know, the theory which permits one to calculate the neutron densities or fluxes is an essentially statistical theory and is called transport theory.   This theory goes back to the last century and Boltzmann's book on the kinetic theory of gases [1] can still be read to advantage.   The fundamental concept is the so-called neutron flux $\Phi(x,E,\Omega;t)$.   This quantity gives for the time t the number of neutrons, multiplied with their speed, which satisfy the following conditions :

(a) They are in unit volume at x.

(b) Their energy is in unit range at E.

(c) The direction of their velocity lies within a unit solid angle about the direction defined by the unit vector Ω.

Naturally, Φ can be replaced by other, equivalent functions, such as $E\Phi$, etc.   Similarly, the variables x, E, Ω can be replaced by equivalent variables. In particular, E and Ω can be replaced by the three components of the velocity.   It may be proper, at this point, to summarize the basic assumptions of transport theory as used in reactor theory.

(1) The number of neutrons is so large that they can be considered to form a statistical assembly and a continuous flux function Φ can describe them adequately.   The basis of this assumption is that even the number of those neutrons is very large which satisfy the aforementioned conditions a, b, c, where the "unit volume", "unit energy", "unit solid angle" are so small that the behavior of neutrons within these unit ranges is undistinguishably similar.   It is assumed, furthermore, that if Φ is given for one time t_0, it is uniquely determined for all later times by a suitable "transport equation". Since the behavior of individual neutrons is subject to statistical laws, this appears to imply such high neutron densities that the time behavior of Φ is governed by the law of averages.   This, incidentally, raises the first mathematical problem : the actual neutron fluxes are not nearly high enough to justify these assumptions.   That the results obtained on the basis of the transport equations are adequate seems to be a consequence not of the large

PROBLEMS OF NUCLEAR REACTOR THEORY   91

number of neutrons which participate in any given process, but rather of the relatively large fluctuations of individual processes which cancel each other because some of these processes yield more, others less, neutrons than can be expected on the average.   However, this point has not been elaborated in detail to my knowledge and the average deviation of the neutron fluxes, calculated on the basis of the transport equations, from the actually fluctuating fluxes has been discussed only in the opposite case of very small neutron numbers [2].   The problem of fluctuations will be discussed at our Symposium by H. Soodak.

(2) The neutron fluxes do not affect the medium of the reactor so that the transport equation for the flux is *linear*.   This also implies the neglect of neutron-neutron collisions.   It is the assumption of linearity which makes the neutron transport equations so much simpler than the transport equations with which the kinetic theory of gases deals.   Of the two assumptions which are necessary to establish the linearity of the equations, the neglect of neutron-neutron collisions is always justified.   The effect of the neutrons on the medium, in particular the heat production by the fission process, cannot be always neglected.   Non-linear equations have to be discussed, in particular, when judging the stability of a reactor or of a neutron distribution and we shall hear more about this question from Drs. Brooks and Welton. Nevertheless, the linear theory is perfectly adequate in the overwhelming majority of all applications and the rest of this report will be based on the linear theory.   This theory also invites comparisons and analogies with the quantum mechanical equations and reference to the methods and concepts used in quantum mechanics will be made repeatedly.

(3) Although the elementary laws of the interaction between neutrons and the medium of the reactor can be calculated only on the basis of quantum mechanical theories, the wave nature of the neutrons can be disregarded and classical mechanics forms the basis of the transport equations.   This is evident already from the simultaneous specification of energy and position in the flux.   There is no reason to doubt this assumption; the only case in which the wave nature of the neutrons plays a macroscopic role is the diffraction in crystalline media.   Even this can be taken into account within the framework of classical transport equations by the use of anisotropic cross sections.

(4) As was just mentioned, quantum theory limits the accuracy with which the classical variables of position and velocity can be specified.   On the other hand, it introduces a new characteristic for particles: their spin. Properly speaking, this should also be one of the variables of the flux Φ, or, more concretely, two flux functions are actually needed to specify the neutron distribution completely.   One of these, Φ_r, would describe the flux due to neutrons of right helicity (spin parallel to velocity), the other, Φ_l, would describe the flux due to neutrons of left helicity (spin antiparallel to velocity).   There are transitions in which the helicity of neutrons changes so

that the equations for Φ_r and Φ_l are coupled. The two types of neutrons behave differently because the cross sections depend on the helicity.[1] Again, it would be possible to use another characteristic instead of the helicity. The unspoken attitude of the reactor scientist toward the introduction of this new variable is that the equations are difficult enough to solve without this added complication. Since the difference between the cross sections for the two helicities is, in general, not very great, and since transitions between the two helicities are so frequent that the neutrons in reactors are practically unpolarized ($\Phi_r = \Phi_l$), one is inclined to agree with this point of view.

The two methods of transport theory. The great variety of methods used to solve the transport equations can be divided, from the point of view of the mathematician, into two groups. These are in the same relation to each other as Riemann and Lebesgue integration, or Euler's and Lagrange's equations of hydrodynamics, or as the story of the man, who described the contents of his chest of drawers by starting at the left side of his top drawer and going down to the right side of the bottom drawer, is related to the story of the man who told where his shirts are, where his socks are, and so on. Similarly, one can focus one's attention on a volume element and an energy-direction range, and obtain the change of the number of neutrons in this

---

[1] The effect of neutron polarization on diffusion was recognized ten years ago by S. Borowitz and M. Hamermesh. See [3]. Actually, the situation is more complicated than represented in this article, or in the text. The reason is that quantum mechanics defines amplitudes rather than intensities for the two helicities and that these amplitudes are complex rather than real. However, one can define, from the two amplitudes, four real quantities, the so-called statistical matrix. One of the real quantities, $\Phi_0(x, E, \Omega, t)$ is the total flux of both polarizations; the three other quantities Φ_x, Φ_y, Φ_z, (which depend on the same variables) describe the state of polarization of the neutrons with energy E, velocity-direction Ω, and position x. If these neutrons are unpolarized $\Phi_x = \Phi_y = \Phi_z = 0$; if the state of polarization is complete, $\Phi_x^2 + \Phi_y^2 + \Phi_z^2 = \Phi_0^2$. There are four Boltzmann equations for the four quantities $\Phi_0, \Phi_x, \Phi_y, \Phi_z$. The streaming terms (1a) are the same in all of them. However, the transference function $\Sigma(x, E' \rightarrow E, \Omega' \rightarrow \Omega)$ becomes a four by four matrix, with rows and columns labeled by $0, x, y, z$. The 0-0 component of this matrix has the same structure as the Σ of (1b); in particular it depends in an isotropic medium only via the scalar product $\Omega \cdot \Omega'$ on the velocity-directions before and after the collision process. The 0-x component of the matrix Σ is also a function of $\Omega \cdot \Omega'$, multiplied, however, with the x component of the vector product $\Omega \times \Omega'$ and the x-0 component has the same form. Similar statements apply to the 0-y, 0-z, y-0, z-0 components of the matrix Σ. Finally, the ij components (where i and j may be x, y, or z) contain five terms:

$$\Sigma_{ij} = \Omega_i \Omega_j g_{00} + \Omega_i' \Omega_j g_{10} + \Omega_i \Omega_j' g_{01} + \Omega_i' \Omega_j' g_{11} + (\Omega \times \Omega')_i (\Omega \times \Omega')_j g.$$

All g are, in general, functions of E, E' and $\Omega \cdot \Omega'$ and, in inhomogeneous media, of x. The form of the transference matrix can be deduced from the theory of polarization in collisions, as given by L. Wolfenstein and J. Ashkin, Phys. Rev. vol. 85 (1952) p. 947. See also L. Waldmann, Z. Naturf. vol. 12a (1957) p. 660 and vol. 13a (1958) p. 609.

PROBLEMS OF NUCLEAR REACTOR THEORY   93

volume element and energy-direction range in unit time. This is Boltz-mann's method. Alternately, one can focus one's attention on a neutron and follow it through its collisions until it is absorbed. Evidently, this latter method is more cumbersome, principally probably because it gives more information than is actually desired. It gives not only the number of neutrons absorbed at a given point but also the origin of these neutrons. Because of its greater complexity, this second method has been pretty much displaced in practical work by Boltzmann's method. Nevertheless, the second method remains useful and a powerful aid in visualizing the changes of the flux. It has also yielded a number of new and interesting results [4].

The Boltzmann equation. Boltzmann's equation is based on the observation that the flux $\Phi(x,E,\Omega)$ changes for two reasons. First, the neutrons move out of the volume element at x. As a result of this, Φ changes by

(1a)
$$\left(\frac{\partial \Phi}{\partial t}\right)_1 = -v\left(\Omega_x \frac{\partial \Phi}{\partial x} + \Omega_y \frac{\partial \Phi}{\partial y} + \Omega_z \frac{\partial \Phi}{\partial z}\right),$$

where $v = \sqrt{2E/M}$ is the speed which corresponds to the energy E. Second, the neutrons change their velocity—but not their position—as a result of collisions. The collision of neutrons with energy E and velocity-direction Ω decreases $\Phi(x,E,\Omega)$; the collision of the neutrons at the point x with energy values different from E and other velocity directions Ω' may increase $\Phi(x,E,\Omega)$ because these neutrons may acquire the energy E and the velocity-direction Ω as a result of a collision. Hence, as a result of the collisions, Φ changes by

(1b)
$$\left(\frac{\partial \Phi(x,E,\Omega)}{\partial t}\right)_2 = -v\,\Sigma(x,E)\Phi(x,E,\Omega)$$
$$+ v \int dE' \int d\Omega'\, \Sigma(x,E' \rightarrow E,\, \Omega' \rightarrow \Omega)\Phi(x,E',\Omega').$$

The total change $\partial \Phi/\partial t$ is the sum of the two expressions of the right sides of (1a) and (1b)

(2)
$$\frac{\partial \Phi}{\partial t} = B\Phi,$$

(2a)
$$B\Phi(x,E,\Omega) = -v\left(\Omega_x \frac{\partial}{\partial x} + \Omega_y \frac{\partial}{\partial y} + \Omega_z \frac{\partial}{\partial z} + \Sigma(x,E)\right)\Phi(x,E,\Omega)$$
$$+ v \int dE' \int d\Omega'\Sigma(x,E' \rightarrow E,\, \Omega' \rightarrow \Omega)\Phi(x,E',\Omega').$$

The removal cross section $\Sigma(x,E)$ is independent of Ω unless the crystal effect, mentioned before, is important; similarly, the transference function $\Sigma(x,E' \rightarrow E,\, \Omega' \rightarrow \Omega)$ depends only on the angle between Ω' and Ω, not on the directions of Ω and Ω' separately. Both Σ are, in the sense of our second postulate, independent of t. They also are, as a rule, independent of x except at the interfaces between different media where they change discontinuously to remain again constant throughout the second medium, and so on. The sizes and shapes of the regions in which the Σ are constant depend on the reactor; in the simplest case of a bare homogeneous reactor, both Σ are independent of x within the region filled by the medium, both vanish in the empty space outside. Both Σ may show, as functions of E, or of E and E', rather violent fluctuations; Dr. Nordheim has referred to these already. The transference function is usually a rather smooth function of $\Omega \cdot \Omega'$. If there is no absorption or multiplication, the integral over the transference function is equal to the removal cross-section, that is,

$$(3) \qquad \int dE \int d\Omega \, \Sigma(x, E' \to E, \Omega' \to \Omega) = \Sigma(x, E').$$

However, if absorption predominates over multiplication, the right side of (3) is larger than the left side while the reverse is true if the contribution of the neutrons of energy E' to neutron multiplication predominates. Thus, by suspending the validity of (3), the expression (1b) can account also for absorption and for the multiplicative processes such as fission.

As pointed out before, the transport theory of neutrons is simpler than the transport theory of gases because its equations (1) and (2) are linear. This greater simplicity is to some degree compensated by the larger mean free path of neutrons as compared with the mean free path of the atoms of a gas. The latter is of the order of 10^{-5} cm at ordinary pressure so that one can assume that the properties of the gas change very little within a mean free path. The mean free path $1/\Sigma$ of neutrons is of the order of a few cm even in condensed material. Hence, a single free path can start and end in media of drastically different properties. The principal approximation of the kinetic theory of gases, which permits the elimination of the direction of motion of the atoms from the equations, is not valid in neutron transport theory and all six variables of Φ have to be taken seriously. The approximation used in kinetic theory, assuming that Φ depends only little on Ω, leads to the simplified equations of diffusion theory and this often can be used only for purposes of orientation for neutrons.

Even if one considers the stationary case, Φ depends in general on six variables and functions of six variables are not only difficult to calculate, they are even difficult to register or to visualize. It is for this reason that the consideration of situations with a high degree of symmetry is so important. As was mentioned before, the reactor equation is almost always invariant under time displacements. Furthermore, since the Σ are regionwise constant, the whole Φ can be pieced together from the general solutions of the transport equation (2) in the various homogeneous regions, by fitting these solutions together so that the whole Φ be continuous at the interfaces. This piecing together is much simplified if the regions have symmetric shapes.

Characteristic value problems. The way in which the displacement symmetry can be used to eliminate one variable is well known. One writes,

PROBLEMS OF NUCLEAR REACTOR THEORY   95

as we have done already, $\partial\Phi/\partial t = B\Phi$ and solves the characteristic value problem

$$(4) \qquad B\Psi_\kappa(x,E,\Omega) = \nu_\kappa\Psi_\kappa(x,E,\Omega),$$

in which the time does not occur any more. Then, $\Psi_\kappa \exp \nu_\kappa t$ is one solution of the reactor equation and, because of the linearity of this equation, one can find a solution

$$(5) \qquad \Phi(x,E,\Omega,t) = \Sigma\, a_\kappa\Psi_\kappa(x,E,\Omega)e^{\nu_\kappa t}$$

with a set of arbitrary constants $a_1, a_2, \cdots$. If the characteristic functions $\Psi_1, \Psi_2, \cdots$ form a complete set, one can choose the constants a_κ to reproduce an arbitrary initial condition, so that (5) is the most general solution of the problem.

This last "if" does uncover, however, a very serious problem. The operator B of (2a) does not belong to the class of normal operators although the first part of it, given in (1a), is skew-hermitean, just as is the corresponding operator of quantum mechanics. The second part, contained in (1b), has no similar property. Hence, the character of the spectrum of B, and the completeness of its characteristic functions Ψ_κ, does not seem to be established. Dr. Weinberg and I, when we wrote our book, realized that the characteristic function of the largest characteristic value of B must have certain properties and made some statements about it [5] which were supported only by physical reasoning. At that time, Birkhoff and Varga [6] had already noted that B belongs to a class of operators already investigated by Frobenius [7], Perron, and Jentzsch; they have extended Frobenius' results [7] and proved about the largest characteristic value and the corresponding function all that we had surmised and a great deal more. The results of these and some subsequent investigations, which form some of the most beautiful mathematical investigations stimulated by reactor theory, will be reported later at this Symposium, and I will not further dwell on the subject. However, the theorems to be reported answer only some of the significant questions: those pertaining to the largest characteristic value. The character of the spectrum, the extent of its discrete and continuous parts and, above all, the possibility and uniqueness of the expansion of an arbitrary $\Phi(x,E,\Omega,0)$ into a series (5), have not been solved to date. There are two questions which should be mentioned in this connection. First, the question of uniqueness which, because of the orthogonality relations, does not enter the similar problem in quantum mechanics, but as will be demonstrated later, does enter here. Second, the need for a practical criterion to decide whether a singular solution of the characteristic value equation (4) belongs to the continuous spectrum. I have often asked the late Dr. von Neumann for such a criterion for the quantum mechanical equations. The theory of distributions should be a more solid foundation on which to base a more tractable criterion than was available at the time of my inquiries.

96   E. P. WIGNER

The simplest example in which the singular solutions of (4) form a continuous spectrum, but can be easily overlooked, concerns one of the simplest
problems of transport theory: the diffusion of monoenergetic neutrons in
plane geometry.   In this case, t, as well as y, z, cease to be variables because
the flux is assumed to be independent of them.   Furthermore, since the
flux is zero except for a single value of E, this is not a significant variable
either.   Finally, the double variable Ω can be replaced by the direction
cosine $\mu = \Omega_x$ of the velocity with respect to the x axis because the flux does
not depend on the azimuthal angle either.   If the further simplification is
made that the scattering is spherically symmetric and if the total cross section is measured in appropriate units, the transport equation assumes the
form

$$(6) \qquad -\mu \frac{\partial \Phi(x,\mu)}{\partial x} - \Phi(x,\mu) + \frac{1}{2}s \int_{-1}^{+1} \Phi\,(x,\mu)d\mu = 0,$$

s being the ratio of scattering to total cross section.   This is, probably, the
simplest transport equation.   It is still invariant under displacements in the
x direction and gives rise, therefore, to the characteristic value problem

$$(7) \qquad -\mu\nu\psi(\mu) - \psi(\mu) + \frac{1}{2}s \int_{-1}^{1} \psi(\mu)d\mu = 0.$$

It is well known that the discrete spectrum consists of a single pair of (oppositely equal) points; the corresponding characteristic functions are smooth
regular functions of μ, easily obtained and recognized.   However, these two
functions certainly do not constitute a complete set of functions of μ in terms
of which all functions of μ, between -1 and 1, could be expressed linearly.

The last term of (7) is independent of μ so that one can set

$$(8) \qquad \psi(\mu) = \frac{c}{1 + \mu\nu} = \frac{-c\lambda}{\mu - \lambda}; \quad \lambda = -\frac{1}{\nu},$$

$$(8a) \qquad c = \frac{1}{2}s \int_{-1}^{1} \psi(\mu)d\mu.$$

For $-1 < \nu < 1$ or for complex ν, the $\psi(\mu)$ of (8) is a regular function of μ in
the relevant interval $(-1 \leq \mu \leq 1)$ but the second equation can be satisfied
only for two real values of ν, to be called $\pm \nu_0$.   These then form the discrete
spectrum.   If there is a continuous spectrum, it must be in one of the real
regions $\nu < -1$ or $\nu > 1$, or $-1 < \lambda < 1$.   For these values of λ, the solution (8) becomes singular but it is tempting to substitute for

$$(9) \qquad \psi_\lambda(\mu) = \frac{c_1}{\mu - \lambda + i\epsilon} + \frac{c_2}{\mu - \lambda - i\epsilon}$$

and consider the ψ of (8) as a limit of the ψ of (9) for $\epsilon \to 0$.   Equation (8a)
then reads

$$(10) \qquad -\frac{1}{\lambda}(c_1 + c_2) = \frac{1}{2}s\left\{c_1\left(\ln\frac{1-\lambda}{1+\lambda} - i\pi\right) + c_2\left(\ln\frac{1-\lambda}{1+\lambda} + i\pi\right)\right\}$$

PROBLEMS OF NUCLEAR REACTOR THEORY   97

and can be satisfied by setting

(10a) $c_1 = \dfrac{i}{\lambda} - \dfrac{1}{2} is \ln \dfrac{1 + \lambda}{1 - \lambda} - \dfrac{1}{2} \pi s = -i\nu + \dfrac{1}{2} is \ln \dfrac{\nu + 1}{\nu - 1} - \dfrac{1}{2} \pi s,$

(10b) $c_2 = -\dfrac{i}{\lambda} + \dfrac{1}{2} is \ln \dfrac{1 + \lambda}{1 - \lambda} - \dfrac{1}{2} \pi s = i\nu - \dfrac{1}{2} is \ln \dfrac{\nu + 1}{\nu - 1} - \dfrac{1}{2} \pi s.$

The total ψ_λ then becomes

(11) $\psi_\lambda(\mu) = \lim_{\epsilon \to 0} \pi s \dfrac{\lambda - \mu}{(\mu - \lambda)^2 + \epsilon^2} + \left[\dfrac{1}{\lambda} - \dfrac{1}{2} s \ln \dfrac{1 + \lambda}{1 - \lambda} \right] \dfrac{2\epsilon}{(\mu - \lambda)^2 + \epsilon^2}.$

If one forms expressions of the form

(12) $$\varphi = \int g(\lambda) \psi_\lambda(\mu) d\lambda,$$

the first fraction of (11) becomes a principal value integral, the second fraction 2π times Dirac's δ-function.   Hence, (12) suggests that

(13) $\Phi(x,\mu) = -\pi s P \displaystyle\int_{-1}^{1} \dfrac{g(\lambda)}{\mu - \lambda} e^{-x/\lambda} d\lambda + 2\pi \left[\dfrac{1}{\mu} - \dfrac{1}{2} s \ln \dfrac{1 + \mu}{1 - \mu} \right] g(\mu) e^{-x/\mu}$

is a solution of (6) where $g(\lambda)$ is an arbitrary function of λ which goes to zero sufficiently well at $\lambda = 0$.   The total flux which corresponds to the $\Phi(x,\mu)$ of (13) is (the first and the last terms of (13) just cancel upon integration over μ)

(13a) $\Phi(x) = \dfrac{1}{2} \displaystyle\int_{-1}^{1} \Phi(x,\mu) d\mu = \pi \int_{-1}^{1} \dfrac{d\mu}{\mu} g(\mu) e^{-x/\mu}.$

The general solution of (6) then would be (13) plus the two characteristic functions of the discrete spectrum with arbitrary coefficients.   It is surely unnecessary to state here that the preceding argument which, incidentally, is not the original argument that led to (13), is not rigorous.   However, one can verify *a posteriori* that (13) is indeed a solution of (6).   The question of the completeness has not been settled to date.[2]

It is interesting to note that the usual expression for the total flux due to an isotropic plane source (see for instance Equations (9.26), (9.27), and (9.29) of reference [5]) can be decomposed into an asymptotic part, which corresponds to the discrete spectrum, and an integral over exponentials $e^{-\nu x}$ of x where ν (denoted by $1 + \eta$ in [5, Equation (9.29)]), runs over the range covered by the continuous spectrum.   It may be well to remark, however, that the possibility of such a decomposition of the total flux into exponentials of x does not prove that there is a solution of the homogeneous equation which corresponds to each exponential separately.   In fact, a different deformation

---

[2] See, however, K. M. Case, Annals of Physics vol. 9 (1960) p. 1; Case has found the continuous spectrum independently.   (Note added in proof.)

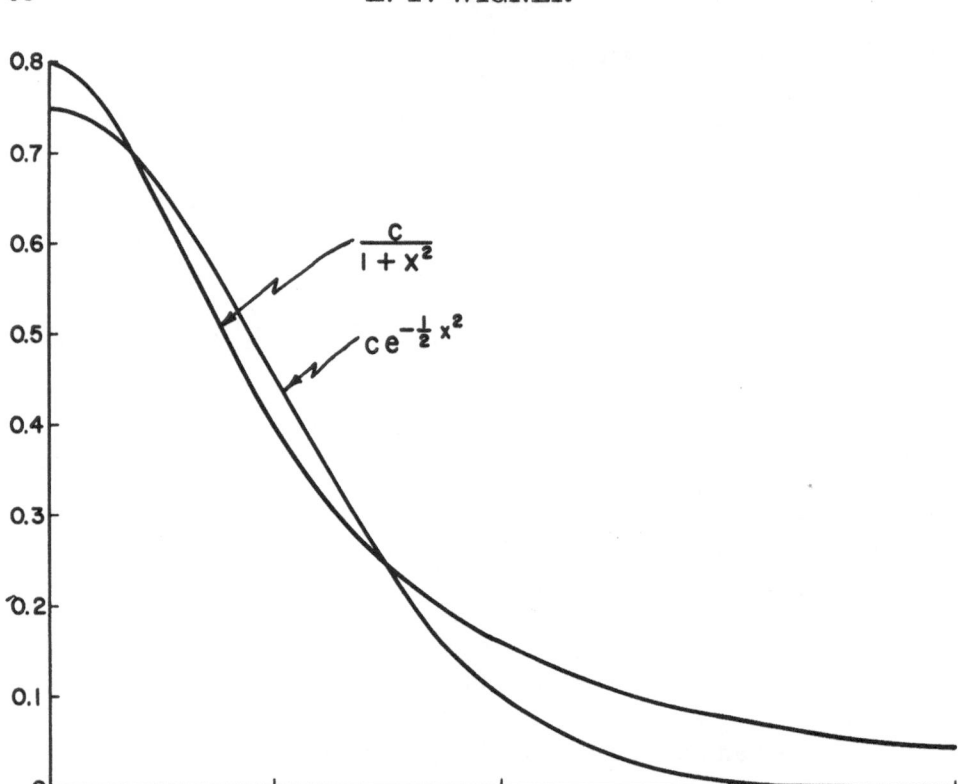

FIGURE 1. Graphs of two functions (both normalized) which are very different analytically but whose vectors in Hilbert space are almost parallel. The cosine of the angle between the two vectors differs from 1 by one half of the integral over the square of the difference of the two functions.

of the path of integration than the one which led to (9.29) represents the total flux as an integral of exponentials $e^{-\nu x}$ with ν which do not belong to the spectrum.

The preceding example was presented because it shows the difficulty of recognizing the continuous spectrum even in a case of almost trivial simplicity. One can find in the literature a variety of interesting cases in which operators related to the B of the Boltzmann equation partly have, partly do not have, continuous spectra [8]. Similarly, the spectrum, which is confined in the preceding case to the real axis, may or may not have points in the complex plane. The spectral problem will be discussed, in the present Symposium, by Drs. Habetler and Martino and Dr. Wing. We shall also hear, I believe, about the approach of the solutions of the various approximation methods,

PROBLEMS OF NUCLEAR REACTOR THEORY   99

such as the spherical harmonics and the S_n method, to the continuous spectrum. In general, the question of the completeness of the characteristic functions and the uniqueness of expansions has not been settled satisfactorily. The situation is even more obscure with respect to the approximations to the Boltzmann equation. Thus, for instance, the appearance of negative neutron fluxes in the spherical harmonic solution of Milne's problem shows that the positivity theorem of Birkhoff and Varga does not hold in this case for the approximate solutions.[3]

Generalization of the characteristic value problem. The characteristic value problem can be formulated as the quest for the irreducible linear manifolds which are invariant under an operator. The principal result of the spectral theory of normal operators can be formulated, from this point of view, as the statement that all irreducible linear manifolds of normal operators are one-dimensional. Similarly, one can ask for irreducible closed linear manifolds which are invariant under a set of operators. Since a closed linear manifold which is invariant under a set of operators is also invariant under the group or algebra generated by these operators, one is naturally led in this way to a linear manifold which belongs to an irreducible representation of a group or an algebra.

The significance of such a generalization of the characteristic value problem is due to the fact that the transport Equations (2), (2a) are, for any of the homogeneous media in the reactor, invariant not only under time displacements but also under the Euclidean group of three dimensions. Thus, the whole group of the transport equation has seven parameters, four parameters describing the displacement in time and space, and three which specify a rotation. Since the number of variables of Φ is also seven, and since every parameter of the invariance group permits in general the elimination of one variable [9], it might appear that a complete set of solutions of the transport equations in a homogeneous medium can be obtained on the basis of invariance considerations. This is, unfortunately, not quite true because, as the theory of the partial determination of the characteristic functions from their transformation properties under a group shows, the decrease in the number of variables is partially compensated by an increase in the number of unknown functions. The number of these becomes equal to the dimension of the representation which determines the transformation properties of the linear manifold considered. If the invariance group is compact, the dimension of all representations is finite. However, "most" representations of the Euclidean group have infinitely many dimensions so that one has an infinity of functions or, more precisely, a two-parametric manifold of functions, instead of the single original function. It follows that the original problem with seven variables can be reduced only to the extent that one has to solve a problem with two variables. In fact, had we not restricted ourselves to

[3] See, for instance, reference [5, Figure 9.6, p. 262].

E. P. WIGNER

the monoenergetic case, (6) would have had, in addition to μ, another essential variable, the energy.

It is probably unnecessary to emphasize for anyone who has considered, for instance, Milne's problem that the knowledge of the general solution of the transport equations in all the distinct media contained in the reactor does not yet constitute a solution of the reactor equations. The adjustment of the boundary conditions at the interfaces between media can be, and often is, more difficult than the determination of the general solution in the homogeneous media. It is the adjustment of boundary conditions which is rendered much simpler if the homogeneous parts of the reactor have shapes of high symmetry.

The considerations sketched in the preceding paragraphs are of a formal nature and do not answer the very important question whether the whole Hilbert space of functions $\Phi(x,E,\Omega,t)$ can be decomposed into irreducible closed linear manifolds which are invariant under the Euclidean plus time-displacement groups. This question is the question of the completeness of the irreducible representations of this group. As to the uniqueness of the decomposition, the example of the $2 + 1$ dimensional Lorentz group shows [10] that this is not assured for non-compact groups. It is the impression of the present writer, nevertheless, that it is correct for the group under consideration. Because two of the variables of the original problem are not eliminated by the symmetry group, the complete decomposability of the symmetry group, even if it should prove correct, does not guarantee that, for instance, the operator B of (2a) has a complete set of characteristic functions which permits a unique expansion of all functions of x,E,Ω. It only reduces B to an operator with two variables which then may, or may not, have these properties.

Approximation methods. The preceding sections dealt either with the mathematical theory of the equations or with their rigorous solutions. In practice, approximation methods play as important a role as accurate solutions and it seems worth while to make a few remarks on this subject. Because of the great variety of approximation procedures, these remarks are bound to remain even more sketchy than those of the preceding sections.

Whether we like it or not, it is clear that computing machines will play an increasing role in solving reactor equations. However, the role which the solutions obtained with computing machines will play in our thinking is less clear. In some respects, computing machines give us the same kind of information as an experiment, for instance, a critical experiment. Computing machines give us the results with a smaller effort and perhaps in a shorter time than critical experiments, but they do not help us in visualizing the factors which determined the final result much more than the critical experiments. For this reason, the results obtained with calculating machines are less able to direct us toward an improved arrangement than the analytical

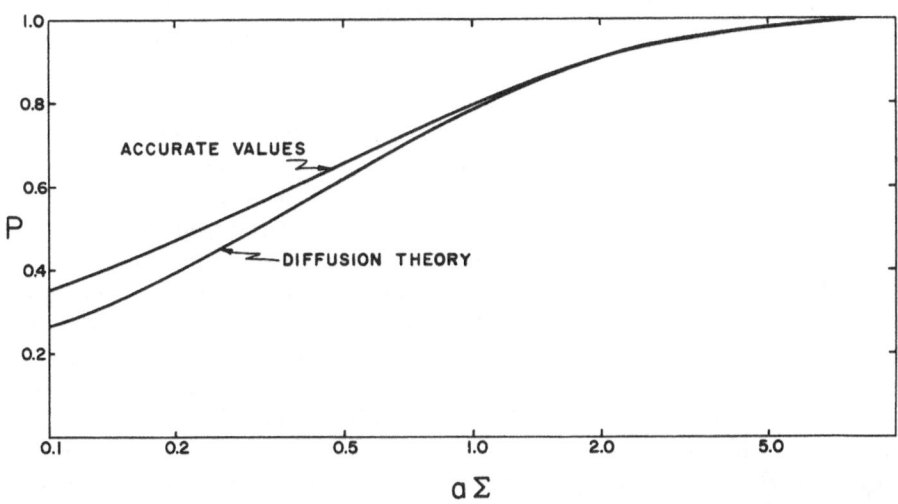

FIGURE 2. Accurate values and Inönü's diffusion theoretic values for the nonescape probability from an infinite slab of thickness 2a, in which the scattering is isotropic and has a macroscopic cross section Σ.

methods of calculation, even less able than less-mechanized methods of numerical calculations. The role which the computing machines will play in the field of reactors, as well as in many other fields, will depend on the extent to which we shall learn to use them not only to give a final result, but also to explain the final result. Surely, very much can be done to use calculating machines more effectively in this regard.

My own experience with calculating machines is so limited that it would be difficult for me to deal with their use in any detail. This would also be unnecessary, since the use of computers will be discussed by Drs. Ehrlich, Varga, Richtmyer and Carlson along with the mathematical models, such as multigroup theories, which the availability of computers inspired.

Among all the non-numerical approximation methods, the effectiveness of the variational methods is perhaps most surprising [11]. This method serves to determine characteristic values of linear operators. Since the time variable can be eliminated from almost every reactor equation by transforming it into a characteristic value problem, the variational method should have wide applications in reactor theory. Its use has been limited, so far, because Boltzmann's operator is not self adjoint or normal. Whether this limitation is a necessary one, remains to be seen. The reason for the great accuracy of the variational principle in simple problems of quantum mechanics is that any function which is positive everywhere and has a single maximum can be so well approximated by any other similar function. Thus

$1/(1 + x^2)$ can be approximated by a function $e^{-\alpha x^2}$ with $\alpha = 0.5$ better than 98 per cent. By this I mean that the cosine of the angle which these two functions include in Hilbert space is more than 0.98. The variational principle becomes less effective in quantum mechanics for problems involving several particles because, on account of the exclusion principle, the wave function cannot be positive everywhere. In reactor theory, on the other hand, the flux for the largest characteristic value is, according to the Birkhoff-Varga theorem, always everywhere positive, and the same applies to the adjoint flux. In this regard, at least, the variational principle is more adapted to problems of transport theory than to those of quantum mechanics. However, this seems to apply only to the Boltzmann equation itself, not to the various approximations to it. As has been mentioned before, the solutions of the approximate equations are in general not everywhere positive.

Dr. Brooks informed me of recent considerations, by himself and Mr. Calame, which, unfortunately, he could not present at the Symposium. They used the variational principle as applied to the Boltzmann equation directly, rather than to equations of approximation methods. This way, the advantages of the positivity of the true solutions are preserved.

The second method which I wish to mention is the diffusion theory. It is well known that this theory gives the spatial variation of the flux quite accurately in regions well removed from interfaces. The trouble is only that in a heterogeneous reactor virtually every point is close to an interface. It seems reasonable, however, that a proper change of the usual boundary conditions will permit not only a fair representation of the asymptotic fluxes far from the interfaces, but will reproduce even the transients, near the interfaces. A similar suggestion was made, before, by Davison [12]. In order to show on an example how this may come about, I wish to call attention to Inönü's calculation [13] of the so-called fast effect. The fast effect was always considered par excellence the process to which diffusion theory is inapplicable because it takes place in a medium which is small as compared with the mean free path. It is indeed true that the calculation of the fast effect by means of diffusion theory and the most simple boundary conditions leads to grossly inaccurate values. However, when Inönü introduced 0.71 mean free paths as the *linear* extrapolation distance, the picture changed completely. It is true that the 0.71 mean free paths is an element foreign to diffusion theory. However, the proposition is just to introduce such foreign elements into diffusion theory, on the basis of a knowledge of the accurate solution in simple cases. The prescription found by Inönü applies to the case of a multiplying medium surrounded by a completely absorbing medium, as the moderator around the fuel element is supposed to rob a fast neutron in a single collision of its ability to induce further fast fission. A short calculation will show that Inönü's boundary condition is equivalent to covering the fuel element with an infinitely thin skin the scattering cross section of which is, however, infinitely high so that the product of

PROBLEMS OF NUCLEAR REACTOR THEORY   103

thickness and macroscopic scattering cross section is finite, namely $0.71 - 0.33 = 0.38$. The solution of the diffusion problem in which there is such a scatterer between the multiplying medium and the perfect absorber surrounding it gives, for the inside of the multiplying medium (the fuel element) the same flux as calculated by Inönü. If one thinks of the transients in the neutron flux in the neighborhood of the interface between two media one will find it most unlikely that the introduction of a surface scatterer will adequately account for the perturbations in the flux in the neighborhood of the interface in all cases. Rather, it will be necessary, in general, to introduce, in addition to the surface scatterer, a surface absorber, positive or negative. In fact, in order to divide the absorption of the transients between the two media, it will be necessary to attribute part of the surface absorption to one, part of it to the other medium, that is, it will be necessary to introduce *two* surface absorbers. Together with the surface scatterer, one should be able to reproduce in this way the effect of the perturbations near interfaces within the framework of diffusion theory.

The purpose of my remarks was to convey not only the impression that the science of reactors could be much helped by the attention of mathematicians interested in its problems, but also that mathematicians can find many interesting and challenging problems in reactor science. Of the two broad fields which were mentioned, the theory of the reactor equations, and the methods of their solutions, I consider the first one even more interesting and even more important than the second one.

It is a pleasure to express my appreciation to Drs. A. M. Weinberg and H. Brooks for several useful comments.

REFERENCES

1. L. Boltzmann, *Vorlesungen über Gastheorie*, Leipzig, J. A. Barth, 1896.

2. D. Hawkins and S. Ulam, MDDC 287 (1944), C. J. Everett and S. Ulam, LADC-533–534; LA-683, LA-690, and LA-707 (1948). For a very brief summary, see F. de Hoffmann, *The science and engineering of nuclear power*, New York, Addison-Wesley Press, 1949, vol. 2, pp. 116–119. Also R. P. Feynman, F. de Hoffmann and R. Serber, J. Nucl. En. vol. 3 (1956) p. 64; and L. D. Pal, Paper 1710, Second United Nations Conference on the Peaceful Uses of Atomic Energy, 1958 ("Second Geneva Conference").

3. S. Borowitz and M. Hamermesh, Phys. Rev. vol. 74 (1948) p. 1285.

4. See in particular C. C. Grosjean, Nuovo Cim. vol. 3 (1956) p. 1262 and Papers 1691–1692 of the Second Geneva Conference, 1958. See also E. P. Wigner, Phys. Rev. vol. 94 (1954) p. 17.

5. A. M. Weinberg and E. P. Wigner, *The physical theory of neutron chain reactors*, Chicago, University of Chicago Press, 1958, p. 406ff.

6. Garrett Birkhoff and R. S. Varga, *Reactor criticality and non-negative matrices*, Westinghouse Atomic Power Division Report no. 166 (1957).

7. G. Frobenius, S-Ber. Akad. Wiss. Berlin, 1912, p. 456.

8. G. M. Wing, J. Math. Mech. vol. 7 (1958) p. 757; J. Lehner and G. M. Wing, Comm. Pure Appl. Math. vol. 8 (1955) p. 217. See also other articles quoted by

104   E. P. WIGNER

Wing at the present Symposium.   Chapter VII of *Linear operators* (Part I) by N. Dunford and J. T. Schwartz, New York, Interscience Publishers, 1958, reviews most of the known general theorems concerning the spectra of linear, not necessarily normal, operators; it deals mostly with bounded operators.   See also K. Friedrichs, Comm. Pure Appl. Math. vol. 1 (1948) p. 361, and F. Rellich, Proceedings of the International Congress of Mathematics, Cambridge, Massachusetts, vol. 1 (1950) p. 606.

9. E. P. Wigner, *Gruppentheorie, und ihre Anwendung auf die Quantenmechanik der Atomspektren*, Braunschweig, Friedr. Vieweg, 1931, Chapter 19.

10. V. Bargmann, Ann. of Math. vol. 48 (1947) p. 568.

11. E. Greuling and C. Marvin, *Graphical method of obtaining critical masses of water-tamped boilers*, Los Alamos Report 493.   For a general discussion, see reference [5], p. 531ff and R. T. Ackroyd, Paper 36, Second Geneva Conference, 1958 and N. C. Francis, J. C. Stewart, L. S. Bohl, ibid., Paper 627.

12. B. Davison, *Neutron transport theory*, Oxford, Clarendon Press, 1957, p. 98.

13. E. Inönü, J. Nucl. Sci. Eng. vol. 5 (1959) p. 248.

PRINCETON UNIVERSITY,
    PRINCETON, NEW JERSEY

29.

Resonance Escape Probability in Lattices

R. F. Christy, A. M. Weinberg, and E. P. Wigner

August 19, 1944

Abstract. The resonance escape probability, p, in a lattice is expressed in terms of the resonance absorption integral $(\int \sigma_a \, dE/E)_{\text{eff}}$ and the spatial distribution of resonance neutrons. It is shown that the dependence on the spatial distribution is very nearly the same whether the slowing down in the moderator is Gaussian or exponential. The dependence on $(\int \sigma_a \, dE/E)_{\text{eff}}$ is rigorous only for a hydrogen moderator; for a heavy moderator the formula overestimates the resonance escape.

Resonance Escape Probability in Lattices

1. The purpose of the present report is to describe the method which has been used during the past two years for estimating the probability that a neutron will escape resonance capture in an infinite lattice. This quantity, which is usually denoted[1] by p, is one of the four factors in the expression for k, the multiplication constant of an infinite system; the other factors are η, the number of primary neutrons produced per thermal neutron absorbed in the fissionable material; ε, the number of neutrons reaching resonance energy per fast neutron produced by thermal fission; and f, the thermal utilization[2] – that is, the fraction of thermal neutrons absorbed in the fissionable material. An exposition of methods for calculating f appears in CP-103, 104, 161 by Plass and the present writers, in CPA-1 by Fermi, Szilard and Weil at Columbia, and in unpublished work by Eckart at Chicago. The calculation of the fast effect, ε (the importance of which was first pointed out by Szilard) is given in CP-644. These reports together with the present one will therefore form a more or less complete description of how the multiplication constant in a chain reacting medium can be estimated. A large number of reports have already appeared in which these methods have been applied to systems both heterogeneous and homogeneous (CP-668) moderated by graphite (CP-372), P-9 (CP-923), H_2O (CP-834), H_2O and P-9 (CP-1834), and Be (CP-1231).

Considering the crude nature of these calculations, the order of accuracy attainable in them is surprisingly high especially as regards the optimal dimensions (i.e. those giving the highest multiplication constant). Thus the optimal

---

[1] Denoted by p_1 in early reports.
[2] Denoted by p_2 in early reports.

dimensions for spheres in a graphite moderated system found even in the earliest calculations (winter 1941–42) were

	U_3O_8	U_3O_8	U	U
ρ	4	6	6	18
r_0	4.5 cm	3.5 cm	4.2 cm	1.7 cm
M_0	1500 gr	1100 gr	1900 gr	360 gr
M_1/M_0	9.2	7.1	7.1	6.6
a	20.5 cm	17 cm	20.5 cm	11.5 cm

Herein ρ is the density of the U_3O_8 or U, r_0 is the radius of the metal or oxide sphere, M_0 the mass of the lump, M_1/M_0 the ratio of graphite and U masses, a the lattice constant of the cubic cell. These results were obtained without taking the fast effect into account. For this reason the optimal metal mass is too low, entailing a loss of less than 1% in the multiplication factor; nevertheless the volume ratios are correctly given. The lattice arrangements quoted above are chain reacting by a comfortable margin.

2. The calculation of the resonance absorption follows. This was, in principle, given in reports A-12, C-1, C-4 and CP-1589, which showed that the resonance absorption consists of two parts: a mass absorption and a surface absorption. In C-4, in order to make the considerations more concrete, two guesses were made concerning the magnitude of these. However, there is no reason to expect that any of these guesses is correct except for order of magnitude. The mere fact that two rather different guesses were made should indicate that neither could have been well founded and the approximate agreement between the first guess and the measurements is more surprising than the relatively small disagreement between them. In all calculations, as also in the present report, the measured values for mass and surface absorptions are used rather than the guessed ones.

The absorption of resonance neutrons depends both on the structure and number of the resonance lines and on the spatial distribution of the resonance neutrons in the cell. The reports C-4 and CP-1589 deal with the former question. They show that it is a reasonable approximation to write for the resonance absorption of the neutrons in unit energy interval

$$A = n_0 N_0 V_0 a + n_s S \beta' . \tag{1}$$

Herein n_0 and n_s are average densities of the neutrons in unit energy interval in the uranium and on its surface, respectively, both multiplied with the corresponding velocity of the neutrons. N_0 is the number of uranium atoms per cm^3, V_0 the volume of the lump, S its surface. a and β' are constants, independent (in our approximation) of the size and shape of the lump but dependent on the chemical constitution thereof (e.g., different for uranium metal and for U_3O_8). They also depend on the temperature of the lump. In addition, a and β' are functions of the energy, just as n_0 and n_s are. The above formula gives (together with the proper expressions for a and β' given in C-4 and CP-1589) the dependence of the resonance absorption on the number and structure of the resonance lines. On the other hand, the calculation of the density distribution of the resonance neutrons, i.e., of n_0 and n_s, is much more complicated and we shall assume for the time being that $n_0(E)$ and $n_s(E)$ are known as function of the energy.

The total number of neutrons slowed down per second below the energy value E in the cell is

$$Q(E) = \sum n_1(E) N_1 V_1 E \xi_1 \sigma_{s1}. \tag{2}$$

Herein n_1 is the average density of resonance neutrons in the moderator of unit energy interval at E, multiplied with their velocity. N_1 is the number of atoms of the moderator per unit volume, V_1 is the volume of the moderator, ξ the logarithmic energy loss of a neutron upon collision with an atom of the moderator (.158 for C), σ_s the scattering cross section of the same. The $\sum$ appears in (2) because the moderator can consist of several constituent atoms (e.g., D_2O), or several parts (e.g., H_2O and graphite, as at W).

The number of resonance neutrons slowed below E decreases as E decreases because of the absorption A. Hence

$$\frac{dQ}{dE} = A. \tag{3}$$

The quantity $\xi \sigma_{s1} E$ is the cross section for the process of moderating neutrons through unit energy interval. This "slowing down" cross section may be called an "absorption" cross section since it measures the probability that a neutron will be slowed down, i.e., "absorbed" out of a unit energy interval. If we adopt this terminology, (2) says that $Q(E)$ is equal to the total "absorption" by the moderator in unit energy interval. Hence, dividing (3) by (2) we obtain

$$\frac{1}{Q}\frac{dQ}{dE} = \frac{d\ln Q}{dE} = \frac{\text{absorption by U}}{\text{"absorption" by moderator}} = \frac{f_R}{1 - f_R} \tag{4}$$

if we denote by f_R the ratio of the absorption by the U per unit energy interval to the absorption by U plus "absorption" by moderator, per unit energy interval. If f_R were known as function of energy, we could write

$$Q = \exp \int \frac{f_R}{1 - f_R} dE.$$

Since the chance of escaping resonance capture, p, is the ratio of Q below the resonance region to Q above the resonance region, this gives

$$p = \exp - \int_{\substack{\text{res.} \\ \text{region}}} \frac{f_R}{1 - f_R} dE. \tag{4}$$

Less abstractly, we obtain if we divide (3) by (2) and use (1) for expressing A

$$\frac{1}{Q}\frac{dQ}{dE} = \frac{d\ln Q}{dE} = \frac{N_0 V_0}{N_1 V_1 \xi_1 \sigma_{s1}} \left(\frac{n_0}{n_1} \frac{a}{E} + \frac{S}{N_0 V_0} \frac{n_s}{n_1} \frac{\beta'}{E} \right) . \tag{5}$$

It is assumed here that the moderator is homogeneous and consists of only one part; if the moderator consists of several parts the quantity $N_1 V_1 \xi_1 \sigma_{s1} n_1$ must be summed. For p, (5) gives as above

$$p = \exp -\frac{N_0 V_0}{N_1 V_1 \xi_1 \sigma_{s1}} \left(\int \frac{a}{d} \frac{dE}{E} + \int \frac{S}{M_0} \frac{\beta}{d_s} \frac{dE}{E} \right) . \tag{6}$$

Herein $d = n_1/n_0$ is called the overall disadvantage factor; it is the ratio of the average neutron densities in moderator and U. Similarly, $d_s = n_1/n_s$ is the disadvantage factor of the surface of the lump. M_0 is the total mass of the lump, and

$$\beta = \beta' M_0 / N_0 V_0 . \tag{6a}$$

Very crudely, one can write the expressions of C-4 and CP-1589 for a and β'.

$$a = \frac{\overline{\sigma_a \sigma_s}}{\sigma_a + \sigma_s} ; \quad \beta' = \frac{1}{4} \overline{\frac{\sigma_a^2}{(\sigma_a + \sigma_s)^2}} \tag{7}$$

where σ_a and σ_s are absorption and scattering coefficients of the lump per U atom. σ_a is a rather rapidly fluctuating function of the energy. The averaging is to be extended over several resonance levels. Hence

$$\frac{\beta}{a} = \frac{\beta' M_0}{a N_0 V_0} \cong \frac{m_u}{4\sigma_s} \tag{7a}$$

where m_u is the mass associated with a U atom (it is the mass of the atom if the lump consists of the pure metal; it is the mass of one atom of U and 8/3 atoms of O if the lump consists of $U_3 O_8$). Similarly σ_s is the scattering cross section associated with a U atom. Equations (7) and (7a) represent extremely crude approximations; they are given here only to give some illustration of the significance of a and β.

The above discussion of the resonance absorption is not quite rigorous and the exponential formula for p underestimates the absorption for heavy moderators; this was pointed out in CP-668. To calculate rigorously the dependence of the resonance escape probability on the energy distribution of absorption lines would require even in the homogeneous case a solution of the Boltzmann equation with energy loss and with variable absorption. In the case of hydrogen this can be done (Bethe, Rev. Mod. Phys. 9, 122, 1937) and the results agree with those quoted above. However, for a heavier moderator we have been able to find a rigorous solution of the integral equation only if the absorption is constant in energy.

Equations (6) or (4) would permit the calculation of the resonance escape probability, assuming that a and β are known, if d and d_s, i.e., the density distribution of resonance neutrons were known. Unfortunately, this is really known only in one case: if the moderator and U are intimately mixed, i.e., in a homogeneous system. In this case, $d = d_s = 1$ and (6) reduces to the expression given already in CP-668. In all other cases one must resort to rather crude approximations in order to obtain d and d_s. These involve, almost always, the assumption of energy independent d and d_s so that these quantities can be taken out of the integral. It may be appropriate, therefore, to give at this point the experimental values of Creutz, Jupnik and Snyder (C-116).

$$\int \frac{adE}{E} = 9.25 \times 10^{-24} \text{ cm}^2, \quad \int \frac{\beta dE}{E} = \frac{8}{3} \frac{\text{cm}^2}{\text{gr}} \times \int \frac{adE}{E}. \quad (8)$$

The sum $\int (a + (S\beta/M_0))(dE/E)$ is generally denoted by $(\int \sigma_a dE/E)_{\text{eff}}$. The values are valid for the metal. For U_3O_8 the first quantity is according to these authors 12×10^{-24} cm^2 and the 8/3 cm^{2}gr^{-1} should be replaced by 5/3 cm^{2}gr^{-1}. The temperature coefficient of the first quantity is a little larger than 10^{-5} per °C, that of the second one is not known. The above values are based on the value

$$\int \frac{adE}{E} = 240 \times 10^{-24} \text{ cm}^2 \quad (9)$$

for infinitely diluted U which is the value obtained for this quantity by Anderson, Fermi and collaborators. If (9) should prove too large (for which there are indications), the integrals in (9) should be decreased correspondingly. In addition, A. C. G. Mitchell, et al., repeated the measurements of C-116 and obtained (CP-1676), still on the basis of (9), 20% lower values for $\int adE/E$. According to CP-1676, therefore, one should write for the metal

$$\int adE/E = 7.8 \times 10^{-24} \text{ cm}^2. \quad (8\,\text{a})$$

Mitchell also gives $\int adE/E$ for U mixed with various scatterers and shows that it depends, within the limits of the experimental accuracy, only on the scattering cross section of the admixture. This would be strictly true if the first equation of (7) were strictly valid.

Unfortunately, the present experimental data for $\int adE/E$ and $\int \beta dE/E$ are not too accurate.

3. The above experiments and calculations refer only to the resonance absorption of 28. There are, however, two other known effects on the resonance neutrons: the resonance absorption of 25 and the fission of 25, induced by resonance neutrons. The first of these can easily be taken into account by adding the a and β of 25 to the corresponding quantities for 28. The 25 fission process, on the other hand, increases the number of neutrons by $\nu - 1$ instead of decreasing it by 1 as the resonance absorption does ($\nu \approx 2.35$ is the number of secondary neutrons per fission). Neglecting second order effects, one should add therefore to $\int a_{28} dE/E$

$$\int a_{25} \, dE/E - (\nu - 1) \int \sigma_f \, dE/E \quad (10)$$

where σ_f is the part of the fission cross section per U atom which is in excess of the $1/v$ law. Since $\nu - 1 \approx 1.35$ and since, apparently, $a_{25} = .17\sigma_f$, (10) becomes

$$-1.18 \int \sigma_f \, dE/E = -\frac{13 \times 10^{-24}}{139} = -.093 \times 10^{-24} \text{cm}^2 . \tag{10a}$$

The value of $\int \sigma_f \, dE/E$ has been obtained from measurements carried out elsewhere; the factor 139 is the isotope ratio of 25 to 28 in normal metal. One sees that (10a) gives a negligible correction to p.

4. The calculation of the disadvantage factors d and d_s which enter in the expression (6) for the resonance escape probability can be performed in principle if the details of the resonance absorption energy spectrum are known. However, such a calculation is impossible at present because the energy distribution and widths and depths of the metal resonance lines are unknown although Dancoff's work (CP-1589) already gives valuable clues in this direction. For this reason we shall present two approximate methods of calculating the disadvantage factors based on simplified models; the results of the two methods are in good agreement. Fortunately a 1% error in d means, near the optimum, only a 0.1% error in p, since d enters only in the exponent (equation 6) which usually is itself 0.1.

The simplest scheme for estimating the disadvantage factor (and the one which has almost always been used in practice), is to treat the slowing down in the same way as in the "two-group" pile model (CP-1461), and to consider all of the resonance absorption to be a constant *volume* absorption inside a sharply defined resonance energy range, zero outside this region. In the two group slowing down model, slowing of a neutron out of the resonance region is likened to an absorption process. The probability that a neutron is knocked out of the resonance region on its next collision is the same regardless of the energy of the neutron. The "absorption" cross section for this process, chosen so as to give correctly the second moment of the spatial distribution of the slowed down neutrons, is $\sigma_{s1}\xi/$(logarithmic width of resonance band).

The width of the energy region in which resonance absorption occurs is not really known experimentally but must be taken from the theory of resonance absorption as developed in C-4 or more recently by Dancoff in CP-1589. It has also been estimated by comparing the diffusion length of resonance neutrons in U_3O_8 as observed by Creutz with the known transport mean free path and total volume absorption. The logarithmic width of the resonance region, $\ln E_0/E_1$ [3] is then the ratio of the total volume absorption to the average volume absorption cross section calculated from the observed diffusion length and the transport mean free path. Both methods of estimating the width of the region lead to about the same result, namely,

---

[3] E_0 and E_1 are the energies of the top and bottom of the resonance band, respectively.

$$\ln E_0/E_1 = 7.3 \qquad \text{for oxide (U}_3\text{O}_8)$$

$$= \frac{385}{500} \times 7.3 = 5.63 \text{ for metal.}$$

The ratio 385/500 is just the ratio of the volume absorption in metal and oxide. We have thus assumed that the difference in volume absorptions in metal and oxide is entirely attributable to a decrease in size of the resonance absorption region, the widths and depths of the resonance lines remaning unchanged.

Once the size of the resonance region has been determined, the average absorption cross section and metal diffusion length are calculated from

$$\sigma_{a0} = \frac{\left(\int \sigma_a \dfrac{dE}{E} \right)_{\text{eff}}}{\ln E_0/E_1} \tag{11}$$

and

$$\kappa_0 = N_0 \sqrt{3\sigma_{t0}\sigma_{a0}} \left(1 - \frac{2}{5}\frac{\sigma_{a0}}{\sigma} \right) \tag{12}$$

where σ_{t0} is the average metal transport cross section and σ is the total (scattering + absorption) cross section. The definition of σ_{a0} involves $(\int \sigma_a \, dE/E)_{\text{eff}}$ instead of just the volume part of the integral, this means that we purposely confound the surface and volume absorption in calculating the disadvantage factor. The metal neutron density $\times$ velocity averaged over the resonance energy region is assumed to satisfy

$$\Delta Q_0 - \kappa_0^2 Q_0 = 0 \tag{13}$$

where the reciprocal diffusion length, κ_0, is an average over the whole energy range.

The distribution in the moderator is calculated, as in the two group model, by assuming that a neutron is "absorbed" when it is thrown out of the resonance energy range. The "absorption" cross section for this process is

$$\sigma_{a1} = \sigma_{s1}\xi/(\ln E_0/E_1). \tag{14}$$

The corresponding inverse diffusion length is

$$\kappa_1 = N_1 \sqrt{3\sigma_{t1}\sigma_{a1}} \left(1 - \frac{2}{5}\frac{\sigma_{a1}}{\sigma_{s1}} \right) \tag{15}$$

where we compute the transport cross section, σ_{t1}, from the scattering cross section, σ_{s1}, by the formula

$$\sigma_{t1} = \sigma_{s1}(1 - \overline{\cos\theta}) = \sigma_{s1}\left(1 - \frac{2}{3A_1} \right), \tag{16}$$

A_1 being the atomic weight of the moderator. It should be noted that σ_{a1} and hence κ_1 depends on the logarithmic width of the resonance energy regions and therefore is different for metal and oxide. The values which we use for $N_1\sigma_{a1}$ and κ_1 for various moderators and U *metal* are given in the following table:

Table. Values of $N_1\sigma_{a1}$ and κ_1 for resonance neutrons (U metal)

Moderator	Density	$N_1\sigma_{a1}(\mathrm{cm}^{-1})$	$\kappa_1(\mathrm{cm}^{-1})$
H_2O	1	.241	.583
D_2O	1.1	.0313	.155
Be	1.85	.0276	.237
BeO	2	.0150	.138
C	1.60	.0108	.1075

The neutrons are assumed to be produced uniformly throughout the moderator; we can take the production rate to be $N_1\sigma_1$ per unit volume per second. The equation describing the density of resonance neutrons in the moderator is therefore

$$\Delta Q_1 - \kappa_1^2 Q_1 = -1 \tag{17}$$

with the boundary conditions:

(a) no net flow on cell boundary, and
(b) neutron density and net neutron flow continuous across metal-moderator surface.

It is evident that the system defined by (13) and (17) is identical with the set of equations used to calculate the thermal utilization in a cell (CP-103, 104, 161). Consequently, the standard formulas given for the thermal utilization apply here, namely

$$\frac{1}{\bar{f}_R} - 1 = \frac{dN_1\sigma_{a1}V_1}{N_0\sigma_{a0}V_0} = \frac{N_1\sigma_{a1}V_1}{N_0\sigma_{a0}V_0}F(\kappa_0 r_0) + E(\kappa_1 r_1, \kappa_1 r_0) - 1 \tag{18}$$

where $\bar{f}_R$ is the "utilization" of resonance neutrons by metal. In (18), V_1/V_0 is the volume ratio of moderator to metal, r_1 and r_0 are inner and outer radii of the moderator, F is the metal disadvantage factor (ratio of density on metal surface to average density inside), d is the overall disadvantage factor and $E-1$ is the ratio of "excess absorption" in the moderator to flow out of moderator, i.e., the ratio of actual moderator to metal absorption minus the value of this ratio if the moderator mean free path is infinite. The functions F and E have been tabulated by J. A. Wheeler and are given in the Project Handbook (Chap. IV E). Expressions for them are also given in CP-103, 104, 161 for spheres, cylinders and slabs.

The overall disadvantage factor, d, and the surface disadvantage factor, d_s, are related by (F is the metal disadvantage factor)

$$d = Fd_s \approx d_s \,; \tag{19}$$

the last equality is correct to about 5% for metal lumps near the optimum size. We can therefore use the expression (18) for d to calculate the resonance escape probability if, in (6), we assume d and d_s are constant and can be taken out of the integral. The result, upon substituting (18) and (19) into (6) is

$$p = \exp - \frac{N_0 V_0}{N_1 \sigma_{s1} \xi_1 V_1 d} \left(\int \sigma_a \frac{dE}{E} \right)_{\text{eff}} = \exp - \frac{\sigma_{a1} \left(\int \sigma_a \dfrac{dE}{E} \right)_{\text{eff}}}{\sigma_{a0} \sigma_{s1} \xi \left(\dfrac{1}{\bar{f}_R} - 1 \right)}. \qquad (20)$$

Equation (20) is the formula which has been used in almost all of the reports which give k calculations. For simplicity it has been usual to use some intermediate value of σ_{a0} for all metal sizes instead of computing it anew from equation (11) for each size.

The last formula in (20) can be somewhat simplified if we substitute the definitions of σ_{a0} from (11) and σ_{a1} from (14). We then obtain, as was pointed out by H. L. Anderson,

$$p = \exp - \frac{\bar{f}_R}{1 - \bar{f}_R}. \qquad (21)$$

The constant σ_{a0} used in the computation of $\bar{f}_R$ must be calculated for each metal size from equation (11) if (21) is used. This formula is similar to (4) except that the integral is eliminated because $\bar{f}_R$ now means the relative absorption by U over the entire resonance region instead of over a unit energy interval.

5. We shall now outline a calculation of the resonance escape probability based on a somewhat more complicated model in which the slowing down in the moderator is Gaussian (instead of exponential as in section 4), and the resonance absorption in the metal consists of a surface and volume term, both of which are constant over the entire resonance energy range. If we assume no energy loss in the metal, then the equation for q_0, the number of neutrons × velocity per cc and unit $\ln E$ in the metal is the same as equation 13:

$$\Delta q_0 - \kappa_0^2 q_0 = 0. \qquad (22)$$

This equation can be solved (e.g., $q_0 = I_0(\kappa_0 r)$ in a cylindrical system) and the metal disadvantage factor, F, read off from the tables in the Project Handbook (Chap. IV E). The absorption in the metal per unit logarithmic energy interval can be written

$$\frac{\text{absorption}}{\text{unit } \ln \text{ energy}} = q_0(r_0)(V_0 N_0 \, a/F + S\beta') \qquad (23)$$

which is essentially another way of writing (1). In the moderator the neutron density is assumed to satisfy the Fermi slowing down equation:

$$\Delta q_1 = \alpha^2 \frac{\partial q_1}{\partial \tau} \qquad (24)$$

where $\tau = \ln E_0/E$ and $\alpha^2 = 3N_1^2 \sigma_{s1} \sigma_{t1} \xi_1$. The boundary conditions for q_1 are: (a) no net flow on the surface of the cell and (b) on the boundary between lump and moderator ($r = r_0$), the flow of neutrons into the lump must correspond to the absorption in the corresponding energy range within the lump. This gives

$$\frac{S}{3N_1\sigma_{t1}} \frac{\partial q_1(r_0,\tau)}{\partial r} = q_1(r_0,\tau)(V_0 N_0\, a/F + S\beta'). \tag{25}$$

Initially ($\tau = 0$), q_1 is a constant ($= 1$) over the moderator.

The equation for the neutron density in the moderator can be solved by developing it with respect to the characteristic functions Z_i which are solutions of

$$\frac{1}{3N_1\sigma_{t1}} \Delta Z_i(r) = -N_1\sigma_{s1}\xi_1\lambda_i Z_i(r). \tag{26}$$

The boundary conditions for Z_i at the interface between the U and the moderator and on the surface of the cell are the same as those for q_1. For a cylindrical system, the Z_i are Bessel functions. In terms of them, q_1 is given by the equation

$$q_1(r,\tau) = \sum_i a_i Z_i(r) e^{-\lambda_i\tau}. \tag{27}$$

The a_i must be determined in such a way that $q_1(r,0) = 1$.

We can obtain an expression for the characteristic numbers λ_i in terms of the disadvantage factors by integrating (26) over the moderator and applying Green's theorem to the left side. This gives, since the gradient of Z_i vanishes over the cell boundary,

$$\frac{S}{3N_1\sigma_{t1}} \frac{\partial Z_i(r_0)}{\partial r} = -N_1\sigma_{s1}\xi_1\lambda_i \int Z_i(r)\, dV. \tag{28}$$

Upon introducing the boundary condition between the U and moderator (equation 25) we find

$$V_0 N_0\, a/F + S\beta' = N_1\sigma_{s1}\xi_1\lambda_i d_{si} \tag{29}$$

where d_{si} is the surface disadvantage factor corresponding to the i-th characteristic function; i.e., it is the average value of Z_i over the moderator divided by the value of $Z_i(r_0)$. Equation (29) is the characteristic equation of the problem; its solutions λ_i form an infinite, discrete set of numbers. The physical significance of the λ_i is suggested by (29); the exponentials $e^{-\lambda_i\tau}$ take care of the diminution of the number of neutrons on account of their streaming into the metal, where they are absorbed.

The resonance escape probability is the number of neutrons leaving the bottom of the resonance energy region per second divided by the number which enter the top of the region per second. This quantity is therefore ($\tau_1 = \ln E_0/E_1$)

$$p = \frac{\int q_1(r,\tau_1)\, dV}{\int q_1(r,0)\, dV} = \frac{\sum_i a_i \int Z_i(r)\, dV\, e^{-\lambda_i\tau_1}}{\sum_i a_i \int Z_i(r)\, dV}. \tag{30}$$

The λ_i can be expressed in terms of the metal absorption and the i-th disadvantage factor according to (29). A further simplification is possible if we normalize the characteristic functions to $\int Z_i^2\, dV = 1$, the integral to be extended over the moderator. Each "Fourier" coefficient a_i is then equal to $\int Z_i \, dV$. The equation for p now takes the form

$$p = \frac{\sum_i a_i^2 e^{-\lambda_i \tau_1}}{\sum_i a_i^2}$$

$$= \sum_i a_i^2 \exp \left\{ -\frac{V_0 N_0}{V_1 N_1 \sigma_{s1} \xi_1} \int \left(\frac{a}{d_i} + \frac{\beta' S}{V_0 N_0 d_{si}} \right) \frac{dE}{E} \right\} \Bigg/ \sum_i a_i^2 \qquad (31)$$

where $d_i = d_{si} F$ is the i-th overall disadvantage factor.

For systems of interest, a_1^2 represents about 99% of the total $\sum a_i^2$. Hence we can replace the sums in (31) by just the first term and the error will be 1% in the exponent or about 0.1% in p; i.e.

$$p = \exp -\frac{V_0 N_0}{V_1 N_1 \sigma_{s1} \xi_1} \int \left(\frac{a}{d_1} + \frac{\beta' S}{V_0 N_0 d_{s1}} \right) \frac{dE}{E} = \exp e^{-\lambda_1 \tau_1} . \qquad (32)$$

The higher terms in the expansion of q_1 represent transients which die away much more rapidly than the fundamental, $i = 1$. Hence the fact that d_1 (a constant) is a good approximation to the true average disadvantage factor means that the neutron distribution reaches its stationary shape very little below the top of the resonance band. The actual computation of the d_i, or equivalently, the computation of λ_i, is performed by solving the characteristic equation (29). The form of this equation will depend on the particular function Z_i, and these functions depend, in turn, on the geometry of the system.

6. In order to compare the calculation of the disadvantage factor based on a Gaussian slowing down model (section 5) with its calculation based on an exponential model (section 4) we shall recalculate the neutron distributions of section 4 by expanding them in terms of the characteristic functions of section 5.

The distribution of "resonance" neutrons in the metal will again satisfy the differential equations (13) and the density will be continuous everywhere. The flow into the metal is given by (25).

In the moderator the neutron density should satisfy (17) with the κ_1 determined from (15). The expression for κ_1 represents a higher order of approximation to the true solution of the transport equation than is justified by the simple slowing down picture of the previous section. We therefore shall use the simplified, first-order expression for κ_1 (not involving the $2/5 \, \sigma_a/\sigma$ correction) in our comparison of the two calculations. In this approximation $\kappa_1^2 = 3 N_1^2 \sigma_{t1} \sigma_1 \xi_1 / \tau_1$, and we have in the moderator,

$$-\frac{1}{3 N_1 \sigma_{t1}} \Delta Q_1 + \frac{N_1 \sigma_1 \xi_1}{\tau_1} Q_1 = 1 . \qquad (33)$$

The differential equations (13), (33) and the corresponding boundary conditions can be easily solved directly as in section 3. However, in order to make the comparison with the calculation of the preceding section, it is more appropriate to solve the present problem by developing Q_1 into a series of characteristic functions of the Gaussian problem:

$$Q_1 = \sum b_i Z_i(r). \tag{34}$$

If we introduce this into (33), we must also develop the right side, i.e., 1 into a series of Z_i. The coefficients of this series are, as we have seen before, equal to the a_i. We can write therefore,

$$N_1 \sigma_{s1} \xi_1 (\lambda_i + 1) b_i = a_i. \tag{35}$$

Since, as we have seen, all a_i are small except a_1, this will hold also for the b_i, particularly because the λ_i also increases with increasing i. Hence the ratio of the average value of Q_1 to its value at r_0 will be very nearly equal to the same ratio for Z_1, i.e., equal to d_{s1}. In this approximation, the ratio of the average value of Q_1 in the graphite to its value at the surface of the U is F times larger than its average in the volume occupied by the U since this is the corresponding ratio for the solution of (13). Hence the ratio of the average values of Q_1 in graphite and U becomes $d_{s1}F$ and this is the disadvantage factor which applies for the mass (volume) absorption. The chance to escape resonance in this approximation is, therefore, again given by (32). This can be again generalized to (31).

It is worth while to remark that practically the only property of the Q_1 of (34) which has been used in the foregoing is that it satisfies the same boundary conditions as the Q_1 of (24) and that it satisfies an equation of the *type* (33). The exact form of (33) is not of great importance as long as the boundary conditions remain unchanged.

One can obtain an accurate expression for the surface disadvantage factor d_s in terms of the a_i. It is, per definition

$$d_s = \frac{\sum b_i \int Z_i(r) \, dV_1}{V_1 \sum b_i Z_i(r_0)} = \frac{\sum \dfrac{a_i^2}{\lambda_i + 1}}{\sum \dfrac{a_i^2}{\lambda_i + 1} \dfrac{1}{d_{si}}}. \tag{36}$$

The second part follows since, as was remarked before, the a_i are equal to the integral of Z_i over the graphite. With (36), the chance to escape resonance absorption would become

$$p = \exp \left\{ -\frac{V_0 N_0}{V_1 N_1 \sigma_{s1} \xi_1} \frac{\sum \dfrac{a_i^2}{\lambda_i + 1} \dfrac{1}{d_{si}}}{\sum \dfrac{a_i^2}{\lambda_i + 1}} \int \left(\frac{a}{F} + \frac{\beta' S}{V_0 N_0} \right) \frac{dE}{E} \right\}. \tag{37}$$

A discussion of this expression can be carried out rather easily and it can be shown that its deviation from (32) is of the same sign and order of magnitude as that of (31). It is to this extent that the pictures of sections 4 and 5 are equivalent.

30.

On the Utilization of Thermal Neutrons – I

E. P. Wigner and G. N. Plass

Probably 1944

1. The present report contains calculations on the proportion of the thermal neutrons which are absorbed, in the usual lattice arrangement, in U rather than in the damper (C). It is assumed that the thermal neutrons are created, with a uniform density, throughout the carbon and none created in the U or U_3O_8. In case of the metal, this assumption is well realized, and even in the case of the oxide the number of neutrons slowed down in the U_3O_8 remains small ($.0875\ M_O/M_C$ where M_O and M_C are the masses of oxide and of carbon per cell). It will be further assumed that one can attribute to the thermal neutrons definite absorption and scattering cross sections both in the damper and the U or U_3O_8, just as if they had one definite energy. This assumption is hardly justifiable, since Maxwell's energy distribution covers a wide energy range. This is particularly significant for the metal, in which there hardly is any energy exchange between neutrons and scatterer. Thus the low energy neutrons, which are most easily absorbed in the outer layers of the U, will not be replenished as the neutrons penetrate further inward and one must expect, even on the average, a higher than thermal energy for the neutrons in the inside of the metal. However, it would be very difficult to take the polyenergetic nature of the thermal neutrons into account rigorously and, at least for the small spheres which seem to offer most prospects, assuming a single energy value for all thermal neutrons will not introduce a great error. This may be particularly true if the constants entering the calculation are obtained under experimental conditions closely approximating those in the arrangement in question. The last relevant approximation which we shall make is the application of diffusion equations inside both damper and U, with proper boundary conditions on the surface of the latter. This approximation is again better for smaller spheres.

Calculations as will be described below have been made, independently, at various places: by C. Eckart in Chicago; Fermi, Szilard, and Weil in New York; and G. Plass and Wigner in Princeton. The approximations made were, in all cases, those enumerated above and the results are therefore essentially identical. The present Report is being written to make a reference to these calculations easier and also to present a discussion of and corrections for the above approximations. (Cf. Appendices I, II, III for the above approximations.)

2. The above approximations make the calculation very simple. At least in the case of a face centered or a body centered lattice, it is a very good approximation to replace the cell by a sphere of equal volume and to assume that the neutron density has zero radial derivative on the surface of the sphere. This is a less good approximation for a simple cubic lattice and the corresponding correction will be given below. This correction constitutes, from the point of view of thermal utilization, the only difference between the simple cubic and face and body centered lattice. For a finite lattice, corrections must be introduced which decrease the thermal utilization. These will be also discussed below.

If we disregard all these corrections, the neutron density in the inside of the U or oxide will be given by

$$n = a\frac{Sh\kappa_0 r}{r}.$$ (1)

The constants with index 0 will all refer to the inner part of the cell; i.e., the U or the oxide. The constants of the damper will have to carry an index 1. The κ_0 is the macroscopic absorption coefficient of the inside, given by

$$\kappa_0 = N_0\sqrt{3\sigma_0\sigma_{a0}}\left(1 - \frac{2}{5}\frac{\sigma_{a0}}{\sigma_0}\right)$$ (2)

where N_0 is the number of molecules per cm^3, σ_0 and σ_{a0} their scattering and absorption coefficients.

The radius of the U or oxide sphere will be denoted by r_0, the radius of the sphere representing the cell by $r_0 + r_1$ so that r_1 is the thickness of the layer of damper which surrounds every U or oxide lump. The density of the neutrons in the damper will be

$$n = \frac{1}{r}(\alpha Ch\kappa_1(r - r_0) + \beta Sh\kappa_1(r - r_0)) + \frac{1}{N_1 v\sigma_{a1}}.$$ (3)

The significance of N_1, κ_1, and σ_{a1} is the same as that of N_0, κ_0, and σ_{a0}, except that they refer to the damper. The first two terms of (3) give the general spherically symmetric solution of the homogeneous part of the differential equation valid in the damper

$$-\Delta n + \kappa_1^2 n = \frac{3\sigma_1 N_1}{v}q.$$ (4)

Herein v is the velocity of the thermal neutrons, q the number of thermal neutrons created per cm^3 and sec, which was assumed to be 1 for (3). Since the absorption is very small in the damper, $\kappa_1^2 = 3N_1^2\sigma_1\sigma_{a1}$ holds therein. The last term of (3) gives a particular solution of (4) so that (3) is the general spherically symmetric solution of (4).

The arbitrary constants a, α, β in (1) and (3) are determined by the boundary conditions at the surface and between the two regions of the cell. These postulate a) that the neutron density have zero radial derivative at the boundary which gives

$$\alpha(-Ch\kappa_1 r_1 + \kappa_1(r_0 + r_1)Sh\kappa_1 r_1) + \beta(-Sh\kappa_1 r_1 + \kappa_1(r_0 + r_1)Ch\kappa_1 r_1) = 0,$$ (5a)

b) that the neutron density be continuous at r_0

$$\alpha - aShκ_0r_0 = -r_0/N_1vσ_{a1} \tag{5b}$$

and c) that the flow

$$n\underline{v} = (Nσ_av/κ^2)\,\mathrm{grad}\,n \tag{6}$$

be also continuous at the boundary. This gives

$$
\begin{aligned}
&- αN_1σ_{a1}/κ_1^2 + βκ_1r_0N_1σ_{a1}/κ_1^2 \\
&+ a(Shκ_0r_0 - κ_0r_0Chκ_0r_0)N_0σ_{a0}/κ_0^2 = 0\,.
\end{aligned}
\tag{5c}
$$

One can eliminate α from (5b) and (5c) and then eliminate both α and β from the resulting equation and (5b) by means of (5a). One obtains for

$$a = \mathcal{N}/\mathcal{D}$$

$$
\begin{aligned}
\mathcal{N} =\ & [κ_1r_1Chκ_1r_1 + (-1 + κ_1^2r_0(r_0 + r_1))Shκ_1r_1]/vκ_1^2 \\
\mathcal{D} =\ & Shκ_0r_0Shκ_1r_1[N_1σ_{a1}(r_0 + r_1) + N_0σ_{a0}/κ_0^2r_0 - N_1σ_{a1}/κ_1^2r_0] \\
& + Shκ_0r_0Chκ_1r_1[N_1σ_{a1}r_1/κ_1r_0 - N_0σ_{a0}κ_1(r_0 + r_1)/κ_0^2r_0] \\
& + (N_0σ_{a0}/κ_0)Chκ_0r_0[-Shκ_1r_1 + κ_1(r_0 + r_1)Chκ_1r_1]\,.
\end{aligned}
$$

The total absorption in the U or oxide sphere is the integral of $nvN_0σ_{a0}$ over the inner sphere. If we divide this by the total production of thermal neutrons, we obtain the utilization factor p_2 for thermal neutrons (f in Fermi's notation). The total production of thermal neutrons is simply the volume of the carbon; i.e. $(4\pi/3)((r_0 + r_1)^3 - r_0^3)$, since we assumed unit production of neutrons per unit time and volume. We obtain in this way for

$$
\frac{1}{p_2} = \frac{κ_0^2}{3N_0σ_{a0}v}\frac{r_1^3 + 3r_1^2r_0 + 3r_1r_0^2}{κ_0r_0Chκ_0r_0 - Shκ_0r_0}\frac{\mathcal{D}}{\mathcal{N}}\,. \tag{7}
$$

In addition to $1/p_2$, it is interesting to note the factor by which the absorption in the carbon is increased by having the uranium concentrated in lumps, rather than spread out uniformly. Since the U has a higher absorption coefficient than the carbon, the neutron density will be larger in the carbon than in the U and we shall call the ratio of these densities the disadvantage factor d. It can be calculated from p_2 as follows.

If the density were the same all over the cell, the ratio of carbon to U absorption would be $(r_1^3 + 3r_1^2r_0 + 3r_1r_0^2)N_1σ_{a1} : r_0^3N_0σ_{a0}$. Actually, this ratio is $1/p_2 - 1$. Hence the disadvantage factor is

$$
d = \frac{r_0^3N_0σ_{a0}}{(r_1^3 + 3r_1^2r_0 + 3r_1r_0^2)N_1σ_{a1}}\left(\frac{1}{p_2} - 1\right)\,. \tag{8}
$$

3. The formulae (7) and (8) are not only rather complicated but also contain small differences between relatively large quantities which makes their numerical evaluation awkward. They will be simplified henceforth. This can be done most easily by assuming that $κ_1r_1$ is small – an assumption which is justified

for the thermal utilization in all practical cases since the thermal utilization is necessarily poor if $\kappa_1 r_1$ is large. We shall proceed, however, first without making this assumption.

With the abbreviation

$$J(x) = xChx - Shx \tag{9}$$

one can write

$$v\mathcal{N} = J(\kappa_1 r_1)/\kappa_1^2 + r_0(r_0 + r_1)Sh\kappa_1 r_1 \tag{9a}$$

and

$$\begin{aligned}
\kappa_0^2 \mathcal{D} = {} & (N_0\sigma_{a0}/r_0)J(\kappa_0 r_0)J(\kappa_1 r_1) + N_0\sigma_{a0}\kappa_1 J(\kappa_0 r_0)Ch\kappa_1 r_1 \\
& + (N_1\sigma_{a1}\kappa_0^2/\kappa_1^2 r_0)Sh\kappa_0 r_0 J(\kappa_1 r_1) \\
& + N_1\sigma_{a1}\kappa_0^2(r_0 + r_1)Sh\kappa_0 r_0 Sh\kappa_1 r_1 \, .
\end{aligned} \tag{9b}$$

If we now further abbreviate

$$\mathcal{G}(x) = \frac{Shx}{J(x)} = \frac{Shx}{xChx - Shx} \tag{10}$$

we can write, after dividing both $\mathcal{D}$ and $\mathcal{N}$ by $J(\kappa_1 r_1)$

$$\begin{aligned}
\frac{\kappa_0^2 \mathcal{D}}{v N_0 J(\kappa_0 r_0)\mathcal{N}} = {} & \left[\frac{\kappa_1^2\sigma_{a0}}{r_0} + \frac{\kappa_1^2\sigma_{a0}}{r_1} + \frac{\kappa_1^2\sigma_{a0}}{r_1}\mathcal{G}(\kappa_1 r_1) + \frac{N_1\sigma_{a1}\kappa_0^2}{N_0 r_0}\mathcal{G}(\kappa_0 r_0) \right. \\
& \left. + \frac{N_1\sigma_{a1}\kappa_1^2\kappa_0^2}{N_0}(r_0 + r_1)\mathcal{G}(\kappa_0 r_0)\mathcal{G}(\kappa_1 r_1) \right] \\
& : [1 + \kappa_1^2 r_0(r_0 + r_1)\mathcal{G}(\kappa_1 r_1)] \, .
\end{aligned}$$

This gives with (7)

$$\begin{aligned}
\frac{1}{p_2} = {} & \frac{\frac{1}{3}r_1^3 + r_1^2 r_0 + r_1 r_0^2}{1 + \kappa_1^2 r_0(r_0 + r_1)\mathcal{G}(\kappa_1 r_1)} \left[\frac{\kappa_1^2}{r_1} + \frac{\kappa_1^2}{r_0} + \frac{\kappa_1^2}{r_1}\mathcal{G}(\kappa_1 r_1) \right. \\
& \left. + \frac{N_1\sigma_{a1}\kappa_0^2}{N_0\sigma_{a0}r_0}\mathcal{G}(\kappa_0 r_0) + \frac{N_1\sigma_{a1}\kappa_1^2\kappa_0^2}{N_0\sigma_{a0}}(r_0 + r_1)\mathcal{G}(\kappa_0 r_0)\mathcal{G}(\kappa_1 r_1) \right] \, .
\end{aligned} \tag{11}$$

This is still rigorous; i.e., a consequence of equs. (5). It is convenient to use the function $\mathcal{G}$ instead of the hyperbolic functions because it can be expressed rather accurately by elementary functions. Thus

$$\mathcal{G}(x) = \frac{3}{x^2} + \frac{1}{5} \quad \text{for } x < 2.5 \tag{11a}$$

$$\mathcal{G}(x) = \frac{1}{x - 1} \quad \text{for } x > 2.5 \tag{11b}$$

is, as the following table shows, sufficiently accurate for practical purposes.

x	.5	1	1.5	2	2.5	3	3.5
$\mathcal{G}(x)$	12.199	3.195	1.522	.9307	.6520	.4965	.3990
$3x^{-2} + 1/5$	12.200	3.200	1.534	.9500	.6800		
$(x-1)^{-1}$					.6667	.5000	.4000

It is evident, in the case of the thermal utilization, that $\kappa_1 r_1$ must be smaller than 1, since, otherwise, the neutrons which become thermal at the surface of the cell have only a very small chance for reaching the U. Hence, we can write $\mathcal{G}(\kappa_1 r_1) = (3/\kappa_1^2 r_1^2) + .2$ and the first factor of (11) becomes $r_1^3/3$ divided by a quantity which is very nearly 1. One can write

$$1 - \frac{\kappa_1^2 r_1^2 r_0^2 (r_0 + r_1)}{5(r_1^2 + 3r_0 r_1 + 3r_0^2)} \tag{12 a}$$

times $r_1^3/3$ for the first factor of (11). The third term is the biggest in the square bracket of (11) and this gives again nearly 1 if multiplied by $r_1^3/3$. If we multiply the whole square bracket with $r_1^3/3$ and again use (11 a) for $\mathcal{G}(\kappa_1 r_1)$ it becomes

$$(\kappa_1^2 r_1^2/3r_0)(r_0 + r_1) + 1 + \kappa_1^2 r_1^2/15$$
$$+ (N_1 \sigma_{a1} \kappa_0^2/N_0 \sigma_{a0}) \mathcal{G}(\kappa_0 r_0)(r_1^3/3r_0 + r_1(r_0 + r_1) + \kappa_1^2 r_1^3 (r_0 + r_1)/15). \tag{12 b}$$

The whole $1/p_2$ is the product of (12 a) and (12 b).

Evidently, the second term of (12 a) is smaller than $\kappa_1^2 r_1^2/15$ and (12 b) is close to $1/p_2$. Hence, if the formula is to be of interest for us, all the terms of (12 b) apart from 1, must be less than .2 and the product of these terms with the second term of (12 a) can be neglected. Furthermore, the last term of (12 b) is $\kappa_1^2 r_1^2/15$ times the preceding term and must be also negligible. This gives, finally,

$$\frac{1}{p_2} = 1 + \frac{N_1 \sigma_{a1} V_C \kappa_0^2 r_0^2}{3 N_0 \sigma_{a0}} \mathcal{G}(\kappa_0 r_0) + \frac{\kappa_1^2 r_1^2}{3r_0} \left(r_0 + r_1 + \frac{r_1^3}{5 V_C r_0^2} \right) \tag{13}$$

where

$$V_C = (r_1^3 + 3r_1^2 r_0 + 3r_1 r_0^2)/r_0^3 \tag{13 a}$$

is the ratio of the volume of carbon to that of the U. The last term of (13) is almost always negligible.

If $\kappa_0 r_0 < 2.5$, which is practically certain, (13) can be further simplified by (11 a). One obtains

$$\frac{1}{p_2} = 1 + \frac{N_1 \sigma_{a1} V_C}{N_0 \sigma_{a0}} \left(1 + \frac{\kappa_0^2 r_0^2}{15} \right) + \frac{\kappa_1^2 r_1^2}{3r_0} \left(r_0 + r_1 + \frac{r_1^3}{5 V_c r_0^2} \right)$$
$$- \frac{\kappa_1^4 r_1^5}{15 V_C r_0} \left(\frac{r_0 + r_1}{r_0} \right)^2 . \tag{14}$$

The last term is obtained if one retains even smaller terms than in the above calculation. If both $\kappa_1 r_1$ and $\kappa_0 r_0$ are smaller than about 2 and if $1/p_2$ does not exceed 1.2, equ. (14) differs from (11) by only about .1 per cent. Usually, it will be quite sufficient to keep only the first, second and fourth and fifth term of (14).

If only the first two terms of (14) were present, the disadvantage factor would be just 1, i.e., the thermal utilization would be just as good as in a homogeneous mixture. Actually, however,

$$d = 1 + \frac{N_0 \sigma_{a0}}{N_1 \sigma_{a1} V_C} \frac{\kappa_1^2 r_1^2}{3 r_0} \left(r_0 + r_1 + \frac{r_1^3}{5 r_0^2 V_C} \left(\frac{r_0 + r_1}{r_0} \right)^2 \right) + \frac{\kappa_0^2 r_0^2}{15}. \qquad (15)$$

Usually, the first three terms, together with the last term, give a sufficiently close approximation. If one replaces, furthermore, κ_1^2 by $3 N_1^2 \sigma_1 \sigma_{a1}$, one obtains the approximate expression

$$d \approx 1 + \frac{N_0 \sigma_{a0} N_1 \sigma_1 r_1^2}{V_C} \frac{r_0 + r_1}{r_0} + \frac{\kappa_0^2 r_0^2}{15}. \qquad (15\,a)$$

4. In the case of the resonance absorption, p_2 is very far from 1 (about .16) and the above approximation becomes very poor. It continues to be possible, however, to use (11 a) for both $\mathcal{G}(\kappa_2 r_2)$ and $\mathcal{G}(\kappa_0 r_0)$ and obtain in this way a manageable expression for the disadvantage factor d. One obtains

$$d = \left[1 + \frac{\kappa_1^2 r_1^3 (r_0 + r_1)}{5 r_0^2 V_C} \right]^{-1} \left[1 + \frac{\kappa_0^2 r_0^2}{15} + \frac{\kappa_1^2 r_1^3 (r_0 + r_1)}{5 r_0^2 V_C} \right. \qquad (16)$$
$$\left. + \frac{N_0 \sigma_{a0} \kappa_1^2 r_1^2}{N_1 \sigma_{a1} V_C} \left(\frac{r_1}{3 r_0} + \frac{2}{5} - \frac{r_1 (r_0 + r_1)}{5 r_0^2 V_C} \right) \right].$$

This can be further simplified if $\kappa_1^2 r_1^2 \ll 1$ to

$$d = 1 + \frac{\kappa_0^2 r_0^2}{15} + \frac{N_0 \sigma_{a0} \kappa_1^2 r_1^2}{N_1 \sigma_{a1} V_C} \left(\frac{r_1}{3 r_0} + \frac{2}{5} - \frac{r_1 (r_0 + r_1)}{5 r_0^2 V_C} \right) \qquad (16\,a)$$

which is identical with (15) except that the fifth (small) term of the latter is missing in (16 a).

Table 1 gives the thermal utilization for the oxide of density 4 as function of the radius r_0 of the oxide sphere and the cube root $\sqrt[3]{V_C + 1}$ of the ratio of cell[4] and oxide volumes. The $1/p_2$ given in this table were calculated on the basis of (11) or (14) and the following constants (ρ is the density of the oxide)

$$\sigma_{a0} = 6.1 \times 10^{-24} \text{ cm}^2 \qquad \sigma_{a1} = 4.5 \times 10^{-27} \text{ cm}^2$$
$$\sigma_0 = 30 \times 10^{-24} \text{ cm}^2 \qquad \sigma_1 = 4.8 \times 10^{-24} \text{ cm}^2$$
$$\kappa_0 = .0463 \rho \text{ cm}^{-1} \qquad \kappa_1 = 2.08 \times 10^{-2} \text{ cm}^{-1}.$$

[4] Original text reads "carbon" instead of "cell". (A.M.W. & A.M.P.)

Table 1. Thermal utilization for the oxide of density 6 [5]

$\sqrt[3]{V_C+1}$	2.0	2.2	2.4	2.6	2.8	3.0	3.2	3.4	3.6
Radius of the oxide sphere									
16	1.137	1.197	1.281	1.409					
14	1.116	1.163	1.241	1.327					
12	1.093	1.137	1.193	1.266					
10	1.076	1.104	1.157	1.210					
8	1.063	1.091	1.125	1.167		1.276			1.533
6	1.051	1.071	1.097	1.128	1.165	1.209	1.260	1.320	1.400
5					1.146	1.183	1.230	1.278	1.329
4	1.041	1.056	1.076	1.100	1.129	1.161	1.199	1.243	1.305
3				1.091	1.116	1.144	1.178	1.215	1.271
2	1.035	1.049	1.065	1.084	1.107	1.133	1.164		

The graphite density was taken to be 1.63. Table 2 gives the disadvantage factors in the same way.

Table 2. Disadvantage factor for the thermal region with oxide spheres of density 6

$\sqrt[3]{V_C+1}$	2.0	2.2	2.4	2.6	2.8	3.0	3.2	3.4	3.6
Radius of the oxide Sphere									
16	4.14	4.30	4.63	5.20					
14	3.45	3.60	3.87	4.16					
12	2.81	2.98	3.18	3.39					
10	2.30	2.39	2.58	2.68					
8	1.90	1.98	2.06	2.12		2.24			2.46
6	1.54	1.56	1.60	1.63	1.66	1.70	1.73	1.76	1.81
5					1.47	1.49	1.51	1.53	1.55
4	1.24	1.24	1.25	1.27	1.30	1.31	1.32	1.34	1.37
3				1.16	1.17	1.17	1.18	1.19	1.20
2	1.06	1.07	1.07	1.07	1.08	1.08	1.09		

Figure 1 gives the thermal utilization for a lattice cell which contains 21300 gr C and 2660 gr U, either in the metal or the oxide form. In the latter case, the cell contains 3140 gr U_3O_8. The abscissa is the density of the U, which is,

---

[5] Entries in Table I are for $1/p_2$, not p_2. (A.M.W. & A.M.P.)

in the case of the U_3O_8, equal to 238/281 times the density of the oxide. In case of the metal, we assumed the following values of the constants

$$\sigma_{a0} = 6.1 \times 10^{-24} \text{ cm}^2 \quad \sigma_0 = 20 \times 10^{-24} \text{ cm}^2 \quad \kappa_0 = .0424\rho \text{ cm}^{-1}.$$

One sees that the thermal loss increases with increasing density. This is natural since, if the U sphere is small, the thickness of the graphite shell is large, and the neutron has to diffuse, on the average, over a larger distance. At the same time, it has to find a smaller sphere. The thermal loss is somewhat larger for the oxide than for the metal because a neutron, even if it has reached the sphere, can be reflected out of it again by the oxygen which therefore has a disadvantageous effect.

Table 3 gives the disadvantage factors for the resonance absorption in an oxide sphere of density 6.

Table 3. Disadvantage factor for the resonance absorption in oxide spheres of density 6

$\sqrt[3]{V_C + 1}$	2.0	2.2	2.4	2.6	2.8	3.0	3.2	3.4	3.6
Radius of the oxide sphere									
16	2.19	2.27	2.34	2.39					
14	1.923	1.990	2.06	2.10					
12	1.692	1.738	1.791	1.831					
10	1.480	1.529	1.568	1.600					
8	1.324	1.353	1.373	1.398		1.464			1.544
6	1.177	1.195	1.204	1.223		1.235		1.252	1.292
5					1.146	1.153			
4	1.057	1.067	1.072		1.100	1.111	1.124		
3					1.054	1.061	1.068		
2	1.007	1.010	1.013		1.021	1.025			

The disadvantage factor can be factored into two factors: the average density $\overline{\rho_c}$ in the C, divided by the average density in the U can be written as

$$d = \frac{\overline{\rho_c}}{\overline{\rho_u}} = \frac{\overline{\rho_c}}{\rho_b} \frac{\rho_b}{\overline{\rho_u}} \tag{17}$$

where ρ_b is the density of neutrons at the boundary between U and C. The last factor can be easily calculated; it is with sufficient accuracy

$$\frac{\rho_b}{\overline{\rho_u}} = 1 + \frac{1}{15}\kappa_0^2 r_0^2 . \tag{17a}$$

One can recognize (17 a) in all three expressions for the disadvantage factor.

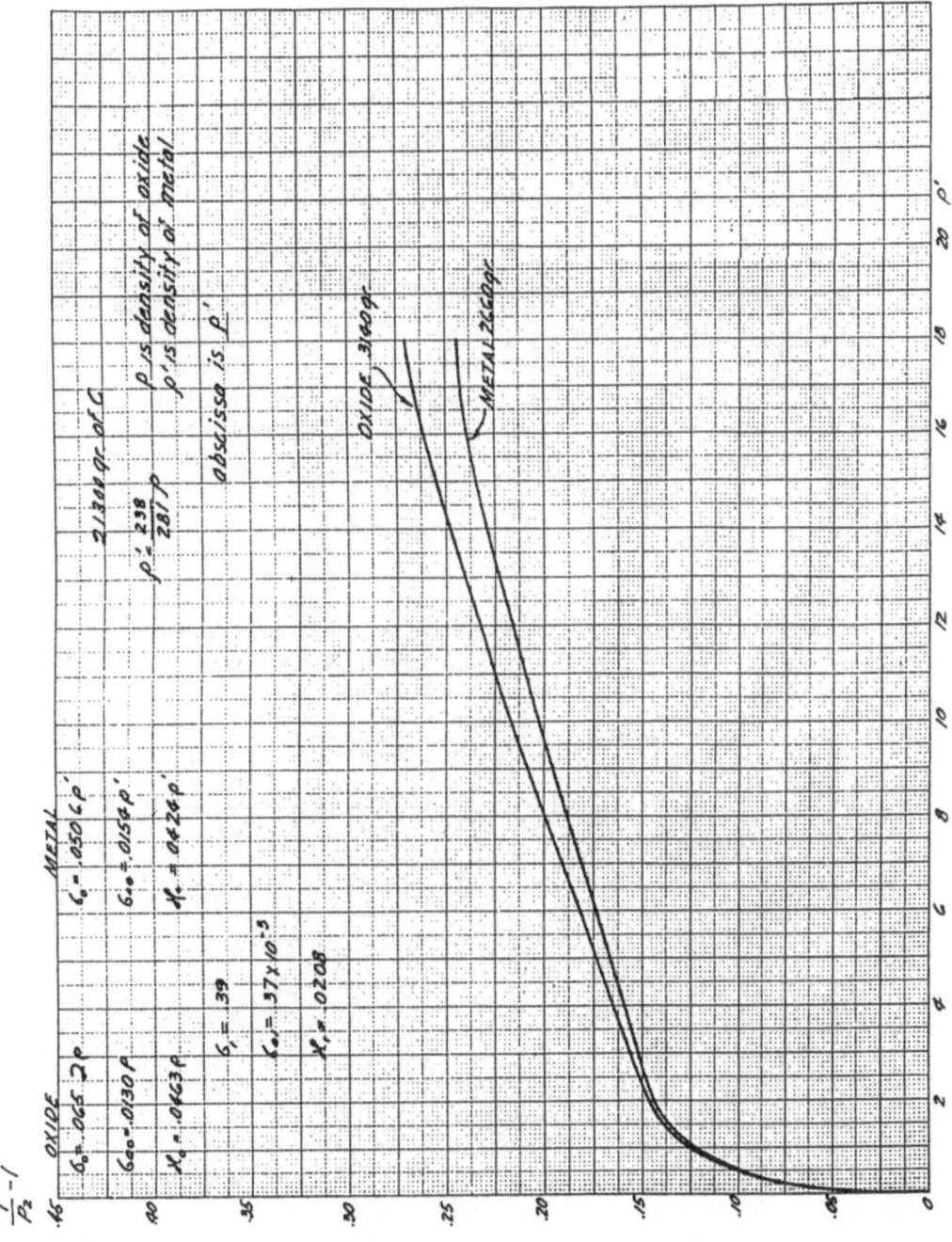

Fig. 1

5. We shall now test the accuracy of our "cellular method", i.e., the replacing of the cell-polyhedra by spheres. We shall do this by estimating the error in case of a simple cubic lattice – the cellular method is evidently much more accurate for face or body centered lattices. The correction which we shall obtain gives, at the same time, the difference between a simple cubic and a centered lattice from the point of view of thermal utilization.

The neutron density in the carbon can be evidently represented by an expression of the form

$$\frac{1}{N_1 \sigma_{a1} v} \left\{ 1 - \alpha \left(\frac{r_s Ch\kappa_1(r_s - r)}{r} - \frac{Sh\kappa_1(r_s - r)}{\kappa_1 r} \right) \right\} \tag{18}$$

since (18) is the most general solution of (4) with zero derivative at $r = r_s = r_0 + r_1$. The constant α is determined by the conditions at $r = r_0$ but we shall not insert its value ($\sim p_2$) for the time being. In the neighborhood of $r = r_s$ we can write for (18) also

$$\frac{1}{N_1 \sigma_{a1} v} \left\{ 1 - \alpha \left(1 + \frac{\kappa_1^2}{2}(r - r_s)^2 \right) \right\} . \tag{18a}$$

Although (18a) evidently satisfies the condition to have zero normal derivative at the surface of a sphere, its normal derivative does not vanish where it really should, i.e., on the surface of the cube-shaped cell. We shall denote the length of the edge of the cube by $2a$ so that

$$8a^3 = 4\pi r_s^3/3; \quad r_s = 1.24a . \tag{19}$$

The normal derivative of (18) will be given, at various points of a side of the cube by the product of $\alpha \kappa_1^2 a / N_1 \sigma_{a1} v$ and the numbers of Table 4.

Table 4

$(a,0,0)$	$(a,\frac{1}{2}a,0)$	$(a,\frac{1}{2}a,\frac{1}{2}a)$	$(a,a,0)$	$(a,a,\frac{1}{2}a)$	(a,a,a)
.24	.11	.01	−.12	−.17	−.28

The integral, over one side of the cube, of the square of the normal derivative is .33 $(\alpha \kappa_1^2 a / 2 N_1 \sigma_{a1} v)^2$. Mathematically, our problem is to find a solution of the differential equation (4), and the equation $\Delta n = \kappa_0^2 n$ inside the U sphere, which satisfies the boundary conditions both at $r = r_0$ and also on the surface of the cell. In this paragraph, we shall call a "solution" of the equations any function which satisfies the differential equations and the boundary condition at $r = r_0$, even if it has not zero normal derivative on the surface of the cube. In order to find that "solution" which does have zero normal derivative on the cube surface, we could proceed in the following way: We could take any "solution", e.g., the one corresponding to (18). Then, we could add to this a "solution" of the homogeneous problem (set $q = 0$ in (4)) the normal derivative of which on the surface of the cube is oppositely equal to the normal derivative of (18).

Such a solution of the homogeneous problem will be, in general, a super-position of solutions which depend on the polar angles in all possible ways. Instead of this, we shall choose one single function, which depends on the angle in one definite way (in the same way as a g-wave) because we shall see that one can eliminate the greatest part of the normal derivative of (18) by such a single function. The angular dependence of this function is the same as that of

$$r^4 P_4 = x^4 + y^4 + z^4 - 3x^2 y^2 - 3y^2 z^2 - 3x^2 z^2 . \tag{20}$$

Near the surface of the cube, it has the form

$$-\varepsilon \frac{\alpha \kappa_1^2}{N_1 \sigma_{a1} v a^2} r^4 P_4 (1 + \gamma r_0^9 / r^9) \tag{20 a}$$

where ε and γ are constants. In the U sphere, the g-solution has, apart from a constant, very nearly the form (20) and from the boundary condition at $r = r_0$ one obtains

$$\gamma = \frac{4\sigma_0 - 4\sigma_1}{5\sigma_0 + 4\sigma_1} \tag{20 b}$$

so that the second term in (20 a) can be neglected near the surface of the cube.

If we take for the neutron density the sum of (18) and (20 a), the normal derivative will be given, on a side of the cube, by the numbers of Table 5 multiplied again by $\alpha \kappa_1^2 a / N_1 \sigma_{a1} v$.

Table 5

$(a,0,0)$	$(a,\frac{1}{2}a,0)$	$(a,\frac{1}{2}a,\frac{1}{2}a)$	$(a,a,0)$	$(a,a,\frac{1}{2}a)$	(a,a,a)
$.24 - 4\varepsilon$	$.11 - 2.5\varepsilon$	$.01 - \varepsilon$	$-.12 + 2\varepsilon$	$-.17 + 3.5\varepsilon$	$-.28 + 8\varepsilon$

If we choose

$$\varepsilon = .043 \tag{20 c}$$

the integral of the square of the normal derivative over one side of the cube will become $.017 \, (\alpha \kappa_1^2 a^2 / 2 N_1 \sigma_{a1} v)^2$, i.e., 20 times smaller than it is for $\varepsilon = 0$. This shows that the sum of (18) and (20 a) satisfies the boundary conditions much more accurately than (18) alone does.

The average of (18) over the carbon, if multiplied by $N_1 \sigma_{a1} v$ gives the thermal loss for a face centered or body centered lattice. Similarly, $N_1 \sigma_{a1} v$ times the average of the sum of (18) and (20 a) will give the thermal loss in a simple cubic lattice with the same C/U ratio.

We shall first integrate the sum of (18) and (20 a) over the whole cube and (18) over the whole sphere.

The integral of (20 a) times $N_1 \sigma_{a1} v$ over the cube is

$$3.2 \varepsilon \alpha \kappa_1^2 a^5 = .137 \alpha \kappa_1^2 a^5 \tag{21}$$

if the term with γ is neglected. For the integral of (20 a) we need

$$\int \int_{-a}^{a} \int r \, dx \, dy \, dz = 4a^4 \ln(2 + \sqrt{3}) - \pi a^4/3 + 2\sqrt{3}a^4 = 7.685a^4 \,.$$

This gives for the integral of (20 a) times $N_1 \sigma_{a1} v$

$$8a^3(1 - \alpha) - 4\alpha\kappa_1^2 r_s^2 a^3 - 4\alpha\kappa_1^2 a^5 + 7.685\alpha\kappa_1^2 r_s a^4 \,. \qquad (21\,\text{a})$$

The integral of (18) over the sphere is

$$4\frac{\pi}{3}r_s^3(1 - \alpha) - \pi\alpha\kappa_1^2 r_s^5/15 \,. \qquad (22)$$

The difference between (21) + (21 a) and (22), divided by the volume of the cell becomes, if we remember (19)

$$-\Delta p_2 = \frac{1}{8}(.137\alpha\kappa_1^2 a^2 - .014\alpha\kappa_1^2 a^2) = .005\alpha\kappa_1^2(2a)^2 \,; \qquad (23)$$

$2a$ is the lattice constant of the cubic lattice.

In order to estimate α,. we can note that, by the presence of the U, the density of neutrons has decreased about in the ratio $1 : 1 - \alpha$. The absorption of the carbon has decreased in the same ratio. Since this absorption is $1 - p_2$ in the presence of the U, and 1 without it, evidently $\alpha \approx p_2$.

Strictly speaking, a correction should be applied to (23). The increase of the absorption at the surface of the cell must be compensated somehow by a decreased absorption somewhere else. This shows that our expression (18) + (20 a) for the density cannot be strictly speaking correct because it does not account for this decreased absorption in the inside of the cell. This is not surprising since we tried to satisfy the boundary conditions at the surface of the cell by a very special function. A somewhat more detailed consideration of the boundary effects on the cell surface shows that we should add to (18) not only (21 a) but also a spherically symmetric "solution" of the homogeneous equations. This corresponds to a neutron distribution which does not differ very greatly from the neutron distribution described by (18) and thus leads to absorptions in U and C in the ratio $p_2 : 1 - p_2$. Its total effect is to compensate for the added neutron absorption given by (23) and gives, therefore $-.005\alpha\kappa_1^2(2a)^2 p_2$ in the U and $-.005\alpha\kappa_1^2(2a)^2(1 - p_2)$ in the C. On the whole the change in the thermal utilization becomes

$$-\Delta p_2 = .005\alpha\kappa_1^2(2a)^2 p_2 = .005p_2^2\alpha\kappa_1^2(2a)^2 \qquad (24)$$

which differs only slightly from (23).

Equation (23) shows that the difference between simple cubic lattice and the face or body centered lattices is very small from the point of view of thermal utilization; $2\kappa_1 a$ is smaller than 1 for even for the largest structures that have been proposed. Furthermore, there is an opposite effect of the same order of magnitude in the resonance absorption.

31.

Effect of the Temperature of the Moderator on the Velocity Distribution of Neutrons with Numerical Calculations for H as Moderator

E. P. Wigner and J. E. Wilkins, Jr.

September 14, 1944

Abstract. In this paper we set up an integral equation governing the energy distribution of neutrons which are being slowed down uniformly thoughout the entire space by a uniformly distributed moderator whose atoms are in motion with a Maxwellian distribution of velocities. The effects of chemical binding and crystal reflection are ignored. When the moderator is hydrogen, the integral equation is reduced to a differential equation and solved by numerical methods. In this manner we obtain a refinement of the dv/v^2 law.

1. There are two particularly simple problems in connection with the energy distribution of neutrons which are present in a medium of finite temperature. In the first problem the slowing down is uniform throughout the entire space which is itself uniformly filled with the slowing down material. In this case the neutron distribution is evidently the same all over space. In the second problem the neutrons enter a half space from one side with uniform intensity and diffuse into it. The question in this case is the density distribution of neutrons at large distances from the boundary plane of the half space and the exponential relaxation length of the neutron density. We shall be interested only in the first problem.

There are three phenomena which make the calculation of the energy distribution of the neutrons somewhat complicated. These are the finite velocity of the particles with which they collide, the effect of the chemical binding on the scattering cross section, and finally the effects of crystal reflection. We shall disregard the last two effects and the calculation to be given will therefore be valid only in a monoatomic gas. Messrs. Seitz and Goldberger are engaged in the study of the crystal effect and Mr. Teller has made considerations on the effect of the chemical binding.

To derive an equation satisfied by $N(v)$, where $N(v)\, dv$ is the number of neutrons per cm^3 whose velocity lies between v and $v + dv$, we proceed as follows. Let us denote the probability that a neutron with velocity v_1 acquire in unit length of time a velocity between v and $v + dv$ by collision with an atom by $N\sigma_s \mathcal{P}(v, v_1)\, dv$. Here N is the number of atoms per cm^3 and σ_s is the scattering cross section of the moderator. Hence the number of neutrons acquiring a velocity between v and $v + dv$ in unit length of time is

$$N\,dv \int_0^\infty \sigma_s \mathcal{P}(v,v_1)N(v_1)\,dv_1\,.$$

Let $N\sigma_s V(v)$ be the probability that a neutron with velocity v will be scattered in unit time, and let $N\sigma_s\gamma$ be the probability that a neutron with velocity v will be absorbed in unit time. If we adopt the $1/v$ law, this second probability will be independent of the velocity of the neutron. We further assume that the scattering cross section is independent of the velocity of the neutron. This is a valid assumption for a moderator gas, since our results will be applied only in a reasonably close neighborhood of thermal energies. It is now clear that $N(v)$ satisfies the following integral equation:

$$\int_0^\infty \mathcal{P}(v,v_1)N(v_1)\,dv_1 = [V(v)+\gamma]N(v)\,. \tag{1}$$

Before calculating $\mathcal{P}$ and V we shall derive a property of the above equation which simplifies calculations greatly and also can serve as a check. Evidently equation (1) is satisfied by the Maxwell distribution if $\gamma = 0$. This together with the principle of detailed balance permits us to give equation (1) a symmetric form. If we denote the Maxwell distribution of velocity by $M(v)$

$$M(v)\,dv = \frac{4\beta^3}{\sqrt{\pi}}v^2 e^{-\beta^2 v^2}\,dv\,,\quad \beta^2 = \frac{1}{2kT} \tag{2}$$

the principle of detailed balance reads

$$\mathcal{P}(v,v_1)M(v_1) = \mathcal{P}(v_1,v)M(v)\,. \tag{3}$$

Herein the mass of the neutron is used as the unit of mass. It follows that

$$\frac{1}{\sqrt{M(v)}}\mathcal{P}(v,v_1)\sqrt{M(v_1)} = \frac{1}{\sqrt{M(v_1)}}\mathcal{P}(v_1,v)\sqrt{M(v)}$$
$$= S(v,v_1) = S(v_1,v) \tag{3a}$$

is a symmetric kernel. Thus if we introduce the new quantity $\nu(v) = N(v)/\sqrt{M(v)}$ into the equation (1), we get the equation

$$\int_0^\infty S(v,v_1)\nu(v_1)\,dv_1 = [V(v)+\gamma]\nu(v) \tag{4}$$

which has a symmetric kernel.

2. The next task is to calculate V and the kernel $\mathcal{P}$. Both quantities depend of course on the temperature of the gas and also on the mass of its atoms. The calculation, although quite laborious, is entirely straightforward and gives for $v_1 < v$

$$\mathcal{P}(v,v_1) = \frac{2\theta^2 v}{\sqrt{\pi}v_1}e^{\beta^2(v_1^2-v^2)}[I(\beta\theta v_1 - \beta\zeta v) + I(\beta\theta v_1 + \beta\zeta v)]$$
$$+ \frac{2\theta^2 v}{\sqrt{\pi}v_1}[I(\beta\theta v - \beta\zeta v_1) - I(\beta\zeta v_1 + \beta\theta v)]\,. \tag{5}$$

For $v_1 > v$ it gives

$$P(v,v_1) = \frac{2\theta^2 v}{\sqrt{\pi}v_1}[I(\beta\theta v - \beta\zeta v_1) + I(\beta\theta v + \beta\zeta v_1)]$$

$$+ \frac{2\theta^2 v}{\sqrt{\pi}v_1}e^{\beta^2(v_1^2-v^2)}[I(\beta\theta v_1 - \beta\zeta v) - I(\beta\zeta v + \beta\theta v_1)]. \tag{5a}$$

Herein

$$\theta = \frac{m+1}{2\sqrt{m}}, \quad \zeta = \frac{m-1}{2\sqrt{m}}, \quad \beta^2 = \frac{1}{2kt} \tag{5b}$$

where m is the mass of the atoms of the moderator in units of the neutron mass. I is the odd function

$$I(x) = \int_0^x e^{-x^2}\,dx. \tag{5c}$$

For V we obtain[6]

$$V(v) = \frac{1}{\sqrt{\pi}}(2v + 1/\beta^2 mv)I(\beta\sqrt{m}v) + e^{-\beta^2 mv^2}/\beta\sqrt{m\pi}. \tag{6}$$

Equations (5) and (6) determine the quantities which occur in the integral equation (1). In the derivation it was assumed that the scattering cross section σ_s is independent of the velocity of the neutrons. This is a valid assumption for a moderator gas since (1) will be applied only in a reasonably close neighborhood of thermal energies. It was further assumed that the scattering is spherically symmetric in the center of mass coordinate system. The physical significance of γ is $\gamma = \sigma_a v/\sigma_s$ where σ_a is the absorption cross section for the relative velocity v between neutron and atom; γ is independent of the latter.

The equations (5) and (6) can be simplified considerably by measuring the velocity in terms of the thermal velocity. Let us denote therefore

$$\beta v = x, \quad \beta v_1 = x_1 \tag{7}$$

and

$$\Gamma = \beta\gamma = \frac{\sigma_a v}{\sigma_s\sqrt{2kT}}. \tag{7a}$$

Then Γ is the probability of absorption of the neutron while it goes one mean free path with the velocity corresponding to the energy kT.

We then have

$$P(x,x_1) = \frac{2\theta^2}{\sqrt{\pi}}\frac{x}{x_1}e^{(x_1^2-x^2)}[I(\theta x_1 - \zeta x) + I(\theta x_1 + \zeta x)]$$

$$+ \frac{2\theta^2}{\sqrt{\pi}}\frac{x}{x_1}[I(\theta x - \zeta x_1) - I(\zeta x_1 + \theta x)] \text{ for } x_1 < x \tag{8}$$

[6] V is also given by G. Jäger in Winkelmann's Handbuch der Physik (Barth, 1906), vol. 3, p. 698.

and

$$P(x, x_1) = \frac{2\theta^2}{\sqrt{\pi}} \frac{x}{x_1} [I(\theta x - \zeta x_1) + I(\theta x + \zeta x_1)]$$
$$+ \frac{2\theta^2}{\sqrt{\pi}} \frac{x}{x_1} e^{(x_1^2 - x^2)} [I(\theta x_1 - \zeta x) - I(\zeta x + \theta x_1)] \tag{8a}$$

for $x_1 > x$. We further have

$$V(x) = \frac{1}{\sqrt{\pi}} (2x + 1/mx) I(m^{\frac{1}{2}} x) + e^{-mx^2}/\sqrt{m\pi} \tag{8b}$$

and the integral equation becomes

$$\int_0^\infty S(x, x_1) \nu(x_1) \, dx_1 = [V(x) + \Gamma] \nu(x). \tag{9}$$

Although the derivation of the expressions for P and V is quite straightforward, it may be worthwhile to give a few of the intermediate steps to facilitate checking the equations.

$V(v)$ is the probability of a collision of the neutron with an atom of unit cross section and the velocity distribution

$$M_m(v_2) \, dv_2 = \frac{4m^{3/2}\beta^3}{\sqrt{\pi}} v_2^2 e^{-\beta^2 m v_2^2} \, dv_2 . \tag{10}$$

The probability for a collision with an atom of velocity v_2 is, if the neutron velocity is v and the cosine of the angle between the directions of motion is μ, simply

$$v_r = \sqrt{v^2 + v_2^2 - 2vv_2\mu} \tag{10a}$$

which is the relative velocity of the two particles. The number of atoms within unit velocity range at v_2 is given by (10), the probability that μ lie between μ and $\mu + d\mu$ is $\frac{1}{2} d\mu$ so that

$$V(v) = \int_{-1}^1 \frac{1}{2} \, d\mu \int_0^\infty dv_2 M_m(v_2) \sqrt{v_2 + v_2^2 - 2vv_2\mu} . \tag{10b}$$

The integration over μ is elementary and gives

$$\frac{1}{2} \int_0^\infty \frac{M(v_2)}{3vv_2} [(v + v_2)^3 - |v - v_2|^3] \, dv_2 . \tag{10c}$$

When integrating this expression one must proceed with the two cases $v_2 < v$ and $v_2 > v$ separately. This kind of disjunction is characteristic also for the calculation of P. With this proviso, the integration of (10c) can be carried out easily and gives (6).

The calculation of P is more cumbersome. Collision of a neutron with the velocity v_1 with an atom of velocity v_2 gives a velocity between v and $v + dv$ for the neutron with the probability

$$\mathcal{P}(v; v_1, v_2, \mu) = 0 \qquad \text{if } v < v_{\min}$$

$$= \frac{2v}{v_{\max}^2 - v_{\min}^2} \qquad \text{if } v_{\min} < v < v_{\max}$$

$$= 0 \qquad \text{if } v > v_{\max}.$$

These equations express the familiar fact that, after the collision, the probability is uniform in energy scale between the minimum and maximum energies. Herein, μ is again the cosine between the directions of motion of neutron and atom before the collision,

$$v_{\min} = v_c - mv_r/(m+1),$$
$$v_{\max} = v_c + mv_r/(m+1), \tag{11a}$$

and

$$v_c = \frac{1}{m+1}\sqrt{v_1^2 + m^2 v_2^2 + 2m v_1 v_2 \mu} \tag{11b}$$

is the velocity of the center of mass of neutron plus atom, $mv_r/(m+1)$ is the velocity of the neutron in the center of mass coordinate system. One derives (11) most easily by a geometrical argument. Now

$$\mathcal{P}(v, v_1) = \int_{-1}^{1} \frac{1}{2} d\mu \int_0^\infty dv_2 M_m(v_2) v_r \mathcal{P}(v; v_1, v_2, \mu)$$

$$= \frac{(1+m)^2 v}{4m} \int \int d\mu \, dv_2 M_m(v_2)(v_1^2 + m^2 v_2^2 + 2m v_1 v_2 \mu)^{-\frac{1}{2}} \tag{11c}$$

where the integration is to be extended over the region limited by $-1 < \mu < 1$ and that $v_{\min} < v < v_{\max}$. The integration itself is quite elementary. However, the domain of integration is quite complicated and is bounded in general by two straight lines and two parts of a hyperbola if one introduces the variable $2m v_1 v_2 \mu$ for μ. A number of cases has to be distinguished and the integration carried out separately over different parts of the domain of integration before one obtains (5) and (5a).

3. The kernel S is greatly simplified in the case $m = 1$, i.e. when the moderator is hydrogen gas. We then have

$$\sqrt{\pi} S(x, x_1) = 4I(x) e^{\frac{1}{2}(x^2 - x_1^2)} \quad (x < x_1), \tag{12a}$$
$$\sqrt{\pi} S(x, x_1) = 4I(x_1) e^{\frac{1}{2}(x_1^2 - x^2)} \quad (x > x_1), \tag{12b}$$
$$\sqrt{\pi} V(x) = (2x + 1/x)I(x) + e^{-x^2} \tag{13}$$

and the integral equation becomes

$$4e^{-\frac{1}{2}x^2} \int_0^x dx_1 I(x_1) e^{\frac{1}{2}x_1^2} \nu(x_1) + 4I(x) e^{\frac{1}{2}x^2} \int_x^\infty e^{-\frac{1}{2}x_1^2} \nu(x_1)\, dx_1$$

$$= \left[\left(2x + \frac{1}{x}\right) I(x) + e^{-x^2} + \sqrt{\pi}\Gamma\right] \nu(x). \tag{14}$$

The $S(x, x_1)$ of (12) has the form of a Green's function of an ordinary second order differential equation. The integral operator of (14) is therefore the reciprocal of a second order differential operator. If one applies this operator to equation (14), it will itself be transformed into a second order differential equation.

Without using the theory of Green's function, one can proceed as follows. Let us find a second order differential operator

$$L_1 = \frac{d^2}{dx_2} + a(x)\frac{d}{dx} + b(x) \tag{15}$$

which gives zero if applied either to $I(x)e^{1/2x^2}$ or to $e^{-1/2x^2}$. This condition gives two linear equations for the two unknowns a and b. One obtains

$$a = \frac{-2I(x)}{e^{-x^2} + 2xI(x)}, \quad b = \frac{e^{-x^2}}{e^{-x^2} + 2xI(x)} - x^2. \tag{15a}$$

In order to bring L_1 into a self adjoint form, one may multiply it with P^{-1} where

$$P = e^{-x^2} + 2xI(x). \tag{16}$$

This gives

$$L = \frac{1}{P}\frac{d^2}{dx^2} - \frac{2I}{P^2}\frac{d}{dx} + \frac{e^{-x^2}}{P^2} - \frac{x^2}{P} = \frac{d}{dx}\frac{1}{P}\frac{d}{dx} + \frac{e^{-x^2}}{P^2} - \frac{x^2}{P}. \tag{16}$$

If one applies either L_1 or L to (14), the integral will give rise to two types of terms. Differentiation of the function before the integral sign gives zero. The other terms will contain no integral. Hence the integral equation reduces to a second order differential equation, the second derivative coming from applying L_1 or L to the right side of (14). By virtue of these remarks, one sees that any solution of the integral equation (14) also satisfies the differential equation

$$-\frac{d}{dx}\frac{1}{P}\frac{d}{dx}(V + \Gamma)\nu(x) + [W(V + \Gamma) - 4/\pi^{\frac{1}{2}}]\nu(x) = 0 \tag{17}$$

where

$$W(x) = x^2/P - e^{-x^2}/P^2. \tag{18}$$

Since $S(0, x_1) = 0$, we also have $\nu(0) = 0$. Conversely, any solution $\nu(x)$ of the differential equation (17) such that $\nu(0) = 0$ is a solution of the integral equation (14). The other solutions of (17) are, however, not solutions of (14). The two solutions of (17) behave at $x = 0$ like a constant and x itself. The former solution would give an $N(v)$ which goes to zero as v, the latter – the one which we have to use – gives an $N(v)$ which goes at $v = 0$ to zero as v^2 which is the same way as the Maxwell distribution $M(v)$ goes to zero.

4. It would be possible to discuss (17) in a general way and to use it directly to obtain numerical results. However, we found it more expedient to proceed as follows. We place

$$\mu(x) = [V(x) + \Gamma]\nu(x), \tag{19}$$

so that we have

$$-(\mu'/P)' + \{W(x) - 4\pi^{-\frac{1}{2}}[V(x) + \Gamma]^{-1}\}\mu = 0, \ \mu(0) = 0, \tag{19a}$$

observing that for $x \geq 0$, $V(x) > 0$, so that $V(x) + \Gamma \neq 0$. We reduce the second order linear differential equation (19 a) to a Riccati equation by making the substitution

$$y = \mu'/\mu P, \tag{20}$$

which leads to

$$y' = W(x) - 4\pi^{-\frac{1}{2}}[V + \Gamma]^{-1} - Py^2. \tag{20a}$$

The boundary condition $\mu(0) = 0$, together with $\mu(x) \not\equiv 0$, implies that we have

$$\lim_{x=0}\{y(x) - x^{-1}\} = 0. \tag{20b}$$

Conversely we obtain a solution $\mu(x)$ from a solution $y(x)$ of (20 a) by setting

$$\mu(x) = \mu'(0)x \exp \int_0^x [P(t)y(t) - t^{-1}]\,dt.$$

The solution $y(x)$ of (20 a) and (20 b) may be written in the form

$$y(x) = \frac{1}{x} - \frac{4(1+\delta)x}{3(1+2\delta)} + \frac{(103 + 380\delta + 364\delta^2)x^3}{90(1+2\delta)^2}$$
$$\frac{-(1163 + 6666\delta + 13200\delta^2 + 8864\delta^3)x^5}{945(1+2\delta)^3} + \cdots, \tag{21}$$

where $\delta = \frac{1}{4}\sqrt{\pi}\Gamma$. We used this series to compute the values of $y(x)$ for $\delta = .1$, .2 and 1 and for $x = .1$, .2, .3, .4 and .5. With these values of y we compute successively μ, ν and N, normalizing the functions so that $N''(0) = 2$.

With the value of $y(\frac{1}{2})$ thus obtained we continued the function $y(x)$ by numerical integration of (20 a). In order to decide how far to carry the numerical integration, we now investigate the behavior of the solutions y of (20 a) as x approaches ∞. We use the following expansion of $I(x)$

$$I(x) = \frac{\sqrt{\pi}}{2} - \frac{e^{-x^2}}{2x}\left(1 - \frac{1}{2x^2} + \cdots\right).$$

Neglecting terms involving e^{-x^2} we see that

$$P(x) \approx \sqrt{\pi}x, \ V(x) \approx x + \frac{1}{2x}, \ W(x) \approx \frac{x}{\sqrt{\pi}}. \tag{22}$$

Hence (20 a) becomes

$$\sqrt{\pi}y' = x - \frac{4x}{x^2 + \frac{1}{2} + \Gamma x} - \pi xy^2. \tag{22a}$$

This has two solutions which are finite at ∞ and which can be expanded in terms of inverse powers of x:

$$\sqrt{\pi}\,y_1 \approx 1 - \frac{2}{x^2} + \frac{2\Gamma}{x^3} - \frac{(2\Gamma^2+3)}{x^4} + \frac{(2\Gamma^3+5\Gamma)}{x^5} - \cdots , \qquad (23\,\mathrm{a})$$

$$\sqrt{\pi}\,y_2 \approx -1 + \frac{2}{x^2} - \frac{2\Gamma}{x^3} + \frac{(2\Gamma^2-1)}{x^4} - \frac{(2\Gamma^3-\Gamma)}{x^5} + \cdots . \qquad (23\,\mathrm{b})$$

These two series give the asymptotic expansions of two linearly independent solutions of (19 a):

$$\mu_1 = \exp\int y_1 P\, dx \approx \exp\left(\frac{1}{2}x^2 - 2\ln x - 2\Gamma/x + \cdots\right), \qquad (24\,\mathrm{a})$$

$$\mu_2 = \exp\int y_2 P\, dx \approx \exp\left(-\frac{1}{2}x^2 + 2\ln x + 2\Gamma/x - \cdots\right). \qquad (24\,\mathrm{b})$$

For the asymptotic behavior of the two linearly independent solutions of (17) this gives because of (19)

$$\nu_1 \approx \frac{1}{x^3}\frac{e^{(\frac{1}{2}x^2 - 2\Gamma/x)}}{1 + \Gamma x^{-1} + \frac{1}{2}x^{-2}} \qquad (25\,\mathrm{a})$$

$$\nu_2 \approx x\frac{e^{(-\frac{1}{2}x^2 + 2\Gamma/x)}}{1 + \Gamma x^{-1} + \frac{1}{2}x^{-2}} \qquad (25\,\mathrm{b})$$

and for N

$$N_1 \approx \frac{1}{x^2}\frac{e^{-2\Gamma/x}}{1 + \Gamma/x \ldots} \qquad (26\,\mathrm{a})$$

$$N_2 \approx x^2\frac{e^{-x^2 + 2\Gamma/x}}{1 + \Gamma/x \ldots}. \qquad (26\,\mathrm{b})$$

It is evident from physical considerations that if $\gamma > 0$ the slowing down at high energies, Nv^2, cannot be zero. Hence in the asymptotic expansion of N at $x = \infty$ the coefficient of N_1 cannot be zero. Hence the coefficient of μ_1 in the asymptotic expansion of μ also is not zero. As a result, μ_1 gives the asymptotic behavior of μ because for large x we have $\mu_2 \ll \mu_1$. It follows that it is necessary to carry out our numerical integration only to a point x_0 at which the series (23 a) gives the same value as the numerical integration. For $x \geq x_0$ the series (23 a) may be used to obtain values of y. In practice it turns out that $x_0 = 5$ is a suitable choice.

Having computed y in this manner for $x \geq .5$ we can compute $\mu(x)$ by the formula

$$\mu(x) = \mu\left(\tfrac{1}{2}\right)\exp\int_{\frac{1}{2}}^{x} P(t)y(t)\, dt$$

and then we can compute $\nu(x)$ and $N(x)$.

The asymptotic form of N at large x or large v is given by (26 a). Replacing $e^{-2\Gamma/x}$ in this by $(1+\Gamma/x)^{-2}$ – which is within the accuracy of the expression in the denominator, we obtain reintroducing σ_s and the life time τ of the neutrons

$$N(v)\, dv \approx \frac{Cv\, dv}{(N\sigma_s\tau v + 1)^3} \qquad (27)$$

where C is constant. This shows that the deviation from the dv/v^2 law arises at high energies not from the positive temperature of the atoms but from the absorption. It causes $N(v)$ to become lower than the dv/v^2 law would indicate. At lower energies the figures show that $N(v)$ becomes higher than $1/v^2$. This is the result of the finite temperature of the atoms. For $T = 0$, (27) is correct for all v. The constant C occurring in (27) is of some importance and may be evaluated directly from the values of $N(x)$ previously calculated. If this be done it is possible to obtain a check on the computations used to find $N(x)$ by calculating C in another manner. The number of neutrons which are slowed down to velocities below v_0 is given, when v_0 is very large, by the formula

$$N\sigma_s \int_{v_0}^{\infty} \frac{vN(v)v_0^2}{v^2}\, dv \sim \frac{N\sigma_s}{2} \lim_{v=\infty} v^2 N(v) = \frac{C}{2N^2\sigma_s^2\tau^3}. \qquad (28)$$

Since v_0 is large this is approximately equal to the number of neutrons that are absorbed at velocities below v_0, and in the limit as v_0 approaches infinitely we have equality. Thus

$$N\sigma_s\gamma \int_0^{\infty} N(v)\, dv = \frac{C}{2N^2\sigma_s^2\tau^3}. \qquad (28\,\text{a})$$

Hence the constant C of the asymptotic expression for $N(v)$ can also be obtained by integrating $N(v)$. The integral was evaluated as follows. Replace v by $x = \beta v$, so that

$$\Gamma \int_0^{\infty} N(x)\, dx = \frac{C\beta^2}{2N^3\sigma_s^3\tau^3}.$$

The integral $\int_0^6 N(x)\, dx = A$ may be computed by numerical integration of the values of $N(x)$ previously computed. To compute $B = \int_6^{\infty} N(x)\, dx$, we replace $N(x)$ by an asymptotic expansion of the form

$$N(x) = \frac{C\beta^2}{2N^3\sigma_s^3\tau^3}\left[\frac{1}{x^2} + \cdots\right] \qquad (29)$$

(which may be obtained by making more precise the reasoning which leads to 26 a) and integrating directly. In this manner we get a linear equation

$$A + B = \frac{C\beta^2}{2N^3\sigma_s^3\tau^3}, \qquad (30)$$

which may be solved for C. The constant C was computed in this manner for the three values of γ previously mentioned, and it was found that they agreed satisfactorily (about 1%) with the values of C computed directly.

Our numerical results are summarized in the attached diagram which shows the graphs of $M(x) = x^2 e^{-x^2}$ $(\gamma = 0)$ and $N(x)$ for $\delta = \frac{1}{4}\sqrt{\pi}\beta\gamma = .1, .2$ and 1 respectively. This corresponds to σ_a/σ_s values of 0.226, 0.452, 2.257 at kT energy. In the three latter cases, the straight line graph of $1/x^2 \lim_{x=\infty} x^2 N(x)$ is also included, so that the deviation from the dv/v^2 law may be visualized.

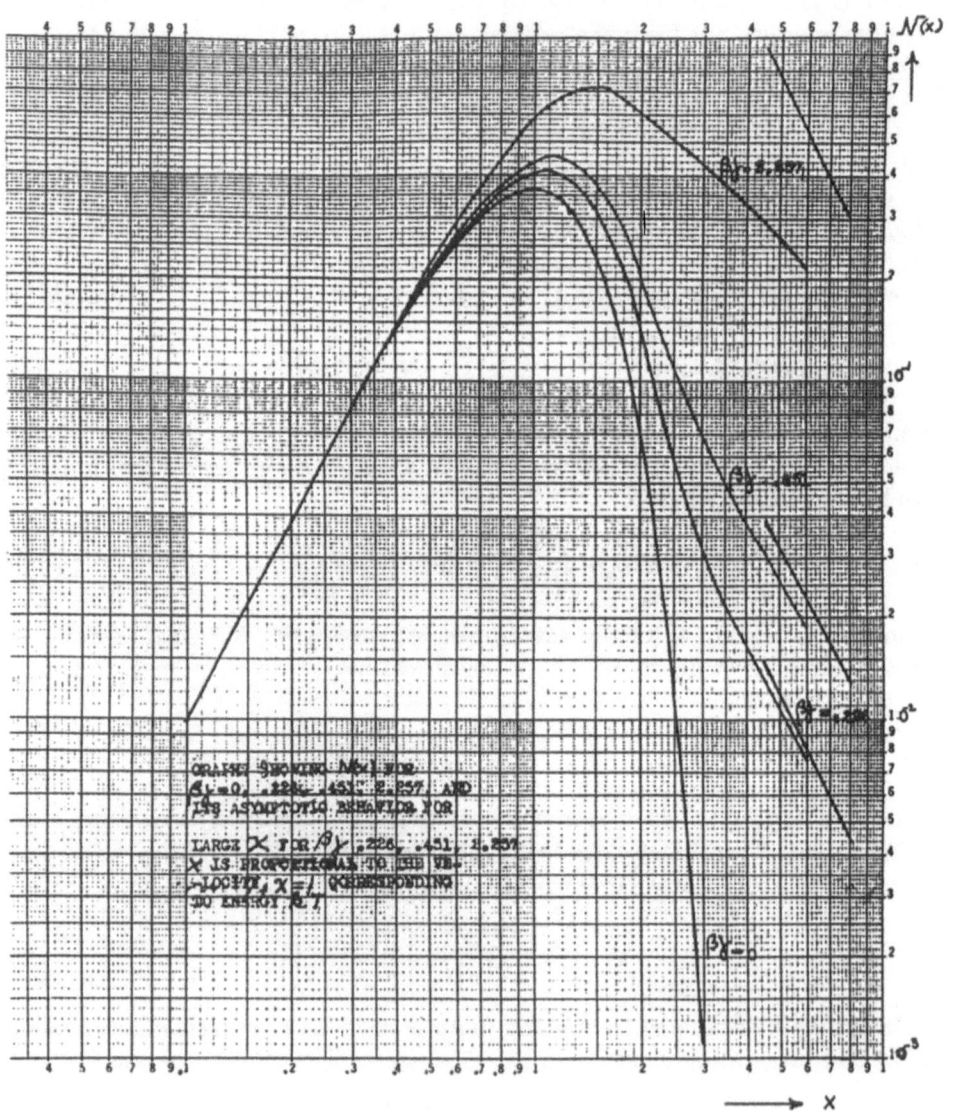

Fig. 1

32.

Approximate Boundary Conditions for Diffusion Equation at Interface Between Two Media

L. Szilard, A. M. Weinberg, E. P. Wigner, and R. F. Christy

July 10, 1942

Ibser and Wheeler (Report C-88) gave a rigorous treatment of the boundary conditions at the interface between two homogeneous multiplying media. This treatment, while it is capable of giving accurate results, is rather laborious to apply. For this reason, it appears worthwhile to describe here a method of adjusting boundary conditions which was used, before Ibser and Wheeler's work became available, for the calculation of heavy water seeds. While this method does not have the accuracy of Ibser and Wheeler's method, more exact formulae were available. It is more exact than the treatment underlying the Report C-11 by Breit and Fermi, which is essentially identical to the one used earlier by V. Halban and Kowarski.

The number of neutrons passing from the first into the second medium per cm^2 and sec. is $\frac{1}{4}\sum n_i v_i$, where n_i is the density of neutrons of velocity v_i in the first medium. If $\frac{1}{4}\sum n_i v_i$ were not equal to $\frac{1}{4}\sum n_i^* v_i$ (where n_i^* is the density of neutrons of velocity v_i in the second medium), there would be a flow of neutrons changing the total density of neutrons in both media by an amount which can be carried away by diffusion only if the gradient of n satisfied the equation

$$\sum \frac{\lambda_i v_i}{3} \operatorname{grad} n_i = \frac{1}{4}\sum n_i v_i - \frac{1}{4}\sum n_i^* v_i \tag{1}$$

i.e., if $\operatorname{grad} n/n$ were of the order $1/\lambda$ (in (1) λ_i is the mean free path of neutrons with velocity v_i). A gradient of this order of magnitude is incompatible with the ideas of diffusion theory and we write, therefore, as the first boundary condition that

$$\sum n_i v_i = \int n(E) v\, dE = \int N(E) \lambda\, dE \tag{2}$$

be continuous i.e. that it has the same value on both sides of the interface. In (2) $n(E)$ is the density of neutrons per unit energy range, $N(E)$ is the number of collisions the neutrons of unit energy range suffer per cm^3 and sec. Both integrals in (2) are really an integral plus another term, referring to the thermal neutrons.

The second boundary condition expresses the fact that the flow to the boundary equals the flow away from the boundary. This gives

$$\frac{1}{3}\sum \frac{\lambda_i v_i \operatorname{grad} n_i}{1-\cos\theta} = \frac{1}{3}\operatorname{grad}\int \frac{\lambda v n(E)}{1-\cos\theta}\,dE = \frac{1}{3}\operatorname{grad}\int \frac{\lambda^2 N(E)}{1-\cos\theta}\,dE \quad (3)$$

be continuous, as above. Again, the integrals in (3) contain, in addition to the integral, an extra term corresponding to the thermal neutrons. The divisor $1-\cos\theta$ enters because the diffusion constant is $\lambda v/3(1-\cos\theta)$ (rather than $\lambda v/3$) where $\cos\theta$ is the average cosine of the angle between the velocities of the neutron before and after the collision.

We must assume now that the energy spectrum in each medium is the same at the boundary as it is in the inside of this medium. This is an arbitrary assumption which is, strictly speaking, incorrect and causes an inaccuracy of our conditions. It is this inaccuracy that has been eliminated by the above mentioned treatment by Ibser and Wheeler. If we make this assumption, $n(E)$ becomes $n\nu(E)$ where $\nu(E)$ is the ratio of neutrons in unit energy range at E and n the total density of neutrons. Similarly $N(E) = qc(E)$ where $c(E)$ is the number of collisions which a neutron suffers while its energy is in a unit interval at E; and q is the number of neutrons created per cm^3 and sec. Our two conditions now become that

$$n\int \nu(E)v\,dE = q\int c(E)\lambda\,dE \tag{4}$$

as well as

$$\frac{1}{3}\int \frac{\lambda v \nu(E)}{1-\cos\theta}\,dE\operatorname{grad}n = \frac{1}{3}\int \frac{\lambda^2 c(E)}{1-\cos\theta}\,dE\operatorname{grad}q \tag{5}$$

are continuous at the interface.

Again, (4) and (5) contain two terms. The first corresponds to fast neutrons and in this $c(E) = 1/(\xi E)$ where ξ is the slowing down power of the material (.158 for carbon). The second term refers to thermal neutrons. In this $\int c(E)\,dE$ is the ratio of total scattering cross section and total absorption cross section per unit cell, multiplied by the thermal disadvantage factor (about 1.4 in the usual arrangement). (This means, of course, that the $\int c(E)\,dE$ of this term is the number of collisions a neutron suffers while it is thermal.)

The second of these equations is capable of a very simple interpretation. The factor of $\operatorname{grad}q$ in (5) is evidently the mean square of the total distance traveled by a neutron from creation to absorption, i.e. the square of the migration length l. We can use a picture in which one "free path" is the migration length. Then q is the number of "collisions" per cm^3 and sec and (5) becomes

$$\frac{1}{3}l^2\operatorname{grad}q. \tag{6}$$

This is, however, equivalent to the usual formula for diffusion flow in which the density is expressed by the number of collisions per cm^3 and sec. If n is the density, q the number of collisions, l the mean free path, $nvl = q$ and $1/3\,lv\operatorname{grad}n = 1/3\,l^2\operatorname{grad}q$. One sees that the equality of the expression (5) for both sides of the interface is only another expression of the equality of the expressions for the flow in our picture.

2. As pointed out before, the preceding considerations, just as those of Ibser and Wheeler, apply only at the interface of two homogeneous media. There is always a considerable ambiguity when they must be applied to the interface between two lattices. In fact, it is evident that the composition of the lattices alone does not suffice to determine the boundary conditions since the diffusion constant of the lattice depends quite considerably on the arrangement of the lumps, etc. within the damper. The following, therefore, involves a further approximation (inherent also in the work of Ibser and Wheeler).

We can express, first $n(E)$ in terms of q in case of a lattice arrangement

$$Vq = n(E)\sum N_iV_i\sigma_{si}v\xi_iE = n(E)\overline{N\sigma_s}\xi VvE. \tag{7}$$

Herein, V is the total volume of the cell, V_i the volume of region i of the cell which contains N_i atoms per cm^3 with the scattering cross section σ_{si} and the slowing down power ξ_i ($\xi_i = 1$ for H, .725 for D, .158 for C, etc.). The relative densities of the neutrons in the different parts of the lattice introduce a correction into (7) but we will disregard this because the density of resonance neutrons is nearly constant over the cell. Similarly, we assume that q is independent of energy. The density of thermal neutrons is given by the equation

$$qV = \sum n_iN_iV_i\sigma_{ai}v_t \tag{8}$$

where n_i is the density of thermal neutrons in the i'th region, σ_{ai} the absorption cross section of the atoms of the i'th region for thermal neutrons and v_t the velocity of thermal neutrons. We can define relative densities ρ_i so that

$$\sum \rho_iV_i = V. \tag{8a}$$

Then $n_i = n\rho_i$ and n is the average number of thermal neutrons per cm^3

$$n = \frac{\sum n_iV_i}{V}.$$

Introducing n into (8) gives

$$q\sum \rho_iV_i = n\sum \rho_iN_iV_i\sigma_{ai}v_t = n\overline{N\sigma_a}v_t\sum V_i\rho_i. \tag{8b}$$

Evidently, when calculating $\overline{N\sigma_a}$ from (8b), it is not necessary to use the ρ normalized by (8a) because they occur linearly on both sides. We can now write for (2)

$$q\left(\int \frac{1}{\overline{N\sigma_s}\xi}\frac{dE}{E} + \frac{1}{\overline{N\sigma_a}}\right). \tag{9}$$

q in this is an average, i.e. the number of neutrons created per second in a cell, divided by the cell volume. The averages $\overline{N\sigma_s}\xi$ and $\overline{N\sigma_a}$ are defined by (7) and (8b); the scattering cross section σ_s refers in the first one to fast, the absorption cross section σ_a in the second to thermal neutrons. The first boundary condition is that (8) be continuous.

For (3) it is reasonable to write if the medium contains several regions

$$\frac{1}{3}\operatorname{grad}\sum\int\frac{\lambda_i v n_i(E)V_i}{(1-\cos\theta)V}\,dE\,.$$

One can write for this by (7) and (8b)

$$\operatorname{grad}q\sum\frac{V_i}{V}\left[\int\frac{1}{3N\sigma_{si}(1-\cos\theta_i)}\frac{dE}{N\sigma_s\xi E}+\frac{D_i}{N\sigma_a v_t}\right]\tag{10}$$

where D_i is the diffusion constant for thermal neutrons in region i. The first term of (10) corresponds to the age, the second to the square of the diffusion length. Their sum is the square of the migration length, and (10) is identical with (6).

One can use, instead of the two boundary conditions (9) and (10) the single condition that their ratio shall be continuous. If we assume a single mean free path λ_f for fast neutrons and another λ_t for thermal neutrons, this can be written as

$$\frac{\operatorname{grad}q}{q}\frac{l^2/3}{\int\frac{dE}{N\sigma_s\xi E}+\frac{1}{N\sigma_a}}=\frac{\operatorname{grad}q}{q}\frac{l^2/3}{c_f\lambda_f+c_t\lambda_t}\tag{11}$$

where c_f is the number of collisions a neutron suffers while it is fast, c_t the number of collisions it suffers while it is thermal, λ_f and λ_t are the two mean free paths. One can obtain (11) directly by dividing (6) by the right side of (4) and postulating continuity for the quantity obtained in that way.

33.

On the Boundary Condition
Between Two Multiplying Media

F. L. Friedman and E. P. Wigner

April 19, 1944

Abstract. The transition region between two parts of a pile which have different compositions is investigated. In the case where the moderator is the same in both parts of the pile, it is found that the diffusion constant times fast neutron density satisfies the usual pile equations everywhere, right to the boundary. More complicated formulae apply in a more general case.

1. The pile equations of Fermi, as well as all the similar equations which describe the densities of neutrons with different energies in chain reacting units have many solutions. All the solutions can be written as superpositions of exponential functions of the position. The relaxation distances of these exponentials are different. Ordinarily, one relaxation distance is imaginary and only exponentials with this relaxation distance occur in the actual solution. We shall call these solutions the "regular" solutions. However, there are other solutions of the equations with real relaxation distances. Since these relaxation distances are all much smaller than the absolute value of the imaginary relaxation distance, they will be called transient solutions. In these, the ratio of the densities of neutrons with different energies cannot be constant throughout the pile. This is the case in the neighborhood of disturbing centers, such as e.g. control rods which absorb only thermal neutrons. Another example is that of a pile which contains two parts, with different compositions. In the latter case, e.g. the ratio of fast and thermal neutrons at the boundary between the two piles will be intermediate between the ratios in the "regular" solutions for both sides and go over into the regular ratio further away from the boundary. These facts have been brought out before on various occasions, most completely by Ibser and Wheeler in C-88.

2. We shall first treat a case in which the moderator is the same all over the pile so that the diffusion constant for both fast and thermal neutrons remains constant throughout the whole system. It has been stated in CP-455 that, for such a system, Fermi's equations can be solved with the two group theory as described in CP-1461 and CP-1554. It will be shown that in this case, the two methods of calculation give the same result. The general case of different moderators in the two parts of the pile will be treated finally by means of the two group theory.

We shall use Fermi's equations in the form

$$\frac{1}{3\sigma}\Delta q + \xi\frac{\partial}{\partial\tau}(\sigma q) + \sigma_a p_2 f(\tau)n = 0$$

$$\frac{1}{3\sigma_0}\Delta n - \sigma_a n + \xi\sigma_0 p_1 q(0) = 0. \tag{1}$$

Here, $\tau = \ln(E/E_{\text{thermal}})$ has a somewhat different definition from the usual one: it is zero for neutrons, the energy E of which is thermal, $\sigma(\tau)$ is the transport cross section which may depend on the energy but does not depend on the position; σ_0, the value of this quantity for $\tau = 0$, i.e. for thermal neutrons; ξ, the average logarithmic energy loss (independent of position); σ_a, the absorption cross section for thermal neutrons which depends on the position. $q(\tau)$ is the density of fast neutrons per unit τ (it is not Fermi's slowing down density Q), multiplied with the velocity, n the density of thermal neutrons times their velocity. $f(\tau)\,d\tau$ is the number of fission neutrons per slow neutron captured in U, for which τ is between τ and $\tau + d\tau$. Finally p_1 is the chance of escaping resonance absorption and p_2 the thermal utilization. The multiplication constant is, here

$$k = p_1 p_2 \int_0^\infty f(\tau)\,d\tau . \tag{2}$$

The advantage of the above way of writing the ordinary pile equations is that it does not assume that the fission neutrons are monochromatic.

If σ_a depends on the position, the ratio of n and $q(\tau)$ will not be constant over the pile and no simple equation will hold for either of them separately. This is natural, since "transients" occur both in n and in q which make their behavior quite complicated. The method of solution to be given consists of finding a quantity ν in which the transients just cancel so that a simple equation shall hold for it.

One can solve the first of the above equations by writing

$$\xi\sigma(\tau)q(\tau) = \left[\int_\tau^\infty f(\tau')e^{t(\tau')\Delta - t(\tau)\Delta}\,d\tau'\right]p_2 n\sigma_a . \tag{3}$$

In this

$$t(\tau) = \int_0^\tau \frac{d\tau}{3\xi\sigma^2} \tag{3a}$$

is Fermi's age for neutrons of an energy E characterized by $\ln(E/E_{\text{thermal}}) = \tau$. The Δ is an operator, to operate on $p_2 n\sigma_a$ and the exponential of an operator is, as usual, the infinite series of operators obtained by expanding it. ($f(\tau)$ and $t(\tau)$ are independent of position.) One can convince oneself easily that the above σq indeed gives at least a formal solution of the first pile equation.

It also gives

$$\xi\sigma_0 q(0) = \left[\int_0^\infty f(\tau')e^{t(\tau')\Delta}\,d\tau'\right]p_2 n\sigma_a \tag{4}$$

and this introduced into the second pile equation gives

$$\Delta\left(\frac{n}{3\sigma_0}\right) - \sigma_a n + p_1\left[\int_0^\infty f(\tau')e^{t(\tau')\Delta}\,d\tau'\right]p_2 n\sigma_a = 0. \tag{5}$$

This is the generalization of the usual equation connecting the multiplication constant and the Laplacian. In the present case, it is useful only if one can neglect the second and higher powers of Δ in the expansion of the exponential.

$$\Delta\left(\frac{n}{3\sigma_0}\right) + p_1 S\Delta(p_2 n\sigma_a) = \sigma_a n\left[1 - p_1\int_0^\infty f(\tau)\,d\tau p_2\right] \tag{6}$$

$$= \sigma_a n(1-k)$$

where

$$S = \int f(\tau')t(\tau')\,d\tau' \tag{6a}$$

is the average age of fission neutrons multiplied by their number. It is independent of position. If the same holds for p_1, we can put it behind the operator Δ and obtain

$$\Delta\left(\frac{n}{3\sigma_0} + p_1 p_2 S n\sigma_a\right) = \frac{1-k}{M^2}\sigma_a n\left[\frac{1}{3\sigma_0\sigma_a} + p_1 p_2 S\right] \tag{7}$$

where

$$M^2 = \frac{1}{3\sigma_0\sigma_a} + p_1 p_2 S = \frac{1}{3\sigma_0\sigma_a} + k\bar{t}. \tag{7a}$$

In this, $\bar{t}$ is the average age of fission neutrons as follows from (2). This shows that

$$\nu = n\sigma_a M^2 \tag{8}$$

satisfies the usual pile equation

$$\Delta\nu = \frac{1-k}{M^2}\nu \tag{9}$$

and has no transients.

The derivation presupposes that not only $\sigma(\tau)$ but also p_1 be constant throughout the pile. It further assumes that $|k-1| \ll 1$, as a consequence of which one can assume that $t(\tau')\Delta \ll 1$. One can discard the second assumption if one is willing to assume that the Laplacian Δ is constant throughout the pile. In this case

$$\nu = \frac{n}{\sigma_0} + p_1\int_0^\infty q(\tau)e^{Bt(\tau)}\frac{d\tau}{\sigma(\tau)} \tag{10}$$

and the calculation is quite elementary giving for B the implicit equation

$$B = 3\sigma_0\sigma_a\left[1 - p\int_0^\infty f(\tau)e^{Bt(\tau)}\,d\tau\right]. \tag{10a}$$

The equation for ν is

$$\Delta\nu = B\nu. \tag{10b}$$

However, the last assumptions are so specialized that they apply in the best case, if the two parts of the pile differ only in temperature which does not affect the Laplacian. The assumptions first made apply reasonably well for a pile which has the same moderator throughout although the amount of metal is different in different parts of it.

3. We now turn to the same problem, except that we shall assume a sharp boundary (at $x = 0$) between the two parts of the pile and use the two group theory described in reports CP-1461 and CP-1554. The notation will be the same as that adopted in CP-1461.

As in CP-1461 n_t is the sum of the velocities of all the thermal neutrons which are present in a cubic centimeter and n_f is the sum of the velocities of the fast neutrons present in a cubic centimeter. The pile equations then are

$$\Delta n_t - \kappa_t^2 n_t + \kappa_t^2 a n_f = 0 \qquad (11\,\text{a})$$

$$\Delta n_f - \kappa_f^2 n_f + \kappa_f^2 (k/a) n_t = 0. \qquad (11\,\text{b})$$

These equations already were used in the above mentioned C-88 by Ibser and Wheeler. In it, κ_t is the reciprocal diffusion length of thermal neutrons in the pile (that takes into account the presence of the metal), κ_f^2 is the reciprocal age, k is the multiplication constant, and

$$a = p_1 \sigma_{af} / \sigma_{at}. \qquad (11\,\text{c})$$

Herein σ_{at} is the absorption cross section for thermal neutrons per cubic centimeter, p_1 is the probability for a neutron to escape resonance absorption, and

$$\sigma_{af} = \frac{\sigma_f \xi}{\ln(E_f / E_t)}. \qquad (11\,\text{d})$$

Herein again σ_f is the total cross section for a fast neutron per cubic centimeter, ξ is the average logarithmic energy loss of a neutron upon collision with an atom of the moderator, E_f and E_t are the energies of fission and thermal neutrons respectively.

The above notation holds for $x < 0$, i.e., on the left side of the boundary. On the right side of the boundary a similar notation will be used from equation (16) on, except that every small letter will be replaced by a capital.

On the left side of the pile the neutron densities are linear combinations of the four expressions

$$e^{\kappa_1 x}, \ e^{-\kappa_1 x}; \quad e^{\kappa_2 x}, e^{-\kappa_2 x}. \qquad (12)$$

It should be remarked that κ_1^2 is smaller than 0 if the pile is to be chain reacting; it is the Laplacian of the neutron density far from the boundary. For the part of the neutron density which is composed of the first two exponentials of (12) the ratio of thermal and fast neutron densities is

$$n_t / n_f = a \kappa_t^2 / (\kappa_t^2 - \kappa_1^2) \approx a. \qquad (12\,\text{a})$$

The last part of the equation with the $\approx$ sign holds if $|k - 1| \ll 1$ in which case $\kappa_1^2 \ll \kappa_t^2$.

For the part of the density proportional to the last two exponentials of (12) the ratio is

$$n_t/n_f = -a\kappa_t^2/(\kappa_f^2 - \kappa_1^2) \approx -\frac{a\kappa_t^2}{\kappa_f^2}.$$ (12b)

We have

$$\kappa_2^2 = \kappa_t^2 + \kappa_f^2 - \kappa_1^2.$$ (12c)

4. Let us first consider a special case in which the ratio of the fast diffusion constants σ_{af}/κ_f^2 of the two sides is the same as the ratio of the slow diffusion constants σ_{at}/κ_t^2 of the two sides. In this case p_1 times the ratio of the two diffusion coefficients, $p_1\sigma_{af}\kappa_t^2/\sigma_{at}\kappa_f^2 = a\kappa_t^2/\kappa_f^2$, is also approximately the same on both sides, and this also holds, if $k - 1 \ll 1$, for the ratio of the coefficients of the second two exponentials of (12) in the slow and fast densities:

$$a\frac{\kappa_t^2}{\kappa_f^2 - \kappa_1^2} = p_1\frac{\sigma_{af}\kappa_t^2}{\sigma_{at}(\kappa_f^2 - \kappa_1^2)}.$$ (13)

The equality of this last quantity for both sides – whether or not it is a consequence of the first assumption – will be assumed in the following.

The second two exponentials of (12) drop out of the expression

$$\nu = n_t + \frac{a\kappa_t^2}{\kappa_f^2 - \kappa_1^2}n_f$$ (14a)

so that this satisfies the equation $\Delta\nu = \kappa_1^2\nu$ and has, therefore, no transients. Since n_f and n_t are both continuous at the boundary, this holds also for ν. Furthermore, the condition for the equality of the fluxes means the ratio of the derivatives of n_f shall be inversely proportional to the ratio of the fast diffusion coefficients and this holds, according to the second assumption, also for n_t. Hence

$$\frac{\sigma_{af}}{\kappa_f^2}\frac{d\nu}{dx} \quad \text{or} \quad \frac{\sigma_{at}}{\kappa_t^2}\frac{d\nu}{dx}$$ (14b)

is also continuous.

The most usual case in which the above conditions are satisfied is that of a common moderator for both regions. In this case the thermal diffusion coefficient σ_{af}/κ_t^2 is also the same on both sides so that not only ν but

$$\frac{\sigma_{at}}{\kappa_t^2}\nu = \frac{\sigma_{at}}{\kappa_t^2}n_t + \frac{a\sigma_{at}}{\kappa_f^2 - \kappa_1^2}n_f$$ (15)

is also continuous. Furthermore, since with our first assumption the fast diffusion coefficients are also the same on both sides, the derivative of the above expression is also continuous. The present n_t was denoted by n in section 1 and $n_f = n_t/a$, so that if one neglects κ_1^2 compared with κ_f^2, this becomes

$$n\sigma_{at}\left(\frac{1}{\kappa_t^2} + \frac{1}{\kappa_f^2}\right).$$ (15a)

The expression in the bracket is the migration area denoted by M^2. We see that (15 a) is the same as (8). In (7 a) $M^2 = (1/\kappa_t^2) + (k/\kappa_f^2)$ but $|k - 1| \ll 1$ is assumed in both derivations and (15 a) is obtained from (15) by replacing k by 1.

The physical interpretation of (15) is quite simple. If the neutrons did not diffuse while in the thermal region but all the diffusion took place in the fast neutron region evidently n_f would be continuous and likewise the product of the derivative of n_f with the diffusion constant. On the other hand, if all the neutron diffusion took place while the neutrons are thermal, n_t would be continuous and so would the thermal diffusion constant times the derivative of n_t. If the diffusion takes place for both thermal and fast neutrons one must expect that a linear combination of the above quantities will be continuous with coefficients which are proportional to the amount of diffusion in the corresponding regions. This is exactly what the above equations show.

5. We now go over to the general case. It will be assumed, however, that $|k - 1| \ll 1$, i.e. $|\kappa_1^2| \ll \kappa_2^2$, $|K_1^2| \ll K_2^2$. In the neighborhood of the boundary we shall use for the neutron density the following expressions

$$n_f = \beta_0 + \beta_1 x + \gamma \kappa_f^2 e^{\kappa_2 x} \tag{16 a}$$

$$n_t = a(\beta_0 + \beta_1 x) - \gamma a \kappa_t^2 e^{\kappa_2 x} \tag{16 b}$$

$$N_f = B_0 + B_1 x + \Gamma K_f^2 e^{-K_2 x} \tag{16 c}$$

$$N_t = A(B_0 + B_1 x) - \Gamma A K_t^2 e^{-K_2 x} . \tag{16 d}$$

The first two of these refer to $x < 0$; the latter two to $x > 0$, i.e. to the right side of the boundary. It would appear that replacing the first two exponentials of (12) by linear functions involves an approximation. This, however, is not the case as we shall use equations (13) only in the immediate neighborhood of $x = 0$. The reason for omitting the $e^{-\kappa_2 x}$ for $x < 0$ and $e^{K_2 x}$ in (13) is that the transient solutions must drop to zero far away from the boundary. The quantities β_0 and B_0 are the extrapolated neutron densities at the left and right side of the boundary if one neglects the transients when making the extrapolations, i.e., uses the equation $\Delta n = \kappa_1^2 n$ at the left side and $\Delta N = K_1^2 N$ at the right side of the boundary. β_1 and B_1 are the values of the derivatives of the extrapolated neutron densities.

The boundary conditions which are valid for the actual densities are

$$n_t(0) = N_t(0) \qquad n_f(0) = N_f(0) \tag{17 a}$$

$$d_t n_t'(0) = D_t N_t'(0) \qquad d_f n_f'(0) = D_f N_f'(0) . \tag{17 b}$$

In this we have introduced the notation

$$\sigma_{at}/\kappa_t^2 = d_t \qquad \sigma_{af}/\kappa_f^2 = d_f \tag{18}$$

for the diffusion constant for thermal and fast neutrons in the left side of the pile and the notation D_t and D_f for the same quantities in the right side of the pile. Equations (17) expressed in terms of the β and γ are

$$\beta_0 + \gamma \kappa_f^2 = B_0 + \Gamma K_f^2 \tag{19a}$$

$$a\beta_0 - a\gamma\kappa_t^2 = AB_0 - A\Gamma K_t^2 \tag{19b}$$

$$d_f(\beta_1 + \gamma\kappa_f^2\kappa_2) = D_f(B_1 - \Gamma K_f^2 K_2) \tag{19c}$$

$$d_t a(\beta_1 - \gamma\kappa_t^2\kappa_2) = D_t A(B_1 + \Gamma K_t^2 K_2). \tag{19d}$$

If one wants the conditions for the extrapolated density one must write down the condition that these four equations for γ and Γ have a solution. The condition for this is that all three-rowed determinants of

$$\begin{vmatrix} \kappa_f^2 & -K_f^2 & \beta_0 - B_0 \\ -a\kappa_t^2 & AK_t^2 & a\beta_0 - AB_0 \\ d_f\kappa_f^2\kappa_2 & D_f K_f^2 K_2 & d_f\beta_1 - D_f B_1 \\ -d_t a\kappa_t^2\kappa_2 & -D_t AK_t^2 K_2 & d_t a\beta_1 - D_t AB_1 \end{vmatrix} = 0 \tag{20}$$

shall vanish. This condition is then the solution of our problem in the general case. It is contained, implicitly, already in C-88.

It may be worthwhile to calculate the magnitude of the transient solutions at the boundary. The sum of the two transients must be equal of course to the jump in the extrapolated neutron density. An easy calculation gives for the fast and thermal n at the boundary ($x = 0$)

$$n_f = \frac{d_f(\kappa_2\beta_0 - \beta_1) + D_f(K_2 B_0 + B_1)}{d_f\kappa_2 + D_f K_2} \tag{21a}$$

$$n_t = \frac{d_t a(\kappa_2\beta_0 - \beta_1) + D_t A(K_2 B_0 + B_1)}{d_t\kappa_2 + D_t K_2}. \tag{21b}$$

These equations are quite symmetric. One must remember that β_0 and B_0 are the extrapolated fast neutron densities, $a\beta_0$ and AB_0 the extrapolated thermal neutron densities, and the significance of β_1 and B_1 is similar. Except if the boundary between the two regions is quite close to the surface of the pile, the terms with β_1 and B_1 in (21) can be neglected because β_1/β_0 is of the order of magnitude κ_1. If one does this, the interpretation of (21) is very simple. It shows that the neutron densities at the boundary are weighted means between the extrapolated densities β_0 and B_0 for fast and $a\beta_0$ and AB_0 for thermal neutrons. The weighting factor is $d_f\kappa_2$ and $D_f K_2$ in case of fast, $d_t\kappa_2$ and $D_t K_2$ in case of thermal neutrons. The weighting factors for fast and thermal neutrons become equal in the special case treated in the fourth section.

34.

On Variations of the Power Output in a Running Pile

E. P. Wigner

November 11, 1942

Abstract. The variation of the neutron intensity is studied under various conditions. These include a sudden increase or decrease of the multiplication constant after steady operation of some length, either under the influence of spontaneous fission or because the multiplication constant was 1 for some time. The expected operation of the "limits control" is also described.

I. The importance of the delayed neutrons for the steady operation of the pile has been early recognized by Allison, Fermi and Szilard (Conferences on the Power Plant, February 1942). The intensities and periods of the different groups of delayed neutrons were measured by Snell, Nedzel and Ibser (C-81). The solution of the equations which give the power output of the pile as function of time has been given by Ibser, Manley and Wheeler (C-65) in their discussion of the effect of a sudden neutron burst. They found in this report that, even if the effective multiplication constant is larger than 1, the total neutron density first decreases after the neutron burst, passes through a minimum and only after a relatively long time begins to increase again. We shall see that in the case in which we are interested, i.e., that of small variations of the multiplication constant around 1, the situation is just the opposite. An initial fast change in the neutron density is followed by a more gradual, slow variation. Thus, e.g., if we start with a steady state with an effective multiplication constant 1 or smaller than 1 (in which latter case the neutrons owe their existence to the spontaneous fission) and the effective multiplication constant suddenly jumps to $1 + k_e$, the neutron density will first be multiplied by about $1 + k_e$ every 1.5×10^{-3} sec. This increase will continue until the neutron density reaches about $1 + 100k_e$ times its original value. After that, the increase in density will be more gradual and, after a long time, will be multiplied by about $1 + 7k_e$ per second. If now the effective multiplication constant drops below 1, i.e., k_e becomes negative the opposite sequence of events will take place: the neutron density will first decrease rapidly, then more gradually until it reaches the value given by the spontaneous fission activity.

II. The following analysis follows quite closely that given by Ibser, Manley and Wheeler. The probability that a neutron yield, when absorbed, a radioactive atom with a decay constant A_j will be denoted by f_j. As Snell, Nedzel and Ibser have shown, the delayed activity can be represented by four periods, with probabilities ranging from a third to two per mille.

Table 1

$A_1 = 1.22 \times 10^{-2} \text{ sec}^{-1} = 1.83 \times 10^{-5}$	$f_1 = .31 \times 10^{-3}$
$A_2 = 2.9 \times 10^{-2} \text{ sec}^{-1} = 4.35 \times 10^{-5}$	$f_2 = 2.30 \times 10^{-3}$
$A_3 = 9.9 \times 10^{-2} \text{ sec}^{-1} = 14.85 \times 10^{-5}$	$f_3 = 2.76 \times 10^{-3}$
$A_4 = 28 \times 10^{-2} \text{ sec}^{-1} = 42 \times 10^{-5}$	$f_4 = 2.76 \times 10^{-3}$

The third column gives the decay constants in units of $(1.5 \times 10^{-3} \text{ sec})^{-1}$, the time 1.5×10^{-3} having been chosen as the unit of time because it represents the time between the creation of a neutron by fission and its absorption. It is unnecessary to mention that the values of Table I are subject to a considerable error. In order to have definite equations to discuss they will be adopted at face value.

It we denote the number of neutrons by n, by s_j the number of radioactive nuclei which emit a delayed neutron of the kind j, we have the equations

$$\dot{s}_j = -A_j s_j + f_j(n + S) \tag{1a}$$

$$\dot{n} = \sum A_j s_j + n(k_e - \sum f_j) + S(1 - \sum f_j). \tag{1b}$$

The unit of time is 1.5×10^{-3} sec, S is the number of spontaneous fission neutrons during this time. Since the number of spontaneous fission neutrons is, according to Fermi and Goldhaber, about 10 per kg U and sec, the value of S is in a 50 ton pile $50000 \times 10 \times 1.5 \times 10^{-3} = 750$. On the other hand, n is about 4.5×10^{10} in a 1 kW pile, S can be neglected except at extremely low power operation.

At any rate, a particular solution of the inhomogeneous equations (1) is

$$s_j = \frac{f_j}{A_j}\left(1 - \frac{1}{k_e}\right)S; \quad n = -\frac{S}{k_e}. \tag{2}$$

If k_e is negative (non operating condition), the particular solution (2) corresponds to the steady state. If k_e is positive, the solution (2) has, of course, no direct physical significance (n is negative) but must be added to solutions of the homogeneous equations which are chosen so that the inital conditions are satisfied.

There are five linearly independent solutions of the homogeneous equations (i.e., of (1) in which $S = 0$ is set), all of which depend exponentially on time. These five solutions ($\rho = 0, 1, 2, 3, 4$) have the form

$$s_j = S_j^\rho \exp(\lambda_\rho t) \quad n = N^\rho \exp(\lambda_\rho t). \tag{3}$$

One sees from (1 a) that

$$S_j^\rho = \frac{f_j}{A_j + \lambda_\rho} N^\rho \tag{3a}$$

and from (1 b) that the λ_ρ satisfy the equation

$$\lambda_\rho = \sum_j \frac{A_j f_j}{A_j + \lambda_\rho} + k_e - \sum f_j \tag{4}$$

or

$$1 + \sum_j \frac{f_j}{A_j + \lambda_\rho} = \frac{k_e}{\lambda_\rho}.$$

The solutions of this equation have been given graphically in C-65 for arbitrary k_e. We shall need the values of λ_ρ for very small k_e but these we need more accurately than they can be obtained from C-65. They have been calculated, for $k_e = 0$, numerically. For small k_e, they have been obtained from the equation

$$-\sum_j \frac{f_j}{(A_j + \lambda_\rho)^2} \frac{d\lambda_\rho}{dk_e} = \frac{k_e}{\lambda_\rho^2} \frac{d\lambda_\rho}{dk_e} + \frac{1}{\lambda_\rho} \tag{4a}$$

which is obtained by differentiating (4 b) with respect to k_e. The solutions of (4) are

$$\lambda_0 = \frac{k_e}{96} + 1.61 k_e^2 \tag{5.0}$$

$$\lambda_1 = -2.068 \times 10^{-5} - .795 \times 10^{-3} k_e \tag{5.1}$$

$$\lambda_2 = -8.86 \times 10^{-5} - 5.68 \times 10^{-3} k_e \tag{5.2}$$

$$\lambda_3 = -31.19 \times 10^{-5} - 8.60 \times 10^{-3} k_e \tag{5.3}$$

$$\lambda_4 = -350 \times 10^{-5} - .994 k_e \tag{5.4}$$

If $k_e = 0$, the solution $\rho = 0$ will represent the steady state; the others are transient solutions. They never can be present alone since they represent states in which some of the s are negative. The same holds also for $k_e > 0$: the solution $\rho = 0$ will outgrow all others after some time. It may be noted that this steady solution increases much less rapidly than one might expect; (5.0) shows that the intensity increases to $1 + k_e$ fold only in 96 generations. The reason for this is (C-65) that the number of radioactive nuclei s_j does not have the natural value $s_j = f_j n / A_j$ as it has for $k_e = 0$, $\lambda_\rho = 0$. It trails somewhat behind the neutron density and (3 a) shows that it is too small by the factor $\sim (1 - \lambda_0 / A_j) = 1 - k_e / 96 A_j$. For this reason the neutron density does not increase to its $1 + k_e$ fold in one generation but only as if the multiplication constant were

$$1 + k_e - \sum_j f_j + \sum_j f_j \left(1 - \frac{k_e}{96 A_j}\right) = 1 + k_e - \frac{95}{96} k_e = 1 + \frac{k_e}{96}$$

(since $\sum_j f_j / A_j = 95$). The expressions (5) are valid only if k_e is substantially smaller than 0.01, which we hope to be true in the operating piles. One obtains the general solution of (1) as a linear combination of the five solutions (3) to which the solution (2) must be added. Except for extremely low power outputs, (2) can be omitted.

It may be for the reader's convenience to repeat here the orthogonality relation of the homogeneous solutions (C-65).

$$N^\rho N^\sigma + \sum_j \frac{A_j}{f_j} S_j^\rho S_j^\sigma = \delta_{\rho\sigma}. \tag{6}$$

From this and (3 a) one obtains for N^ρ

$$\left(\frac{1}{N^\rho}\right)^2 = 1 + \sum_j \frac{f_j A_j}{(f_j + A_j)^2} = 1 + \sum_j \frac{f_j}{A_j + \lambda_\rho} - \lambda_\rho \sum_j \frac{f_j}{(A_j + \lambda_\rho)^2}. \tag{6 a}$$

This gives with (4)

$$\left(\frac{1}{N^\rho}\right)^2 = \frac{k_e}{\lambda_\rho} - \lambda_\rho \sum_j \frac{f_j}{(A_j + \lambda_\rho)^2}. \tag{6 b}$$

The values of this last sum are for $k_e = 0$

$\rho =$	0	1	2	3	4
$\sum_j \dfrac{f_j}{(A_j + \lambda_\rho)^2} =$	2.28×10^6	60.2×10^6	1.99×10^6	$.364 \times 10^6$	$118.$

III. If the pile has been steadily operating for some time at $k_e = 0$, the solutions (3) with $j = 1, 2, 3, 4$ will have disappeared because the corresponding λ_ρ are negative. In this "natural state",

$$s_{j0} = \frac{f_j}{A_j} n_0. \tag{7}$$

If now k_e suddenly becomes positive, the neutron density will become

$$n = n_0 \left\{ e^{k_e t/96}(1 + 248.5 k_e) - 40 k_e e^{-2.01 \times 10^{-5} t} - 64 k_e \right.$$
$$\left. \times e^{-8.86 \times 10^{-5} t} - 27.5 k_e e^{-31.2 \times 10^{-5} t} - 117 k_e e^{-850 \times 10^{-5} t} \right\}$$
$$= n_0 \left\{ (1 + 248.5 k_e) e^{7 k_e t'} - 40 k_e e^{-.0137 t'} - 64 k_e e^{-.058 t'} \right.$$
$$\left. - 27.5 k_e e^{-.208 t'} - 117 k_e e^{-5.66 t'} \right\} \tag{8}$$

where t' is the time in seconds. The neutron density first increases rapidly, corresponding to the disappearance of the negative transients. At the very beginning, it increases to the $1 + k_e$ fold in one generation ($t = 1$ or $t' = 1.5 \times 10^{-3}$ sec). This initial behavior can be seen most directly from the differential equations (1), by substituting (7). After the transients have disappeared, the increase is slower, it is to $1 + k_e$ fold in $1/7$th of a second.

Let us now consider the case that initially k_e was negative, $k_{ei} < 0$, and the neutron intensity given by the spontaneous fission $n_i = -S/k_{ei}$. If k_e now becomes positive and equal to k_{ef}, the neutron density will be given by (8) with

$$n_0 = \frac{S}{k_{ef}} - \frac{S}{k_{ei}} \tag{8a}$$

to which the constant negative "spontaneous activity" $-S/k_{ef}$ must be added (i.e., S/k_{ef} subtracted). It is

$$n = \left(\frac{S}{k_{ef}} - \frac{S}{k_{ei}}\right)\left\{(1 + 248.5k_{ef})e^{7k_{ef}t'} - 40k_{ef}e^{-.0137t'} - 64k_{ef}e^{-.059t'}\right.$$
$$\left. - 27.5k_{ef}\,e^{-.208t'} - 117k_{ef}e^{-5.66t'}\right\} - \frac{S}{k_{ef}}. \tag{8b}$$

The initial derivative is even higher (by $1 - k_{ei}/k_{ef}$) than if we had started from a running pile (8). However, the rate of increase soon settles down to the same asymptotic value.

If e.g., the k_e was increased from $k_{ei} = -.001$ to $k_{ef} = +.001$, the n_0 in (8b) is, in a pile containing 60 tons of U, $n_0 = 375 \times 2000 = 7.5 \times 10^5$. After some time (about 2 minutes) the spontaneous activity and all terms of (8) except the first one can be neglected and the activity becomes

$$7.5 \times 10^5 e^{.007t'} \quad (t' \text{ in seconds}).$$

The activity 4.5×10^{10}, corresponding to 1 kW, will be reached at $t' = 1570$ sec. If k_e was increased from $-.0005$ to $.0005$ the n_0 in (8) is 15×10^5 in the same pile, and, after some time, the activity will be given by

$$15 \times 10^5 \times e^{.0035t'}$$

and the 1 kW activity will be reached in 2940 sec.

The bracket in (8) is represented in Fig. 1.

IV. Let us now consider an earlier stage, in which k_e is increased but still remains negative. This corresponds to the initial stages of the pulling out of the control rod. The neutron activity is still given by (8b), except that k_{ef} is also negative. The initial rate of increase is such that the neutron density increases to its $1 + k_{ef} - k_{ei}$ fold in 1.5×10^{-3} sec. This initial increase is soon checked (after about .2 sec) and after about $1/7\,k_{ef}$ seconds the density begins to approach its asymptotic value $-S/k_{ef}$. A few typical curves of this kind are shown in Fig. 2.

By observing the neutron density as function of time, one may arrive at an estimate of the prevailing k_{ef}. However, one should not trust (8b) too much since the underlying experimental values of Table I may not be too accurate. It is, probably, better to plot the reciprocal neutron density in the steady state against the position of the control rod. This plot should give a straight line, the intercept of which with the position axis gives the $k_e = 0$ position of the rod. If one does this, it is advisable to measure the neutron intensity not too close to the control rod, because the shape of the first harmonic is changing too

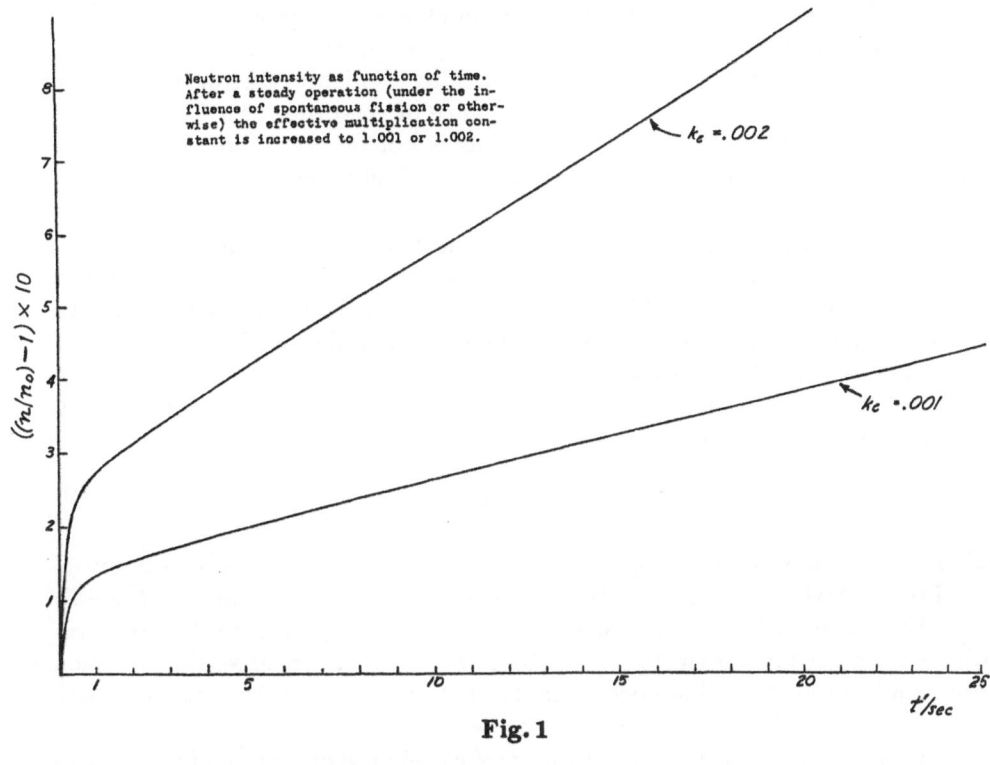

Fig. 1

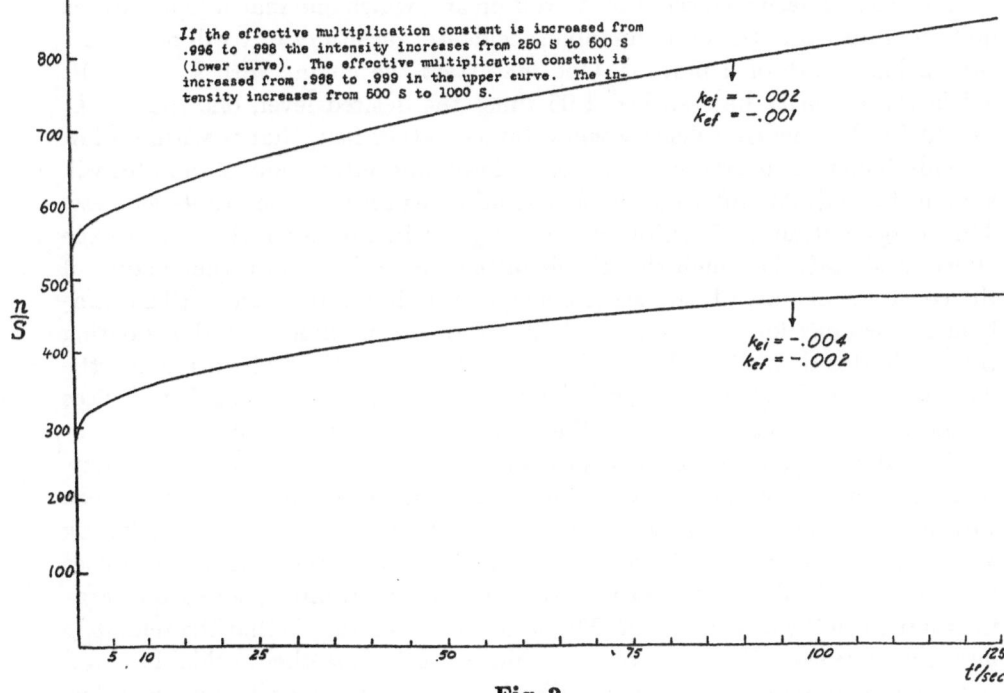

Fig. 2

much in that neighborhood. On the other hand, one should not be too close to the edge of the pile because one is interested only in the intensity of the first harmonic which is weakest and most subject to accidental fluctuations at the edge. If k_e is of the order of $-.003$, a distance of the order of a foot or a few feet inside may be quite satisfactory. If this is not feasible, then the neutron stream outside could be measured and calculated as function of k_e.

V. At last the case should be described in which k_e is originally positive, $k_{ei} > 0$, and the neutron density increases steadily until, suddenly, the control is pushed in again deeper. This will happen if the operation is started and the neutron intensity has approximately reached the desired level. The spontaneous fission activity can be neglected in this case.

The initial conditions are, according to (3 a), since only the solution $\rho = 0$ remains after a long time,

$$s_j = \frac{f_j}{A_j + \lambda_{0i}} n_0 \tag{9}$$

where λ_{0i} is the value of λ_0 for $k_e = k_{ei}$. In the solution of (1) with the initial conditions (9) the number of neutrons is again given by (8), except that $k_{ef} - k_{ei}$ must be substituted for k_e everywhere apart from the exponent in the first term where k_{ef} must be substituted for it. Since $k_{ef} - k_{ei}$ is negative, the situation will be just opposite to that encountered at the discussion of (8): a fast decrease of n will be followed by a more gradual one.

The above permits one to describe more closely the operation of the control rods which has been suggested by V. Wilson and which one may adopt if one is quite confident that the unpredictable, accidental variations of k will be smaller than a hundredth of a percent. One may start the engine with $k_e = .0001$. When the intensity has reached 1.06 times the desired level, one reduces k_e to $-.0001$. The neutron density will drop 3% rather fast, that is within a few seconds. Then the decrease will become slower and after about 2 minutes will amount to 12%. At this time the k_e should be again increased to $+.0001$ etc. The energy output as function of time is given in Fig. 3 which refers to the above ideal case. Although the details of this curve depend on the validity of the constants of Table I, the general shape of it should be correct. (The same holds, of course, for Figs. 1 and 2). If the operator finds that the inflating time is shorter than the deflating time, he will lower both the inflating and the deflating positions of the rod, and vice versa. Naturally, various safety devices should be introduced for the case that something should go wrong.

The above operation of the controls has the great advantage of simplicity. It has two disadvantages: First, it will operate smoothly only if the unforeseen variations of k are smaller than the difference between inflating and deflating k_e. These cannot be too large if one wants to avoid too frequent switches from inflation to deflation and vice versa and if one wants to avoid too large fluctuations in the power output. The second disadvantage is that the relatively large fluctuations of the power output which are not altogether avoidable, entail relatively large temperature changes in the U, except for a pile with very low

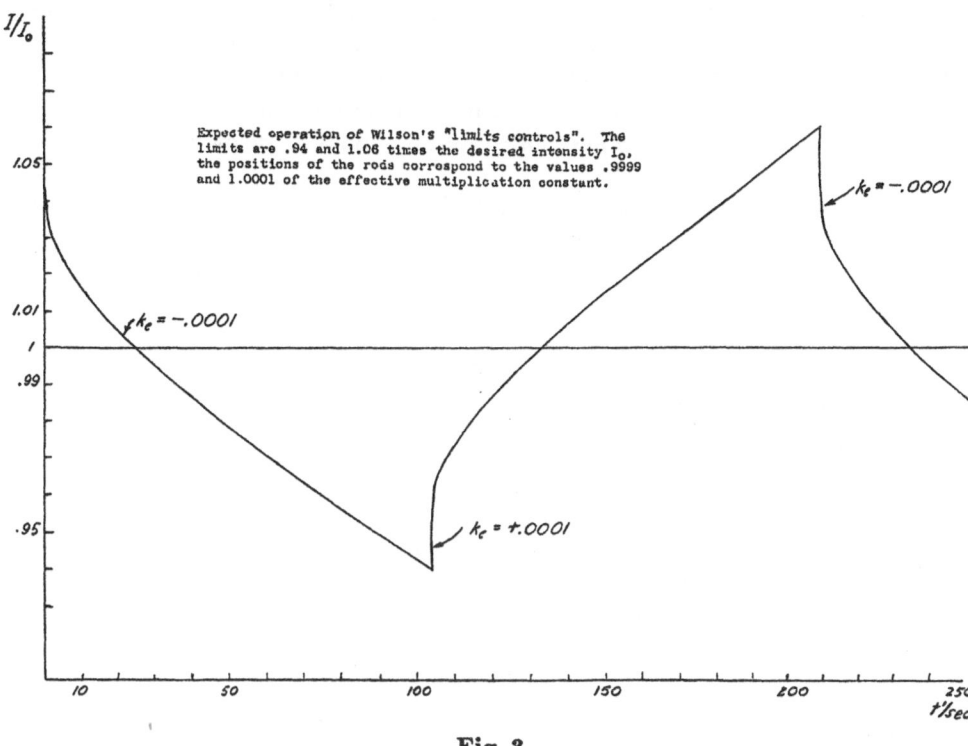

Expected operation of Wilson's "limits controls". The limits are .94 and 1.06 times the desired intensity I_0, the positions of the rods correspond to the values .9999 and 1.0001 of the effective multiplication constant.

Fig. 3

power output (of the order of 100 kW). For reasonably high power output, the difference between the temperatures of the U and the cooling agent will be proportional to the power output. This means, even in the above sketched very favorable case, an about 20 °C temperature variation takes place every 2 minutes in a substantial part of the U in the He cooled plant of report CE-277.

It may be desirable to avoid these disadvantages and Mr. Christy will shortly report on other controlling mechanisms which try to avoid these defects. More generally, if one wants to use the "limits control" in such a way that the intensity varies between $n_0(1-v)$ and $n_0(1+v)$ and if one uses upper and lower positions of the control rods which give effective multiplication constants $1 + k_e$ and $1 - k_e$, the time t' during which the control rod will remain in the upper or lower position will be given by the equation

$$2v = 497k_e + 7k_e t' \tag{10}$$

(t' in seconds). Evidently the position of the rod will change less frequently if v is permitted to be larger and if the two positions of the control rod are close to each other (k_e small). The above applies only if the t' obtained from (10) is more than a minute, otherwise the $-40k_e^{-.0137t'}$ term can not be neglected in (8) and this must be used more directly.

If the k_e which gives a sufficiently large t' should turn out to be so low that one must fear accidental variations of the effective multiplication constant which are of the same order of magnitude, additional controls should be introduced with a more drastic action on the multiplication constant. These should operate if the neutron density goes substantially above or below the permitted limits $n_0(1 + v)$ or $n_0(1 - v)$.

The above is a more detailed account of work which has been summarized in the Report of the Theoretical Group for the month ending September 15, 1942.

35.

Use of the Pile Oscillator
for the Measurement of Pile Constants

E. P. Wigner

June 23, 1945

Abstract. The pile oscillator is an instrument which causes periodic variations of the neutron intensity in a pile of critical or smaller than critical size. A realization of the pile oscillator is for instance an absorbing material which can be moved periodically back and forth in the pile. The effect of such a motion on the neutron density has been investigated before (Cahn, Monk and Weinberg, CP-2907) by means of a theory which recognizes only one kind of neutrons. This theory is extended in the present report by taking the polyenergetic nature of the neutrons into account. It is shown that the spatial variation of the oscillating part of the neutron density is proportional to a function ψ which obeys an equation $\Delta\psi = B\psi$ where B is a complex number depending on the period of the oscillations. An expression is found for B in terms of the constants of the pile. B on its turn determines the wave length and apparent absorption coefficient of the pile for oscillations of a given period.

The maxima and minima of the epicadmium neutron intensity at a given point do not occur simultaneously with the maxima and minima of the thermal neutron intensity at the same point because there is a phase shift between the oscillations of the intensities of the different kinds of neutrons. The thermal neutron intensity is delayed with respect to the epicadmium neutron intensity by the product of the life time of the neutrons (excluding delayed neutrons) and the ratio of age to migration area.

Introduction

For the purpose of measuring danger coefficients in unsteady piles, A. Cahn, A. Monk and A. Weinberg developed a theory of the pile oscillator (CP-2907). The pile oscillator is a device, considered by the theoretical group as early as the spring of 1942, which sets the neutron density in the pile into periodic oscillations. This can be done, e.g., by setting an absorber situated in the pile into periodic motion. As a result of this motion, the neutron density in the pile will undergo oscillations with the same period and with an amplitude which is proportional to the absorbing power of the body in motion. The oscillations in neutron density can be detected and their amplitude and phase measured by means of an ionization chamber-amplifier system. This measurement is not influenced to any considerable extent by accidental fluctuations in the pile since these will not have, in general, the same period as the pile oscillator – hence the application of this device to measuring danger coefficients.

The problem of the pile oscillator will be taken up here again for two reasons. First, CP-2907 recognizes only one kind of neutrons while, as will be seen, it should be possible to observe some interesting effects due to the non-monochromatic nature of the same. This is particularly true if the pile oscillator is used to determine the pile constants – a purpose for which, as will be shown, it is well adapted. Second, the mathematical procedure to be employed here appears simpler than that used in CP-2907 and, in spite of recognizing the polyenergetic nature of the neutrons, gives the results in a form which is more easily visualizable than that given before. It starts from a discussion of the periodic solutions of the homogeneous pile equations, i.e., the pile equations without the oscillator. These equations must be valid everywhere in the pile, except the region of the oscillator. The effect of the oscillator in this treatment is to specify the boundary conditions for the above periodic solution in the neighborhood of the oscillator. One will easily recognize the analogy between the two treatments of the pile equations and the two possible treatments of light emission.

The Pile Equations

The pile equations will be used here in the form given in CP-3048, except that the effect of the delayed neutrons will also be taken into account. We have

$$\operatorname{div} D \operatorname{grad} q + \frac{\partial}{\partial \tau} Sq - r(\tau)q + \zeta \sigma_a (1 - \beta) f(\tau) n$$
$$+ \sum \lambda_j c_j g_j(\tau) = \frac{1}{v_0} e^{-\frac{1}{2}\tau} \frac{\partial q}{\partial t}. \tag{1}$$

In this, E_0 is the energy of thermal neutrons, $\tau = \ln E/E_0$ where E is the neutron energy, D, which may be a function of τ, is one-third of the transport mean free path, $q(\tau, x, y, z)$ is the sum of the velocities of those neutrons for which $\ln E/E_0$ is in unit interval at τ and which are in unit volume at x, y, z. $S = \sum \sigma \xi$ is the slowing down power, σ being the scattering cross section per unit volume of the pile and ξ the average loss in $\tau = \ln E/E_0$ in a collision. The $r(\tau)$ is the effective resonance absorption cross section per unit volume, $\zeta = \eta \varepsilon f = k/p$. The sum of the velocities of the thermal neutrons contained in unit volume is denoted by n. The σ_a is the absorption cross section for thermal neutrons per unit volume of the pile, β is the fraction of the delayed fission neutrons, $1 - \beta$ the fraction of the prompt fission neutrons. The $f(\tau)$ gives the energy distribution of the latter, its integral $\int_0^\infty f(\tau)\, d\tau = 1$. The c_j are the densities of the different fission products which emit neutrons, $g_j(\tau)$ characterizes the energy distribution of the delayed neutrons emitted by the j'th type of fragment, λ_j its decay constant. We have

$$\frac{dc_j}{dt} = \zeta \beta_j \sigma_a n - \lambda_j c_j; \tag{1a}$$

β_j is the fraction of delayed neutrons coming from the j'th type, $\sum \beta_j = \beta$. It is possible that the fragments which emit the delayed neutrons are not immediately formed by the fission in which case the above equations would not hold. It will be seen, however, that the detailed mechanism of the emission of the delayed neutrons does not affect the results. The $g_j(\tau)$ represent the energy distribution of the neutrons emitted by the fragment of type j. This also will play only a subordinate role in the following.

The equation for the thermal neutrons is

$$\operatorname{div} D_0 \operatorname{grad} n + S_0 q(0) - \sigma_a n = \frac{1}{v_0} \frac{\partial n}{\partial t}. \tag{1b}$$

In this D_0 is one-third of the transport mean free path for thermal neutrons, S_0 the value of the slowing down power S for $\tau = 0$, i.e., for thermal neutrons.

Solution of the Pile Equations

For the stationary case, in a simple bare pile, equations (1) have been solved before (CP-3048), except that the existence of delayed neutrons was not taken into account there. As pointed out there, this does not introduce any difficulties. In order to save some space, the stationary problem will not be treated separately but we go over at once to periodic solutions, i.e., assume that all quantities q, c_i, n depend on time by means of a factor $e^{i\omega t}$. This means that

$$\frac{\partial q}{\partial t} = i\omega q; \quad \frac{\partial c_j}{\partial t} = i\omega c_j; \quad \frac{\partial n}{\partial t} = i\omega n. \tag{2}$$

From (1 a) and (2) we have at once

$$c_j = \frac{\zeta \beta_j \sigma_a n}{\lambda_j + i\omega}. \tag{3}$$

This can be reintroduced into (1) and gives

$$\operatorname{div} D \operatorname{grad} q + \frac{\partial}{\partial \tau} S q - r(\tau) q + \zeta \sigma_a \bar{f}(\tau) n = \frac{i\omega}{v_0} e^{-\frac{1}{2}\tau} q \tag{4}$$

where

$$\bar{f}(\tau) = (1 - \beta) f(\tau) + \sum \frac{\lambda_j \beta_j}{\lambda_j + i\omega} g_j(\tau). \tag{4a}$$

For the stationary case, (4) now has exactly the same form as in CP-3048. In the non-stationary case, $\bar{f}(\tau)$ can be replaced by $(1 - \beta) f(\tau)$ if the period of the oscillation to be considered is a reasonably short fraction of a second.

For a simple bare pile, all quantities D, S, r, ζ, σ_a, D_0, S_0 are independent of position. As a result, div D grad and div D_0 grad can be replaced by $D\Delta$ and $D_0\Delta$ and q, c_j, n by

$$q(\tau, x, y, z) = a(\tau)\psi(x, y, z) \tag{5}$$
$$c_j(x, y, z) = b_j\psi(x, y, z); \quad n(x, y, z) = a_0\psi(x, y, z)$$

where $a(\tau)$, b_j and a_0 are independent of position. The ψ will satisfy an equation of the form

$$\Delta\psi = B\psi \tag{6}$$

with the boundary condition that ψ vanish at the surface of the pile. All this is a repetition from CP-3048. The only difference is that B will depend in the present case on ω, i.e., we have sort of a frequency dependent refraction index which, in addition, will turn out to be complex. For the present, we shall disregard the boundary conditions and determine B as function of ω.

Introducing (5) and (6) into (4) we have

$$DBa(\tau) + \frac{\partial}{\partial\tau}Sa(\tau) - ra(\tau) + \zeta\sigma_a\bar{f}(\tau)a_0 = \frac{i\omega}{v_0}e^{-\frac{1}{2}\tau}a(\tau). \tag{7}$$

The solution of this is

$$a(\tau) = \frac{\zeta\sigma_a a_0}{p(\tau)S(\tau)}e^{-Bt(\tau)}\int_\tau^\infty \bar{f}(\tau')p(\tau')e^{Bt(\tau')-i\omega s(\tau,\tau')}\,d\tau' \tag{8}$$

where t is Fermi's "age" (has nothing to do with the time)

$$t(\tau) = \int_0^\tau \frac{D(\tau')}{S(\tau')}\,d\tau', \qquad \ln p(\tau) = -\int_0^\tau \frac{r(\tau')}{S(\tau')}\,d\tau' \tag{8a}$$

and

$$s(\tau,\tau') = \frac{1}{v_0}\int_\tau^{\tau'} e^{-\frac{1}{2}\tau''}\frac{d\tau''}{S(\tau'')}. \tag{8b}$$

No approximation has been made so far and, for $\omega = 0$, this is identical with formulae given in CP-3048.

For (1b) we shall need $q(0)$. In order to obtain that, $s(0,\tau')$ shall be calculated first. This is easily done if one takes the definition of S at face value. Since the scattering cross sections do not change in the thermal region, the $e^{-\frac{1}{2}\tau''}$ factor in (8b) permits one to replace S in (8b) by S_0 if $\tau = 0$. This gives

$$s(0,\tau') = \frac{2}{v_0 S_0}(1 - e^{-\frac{1}{2}\tau'}). \tag{9}$$

Actually, this is not altogether justified since the slowing down power does change at low energies because of the effects of chemical binding and lattice structure. These effects are not properly within the scope of the pile equations which we use, just as the whole notion of thermal neutrons is inaccurate. Since, however, there is no better theory available at present, this difficulty will be disregarded.

If (9) is inserted into (8) the $e^{-\frac{1}{2}\tau'}$ term can be neglected because $f(\tau')$ vanishes unless τ' is quite large. $p(0) = 1$, $S(0) = S_0$, $t(0) = 0$ and we have

$$a(0) = \frac{\zeta\sigma_a a_0}{S_0}\int_0^\infty \bar{f}(\tau)p(\tau)e^{Bt(\tau)}\,d\tau e^{-2i\omega/v_0 S_0}. \tag{10}$$

Inserting (5) and (6) into (1b) gives an equation for a_0 and $a(0)$. Using (10) for the latter and cancelling a_0 from the equation gives

$$D_0 B + \zeta\sigma_a e^{-2i\omega/v_0 S_0}\int_0^\infty \bar{f}(\tau)p(\tau)e^{Bt(\tau)}\,d\tau - \sigma_a = i\omega/v_0. \tag{11}$$

This is the equation defining B. For $\omega = 0$ it goes over into the customary pile equation. Usually $p(\tau)$ has assumed its asymptotic value p for such τ for which $\bar{f}(\tau)$ begins to be appreciable. If this is the case, p can be brought out of the integral and ζp replaced by k. Dividing (11) by σ_a permits the introduction of the average lifetime of thermal neutrons $T_0 = 1/\sigma_a v_0$, the average lifetime of fast neutrons $T_f = 2/v_0 S_0$ and of the thermal diffusion length in the pile $L_0 = \sqrt{D_0/\sigma_a}$. With these notations, (11) becomes

$$L_0^2 B + k e^{-i\omega T_f} \int_0^\infty \bar{f}(\tau) e^{Bt(\tau)} \, d\tau - 1 = i\omega T_0 . \tag{11 a}$$

For $\omega = 0$, i.e., for the stationary solution, $\bar{f}(\tau)$ is the energy distribution of all prompt and delayed neutrons. For periods of .1 sec or less, ω becomes 60 or more and the sum in (4a) can be neglected. If this is true

$$\bar{f}(\tau) = (1 - \beta) f(\tau), \qquad \omega > 60 . \tag{11 b}$$

Equation (11 a) is a transcendental equation for B. It can be simplified to some degree as follows (CP-2881). Let us denote the average age of prompt fission neutrons by

$$\bar{t}_p = \int_0^\infty f(\tau) t(\tau) \, d\tau . \tag{12 a}$$

This is not significantly different from the average age which is used commonly. The integral of (11 a) can then be written as

$$e^{B\bar{t}_p} \int_0^\infty f(\tau) \left[1 + B(t(\tau) - \bar{t}_p) + \frac{1}{2} B^2 (t(\tau) - \bar{t}_p)^2 + \dots \right] d\tau$$
$$= e^{B\bar{t}_p} \left(1 + \frac{1}{2} B^2 m_p + \dots \right) \tag{12 b}$$

where

$$m_p = \int \bar{f}(\tau) (t(\tau) - \bar{t}_p)^2 \, d\tau . \tag{12 c}$$

In order to obtain the right side of (12 b), one must notice that the term with the first power of B in the square bracket cancels because of (12 a). The quantity (12 c) has been calculated by Plass and found to be very small for C and $D_2 O$ as moderator but relatively large in a water moderated system. It will be omitted here in order to avoid complications. This gives with (11 b)

$$L_0^2 B + k(1 - \beta) e^{-i\omega T_f + B\bar{t}_p} = 1 + i\omega T_0 . \tag{13}$$

For $\omega = 0$, this is Fermi's equation for the Laplace factor B. This is real for $\omega = 0$ but not otherwise. Since T_f is very small ($\sim 10^{-4}$ sec in a graphite pile), for small $k - 1$, i.e., small $B\bar{t}_p$, (13) can be simplified to (the migration area will be denoted by $M^2 = L_0^2 + \bar{t}_p$)

$$M^2 B = (L_0^2 + \bar{t}_p) B = 1 - k(1 - \beta) + i\omega (T_0 + T_f) . \tag{13 a}$$

This, of course, holds only if $k-1 \ll 1$. It permits one to see more easily than the general equation (13) does, which quantities a determination of B would yield. These are, clearly $(1-k(1-\beta))/M^2$ and $(T_0+T_f)/M^2$. Since $(1-k)/M^2$ is the Laplace factor for $\omega = 0$ (in which case the delayed neutrons are fully utilized), knowledge of the first quantity (i.e., of the real part of B) gives one β/M^2, i.e., the migration area if the fraction β of delayed neutrons is known and the fraction β if the migration area is known. However, this determination is likely to be very inaccurate because it involves taking the difference between two not too different quantities which are determined by very different methods (measurements of Laplace factor in steady pile and measurement or real part of B, to be discussed below). On the other hand, measurement of the imaginary part of B should give the ratio of the total neutron lifetime $T_0 + T_f$ to the migration area $M^2 = L_0^2 + \bar{t}_p$. The accuracy of this measurement could be increased by carrying it out for several ω.

Cd Ratio

Before going over to the measurement of B, we shall calculate the time difference, d, between the maxima (and minima) of the densities of epicadmium and thermal neutrons at a given point. The reason that the maxima of thermal neutron intensity do not coincide in time with the maxima of the epicadmium neutron intensity is only partly that the epicadmium neutron maxima coincide only with the maxima of the *production* of thermal neutrons (and conversely). Another factor which enters is that the thermal neutrons, and also the fast neutrons, move and the thermal neutrons at one position may have been fast at a rather different position.

The calculation itself is simply the calculation of the relative complex phases of fast and thermal neutrons. We assume that the fast neutrons are detected with a $1/v$ detector the sensitivity of which has, however, a lower limit. This should correspond to a Cd covered B or 25 chamber. Its response will be proportional to

$$\int_{\tau_e}^{\infty} q(\tau)e^{-\frac{1}{2}\tau}\,d\tau \tag{14}$$

where $\tau_e \approx \ln(.45/.04) \approx 2.4$ corresponds to the Cd limit. By dividing (14) with the density n of thermal neutrons, one obtains a complex number the absolute value of which is proportional to the ratio of the responses of Cd covered and not covered chambers, the complex phase of which is, on the other hand, equal to the product of ω and the time delay.

Because of (5), the ratio of (14) and n is equal to a similar expression involving the a instead of the q and (8) permits one to calculate this

$$
\begin{aligned}
e^{i\omega d} &= R \int_{\tau_e}^{\infty} a(\tau)e^{-\frac{1}{2}\tau}\,d\tau/a_0 \\
&= R \int_{\tau_e}^{\infty} \frac{\zeta\sigma_a e^{-\frac{1}{2}\tau - Bt(\tau)}}{p(\tau)S(\tau)} \int_{\tau}^{\infty} \bar{f}(\tau')p(\tau')e^{Bt(\tau')-i\omega s(\tau,\tau')}\,d\tau'\,d\tau .
\end{aligned}
\tag{14a}
$$

R is a real number to make the absolute value of the right side 1. Because of the $e^{-\frac{1}{2}\tau}$ factor, in all other quantities $\tau = \tau_e$ can be substituted on the right side of (14 a). If the Cd limit is below all resonance lines $p(\tau_e) = 1$ and one can also write $S(\tau_e) = S_0$. This assumes that there is no very low energy resonance absorption present but is sufficiently general for our purposes. The integral over τ' can be extended as well from 0 to ∞ as from τ_e to ∞ since $f(\tau')$ certainly vanishes below τ_e. We then have, carrying out the integration over τ

$$e^{i\omega d} = R\frac{2\zeta\sigma_a}{S_0}e^{-Bt(\tau_e)}\int_0^\infty \bar{f}(\tau')p(\tau')e^{Bt(\tau')-2i\omega/v_e S_0}\,d\tau'. \qquad (14\,\text{b})$$

We have substituted for $s(\tau_e, \tau')$ an expression similar to (9); $v_e = v_0 e^{\frac{1}{2}\tau_e}$ is the velocity of the neutrons at the Cd limit. Since already $\omega T_f = 2\omega/v_0 S_0$ was negligible except for particularly large ω, the factor involving the exponential of $2\omega/v_e S_0 = 2e^{-\frac{1}{2}\tau_e}\omega T_f$ can be safely neglected. The value of the integral in (14 b) can be obtained from (11), all real factors can be omitted if one replaces the real constant R by another constant R'. Hence

$$e^{i\omega d} = R'e^{-Bt(\tau_e)+i\omega T_f}(1 - BL_0^2 + i\omega T_0). \qquad (14\,\text{c})$$

If we again go over to the case of $|k-1| \ll 1$, (13 a) can be used for B. It gives

$$e^{i\omega d} = R''e^{-i\omega(T_0+T_f)t(\tau_E)/M^2+i\omega T_f}$$
$$\times\left[1 - \frac{L_0^2}{M^2}(1-k(1-\beta)) + i\omega T_0\frac{t_p}{M^2} - i\omega T_f\frac{L_0^2}{M^2}\right]. \qquad (15)$$

Since, under ordinary conditions the 1 in the square bracket is very much larger than the other quantities herein, one obtains

$$d = (T_0 + T_f)(\bar{t}_p - t(\tau_e))/M^2. \qquad (16)$$

This may not be quite accurate under unusual conditions (very large ω, large $k-$ 1, etc.) but it is sufficient to show what the measurement of d essentially gives. Since the factor $(T_0+T_f)/M^2$ is just the imaginary part of B, a combination of the measurement of (16) with that of B should give directly $\bar{t}_p - t(\tau_e)$, i.e., the age difference of fission and epicadmium neutrons. Again, the accuracy of the measurement could be increased by carrying it out for different frequencies.

Altogether, we obtained expressions for $\bar{t}_p - t(\tau_e)$, for $(T_0+T_f)/M^2$, and for $(1-k(1-\beta))/M^2$. If $t(\tau_e)$ is known – it will be ordinarily a rather small quantity as compared with $\bar{t}_p$ – one has $\bar{t}_p$. If one further succeeds to obtain the life time $T_0 + T_f$ of neutrons, one also has M^2 and $1 - k(1-\beta)$. Some information about the quantities $t(\tau_e)$ and $T_0 + T_f$ can be obtained by going over to frequencies ω so high that $\omega(T_0+T_f)$ becomes of the order of magnitude 1. If this is done, the equations (13 a) and (16) lose their validity even if $|k-1| \ll 1$ so that one has to use the accurate expressions (13) and (14 c). As a result, B does not remain linear in ω, nor d independent of ω. The deviation from the linear law in the first case and the dependence of d on ω in the second case give further information about the quantities $T_0 + T_f$ and $t(\tau_e)$ which could not be determined as long as one was restricted to relatively low ω. It is questionable, however, to what extent these deviations could be determined experimentally and for this reason the question will not be pursued further here.

Measurement of B

We found that the neutron densities, as far as their dependence on the position is concerned, are proportional to a function ψ which satisfies the equation (6) $\Delta\psi = B\psi$. This, of course, does not completely determine ψ as one has to adjust the boundary conditions both at the surface of the pile and also at the pile oscillator.

The adjustment of the boundary conditions at the surface of the pile is simplest for a parallelepipedic or for a very large pile. In the first case, the well known theory of images leads to the goal; in the second case, they will be satisfied automatically. One chooses in both cases solutions ψ which have singularities at the pile oscillator but go to zero exponentially at very large distances. If the pile is very large, this kind of solution will be practically zero at the pile boundaries and the boundary condition is automatically satisfied. Otherwise, one adds to the original ψ its "images" with respect to the pile boundary, with opposite sign, as if the whole pile were infinite and one had negative pile oscillators at the image points of the actual pile oscillator with respect to the pile boundary. This procedure is illustrated schematically in the figure. The actual pile is shaded in this figure and the oscillator in it is indicated by an arrow.

An exponentially decreasing solution is found with the arrow as its origin; to this are added similar exponentially decreasing solutions with the other arrows as origin. These other solutions will have the same sign as the original one if the arrow is directed to the left. The resulting function will satisfy equation (6) everywhere and its only singularity in the shaded region will be at the arrow. At the boundary of the shaded region the positive and negative solutions will just cancel so that the boundary conditions are satisfied. Except if the absolute value of B is very much smaller than the square of the pile edge, only the solutions originating at the arrows very close to the shaded area will have an appreciable contribution within the shaded area. With other words, the method converges well. The method of images has been described in some detail because it enables one to find the solution of (6) even for a pile as small as used in an exponential experiment.

The preceding discussion shows that it is possible to compose the solution ψ from solutions which are decreasing exponentially with increasing distance from their respective origins. It remains to adjust the boundary conditions at the pile oscillator, i.e., find the exponentially decreasing solution which the oscillator would excite in an infinite pile. This will be called the fundamental solution ψ_0. It will depend, in general, on the type of oscillator one uses. If the oscillator has spherical symmetry (i.e., it is an absorber of variable cross section but constant position such as may be approximated by two co-axial hoops rotating in opposite directions), the fundamental solution will also have spherical symmetry and will be, hence, proportional to

$$\psi_0 = \frac{c}{r}e^{\kappa r} \tag{17}$$

where

$$\kappa = \sqrt{B} \tag{17a}$$

is the square root of B the real part of which is negative.

In the more practical case that the oscillator is an absorber moved back and forth along a line, the fundamental solution will have axial symmetry. It will be, therefore, a superposition of axially symmetric waves

$$\psi_0 = \left(c_0 + c_1\frac{\partial}{\partial z} + c_2\frac{\partial^2}{\partial z^2} + \dots\right)\frac{1}{r}e^{\kappa r}. \tag{17b}$$

The coefficients c of (17) can be determined in the following way, which is partially given already by Cahn, Monk and Weinberg in CP-2907. The absorber can be considered to be a moving and variable negative source of neutrons of the kind it absorbs. It will be assumed that these are thermal neutrons. The intensity of the source is proportional to the steady part of the intensity of neutrons which it absorbs. Its position is the instantaneous position of the absorber.

Let us place our coordinate system at the center of the oscillator and its x axis in the direction of its motion. Then the source of neutrons is

$$s(x, y, z, t) = -sn_s(x, y, z)\delta(y)\delta(z)\delta(x - a\sin\omega t). \tag{18}$$

In this s is the absorption cross section of the oscillating absorber, n_s the steady part of the thermal neutron density, a the amplitude of the oscillation. Because of the δ functions, one can replace in (18) $n_s(x, y, z)$ by

$$n_s + a\sin\omega t\, n_s' + \frac{1}{2}a^2\sin^2\omega t\, n_s'' + \dots \tag{18a}$$

where n_s, n'_s, n''_s,... are the zero'th, first, second, etc. derivatives of n_s at $x = y = z = 0$. One can, furthermore, write formally

$$\delta(x - a \sin \omega t) = \delta(x) - a \sin \omega t \, \delta'(x) + \frac{1}{2}a^2 \sin^2 \omega t \, \delta''(x)\dots \quad (18\,\mathrm{b})$$

Inserting (18 a) and (18 b) into (18) one obtains s as a power series of $\sin \omega t$. This can be written also as a Fourier series of $\sin n\omega t$. Since the detecting equipment will discover only fluctuations which have the simple period $2\pi/\omega$, only the terms proportional to $\sin \omega t$ need to be retained. These are

$$s_1(x,y,z,t) = -s\delta(y)\delta(z) \sin \omega t \left[an'_s\delta(x) - an_s\delta'(x) + \frac{1}{8}a^3 n'''_s\delta(x) \right.$$

$$\left. - \frac{3}{8}a^3 n''_s\delta'(x) + \frac{3}{8}a^3 n'_s\delta''(x) - \frac{1}{8}a^3 n_s\delta'''(x)\dots \right]. \quad (18\,\mathrm{c})$$

The function ψ_0 which has the singularities produced by this source is

$$\psi_0 = \frac{s}{4\pi D_0} \sin \omega t \left\{ an'_s + \frac{1}{8}a^3 n'''_s - \left(an_s + \frac{3}{8}a^3 n''_s \right) \frac{\partial}{\partial x} + \frac{3}{8}a^3 n'_s \frac{\partial^2}{\partial x^2} \right.$$

$$\left. - \frac{1}{8}a^3 n_s \frac{\partial^3}{\partial x^3} \right\} \frac{1}{r} e^{\kappa r}. \quad (19)$$

The definition of this ψ_0 is different from the one used so far because it already contains the $\sin \omega t$ factor. The terms with the first power of the amplitude a were found already by Cahn, Monk and Weinberg; there are no terms with even powers of a as could have been foreseen. If the size of the pile is large as compared with the amplitude of the oscillations, the terms containing higher derivatives of n_s will be negligible. Similarly, if both $a\kappa \ll 1$ and $a/r \ll 1$ the higher derivatives of $r^{-1}e^{\kappa r}$ can be neglected, leaving only the terms considered already in CP-2907. Since this last pair of conditions may not be always satisfied, it may be better to write

$$\psi_0 \approx \frac{s}{4\pi D_0} \sin \omega t \left\{ an'_s - an_s \frac{\partial}{\partial x} + \frac{3}{8}a^3 n'_s \frac{\partial^2}{\partial x^2} - \frac{1}{8}a^3 n_s \frac{\partial^3}{\partial x^3} \dots \right\} \frac{e^{\kappa r}}{r}. \quad (19\,\mathrm{a})$$

If the size of the pile is very large, even the first and third terms can be omitted. Alternately, one can omit the second and fourth term, if one takes all measurements in the plane through the oscillator perpendicular to its direction. Two terms remain in each case and, unless $a\kappa \ll 1$, $a/r \ll 1$, one has to make a correction for the second.

The methods for obataining the real and imaginary parts of κ by measuring the dependence of the intensity on the distance from the oscillator and phase differences between different points are too well known in principle to bear repetition. If $\kappa = \kappa_r + i\kappa_i$, κ_r and κ_i being real and the former negative, the square of the amplitude of the neutron density oscillation is proportional to

$$\left[\left(\frac{1}{r} - \kappa_r \right)^2 + \kappa_i^2 \right] e^{2\kappa_r r} \quad (19\,\mathrm{b})$$

in the direction of the oscillator if only the second term of (19 a) needs to be retained, i.e., if all quantities an_s'/n_s, a/r, and $\sqrt{\kappa_r^2 + \kappa_i^2}a$ are small. The phase is proportional to

$$\kappa_i r - \text{arctg} \frac{\kappa_i r}{1 - \kappa_r r} \tag{19 c}$$

under the same conditions. By measuring phase and amplitude at more than two points one may be able to obtain direct experimental evidence on the sufficiency of a single term of (19 a).

The above formulae (19), (19 a), (19 b), (19 c) assume, of course, that the images of the original neutron oscillator can be neglected. This may not be correct in a small pile and for low κ_r. (An estimate of κ_r and κ_i will follow.) A further complication may enter on account of the lattice structure which will make it necessary in general to take all measurements at corresponding points in the lattice.

The values of κ_r and κ_i can be calculated as follows. Denoting the real and imaginary parts of B by B_r and B_i we have

$$\kappa_r^2 - \kappa_i^2 = B_r ; \quad 2\kappa_r \kappa_i = B_i \tag{20}$$

whence

$$\kappa_r = \pm\sqrt{\frac{1}{2}B_r + \frac{1}{2}\sqrt{B_r^2 + B_i^2}} ; \quad \kappa_i = \pm\sqrt{-\frac{1}{2}B_r + \frac{1}{2}\sqrt{B_r^2 + B_i^2}} . \tag{20 a}$$

Let us assume now that we can use (13 a) for B, i.e., that $B\bar{t}_p$ is small. Then, in a graphite lattice $B_r = [1 - k(1 - \beta)]/M^2$ will be about $-.059/710$ cm^2 = -83×10^{-6} cm^2. T_0 is equal to $1/v\sigma_a$ for graphite, multiplied with $1 - f$ where f is the thermal utilization. $T_0 \approx .106/2.2 \times 10^5 \times 3.6 \times 10^{-4} = 1.34 \times 10^{-3}$ sec. T_f is much smaller than T_0 so that $B_i \approx 1.34 \times 10^{-3}w/M^2 = 1.89 \times 10^{-6}w$ sec/cm^2 (the actual figures used are the Handbook values for the X pile). As a result one obtains the following table for κ_r and κ_i which can serve as an orientation.

$\omega(\text{sec}^{-1})$	$B_r(\text{cm}^{-2})$	$B_i(\text{cm}^{-2})$	$\kappa_r(\text{cm}^{-1})$	$\kappa_i(\text{cm}^{-1})$
6.28	-83×10^{-6}	11.9×10^{-6}	$- .65 \times 10^{-3}$	$- 9.1 \times 10^{-3}$
31.4	-83×10^{-6}	59.4×10^{-6}	-3.1×10^{-3}	$- 9.6 \times 10^{-3}$
62.8	-83×10^{-6}	119×10^{-6}	-5.6×10^{-3}	-10.7×10^{-3}
126	-83×10^{-6}	238×10^{-6}	-9.5×10^{-3}	-12.9×10^{-3}

36.

Effect of Small Perturbations on Pile Period

E. P. Wigner

June 13, 1945

Abstract. A perturbation theory is developed for calculating the change of the pile period caused by the introduction of small amounts of extraneous materials into the pile, such as impurities, additional fissionable material, etc. It is shown that the effect can be calculated in first approximation most easily by solving the adjoint of the pile equation. This leads to the definition of an "adjoint neutron density", also depending on the position and the energy of the neutron. The weight factor for a perturbation is the product of the ordinary neutron density and the adjoint density. The adjoint equation is developed both for the two group and the ordinary pile theory and solved in both cases for a simple pile without reflector. The purpose of the work is to find a method for calculating the effect of small perturbations on composite piles and to separate the effects on fast, resonance and slow neutrons.

Introduction

The effect of small perturbations (localized impurities, etc.) on the pile is important not only because of the theory of danger coefficient measurements but also because the effect of certain influences cannot be taken into account easily by any other than the perturbation method. The present report endeavors to go beyond Fermi's simple theory which applies only to a uniform bare pile and recognizes only one kind of neutrons (thermal).

It was customary so far to express the effect of small perturbations as a change in the multiplication constant. However, this is a possible procedure only in the case of a uniform, bare pile. A composite pile, or even a simple pile with a reflector, has no single multiplication constant and it was thought best to express the effect of perturbations in terms of the change in reciprocal pile period. However, in order to avoid unnecessary complications, the effect of the delayed neutrons was omitted. Their inclusion would not cause any fundamental difficulty but would complicate the formulae and make them more cumbersome. They were omitted for this reason and the results for the change in period apply as they stand, only for very short perturbations (short as compared with the period of the delayed neutrons) or if the "infinitely small" perturbations cause a very much larger change in some sort of multiplication constant than the about 1% for which the delayed neutrons are responsible. However, as was stated before, it would not be difficult at all to include the delayed neutrons into the considerations and one can in most cases guess even without a calculation what their effect would be.

In many of the calculations one is interested not so much in the change of the pile period caused by one perturbation as in the amount of another type of perturbation which would counteract the perturbation of the first kind. This can be calculated most simply by determining the joint effect of both perturbations and choosing the second one so large that the joint effect should vanish. The delayed neutrons do not play ordinarily any role in such calculations.

General Perturbation Theory

All pile equations considered heretofore have the form

$$\frac{\partial n}{\partial t} = Mn. \tag{1}$$

In this, the neutron density n is a function of the position and, in most theories, also of the energy. It could be made to be a function of the direction of the velocity also without the present section losing its applicability.

One usually solves (1) by considering solutions which depend on time exponentially. These solutions satisfy characteristic value equations of the form

$$Mn = \lambda n \tag{1a}$$

where λ is a constant. Its reciprocal is often called the period of the pile. The steady solution of the critical pile satisfied the equation $Mn = 0$ corresponding to $\lambda = 0$. What we shall be interested in is always the largest λ, which is usually in the neighborhood of zero, all other λ of the pile being negative. We wish to calculate the change of this λ due to a change V in the pile equation

$$(M + V)n' = \lambda'n'. \tag{2}$$

The change V may be caused by the introduction of some additional absorber into the pile, or of a fissionable element or something similar. V causes a change in both the neutron distribution n which will become n' and also in the period which changes from λ to λ'.

The above problem is very much the same as the one occurring, e.g., in quantum mechanics where it is solved by the Rayleigh-Schrödinger perturbation theory. The only difference is that M is not i times a self adjoint (Hermitian) operator. For this reason, it is necessary to consider, in addition to (1 a), the adjoint equation

$$M^+N = \lambda N \tag{3}$$

the operator M^+ is defined by the equation

$$(\phi, M\psi) = (M^+\phi, \psi). \tag{4}$$

In this ϕ and ψ are two arbitrary functions satisfying the boundary conditions of the problem and M^+ is defined so that (4) is valid for any such ϕ and ψ.

The (,) denotes the scalar product of the function before and after the comma, i.e., the integral over the product of both. Examples will be given later which show the way to find the adjoint M^+ to a given M. If M is real, its characteristic values are in general the same as those of M^+. If M is complex, the characteristic values of M^+ are conjugate complex to the characteristic values of M. In this case, with which we shall not be concerned, it is better to define the scalar product as the integral over the product of the second factor and the conjugate complex of the first factor.

It is impossible to derive formulae for n' and λ' of (2) which are entirely analogous to the formulae of the Rayleigh-Schrödinger theory, except that in all scalar products N occurs before the comma, instead of n. This will not be done here as we can restrict ourselves to the first approximation. We find that

$$\lambda' = \frac{(N, (M + V)n')}{(N, n')} = \frac{(M^+N, n') + (N, Vn')}{(N, n')}$$

$$= \frac{\lambda(N, n') + (N, Vn')}{(N, n')} = \lambda + \frac{(N, Vn')}{(N, n')} \,. \tag{5}$$

No approximation has been made to obtain (5) which gives an expression for $\lambda' - \lambda$. This is supposed to be a small quantity so that in the expression therefore one can replace n' by its unperturbed value n to obtain

$$\lambda' - \lambda = \frac{(N, Vn)}{(N, n)} \,. \tag{6}$$

This is the equation we shall use in the rest of this report.

It may be worth while to remark, for the sake of rigorous mathematics, that the extension of the full scope of the Rayleigh-Schrödinger theory is probably unjustified in general for arbitrary (e.g., not self-adjoint) operators M because it involves the assumption that the characteristic functions of M form a complete system. This is not true in general. On the other hand, the derivation of the first approximation (6) involved no such assumption. The only two assumptions that had to be made were that (N, n) shall not vanish and that n' approach n if V is made very small. The first assumption breaks down if λ is an elementary divisor and in this case, in fact, (6) is incorrect. One would notice this occurrence, however, when trying to use (6). For certain operators V, n' does not converge to n even if the coefficient of V converges to zero. This can occur for self-adjoint M as well as in our case and examples for it are well known in quantum mechanics (action of weak uniform electric field on an atom). This possibility will be disregarded here, as it is not believed that it plays any role, just as it is quite customary to disregard it in quantum theory.

There may be some question as to whether the change in λ (reciprocal pile period) is the most important quantity, in the changes of which one is interested foremost. This is, of course, a matter of opinion. It may be worth while to remark that there is, of course no such ambiguity if one merely calculates the ratio of two kinds of perturbations the joint effect of which is zero.

Two Group Theory

We shall use the pile equations in the form

$$\operatorname{div} D \operatorname{grad} q + \frac{\partial}{\partial \tau} S q + \zeta \sigma_a f(\tau) n_0 - r(\tau) q = \frac{e^{-\tau/2}}{v_0} \frac{\partial q}{\partial t} \qquad (7)$$

$$\operatorname{div} D_0 \operatorname{grad} n_0 + S_0 q(0) - \sigma_a n_0 = \frac{1}{v_0} \frac{\partial n_0}{\partial t} . \qquad (7\,\mathrm{a})$$

In these, $\tau = \ln E/E_0$ is the logarithm of the energy of the neutron in units of the thermal energy E_0. The $q(\tau)$ is the number of neutrons per unit volume in unit τ range, multiplied with their velocity; n_0 is the density of thermal neutrons, multiplied with their velocity. D and D_0 are the diffusion constants (divided by the velocity) for fast and thermal neutrons; the former is a function of τ. S is the slowing down per unit volume, S_0 is the value of S for thermal energies, i.e., $\tau = 0$,

$$D(\tau) = \frac{1}{3 \sum \sigma(1 - \cos \theta)} ; \quad S = \sum \sigma \xi . \qquad (8)$$

In these and the above equations, the σ are cross sections per unit volume, the σ in (8) is scattering cross section, ξ the average loss in τ per collision. The $\sum$ are used because the material may contain different types of atoms. The σ_a is the thermal absorption cross section; $r(\tau)$ the resonance absorption cross section per unit volume. The $\zeta = \eta \varepsilon f$ is the multiplication constant divided by the resonance escape probability. The product of thermal utilization f and σ_a is the effective cross section of uranium per unit volume, i.e., its cross section per unit volume multiplied by the thermal neutron density in it and divided by the average thermal neutron density. One can write, therefore, σ_u for $f\sigma_a$. If one multiplies this with η the result is the same as $\sigma_f \nu$ where σ_f is the effective fission cross section for thermal neutrons per unit volume, ν the number of fast neutrons per fission. As a result, the third term in (7) can be written also as (ε is the multiplication by fast effect)

$$\int \sigma_a f(\tau) n_0 = \nu \varepsilon \sigma_f f(\tau) n_0 . \qquad (8\,\mathrm{a})$$

The $f(\tau)$ is the fraction of fission neutrons for which τ is in unit interval at τ so that

$$\int_0^\infty f(\tau)\,d\tau = 1; \qquad (8\,\mathrm{b})$$

v_0 is the velocity of thermal neutrons.

One obtains the first equation of the two group theory by integrating (7) with respect to τ and writing

$$\int_0^\infty q(\tau)\,d\tau = n_f . \qquad (9)$$

In addition, one has to assume that D is independent of τ or, otherwise, substitute a suitable average for it. The last term on the left side causes a decrease in the total number of neutrons. This term is omitted in the two group theory and will be replaced by inserting a factor p in the transcription of the second equation (7 a). Writing τ_f for the average τ of fission neutrons, one has, approximately

$$n_f/q(0) = \int_0^\infty q\, d\tau/q(0) = \tau_f. \qquad (9\,\mathrm{a})$$

Finally, in order to integrate the right side of (7), it is necessary to assume that, around $\tau = 0$, q does not depend strongly on τ. If this is done, $\partial q(0)/\partial t$ can be written for $\partial q/\partial t$ and (9 a) used for evaluating this term. One obtains in this way

$$\operatorname{div} D \operatorname{grad} n_f - \sigma_{af} n_f + \zeta \sigma_a n_0 = \frac{2}{v_0 \tau_f} \frac{\partial n_f}{\partial t} \qquad (10)$$

where

$$S_0/\tau_f = \sigma_{af} \qquad (9\,\mathrm{b})$$

was substituted.

Instead of (7 a), one has

$$\operatorname{div} D_0 \operatorname{grad} n_0 - \sigma_a n_0 + \sigma_{af} p n_f = \frac{1}{v_0} \frac{\partial n_0}{\partial t}. \qquad (10\,\mathrm{a})$$

The factor p is the resonance escape probability for the inclusion of which the reason was given above. Equations (10), (10 a) go over into (5 a), (5 b) of CP-1461, if one restricts oneself to the stationary case, i.e., sets the right sides of (10), (10 a) equal zero, except that the present equations apply also if the diffusion coefficients are not constant throughout the pile. The purpose of rederiving (10) and (10 a) was not to obtain the latter, rather trivial generalization but to obtain the proper factor for the right side of (10). Cf. also CP-1554.

The operator appearing in (10) can be written in a matrix form

$$
M = \left\|
\begin{array}{cc}
\frac{1}{2} v_0 \tau_f (\operatorname{div} D \operatorname{grad} -\sigma_{af}) & \frac{1}{2} v_0 \tau_f \zeta \sigma_a \\
v_0 \sigma_{af} p & v_0 (\operatorname{div} D_0 \operatorname{grad} -\sigma_a)
\end{array}
\right\|
$$
$$
= \left\|
\begin{array}{cc}
M_{ff} & M_{f0} \\
M_{0f} & M_{00}
\end{array}
\right\|. \qquad (11)
$$

The last line serves as the definition of M_{ff}, M_{f0}, etc.: for instance, $M_{f0} = \frac{1}{2} v_0 \tau_f \zeta \sigma_a$. The operator M is applied to the pair of functions n_f, n_0 in the usual way: the fast component $(Mn)_f$ of Mn is $(Mn)_f = M_{ff} n_f + M_{f0} n_0$ and the thermal component of Mn is $(Mn)_0 = M_{0f} n_f + M_{00} n_0$. The equation for the rate of change of the n can be written as

$$\frac{\partial n}{\partial t} = Mn \qquad (11\,\mathrm{a})$$

meaning that $\partial n_f/\partial t = (Mn)_f$ and $\partial n_0/\partial t = (Mn)_0$.

In order to obtain the operator adjoint to M one has to carry out two operations: first replace all four M_{ff}, M_{f0}, etc., with their adjoint operators and, second, interchange rows and columns. The first step does not introduce any change, since all M_{ff}, M_{f0} etc., are self adjoint. This is least evident for the M_{ff} and the M_{00}. In this case $\int N_f \frac{1}{2} v_0 \tau_f \operatorname{div} D \operatorname{grad} n_f \, dV$ can be transformed in the following way: v_0 and τ_f are assumed to be constants and can be brought before the integration sign. Then, one can write

$$
\begin{aligned}
N_f \operatorname{div}(D \operatorname{grad} n_f) &= \operatorname{div}(N_f D \operatorname{grad} n_f) - D(\operatorname{grad} N_f \operatorname{grad} n_f) \\
&= \operatorname{div}(N_f D \operatorname{grad} n_f) + n_f \operatorname{div}(D \operatorname{grad} N_f) \\
&\quad - \operatorname{div}(n_f D \operatorname{grad} N_f).
\end{aligned}
$$

Integrating over both sides of the equation and applying Gauss' theorem to the right side gives

$$
\begin{aligned}
\iiint N_f \operatorname{div}(D \operatorname{grad} n_f) \, dV &= \iiint n_f \operatorname{div}(D \operatorname{grad} N_f) \, dV \\
&+ \iint (N_f D \operatorname{grad} n_f - n_f D \operatorname{grad} N_f)_{\text{normal}} \, dS.
\end{aligned}
\tag{12}
$$

The second line is a surface integral which vanishes if n_f and N_f vanish on the surface of the domain of integration, i.e., the surface of the pile. If D has discontinuities in the pile, the above transformation must be applied separately to all parts of the pile in which D is continuous. In this case, the surface integrals of (12) will not vanish for the separate domains but their sum will be zero if the boundary conditions for n_f and N_f are

$$
n_f \text{ and } N_f \text{ continuous}
$$
$$
D \operatorname{grad} n_f \text{ and } D \operatorname{grad} N_f \text{ continuous}.
\tag{12a}
$$

These boundary conditions for n_f are the usual ones; they will be chosen to be valid for N_f also. In any case, then, the surface integrals can be omitted in (12) and one sees that $M_{ff} = \frac{1}{2} v_0 \tau_f (\operatorname{div} D \operatorname{grad} -\sigma)$ is a self adjoint operator. One sees in a similar way that $M_{00} = v_0(\operatorname{div} D_0 \operatorname{grad} -\sigma_a)$ is self adjoint, provided that the boundary conditions at a discontinuity of D_0 are

$$
n_0 \text{ and } N_0 \text{ are continuous}
$$
$$
D_0 \operatorname{grad} n_0 \text{ and } D_0 \operatorname{grad} N_0 \text{ are continuous}.
\tag{12b}
$$

This, and trivial equations like

$$
\iiint N_o(v_0 \sigma_{af} n_f) \, dV = \iiint n_f(v_0 \sigma_{af} N_0) \, dV
$$

show that all four M_{ff}, M_{f0}, M_{0f}, M_{00} are self adjoint. As a result, the adjoint to M can be obtained by interchanging the rows and columns of M:

$$
M^+ = \left\| \begin{array}{cc} \frac{1}{2} v_0 \tau_f (\operatorname{div} D \operatorname{grad} -\sigma_{af}) & v_0 \sigma_{af} p \\ \frac{1}{2} v_0 \tau_f \zeta \sigma_a & v_0(\operatorname{div} D_0 \operatorname{grad} -\sigma_a) \end{array} \right\|.
\tag{13}
$$

Examples for the Application of (6)

The purpose of the present report is not to solve any specific problems and only a very simple application of equation (6) will be given. It will be assumed that the pile is a simple one, i.e., that all quantities D, σ_{af}, p, D_0, σ_a which are, in general, functions of the position, are constant throughout the pile. In this case the div D grad can be replaced by $D\Delta$ and, as will be seen, n_f, n_0, N_f, and N_0 are all multipla of a common space function $\psi(x, y, z)$:

$$n_f = a_f\psi; \quad n_0 = a_0\psi; \quad N_f = A_f\psi; \quad N_0 = A_0\psi. \tag{14}$$

The equation $Mn = 0$ then reads

$$a_f(D\Delta\psi - \sigma_{af}\psi) + a_0\zeta\sigma_a\psi = 0$$
$$a_f\sigma_{af}p\psi + a_0(D_0\Delta\psi - \sigma_a\psi) = 0,$$

whence

$$\Delta\psi = D^{-1}(\sigma_{af} - \zeta\sigma_a a_0/a_f)\psi = D_0^{-1}(\sigma_a - p\sigma_{af}a_f/a_0)\psi. \tag{14a}$$

It follows from this, as is well known (CP-1461), that

$$\Delta\psi = B\psi \tag{15}$$

where B is the negative solution of the equation ($k = p\zeta = p\eta\varepsilon f$)

$$(DB - \sigma_{af})(D_0B - \sigma_a) - k\sigma_a\sigma_{af} = 0 \tag{15a}$$

$$\frac{a_f}{a_0} = \frac{\zeta\sigma_a}{-DB + \sigma_{af}} = \frac{-D_0B + \sigma_a}{p\sigma_{af}}. \tag{15b}$$

The equation for A_f and A_0 can be obtained in a similar way

$$\frac{A_f}{A_0} = \frac{p\sigma_{af}}{\frac{1}{2}\tau_f(-DB + \sigma_{af})} = \frac{-D_0B + \sigma_a}{\frac{1}{2}\tau_f\zeta\sigma_a}. \tag{15c}$$

One obtains from this the same equation (15 a) for B which (15 b) gave – which is only a demonstration of the fact that if the equation $Mn = 0$ has a solution, the equation $M^+N = 0$ has one also. The above equations determine n and N except for a constant factor which is arbitrary anyway.

It is easy now to obtain the effect of different perturbations on the period. Let us consider a few cases:

(a) Impurity with unit cross section for thermal neutrons at a definite position $\vec{r}_0$. This, contrary to appearance, does not change M_{f0} of (11) because, as (8 a) shows, this depends only on σ_f. (This may not be quite true in an actual lattice because σ_f is the effective fission cross section, i.e., the fission cross section multiplied with the neutron density ratio in the U over the average neutron density in the lattice. This point will be disregarded, however.) The only quantity in (11) that is changed is M_{00} so that V is the matrix

$$V = \begin{Vmatrix} 0 & 0 \\ 0 & -v_0 \delta(\vec{r} - \vec{r}_0) \end{Vmatrix}. \tag{16}$$

As a result, the change in λ is according to (6)

$$-\frac{\iiint A_0 \psi(\vec{r}) v_0 \delta(\vec{r} - \vec{r}_0) a_0 \psi(\vec{r}) \, dV}{\iiint (A_f a_f + A_0 a_0) \psi(\vec{r})^2 \, dV} = -\frac{A_0 a_0 v_0}{A_0 a_0 + A_f a_f} \frac{\psi(\vec{r}_0)^2}{\iiint \psi(\vec{r})^2 \, dV}$$

$$= -\frac{1}{1 + \frac{2}{\tau_f} \frac{-D_0 B + \sigma_a}{-DB + \sigma_{af}}} \frac{v_0 \psi(\vec{r}_0)^2}{\iiint \psi^2 \, dV}. \tag{16a}$$

The last factor of this is the one commonly used. The existence of fast neutrons is responsible for the first factor and the fact that the normalization contains the integral over $N_f n_f$ in addition to the integral over $N_0 n_0$. The former integral is, however, in the present case much smaller than the latter: τ_f is of the order of 20 and σ_a/σ_{af} is about 1/20 in a graphite pile. The $D_0 B$ and DB are negligible. As a result, the first factor on the right side of (16a) is very close to 1 and

$$\Delta \lambda = -\frac{v_0 \psi(\vec{r}_0)^2}{\iiint \psi^2 \, dV} \tag{16b}$$

is the change in reciprocal period due to the introduction of a thermal absorber of unit cross section at the point $\vec{r}_0$.

(b) As a second example we consider a change in η all over the pile. Since we are dealing with a simple pile, a multiplication constant can be defined and a change of η by $\eta\delta$ corresponds to a change in k by $\delta k = k\delta$. Of course, η cannot be changed easily over the whole pile but the calculation of the effect of such a change will tie in the present work with earlier considerations.

If η is changed by $\eta\delta$, ζ will be changed by $\zeta\delta$ and V will become

$$V = \begin{Vmatrix} 0 & \frac{1}{2} v_0 \tau_f \zeta \sigma_a \delta \\ 0 & 0 \end{Vmatrix}. \tag{17}$$

As a result, the fast component of Vn is $\frac{1}{2} v_0 \tau_f \zeta \sigma_a \delta n_0$, the thermal component is zero. We have

$$\Delta \lambda = \frac{1}{2} v_0 \tau_f \zeta \sigma_a \delta \frac{A_f a_0}{A_f a_f + A_0 a_0}. \tag{17a}$$

The integral over ψ^2 cancels out in this case. Neglect of $A_f a_f$ in the denominator gives with (15c) if $-D_0 B$ is neglected in the latter

$$\Delta \lambda = v_0 \sigma_a \delta. \tag{17b}$$

Considering that $1/v_0 \sigma_a$ is the life time of a thermal neutron and that the life time of a fast neutron is much shorter, this is the expected result. Setting the sum of (16b) and (17b) equal to zero gives for the fractional change in η (or k) which compensates the unit absorption at r_0

$$\delta = \frac{\delta k}{k} = \frac{\psi(\vec{r}_0)^2}{\sigma_a \iiint \psi(\vec{r})^2 \, dV}. \tag{17c}$$

This is the commonly used formula.

(c) As a last example, we consider the introduction of some additional, non-absorbing moderator, such as a small amount of D_2O. This will change D by decreasing the mean free path, it will change p by decreasing the resonance loss and increase σ_{af} by increasing S (cf. 9 b). It also will change D_0 by changing the thermal mean free path as well as the fast mean free path and it may change even $\sigma_a \zeta = \nu \varepsilon \sigma_f$ (cf. 8 a) because it may change the disadvantage factor. Of all these changes, only that corresponding to the change of σ_{af} will be calculated. This is greatest if the average logarithmic energy loss ξ of the newly introduced substance is greater than the same quantity in the pile.

If S is increased by unit area at the point $\vec{r}_0$, the σ_{af} is increased by $1/\tau_f$ and V becomes

$$V = \left\| \begin{matrix} -\frac{1}{2} v_0 \delta(\vec{r} - \vec{r}_0) & 0 \\ v_0 p/\tau_f \delta(\vec{r} - \vec{r}_0) & 0 \end{matrix} \right\| . \tag{18}$$

The change in reciprocal period is

$$\Delta \lambda = \frac{-\frac{1}{2} A_f a_f v_0 + A_0 a_f v_0 p/\tau_f}{A_f a_f + A_0 a_0} \frac{\psi(\vec{r}_0)^2}{\int \int \int \psi^2 \, dV} . \tag{18 a}$$

Neglecting $A_f a_f$ in the denominator gives with (15 b) and (15 c)

$$\Delta \lambda = \frac{\psi(\vec{r}_0)^2}{\int \int \int \psi^2 \, dV} \frac{a_f v_0}{a_0} \left(-\frac{1}{2} \frac{A_f}{A_0} + \frac{p}{\tau_f} \right)$$

$$= \frac{\psi(\vec{r}_0)^2}{\int \int \int \psi^2 \, dV} \frac{\sigma_a v_0}{p \sigma_{af}} \left(\frac{-\sigma_{af}}{\sigma_{af} - DB} + 1 \right) \frac{p}{\tau_f}$$

$$= \frac{\psi(\vec{r}_0)^2}{\int \int \int \psi^2 \, dV} \sigma_a v_0 \frac{-DB}{\sigma_{af}^2 \tau_f} . \tag{18 b}$$

All these results can be obtained, of course, also with standard methods. This is not true, however, if one has to deal with composite piles, with piles with reflectors, control rods, etc. In all these cases in which the equations of the two group theory can be solved, the adjoint equation for the "adjoint density" can be solved equally easily and one obtains the effect for arbitrary perturbations by the simple equation (6).

Pile Equations (7), (7 a)

The discussion of the two group equations (10), (10 a) leads only to rather obvious results in the case of a simple, bare pile and their usefulness becomes apparent only in less simple cases. The situation is rather the opposite for the modified Fermi equations (7), (7 a) which can be solved easily only for a simple bare pile. Even finding the adjoint operator to

$$M = \left\| \begin{matrix} v_0 e^{\tau/2} \left(\operatorname{div} D \operatorname{grad} + \frac{\partial}{\partial \tau} S - r \right) & v_0 e^{\tau/2} \zeta \sigma_a f(\tau) \\ v_0 S_0 \int d\tau \delta(\tau) & v_0 (\operatorname{div} D_0 \operatorname{grad} - \sigma_{at}) \end{matrix} \right\| \tag{19}$$

is not quite trivial and we shall first proceed with this task.

Formal application of the rule for obtaining the adjoint gives

$$M^+ = \left\| \begin{array}{cc} v_0 \left(\operatorname{div} D \operatorname{grad} + S\frac{\partial}{\partial\tau} - r \right) e^{\tau/2} & v_0 S_0 \delta(\tau) \\ v_0 \zeta \sigma_a \int d\tau\, f(\tau) e^{\tau/2} & v_0(\operatorname{div} D_0 \operatorname{grad} - \sigma_a) \end{array} \right\| .$$

This seems to give for the "adjoint densities" Q and N_0 the first equation

$$v_0 \left(\operatorname{div} D \operatorname{grad} + S\frac{\partial}{\partial\tau} - r \right) e^{\tau/2} Q + v_0 S_0 \delta(\tau) N_0 = \lambda Q$$

which is singular at $\tau = 0$. Around this point, the dominant terms are $-S\partial/\partial\tau\, e^{\tau/2} Q + S_0 \delta(\tau) N_0 = 0$. Since the value of S for $\tau = 0$ is S_0 this factor drops out and one sees that $e^{\tau/2} Q$ has a discontinuity of the amount N_0. It is natural to assume that Q vanishes below this discontinuity and the above shows then that Q is equal to N_0 immediately above $\tau = 0$. As a result, we can omit the $v_0 S_0 \delta(\tau)$ term and replace it with the boundary condition

$$Q(0) = N_0 . \tag{20}$$

If this boundary condition is used instead of $Q(0) = 0$, the $v_0 S_0 \delta(\tau)$ term must be omitted. As a result we have

$$M^+ Q(\tau) = v_0 e^{\tau/2} \left(\operatorname{div} D \operatorname{grad} - r - S\frac{\partial}{\partial\tau} - \frac{1}{2}S \right) Q(\tau) \tag{20 a}$$

and

$$M^+ N_0 = v_0 \zeta \sigma_a \int_0^\infty f(\tau) e^{\tau/2} Q(\tau)\, d\tau + v_0(\operatorname{div} D_0 \operatorname{grad} - \sigma_a) N_0 . \tag{20 b}$$

With this definition of $M^+ Q$ and $M^+ N_0$ one can easily convince oneself that, for any Q, q, N_0, n_0 satisfying the boundary conditions

$$\iiint dV \int d\tau\, Q(Mq) + \iiint dV\, N_0(Mn_0)$$

$$= \iiint dV \int d\tau\, (M^+ Q)q + \iiint dV\, (M^+ N_0)n_0 \tag{21}$$

so that (20 a), (20 b) really define the adjoint operator to M, provided Q satisfies the boundary condition (20), both Q and q vanish for $\tau = \infty$ and all quantities are zero at the surface of the pile. At discontinuities of D and D_0, the quantities q, n_0, Q, N_0, $D \operatorname{grad} q$, $D_0 \operatorname{grad} n_0$, $D \operatorname{grad} Q$, $D_0 \operatorname{grad} N_0$ must be continuous. The definition of Mq and Mn is already given in (7), (7 a) as well as in (9)

$$Mq = v_0 e^{\tau/2} \left(\operatorname{div} D \operatorname{grad} - r + \frac{\partial}{\partial\tau}S \right) q + v_0 e^{\tau/2} \zeta \sigma_a f(\tau) n_0 \tag{19 a}$$

$$Mn_0 = v_0 S_0 q(0) + v_0(\operatorname{div} D_0 \operatorname{grad} - \sigma_a) n_0 . \tag{19 b}$$

The equations $Mq = Mn_0 = M^+ Q = M^+ N_0 = 0$ for q, n_0, Q, N_0 will now be solved for a simple, bare pile.

Adjoint Neutron Density for Simple, Bare Pile

In case of a simple bare pile div D grad can be replaced by $D\Delta$ and div D_0 grad by $D_0\Delta$. The dependence of all quantities q, n_0, Q, N_0 on the position will be the same so that one can write, in analogy to (14)

$$q(x,y,z,\tau) = a(\tau)\psi(x,y,z) \quad n_0(x,y,z) = a_0\psi(x,y,z) \tag{22 a}$$
$$Q(x,y,z,\tau) = A(\tau)\psi(x,y,z) \quad N_0(x,y,z) = A_0\psi(x,y,z) \tag{22 b}$$

and ψ will satisfy the equation

$$\Delta\psi = B\psi \tag{22}$$

in which B is a constant which will be defined later (25). As a result, we have for (19 a) and (20 a)

$$(BD - r)a(\tau) + \frac{\partial}{\partial\tau}Sa(\tau) + \zeta\sigma_a f(\tau)a_0 = 0 \tag{23 a}$$

and

$$\left(BD - r - \frac{1}{2}S\right)A(\tau) - S\partial A(\tau)/\partial\tau = 0. \tag{23 b}$$

Both of these can be solved by elementary methods although D, r and S are functions of τ. One obtains with (20)

$$a(\tau) = \frac{a_0\zeta\sigma_a}{p(\tau)S(\tau)}e^{-Bt(\tau)}\int_\tau^\infty f(\tau')p(\tau')e^{Bt(\tau')}\,d\tau' \tag{24 a}$$

$$A(\tau) = A_0 e^{-\tau/2 + Bt(\tau)}p(\tau). \tag{24 b}$$

In these, $t(\tau)$ has nothing to do with time but denotes the "age" of neutrons with the energy $E_0 e^\tau$

$$t(\tau) = \int_0^\tau \frac{D(\tau')}{S(\tau')}\,d\tau'; \quad \ln p(\tau) = -\int_0^\tau \frac{r(\tau')}{S(\tau')}\,d\tau'. \tag{24}$$

$p(\tau)$ is the probability that a neutron of energy $E_0 e^\tau$ will not be absorbed at resonance before reaching thermal energy. Usually, for energies as high as that of fission neutrons, i.e. if $f(\tau')$ is not zero any more, $p(\tau')$ has assumed its asymptotic value. The reason for this is that r decreases with increasing energy and is practically zero at the energy of fission neutrons. If this is true – and it is ordinarily – the $p(\tau')$ in the integral of (25 a) can be replaced by $p(\infty) = p$, the energy independent chance for a fission neutron to escape resonance capture.

Equations (19 b) and (20 b) read with (22), (22 a), (22 b)

$$S_0 a(0) + (BD_0 - \sigma_a)a_0 = 0 \tag{25 a}$$

$$\zeta\sigma_a \int_0^\infty f(\tau)e^{\tau/2}A(\tau)\,d\tau + (BD_0 - \sigma_a)A_0 = 0. \tag{25 b}$$

Inserting into (25 a) its value from (24 a), we have

$$-a_0(BD_0 - \sigma_a) = a_0 \zeta \sigma_a \int_0^\infty f(\tau')p(\tau')e^{Bt(\tau')} d\tau' \qquad (25)$$

which is the equation defining B and, therefore, the size of the critical pile by (22). It assumes the usual form (CP-1662) if, as explained after equation (24), we replace $p(\tau')$ by p, cancel $\sigma_a a_0$, write k for ζp

$$-BD_0/\sigma_a + 1 = k \int_0^\infty f(\tau)e^{Bt(\tau)} d\tau . \qquad (25\,c)$$

D_0/σ_a is the square of the thermal diffusion length.

One obtains the same equation (25) from (25 b) and (24 b) which shows only that the condition for 0 to be a characteristic value of M is the same as the condition for it to be a characteristic value of M^+.

Equations (22), (22 a), (22 b), (24 a) and (24 b) give the complete set of solutions of the equations resulting from setting (19 a), (19 b), (20 a), (20 b) equal to zero. The a_0 and A_0 remain arbitrary constants. These solutions permit one to calculate by (6), the effect of any small perturbation on a simple bare pile. We shall consider only two examples: (a) that of a small amount of thermal absorber, and (b) that of a resonance absorber.

(a) A thermal absorber of unit cross section at the point $\vec{r}_0$ will not change Mq of (19 a) for the same reason which was discussed before (16). As a result $Vq = 0$, while $Vn_0 = -v_0\delta(\vec{r}-\vec{r}_0)n_0$. The perturbation in the reciprocal period becomes

$$\Delta\lambda = \frac{-\iiint N_0 v_0 \delta(\vec{r}-\vec{r}_0)n_0\, dV}{\iiiint Q(\tau)q(\tau)\, d\tau\, dV + \iiint N_0 n_0\, dV} . \qquad (26)$$

The second integral in the denominator is clearly $A_0 a_0 \iiint \psi^2\, dV$. The integral in the numerator is $-A_0 a_0 v_0 \psi(\vec{r}_0)^2$. The first integral in the denominator is $\int A(\tau)a(\tau)\, d\tau \iiint \psi^2\, dV$. For the former of this we have from (24 a) and (24 b)

$$\int_0^\infty A(\tau)a(\tau)\, d\tau = A_0 a_0 \zeta \sigma_a \int_0^\infty \frac{d\tau}{S(\tau)} e^{-\tau/2} \int_\tau^\infty f(\tau')p(\tau')e^{Bt(\tau')} d\tau' .$$

Because of the factor $e^{-\tau/2}$, in the rest of the integrand $\tau = 0$ can be substituted. Hence, (25) gives for the above expression $2(A_0 a_0/S_0)(-BD_0 + \sigma_a)$. Consequently,

$$\Delta\lambda = -\frac{v_0}{1 + 2(-BD_0 + \sigma_a)/S_0} \frac{\psi(\vec{r}_0)^2}{\iiint \psi^2\, dV} . \qquad (26\,a)$$

This differs from (16 a) only inasmuch as the $-DB$ term must be omitted in the latter. If this is done, (26 a) goes over into (16 a) by virtue of (9 b). As is well known, the two group theory's validity depends on a relatively small k and $-DB$ is small if k is close to 1. Just as in (16 a), the term $2(-BD_0 + \sigma_a)/S_0$ can ordinarily be neglected against 1.

(b) Let us denote the effective resonance absorption of the added absorber by $r'(\tau)$ and let us assume that it is uniformly distributed over the pile. Then $Vq = -v_0 e^{\tau/2}r'q$ and $Vn = 0$. As a result

$$\Delta\lambda = \frac{-\iiint\int Qv_0 e^{\tau/2} r' q \, d\tau \, dV}{\iiint\int Qq \, d\tau \, dV + \iiint Nn \, dV}. \qquad (27)$$

The denominator was calculated before and is $A_0 a_0$ times that of (26 a). For the numerator we have

$$\int_0^\infty \frac{A_0 a_0 v_0 \zeta \sigma_a}{S(\tau)} r'(\tau) \int_\tau^\infty f(\tau') p(\tau') e^{Bt(\tau')} \, d\tau' \, d\tau \iiint \psi^2 \, dV.$$

As pointed out after (24), $p(\tau')$ can be replaced by p and ζp then by k. The result then is

$$\Delta\lambda = -\frac{v_0 k \sigma_a}{1 + 2(\sigma_a - BD_0)/S_0} \int_0^\infty \frac{r'(\tau)}{S(\tau)} \int_\tau^\infty f(\tau') p(\tau') e^{Bt(\tau')} \, d\tau' \, d\tau. \qquad (27\,a)$$

The integral over τ' gives the fraction of neutrons of energy $E_0 e^\tau$ which become thermal in the pile, i.e., do not escape from the pile during the slowing down process (B is negative). The fraction $r'(\tau)/S(\tau)$ of them is absorbed so that the integral over τ gives the fractional loss in the production of thermal neutrons. The factors before the integral convert this loss into change of reciprocal period. If the resonance absorber had not been uniformly distributed over the pile but concentrated at a point r_0, the result would have contained the $\psi(\vec{r}_0)^2 / \iiint \psi^2 \, dV$ factor.

For sake of reference, the change of λ caused by a change $\eta\delta$ of η (or $k\delta$ of k) may be written down. In this case $Vq = v_0 e^{\tau/2} \zeta \sigma_a f(\tau) n_0 \delta$ and

$$\Delta\lambda = \delta \frac{v_0 \zeta \sigma_a \int e^{Bt(\tau)} f(\tau) p(\tau) \, d\tau}{1 + 2(\sigma_a - BD_0)/S_0}.$$

Because of (25) this again goes over into

$$\Delta\lambda = \frac{v_0(\sigma_a - BD_0)\delta}{1 + 2(\sigma_a - BD_0)/S_0}. \qquad (27\,b)$$

This is again identical with (17 a) if we make the same approximation in the denominator of the latter which was mentioned at (26 a).

37.

Efficiency of Control Rods
Which Absorb Only Thermal Neutrons

E. P. Wigner, A. M. Weinberg, and R. R. Williamson

February 24, 1944

Abstract. Simplified pile equations are derived under the assumption that the fast neutrons have at every collision with the moderator the same probability of reaching thermal energy. The approximation involved is much less accurate than Fermi's whose pile equations assume that the slowing down occurs in a definite number of collisions. However, the resulting equations are much more easily solved than Fermi's since they are ordinary instead of partial differential equations.

Solutions are given for the case of a control rod at the center of a cylindrical pile and the result compared with Murray's calculation, carried out on the basis of Fermi's equations. Second, the efficiency of a regular array of control rods is calculated and compared with older work and with an experimental value. The agreement is good in each case.

1. If one replaces the lattice arrangement by a homogeneous mixture one can describe the pile by the equations

$$\Delta n_t - \kappa_t^2 n_t + \kappa_t^2 q(t_0) = 0 \qquad (1\,a)$$

$$\Delta q(t) = \frac{\partial q(t)}{\partial t}; \quad q(0) = kn_t. \qquad (1\,b)$$

In these n_t is the density of thermal neutrons multiplied by their velocity, $q(t)$ is proportional to the slowing down density; t is the age of neutrons, t_0 the age of thermal neutrons, k the multiplication constant, κ_t the reciprocal diffusion length of thermal neutrons in the mixture. The above equations were derived by Fermi on the assumption that the number of collisions which a fission neutron suffers before it becomes thermal is the same for all neutrons and that the distance between two collisions is also a well-defined quantity depending only on the energy.

Actually the lengths of the paths between two collisions are distributed according to the exponential law and the energy loss depends on the angle of deflection for the collision. As a result of this and, in particular, of the variation in the initial (fission) energy, the number of collisions which bring a neutron down to thermal energy is not the same for all neutrons but shows considerable fluctuation. This causes the second of the above equations to be somewhat inaccurate, particularly if the mean free path changes strongly with energy as is the case, for instance, in water.

One may try to replace the definite number of collisions underlying Fermi's equation with an average number of collisions. This picture is, for a graphite lattice much less accurate than Fermi's picture because the number of collisions which bring a say 1 MeV neutron down to thermal energies is always about 100 and certainly there is no chance that during the first 50 collisions the energy should drop to 1/40 eV. Nevertheless, it seems worth while to try to use the above picture because of its mathematical convenience.

The pile equations become, if one uses the above simplified picture in which a fast neutron has, at every collision, an equal chance σ_{af}/σ_f to become thermal,

$$\Delta n_t - \kappa_t^2 n_t + (\kappa_t^2/\sigma_{at})P_t = 0 \tag{2a}$$
$$\Delta n_f - \kappa_f^2 n_f + (\kappa_f^2/\sigma_{af})P_f = 0; \tag{2b}$$

herein n_f is the density of fast neutrons multiplied with their velocity,

$$\kappa_f^2 = 1/t_0 \tag{3a}$$

is six times the reciprocal mean square distance of travel of a neutron while it is being slowed down. (Just as κ_t^2 is six times the reciprocal mean square distance of travel of a thermal neutron.) Plass has shown (CP-992) that, in case of a pile, one should write

$$\kappa_t^2 = \frac{1}{L^2(1 - p_2)} \tag{3b}$$

where L is the diffusion length in the moderator and p_2 the thermal utilization. σ_{at} and σ_{af} are "absorption cross sections" per cm^3 for thermal and fast neutrons. In the latter case, the "absorption" is slowing down to thermal energy:

$$\sigma_a/\sigma = \xi(\ln E_f/E_t)^{-1}$$

where σ is the total scattering cross section, ξ the logarithmic energy loss upon a collision, E_f/E_t the ratio of fission and thermal energies. Both will drop out of the final results. P_t and P_f are the production of thermal and fast neutrons

$$P_t = p_1\sigma_{af}n_f \qquad P_f = p_2\varepsilon\eta\sigma_{at}n_t \tag{4}$$

where p_1 is the probability of escaping resonance absorption during the slowing down process, $p_2 = f$ the thermal utilization, ε the fast effect, η the number of fission neutrons per neutron absorbed in uranium.

Introducing (4) into (2) one obtains

$$\Delta n_t - \kappa_t^2 n_t + \kappa_t^2 A n_f = 0 \tag{5a}$$
$$\Delta n_f - \kappa_f^2 n_f + \kappa_f^2 B n_t = 0 \tag{5b}$$
$$A = p_1\sigma_{af}/\sigma_{at}; \ B = p_2\varepsilon\eta\sigma_{at}/\sigma_{af}; \ AB = \eta\varepsilon p_1 p_2 = k. \tag{5c}$$

Equation (5b) implies that the production of thermal neutrons from a point source of fast neutrons has the form of an exponential

$$(e^{-\kappa_f r})/r$$

while it is a Gaussian if Fermi's picture is correct. As said before, in the case of a graphite lattice Fermi's assumption is more correct, while equation (5 b) is more accurate, for instance, in the case of water. The picture underlying the latter equations has been used repeatedly before, for instance in the calculation of the resonance disadvantage factor, also by Teller in his theory of the exponential pile, and more recently by S. Lloyd and F. Friedman in their theory of a pile reflector. In the first of these cases, Christy has shown (not yet reported) that the solutions of (2 b) and (1 b) agree with each other extremely closely.

The advantage of equations (5) is that none of them is a partial differential equation involving the energy explicitly so that the number of variables is reduced by one. One can always solve them if one can solve the equation

$$\Delta n = \kappa^2 n$$

with suitable boundary conditions.

2. If one eliminates n_t or n_f from the above equations one obtains for either of them

$$\Delta \Delta n - (\kappa_t^2 + \kappa_f^2)\Delta n - (k-1)\kappa_t^2\kappa_f^2 n = 0 \,. \tag{6}$$

This shows that n_f and n_t have the form

$$n_f = a_f Z_1(\underline{r}) + b_f Z_2(\underline{r}) \qquad n_t = a_t Z_1(\underline{r}) + b_t Z_2(\underline{r}) \tag{7}$$

where Z are solutions of the equations

$$\Delta Z_1 = \kappa_1^2 Z_1 \qquad \Delta Z_2 = \kappa_2^2 Z_2 \tag{8a}$$

$$\kappa_1^2 = \frac{1}{2}(\kappa_t^2 + \kappa_f^2) - \sqrt{\frac{1}{4}(\kappa_t^2 + \kappa_f^2)^2 + (k-1)\kappa_t^2\kappa_f^2} \tag{8b}$$

$$\kappa_2^2 = \frac{1}{2}(\kappa_t^2 + \kappa_f^2) + \sqrt{\frac{1}{4}(\kappa_t^2 + \kappa_f^2)^2 + (k-1)\kappa_t^2\kappa_f^2} \,. \tag{8c}$$

Inserting the above again into (5) one obtains

$$
\begin{aligned}
(\kappa_t^2 - \kappa_1^2)a_t = A\kappa_t^2 a_f \text{ or } (\kappa_f^2 - \kappa_1^2)a_f = B\kappa_f^2 a_t \\
(\kappa_t^2 - \kappa_2^2)b_t = A\kappa_t^2 b_f \text{ or } (\kappa_f^2 - \kappa_2^2)b_f = B\kappa_f^2 b_t \,.
\end{aligned}
\tag{9}
$$

If $k - 1 \ll 1$, which is true in all cases of interest here

$$\kappa_1^2 \approx -\frac{k-1}{L^2(1-p_2)+t_0} = -\frac{k-1}{M^2} \,; \tag{10a}$$

it is negative if $k > 1$. On the other hand, κ_2^2 is always positive and much larger than κ_1^2,

$$\kappa_2^2 \approx \kappa_t^2 + \kappa_f^2 \,. \tag{10b}$$

Introducing the above expressions into (9), one has $AB \sim 1$

$$a_t = A a_f$$
$$-\kappa_f^2 b_t = A \kappa_t^2 b_f . \tag{11}$$

The Z_1, Z_2, a, b are only partially determined by the equations (8 a) and (9) and are more carefully determined by the boundary conditions which are imposed on n_t or n_f. If these boundary conditions are the same for n_t and n_f, as is the case e.g. for an ordinary pile without control rods, one can set $b = 0$ and one obtains the usual pile equation. In the following the above equations will be applied to the case of a single rod in the center of a cylindrical pile to calculate the reduction in the effective multiplication constant. It will also be applied for the same purpose in case of a regular array of control rods as is the case for instance in the central region of the W pile.

3. The problem of a cylindrical pile with one or more control rods parallel to its axis can be treated as a two-dimensional problem by allotting part of the multiplication constant to the escape of neutrons parallel to the axis of the cylinder. If one does this, the boundary conditions for the variation of the neutron density perpendicular to the axis are: (1) both n_t and n_f vanish at the boundary of the pile; (2) n_t also vanishes at the surface of the control rods; (3) grad n_f is constant on the surface of the control rod. In the cases to be treated here, grad $n_f = 0$ will hold for reasons of symmetry.

We begin with a single rod at the center of the pile. It is possible to solve this problem rigorously. However, an approximate solution which assumes that $\kappa_t r_0 \ll 1$, $\kappa_f r_0 \ll 1$, $\kappa_t R \gg 1$, $\kappa_f R \gg 1$ where r_0 is the effective radius of the rod and R the radius of the pile is more easily discussed.

Without the control rod, we can assume that

$$n_t = A n_f = J_0(\bar{\kappa}_0 r)$$

where $\bar{\kappa}_0 R = 2.405$ and R is the radius of the pile, $\bar{\kappa}_0^2 = -\kappa_0^2 = (k_0 - 1)/M^2$. If one inserts a control rod and still wishes to keep the pile chain reacting, one must increase k_0 to k and hence $\bar{\kappa}_0^2$ to $\bar{\kappa}_1^2 = -\kappa_1^2 = (k - 1)/M^2$. In order to satisfy the boundary condition $Z_1(R) = Z_2(R) = 0$, one must write (the notation for the Bessel functions is that of Jahnke-Emde[7])

$$\begin{aligned} Z_1(r) &= J_0(\bar{\kappa}_1 r) - \frac{J_0(\bar{\kappa}_1 R)}{N_0(\bar{\kappa}_1 R)} N_0(\bar{\kappa}_1 r) \\ &= J_0(\bar{\kappa}_1 r) + \frac{\bar{\kappa}_1 - \bar{\kappa}_0}{\bar{\kappa}_0} \frac{2.405 \times .519}{.514} N_0(\bar{\kappa}_1 r) \end{aligned} \tag{12 a}$$

and

---

[7] The equivalent British Association Tables notation is $J_n(x) \cong J_n(x)$, $N_n(x) \cong Y_n(x)$; $J_0(ix) \cong I_0(x)$, $-i J_1(ix) \cong I_1(x)$; $i H_0^{(1)}(ix) \cong \frac{2}{\pi} K_0(x)$, $-H_1^{(1)}(ix) \cong \frac{2}{\pi} K_1(x)$

$$Z_2(r) = iH_0^{(1)}(i\kappa_2 r) - \frac{iH_0^{(1)}(i\kappa_2 R)}{J_0(i\kappa_2 R)} J_0(i\kappa_2 r) \tag{12}$$

$$= iH_0^{(1)}(i\kappa_2 r) - 2e^{-2\kappa_2 R} J_0(i\kappa_2 r).$$

The second term is appreciable only near the surface of the pile.

Assuming $a_f = 1$ (which is possible since a multiplicative constant is free in the neutron densities), we have for

$$n_f = Z_1(r) + b_f Z_2(r). \tag{13}$$

The condition $\operatorname{grad} n_f(r_0) = 0$ is equivalent, for small r_0, to the condition that n_f go to a constant and have no singularity at $r = 0$. Using the asymptotic expressions for the Bessel functions in (12), this gives

$$b_f = 2.43\frac{\bar{\kappa}_1 - \bar{\kappa}_0}{\bar{\kappa}_0}. \tag{13a}$$

Hence we have, for (11)

$$a_t = A; \quad b_t = -2.43A\frac{\kappa_t^2}{\kappa_f^2}\frac{\bar{\kappa}_1 - \bar{\kappa}_0}{\bar{\kappa}_0} \tag{14}$$

and n_t/A becomes, in the neighborhood of $r = 0$,

$$\frac{n_t}{A} \sim 1 - 2.43\frac{\bar{\kappa}_1 - \bar{\kappa}_0}{\bar{\kappa}_0}\frac{2}{\pi}\ln\frac{2}{\gamma\bar{\kappa}_1 r} - 2.43\frac{\kappa_t^2}{\kappa_f^2}\frac{\bar{\kappa}_1 - \bar{\kappa}_0}{\bar{\kappa}_0}\frac{2}{\pi}\ln\frac{2}{\gamma\kappa_2 r}. \tag{14a}$$

The boundary condition for n_t at the surface of the rod will be satisfied if (14a) is zero for $r = r_0$. This gives, together with

$$\frac{\bar{\kappa}_1 - \bar{\kappa}_0}{\bar{\kappa}_0} \approx \frac{\bar{\kappa}_1^2 - \bar{\kappa}_0^2}{2\bar{\kappa}_0^2} = \frac{\Delta k}{2(k-1)}$$

and finally, with $\ln 2/\gamma = .116$,

$$\frac{\Delta k}{k-1} = \frac{\pi}{2.43}\left[.116\left(1 + \frac{\kappa_t^2}{\kappa_f^2}\right) - \ln\bar{\kappa}_1 r_0 - \frac{\kappa_t^2}{\kappa_f^2}\ln\kappa_2 r_0\right]^{-1}. \tag{15}$$

The simple theory, which takes into account only one kind of neutrons, gives in the same approximation

$$\frac{\Delta k}{k-1} = \frac{\pi}{2.43}[.116 - \ln\bar{\kappa}_1 r_0]^{-1} \tag{15a}$$

which is obtained from (15) by setting $\kappa_f = \infty$. It must be remembered that $k - 1$ in (15) involves only that part of the multiplication constant which is available for the density variation perpendicular to the axis. In other words

$$k - 1 = M^2\frac{2.405^2}{R^2} = \frac{2.405^2}{R^2}\left(\frac{1}{\kappa_t^2} + \frac{1}{\kappa_f^2}\right).$$

Similarly, the approximations involve $\bar{\kappa}_1 R_0 = 2.405$, $\kappa_2 = \sqrt{\kappa_t^2 + \kappa_f^2}$. Hence

$$\Delta k = \frac{7.5}{R^2}\left(\frac{1}{\kappa_t^2} + \frac{1}{\kappa_f^2}\right)\left[.116\left(1 + \frac{\kappa_t^2}{\kappa_f^2}\right) - \ln\frac{2.405 r_0}{R} - \frac{\kappa_t^2}{\kappa_f^2}\ln\sqrt{\kappa_t^2 + \kappa_f^2}\,r_0\right]^{-1}.$$

(16)

The ratio of (15) and (15 a) is plotted in Fig. 1 against r_0 for $\kappa_t = \kappa_f$; $k - 1 = .05$, hence $\bar{\kappa}_1 = .158\kappa_t$; $\kappa_2 = \sqrt{2}\kappa_t$. It also contains calculations for $k - 1 = .019$, both on the basis of the present method and on the basis of Fermi's equations (1). The latter are due to F. Murray (CP-742) and the agreement between the two types of calculations is seen to be very good.

For the very small radii of the control rod, the ratio of the efficiency of the rod which absorbs only thermal neutrons is only $\kappa_t^{-2}/(\kappa_t^{-2} + \kappa_f^{-2})$ times as great as the efficiency of a rod which absorbs both fast and thermal neutrons. This is the ratio which has been used occasionally for all rod sizes. As r_0 increases, the ratio of the efficiencies of the two types of rods tends to one. This is seen from the Fig. 1 but is somewhat obscured in the formulae on account of the approximations used for the Bessel functions. The Fig. 1 shows that the ratio of efficiencies of the two types of control rods depends essentially only on their absolute radius, i.e. on $\kappa_f r_0$. It is practically independent of the multiplication constant, i.e. $\kappa_1 r_0$.

The fact that the efficiency of a thermally black rod depends very little on its blackness to fast neutrons, provided the rod is large enough, may be explained by referring to Fig. 3. In this figure, the fast and thermal neutron densities are plotted for three rods of radii 1 cm, 5 cm and 10 cm, respectively. In all three cases the thermal density dips sharply in the neighborhood of the rod; on the other hand, the dip in the fast distribution is practically smoothed out for the small rod but is almost as great as the thermal neutron dip for the large rod. The reason is that if the rod size is small compared to the age ($\kappa_f r_0 \ll 1$) any initial variation in the nascent neutron density will be smoothed during the slowing down process. As the rod size increases, the original dip tends to persist more strongly, until, for a rod very large compared to the slowing down length, the fast distribution remains very similar to the slow neutron distribution. In this case the distribution of fast neutrons, and therefore the overall effectiveness, is practically the same as it is for a rod black to both kinds of neutrons.

The effect of a control rod is partly due to leakage of neutrons into the rod, and partly to increased leakage of neutrons to the outside. It is of some interest therefore to compare the ratio (ρ) of excess outward leakage to rod leakage. The excess leakage outward, per neutron absorbed in the pile, is

$$\frac{-\frac{\sigma_{a_t}}{\kappa_t^2}R\frac{dn_t}{dr}\Big|_R - \frac{\sigma_{a_f}}{\kappa_f^2}R\frac{dn_f}{dr}\Big|_R}{\sigma_{a_t}\int_{r_0}^{R} n_t r\,dr} - \left(\frac{1}{\kappa_t^2} + \frac{1}{\kappa_f^2}\right)\frac{\overline{2.405}^2}{R^2}$$

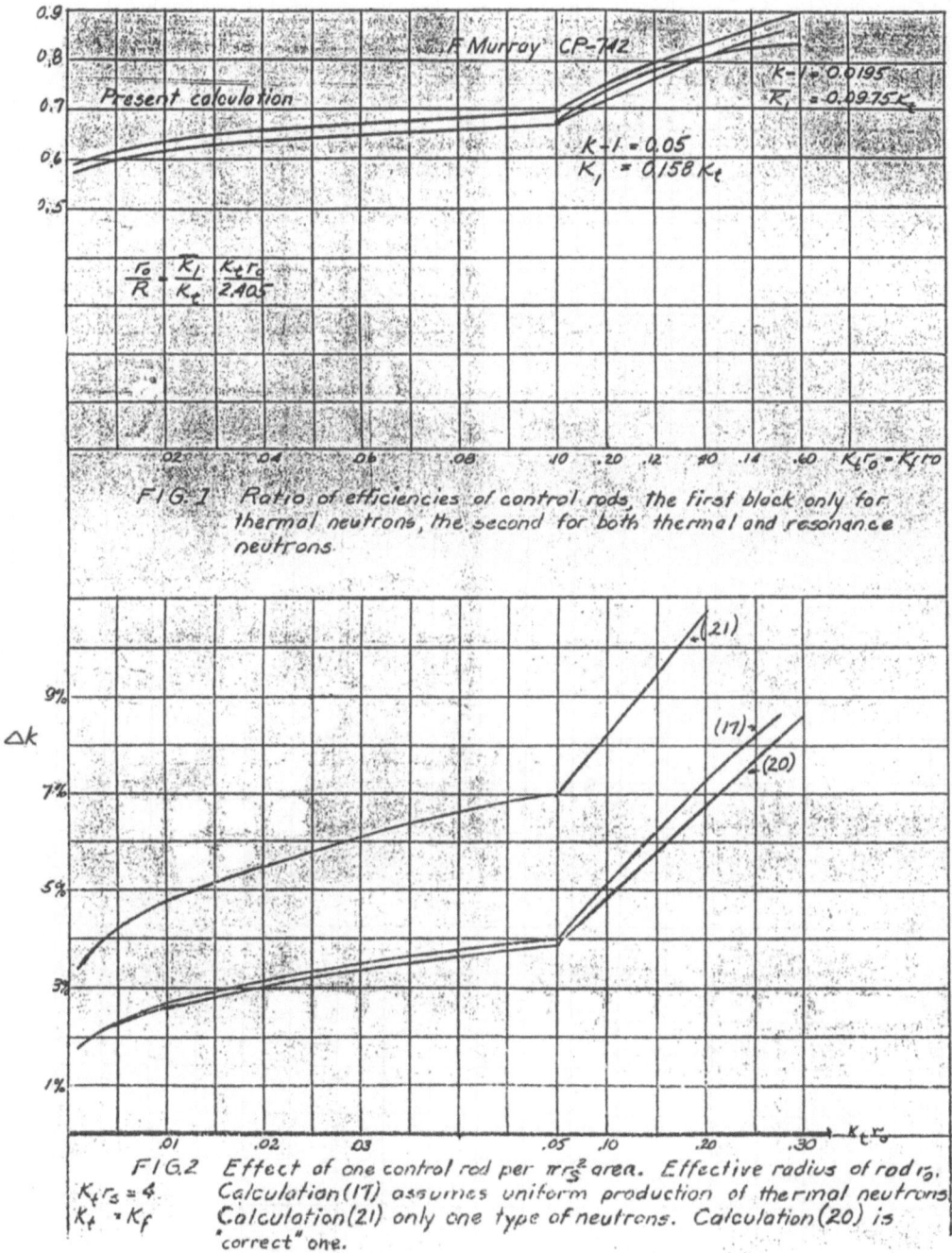

FIG.1   Ratio of efficiencies of control rods, the first block only for thermal neutrons, the second for both thermal and resonance neutrons.

FIG.2   Effect of one control rod per πr_3^2 area. Effective radius of rod r_3.
$K_t r_3 = 4$.   Calculation (17) assumes uniform production of thermal neutrons.
$K_t = K_f$.   Calculation (21) only one type of neutrons. Calculation (20) is "correct" one.

Figs. 1 and 2

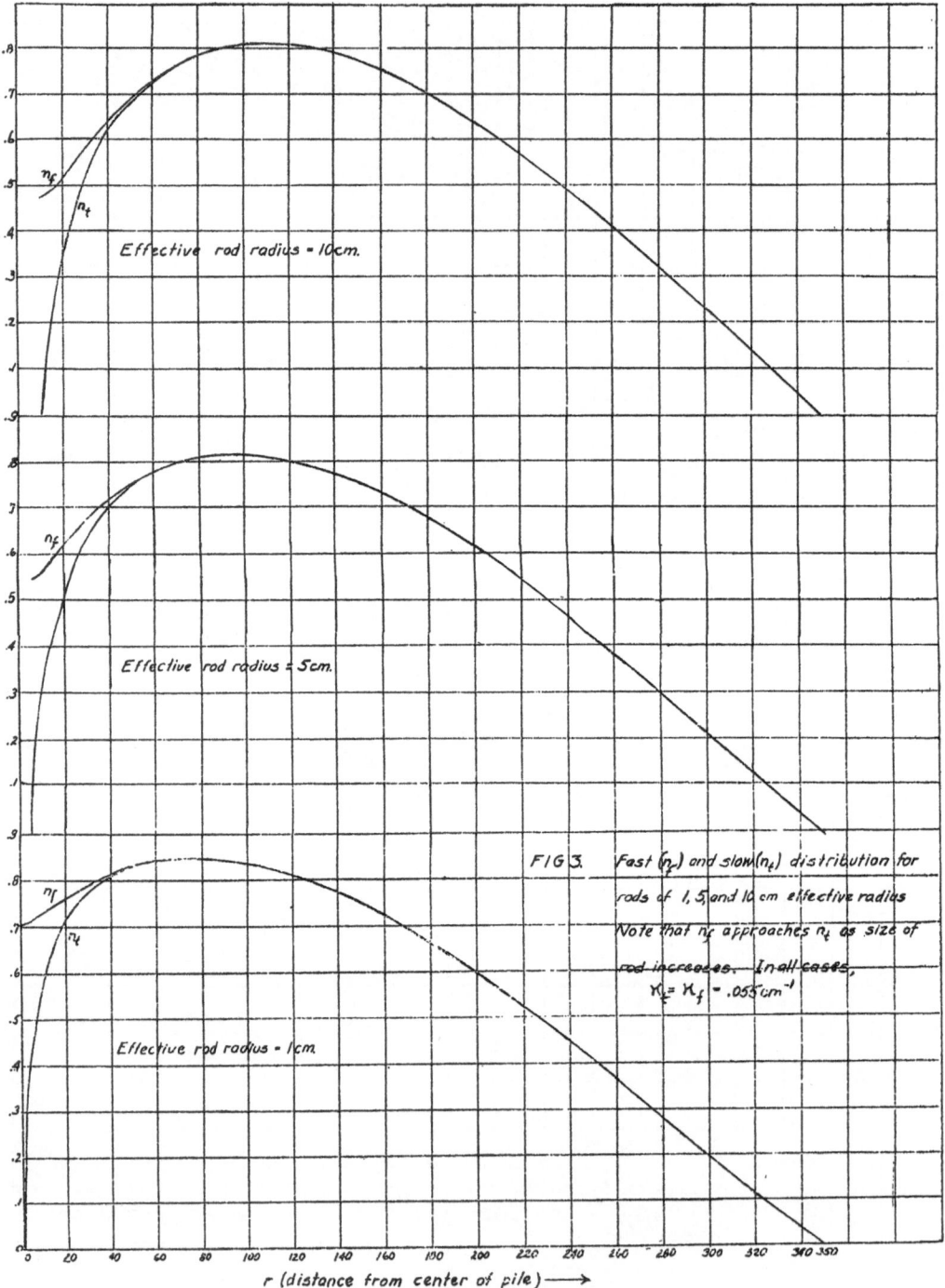

Fig. 3

while the rod leakage, per neutron absorbed in the pile, is

$$\frac{-\frac{\sigma_{a_t}}{\kappa_t^2} r_0 \frac{dn_t}{dr}\Big|_{r_0}}{\sigma_{a_t} \int_{r_0}^R n_t r \, dr}.$$

Substituting for n_t and n_f the expressions found previously, we obtain

$$\frac{\text{excess leakage outward}}{\text{leakage into rod}} = -\frac{N_0(\bar{\kappa}_1 r_0)}{N_0(\bar{\kappa}_1 R)} \frac{\left(1 - \frac{\bar{\kappa}_0^2}{\bar{\kappa}_1^2}\right)}{1 + \frac{\pi}{2} N_1(\bar{\kappa}_1 r_0)\bar{\kappa}_1 r_0 Z_1(r_0)}$$

$$- \frac{\bar{\kappa}_0^2}{\bar{\kappa}_1^2}\left(1 - \frac{\bar{\kappa}_1^2}{\bar{\kappa}_2^2}\frac{\kappa_t^2}{\kappa_f^2}\right)$$

which reduces, in the case of only one kind of neutron ($\kappa_f = \infty$), to

$$\rho = -\frac{N_0(\bar{\kappa}_1 r_0)}{N_0(\bar{\kappa}_1 R)}\left(1 - \frac{\bar{\kappa}_0^2}{\bar{\kappa}_1^2}\right) - \frac{\bar{\kappa}_0^2}{\bar{\kappa}_1^2}.$$

The ratio ρ decreases as the size of the rod decreases and approaches the limit

$$\lim_{r_0 \to 0} \rho = 1.608 - \frac{\bar{\kappa}_0^2}{\bar{\kappa}_1^2}\left(1 - \frac{\kappa_t^2}{\kappa_f^2}\frac{\bar{\kappa}_1^2}{\bar{\kappa}_2^2}\right) \approx .608$$

for a rod of zero radius. This ratio is useful in calculating, for example, the amount of boron which is transmuted by n, α processes in a boron coated rod which takes up a certain amount of k.

4. Before going into the details of the calculation of the efficiency of an array of control rods a few remarks should be made. The effect of control rods can always be considered in two ways: In the first picture, the rods change the effective size of the pile, i.e., change the boundary conditions for the neutron density. In the second picture, the control rods belong to the pile and form sort of a poisoning material thus decreasing the multiplication constant. It is interesting to note that in the first picture one calculates an increase in the multiplication constant which is necessary to keep the pile chain reacting in spite of the changed boundary conditions. In the second picture, one calculates the decrease of the multiplication constant due to the rods. In order to keep the pile chain reacting one again would have to increase k in the material of the pile by the same amount. Thus, although in different ways, both pictures give the amount of increase in k which could compensate for the insertion of the control rods.

If there is only one control, only the first picture can be used, and the calculations described in this report are based in every case on that picture. The same holds for the early calculations in which only one type of neutrons was considered. However, the calculations for the safety rods in the W system were based on the second picture and a comparison between the two methods will be given here.

In the second picture, the effect of the control rods is to absorb some of the thermal neutrons. If we continue to consider the material of the pile proper as a homogeneous mixture the control rods can be considered to form a lattice just as the uranium rods form a lattice in graphite. The difference between the two cases is, however, that the neutrons which are absorbed in the uranium are responsible for the further multiplication while the neutrons absorbed in the control rod are excluded from it. Thus if one calculates the fraction p_2 of neutrons absorbed in the control rod, one has calculated the number of neutrons which are excluded from further multiplication. Hence p_2 is the change in the multiplication constant caused by the presence of the control rods.

It appears reasonable to apply the ordinary formulae for calculating p_2 such as given in report (CP-104) by Christy and Monk. The only objection to this is that the cell is so great in the present case that the production of thermal neutrons is not uniform any more over the cell but shows a dip in the neighborhood of the rod. For this reason the formulae of (CP-104) give a too high thermal utilization p_2 and hence tend to exaggerate the effect of the control rods. The equation (πr_s^2 is the area of a cell in which there is one control rod, $\sigma_{a0} = \infty$ must be used in CP-104 since the rod is black)

$$\Delta k = p_2 = \frac{2r_0}{\kappa_t^2(r_s^2 - r_0^2)} \frac{H_1^{(1)}(i\kappa_t r_s)iJ_1(i\kappa_t r_0) - iJ_1(i\kappa_t r_s)H_1^{(1)}(i\kappa_t r_0)}{H_1^{(1)}(i\kappa_t r_s)J_0(i\kappa_t r_0) + iJ_1(i\kappa_t r_s)iH_0(i\kappa_t r_0)} \quad (17)$$

gives therefore an upper limit for the change of the multiplication constant. It is possible to simplify this equation somewhat without sacrificing accuracy by using the asymptotic formulae valid if $\kappa_t r_0$ is small and this gives

$$\Delta k = p_2 = \frac{2}{\kappa_t^2(r_s^2 - r_0^2)} \left[\ln \frac{2}{\gamma\kappa_t r_0} + \frac{\pi}{2} \frac{H_1^{(1)}(i\kappa_t r_s)}{iJ_1(i\kappa_t r_s)} \right]^{-1} \quad (17\,\text{a})$$

$$\sim \frac{2}{\kappa_t^2 r_s^2} \left[\ln \frac{2}{\gamma\kappa_t r_s} \right]^{-1} .$$

As said above, this is an upper limit for the efficiency of the system of control rods. However, with Plass' help, we applied a small correction to take into account the non-uniform production of thermal neutrons before we used it for the W system. The calculation of the correction is somewhat complicated and will not be reported here.

We now return to the picture in which the effect of the control rod modifies the boundary conditions. We must first calculate the change in the Laplacian which will keep this pile chain reacting. In the present case it appears simplest to consider only the central portion of the pile which is filled by the array of control rods and impose upon it the boundary condition that the derivative of the neutron densities vanishes at its surface. Without control rods this would correspond to a zero Laplacian and $k = 1$. In order to maintain this boundary condition with the control rods inserted the Laplacian must be negative κ_1^2 and the multiplication constant therefore larger than one. The above boundary condition for the whole center portion of the pile can be replaced by the same boundary condition for each cell. Hence the problem which we are going to solve is that of equations (5) with the boundary conditions that both n_t and n_f have zero derivatives at $r = r_s$. The same holds for n_f at $r = r_0$ while n_t vanishes at this point. The condition that $\mathrm{grad}\, n_f$ vanish at $r = r_0$ will be replaced again by the condition that it be regular and hence have zero gradient at $r = 0$. This is a good approximation if r_0 is small. The calculation can be carried out equally easily without using this assumption but the final formulae get much more involved.

The calculation can be carried out as before. We have

$$Z_1 = J_0(\bar{\kappa}_1 r) - \frac{J_1(\bar{\kappa}_1 r_s)}{N_1(\bar{\kappa}_1 r_s)} N_0(\bar{\kappa}_1 r) \tag{18a}$$

and

$$Z_2 = i H_0^{(1)}(i\kappa_2 r) + \frac{H_1^{(1)}(i\kappa_2 r_s)}{i J_1(i\kappa_2 r_s)} J_0(i\kappa_2 r)$$
$$= i H_0^{(1)}(i\kappa_2 r) - 2 e^{-2\kappa_2 r} J_0(i\kappa_2 r). \tag{18b}$$

We assume again n_f in the form of (13). The condition that it have no singularity at $r = 0$ yields this time

$$b_f = -\frac{J_1(\bar{\kappa}_1 r_s)}{N_1(\bar{\kappa}_1 r_s)}; \tag{19}$$

and

$$n_t/A = Z_1 + \frac{\kappa_t^2}{\kappa_f^2} \frac{J_1(\bar{\kappa}_1 r_s)}{N_1(\bar{\kappa}_1 r_s)} Z_2. \tag{19a}$$

This vanishes at r_0 if

$$\ln \frac{2}{\gamma \bar{\kappa}_1 r_0} + \frac{\pi}{4} \frac{N_1(\bar{\kappa}_1 r_s)}{J_1(\bar{\kappa}_1 r_s)} - \frac{1}{2} \ln \frac{\kappa_2}{\bar{\kappa}_1} = 0. \tag{20}$$

The above equation gives the effective radius of the control rod which permits satisfying the boundary condition at the surface of the central portion of the pile with a Laplacian $\bar{\kappa}_1^2$. The corresponding $k = 1 + M^2 \bar{\kappa}_1^2$ gives the increase in multiplication constant necessitated by the insertion of the control rods.

In order to compare this with the change in multiplication constant given by the previous picture one can assume that $\bar{\kappa}_1 r_s$ is small which then leads to

$$-\ln \frac{2}{\gamma \bar{\kappa}_1 r_0} + \frac{1}{\bar{\kappa}_1^2 r_s^2} + \frac{1}{2} \ln \frac{\kappa_2}{\bar{\kappa}_1} = 0$$
$$\bar{\kappa}_1^2 = \frac{1}{r_s^2} \left[\ln \frac{2}{\gamma \bar{\kappa}_1 r_0} - \frac{1}{2} \ln \frac{\kappa_2}{\bar{\kappa}_1} \right]^{-1}. \tag{20a}$$

This change in the Laplacian corresponds to the change in the multiplication constant

$$\Delta k = \frac{2}{\kappa_t^2 r_s^2} \left(1 + \frac{\kappa_t^2}{\kappa_f^2} \right) \left[\ln \frac{2}{\gamma \bar{\kappa}_1 r_0} + \frac{\kappa_t^2}{\kappa_f^2} \ln \frac{2}{\gamma \kappa_2 r_0} \right]^{-1} . \tag{20 b}$$

In the same way, the simplified picture which considers only one type of neutrons leads to

$$\ln \frac{2}{\gamma \bar{\kappa}_1 r_0} + \frac{\pi}{2} \frac{N_1(\bar{\kappa}_1 r_s)}{J_1(\bar{\kappa}_1 r_s)} = 0 \tag{21}$$

which under the same assumption goes over into

$$\Delta k = \frac{2}{\kappa_t^2 r_s^2} \left(1 + \frac{\kappa_t^2}{\kappa_f^2} \right) \left[\ln \frac{2}{\gamma \bar{\kappa}_1 r_0} \right]^{-1} . \tag{21 a}$$

This last equation gives an overestimate for the efficiency of the control rod. The expression (20) must go over into (17) if $\kappa_f = 0$, into (21) if $\kappa_f = \infty$. This can be seen to hold for the approximate expressions (20 b), (17 a), (21 a) which permit an easier comparison. For finite κ_f, the present calculation [(20) or (20 b)] gives a lower Δk than either of the other methods and is considered to give the correct result.

Figure 2 gives again for $\kappa_t = \kappa_f$, $k - 1 = .05$ and $\kappa_t r_s = 3$ the reduction of the multiplication constant as calculated by the three formulae. These curves were calculated without the approximations which led to (17 a), (20 b), and (21 a).

It is worthwhile to remark that pile radius R and rod radius r_0 are "effective radii". They denote the radii at which the thermal neutron density, if extrapolated from the pile, would become zero. This is particularly important as far as the rod radius r_0 is concerned. As Placzek and Seidel have shown (MT-5) the thermal neutron density does not vanish at the surface of an absorbing rod but obeys the condition $n_t = .710\lambda \, dn_t/dr$ where λ is the mean free path. It should be remembered, however, that even with this correction, the whole treatment becomes invalid if r_0 is small as compared with the mean free path.

5. An experimental test of control rod theory is afforded by the recent West Stands experiments of Morrison and his group in which the change in ζ, the relaxation length, was measured when a 2.08 cm radius Cd covered rod was inserted along the axis of an exponential pile. The pile was square with 272.2 cm sides; this introduces a harmonic perturbation (since a single harmonic represents only a circular pile) which, however, we neglect in the following.

Since the effectiveness of a black rod varies only logarithmically with the radius of the rod, small errors in the observed relaxation distance would have a very great effect on the effective rod radius calculated from the observed change in ζ. It was therefore thought more advisable to assume the Placzek rod boundary conditions (used also in the design of the W safety rods) and, on the basis of the present theory, to calculate a ζ which could be compared directly with the experimental ζ. By this procedure we can arrive at an estimate of the adequacy of the W safety rod design, since the theory used for W is essentially identical to the present 2-group theory.

The theory of the exponential pile with a control rod follows very closely the theory of section 2 for a going pile. The fast and slow distributions are given by [cf. (7) and (13)]

$$n_t = [Z_1(\kappa_1 r) + b_t Z_2(\kappa_2 r)]e^{-z/\zeta}$$

$$n_f = \left[Z_1(\kappa_1 r) - b_t \frac{\kappa_f^2}{\kappa_t^2} Z_2(\kappa_2 r) \right] e^{-z/\zeta}$$

where the κ_1^2 of (10 a) is now replaced by

$$\kappa_1^2 = -\frac{k-1}{M^2} - \frac{1}{\zeta^2} \tag{22}$$

$(k-1)/M^2$ being the Laplacian before insertion of the rod. As boundary conditions on the rod surface, r_0, we require $dn_f/dr = 0$ and $1.917(dn_t/dr) = n_t$, the thermal neutron condition coming from Placzek and Seidel's .710λ extrapolation distance. Since we cannot satisfy the condition $n_t, n_f = 0$ on a square boundary with only one harmonic, we replace the square pile by a circular one which has the same transverse buckling.

Since the transverse buckling in a square pile of side W is $2\pi^2/W^2$ and of a circular pile is $(2.4048)^2/R^2$, we find for the circular pile equivalent to a square one of side 272.2 cm a radius $R = 147.3$ cm. It should be pointed out that this procedure is not completely unambiguous since an accurate solution of the problem requires an infinite series of Bessel functions of order 4ν; however, it seems reasonable that such a choice of R gives the best one harmonic representation of this sum. An estimate of the error made in neglecting the fourth harmonic is given at the end of this report.

The expressions for n_t and n_f involve one arbitrary constant, b_t, and a characteristic value κ_1^2 to be determined from the two boundary conditions on the rod. Substituting into the boundary conditions we obtain

$$\frac{N_0(\mu x)}{J_0(\mu x)} = \frac{N_0(\mu) + \Phi\mu N_1(\mu)}{J_0(\mu) + \Phi\mu J_1(\mu)} \tag{23}$$

where $\mu = \bar{\kappa}_1 r_0$, $x = R/r_0$ and

$$\Phi = \frac{1.917}{r_0}\left(1 + \frac{\kappa_t^2}{\kappa_f^2}\right) - \frac{\kappa_t^2}{\kappa_f^2}\frac{Z_2(\kappa_2 r_0)}{\kappa_2 r_0 Z_2'(\kappa_2 r_0)} = 4.875.$$

In evaluating Φ we have taken $\kappa_t^2/\kappa_f^2 = 1.365$; this is based on ·a thermal utilization of 0.9, a diffusion length of 50.7 cm, and an age of 350 cm². The lowest root of (23), which is easily obtained by plotting the left and right hand sides as functions of μ, is found to be $\mu = .03657$. This gives a $\kappa_1^2 = -309 \times 10^{-6}$ cm⁻², a ζ (equation 22) of 72.7 cm since the original no rod $(k-1)/M^2$ is 120×10^{-6} cm⁻², and an effective Laplacian $(= -(2\pi^2/W^2) + 1/\zeta^2)$ of -78×10^{-6} cm⁻². The observed ζ is 71.6 cm which corresponds to a Laplacian of -72×10^{-6} cm⁻². Thus the actual rod takes up 48×10^{-6} cm⁻², while the calculated rod takes up only 42×10^{-6} cm^{-2}. In other words, the effectiveness of a control rod is about 14% greater than calculated here. Since the Hanford safety rods were calculated on this same basis, one might conclude from the present experiment that the W safety system is overdesigned by 14%.

There are two factors, however, which make this statement somewhat misleading:

(1) The discrepancy of 6×10^{-6} cm^{-2} in the calculated and observed rod effect is well within experimental error since the results of *two* exponential experiments (with and without the rod) are needed in the comparison, and each experiment may be off by 5×10^{-6} cm^{-2}. Thus, if the no rod Laplacian were only 117×10^{-6} instead of 120×10^{-6} and the rod were observed to be 72.2 cm instead of 71.6, the agreement between theory and experiment would have been perfect.

(2) The resonance absorption in the rod is neglected in the theory. Since this is different for a Cd rod and a W boron steel rod, it is unsafe to assume that the W safety rods have been proved to be overdesigned by the present experiments.

6. To estimate the error made in the previous section when we replaced the square exponential pile by an equivalent circular pile we shall calculate the effect of the fourth harmonic correction for a pile which produces only thermal neutrons ($\kappa_f = \infty$). The thermal neutron density, n_t, in a square pile with a centered rod is a function of both r and the azimuthal angle θ; because of the 90° symmetry of the problem, n_t is a sum of cylindrical harmonics of order 4ν, ν being an integer. It is sufficiently accurate, however, to take only the zeroth and fourth harmonic; this allows us to satisfy the boundary condition $n_t = 0$ at four points along each side of the square, instead of only two points as is the case if only the zeroth harmonic is used. The curve along which the density vanishes will then approximate a square (see adjoining figure).

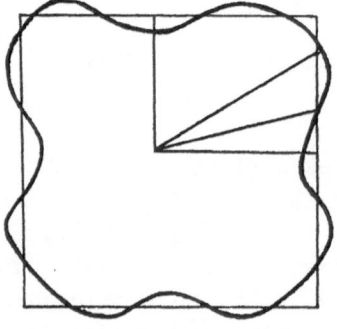

The thermal density, in this approximation, is

$$n_t = J_0(\bar{\kappa}_1 r) - \phi_0(\bar{\kappa}_1 r_0) N_0(\bar{\kappa}_1 r) + C[J_4(\bar{\kappa}_1 r) - \phi_4(\bar{\kappa}_1 r_0) N_4(\bar{\kappa}_1 r)] \cos 4\theta \quad (24)$$

where, in order to satisfy the Placzek boundary condition $\alpha (dn_t/dr) = n_t$ at $r = r_0$, we must set

$$\phi_i(\kappa_1 r_0) = \frac{J_i(\kappa_1 r_0) - \alpha \bar{\kappa}_1 J_i'(\bar{\kappa}_1 r_0)}{N_i(\bar{\kappa}_1 r_0) - \alpha \bar{\kappa}_1 N_i'(\bar{\kappa}_1 r_0)} .$$

Since (24) contains *two* parameters (the constant C and the characteristic number $\bar{\kappa}_1$), the condition $n_t = 0$ on the outer boundary can be satisfied at *two* points in each half side or 16 points on the whole boundary. We may space these 16 points at equal intervals along the sides of the square; then if (r_1, θ_1) and (r_2, θ_2) are the coordinates of the two points in a given half-quadrant at which n_t vanishes (see figure), we find for the characteristic equation which determines the value of $\bar{\kappa}_1$,

$$\frac{J_0(\bar{\kappa}_1 r_1) - \phi_0(\bar{\kappa}_1 r_0) N_0(\bar{\kappa}_1 r_1)}{J_4(\bar{\kappa}_1 r_1) - \phi_4(\bar{\kappa}_1 r_0) N_4(\bar{\kappa}_1 r_1)} \cos 4\theta_1 = \frac{J_0(\bar{\kappa}_1 r_2) - \phi_0(\bar{\kappa}_1 r_0) N_0(\bar{\kappa}_1 r_2)}{J_4(\bar{\kappa}_1 r_2) - \phi_4(\bar{\kappa}_1 r_0) N_4(\bar{\kappa}_1 r_2)} \cos 4\theta_2 .$$
$$(25)$$

The value of $\bar{\kappa}_1 r_0$ found from (25) is 0.03880 (assuming the geometry of Morrison's exponential pile); this may be compared with $\bar{\kappa}_1 r_0 = 0.03721$ calculated using one harmonic and an "equivalent" circular pile of the same area, and with $\bar{\kappa}_1 r_0 = .03883$ calculated as in section 5, by using one harmonic and an equivalent circular pile of the same radial buckling. It is seen that the pile with equivalent buckling is an exceedingly good approximation to the true square pile, the error in Laplacian being about 0.1%. We therefore conclude that the procedure used in section 5 for reducing the experimental observations is justified.

General Articles on Nuclear Energy and Energy Policy

Annotation by Alvin M. Weinberg

38. Wigner, E. P., "Atomic Energy", *Science* Vol. 108, Nov. 12, 1948.

Argues that nuclear energy may eventually have to compete with solar energy.

39. Wigner, E. P., "Impact of the Developments in Atomic Energy on the Sciences", *Bulletin of the Atomic Scientists 7*, 66 (1951).

40. Wigner, E. P., "Ziele und Probleme von Reaktoren", *Acta Physica Austriaca 11*, 410, 1958.

41. Wigner, E. P. and Weinberg, A. M., "Longer Range View of Nuclear Energy", *Bulletin of the Atomic Scientists 16*, 10, 1960.

42. E. P. Wigner, "Our Needs for Energy and Ways to Satisfy Them", *Revista Interamericana VI*, 485 (1976).

T. E. P. Wigner, "Roots of the Atomic Age", in *One World or None*, McGraw Hill, New York, 1946.

T. E. P. Wigner, Commentary: Atomic Energy *Research 1*, 577 (1948).

T. Farrington Daniels, M. K. Hubbert and E. P. Wigner, "Our Energy Resources", *Physics Today 2*, 19 (1949).

T. F. Seitz and E. P. Wigner "On the Geneva Conference, A Dissenting Opinion", *Bulletin of the Atomic Scientists*, January 1956, p. 23.

T. E. P. Wigner, "Review of Progress in Nuclear Energy", Vol. II *Phys. Today*, March 1957.

T. E. P. Wigner, "Review of Resonance Capture in Lumps", *Proceeding of Brookhaven Conference on Resonance Absorption in Nuclear Reactors*, p. 67, July 1957, Office of Tech. Services, Dept. of Commerce, Washington, D.C.

T. E. P. Wigner, "Il y a Vingt Ans: Le Premiere Reacteur en Chaine: Reflexion Sur Un Anniversaire", *Agence Int. de l'Energie Atomique Bulletin*, Dec. 2, 1962.

T. E. P. Wigner, "20th Birthday of Atomic Age", *NY Times Magazine*, Dec. 2, 1962, p. 34.

T. E. P. Wigner, "Nuclear Reactors Offer The Surest, Safest Way of Meeting Energy Needs", *U.S. Information Service*, 9/75 #3.

T. E. P. Wigner, "Mit Atom für den Frieden", *Budapester Rundschau*, Nov. 28, 1977.

T. E. P. Wigner, "Weighing our Energy Options", *Prism*, Jan. 1976, p. 51, Vol. VI, 485 (1976).

T. E. P. Wigner, "Nuclear Energy and Alternatives", in B. Kursunoglu, et. al. *Nuclear Energy and Alternatives*, pp. XXII–XXIII, Ballinger Co., Cambridge, MA, 1977.

T. E. P. Wigner, "Nuclear Reactor Operations", Letter to the Editor, *Science* **205** 148 (1979)

T. E. P. Wigner, "Nuclear Plant Safety", Letter to the Editor, *Morning Advocate*, Baton Rouge, LA, April 10, 1979.

T. E. P. Wigner, "The Beginning of the Atomic Age" (in Hungarian) Rutgers University Publication (Tanuk Korukol Series).

T. E. P. Wigner, "The Story of the 1st Chain Reaction", Address at the American Nuclear Society Meeting of November 16, 1982.

T. E. P. Wigner, "My Work for Nuclear Reactors", *Princeton Packet*, July 22, 1985.

T. E. P. Wigner and A. M. Weinberg, "New Light on the Einstein Letter", *The Oak Ridger*, March 11, 1986, p. 1.

T. E. P. Wigner, "Second Thoughts to the Problem of Nuclear Energy", *Science*, Nov. 1948, Vol. 108.

T. E. P. Wigner, "Why Reactor Development is Economical", Letter to the Editor, *Bull. At. Sci.*, Vol. XVII, 120, (1961).

T. E. P. Wigner, "Advantages of Reprocessing Nuclear Fuel vs. Proliferation". Address on June 26, 1980 Stanford U.

T. "Interview: Eugene Paul Wigner", *New Engineers* Vol. I, Number 2, Nov. (1971) p. 4.

T. E. P. Wigner, "Resonance Reactions", *Proc. Am. Phil. Soc.* Vol. 90, No. 1, P. 25, (1946).

T. E. P. Wigner, "Conference Summary and Overview", Lowell University, July 9, 1976.

T. E. P. Wigner, "Roundtable Discussion of the Significance and Accomplishments of the I.U.P.A.P. Conference", Washington, D.C., March 3–7, 1975.

38.

Atomic Energy

E. P. Wigner

Science *108*, 517–521 (1948)

THE YEARS THAT HAVE PASSED since the discovery of the nuclear chain reaction have not damped our high expectations in the future usefulness of atomic energy for peaceful pursuits. They have, furthermore, helped us to recognize, in addition to the size of this giant, his special skills. However, they have also helped us to realize, perhaps more clearly than we first did, that much hard and persevering work will be necessary before any of the benefits of atomic energy will be really ours.

During the period of abundance of the sources of energy which are now in use there will be two ways in which atomic energy can prove its significance. It may compete with our current sources of fossil energy and, second, it may open up new fields. As to the

Real success will therefore come to atomic energy in the near future only on the second path—by the discovery of new needs which atomic energy is able to satisfy better than existing sources can, by opening up new possibilities which it would be difficult or even impossible to realize with the sources of energy which are now in use. This task atomic energy has not yet achieved or even tackled. In fact, research on nuclear energy has to be so sheltered and separated from other industrial and economic problems that it will require extraordinarily keen vision to discover those needs which it is particularly suited to satisfy.

Some time hence, when the currently used sources of energy will near exhaustion, the situation will be different. Then nuclear energy may become the savior of our abundant life. But even then, nuclear

TABLE 1

Source	Coal	Oil	Atomic energy	Solar energy
Energy available in U.S.A. 10^{15} kcal	18,000	25 high-grade 300 low-grade	100 high-grade ores 3×10^{10} very low-grade ores and rocks	20,000 per year
Consumption per year	3.6	2.5 high-grade 0.5 low-grade	?	
Investment per/kw power plant	$110		$250	
Investment for producing 1 kjoule fuel/sec		$110 from high-grade $150 from low-grade sources		

former, our industrial and even our everyday life has adapted itself to the possibilities of chemical fuels to a degree of which we are rarely conscious. The transition to a new source of energy would involve a reorientation of many methods of manufacturing and also cause a shift in the character of many of the commodities and services which industry can make available. It will not be easy, therefore, for atomic energy to woo away very much territory from the chemical fuels in the near future. Even if it did, its success along this line would be quite comparable with the success of the turbine—which is great, but not decisive for the over-all economic or social life.

"Atomic Energy" was one of the addresses delivered at the Symposium on Sources of Energy, held in Washington, D. C., on September 15, during the Centennial Celebration of the AAAS.

energy will not be the only one in the field; it will have to compete at least with solar energy, of which there is an immense abundance.

Table 1,[1] which I am sure you have seen before in this or another form but the contents of which are well

[1] The following publications were used to obtain the figures of Table 1: (a) "Geochemische Verteilungsgesetze der Elemente," by V. M. Goldschmidt. Norske Videnskaps Akademi i Oslo, Mat. Naturv. Klasse, 1937; (b) "Power and Fuel Data," by Gale Young, December 1945 (unpublished); (c) "Nuclear Power," Scientific Information Transmitted to the United Nations Atomic Energy Commission by the United States Representative, Vol. IV, September 1946 (by C. A. Thomas, *et al.*). Also "Non-Military Uses of Atomic Energy," by C. A. Thomas. *Chem. eng. News*, 1946, **24**, 2480, and "Atomic Energy: Its Future in Power Production," by J. B. Condliffe, *et al. Chem. Eng.*, 1946, **53**, 125; (d) "The New Power," by Gale Young. Chap. 4 in *One world or none*. New York: McGraw-Hill, 1946; (e) "Natural Gas, Coal, Oil Shale as Sources of Liquid Fuels," by E. V. Murphree. *Oil and Gas J.*, April 1948.

kept in mind, illustrates this situation. It gives, for the different energy resources—coal, oil, atomic energy, and sunshine—the magnitude of the reserves and the yearly consumption. With respect to coal we have enough for 5,000 years at the present consumption. The situation with respect to oil is more precarious. As to atomic energy, you see that there is not too much of it in the form of high-grade ores. The supply in low-grade ores is practically inexhaustible. The magnitude of the solar energy is obviously great.

There are three points in Table 1 which I want to emphasize particularly. First, the total amount of coal under the ground in the United States has somewhat less heat content than the United States receives as sunshine during a single year. The over-all situation for the whole earth is even worse. Paradoxical as it may sound, the sunshine which falls on an acre of land during a single year would have, in the form of coal, a value of about $5,000. Second, if we look a little further ahead than a few hundred years, the chemical sources of energy are surely insufficient, and some of the new sources of energy will have to be utilized. Only two such sources are now known: nuclear energy from low-grade ores and solar energy. The question to which of these belongs the future will probably be decided by the relative convenience with which these two sources of power can be utilized and by the magnitude of the effort needed to exploit low-grade ores, on the one hand, and to concentrate solar energy, on the other. The last point which I wish to make is that oil or gasoline consumption is, in spite of the higher price of this fuel, almost as great as that of coal. This shows that the price of fuel is not always the decisive consideration; its adaptability and concentration are often more important.

The figures of our table clearly show that coal and oil cannot remain very long the predominant fuels. Nuclear energy may eventually replace them, but the above figures do not do more than to leave this possibility open. As for the present, a number of independent studies show, first, that nuclear energy is on the verge of competing with coal and, second, that a cheapening of power may have a stimulating influence on our economy, which could go far beyond the direct benefits calculable on a dollar-and-cent basis. The stimulating influence on more backward countries may be even greater.[2]

[2] Cf. in particular the Special Papers of the Cowles Commission, "Nuclear Fission as a Source of Power," by John R. Menke, and "Economic Aspects of Atomic Power," by Jacob Marschak, Sam H. Schurr, and Philip Sporn. Chicago: Univ. Chicago Press, 1947. Also, "Some Economic Implications of Atomic Energy," by Walter Isard. *Quart. J. Econ.*, 1948, **72**, 202. I am also personally indebted to Prof. Marschak, Dr. Schurr, and their collaborators for communicating to me a vast amount of unpublished material.

My personal impression would be that the emphasis on the stimulation of economic life is perhaps somewhat exaggerated. A similar and even more intense stimulation could be expected from the easier availability of many other types of goods—for instance, ingredients of housing. On the other hand, it seems to me that most price estimates disregard the ability of a stationary power plant, which uses the raw materials uranium and thorium not only to furnish heat and electricity but also to manufacture a pure fissionable material which is bound to occupy the role of a high-grade fuel (such as gasoline). The investment cost for nuclear energy, given in the last row, which militates so strongly against the economic attractiveness of nuclear energy, should be compared, therefore, not with the investment cost of a stationary power plant but with the joint investment costs of a power plant plus an oil refinery. This would improve considerably the economic attractiveness of atomic energy, while the first point I made would tend to decrease the importance of energy sources in general for our present economy. Perhaps even more important than these factors, which can be reduced to a dollar-and-cent basis, will be the relative convenience and safety with which the different types of plants can be operated. The full impact of the enormously dangerous radioactivity accompanying all nuclear energy operations is being felt increasingly, and the need of training a large number of people in new techniques involves an additional investment, the magnitude of which is difficult to estimate.

Let me now go over somewhat to the technical side, and, although this has been done on many occasions before, describe once more the broad features of the arrangements in which uranium can be used for the generation of energy.

Just as a single log cannot burn in our fireplace, in a similar way there is a minimum amount of uranium which is necessary to produce power. This minimum amount is called the critical amount. Once this critical amount is assembled in the so-called reactor space, it undergoes fission, and the energy of the fission fragments is converted into heat. This heat can be transferred by means of a heat transfer medium, which circulates through the reactor space, to a conventional heat engine.

Nothing could be simpler in principle than this, and there are only two problems which are not encountered in conventional engineering. These are the limitations of the heat transfer medium to substances which do not stop the chain reaction and the need to surround most of the equipment with a tight and thick shield. This shield has to protect the environment from the deadly radiation of the reactor and of the heat transfer medium which becomes radioactive within it.

The energy which can be liberated from uranium is about 3,000,000 times greater than that contained in the same mass of coal. The ratio is 10,000,000 if we add to the weight of coal the weight of oxygen which it needs for burning. This establishes the most im-

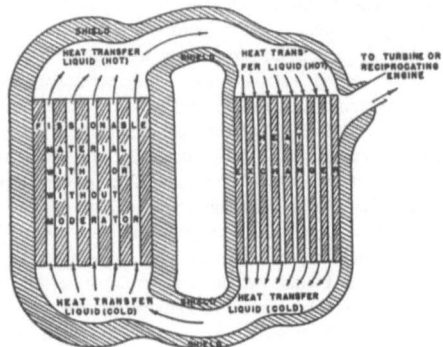

FIG. 1

portant characteristic of uranium as a fuel: it is practically weightless. This is, of course, not true of the whole power-generating equipment. In particular, the weight of the shield in many, if not most, cases overbalances the saving in fuel weight. This is particularly true in small engines and when refueling is easy. A serious disadvantage of the nuclear fuel is, furthermore, that any accident which breaks the shield is likely to liberate a vast amount of radioactivity and thus develop into a calamity much beyond the calamity which may result from an accident in the operation of the conventional sources of power.

Primarily, nuclear energy appears as the kinetic energy of fission fragments. The velocity of these corresponds to a temperature of about 600,000,000,-000° C, and one feels that it is a pity to degrade this high temperature to a pittance of a couple of thousand degrees. For this reason, a good deal of thought has been spent on methods for a direct utilization of the energy of fission. Electric, electromagnetic, thermoelectric, and chemical methods have been discussed in some detail.[3] To date, none of these methods has proved attractive, and it is at least temporarily conceded that the fission energy will have to be converted into heat at a tractable temperature before it is further utilized. For land-based power plants, in which the rejected heat can be easily discarded at a few hun-

[3] Much of the material referred to remains unpublished. Cf., however, Marschak, Schurr, and Sporn, footnote 2.

dred degrees, this is not a major disadvantage, since the efficiency in this case is already close to its optimal value if the prime heat is delivered above 1,000°. However, the need for converting the energy of the fission fragments into heat becomes more of a drawback if one tries to exploit the most outstanding feature of nuclear energy—its enormous concentration.

Figs. 1 and 2 show the by now conventional arrangements to generate power and thus illustrate what I have previously called the competitive uses of nuclear energy. In the arrangement of Fig. 1 the heat transfer medium first traverses the fissionable material through a number of channels, gathering up the heat generated, and then flows to a heat exchanger. In this heat exchanger the heat of the primary coolant is transferred to another medium which, in its turn drives a turbine or a reciprocating engine. In the arrangement of Fig. 2 the primary coolant drives the turbine directly. This arrangement has fewer parts, but a larger shield than the former, and a turbine which is,

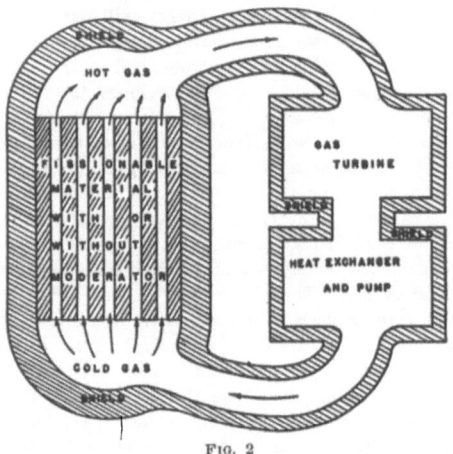

FIG. 2

because of the radioactivity of the primary coolant, inaccessible. It is not yet possible to say with certainty which of the two arrangements is more advantageous and under what conditions.

The time scale for the development of nuclear energy on a substantial scale naturally comes up at this point, but it is a question most difficult to answer. Our uncertainty concerning this point not only has its origin in our inability to answer several technical and scientific questions, but is caused, to an equal degree, by the circumstance that the answer is bound to depend on the strength of our desire to see nuclear energy prove itself soon, on our courage, and on our confidence in our technical judgment and foresight.

In other words, the human element strongly enters the picture.

Disregarding this human element, M. H. L. Pryce gave a tentative answer in a most thoughtful article in a recent issue of the *Bulletin of Atomic Scientists* (1948, **4**, 245). He estimates that nuclear energy may begin to replace coal in about 30 years. The number 30 is uncertain, but it is not likely to be less than 5 or more than a few hundred.

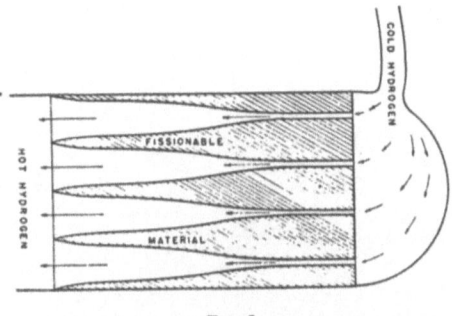

Let me now go over to the more speculative uses of nuclear energy. The high concentration of nuclear energy would seem to make it the ideal fuel for providing power for transportation. As long as one considers the most conventional types of transportation— land and sea routes—the rejection of part of the energy still remains a subordinate difficulty, and it is, in fact, in powering ships, in which the problem of radioactivity can be mastered more easily, that the first application of nuclear energy may come. In long-range aircraft, flying at high altitudes, the rejection of the waste heat is already much more difficult, unless one is willing to take higher temperatures of rejection into the bargain and thus reduce thermodynamic efficiency. If one considers, finally, travel outside the gravitational sphere of the earth, the problem of the rejection of waste heat becomes dominant.

In order to escape the gravitational field of the earth, one needs about 15,000 kcal/kg of escaping material. Since the energy content of a fissionable material is more than 1,000,000 times greater than this, the energy requirement is not, in itself, prohibitive even if one assumes a relatively low efficiency, η, for the process which furnishes the needed energy. However, for an efficiency, η, the waste heat amounts to 15,000 $(1-\eta)/\eta$ kcal/kg, and unless one can dispose of this, it will surely vaporize the body of the ship. As we discussed it before, the problem of elimination of the waste heat can easily be solved on the sea; it can also be solved in the air, but if the ship

is to have power also outside the atmosphere, it can keep cool only either by throwing off hot parts or by radiation. The first alternative is the one which is discussed most commonly,[4] but it has its definite limitations. Current opinion is that it may be barely sufficient to achieve the purpose: to raise a rocket off our planet. What runs out first is, characteristically, not the energy of the uranium but the hydrogen.

The second alternative, discarding the waste heat by radiation, also has clear limitations. The efficiency decreases very strongly if the time of ascent is much more than 1,000 sec. Taking this into account, one finds that, for a radiating temperature of 200° C, a radiating area of about 20$(1-\eta)/\eta$ m²/kg of the vessel is needed—a practical impossibility. For a radiating temperature of 1,000° C the radiating area becomes more manageable: about 0.4$(1-\eta)/\eta$ m². At this temperature of the radiator, however, the thermodynamic efficiency is necessarily rather low in any conventional heat engine. This example shows again how problems of an apparently secondary nature can push themselves in a most disappointing fashion into the center of the picture.

Breaking the gravitational prison of the earth is so challenging a problem that I wanted to say a few words about it, even though it would be clearly premature to discuss it in detail. Furthermore, it is not the direction in which nuclear energy has so far proved itself most decisively. That field is indeed an application of nuclear energy in which a new need has been discovered. It is the procurement of research facilities for biology, chemistry, and physics by radioactive tracers, by new and more intense types of radiation. Even though this subject is the last one on my list, it is at present the most important one, and it is quite possible that it will maintain this position for a long time. The subject, which has received adequate treatment on several occasions,[5] lies outside the scope of our symposium. If we could divest ourselves from our admiration of the spectacular, we might easily find that the nuclear research facilities are for the present more important than nuclear energy. The success of the research which they support is a more real and more truly human need than is the need for additional energy and power.

However, there is good reason to look forward with confidence also to the more direct applications of nuclear energy. In order to be fully successful, these

[4] Cf. e.g. "Atomic Power for Airplanes and Rockets" article in the March 1947 issue of Atomic Information based on L. Alvarez's address.

[5] See, for example, (a) *Radioactive tracers in biology*, by M. D. Kamen. New York: Academic Press, 1947; (b) *The use of isotopes in medicine and biology* (Symposium Report). Madison: Univ. Wisconsin Press, 1948; (c) various articles in *Nucleonics*, 1948.

applications will require more of the undeviating interest which is so necessary for technical success but not enough of which they have received so far. They will surely receive this interest in the future, and we may hope that they will receive it from us—not only from our neighbors and children. And we may even dare to hope that the success may be so overwhelming that the first application of nuclear energy will appear just as insignificant, in comparison, as the first and still most efficient heat engine, the cannon, is in comparison with our generators of electricity and industrial power.

39.

Impact of the Developments in Atomic Energy on the Sciences

E. P. Wigner

Bulletin of the Atomic Scientists *VII*, 66–70 (1951)

The author of the following article is professor of mathematical physics at Princeton University. He has contributed greatly to the development of uranium reactors, being directly involved in the design of the reactors at Hanford. Dr. Wigner recently received the Franklin Medal of the Franklin Institute for his work in atomic physics.

THE importance of the unhindered and vigorous development of science for technical progress has often been discussed. I wish to deal, on this occasion, with the reverse side of the picture, with the stimulating influence which so many branches of science received from technological accomplishments in the field of atomic energy.

The time to bring up this subject is a particularly propitious one. On December 4, the Atomic Energy Commission declassified the last pieces of scientific information which will enable "any competent group of scientists or engineers" to proceed with the design and construction of nuclear reactors. Few still entertain the enthusiastic attitude of the early days of the Uranium Project, when it was widely believed that the release of this information would be the day for scores of privately owned atomic energy installations to sprout throughout the country. Our nation is too much occupied with tasks of more immediate urgency to permit this and, furthermore, many who were looking forward to such activity have long ago given up waiting and have turned their attention to other matters. In addition, it is to be feared that the astronomical cost of the installations of the Atomic Energy Commission are an effective deterrent. Yet this need not be so. We must realize that a very large part of the high cost of nuclear reactors has been due to the secrecy which had to surround their construction and which, in some cases, necessitated the building of whole new villages to house the builders and operators of the planned reactor. In addition, the emphasis on secrecy and certain other considerations have made the administration of the work on nuclear reactors so difficult that economic considerations had often to be sacrificed. This point is not brought up here as a criticism of past procedures, but as a circumstance which should give the recent release of hitherto secret information more significance than may meet the eye. While it cannot be expected that nuclear reactors will henceforth be standard equipment for our universities and industrial laboratories, some work not wholly financed by the government will begin in the field, and one project has, in fact, already been started. In the absence of secrecy, this work may proceed with greater emphasis on economic considerations and exert an invigorating influence on all atomic energy work in the country. In other words, one may hope that the recent release of information, and further releases to come, will create a more general interest in reactors and their accessories. This broader interest

Based on an address given at the annual meeting of the American Association for the Advancement of Science at Cleveland, Ohio, on December 28.

may stimulate a competition in ideas and may stimulate the desire to do better than is done now—the desire which is the basis of American efficiency. It can be hoped, finally, that now that the last element of mystery is removed from them, nuclear reactors, instead of being something to be shied away from, will become one of the instruments to which a scientist is ready to resort in case of need, and will thus increasingly fulfil their function as tools, and sources of tools, for a great variety of scientific and technical investigations.

This brings us back to our subject proper—the effects of the developments in atomic energy on the sciences. These effects are remarkable not so much because of their spectacular nature as because of their great variety, their extension to so many fields. One cannot point to any single discovery made in the course of atomic energy development, which had as profound an influence on the fundamental concepts of physics as the "Lamb shift"[1]—a discovery made possible by wartime research on radar; but, on the other hand, there are so many fields in which progress was greatly enhanced by the developments in the atomic energy field that it would be impossible to deal with all of them in a short article.

I will attempt to review some of the influence of the developments in nuclear energy on the sciences under four headings: provision of research tools, progress of nuclear physics itself, stimulation of interest in new problems in other fields, and, last, but not least, the effect on the thinking of so many of us on certain fundamental questions of science.

TOOLS FOR RESEARCH

The most important research tools with which large-scale nuclear energy

has provided us are unquestionably the radioisotopes. The radioisotope program of the Atomic Energy Commission, originated by a few chemists and biochemists of vision in Oak Ridge, under the leadership of W. E. Cohn, became, under able administration, an institution of more than national importance. It would be quite impossible to enumerate all the research problems in which radioisotopes have been successfully employed or even to mention all the significant scientific results to which they have contributed. Fortunately, this is not necessary because the facts are widely known and there are excellent and comprehensive publications on the subject. Let me simply mention the fact that since the inception of the radioisotope program, in the summer of 1946, the number of shipments of tracer materials to installations outside the Atomic Energy Project has reached almost 20,-000 and that the number of investigations in which these tracers have been used runs into the thousands.

The radioisotope program has been supplemented lately by the stable isotope program and by the facilities for neutron irradiation either in one of the reactors or, by means of a photoneutron source,[2] in any properly equipped laboratory.

A very interesting, although not at all typical example of the use of stable isotopes is the investigation, carried out in Los Alamos, of the cause of the "superfluidity"[3] of liquid "helium 2." The London brothers, and other physicists interpreted this phenomenon as the "Bose condensation" of helium. Ordinarily, gases condense at low temperatures because their molecules attract each other. However, as was first shown by Einstein, a gas obeying Bose

---

[1] Small difference between two sets of levels in the spectrum of the hydrogen atom, whose coincidence was required by the previously accepted theory.

[2] Photoneutrons are neutrons produced by some nuclei during or after irradiation with γ-rays.

[3] Liquid helium exists in two forms—one with the properties of an ordinary liquid, and another "superfluid" one, which shows no cohesion at all.

statistics,[4] such as He 4, should condense even if there were no attraction between its molecules. According to the Londons, this phenomenon must play an important if not decisive role in the conversion of ordinary helium into the superfluid liquid. The London hypothesis could be confirmed by the liquefication of He 3, which is an isotope of ordinary helium. Having an odd mass, He 3 can be expected to obey the Fermi rather than the Bose statistics. The observation that He 3 could not be brought into the superfluid state gave a striking confirmation of the London theory. Further investigations on the properties of mixtures of the two isotopes can be expected to illuminate the mechanism of the phenomenon more closely. Incidentally, this work illustrates our Atomic Energy Commission's interest in pure science. Not only was the necessary stable isotope, He 3, furnished by the commission, but the whole investigation was carried out in one of the Atomic Energy Commission's laboratories.

I may mention in passing another important discovery in the low temperature field which was made possible by the stable isotope program: the isotope effect on the temperature at which certain metals become "superconducting."[5] This discovery reveals quite clearly that this long misunderstood phenomenon is caused by the interaction between the motion of the electrons that carry the electric current, and the vibrations of the crystal lattice of the metal.

As a possible application of irradiation techniques, one may mention the determination of certain impurities in almost unbelievably small quantities. It should be possible, for instance, to

determine in this way, with an accuracy of ± 1 per cent in a sample of only a few grams, arsenic in amounts as low as one part in 10^{10}. For most other elements, the accuracy is less spectacular, but still very high, and this includes some cases in which ordinary chemical assay is very difficult. All that is necessary to carry out the analysis is to irradiate the sample by placing it into the reactor for a few hours, remove and dissolve it, add a small amount of a solution containing inactive arsenic as a "carrier," and to precipitate the latter chemically free, as far as possible, from impurities. The active arsenic originally present in the sample, no matter how small its amount, will go into the precipitate along with the inactive arsenic. If the precipitation reaction is suitably chosen, of all the radioactive elements present in the original sample as a result of the irradiation, arsenic alone will be carried into the precipitate. Thus, the radioactivity of the precipitated arsenic will be a direct measure of the amount of this contaminant in the original sample. Unfortunately, neither hydrogen nor oxygen can be determined in solid samples by the irradiation technique. For the determination and tracing of these—which are vital elements in so many applications —one still has to resort to stable isotopes or, in the case of hydrogen, to the new artificial radioactive isotope, tritium.

The production and the preparation for more ready use of the radioisotopes and stable isotopes, along with counters, scalers, etc., has become a veritable industry serving research and development work; several companies specializing in these activities. Instruction in the use of these tools is one of the principal activities of the Oak Ridge Institute of Nuclear Studies.

A very different type of research instrument are the powerful neutron beams which emerge from the shields of reactors wherever one opens a window. These, of course, cannot be separated from the reactors proper. They can be used for most of the purposes for which X-rays are more commonly

[4] The kind of statistical treatment to be used, according to theoretical physics, for nuclei consisting of an even number of elementary particles, such as He 4. Fermi statistics is the kind applicable to single protons, neutrons or electrons and nuclei containing an odd number of these particles.

[5] That means, lose their resistance to the passage of an electric current.

used, and for a few others. Their main
advantages, as compared with X-rays,
stem from the fact that the scattering
of neutrons is produced by atomic
nuclei, while that of X-rays is caused
by the electronic cloud surrounding
the nuclei in the atoms. As a result, the
neutron-scattering properties of the
crystal are almost exactly those of a
geometrical point lattice. This permits,
in principle, a much more accurate interpretation of the diffraction pattern
produced by the scattering of neutrons
than is possible for X-rays since, in the
latter case, the scattering power is described by the somewhat nebulous concept of "effective electron density."
Furthermore, the scattering power of
various nuclei differs much less than
that of their electronic clouds. It is thus
possible, for instance, as Shull and
Wollan have shown, to determine, from
neutron scattering, the position of hydrogen atoms in a crystal lattice—a task
which X-ray crystal analysis was unable
to accomplish. It would be impossible
in a short article to describe the results
of all the interesting experiments which
have been carried out with neutron
beams and which have greatly enriched
our knowledge of crystals. In spite of
repeated atempts, even a qualitative
demonstration of neutron diffraction
was never fully successful before reactors became available as neutron
sources. With these new sources, barely
seven years old, neutron diffraction is
already being used to solve intricate
problems of crystal physics, and its
use is sure to grow as reactors producing more powerful neutron beams become available to overcome the still
troublesome problem of insufficient intensity.

PROGRESS IN NUCLEAR PHYSICS

It certainly is not surprising that
work in nuclear physics received a
strong impetus from the development
of atomic energy, and the reviewer's
task in this case must consist of selecting a few of the most important
items for discussion. Not all the effects
of the discovery of atomic energy

were favorable for the development
of nuclear physics: the secrecy which
had to surround some of the results
created some deep gaps in the otherwise coherent structure of our knowledge in the field.

The most important development
which nuclear physics owes to the
atomic energy program is, in my opinion, the vast amount of numerical data
which has been accumulated in the last
decade. As compared with our present
knowledge of nuclear events, the information available ten years ago consisted of a few scattered data. Bethe
and his collaborators, in a well-known
series of articles in the *Reviews of
Modern Physics*, not only summarized
all the accepted theories up to 1937,
but also gave a comprehensive and almost complete review of the then
available experimental information.
Since then, stimulated by the need for
materials with specific nuclear characteristics, we have accumulated a great
deal of knowledge on radioactivities,
excited states of nuclei, their cross-
sections, and so on, and the new data
are much more reliable and accurate.
No well-measured nuclear absorption
curve for neutrons in the neighborhood
of a nuclear resonance was known before the war; we now have enough to
make and verify statistical statements
on the density, width, and other characteristics of nuclear resonance levels.
The same holds for excited states. It
may be true that very few of these data,
taken by themselves, are of fundamental significance; but their accumulation
has already served to clarify our picture
of nuclear structure in at least one
fundamental respect: by adducing evidence for the nuclear shell model. Such
models had already been proposed before the war; in fact, the "magic" nature of the nucleon numbers 20, 50, 82,
etc., had been noticed long ago by V.
M. Goldschmidt.[6] However, the scarc-

[6] "Magic numbers" are numbers of protons or neutrons at which the stability of
the nuclei changes suddenly. In analogy to
Bohr's theory of the periodic table of elements, these numbers are supposed to cor-

ity of experimental information did not permit the collection of enough evidence to prove these theories. They remained interesting, but not adequately supported surmises, until a few years ago Mrs. Mayer again brought the idea up, at a time when experimental evidence in its favor piled up almost spontaneously. Her discovery was connected with the development of nuclear energy not only, and perhaps not mainly, because Mrs. Mayer worked in an atomic energy laboratory, but because the material on which she based her conclusions would not have been collected without the practical success of the atomic energy project.

Turning now to a somewhat different example, let us look at the nuclear data just released by the Atomic Energy Commission. The very first of these tells us that for U 235 the usual n-γ reaction, the capture of a slow neutron, is only six times less probable than fission. Since the γ widths in heavy odd elements are known to be smaller than one-tenth of a volt, and since the six times larger probability of fission than of capture indicates that the fission width is only six times greater than the γ width, this must be under 1 eV. The original theoretical estimates had been many times greater. The third item of the same release gives the number of neutrons released per fission as 2.5. Before the war the very existence of fission neutrons (apart from the delayed ones) was doubtful. Examples of this kind, which this single release would permit us to multiply, show not only how important theories have to be adjusted constantly to the growing body of experimental information, they also show how conceptual progress can be limited by the requirements of secrecy. It is certain that, had it been possible to release the above-quoted information at an earlier date, the theory of

respond to the completion of "closed shells" of protons or neutrons in the nucleus; the beginning of a new shell is marked by relative instability of the nuclei which follow the "magic ones."

fission would have been by now more closely adjusted to the experimental facts.

Before leaving the field of nuclear physics, let me mention two important experiments which would not have been possible without operating reactors. The first of these is Wollan's determination of an upper limit to the collision cross-section between neutrinos[7] and heavy nuclei.

Very little information was available on this question before the war—quite naturally so since there were no neutrino sources to speak of. On the other hand, a reactor with only 1,000 kilowatt heat production—a very small reactor indeed—emits a constant stream of 50 kw, that is almost 70 h.p. of neutrinos! There was some apprehension, at one time, that these neutrinos might constitute a danger to the operating personnel. Wollan's measurement, which gave us an upper limit 2×10^{-30} cm^2 for the collison cross-section between a neutrino and a proton, in addition to being of considerable scientific interest, dispelled these fears by showing that even a very intense beam of neutrinos will pass through the human body without scoring a dangerous number of "hits" on the atoms in it.

The other experiment which I wish to mention is the proof, first furnished by Snell in Oak Ridge, of the instability of the neutron and its decay into a proton and an electron. This was probably the first case in which spontaneous disintegration of what is usually called an "elementary particle," i.e., the fact that it disintegrates even if under no influence from other particles, was proved experimentally. This goes far toward proving the complex nature of the concept of "elementary particle." Unfortunately, neither Snell nor

[7] Neutrinos are extremely elusive particles with a very small mass and no electric charge, the existence of which had been postulated to permit the laws of conservation of energy and momentum to remain valid in such phenomena as emission of beta particles by radioactive atoms.

Robson have been successful so far in establishing accurately the average life of the neutron, even though Robson worked with the Canadian pile which has the highest neutron density. This enabled him, however, to measure the energy spectrum of the electrons released by disintegrating neutrons.

STIMULATION OF OTHER FIELDS

Let me begin my review here with a story, even though it is probably apocryphal. Dr. Compton, the leader of the plutonium project, discussed the problems and difficulties of his project with another leader, himself one of the country's most distinguished chemists. "Even if you get the chain reaction established, there still remains the big problem of the chemical separation of plutonium on an industrial scale," remarked the latter. "Seaborg thinks he will have a process worked out in two years," countered Compton. "He is good, but not that good," was the answer. He was, however; in fact, he beat the deadline. His and his colleagues' work lead to a revival of inorganic chemistry—a reversal of the conviction so prevalent with many before the war, that problems of inorganic chemistry were uninteresting because they were all already "solved in principle." They may have been solved in principle, but no one was able to apply those principles in all the important cases, and the exploration of many a neglected corner of inorganic chemistry proved richly rewarding for the explorer. Thus, inorganic chemistry again became an experimental discipline, much of it in the old spirit, though not unmindful of the great accomplishments of the thermodynamic studies of the last half-century. The work of those of Seaborg's colleagues whose ideas have not been used in the process actually employed for plutonium separation contributed equally to the rebirth of some branches of chemistry. One of these processes, in particular, led to the development of a successful method for the separation of the rare earth metals from each other, an objective which was never before fully attained, and to the

separation of nucleic acids which heralds new progress in biology. This process is the ion exchange process, so successfully employed by Boyd, Cohn, Spedding,[8] and others.

Leaving aside the work on the effects of radiation on chemicals and on living organisms—fields in which progress was to be expected—we turn to a subject somewhat less obviously connected with atomic energy, but which has also greatly benefited from it. Problems of metallurgy and questions concerning solids arose because most ordinary structural materials absorb too many neutrons to be used in a reactor.

The requirement of low neutron absorption cross-section excludes all but a handful of elements from use in reactors. In addition, reactor materials must have high thermal conductivities and good ductility. Otherwise, they will rupture as a result of the intense heat flow produced in a reactor operating at a high energy level. Finally, the characteristics of reactor materials must be affected as little as possible by the intense radiation to which they are exposed. Metallurgy has to solve a very difficult problem indeed.

As a by-product, the investigation of the effect of radiation on solids has already yielded much interesting information about the solid state. The effect is twofold: the atoms which scatter the neutron elastically are thrown out of their regular positions in the lattice, leave a vacant position behind, and either get stuck in the lattice as "interstitial" atoms or wander off to the

[8] See F. H. Spedding, "Chemical Aspects of the Atomic Energy Problem," *Bulletin*, V (February 1949), 48–50.

surface. The ratio of the two gives direct information on the facility with which atoms can move about in the crystal lattice. Even more important is the possibility of artificially creating a controlled number of vacancies and interstitial atoms in a crystal lattice and of studying their physical and chemical behavior as function of temperature. These studies opened up a new experimental technique which is as yet hardly exploited.

The second type of radiation-induced change in solids is caused by the nuclei which absorb a neutron or undergo some other nuclear reaction. As a result of these processes a new element replaces the old one in the lattice; an artificial "impurity" has been created. Such impurities are responsible for many interesting and important properties of solids. For example, the conductivity of germanium, the miracle element of transistors, is governed by the type of impurity present in it, and Lark-Horovitz and his collaborators, at Purdue and Oak Ridge, were able to change the type of electric conductivity shown by this material by irradiating it. This is one of two ways in which reactors helped transistor research; the other is due to the fact that without the radioactive tracer technique, the measurement of "natural" impurities responsible for the transistor action could hardly have been carried out with the required precision.

Before leaving this subject—the stimulation of various sciences by the problems and facilities created by reactors—I should apologize for not having more than mentioned the ways in which reactors have influenced the life sciences. Space limitations naturally imposed on an article of this type provide a welcome excuse to restrict myself to subjects of which I have some comprehension.

PHILOSOPHY OF THE SCIENTIST

We are still too close to the changes which the Uranium Project has evoked in our thinking to assess them in any but a most tentative fashion. The two most obvious general effects of the advent of atomic energy on the scientist's way of life are clear enough: he has more financial assistance in his scientific endeavors than he ever had before, perhaps more than is good for him, and he has to take himself more seriously than he had before, perhaps more seriously than is good for him. These changes are so obvious that they are hardly worth a comment. However, the scientist's experiences while working on the atomic energy program also had effects on his general evaluation of the role and essence of science which are more subtle.

The work on atomic energy gave those who participated in it a very deep sense of the power and scope of our science. The experience is, in a sense, similar to that of Newton's lad picking up pebbles on the seashore, except that we recognize so many of the pebbles. The concerted efforts of large groups of scientists enabled them to foresee essentially all the behavior and all the complications in the functioning of two such novel types of machines as the nuclear reactor and the fission bomb. Even some of the most courageous and most far-sighted scientists sincerely doubted that this would be possible; yet it came to pass. This gave us a very deep conviction of the magnitude of our knowledge.

While the experience just described might have given us an exaggerated sense of the power of our knowledge, the realization of the many parts of the knowledge which went into that foresight, and of the very small fraction of that knowledge which anyone has mastered individually, restored some of our humbleness—even made some of us despair. The reaction of the scientists to these impacts covered a wide spectrum. On one side of the spectrum was the trend toward the attitude that it is hopeless and purposeless to be physicists in the sense of knowing physics, that all we can hope to accomplish is to be specialized researchers with interest in a very narrow field. There was a tendency, soon after the termination of the hostilities, to make this field so narrow that even nuclear physics was excluded from it—a tendency which fortunately has not prevailed. However, the opinion was not uncommon that as large parts of physics as possible should be forgotten, that we should have as little as possible to do with anything but research on the questions which appear most fundamental and that, in particular, all should be scorned that smacked of engineering or applied science.

The trend on the other end of the spectrum was set by those who were deeply impressed by the multitude of the branches of science which they had seen to function and by the realization that it would always remain impossible for them to master more than an infinitesimal part of even their own discipline. Nevertheless, they felt it to be a worth-while purpose to maintain an acquaintance and an interest in as large a part of their discipline as they could, the parts of particular interest being determined, to some degree, by their usefulness. They remembered that the work of many scientists, from Lavoisier to Einstein, has maintained at least some contact with practical life.

It is possible that the increasing growth of science and the consequent need for specialization will not permit the ultimate survival of the trend at this latter end of the spectrum. That followers of it will be badly needed in the near future seems unquestionable.

40.

Ziele und Probleme von Reaktoren

E. P. Wigner

Acta Physica Austriaca *11*, 410–421 (1958)

Zusammenfassung

Reaktoren haben in der Vergangenheit hauptsächlich Hilfsmittel für Forschungsarbeiten geliefert. In der Zukunft erwarten wir, daß sie die fossilen Energiequellen ersetzen. Die künftige Notwendigkeit von neuen Energiequellen wird dargelegt und die Rolle der Kernenergie in diesem Zusammenhang untersucht. Der Bedarf für neue Energiequellen ist in Europa dringender als in den Vereinigten Staaten. Dies rechtfertigt das Interesse europäischer Forscher in fast unmittelbar realisierbaren Reaktoren und das Vorwiegen des amerikanischen Interesses an dem mehr fortgeschrittenen Typus von Brütern.

Reaktoren sind Gebilde, in denen die Uranspaltungsreaktion auf großem Maßstabe vor sich geht. Die erste Frage, die sich dem Laien in diesem Zusammenhange aufdrängt ist „wozu sind solche Reaktoren gut" und es ist diese Frage, die ich zuerst diskutieren will.

Außer ihren militärischen Anwendungen haben uns die Reaktoren in der Vergangenheit hauptsächlich in unserer Forschungsarbeit unterstützt. In der Zukunft erhoffen wir von den Reaktoren eine neue Energiequelle. Ich hoffe und glaube aber, daß wir die Rolle des Reaktors als Forschungsinstrument auch weiter im Auge behalten werden. In einem gewissen Sinne ist die Unterstützung der Forschung eine viel höhere Funktion als die Energieerzeugung; sie hilft uns in unserem höchsten Bestreben, unsere Umwelt, unsere Nachbarn und auch uns selber besser zu verstehen. Eine neue und ergiebigere Energiequelle dient nur unseren irdischen Bedürfnissen und unserer Bequemlichkeit.

Wenn man über die Hilfe spricht, die Reaktoren der Forschungstätigkeit angedeihen ließen, so denkt man in erster Linie an radioaktive Indikatoren. Diese werden in fast jedem Forschungsgebiet heute so allgemein verwendet, sind so leicht zu verschaffen, daß wir geneigt sind zu vergessen, daß sie in erheblichem Umfange erst seit ungefähr zehn Jahren zugänglich sind. Ich habe einmal im Physical Review nachgezählt und gefunden, daß in etwa jeder dritten experimentellen Arbeit radioaktive Indikatoren eine Rolle spielten. Vielleicht erklärt diese weite Verwendung der Indikatoren, ihre große Zugänglichkeit, daß wir nicht mehr über sie reden. Wir reden ebensowenig darüber, daß die Luft, die wir zum Atmen und für unsere Verbrennungsmaschinen brauchen, auch immer und frei vorhanden ist.

Die Indikatoren sind aber nicht das einzige Forschungsmittel womit uns die Reaktoren bereichert haben; sie stellen auch Strahlungsquellen zu unserer Verfügung, sowohl für β- und γ-Strahlen, wie auch für Neutronen. Diese letzteren haben die Erforschung der Struktur mehrerer Festkörper gefördert, teils durch Interferenzexperimente, teils durch die Erzeugung von Fehlstellen. Die Erforschung der durch Neutronenstrahlung geschaffenen Veränderungen im Festkörper beschäftigt viele unserer besten Forscher und Laboratorien, auch der Vorsitzende Ihrer Gesellschaft ist daran tief interessiert. Da ist ein neues Wissensgebiet entstanden als Nebenprodukt der Reaktorforschung.

In den Beispielen, die bisher genannt worden sind, wurde der Reaktor nur als Hilfsmittel verwendet. In einer anderen Reihe von Verwendungen ist der Reaktor das Versuchsobjekt selber. Ich denke an die Messung von Wirkungsquerschnitten, an die Eigenschaften des Neutrons selber, seine Spaltung und Lebensdauer, an die Versuche, in denen die Intensität der Kernreaktion in sowohl zeitliche, wie auch räumliche Schwingungen versetzt wird.

Die Rolle, die Reaktoren in unserer Forschungsarbeit spielen, steht mir sehr nahe und ich könnte viele Stunden darüber reden. Der Mensch ist aber so beschaffen, daß ihn die Zukunft mehr als die Gegenwart interessiert. Deshalb will ich zu einem anderen Punkt übergehen und über den Reaktor als Energiequelle sprechen. Es sind da zwei Fragen, die mir besonders wichtig erscheinen: wie notwendig ist eine neue Energiequelle für unsere Zivilisation und unser Weiterkommen und inwiefern können wir hoffen, daß Kernreaktionen unsere Bedürfnisse in dieser Hinsicht befriedigen können? Die zweite Frage bezieht sich auf den Unterschied der Wege, die die verschiedenen Reaktorprogramme gegangen sind, die Ursachen und Berechtigung für diese Unterschiede und schließlich die Probleme und Hindernisse des amerikanischen Programmes.

Als ich mir die Statistiken zum letztenmal angesehen habe, haben die Energiequellen in den Vereinigten Staaten 3% des Nationaleinkommens in Anspruch genommen. Dies bedeutet, daß wir, wenn die Energie frei wäre wie die Luft, unsere anderen Ausgaben um 3% erhöhen könnten. Es ist klar, daß Reaktoren als Energiequellen kein goldenes Zeitalter erzeugen werden. Wenn sie überhaupt ein goldenes Zeitalter erzeugen, so werden sie das als Forschungsmittel tun und selbst das ist fragwürdig, da sie bisher die Psychologie nur sehr indirekt beeinflußt haben.

Was ich eben sagte ist gleichbedeutend damit, daß die Energie, wenigstens in den Vereinigten Staaten, jetzt fast so frei ist wie die Luft. Ist das überall so? Wird das auch in den Vereinigten Staaten immer so bleiben? Die Antwort auf beide Fragen ist, wenn wir von neuen Energiequellen absehen, zweifellos Nein. Die meisten Länder haben weniger Kohle und Öl als Amerika und auch in Amerika können die Vorräte

nicht unbegrenzt aushalten. Die Atomenergie, wenn sie auch kein
goldenes Zeitalter einleiten kann, mag dazu berufen sein, unser eisernes
Zeitalter weiter möglich zu machen.

Tabelle 1. Hypothetischer Energiehaushalt

Brennstoff	Vorrat	Verbrauch per Jahr	Dauer des Vorrats
Kohle	60×10^{18}	75×10^{15}	400 Jahre
Petroleum	2×10^{18}	75×10^{15}	30 Jahre
ges. Atomenergie	16×10^{27}		
Atomenergie in Erzen	weniger als Kohle	Alle Zahlen in kgcal.	
Sonnenlicht	5×10^{21} per Jahr		

Wie lange die Kohlenvorräte ausreichen ist eine viel umstrittene
Frage; die Schätzungen variieren zwischen 50 und 3000 Jahren. Vielleicht können wir uns ein besseres Bild über die Situation verschaffen,
wenn ich zwei Zahlen anführe, die keine Hypothesen enthalten[1]. Wenn
Amerikas Energiebedarf von heute ab konstant bleibt und der der
übrigen Welt auf die amerikanische Ziffer steigt, und wenn Kohle und
Öl mit derselben Wirksamkeit wie heute gefördert werden, so haben wir
genug Kohle für 400 Jahre und genug Öl oder Petroleum für etwa
30 Jahre. Dies steht ziemlich fest; die großen Meinungsunterschiede
beruhen darauf, daß man auf eine weitere Zunahme der Bevölkerung
der Erde rechnet, daß man annimmt, die Energieanforderung würde
selbst in Amerika und selbst bei konstanter Bevölkerungszahl weiter an-
wachsen. Diese Annahmen sind sicher alle berechtigt, aber ihr Ausmaß
ist wohl schwer vorauszusagen. Andererseits wird darauf hingewiesen,
daß Öl sicher aus Kohle produziert werden kann (was allerdings die
Lebenszeit der Kohlevorräte verringert, vgl. die Zahl 400 Jahre in der
Tabelle) und daß die Technik der Kohleförderung sicher Ver-
besserungen zugänglich ist, so daß es möglich sein wird Vorräte abzu-
bauen, die heute als unbrauchbar angesehen werden. Wiederum sind
beide Annahmen sicher richtig, ihr Ausmaß wieder umstritten. Zu
alledem soll noch ein Punkt erwähnt werden, der selten beachtet wird:
es wäre ratsam mit unseren Kohlevorräten sparsam zu sein, weil Kohle

---

[1] Vgl. E. Ayres und C. A. Scarlott, Energy Sources — The Wealth of the
World. McGraw-Hill Co., New York 1952. Eine kürzere Übersicht über den-
selben Gegenstand findet man bei J. Daniels, M. K. Hubbert und E. Wigner,
Physics Today, 2, No. 4 (1949). Vgl. auch H. Thirring, Power Production, the
Practical Application of World Energy. G. G. Harrap and Co., London 1956.
Merkwürdigerweise scheint die Größe der Petroleumvorräte ziemlich unsicher
zu sein.

nicht nur eine Energiequelle sondern auch ein chemisches Reduktionsmittel ist. Die neuen Energiequellen, auf die wir gleich kommen werden,
bieten kein solches Reduktionsmittel. Wenn wir keine Kohle hätten,
so müßten wir unsere Eisenerze anderswie — vielleicht mit Aluminium —
reduzieren, was sicher viel kostspieliger wäre als die Reduktion mit
Kohle. Aluminium selber kann durch elektrische Energie produziert
werden, die Reduktion auf diesem Wege beansprucht aber sehr viel
mehr Energie als die Reduktion des Eisens mit Kohle.

Was sind nun die neuen Energiequellen, die Kohle und Öl ersetzen
werden? Man denkt, außer an Atomenergie, auch an eine effektivere Ausnützung des Sonnenlichtes. Obwohl in unseren Kreisen die effektivere
Ausnutzung des Sonnenlichtes wenig diskutiert wird, bin ich über-
zeugt, daß die Frage ob das Atom oder die Sonne die Energie der
Zukunft liefern wird, noch keineswegs entschieden ist[2]. Das Sonnen-
licht, das auf etwa 50 m² fällt, könnte den Energiebedarf einer Person,
selbst einer Person in den Vereinigten Staaten, liefern. Die Frage ist nur,
wie wir diese Energie sammeln und etwa in Elektrizität umwandeln
können. Pflanzen haben eine ganz löbliche Wirksamkeit und eine ge-
wisse Art von Fichtenholz speichert ungefähr ein Prozent der Sonnenenergie als chemische Energie. Die meisten Pflanzen sind aber viel
weniger effektiv und im Durchschnitt nehmen die Pflanzen nur etwa
ein hundertstel Prozent der Sonnenenergie auf. In Japan experimentiert
man mit Algen (Chlorella), die im Wasser leben und hofft mit Hilfe
dieser den Energiemangel zu überwinden. All dies liegt in der weiten
Zukunft und die Mittel, die zur Erforschung dieser Energiequelle zur
Verfügung stehen, sind Zwerge, wenn man sie mit den Mitteln, die in
der Atomforschung verwendet werden, vergleicht. In Amerika beschäftigt sich, soweit mir bekannt ist, nur FARRINGTON DANIELS, an der
Universität von Wisconsin, intensiv mit dieser Frage. Er hat gezeigt,
daß die maximale Ausbeute des Sonnenlichtes, die man von Pflanzen
erwarten kann, ungefähr 3% beträgt.

Was Sie soeben gehört haben, läuft darauf hinaus, daß die Sonnen-
strahlung, die auf die Erde fällt genügend Energie für alle voraussehbaren
Zwecke enthält, daß sie nur gesammelt und in elektrische Energie um-
gesetzt werden müßte, um alle unsere Bedürfnisse zu befriedigen. Gilt
das auch für die Atomenergie? Diese Frage ist zweifellos zu bejahen.
Man kann diese Tatsache besonders plastisch illustrieren, indem man
bemerkt, daß die Energie im Uran, die in einer Tonne gewöhnlichen
Gesteins enthalten ist, zwei bis dreimal größer ist als die Energie, die in der
Tonne Kohle verborgen ist. Nur ist sie, leider, viel besser verborgen im
Uran des Granits als in der Kohle. Die Erdrinde enthält zudem etwa
viermal soviel Thorium als Uran. Es ist also nicht die Energiemenge
selber, die knapp ist, die Schwierigkeit besteht in der Ausnützung und
Umwandlung dieser Energie. Dies ist eine Tatsache, die wir uns auch vor

[2] Vgl. Solar Energy Research, Farrington Daniels, Editor. University of
Wisconsin Press. Madison 1955.

Augen halten müssen, wenn wir die Atomenergie von der Uran-Spaltung
mit der Atomenergie der thermonuklearen Reaktion, also des Element-
aufbaues, vergleichen. Zur Zeit scheinen mir die Schwierigkeiten des
letzteren Prozesses so viel größer als die der Atomspaltung, daß ich
mich im weiteren nur mit dem Spaltungsprozeß beschäftigen will.

Die Schwierigkeit in der Ausnützung des Sonnenlichtes besteht
zweifellos in der großen Verdünnung, in der diese Energie zu uns kommt
und in der hierdurch verursachten Notwendigkeit, diese Energie zu
sammeln. Worin bestehen die prinzipiellen Schwierigkeiten der Verwendung der Uranspaltungsenergie?

Ich muß hier auf die Unterschiede in den Orientierungen der verschiedenen Reaktorprogramme kommen, weil die Schwierigkeiten und
auch die Ziele der Reaktorprogramme tief mit ihren Orientierungen ver-
knüpft sind. Wie schon früher erwähnt wurde, sind die Vereinigten
Staaten verhältnismäßig reich an fossilen Brennstoffen, Kohle und
Petroleum. Dies ist nicht der Fall im westlichen Europa; Österreich,
wie immer, bildet eine glückliche Ausnahme. Deshalb ist die Notwendigkeit einer neuen Energiequelle in Europa dringender als in
Amerika, deshalb sind die Pläne, die unsere Kollegen in Westeuropa
verfolgen, mehr auf das unmittelbar Mögliche gerichtet. Auch in einer
anderen Hinsicht passen sich diese Pläne an die vorliegenden Verhältnisse an: der Preis von Kohle und Öl in Westeuropa ist fast doppelt
so hoch wie in den Vereinigten Staaten.

Diese Pläne werden im Laufe unserer Konferenz später ausführlicher
besprochen werden. Sie beruhen fast ausschließlich auf der Verwendung
von natürlichem Uran, dessen Eigenschaften und Verhalten unter der
Bestrahlung die für natürliches Uran möglich ist, besser bekannt sind
als die der mehr esoterischen Substanzen, die die amerikanischen
Forscher beschäftigen. Die Reaktionsfähigkeit von Uran vermindert
sich ziemlich rasch, sobald sich sein Gehalt am aktiven Isotop, U-235,
vermindert. Deshalb wird in den in Westeuropa geplanten Reaktoren
nur ein kleiner Teil des Urans verbrannt, der größte Teil bleibt übrig,
verarmt an U-235. Man könnte zwar das im Uran gebildete Plutonium
von diesem chemisch isolieren und damit das erschöpfte Uran wieder
reaktionsfähig machen. Ob und inwieweit dies geplant wird, ist uns in
Amerika nicht völlig bekannt, vielleicht wird diese Frage von anderen
ausführlicher behandelt. Wenn das Uran nur solange abgebrannt wird,
bis sich seine Reaktionsfähigkeit erschöpft, so braucht man sich um die
chemische Separation nicht kümmern, was eine große Erleichterung ist.

Der erste Eindruck der amerikanischen Forscher über die westeuropäischen Pläne ist ein Eindruck der Freude. Wir wissen, daß wir
Reaktoren mit natürlichem Uran vernachlässigt haben, wir wissen,
daß es möglich ist, daß die Zukunft diesen gehört. Wir haben bereits
sehr viel von unseren europäischen Kollegen über Reaktoren mit natürlichem Uran gelernt und damit auch über die Theorie der Reaktoren im
allgemeinen. Ob diese Reaktoren auf die Dauer den Energiebedarf
der Welt decken können, wird von zwei Umständen abhängen. Der

erste betrifft die Gesamtmenge von Uran, die verfügbar ist. Wenn man auf alle Uranvorkommnisse rechnen kann, so sind unsere Zweifel in dieser Hinsicht grundlos. Wenn man aber nur die an Uran reicheren Erze aufschließen kann — mit einem Urangehalt von wenigstens zwei hundertstel Prozent —, so kann die Gesamtenergie, die natürliche Uranreaktoren liefern können, auf die Dauer bestenfalls nur genügen, wenn man das in den Reaktoren gebildete Plutonium auch zur Energieerzeugung mit heranzieht. Diese Notwendigkeit würde die Konstruktion der natürlichen Uranreaktoren stark komplizieren.

In den gegenwärtigen natürlichen Uranreaktoren ist die Plutoniummenge, die produziert wird, ungefähr 80% der Menge des verbrauchten U-235. Wenn das U-235 durch Plutonium ersetzt wird, so würde das Verbrennen dieses Plutoniums neues Plutonium erzeugen, aber nur etwa 70% dessen was verbrannt wird. Die Ursache für diese niedrigere Zahl liegt in den kernphysikalischen Eigenschaften des Plutoniums, insbesondere in der niedrigen Neutronemission der Zerfälle, die durch die Resonanzlinie bei 0,3 Elektronenvolt verursacht sind. Man berechnet leicht, daß man mit den heutigen natürlichen Uranreaktoren bestenfalls 4,5% der Energie des Urans erhalten könnte. Ob das hinreicht, hängt, wie gesagt, von der Gesamtmenge des abbaubaren Urans, also von unseren Mineralogen, Bergwerksingenieuren und von unseren Metallurgen ab.

Selbst wenn die Uranvorkommnisse niedrigen Inhaltes — in denen das meiste Uran der Erdkruste verborgen liegt — sich widerstandsfähiger erweisen sollten als es unsere europäischen Kollegen hoffen, bleibt der Weg offen, die Konversionsziffer der natürlichen Uranreaktoren zu erhöhen. Die 80% der heutigen Reaktoren ist keinesfalls die höchste erreichbare Ziffer und jede Erhöhung dieser Ziffer wirkt sich ganz mächtig in einer Erhöhung des verbrennbaren Anteils des Urans aus. Wenn sich die Ziffer auf 85% erhöhen sollte, statt der heutigen 80%, so würde sich die verbrennbare Menge von $4\frac{1}{2}\%$ auf $5\frac{1}{4}\%$, also um 17% erhöhen. Leider sind aber diese Zahlen nicht realistisch, weil sie den sogenannten kleinen Verlusten keine Rechnung tragen. Ich will hierauf später noch zurückkommen. Jedenfalls wird, glaube ich, jeder der Behauptung zustimmen, daß die heutigen natürlichen Uranreaktoren noch verbesserungsfähig sind.

Von dieser Seite aus betrachtet, unterscheiden sich die Einstellungen auf den beiden Ufern des Atlantischen Ozeans mehr in den Wegen, die sie einschlagen wollen, als in den Zielen. Die amerikanischen Forscher wollen das Ziel — die Breeder Reaktoren (Brüter) direkt erreichen, unsere Kollegen hier hoffen durch eine ständige Verbesserung der jetzt möglichen Reaktoren hinreichend nahe an die Breeder zu kommen. Ob und inwieweit dies möglich sein wird, läßt sich heute wohl ebensowenig voraussehen, wie es sich voraussehen läßt, ob die Kühnheit der amerikanischen Forscher berechtigt ist.

Ich komme jetzt zu den Plänen der Forscher in den Vereinigten Staaten und dies ist natürlich der Gegenstand, bei dem ich mich am meisten zu Hause fühle. Die Untersuchungen, über die ich zunächst berichten möchte, betreffen den Breeder, das ist einen Reaktor der nicht nur Energie sondern auch spaltbares Material erzeugt. Man kann die Spaltungsreaktion in der Form:

$$n + \mathrm{Pu}^{239} = \frac{2}{1+\alpha} \text{ Spaltprodukte} + \frac{\alpha}{1+\alpha} \mathrm{Pu}^{240} + \eta \text{ Neutronen}$$

$$+ \, \beta \text{ und } \gamma \text{ Strahlung} + \frac{200}{1+\alpha} \text{ MeV Energie}$$

schreiben; α ist das Verhältnis der Anzahl der Prozesse, die zur Bildung von Pu-240 führen, zu der Zahl der Prozesse die in der Spaltreaktion resultieren; η ist die mittlere Zahl der Neutronen, die nach der Absorption eines Neutrons emittiert werden. Die hingeschriebene Gleichung bedeutet also eine Zunahme der Neutronenzahl mit $\eta - 1$, wogegen ein spaltbarer Kern verloren geht. Da, wenn man Nebenprozesse außer acht läßt, ein Neutron wiederum zur Erzeugung eines spaltbaren Kernes benützt werden kann, besagt unsere Gleichung, daß das Verhältnis von neuen spaltbaren Kernen zu verbrauchten spaltbaren Kernen $\eta - 1$ ist. Unter der Voraussetzung, daß man Nebenreaktionen vernachlässigen kann, folgt dann, daß die Voraussetzung für den Breeder $\eta - 1 > 1$ lautet, oder daß η größer sein muß als zwei.

Tabelle 2

Energie der Neutronen	α	η
Thermisch	0,416	2,02
0,296 eV	0,70	1,7
(7,1 eV)	0,55	(1,8)
7,85 eV	1,35	1,2
10,95 eV	0,40	2,05

Die Zahl η hängt sowohl vom Material ab, das mit den Neutronen reagiert, wie auch von der Energie der Neutronen. Sie ist verschieden für verschiedene Resonanzlinien. Ihr Wert für thermische Neutronen und die ersten vier Resonanzlinien im Pu-239 ist in Tab. 2 gegeben. Die Werte für η erscheinen zunächst nicht so sehr verschieden voneinander, aber man erhält ein anderes Bild, wenn man $\eta - 2$ betrachtet, die Zahl, die für die Möglichkeit des Brütens maßgebend ist. Sie sehen zugleich, daß diese Zahlen alle niedriger sind als die entsprechende Zahl für U-235, nämlich 2,09. Dies erklärt die Verminderung der Konversionsziffer des natürlichen Uranreaktors, wenn das U-235 darin durch Plutonium ersetzt wird. Die Klammer für die dritte Resonanzlinie soll andeuten, daß diese so schwach ist (wegen der geringen Neutronenbreite), daß sie außer acht gelassen werden kann.

Es hat sich schon sehr früh die Auffassung durchgesetzt, daß die beiden Verhältnisse, bei denen man am ehesten hoffen kann einen Breeder zu erhalten, bestehen werden, wenn man entweder thermische Neutronen auf U-233 einwirken läßt, oder sehr rasche Neutronen auf Pu-239. Im ersten Fall beträgt η ungefähr 2,30, das Verhältnis von erzeugten zu verbrauchten spaltbaren Kernen ist also ungefähr 1,30.

Im zweiten Falle ist η wesentlich höher, es beträgt ungefähr 2,9, das vorerwähnte Verhältnis also 1,9. Dies ist mehr als das Quadrat der entsprechenden Zahl für thermische Neutronen und U-233.

Das Verhältnis der beiden η-Werte vergrößert sich weiter, wenn man die Nebenprozesse betrachtet. Um diese völlig zu verstehen, müssen wir bedenken, daß die vorerwähnte Reaktion die Kette noch nicht schließt. Um sie zu schließen, müssen wir die Neutronen absorbieren lassen. Im ersten Falle muß das in Thorium geschehen, dessen einziges Isotop Th-232 durch Neutronabsorption in Th-233 übergeht, das nach zwei radioaktiven Zerfällen wieder U-233 liefert. Die zweite Reaktion ist also im Falle des U-233 Kreisprozesses

$$\text{Th}^{232} + n = \text{Th}^{233} \xrightarrow{23 \text{ Min.}} \text{Pa}^{233} \xrightarrow{27 \text{ Tage}} \text{U}^{233}.$$

Im Prozeß, der schnelle Neutronen verwendet, schließt sich die Kette durch die wohlbekannten Glieder

$$\text{U}^{238} + n = \text{U}^{239} \xrightarrow{23 \text{ Min.}} \text{Np}^{239} \xrightarrow{2,3 \text{ Tage}} \text{Pu}^{239}.$$

Man wird also den Prozeß im ersten Falle so leiten, daß die Neutronen, die bei der Spaltung des U-233 emittiert werden, durch Thorium absorbiert werden. Am zweckmäßigsten bewirkt man dies so, daß man den eigentlichen Reaktor mit einer Hülle umgibt, in der Thorium der wirksame Neutronenabsorber ist. Das sich langsam in der Hülle ansammelnde U-233 muß dann von Zeit zu Zeit vom Thorium chemisch geschieden und in den eigentlichen Reaktor an Stelle des dort verbrauchten Urans eingeführt werden. Funktioniert der Breeder wie beabsichtigt, so wird die chemische Separation mehr Uran liefern, als im eigentlichen Reaktor verbraucht wurde und dieser Überschuß ist eben das neu erzeugte spaltbare Material.

All dies zeigt, daß der Brüter nicht so einfach funktioniert wie etwa ein natürlicher Uranreaktor, den man einfach laufen läßt bis sein Gehalt an U-235 soweit gefallen ist, daß er nicht mehr reagieren kann. Jede Komplikation, die in den Breeder hineingebaut werden muß, beansprucht nicht nur Kapital, sie stört auch den Neutronenhaushalt. Ich habe einmal 16 solche Neutronenverluste aufgezählt — eine sehr hohe Zahl, wenn man bedenkt, daß nur 30% Überschuß an Neutronen zur Verfügung steht. Ich will einige der Prozesse, die Neutronen verzehren, nennen.

1. Da unser Reaktor mit thermischen Neutronen arbeitet, muß er einen Moderator haben, der aber auch Neutronen nutzlos verschlingen wird.

2. Die Spaltprodukte, die nicht sofort vom Reaktor entfernt werden können, absorbieren auch Neutronen. Besonders gefährlich unter diesen

ist das berühmte Xe-135 mit einem Querschnitt von $3\frac{1}{2}$ Millionen Barn ($3,5 \times 10^{-18}$ cm^2).

3. Um diese Verluste möglichst niedrig zu halten, wird man das Uran samt den Spaltprodukten zirkulieren lassen und die gasförmigen Spaltprodukte bei jedem Umlauf wenigstens teilweise entfernen. Dies führt zwar zu einer Verringerung der eben erwähnten Verluste, verursacht aber eine dritte Verlustquelle, weil ein Teil der Neutronen erst einige Zeit nach der Spaltung, also außerhalb des Reaktors ausgesandt wird.

4. Eine vierte Verlustquelle ergibt sich aus der Tatsache, daß kein Moderator wirksam genug ist, um die Spaltung durch Resonanzneutronen gänzlich zu vermeiden. Es scheint aber, daß η für Resonanzneutronen niedriger liegt als für thermische Neutronen (vgl. die Tab. 2, die sich allerdings auf Pu-239 bezieht), so daß die Spaltungsprozesse, die durch Resonanzneutronen eingeleitet werden, eine Verminderung der Anzahl der Sekundärneutronen mit sich bringen.

Dies sind die Verluste im Reaktor selber. Dazu kommen Verluste bei dem Übergang der Neutronen in die Hülle, und in der Hülle selber. Besonders gefährlich unter diesen ist die Absorption des verhältnismäßig langlebigen Pa, welche nicht nur Neutronen, sondern auch spaltbares Material verzehrt durch Unterbrechung der Prozesse, die zur Bildung des U-233 führen. Schließlich fängt keine Hülle alle Neutronen, die in sie eintreten, und die chemische Separation des gebildeten Urans vom überschüssigen Thorium kann auch nie vollkommen und verlustfrei sein. Es erscheint mir immer als ein Wunder, daß man trotz all dieser Schwierigkeiten mit Recht hoffen kann, daß der U-233 Breeder als Breeder funktionieren wird.

Die Verhältnisse bei dem schnellen Brüter, bestehend aus Plutonium und umgeben von einer Uranhülle, stehen weit günstiger. Erstens stehen einem in diesem Fall nicht 30%, sondern 90% Überschuß an Neutronen zur Verfügung. Zweitens ist aber der wichtigste Nebenprozeß nicht schädlich, sondern nützlich vom Standpunkt des Neutronenhaushalts. Ich beziehe mich auf den Spaltprozeß im U-238, den die schnellen, vom eigentlichen Reaktor entfliehenden Neutronen sehr wohl einleiten können und der Neutronen liefert, ohne spaltbares Material zu zerstören. Dieser Prozeß ist so wichtig, daß man zunehmend dazu neigt, wenigstens einen Teil des U-238 im Reaktor selber unterzubringen, das heißt Reaktor und Hülle wieder zu vereinigen. Dies wäre ein Schritt in der Richtung des natürlichen Uranreaktors mit dem Unterschied, daß das spaltbare, also wirksame, Material nicht U-235, sondern Plutonium ist. Ein zweiter und vielleicht noch wichtigerer Unterschied ist, daß man stillschweigend annimmt, wenn man vom natürlichen Uranreaktor spricht, daß er thermische Neutronen verbraucht. Dagegen wird im eben besprochenen Reaktor die Mehrzahl der Spaltungen von raschen Neutronen eingeleitet. Wenn man aber in natürlichen Uranreaktoren, oder vielmehr in Reaktoren, in denen das U-235 des natürlichen Urans durch Plutonium ersetzt ist, zu höheren Neutronenenergien übergehen würde, so erreichte man dieselbe Anordnung, die wir eben für einen Breeder als eine günstige Abart erkannt haben. Dies zeigt erneut, daß am Ende die beiden Orientierungen der Reaktorpläne, auf beiden Seiten des Atlantiks, wohl zu derselben Maschine konvergieren könnten.

Der vorangehende Vergleich legt die Frage nach der Berechtigung des thermischen U-233 Reaktors nahe. Was ist der Vorteil dieses Reaktors, wenn der rasche Plutonium-Reaktor ein so viel größeres Konversionsverhältnis hat? Man wird uns, ich hoffe, nicht als zu pessimistisch ansehen, wenn wir das Konversionsverhältnis des thermischen U-233 Reaktors als 1,1 ansetzen, das heißt annehmen, daß er 110 U-233 Atome für 100 verbrannte U-233 Atome produziert. Bei dem Plutonium-Reaktor spricht man oft über ein Konversionsverhältnis von 2. Die Ursache, daß der U-233 Reaktor im Rennen ist, ist, daß er einen so viel höheren Umsatz des Materials erlaubt. Das Hauptproblem, dem man mit dem raschen Plutoniumreaktor begegnet, besteht gerade in der Schwierigkeit, das Energiespektrum der Neutronen hoch zu erhalten trotz der Notwendigkeit, die durch die Kernreaktion erzeugte Wärme aus dem Reaktor abzuführen. Fast jedes Material, auch das zur Verdünnung des Plutoniums benutzte, auch das U-238, das in den Reaktor eingeführt wird, verlangsamt die Neutronen. Die schweren Elemente tun dies durch inelastische Streuung, die leichten Elemente durch den elastischen Rückstoß, durch den sie Energie den Neutronen entziehen. Wenn aber die mittlere Energie der Neutronen, die unmittelbar nach dem Spaltungsprozeß etwa 2 MeV beträgt, auf etwa 0,1 MeV fällt, so fällt η schon von 2,9 auf 2,5, also $\eta - 2$ von 0,9 auf 0,5. Eine weitere Degradierung der Neutronenenergie würde η sogar unter 2 bringen. Demgemäß ist man in der Konstruktion der raschen Reaktoren sehr beschränkt, sowohl was Menge wie auch was das Material des Wärmetransportmittels und des Verdünnungsstoffes des spaltbaren Materials betrifft. All dies begrenzt den Wärmeumsatz. Dagegen ist der thermische U-233 Reaktor von solchen Beschränkungen im wesentlichen frei. Es ist sogar möglich, daß die Umsatzgeschwindigkeit dieses Reaktors hauptsächlich durch die Langlebigkeit des Zwischenprodukts Pa beschränkt wird.

Aus diesem Grunde wird die Zeit, während der sich die Menge des spaltbaren Materials verdoppelt, für beide Reaktorarten als ungefähr gleich lang angenommen. Man schätzt diese Zeit oft auf sieben Jahre, was aber sicher reine Spekulation ist. Wir wissen ja noch gar nicht, ob Brüter wirklich physikalisch möglich sind.

Sieben Jahre sind eine verhältnismäßig lange Zeit, so daß es bestenfalls vieler Jahre bedürfen wird, bevor man soviel spaltbares Material angesammelt hat, daß man eine für den Energiebedarf der Welt genügend große Anzahl von Reaktoren bauen kann. Andererseits kann ich der Meinung nicht zustimmen, daß es fast gleichgültig ist, ob das Konversionsverhältnis 1,1 oder 0,9 ist. Im ersten Falle erhöht sich der Bestand an spaltbarem Material in 50 Jahren auf das hundertfache, im zweiten Fall bleibt nur ein Prozent davon übrig.

Vom Reaktorprogramm der Vereinigten Staaten habe ich bisher nur das Breederprogramm berücksichtigt. Dies ist, meiner Ansicht nach, der wichtigste und kühnste Teil des Programms, es ist aber nicht der einzige. Wir wissen, daß Energie nicht überall so reichlich vorhanden ist wie bei uns und selbst in Amerika ist Energie nicht überall gleich billig. In Alaska zahlen wir höhere Preise für Elektrizität als hier in Wien und ich höre eben, daß man Bauxit von Jamaica nach Kitimat in Britisch Columbien, 9000 km weit, transportieren will, weil dort die elektrische Energie am billigsten ist. Der Transport elektrischer Energie ist überraschend teuer und man leitet sie ungern über mehr als 100 km. Atomenergie hat unter diesen Umständen eine wichtige Rolle zu erfüllen, die Transportkosten des Brennstoffes sind vernachlässigbar und die der Asche, des erschöpften spaltbaren Materials, trotz seiner Radioaktivität noch sehr viel niedriger als die von Kohle mit dem gleichen Energiegehalt. Deshalb haben wir auch, unter anderem, Pläne, in Alaska einen Reaktor zu errichten und ein leicht transportabler Reaktor, der sich in abgelegenere Teile der Erde schicken läßt, wird bald in größerem Maßstab produziert. Aber selbst auf diesem Gebiet, des kleinen und kostspieligen Reaktors, stehen wir nicht in direktem Wettbewerb mit den westeuropäischen Reaktoren. Keiner dieser Reaktoren verwendet natürliches Uran. Die Verwendung des an U-235 angereicherten Materials läßt sich in abgelegenen Stellen dadurch rechtfertigen, daß dieses Material die Errichtung kleinerer Einheiten gestattet, als das natürliche Uran. Der Energiebedarf an diesen Stellen ist aber im allgemeinen wenigstens vorläufig so niedrig, daß sich die Installation eines großen Reaktors nicht rechtfertigen läßt. Ob sich die Verwendung von angereichertem Uran auch anderswo, etwa in Pittsburgh, als berechtigt erweisen wird, ist eine mehr anfechtbare These, deren Richtigkeit nur die Zukunft wird entscheiden können.

Der Entwurf und die Konstruktion der Reaktoren, über die ich soeben sprach und die im Prinzip unmittelbar realisierbar sind, ist viel mehr dezentralisiert, als der Entwurf der Brüter. Der Reaktor in Alaska wird von einer Gesellschaft entworfen, die vor ein paar Jahren nur wenige tausend Dollar ihr eigen nennen konnte. Dies hat den Vorteil größerer Anpassungsfähigkeit an die spezialen lokalen Verhältnisse und Bedürfnisse. Trotzdem leidet auch dieser Teil des amerikanischen Reaktorprogramms an einem Mangel an Talent und ausgebildetem Personal. Ich werde nie vergessen, daß als ich vom Uranprojekt in Chicago nach Princeton zurückkehrte, die elektrische Schaltung in einem Zimmer versagte. Ich zerlegte den Schalter um die Ursache des Versagens zu finden und war überwältigt von der Genialität, die in die Konstruktion dieses kleinen Instruments ging. Er hatte weniger Teile als ich dies für möglich gehalten hätte und diese konnten offenbar leicht fabriziert und vereinigt werden. Ich habe damals die Hoffnung ausgedrückt, daß ich einmal etwas ähnlich Geschicktes an einem Reaktor sehen möge. Diese Hoffnung ist bisher nicht in Erfüllung gegangen. Vielleicht haben wir in Amerika mehr angebissen als wir verdauen können.

Es ist nicht leicht sich eine Welt vorzustellen, in der Atomenergie an die Stelle von Kohle und Öl getreten ist. Ich möchte aber den Gegenstand nicht verlassen, bevor ich diesen Umstand etwas illustriert habe. Die Energieproduktion, die als wahrscheinlich hingestellt wurde, entspricht der Verbrennung von 7500 Tonnen spaltbarem Material im Jahre und dem, daß ständig ungefähr ebensoviel radioaktives Strontium irgendwo auf der Erde vorhanden ist. Die Anzahl der Megaton-Waffen, deren Explosion dieselbe Energie und demgemäß dieselbe Radioaktivität liefern, wird man nicht gerne angeben. Die Kontrolle und Speicherung dieser unglaublich großen Menge radioaktiven Materials wird selber ein technisches und administratives Problem großer Dimension werden. Es ist klar, daß wir in der Zukunft viel mehr Aufmerksamkeit diesem und *vielen anderen* ähnlichen Problemen widmen müssen, wenn wir den Anschein, ernste und verantwortungsbewußte Menschen zu sein, aufrecht erhalten wollen.

Damit will ich meinen Vortrag schließen. Ich erinnere wieder daran, daß ich hauptsächlich über die Zukunft sprach, als ich die Erzeugung nützlicher Energie diskutierte, daß wir aber schon jetzt in vielen Gebieten eine mächtige Bereicherung unseres Wissens den Reaktoren und ihren verschiedenen Nebenprodukten verdanken. Mögen sie unser Interesse auch weiter behalten und zum Wohle der Wissenschaft und von uns allen gedeihen.

41.
Longer Range View of Nuclear Energy

A. M. Weinberg and E. P. Wigner

Bulletin of the Atomic Scientists *XVI*, 400–403 (1960)

Alvin M. Weinberg is the Director of Oak Ridge National Laboratory. Eugene P. Wigner, who headed the scientific team that designed the first nuclear reactor, is now professor of mathematical physics at the Palmer Physical Laboratory, Princeton University. Both Dr. Weinberg and Dr. Wigner recently received the Ford Motor Company's "Atoms for Peace" award in a presentation at the National Academy of Sciences.

IT MAY be startling to realize that only about 3 per cent of the U.S. national income is spent to furnish the energy we use. From this, it is evident that no golden age can be created by reducing this expenditure —not even by reducing it to nothing.

On the other hand, a substantial increase in the effort now necessary to provide our energy requirements might be a serious calamity. Thus, the true objective of reactor research may well be the preservation of our present iron age. This point underlies the following discussion[1] of whether reactor research ought to have as its primary aim the development of "burners" or of "breeders."

Breeders and Burners

Natural uranium contains only 0.7 per cent of the fissionable isotope U-235. In "burners" (burner reactors) only this 0.7 per cent of the total energy content of the uranium is utilized, and even this fraction cannot be utilized completely. "Converter" reactors are somewhat more efficient; in their present-day embodiments, they produce enough plutonium to permit, in principle, the doubling of the energy content of the U-235.

Breeder reactors, on the other hand, can convert virtually all of U-238 into fissionable plutonium, producing about 140 times more energy than burners, and 70 times more than low-conversion ratio reactors.

The breeder vs. burner argument, therfore, revolves around the question: How necessary is it to utilize the bulk of the energy content of uranium and thorium, and to what extent can we be satisfied with the utilization of a small fraction of this energy content. Through-

out this article we will assume that the future energy economy will be based on fission, i.e., on the energy content of uranium and/or thorium. The discussion will also be based on the supply and consumption as they can be expected to develop in the United States. The conditions in most other parts of the world will probably become quite comparable to those in this country. Hence, it appears justifiable to base our conclusions on the conditions which we know best.

On the basis of these assumptions, the problem of breeder vs. burner reduces to the question of the supply of uranium (and thorium) available, vs. future energy requirements.

Supply and Demand of Uranium

The amount of uranium available, in the form of uranium oxide (U_8O_8), depends on the price one is willing to pay to obtain it. Thus, the cost of the energy obtained, whether from a burner, a converter, or a breeder, may depend on the cost of the uranium. This is shown in Table 1. The figures of this Table are,

---

[1] See also E. P. Wigner, *Acts Physical Austriaca 11,* 410 (1958).

of course, only estimates. This applies particularly to the magnitude of the supplies of more highly priced material, which should actually be tripled if breeders come into general use, because of the increased usability of thorium.

It may be mentioned at this point that the total coal reserves of the United States[2] amount to about 2×10^{12} tons, with an energy content of 50×10^{18} BTU. At present, about one-third of all the energy produced in the United States is derived from coal. The price of bituminous coal is about $4.50 per ton, giving a price of about 17¢ per 10^6 BTU. Even though used at present to a similar extent, the total energy content of the oil and gas supplies is much lower than that of coal. Furthermore, about one-half of the total coal reserves can be obtained only at more than twice the present cost. Hence, if only fuel costing less than twice the present price is considered to be available, the total fossil fuel of the U.S. amounts to 25×10^{18} BTU.

On the basis of Table 1, breeding appears to have an enormous advantage over burning. It must be remembered, however, that the price of the fuel is only one of the cost items of a reactor (around 20 per cent) and it is believed that, as a rule, the capital cost of a breeder may be higher than that of a burner. The present tendency to build burners clearly shows that it is at least easier to build a burner than a breeder. Finally, beyond a certain point, the reduction of the price of the fuel influences the cost of power very little.

Similarly, the need for energy has its limits and the availability of further energy in the form of fuel is, beyond a certain point, more of academic than of real interest.[3] On the other hand, the fact that the energy content of low-price uranium, used in a low-conversion reactor, is only a small fraction of the energy content of fossil fuels, appears most significant. This will become evident when the power requirements are discussed.

Future Power Requirements

The magnitude of the energy requirements will not be based on the production of electrical power for two reasons. First, the electrical power produced in the U.S. is less than 6 per cent of our total energy consumption. Even after making generous allowance for the relatively low thermal efficiency of the production of electrical power, it is clear that it accounts for only a fraction of our energy requirements. Second, the energy requirements increase only slowly, whereas the increase in the production of electrical energy is relatively fast. The

[2] See, for instance, Statistical Abstract of the U.S., 1957, Government Printing Office, Washington, D.C., pages 733 and 730; Also Sam H. Schurr's statement before congressional subcommittee, Reprint 14, Resources for the Future, Inc., Washington, 1959.

[3] See, however, Alvin M. Weinberg, Physics Today, 12 (1959), 18.

TABLE 1

Cost in $/lb.	10^6 tons U_3O_8	Cost in cents/10^6 BTU			Energy content in 10^{18} BTU		
		only U^{235}	low conv.	breeding	only U^{235}	low conv.	breeding
10	0.9	5.5	2.7	0.039	0.39	0.78	56
20	3.5	11	5.5	0.077	1.35	2.7	190
30	8	16.5	8.2	0.12	3.5	7	500
50	30	27	13.5	0.19	13	26	1850
100	80	55	27	0.39	35	70	5000

The second column gives the amout of U_3O_8 that is available at the price given in the first column. The data are based on An Analysis of the Current and Long-Term Availability of Uranium and Thorium Raw Materials, TID-8201, Technical Information Service, Atomic Energy Commission, and on Sornein's estimate (article in The Industrial Challenge of Nuclear Energy, Office of European Economic Cooperation, June 1957) according to which the available amounts of uranium are roughly proportional to the square of the price one is willing to pay. This assumption is consistent with the data of the Technical Information Service. Furthermore, the numbers of Table 1 are very nearly equal to those given in a recent study by C. S. Starr and R. A. Laubenstein (The Availability of Uranium for a Nuclear Power Industry, Report AI-4945, Atomics International, 1960).

The cost of the fuel per unit amount of energy depends on the fraction of the fuel that is used. Hence, the price of the fuel in cents per million BTU, given in columns three, four and five, is different for the three types of reactors considered. Column three refers to a pure burner, column four to a reactor with conversion ratio 0.5, column five to a breeder. The figures are somewhat optimistic because no fuel can be used completely. The last three columns give the total energy obtainable in the types of reactors just enumerated. The unit of energy is 10^{18} BTU (1 BTU = 0.252 kcal = 1,055 Joules).

former slow increase can be expected to remain more nearly steady over extended periods of time. It may be worth noting, however, that the conclusions to be arrived at would be modified only little if the calculation were based on the electrical power production. In fact, at least for the more distant future, the calculation used here gives lower energy requirements.

The total energy consumption in the United States was 40×10^{15} BTU in 1955, and the consumption appears to double in a little less than 25 years.[4] Table 2 gives the total consumption in 10^{18} BTU estimated, on the basis of the permanence of this increase, up to the year 2080, when it will be about thirty times the 1955 level.

TABLE 2

Year	1980	2005	2030	2055	2080	
Cumulative consumption	1.6	4.7	11	23	48	$\times 10^{18}$ BTU

The objection may be raised that it is very dangerous to make any forecast for the distant future, and this objection is quite justified. Nevertheless, the figures of Table 2 seem to be the most reasonable ones—one may even expect them to be underestimates.[5] They show,

[4] Reference 2, page 526; also Sam H. Schurr, loc cit.

first, that the supply of fossil fuels (around 25×10^{18} BTU) should last for not more than 100 years. Serious local shortages are bound to develop much before. Second, they show that by the time there will be a real need for a substitute for the fossil fuels, low-conversion reactors would consume all the uranium supply of the country in about 25 years, as long as one does not wish to pay a higher price for the energy contained in uranium than one now pays for the energy contained in coal (line before last of Table 1). Hence, low-conversion-ratio reactors would provide only a stopgap for a short period when an urgent need for a new energy source will arise.[6]

On the contrary, the energy supply can be expected to be ample for many hundreds of years if uranium and thorium are used in a breeding cycle. The reason for this is not only that a pound of these materials gives about 70 times more energy if used in the breeding cycle; even more important is the fact that the exploitation of the ample supply of low concentration ores becomes possible without an increase in the price of the energy content of the fuel above the present price. As Table 1 shows, the cost of a certain amount of energy in the form of uranium remains about 140 times lower if this uranium is used in a breeding cycle, that is completely, than if only the skim of it—the U-235—is used. The ratio is 70 against a low-conversion reactor. As a result, according to Sornein's rule, the amount obtainable at a given price is $140^2 = 20,000$ times greater than for burners, and $70^2 = 5,000$ times greater than for low-conversion-ratio converters. It is this disparity between both price and availability of energy, through the use of burners and of low-conversion reactors on the one hand, and breeders on the other, which induces so many students of these questions to emphasize the advantages of the breeders over the burners.

Why Develop Nuclear Power?

In view of the sufficiency of the fossil fuels for about 100 more years, one may doubt the present need for the development of any new source of energy. It should be pointed out, however, that the situation may be rather worse than indicated, principally because fossil fuels are needed for many purposes other than energy production—for instance, for the reduction of iron ores. Hence, it would be irresponsible to exhaust the supply of fossil fuels to a very large extent. We must not forget that what we are after is the preservation of our iron age. This point has been brought out with particular clarity by G. Young; he even proposes the name "fossil organics" for coal, oil, and perhaps also for natural gas. As a

[5] The picture presented in Palmer C. Putnam's book, *Energy in the Future* (Van Nostrand & Co., New York, 1953) is much more pessimistic.

[6] This conclusion remains valid under the assumptions made in Palmer C. Putnam's book (reference 5).

result of these circumstances, the need for new energy sources may appear earlier than in a hundred years, even if the relatively low estimates of the energy requirements given in Table 2 should prove to be valid.

Second, the rapidity of our progress in science should not blind us to the fact that far-reaching changes in our methods of production continue to take a long time. Power plants, in particular, have a useful life of about 50 years, so that the conversion to nuclear economy will surely extend over a long period of time. It should not be forgotten, either, that all present thinking is directed toward replacing fossil fuels by fission energy in the electrical power industry; the problem of the replacement of fossil fuels for other purposes is still essentially untouched. As mentioned before, at present only a relatively small, though increasing, fraction of our energy is used for the generation of electricity.

Third, the transition to nuclear economy will be possible only if vast amounts of fissionable material will be available when needed. Some such fissionable material will come from isotope separation plants. However, the large inventories necessary for the huge nuclear fueled power industry of the future can be conveniently obtained only from breeders which produce fissionable material well ahead of the time when these materials will be needed for power production. In fact, several generations of breeders fueled by materials produced by earlier breeders will be necessary to produce the inventories for the critical period.

Need for Burners

There is, finally, a fourth and perhaps most important factor which renders a deliberate approach toward the use of fission energy imperative: the attendant production of vast amounts of radioactive materials. The storage and disposal of these materials will present many problems, the solution of which will have to be based on protracted practical experience. Not many mistakes can be allowed while the proposed solutions are tried out. If there were no other reasons for developing some nuclear power fast, the difficulty and urgency of the problem of the disposal of radioactive wastes would be an adequate reason.

There are, however, several other important reasons for producing nuclear power on a reasonable scale in the near future, and it is hoped that the present analysis will not be interpreted as a blanket rejection of the development and use of burners. Three of these reasons are most cogent from our point of view:

(a) As has been emphasized it is quite uncertain whether the problems of the breeder will be solved by a frontal assault, or by successive improvements on burners.

(b) Nuclear power, even before it becomes economical on a large scale, will have many specialized applications in which it may be not only economical but almost irreplaceable. The Army Package Power Reactor may not fall into this latter category, but the nuclear batteries for rockets may. Once low-weight high-capacity batteries are easily available, many uses will be found for them.

(c) Burners will make it possible to tackle the waste disposal problem at an early date.

Conclusions

The preceding analysis suggests first, more emphasis on the search for a practical breeder; second, more emphasis on the improvement of the conversion ratio of burners; and third, more emphasis on solving the long-term waste disposal problem. As of today, nuclear energy for large power stations is uneconomical. Hence, it makes little difference whether it is very uneconomical. The large-scale use of money and of scientific manpower for the development of nuclear energy cannot be justified on the basis that it leads only to a small economical disadvantage. It can be justified only if it is directed toward the use of nuclear energy when it will be needed, and if the use of nuclear energy will indeed satisfy the need which will arise. This will be true only if breeders will be available. Short-term economic advantages, or rather, diminution of the disadvantages, are, in this connection, largely irrelevant.

42.

Our Needs for Energy and Ways to Satisfy Them

E. P. Wigner

Revista Interamericana *VI*, 485–496 (1976)

I. Observations of a Physicist

Our subject is not only a very important one, it is also very large and complex. As a physicist, I should first point out that the problem is not really one of energy. We have plenty of energy: the atmosphere has a temperature of around 300° absolute and so have our seas (this corresponds to about 50°F). If we could bring only a tiny fraction of their heat content into our houses in the winter, we would need no heating. Unfortunately, heat does not move from a colder region to a warmer one, that is it does not move from the outside of our houses in the winter to the warmer region inside—it flows in the other direction. When we speak about energy requirements, we really mean requirement of *available* energy and this is limited, even in principle, by the second law of thermodynamics. It is limited also by our lack of skill to make full use of even that part of the energy which seems to be available even if the second law of thermodynamics is taken into account. An expression for this latter amount was already given by Gibbs: if a container is available which can act both as a source and a sink of heat at a definite temperature T_o—and our atmosphere or our seas can furnish or accept heat at a temperature around 300°K—the amount of available energy is given by the expression $G = U - T_o S$ where U is the actual energy difference between initial and final state of the material furnishing us with what we call energy in this discussion, and S is the difference between the entropies of the two states of this material, the difference of the entropies of its initial and its used-up state. If we apply Gibbs' law to an inert material, such as water, it follows that it contains some available energy if its temperature is above T_o, none if its temperature is T_o, but some also if its temperature is below T_o,

EUGENE P. WIGNER is a Nobel Prize Winner in Physics. The above remarks were made at the Louisiana State University Nuclear Energy Symposium held on April 23, 1976. [They have been edited for print by removing references to other speakers and by minor textual emendations.—Ed.]

For very energy-rich substances, such as coal, oil, or nuclear fuel, the T_0S term is very small as compared with the term of true energy U, so that practically all the energy content of these materials is "available."

Unfortunately, or fortunately, there is no general formula giving the amount of energy which we could extract from a substance with the skill we possess, and most of our discussion will be devoted to this subject.

II. Some General Observations on Energy Consumption

Having discharged my obligation as a physicist by calling attention to the difference between total and available energy, I will now turn to our more immediate problems. The first, and perhaps most surprising observation I'll make is that man, as a machine, is a one-tenth kilowatt machine. The food we consume has the energy which could run such a machine or burn a 100 watt lamp. In contrast, the "available energy" which the machines of the United States, our lighting and heating systems, consume, if divided by our population, amounts to 10 kilowatts. This is a hundred times more than what we consume in the form of food. It may be useful to call attention to the fact that this latter number is enormously large. There are only two other countries which consume similarly large amounts of energy per person and Canada is one of them, Monaco the other. All other countries, with the exception of Sweden and perhaps England, consume less than one half as much energy per person as does the U.S. This includes France, Germany, Italy, Holland, etc. Just the same, these countries are getting along quite well and do not suffer privations. This already indicates something I will emphasize later, namely that we could economize a bit in our energy consumption.

The next fact I want to mention is that the sun rays' energy falling on our country amounts to 10,000 kilowatts per person. This is, of course, very much more than in the other countries I mentioned because the density of our population is hardly more than one tenth of the average density of the populations of these countries. Nevertheless, it shows that if the sun rays' energy could be made "available" with reasonable efficiency, the energy problem would be close to its solution everywhere.

How much does it cost to provide the energy we use? Its cost amounts to about 4 percent of the gross national product, that is about $250 per person a year. Naturally, only a relatively small part of this is spent directly by us for our heating and lighting. Most of the energy is not used in our houses. The total energy usage can be divided into five categories and the consumption of the various categories is given in the Table that follows:

Transportation	25%
Industrial (manufacturing)	30%
Commercial (stores, etc.)	6%
Electricity production	25%
Residential	15%

Eugene P. Wigner   487

III. COULD OUR ENERGY CONSUMPTIONS OF THE VARIOUS CATEGORIES BE REDUCED?

The consumption for transportation plays, in our country, a much more significant role than in the other developed countries, except Canada, the situation of which is very similar to ours. Part of the reason for our high consumption for transportation is that our country is so large; if we want to visit a place in Washington state, we have to travel a great distance. In addition, our travels per mile consume much more energy than they do in the other developed countries. We travel much more by automobile and these are, on the whole, larger than those used in other countries and we drive them faster than they do. The reason for all this is, partly, that our gasoline is so inexpensive—its cost is about one third of that in other developed countries, excepting again Canada. Partly as a result of this, our public transportation system is inferior to those of Western Europe. It is not pleasant to go into the other reasons for this and I will not do it. It is evident, however, that the energy consumption for transportation could be much reduced if it were truly important to do so. It is pleasant, however, to record that the number of people who take pleasure in bicycle rides, rather than automobile excursions, has much increased. I do believe that this is a step in the right direction taken by people who increasingly realize the need to reduce our energy consumption.

Whether the industry's consumption of energy could be much reduced is difficult to tell, and I cannot. I recall that once when I complained to an Atomic Energy Commissioner about the large size of our energy consumption, he countered by pointing out that our energy consumption per gross national product is not very different from that of the other developed nations. This seems to indicate that if we want to maintain our industrial production the energy consumption of our industry would be difficult to reduce substantially.

The consumption of our commercial enterprises, in particular stores, is relatively small. People do like to have the stores well illuminated and do take pleasure in finding a luxurious atmosphere there. We could, but it is not clear whether we should, decrease this pleasure. It does not consume a large fraction of all the energy we use.

Much of our electricity is, unquestionably, used wastefully. One very often sees street lamps burning in broad daylight and I know that the lights in the houses are left on when not needed much more than in Europe. Part of the reason is again that the cost of electricity is, as a fraction of the income, much lower than in other countries—just as that of gasoline is. One may well hope that the attitude which diverted so many young people from automobiles to bicycles will also have an effect on electricity consumption in lodgings and that people will be more inclined to turn off the light if they leave a room.

The residential energy consumption, around 15 percent, refers to heating and also to air conditioning. It includes also cooking. As to the heating of the lodgings, the energy requirement for this could be reduced significantly by not using the heat of some fuel directly, but "pumping" heat from the outside into the lodging. This is being used in some countries—it is just the opposite of air conditioning which pumps heat from the (hopefully) cooler interior into the warmer outside. The heat pump brings heat from the cooler outside into the warmer inside. This requires, of course, "available energy," usually in the form of electricity, but requires much less "available energy" than the amount of energy pumped. This is now provided by the heating material, most of the energy of which is "available energy." The heat pumps installation would require, however, considerable investment and has not been considered seriously in our country on a significant scale.

How much energy could we save without changing our modes of life, without noticeably decreasing our comforts? This is very difficult to estimate and I can not recall having seen estimates. Our energy consumption doubled in the past 20 years which means that the per capita consumption increased in that period by a factor of 1.6. In addition, the consumption shifted heavily toward electrical energy. Some of the increase of our consumption would have been avoided had there been more incentive. I am convinced that the energy consumption will not double again in the next 20 years, but will approach a final value—we have no use for much more energy than we now use. In fact, I believe that our consumption *could* be reduced by about 20 percent without affecting our lifestyles. The conviction that we do approach a final value in our energy consumption finds some support in the fact that its increase in the past 20 years was much slower than that of countries with lower per capita consumption. In Western Europe the doubling time was 12 years, in Japan around 6 years. At our level of consumption we may be approaching a steady state. I believe we have, essentially, enough energy.

More generally, I believe that the problem of our society is shifting from problems of poverty to those of affluence. We do not know where to direct our efforts, our ambitions. The food and lodging, the procurement of which for ourselves and our loved ones used to be the purpose of the efforts of men, is now essentially assured. If I compare our affluence with the poverty which I saw in my early years in Hungary, I become speechless. The farmer or agricultural worker earned 25¢ a day for twelve hour's hard work in the summer, 25¢ a day. In the winter he earned less. Those 25¢ were worth as much as a full dollar now, perhaps as much as $1.25. But $1.25 for a twelve hour day! Truly, we live in affluence and I do not believe that we need more of it, or that we will continue to strive for much more of it with the intensity of past years. Hence, I also believe that the quest for more energy will subside. I realize that many of my friends disagree with me on this point, but I have to give my own convictions and surprisingly many people actually agree with me.

However, we surely want to maintain our present freedom of using energy for our needs and some of our extravagances and the sources which can satisfy our future energy requirements will be our next subject.

IV. Our Energy Resources

For the 10 kw energy consumption per person, coal furnishes 2.5 kw, oil, 6 kw, natural gas, 0.5 kw, nuclear reactors about 0.4 kw and around 1 kw comes from other sources, including hydroelectric power. The question which naturally arises is: how long will it be possible to continue this consumption, how long will our supplies last even if the energy consumption does not double again in the next 20 years, and so on. This is again difficult to estimate even if we assume that the consumption will remain constant. One reason for this is that what we mean by available supplies is the amount of the material that can be recovered at the present cost, using presently available technical means. Both the price and the technical methods of recovery change—the price, as a rule, increases, the techniques of recovery improve. As a result, the estimates for the magnitude of the energy resources show, as a rule, a tendency to increase. It must be realized, on the other hand, that at least some of the resources are needed also for other purposes, not only for energy production. At present, we cannot produce steel without coal, this is needed for the reduction of the mineral iron oxide to metallic iron. The reduction could be done also with electricity, but this appears at present impractical and, if the electricity is produced by burning coal, may consume more coal than the present reduction process requires. Of course, aluminum oxide is reduced to the metal by electrolysis—but that is a different story.

For what length of time do we have enough coal at the present rate of consumption? The very lowest estimate is a few hundred years, the highest about a thousand. There is a certain constancy in this estimate—it is the same as it was 40 years ago even though the consumption increased considerably in the meantime.

There is at least one additional reason for the uncertainty in the estimates of the size of our coal supplies. Most coal contains a certain amount of sulphur and this appears as sulphur dioxide in the combustion products and is discharged as such through the smoke stacks into the atmosphere. Sulphur dioxide is poisonous, its discharge pollutes the air badly. Hence, there is a valid objection to the burning of coal with a high sulphur content. I recall a discussion with a group of nuclear energy opponents in the office of the governor of Vermont, the Honorable Mr. Davis. I pointed to the fact that most coal-fired plants discharge more radioactivity than most nuclear plants and that, in addition, the former produce sulphur dioxide in amounts which cause more danger than the radioactivity. The opponents of nuclear energy said, in reply, that "as to the sulphur dioxide, we eliminate it!" When I asked, how do you eliminate it, the answer was "with machines!" and the methods of eliminating the sulphur dioxide have not assumed a much more concrete form since.

In addition to the requirement of using coal with a low sulphur content, there is also the objection to strip mining this material. The undesirability of this also contributes to the uncertainty in the estimates of the true availability of coal.

The situation is even worse with respect to oil. In the 1940s, when I became first interested in the energy question, it was estimated that we had enough oil for 10 years. In the meantime, the oil consumption increased so much that we now consume every year about one half of what was estimated then to be the total supply. The present estimate is that, with the present consumption, our supplies can last 13 years. But that is all—I am afraid that the present estimates are more realistic than were those of the '40s. But even these are uncertain; even when an oil field is considered "exhausted," more than half of the oil is still in the ground. Hence, the present 13 year estimate may also be too low, but, I fear, not very low.

Natural gas is, even at the present consumption, in lowest supply. It is not expected to last more, at the present consumption, than five or six years. Yet we depend a great deal on it—in Princeton where I come from, practically every house is heated by natural gas, and we also use it for our cooking.

Before entering the discussion of the energy which I'll take up last, let me mention a common property of those we just discussed. They are fossil fuels, derived from the remains of living organisms which, in their turn, derived their energy from the sun's rays. Their energy is freed by combustion, and the two most abundant, and surely unavoidable, combustion products are water and carbon dioxide. Water vapor, surely, appears harmless and carbon dioxide is not really poisonous either (carbon monoxide, which however is produced in very small quantities in a well-burning furnace, is very poisonous). However, carbon dioxide may have an effect on our environment by absorbing the infrared radiation emitted by the Earth while transmitting the visible radiation coming from the sun. This is called the "greenhouse effect" and may increase the temperature of our surroundings much more than the direct effect of our energy production. The temperature increase caused by the latter amounts to about 1/20 of a degree (centigrade) in our country though it is a bit larger in more densely populated ones. The greenhouse effect may amount to 2 degrees centigrade (about 4° Fahrenheit) and an increase of this magnitude may affect the vegetation significantly. A temperature increase in Vermont in the last 40 years has decreased the prevalence of maple greatly. I'll now turn to the consideration of the energy source which is of most interest for the present audience, that is of nuclear energy.

V. Availability of Nuclear Energy

There are two elements the nuclear energy of which we can use "in principle": thorium and uranium. Both contain about an equal amount of energy, about 2.5 million times more than an equal amount of coal. Thorium is about three times more abundant in the Earth's crust than uranium, but we use at present only the latter and of this principally only a small fraction, the isotope U^{235} which constitutes only 0.7 percent of the whole uranium. Actually, one of the prime arguments against the further installation of nuclear energy plants is that the nuclear fuel is close to exhaustion and this is, of course, more nearly true if only uranium, and only the small fraction of 0.7 percent of it is used. In fact, the cost of uranium has greatly increased in the past few years; the price of a pound increased from \$7 to about \$31. The cost of coal is about this much *per ton*. This means that, as fuel, if only U^{235} is used, its energy is only about 8 times less expensive than that of coal—of course, if all were used it would be about a thousand times less expensive.

In my opinion, the principal reason for the cost increase of uranium is that the people who own the uranium mines realize that its present price is low and that it may be worthwhile to hang on to it and to give it away only at a price which is not terribly far, per unit amount of energy, from that of coal. The actual amount of uranium in the Earth's crust is about 4 parts per million, which means that an average ton of rock contains about 10 times more energy in the form of uranium than does a ton of coal. This is no longer true if we take only the U^{235} into account—in that case the average rock contains about 14 times *less* energy than the same weight of coal. Even this is a great deal, but, of course, it would be quite expensive to break up the average rock to get the uranium out of it.

The uranium ores used at present contain about 1000 parts of uranium per million parts of ore, *i.e.*, the weight percentage of uranium is about 0.1. The amount of ore of this quality available in our country contains about 1,000,000 tons of uranium. If all of this could be used, it would cover all our energy requirements for about 1250 years, if only the U^{235} is burned, for only 10 years. But, of course, at present only a few percent of our energy derives from nuclear power, and even if this should double in the next few years as it is expected to do, the present grade ores can supply the consumption for some decades. The uranium ores with more than 500 parts per million uranium content are supposed to contain twice more uranium than those now used, those with 200 parts per million about 4 times more, about 4,000,000 tons. The cost of the extraction of the uranium from these ores would be more expensive than the extraction from the ores used at present, but the cost per unit amount of energy would still remain much below that of coal. If not only the U^{235} but the whole uranium could be burned, the cost, as compared with those of fossil fuels, would be quite negligible. Of course, the uranium in the average rock constitutes an essentially inexhaustible supply, but a probably very expensive one, and an unnecessary one.

VI. The Availability and Cost of Nuclear Energy

Why is it then that only such a small fraction of our energy requirements is furnished by nuclear power? This has many reasons and I do not believe I could enumerate and properly value all of them under any conditions. The first and perhaps most potent reason is that it took a rather long time to develop the technology of nuclear power, to make detailed plans for nuclear reactors furnishing electrical energy. Right after the Second World War those of us who collaborated on the design of nuclear reactors thought that in a few years nuclear energy will be preponderant. Well, there were some wiser people, particularly the physicist M. H. L. Pryce, who realized that it will take 30 years before nuclear energy will become economical and attractive, that it will take a lot of planning, a lot of engineering development, to make nuclear energy practical.

The 30 years have passed by now and nuclear energy is indeed practical and its contribution to our energy requirements is vigorously increasing. However, it is good to realize that its economic advantages are by no means as great as the comparison of the raw material prices for a unit amount of energy may indicate. As was mentioned before, this is about 8 times lower than that of coal and the comparison with oil is even more advantageous. One thousand kilowatt-hours of nuclear "energy" would cost, if all of it were "available," 0.3¢ at the present price of $31 per pound of uranium. Since with the present skill only the U^{235} is used and since we can use this only with an efficiency of about 32 percent, the actual contribution of the raw fuel cost to the cost of one kilowatt of electricity is about 0.13¢. The corresponding number for coal is about 0.85¢ per kwhr (the thermal efficiency of coal-fired plants is somewhat higher than that of nuclear plants). The ratio is still enormous, but there are several factors which decrease it considerably.

The first of these is that coal can be burned directly, the uranium must be put into the form of fuel-elements. First, its U^{235} content has to be increased—by a factor about 4—second it must be reduced to a metal, given a definite form and "clad" so that the heat transfer liquid (water in the present reactors) does not affect it chemically. These operations increase the price by a factor almost two. Second, the nuclear plants are more expensive to build than coal or oil-fired plants and the capital investment must be amortized. The amortization cost, according to a very informative article by D. J. Rose in the 1974 *Science*, is 0.78¢ per kwhr for oil-fired plants, 1.18¢ per kwhr for coal-fired plants, and 1.40¢ per kwhr for nuclear plants. This still leaves a considerable advantage for the nuclear plants, but surely not by a factor of 8 as the raw material prices would indicate. It so happens that I received in the past few days statements by two power companies which illustrate this: the Long Island Light Co. bought some of its power from a nuclear plant and this led to a saving of $225,000 a month which was used to reduce the bills of its customers. The Carolina Power and Light Co. saved $52 million in the last year—the power that would have cost $62 million was produced for $10.5 million—a fantastic decrease. Some people accuse the power companies of slanting their figures to favor nuclear energy. I cannot see any advantage that they could gain thereby—in fact, the building and planning of nuclear plants runs into so many bureaucratic obstacles that it seems to me that they would favor fossil-fueled plants if these could be run at the same price as nuclear ones. It may be worth remarking that according to David Rose's article nuclear reactors produce energy (electricity) at a price 25 percent lower than coal or oil-fired plants. Naturally, this ratio does depend on the distance of the plant from the coal mines, from the magnitude of the fluctuations of the energy demand, and probably some other circumstances.

It was mentioned before that, though uranium, even at its present price, is inexpensive as an energy source, the availability of the ore of the presently used quality is limited. If we used the uranium of this ore for all our energy requirements, and used only the U^{235} contained in it, we would run out of the ore in 10 years. It would be, therefore, clearly good if we could use all the uranium, not only 0.7 percent of it. This is made possible by the so-called breeding process which makes it possible to use, eventually, the energy of all the uranium—it increases the effective energy content of the ore by a factor of around 140! There are several ways to accomplish this—the arrangement which seems most popular at present is called liquid metal cooled fast breeder reactor. I will not describe this reactor in detail, nor the other breeding processes—it would lead us too far. I wish to mention though that there is a liquid metal cooled fast breeder reactor, called Phenix, in operation in France since the end of 1973. There are breeders also in the USSR and England. Even though the idea of the liquid metal cooled fast breeder originated in our country, we are quite far from having one in operation. This bothers me a great deal. Do we have too much red tape? Do we listen too much to people who object to all innovations which would make our lives easier, which would strengthen our country? Are other nations far superior to us in these regards? A second breeder, even larger than the Phenix, is being built in France and it will be in operation before any of ours.

Actually, breeders increase the energy supply not only by increasing the energy content of the ores now in use by a factor 140. They would also render the mining of ores, with much smaller uranium content than the presently mined ores, economical. After all, the "available" energy content of an ore with 140 times less uranium would be the same as that of our presently mined ores. This means that breeders would provide an essentially inexhaustible supply of energy—a point well worth remembering.

The red tape to be overcome in order to obtain permission even for the operation of a reactor of a type of which many are working is quite enormous. It takes about 8 years from decision to operation. This, in spite of the fact that only about 7% of the people living in the close neighborhood of reactors oppose these. Of the 8 years, almost 3 years are needed to convince the authorities that the reactor will not explode, that it will cause no real harm to the environment. The reactors which are functioning in our country have a potential output of 40 million electrical kilowatts. Those planned and in construction have a 5 times larger capacity. Much of the delay in getting these into operation is needed for the settlement of objections against building and operating them and a cursory discussion of these objections—whether reasonable or unreasonable—will be our next subject.

VII. Objections to Nuclear Energy

Before entering the discussion of the objections to nuclear energy, let me mention that I once read a small book on the history of railroads. When railroads were ready to be introduced, there was tremendous opposition to them. It was claimed that people living in the neighborhood of tracks would go crazy because of the noise, people inside the cars will be dumped on each other—it will be just terrible. There were also objections against the introduction of electricity into lodgings—people will be electrocuted by it. And, of course, there was some truth in these objections—just about two years ago, 106 people were killed in a railroad accident in Chicago. But if we compare the dangers caused by the technical developments with the benefits arising from them, the comparison favors these developments overwhelmingly. The life expectancy in the United States increased during the present century from 47 to 71 years in spite of the railroad accidents and the even much more numerous (around 50,000 per year) automobile accident caused fatalities. It seems that even though technical innovations do cause new types of accidents, their total life-saving capacity has, at least in the past, far outweighed their harmful effects. The question is, therefore, only: will this be true also for nuclear energy?

The two most common objections to nuclear energy are easily disposed of. The radioactivity released by most nuclear plants is smaller than the radioactivity released by a coal fired plant with the same energy output. The amounts of radioactive materials contained in coal are very small but 2.5 million times more coal has to be burned than uranium consumed to produce the same amount of energy. The sulphur dioxide problem of coal-fired plants was mentioned before. The second often heard objection is that nuclear plants may explode as bombs do. The U^{235} enrichment in the uranium used in our plants is far too small to render an explosion possible.

The existence of unreasonable objections to nuclear plants should not close our eyes to possible real dangers—after all railroads do have accidents and it is good to do all one can to minimize the real dangers. The first of these is that some error is made—there are scores of them which could cause trouble. It is important to employ highly competent people in supervisory positions and every effort should be made in this direction. This, if properly implemented, will also minimize the danger from sabotage, and I must admit that I am not fully satisfied with the efforts to minimize this danger. Guards often get bored and pay little attention to their duties and it is not impossible to sneak into "guarded" areas. I hope that more attention will be paid to the problem to keep the guards interested in their jobs and convinced of the importance thereof.

Bombings can destroy nuclear plants and scatter their radioactive materials over wide areas. In this way, the effect of a precisely aimed non-nuclear bomb can be increased. As to nuclear weapons, their direct effect is greater than the one they can cause by destroying a nuclear reactor. Nevertheless, it has been proposed to place the nuclear reactors underground and, in Sweden at least, this is done.

Chemical separation plants, dealing with vast amounts of radioactive materials, may present a problem. There is good reason to believe that they can be designed well enough to avoid calamities. The ultimate disposal of the radioactive waste products presents a problem which has been very much emphasized. Indeed, no final decision has been made about it so far. However, I am not very concerned about this problem. There are several ways to dispose of the waste products safely and the fact that no final and detailed choice between the methods has been made in no way indicates that these are unsafe. In addition, in ten years or so, it will be possible to shoot them up into empty space in rockets and they will be away from us for good. For the present, I do not recommend this even though we have so far had no accidents with rockets, because I consider, for the time being, the other disposal methods even safer. In ten years, rockets will be entirely safe.

Let me recall, in conclusion, that in spite of many years of operations of dozens of nuclear energy plants (we now have about 50 in operation) so far no accident involving a person was caused by them. Surely, safety measures had to be put into operation, just as semaphores operate to prevent collisions of trains. However, no health has been impaired so far—we have *nothing* to compare with the 100,000 coal miners affected by the black lung disease. This does not indicate, by any means, that we should relax our attention to safety problems, but it surely does speak against scare stories opposing nuclear power plants.

VIII. Other Sources of Energy

Let me now come to the discussion of possible future sources of energy. Even if those presently known were assured to be entirely satisfactory, we should not neglect their exploration.

Fusion is the most abundant alternative to present sources which is now in sight. This is a nuclear reaction also, but its burning products contain much less radioactivity by far than those of uranium or thorium. Fusion has been under development for a long time and, personally, I have been very optimistic about its technical feasibility. I continue to believe that its feasibility will be demonstrated in the reasonably near future. Since, however, this has not been done so far, the question of the economic problem can not yet be seriously attacked.

The next energy source on my list is the geothermal energy. If one digs about 90 feet underground, the temperature increases by 1°F, and so on. This means, as I have mentioned earlier, that there is some "available" energy present. In Iceland this is used to a considerable extent. It is, of course, a very meager source of energy. There is yet another energy source;

it is based on the fact that the temperature *decreases* as one sinks deeper and deeper into the oceans.

Finally, I want to mention solar energy. As I have already said earlier, the solar radiation coming to our country contains a thousand times more energy than the artificial energy we use. It nourishes, of course, the trees of our forests and the plants of our crops and in this way we do make use of it, even though, as far as energy is concerned, only a minute fraction of it. Some trees, however, such as the southern pine, can convert a few percent of the sun's rays energy into chemical energy and, as well known, the chemical energy of trees was the only source of artificial energy until a few hundred years ago. In Israel, and probably many other places, solar heat is used to heat houses and there are many interesting plans and ideas for using it for other purposes also, in particular also for the generation of electricity. These ideas are not only interesting, they may well turn out also to be very useful. We do not know.

Let me say in summary that there are probably many alternatives to the present sources of energy, fossil and nuclear. But we should not neglect the present ones—quite likely, objections can be raised also against the energy sources of the future, particularly if one wants to stop all technical developments. For the present, we should try to reduce the unquestionable dangers from the consumption of coal and reduce also our, and our friends', tragic dependence on the oil of other countries—this can work badly against ourselves and our freedoms. It would be very desirable also not to rely on a single source of energy, but have a variety available—perhaps fossil, nuclear and some future ones.

Comments on Dr. Wigner's Article*

A. Cobas

Revista Interamericana *VI*, 505–511 (1976)

In his article, "Our Needs for Energy and Ways to Satisfy Them," Dr. Eugene P. Wigner points out that the energy problem is not one of the total amount of energy in the world—since that is practically unlimited—but that only a very tiny fraction of the total is *available* for our use, *available* in the sense that such energy may be made use of through some appropriate technology. Dr. Wigner gives, as an example, the atmosphere, in which the enormous amount of existing energy is not available. Fossil and nuclear fuels, on the other hand, are substances with a high concentration of available energy, and the problem is that quantities of these substances are limited.

Dr. Wigner points out that the human body is a system which, under normal conditions, consumes the same amount of energy as a 100 watt bulb—which is fairly little. He indicates that the per capita consumption of energy in the United States is equivalent to that of a 10,000 watt machine, which is one hundred times greater than that used by the body. In all the other countries of the world, the per capita consumption of energy is less than in the United States, and in some countries it is very considerably less.

Though these statistics indicate the fact, Dr. Wigner does not point out that in the United States energy is wasted unnecessarily. Stewart Udall, in his book *The Energy Balloon*, calls attention to this wastage, giving as an example the many families with two or more automobiles, with their enormous engines functioning at 20% efficiency; ornamental lamps burning natural gas 24 hours a day; hermetically sealed skyscrapers which require constant artificial temperature control; buildings which often require more energy to operate than a town of 150,000 inhabitants, and all kinds of huge structures such as airports, sports facilities, shopping centers, etc. It is estimated that the per capita consumption of energy in the United States is the equivalent of each person having 200 servants.

Dr. Wigner thinks it is very difficult to know whether it is possible to

* Translated by Marshall Morris

reduce the consumption of energy by industry. However, the document "Potential for Effective Use of Fuel in Industry," published by the Thermo Electron Corporation and sponsored by the Ford Foundation Energy Project, indicates that by the use of technology it is possible to reduce energy consumption by 33% in one sector of industry, and that greater savings will be possible using industrial processes now on the technological horizon.

Dr. Wigner says that all indications are that in the United States it would be difficult to reduce the consumption of energy and at the same time maintain industrial production. However, in the report "A Time to Choose—America's Energy Future," published by the Energy Project of the Ford Foundation, we are informed of an analysis of the relation between the consumption of energy and the Gross National Product, using an econometric model designed by Data Resources, Inc., which concludes that a gradual reduction in energy consumption would have a surprisingly small effect. This is not the case when there is a sudden, unexpected interruption in the supply of fuels such as occurred in the case of the oil embargo by the Arabs.

In regard to the consumption of energy in the residential sector, Dr. Wigner asks how much energy we could save without changing the life style of the United States. It seems to me that given the rapidity with which we are consuming the fossil fuels which were produced millions of years ago, and cannot be replaced, the question ought not be how much we can save without changing our life style, but what changes we ought to make in our life style in order not to exhaust the fossil fuels which are the heritage of all humanity. There is no justification at all for generations—recent, present and in the immediate future—to use them up, forgetting the right of future generations also to enjoy their use. Dr. Wigner then arrives at the conclusion that in the residential sector we are coming to a stable level of energy use, but he does not give convincing data to support his thesis. He concludes, inexplicably, that the search for more energy will subside.

Dr. Wigner enumerates the energy resources of the inhabitants of the United States, indicating that oil accounts for 60% of these, coal for 25%, natural gas 5%, and nuclear reactors 4%, the rest being produced by hydro-electric and other sources. He also gives the following estimates of the time these will last: oil, 13 years (which I consider very low); coal, 100 to 1,000 years (which seems extremely high); natural gas, 5 years (very low). He calls our attention to the fact that coal mining causes serious ecological damage, and the burning of coal produces great quantities of sulphur dioxide, a harmful pollutant. He goes on to say that the burning of fossil fuels increases the carbon dioxide content of the atmosphere, which has grave consequences for the ecology.

Given this serious situation, Dr. Wigner offers nuclear energy as an ideal solution for the energy problem. According to his analysis, the use of nuclear energy is cheaper and less dangerous ecologically than the burning of fossil fuels; that, by making use of the breeder reactor, the supplies of uranium become an unlimited source of energy. In his enthusiasm for nuclear energy,

Dr. Wigner says: (a) that, due to the uranium content, a ton of any kind of rock contains ten times as much energy as a ton of coal, which is true, but at the present it is not possible technologically to make use of it; (b) that the electrical energy produced by nuclear reactors is cheaper than that produced by fossil fuel burning generators, which is true; (c) that there are breeder reactors functioning in France, England and Russia, and that their installation in the United States has been prevented by red tape; (d) that objections to the use of breeder reactors are not justified.

It is curious that Dr. Wigner should place such great emphasis on the pollution produced by the burning of fossil fuels while he gives very little importance to radioactive contamination produced by nuclear reactors or to the danger of possible catastrophic accidents. At the same time, he admits that he is not satisfied with the measures taken to prevent the sabotage of nuclear installations, and that it is quite possible to enter "protected" areas illegally. As to nuclear wastes, which involve the gravest dangers to mankind for periods of over a thousand years, Dr. Wigner thinks that this problem will be resolved when such wastes can be sent off into space by using rockets. This is not done now because present-day rockets are not safe enough for the task.

On October 1, 1976, a news item was published in the *New York Times* regarding the decision of the Atomic Safety and Licensing Appeals Board to stop construction indefinitely on a $6,000,000 nuclear center in Seabrook, New Hampshire, due to unsatisfactory answers to questions about the disposal of nuclear wastes. On October 9, in the same newspaper, it was reported that a fourth-year physics student at Princeton had designed an atomic bomb costing about $2,000 to build. The student pointed out that any physics major who pays attention to his teachers can design and build an atomic bomb if he can get the necessary materials. The bomb designed by this student is one-third the strength of the Hiroshima bomb exploded at the end of World War II.

In the same paper, on October 8, there was an item concerning the Stockholm International Peace Research Institute, indicating that within nine years 35 nations will be capable of manufacturing atomic bombs. Until now only six nations have exploded bombs: the United States, the Soviet Union, England, France, China and India. In May of 1974 India exploded an atomic bomb made of materials and equipment provided for peaceful purposes by the United States and Canada.

Dr. Wigner concludes his article with a brief discussion of other energy sources. He begins by mentioning the energy obtainable from the process of fusion, and indicates that this is the most abundant source of energy available today, and that it will soon be shown to be viable. I believe that Dr. Wigner is very optimistic. Then he makes mention of geothermal energy, energy obtainable from the conversion of thermal energy from the sea, and finally solar energy—to which only a few lines are given. It is inexplicable that Dr. Wigner should give so little importance to solar energy considering that this is the most abundant source of energy we have and that it is virtually inexhaustible. To give an idea of just how enormously abundant solar energy is, it is sufficient to say that the solar energy which reaches Puerto Rico in a single day is *five times* what the Autoridad de Fuentes Fluviales (the government power authority) generates in an entire year. This means that *if* we could convert the twentieth part of one percent of the available solar energy to electrical energy, we would have as much energy as Fuentes Fluviales now produces. This is not technologically feasible at present, but it makes it clear that if even a tiny fraction of the available solar energy were used, it would allow us to reduce petroleum imports by several millions of barrels per year.

In the October 1976 edition of *Foreign Affairs** Dr. Amory B. Lovins, Consulting Physicist and Representative for England to Friends of the Earth, Inc., published an article entitled "Energy Strategy: The Road Not Taken?" in which he outlines and contrasts two possible paths that we can follow in trying to resolve the energy problem. The first of these is to follow along the lines traced in the recent past, making use of advanced, centralized technology and increasing energy supplies. The second path combines a firm commitment to efficient use of energy, a rapid development of renewable energy sources, making sure that the quantity and the quality of the energy are adequate to necessities, and a transitional technology using fossil fuels. These two paths are mutually exclusive since a commitment to one makes it impossible to follow the other. According to Dr. Lovins we must make our decision soon.

The first path requires rapid expansion in three sectors: coal (principally by open-pit mining), oil and gas (largely from Arctic and off-shore wells), and nuclear fission (eventually using breeder reactors). Along this path, energy conservation is achieved more by price than by policy; that is, it is more rhetorical than real. Non-traditional energy production is relegated to an unimportant role and its contribution is postponed until the year 2000. It is held that this path can be maintained in the long run by using a combination of both fission and fusion breeder reactors. According to studies already made, this path would require the following in the next 10 years: 900 additional off-shore oil wells, and 170 additional open-pit coal mines which would have to produce 320 million tons per year. The nuclear reactors would require 100 additional uranium mines, one additional unit for enriching uranium 235, 40 additional factories to produce the fuel elements for the reactors and three installations for reprocessing. The electrical system required for this would include 180 coal-burning electrical generating plants of 800 megawatts, 140 reactors of 1,000 megawatts, 160 hydroelectric plants and 350 gas turbines. To construct these units 100,000 engineers, 420,000 skilled workers and 140,000 unskilled workers would be required. By the year 2000 there would be 450 to 800 reactors (including 80 breeder reactors

* Vol. 55, No. 1, pp. 65–96.

each using 2.5 metric tons of plutonium), from 500 to 800 coal-fueled electrical generating plants, from 1,000 to 1,600 additional coal mines and 15 million electric automobiles. Massive electrification would produce a volume of waste heat that would increase the temperature of the rivers of the United States, excluding Hawaii and Alaska, from 34° to 49° Farenheit. The volume of carbon dioxide in the atmosphere would double, producing substantial and irreversible changes in the climate of the globe.

The second path involves technical modifications in the ways energy is used. In the first place, energy-saving technologies would be employed, permitting us to produce the same volume of products and services as at present but with an expenditure of much less energy. The means include the following: thermal insulation, more efficient motors, reduction of excessive illumination and ventilation in commercial buildings, and the use of devices to recuperate waste heat in industry. With these technical modifications, efficiency in the use of energy could be doubled, and we could keep the same level of economic activity with primary energy consumption approximately fixed for the next decades, and in this way stretch our present supplies rather than have to augment them massively. It is well known that Americans could considerably reduce energy consumption if they were as efficient as the Swedes.

Some of these changes have already come about and are functioning. The 1977 cars will get 27% more mileage per gallon of gas than those produced in 1974. Kitchen utensils waste great amounts of energy. The design and construction of more efficient utensils would produce twice the number of jobs as the construction of electrical generating plants—which is the least labor intensive of industries.

It is estimated that the capital investment required to reduce the consumption of oil by a barrel a day varies from zero to $25,000, in comparison with estimates of from $100,000 to $300,000 to produce an increase in the energy generated by the equivalent of a barrel of oil per day in generators that burn coal or other fossil fuels.

Another area of energy saving along the second path is what is called co-generation which consists of the generation of electrical energy as a by-product of steam produced in manufacturing. It is estimated that by 1985 American industry could supply half its own needs for electrical energy by this technique. Co-generation could save $20 to $50 billion in capital investment, and from 2 to 3 million barrels of oil per day, and it would eliminate the necessity to construct 50 large, nuclear generating plants. At present, co-generation provides 29% of the electrical energy produced in West Germany but only 4% in the United States. These modifications, besides being permanent, would cost less than increasing the supplies of energy.

The barriers to the efficient use of energy are neither technical nor, for the most part, economic. They are the product of 3,000 conflicting and obsolete building codes, of a construction industry resistant to change, of a lack of mechanisms to facilitate the transition from kinds of jobs we no longer need to those that we now do, of the way engineers are paid (by fixing a percentage of the cost of equipment installed), of a fragmentation of government responsibility, etc.

The technologies required for the first path are called *hard technologies,* and those of the second *soft technologies.* The characteristics of the soft technologies are these:

1. They depend on renewable energy sources which will always be available, whether we use them or not, such as solar, wind and vegetable energy. It is a matter of using energy "income" that will last indefinitely rather then energy "capital" which can be exhausted.
2. There are many forms, and the energy supply is the product of many small, individual contributions.
3. They are flexible and involve a technology which is simple to use without esoteric skills.
4. They adapt in scale and geographic distribution to the purposes to which they will be put.
5. They adapt in regard to the quality of the energy produced to the purposes to which they will be put.

People do not want either electricity or oil as such; they want comfortable homes, light, vehicles, food and other things which are real and palpable. In the United States 58% of the energy is needed, in its final form, as heat—about half above and half below the boiling point of water. In addition, about 38% is used for driving machines: 31% for vehicles, 3% for tubing, and 4% for industrial engines. The remaining 4% is used for illumination, electronic equipment, telecommunications, electrometallurgy, electrochemistry, arc-welding and electric motors for kitchen utensils. From all this one comes to the conclusion that we use fuels of high quality and electricity for many purposes not requiring such high quality. Where we need to produce heat on a scale of tens of degrees we ought to use an energy source capable of producing tens or hundreds of degrees, not a flame whose temperature is thousands of degrees or a nuclear temperature of millions of degrees. This is like cutting butter with a mechanical saw.

Soft technologies can produce savings in the following ways:

1. They can reduce overhead. Approximately half of the charges that you pay for the electrical energy that you consume are due to overall costs: transformers, cables, meters, and the people necessary to read them, the planners and computers, etc.
2. They eliminate the losses of energy involved in distribution.
3. They eliminate the need for large reserve units to supply sudden energy requirements.
4. They eliminate the costs of errors in the calculation of demands for energy.
5. They permit the construction, on a massive scale, of small units with the consequent savings in production costs.

The use of fusion energy involves the following dangers:
1. The neutrons could be used to make materials for nuclear bombs.
2. If it turns out to be a "dirty" source of energy as many experts think it will, we will use it anyway; and if it turns out to be "clean," the volume of waste heat produced would alter the temperature of the earth dangerously.
3. The use of a complex, costly and centralized technology to produce electricity takes us in the direction opposite to the one we ought to follow.

Hard technology and soft technology both carry us into a post-petroleum, post-gas era. Soft technology is simple, inexpensive and uncomplicated; hard technology is extremely costly and complex. The investment costs of soft technology are small and the user himself covers them; hard technology involves enormous costs and the government has to cover them. Risks are consequently greater for hard technology. Hard technology depends on a few, highly complex technologies while soft technology depends on a great variety of simple technologies.

All that has been said here with reference to the United States applies to Puerto Rico, but with the added, weighty factor that Puerto Rico does not have its own domestic sources of energy. Almost 100% of the energy consumed in Puerto Rico comes from imported oil. The possibilities of getting oil from the fields which are said to exist off the coast of Puerto Rico are very remote. In contrast to this uncertainty we have the sure fact of the enormous amount of solar energy which reaches us daily and which we do not use. The task before us is simple and clear: *our future is in the use of solar energy*.

Response to Dr. Cobas' Comments*

E. P. Wigner

Revista Interamericana *VI*, 512–514 (1976)

Several of Dr. Cobas' statements are in agreement with my views. For example, it surely would be useful to economize and not to allow our energy consumption to increase much more, and certainly not to the extent that Dr. Cobas assumes it will. I also heartily concur with his suggestion to reduce the consumption of fossil fuels. As we both observed, these will continue to be needed for a variety of other functions, including the reduction of iron ores to iron. I also mentioned the danger of the "greenhouse effect," an increase in the temperature of our environment by the accumulation of carbon dioxide produced by the burning of coal, oil, or natural gas. I concur too with the suggestion to provide more effective insulation for our homes and more efficient motors for our transportation media. Dr. Cobas does not mention the possibility of using heat pumps in our homes, but this too would be desirable. Also, I am wholeheartedly in favor of the further exploration of the use of solar energy.

Dr. Cobas suggests that the power consumption of industry should be reduced, but proposes no concrete measures towards achieving this goal. It should be pointed out that, at present, U.S. industry hardly consumes more power per unit gross national product than the industries of other nations.

Some of Dr. Cobas' proposals are not very clearly defined. He proposes the production of "electrical energy as a by-product of the steam produced in factories." Which factories produce significant amounts of steam and how is that steam to be used? Does he mean, conversely, that the heat rejected when electricity is generated can be used for the heating of homes? This would be a worthwhile proposal, even though it would require considerable investment to make it operational.

Dr. Cobas criticizes the relatively low estimate given in my article for U.S. power requirements. These estimates are based on the belief that we shall

* I am much indebted to Dr. Bernard Lowy for helping me to formulate these comments on Dr. Cobas' article. [Dr. Lowy is an ethnomycologist, Director of the Louisiana State University Herbarium, and Consulting Editor of the *Revista/Review Interamericana*. Ed.]

Eugene P. Wigner   513

not want to consume much more energy in the future than we are doing now. Similarly, I do not believe we want to consume more food than we now consume. That some estimates of our future energy consumption are grossly exaggerated is supported by the observation that our per capita energy consumption increased only by a factor of 1.6 in 20 years, whereas that of Western Europe doubled in 12 years, that of Japan in 6 years. Does this not indicate that we are approaching the fulfillment of our energy demands? Dr. Cobas does not comment on this question.

Another of Dr. Cobas' points that is not very clear is his doubt that an increase in the cost of energy will induce the more economical use thereof. He does not believe that "energy conservation is achieved more by price than by policy; that is, it is more rhetorical than real." Is it not reasonable to assume that a higher cost of energy would induce individuals, and industries probably to an even greater extent, to adopt measures to reduce their energy consumption? Finally, what is meant by "soft technologies"?

Let us now turn to those points on which I disagree with Dr. Cobas. The most important of these is associated with his fear that unless we discontinue the operation and construction of nuclear reactors, terrorists will acquire nuclear explosives and create havoc. We are as apprehensive of terrorist acts as he is and in my article I suggest that more attention be given to this problem. However, abolishing nuclear power would give very little assurance against the acquisition of nuclear materials by a terrorist, as long as other nations have nuclear reactors. Nuclear materials can be stolen from, or even furnished by, other countries, then clandestinely imported. In fact, it is highly probable that the perpetrators of an explosion could be more easily traced, had they previously visited one of our nuclear installations. Consequently, the terrorists would do better acquiring their explosives in another country. Even more important is the fact that other terrorist weapons such as pathogenic bacteria can be used to contaminate a water supply. These organisms can be acquired with relative ease and a saboteur is far more likely to escape detection. Altogether, I do not believe that the operation of nuclear reactors adds significantly to the danger of terrorist acts.

I should like to add a remark on the bomb design made by a Princeton University student mentioned by Dr. Cobas. As it happens, I was one of the readers of that student's thesis. I thought it showed considerable skill and understanding and it was therefore recommended for acceptance. Nonetheless I must emphasize that the equipment designed by that student would never have exploded. This is not surprising, since the design of the nuclear bomb which eventually did explode took the concentrated effort over a considerable period of time of many of our most able physicists. Their work could hardly be duplicated by a single individual's sideline work.

Dr. Cobas criticizes the author, who "places such great emphasis on the pollution produced by the burning of fossil fuels while he gives very little importance to radioactive contamination produced by nuclear reactors or to the danger of possible catastrophic accidents." It was pointed out, however, that the burning of most coal causes more radioactivity to escape into the atmosphere than is emitted by a nuclear reactor with the same power production. This is, perhaps, difficult to believe but true nonetheless. Furthermore, most coal contains sulphur, and its oxidized product, sulphur dioxide, creates a much greater health hazard than its radioactivity. We hope that the means will be discovered to reduce the emission of sulphur dioxide by coal burning plants, and we expect likewise, that the safety of nuclear reactors will be still further perfected. This remains desirable even though, in contrast to the record of some 100,000 cases of black lungs among coal miners, nuclear reactors have *not caused a single known accident* to date.

Regarding the employment question, it would be interesting to know the reason for Dr. Cobas' objection to the large number of engineers (100,000), skilled workers (420,000), and unskilled workers (140,000) which the construction of nuclear reactors would require, since he strongly approves of the fact that fossil fuel plants would "provide work" for many people? Is there not a "slight" contradiction between these views? In any case, I do not think it likely that the design, construction, and supervision of 830 units of machinery, many of them identical, would require the work of 100,000 engineers for 10 years.

The refutation of some of Dr. Cobas' arguments should not obscure the fact that in spite of our different interpretations regarding the uses of nuclear power, I am grateful to him for having stated his views. Only by considering such an exchange of ideas will it be possible for the layman to reach an informed decision. As to myself, the reading of the points he makes, which I assume to be a fair summary of the objections to nuclear power, strengthens my belief that an intelligent and careful use of this power will be beneficial in the long run.

Books on Reactor Theory
and Nuclear Physics

Eisenbud, L. and Wigner, E. P., *Nuclear Structure*, Princeton University Press, Princeton, New Jersey, 1958.

Weinberg, A. M. and Wigner, E. P., *The Physical Theory of Neutron Chain Reactors*, University of Chicago Press, Chicago, Illinois, 1958.

PART IV

The Wigner Patents

The Wigner Patents

Annotation by Alvin M. Weinberg

Wigner holds 37 U.S. patents on nuclear reactors. The 12 most important patents are reproduced in their entirely; derivative patents are listed only by title (T).

Graphite-Moderated

Hanford Plutonium-Producing Reactor

	Date of filing	U.S. Number	Inventors	Title
1P	1/23/45	2,910,418	E. Creutz L. A. Ohlinger A. M. Weinberg E. P. Wigner G. J. Young	Neutronic Reactor

1P is the original Hanford patent, based upon paper # 16, CE-407.

	Date of filing	U.S. Number	Inventors	Title
2P	5/28/45	2,856,339	E. P. Wigner G. J. Young	Neutronic Reactors
3P-T	8/28/45	2,830,944	E. P. Wigner A. M. Weinberg G. J. Young	Neutronic Reactor

2P is the earliest version of Hanford; 3P-T followed by a few months. In both designs the fuel is disposed vertically, and is cooled internally.

	Date of filing	U.S. Number	Inventors	Title
4P-T	12/19/44	2,890,158	L. A. Ohlinger E. P. Wigner A. M. Weinberg G. J. Young	Neutronic Reactor
5P-T	5/22/45	2,850,447	A. M. Weinberg G. J. Young L. A. Ohlinger E. P. Wigner	Neutronic Reactor

4P-T and 5P-T describe methods of unloading the radioactive fuel from Hanford.

	Date of filing	U.S. Number	Inventors	Title
6P-T	2/20/46	2,886,503	L. Szilard E. P. Wigner E. Creutz	Jacketed Fuel Elements for Graphite Moderated Reactors
7P-T	5/8/46	2,872,401	E. Creutz L. Szilard E. P. Wigner	Jacketed Fuel Elements

6P-T and 7P-T describe the design of the jacketed uranium slugs that was actually used at Hanford. These ideas were crucial since jacketing the slugs was one of the most difficult problems at Hanford.

	Date of filing	U.S. Number	Inventors	Title
8P-T	5/22/45	2,853,624	E. P. Wigner G. J. Young	Radiation Shielding Device
9P-T	9/18/45	2,856,339	E. P. Wigner L. Ohlinger	Neutronic Reactor Shield and Spacer Construction

8P-T and 9P-T give details of this "wet" shield (water and iron) for Hanford. DuPont substituted masonite for water in the actually built reactors.

	Date of filing	U.S. Number	Inventors	Title
10P-T	5/14/45	2,990,357	A. M. Weinberg E. P. Wigner G. J. Young	Method and Apparatus for Controlling Neutron Density

10P-T is the first description of a method of flattening the power distribution in Hanford-type reactors by enlarging the water cooling annulus around the slugs at the center of the reactors.

	Date of filing	U.S. Number	Inventors	Title
11P-T	8/18/47	2,806,820	E. P. Wigner	Neutronic Reactor

11P-T is a version of Hanford in which the coolant is superheated steam, not unlike the Chernobyl RMBK reactor.

Air-Cooled Graphite-Moderated

	Date of filing	U.S. Number	Inventors	Title
12P	6/15/52	2,928,781	E. P. Wigner E. Creutz	Cooled Neutronic Reactor

12P is essentially the Brookhaven split-flow, air-cooled reactor.

D$_2$O-Moderated

These patents arose from the work of the P-9 project during 1944.

	Date of filing	U.S. Number	Inventors	Title
13P	8/29/45	2,736,696	E. P. Wigner L. Ohlinger G. Young A. M. Weinberg	Reactor

13P is the original description of D$_2$O-moderated reactors, including homogeneous slurry as well as heterogeneous systems.

	Date of filing	U.S. Number	Inventors	Title
14P-T	5/3/46	2,770,591	E. P. Wigner L. Ohlinger G. Young A. M. Weinberg	Heavy Water Moderated Neutronic Reactor
15P-T	11/6/45	2,810,689	E. P. Wigner L. Ohlinger G. Young A. M. Weinberg	Fluid Moderated Reactor

14P-T and 15P-T are refinements of the heavy water moderated heterogeneous reactors described in 13P. In 14P-T, the coolant is light water. In 15P-T, the heavy water moderator also serves as coolant.

	Date of filing	U.S. Number	Inventors	Title
16P-T	8/28/45	2,961,362	E. P. Wigner	Neutronic Reactors

16P-T describes a reactor in which the moderator is a combination of graphite and heavy water.

	Date of filing	U.S. Number	Inventors	Title
17P	8/29/45	3,052,613	E. P. Wigner L. Ohlinger G. Young A. M. Weinberg	Method and Apparatus for Conducting a Nuclear Chain Reaction

17P is the basic patent on the D$_2$O-moderated, homogeneous slurry reactor.

	Date of filing	U.S. Number	Inventors	Title
18P-T	3/01/45	3,070,529	E. P. Wigner	Neutronic Reactor

18P-T describes a boiling homogeneous slurry reactor.

Breeders and Converters

	Date of filing	U.S. Number	Inventors	Title
19P	2/04/46	2,954,335	E. P. Wigner	Neutronic Reactor

Date of filing	U.S. Number	Inventors	Title

19P is the first detailed description of a heterogeneous D_2O or H_2O moderated thermal reactor for converting Pu^{239} into U^{233}. This design was Wigner's response to the discovery in 1943 at Los Alamos that a gun-type Pu^{239} bomb would not work. Many of the ideas now embodied in light water reactors for power production and research derive from this original design.

| 20P-T | 5/14/46 | 2,986,510 | L. Szilard
R. F. Christy
F. L. Friedman | Massive Leakage
Radiator |

20P-T is a very early generalized description of a converter in which the isotope to be bombarded with neutrons surrounds the chain reacting core. Both heterogeneous and homogeneous embodiments of the core, and various moderators, are covered.

| 21P | 11/13/45 | 2,815,321 | E. P. Wigner
L. Ohlinger
G. Young | Isotope Conversion
Device |

21P is the first description of a circulating $D_2O\text{-}UO_2$ slurry converter in which the fertile ThO_2 is also circulated as a slurry in the blanket.

| 22P-T | 8/27/46 | 2,743,225 | E. P. Wigner
A. M. Weinberg
G. Young
L. Ohlinger | Reactor |
| 23P-T | 11/02/45 | 2,990,355 | L. W. Nordheim
E. P. Wigner | Means for Controlling
Reactions |

22P-T and 23P-T are refinements on the basic slurry system of 21P: 22P-T introduces a neutron-transparent tank between core and blanket; 23P-T describes a control rod in the blanket of a homogeneous converter in which the rod's effectiveness is increased by immersing it in pure heavy water.

| 24P | 11/13/45 | 2,860,093 | E. P. Wigner
L. Ohlinger | Isotope Conversion
Device and Method |

24P describes a homogeneous converter in which the slurry is pumped back and forth through the heat exchanges by means of pulsating gas pressure. Wigner invented this scheme as a way of avoiding radioactive contamination of circulating pumps, but the idea was never tried in practice.

| 25P | 3/19/47 | 3,102,851 | E. P. Wigner | Neutronic Reactor
System |

Date of filing	U.S. Number	Inventors	Title

25P, based on work at Clinton laboratories in 1946, describes a homogeneous solution breeder. The Homogeneous Reactor Experiment and the Homogeneous Reactor Test, both built at Oak Ridge, embodied the ideas of 25P.

26P-T	10/08/46	2,969,311	E. P. Wigner A. M. Weinberg	Means for Producing Plutonium Chain Reactions

26P-T is the first description of a heterogeneous converter moderated and cooled by H_2O and operating with resonance energy neutrons (0.5 to 100 e.v.)

27P	6/14/48	2,993,850	H. Soodak E. P. Wigner	Fast Neutron Breeder

27P is the basic patent on the liquid metal fast breeder (LMFBR). The coolants mentioned were liquid Na, Na-K, Bi, and Pb-Bi.

Materials Testing Reactor (High Flux Pile)

	Date of filing	U.S. Number	Inventors	Title
28P	10/14/52	2,831,806	E. P. Wigner	Neutronic Reactor
29P-T	6/08/53	2,832,732	E. P. Wigner	Fuel Assembly for Neutronic Reactor
30P-T	5/25/53	2,985,574	H. W. Newson E. P. Epler T. E. Cole E. P. Wigner	Overall Control System for High Flux Pile

28P, 29P-T, and 30P-T are the basic patents on the Materials Testing Reactor which was largely designed while Wigner was at Clinton Laboratories in 1947. MTR was the first high flux reactor (30,000 megawatts, maximum thermal flux 4×10^{14} neutrons/cm^2/sec). 28P describes the curved fuel plate invented by Wigner, 29P-T focuses on the flux trap, which was embodied in MTR, and 30P-T gives details of the control system. Since the MTR was a highly enriched reactor, too rapid withdrawal of control rods could lead to a prompt criticality excursion. 30P-T shows how this problem was dealt with.

Miscellaneous Patents

	Date of filing	U.S. Number	Inventors	Title
31P	8/04/47	2,781,307	E. P. Wigner	Apparatus for the Measurement of Neutron Absorption

Date of filing	U.S. Number	Inventors	Title

31P describes the pile oscillator method of measuring neutron absorption cross sections.

32P-T 5/28/45	2,833,618	E. Creutz E. P. Wigner	Separating Uranium- Containing Solids Suspended in a Liquid
33P-T 3/29/45	2,849,284	E. P. Wigner W. A. McAdams	Method of Separating Uranium Suspensions

32P-T describes the use of an organic liquid that preferentially wets the uranium oxide and forms a separate liquid phase. 33P-T adds H_2O_2 to the uranium slurry to convert the colloidal UO_2 to non-colloidal U_2O_4 which precipitates out of the original slurry. These two patents illustrate Wigner's grasp of chemistry.

34P-T 6/15/45	2,856,340	E. P. Wigner R. R. Williamson	Shaped Fissionable Bodies

34P-T describes a fuel element with fins to improve heat transfer.

35P-T 5/14/46	2,975,115	L. Ohlinger E. P. Wigner G. Young	Bellows Seal on Reactor Coolant Tube

35P-T describes a bellows seal for the Hanford reactors.

36P-T 10/08/46	2,873,243	E. P. Wigner L. Ohlinger G. Young A. M. Weinberg	Means for Shielding and Cooling Reactors
37P-T 1/09/47	2,874,307	E. P. Wigner L. Ohlinger G. Young A. M. Weinberg	Reactor Shield

These two patents describe variants of shields based on hydrogen-containing liquids. 36P-T describes a shield consisting only of a thick layer of hydrocarbon liquid that floats on top of the water coolant. 37P-T is an iron-water shield.

Oct. 27, 1959   E. C. CREUTZ ET AL   2,910,418

NEUTRONIC REACTOR

Filed Jan. 23, 1945   14 Sheets—Sheet 1

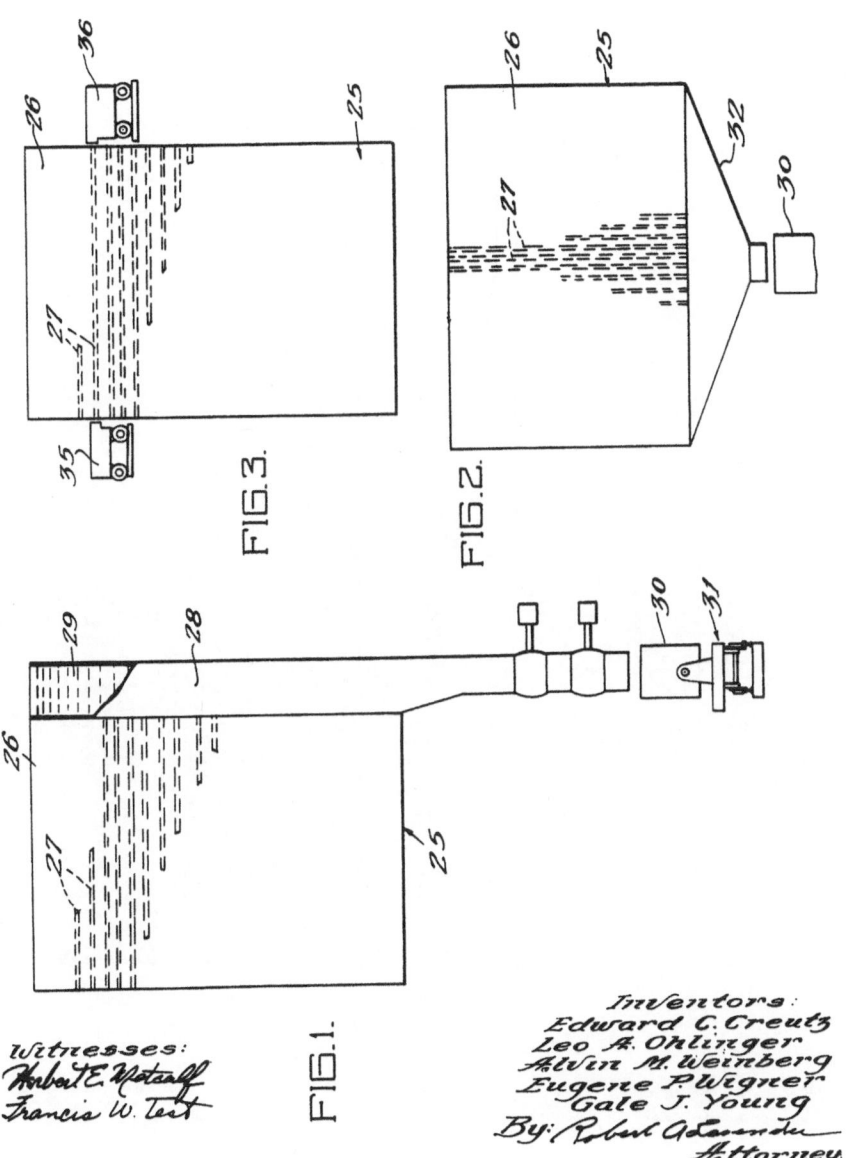

FIG. 3.

FIG. 2.

FIG. 1.

Witnesses:
Herbert E. Metcalf
Francis W. Test

Inventors:
Edward C. Creutz
Leo A. Ohlinger
Alvin M. Weinberg
Eugene P. Wigner
Gale J. Young
By: Robert A. Lavender
Attorney

Oct. 27, 1959   E. C. CREUTZ ET AL   2,910,418

                            NEUTRONIC REACTOR

Filed Jan. 23, 1945   14 Sheets—Sheet 2

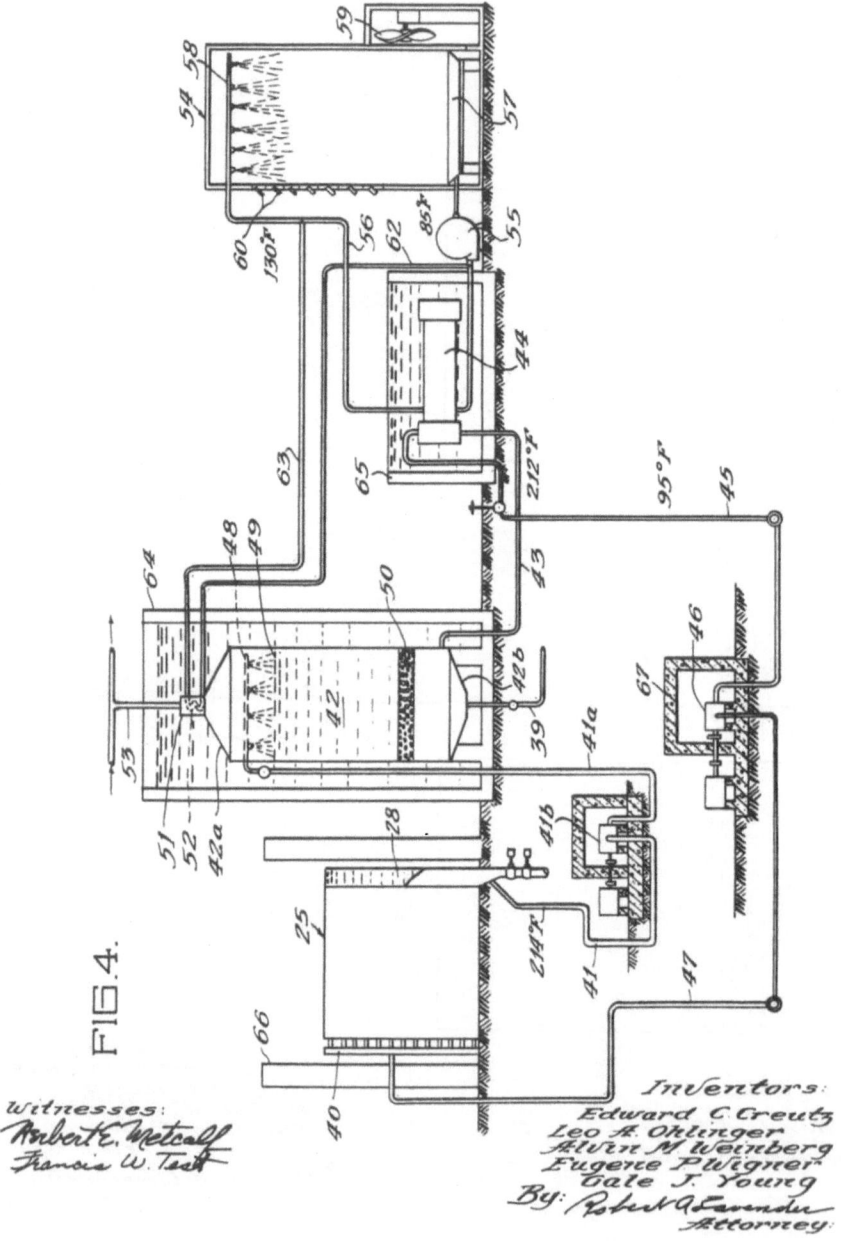

FIG. 4.

Witnesses:
Robert E. Metcalf
Francis W. Test

Inventors:
Edward C. Creutz
Leo A. Ohlinger
Alvin M. Weinberg
Eugene P. Wigner
Gale J. Young
By: Robert A. Lavender
Attorney.

Oct. 27, 1959   E. C. CREUTZ ET AL   2,910,418
                                NEUTRONIC REACTOR

Filed Jan. 23, 1945   14 Sheets-Sheet 3

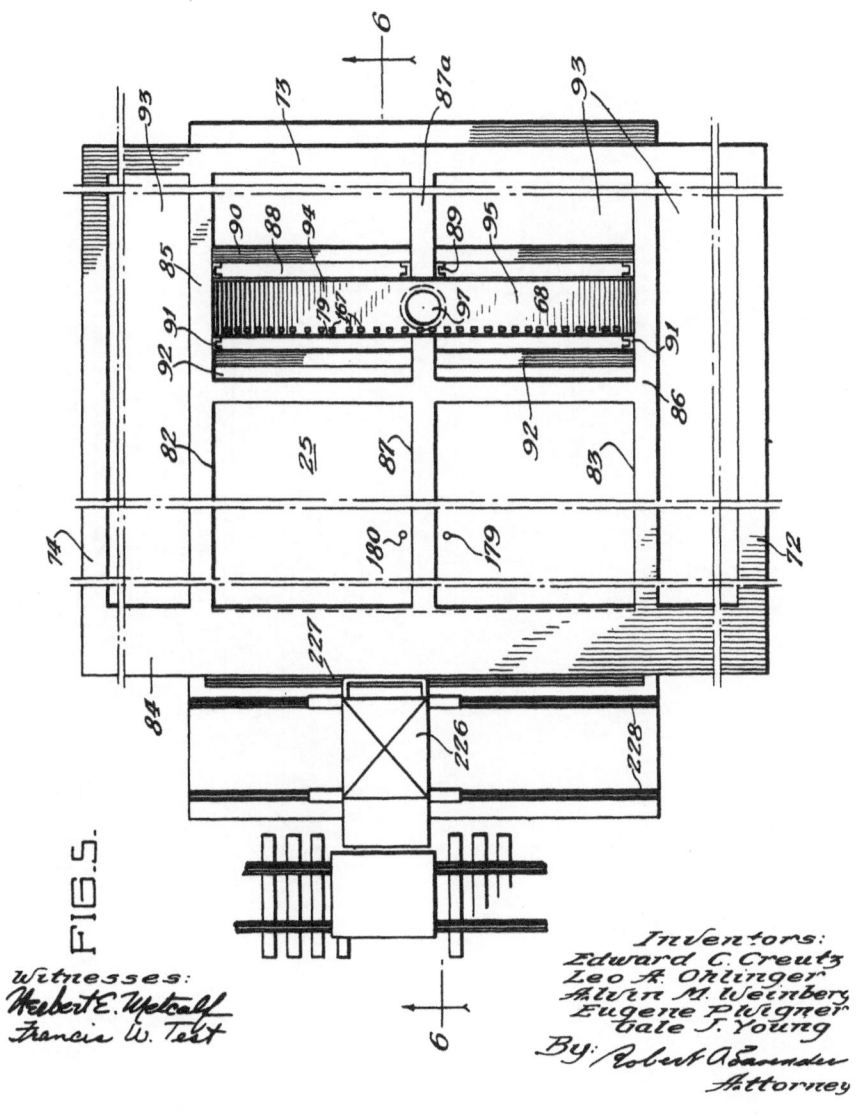

FIG. 5.

Witnesses:
Herbert E. Metcalf
Francis W. Test

Inventors:
Edward C. Creutz
Leo A. Ohlinger
Alvin M. Weinberg
Eugene P. Wigner
Gale J. Young
By: Robert A. Lavender
        Attorney

Oct. 27, 1959   E. C. CREUTZ ET AL   2,910,418

NEUTRONIC REACTOR

Filed Jan. 23, 1945   14 Sheets—Sheet 4

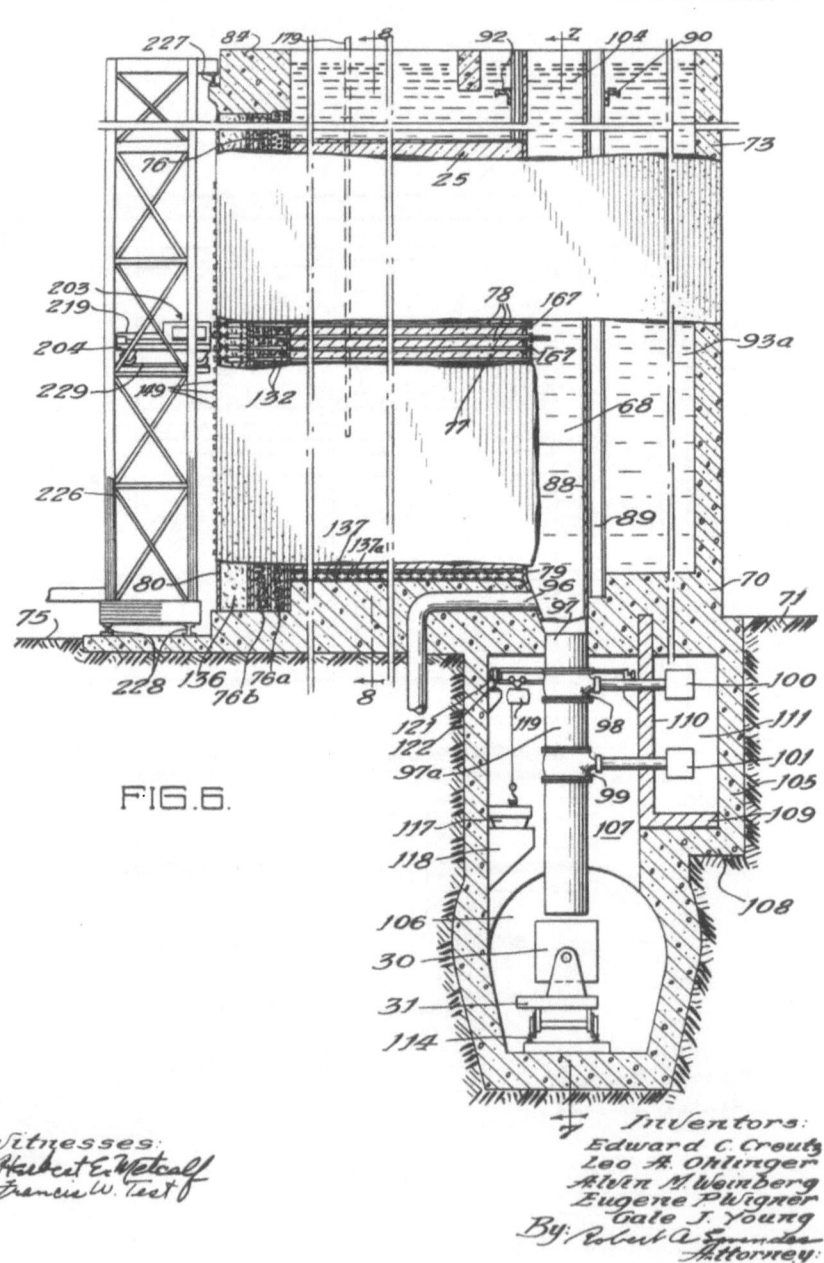

FIG.6.

Inventors:
Edward C. Creutz
Leo A. Ohlinger
Alvin M. Weinberg
Eugene P. Wigner
Gale J. Young
By: Robert A. Spencer
Attorney.

Oct. 27, 1959   E. C. CREUTZ ET AL   **2,910,418**

NEUTRONIC REACTOR

Filed Jan. 23, 1945   14 Sheets—Sheet 5

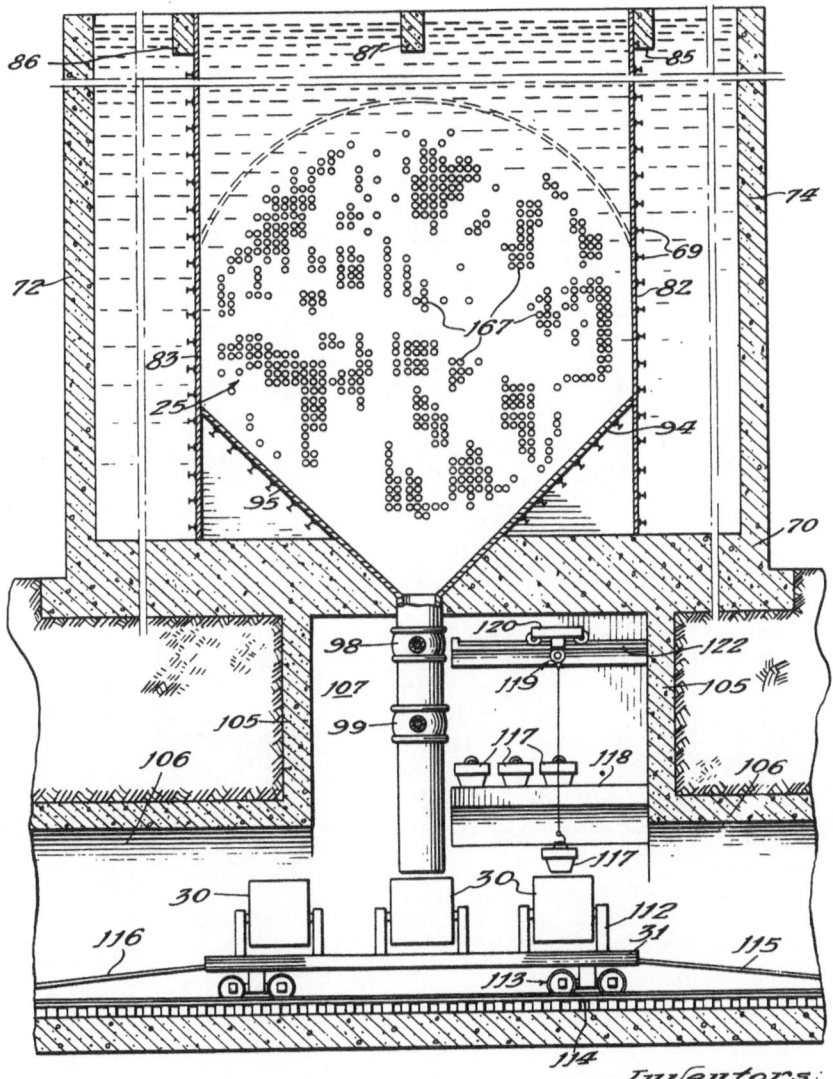

FIG.7.

Witnesses:
Herbert E. Metcalf
Francis W. Test

Inventors:
Edward C. Creutz
Leo A. Ohlinger
Alvin M. Weinberg
Eugene P. Wigner
Gale J. Young
By: Robert A. ___
Attorney

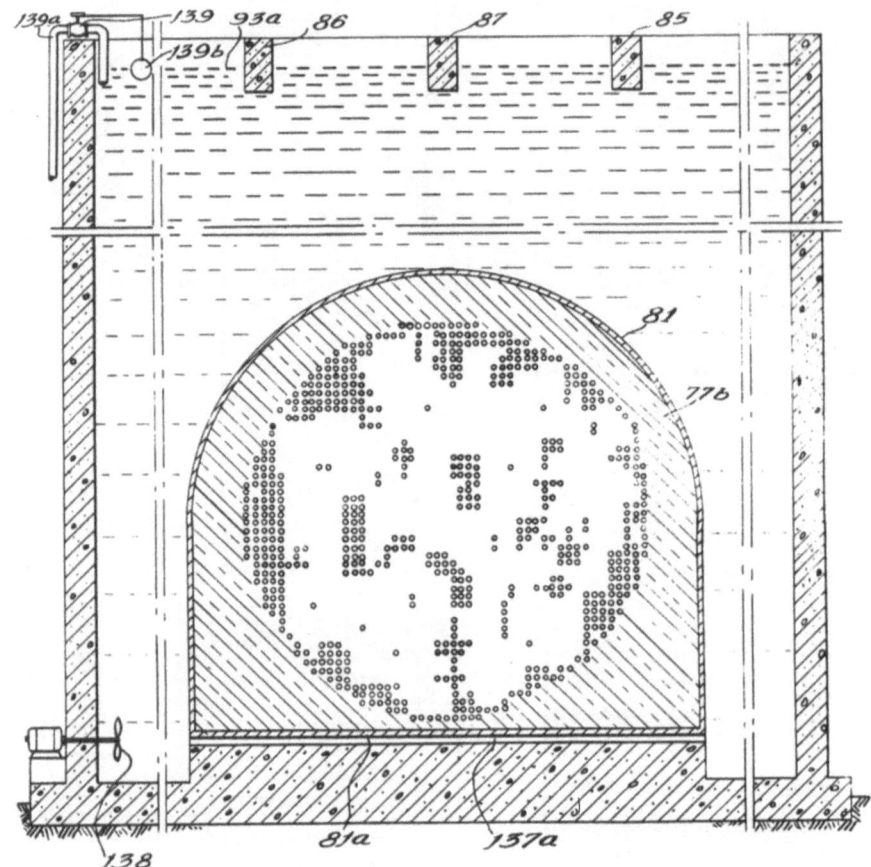

Oct. 27, 1959   E. C. CREUTZ ET AL   2,910,418

NEUTRONIC REACTOR

Filed Jan. 23, 1945   14 Sheets-Sheet 6

FIG. 8.

Witnesses:

Herbert E. Metcalf
Francis W. Test

Inventors:
Edward C Creutz
Leo A. Ohlinger
Alvin M. Weinberg
Eugene P. Wigner
Gale J. Young
By: Robert A Summder
Attorney

Oct. 27, 1959   E. C. CREUTZ ET AL   2,910,418

NEUTRONIC REACTOR

Filed Jan. 23, 1945   14 Sheets-Sheet 7

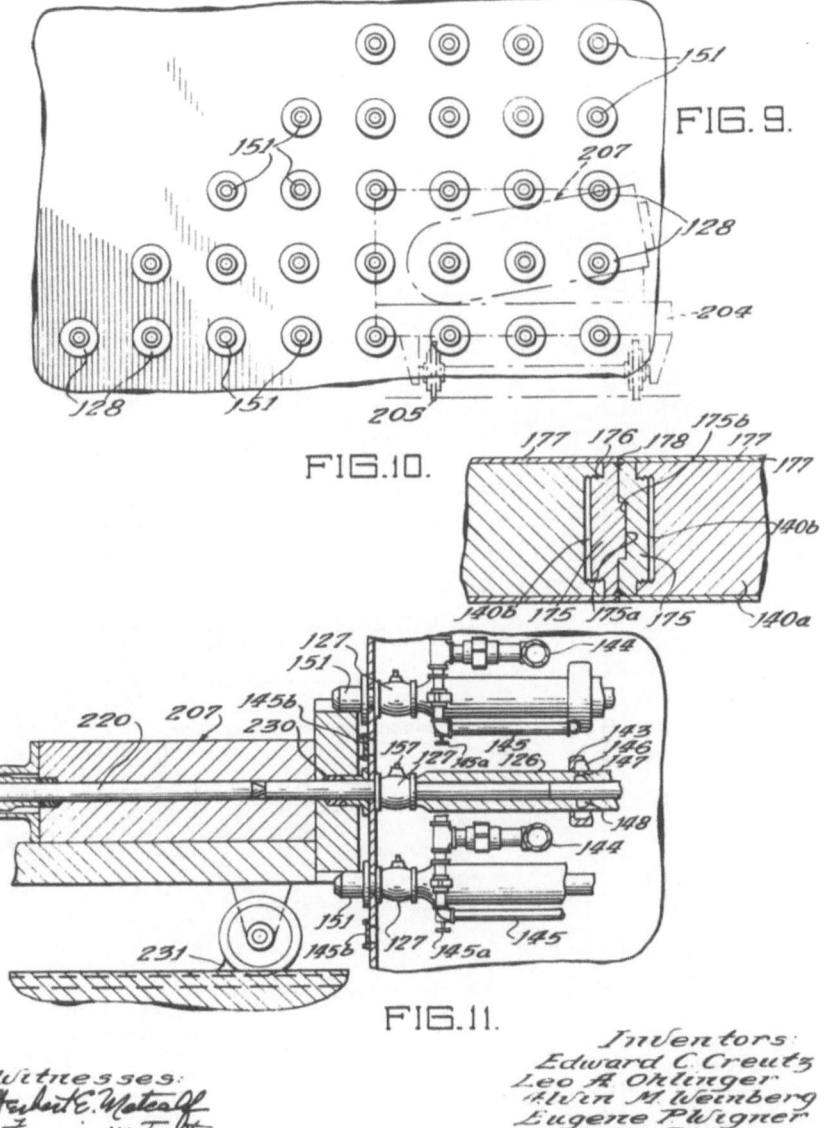

FIG. 9.

FIG. 10.

FIG. 11.

Witnesses:
Herbert E. Metcalf
Francis W. Tett

Inventors:
Edward C. Creutz
Leo A. Ohlinger
Alvin M. Weinberg
Eugene P. Wigner
Gale J. Young
By Robert A. Savender
Attorney.

The Wigner Patents

638

Oct. 27, 1959   E. C. CREUTZ ET AL   2,910,418

NEUTRONIC REACTOR

Filed Jan. 23, 1945   14 Sheets—Sheet 8

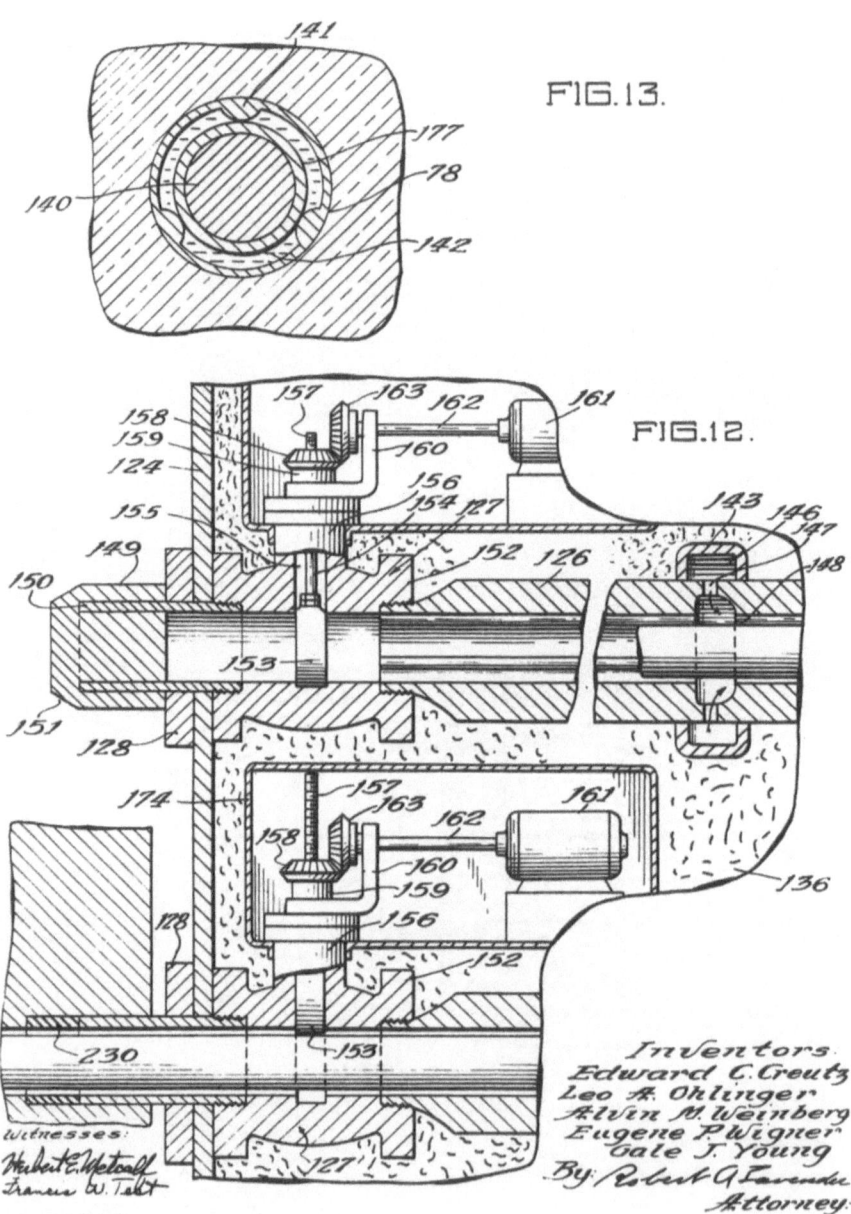

FIG.13.

FIG.12.

Inventors.
Edward C. Creutz
Leo A. Ohlinger
Alvin M. Weinberg
Eugene P. Wigner
Gale J. Young
By Robert G Lavender
Attorney.

Witnesses:
Herbert E. Metcalf
Francis W. Tabt

Oct. 27, 1959   E. C. CREUTZ ET AL   2,910,418

NEUTRONIC REACTOR

Filed Jan. 23, 1945   14 Sheets—Sheet 9

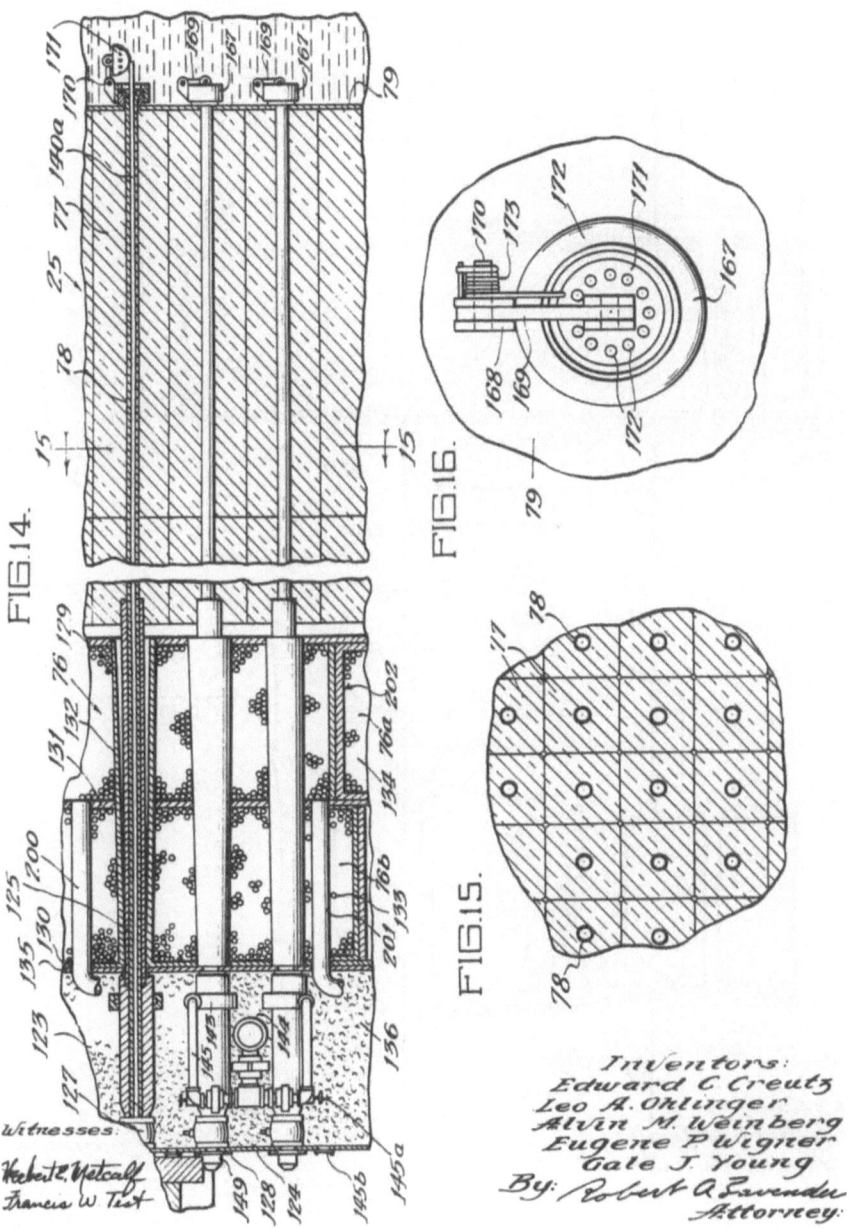

FIG.14.

FIG.15.

FIG.16.

Witnesses:
Herbert E. Metcalf
Francis W. Test

Inventors:
Edward C. Creutz
Leo A. Ohlinger
Alvin M. Weinberg
Eugene P. Wigner
Gale J. Young
By: Robert A. Zwierlein
Attorney.

Oct. 27, 1959   E. C. CREUTZ ET AL   2,910,418

NEUTRONIC REACTOR

Filed Jan. 23, 1945   14 Sheets—Sheet 10

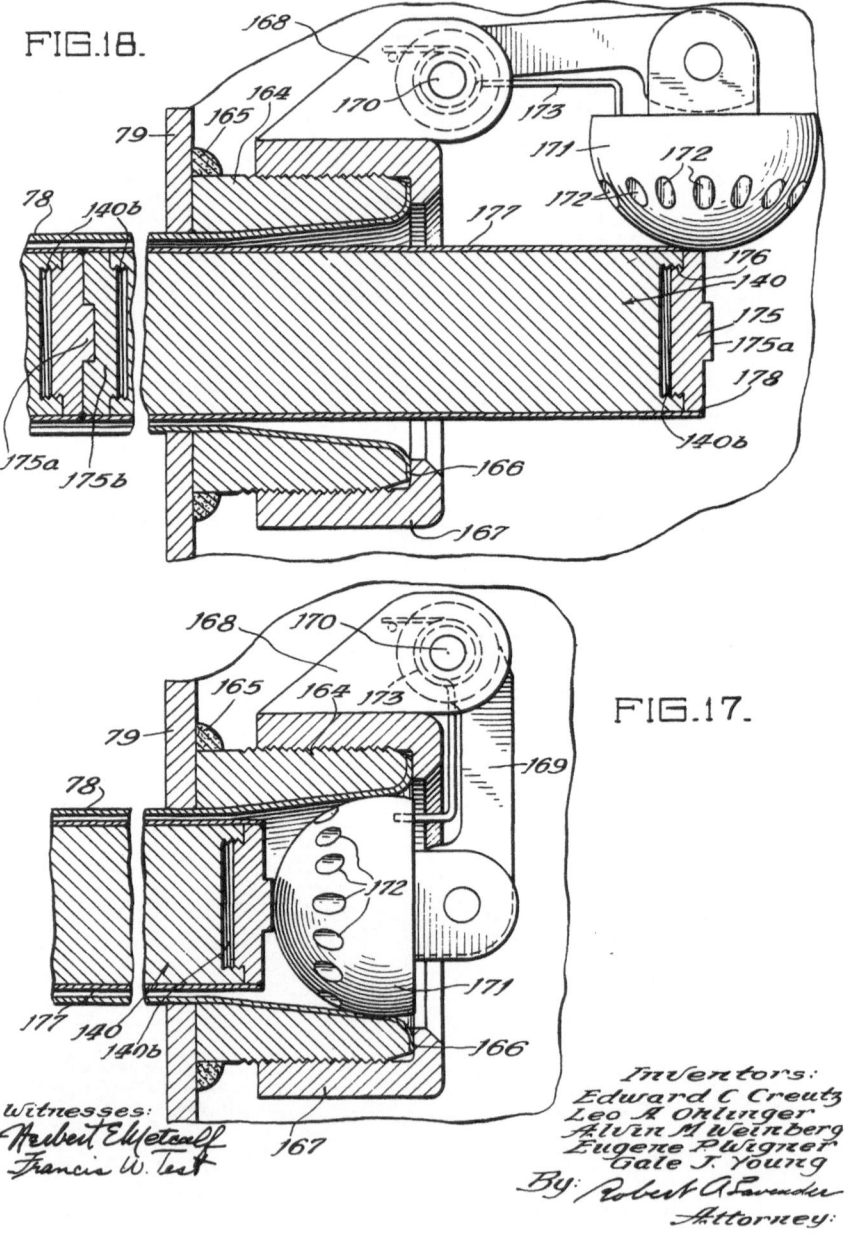

FIG.18.

FIG.17.

Inventors:
Edward C Creutz
Leo A Ohlinger
Alvin M Weinberg
Eugene P Wigner
Gale J Young
By Robert A Lavender
Attorney:

Oct. 27, 1959   E. C. CREUTZ ET AL   2,910,418
NEUTRONIC REACTOR

Filed Jan. 23, 1945   14 Sheets-Sheet 11

FIG.19.

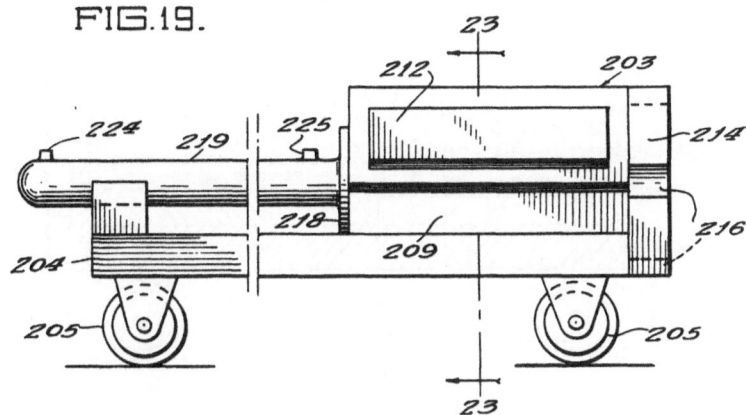

FIG.20.

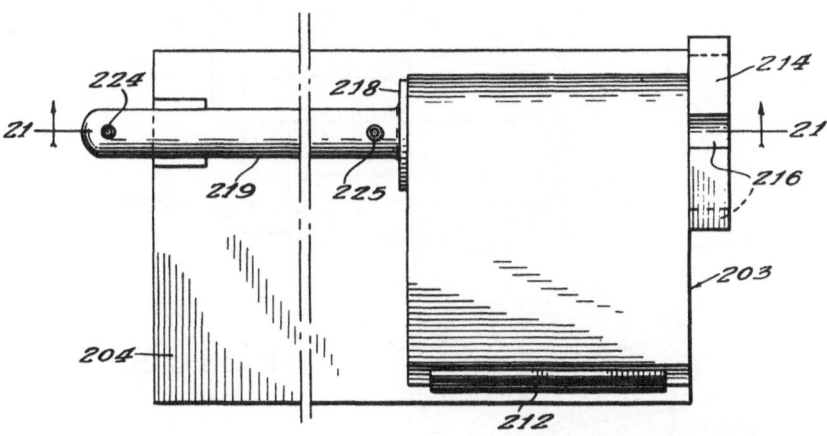

Witnesses:
Herbert E. Metcalf
Francis W. Test

Inventors:
Edward C. Creutz
Leo A. Ohlinger
Alvin M. Weinberg
Eugene P. Wigner
Gale J. Young
By: Robert A. Lavender
Attorney.

Oct. 27, 1959   E. C. CREUTZ ET AL   2,910,418
                                   NEUTRONIC REACTOR

Filed Jan. 23, 1945   14 Sheets-Sheet 12

FIG.21.

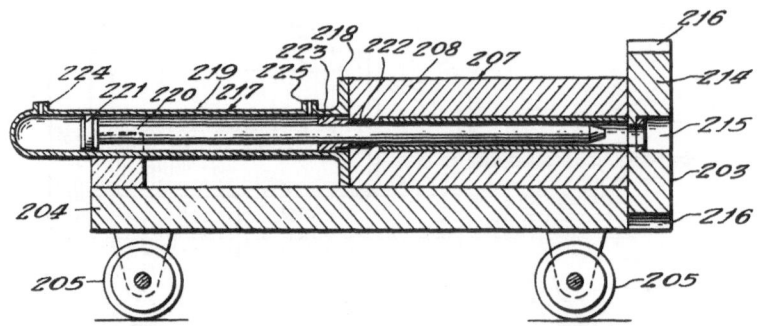

FIG.22.

FIG.23.

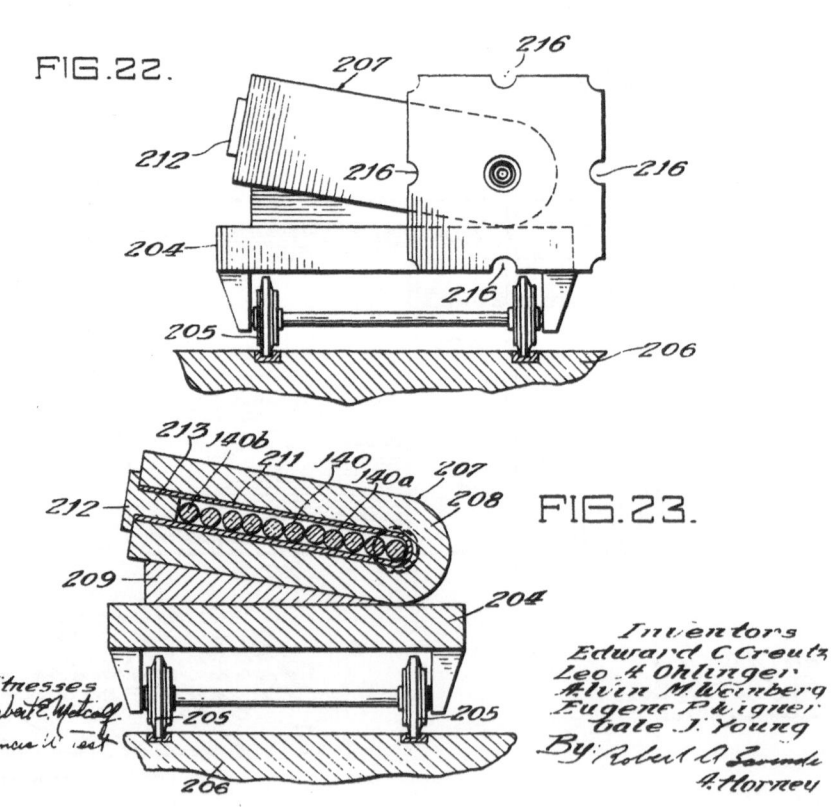

Inventors
Edward C Creutz
Leo A Ohlinger
Alvin M Weinberg
Eugene P Wigner
Gale J Young
By Robert A Lavender
A Horney

Witnesses
Hubert E Metcalf
Francis W West

Oct. 27, 1959   E. C. CREUTZ ET AL   2,910,418

NEUTRONIC REACTOR

Filed Jan. 23, 1945   14 Sheets—Sheet 13

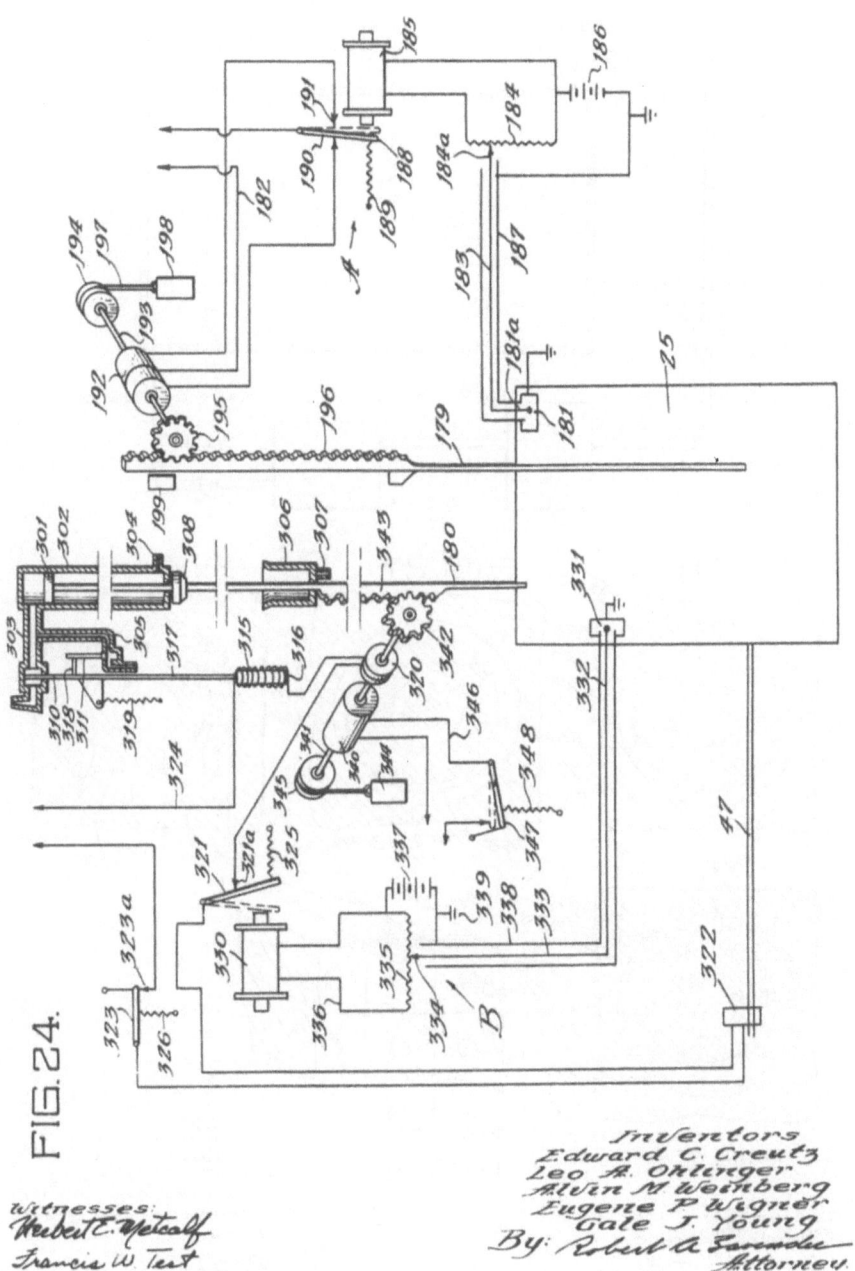

FIG. 24.

Inventors
Edward C. Creutz
Leo A. Ohlinger
Alvin M Weinberg
Eugene P Wigner
Gale J. Young
By: Robert A. Lavender
Attorney.

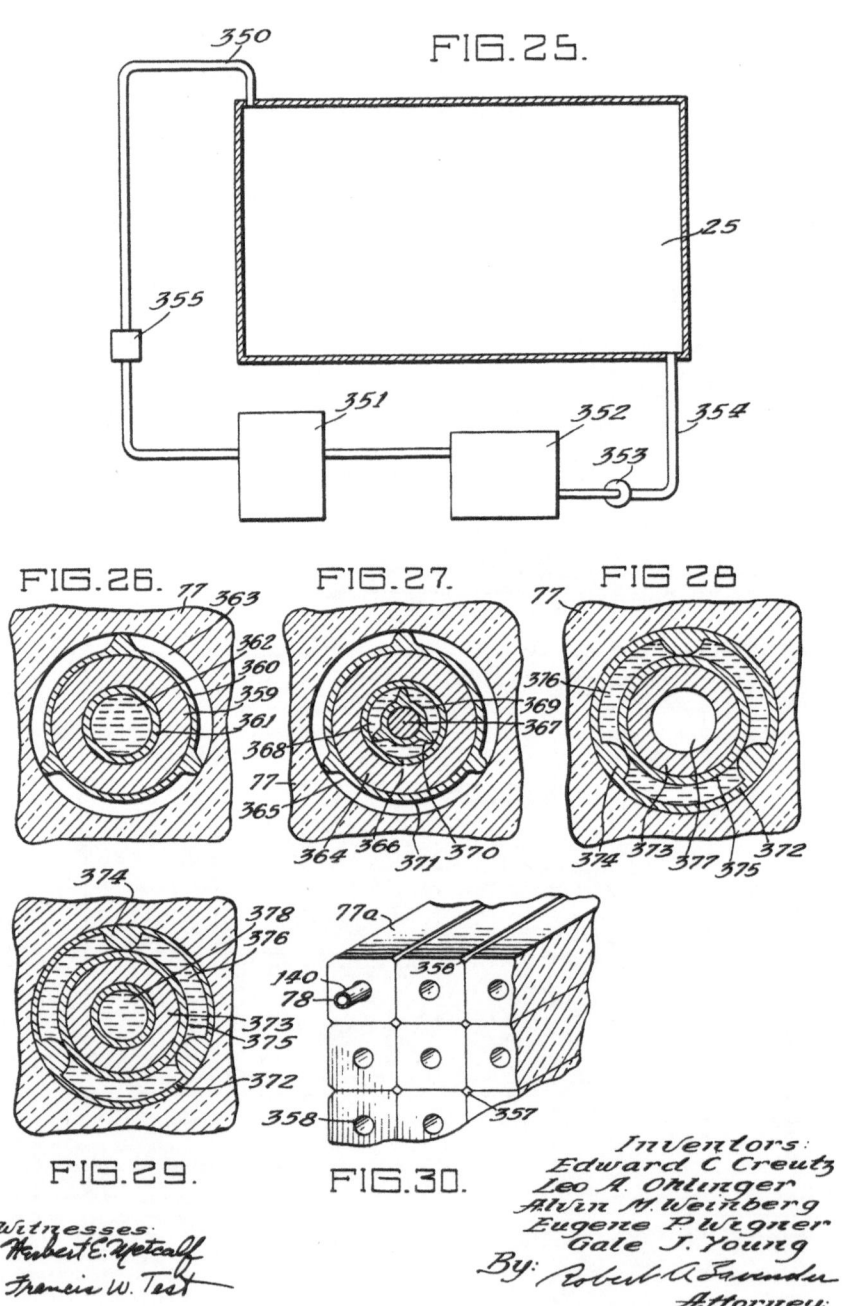

Oct. 27, 1959   E. C. CREUTZ ET AL   2,910,418

NEUTRONIC REACTOR

Filed Jan. 23, 1945   14 Sheets—Sheet 14

FIG. 25.

FIG. 26.

FIG. 27.

FIG 28

FIG. 29.

FIG. 30.

Witnesses:

Inventors:
Edward C Creutz
Leo A. Ohlinger
Alvin M. Weinberg
Eugene P. Wigner
Gale J. Young
By: Robert A. Zwenda
Attorney.

United States Patent Office

2,910,418
Patented Oct. 27, 1959

1

2,910,418

NEUTRONIC REACTOR

Edward C. Creutz, Leo A. Ohlinger, Alvin M. Weinberg, Eugene P. Wigner, and Gale J. Young, Chicago, Ill., assignors to the United States of America as represented by the United States Atomic Energy Commission

Application January 23, 1945, Serial No. 574,153

10 Claims. (Cl. 204—193.2)

The present invention relates to the subject of neu- 15 tronics, and more particularly to a liquid cooled neutron chain reacting system, also referred to as a neutronic reactor, or pile, the latter name having been originally adopted for the active portions of systems employing uranium or other fissionable bodies geometrically ar- 20 ranged in graphite or other moderator in the form of lattice structures. More specifically the invention has to do with liquid cooled neutronic reactors in which a coolant passes over bodies of fissionable material disposed in channels inside the reactor and this coolant 25 may be maintained under pressure which can readily be varied. The flow of coolant through any one or more of the tubes or channels containing fissionable material may be independently regulated and if desired may be entirely shut off so that a defective tube or channel may 30 be blocked off and removed from the reactor without permanently disabling the remainder of the reactor.

As a result of the chain reaction, when U^{238} is present (as in natural uranium), transuranic element 94^{239}, known as plutonium, is produced. This material is fis- 35 sionable and is valuable when added to natural uranium for use in a chain reacting system, as a fissionable body in lieu of or conjunction with natural uranium.

Natural uranium contains both uranium isotopes U^{235} and U^{238} in the ratio of 1 to 139. The U^{235} is the 40 isotope fissionable by slow neutrons.

When fission occurs in the U^{235} isotope, the following reaction takes place:

$$_{92}U^{235} + \text{neutron} \rightarrow A + B + \text{about 2 neutrons (average)}$$

where "A" represents "light" fission fragments having 45 atomic masses ranging from 83 to 99 inclusive and atomic numbers from 34 to 45 inclusive; for example, Br, Kr, Rb, Sr, Y, Zr, Cb, Mo, Ma, Ru, and Rh; and "B" represents "heavy" fission fragments having atomic masses ranging from 127 to 141 inclusive, and atomic numbers 50 from 51 to 60 inclusive; for example, Sb, Te, I, Xe, Cs, Ba, La, Ce, Pr, and Nd.

The elements resulting from the fissions are unstable and radioactive, with half-lives varying in length in accordance with the element formed. 55

The absorption of thermal or resonance neutrons by the U^{238} isotope gives rise to the conversion of U^{238} to U^{239} which ultimately decays to transuranic element 94^{239}. The reaction is as follows: 60

$$92^{238} + n \longrightarrow 92^{239} \text{ [plus 6 mev. of } \gamma \text{ rays, not necessarily all of one frequency]}$$

$$92^{239} \xrightarrow{\text{23 min.}} 93^{239} + \beta^- \text{[1 mev. } \beta^-, \text{ no } \gamma \text{ rays]}$$

$$93^{239} \xrightarrow{\text{2.3 days}} 94^{239} + \beta^- \text{[600 kv. upper } \beta^- \text{ energy limit. Also 2 } \gamma \text{ rays, 400 kv. and 270 kv., about } \frac{1}{2} \text{ of which are converted to electrons]}$$

Most of the neutrons arising from the fission process 70 are set free with the very high energy of above one mil-

2

lion electron volts average and are therefore not in condition to be utilized efficiently to create new thermal neutron fissions in a fissionable body such as U^{235} when it is mixed with a considerable quantity of U^{238}, particularly as in the case of natural uranium. The energies of the fission-released neutrons are so high that most of the latter would tend to be absorbed by the U^{238} nuclei, and yet the energies are not generally high enough for production of fission by more than a small fraction of the neutrons so absorbed. For neutrons of thermal energies, however, the absorption cross section of U^{235}, to produce fission, is a great deal more than the simple capture cross section of U^{238}; so that under the stated circumstances the fast fission neutrons, after they are created, must be slowed down to thermal energies before they are most effective to produce fresh fission by reaction with additional U^{235} atoms. If a system can be made in which neutrons are slowed down without excessive absorption until they reach thermal energies and then mostly enter into uranium rather than into any other element, a self-sustaining nuclear chain reaction can be obtained, even with natural uranium. Light elements, such as deuterium, beryllium, oxygen or carbon, the latter in the form of graphite, can be used as slowing agents. A special advantage of the use of the light elements mentioned for slowing down fast fission neutrons is that fewer collisions are required for slowing than is the case with heavier elements, and furthermore, the above-enumerated elements have very small neutron capture probabilities, even for thermal neutrons. Hydrogen would be most advantageous were it not for the fact that there may be a relatively high probability of neutron capture by the hydrogen nucleus. Carbon in the form of graphite is a relatively inexpensive, practical, and readily available agent for slowing fast neutrons to thermal energies. Recently, beryllium has been made available in sufficiently large quantities for test as to suitability for use as a neutron slowing material in a system of the type to be described. It has been found to be in every way as satisfactory as carbon. Deuterium while more expensive is especially valuable because of its low absorption of neutrons and its compounds such as deuterium oxide have been used with very effective results.

However, in order for the premise to be fulfilled that the fast fission neutrons be slowed to thermal energies in a slowing medium without too large an absorption in the U^{238} isotope of the uranium, certain types of physical structure should be utilized for the most efficient reproduction of neutrons, since unless precautions are taken to reduce various neutron losses and thus to conserve neutrons for the chain reaction the rate of neutron reproduction may be lowered and in certain cases lowered to a degree such that a self-sustaining system is not attained.

The ratio of the number of fast neutrons produced by the fissions, to the original number of fast neutrons creating the fissions, in a system of infinite size using specific materials is called the reproduction or multiplication factor of the system and is denoted by the symbol K. If K can be made sufficiently greater than unity to create a net gain in neutrons and the system made sufficiently large so that this gain is not entirely lost by leakage from the exterior surface of the system, then a self-sustaining chain reacting system can be built to produce power by nuclear fission of natural uranium. The neutron reproduction ratio, r, in a system of finite size differs from K by the leakage factor and by localized neutron absorbers such as control rods, and must be sufficiently greater than unity to permit the neutron density to rise exponentially. Such a rise will continue indefinitely if not controlled at a desired density corresponding to a desired power output.

3

During the interchange of neutrons in a system comprising bodies of uranium of any size in a slowing medium, neutrons may be lost in four ways, by absorption in the uranium metal or compound without producing fission, by absorption in the slowing down material, by absorption in impurities present in the system, and by leakage from the system. These losses will be considered in the order mentioned.

Natural uranium, particularly by reason of its U^{238} content, has an especially strong absorbing power for neutrons when they have been slowed down to moderate energies. The absorption in uranium at these energies is termed the uranium resonance absorption or capture. It is caused by the isotope U^{238} and does not result in fission but creates the isotope U^{239} which by two successive beta emissions forms the relatively stable nucleus 94^{239}. It is not to be confused with absorption or capture of neutrons by impurities, referred to later. Neutron resonance absorption in uranium may take place either on the surface of the uranium bodies, in which case the absorption is known as surface resonance absorption, or it make take place further in the interior of the uranium body, in which case the absorption is known as volume resonance absorption. It will be appreciated that this classification of resonance absorptions is merely a convenient characterization of observed phenomena, and arises, not because the neutron absorbing power of a U^{238} nucleus is any greater when the nucleus is at the surface of a body of metallic, or combined uranium, but because the absorbing power of U^{238} nuclei for neutrons of certain particular energies is inherently so high that practically all neutrons that already happen to have those energies, called resonance energies as explained above, are absorbed almost immediately upon their arrival in the body of uranium metal or uranium compound, and thus in effect are absorbed at the surface of such body. Volume resonance absorption is due to the fact that some neutrons make collisions inside the uranium body and may thus arrive at resonance energies therein. After successfully reaching thermal velocities, about 40 percent of the neutrons are also subject to capture by U^{238} without fission, to produce U^{239} and eventually 94^{239}.

It is possible, by proper physical arrangement of the materials, to reduce substantially uranium resonance absorption. By the use of light elements as described above for slowing materials, a relatively large increment of energy loss is achieved in each collision and therefore fewer collisions are required to slow the neutrons to thermal energies, thus decreasing the probability of a neutron being at a resonance energy as it enters a uranium atom. During the slowing process, however, neutrons are diffusing through the slowing medium over random paths and distances so that the uranium is not only exposed to thermal neutrons but also to neutrons of energies varying between the emission energy of fission and thermal energy. Neutrons at uranium resonance energies will, if they enter uranium at these energies, be absorbed on the surface of a uranium body whatever its size, giving rise to surface absorption. Any substantial reduction of overall surface of the same amount of uranium relative to the amount of slowing material (i.e. the amount of slowing medium remaining unchanged) will reduce surface absorption, and any such reduction in surface absorption will release neutrons to enter directly into the chain reaction, i.e., will increase the number of neutrons available for further slowing and thus for reaction with U^{235} to produce fission.

For a given ratio of slowing material to uranium, surface resonance absorption losses of neutrons in the uranium can be reduced by a large factor from the losses occurring in a mixture of fine uranium particles and a slowing medium, if the uranium is aggregated into substantial masses in which the mean radius of the aggregates is at least 0.25 centimeter for natural uranium metal and when the mean spatial radius of the bodies is at least

4

0.75 centimeter for the oxide of natural uranium (UO_2). Proportionate minimums exist for other uranium compounds the exact minimum value being dependent upon the uranium content and the denstiy of the product. An important gain is thus made in the number of neutrons made directly available for the chain reaction. A similar gain is made when the uranium has more than the natural content of fissionable material. Where a maximum K factor is to be desired we place the uranium in the system in the form of spaced uranium masses or bodies of substantial size, preferably either of metal, oxide, carbide, or other compound or combinations thereof. The uranium bodies can be in the form of layers, rods or cylinders, cubes or spheres, or approximate shapes, dispersed throughout the graphite, preferably in some geometric pattern. The term geometric is used to mean any pattern or arrangement wherein the uranium bodies are distributed in the graphite or other moderator with at least either a roughly uniform spacing or with a roughly systematic non-uniform spacing, and are at least roughly uniform in size and shape or are systematic in variations of size or shape to produce a volume pattern conforming to a roughly symmetrical system. If the pattern is a repeating or rather exactly regular one, a system embodying it may be conveniently described as a lattice structure. Optimum conditions are obtained with natural uranium by using a lattice of metal spheres.

The number of neutrons made directly available to the chain reaction by aggregating the uranium into separate bodies spaced through the slowing medium is a critical factor in obtaining a self-sustaining chain reaction utilizing natural uranium and graphite. The K factor of a mixture of fine uranium particles in graphite, assuming both of them to be theoretically pure, would only be about .785. Actual K factors as high as 1.07 have been obtained using aggregation of natural uranium in the best known geometry, and with as pure materials as it is presently possible to obtain.

Assuming theoretically pure carbon and theoretically pure natural uranium metal, both of the highest obtainable densities, the maximum possible K factor theoretically obtainable is about 1.1 when the uranium is aggregated with optimum geometry. Still higher K factors can be obtained by the use of aggregation in the case of uranium having more than the naturally occurring content of fissionable elements. Adding such fissionable material is termed enrichment of the uranium.

It is thus clearly apparent that the aggregation of the uranium into masses separated in the slowing material is one of the most important, if not the most important factor entering into the successful construction of a self-sustaining chain reaction system utilizing relatively pure natural uranium in a slowing material such as graphite in the best geometry at present known, and is also important in obtaining high K factors when enrichment of the uranium is used.

Somewhat higher K factors are obtainable where moderators such as deuterium oxide or beryllium are used. Thus with beryllium it is possible to secure a K factor as high as 1.10 with optimum geometry and absolute purity. Moreover with deuterium oxide K factors of about ~ 1.27 may be obtained. When such moderators are used the problem of aggregation may be somewhat less important although it is an essential factor if maximum K factors and minimum size reactors are to be obtained.

The thermal neutrons are also subject to capture by the slowing material. While carbon and beryllium have very small capture cross sections for thermal neutrons, and deuterium still smaller, an appreciable fraction of thermal neutrons (about 10 percent of the neutrons present in the system under best conditions with graphite) is lost by capture in the slowing material during diffusion therethrough. It is therefore desirable to have the neutrons reaching thermal energy promptly enter uranium.

2,910,418

5

In addition to the above-mentioned losses, which are inherently a part of the nuclear chain reaction process, impurities present in both the slowing material and the uranium add a very important neutron loss factor in the chain. The effectiveness of various elements as neutron absorbers varies tremendously. Certain elements such as boron, cadmium, samarium, gadolinium, and some others, if present even in a few parts per million, could prevent a self-sustaining chain reaction from taking place. It is highly important, therefore, to remove as far as possible all impurities capturing neutrons to the detriment of the chain reaction from both the slowing material and the uranium. If these impurities, solid, liquid, or gaseous, and in elemental or combined form, are present in too great quantity, in the uranium bodies or the slowing material or in, or by absorption from, the free spaces of the system, the self-sustaining chain reaction cannot be attained. The amounts of impurities that may be permitted in a system, vary with a number of factors, such as the specific geometry of the system, and the form in which the uranium is used—that is, whether natural or enriched, whether as metal or oxide—and also factors such as the weight ratios between the uranium and the slowing down material, and the type of slowing down or moderating material used—for example, whether deuterium, graphite or beryllium. Although all of these considerations influence the actual permissible amount of each impurity material, it has fortunately been found that, in general, the effect of any given impurity or impurities can be correlated directly with the weight of the impurity present and with the K factor of the system, so that knowing the K factor for a given geometry and composition, the permissible amounts of particular impurities can be readily computed without taking individual account of the specific considerations named above. Different impurities are found to affect the operation to widely different extents; for example, relatively considerable quantities of elements such as hydrogen may be present, and, as previously suggested, the uranium may be in the form of oxide, such as UO_2 or U_3O_8, or carbide, although the metal is preferred. Nitrogen may be present to some extent, and its effect on the chain reaction is such that the neutron reproduction ratio of the system may be changed by changes in atmospheric pressure. This effect may be eliminated by enclosing or evacuating the system if desired, or may be utilized by determining changes in a particular system in the reproduction ratio as changes occur in the atmospheric pressure. A sensitive barometer is thus obtained. In general, the inclusion of combined nitrogen is to be avoided.

The effect of impurities on the optimum reproduction factor K may be conveniently evaluated to a good approximation, simply by means of certain constants known as "danger coefficients" which are assigned to the various elements. These danger coefficients for the impurities are each multiplied by the percent by weight of the corresponding impurity, and the total sum of these products gives a value known as the total danger sum. This total danger sum is subtracted from the reproduction factor K as calculated for pure materials and for the specific geometry under consideration.

The danger coefficients are defined in terms of the ratio of the weight of impurity per unit mass of uranium and are based on the cross section for absorption of thermal neutrons of the various elements. These values may be obtained from physics textbooks on the subject and the danger coefficient computed by the formula

$$\frac{\sigma_i}{\sigma_u} \cdot \frac{A_u}{A_i}$$

wherein σ_i represents the cross section for the impurity and σ_u the cross section for the uranium, A_i the atomic weight of the impurity and A_u the atomic weight for uranium. If the impurities are in the carbon, they are computed as their percent of the weight of the uranium of the system.

6

Presently known values for danger coefficients for some elements are given in the following table, wherein the elements are assumed to have their natural isotopic constitution unless otherwise indicated, and are conveniently listed according to their chemical symbols:

Element	Danger Coefficient
He	0
Li	310
B	2,150
N	4.0
F	0.02
Na	0.65
Mg	0.48
Al	0.30
Si	0.26
P	0.3
S	0.46
Cl	31
K	2.1
Ca	0.37
Ti	3.8
V	4
Cr	2
Mn	7.5
Fe	1.5
Co	17
Ni	3
Cu	1.8
Zn	0.61
Ga	~1
As	2
Se	6.3
Br	2.5
Rh	50
Ag	18
Cd	870
In	54.2
Sn	0.18
Sb	1.6
I	1.6
Ba	0.30
Sm	~1,430
Eu	435
Gd	~6,320
Pb	0.03
Bi	0.0025
Th	1.1

Where an element is necessarily used in an active part of a system, it is still to be considered as an impurity; for example, in a structure where the uranium bodies consist of uranium oxide, the actual K factor would ordinarily be computed by taking that fact into account using as a base K a value computed for theoretically pure uranium.

As a specific example, if the materials of the system under consideration have .0001 part by weight of Co and Ag, the total danger sum in K units for such an analysis would be:

$$.0001 \times 1 + .0001 \times 18 = .0035 \text{ K units}$$

This would be a rather unimportant reduction in the reproduction factor K unless the reproduction factor for a given system, without considering any impurities, is very nearly unity. If, on the other hand, the impurities in the uranium in the previous example had been Li, Co, and Rh, the total danger sum would be:

$$.0310 + .0017 + .0050 = .0377 \text{ K units}$$

This latter reduction in the reproduction factor for a given system would be serious and might well reduce the reproduction factor below unity for certain geometries and certain moderators so as to make it impossible to effect a self-sustaining chain reaction with natural uranium and graphite, but might still be permissible when using enriched uranium in a system having a high K factor.

This strong absorbing action of some elements renders a self-sustaining chain reacting system capable of control. By introducing neutron absorbing elements in the form of rods or sheets into the interior of the system, for instance in the slowing material between the uranium masses, the neutron reproduction ratio of the system can be changed in accordance with the amount of absorbing material exposed to the neutrons in the system. A sufficient mass of the absorbing material can readily be inserted into the system to reduce the reproduction ratio

2,910,418

7

of the system to less than unity and thus stop the reaction. Consequently, it is another object of our invention to provide a means and method of controlling the chain reaction in a self-sustaining system.

When the uranium and the slowing material are of such purity and the uranium is so aggregated that fewer neutrons are parasitically absorbed than are gained by fission, the uranium will support a chain reaction producing an exponential rise in neutron density if the overall size of the system is sufficiently large to overcome the loss of neutrons escaping from the system. Thus the overall size is important.

The size of the system will vary, depending upon the K factor of the system, and upon other things. If the reproduction factor K is greater than unity, the number of neutrons present will increase exponentially and indefinitely, provided the structure is made sufficiently large. If, on the contrary, the structure is small, with a large surface-to-volume ratio, there will be a rate of loss of neutrons from the structure by leakage through the outer surfaces, which may overbalance the rate of neutron production inside the structure so that a chain reaction will not be self-sustaining. For each value of the reproduction factor K greater than unity, there is thus a minimum overall size of a given structure known as the critical size, above which the rate of loss of neutrons by diffusion to the walls of the structure and leakage away from the structure is less than the rate of production of neutrons within the system, thus making the chain reaction self-sustaining. The rate of diffusion of neutrons away from a large structure in which they are being created through the exterior surface thereof may be treated by mathematical analysis when the value of K and certain other constants are known, as the ratio of the exterior surface to the volume becomes less as the structure is enlarged.

In the case of a spherical structure employing uranium bodies imbedded in graphite in the geometries disclosed herein and without an external reflector the following formula gives the critical overall radius (R) in feet:

$$K-1=\frac{C}{R^2}$$

where C is a constant that varies slightly with geometry of the lattice and for normal graphite lattices may have a value close to 7.2.

For a rectangular parallelepiped structure rather than spherical, the critical size can be computed from the formula

$$K-1=C\left(\frac{1}{a^2}+\frac{1}{b^2}+\frac{1}{c^2}\right)$$

where a, b, and c are the lengths of the sides in feet. The critical size for a cylindrical structure is given by the formula, irrespective of the shape of the uranium bodies, cylinder height h ft., radius R ft.

$$K-1=C\left(\frac{1}{h^2}+\frac{.59}{R^2}\right)$$

However, when critical size is attained, by definition no rise in neutron density can be expected. It is therefore necessary to increase the size of the structure beyond the critical size but not to the extent that the period for doubling of the neutron density is too short, as will be explained later. Reactors having a reproduction ratio (r) for an operating structure with all control absorbers removed and at the temperature of operation up to about 1.005 are very easy to control. Reproduction ratio should not be permitted to rise above about 1.01 since the reaction will become difficult to control. The size at which this reproduction ratio can be obtained may be computed from modifications of the above formulae for critical size. For example, for spherical active structures the formula

$$K-r=\frac{C}{R^2}$$

8

may be used to find R when K is known and r is somewhat over unity. The same formula will, of course, give r for given structures for which K and R are known.

Critical size may be attained with a somewhat smaller structure by utilizing a neutron reflecting medium surrounding the surface of the active structure. For example, a 2 foot thickness of graphite having low impurity content, completely surrounding a spherical structure is effective in reducing the diameter of the uranium bearing portion by almost 2 feet, resulting in a considerable saving of uranium or uranium compound.

The rate of production of element 94^{239} will depend on the rate of neutron absorption by U^{238} and is also proportional to the rate at which fissions occur in U^{235}. This in turn is controlled by the thermal neutron density existing in the reactor while operating. Thus for maximum production of element 94^{239}, it is essential that the thermal neutron density be at a maximum value commensurate with thermal equilibrium.

Considerable heat is generated during a neutronic reaction primarily as the result of the fission process. Following are tables showing more specifically the type of heat generated in the reactor.

SUMMARY BY TYPE

	Mev./ fission	Percent
Gamma radiation	18	9
Beta radiation	16	8
Kinetic energy of fission fragments	160	80
Kinetic energy of neutrons	6	3
	200	100

SUMMARY BY LOCALE WHERE HEAT IS GENERATED

	Mev./ fission	Percent
In uranium	174	87
In moderator	16	8
Outside pile	10	5
	200	100

SUMMARY BY TYPE AND LOCALE

	Mev. per fission	Percent in U	Percent in C	Percent Outside
Kinetic energy of fission fragments	159	100		
Kinetic energy of neutrons	6		99	1
Gamma radiation from fission products	5	50	45	5
Beta radiation from fission products	6	100		
Nuclear affinity of neutrons (gamma radiation)	12	70	25	5

When the system is operated for an extended period of time at a high production output of element 94^{239}, the large amount of heat thus generated must be removed in order to stabilize the chain reaction. Most of the heat in an operating device is generated as the result of the nuclear fissions taking place in the U^{235} isotope. Thus, the rate of heat generation is largely proportional to the rate at which the fissions take place. In other words, if the rate of generation of neutrons is increased, a greater amount of coolant must be passed through the reactor in order to remove the heat thus generated to avoid damage, particularly at the central portion of the pile, by excessive heat. Thus, the highest obtainable neutron density at which a system can be operated for an extended period of time is limited by the rate at which the generated heat can be removed. That is to say, the maximum power output of a system is limited by the capacity of the cooling system. An effective cooling system is therefore a primary requirement for high power operation of a neutronic reactor and it has been found

2,910,418

9

that this cooling may be accomplished most effectively by passage of the coolant in contact with or in close proximity to the uranium.

After the neutronic system has operated for a period of time sufficient to cause a quantity of element 94^{239} to be produced, it may be desirable to remove at least some of the uranium rods from the reactor in order to extract element 94^{239} and the radioactive fission products, both being formed in the uranium rods or for other purposes.

In many neutronic reactors, a neutron density variation occurs across the reactor; that is, the neutron concentration at the periphery is relatively small and increases to a maximum value at the center. Actually, therefore, since the rate of production of element 94^{239} is dependent upon the neutron density, the reactor will have zones which may be likened to three dimensional shells, the average concentration of element 94^{239} being uniform throughout any given zone. In a reactor built in the form of a sphere these would, of course, be in the shape of concentric spheres of different diameters, while one built in the shape of a cylinder would have similar zones but of different shapes.

Where this variation in concentration exists in a reactor it is often desirable to resort to a systematic schedule of removal depending upon the time of operation and the location of the uranium for removing and discharging uranium metal that has been subjected to neutron bombardment. In the case of a new system of this character the operation would normally continue until the metal in the center portion of the reactor reaches a desired content of element 94^{239}, at which time this metal would be removed and replaced with fresh metal. The next removal then would be from the section next adjacent to the center section of the reactor where the desired content of element 94^{239} is reached after further operation. The process would then proceed with the removal of the metal at various times until the metal recharged at the center of the reactor has reached the desired content of element 94^{239}. This would then be replaced and the process of progressing towards the periphery continued with periodic return to more central areas. Since the neutron density in the central areas of such a reactor would, ordinarily, greatly exceed the neutron density near the periphery, the metal in the central areas may be replaced several times for each replacement of the metal near the periphery. A removal schedule can be developed by calculation and checked by actual experience after the system has been placed in operation.

Different schedules may be developed with other reactors having different reactivity curves. For example, certain reactors are constructed in a manner such that the neutron concentration is substantially uniform throughout a large volume of the reactor. In such a case the schedule for removal of uranium bodies may be modified accordingly.

Since the heat generated in the reactor results from fissions in the uranium, it is evident that this heat is not formed uniformly throughout the reactor but that it must vary across the reactor with the local rate at which fissions occur and element 94^{239} formed. Consequently, the relative values for the production of element 94^{239} apply also to heat distribution; that is, the heat generated may increase from a minimum at the outer surface of the reactor to a maximum at the center in certain reactors.

As the total weight of the radioactive fission elements is proportional to that of the 94^{239} at the time of fissions, it might be assumed that the amounts of these radioactive fission elements and of 94^{239} present in metal removed from the reactor are also of the same proportion. This is not true, however, as the fission elements when produced are highly radioactive and immediately start to decay, some with short half-lives and others with longer half lives until, through loss of energy, these unstable fission elements arrive at a stable non-radioactive element or iso-

10

tope and no longer change. The 94^{239} on the other hand is a relatively stable element when formed, having a radioactive half-life of about 2×10^4 years.

At the start of the reaction in new metal the radioactive fission elements and the 94^{239} both increase in amounts. After a certain period of operation during which time the metal is subjected to intense neutron bombardment the radioactive fission elements will reach a state of equilibrium and from that time on the amounts of these radioactive elements remain constant, as the fission elements with shorter half lives are reaching a stable condition at the same time new ones are being produced. The amount of the stable end products of fission, however, continues to increase with the increase in element 94^{239}. Consequently, the rate of formation of the fission end products is dependent upon the location of any particular metal in the reactor, and the power at which the system operates controls the maximum radioactive fission element content regardless of the length of time the system operates after equilibrium occurs. The quantity of element 94^{239} on the other hand, and of the final and stable end products of fission continue to increase as the operation of the system continues. The amounts of both 94^{239} and fission end products present are controlled only by the location of the metal in the reactor and the time and power of operation. The highly radioactive fission elements may, therefore, vary from a substantial percentage of the weight of element 94^{239} present in the metal at the center of the reactor after a short period of operation, to a very small percentage in metal from a position near the periphery of the reactor after an extended operating period at a given power.

It is not to be assumed, however, that the fact that equilibrium can be obtained between the original highly radioactive fission elements and the stable fission end products that all radioactivity will cease when the original fission elements have been permitted to decay for a time equal to the equilibrium period, for example. Many of the original fission elements have long half lives that, taken together with their successive radioactive distintegration products existing long after the fission elements having a shorter half life have decayed, renders the uranium still radioactive especially after prolonged bombardment at high neutron densities. In addition, the successive radioactive disintegration products of the original shorter lived fission elements may still be present.

The equilibrium radioactivity is so intense that metal taken from the reactor for the recovery of element 94^{239} and fission products immediately after bombardment at high neutron densities will heat spontaneously due to self absorption of the intense radioactivity of the remaining radioactive fission products. The amount of heat generated as the result of the spontaneous heating will depend particularly on three factors: (1) the concentration of element 94^{239} and fission products in the metal; (2) the period of time for continuous operation required to reach this concentration; and (3) the elapsed time since the reactor was shut down and the metal was removed.

The metal from the center of the reactor in a system operating at a high power output, for example, at a 94^{239} concentration of 1 to 2,000, if not cooled, can increase in temperature at the rate of about $2000°$ C. per hour one day after the neutron activity of the system has been shut down. After 30 days shut down following an operation of 100 days at an output of 500,000 kilowatts, the average temperature rise can be approximately $572°$ C. per hour. The uranium metal of the type used in the chain reacting systems herein under consideration melts at about $1100°$ C.

Under these conditions uranium bombarded with neutrons for an extended period of time at high rates of power output can be safely removed from the reactor under one of the following methods:

(1) The neutron activity of the system is shut down

2,910,418

11

and the uranium is kept in the reactor and continuously cooled until the radioactivity decays to a point where the metal can be removed without melting in ambient air. This procedure may require that the metal remain in the reactor for a period of from 30 to 50 days after the neutron bombardment has ceased.

(2) The neutron activity of the system is shut down and the uranium is kept in the reactor with the cooling system in operation for only a few days to permit the most violent radioactivity to subside and then the metal is removed from the reactor with the cooling discontinued during the removal except for cooling by the atmosphere or by water spray. The metal is then promptly placed under more efficient cooling conditions before the temperature of the uranium has become excessive.

(3) The neutron activity of the system is shut down and the uranium is removed while cooling the uranium body at least to an extent sufficient to prevent the temperature from becoming excessive. This modification of the present invention is particularly effective.

It is also important, of course, from the point of view of biological safety of operating personnel that adequate shielding be provided to absorb the strong gamma radiations from the fission products present in the active uranium while being removed from the reactor. The neutron activity in the reactor completely ceases within 30 minutes after shut down of the neutronic reaction during which period delayed neutrons are being emitted. In no case then should the uranium be removed from the reactor immediately following shut down of the neutronic reaction, but sufficient time should be given to permit all delayed neutrons to be omitted. Thus, the shielding required during the removal of the uranium rods from the system is primarily intended to protect personnel from gamma radiations. As stated above, immediately following shut down of the neutronic reaction, there are many short lived radioactive fission elements in the uranium causing the gamma radiation to be very intense. Many of these elements decay into more stable products within the first thirty minutes following shut down of the reaction. Thus, the fission products lose a large amount of their radioactivity during this period.

While the method of extracting the fission products and element 94^{239} from the bombarded uranium taken from the reactor forms no part of the present invention, the fission products and element 94^{239} are removable and when removed are extremely useful. The radioactive fission products are valuable for use as radiation sources, many having long half lives with high energy gamma radiation sufficient for radiography of even heavy metal castings. In addition, some of the fission products are useful as radioactive tracers in biological and physiological research.

Element 94^{239} is exceptionally useful because it is fissionable by slow neutrons in the same manner as the uranium isotope 92^{235} contained in natural uranium. The separation of 92^{235} from 92^{238} in natural uranium is extremely difficult since both are isotopes of the same element and these isotopes vary only a small percentage in comparative weight. Element 94^{239} on the other hand, is a different element from uranium, having different chemical properties than uranium, and therefore can be chemically separated from uranium. After separation, for example, element 94^{239} can be added to natural uranium to supplement the 92^{235} content, thus increasing the amount of fissionable material in the uranium. This enriched uranium can then be used in neutronic systems making it possible to provide more cooling facilities, for example, than can be used in a system of the same geometry employing only natural uranium. Thus, an enriched neutronic system may provide a greater power output than would be possible in a natural uranium system having the same geometry.

12

To summarize, the present invention is concerned with a liquid cooled neutronic reactor capable of generating large quantities of heat and of producing element 94^{239} and radioactive fission products and is well adapted for long and continuous operation. The pressure of the coolant flowing through the reactor can be regulated either generally or by zones inside the reactor and any one or more portions of the reactor may be blocked off without in any way disabling the entire reactor.

Other utilities of neutronic reactors are disclosed in assignee's copending application, Serial No. 568,904, filed on December 19, 1944, by Enrico Fermi and Leo Szilard, now Patent No. 2,708,656, dated May 17, 1955, which discloses and claims neutronic reactors wherein surface resonance absorption losses of neutrons in the uranium are minimized by aggregating uranium bodies into substantial masses and by appropriate spacing as described above.

The foregoing constitute some of the principal objects and advantages of the present invention, others of which will become apparent from the following description read in conjunction with the drawings, in which:

Fig. 1 is a diagrammatic view of a preferred embodiment of the present invention showing horizontally disposed tubes in a graphite moderator and further illustrating a water filled chute into which uranium rods are ejected;

Fig. 2 is a diagrammatic view of a second embodiment of the present invention similar to Fig. 1 but showing the tubes disposed vertically in the graphite moderator;

Fig. 3 is a diagrammatic view of a third embodiment of the present invention showing tubes arranged horizontally in a graphite moderator and further illustrating shielded cars for charging and discharging uranium rods into and from the reactor;

Fig. 4 is a schematic diagram showing the external circulating system for the coolant;

Fig. 5 is a plan view of the power unit forming the preferred embodiment of the invention;

Fig. 6 is a vertical sectional view taken on the line 6—6 of Fig. 5, the view being shown partially in elevation;

Fig. 7 is a vertical sectional view taken on the line 7—7 of Fig. 6, the view being shown partially in elevation;

Fig. 8 is a vertical sectional view taken on the line 8—8 of Fig. 6, the tubes being only partially indicated and the graphite between the tubes being shown in elevation;

Fig. 9 is an enlarged fragmentary front elevational view showing the ends of the tubes at the loading end of the reactor;

Fig. 10 is an enlarged fragmentary longitudinal sectional view taken through two segments of uranium rods showing the interlock between rod segments;

Fig. 11 is an enlarged fragmentary view at the loading end of the reactor showing partially in section and partially in elevation, the relationship between the uranium rod loading car and the horizontally disposed tubes in the reactor during the loading of uranium rods into the reactor;

Fig. 12 is an enlarged fragmentary detailed sectional view corresponding to Fig. 11 but showing in particular the valve arrangement at the loading end of the reactor;

Fig. 13 is an enlarged transverse sectional view through one of the horizontal tubes in the reactor showing a uranium rod in position with the cooling water film filling the space between the rod and the walls of the tube, the view further showing portions of the graphite moderator surrounding the tube;

Fig. 14 is a fragmentary vertical view through the reactor shown partially in section and partially in elevation;

2,910,418

13

Fig. 15 is a vertical sectional view taken on the line 15—15 of Fig. 14;

Fig. 16 is an enlarged end elevational view at the emerging end of one of the horizontal tubes showing the rod retainer in closed position;

Fig. 17 is an enlarged fragmentary vertical sectional view through the discharge end of one of the tubes showing in elevation the retainer in closed position;

Fig. 18 is a view corresponding to Fig. 17 but showing a uranium rod projecting from the end of the tube with the retainer held in open position by the rod;

Fig. 19 is a side elevational view of a uranium rod loading car used in charging uranium rods into the reactor, a portion of the car being broken away;

Fig. 20 is a plan view of the car shown in Fig. 19;

Fig. 21 is a longitudinal sectional view taken on the line 21—21 of Fig. 20, the plunger being shown in elevation;

Fig. 22 is a front elevational view of the car shown in Fig. 19;

Fig. 23 is a vertical sectional view taken on the line 23—23 of Fig. 19;

Fig. 24 is a schematic line diagram of a control system for the neutron reactor, the electrical circuit operating this control system being reduced to lowest terms for purposes of illustration;

Fig. 25 is a diagrammatic view of the reactor showing a circuit for the helium in the reactor;

Fig. 26 is a transverse sectional view through a modified form of assembly of uranium, aluminum, and coolant;

Fig. 27 is a sectional view corresponding to Fig. 26 but showing a second modification;

Fig. 28 is a sectional view corresponding to Fig. 26 but showing still another modified arrangement;

Fig. 29 is a sectional view corresponding to Fig. 26 but showing a fourth modification; and

Fig. 30 is an enlarged fragmentary perspective view of the graphite blocks employed in the reactor showing one uranium rod and tube in position.

General organization

An entire neutronic system accomplishing the objects hereinbefore set forth comprises broadly a power unit including a neutronic reactor, a complete heat extracting or cooling circuit adapted to remove the reactor heat generated as a result of the neutronic reaction, an effective control system regulating the operation of the neutronic reaction to conform to selected conditions of operation, and provisions for loading bodies containing fissionable material into the reactor and discharging the bodies from the reactor after they have been subjected to neutron bombardment for a predetermined period of time.

The reactor forming a part of the power plant can be one of a variety of types and for purposes of illustration, one employing uranium rods disposed in graphite has been selected.

Referring to Figs. 1 to 3 inclusive, three embodiments of the present invention are diagrammatically shown. In Fig. 1, the reactor is generally indicated at 25 and comprises broadly a graphite portion 26 serving as the moderator and a plurality of horizontally disposed tubes 27 extending throughout the width of the graphite. Uranium rods are disposed in the tubes 27 as will be brought out hereinafter and these rods are adapted to be discharged into a vertical chute 28 filled with water 29 and discharging at its lower end into a shielded receptacle 30 mounted on a car generally indicated at 31. This embodiment of the invention will be illustrated more completely hereinafter.

Referring to Fig. 2, a second form of the invention is shown wherein the tubes 27 are disposed vertically in the graphite 26. In this form of the invention, the uranium rods are discharged from the tubes 27 into a

14

chute 32 positioned underneath the reactor 25. This chute 32 discharges into the waiting shielded receptacle 30.

A third embodiment of the invention is illustrated in Fig. 3. In this form the tubes 27 are again disposed horizontally in the graphite 26. Uranium rods, not shown, are fed from a car 35 disposed adjacent to one side of the reactor 25 into the tubes 27 and the active uranium rods are ejected from the tubes 27 at the opposite side of the reactor 25 into a waiting shielded car generally indicated at 36. This shielded car 36 is filled with water and some means for constantly replenishing this water is provided.

The cooling circuit

Irrespective of the form the reactor may take, the power may be extracted by circulating a coolant through the reactor in heat exchange relationship with the uranium rods. If the power unit is located near a relatively pure body of water such as a river of sufficient size to supply the necessary quantity of water to extract the required amount of heat, then this river water can be passed through the reactor after first being processed to obtain the required purity. Under these conditions the water after passing through the reactor is returned to the river.

In the event a natural body of water is not available, then a supply of coolant must be provided. In this event, it is desirable to recirculate the coolant through the reactor many times and thus reduce to a minimum the total quantity required. The coolant leaving the reactor may carry with it certain free gases such as hydrogen and oxygen if water is used, which should be eliminated from the cooling circuit. For this purpose, the coolant leaving the reactor is passed through a flash tank where these gases are removed. The coolant is then cooled and finally pumped back through the reactor and recirculated.

For purposes of illustration an external circulating system for the coolant is shown in Fig. 4, wherein water, as the coolant, after passing through the reactor 25 and extracting heat, is treated and filtered and then passed in heat exchange relationship with a secondary coolant and finally is recirculated through the reactor.

In Fig. 4, the reactor is diagrammatically shown at 25, the water inlet header being illustrated at 40. The cooling water is discharged from the reactor into the vertical chute 28 and then enters the pipe 41 through which it is conveyed by pump 41b to a flash tank 42, after passing through a throttling valve 41a. The water leaves the flash tank 42 through pipe 43, passes through heat exchanger 44 where it is cooled by flowing in heat exchange relationship with a cooler liquid, and then flows through pipe 45 and is returned to the reactor 25 by the pump 46 through pipe 47.

The water entering the heat exchanger 44 has a temperature only slightly less than the boiling point at the existing pressure. The water leaving the heat exchanger 44 has a temperature of about 95° F., this cooling being accomplished by transferring the heat to the cooler fluid in a secondary cooling system.

This secondary system includes a cooling tower generally indicated at 54, a pump 55, the heat exchanger 44 and suitable piping 56. The cooling fluid in the cooling tower 54, and for purposes of illustration water has been selected for this fluid, is collected in a reservoir 57 at the bottom of the tower from which the water is withdrawn by pump 55 and passed to the heat exchanger 44. This water entering the heat exchanger 44 is at a temperature of about 85° F. and leaves the heat exchanger at a temperature of about 130° F. This hot water passes through pipe 56 into spray head 58 disposed adjacent to the top of the cooling tower 54. The hot water is sprayed in a fine mist into the cooling tower 54, and mixes with air circulated through the cooling tower by

2,910,418

15

a blower 59. Evaporation takes place resulting in the cooling of the water in the cooling tower 54 so that the water collected in the reservoir 57 has been effectively cooled by this process of evaporation. Losses due to vaporization may be replaced as will be understood in the art. Louvered openings 60 are provided in a wall of the cooling tower 54 through which air is discharged.

A portion of the cold water leaving pump 55 is bypassed from the secondary cooling system through a pipe 62 feeding condenser coils 52 and is returned to the secondary cooling system through pipe 63.

For purposes of illustration, distilled water is used in both cooling circuits and conventional means (not shown) may be provided for replenishing water lost by evaporation or leakage. Inhibitors may be added to the water to reduce the corrosive effect on the metal in the system. A mixture of potassium or ammonium phosphate and silicate is satisfactory for this purpose.

The entire primary water circulating system is provided with radiation shielding to protect personnel from the harmful effects of gamma radiations. As shown, the flash tank 42 is surrounded by a cylindrical concrete wall 64 extending above the top of the flash tank 42 and forming an enclosure which is completely filled with water. A similar wall 65 surrounds the heat exchanger 44 and the tank formed thereby likewise is filled with water.

For purposes of illustration only a diagrammatic showing is made at 66 of a shield surrounding the reactor 25. More complete details of this latter shield are brought out elsewhere. Similarly a diagrammatic showing of a shield surrounding the pump 46 is illustrated at 67.

The power unit

Since energy is extracted in the form of heat, the source of the heat may be termed the power unit. Referring to Figs. 5 to 7 inclusive, the power unit, including the reactor 25, as shown. The reactor 25 is mounted on a concrete base 70 resting on the ground 75. Upstanding walls 72, 73, and 74 are disposed on three sides of the reactor 25, extending upwardly from the base 70 to a position substantially above the top of the reactor 25. The fourth side of the reactor 25 is enclosed by a steel shot and water shield, generally indicated at 76 and forming the fourth side of a large tank.

Referring paricularly to Figs. 6 and 7, the reactor 25 comprises a graphite moderator 77 in the form of a horizontally disposed cylinder through which are arranged a plurality of horizontally disposed parallel tubes 78, this entire assembly being surrounded by a graphite reflector 77b, as shown in Figure 8, of about 50 centimeters in thickness. As shown in Fig. 7, the tubes 78 are arranged in a square geometry (i.e. in uniformly spaced vertical and horizontal rows) throughout the cross sectional area of the graphite moderator 77.

The graphite moderator 77 may be built of graphite blocks 77a (Fig. 30) cut to a size convenient for handling. The tubes 78 may be disposed in holes drilled through the graphite blocks that are aligned to form continuous channels or passages throughout the length of the cylindrical graphite moderator 77.

The reactor 25 is enclosed in a gas tight shell comprising two steel end sheets 79 and 80, a steel side sheet 81 (see Fig. 8) and a steel bottom sheet 81a all welded together.

Spanning the distance between the concrete walls 72 and 74 at the front of the reactor 25 is a concrete beam 84 disposed above the shot and water shield 76. A pair of spacing beams 85 and 86 extend from beam 84 to the rear upstanding concrete wall 73. An intermediate spacing beam 87 extends from the beam 84 to the steel sheet 79.

Spaced from the sheet 79 is a parallel sheet 88 and extending from this sheet to the concrete wall 73 is an extension 87a of the intermediate spacing beam 87. Two

16

reinforcing channel members 89 extend from the concrete base 70 to the plane of the top edges of the concrete walls 72, 73 and 74 and the sheet 88 is fastened to these channel members and is reinforced thereby. A horizontally disposed channel beam 90 is fastened to the upstanding channel members 89 and serves as additional support for the spacing beams 85, and 86 and 87a. Spaced apart, upstanding sheets 82 and 83 together with sheets 79 and 88 form the sides of a tank 68 at the discharge ends of the tubes 78.

A plurality of upstanding vertical channel backing members 91 reinforces the sheet 79 in the region above the reactor 25. The horizontally disposed channel beam 92 is fastened to the channel members 91 and serves as additional support for the horizontal spacing beams 85, 36 and 87.

As shown in Fig. 5, the steel sheets 82, 83 and 88 form three sides of a box-like enclosure spaced from the concrete walls 72, 73 and 74 so as to provide a shield 93 in the form of a tank surrounding the three sides of this box-like enclosure. Water indicated at 93a in Fig. 6, fills this tank forming part of the shield 93. Thus, except for the front face, the reactor 25 is immersed in water as best shown in Fig. 8.

Reactor 25 is supported on a plurality of spaced stringers 137 in the form of I-beams resting at their ends on concrete. Between the stringers 137 are a plurality of passages 137a (see Fig. 6) through which water 93a is circulated by a motor driven propeller 138 (Fig. 8). As shown in Fig. 8, the level of the water 93a above the reactor 25 is maintained at a selected position by an automatically controlled valve 139 in water supply pipe 139a. A float 139b in the water 93a controls the valve 139 in any one of several well known manners.

Projecting diagonally downward from the sheets 82 and 83 are walls 94 and 95 secured along one side to sheet 88 (Fig. 6) and along the other side to sheet 79. A plurality of I-beams are spaced throughout the extent of the sheets 82 and 83 and the diagonal walls 94 and 95 for reinforcement (see Fig. 7). The sheet 79 adjacent to its bottom is sloped toward the rear sheet 88 and together with the walls 94 and 95 and the rear sheet 88 form a funnel-like portion merging into a vertical discharge chute 97 that projects downwardly below the concrete base 70. This discharge chute 97 is provided with a pair of gate valves 98 and 99 having operating means generally indicated at 100 and 101, respectively. The vertical chute 97 is open at the bottom and directly below the opening is the car 31 supporting the container 30 for receiving materials discharged from the vertical chute 97.

The tank 68 that merges into the vertical chute 97 is filled with water, the upper level 104 of which is approximately adjacent to the tops of the concrete walls 72, 73 and 74.

As previously mentioned, the chute 97 projects through the concrete base 70 to a position below the ground level 71. Concrete walls 105 (Figs. 6 and 7) project downwardly below the concrete base 70 and form a discharge chamber generally indicated at 107 communicating with a tunnel 106. As best shown in Fig. 6, the walls 105 are so formed as to provide an off-set portion or floor 108 on which is mounted a lead angle member 109 having an upstanding flange 110 anchored in the concrete base 70. The flange 110 forms one wall of a room 111. The other walls of the room 111 are formed by the concrete walls 105 and the ceiling is formed by the concrete base 70. This lead angle member 109 is of sufficient thickness to provide adequate shielding to protect personnel who may be located in room 111 from the harmful effects of radioactive radiations emanating from fission products discharged through the vertical chute 97. The concrete base 70 is of sufficient thickness to absorb any neutrons directed towards the room 111 from the reactor 25.

2,910,418

17

The container 30 is mounted on the car 31 by means of a cradle 112 and as shown in Fig. 7; three such containers are thus carried on each car 31. Car 31 is mounted on wheeled trucks 113 that operate on rails 114 and a cable 115 is attached to one end of the car for pulling the car in one direction through tunnel 106. Cable 116 is attached to the opposite end of the car for pulling the car in the opposite direction through tunnel 106. Each container 30 is provided with thick lead walls, not shown, adapted to absorb gamma radiations, and lead cover 117 fits in an opening, not shown, in the top of each container. A shelf 118 is mounted on one of the walls 105 on which are placed the covers 117 for the containers 30. A conventional overhead crane is diagrammatically shown at 119 and is mounted on a movable track 120. The movable track 120 is provided with wheels 121 at each end; these in turn are mounted on fixed tracks 122.

Referring to Fig. 14, the arrangement of the tubes 78 in the graphite moderator 77 is clearly illustrated. The moderator 77 is broken away so that only a fragment of the graphite is illustrated but the tubes 78 extend throughout the entire length of the cylindrical graphite moderator 77. These tubes terminate adjacent to the sheet 79 and at the opposite side of the graphite 77 project through the sheet 79, the shield generally indicated at 76, and finally through a chamber 123 terminating after passing through front wall 124 of the power unit. A cylindrical sleeve 125 surrounds the tube 78 in the region from the graphite reactor 77 to the chamber 123 and a second sleeve 126 surrounds the portion of the tube passing through the chamber 123. Sleeve 126 is threaded on to the end of sleeve 125. A gate valve 127 is disposed on the tube 78 adjacent to the inner face of the front wall 124 of the power plant. A washer 128 is mounted on the free end of the tube 78 and against the outer face of wall 124.

The shield 76 comprises a plurality of tanks formed by spaced inner and outer walls 129 and 130 and intermediate wall or partition 131. A sleeve 132 in the form of a truncated cone surrounds each of the sleeves 125 and is welded to the walls 129, 130 and 131. Lead or iron shot 133 together with water indicated at 134 are contained in the shields 76. A tube sheet 135 is disposed against the shield wall 130 and is perforated to receive the tubes 78 with their surrounding sleeves 125. This tube sheet 135 together with the front wall 124 of the power unit form two walls of the chamber 123. Lead wool or shot 136 fills this chamber 123.

Water system in reactor

As shown in Fig. 13, the tubes 78 are adapted to receive uranium rods 140 and are provided with internal longitudinally disposed ribs 141. These ribs 141 support and center the uranium rods 140 and properly space them from the walls of the tube 78 so as to form passages 142 through which cooling medium passes.

Referring to Figs. 11, 12, and 14, a ring header 143 is provided around each of the sleeves 126 and these headers 143 are supplied with a liquid coolant from header pipes 144 through a piping system generally indicated at 145 having control valves 145a therein controlling the flow of coolant into the tubes 78. A removable cover plate 145b (Fig. 11) provides access to the valve 145a. The ring header 143 (see Fig. 12) has an annular passage 146 into which the coolant is fed from the piping system 145 (see Fig. 14). The sleeve 126, in turn, is provided with a plurality of passages 147 feeding from the annular passage 146 into a clearance space 148 from which the coolant passes into the tubes 78 and flows in the direction left to right in Fig. 14. This water passing through the tubes 78 discharges through the open ends of tubes 78 into the water filled tank 68 (Fig. 6). Thus, the ends of the tubes 78 adjacent to the coolant inlet headers 143 will hereinafter be designated the inlet ends of the tubes, whereas the opposite ends will be designated the outlet or discharge ends of the tubes.

18

At the inlet end each of the tubes 78 is provided with a plug 149, best shown in Fig. 12, provided with a shank portion 150 adapted to fit inside the end of the tube and a cap portion 151. The shank portion 150 should be of a material such as lead of adequate thickness to adsorb neutrons and gamma rays directed through the tube from the interior of the reactor.

Each of the valves 127 comprises a valve casing 152 housing a movable valve member 153 fastened to a valve stem 154 projecting upwardly from the movable valve member 153 through a passage 155 in an upstanding boss 156.

The upper end of the valve stem 154 is threaded as shown at 157 to receive a bevelled gear 158 rotatably supported and held against axial displacement on the top of the boss 156 by means of an intermediate collar 159 and on a flange of an angular bracket 160. The bevelled gear 158 is adapted to rotate about the valve stem 154 and while so doing moves the stem upwardly or downwardly due to the interaction between the threaded portion 157 of the valve stem 154 and the threaded bore of the bevelled gear 158. Thus, by rotation of bevelled gear 158 the movable valve member 153 is moved upwardly or downwardly to open or close the passage through the valve housing 127.

As shown in Fig. 12, an electric motor 161 is provided for operating each of the valves 127. Each motor 161 is drivingly connected to a shaft 162 journalled in the upstanding flange of the annular bracket 160 and has on its free end a bevelled gear 163 meshing with bevelled gear 158 to rotate bevelled gear 158 in response to the operation of the motor 161. A conventional electric control system, not shown, operates the motor 161. A casing 174 encloses the assembly operating the valve 153; i.e., the motor 161, shaft 162, gears 158 and 163, and the related parts and supports of the motor 161.

At the discharge end each tube 78 passes through sheet 79, and then terminates. As shown in Fig. 18, sleeves 164 are welded or otherwise secured as at 165 to the sheet 79. Each tube 78 passes through a sleeve 164. The sleeve 164 is threaded on its peripheral face to receive a cap 167. It is important that water 104 is not permitted to pass between the tube 78 and the sleeve 164 and for this reason each tube 78 is flanged at its end at 166 over the outer edge of the sleeve 164, and cap 167 applies a pressure seal against this flange 166 of tube 78.

The cap 167 is provided with a bifurcated bracket 168 adapted to receive an arm 169 pivotally mounted at 170 on the bifurcated bracket 168. A retainer plug 171 is mounted on the free end of the arm 169 and this plug is substantially hemispherical in shape and adapted to fit into the flared end of the tube 78 as best shown in Fig. 17. The retainer plug 171 is provided with a plurality of perforations 172 to permit the passage of cooling fluid from the tube 78 through plug 171. A torsion spring 173 (see Fig. 16) normally holds retainer plug 171 in its closed position shown in Fig. 17 so that force is required to open the plug to the position shown in Fig. 18 and to read it in this open position.

Each uranium rod 140 comprises a plurality of rod segments 140a as best shown in Figs. 10 and 18 interlocked together to form in effect the continuous rod 140. On each end of each segment 140a is threaded as at 176 an aluminum end cap 175. A space 140b is provided between each end cap 175 and the uranium body to serve as an insulating dead air space to keep the end cap cool. In this way corrosion and alloying problems with respect to the end caps are reduced to a minimum. An aluminum sheath 177 closely fits around the periphery of each rod segment 140a and is welded to the end cap 175 as at 178, thus completely sealing the uranium rod segment 140a within aluminum. The rod segments 140a are arranged end to end in each tube 78, and the end

2,919,418

19

caps 175 of adjacent ends of adjoining rod segments 140a are provided respectively with male and female portions 175a and 175b cooperating to form an effective interfit between adjacent rod segments.

As shown in Fig. 14, the rod segments 140a when interfitted form, in effect, a continuous uranium rod extending from the valve 127 to the discharge end of the tube 78 where the rod retainer 171 is disposed.

The cooling medium, that for purposes of illustration is water, enters the tubes 78 through ring headers 143 and then passes lengthwise through the tubes in the annular space 142 between the walls of tubes 78 and the rods 140. Thus, the water passes over the surface of the aluminum sheath 177 surrounding the uranium in each tube and thereby extracts heat that is conducted to the surface of the aluminum sheath 177 from the uranium rod 140. The water thus passing through each tube 78 flows through the openings 172 in the normally closed rod retainer 171 into the water filled tank 68.

Diphenyl as coolant

In this description, reference has been made primarily to light water as the coolant for the system. As previously mentioned, diphenyl, also known as biphenyl or phenylbenzene, is also a satisfactory liquid coolant. This substance has the chemical formula $C_6H_5C_6H_5$ and is in the form of a solid at atmospheric temperature, melts at 70° C. and has a boiling point of 225° C. Thus a sysem employing diphenyl as the coolant must be operated at a temperature such that the coolant at all times maintains a temperature above 70° C.

Diphenyl has a lower absorbing characteristic for thermal neutrons than has water; for example, a loss of neutrons by absorption due to a diphenyl cooling layer of 4 millimeters corresponds to the loss due to a 2.2 millimeter layer of light water. Thus for the same reproduction ratio almost twice as much diphenyl as light water can be circulated through the reactor. The cooling passages for a system employing diphenyl can then be almost twice as wide as those for light water in systems having the same reproduction ratio; the pressure drops will therefore be less. About 10 percent to 15 percent more pumping power is required to circulate the diphenyl due to its greater viscosity. Since diphenyl solidifies at 70° C., special measures are taken to prevent solidification of the coolant while it is disposed in the portion of the cooling circuit outside of the reactor. This, of course, can be accomplished by merely maintaining the temperature of diphenyl at all times safely above 70° C. as by regulating the amount of heat extracted from the diphenyl.

Whichever coolant is used, it must be maintained in liquid form and not be allowed to evaporate into steam. If steam is produced in the tubes 78 of the reactor 75, the reduction in water density resulting therefrom immediately increases the value of the reproduction ratio for the reactor. This, of course, upsets the control balance and might prove dangerous as the exponential rise in neutron density might become too rapid to control. Thus it is important to guard against this condition by preventing the formation of steam in the tubes 78. Proper control of the neutron density in the reactor and the rate of water flow through the reactor will accomplish this result.

Helium system

In accordance with a further modification helium, air or similar gaseous agent may be used to cool the reactor. The use of helium as a coolant in the reactor offers many advantages. The neutron absorption of pure helium is negligible and it has the characteristic of readily conducting heat. For this reason a charge of helium in the reactor tends to promote heat exchange throughout the reactor. The use of helium has an additional advantage in that it replaces the nitrogen that normally would permeate a reactor exposed to the atmosphere. Nitrogen has a much higher neutron absorption factor and conse-

20

quently any nitrogen absorption of neutrons to the detriment of the neutronic reaction is prevented when helium is used. Furthermore, the nitrogen density in the reactor exposed to the atmosphere changes with variations in atmospheric pressure and any such changes cause variations in the reproduction ratio of the reactor and therefore alters control requirements. Consequently, a sealed reactor containing helium is immune to changes in atmospheric pressure.

Another advantage in replacing the air content of a reactor with helium is based on the fact that the argon content of ambient air becomes highly radioactive when subjected to neutron bombardment. This radioactive argon diffuses out of the reactor exposed to the atmosphere and thus becomes a biological hazard. Such a condition does not occur in a helium filled reactor.

The air in the reactor can be replaced with helium in various ways, for instance by simple gravitational replacement; i.e., by introducing the helium into the enclosed reactor and allowing it to displace the heavier air which escapes through a suitable outlet. Another method would be first to evacuate the air in the reactor and then replace it with helium.

The helium may be circulated through a circuit as shown in Fig. 25. Here the gas is drawn from the top of the reactor 25 through pipe 350, is circulated through a container 351, driers 352, and is then returned to the reactor 25 by pump 353 through pipe 354. A water detector diagrammatically shown at 355 is placed in the pipe 350 to detect leakage of water which, if it gets into the helium, would likewise get into the graphite. In the event of such leakage, the faulty tube is located by suitable means not shown and if it cannot be repaired or replaced, it is closed off and plugged and thereafter not used.

The graphite moderator 77 may be built up of a plurality of blocks 77a (Fig. 30) bevelled along their edges as at 356 to form passages 357 through which helium gas may circulate. The tubes 78 pass through aligned holes 358 in the graphite blocks.

SPECIFIC DIMENSIONS AND VALUES FOR REACTOR

The size of the reactor required to produce a given amount of power depends upon such considerations as the geometry of the uranium and the graphite, the volume ratio between the uranium and the graphite, and the impurities in the reactor including impurities in the uranium and the graphite as well as other neutron absorbing materials such as the aluminum in the tubes and coatings and the water layer serving as the coolant. In other words, the size of the reactor depends upon the ratio of the fast neutrons produced in one generation by the fissions to the original number of fast neutrons in a theoretical system of infinite size where there can be no external loss of neutrons and where losses due to control rods, limiting rods, and the like are disregarded. This ratio is known as the reproduction or multiplication factor and is referred to by the symbol K.

For a reactor employing uranium rods disposed in graphite in accordance with optimum geometry conditions and utilizing uranium and graphite that are free of impurities, the value of K would be about 1.074. This value is known as the base K for a uranium rod and graphite reactor. The value of K for the structure shown herein is determined as follows:

Base K for uranium rods in graphite	1.074
Actual K loss due to aluminum sheathing and tube	0.013
Actual K loss due to water	0.023
Actual K loss due to heating of rods	0.003
Total K loss	0.041
The value of K for the structure shown	1.035

2,910,418

21

These figures are based on a reactor of infinite size. As the size of the reactor is reduced to a finite size, the loss of neutron from the exterior of the reactor rises so that the reproduction ratio of a reactor of finite size is necessarily less than the value of K given above. As the size becomes finite, there is a size where the ratio between the fast neutrons produced by fission and the fast neutrons present to start the chain reaction becomes unity. The size corresponding to this condition is known as the critical size for the reactor, below which the chain reaction is not self-perpetuating. The finite size of the reactor then for a self-perpetuating system must be slightly over the critical size. It has been found in practice, however, that the size should be such that the reproduction ratio is not greater than about 1.01 preferably not greater than 1.005.

If the reproduction ratio exceeds this value, the exponential rise in neutron density, when the control rods are removed from the reactor, may take place so rapidly that this rise is difficult to stop and may not be possible to stop until serious damage has been done.

The critical and operating sizes for a neutronic reactor radius R can be computed in accordance with the formulae previously mentioned.

The power unit disclosed herein for illustrative purposes may be rated at about 200,000 kilowatts, the heat energy that can be extracted from the reactor during normal operation. For this output, the required value for K with the reflector is 1.024 and without the reflector is 1.028 to give a reactor of practical size. Actually, the value of K in the reactor shown is 1.035. Thus, the requirements are met.

The principal dimensions of the reactor are approximately as follows:

Axial length of active cylinder of reactor 25_____ 7 meters (23 ft.).

Radius of active cylinder of reactor 25 _____ 4.94 meters.
Thickness of graphite reflector 77a_____ 50 centimeters.
Total weight of uranium metal in rods 140 _____ 200 metric tons.
Weight of graphite 77 in reactor (exclusive of reflector)_____ 850 metric tons.
Size of each graphite block_____ 8⅜" x 8⅜" x 50".
Weight of graphite in reflector_____ 315 metric tons.
Radius of uranium metal rods 140_____ 1.7 centimeters.
Length of each rod segment 140a_____ 2 feet.
Thickness of aluminum sheath 177____ 0.5 millimeter.
Thickness of aluminum end cap 175____ 17 gauge.
Thickness of aluminum tube 78_____ 1.5 millimeters.
Thickness of liquid layer 142_____ 2.2 millimeters with water; 4 millimeters with diphenyl.

Number of rods 140 in reactor_____ 1695.
Weight of aluminum in reactor proper__ 8.7 metric tons.
Rod spacing in square array_____ 21.3 centimeters (8⅜ inches).

Surrounding each tube 78 is essentially a graphite shell having a substantially uniform heat production due to neutron absorption in the graphite. This heat must pass through the walls of tubes 78 and be carried away in the water stream.

When the heat generated in the rods 140 has passed through the aluminum sheaths 177 on the uranium rods 140, it must be transferred to the flowing liquid coolant 142 and be carried away. A temperature drop occurs between the wall of aluminum tube 78 and the interior of the liquid stream, this drop being expended across the non-turbulent film of liquid coolant adjacent to the walls of the tube 78. There is also a rise in the temperature

22

of the liquid coolant as the coolant flows through the reactor 25. The wall temperature of the aluminum tube 78 at any point consists of the rise in coolant temperature to that point plus the film drop at that point.

The coolant enters the reactor at considerable pressure and the pressure decreases as the coolant moves along the tube. Thus, the temperature at which the coolant will boil decreases as the liquid moves through the reactor. Local boiling in the coolant film may hamper the flow of heat into the flowing liquid, and for this reason the wall temperature everywhere is kept lower than the boiling temperature of the liquid coolant. By keeping the coolant under some excess pressure as it leaves the reactor the boiling point of the coolant throughout the length of the tube is raised and the coolant can thus be allowed to have a greater temperature rise.

Specific values for the present system are as follows:

Pressure drop along rod 140_____ 115 lbs./in.²
Total flow rate in tube 78 neglecting ribs 141_____ 1790 cm.³/sec.
Velocity of flow_____ 7 meters/sec.; 23 feet/sec.
Allowance for reduction of stream cross section by ribs 141_____ 10 percent.
Flow rate with ribs 141_____ 1600 cm.³/sec.
Output of central rod 140_____ 56,000 cal./sec.
Average output per rod 140_____ 118 kw.; 28,200 cal./sec.
Total volume of water in reactor 25___ 2.75 cubic meters.
Maximum heat from rod 140 per unit length _____ 110 cal./sec./ cm.
Maximum heat into water per unit area _____ 10 cal./sec./ cm.²
Film transfer coefficient_____ 1.2 cal./sec./ cm.²/°C.
Input temperature of water to reactor 25 _____ 35° C.

The foregoing values refer to the water cooled system. If diphenyl is used the only construction change required in the reactor 25 is to increase the thickness of the annular cooling space 142 to 4 millimeters as previously mentioned. The temperature rise of diphenyl is about the same as in the case of water, but the drop in temperature across the non-turbulent film adjacent to the inner walls of each tube 78 is about 5 times that of water.

Again referring to a system wherein water cooling is employed it is seen that the emergent temperature of the central water stream is 123° C. for the input temperature specified, so that the emergent pressure must be at least 20 pounds per square inch to avoid boiling of the coolant water in the tubes 78.

In the arrangement disclosed herein the coolant completely separates uranium metal 140 and the graphite 77 and thus interposes a neutron scattering layer between the moderator and the uranium to slow neutrons passing from the graphite to the uranium metal. This blocking effect forces the density of the thermal neutrons in the graphite to be greater than they otherwise would be and thereby increases the graphite absorption of neutrons. The loss due to the use of a 4 millimeter cooling layer of diphenyl corresponds to that for a 2.2 millimeter layer of water so that the two arrangements disclosed herein, i.e., the one employing water as the coolant and the other diphenyl, have the same multiplication factor.

The foregoing values for the multiplication factors are based on the assumption that the air has been removed from the reactor so as to eliminate the absorption of neutrons by nitrogen. This can be done effectively by

23

charging the reactor with helium gas having a high thermal conductivity to aid in the transfer of heat to the coolant.

Aluminum is a satisfactory material from which the tubes 78 and the sheaths 177 are made. Beryllium also may be used.

Shields

The required thickness of shield 93 depends on how close persons will stand with respect to the reactor and for how long a time they will be thus exposed. For purposes of illustration, a thickness of 12 feet has been selected as the minimum thickness for this shield. An equivalent depth of water exists over the top of the reactor 25.

The water and concrete shield 93 surrounds only three sides and the top of the reactor 25. The fourth side constitutes the inlet or loading side of the reactor where uranium rods 140 are charged into the reactor 25. The shield 76 (see Fig. 14) serves to protect operating personnel from the harmful radiations resulting from the neutronic reaction.

This shield 76 comprises an inner and an outer shield portion 76a and 76b, respectively, separated by the partition 131. This partition 131 maintains a complete separation between the shield portions 76a and 76b. Shield portion 76a is disposed adjacent to the reactor 75 and thus is subjected to neutron bombardment and thereby becomes radioactive, generating considerable heat. The water 134 in the shield portion 76a is therefore circulated out of the shield through pipe 200, is cooled, and then is returned to the shield portion 76a through return pipe 201. Thus, the temperature of shield portion 76a is controlled. The temperature of shield portion 76b may be sufficiently stable without requiring artificial cooling.

By separating the shield 76 into the two portions 76a and 76b as described, an advantage is derived other than the easy maintenance of the proper temperature. Neutrons entering the inner shield portion 76a are absorbed before they reach the outer shield portion 76b. Since there is no intermixing of the water between the two shield portions 76a and 76b, the radioactivity resulting from neutron absorption is limited to the inner shield portion 76a. Thus, the arrangement is effective in protecting operating personnel from any harmful radiations induced in the shield 76.

Referring again to Fig. 14, the shield 76 may be fabricated in the form of a plurality of tanks nesting together and held by means of complementary interlocking joints generally indicated at 202.

On the outside of shield 76 in the space between tube sheet 135 and plate 124 is disposed the lead wool or shot 136 serving as an additoinal absorber of gamma rays.

Control system

The device above described is ideally adapted for automatic control to maintain the neutron density inside the reactor substantially constant, and thus gives a substantially constant rate of production of element 94^239. Due to the fact that large masses of materials are utilized in the reactor, there is a temperature lag therein. Consequently, it is convenient to monitor the structure by means of ionization chambers and equivalent devices responding to the neutron density at the periphery of the reactor. As the rate of neutron diffusion out of the neutronic reactor 25 is always proportional to the rate of release of neutrons within the structure, the ionization chambers can readily be placed at the periphery of the reactor and in fact are preferably so placed that they may not be subjected to the extremely high neutron density existing within the reactor.

For purposes of illustration, one control rod 179 and one emergency or safety rod 180 are shown in position for operation in the neutronic reactor Fig. 24. Obviously many more control rods and safety rods may be employed. For example, two control rods of boron may be required

24

for normal operation and twenty-nine safety rods distributed throughout the reactor may be used. Also seven shim rods may be installed. These rods may project into aluminum lined channels in the moderator, and contain about 85 grams of boron per meter length of rod. The rods are adapted to pass through the length of the reactor.

The safety rods 180 are normally disposed in their extreme upper position out of the reactor and are held there during operation of the neutronic reaction. Their purpose is to stop the reaction either when it becomes too violent, or under normal conditions when the reaction is to be stopped. They function as efficient neutron absorbers, and when in their lowered position in the reactor they absorb a sufficient number of neutrons to lower the reproduction ratio of the reactor to a value well below unity.

Iron and cadmium are relatively good neutron absorbers, and therefore serve as very satisfactory materials for safety rods. The incorporation of boron in the iron increases the efficiency of the rod as a neutron absorber, and therefore is advantageous in the rod composition.

Before proceeding to a detailed description of a control circuit that may be utilized in controlling the neutronic reaction, it is desirable to point out the manner in which the control rods operate to regulate the neutron density.

In any self-sustaining chain reacting structure adapted to produce power or element 94^239, the attainable neutron reproduction ratio and reproduction factor of the system must be sufficiently greater than unity to establish and maintain the desired output. For any value over unity, the chain reaction becomes self-sustaining and the neutron density, without control, would increase exponentially in point of time until the device is destroyed. For proper control, the system must be held in balance by maintaining the neutronic reaction at some point where the production of new neutrons is balanced with the neutrons initiating the chain. Under these conditions, the reaction will continue to maintain the neutron density in the reactor the same as existed when the system was balanced.

However, in order to enable the reactor to reach a desired neutron density, the system must be permitted to rise in neutron density for a period of time until the desired density is reached. It is necessary thereafter only to hold the system in balance.

Inasmuch as the reproduction ratio in any self-sustaining chain reacting system is reduced by the presence of impurities that absorb neutrons, such impurities can be introduced into the active portion of the structure in the form of a control rod of a material such as boron or cadmium capable of absorbing large amounts of neutrons. The depth the rod penetrates into the reactor will determine the amount of neutron absorption and therefore the reproduction ratio of the system. A range can be obtained between a condition providing a neutron reproduction ratio that is greater than unity and a condition at which no chain reaction can be maintained. The exponential rise in neutron density can be made relatively fast or relatively slow in accordance with whether the reproduction ratio is permitted to be much greater than unity or only slightly greater than unity.

There is a small percentage of delayed neutrons emitted in the fission process causing the neutron density in the reactor to rise in a finite time rather than instantaneously. The time required for doubling the neutron density increases as the reproduction ratio approaches unity, and any desired rate of rise in neutron density can be obtained.

As a general rule, the reproduction ratio of a neutronic system with all control rods in their withdrawn position should not be greater than about 1.005. At this value the neutron density in the system will double itself every seven or eight seconds and can therefore be easily controlled.

When initially placing a neutronic system such as that

2,910,418

25

shown in Fig. 8 into operation, the safety rod 180 diagrammatically illustrated in Fig. 24 is withdrawn and then the control rod 179 is withdrawn from the reactor to a point where there is an exponential and preferably slow rise in neutron density. When a desired neutron density has been reached, the control rod 179 is then returned into the reactor to a point where the reaction is balanced. This balance is then maintained so as to keep a constant power output or production output of element 94²³⁹ in the reactor. The maintenance of the balance point with the control rod would be relatively simple were it not for the fact that changes in temperature in the reactor result in changes in the reproduction ratio. It is desirable, therefore, that the control rod be so manipulated that a substantially constant neutron density is maintained within the system. Such a method of control may be accomplished by automatically connecting the control rod 179 through automatic adjusting means controlled by an ionization chamber or similar device responding to neutron density conditions within the reactor. Furthermore, due to the exponential rise in neutron density within the reactor when the reproduction ratio is greater than unity, all possible precautions must be taken to prevent a continued exponential rise in neutron density in the event of failure of the control rod to return to position.

While there are many means by which the control rod can be operated, it is believed sufficient to illustrate and describe one simplified control circuit to the end that fully equivalent circuits will be made apparent to those skilled in the art.

Referring to Fig. 24, there is shown diagrammatically one form of control circuit reduced to lowest terms, that might be used for regulating by means of control rods the output of the plant hereinbefore described, and attention is focused first on the control circuit shown at A. A control ionization chamber 181 is placed at the periphery of the reactor 25 and is filled with boron fluoride. A central electrode 181a is provided within the chamber 181 and connected to a wire 183 leading outside the reactor to a movable slider 184a on a resistor 184. The resistor 184 is connected across a relay coil 185; one side of this coil is connected to a battery 186 and the other side is connected to a shield 187 around the wire 183. The shield 187 is grounded as is the chamber 181.

Alpha ray ionization due to the neutron reaction with the boron within the chamber 181 is proportional to the neutron density in the region of the resistor 184. Thus, the current in the resistor is varied in accordance with neutron density reaching the ionization chamber 181.

The relay coil 185 operates a relay armature 188 that is spring biased by a conventional spring 189 to contact one motor contact 190 and is urged by the electromagnetic attraction in response to a sufficient current passing through the relay 185 to contact a second motor contact 191. Contacts 190 and 191 connect to the outside of a split winding or motor 192, the central terminal 182 being connected through power mains to the relay armature 188.

The motor 192 rotates a shaft 193 having on one end thereof a pulley 194 and on the other end thereof a control rod gear 195. The control rod gear 195 meshes with a rack 196 on the control rod 179.

The pulley 194 has a cable 197 wound thereon that in turn is connected to a counter weight 198 so that the weight of the control rod 179 is substantially balanced by the counter weight 198 so as to permit the motor 192 to run easily in either direction.

During normal operation of the system where the neutronic reaction is controlled only by control rods 179, the slider 184a on the resistor 184, having previously been calibrated in terms of neutron density, is moved to the density position at which the reactor is intended to operate, taking into account the difference in neutron density at the center of the active portion of the reactor and at the periphery thereof during the operation. This

26

difference is a constant ratio at various operative densities. When the reproduction ratio in the reactor is less than unit, as is the case when the control rods 168 are disposed at a sufficient distance in the active portion of the reactor, a neutron density at the position of the ionization chamber 181 is much lower than that required to energize the relay coil 185 sufficiently to overcome the action of spring 189 so as to move the armature 188 away from motor contact 190. Thus, the motor 192 under these conditions is energized to withdraw the control rod 179 from the active portion of the reactor to a point determined by a limit stop 199 where the reproduction ratio of the reactor is just sufficiently greater than unity to permit a slow rise in neutron density in the reactor. The motor 192 will stall when the rod 179 is at the stop 199, and should be of the type permitting stalling for a sufficient time to allow the next succeeding operations to be effected. The reaction at this position of the control rod becomes self-sustaining and the neutron density rises. In consequence, the ionization taking place within the ionization chamber 181 rises. As the ionization increases, the current passing through relay coil 185 also increases until a predetermined field density is reached. When the induced magnetism in the relay coil 185 has reached a predetermined strength the effect of the spring 189 is overcome and the armature 188 is attracted to the adjacent pole of the relay 185 thus connecting the armature 188 to the motor contact 191. This energizes the motor 192 so as to operate in the reverse direction to that of original operation, so as to drive control rod 179 downwardly into the active portion of the reactor thereby causing the neutron density in the reactor to rise less rapidly and eventually to decay. As soon as the neutron density decays to a predetermined lower value that is approximately at or slightly below that corresponding to the setting of the slider 184a, the control rod 179 having then adjusted the reproduction ratio to a value approximately equal to or usually slightly below unity, the magnetic strength of the relay coil 185 becomes weakened to the extent that the spring 189 again acts to move the armature 188 against the contact 190. The control rod 179 will thus hunt between a point above the balance position where the neutron density rises and a point below the balance position where the neutron density decays, thus providing an average neutron density within the reactor as determined by the setting of slider 184a on the resistor 184. Since the mass of the reactor causes temperature change to lag behind neutron density change, the temperature of the reactor is maintained substantially constant. If desired, any of the well-known anti-hunting circuits may be utilized as will be apparent to those skilled in the art.

The main purpose of the control circuit A is to regulate the control rod 179 to balance the neutron density in the active portion of the reactor so as to maintain any desired average temperature within the reactor.

Emergency control system

In the event of emergency, or for other reasons, it may be desirable to discontinue the neutronic reaction. This may be done by rapidly driving into the reactor a sufficient number of safety rods to reduce the reproduction ratio of the reactor below unity.

Referring to Fig. 24, one such emergency or safety rod is shown at 180. At the upper end of the rod is a piston 301 operating in a cylinder 302. At the upper end of the cylinder 302 is a passage 303 connecting the cylinder to a source of fluid pressure, not shown. At the lower end of the cylinder 302 is an exhaust port 304. The piston 301 is adapted to move up and down in the cylinder 302. A second exhaust port 305 is disposed in the passage 303. The function of the fluid pressure created in the cylinder 302 is to force the piston 301 from an upward position in the cylinder 302 to a position

2,910,418

27   28

adjacent to the bottom of the cylinder, as a result of which movement the emergency or safety rod 180 is rapidly inserted into the reactor 25. Inasmuch as the rod 180 is heavy some means is provided to cushion the rod when it reaches its lower position. This cushioning means is illustrated in the form of a dash pot 306 provided with a small exhaust opening 307 and adapted to receive the plunger 308.

During normal operation of the reactor 25 the safety rod 180 is in its upper position as illustrated in the drawings. An inlet valve 310 in the pressure line 303 is normally closed and an exhaust valve 311 in the exhaust passage 305 is normally open.

For purposes of illustration a simple control circuit for operating the safety rod 180 is generally indicated at B. This control circuit includes an electromagnetic solenoid 315 that is normally energized from the power mains to hold a movable armature 316 inside the solenoid. This armature 316 is attached to the valve operating arm 317 connected at its opposite end to the inlet valve 310. The exhaust valve 311 likewise is fastened to the operating arm 317 by means of member 318. A tension spring 319 normally urges the valve 310 toward its open position and at the same time urges valve 311 toward its closed position. As long as solenoid 315 is energized, however, from the power mains, the force of spring 319 is overcome and the armature 316 is held inside the solenoid 315 so that valve 310 is normally closed, whereas valve 311 is normally open.

Solenoid 316 is one element in a holding circuit that also includes a magnetic clutch 320, a circuit breaker 321, a flow switch 322 and a manually operated circuit breaker 323, all connected in series by wiring 324. The circuit breaker 321 normally makes contact with contact 321a, whereas circuit breaker 323 normally makes contact with contact 323a. A spring 325 normally urges the circuit breaker 321 against contact 321a and a spring 326 normally urges a manual circuit breaker 323 against contact 323a. The flow switch 322 is disposed in the cooling water header 47 and is normally maintained in closed circuit position by the normal pressure of the cooling fluid passing through the header 47. If for any reason water ceases to flow through header 47 or the flow drops below a predetermined rate, the flow switch 322 automatically opens thereby breaking the circuit through the solenoid 315.

The circuit breaker 321 serves as an armature in a magnetic relay. The magnetically operated relay coil 330 is disposed in an operating circuit that includes an ionization chamber 331 situated at the periphery of the neutronic reactor 25, a central electrode 332, a conductor 333 connecting the central electrode 332 to a movable contact slider 334 operating along a resistor 335. The relay coil 330 is disposed in the circuit with the resistor 335, the circuit being completed through conductor 336. A battery 337 is in the circuit with the relay coil 330 and is connected to a shield 338 surrounding the conductor 333. The battery 337 is grounded at 339.

After the safety rod 180 has been forced into the reactor it must be raised by some power means to return it to its normal withdrawn position. For this purpose a motor 340 mounted on a shaft 341 is connected to a pinion wheel 342 that in turn engages a rack 343 mounted on the safety rod 180. The weight of the safety rod 180 is counterbalanced by a weight 334 operating over a pulley 345 mounted on the shaft 341. The magnetic clutch 320 is mounted on the shaft 341 between the motor 340 and the pinion 342. Power for the motor 340 is provided from the power mains through conductors 346 and a manually operated circuit breaker 347 opens and closes the circuit depending upon the requirements for the motor 340. A tension spring 348 normally holds the circuit breaker 347 open.

The operation of circuit B in controlling the position of the safety rod 180 will now be explained. When the neutronic reactor 25 is normally operating, the safety rod 180 is in its raised position outside the reactor as shown in Fig. 24. Under this condition the circuit breakers 321 and 323 as well as the flow switch 322 are closed so that the solenoid 315 is energized from the power mains. The magnetic clutch 320 also is energized through the same circuit so that the safety rod 180 is held in its raised position by the counterweight 344. As shown, the solenoid 315 will be de-energized in the event of four different conditions or any combinations thereof as follows: (1) power from the mains fails; (2) the neutron density in the reactor 25 exceeds a predetermined value so as to cause ionization in the chamber 331 to the extent that the magnetic relay 330 becomes energized sufficiently to break the circuit through the circuit breaker 321; (3) the flow switch 322 opens the circuit due to complete or partial failure of the cooling medium to flow through the header 47; and (4) the circuit breaker 323 is opened manually.

As soon as the solenoid 315 becomes de-energized by any one or combination of the foregoing events the armature 315 moves in response to the action of spring 319 to open the valve 310 and simultaneously close valve 311, thus introducing into cylinder 302 fluid through the fluid pressure line 303. Simultaneously with the de-energization of the cylinder 315, the magnetic clutch 320 also becomes de-energized. Thus, the safety rod 180 is free to drop by gravity into reactor 25, but in addition to the force of gravity the fluid pressure against the piston 301 rapidly forces the rod into the reactor 25. When the rod approaches its lower position in the reactor 25, the plunger 308 enters the dash pot 306 compressing the air therein to cushion the final stages of the fall. In this manner the neutronic reaction in the reactor 25 is stopped.

When it is again desired to place the neutronic reactor in operation and the various circuit breakers and the flow switch 322 are in position so that this can be done the circuit can be closed to return armature 316 to its position inside the solenoid 315, thus closing the fluid pressure line and opening the discharge port 305 on the fluid inlet side of the piston 301. At the same time, the magnetic clutch 320 becomes energized so that on closing the manually operated circuit breaker 347, the motor 340 becomes energized and operates to raise the safety rod 180 to its withdrawn position illustrated in the drawings.

The system described has been reduced to lowest terms for illustrative purposes and is not deemed to be limiting, as equivalents will readily be suggested to those skilled in the art.

Other embodiments

Referring to Figs. 26 to 29 inclusive various different arrangements are shown for cooling the uranium in the reactor. In Fig. 26, the uranium is in the form of a hollow tube 359 surrounded by an aluminum sheath 360 and lined with an aluminum liner 361. In the arrangement shown in Fig. 26, a thin coating of non-fissionable material may be used instead of the sheath 360, since as shown its purpose is to prevent the escape of fission products. The water passes through the uranium tube 359 as shown at 362, the liner 361 serving as a protective coating to prevent corrosion of the uranium tube 359 by the water 362. A gap 363 is provided between the uranium tube 359 and the surrounding graphite moderator 77. This gap is needed to a greater extent in this arrangement than in the case where the water film is disposed between the uranium rod and the graphite due to the greater differential in heat between the uranium and the graphite. A gas, such as helium, may be passed through this gap 363.

Referring to Fig. 27, another arrangement is shown wherein a uranium tube 364 is covered by an aluminum sheath 365 and lined with an aluminum liner 366; and a

2,910,418

29

uranium rod 367 is disposed in the hollow portion of the uranium tube 364 but is of less diameter than the inside diameter of the uranium tube 364 so as to provide a space between the uranium rod 367 and the inner face of the liner 366 through which space water 368 circulates. This uranium rod 367 is covered by an aluminum sheath 369 that is provided with longitudinal ribs 370 serving as spacers to properly center the uranium rod 367 inside the uranium tube 364. Again a gap 371 is provided between the uranium tube 364 and the graphite moderator 77.

Still another modification is shown in Fig. 28. Here an aluminum tube 372 is disposed in the graphite 77 and inside this tube is a uranium tube 373 of lesser diameter than the inside diameter of the aluminum tube 372, and longitudinal ribs 374 are provided along the inner face of the aluminum tube 372 cooperating to properly center the uranium tube 373 in the aluminum tube 372. The uranium tube 373 is covered on its outer surface with an aluminum sheath 375. Water is circulated at 376 in the annular passages betwen the aluminum sheath 375 and the aluminum tube 372. The center portion 377 of the uranium tube 373 may be hollow and left empty or as shown in Fig. 29 may serve as a second passage through which water or gas 378 circulates. Thus, in the last embodiment shown in Fig. 29, the uranium tube 373 is externally cooled by water stream 376 and is internally cooled by water or gas stream 378 or by a different fluid such as diphenyl, air etc.

Auxiliaries

Flash tank.—The flash tank 42 shown in Fig. 4 comprises a steel cylindrical tank approximately 22 feet in diameter and 40 feet high with conical top and bottom 42a and 42b, respectively, and contains about 13,300 cubic feet of water. The water leaving the reactor 25 has a temperature of about 2° F. above the boiling point at the existing atmospheric pressure and is at a pressure of about 20 pounds per square inch gauge. This water before entering the flash tank 42 first passes through the throttling valve 41a, where the pressure is reduced to about atmospheric pressure. As the result of this reduction in pressure, some of the water vaporizes and most of the free oxygen and hydrogen in the water is carried into the water vapor. The water at this reduced pressure enters the tank through the spray head 48 and additional steam is formed inside the flash tank 42. The liquid water leaving the spray head 48 has a temperature approximately corresponding to the boiling point at the existing pressure and passes through approximately a 5 foot spray space striking the top surface, indicated at 49 of the water contained in the flash tank 42. The hot water in the tank passes downwardly through a gravel filter 50, of about 2 feet in depth, where suspended solids such as iron rust and aluminum hydroxide are removed from the circulating system. The water passes through the filter 50 at the rate of about 0.17 foot per second. The pressure drop through the gravel is about 1.4 pounds per square inch. The water leaves the tank through the pipe 43. A discharge pipe 39 provides a passage through which the flash tank 42 can be drained.

Flash tank condenser 52.—The mixture of steam, hydrogen and oxygen in the flash tank 42 passes off through an opening 51 in the top of the flash tank, over flash chamber condenser coils 52 where some of the steam condenses and drops back into the flash tank 42. The remaining gases, hydrogen, oxygen and some water vapor pass out through stack 53 and are diluted with air to reduce them below an explosive mixture and are then exhausted into the air. Because of the high oxygen content of the atmosphere around the condenser coils 52 and the severe corrosive action caused thereby, the coils 52 are preferably made of brass while the tube sheets are made of steel.

30

Specific data for a satisfactory condenser is as follows:

FLASH CHAMBER CONDENSER

Water vapor condensed	29,500 lb./hr.
Heat transferred	28,600,000 B.t.u./hr.
Temp. condensate	212° F.
Temp. cooling water in	85° F.
Temp. cooling water out	130° F.
Heat transfer coefficient	150 B.t.u./hr. ft.^{2} °F.
Heat transfer area	1900 ft.^{2}
Tube diameter	¾ in.
Wall thickness tubes	14 BWG.
Tube length	4 ft.
No. tubes	2520.
Shell diameter	60 ins.
Tube material	Brass.
Tube sheet material	Brass.
Shell and heads	Iron.

Heat exchangers 44.—Though only one heat exchanger 44 is shown, more than one may be used. The heat exchangers 44 per se form no part of the present invention. As an example of a specific type, however, that is satisfactory for the system disclosed, heat exchangers of the shell and tube type are suggested with the water from the reactor 25 disposed inside the tube and the cooling water from the cooling tower 54 being on the outside of the tubes. If steel tubes are used in the heat exchangers corrosion inhibitors in both water streams would be used.

As a specific example, for a heat exchanger, the following data are given:

Heat transfer rate	1.679×10^6 B.t.u./hr.
Temperature dist. water in	212° F.
Temp. dist. water out	95° F.
Temp. cooling water in	85° F.
Temp. cooling water out	130° F.
Distilled water flow rate	28,700 g.p.m.
Cooling water flow rate	74,600 g.p.m.
Log mean temperature difference	34.2° F.
Heat transfer area, outside of tubes including spares	329,000 ft.^{2}.
Overall heat transfer coefficient	186 B.t.u./hr. ft.^{2} ° F.
Tube fluid	Distilled water.
Shell fluid	Cooling tower water.
Tube length	20 ft.
Tube O.D.	0.75″.
Tube I.D.	0.584″.
No. tube passes per exchanger	2.
No. shell passes per exchanger	2.
Shell I.D.	6 ft.
No. tubes per exchanger	3500.
No. exchangers, total	24.
No. exchanger banks, in parallel including 1 spare bank	6.
No. exchangers per bank, in series	4.
No. banks in normal operation	5.
No. stand-by banks	1.
Pressure drop on shell side, total for battery	30 p.s.i.
Pressure drop on tube side, total for battery	13 p.s.i.
Material	Steel and iron.

Pumps 46.—Although only one pump 46 is shown in Fig. 4, of the drawings, more than one actually would be provided to assure circulation of water through the reactor 25 in the event of failure of one of the pumps. The combined normal capacity of the pumps handling the flow of the cooling water through the reactor 25 must be about 28,700 gallons per minute. It is suggested that eight pumps be provided, any six

2,910,418

31

of which can carry the normal load. With this arrangement, two pumps are available for standby service in the event of failure of one or two of the six normally carrying the load. Again for the purpose of illustration, specific data for the water pumps 46 are as follows:

Pumping rate _____ 28,700 g.p.m.
Minimum No. units desired _____ 6.
No. spare units desired _____ 2.
Discharge pressure _____ 150 p.s.i. gage.
Suction pressure _____ 0 p.s.i. gage.
Suction temperature _____ 95° F.
Power required _____ 3300 kw.
Power material _____ Stainless steel

Cooling tower 54.—The cooling towers 54 can be of conventional design. Though in Fig. 4 the one illustrated is disposed relatively close to the remaining portions of the system, it is possible to locate the cooling towers at some distances.

As a specific example of a satisfactory cooling tower and cooling water for the system, specifically disclosed herein the following data are given:

Temperature water in _____ 130° F.
Temperature water out _____ 85° F.
Loading density _____ 2 g.p.m./ft.².
Hot water rate _____ 75,800 g.p.m.
Tower area required _____ 37,900 ft.².
Water evaporation rate _____ 3,300 g.p.m.
Maximum probable water draw-off rate___ 165 g.p.m.
Windage water loss rate _____ 760 g.p.m.
Treated water make-up rate _____ 4,225 g.p.m.
Power required for cooling tower fans __ 1,000 kw.
Tower material _____ Wood.

Secondary cooling water pump 55.—Although only one pump 55 is shown in Fig. 4, more than one can be used. The design of the pump disclosed is conventional and again for purposes of illustration, the following specific data for a satisfactory pump are given:

Pumping rate _____ 75,800 g.p.m.
Minimum No. units desired _____ 7.
No. spare units recommended _____ 1.
Discharge pressure _____ 50 p.s.i. gage.
Suction pressure _____ 0 p.s.i. gage.
Suction temperature _____ 85° F.
Power required _____ 3000 kw.
Pump material _____ Iron or steel.

ADDITIONAL ACCESSORIES

It may be necessary to discard the water after it is passed through the reactor 25 because of contamination with certain impurities, some of which may be radioactive. It may be desirable further to employ inhibitors, in which case, the inhibitor concentration may gradually increase and for this reason the water in the primary cooling circuit passing through the reactor 25 may have to be discarded. The disposal of this discarded water must be done with considerable care because of the likelihood of radioactive materials being present in the water. It may be desirable to provide an evaporator or still, not shown, to which the water from the reaction 25 can be directed. The purpose of this evaporator would be to concentrate the waste material in the water to a relatively small volume to permit discarding it into deep wells.

Construction

When building the plant herein described, the necessary excavations are made, all the concrete is poured including the concrete walls forming the shields surrounding the reactor, the metal work is installed, the graphite and aluminum tubes included in the reactor are installed and the entire plant is assembled including all of the external equipment except that the uranium is not inserted in the aluminum tubes.

32

During the operation of the neutronic reactor, particularly at high neutron densities, radioactive elements of exceedingly high capture cross section for neutrons may be formed in the uranium as an intermediate element in the decay chains of fission fragments and this formation will reduce the value of the reproduction factor for the system. Radioactive xenon[135] is an example of such an intermediate element, this product having a half life of about 9 hours and being formed mostly from radioactive iodine which has a half life of about 6.6 hours and ultimately decays to barium through cerium. When building the neutronic reactor there should be sufficient size to produce an excess in the reproduction ratio sufficient to overcome this reduction in the reproduction factor. It may be desirable initially to construct the reactor sufficiently oversized to supply this excess reproduction ratio when needed, but to leave these excess uranium bodies out of the reactor until they are needed to overcome the effect of xenon. It may be more desirable, however, initially to construct the reactor complete with uranium sufficiently oversized to supply the excess reproduction ratio when needed in which case removable impurities, for example in the form of additional shim, limiting or control rods, may be initially placed in the reactor and kept there at all times until the reproduction ratio commences to fall, while the reactor is in operation, as the result of the formation of these intermediate decay elements. To compensate for this fall in the reproduction ratio these rods would then be withdrawn from the reactor sufficiently to maintain the reproduction ratio at a value of unity or at any other value desired.

Thus, before the uranium is charged into the reactor, the entire plant is assembled and all the shields are fitted into place and filled with water, or water and shot, as the case may be and all control rods are installed. The electrical control circuit is installed and placed into operating condition.

After the entire plant has thus been assembled and it is in readiness for operation with the exception it lacks the uranium, the time is then proper for charging the uranium into the reactor. Before this is done, however, all of the control rods and safety rods are inserted their full distance into the reactor. The loading operation is then started.

Loading and unloading reactor

Referring to Fig. 14, the uranium rods are loaded or charged into the reactor at the inlet side of the reactor shown at the extreme left-hand side of the figure and are discharged at the outlet side of the reactor shown at the right-hand side.

Referring to Figs. 19 to 23, inclusive, a loading car generally indicated at 203 is illustrated and comprises a flat car 204 mounted on wheeled axles generally indicated at 205 operating on tracks 206, the flat car 204 carrying a loading device generally indicated at 207 in Fig. 21.

The loading device 207 comprises a magazine 208 mounted on a wedge shaped support 209 (see Fig. 23). The magazine 208 is generally U shape in cross section and is adapted to receive a plurality of uranium rods 140. Thick walls of lead are provided on the magazine 208 and they are lined at 211 with a steel plate. A cover 212 closes the open side of the U-shaped magazine 208 and this cover is in the form of a plug having a shank portion 213 fitting into the open portion of the magazine.

The front of the car 203 is provided with a lead plate shield 214 having a central opening 215 and cut-out semicircular portions 216 at each edge. The central opening 214 is alined with the bottom of the open portion of the U-shaped magazine 208.

Behind the magazine portion 208 is a plunger assembly 217 secured to the magazine portion 208 by flanges

2,910,418

33

218 and comprising a cylinder 219 having mounted therein a plunger 220 carrying a piston 221. The plunger 220 projects into the magazine 208 in alinement with the opening 215 in the end shield 214 and a conventional packing gland 222 held in place by a bushing 223 seals the opening between the cylinder 219 and the magazine 208. The plunger 220 may be operated hydraulically, in which case inlet and outlet ports 224 and 225 respectively are disposed in the cylinder 219 on opposite sides of the piston 221.

Referring to Figs. 5 and 6, an elevator structure 226 mounted on upper and lower trackways 227 and 228, respectively is disposed at the loading side of the reactor 25 and is adapted to move transversely across the loading end of the tube 78. An elevator platform 229 (see Fig. 6) is mounted on the elevator structure 226 and is adapted to move vertically with respect to the tubes 25.

A loading car 203 is carried on the elevator platform 229. Thus it is possible to select any tube 78 and place the car 203 with the opening 215 in its front face directly alined with the selected tube. The cap 151 (Fig. 12) on the end of the selected tube is removed and the car 207 is moved so that the free end of the selected tube 78 enters the opening 215 in the front face of the car and the tube is then sealed against a gasket 230 (Figs. 10 and 12). The wheels of the car 207 may then be blocked as at 231 to lock the car in place. The car is now in position to commence loading uranium rods into the reactor.

The magazine 208 may contain just the required number of rods 140a and dummy rods 140b to fill one tube 78. The strength of coil spring 173 on the rod retainer 171 holds the rods in place in the reactor.

In the instance where the rods 140 only fill those portions of the tubes disposed in the reactor 25, it is necessary that the plunger 220 (Figs. 11 and 21) be of sufficient length to force the uranium rod sections 140a into their proper positions in the reactor 25.

After a tube 78 has been loaded with uranium rods in the manner described, the plunger 220 is withdrawn from the tube into the car 207 a sufficient distance to clear the end of the tube 78. The valve 127 is then closed and the car 207 is backed away from the ends of the tubes. The cap 151 (Fig. 12) is replaced on the end of the newly loaded tube 78 and a freshly loaded car 207 is alined with the next selected tube 78 by moving the elevator 229. The loading process is then repeated until all of the desired tubes are filled with uranium rods. The steps for replenishing the tubes with fresh rods include essentially the procedure of the initial charging. During this period of loading and unloading or at least during a major portion of such period, circulation of water or other coolant through the system frequently is continued at a rate sufficient to prevent overheating of the fissionable material (uranium) and to remove heat efficiently.

During all of the preparations for replenishing, the valve 127 on the selected tube 78 has been closed and the cooling water has been circulated through the reactor. At the actual time unloading is to be begun uranium rods are disposed throughout the entire length of the reactor in the selected tube 78 but the neutronic reaction should have been shut down for a sufficient length of time to permit the delayed neutron emission to cease to such an extent as to be no longer dangerous. Approximately a period of one-half hour following the insertion of the control rods is required for this fission process to drop below the danger level.

Following this thirty minute period, there is no further need to circulate the water through the reactor at the same rapid rate of flow as was required while the neutronic reaction was in effect. The heat generation taking place following the termination of the fission process is due to radioactive decay of the fission products in the uranium. This heating is commonly referred to as "self-heating" and diminishes rapidly following shut-down of a neutronic reaction. It is thus possible after this 30

34

minute period to diminish the rate of flow of water through the reactor although it is not possible to stop the flow of water since a substantial quantity of heat continues to be evolved. Thus the water flow through the selected tube 48 is reduced to some lower rate required to extract the heat generated by self-heating in the uranium rods.

A valve 127 is now opened by means of closing the circuit to the motor 151 thus placing the selected tube 78 in communication with the interior of the loading magazine 208 on the loading car 207. The plunger 220 is now withdrawn by admitting hydraulic fluid through port 225 into the cylinder 219 until a uranium rod section 140a drops by gravity into the lowest position in the loading magazine 208 opposite the opening 215 in the front wall 214 of the loading car 207. The plunger 220 is then forced against the uranium rod section 140a thus moved into position by admitting hydraulic fluid through the port 224 into the cylinder 219. The rod section 140a is thus forced through the open tube 78. Since the tube 78, as previously mentioned, is already completely full of uranium rod sections 140a, the charging of the new rod sections into the tube 78 causes the old rods to be forced toward the discharge side of the reactor, forcing the last rod section 140a against the rod retainer 171, thereby opening the retainer and the last rod is thus ejected from the tube 78 into the water filled tank 68. The rod thus discharged immediately drops to the bottom of tank 65 through the chute 97, past open valve 98 and comes to rest against the closed valve 99. The plunger 220 forces the first fresh rod unit clear of the loading magazine 208 and then the plunger 220 is again withdrawn to allow the second rod 140 to drop to the lowest position in the magazine 208. The operation of charging this second rod section 140a into the reactor is the same as that described with respect to the first rod section and once again an old rod section is ejected at the discharge end of the tube 78 and comes to rest directly above the valve 99 in the discharge tube 97. This operation is repeated until the selected tube 78 is completely filled with fresh rods, or at least is filled with fresh rod sections 140a throughout the part of the tube 78 that traverses the length of the reactor 25. The remainder of the tube 78, that is that portion projecting through the shield 76 may be filled with dummy plugs, 140b, for example of aluminum, but the charging operation even in this event is the same as though the entire tube 78 were filled with uranium rods. The top two or three rod sections 140a in a magazine 208 are the aluminum dummy plugs 140b (Fig. 23) and as the rods are ejected, the aluminum dummy plugs are forced out along with the uranium.

Referring to Figs. 6 and 7, after all of the rods in one tube 78 have been ejected in the manner just described and are disposed in the discharge chute 97 directly above the closed valve 99, valve 98 is closed to form water lock 97a. Valve 99 is then opened allowing the rods together with the charge of water in lock 97a to drop into the container 30 disposed directly below the chute and mounted on the car 31, as hereinbefore described. The valve 99 is then again closed and the car 31 moved to a position shown in Fig. 7, where a cover 117 can be taken from the shelf 118 by the hoist generally indicated at 119 and placed on the container 30 to close the opening in the container.

As shown in Fig. 7, the car 31 is provided with three containers 30, and 3 tubes 78 can be emptied before the car 31 is fully loaded. The car 31 when fully loaded is moved through the tunnel 106 by the cable 116 or 115 to a soaking pit where the uranium rods are allowed to remain for a period of time to allow the radioactivity to decay to safe limits to permit handling of the rods. Following this, the uranium rods are processed for the purpose of extracting fission products and

2,910,418

35

element 94²³⁹. These later steps form no part of the present invention and are therefore not detailed here. Frequently a substantial amount of heat is developed in the uranium or similar material during this soaking period. If desired the rods may be immersed in a body of water and steam generated directly or indirectly by the heat transferred to the water. For example the receptacle containing the water and the fissionable material may be used as a boiler to generate high or low pressure steam.

Summary

It can thus be seen that the uranium rods 140 can be loaded into the reactor 25 at one side and discharged from the other. The length of time the rods are left in the reactor, of course, will depend upon the conditions of operation and the ultimate objective sought. When it is desired to remove the rods from the reactor the neutronic reaction is first discontinued. The rods then can be left in the reactor for a period of from thirty to fifty days under continuous cooling, during which period the fission products in the rods are allowed to decay to the point where the rods may be removed without further cooling. Under these conditions, the flow of cooling water through the tube from which a rod is to be removed can be completely shut off before the valve 127 is opened. The problem of removing the rod under these conditions is very simple.

It may be desirable to allow the rods to remain in the reactor following cessation of the neutronic reaction for the period of time only sufficient to permit the most violent radioactivity in the fission products to subside. Under these conditions the flow of cooling water through the tube from which a rod is to be removed can be stopped long enough to allow the rod to be discharged from the tube 78, with cooling of the rod thus discharged then being continued outside the reactor for example by discharging the rods into a body of water as previously described.

Under either of the foregoing conditions the water level in the tank 68 at the discharge end of the tubes 78 should be below the tubes from which rods are to be rejected for otherwise the water in this tank 68 will back up through the open ends of tubes 78.

The primary advantage of the arrangement shown herein is the fact that rods may be removed from the reactor promptly following the cessation of the neutron reaction. Only sufficient time need be allowed to permit all delayed neutrons to disappear. Provisions are made for continuous and adequate cooling of the rods throughout the entire removal operation, thus making the subject matter of the invention particularly well suited for the continuous production of element 94²³⁹.

While the invention has been described with reference to the use of water as the cooling media it is to be understood that this is merely illustrative since other liquid coolants such as diphenyl, deuterium oxide, or other liquid which does not absorb neutrons to an excessive degree may be used. Moreover the invention herein contemplated may be used in connection with other moderators such s beryllium or liquid moderators such as deuterium oxide. Where liquid moderators are used the channels for the uranium should possess sufficient rigidity to be self-sustaining.

In the specific embodiments illustrated herein the vertical chute 28 is shown filled with water 29 to a level considerably above the discharge ends of the tubes 27. It is of course understood that this level may be altered and if desired may be lowered even to a position below the discharge ends of the tubes. In this latter instance the uranium rods discharged from the tubes would first drop through air but they would be received by the lower portion of the chute 28 containing water. The lowering of the water level in the vertical chute 28 would decrease the back pressure against the coolant in tubes 27.

36

From this it can be seen that at the time it is desired to open the charging ends of the tubes, the water pressure inside the tubes can be reduced by merely lowering the level of the water in the chute 28. Separate control valves instead could be used in the water pipes or headers 144 to regulate the flow of water through each pair of tubes 27, or if desired each tube 27 itself can be supplied with an individual control valve and this, for example, can be accompanied by placing the valve 145a in pipe 145 (see Fig. 11).

While the invention has been described with reference to uranium it should be noted that numerous compositions capable of fission to yield neutrons in greater amount than are consumed in the fission may be used. Natural uranium is an effective composition of this type since it contains U^{238} and U^{235} in the proportion of 139 to 1. Enriched compositions containing higher concentrations of U^{295} also may be used. Generally speaking a neutronic chain reaction may be conducted with various materials which yield a surplus of neutrons upon fission by neutrons such as U^{233}, U^{235}, 94²³⁹ or other fissionable isotopes. These fissionable materials may be mixed or otherwise combined with materials which absorb neutrons to yield a fissionable isotope or which yields an isotope decaying to a fissionable isotope such as Th^{232} or U^{238}.

While the theory of the nuclear chain fission mechanism in uranium set forth herein is based on the best presently known experimental evidence, we do not wish to be bound thereby, as additional experimental data later discovered may modify the theory disclosed. Any such modification of theory, however, will in no way effect the results to be obtained in the practice of the invention herein described and claimed.

Obviously, many modifications may be made in the specific embodiments disclosed without departing from the intended scope of the invention.

We claim:

1. In a neutronic reactor, a neutron moderator capable of slowing fast neutrons to thermal energy, a tube through the moderator, a rod in the tube and containing a thermal neutron fissionable material, the rod having an area in cross section less than that of the tube, longitudinal ribs between the rod and the tube wall forming passages disposed lengthwise with respect to the tube, and a coolant in the passages, sufficient of thermal neutron fissionable material being used in said neutronic reactor to provide a neutron reproduction ratio greater than unity.

2. In a neutronic reactor, a neutron moderator capable of slowing fast neutrons to thermal energy, a tube through the moderator, a uranium rod in the tube and having a cross sectional area less than that of the tube so as to form a space between the rod and the inner face of the tube, means inside tube for spacing the uranium rod away from the tube wall, and a coolant in said space, sufficient uranium being used in said neutronic reactor to provide a neutron reproduction ratio greater than unity.

3. In a neutronic reactor, a neutron moderator capable of slowing fast neutrons to thermal energy, a tube through the moderator, a rod in the tube and comprising a thermal neutron fissionable material, said rod having a cross sectional area less than that of the tube so as to form a passage between the rod and the inner face of the tube, means inside the tube for spacing the rod away from the tube wall and a coolant in the passage, sufficient of thermal neutron fissionable material being used in said neutronic reactor to provide a neutron reproduction ratio greater than unity.

4. In a neutronic reactor, a neutron moderator in the form of a solid mass capable of slowing fast neutrons to thermal energy having a passage therethrough, a tube in the passage and containing thermal neutron fissionable material, the cross sectional area of the tube being less

2,910,418

37

than that of the passage so as to form a space between the tube and the walls of the passage, means for spacing the tube away from the walls of the moderator providing the passage, and a coolant in the passage, sufficient of thermal neutron fissionable material being used in said neutronic reactor to provide a neutron reproduction ratio greater than unity.

5. In a neutronic reactor, a mass of graphite having a passage therethrough, a tube in the passage containing uranium, the over-all cross sectional area of the tube being less than the area of the passage so as to provide space between the tube and the walls of the passage a coolant in said space, means for spacing the tube away from the walls of the graphite providing the passage, and a coolant in the tube, sufficient uranium being used in said neutronic reactor to provide a neutron reproduction ratio greater than unity.

6. In a neutronic reactor, a neutron moderator capable of slowing fast neutrons to thermal energy, a plurality of tubes through the moderator, a second tube containing thermal neutron fissionable material disposed in each of the first mentioned tubes, the over-all cross sectional area of each of the second tubes being less than the cross sectional area between the inner walls of the corresponding first mentioned tubes, longitudinal ribs between each first mentioned tube and its corresponding inner tube so as to form a space between each first mentioned tube and its corresponding inner tube, and a coolant in the space, sufficient of thermal neutron fissionable material being used in said neutronic reactor to provide a neutron reproduction ratio greater than unity.

7. In a neutronic reactor, a neutron moderator capable of slowing fast neutrons to thermal energy, a plurality of tubes through the moderator, bodies in the tubes containing thermal neutron fissionable material, the cross section area of the bodies being less than that of the corresponding tubes so as to allow space between the bodies and the walls of the tubes, means for spacing the bodies of thermal neutron fissionable material away from the walls of the tubes, a separate coolant pipe connected to each tube and communicating with said space, a coolant in the pipe and the space, means circulating the coolant through the pipe and the space, and a shut-off valve in each pipe adapted to regulate the flow of coolant in the pipe, sufficient of thermal neutron fissionable material being used in said neutronic reactor to provide a neutron reproduction ratio greater than unity.

8. A neutronic system comprising a neutronic reactor having a neutron moderator capable of slowing fast neutrons to thermal energy, and having a plurality of coolant passages disposed therethrough, rods containing thermal

38

neutron fissionable material in said passages, means for spacing said rods away from the surfaces of the moderator providing the passages, a coolant supply header connected to each passage, a supply of coolant, and means selectively regulating the flow of coolant from the supply header through each passage, sufficient of thermal neutron fissionable material being used in said neutronic reactor to provide a neutron reproduction ratio greater than unity.

9. In a neutronic reactor, a neutron moderator having passages therethrough, a tube in each passage, longitudinally extending ribs projecting inwardly on the inner face of each tube, and bodies of thermal neutron fissionable material in each tube spaced from the tube walls by the ribs, the free spaces between the bodies and the tube walls forming coolant passages, sufficient of thermal neutron fissionable material being used in said neutronic reactor to provide a neutron reproduction ratio greater than unity.

10. In a neutronic reactor, a mass of graphite having passages therethrough, a tube in each passage, longitudinally extending ribs projecting inwardly on the inner face of each tube, bodies of uranium in each tube spaced from the tube walls by the ribs, the free spaces between the bodies and the tube walls forming coolant passages, coolant in the said passages, and means for moving the coolant through the passages, sufficient uranium being used in said neutronic reactor to provide a neutron reproduction ratio greater than unity.

References Cited in the file of this patent

UNITED STATES PATENTS

2,206,634	Fermi et al. _____	July 2, 1940
2,890,158	Ohlinger et al. _____	June 9, 1959

FOREIGN PATENTS

114,150	Australia _____	May 2, 1940
114,151	Australia _____	May 3, 1940
861,390	France _____	Feb. 7, 1941
233,011	Switzerland _____	Oct. 2, 1944

OTHER REFERENCES

Anderson et al.: "Neutron Production and Absorption in Uranium," Physical Review, vol. 56, pp. 284–286, Aug. 1, 1939.

Power, July 1940, page 58.

Smyth: "Atomic Energy for Military Purposes," August 1945. (Copy may be purchased from Supt. of Documents, Washington 25, D.C.)

Kelly et al.: Phy. Rev. 73, 1135–9 (1948).

Oct. 14, 1958   E. P. WIGNER ET AL   2,856,339
                           NEUTRONIC REACTORS

Filed May 28, 1945   9 Sheets—Sheet 1

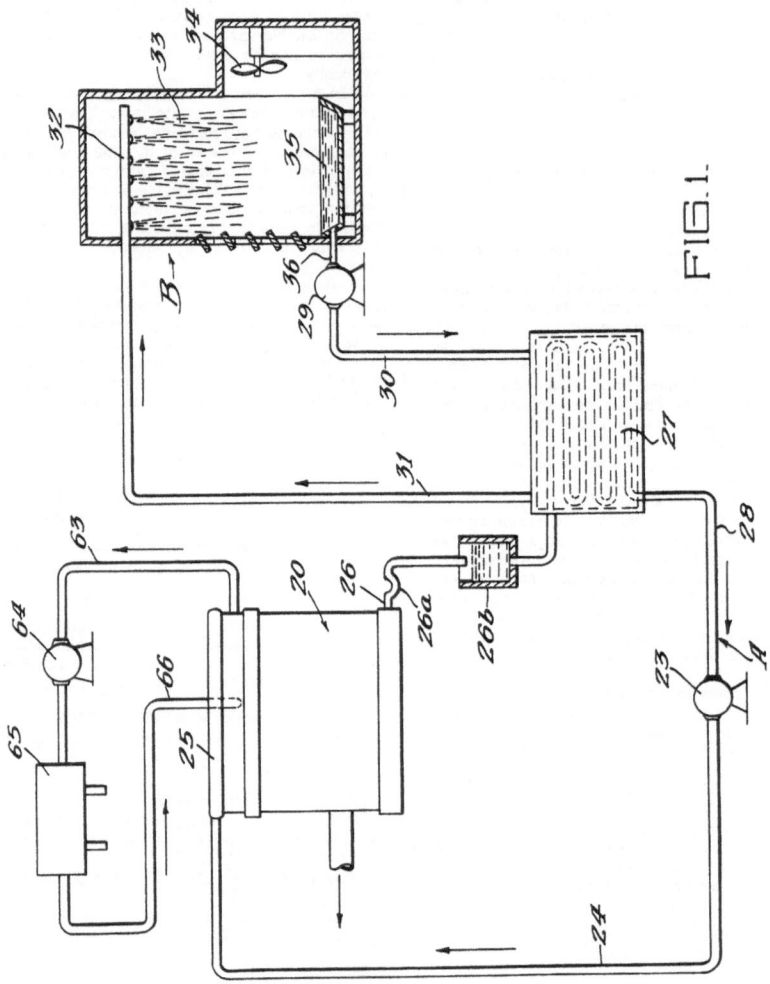

FIG. 1.

Oct. 14, 1958   E. P. WIGNER ET AL   2,856,339

NEUTRONIC REACTORS

Filed May 28, 1945   9 Sheets—Sheet 2

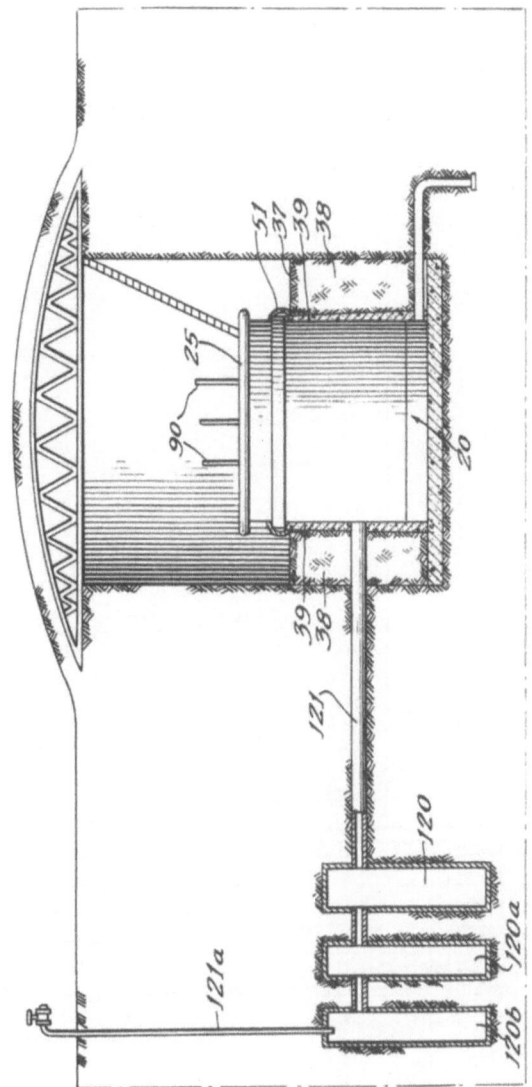

FIG.2.

Witnesses:
Herbert E. Metcalf
Richard E. Burn

Inventors:
Eugene P. Wigner
Gale J. Young

By:
Robert A. Commenter
Attorney.

Oct. 14, 1958   E. P. WIGNER ET AL   2,856,339
                            NEUTRONIC REACTORS

Filed May 28, 1945   9 Sheets—Sheet 3

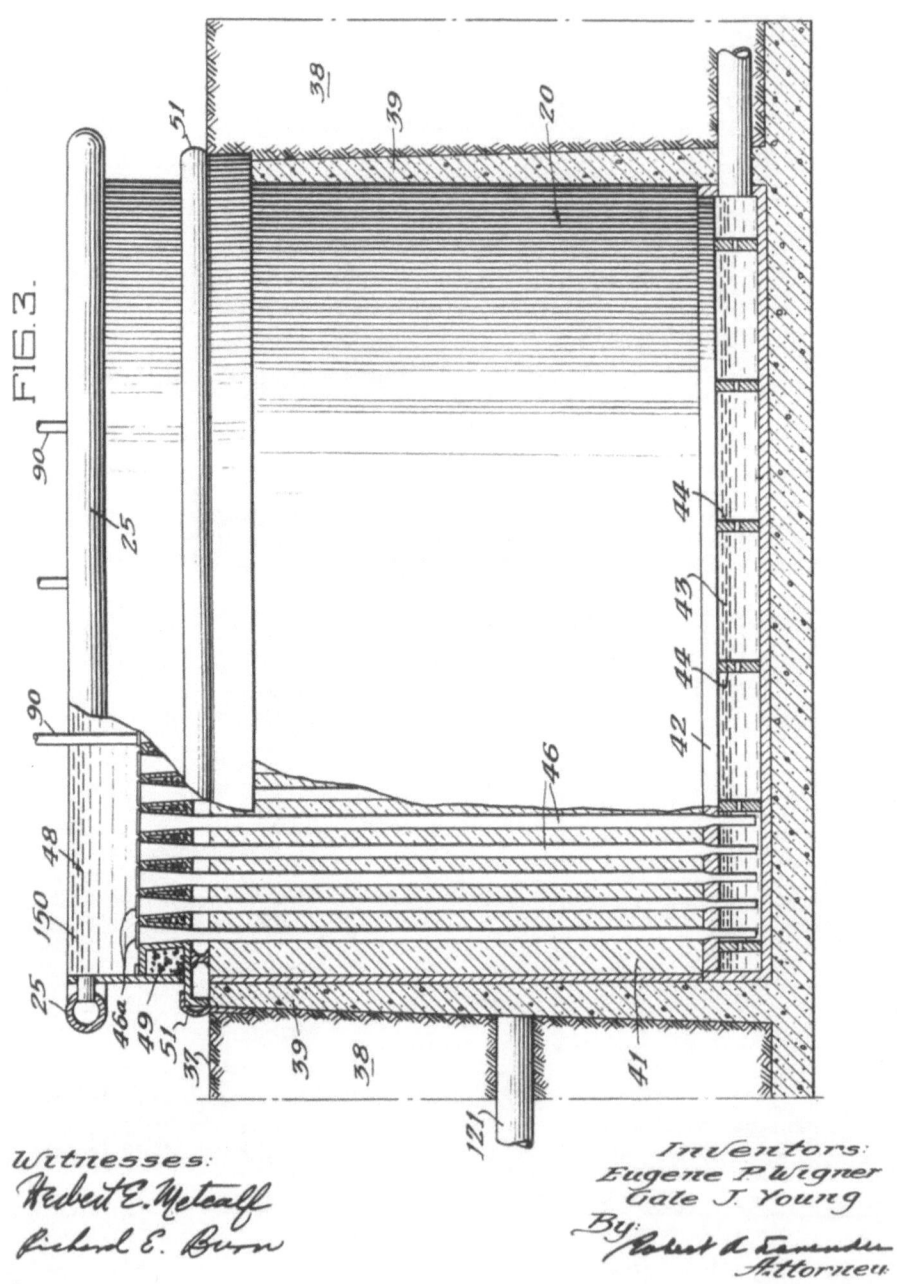

FIG. 3.

Oct. 14, 1958   E. P. WIGNER ET AL   2,856,339
NEUTRONIC REACTORS

Filed May 28, 1945   9 Sheets—Sheet 4

FIG.4.

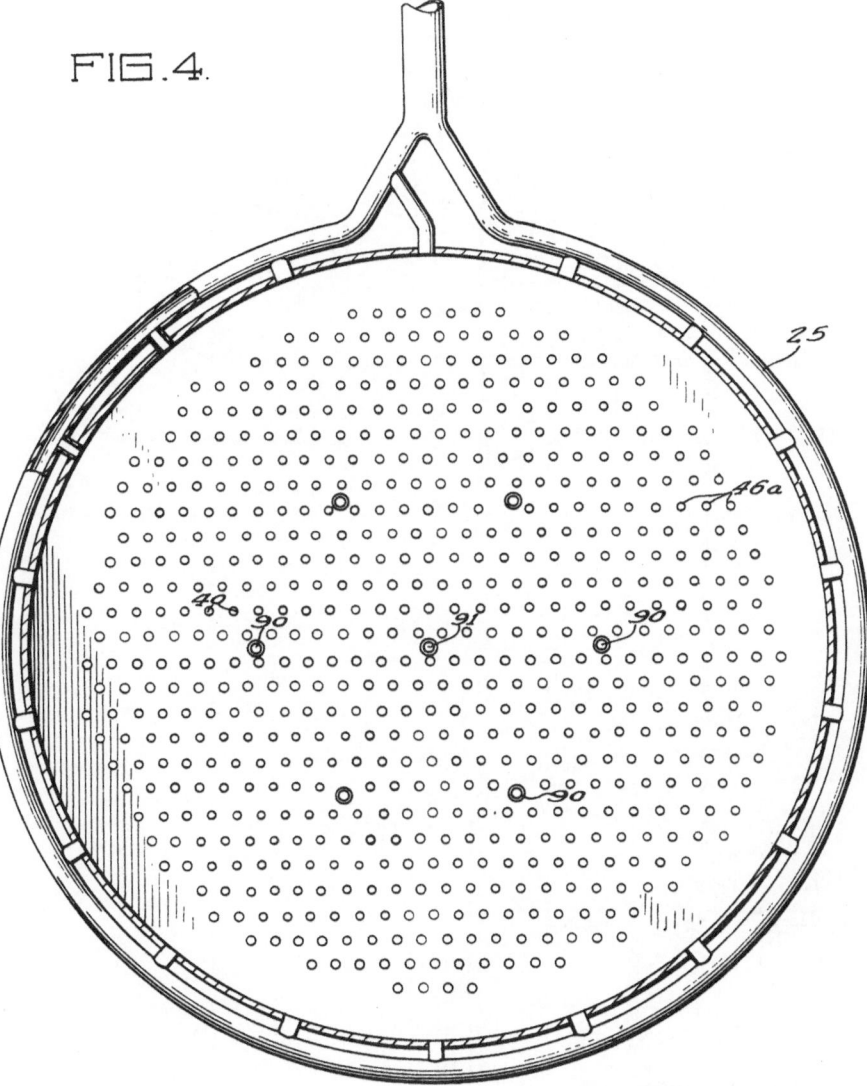

Oct. 14, 1958   E. P. WIGNER ET AL   2,856,339
                         NEUTRONIC REACTORS

Filed May 28, 1945   9 Sheets—Sheet 5

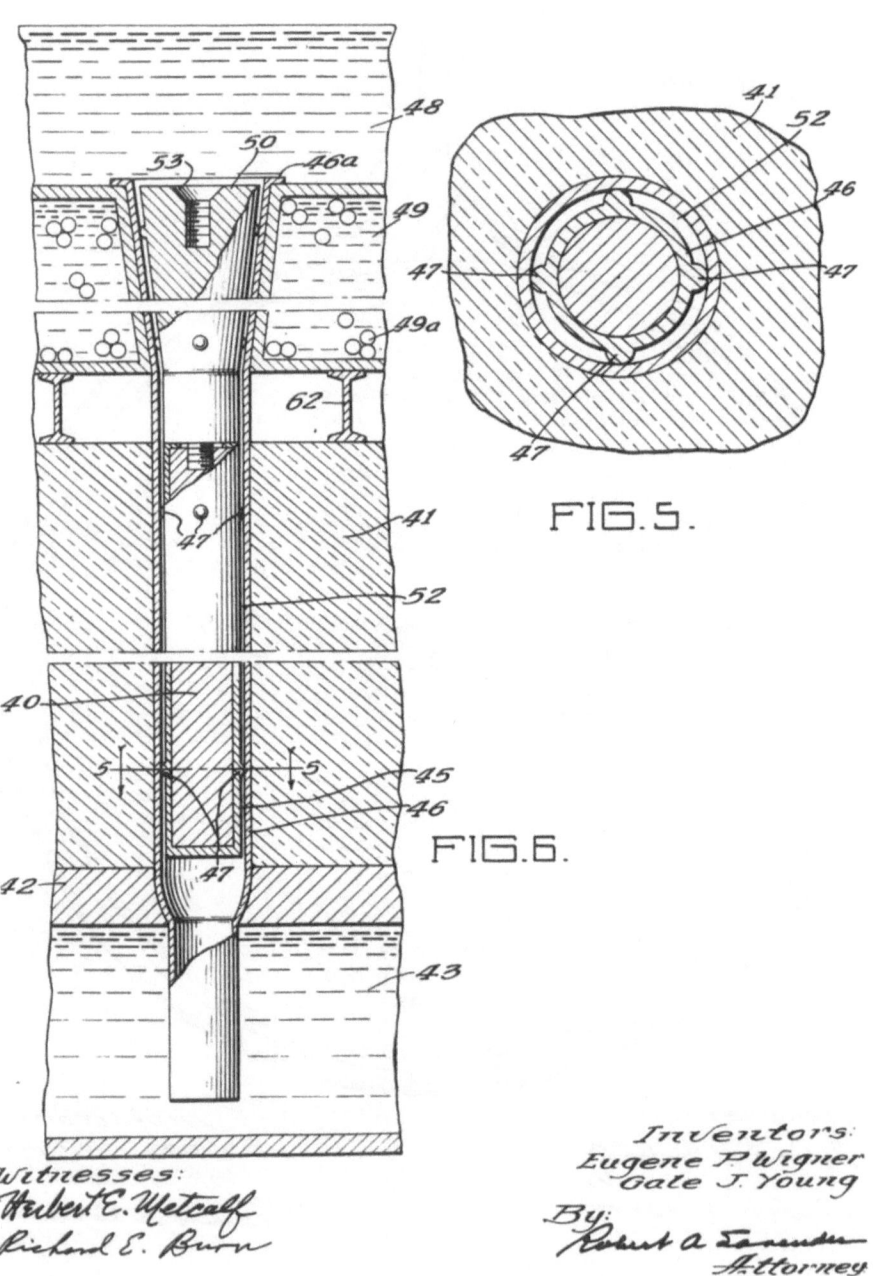

FIG.5.

FIG.6.

Inventors:
Eugene P. Wigner
Gale J. Young

By:
Robert A. Sanerson
Attorney

Witnesses:
Herbert E. Metcalf
Richard E. Burn

Oct. 14, 1958   E. P. WIGNER ET AL   2,856,339
NEUTRONIC REACTORS

Filed May 28, 1945   9 Sheets—Sheet 6

FIG.7.

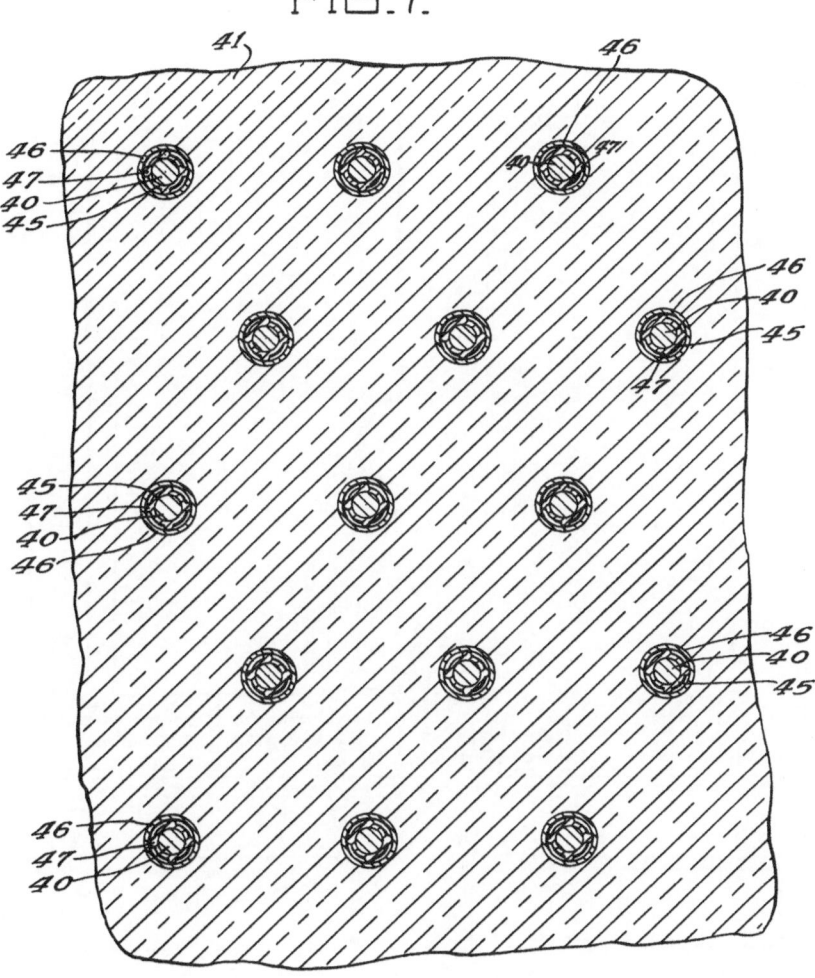

Witnesses:
Herbert E. Metcalf
Richard E. Burr

Inventors:
Eugene P. Wigner
Gale J. Young

By:
Robert A. Lavender
Attorney.

Oct. 14, 1958   E. P. WIGNER ET AL   2,856,339

NEUTRONIC REACTORS

Filed May 28, 1945   9 Sheets—Sheet 7

FIG.8.

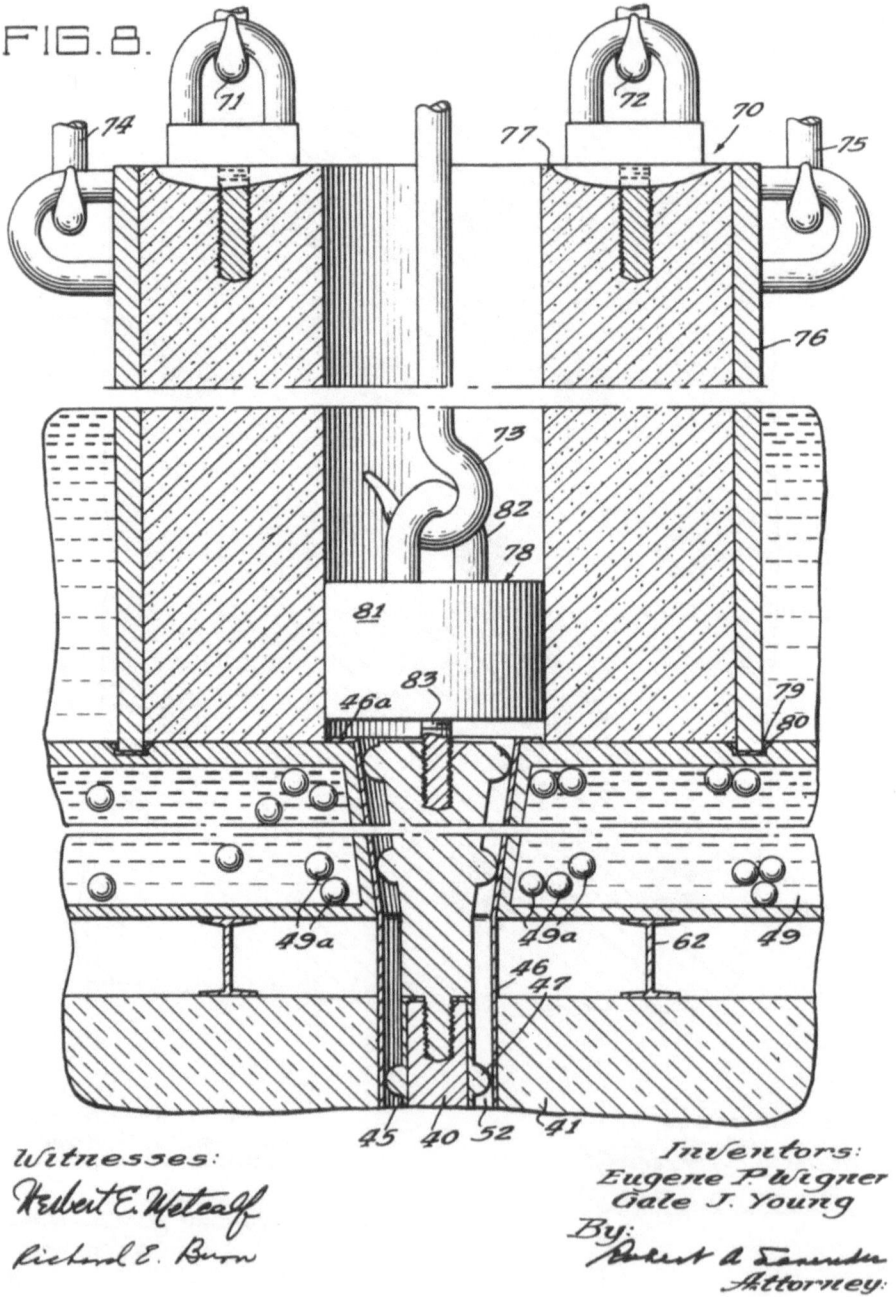

Witnesses:

Herbert E. Metcalf

Richard E. Burn

Inventors:
Eugene P. Wigner
Gale J. Young
By:
Robert A. Lavender
Attorney.

Oct. 14, 1958   E. P. WIGNER ET AL   2,856,339
                                    NEUTRONIC REACTORS

Filed May 28, 1945   9 Sheets—Sheet 8

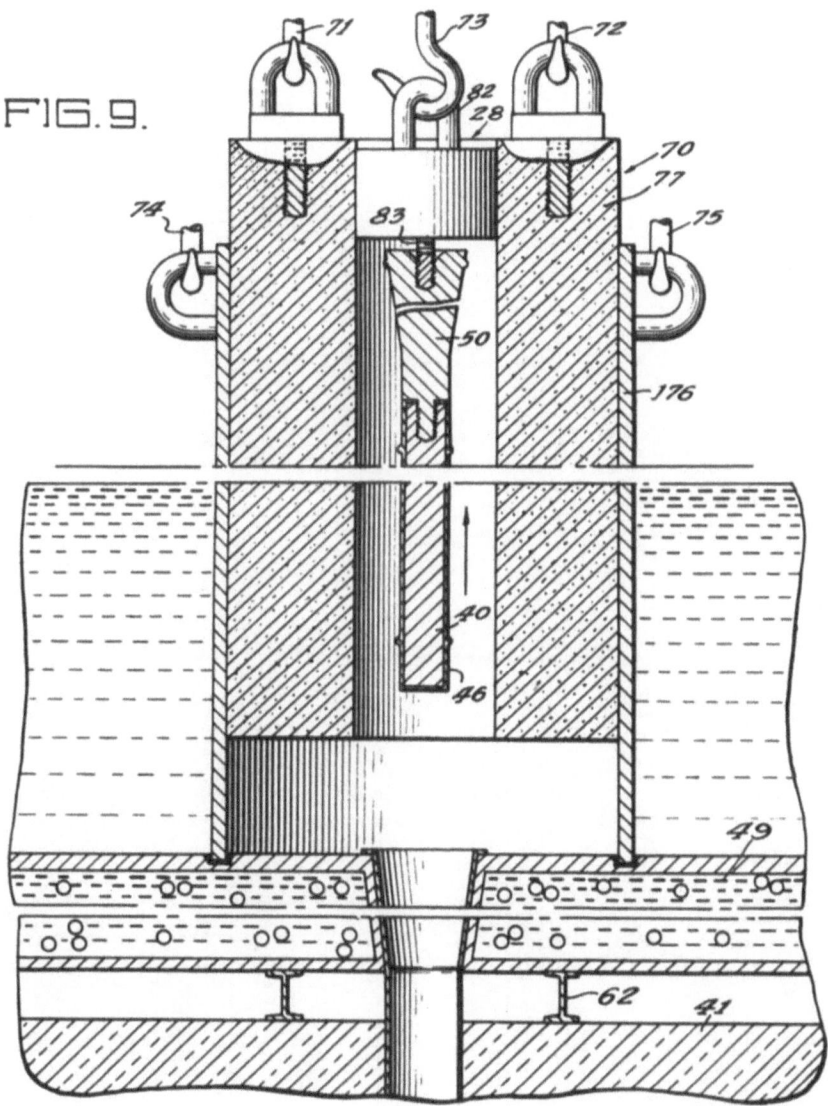

FIG. 9.

Oct. 14, 1958   E. P. WIGNER ET AL   2,856,339

NEUTRONIC REACTORS

Filed May 28, 1945   9 Sheets—Sheet 9

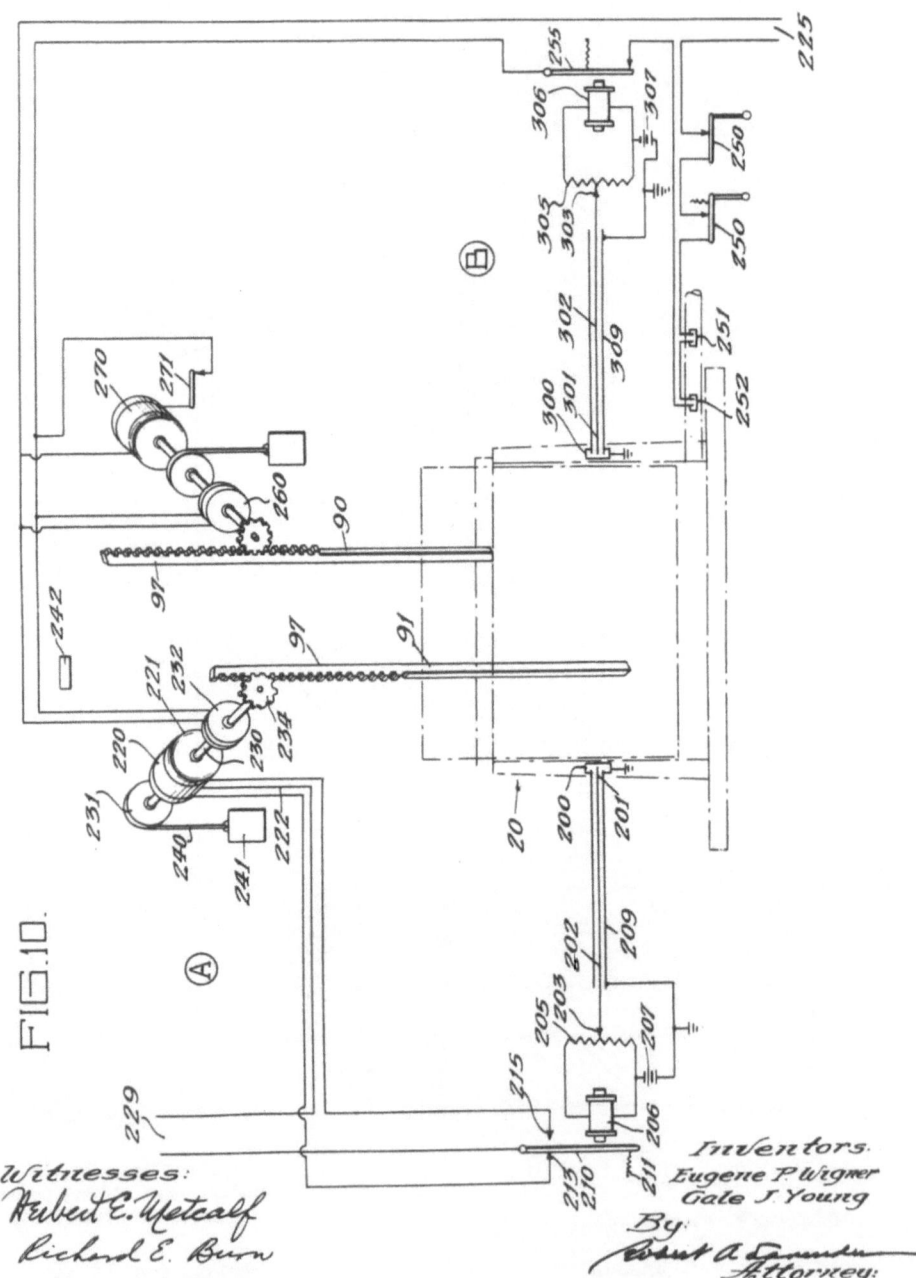

FIG. 10.

Inventors
Eugene P. Wigner
Gale J. Young
By:
Attorney:

Witnesses:
Hubert E. Metcalf
Richard E. Burn

United States Patent Office

2,856,339

Patented Oct. 14, 1958

1

2,856,339

NEUTRONIC REACTORS

Eugene P. Wigner and Gale J. Young, Chicago, Ill., assignors to the United States of America as represented by the United States Atomic Energy Commission

Application May 28, 1945, Serial No. 596,219

2 Claims.   (Cl. 204—193.2)

The present invention relates to the removal of bodies of fissionable material from a neutron chain reacting system, also referred to as a neutronic reactor system. In the form of the invention shown bodies of uranium are disposed in passages arranged in a graphite moderator, the said bodies having a cross sectional area less than the cross sectional area of the passages in which they are disposed so as to provide spaces around the bodies through which the coolant can pass. The bodies are removable from the reactor through openings in the passages at one side of the reactor, the openings normally being submerged in water.

As a result of the chain reaction, when U^{238} is present (as in natural uranium), transuranic element 94^{239}, known as plutonium, is produced. This material is fissionable and is valuable when added to natural uranium for use in a chain reacting system, as a fissionable body in lieu of or conjunction with natural uranium.

Natural uranium contains both uranium isotopes U^{235} and U^{238} in the ratio of 1 to 139. The U^{235} is the isotope fissionable by slow neutrons.

When fission occurs in the U^{235} isotope, the following reaction takes place:

$$_{92}U^{235} + \text{neutron} \quad A + B + \text{about 2 neutrons (average)}$$

where "A" represents "light" fission fragments having atomic masses ranging from 83 to 99 inclusive and atomic numbers from 34 to 45 inclusive; for example, Br, Kr, Rb, Sr, Y, Zr, Cb, Mo, Ma, Ru, and Rh; and "B" represents "heavy" fission fragments having atomic masses ranging from 127 to 141 inclusive, and atomic numbers from 51 to 60 inclusive; for example, Sb, Te, I, Xe, Cs, Ba, La, Ce, Pr, and Nd. The elements resulting from the fissions are unstable and radioactive, with half-lives varying in length in accordance with the element formed.

The absorption of thermal or resonance nuetrons by the U^{238} isotope gives rise to the conversion of U^{238} to U^{239} which ultimately decays to transuranic element 94^{239}. The reaction is as follows:

$$_{92}U^{239} + n \longrightarrow {}_{92}U^{239} \quad \text{[plus 6 m. e. v. of } \gamma \text{ rays, not necessarily all of one frequency.]}$$

$$_{92}U^{239} \xrightarrow{\text{23 min.}} {}_{93}U^{239} + \beta^- \text{[1 m. e. v. } \beta^-, \text{no } \gamma \text{ rays.]}$$

$$_{93}U^{239} \xrightarrow{\text{2.3 days}} {}_{94}U^{239} + \beta^- \text{[600 kv. upper } \beta\text{-energy limit. Also 2 } \gamma \text{ rays, 400 kv, and 270 kv, about } \frac{1}{2} \text{ of which are converted to electrons.]}$$

Most of the neutrons arising from the fission process are set free with the very high energy of above one million electron volts average and are therefore not in condition to be utilized efficiently to create new thermal neutron fissions in a fissionable body such as U^{235} when it is mixed with a considerable quantity of U^{238}, particu-

2

larly as in the case of natural uranium. The energies of the fission-released neutrons are so high that most of the latter would tend to be absorbed by the U^{238} nuclei, and yet the energies are not generally high enough for production of fission by more than a small fraction of the neutrons so absorbed. For neutrons of thermal energies, however, the absorption cross section of U^{235}, to produce fission, is a great deal more than the simple capture cross section of U^{238}; so that under the stated circumstances the fast fission neutrons, after they are created, must be slowed down to thermal energies before they are most effective to produce fresh fisssion by reaction with additional U^{235} atoms. If a system can be made in which neutrons are slowed down without excessive absorption until they reach thermal energies and then mostly enter into uranium rather than into any other element, a self-sustaining nuclear chain reaction can be obtained, even with natural uranium. Light elements, such as deuterium, beryllium, oxygen or carbon, the latter in the form of graphite, can be used as slowing agents. A special advantage of the use of the light elements mentioned for slowing down fast fission neutrons is that fewer collisions are required for slowing than is the case with heavier elements, and furthermore, the above-enumerated elements have very small neutron capture probabilities, even for thermal neutrons. Hydrogen would be most advantageous were it not for the fact that there may be a relatively high probability of neutron capture by the hydrogen nucleus. Carbon in the form of graphite is a relatively inexpensive, practical, and readily available agent for slowing fast neutrons to thermal energies. Recently, beryllium has been made available in sufficiently large quantities for test as to suitability for use as a neutron slowing material in a system of the type to be described. It has been found to be in every way as satisfactory as carbon. Deuterium while more expensive is especially valuable because of its low absorption of neutrons and its compounds such as deuterium oxide have been used with very effective results.

However, in order for the premise to be fulfilled that the fast fission neutrons be slowed to thermal energies in a slowing medium without too large an absorption in the U^{238} isotope of the uranium, certain types of physical structure should be utilized for the most efficient reproduciton of neutrons, since unless precautions are taken to reduce various neutron losses and thus to conserve neutrons for the chain reaction the rate of neutron reproduction may be lowered and in certain cases lowered to a degree such that a self-sustaining system is not attained.

The ratio of the number of fast neutrons produced by the fissions, to the original number of fast neutrons creating the fissions, in a system of infinite size using specific materials is called the reproduction or. multiplication factor of the system and is denoted by the symbol K. If K can be made sufficiently greater than unity to create a net gain in neutrons and the system made sufficiently large so that this gain is not entirely lost by leakage from the exterior surface of the system, then a self-sustaining chain reacting system can be built to produce power by nuclear fission of natural uranium. The neutron reproduction ratio, r, in a system of finite size differs from K by the leakage factor and by localized neutron absorbers such as control rods, and must be sufficiently greater than unity to permit the neutron density to rise exponentially. Such a rise will continue indefinitely if

2,856,339

not controlled at a desired density corresponding to a desired power output.

During the interchange of neutrons in a system comprising bodies of uranium of any size in a slowing medium, neutrons may be lost in four ways, by absorption in the uranium metal or compound without producing fission, by absorption in the slowing down material, by absorption in impurities present in the system, and by leakage from the system. These losses will be considered in the order mentioned.

Natural uranium, particularly by reason of its U^{238} content, has an especially strong absorbing power for neutrons when they have been slowed down to moderate energies. The absorption in uranium at these energies is termed the uranium resonance absorption or capture. It is caused by the isotope U^{238} and does not result in fission but creates the isotope U^{239} which by two successive beta emissions forms the relatively stable nucleus 94^{239}. It is not to be confused with absorption or capture of neutrons by impurities, referred to later. Neutron resonance absorption in uranium may take place either on the surface of the uranium bodies, in which case the absorption is known as volume resonance absorption. It will be appreciated that this classification of resonance absorptions is merely a convenient characterization of observed phenomena, and arises, not because the neutron absorbing power of a U^{238} nucleus is any greater when the nucleus is at the surface of a body of metallic, or combined uranium, but because the absorbing power of U^{238} nuclei for neutrons of certain particular energies is inherently so high that practically all neutrons that already happen to have those energies, called resonance energies as explained above, are absorbed almost immediately upon their arrival in the body of uranium metal or uranium compound, and thus in effect are absorbed at the surface of such body. Volume resonance absorption is due to the fact that some neutrons make collisions inside the uranium body and may thus arrive at resonance energies therein. After successfully reaching thermal velocities, about 40 percent of the neutrons are also subject to capture by U^{238} without fission, to produce U^{239} and eventually 94^{239}.

It is possible, by proper physical arrangement of the materials, to reduce substantially uranium resonance absorption. By the use of light elements as described above for slowing materials, a relatively large increment of energy loss is achieved in each collision and therefore fewer collisions are required to slow the neutrons to thermal energies, thus decreasing the probability of a neutron being at a resonance energy as it enters a uranium atom. During the slowing process, however, neutrons are diffusing through the slowing medium over random paths and distances so that the uranium is not only exposed to thermal neutrons but also to neutrons of energies varying between the emission energy of fission and thermal energy. Neutrons at uranium resonance energies will, if they enter uranium at these energies, be absorbed on the surface of a uranium body whatever its size, giving rise to surface absorption. Any substantial reduction of overall surface of the same amount of uranium relative to the amount of slowing material (i. e. the amount of slowing medium remaining unchanged) will reduce surface absorption, and any such reduction in surface absorption will release neutrons to enter directly into the chain reaction, i. e., will increase the number of neutrons available for further slowing and thus for reaction with U^{235} to produce fission.

For a given ratio of slowing material to uranium, surface resonance absorption losses of neutrons in the uranium can be reduced by a large factor from the losses occurring in a mixture of fine uranium particles and a slowing medium, if the uranium is aggregated into substantial masses in which the mean radius of the aggregates is at least 0.25 centimeter for natural uranium metal and when the mean spatial radius of the bodies is at least 0.75 centimeter for the oxide of natural uranium (UO_2). Proportionate minimums exist for other uranium compounds the

exact minimum value being dependent upon the uranium content and the density of the product. An important gain is thus made in the number of neutrons made directly available for the chain reaction. A similar gain is made when the uranium has more than the natural content of fissionable material. Where a maximum K factor is to be desired we place the uranium in the system in the form of spaced uranium masses or bodies of substantial size, preferably either of metal, oxide, carbide, or other compound or combinations thereof. The uranium bodies can be in the form of layers, rods or cylinders, cubes or spheres, or approximate shapes, dispersed throughout the graphite, preferably in some geometric pattern. The term geometric is used to mean any pattern or arrangement wherein the uranium bodies are distributed in the graphite or other moderator with at least either a roughly uniform spacing or with a roughly systematic non-uniform spacing, and are at least roughly uniform in size and shape or are systematic in variations of size or shape to produce a volume pattern conforming to a roughly symmetrical system. If the pattern is a repeating or rather exactly regular one, a system embodying it may be conveniently described as a lattice structure. Optimum conditions are obtained with natural uranium by using a lattice of metal spheres.

The number of neutrons made directly available to the chain reaction by aggregating the uranium into separate bodies spaced through the slowing medium is a critical factor in obtaining a self-sustaining chain reaction utilizing natural uranium and graphite. The K factor of a mixture of fine uranium particles in graphite, assuming both of them to be theoretically pure, would only be about .785. Actual K factors as high as 1.07 have been obtained using aggregation of natural uranium in the best known geometry, and with as pure materials as it is presently possible to obtain.

Assuming theoretically pure carbon and theoretically pure natural uranium metal, both of the highest obtainable densities, the maximum possible K factor theoretically obtainable is about 1.1 when the uranium is aggregated with optimum geometry. Still higher K factors can be obtained by the use of aggregation in the case of uranium having more than the naturally occurring content of fissionable elements. Adding such fissionable material is termed enrichment of the uranium.

It is thus clearly apparent that the aggregation of the uranium into masses separated in the slowing material is one of the most important, if not the most important factor entering into the successful construction of a self-sustaining chain reacting system utilizing relatively pure natural uranium in a slowing material such as graphite in the best geometry at present known, and is also important in obtaining high K factors when enrichment of the uranium is used.

Somewhat higher K factors are obtainable where moderators such as deuterium oxide or beryllium are used. Thus with beryllium it is possible to secure a K factor as high as 1.10 with optimum geometry and absolute purity. Moreover with deuterium oxide K factors of about 1.27 may be obtained. When such moderators are used the problem of aggregation may be somewhat less important although it is an essential factor if maximum K factors and minimum size reactors are to be obtained.

The thermal neutrons are also subject to capture by the slowing material. While carbon and beryllium have very small capture cross sections for thermal neutrons, and deuterium still smaller, an appreciable fraction of thermal neutrons (about 10 percent of the neutrons present in the system under best conditions with graphite) is lost by capture in the slowing material during diffusion therethrough. It is therefore desirable to have the neutrons reaching thermal energy promptly enter uranium.

In addition to the above-mentioned losses, which are inherently a part of the nuclear chain reaction process,

2,856,339

5

impurities present in both the slowing material and the uranium add a very important neutron loss factor in the chain. The effectiveness of various elements as neutron absorbers varies tremendously. Certain elements such as boron, cadmium, samarium, gadolinium, and some others, if present even in a few parts per million, could prevent a self-sustaining chain reaction from taking place. It is highly important, therefore, to remove as far as possible all impurities capturing neutrons to the detriment of the chain reaction from both the slowing material and the uranium. If these impurities, solid, liquid, or gaseous, and in elemental or combined form, are present in too great quantity, in the uranium bodies or the slowing material or in, or by absorption from, the free spaces of the system, the self-sustaining chain reaction cannot be attained. The amounts of impurities that may be permitted in a system, vary with a number of factors, such as the specific geometry of the system, and the form in which the uranium is used—that is, whether natural or enriched, whether as metal or oxide—and also factors such as the weight ratios between the uranium and the slowing down material, and the type of slowing down or moderating material used—for example, whether deuterium, graphite or beryllium. Although all of these considerations influence the actual permissible amount of each impurity material, it has fortunately been found that, in general, the effect of any given impurity or impurities can be correlated directly with the weight of the impurity present and with the K factor of the system, so that knowing the K factor for a given geometry and composition, the permissible amounts of particular impurities can be readily computed without taking individual account of the specific considerations named above. Different impurities are found to affect the operation to widely different extents; for example, relatively considerable quantities of elements such as hydrogen may be present, and, as previously suggested, the uranium may be in the form of oxide, such as UO_2 or U_3O_8, or carbide, although the metal is preferred. Nitrogen may be present to some extent, and its effect on the chain reaction is such that the neutron reproduction ratio of the system may be changed by changes in atmospheric pressure. This effect may be eliminated by enclosing or evacuating the system if desired, or may be utilized by determining changes in a particular system in the reproduction ratio as changes occur in the atmospheric pressure. A sensitive barometer is thus obtained. In general, the inclusion of combined nitrogen is to be avoided.

The effect of impurities on the optimum reproduction factor K may be conveniently evaluated to a good approximation, simply by means of certain constants known as "danger coefficients" which are assigned to the various elements. These danger coefficients for the impurities are each multiplied by the percent by weight of the corresponding impurity, and the total danger sum. This total danger sum is subtracted from the reproduction factor K as calculated for pure materials and for the specific geometry under consideration.

The danger coefficients are defined in terms of the ratio of the weight of impurity per unit mass of uranium and are based on the cross section for absorption of thermal neutrons of the various elements. These values may be obtained from physics textbooks on the subject and the danger coefficient computed by the formula

$$\frac{\sigma_i}{\sigma_u}\frac{A_u}{A_i}$$

wherein σ_i represents the cross section impurity and σ_u the cross section for the uranium, A_i the atomic weight of the impurity and A_u the atomic weight for uranium. If the impurities are in the carbon, they are computed as their percent of the weight of the uranium of the system.

Presently known values for danger coefficients for some elements are given in the following table, wherein the elements are assumed to have their natural isotopic

6

constitution unless otherwise indicated, and are conveniently listed according to their chemical symbols:

Element	Danger Coefficient	Element	Danger Coefficient
He	0	Cu	1.8
Li	310	Zn	0.61
B	2,150	Ga	1
N	4.0	As	2
F	0.02	Se	6.3
Na	0.65	Br	2.5
Mg	0.48	Rh	50
Al	0.30	Ag	18
Si	0.26	Cd	870
P	0.3	In	54.2
S	0.46	Sn	0.18
Cl	31	Sb	1.6
K	2.1	I	1.6
Ca	0.37	Ba	0.30
Ti	3.8	Sm	1,430
V	4	Eu	435
Cr	2	Gd	6,320
Mn	7.5	Pb	0.03
Fe	1.5	Bi	0.0025
Co	17	Th	1.1
Ni	3		

Where an element is necessarily used in an active part of a system, it is still to be considered as an impurity; for example, in a structure where the uranium bodies consist of uranium oxide, the actual factor K would ordinarily be computed by taking that fact into account using as a base K a value computed for theoretically pure uranium.

As a specific example, if the materials of the system under consideration have .0001 part by weight of Co and Ag, the total danger sum in K units for such an analysis would be:

$$.0001 \times 17 + .0001 \times 18 = .0035 \text{ K units}$$

This would be a rather unimportant reduction in the reproduction factor K unless the reproduction factor for a given system, without considering any impurities, is very nearly unity. If, on the other hand, the impurities in the uranium in the previous example had been Li, Co, and Rh, the total danger sum would be:

$$.0310 + .0017 + .0050 = .0377 \text{ K units}$$

This latter reduction in the reproduction factor for a given system would be serious and might well reduce the reproduction factor below unity for certain geometries and certain moderators so as to make it impossible to effect a self-sustaining chain reaction with natural uranium and graphite, but might still be permissible when using enriched uranium in a system having a high K factor.

This strong absorbing action of some elements renders a self-sustaining chain reacting system capable of control. By introducing neutron absorbing elements in the form of rods or sheets into the interior of the system, for instance in the slowing material between the uranium masses, the neutron reproduction ratio of the system can be changed in accordance with the amount of absorbing material exposed to the neutrons in the system. A sufficient mass of the absorbing material can readily be inserted into the system to reduce the reproduction ratio of the system to less than unity and thus stop the reaction. Consequently, it is another object of our invention to provide a means and method of controlling the chain reaction in a self-sustaining system.

When the uranium and the slowing material are of such purity and the uranium is so aggregated that fewer neutrons are parasitically absorbed than are gained by fission, the uranium will support a chain reaction producing an exponential rise in neutron density of the overall size of the system is sufficiently large to overcome the loss of neutrons escaping from the system. Thus the overall size is important.

The size of the system will vary, depending upon the K factor of the system, and upon other things. If the reproduction factor K is greater than unity, the number

7

of neutrons present will increase exponentially and indefinitely, provided the structure is made sufficiently large. If, on the contrary, the structure is small, with a large surface-to-volume ratio, there will be a rate of loss of neutrons from the structure by leakage through the outer surfaces, which may overbalance the rate of neutron production inside the structure so that a chain reaction will not be self-sustaining. For each value of the reproduction factor K greater than unity, there is thus a minimum overall size of a given structure known as the critical size, above which the rate of loss of neutrons by diffusion to the walls of the structure and leakage away from the structure is less than the rate of production of neutrons within the system, thus making the chain reaction self-sustaining. The rate of diffusion of neutrons away from a large structure in which they are being created through the exterior surface thereof may be treated by mathematical analysis when the value of K and certain other constants are known, as the ratio of the exterior surface to the volume becomes less as the structure is enlarged.

In the case of a spherical structure employing uranium bodies imbedded in graphite in the geometries disclosed herein and without an external reflector the following formula gives the critical overall radius (R) in feet:

$$K-1=\frac{C}{R^2}$$

where C is a constant that varies slightly with geometry of the lattice and for normal graphite lattices may have a value close to 7.2.

For a rectangular parallelepiped structure rather than spherical, the critical size can be computed from the formula

$$K-1=C\left(\frac{1}{a^2}+\frac{1}{b^2}+\frac{1}{c^2}\right)$$

where a, b, and c are the lengths of the sides in feet. The critical size for a cylindrical structure is given by the formula, irrespective of the shape of the uranium bodies

Cylinder height h ft.
Radius R ft

$$K-1=C\left(\frac{1}{h^2}+\frac{.59}{R_2}\right)$$

However, when critical size is attained, by definition no rise in neutron density can be expected. It is therefore necessary to increase the size of the structure beyond the critical size but not to the extent that the period for doubling of the neutron density is too short, as will be explained later. Reactors having a reproduction ratio (r) for an operating structure with all control absorbers removed and at the temperature of operation up to about 1.005 are very easy to control. Reproduction ratio should not be permitted to rise above about 1.01 since the reaction will become difficult to control. The size at which this reproduction ratio can be obtained may be computed from modifications of the above formulae for critical size. For example, for spherical active structures the formula

$$K-r=\frac{C}{R^2}$$

may be used to find R when K is known and r is somewhat over unity. The same formula will, of course, give r for given structures for which K and R are known.

Critical size may be attained with a somewhat smaller structure by utilizing a neutron reflecting medium surrounding the surface of the active structure. For example, a 2 foot thickness of graphite having low impurity content, completely surrounding a spherical structure is effective in reducing the diameter of the uranium bearing portion by almost 2 feet, resulting in a considerable saving of uranium or uranium compound.

The rate of production of element 94^{239} will depend on the rate of neutron absorption by U^{238} and is also pro-

8

portional to the rate at which fissions occur in U^{235}. This in turn is controlled by the thermal neutron density existing in the reactor while operating. Thus for maximum production of element 94^{239}, it is essential that the thermal neutron density be at a maximum value commensurate with thermal equilibrium.

Considerable heat is generated during a neutronic reaction primarily as the result of the fission process. Following are tables showing more specifically the type of heat generated in the reactor.

Summary by type

	M. e. v./ fission	Percent
Gamma radiation	18	9
Beta radiation	16	8
Kinetic energy of fission fragments	160	80
Kinetic energy of neutrons	6	3
	200	100

Summary by locale where heat is generated

	M. e. v./ fission	Percent
In uranium	174	87
In moderator	16	8
Outside pile	10	5
	200	100

Summary by type and locale

	M. e. v. per fission	Percent in U	Percent in C	Percent Outside
Kinetic energy of fission fragments	160	100		
Kinetic energy of neutrons	6		99	1
Gamma radiation from fission products	6	50	45	5
Beta radiation from fission products	16	100		
Nuclear affinity of neutrons (gamma radiation)	12	70	25	5

When the system is operated for an extended period of time at a high production output of element 94^{239}, the large amount of heat thus generated must be removed in order to stabilize the chain reaction. Most of the heat in an operating device is generated as the result of the nuclear fissions taking place in the U^{235} isotope. Thus, the rate of heat generation is largely proportional to the rate at which the fissions take place. In other words, if the rate of temperature of neutrons is increased, a greater amount of coolant must be passed through the reactor in order to remove the heat thus generated to avoid damage, particularly at the central portion of the pile, by excessive heat. Thus, the highest obtainable neutron density at which a system can be operated for an extended period of time is limited by the rate at which the generated heat can be removed. That is to say, the maximum power output of a system is limited by the capacity of the cooling system. An effective cooling system is therefore a primary requirement for high power operation of a neutronic reactor and it has been found that this cooling may be accomplished most effectively by passage of the coolant in contact with or in close proximity to the uranium.

After the neutronic system has operated for a period of time sufficient to cause a quantity of element 94^{239} to be produced, it may be desirable to remove at least some of the uranium rods from the reactor in order to extract element 94^{239} and the radioactive fission products, both being formed in the uranium rods or for other purposes.

In many neutronic reactors, a neutron density variation occurs across the reactor; that is, the neutron concentration at the periphery is relatively small and in-

9

creases to a maximum value at the center. Actually, therefore, since the rate of production of element 94²³⁹ is dependent upon the neutron density, the reactor will have zones which may be likened to three dimensional shells, the average concentration of element 94²³⁹ being uniform throughout any given zone. In a reactor built in the form of a sphere these would, of course, be in the shape of concentric spheres of different diameters, while one built in the shape of a cylinder would have similar zones but of different shapes.

Where this variation in concentration exists in a reactor it is often desirable to resort to a systematic schedule of removal depending upon the time of operation and the location of the uranium for removing and discharging uranium metal that has been subjected to neutron bombardment. In the case of a new system of this character the operation would normally continue until the metal in the center portion of the reactor reaches a desired content of element 94²³⁹, at which time this metal would be removed and replaced with fresh metal. The next removal then would be from the section next adjacent to the center section of the reactor where the desired content of element 94²³⁹ is reached after further operation. The process would then proceed with the removal of the metal at various times until the metal recharged at the center of the reactor has reached the desired content of element 94²³⁹. This would then be replaced and the process of progressing towards the periphery continued with periodic return to more central areas. Since the neutron density in the central areas of such a reactor would, ordinarily, greatly exceed the neutron density near the periphery, the metal in the central areas may be replaced several times for each replacement of the metal near the periphery. A removal schedule can be developed by calculation and checked by actual experience after the system has been placed in operation.

Different schedules may be developed with other reactors having different reactivity curves. For example, certain reactors are constructed in a manner such that the neutron concentration is substantially uniform throughout a large volume of the reactor. In such a case the schedule for removal of uranium bodies may be modified accordingly.

Since the heat generated in the reactor results from fissions in the uranium, it is evident that this heat is not formed uniformly throughout the reactor but that it must vary across the reactor with the local rate at which fissions occur and element 94²³⁹ formed. Consequently, the relative values for the production of element 94²³⁹ apply also to heat distribution; that is, the heat generated may increase from a minimum at the outer surface of the reactor to a maximum at the center in certain reactors.

As the total weight of the radioactive fission elements is proportional to that of the 94²³⁹ at the time of fissions it might be assumed that the amounts of these radioactive fission elements and of 94²³⁹ present in metal removed from the reactor are also of the same proportion. This is not true, however, as the fission elements when produced are highly radioactive and immediately start to decay, some with short half-lives and others with longer half lives until, through loss of energy, these unstable fission elements arrive at a stable non-radioactive element or isotope and no longer change. The 94²³⁹ on the other hand is a relatively stable element when formed, having a radioactive half-life of about 2×10⁴ years.

At the start of the reaction in new metal the radioactive fission elements and the 94²³⁹ both increase in amounts. After a certain period of operation during which time the metal is subjected to intense neutron bombardment the radioactive fission elements will reach a state of equilibrium and from that time on the amounts of these radioactive elements remain constant, as the fission elements with shorter half lives are reaching a

10

stable condition at the same time new ones are being produced. The amount of the stable end products of fission, however, continues to increase with the increase in element 94²³⁹. Consequently, the rate of formation of the fission end products is dependent upon the location of any particular metal in the reactor, and the power at which the system operates controls the maximum radioactive fission element content regardless of the length of time the system operates after equilibrium occurs. The quantity of element 94²³⁹ on the other hand, and of the final and stable end products of fission continue to increase as the operation of the system continues. The amounts of both 94²³⁹ and fission end products present are controlled only by the location of metal in the reactor and the time and power of operation. The highly radioactive fission elements may, therefore, vary from a substantial percentage of the weight of element 94²³⁹ present in the metal at the center of the reactor after a short period of operation, to a very small percentage in metal from a position near the periphery of the reactor after an extended operating period at a given power.

It is not to be assumed, however, that the fact that equilibrium can be obtained between the original highly radioactive fission elements and the stable fission end products that all radioactivity will cease when the original fission elements have been permitted to decay for a time equal to the equilibrium period, for example. Many of the original fission elements have long half lives that, taken together with their successive radioactive disintegration products existing long after the fission elements having a shorter half life have decayed, renders the uranium still radioactive especially after prolonged bombardment at high neutron densities. In addition, the successive radioactive disintegration products of the original shorter lived fission elements may still be present.

The equilibrium radioactivity is so intense that metal taken from the reactor for the recovery of element 94²³⁹ and fission products immediately after bombardment at high neutron densities will heat spontaneously due to self absorption of the intense radioactivity of the remaining radioactive fission products. The amount of heat generated as the result of the spontaneous heating will depend particularly on three factors: (1) the concentration of element 94²³⁹ and fission products in the metal; (2) the period of time for continuous operation required to reach this concentration; and (3) the elapsed time since the reactor was shut down and the metal was removed.

The metal from the center of the reactor in a system operating at a high power output, for example, at a 94²³⁹ concentration of 1 to 2,000, if not cooled, can increase in temperature at the rate of about 2000° C. per hour one day after the neutron activity of the system has been shut down. After 30 days shut down following an operation of 100 days at an output of 500,000 kilowatts, the average temperature rise can be approximately 572° C. per hour. The uranium metal of the type used in the chain reacting systems herein under consideration melts at about 1100° C.

Under these conditions uranium bombarded with neutrons for an extended period of time at high rates of power output can be safely removed from the reactor under one of the following methods:

(1) The neutron activity of the system is shut down and the uranium is kept in the reactor and continuously cooled until the radioactivity decays to a point where the metal can be removed without melting in ambient air. This procedure may require that the metal remain in the reactor for a period of from 30 to 50 days after the neutron bombardment has ceased.

(2) The neutron activity of the system is shut down and the uranium is kept in the reactor with the cooling system in operation for only a few days to permit the most violent radioactivity to subside and then the metal

11

is removed from the reactor with the cooling discontinued during the removal except for cooling by the atmosphere or by water spray. The metal is then promptly placed under more efficient cooling conditions before the temperature of the uranium has become excessive.

(3) The neutron activity of the system is shut down and the uranium bodies from one passage are removed while cooling of the remaining uranium bodies in the reactor is continued at least to an extent sufficient to prevent the temperature from becoming excessive.

It is also important, of course, from the point of view of biological safety of operating personnel that adequate shielding be provided to absorb the strong gamma radiations from the fission products present in the active uranium while being removed from the reactor. The neutron activity in the reactor completely ceases within 30 minutes after shut down of the neutronic reaction during which period delayed neutrons are being emitted. In no case then should the uranium be removed from the reactor immediately following shut down of the neutronic reaction, but sufficient time should be given to permit all delayed neutrons to be emitted. Thus, the shielding required during the removal of the uranium rods from the system is primarily intended to protect personnel from gamma radiations. As stated above, immediately following shut down of the neutronic reaction, there are many short lived radioactive fission elements in the uranium causing the gamma radiation to be very intense. Many of these elements decay into more stable products within the first thirty minutes following shut down of the reaction. Thus, the fission products lose a large amount of their radioactivity during this period.

While the method of extracting the fission products and element 94^{239} from the bombarded uranium taken from the reactor forms no part of the present invention, the fission products and element 94^{239} are removable and when removed are extremely useful. The radioactive fission products are valuable for use as radiation sources, many having long half lives with high energy gamma radiation sufficient for radiography of even heavy metal castings. In addition, some of the fission products are useful as radioactive tracers in biological and physiological research.

Element 94^{239} is exceptionally useful because it is fissionable by slow neutrons in the same manner as the uranium isotope 92^{235} contained in natural uranium. The separation of 92^{235} from 92^{238} in natural uranium is extremely difficult since both are isotopes of the same element and these isotopes vary only a small percentage in comparative weight. Element 94^{239} on the other hand, is a different element from uranium, having different chemical properties than uranium, and therefore can be chemically separated from uranium. After separation, for example, element 94^{239} can be added to natural uranium to supplement the 92^{235} content, thus increasing the amount of fissionable material in the uranium. This enriched uranium can then be used in neutronic systems making it possible to provide more cooling facilities, for example, than can be used in a system of the same geometry employing only natural uranium. Thus, an enriched neutronic system may provide a greater power output than would be possible in a natural uranium system having the same geometry.

A more complete description of the design construction and operation of neutronic reactors will be found in Fermi et al. Patent 2,708,656, dated May 17, 1955.

Among the objects of the present invention are the following:

To provide a liquid-cooled neutronic reactor arrangement operating by virtue of nuclear fission of fissionable material in the form of bodies over which the liquid coolant passes and in which the bodies may be readily removed from the reactor and replaced;

To provide a neutronic reactor arrangement wherein the liquid coolant is passed between the moderator and

12

the bodies containing fissionable material wherein the bodies are readily removed from the reactor and replaced;

To provide a neutronic system employing a reactor at least partially covered with water wherein the bodies of fissionable material are readily removed from the reactor through the water; and

To provide a neutronic system wherein the bodies of fissionable material are made available for removal from the system, requiring only a limited period following cessation of the neutronic reaction for the decay of short-life radioactivity of certain fission products to take place before the removal operation is commenced.

The foregoing constitute some of the principal objects and advantages of the present invention, others of which will become apparent from the following description and the drawings, in which:

Fig. 1 is a schematic drawing of one embodiment of the complete system;

Fig. 2 is a schematic drawing of the reactor shown disposed below the level of the ground;

Fig. 3 is an enlarged vertical sectional view taken through the reactor shown in Fig. 2 and illustrating the arrangement of the uranium and graphite and the disposition of the water and tubes forming the cooling system, a portion of the reactor being shown in elevation, at least portions of said drawing not being drawn to scale;

Fig. 4 is a horizontal sectional view not to scale taken through the water tank at the top of the reactor showing the arrangement of the tubes extending downwardly from the tank;

Fig. 5 is an enlarged, fragmentary, sectional view taken on line 5—5 of Fig. 6;

Fig. 6 is an enlarged, fragmentary, detailed sectional view taken vertically through the reactor showing one uranium rod and its relationship to the graphite and other elements making up the reactor;

Fig. 7 is an enlarged fragmentary sectional view taken horizontally through the reactor showing the arrangement of the uranium rods in the graphite moderator;

Fig. 8 is an enlarged fragmentary vertical sectional view taken through the reactor and showing partly in section and partly in elevation a cofferdam and lead shield mounted over one vertical tube;

Fig. 9 is a view partly in elevation and partly in vertical section showing the cofferdam in place and illustrating the removal shield and one rod withdrawn from the reactor; and

Fig. 10 is a diagrammatic view showing the control system for the reactor electrical circuit being reduced to the lowest terms.

In the reactor forming the subject matter of the present invention, the coolant is passed over the exterior surfaces of the uranium. For purposes of convenience, such an arrangement will be identified as an externally cooled structure. This will serve to distinguish from the internally cooled arrangement where the coolant passes through a central longitudinal passage in the uranium bodies.

As shown, the uranium rods are disposed in tubes or liners disposed in the graphite moderator. The diameter of each rod is less than the inside diameter of its corresponding tube so that there is provided a passage between the rod and the tube walls through which the coolant may pass. Heat generated in the rods as the result of the neutronic reaction, or due to the intense radioactivity of the short-lived fission elements in the rods resulting from the neutronic reaction, passes to the outer surfaces of the rods and is carried away by the stream of coolant. Likewise, heat originating in the graphite is conducted through the graphite mass to the cooling passages where it is picked up by the coolant and carried away.

Since the coolant used has a high neutron capture cross section, the film of coolant between the graphite moderator and the uranium rods absorbs some of the thermal neutrons that would otherwise pass from the

13

graphite into the rods and produce fissions. Other thermal neutrons passing from the graphite toward the rods are reflected back into the graphite by the film of coolant. Thus there is a greater opportunity for thermal neutrons to be absorbed parasitically in the graphite and the coolant than would be the case if the coolant were not interposed between the rods and the graphite. This is known as the blocking effect of the coolant and because of this effect, the reproduction factor for a given geometry, volume ratio and purity may be less for externally cooled systems than for internally cooled systems. Thus the thickness of the water layer surrounding the uranium rods must be so proportioned that the reproduction factor for the neutronic reactor is above unity, for otherwise the chain reaction would not be self-sustaining.

The system disclosed herein includes generally a power unit, herein referred to as the neutronic reactor, a complete heat extracting or cooling circuit adapted to remove from the reactor heat generated as a result of the neutronic reaction, and an effective control system regulating the operation of the neutronic reaction to conform to selected conditions of operation.

The reactor can be one of a variety of types and for purposes of illustration, two embodiments have been selected wherein uranium rods are disposed in graphite. Referring to Fig. 1 one embodiment of the invention is shown. The neutronic reactor is generally indicated at 20 and includes bodies containing uranium geometrically spaced in graphite blocks piled to form the reactor. The specific details of the reactor will be explained presently. The heat generated in the reactor as the result of the chain reaction is removed from the reactor by means of water in a cooling system which may be divided into two circuits; viz. a primary cooling circuit indicated at A and a secondary cooling circuit indicated at B in heat exchange relationship with the primary circuit.

The primary cooling circuit A comprises a pump 23 circulating water through an inlet pipe 24 into the reactor 20 through a ring header 25 at the top of the reactor 20. The coolant enters the reactor 20 and passes vertically down through the reactor, as shown in Figs. 1 and 3 and as will be explained later, and leaves the reactor through outlet pipe 26, passing through a water trap 26a to prevent gas escape from the outlet water tank, an outlet sump 26b, through a heat exchanger 27, and finally through a pipe 28 to the pump 23 for recirculation. Thus the outlet water of the reactor is controlled by gravity alone.

The secondary cooling circuit B serves to cool the water in the primary circuit and includes a pump 29 discharging into a pipe 30, which empties into the heat exchanger 27. The cooling medium leaves the heat exchanger 27 through a pipe 31 and passes to a spray head 32 in a cooling tower 33 wherein the water spray is subjected to currents of air produced by a suitable blower fan 34. By means of evaporation of a certain percentage of the water in the cooling tower 33, the remaining water is cooled and collects in a suitable pan or sump 35 at the bottom of the cooling tower, from which the pump 29 draws the coolant through a pipe 36.

For purposes of illustration, water leaving the reactor 20, may be at a temperature of approximately 200 degrees Fahrenheit, and is cooled in the heat exchanger 27 to a temperature of about 85 degrees Fahrenheit. These conditions illustrate a suitable temperature differential between the cooling water entering and leaving the reactor 20. In the secondary cooling circuit B, the water leaving the heat exchanger 27 through a pipe 31 is at a temperature of approximately 130 degrees Fahrenheit, and this liquid is cooled by a process of evaporation in the cooling tower 33 to a temperature of about 75 degrees Fahrenheit. Thus, the temperature differential between the water entering and leaving the heat exchanger 27 is about 55 degrees

Because intense gamma radiation and both fast and

14

slow neutrons escape from the pile, it is essential to surround the reactor with some suitable protective shielding. As shown in Fig. 2 the reactor 20 may be disposed below the level of the ground indicated at 37. The bottom and sides of the reactor are surrounded by earth 38 so that escaping neutrons and harmful gamma rays are absorbed in the earth, and by this means protection is provided to individuals working about the reactor. Further shielding is required across the top of the chain reacting unit, details of which will be described presently.

The reactor itself may take a variety of shapes and arrangements. Uranium metal is preferred to furnish the neutrons for the chain reaction, though uranium compounds may be used, as, for example, uranium oxide or uranium carbide. Also, combinations of these forms may be employed.

Referring to Fig. 3, a specific form of the invention is illustrated, wherein the active portion of the reactor takes the form of a cylinder about 5½ meters high and 7.6 meters in diameter. Uranium metal is used in the form of rods each having a diameter of about 3.22 centimeters and a height of about 5½ meters. These rods are arranged approximately 25 centimeters apart, and in all there are about 850 rods. This is for a system employing about 68 tons of uranium metal rods of high purity, said rods being geometrically arranged in approximately 450 tons of high quality graphite in the form of rectangular blocks. If a carbon reflector of sufficient thickness is placed around the outside of the active portion, the uranium could be reduced to 50 tons.

The reactor rests on a concrete floor and is surrounded by a concrete wall 39, which completely surrounds the reactor unit on its sides as shown in Fig. 3. The active portion of the reactor 20 consists of the vertically disposed uranium rods 40 surrounded by the moderator, such, for example, as graphite 41.

Referring to Fig. 4, the uranium rods 40 are shown spaced in a triangular arrangement in the graphite 41. It is recognized that other spacing arrangements are equally suitable—such, for example, as a square disposition of the rods.

Again referring to Fig. 3, the rods 40 are disposed vertically and extend from a position substantially at the top of the concrete wall 39 to a floor 42 spaced above the bottom of the foundation. Between the floor 42 and the bottom of the foundation is a discharge tank or header 43 extending over the entire floor area of the foundation and having a depth of about ½ meter. This tank serves as the outlet or discharge header for the water in the cooling circuit. The floor 42 may be supported in any suitable manner, such, for example, by posts 44 resting on the bottom of the discharge header 43.

Referring to Figs. 5 and 6, each of the uranium rods 40 is covered with a continuous metallic sheath 45. The purpose of this sheath is to keep the water from directly contacting the uranium and to keep fission products from entering the water or the carbon. The corrosion of the uranium due to water reaction is particularly serious when the water is at a temperature close to the boiling point.

A zinc coating of 5 mils in thickness may be used, but a 1 millimeter covering of aluminum is more resistant to corrosion. However, the 1 mm. aluminum covering reduces the reproduction factor K by approximately .0055 while the 5 mil. coating of zinc reduces it only .004. The sheath should be bonded to the uranium in order to achieve as good heat transfer as is possible. Each sheathed uranium rod 40 is suspended inside of an aluminum tube or liner 46 that extends through a passage in the moderator from the upper water tank 48 through the reactor floor 42 to the lower discharge tank 43. The tubes 46 are in close contact with the graphite 41 to provide a good heat transfer between water and the graphite. The diameter of each tube 46 is reduced on its lower end. By varying the amount of reduction of

15

the tubes 46 the water flow may be adjusted so that more water passes around the hotter uranium rods in the center of the reactor, than around the cooler rods nearer the edge.

The upper end of each tube 46 is funnel-shaped as shown in Fig. 6, with the extreme upper end formed into a flange 46a that is sealed to the enclosed water tank 49.

Each sheathed uranium rod 40 is suspended inside of a tube 46 by means of a steel plug 50 having the same general outline as the funnel shaped portion of tube 46 and being slightly smaller than said tube. Plug 50 is screwed onto the upper end of the rod 40 and extended to the top of the graphite as shown in Fig. 6. A plurality of rounded protuberances or lugs 47 center the rod 40 in tube 46 and space it therefrom, leaving a 1.7 millimeter annular passage 52 surrounding the rod through which water passes from the upper water tank 48 to the discharge tank 43.

Plug 50 being of steel absorbs a portion of the neutrons and gamma rays escaping up tube 46. The water in tank 48 absorbs most of those that pass around plug 50. A threaded socket 53 is provided in the upper end of plug 50 into which a tool may be inserted for removing rod 40 from the reactor. This operation will be further described hereinafter.

If the reactor is opened to the atmosphere there is considerable neutron absorption by the air. Therefore the reactor is preferably sealed and the air replaced by helium which has a much lower neutron absorption coefficient than oxygen or nitrogen. As illustrated in Fig. 2, gas expansion tank 120, 120a and 120b are connected to the reactor 20 by means of pipe 121. A fill pipe 121a is connected to tank 120b. As the temperature of the reactor rises the helium expands into the gas expansion tanks 120, 120a and 120b.

As shown in Fig. 3, a horizontal combination shield and water tank assembly is generally indicated at 48 and extends entirely over the top of the pile and is supported on the foundation 39. This assembly is made up of a lower enclosed tank 49 disposed horizontally across the top of the pile and serving as a shield, and an upper water inlet tank or header 50 coextensive with the lower tank. A gas seal is effected between the water tank assembly 48 and the wall 39 by means of a sealing member indicated at 51 in Fig. 3. This sealing member 51 is fastened to the assembly 48 and extends downwardly along the outer surface of the foundation 39 to a position below the ground level indicated at 37. By this means the entire reactor is gas sealed to retain all gases inside of said reactor.

The upper tank 150 is fed with water by ring header 25. Tank 150 may be of any desired depth to give the necessary head to achieve the flow desired around the uranium rods 40. Although the tank 150 is shown open on top, it will be understood that if a higher pressure is necessary the top may be closed so that pump 23 may place the water in said tank under any desired pressure. The closed top of the tank is usually divided into segments so that portions of the top may be removed for unloading of uranium.

In the space between the graphite 41 and the shield 49 are suitable beams 62 (Fig. 6) resting on the concrete foundation 39 and supporting the water tank and shield assembly 48.

The shield 49 is filled with iron shot 49a and water 49b, the combination of which provides a satisfactory neutron and gamma ray absorber so as to reduce the escape of these harmful radiations from the reactor to safe biological values for personnel on top of the pile. For a plant having an output of about 100,000 kilowatts, a satisfactory thickness for the shield 49 is 60 centimeters.

The water in the shield 49 must be cooled. For this purpose, a separate cooling system may be employed.

16

Referring to Fig. 1 the water is pumped out of the shield 49 through a pipe 63 by a pump 64, and is passed through a suitable heat exchanger or cooler 65, and then is returned to the shield through a pipe 66. A cooling medium is passed in heat exchange relationship with the water flowing through the cooler 65.

Referring to Figs. 3 and 4, six safety rods 90 are disposed circumferentially about the center of the reactor. A control rod 91 is placed at the center of the reactor. All of these rods extend vertically through and project out the top of the pile as indicated in Fig. 3.

Each safety rod when in its lowered position, shown schematically in Fig. 10 extends from the top of the water tank 150 to a position slightly above the bottom of the graphite 41. The safety rods 90 and control rod 91 operate through stuffing boxes so that they may slide freely without allowing the escape of gas from the reactor.

The safety rods 90 are normally in their extreme upper position and are held there during operation of the pile. Their purpose is to stop the chain reaction in the pile either when the reaction becomes too violent, or under normal conditions when the reaction is to be stopped. They function as an efficient neutron absorber, and when in their lowered positions, they absorb a sufficient number of neutrons so as to lower the reproduction ratio of the pile to a value well below unity.

Iron is a relatively good neutron absorber, and therefore serves as a very satisfactory material for the safety rods. The incorporation of boron in the iron serves to increase the efficiency of the rod as a neutron absorber, and consequently is advantageous in the rod composition. A rack 97 is shown at the top of each safety rod 90, and serves as a convenient member for manipulating the rod.

As shown, the safety rods 90 pass through the iron and water shield 49 so as to break the continuity of the shield across the top of the reactor. Each safety rod, then, serves the additional function of plugging the opening through which it passes, to prevent the escape of neutrons and gamma radiations from the reactor.

The control rod 91 is also shown schematically in Fig. 10. This control rod may take several forms. We prefer to use a sheet of cadmium enclosed in aluminum for strength, because cadmium has a very high neutron absorption coefficient so that a relatively small quantity inserted into reactor will absorb a large quantity of neutrons.

The power plant above described is ideally adapted for automatic control to maintain the neutron density within the reactor substantially constant, and thus give a substantially constant power output. Due to the fact that large masses of materials are utilized in the reacting portion of the structure, there is a temperature lag therein. Consequently, it is convenient to monitor and control the structure by means of ionization chambers, or equivalent devices which will measure the neutron density at the periphery of the lattice portion of the structure. As the rate of neutron diffusion out of a chain reacting system is always proportional to the rate of generation of neutrons within the structure, the ionization chambers can readily be placed at the periphery of the active portion or lattice, and in fact are preferably so positioned in order that they be not subjected to the extremely high neutron densities existing near the center of the reactor.

Before proceeding to a description of one type of control system that may be utilized in controlling the pile described herein, it is desirable to point out the manner in which the control rods operate to regulate the neutron density. In any self-sustaining chain reacting structure adapted to produce power, the attainable neutron multiplication ratio of the system must be greater than unity. For any value over unity, the chain reaction becomes self-sustaining and the neutron density, without control,

2,856,339

<table>
<tr><td>17</td><td>18</td></tr>
</table>

would increase exponentially in point of time, until the device is destroyed. For proper control, the system must be held in balance by maintaining the chain reaction at some point where the production of new neutrons is balanced with the neutrons initiating the chain. Under these conditions, the reacting portion of the structure will continue to maintain the neutron density therein which obtained when the system was balanced.

However, in order to enable the reactor to reach a desired neutron density, the system must be permitted, for a period of time, to rise in neutron density until the desired density is reached. After that it is only necessary to hold the system in balance.

Inasmuch as the reproduction ratio of any pile is reduced by the presence of impurities which absorb neutrons, such impurities can be introduced in the pile in the form of the control rod which can be of a material such as cadmium or boron which will absorb large amounts of neutrons. The depth to which this control rod penetrates into the pile will determine the amount of neutron absorption and therefore the reproduction ratio of the pile. A range can be obtained between a condition providing a neutron reproduction ratio which is greater than unity and a condition at which no chain reaction can be maintained. The exponential rise of neutron density can be made relatively fast or relatively slow, in accordance with whether the multiplication ratio is permitted to be much greater than unity, or only slightly greater than unity. Between 0.05 percent and one percent of the neutrons emitted in the fission process are delayed neutrons. These delayed neutrons cause the neutron density to rise in a finite time rather than instantaneously. The time required for doubling the neutron density increases as the multiplication ratio approaches unity, and any desired rate of rise can be obtained.

The broad method of control preferred is to withdraw all safety rods from the pile and then withdraw the control rod from the structure to a point where there is an exponential, and preferably slow rise in neutron density within the structure. When a desired neutron density has been reached, the control rod is then returned into the pile to a point where the reaction is balanced. This balance is then maintained to maintain a constant power output in the reactor. The maintenance of the balance point with the control rod would be relatively simple were it not for the fact that changes in temperature in the pile change the reproduction ratio of the structure slightly. It is desirable, therefore, that the control rod be so manipulated that a substantially constant neutron density within the system is maintained. Such a method of control may be accomplished by automatic connection of the control rods with an ionization chamber or similar device measuring neutron density, positioned within the reactor close to the active portion of the pile.

Furthermore, due to the exponential rise of neutron density, within the reacting structure when the multiplication ratio is greater than unity, all possible precautions must be taken to prevent a continued exponential rise in neutron density, in case of failure of the control rod to return to the balance position.

While there are many means by which the control rod and the safety rods can be operated, it is believed that by the illustration and description of one simplified circuit, other and fully equivalent circuits will be made apparent to those skilled in the art.

Referring, therefore to Fig. 10, which shows diagrammatically and reduced to lowest terms one form of control circuit that might be used for regulating the output of the power plant hereinbefore described, and referring first to control circuit A, a control ionization chamber 200 is placed within the reacting structure adjacent to the periphery of the pile, and filled with barium fluoride. A central electrode 201 is provided within the chamber 200 and connected to wire 202 leading outside of the re-

actor to a movable contact 203 on a resistor 205. Resistor 205 is connected across a relay coil 206. One side of relay coil 206 is connected to battery 207, the other end of which is connected to shield 209 around wire 202. Shield 209 is grounded, as is chamber 200. Alpha ray ionization due to neutron reaction with the barium within chamber 200 is proportional to the neutron density. Thus the current in resistor 205 is varied in accordance with neutron density reaching the ionization chamber. Relay coil 206 operates a relay armature 210 which is spring biased by spring 211 to contact one motor contact 213, and is urged by current in coil 206 to contact a second motor contact 215. Contacts 213 and 215 connect to the outside of split winding 220 of motor 221, the center connection 222 of which is connected through power mains 225 to armature 210. Motor 221 operates shaft 230 having secured thereon a pulley 231, one end of said shaft being connected through a magnetic clutch 232 to a control rod gear 234. Control rod gear 234 meshes with the rack 97 on the control rod 91. Pulley 231 has a cable 240 wound thereon connected to a counter weight 241 so that the weight of the control rod is substantially balanced by counter weight 241 thus permitting motor 221 to run easily in either direction.

Having described a circuit for controlling the position of a control rod, I will now describe its operation, considering the safety rods withdrawn. Slider 203 on resistor 205, having previously been calibrated in terms of neutron density, is moved to the density position at which it is desired the reactor to operate, taking into account the difference in neutron density at the center of the lattice and at the periphery thereof during operation. This difference is a constant ratio at various operative densities. The reactor, having at best a neutron density much lower than the desired density at which relay coil 206 will receive enough current to operate armature 210, very little ionization takes place in ionization chamber 200 thus causing armature 210 to rest against contact 213. Motor 221 is thus energized to withdraw the control rod 98 from the reactor to a point as determined by a limit stop 242 where the multiplication ratio of the reactor is just sufficiently greater than unity to permit an exponential rise in neutron density within the reactor. The motor 221 will stall when the rod is at stop 242 and should be of a type permitting stalling for the required time. The reaction at this position of the control rod becomes self-sustaining and the neutron density rises. In consequence the ionization within chamber 200 rises. As the ionization in chamber 200 increases, more and more current passes through relay coil 206 until the desired value has been reached. Relay coil 206 then operates to cause armature 210 to connect with contact 215, thus reversing the motor 221 to drive the control rod into the reactor to a point where the neutron density starts to decay. The control rod 90, will thereafter hunt between a point above the balance position where the neutron density rises, and a point below the balance position where the neutron density decays, thus providing an average neutron density within the reactor as determined by the setting of slider 203 on resistor 205. As the mass of the reactor causes any temperature change to lag behind any neutron density change, the temperature of the reactor is maintained substantially constant. If desired, any of the well-known anti-hunting circuits may be utilized, as will be apparent to those skilled in the art.

The main purpose of the control circuit A is to regulate the control rod to substantially balance the neutron density to maintain any desired average temperature within the reactor.

Due to the fact that it might be possible for the control system as described to fail, and thereby leave the control rod in a position where the neutron density would continue to rise indefinitely, both the safety rods 90 and the control rod 98 are preferably to be operated so as to enter the pile immediately upon any failure of the control rod

19

system. One such emergency circuit that illustrates the operation of the safety rods is illustrated in circuit B. The circuit for each of the safety rods is the same.

Circuit B comprises a connection to the power mains 225. One side of the mains passes through a plurality of hand operated emergency switches 250, a flow switch 251 in water outlet pipe 26, a thermostat 252 in the same pipe, and a contact 254 held in connection with a relay armature 255 by a spring 256, all switch connections being normally closed. Both legs of mains 225 then are connected to energize magnetic clutch 232 on the control rod shaft 230 and also a similar magnetic clutch 260 on a safety rod shaft 261.

Safety rod shaft 261, on one side of magnetic clutch 260 ends in a safety rod gear 262 meshing with safety rod rack 97. On the other side of clutch 260 a safety rod pulley 264 has a cable 265 wound thereon connecting with a safety rod counter weight 266. Thus the safety rod, with clutch 260 energized, will stay in any position it is placed.

Safety rod 90 is normally moved in and out of the pile by safety rod motor 270 by connection to mains 255 through hand switch 271.

Relay armature 255 is controlled by an ionization chamber circuit identical in all respects with the circuit previously described for operating the control rod, this safety circuit being given the same numbers plus 100. The circuit may be adjusted, for example, to attract relay armature 255 and open the circuit when the neutron density reaches 5% over the normal operating density. In addition, flow switch 251 opens the circuit when output water flow falls below a safe level. When output water temperature reaches 210° F., for example, thermostat 252 will open the circuit. Hand-switches 250 located in convenient spots can be used for the same purpose by the operating personnel.

Safety rod magnetic clutch 260 is connected in parallel with control rod magnetic clutch 232. Thus, upon any failure of power in the mains or the opening of any of the switches 251, 252, or 250, all of the rods will be relieved of their counter weights and will fall into the pile by gravity. Stops may be provided on the rods, and any suitable braking action used to reduce the impact shock on the reactor.

Irrespective of how the circuit is opened, magnetic clutches 232 and 260 disengage, and remove the influence of the counter weights from the rods. The rods then drop into the system by gravity, stopping the reaction.

To construct the system, a hole is first dug in the earth 15 meters deep and 15 meters in diameter, within a suitable earth retaining wall, if such is needed. A concrete floor of suitable thickness to cover the bottom of the hole is then poured. Concentric within the hole is erected a concrete or reinforced concrete wall of 9 meters inside diameter and 6 meters in height, of suitable thickness and taper, and with proper footings and stiffening columns. Through the wall near the floor pass the pipes for the water system and the gas expansion chambers.

The bottom water tank is then installed. A hole in one side of the tank joins to the water outlet pipe. The water in this tank is not under pressure, and it drains by gravity into sump 26b.

Supports are positioned to rest on the tank bottom and extend to the height of the tank sides. On these supports the floor for the pile is laid, which is also the lid of the outlet water tank. This floor is pierced with openings through which the ends of the cooling tubes will later descend. The reactor shell is then built.

The graphite, preferably in the form of blocks, is then stacked in the space above the floor up to the height of, and out to, the surrounding concrete wall. Between the blocks sufficient clearances are left to allow for graphite expansion when the pile is operating, and for helium diffusion through the pile. Some 850 holes run vertically through the graphite blocks and are kept in vertical align-

20

ment with the holes in the floor. Seven vertical holes of about 12 cm. in diameter are provided in the graphite for the control and safety rods, one in the center for the control rod and six on a circle about the center of some four meters diameter. The pipe leading to the gas expansion chamber is just above floor level, and enough space should be left between the graphite blocks for passage of the helium and for graphite blocks expansion.

The vertical holes must be aligned with some care so that the tubes 46 may be inserted and the uranium and control rods will be able to slide freely up and down. These are checked with dummy rods hung from an overhead frame as the stacking of the graphite progresses. The expansion clearances are differentially patterned about the center in accordance with the expected distribution of temperature in the operating plant.

As the graphite stacking progresses, earth is filled in outside the concrete wall to the sides thereof, and up to a little below the top of the wall. The time at which this is done is immaterial. At that time the helium expansion tanks are installed.

The beams over the pile which rest on the walls and are to support the shield and upper water tank are put in place. The shield sections come next, being careful to align the holes therein with those in the graphite.

The tubes or liners 46 are set in place and the shield sections are sealed together so that they represent a continuous gas tight disc except for the safety, control, and uranium rod openings. Sealed to the shield, and extending outwardly and downwardly from it, there is placed the thin metal gas cover which extends over the outside of the concrete wall. The upper water tank sides and water mains and connections are then installed, and the control and safety rod placed in their receptacles, connected to the exterior circuits and test operated.

With the control and safety rods inserted fully into the pile, the uranium rods are then inserted, as for example, by a crane operating from above ground, lowering each rod into place until all holes are filled.

Preliminary tests are then made of the chain reaction at low power output, with the circulating system in operation. When operating conditions are found to be satisfactory, the roof is closed over and covered with earth. To guard against helium leakage carrying radioactive fission fragments, the space between the reactor and the roof is well ventilated, preferably by blowers.

The system is then ready for operation at high powers. In the present instance, the power output is preferably stepped up a few thousand kilowatts a day, with all operating mechanism carefully checked, until the desired 100,000 kilowatts output is being obtained. The safety circuit B is then set to trip the safety rods at a 5% increase in neutron density over the steady output value, and thermostat 252 is set to trip the safety rods at 210° water temperature.

After 80–100 days of operation, sufficient fission products and 94^{239} have been produced in the center of the pile to warrant separation. The chain reaction is then shut down by inserting the control and safety rods, meanwhile maintaining full water circulation for a period at which time the spontaneous heating of the uranium will have been so reduced by radiation decay of the fission fragments, as to allow removal of the uranium without melting, under such cooling as may be obtained from the contact with the atmosphere.

However, as the radioactivity still remaining in the rods is so great that they cannot be closely approached, they must be unloaded with full precautions for biological safety of the operating personnel. This is accomplished without halting the flow of coolant by means of the apparatus shown in Figs. 8 and 9. The earth and roof over the reactor are first removed, exposing the upper water tank 48. If a closed water tank is used the water flow is reduced and a segment of the top of the tank removed.

2,856,339

21

However, in the described reactor with the open tank, the water flow is only reduced slightly.

One means of removing the rods is a crane, not shown, operated from the ground surface, at one side of the opening. The crane may be provided with a long crane arm and lead shields so that operating personnel are protected from harmful radiation. A coffer dam assembly 70 supported in crane hooks 71, 72, 73, 74 and 75 is lowered over the top of the uranium rod 40 that is to be removed. The assembly 70 comprises a steel coffer dam 76, inside of which slides a lead coffin or removal shield 77 and inside of said coffin a rod removal tool 78. The assembly 70 may be lowered as a unit until the coffer dam 76 is resting on top of water tank 49 and centered in groove 79 that is concentric with the rod opening. A gasket 80 is positioned in groove 79 and the weight of the coffer dam 76 forces the bottom of said dam against said gasket making a water-tight joint.

The annular lead coffin 77 slides inside of coffer dam 76 and also rests on the water tank 49. The rod removal tool 78 has a cylindrical body 81 to the upper side of which is secured a ring 82. On the lower side of body 81 is a threaded member 83 adapted to screw into threaded socket 53 in plug 50. By turning hook 73 that is inserted in ring 82 the removal tool 78 is revolved until the threaded member 83 is screwed into threaded socket 53. At this time hook 73 is lifted upward carrying with it removal tool 78, plug 50 and rod 40, until the rod 40 is entirely surrounded by coffin 77. The coffin 77 and the rod 40 are then lifted together from coffer dam 76 and removed from the reactor to a position where the rod is dropped from the coffin 77 for further processing. A new uranium rod 40 and plug 50 are then pulled inside of coffin 77, the coffin is re-inserted in the coffer dam 76 and removal tool 78 is unscrewed from plug 50 leaving rod 40 suspended in the reactor in operating position. The coffer dam 76, coffin 77 and removal tool 78 are then lifted and moved to the next rod to be replaced and the process repeated. While the coffer dam is in place all the rods not surrounded by the dam are receiving the uninterrupted flow of coolant in normal amount since the coffer dam 76 prevents the evacuated hole or holes from filling with coolant. The use of the lead coffin protects the operating personnel from direct radiation of the "hot" uranium metal. It is of sufficient thickness that enough gamma radiation may be absorbed to make it safe for personnel to handle an irradiated rod in said coffin.

After the required rods are removed and replaced, the system is again placed in operation by replacing all parts and starting up the chain reaction. Systematic removal of rods in various radial zones of the pile is practiced so that the outer rods finally reach the required production of the end products.

Several important features of the system described herein should be noted, particularly the type of cooling arrangement, i. e., circulation of the cooling fluid around the outside of the uranium bodies between the uranium and the moderator. If greater cooling is desired this method may be combined with a method of internal cooling in which the uranium is in the form of tubes and the cooling fluid is circulated through a passage extending lengthwise through the center of the uranium bodies as well as through a passage between the uranium and the graphite.

As has been explained heretofore, the annulus of water around each uranium tube acts as a reflector to return some of the neutrons to the moderator instead of allowing them to enter the uranium. Therefore, a reactor using such cooling must be designed with sufficient excess maximum reproduction ratio r to allow this loss of neutrons.

A cylindrical reactor using materials and geometry of the type described should have a maximum reproduc-

22

tion factor of about 1.073 without such impurities as aluminum covers 45, aluminum tubes 46 or water in the system. The losses in K due to these impurities are as follows:

    1 mm. of Al in tubes 46=.0055W units
    1 mm. of Al in rod coverings=0.12 K units
    Water in the reactor=0.12 K units
    Other miscellaneous losses=.005 K units
        Total losses=0.0280 K units

Thus when these losses are deducted from the maximum K, the value of K for the present reactor becomes 1.0730−0.028=1.045. A reproduction ratio of about 1.045 is considered desirable in neutronic reactor, because it allows ease of control. As is explained in the previous discussion of control rod action less than 1 percent of the neutrons are delayed. Therefore, the reproduction ratio should be kept below 1.1 in order that the control rod need only absorb a percentage of neutrons less than the percentage of delayed neutrons to reduce the reproduction ratio to unity.

Note that the water loss of 0.012 r units is the greatest loss of those listed. However, by means of our invention using materials of high purity and the proper geometry, this loss may be overcome and a neutronic chain reaction is achieved.

In this description reference has been made primarily to water as the coolant for the pile. Diphenyl, also known as biphenyl or phenylbenzene, serves as another satisfactory coolant. This substance has the chemical formula $C_6H_5C_6H_5$ and is in the form of a solid at atmospheric temperature, melts at 70° C., and has a boiling point of 255° C. Thus the operating temperatures of a pile cooled with diphenyl can be higher than that for a pile using water.

Diphenyl has a lower absorption of thermal neutrons than has water. For example, a loss of neutrons by absorption due to a diphenyl cooling layer 4 millimeters thick corresponds to the loss due to a 2.2 millimeter layer of water. Thus for the same reproduction ratio almost twice as much diphenyl as water can be circulated through the pile. The cooling passages then, for a pile employing diphenyl can be almost twice as wide as those set forth for the water passage. About 10 percent to 15 percent more pumping power is required to circulate the diphenyl due to its greater viscosity, and since diphenyl solidifies at 70° C. special measures are taken to prevent freezing of the coolant while it is disposed in the portion of the cooling circuit outside the pile. This can be done by maintaining the temperature of the diphenyl at all times safely above 70° C. while being circulated.

In addition, the use of helium in the pile offers many advantages. The neutron absorption of pure helium is negligible, and as heretofore pointed out it promotes heat exchange in the pile. The use of helium, however, has another advantage as it replaces the nitrogen that normally would permeate a pile exposed to the atmosphere. Nitrogen has a definite neutron absorption factor and consequently any nitrogen absorption of neutrons to the detriment of the chain reaction is prevented when helium is used. Furthermore, the nitrogen density in a pile exposed to the atmosphere changes with atmospheric pressure, and any such change causes change in the reproduction ratio of the pile, requiring correction by the control rods. Consequently, a sealed pile containing helium is immune to changes in atmospheric pressure. Again, the argon content of ambient air becomes highly radioactive with neutron bombardment. Such radioactive argon diffuses out of a pile exposed to the atmosphere and becomes a biological hazard. Such a condition does not occur in the helium filled pile described herein.

Other advantages of the invention set forth in the speci-

2,856,339

23

fication and attached claims will be apparent to those skilled in the art.

During the operation of the neutronic reactor particularly at high neutron densities radioactive elements of exceedingly high capture cross section may be formed in the uranium as an intermediate element in the decay chains of fission fragments and this formation will lower the value of the reproduction factor for the system. Radioactive xenon[135] is an example of such an intermediate element. this product having a half life of about 9 hours and being formed mostly from radioactive iodine which has a half life of about 6.6 hours and decays to barium. There should be sufficient excess in the reproduction ratio of the reactor so that in the event the reproduction factor is reduced as a result of the formation of an intermediate decay element having a high capture cross section for neutrons the control rods may be withdrawn sufficiently to maintain the reproduction ratio at a value of unity while maintaining the power output at the desired level. It might be desirable to initially construct the reactor sufficiently over-sized to supply this excess reproduction ratio when needed and in this event removable impurities for example in the form of additional shim or control rods may be initially placed in the reactor and kept there at all times until the reproduction ratio commences to fall as the result of the formation of these intermediate decay elements.

While the theory of the nuclear chain fission mechanism in the uranium set forth herein is based on the best presently known experimental evidence, it is not intended that the present inventors be bound thereby, as additional experimental data later discovered may modify the theory disclosed. Any such modification of theories, however, will in no way effect the results to be obtained in the practice of the invention herein discovered and claimed.

What is claimed is:

1. In a neutronic reactor having vertically disposed rods of thermal neutron fissionable material and having a chamber of liquid coolant surrounding the top ends of

24

said rods, the improvement comprising a coffer dam sleeve sealable to the floor of the chamber in positions surrounding the ends of the rods, a radiation shield sleeve disposed in and vertically reciprocable through the coffer dam sleeve, and an engaging member disposed in and reciprocable axially of the radiation shield and adapted to engage a rod, whereby the coffer dam may be placed over the end of a rod, the rod withdrawn into the shield, and the shield and rod removed.

2. In a neutronic reactor having rods of thermal neutron fissionable material and having a chamber of liquid coolant surrounding the ends of said rods, the improvement comprising a coffer dam sleeve sealable to the chamber in positions surrounding the ends of the rods, a radiation shield sleeve disposed in and reciprocable through the coffer dam sleeve, and an engaging member disposed in and reciprocable axially of the radiation shield and adapted to engage a rod, whereby the coffer dam may be placed over the end of a rod, the rod withdrawn into the shield, and the shield and rod removed.

References Cited in the file of this patent

UNITED STATES PATENTS

92,324	Lewis	July 6, 1869
367,547	Poetsch	Aug. 2, 1887
860,391	Lea	July 16, 1907

FOREIGN PATENTS

114,150	Australia	May 2, 1940
861,390	France	Oct. 28, 1940
233,011	Switzerland	Oct. 2, 1944

OTHER REFERENCES

Smyth: "Atomic Energy for Military Purposes," pp. 103–104, August 1945. Copy may be purchased from Supt. of Doc., Washington, D. C.

Kelly et al.: Phy. Rev. 73, 1135–9 (1948). Copy in Patent Office Library.

March 15, 1960   E. P. WIGNER ET AL   2,928,781
                        COOLED NEUTRONIC REACTOR

Filed June 12, 1952   4 Sheets–Sheet 1

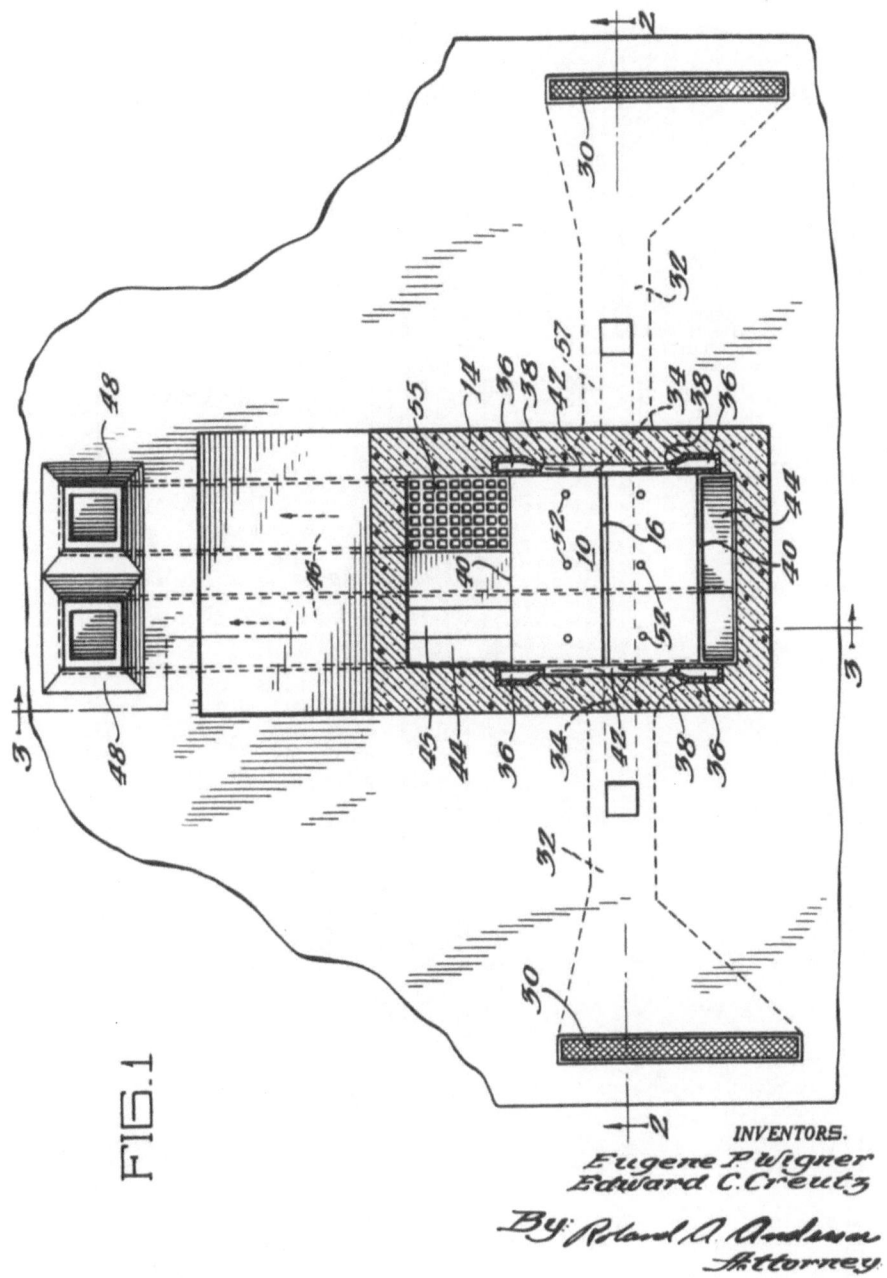

FIG.1

INVENTORS.
Eugene P. Wigner
Edward C. Creutz
By: Roland A. Anderson
Attorney.

March 15, 1960   E. P. WIGNER ET AL   2,928,781

COOLED NEUTRONIC REACTOR

Filed June 12, 1952   4 Sheets—Sheet 2

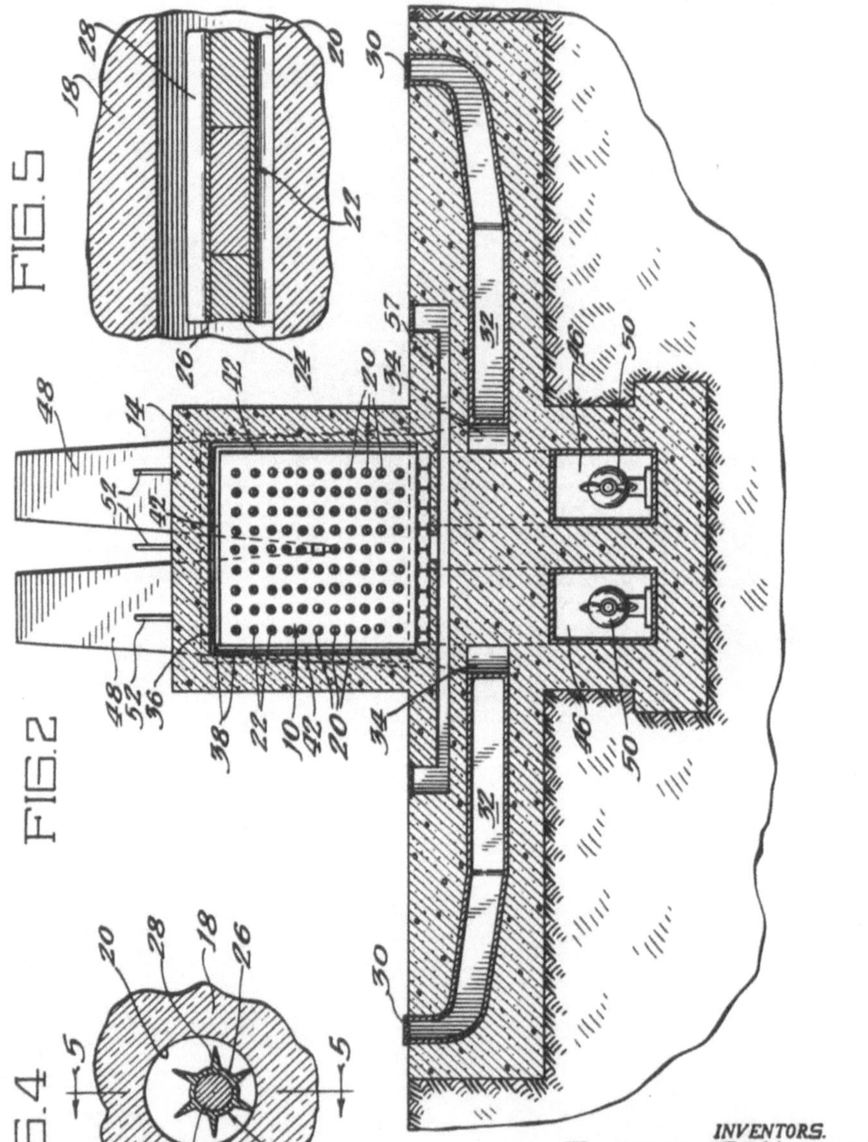

FIG. 5

FIG. 2

FIG. 4

INVENTORS.
Eugene P. Wigner
Edward C. Creutz

By: Roland A. Anderson
Attorney.

March 15, 1960   E. P. WIGNER ET AL   2,928,781

COOLED NEUTRONIC REACTOR

Filed June 12, 1952   4 Sheets—Sheet 3

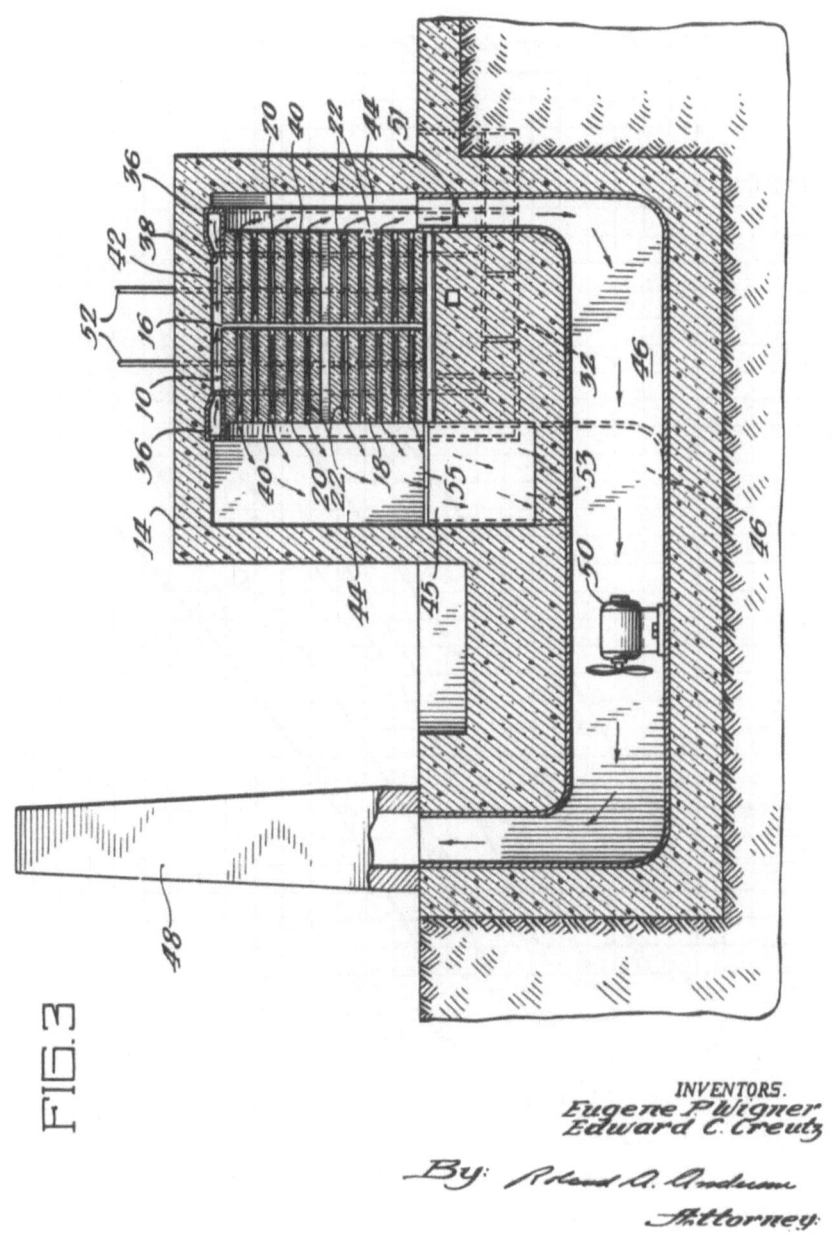

FIG.3

INVENTORS.
Eugene P. Wigner
Edward C. Creutz

By: Roland A. Anderson
Attorney.

March 15, 1960   E. P. WIGNER ET AL   **2,928,781**

COOLED NEUTRONIC REACTOR

Filed June 12, 1952   4 Sheets—Sheet 4

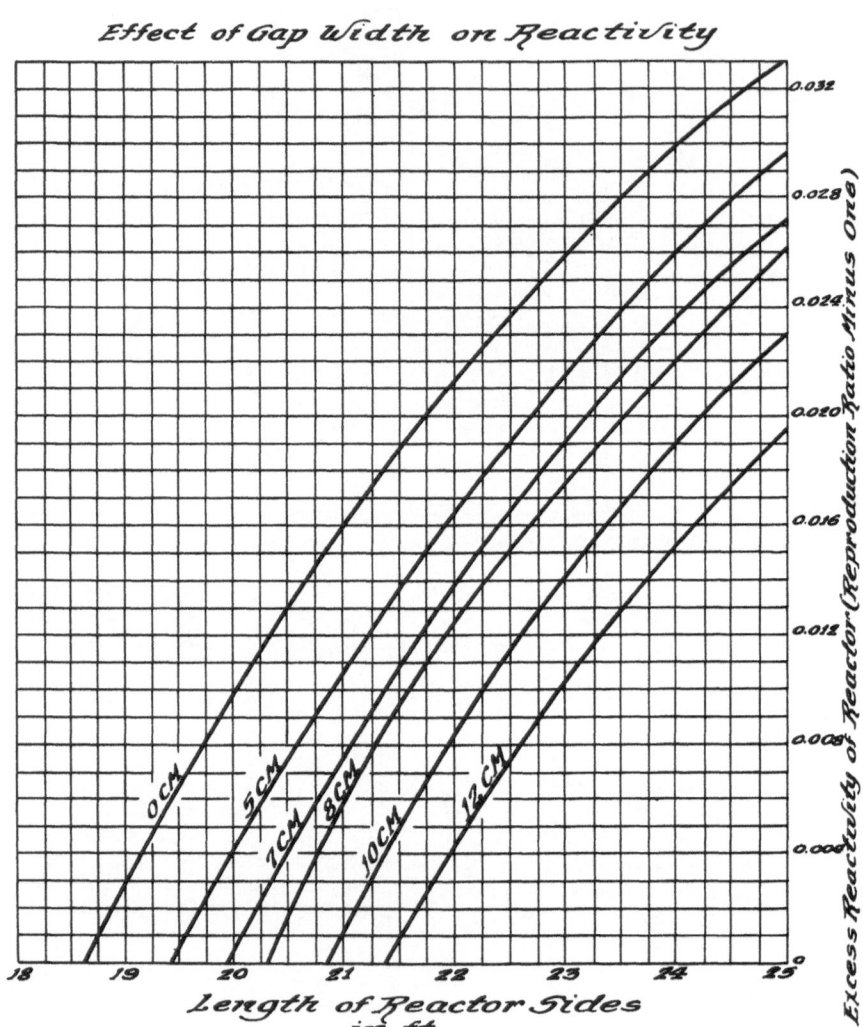

Effect of Gap Width on Reactivity

FIG.6

INVENTORS.
Eugene P. Wigner
Edward C. Creutz

By: Roland A. Anderson
Attorney.

United States Patent Office

2,928,781
Patented Mar. 15, 1960

1

2

2,928,781

COOLED NEUTRONIC REACTOR

Eugene P. Wigner, Princeton, N.J., and Edward C. Creutz, Pittsburgh, Pa., assignors to the United States of America as represented by the United States Atomic Energy Commission

Application June 12, 1952, Serial No. 293,104

1 Claim. (Cl. 204—193.2)

The present invention relates generally to neutronic re- 15 actors, and more specifically to methods and apparatus for cooling neutronic reactors.

It is well known, that the absorption of a neutron by an atom of fissionable material may result in the splitting of the fissionable atom and the liberation of 20 a number of neutrons and considerable energy. Neutronic chain reactors operate by virtue of the fact that more neutrons are liberated by the fissioning of atoms in a chain reacting system than are lost to the reaction by diffusion and absorption of neutrons. The energy liberat- 25 ed by the fission process appears in the form of fission fragments, particles, and radiations. The fission fragments, particles, and a portion of the radiations, are absorbed within the structure of the neutronic reactor, thereby liberating large quantities of heat. It thus be- 30 comes necessary to cool the structure of a neutronic reactor in order to prevent melting of solid materials, the increasing of oxidation rates, and other undesirable effects within the reactor structure.

Not all of the cooling agents known to the art are 35 suitable for cooling a neutron reactor. In addition to the usual properties required of cooling agents, a neutronic reactor coolant must possess nuclear properties which will not present too deleterious an effect upon the neutronic chain reaction. This, in general, means that 40 a neutronic reactor coolant must have a small neutron capture cross section as well as favorable thermal properties, since neutrons absorbed in the coolant will be lost to the neutronic chain reaction. This requirement generally dictates a reactor design utilizing a minimum 45 volume of the coolant within the reactor for heat removal. A good coolant for a neutronic reactor should be highly stable from the point of view of both thermal properties and nuclear properties, since changes in these properties of a coolant will effect the reactivity of a neu- 50 tronic reactor. The coolant should also produce very limited corrosion, and require a minimum power consumption for circulating the coolant throughout the reactor. A number of materials have been used as coolants in neutronic reactors, among these air, water, and heavy water 55 have proven to be most readily adapted to the coolant processes used in reactors in which neutrons, primarily of thermal energy produce fission. Other coolants, such as potassium and sodium potassium alloys, as well as the coolants mentioned, may be used in reactors operating 60 with neutrons of higher energies. The necessary criteria for a coolant material has been widely discussed in the art, the patent of Enrico Fermi and Leo Szilard, Serial No. 568,904, filed December 19, 1944, now Patent No. 2,708,656, being an example. 65

The existing reactors have generally utilized ducts extending through the active portion of a neutronic reactor and forced the coolant through the ducts from one end to the other. As a result, the lowest temperature of the coolant in the reactor occurs at the intake end of the 70 duct, and the highest temperature at the outlet end of the duct. It is also true, that the neutron density within a

neutronic reactor is greatest at the center of the active portion of the reactor, and hence the greatest proportion of the heat liberated by the reaction occurs in this center portion. Also, the heat removed from any structure by 5 a coolant is proportional to the temperature difference between the structure and the coolant, so that less heat is removed from the region of the reactor adjacent to the exhaust end of the ducts than is removed from the region adjacent to the intake end of the ducts. As a re- 10 sult, such a neutronic reactor will have a central portion operating at a higher temperature than the peripheral portion, and also a poorer transfer of heat from the central region to the coolant than would be obtained if the coolant entering the center of the reactor had not 15 already become quite hot. The present invention is directed to eliminating this difficulty.

The maximum power level to which a neutronic reactor may be operated is generally determined by a limiting maximum temperature within the neutronic re- 20 actor structure, although other limiting factors could also determine the maximum power level. In order to obtain maximum cooling, the coolant is generally introduced into the reactor coolant ducts at a high pressure, this pressure decreasing as the coolant traverses through the 25 reactor. The amount of heat carried away by a stream of coolant is a function of the velocity of the coolant. Since the velocity of the coolant is reduced by a pressure drop along a duct, it is desirable to minimize the pressure drop along the length of the coolant duct as the 30 coolant traverses the duct, and the present invention is also directed to this end.

A more complete understanding of the present invention may be had by a further reading of the specification, the invention having been illustrated by means of an air 35 cooled neutronic reactor, although it is to be understood that a water cooled, or other type of neutronic reactor, could also be used to illustrate the present invention. Such an air cooled neutronic reactor is illustrated in the drawings, in which: 40

Figure 1 is a plan view, partly in section, of a neutronic reactor embodying the present invention;

Figure 2 is a sectional view of the neutronic reactor shown in Figure 1 taken along line 2—2 thereof;

Figure 3 is a vertical sectional view of the reactor 45 taken along line 3—3 of Figure 1;

Figure 4 is an enlarged vertical sectional view of a fragment of the active portion of the reactor;

Figure 5 is an enlarged fragmentary view of a portion of the active portion of the reactor taken along line 5—5 50 of Figure 4; and

Figure 6 is a graph showing the effect of different gap widths upon the reactivity and critical size of the reactor shown in Figures 1 through 5.

The neutronic reactor described in Figures 1 through 5 55 has an active portion 10 surrounded by a shield 14. The active portion of a reactor may be defined as the region in which the neutron fissionable material is disposed and in which the reaction produced neutrons are generated. The active portion 10 is in the form of a cube, and is 60 provided with a central gap 16 which traverses the active portion 10 in a vertical plane.

In this embodiment of the invention, the active portion 10 consists of a moderator 18 having two portions and forming a cube at least 20 feet on a side with a gap less 65 than 20 centimeters and preferably approximately 7 centimeters, a plurality of coolant channels 20 extending through the moderator 18 normal to the plane of the central gap 16, and fuel elements 22 disposed within the coolant channels 20. The fuel elements 22 have a body 24 of material fissionable by neutrons of thermal energy, such as natural uranium, or isotopically enriched uranium, or plutonium in a suitable diluent. The fissionable body

3

24 is enclosed within a protective casing 26 which may be constructed of aluminum. The casing 26 is also provided with a plurality of longitudinal fins 28 for supporting the fuel elements 22 in the coolant channels 20.

Air is introduced into the reactor for cooling purposes through filters 30 disposed upon opposite sides of the reactor. The air is transported to the reactor by means of ducts 32 which split into a T-connection 34 adjacent to the sides of the reactor. A manifold 36 is disposed adjacent to the two exhaust faces 40 of the reactor and is attached to the top and sides of the reactor active portion 10. Each of the T-connections 34 connects one of the intake ducts 32 to each of the manifolds 36. The manifolds 36 are provided with a plurality of apertures 38 contiguous to the sides of the active portion 10, but sealed from the exhaust faces 40 of the reactor. The manifolds 36 are imbedded within the concrete shield 14, but an air passage 42 is provided adjacent to the surfaces of the reactor between the apertured sides of the manifolds 36, the passage 42 girdling the active portion 10 of the reactor except for the bottom thereof, as best illustrated in Figures 2 and 3.

Exhaust chambers 44 are disposed adjacent to each of the exhaust faces 40 of the active portion 10 of the reactor. A channel 45 extends downwardly from the floor of one of the exhaust chambers 44, as illustrated in Figures 1 and 2, and may be used to remove irradiated fuel elements 22 from the reactor. Exhaust ducts 46 disposed on a level beneath the active portion 10 of the reactor extend from the exhaust chambers 44 to stacks 48. The exhaust ducts 46 are connected to the exhaust chambers 44 by passages 51 and 53, a grate 55 being disposed at the entrance of passage 53. Exhaust fans 50 are disposed within the exhaust ducts 46 for the purpose of forcing the air through the stacks 48.

The exhaust fans 50, by forcing air from the stacks 48, lower the air pressure in the exhaust ducts 46 and the exhaust chambers 44, thereby drawing air into the coolant system of the reactor through the air filters 30. The coolant air flows from the filters 30 through the intake ducts 32 to the manifolds 36. The air flows from the manifolds 36 across three of the surfaces of the reactor, all of the faces except the exhaust faces 40 and bottom. The air then flows into the central gap 16. At this point the air flow divides, approximately equal portions of air flowing through the coolant channels 20 in each portion of the active portion 10 of the reactor. The coolant air thus flows across the fuel elements 22 within the coolant channels 20 and into the two exhaust chambers 44. From thence the coolant air proceeds down the two exhaust ducts 46 to the stacks 48.

As indicated in Figure 6, the active portion 10 of the reactor is constructed with excess reactivity, so that changes in the nuclear characteristics of the reactor will not reduce the reproduction ratio below unity, and so that the reactor may be used to irradiate non-fissionable objects placed within the active portion 10 of the reactor. The reproduction ratio is the ratio of the number of neutrons in one generation to the number of neutrons in the preceding generation for a given reactor. It is to be noted, that the critical sizes of reactors with the shown gap widths are indicated in Figure 6, since a reactor will be just critical when the excess reproduction ratio vanishes (all reactivity control positioned for maximum reactivity).

A plurality of control rods 52 are positioned within the active portion 10 of the reactor in order to absorb the excess neutrons and reduce the reproduction ratio to unity. These rods 52 are slidably disposed, and may be further inserted into the active portion 10 or withdrawn therefrom the increase or decrease the neutron reproduction ratio of the reactor. The control rods 52 may also be fully inserted into the active portion 10 of the reactor for the purpose of stopping the neutronic chain reaction and shutting down the reactor.

The reactor is provided with access points for irradiat-

4

ting materials. One of these is a duct 57 which extends beneath the active portion 10 of the reactor.

In one particular embodiment of the present invention, the active portion 10 is construced with a graphite moderator 18 to form a cube 25 feet long on each side. The moderator could be constructed of other materials than graphite, any material having suitable chemical and physical properties and a moderating ratio at least as great as that of water being satisfactory, the term moderating ratio being defined by the expression

$$\frac{\sigma_s \xi}{\sigma_c}$$

where σ_s is the neutron scattering cross section for the material, σ_c the neutron capture cross section for the material, and ξ the mean logarithmic energy loss for a collision of a neutron with an atom of the material. The coolant channels 20 are disposed in a rectangular lattice spaced 8 inches from the nearest adjacent channels. The circular area of the coolant channels is 5.6 square inches, and the diameter of natural uranium fissionable bodies 24 disposed within the coolant channels 20 is 1.1 inches. The casing 26 surrounding the bodies 24 of natural uranium is constructed of aluminum, the ratio of the volume of aluminum to the volume of uranium in the reactor active portion 10 being 0.2. The gap 16 at the center of the reactor is 2.75 inches. Thus constructed, it has been found that the 463 coolant channels 20 nearest the center of the active portion 10 of the reactor must be loaded with 37.2 metric tons of uranium in order to make the neutron chain reaction self-sustaining with the control rods 52 totally withdrawn from the active portion 10 of the reactor.

This reactor was also constructed with two types of graphite. The core of the reactor was constructed of 650,000 pounds of the first type of graphite, corresponding to a cube of approximately 18.5 feet on a side, while the remainder of the reactor was constructed of the second type of graphite. The diffusion length for thermal neutrons has been found by test to average 52 centimeters for the first type of graphite and 48 centimeters for the second type of graphite. The thermal neutron diffusion length L is given by the equation

$$L^2 = \frac{\overline{r^2}}{6}$$

where $\overline{r^2}$ is the root mean square distance from source point to absorption point for thermal neutrons in the medium. If the entire reactor were constructed of the first type of graphite, the multiplication factor would be 1.074, as contrasted with 1.057 for the same reactor constructed with the second type of graphite due to the differences in the moderating power of the two types of graphite. The multiplication factor is the ratio of the number of neutrons in one generation to the number of neutrons in the preceding generation in a reactor of infinite size with an active portion identical to the reactor in question. These values for the multiplication factor are attained with the reactor operating with a graphite temperature of about 200° C., a maximum uranium temperature within the reactor of approximately 350° C., and a neutron flux at the center of the reactor of approximately 4×10^{12} neutrons per square centimeter second.

The uranium used in the fuel elements 22 has the isotopic content found in nature and must be purified to remove ordinary impurities, such as common metals and silica found in the uranium in its natural state, and to reduce the impurities which have high neutron capture cross sections to less than 10^{-6} grams per gram of uranium, including such elements as boron, cadmium, indium and some of the rare earths.

Sixteen control rods 52 having square cross sections of 4 square inches each and containing 1.75 percent boron steel have been found to be adequate to control the reactor. While the control rods have been illus-

2,928,781

5

trated as entering the reactor vertically, they could equally well be disposed horizontally or diagonally.

The gap 16 which divides the active portion 10 of the reactor permits the coolant air to be introduced into the center of the active portion 10 where the highest temperatures are experienced. Thus, the maximum heat removal occurs at the point of greatest heat generation, and permits the reactor to operate at higher power levels than would be possible if the air coolant were merely flown through the coolant channels 20 from one end to the other end. Also, by introducing the air through the central gap 16, the pressure drop of the coolant air along the channels 20 has been found to be approximately one-fifth as great as it would be in a similarly constructed reactor without the central gap and in which the coolant air would be flown from one end of the channel to the other end of the channel.

However, the gap 16 also separates the active portion 10 of the reactor on a plane passing through the region of greatest neutron flux. As a result, the neutron reproduction ratio has been appreciably reduced. This is best illustrated by the graph in Figure 6 which shows the decrease in the excess reactivity of the reactor for different width gaps. In order to overcome the decrease in the neutron reproduction ratio caused by the gap at the center of the reactor, it is necessary to increase the physical dimensions of the reactor, as shown in Figure 6.

The man skilled in the art will readily devise many other applications and embodiments of the present invention from a reading of the foregoing specification. For instance, it is clear that the invention is not limited to air cooled neutronic reactors, but could equally well be applied to water cooled or other types of cooled reactors. For this reason, it is intended that the scope of the present invention be limited only by the following claim, and not by the specific disclosure.

What is claimed is:

A neutronic reactor comprising, in combination, a pair

6

of equally dimensioned rectangular graphite blocks confronting each other and forming a cube, said blocks being separated by a gap of approximately 7 centimeters and being provided with 463 annular channels approximately 5.6 square inches in cross section extending from the gap through each of the graphite blocks, the channels within the reactor being spaced by approximately 8 inches in a rectangular lattice, the graphite blocks together forming a cube of approximately 25 feet on a side, a plurality of fuel elements containing cylinders of uranium having a diameter of 1.1 inches and jackets of aluminum, the ratio of the volume of aluminum to the volume of uranium in the active portion of the reactor being approximately 0.2, the 463 cooling channels nearest the center of each of the graphite cubes containing fuel elements containing a total of 37.2 metric tons of uranium, means for introducing air into the gap between the graphite blocks, and means to exhaust the air from the ends of the channels opposite to the gap.

References Cited in the file of this patent

UNITED STATES PATENTS

2,708,656   Fermi et al. _____ May 17, 1955

OTHER REFERENCES

Nucleonics, February 1950, pages 54–60, an article by Ohlinger.

Atomics, February 1951, pages 51–56, an article on "Gleep."

Atomics, June 1951, pages 176–180, an article on "Bepo."

U.S. Atomic Energy Commission, BNL–152, January 1952, I. Kaplan and J. Chernick, pages 6–8, 37, 38, 40, 41, 42, 45, 46, 47, 48, 76, 78, 79. Available from Technical Information Service, Oak Ridge, Tenn.

"Principles of Nuclear Reactor Engineering," by Samuel Glasstone, D. Van Nostrand Co., New York, N.Y., July 1955, pages 797–800.

Feb. 28, 1956   E. P. WIGNER ET AL   2,736,696

REACTOR

Filed Aug. 29, 1945   13 Sheets—Sheet 1

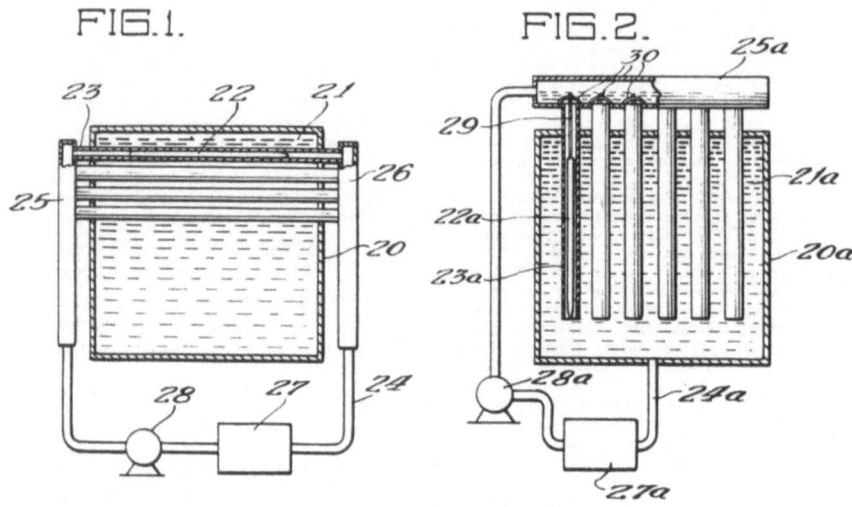

FIG.1.   FIG.2.

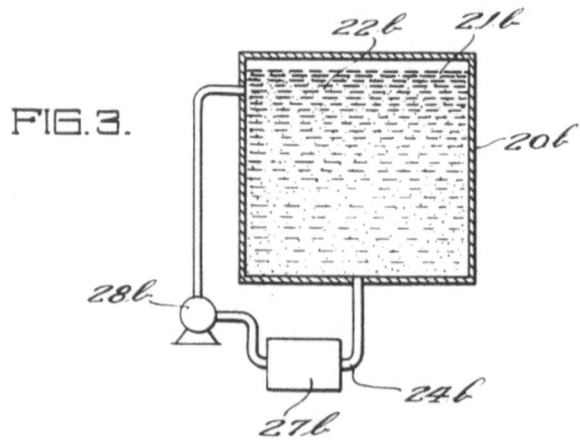

FIG.3.

Inventors:
Eugene P Wigner
Leo A. Ohlinger
Gale Young
Alvin M. Weinberg
By: Robert A. Lavender
        Attorney

Witnesses:
Herbert E. Metcalf
Francis W. Test

Feb. 28, 1956   E. P. WIGNER ET AL   **2,736,696**

REACTOR

Filed Aug. 29, 1945   13 Sheets—Sheet 2

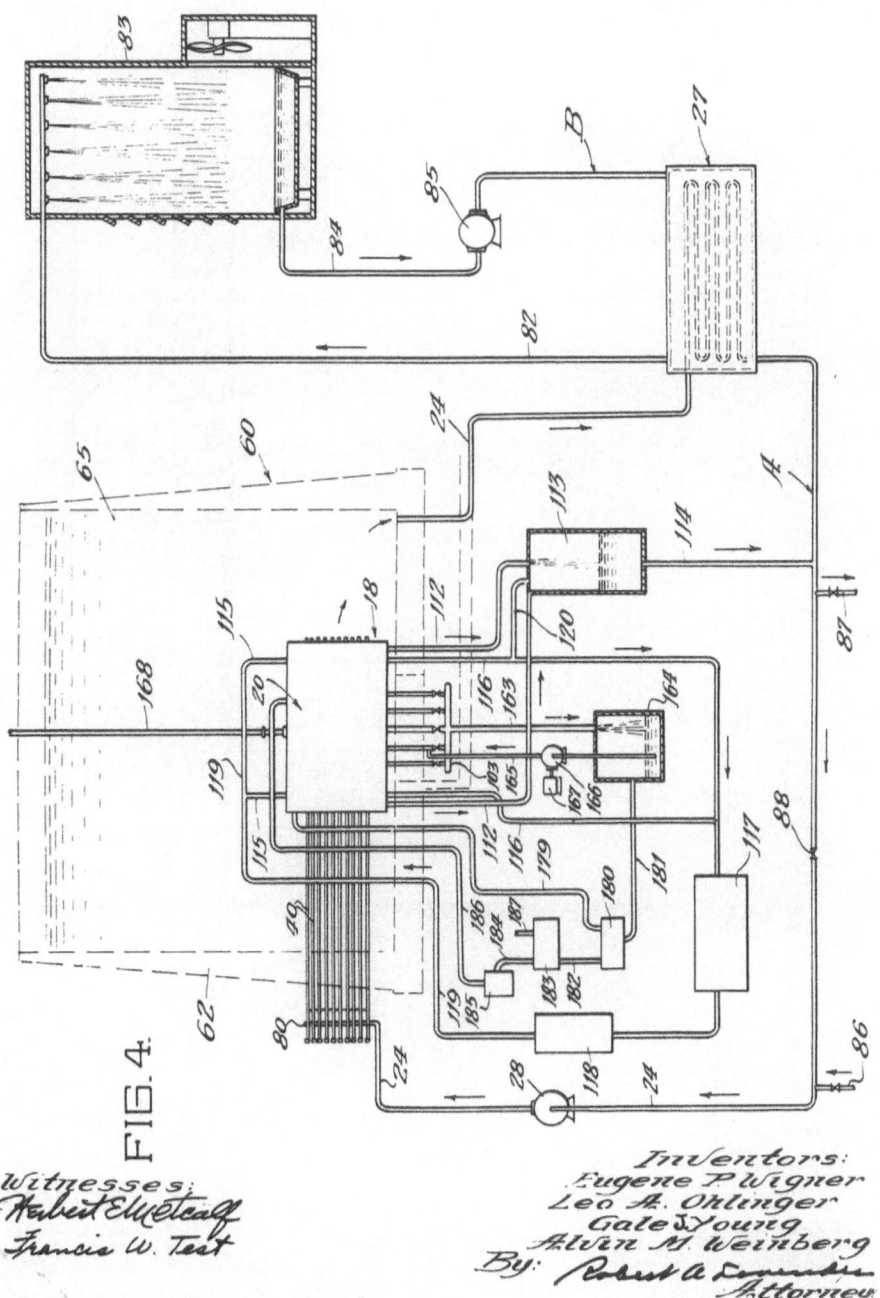

FIG. 4.

Witnesses:
Herbert E. Metcalf
Francis W. Test

Inventors:
Eugene P. Wigner
Leo A. Ohlinger
Gale J. Young
Alvin M. Weinberg
By: Robert A. Commander
    Attorney

Feb. 28, 1956   E. P. WIGNER ET AL   2,736,696

REACTOR

Filed Aug. 29, 1945   13 Sheets—Sheet 3

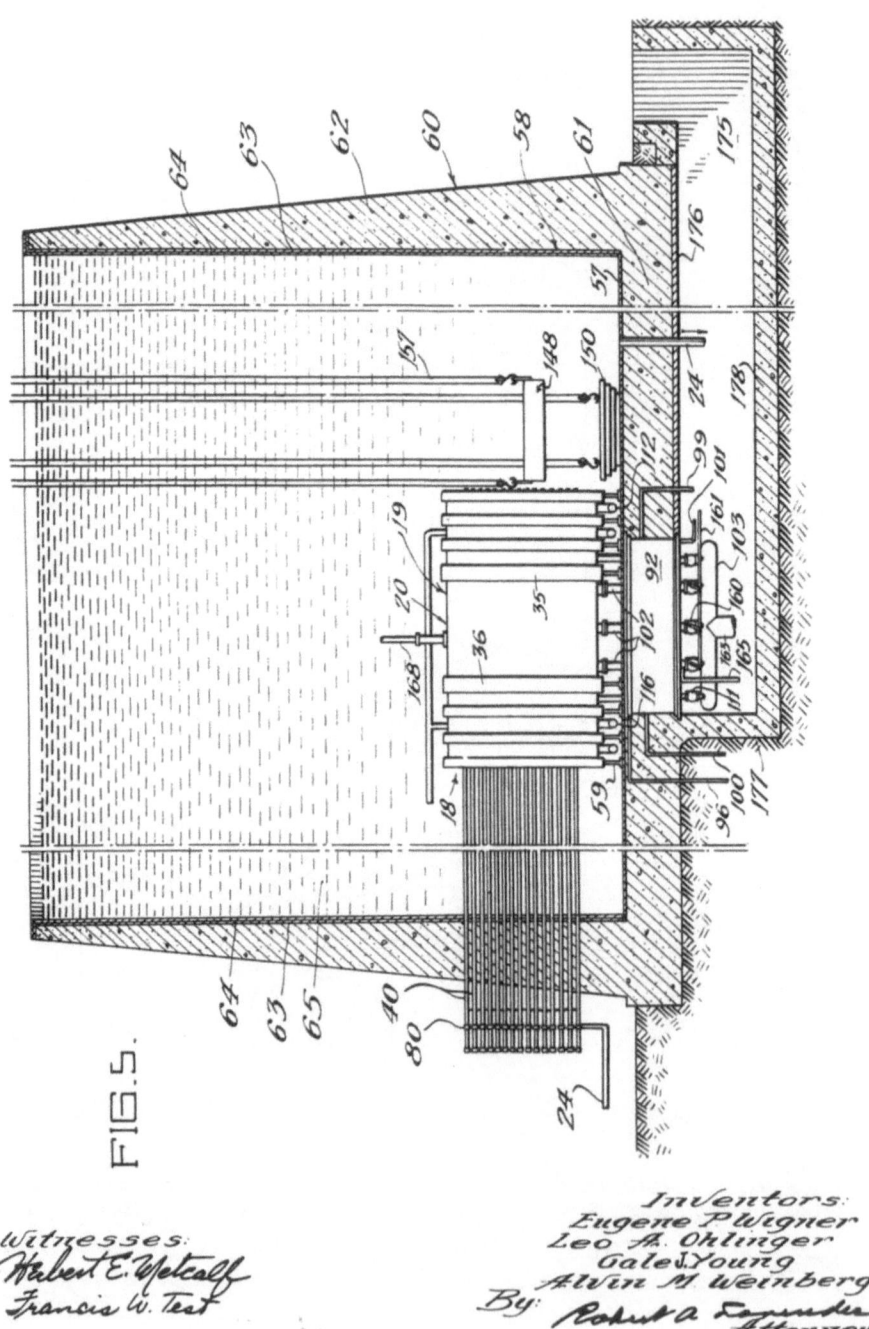

FIG. 5.

Witnesses:
Herbert E. Metcalf
Francis W. Test

Inventors:
Eugene P. Wigner
Leo A. Ohlinger
Gale J. Young
Alvin M. Weinberg
By: Robert A. ___
Attorney.

Feb. 28, 1956   E. P. WIGNER ET AL   **2,736,696**

REACTOR

Filed Aug. 29, 1945   13 Sheets—Sheet 4

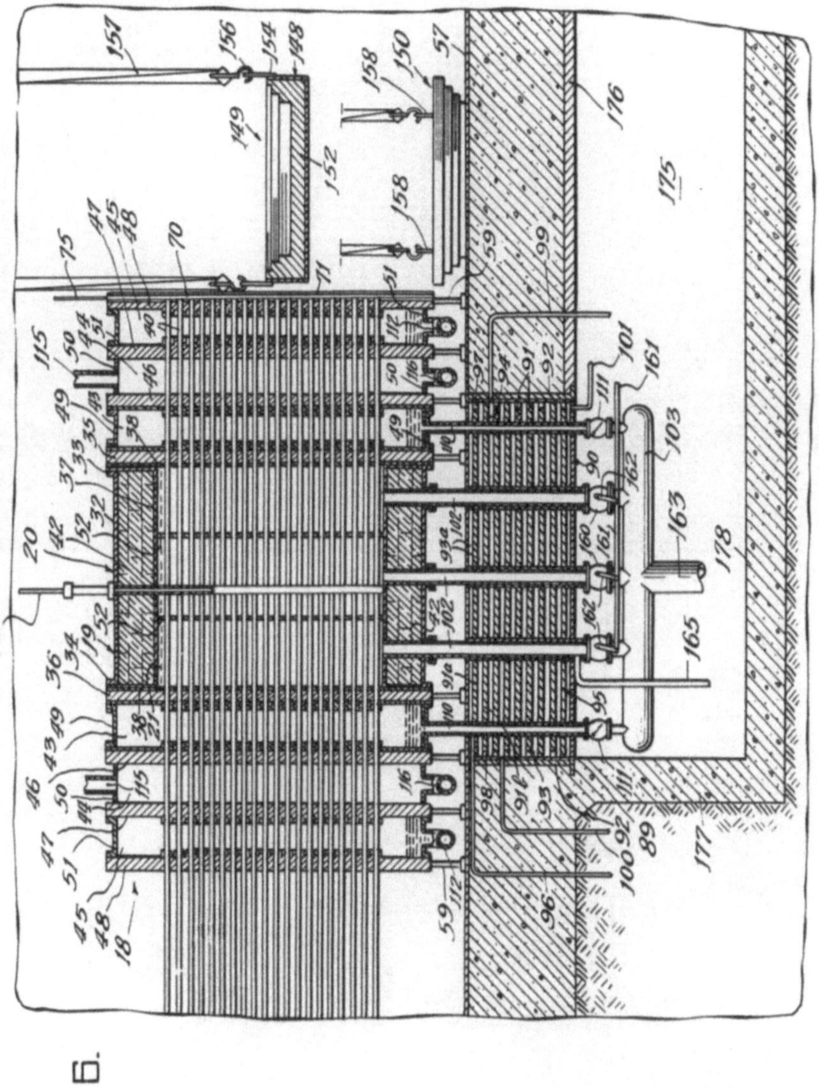

FIG.6.

Witnesses:
Herbert E. Metcalf
Francis W. Test

Inventor
Eugene P.Wigner
Leo A. Ohlinger
Gale J.Young
Alvin M. Weinberg
By: Robert A. Lorenson
Attorney.

Feb. 28, 1956   E. P. WIGNER ET AL   2,736,696

REACTOR

Filed Aug. 29, 1945   13 Sheets—Sheet 5

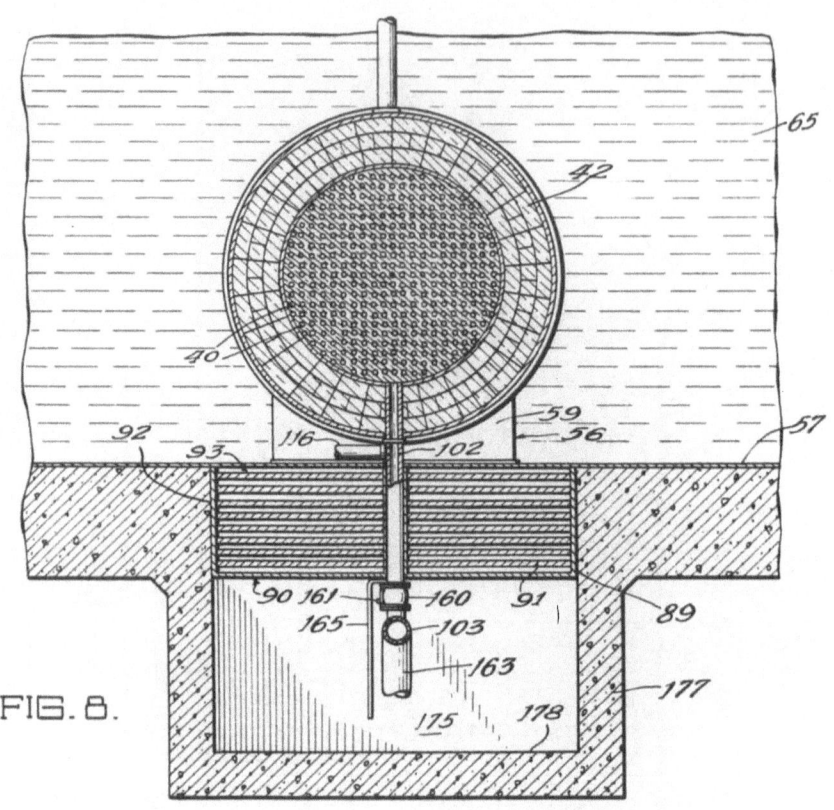

FIG. 8.

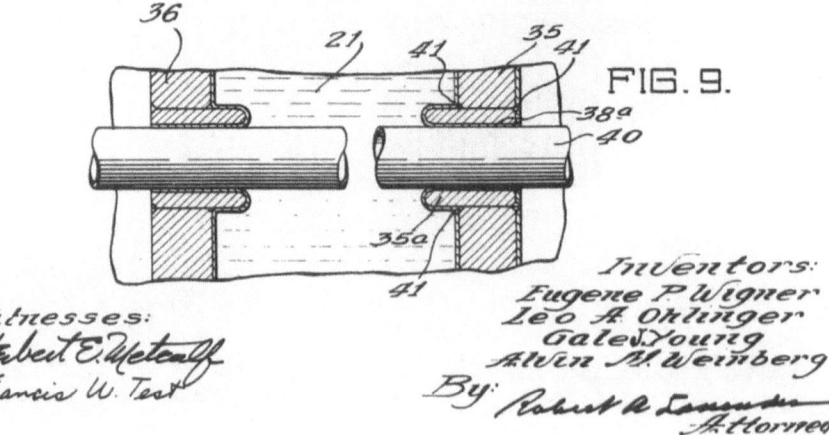

FIG. 9.

Witnesses:
Herbert E. Metcalf
Francis W. Test

Inventors:
Eugene P Wigner
Leo A. Ohlinger
Gale J. Young
Alvin M. Weinberg
By: Robert A. Lamar
Attorney

Feb. 28, 1956   E. P. WIGNER ET AL   2,736,696

REACTOR

Filed Aug. 29, 1945   13 Sheets—Sheet 6

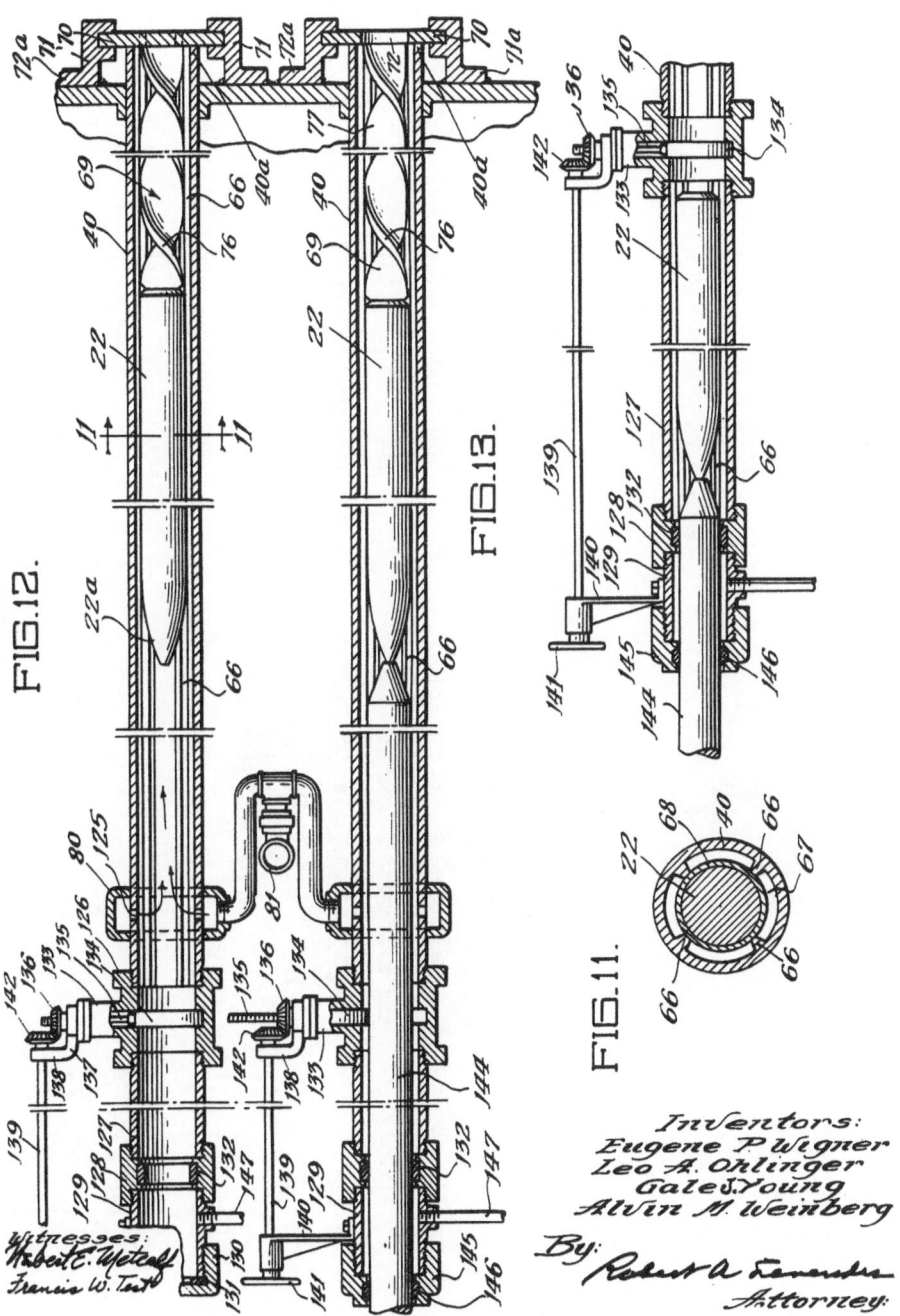

Inventors:
Eugene P. Wigner
Leo A. Ohlinger
Gale J. Young
Alvin M. Weinberg

By
Robert A. Hardester

Attorney

Witnesses:
Hubert E. Metcalf
Francis W. Test

Feb. 28, 1956   E. P. WIGNER ET AL   **2,736,696**

REACTOR

Filed Aug. 29, 1945   13 Sheets—Sheet 7

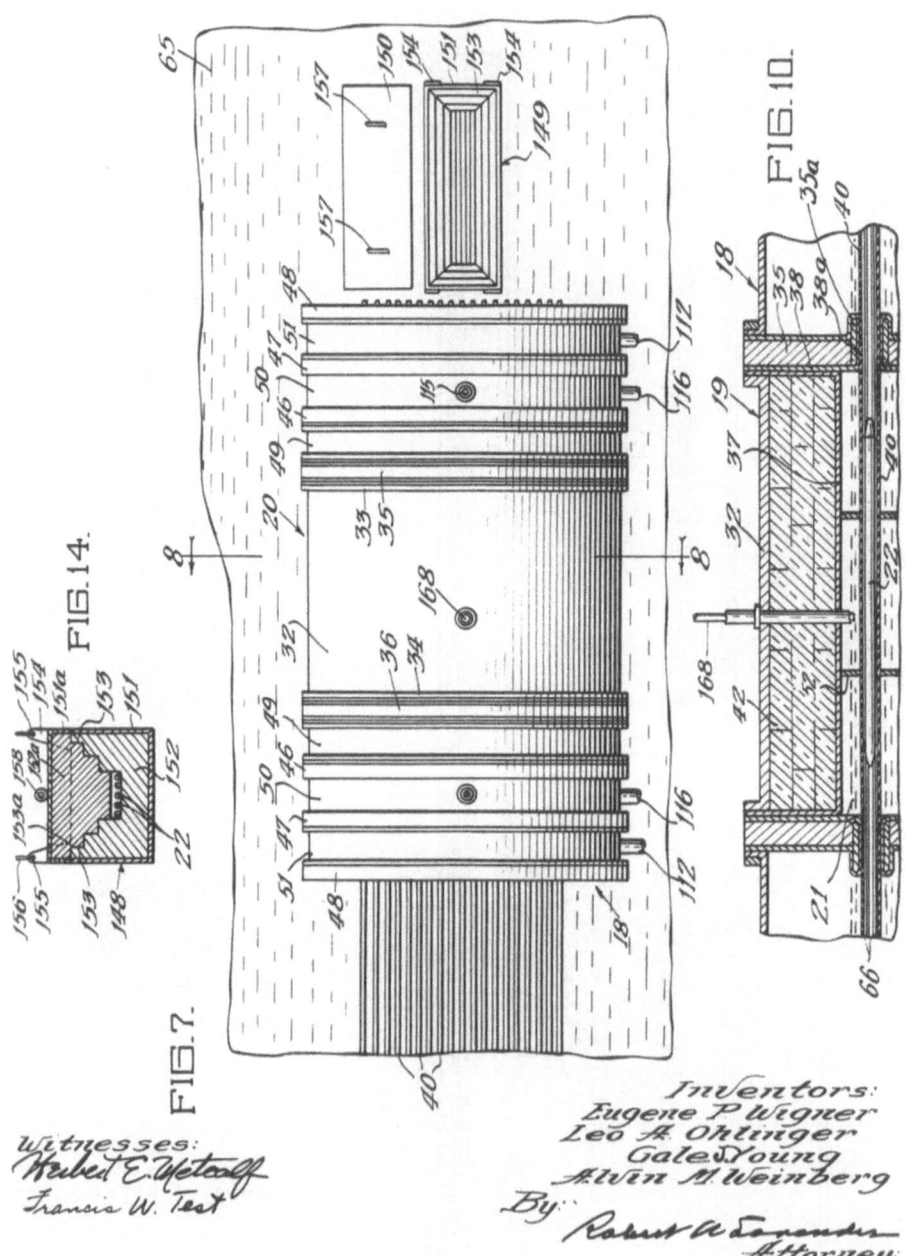

Inventors:
Eugene P. Wigner
Leo A. Ohlinger
Gale Young
Alvin M. Weinberg

By

Robert A. Lorenders
Attorney

Witnesses:
Herbert E. Metcalf
Francis W. Test

Feb. 28, 1956   E. P. WIGNER ET AL   2,736,696

REACTOR

Filed Aug. 29, 1945   13 Sheets—Sheet 8

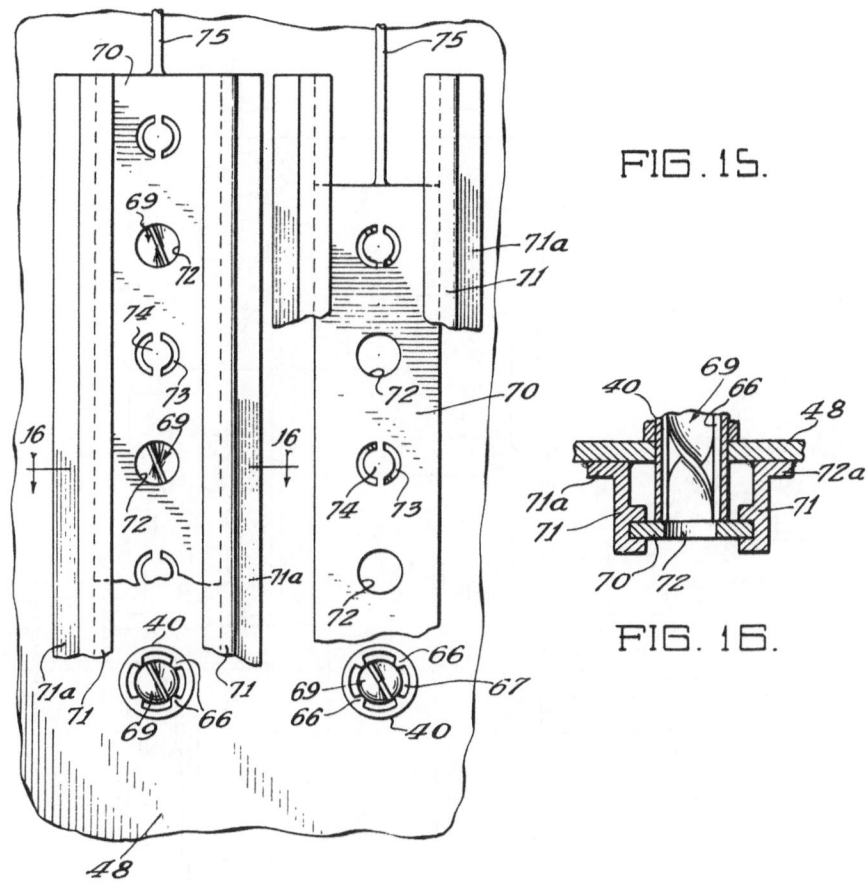

FIG. 15.

FIG. 16.

FIG. 18.

FIG. 17.

Witnesses:
Hubert E. Metcalf
Francis W. Test

Inventors:
Eugene P. Wigner
Leo A. Ohlinger
Gale J. Young
Alvin M. Weinberg
By: Robert A. _____
Attorney.

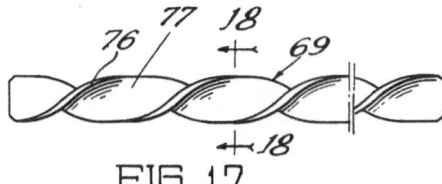

Feb. 28, 1956   E. P. WIGNER ET AL   2,736,696

REACTOR

Filed Aug. 29, 1945   13 Sheets—Sheet 9

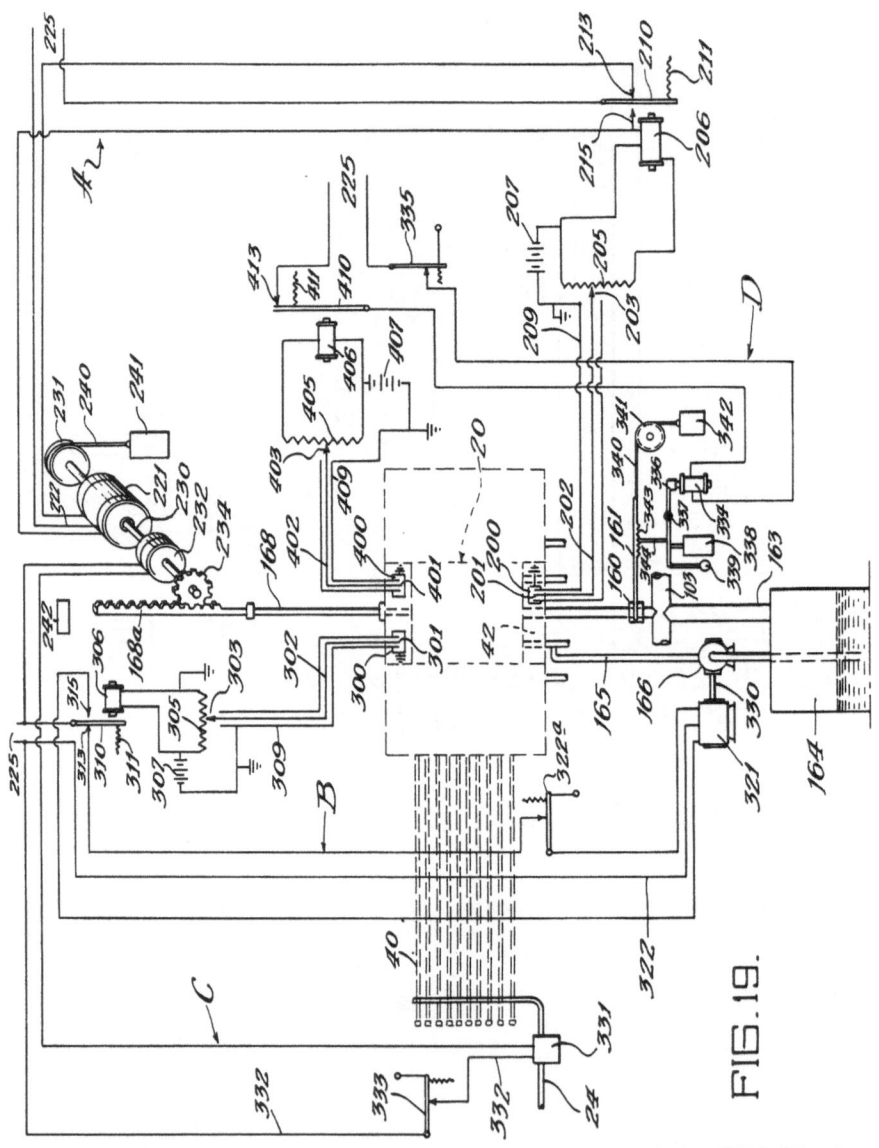

FIG.19.

Witnesses:
Herbert E. Metcalf
Francis W. Test

Inventors:
Eugene P. Wigner
Leo A. Ohlinger
Gale J. Young
Alvin M. Weinberg
By:
Robert A. Fer_____
Attorney

Feb. 28, 1956   E. P. WIGNER ET AL   2,736,696
                                        REACTOR

Filed Aug. 29, 1945   13 Sheets—Sheet 10

FIG.20.

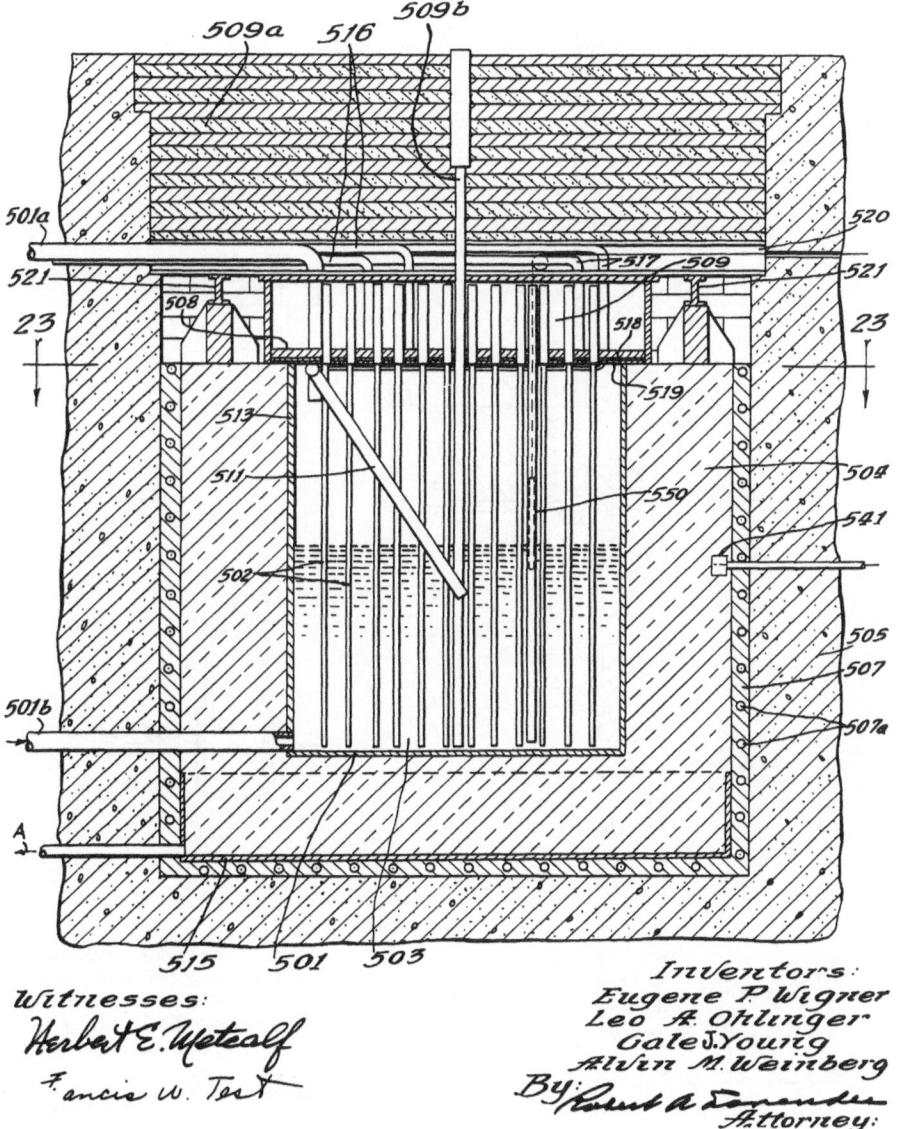

Witnesses:
Herbert E. Metcalf
Francis W. Test

Inventors:
Eugene P. Wigner
Leo A. Ohlinger
Gale J. Young
Alvin M. Weinberg
By Robert A. Levander
Attorney.

Feb. 28, 1956   E. P. WIGNER ET AL   2,736,696

REACTOR

Filed Aug. 29, 1945   13 Sheets—Sheet 11

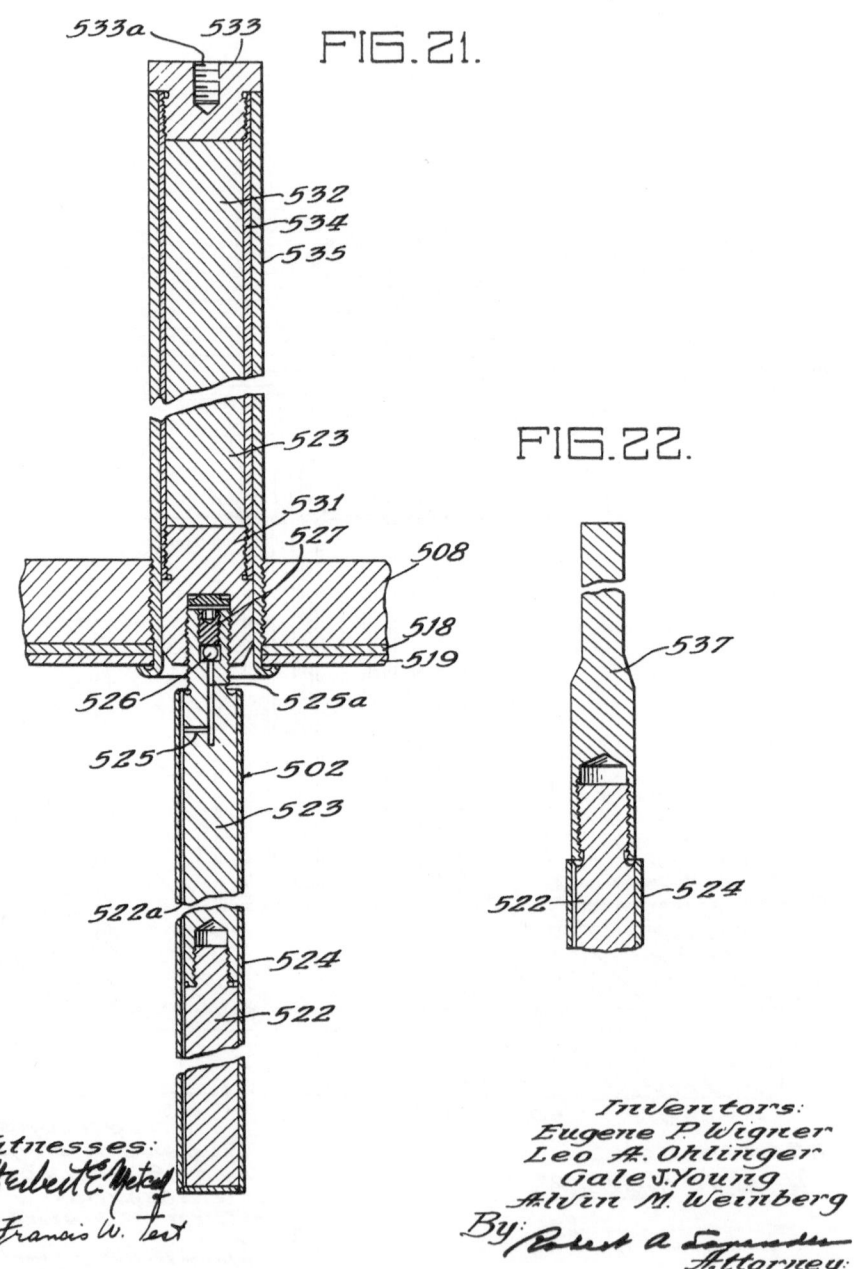

FIG.21.

FIG.22.

Inventors:
Eugene P. Wigner
Leo A. Ohlinger
Gale J. Young
Alvin M. Weinberg
By: Robert A. Spangler
Attorney.

Feb. 28, 1956   E. P. WIGNER ET AL   2,736,696
                                          REACTOR

Filed Aug. 29, 1945   13 Sheets—Sheet 12

FIG.23.

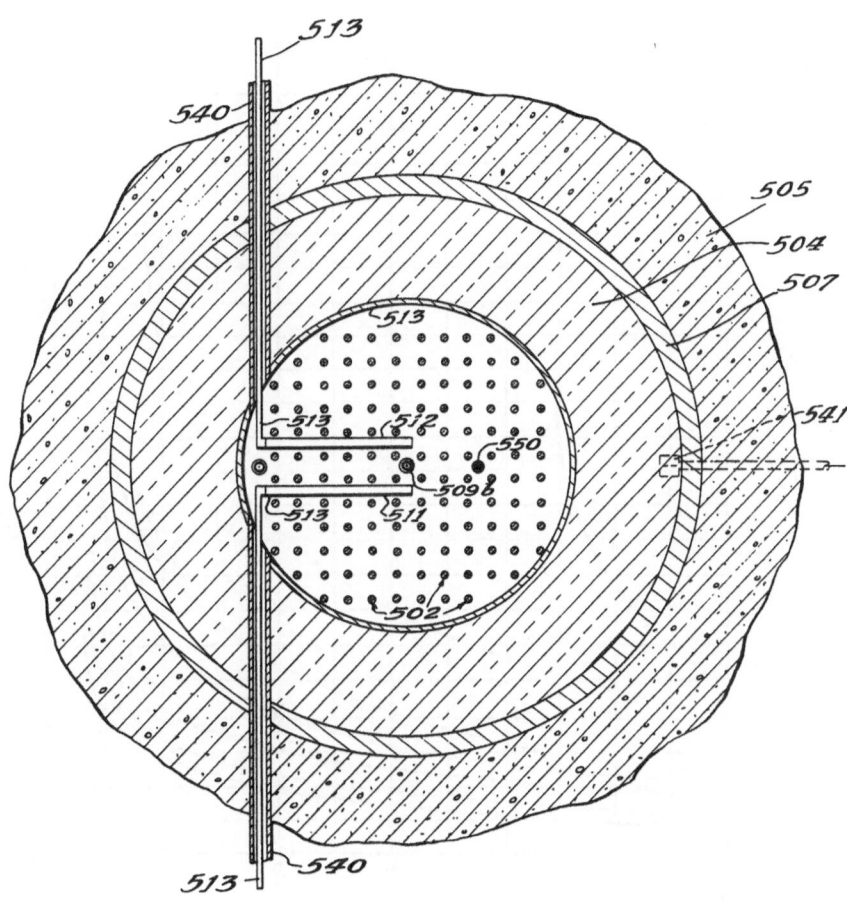

Inventors:
Eugene P. Wigner
Leo A. Ohlinger
Gale J. Young
Alvin M. Weinberg
By: Robert A. Lavender
Attorney.

Feb. 28, 1956   E. P. WIGNER ET AL   2,736,696

REACTOR

Filed Aug. 29, 1945   13 Sheets—Sheet 13

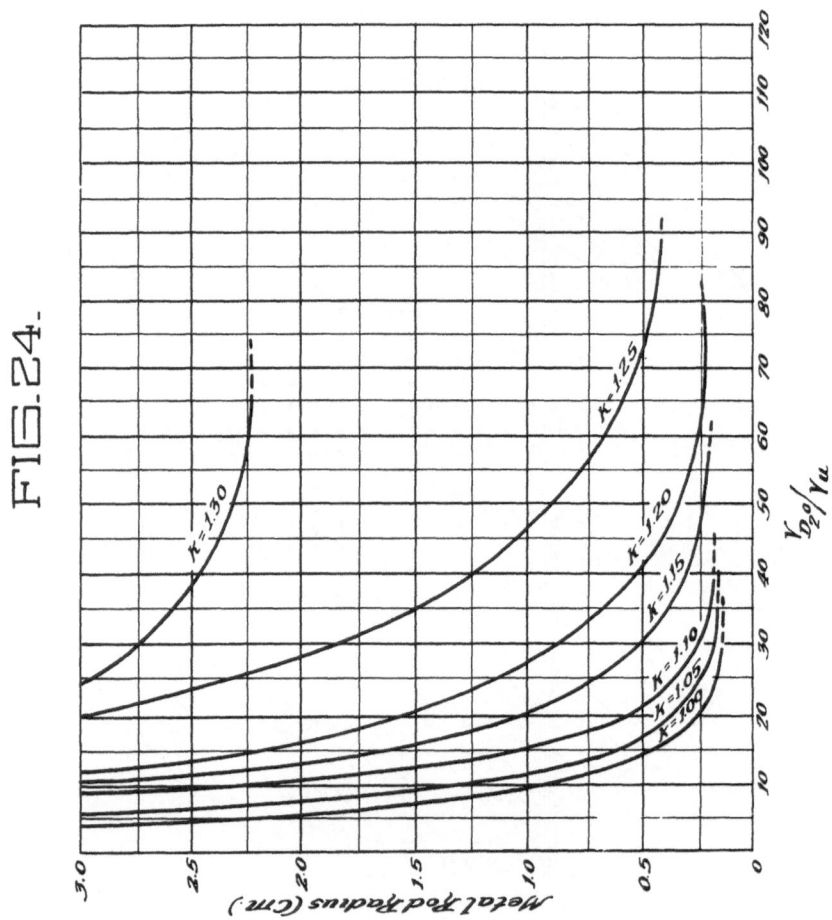

United States Patent Office

2,736,696
Patented Feb. 28, 1956

1

2,736,696

REACTOR

Eugene P. Wigner, Leo A. Ohlinger, Gale J. Young, and Alvin M. Weinberg, Chicago, Ill., assignors to the United States of America as represented by the United States Atomic Energy Commission

Application August 29, 1945, Serial No. 613,355

7 Claims.   (Cl. 204—193)

The present invention relates to the subject of neutronics, and more particularly to a suitably cooled slow neutron chain reacting system, also referred to as a neutronic reactor, or pile, the latter name having been originally adopted for the active portions of systems employing uranium bodies geometrically arranged in graphite in the form of lattice structures. The present invention specifically relates to a neutronic reactor wherein a fissionable material such as the U^{235} content of natural uranium is arranged in a neutron slowing material or moderator of deuterium in some suitable form such as deuterium oxide (D$_2$O), commonly known as heavy water. As a result of the chain reaction, and when U^{238} is present as for example in natural uranium, transuranic element 94^{239}, known as plutonium, is produced. This material is fissionable and is valuable when added to natural uranium for use in a chain reacting system, because it enriches the natural uranium with fissionable material in the system so as to reduce the overall critical size; that is, that size above which the system must be built in order to effect a self-sustaining neutron chain reaction, and below which the system is normally not self-sustaining.

Natural uranium contains both uranium isotopes U^{235} and U^{238} in the ratio of 1 to 139. The U^{235} is the isotope fissionable by slow neutrons, as will be presently explained.

The slow neutron chain reaction system, operating by virtue of nuclear fission produced by absorption of thermal neutrons in uranium is accompanied by the ejection of secondary fast neutrons liberated in a uranium body as the result of fissions of the U^{235} isotope. A small portion of these fast neutrons are absorbed directly by the isotope U^{238} to produce what is termed fast fission of the nuclei of this isotope with the resulting production of a few new fast neutrons. All of these fast neutrons leave the uranium bodies and diffuse through the system; that is, through the heavy water and the uranium. Some of the neutrons escape from the system and are lost. The remaining neutrons, as the result of many elastic collisions with nuclei of the heavy water and uranium, lose energy at each collision and are thereby slowed down. This slowing down process occurs during a diffusion process whereby the neutrons travel in random paths through the heavy water and the uranium bodies, colliding many times with nuclei of the heavy water and the uranium. Some of the neutrons thus may reach a velocity corresponding to a strong so-called resonance energy of uranium. Those neutrons reaching that velocity and contacting a uranium body while at that energy, or those which are slowed to that energy while inside a uranium body, are absorbed in the nuclei of the isotope U^{238} without producing fission. This absorption is known as resonance absorption. As will be explained presently, this resonance absorption of neutrons constitutes one productive source for element 94^{239}, though, in reality, all resonance absorption of neutrons constitutes a loss of neu-

2

trons to the neutronic reaction, since the resonance neutrons absorbed might otherwise be made available to produce new fissions, provided of course that in slowing down to thermal energies the particular neutron was not absorbed in U^{238} at thermal energies. This thermal absorption of neutrons by U^{238} constitutes the other productive source for the element 94^{239}.

Neutrons escaping resonance absorption in uranium continue to diffuse in the moderator and the uranium, and continue to lose energy until they finally reach thermal energy, when they are in thermal equilibrium with their surroundings. The neutrons thus slowed to thermal energy may continue to diffuse in the moderator in random paths, but without losing further energy. Some of these thermal neutrons escape from the system and are lost. A very small proportion of the neutrons while thus diffusing at thermal velocity are absorbed by the heavy water. Others are absorbed by impurities in the slowing material and the uranium. All of these absorptions constitute losses of neutrons to the neutronic reaction.

The thermal neutrons remaining to enter the uranium bodies are absorbed either in the nuclei of the isotope U^{235} to produce fission or in the isotope U^{238} leading to the production of 94^{239}.

When fission occurs in the U^{235} isotope, the reaction which takes place may be represented by the following:

$$92U^{235} + \text{neutron} \rightarrow A + B + \text{about 2 neutrons (average)}$$

where "A" represents "light" fission fragments having atomic masses ranging from 83 to 99 inclusive and atomic numbers from 34 to 45 inclusive; for example, Br, Kr, Rb, Sr, Y, Zr, Cb, Mo, Ma, Ru, and Rh; and "B" represents "heavy" fission fragments having atomic masses ranging from 127 to 141 inclusive, and atomic numbers from 51 to 60 inclusive; for example, Sb, Te, I, Xe, Cs, Ba, La, Ce, Pr, and Nd.

The elements resulting from the fissions appear in general to be unstable and radioactive, with half-lives varying in length in accordance with the element formed.

The absorption of thermal or resonance neutrons by the U^{238} isotope gives rise to the conversion of U^{238} to U^{239} which ultimately decays to transuranic element 94^{239}. The reaction is as follows:

$$92U^{238} + n \longrightarrow 92U^{239} \quad \text{[plus 6 Mev of } \gamma \text{ rays, not necessarily all of one frequency}$$
$$92U^{239} \xrightarrow{\ 23\ \text{min.}\ } 93^{239} + \beta^{-} \quad \text{[1 Mev } \beta^{-}\text{, no } \gamma \text{ rays}$$
$$93^{239} \xrightarrow{\ 2.3\ \text{days}\ } 94^{239} + \beta^{-} \quad \text{[600 kv. upper } \beta^{-} \text{ energy limit. Also 2 } \gamma \text{ rays, 400 kw., and 270 kv., about } \frac{1}{2} \text{ of which are converted to electrons.}$$

The new fast neutrons resulting from the fissions in the isotope U^{235} pass through the same neutronic cycle as just described, there being a certain proportion that will produce fast fission, some that will be lost to the chain reaction, and others that will reach thermal energy and be absorbed in the uranium.

Deuterium has an extremely low neutron absorption characteristic for thermal neutrons; i. e., thermal neutrons can diffuse in the deuterium with relatively little danger of being absorbed by deuterium nuclei. For this reason heavy water is an ideal moderator.

Of course, any neutron absorbing impurities in the heavy water will increase the probability of parasitic capture by such impurities.

By arranging the uranium in bodies or masses of suitable shape and size and selecting the correct volume ratio of uranium to heavy water, and, further, by suitably limiting the impurities in the uranium and the heavy water, and by limiting the escape of neutrons from the system (i. e., the percentage of neutrons escaping) by making the active portion of the system sufficiently large in size, it is

2,736,696

3

possible to produce in each generation more fast neutrons by fission than were present to start the chain so that the chain reaction in the system is perpetuated. The ratio between these amounts is known as the reproduction ratio and for an operative reactor is greater than unity.

After all of the neutron losses that enter into the chain reaction have been evaluated for a specific lattice, with the exception of the loss by leakage from the exterior of the system, the size to which the system is built to effect a self-sustaining chain reaction has to be determined. The size at which the neutrons gained equal the neutrons lost is known as the critical size. The operating size is somewhat larger than the critical size.

There are several ways by which critical and operating sizes are determined, and it is desirable that these sizes be found within a low margin of error prior to the actual building of the full size reactor so that auxiliary equipment such as shields, for example, can be constructed of proper size and not too small to enclose the operating reactor.

One very satisfactory method of determining critical size is to measure the Laplacian (Δ) in an exponential pile, i. e., a structure that is similar in all respects to the full size reactor contemplated, but is considerably smaller than that size required to make the reaction self-sustaining. A detailed discussion of how this is done can be found in the United States Patent of Enrico Fermi et al., No. 2,708,656. In case the reactor is to be built in the form of a spherical structure employing uranium bodies of any shape or size imbedded in a heavy water (D_2O) moderator, the following formula gives the critical overall radius:

$$R = \frac{56.5}{\sqrt{K-r}}$$

where R is the radius in centimeters and r for critical size equals one.

For a rectangular parallelopiped structure rather than spherical, using a heavy water moderator, the critical size can be computed from the formula:

$$K - r = 323\pi^2 \left(\frac{1}{a^2} + \frac{1}{b^2} + \frac{1}{c^2} \right)$$

where a, b, and c are the lengths of the sides in centimeters.

The critical size for a cylindrical structure using a heavy water moderator is given, irrespective of the shape of the uranium bodies by the formula:

$$K - r = \frac{323\pi^2}{H^2} + \frac{777}{R^2}$$

where H is the height in centimeters and R is the radius in centimeters. Operating sizes are also determinable from these formulae by inserting the operating value of r desired.

The following table will show the approximate critical sizes for operative reactors, utilizing D_2O, for different values of K where K is changed due to change in geometry. In evaluating the values given in the table it must be kept in mind that the critical size is that size where the chain reaction just will become self-sustaining.

U Metal-heavy water, K-1	Sphere Radius	Critical Sizes in Feet, Cube Side
.01	18.5	32
.02	12.3	21.5
.05	7.7	13.4
.1	5.45	9.45
.2	3.84	6.65
.3	3.14	5.45

The use of a reflector, through the action of scattering neutrons back into the reactor, raises the density of thermal neutrons throughout the reactor and this, in turn,

4

increases the reproduction ratio of the reactor. Thus critical size ($r=1$) with a reflector is smaller than without a reflector. Consequently the use of a reflector permits a smaller reactor to be constructed for the same K factor.

Since the neutronic reactor, when it is built to operating size, is then capable of producing fast neutrons at a greater rate than neutrons are lost, there would be an exponential rise in the neutron density to infinity (at least in theory) as the system is operated unless the density rise is controlled. This is accomplished by controlling the ratio between the neutrons lost from the chain reaction to those gained from new fissions. This can be accomplished in a variety of ways, some of which will be hereinafter explained. The point at which the exponential rise is stabilized is a matter of choice and will depend on the desired output of element 94^{239} as well as on considerations involving safety, and ability to remove heat from the reactor.

In order to effect a self-sustaining neutronic reaction at all in a system employing uranium and heavy water, a certain minimum amount of heavy water moderator is required. This minimum quantity is about five tons, provided an optimum arrangement of uranium bodies is used.

Referring to Fig. 24, K values for various reactor arrangements are given for metal rod geometry. In other words it can be determined from these contour curves what value of K can be expected for a given geometry specified in terms of uranium metal rods of a specific radius disposed in D_2O with a given volume ratio between the D_2O and the uranium. These K values are determined in accordance with the method disclosed in column 27, line 48 through column 30, line 62 of U. S. Patent No. 2,708,656 of Fermi et al.

It has been found that if the D_2O has any light water (H_2O) in solution, the light water will pick up an extra neutron as a result of the neutron bombardment taking place during the neutronic reaction, transforming the light water into D_2O. Thus, as a result of the neutronic reaction the D_2O moderator will become more pure resulting in a slight increase in the K factor. This factor should be considered in the design of the reactor and can be taken care of by an adjustment in the balanced or equilibrium position of the controls.

The rate of production of element 94^{239} will depend on the rate of absorption of neutrons by U^{238}, which is proportional to the rate at which fissions occur in the U^{235}. This, in turn, is governed by the thermal neutron density in the active part of the reactor. Thus, for maximum production of element 94^{239} for a given system, it is essential that the thermal neutron density be at a maximum.

Considerable heat is generated during the chain reaction, primarily as a result of the fission process. When the system is operated for an extended period of time at a high production output of element 94^{239}, the large amount of heat thus generated must be removed in order to stabilize the chain reaction.

Most of the heat in an operating device is generated as the result of the nuclear fissions taking place in the U^{235} isotope. Thus the rate of heat generation is largely proportional to the rate at which the fissions take place. In other words, as the rate of generation of neutrons is increased, a greater amount of coolant must be passed through the reactor in order to remove the additional heat thus generated. Thus the highest attainable neutron density at which a system can be operated for an extended period of time is limited by the rate at which the generated heat can be removed. That is to say, the maximum power, or element 94^{239}, output of a system is limited by the capacity of the cooling system. An effective cooling system is therefore a primary requirement for high power operation of a neutronic reactor.

Heavy water is by far the most efficient moderator

5

presently known. The outstanding quality of the deuterium nucleus for slowing down neutrons arises first from its low capture cross section for thermal neutrons and second from its small mass which results in large reduction of neutron energy at each collision. In the latter respect it is surpassed only by hydrogen (of mass 1). A neutron requires about 35 per cent more collisions with deuterium, on the average, to undergo a given energy reduction, than with hydrogen, but only about one fifth as many with deuterium as with carbon, and about one fourth as many as with beryllium. The thermal neutron capture cross section of deuterium is much less than that of other suitable light nuclei, being about 5 times smaller than that of carbon, 15 times smaller than that of beryllium, and 500 times smaller than that of hydrogen.

Among deuterium compounds, heavy water is outstanding as a moderator because oxygen has also a very low capture cross section, the number of deuterium nuclei per unit volume is high in heavy water, it has acceptable chemical properties, and it has low viscosity and high specific heat, whereby the heat generated in it by impact of fast neutrons is easily conveyed to the cooling system when the heavy water is used in a cooled neutronic reactor.

The net gain in neutrons provided by the low neutron absorption in the heavy water may be used up, for example, by the insertion of more parasitic neutron absorbers into the system than could be tolerated with other and less efficient moderators. As high power output is clearly desirable, the parasitic neutron absorbers may take the form of cooling tubes and coolant volume. As more coolant can be tolerated in a heavy water system, high powers can be attained without boiling the moderator, and the system can still be made smaller than, for example, a uranium-graphite system of equal power.

It is an object of the present invention, then to provide a chain reacting system having a neutron moderator characterized by very high slowing capability while at the same time having very low neutron absorption so that neutron losses in the slowing medium are reduced to a minimum.

It is another object to combine a neutron moderator with uranium bearing material such that a controllable self-sustaining nuclear chain fission reaction may be obtained with resultant regulated production of neutrons, liberation of heat, the production of radioactive fission products and new elements, both radioactive and stable, caused by the absorption of neutrons.

Heavy water, of course, is a liquid and, therefore, must be contained in a tank. This tank may serve as the reactor within which the neutronic reaction takes place. The heat generated during the reaction may be removed by flowing a coolant through the tank necessitating the use of a plurality of tubes in the tank. These tubes pass through the tank walls and are connected to a circulating system.

During the neutronic reaction the tubes inside the tank and the tank walls become highly radioactive with the result that the maintenance of conventional packing glands around the tubes where they pass through the tank walls is most impractical, if not impossible. It is important, however, that provisions be made to prevent the loss of heavy water from the system and obviously the greatest possibility for this loss to occur is through the perforations in the tank walls. Heavy water as it is produced today is expensive and for that reason even slight losses cannot be tolerated.

Thus it is one of the objects of the present invention to provide a permanently operative neutronic reactor of the heavy water type whereby loss of heavy water from the system is prevented. This is made possible without requiring constant maintenance of parts closely associated with the highly radioactive portions of the system.

6

Among other objects of the present invention are:

To provide a neutronic system employing a fissionable material as the source of fast neutrons and an efficient fluid moderator capable of slowing fast neutrons to thermal energy;

To provide a neutronic system employing a fissionable material as the source of fast neutrons and heavy water as the moderator ;

To provide such a system wherein during its operation for extended periods of time at high neutron densities the temperature is stabilized;

To provide means for preventing losses of the fluid moderator due to leakage from the reactor or contamination by light water;

To provide a neutronic system from which the active fissionable material, directly following extended periods of operation, may be removed from the reactor and replaced;

To provide an effective emergency control for a neutronic system whereby the overall size of the active part of the reactor is rapidly reduced when required to stop the chain reaction; and

To provide a neutronic system having a high power output and one that is safeguarded in the event of failure of the cooling system by having automatic means for continuing proper cooling of the active fissionable material in the system promptly following the failure of the original cooling system.

The foregoing constitute some of the principal objects and advantages of the present invention, others of which will become apparent from the following description read in conjunction with the drawings, in which

Fig. 1 is a diagrammatic view of a preferred form of the invention showing a uranium-heavy water system cooled by means of a coolant circulated through tubes disposed in the active part of the reactor;

Fig. 2 is a diagrammatic view of one modified form of the invention, showing a uranium-heavy water system cooled by circulating the moderator outside the reactor through a heat exchanger and then returning the cooled moderator to the reactor;

Fig. 3 is a diagrammatic view of a second modified form of the invention showing the uranium and heavy water in the form of a slurry cooled by circulation of the slurry outside the reactor over a heat exchanger and then returning it to the reactor;

Fig. 4 is a schematic view of the reactor enclosed in a water and concrete shield showing the light water cooling circuit, the helium circuit, and the heavy and light water leakage circuits;

Fig. 5 is an enlarged side elevational view of the reactor surrounded by the light water and concrete shield (in section) and showing the relationship between the reactor and the dumping coffin;

Fig. 6 is an enlarged vertical longitudinal sectional view taken through the reactor and coffin, shown partially in elevation, and showing the steel and water shield below the reactor and the mechanism employed for effecting an emergency discharge of the moderator from the system;

Fig. 7 is a top plan view of the reactor shown in Fig. 6;

Fig. 8 is a transverse sectional view taken on line 8—8 of Fig. 7, and shown partially in elevation;

Fig. 9 is an enlarged fragmentary detailed sectional view through the tube sheets at both sides of the reactor showing in elevation one cooling tube passing through said sheets;

Fig. 10 is an enlarged fragmentary longitudinal view shown partially in side elevation and partially in section through the top portion of the reactor showing the graphite shield surrounding the reactor and further illustrating one uranium rod disposed in one of the cooling tubes;

Fig. 11 is an enlarged vertical sectional view taken on the line 11—11 of Fig. 12;

Fig. 12 is an enlarged, fragmentary, horizontal view shown partially in side elevation and partially in section

7

through two of the cooling tubes showing primarily the inlet header and further showing in the upper tube a uranium rod in its normal position in the reactor with the header valve closed and the cooling water circulating while in the lower tube showing the header valve open and the plunger in position to remove the uranium rod from the reactor;

Fig. 13 is a fragmentary, horizontal view shown partially in elevation and partially in section and corresponding to a portion of Fig. 12 but showing a uranium rod in position in front of the header valve and about to be charged into the system by the plunger;

Fig. 14 is an enlarged, vertical, transverse sectional view through the coffin shown in Fig. 7, but illustrating uranium rods in place in the coffin with the lid closed;

Fig. 15 is an enlarged, fragmentary, front elevational view of two holding plates, one shown in the discharge position for removal of the uranium rods from the reactor, and the other plate shown in the rod holding position;

Fig. 16 is a horizontal sectional view taken on the line 16—16 of Fig. 15 and showing in plan view the dummy rod;

Fig. 17 is an enlarged fragmentary plan view of a dummy rod;

Fig. 18 is a transverse sectional view of the dummy rod shown in Fig. 17 and taken on the line 18—18 of Fig. 17;

Fig. 19 is a schematic line diagram showing an electrically operated control circuit for the system, the electrical circuit being reduced to lowest terms for illustrative purposes;

Fig. 20 is a vertical sectional view of a modified neutronic reactor having deuterium oxide as the moderator and vertically disposed uranium rods, the outer portions of the shield being broken away;

Fig. 21 is an enlarged fragmentary vertical sectional view through a portion of the reactor of Fig. 20, showing in particular details of a uranium rod;

Fig. 22 is an enlarged vertical sectional view of a portion of a uranium rod equipped with an attached adapter for removing the uranium rod from the reactor;

Fig. 23 is a horizontal sectional view taken on the line 23—23 of Fig. 20; and

Fig. 24 is a graph showing K contour lines for uranium metal rods immersed in D_2O.

Referring to Figs. 1 to 3, inclusive, of the drawings three embodiments of the invention are diagrammatically shown. In the embodiment illustrated in Fig. 1, the heat is removed from the active part 20 of the reactor by means of a separate coolant passed through the reactor in suitable tubes. In Fig. 2, a system is shown wherein the heavy water moderator is circulated out of the reactor, passed through a heat exchanger where the heat is extracted, is then passed over the uranium bodies and finally is returned to the reactor. The heavy water may be returned direct to the reactor after cooling and not over the uranium bodies. Fig. 3 shows a third species of the invention wherein the uranium is in the form of a slurry of small particles held in suspension in the heavy water moderator. In this last arrangement, both the uranium and the heavy water moderator are circulated out of the reactor, passed through heat exchangers for extraction of heat and then are returned to the reactor.

Referring again to Fig. 1, the active portion 20 of the reactor may be termed the reactor tank and contains a quantity of heavy water 21. The cooling tubes 23 may be made of aluminum and extend through the reactor tank 20. The uranium is shown at 22 in the form of horizontally disposed rods contained in the tubes 23. The coolant, passing through the tubes 23, flows in heat exchange relationship with the uranium rods 22 and is circulated in a closed system comprising piping 24, inlet and discharge headers 25 and 26 respectively, a heat exchanger 27 for extracting the heat from the coolant and a pump 28.

8

In Figs. 2 and 3, the parts of the devices corresponding to those shown in Fig. 1 bear identical reference characters followed by the letter "a" in the case of Fig. 2, or "b" in Fig. 3.

In Fig. 2, the uranium 22a is in the form of vertical rods suspended in tubes 23a disposed vertically in the moderator 21a. Conventional means such as cables 29 suspended from spiders 30 supported in header 25a hold the uranium rods 22 in place. The heavy water 21a is withdrawn from the reactor 20a through piping 24a, is passed through a heat exchanger 27a, and is then pumped back into the reactor by pump 28a through tubes 23a. The coolant is thus circulated in heat exchange relationship with the uranium 22a. The tubes 23a are open at the bottom to discharge the heavy water into the reactor tank 20a.

In Fig. 3, the uranium 22b is in the form of small particles held in suspension in the heavy water 21b. Both the uranium and the heavy water as a slurry, are circulated through pipe 24b, heat exchanger 27b, and are returned to the reactor tank 20b by pump 28b.

As a detailed specific example of a suitably cooled uranium heavy water system, the embodiment diagrammatically shown in Figs. 4 to 19 inclusive is selected as a preferred embodiment. As coolants for this type of system, light water and diphenyl, for example, are two satisfactory liquids but there must be a proper relationship between the uranium, the heavy water, the tubes for circulating the light water or diphenyl through the reactor, and the light water or diphenyl in the tubes. This relationship must be such that the reproduction ratio is greater than unity; that is, the ratio between the neutrons gained to those lost to the chain reaction. Heavy water itself may be used as the coolant in this embodiment. This is particularly desirable where extremely high power outputs are sought, because the low neutron absorption characteristic of this material permits the presence of a very large quantity of this coolant in the active part of the reactor without reducing the reproduction ratio of the system below unity.

In the present embodiment, light water is selected as the coolant and it is circulated through the reactor in aluminum tubes. Beryllium is also suitable as tube material, and may be more desirable than aluminum in at least two respects; that is, it has a lower absorption for thermal neutrons, and in itself is effective as a neutron slowing material.

As an example of an operative system, an output of about 50,000 kilowatts, the electrical equivalent of the heat energy removed from the reactor, can be realized in a system employing about ten tons of heavy water and six tons of uranium metal in the form of rods. This presupposes that a peripheral neutron reflector is used and for illustrative puposes a graphite reflector has been selected. A considerable number of neutrons that might otherwise be lost by escaping from the active portion of the system are reflected back into the latter by the scattering or reflecting effect of the neutron-reflecting material, and thus a small reduction in overall size for the active part of the reactor can be effected by using a reflector. About 80 tons of graphite are required and such a system is illustrated in the drawings and the description to follow. Cooling water is circulated through the system at the rate of 13,000 gallons per minute.

It is recognized, of course, that other arrangements can be employed than those shown and the uranium may be in one of a variety of forms. For example, the uranium may vary in size and shape from small particles to larger lumps or bodies in any convenient shape, such as spheres, tubes, or rods. The uranium may be in the form of metal, or it may be in a compound such as UO_2 or U_3O_8. Unless otherwise specified reference to uranium bodies is intended generically to mean bodies including the metal or its compounds. If desired, the natural uranium in such bodies may be enriched as to its isotope U^{235} content, or

2,736,696

9

with other fissionable material, such, for example, as 94^{239}.

Again for illustrative purposes, the uranium rods are shown disposed horizontally in the heavy water moderator. They may be disposed vertically or even at some inclined position with equally satisfactory results.

Since heavy water has a low absorption characteristic for thermal neutrons, the actual spacing of the uranium rods in this moderator is not as critical as in the case where a moderator such as graphite is used. In graphite, for example, because of the probability of its absorbing thermal neutrons, it is important that the uranium bodies be so arranged that neutrons have a high probability of entering a uranium body promptly after being slowed to thermal energy. With heavy water, however, thermal neutrons may diffuse larger distances in the moderator without being absorbed by the heavy water. Thus, generally speaking there is a greater range of body spacings that can be tolerated in a heavy water system than in one wherein graphite is employed as the moderator.

Referring now to Figs. 4, 6, 8 and 10, the neutronic reactor is shown at 18 and consists essentially of an elongated horizontally disposed cylindrical tank or shell. The center portion 19 of the reactor is the active part of the system wherein the neutronic reaction takes place and comprises a cylindrical tank 20 containing a quantity of heavy water 21. This tank 20 may be referred to as the reactor tank.

The reactor tank 20 is a horizontally disposed cylindrical aluminum lined, steel tank about 8 feet in diameter and 9 feet long containing 10 tons of heavy water 21.

In all, there are approximately 460 rods in the reactor arranged in a square geometry as shown in Fig. 8, with the rods spaced 4 inches apart center to center. Each rod is 2 centimeters in diameter and 7 feet long, and in this example comprises natural uranium metal, as explained below.

The side wall or shell of the cylindrical tank 20 is shown at 32 in Fig. 6 and is made of steel and is flanged at its opposite ends at 33 and 34. To these flanges are secured steel end walls 35 and 36 respectively, which also serve as tube sheets as will presently be explained. The tank 20 is lined on its inner face with aluminum lining 37, on the sides or shell of the tank, and 38, on the end walls or tube sheets 35 and 36, as corrosion of aluminum by heavy water is much less than corrosion of steel, for example.

The uranium rods 22 are disposed in tubes 40 (see Fig. 10) which in turn extend through the end walls of the tank 20 and constitute circulating tubes through which the light water may be passed for stabilizing the temperature of the chain reaction as will be brought out hereinafter. The end walls or tube sheets 35 and 36 are perforated to accommodate the tubes 40. The tube sheets are supporting members for the tubes 40 and the uranium rods 22 and thus have a substantial thickness.

As shown in Figs. 9 and 10, each of the perforations is provided with steel sleeve 35a having its inner surface lined with aluminum as shown at 38a. This may be accomplished by making each hole in the tube sheets 35 and 36 somewhat larger than the outer diameter of the tubes 40, then inserting the steel sleeve 35a into this perforation and welding this sleeve to the tube sheet as indicated at 41. The inner diameter of this sleeve is such that the sleeve when provided with aluminum lining 38a fits snugly over the tube 40. The aluminum lining 37 (see Fig. 10) on the inside of the tank sides is spaced from the steel shell 32 to provide space for a graphite neutron reflector 42 extending circumferentially about the tank and having a thickness of about 18 inches.

The reflector 42 serves to reflect back into the heavy water moderator some of the neutrons that normally would escape from the system. As shown in Figs. 6, 8, and 10 graphite in the form of blocks is employed, though other suitable material such as paraffin may be used.

10

Graphite is readily machinable with conventional wood working machine tools and hence the block form is a convenient shape for the graphite.

In assembling a reflector for a liquid moderator pile the positioning of the reflector inside the reactor tank with the liquid in direct contact with the reflector material will, of course, give the highest neutronic efficiency. However, a reflector placed inside of the reactor tank must be able to exclude the moderator from the reflector material and for that reason the reflector if used inside could not be built from porous material, such as graphite, or any material in the form of bricks, because any impurities in the moderator (such as corrosion products) would lodge in the reflector material or between reflector blocks and become more and more radioactive during operation of the reactor thus producing localized "hot spots." To prevent this occurrence, the reflector would have to be one piece, such as, for example, a beryllium casing inside the tank, or a beryllium oxide lining with a glazed surface or similar arrangement and at present such construction is impractical.

It has been found that if the tank wall is not made too thick, and is made of materials having not too high a neutron danger coefficient, then it is possible to place the reflector on the opposite side of the tank wall from the moderator and still have an efficient neutron interchange and reflection back into the liquid moderator in spite of the neutron absorption characteristics of the tank wall, such as the reflector 42. This arrangement is particularly practical from a nuclear physics point of view in the D$_2$O reactors herein described, because they are small, and consequently have a high neutron loss from the exterior thereof. A certain percentage of the escaping neutrons can be absorbed by the tank wall, and still there will be a high percentage of neutrons returned to the reactor by the reflector.

At each end of the cylindrical tank 20 are three end chambers 43, 44 and 45 formed by spaced steel walls or tube sheets 46, 47 and 48 and circumferentially disposed steel shell walls 49, 50 and 51 respectively. The function of these chambers 43, 44 and 45 will be brought out presently.

The tubes 40 extend throughout the length of the tank 20 and continue through all of the end chambers 43 to 45 inclusive, passing through each of the tube sheets 35, 36, 46, 47 and 48. In the tank 20 perforated aluminum diaphragms 52 are interposed between the opposite tube sheets 35 and 36 through which pass the tubes 40. The diaphragms serve as intermediate supports for the tubes. This, of course, could be accomplished by other conventional means, such as by hangers.

Referring to Figs. 4, 5, 6 and 8, the reactor is supported on a cradle generally indicated at 56 on the bottom 57 of a cylindrical steel tank 58. This cradle 56 comprises spaced supporting members 59 which in turn rest on the bottom of the steel tank 58. A supporting member 59 is disposed under each of the tube sheets 35, 36, 46, 47 and 48. The tank bottom 57 rests on a concrete slab 61. A cylindrical concrete wall 62 about 10 feet thick projects upwardly from the concrete slab 61 and completely surrounds cylindrical side wall 63 of the steel tank 58. The wall 63 of the tank 58 is lined with lead as indicated at 64. The steel tank 58 is about 70 to 80 feet in diameter and 40 to 60 feet high and is filled with light water 65 completely submerging the reactor 31 in water. Thus an effective neutron and gamma radiation shield 60 is provided, the concrete serving finally to absorb any radiations which may penetrate the water, lead and steel layers.

The aluminum tubes 40, as shown in Fig. 5, extend continuously from outside the concrete wall 62, through the concrete wall, lead liner 64, steel wall 63, and then project through the light water 65, into the end wall of the adjacent light-water leakage chamber 45 (Fig. 6), and then continue through the intermediate chambers 44 and 43 and the reactor tank 20, terminating after passing

11

through the most remote light-water leakage chamber 45. These tubes are ribbed as indicated at 66 (see Figs. 10 and 11). Between the ribs are light water channels 67. As best shown in Fig. 11, the ribs 66 are disposed on the inner faces of tubes 40 and extend longitudinally of the tubes forming supporting members for the uranium rods 22 and approximately centering the rods in the tubes. Enveloping each uranium rod 22 is an aluminum coating or sheath 68 completely sealing the uranium in aluminum to prevent light water which passes through the channels 67 from attacking the uranium. This sheath 68 is about one-half millimeter in thickness, and the water annulus surrounding the rod 22 is about 2.2 millimeters in thickness. Each ribbed tube 40 is about 3 millimeters in thickness.

The cooling water passing through the tubes 40 enters the pile through ring headers 80 (Figs. 4, 5 and 12) and flows in a direction from left to right through the tubes, as shown in the drawings. Each ring header 80 is supplied with water from a header pipe 81, which in turn is fed by the main supply pipe 24 (Fig. 4).

As shown in Figs. 4 and 5, the cooling water passing through the tubes 40 is discharged directly into the steel tank 58 forming part of the water shield. The shield tank 65 in turn may be circulated through an external primary cooling circuit shown at A in Fig. 4. The water in this circuit may be cooled by being circulated in heat exchange relationship with a secondary cooling circuit B.

Cooling circuit

The primary cooling circuit comprises a pipe system 24 drawing light water 65 from the water shield, and then circulates it through the heat exchanger 27, and by means of pump 28 conveys the water to the pipe headers 81 (Fig. 12) from which the water passes through ring headers 80 into the tubes 40. This water is finally discharged out the open ends of tubes 40 into the shield water 65.

The secondary cooling circuit B comprises the heat exchanger 27 wherein heat is extracted from the hot water in the primary circuit A, a pipe 82 that conveys the hot water in the secondary circuit to a cooling tower 83 where the water is cooled by evaporation, and a return pipe 84 that conveys the cooled secondary water stream back to the heat exchanger 27 in response to a circulating pump 85.

Rather than recirculate the cooling water in the primary circuit, it may be preferred to pass only fresh water through the pile. This may be advantageous where the pile is situated close to a river or a large body of water. The coolant may then be introduced into the cooling circuit at 86 and be withdrawn at 87 and disposed of in some suitable manner. A water purifying and filtering plant may be required in such an arrangement. In the circuit shown, a shut-off valve 88 would be in its "off" position under these circumstances.

During normal operation of the system the uranium rods 22 (Figs. 10 and 12) are disposed in the active portion 19 of the reactor 18. As best shown in Fig. 12, the cooling water entering through the header pipes 81 is introduced into the tubes 40 through ring headers 80. This water, as indicated by the arrows in Fig. 12, passes through openings 125 in the walls of the tubes 40 and then flows in the direction from left to right through the tubes 40. When the cooling water reaches a uranium body 22 it is divided by the tapered end 22a of the uranium rod 22 into separate streams passing through the water channels 67 bounded by the uranium rod 22, the wall of the tube 40, and the inwardly projecting ribs 66 on the inner face of the tube 40. Thus, the water is passed in heat exchange relationship with the coating 68 surrounding the uranium rod 22 and thereby extracts heat from the rod generated as a result of the neutron chain reaction. The cooling water is then passed throughout the remainder of the length of the tube 40 and discharged through an open end 40a of the tube into the shield water 65.

12

Plate shield

Referring to Figs. 6 and 8, directly below the reactor 18 is an opening 89 in the bottom of tank 58 and the concrete slab 61 of rectangular shape in which is disposed a neutron and gamma radiation shield generally indicated at 90. This shield 90 comprises a plurality of horizontally disposed steel sheets or plates 91 extending entirely across the opening 89 in the concrete slab and held at their ends by means of steel end walls 92. The plates 91 are spaced apart to provide passages 93 therebetween.

Referring to Fig. 6, the shield 90 is divided into two segments, that is, an inner hot shield 94 which is treated separately from an outer cold shield indicated as 95. The inner shield 94 is cooled by light water entering through a pipe 96 into the innermost passage 93a, and passes throughout the entire length of the shield until it reaches the opposite end of the shield. The water then passes downwardly through an opening 97 in the steel plate 91a bordering the passage 93a and into the next lower passage 93b wherein the cooling water reverses its direction of flow and passes to the opposite end of the shield. The coolant then passes through opening 98 in the steel plate 91b and into the next lower passage, from which the cooling water is withdrawn at the opposite end through a discharge pipe 99.

The outer or cold shield 95 is similarly cooled by light water entering through a suitable pipe 100 and leaving through a discharge pipe 101 and circulating through the outer shield in a manner similar to that just explained for the inner shield.

By dividing the two shields in the manner just described, any radioactivity induced in the cooling water in the inner or hot shield is removed with the water discharged through the pipe 99 without contaminating the cooling water in the outer or cold shield 95. Thus, the outermost layer of the shield 90 is kept relatively free from radioactivity. Obviously the shield 90 may be divided into more than two segments.

Leakage control

The active portion 19 of the reactor 18 can be completely sealed so as to prevent leakage of heavy water except in the end walls or tube sheets 35 and 36 through which the tubes 40 pass. In order to prevent loss of heavy water due to leakage through these openings, leakage chambers 43 are provided. Thus, any heavy water leaking from the tank 20 into these leakage chambers is collected in the bottom of the chambers and is passed therefrom through pipes 110 into a header 103. Check valves 111 are provided in the pipes 110 and allow the heavy water to flow from the leakage chambers 43 into the headers 103 but serve to prevent the flow of heavy water in the opposite direction.

The light water leakage chambers 45 are provided for the purpose of receiving water passing from the water shield surrounding the reactor 18 through the openings in the tube sheets 48 and this water is collected in the bottom of the leakage chamber 45 from which it is withdrawn through a pipe 112.

Referring to Fig. 4, the light water thus withdrawn from the leakage chambers 45 through pipes 112 is collected in a sump 113 from which it is again returned to the water cooling circuit A through a pipe 114.

As shown in Fig. 6, the third chamber 44 is disposed between the inner heavy water leakage chamber 43 and the outer light water leakage chamber 45 and is referred to as the intermediate chamber. This intermediate chamber 44 is filled with helium entering the chamber through a pipe 115 and leaving through discharge pipe 116. This gas is maintained under a pressure higher than that existing in either the inner heavy water leakage chamber 43 or the outer light water leakage chamber 45 so that in the event of leakage through the tube passages in either of the tube sheets 46 or 47 the direction of flow through such leakage space is from the helium chamber to one

2,736,696

or both of the other chambers rather than in the reverse direction. In this manner contamination of the heavy water in the leakage chamber 43 by light water vapor from the leakage chamber 45 is prevented.

Again referring to Figs. 4 and 6, the helium leaving the intermediate passage 44 through the discharge pipe 116 is passed through conventional driers indicated at 117 and then into a compressor 118 from which the helium is returned to the intermediate chambers 44 under pressure through piping 119.

Since helium is apt to leak into the outer light water leakage chamber 45, provision is made to remove this gas from the water system. Such is shown in Fig. 4 where a pipe 120 enters the sump 113 at the top above the water level therein and conveys helium gas that naturally rises to the top of the sump 113 back to the helium line 116 where the gas is passed through the driers 117 to remove any water vapor present.

Rod removal

As previously mentioned, after the system has operated for a period of time sufficient to cause a quantity of element 94^{239} to be produced, it may be desirable to remove at least some of the uranium rods 22 from the reactor to extract element 94^{239} and radioactive fission products from the uranium.

In the neutronic reactor, a neutron density variation occurs across the active portion 19 of the reactor 18; that is, the neutron concentration at the periphery is relatively small and increases to a maximum value at the center. Actually, therefore, since the rate of production is dependent upon the neutron density, the active portion of the reactor will have three-dimensional shells or shell-like regions and the concentration of element 94^{239} throughout any given shell or region is substantially uniform. In a reactor built in the form of a sphere, these would, of course, be in the shape of concentric spheres, that is, spherical shells, of different diameters, while one built in the shape of a cylinder would have similar zones of equal concentration but of different shapes.

This variation in concentration indicates the desirability of a systematic schedule depending upon the time of operation and the location of the uranium for removing and discharging uranium metal that has been subject to neutron bombardment. In the case of a new system, the operation would normally continue until the metal in the center portion of the reactor reaches a desired content of element 94^{239}, at which time this metal should be removed and replaced with fresh metal. The next removal then would be from the section next adjacent to the center section of the reactor where the desired content of element 94^{239} is reached after further operation. The process would then proceed with the removal of metal at various times until the metal recharged at the center of the reactor has reached the desired content of element 94^{239}. This would then be replaced and the process of progressing toward the periphery continued, with periodic return to more central areas. A removal schedule can be developed by calculation, and checked by actual experience after the system has been placed in operation.

The power output of a system, stated in terms of heat energy, refers to the total heat production for the entire active portion of the reactor. Since the heat results from fissions in the uranium, it is evident that this heat is not formed uniformly throughout the reactor but that it must vary across the active portion of the reactor with the local rate at which fissions occur and element 94^{239} is formed. Consequently, the relative values for the production of 94^{239} apply also to heat evolution; that is, the heat generated increases from a minimum at the outer surface of the reactor to a maximum at the center.

As the total weight of the radioactive fission elements is proportional to that of the 94^{239} at the time of fission, it might be assumed that the amount of these fission elements and of 94^{239} present in metal removed from the

reactor are also of the same proportion. This is not true, however, as the fission elements when produced are highly radioactive and immediately start to decay, some with short half-lives and others with longer half-lives until, through loss of energy, these unstable fission elements arrive at a stable element or isotope, and no longer change. The 94^{239}, on the other hand, is a relatively stable element when formed having a radioactive half-life of the order of 10^4 years.

At the start of the reaction in new metal, the radioactive fission elements and the 94^{239} both increase in amounts. After a certain period of operation during which time the material is subjected to intense neutron bombardment, the radioactive fission elements may reach a state of equilibrium, and from that time on, the amounts of these elements remain constant as the shorter-lived fission elements are reaching a stable condition at the same rate new ones are being produced. The amounts of stable end products of fission, however, continue to increase with the increase in element 94^{239}. Consequently, the rate of formation of the end products, that is dependent upon the location of any particular metal in the reactor and the power at which the system operates, controls the maximum radioactive fission element content regardless of the length of time the system operates. The quantity of element 94^{239} on the other hand, and the final and stable end products of fission, continue to increase as the operation of the system continues. The amounts of the latter present are controlled only by the location of the metal in the reactor and the time and power of operation. The original high radioactive fission elements may, therefore, vary from a substantial percentage of the weight of element 94^{239} present in the metal at the center of the reactor after a short period of operation to a very small percentage in metal from a position near the periphery of the reactor after an extended operating period at a given power.

It is not to be assumed, however, that the fact that equilibrium can be obtained between the original highly radioactive fission elements and the stable fission end products, that all radioactivity will cease when the original fission elements have been permitted to decay for a time equal to the equilibrium period, for example. Many of the original fission elements have long half-lives which, together with their successive radioactive disintegration products existing long after the fission elements having the shorter half-lives have decayed, renders the uranium still radioactive, especially after prolonged bombardment at high neutron densities. In addition, the successive radioactive disintegration products of the original shorter-lived fission elements may still be present.

The equilibrium radioactivity is so intense that metal taken from the reactor for the recovery of element 94^{239} and fission products immediately after bombardment at high neutron densities, will heat spontaneously due to the intense radioactivity of the shorter-lived fission elements. The amount of heat given off as a result of the spontaneous heating will depend particularly on three factors:

(1) The concentration of element 94^{239} and fission elements in the metal; (2) The period of time of continuous operation required to reach this concentration; and (3) The elapsed time since the reactor was shut down and the metal was removed.

The metal from the center of the reactor in a system operating at a high power output, for example, at a 94^{239} concentration of 1 to 2000, if not cooled, might increase in temperature at the rate of about 2000° C. per hour, one day after the neutron activity of the system has been shut down. After 30 days shutdown after operating for 100 days at an output of 50,000 kilowatts, the average temperature rise is approximately 126° C. per hour. Uranium metal of the type used in chain reacting systems melts at about 1100° C.

Under these conditions, uranium metal bombarded

2,786,696

15

with neutrons for an extended period of time at high rates of power output can only be safely removed from the reactor under the following procedures:

A. The neutron activity of the system is shut down, and the uranium metal is kept in the reactor and continuously cooled until the radioactivity decays to a point where the metal can be removed without melting in ambient air. This procedure may require that the metal remain in the reactor for a period of from 30 to 50 days after the neutron bombardment has ceased;

B. The neutron activity of the system is shut down and the metal is kept in the reactor with the cooling system in operation for only a few days to permit the most violent radioactivity to subside and then the metal is removed from the reactor with the cooling discontinued during the removal, except for cooling by the atmosphere or by air or water spray. The metal is then promptly placed under more efficient cooling conditions before the temperature of the metal has had sufficient time to reach the melting point of uranium; or

C. The neutron activity of the system is shut down and the uranium metal is removed under continuous and effective cooling such as is disclosed in the preferred embodiment of the present invention, wherein the uranium is discharged directly into the shield water 65.

Obviously, methods A and B can be performed with all the embodiments of the invention shown, while method C can be performed with the preferred embodiment shown, for example, in Fig. 6.

It is important, of course, from a point of view of biological safety, that adequate shielding be provided to absorb the strong gamma radiations from the fission elements present in the active uranium while being removed from the reactor. The neutron activity in the system ceases within 30 minutes after shutdown, during which period delayed neutrons are being emitted from fission fragments. In no case, then, should the uranium be discharged from the reactor immediately after shutdown of the reactor but sufficient time will be given to permit all delayed neutrons to be emitted. Thus the shielding during the removal of uranium is provided primarily to protect personnel from gamma radiations.

While the methods of extracting the fission products and the 94²³⁹ from the bombarded uranium metal taken from the reactor form no part of the present invention, the fission products and element 94²³⁹ are removable and when removed are extremely useful. The radioactive fission products are valuable for use as radiation sources, many having high energy gamma radiations with long half-lives, sufficient for radiography of even heavy metal castings. In addition, some of the fission products are useful as radioactive tracers in biological and physiological research, and are in demand for these purposes.

Element 94²³⁹ is very useful, as it is fissionable by slow neutrons in the same manner as the uranium isotope 92²³⁵ contained in natural uranium. The separation of 92²³⁵ from 92²³⁸ in natural uranium is extremely difficult since both are isotopes of the same element. Element 94²³⁹, on the other hand, is a different element from uranium, having different chemical properties than uranium, and therefore can be separated from the uranium by employment of chemical processes. After separation, element 94²³⁹ can be added to natural uranium to supplement the 92²³⁵ content, thus increasing the amount of fissionable material in the uranium. This enriched uranium can then be used in chain reacting system making it possible to provide more neutron absorbing cooling facilities, for example, than could be used in a system of similar geometry employing only natural uranium. Thus, an enriched system may provide a greater power output than would be possible in a natural uranium system of the same geometry.

Referring to Figs. 12 and 13, an arrangement is shown whereby the uranium rods 22 are made accessible from a position outside the concrete wall 62 to effect this re-

16

moval. Each of the aluminum tubes 40 terminates at a position near the ring header 80. On the end of each tube is threaded a valve housing member 126. On the opposite end of the valve housing 126 is threaded a tube section 127 on the end of which is threaded a packing gland 128 on the opposite end of which is threaded a length of tubing or tube section 129. This tube section 129 is threaded on its free end to receive a cap 130 that in turn is provided with a gasket 131 operating against the end face of the tube section 129 to seal the tube from the atmosphere. The distance from the packing gland 128 to the valve housing 126 is at least sufficient to receive therebetween a uranium rod 22. The packing gland 128 is provided with a resilient packing ring 132.

The valve housing 126 has an upstanding boss 133 adapted to receive a slidable valve member 134 with a valve stem 135. This valve stem 135 projects beyond the top of boss 133 and is threaded on its free end to receive a beveled gear 136 that is supported on the top of the boss 133 and is adapted to rotate about the valve stem 135, causing the stem to feed upwardly or downwardly by virtue of the screw threads, thereby raising or lowering the valve member 134. An angular bracket 137 is supported on the upstanding boss 133 and is provided with a flange 138 that supports one end of a horizontally disposed shaft 139. The other end of this shaft 139 is journaled in a bracket member 140 that in turn is fastened to the top face of the tube section 129. The shaft 139 projects through the journal on the bracket 140 and is provided with a hand crank or wheel 141 at its free end. At the opposite end of the shaft 139 is a beveled gear 142 adapted to mesh with the first-mentioned beveled gear 136, thus imparting a vertical sliding motion to valve member 134. During normal operation of the system, when the uranium rods 22 are in place in the reactor and the chain reaction is occurring, valve member 134 is disposed in its closed position shown in the upper tube of Fig. 12.

The light water flows through the tubes 40 at a rather high rate of speed so that some means should be provided for holding the uranium rods 22 in place in the reactor to prevent the force of the cooling water from moving the rods toward the discharge ends of the tubes. For this purpose, as shown in Figs. 12 and 15 to 18 inclusive, each uranium rod 22 is held in place in the reactor by means of a dummy rod 69 cooperating with a perforated holding plate 70. As shown in Figs. 12, 15 and 16, the holding plate 70 is disposed vertically at the open ends of a vertical row of tubes 40 and is slidable vertically within guide strips 71. Each guide strip 71 is more or less Z-shaped in cross section and is provided with a flange 71a for fastening the strip at the end wall 48 at the discharge ends of the tubes 40. The guide strips 71 may be welded or otherwise suitably secured to the end wall 48.

The holding plate 70, as best shown in Fig. 15, is provided with two types of spaced perforations 72 and 73, the former constituting circular holes and the latter being in the form of spaced, substantially semi-circular openings separated by an intermediate bridging member 74 integral with the plate 70.

A manipulation rod 75 is fastened to each holding plate 70 and extends upwardly to a position at the top of the shield water 65. Thus, each holding plate 70 may be moved independently in a vertical direction so as to place either the circular holes 72 opposite the ends of each of the corresponding tubes 40 or to place the semicircular openings 73 opposite the ends of said tubes. The circular holes 72 have a diameter at least equal to, although preferably slightly greater than, the inner diameter of the tubes 40. Thus, when the circular openings 72 are disposed opposite to the open ends of tube 40, the passage from the tubes into the shield water 65 is completely unobstructed. The semi-circular openings 73 have an outer diameter approximately equal to the inner

17

diameter of the tubes 40 so that when these semi-circular openings 73 are disposed opposite to the open ends of the tubes 40, the passages from the tubes to the shield water 65 is restricted by the bridging member 74. The semi-circular openings 73 are of sufficient width to accommodate the normal flow of cooling water passing through the tubes 40 during the operation of the system at its maximum production output. Thus, these semi-circular openings in no way restrict the intended output of the system.

Referring to Fig. 17, the dummy rod 69 may be a flat plate twisted to provide flutes 76 disposed in the form of helix about the rod. Thus, helical channels 77 are provided between adjacent flutes 76 so as to provide a continuous helical passage for the flow of cooling water from one end of the dummy rod 69 to the other. Referring to Fig. 12, the dummy rod or plug 69 is disposed between one end of the uranium rod 22 and the holding plate 70.

After the system has operated for a period of thirty or more days at a relatively higher power output, as for example 50,000 kilowatts, the temperature of the system is in a state of equilibrium. If it is desired to remove some or all of the uranium rods in the reactor following the operation of the system for this period of time, the neutron reaction in the system must first be shutdown (as hereinafter described) by inserting the safety or control rods 168 all the way into the active portion of the reactor or by discharging the heavy water from the reactor into the sump 164. The operation of the cooling system is continued, however. Assuming that the system has been operating continuously over this period of time, the decline of heat generated in the active portion of the system immediately following the discontinuance of the chain reaction is approximately 85 per cent of the heat being generated while the system was in operation. In other words, immediately after the reproduction ratio of the system is reduced below unity, the heat generation rate assumes a value approximately equal to 15 per cent of the heat generation rate existing prior to the reduction in the reproduction ratio. Approximately one second after the reproduction ratio is reduced to a value below unity, the heat generation rate has been reduced to approximately 10 per cent of the original value. One minute after the original shutdown of the system, the heat generation rate has been reduced to approximately 4.2 per cent of the original rate, and one hour following the original shutdown the heat generation rate has been reduced to approximately 1.8 per cent of the original rate.

Directly following the shutdown of the system and for a period of approximately thirty minutes thereafter, there are fissions continuing in the active portion of the reactor due to the release of delayed neutrons. Some of the heat generated during this thirty-minute period following the shutting down of the system is due to these delayed neutrons. The remainder of the heat being generated during this period is due to the radioactive decay of the fission products produced during the operation of the system. The delayed neutrons disappear within approximately thirty minutes following the shutdown of the system so that the heat generated after this thirty-minute period is caused by the decay of the radioactive fission products.

It is thus seen that the required amount of cooling after the operation of the system has been shutdown is much less than that needed before shutdown, so that it is possible immediately following the insertion of the control rods 168 or the discharge of the heavy water from the reactor tank 20 to reduce the rate of flow of light water through the cooling tubes 40 by a substantial amount. Thus, it is possible immediately following the shutting down of the system to decrease the rate of flow of the cooling water through the tubes 40 to a rate ap-

18

proximately 25 per cent, more or less, of that originally required during normal operation.

At this decreased rate of flow, there is no further need for holding the uranium rods 22 in place in the reactor against the water pressure so that the holding plates 70 may be moved from their normal position, wherein the bridge-like members 74 obstruct the open ends of the tubes 40, to their open position wherein the circular openings 72 are disposed opposite to the open ends of tube 40, thereby entirely removing all obstructions from the open ends of the tubes 40. It is thus possible under these conditions to remove the dummy plugs 69 and the uranium rods 22 from the reactor by merely applying force on the uranium rods 22 to force them towards the open ends of the tubes 40.

When removing a rod 22 from the reactor, a plunger 144 is employed as shown in Fig. 12 in the lower tube 40. This plunger is inserted in the tube at a time when the movable valve 134 is in its closed position while the cooling water is still being used in the tubes 40 although at a reduced rate of flow. The end cap 130 is removed and a packing gland 145 is threaded onto the free end of the tube section 129. This gland is provided with a packing ring 146 through which the plunger 144 operates. The inner diameters of the packing rings 146 and 132 are such as to provide a tight fit about the plunger 144, thereby reducing to a minimum the leakage of water past these glands. A suction pump, not shown, is normally in operation to withdraw through a tube 147 any cooling water inside the tube section 129.

With the arrangement shown it is possible to remove the uranium rods 22 and to replace them with fresh rods without halting the operation of the system for more than a few hours. This is made possible in the present device because the discharge of the uranium rods is effected under a sufficient thickness of water to effectively reduce to safe limits the neutron and gamma radiations outside the shield, and also cooling of the uranium is continued throughout the removal operations and the uranium is maintained under water. Referring to Fig. 12, a uranium rod 22 which is in its normal position, shown in the upper tube of the figure, is removed by the use of a plunger 144. While the sliding valve member 134 is in its closed position as shown, the end cap 130 is removed. The packing gland 145 is threaded onto the end of tube section 129 and the plunger 144 is passed through both glands 145 and 128 so as effectively to seal the end of the tube. The sliding valve member 134 is then raised by rotating the wheel 141 until the sliding valve 134 is in its elevated position shown in the lower tube of Fig. 12 and then the plunger 14 is moved through the tube 40 until it engages the uranium rod 22.

Force is then applied against a uranium rod 22 to move the rod toward the open end of the tube 40. A basket, not shown, may first be disposed adjacent to the discharge end of the tube 40 to receive the dummy plug 69 that is first ejected. A shielded coffin, generally indicated at 148 in Figs. 5, 6 and 7, is then placed in position adjacent to the discharge end of the tube 40 to receive the uranium rod 22 and the plunger 144, then pushes rod 22 into the waiting coffin 148. While the plunger 144 is in a position wherein the packing rings 146 and 132 cannot limit the leakage of water, there may be some leakage into the tube section 129. This water however is withdrawn from the tube section 129 through the pipe 147, as previously explained, so that there is no leakage of water past the packing gland 145.

The coffin 148 (Fig. 14), as shown, is rectangular in shape (see Figs. 7 and 14) and comprises a bottom portion 149 and a lid 150. The bottom 149 includes a steel shell 151 lined with lead 152 to a substantial thickness. The cover likewise comprises a steel shell 151a lined with lead 152a. The lead lining in the bottom 149 and the cover 150 is stepped at 153 and 153a respectively throughout the periphery of the open side of each so as

2,736,696

19

to provide a tortuous path for any radio-active radiations passing from inside the closed coffin toward the outside. Such radiations are absorbed in the lead and are thus prevented from escaping.

The coffin is provided with ears 154 (Figs. 6, 7 and 14) fastened to the bottom portion 149 having openings 155 (Fig. 7). By means of grab hooks 156 on the end of cables 157 (Fig. 6) the coffin can be handled conveniently from overhead by means of a crane, not shown, located above shield water 65.

The lid 150 of the coffin is provided with eyelets 158 to facilitate handling with conventional hooks and cables shown in Figs. 5 and 6.

During removal of a uranium rod 22 from the reactor, the bottom portion 149 of the coffin is disposed at the discharge end of the reactor as shown in Fig. 6 directly opposite to and below the end of the tube 40 from which the rod 22 is to be discharged. The rod thus discharged is received by the bottom portion of the coffin, and after the coffin has been filled, as shown in Fig. 14, the lid 150 is moved into place on the bottom portion 149.

At the time the active uranium rods are placed in the coffin 148, they are disposed under water and are thus cooled and shielded. After the coffin is filled, it is raised to the top of the shield water 65 and removed to an aging pit, not shown, and placed under water where the radioactive decay continues. Means for cooling the rods must be provided during this removal, and for present purposes, since the coffin, per se, forms no part of the invention, it will suffice to say that the cooling can be effected by circulating cooling water through the coffin from an outside water supply in any well known manner.

The actual separation of element 94²³⁹ and radioactive fission products from the uranium is not started until the radioactivity of the fission products has decayed to limits wherein the material can be handled with the exercise of practical safety precautions.

Heavy water discharge

Extending downwardly from the active portion 19 of the reactor 18 are a plurality of heavy water discharge pipes 102 which project through the shield 90 and terminate in a heavy water discharge header 103 (Figs. 4, 5, and 6). A suitable gate valve 160 is disposed in each pipe 102 and is operatively connected to an operating rod 161 by a suitable arm 162. As shown, all of the valves 160 connected to the rod 161 open or close simultaneously by manipulation of the rod 161.

Normally valves 160 are closed and the level of the heavy water in the tank 20 is such that the overall size of the uranium-heavy water mass is slightyl greater than the critical size for the specific geometry and purity of the system. Thus, the reproduction ratio for the active portion of the reactor is slightly greater than unity (about 1.005) so that a controlled exponential rise in the neutron density can be effected in the system. In the event of some emergency, the gate valves 160 are opened to stop the chain reaction rapidly, as will be explained presently when the details of an emergency control circuit are described.

The discharge header 103 empties into a main discharge pipe 163 which in turn empties its contents into a sump 164 (Fig. 4). This sump 164 is of sufficient size to receive all of the heavy water in the system.

Extending off from one of the discharge pipes 102 is a level control pipe 165 projecting downwardly into the sump 164 and terminating close to the bottom of the sump. A two-way operating pump 166 driven by a reversible motor 167 is provided in the level control pipe 165. The purpose of this level control pipe 165 is to convey heavy water back and forth between the sump 164 and the active portion 19 of the reactor 18 in response to the action of the pump 166 so as to control the level of the heavy water in the tank 20. This may

20

be done in conjunction with an electrical control circuit as will be brought out hereinafter.

Heavy water circuit

A heavy water system of the type disclosed herein is temperature stable. By that is meant that as the neutron density in the reactor increases exponentially at a controlled rate, the temperature of the uranium and the heavy water rises as a result thereof. This rise in temperature of the heavy water causes expansion of this moderator effectively lowering its density and hence the reproduction ratio of the system. This results in a decrease in the neutron density that in turn lowers the rate of heat generation. Thus the temperature of the heavy water is lowered and its density again becomes greater, resulting in a rise in the reproduction ratio of the system. In this manner, the cycle repeats itself and the control of the reaction is automatic, within short limits.

Additional control means may be provided, however, to insure adequate control. One of said means consists of one or more safety rods 168 comprising a material of high neutron absorbing characteristics such as cadmium. This rod per se forms no part of the present invention so a detailed description is believed unnecessary. It will be sufficient to explain that this rod can be moved in and out of the active portion of the reactor and is maintained at a selected position in the reactor to absorb sufficient neutrons to stabilize the chain reaction. If a thin sheet of cadmium is used as the absorbing material in the control rod, only thermal neutrons will be absorbed, the fast neutrons being capable of penetrating the material. Thus, long lived radioactivity in the control rod resulting from fast neutron absorption is prevented. The control rod or rods may be operated manually or by some electrical circuit. An operative circuit will be brought out hereinafter.

The heavy water discharge header 103 and valve mechanism associated with the discharging of the heavy water from the reactor must be readily accessible for maintenance purposes. As shown in Figs. 5, 6 and 8, a tunnel 175 is provided underneath the light water shield 60. A lead plate 176 forms the ceiling for the tunnel and supplements the concrete slab 61 in absorbing neutrons and gamma rays. Concrete walls 177 and floor 178 complete the walls of the tunnel 175. The entrance to the tunnel may be adjacent to the outer face of the concrete wall 62 of the light water shield 60. The steel and light water shield 90 serves as protection against dangerous exposures to neutrons and radioactive radiations directly under the active portion of the reactor 18.

Helium system

The heavy water in the reactor tank 20 does not completely fill the tank, the space above the level of the heavy water constituting a vapor and gas space. It is desirable that this space also be filled with helium under pressure that is preferably circulated outside the reactor tank 20, so as to remove heavy water vapor and gases accumulating in this vapor space above the heavy water. Helium being an inert gas prevents an explosive mixture from forming above the heavy water level.

Under the effect of the dense neutrons in the reactor, the deuterium and the oxygen in the heavy water become dissociated and bubble through the moderator in the form of gases that collect in the vapor space above the heavy water. These gas bubbles reduce the effective density of the heavy water and hence lower the efficiency of the moderator as a neutron slowing agent thus lowering somewhat the reproduction ratio. These two gases are removed from the vapor space along with the helium.

This decomposition takes place in any type of heavy water system, and is particularly prevalent in the so-called "slurry system" shown in Fig. 3 where the uranium is in the form of small particles held in suspension in the heavy water. The primary energies released as the re-

21

sult of fissions in the uranium particles in the case of the slurry system leave the uranium bodies and are spent in the heavy water. The decomposition of the deuterium and the oxygen as a result of these primary energies is much greater than that caused by the bombardment of the heavy water nuclei by neutrons.

In the uranium rod system disclosed herein, the uranium bodies are so large and are so well enclosed in aluminum that the fission fragments cannot leave the uranium bodies. Thus, these primary energies are spent by being converted into heat in the uranium bodies so that the decomposition in such a system is caused primarily by the bombardment of the heavy water nuclei with neutrons. There is sufficient decomposition, however, even in the present system to require some means for recombining the deuterium and the oxygen. For example, in a 50,000 kw. plant with a tank having a volume of 10 cubic meters, gas is liberated from the moderator at the rate of about 10 liters per second. This is equivalent on the average to about 10^{-3} cm.^{3} of gas per second per cm.^{3} of active portion of the chain reacting system which results in a reduction in water density of about 0.8%. An increase of about 1.7% of material required in the active portion of the reactor is needed to compensate for this water density reduction.

Referring to Fig. 4, a pipe 179 conveys the helium together with the heavy water vapor and uncombined gases from the vapor space above the heavy water in the reactor tank 20 to a combiner and condenser 180 located entirely outside the shield 60. This combiner, per se, forms no part of the present invention but includes a hot grid or the like, not shown, against which the uncombined gases, deuterium and oxygen are directed for recombination and condensing into deuterium oxide. The heavy water thus formed is directed into the sump 164 through a pipe 181. The helium gas serving to wash the dissociated gases out of the reactor tank 20 leaves the combiner 180 through a pipe 182 and enters a purifier 183 where it is purified and separated from noncondensible gases, and then passes through a pipe 184 and into a compressor 185 where the helium is compressed and is then returned through a pipe 186 to the vapor space above the heavy water in the reactor tank 20. The non-condensible gases in the purifier 183 are exhausted through a suitable pipe 187.

It is to be understood, of course, that all of the equipment shown in Figure 4 outside of the shield 60 must be suitably shielded either by being disposed below ground or by some other appropriate means to protect the personnel in the vicinity from the harmful radiations.

Controls

The device above described is ideally adapted for automatic control to maintain the neutron density within the reactor substantially constant, and thus give a substantially constant rate of production of element 94²³⁹. Due to the fact that large masses of material are utilized in the active portion of the reactor, there is a temperature lag therein. Consequently, it is convenient to monitor and control the structure by means of ionization chambers, or equivalent devices that will respond to the neutron density at the periphery of the active portion of the reactor. As the rate of neutron diffusion out of a chain reacting system is always proportional to the rate of generation of neutrons within the structure, the ionization chambers can readily be placed at the periphery of the active portion of the reactor, and in fact are preferably so positioned that they may not be subjected to the extremely high neutron density existing within the reactor.

There are disclosed herein two means for normally controlling the chain reaction. The control rod 168 (Fig. 19) may be used alone or with other control rods. The chain reaction may also be controlled either without any other control means or in conjunction with the control rod 168 by means of varying the level of the heavy

22

water in the active portion 19 of the reactor 18. This latter control is effected by means of the reversible pump 166 that selectively draws heavy water from the reactor tank 20 and discharges it into the sump 164, or conversely, pumps water from the sump 164, and discharges it into the reactor tank 20, in both instances passing the water through the level control pipe 165. Either one or both of these devices may constitute the normal control for the system.

Before proceeding to a detailed description of a control circuit that may be utilized in controlling the neutron chain reaction, it is desirable to point out the manner in which the control rods operate to regulate the neutron density. In any self-sustaining chain reacting structure adapted to produce power or element 94²³⁹, the attainable neutron reproduction ratio of the system must be capable of being made greater than unity. For any value over unity, the chain reaction becomes self-sustaining and the neutron density, without control, would increase exponentially in point of time, until the device is destroyed. For proper control, the system must be held in balance by maintaining the neutronic reaction at some point where the production of new neutrons is balanced with the neutrons initiating the chain. Under these conditions, the active portion of the reactor will continue to maintain the neutron density therein that obtained when the system was balanced.

However, in order to enable the reactor to reach a desired neutron density, the system must be permitted to rise in neutron density for a period of time until the desired density is reached. It is necessary thereafter only to hold the system in balance.

Inasmuch as the reproduction ratio in any self-sustaining chain reacting system is reduced by the presence of impurities that absorb neutrons, such impurities can be introduced in the active portion of the structure in the form of a control rod that can be of a material such as boron or cadmium capable of absorbing large amounts of neutrons. The depth to which this control rod penetrates into the active portion of the structure will determine the amount of neutron absorption and, therefore, the reproduction ratio of the system. A range can be obtained between a condition providing a neutron reproduction ratio that is greater than unity and a condition at which no chain reaction can be maintained. The exponential rise in neutron density can be made relatively fast or relatively slow in accordance with whether the reproduction ratio is permitted to be much greater than unity or only slightly greater than unity.

There is a small percentage of delayed neutrons emitted in the fission process. These delayed neutrons cause the neutron density to rise in a finite time rather than instantaneously. The time required for doubling the neutron density increases as the reproduction ratio approaches unity, and any desired rate of rise in neutron density can be obtained.

As a general rule, the reproduction ratio of a neutronic system with all control rods withdrawn should not be greater than about 1.005. At this value the neutron density in the system will double itself every seven or eight seconds and can easily be controlled. Thus at no time will the heavy water level in the reactor tank 20 be permitted to exceed that corresponding to a reproduction ratio of about 1.005.

If the control rods 168 are used alone to control the neutron chain reaction then one method of control would be to withdraw the control rods from the active portion of the reactor to a point where there is an exponential and preferably slow rise in neutron density within the structure. When a desired neutron density has been reached, the control rod is then returned into the active portion of the reactor to a point where the reaction is balanced. This balance is then maintained so as to keep a constant power output or production output of element 94²³⁹ in the reactor. The maintenance of the

2,786,696

23

balance point with the control rod would be relatively simple were it not for the fact that changes in temperature in the reactor result in changes in the reproduction ratio. It is desirable, therefore, that the control rod be so manipulated that a substantially constant neutron density within the system is maintained. Such a method of control may be accomplished by automatic connection of the control rods with an ionization chamber or similar device responding to neutron density and positioned within the reactor close to the active portion thereof.

Furthermore, due to the exponential rise of neutron density within the active portion of the structure when the reproduction ratio is greater than unity, all possible precautions must be taken to prevent a continued exponential rise in neutron density in case of failure of the control rod to return to the balance position.

While there are many means by which the control rod can be operated, it is believed sufficient to illustrate and describe one simplified control circuit to the end that fully equivalent circuits will be made apparent to those skilled in the art.

Referring to Fig. 19 which shows diagrammatically and reduced to lowest terms one form of control circuit that might be used for regulating by means of control rods the output of the plant hereinbefore described, attention is focused first on the control circuit shown at A. A control ionization chamber 200 is placed in the graphite 42 surrounding the active portion 19 of the reactor 20 and is filled with boron fluoride. A central electrode 201 is provided within the chamber 200, and is connected to a wire 202 leading outside the reactor to a movable contact 203 on a resistor 205. The resistor 205 is connected across a relay coil 206, one side of which coil is connected to a battery 207, the other side of which is connected to a shield 209 around the wire 202. The shield 209 is grounded as is the chamber 200.

Alpha ray ionization due to neutron reaction with the boron within the chamber 200 is proportional to the neutron density in the region of the ionization chamber 200. Thus, the current in the resistor 205 is varied in accordance with neutron densities reaching the ionization chamber 200.

The relay coil 206 operates a relay armature 210 which is spring biased by a conventional spring 211 to contact one motor contact 213, and is urged by the electromagnetic attraction of the relay coil 206, in response to a sufficient current passing through the relay 206, to contact a second motor contact 215. Contacts 213 and 215 connect to the outside of a split winding in motor 221, the center connection 222 being connected through power mains 225 to the armature 210.

The motor 221 rotates a shaft 230 having on one end thereof a pulley 231 and on the other end thereof a control rod gear 234. The shaft 230 between the motor 221 and the gear 234 is split and an electromagnetic clutch 232 connects the two split portions of the shaft 230 when the magnetic clutch 232 is energized. The control rod gear 234 meshes with a rack 168a on the control rod 168.

The pulley 231 has a cable 240 wound thereon which in turn is connected to a counterweight 241 so that the weight of the control rod 168 is substantially balanced by the counterweight 241 so as to permit the motor 221 to run easily in either direction.

During normal operation of the system where the neutronic reaction is controlled only by control rods 168, the slider 203 on the resistor 205, having previously been calibrated in terms of neutron density, is moved to the density position at which it is desired the reactor to operate, taking into account the difference in neutron density at the center of the active portion of the reactor, and at the periphery thereof during the operation. This difference is a constant ratio at various operative densities. When the reproduction ratio in the reactor is less than unity, as is the case when the control rods 168 are disposed at a

24

sufficient distance in the active portion of the reactor, a neutron density at the position of the ionization chamber 200 is much lower than that required to energize the relay coil 206 sufficiently to overcome the action of spring 211 so as to move the armature 210 away from motor contact 213. Thus, the motor 221 under these conditions is energized to withdraw the control rods 168 from the active portion of the reactor to a point determined by a limit stop 242 where the reproduction ratio of the reactor is just sufficiently greater than unity to permit a slow rise in neutron density in the reactor. The motor 221 will stall when the rods 168 are at the stop 242, and should be of the type permitting stalling for a sufficient time to allow the next succeeding operations to be effected. The reaction at this position of the control rod becomes self-sustaining and the neutron density rises. In consequence, the ionization taking place within the ionization chamber 200 rises. As the ionization increases, the current passing through relay coil 206 also increases until a predetermined field intensity is reached. When the induced magnetism in the relay coil 206 has reached a predetermined strength the effect of the spring 211 is overcome and the armature 210 is attracted to the adjacent pole of the relay 206 thus connecting the armature 210 to the motor contact 215. This energizes the motor 221 so as to operate in the reverse direction to that of original operation, so as to drive control rods 168 downwardly into the active portion of the reactor thereby causing the neutron density in the reactor to decay. As soon as the neutron density decays to a predetermined lower value that is approximately at or slightly below that corresponding to a reproduction ratio of unity, the magnetic strength of the relay coil 206 becomes weakened to the extent that the spring 211 again acts to move the armature 210 against the contact 213. The control rods 168 will thus hunt between a point above the balance position where the neutron density decays, thus providing an average neutron density within the reactor as determined by the setting of slider 203 on the resistor 205. Since the mass of the reactor causes any temperature change to lag behind any neutron density change, the temperature of the reactor is maintained substantially constant. If desired, any of the well-known anti-hunting circuits may be utilized as will be apparent to those skilled in the art.

The main purpose of the control circuit A is to regulate the control rods 168 to balance the neutron density in the active portion of the reactor so as to maintain any desired average temperature within the reactor.

Instead of employing control rods for monitoring the system, it may be desirable to vary the level of the heavy water in the reactor tank 20 as previously suggested. A system of the present type lends itself particularly well to this type of control because of the ease by which the liquid moderator can be pumped in and out of the active portion of the reactor.

Referring again to Fig. 19, a simplified circuit effecting such a control is generally shown at B. The various parts making up this circuit are practically the same as those employed in circuit A, and for the sake of convenience similar reference characters are employed for corresponding positions of the circuit except that the values of the reference characters have been increased by a factor of 100. Thus, the control ionization chamber is shown at 300, the central electrode at 301, the conductor 302 is shown connecting the central electrode 301 to the movable contact slider 303 operating along resistor 305. A relay coil is shown at 306 in a circuit with battery 307 suitably grounded and connected to the shield 309. An armature 310 is normally urged away from the relay coil 306 by a tension spring 311 to engage a contactor 313. Upon energization of the relay coil 306, the armature 310 is attracted to the pole of the relay coil 306 so as to break the circuit through the contact 313, and to complete the circuit through contact 315. Contacts 313

25

and 315 are connected to the outside of a split winding of motor 321, the central connection 322 of which is connected through power mains 225 to the armature 310. The motor 321 drives the reversible pump 166 through a shaft 330.

Having described the circuit for controlling the level of the heavy water in the reactor tank 20, we shall now describe its operation, considering the control rods 168 withdrawn from the system and having no effect on its operation. The slider 303 on the resistor 305 having previously been calibrated in terms of neutron density is moved as previously explained on the slider 303 to the density position at which it is desired the reactor to operate, again taking into account the difference in neutron density at the center of the reactor tank 20 and at the periphery thereof during operation.

For this description, it is assumed that the level of the heavy water in the reactor tank 20 is below the critical level at which the reproduction ratio for the system is equivalent to unity. In other words, it is assumed that the level of the moderator in the tank 20 is such that the effective overall size of the moderator with the uranium contained therein is less than the critical size at which the reproduction ratio is equivalent to unity. Thus, the chain reaction is not self-sustaining and in order to place the system in operation it is necessary to increase the effective overall size of the active portion 19 of the reactor 18 above the critical size so as to raise the value of the reproduction ratio to slightly above unity. A manual switch 322a is first closed so as to complete the circuit to the motor 321. Since the reproduction ratio of the system is less than unity, the neutron density in the reactor is below that required to energize the relay 306 sufficiently to open the circuit through contact 313. Thus, as soon as the manual switch 322a is closed, the motor 221 operates to pump heavy water from the sump 164 into the reactor tank 20. The heavy water level thereby rises in the reactor tank 20, increasing the effective amount of uranium and moderator and thus increasing the overall effective size of the active part of the reactor. As soon as the heavy water in the reactor tank 20 reaches a level slightly greater than that corresponding to the critical size of the system, the relay 306 responding to the ionization chamber 300 becomes energized sufficiently to move the armature 310 away from the contact 313 and into engagement with contact 315. The motor 321 is thus operated in the reverse direction to return some of the heavy water from the reactor tank 20 to sump 164, thereby again reducing the effective overall size of the neutronic system and hence lowering the reproduction ratio. This reduction will continue until the reproduction ratio corresponds approximately to or is a little less than unity. Thereafter the level of the heavy water in the reactor 20 hunts between positions corresponding to overall effective sizes slightly in excess and slightly less than the critical size.

Emergency control

In the event of emergency, or for other reasons, it may be desirable to discontinue the operation of the system. This may be done by rapidly driving into the active portion of the reactor a plurality of safety rods that may be similar to the control rods 168. These rods in order to serve their intended function must absorb a sufficient number of neutrons to lower the reproduction ratio of the system below unity.

For present purposes, the control rod 168 may also represent a safety rod. It is to be understood, of course, that separate rods may be used. This rod may be driven into the reactor in response to excessive neutron densities in the reactor, or it may respond either separately or additionally to temperature or flow conditions of the light water passing through the cooling system.

As shown in Fig. 19, the rods 168 respond to a flow switch 331 disposed in the light water cooling pipe 24

26

near the inlet ends of the tubes 40. The electric circuit for accomplishing this is shown at C.

In this circuit, the flow switch is connected to one side of power mains 225 by a conductor including a normally closed circuit breaker 333 adapted to be manually operated. The other side of the flow switch is connected to the magnetic clutch 232 and thence to the power main 225. As long as the proper amount of water is passing through the pipe 24, the flow switch 331 maintains a closed circuit through the magnetic clutch 232 so that the control rods 168 operate in response to circuit A to stabilize the chain reaction. If, for any reason, the power source for the power mains 225 should fail, or if the flow of water past the flow switch 331 should stop or diminish below a predetermined rate to break the circuit through the flow switch 331, then the magnetic clutch 232 becomes de-energized so as to free the gear 234 from the counterweight 241, thereby allowing rod 168 to drop by gravity into the active portion of the reactor. There is a sufficient number of rods 168 to thus stop the chain reaction.

Since the moderator disclosed herein is in liquid form, a very effective emergency control for the present system is to withdraw rapidly from the active portion of the reactor the heavy water moderator to reduce the overall size of the active part of the reactor below the critical size thereby stopping the chain reaction. The pressure of the helium on the upper surface of the heavy water aids in discharging the heavy water from the reactor. This discharge is accomplished in response to neutron densities in the reactor, the circuit for this control being shown at D in Fig. 19.

Circuit D is similar to circuits A and B and again for the sake of simplicity similar reference characters for parts corresponding to those in circuit A are used but they are increased by two hundred.

An ionization chamber 400 controls the circuit and varies the current flowing through the relay coil 406. An armature 410 is normally held against contact 413 by spring 411 and is connected on its other end to an electromagnetic relay coil 334, which in turn is connected at its other end to the power line 225 through a normally closed circuit breaker 335 adapted to be operated manually. An armature 336 in the form of an arm pivoted intermediate its ends at 337 is normally held against the pole of the relay coil 334, as shown, on the opposite side of the fulcrum 337 from the relay coil 334 and a weight 338 is adapted to pivot the armature 336 in a counterclockwise direction about the fulcrum 337 when the attraction of the relay coil 334 ceases. A hand grab 339 also is provided for effecting this counterclockwise motion.

Attached to the end of the operating arm 161 is a cable 340 passing over a pulley wheel 341 and having a weight 342 attached to its free end. A rack 343 is fastened to the operating arm 161 and cooperates with a pawl 344, fastened to the armature 336, to hold the operating arm 161 so as to maintain the gate valves 160 closed. Upon removal of the pawl 344 from engagement with the rack 343, the operating arm 161 is released and is moved by the falling weight 342 to open the valves 160 so as to discharge the heavy water from the active part 19 of the reactor 18 into the sump 164.

Normally the armature 336 is disposed as shown in Fig. 19 in the position maintaining the gate valves 160 closed. This is accomplished by maintaining the electromagnetic relay coil 334 energized from the power source 325 through the control circuit D. In the event that the neutron density in the reactor 18 reaches a predetermined high value the ionization chamber 400 responding to the neutron density causes the magnetic reaction of the relay coil 406 to reach a value sufficient to move the armature 410 to the pole of the magnetic relay 406. Thus, circuit D is broken at the contact 413 thereby de-energizing the electromagnetic relay coil 334 causing

2,786,696

27

the armature 336 to move about its fulcrum 337 in response to the falling weight 338. This movement of the armature 336 withdraws pawl 334 from the rack 343 to release the weight 342 thereby causing the operating arm 161 to open the valves 160. The heavy water in the reactor tank 20 is thereby permitted to pass through the valves 160 and empty into the discharge header 103, and thence into the sump 164.

The initial calibration of the ionization chamber 400 is such that the relay coil 406 is energized only in the event that the neutron density in the reactor 18 reaches a value greater than that sufficient to operate either circuits A or B. In other words, circuit D operates to discharge the heavy water from the active portion of the reactor only as a last resort in the case of emergency, and only after circuits A or B have failed to operate or otherwise to prevent the neutron density in the system from rising above that required to operate circuit A or B.

Circuit D, as shown, will also function to open the discharge valves 160 in the event that there is a failure in the power mains 225. On failure of the power source, the electromagnetic coil 334 becomes de-energized thereby releasing the armature 336 to open valves 160 in the manner previously explained.

As an added precaution, a hand operated circuit breaker 335 is placed in series in the circuit D so that the heavy water can be discharged from the reactor tank 20 merely by manually opening the circuit breaker 335.

While the rods 22 are disposed inside the aluminum tubes 40, the cooling water is passed over the rods and heat is thus conveyed away in the form of sensible heat, thus causing the cooling water to increase in temperature. In the system shown herein, it is important that the cooling water is not permitted to boil for otherwise the density of the coolant in the tubes 40 will drop resulting in a rapid increase in the reproduction ratio for the system. Thus, the control is thrown out of balance and it may be difficult, if not impossible again to stabilize the reaction. Consequently, various factors such as the control of the reaction, and the rate of flow of cooling water, are preferably so regulated as to prevent the cooling water from boiling in the tubes when they are surrounded by the heavy water moderator.

Upon discharging the uranium rods from the aluminum tubes 40 into the shield water 65, the rods become immersed in a more or less stationary body of water so that boiling of the water coming in contact with the uranium rods takes place. This is advantageous because the efficiency of the cooling is enhanced by the fact that the latent heat of evaporization is utilized to remove heat from the uranium rods Thus there is no problem of cooling after the rods 22 have once been discharged from the aluminum tubes 40 into the shield water 65.

In the arrangement (Fig. 4) where the water enters the cooling system at 86 and leaves at 87, if for any reason the cooling system should fail while the active uranium rods 22 are disposed inside the active portion 19 of the reactor 18, the embodiment disclosed herein has the additional advantage of providing an automatic cooling system for removing the heat in the rods 22 generated as the result of delayed fissions and radioactive decay of fission products in the uranium so as to forestall boiling in tubes 40. As soon as the flow of cooling water through the tubes 40 ceases, the shield water 65 directed by its 40 foot head enters the open ends of the tubes 40 and passes in reverse direction through these tubes and over the active uranium rods 22 causing the shield water 65 to drain from the tank 58 through pipe 24, pump 28 and finally through pipe 86. Thus cooling of the uranium rods 22 is maintained and can be continued indefinitely assuming some means are furnished for replenishing the shield water 65 in the tank 58. This can be done by the provision of a separate water supply system, not shown, of conventional type.

28

Construction

When constructing the system above described, the fabrication of the reactor is important at least insofar as concerns the manner by which the heavy water is placed in the active portion of the reactor. The entire reactor shell with all of its fittings including the tubes 40 should be completely fabricated and all sealed joints made without either the uranium or the heavy water being in the reactor. The control rods, if used, should be placed into position and the proper seals made. All external circuits should be installed and placed in readiness for use.

After the system has been completely fabricated, the uranium rods can be inserted, together with their dummy rods 69. The holding plates 70 are in position to hold the rods in place in the reactor. Next after checking to be sure the valve members 134 are closed, the shield water 65 is placed in the shield and the flow of cooling water through the tubes 40 is started. Likewise, the cooling water flow through the shield 90 is started. The control rods 168 are inserted entirely into the active portion of the reactor as are any additional emergency rods that may be used. The heavy water is now pumped slowly into the reactor tank 20 by pump 166, withdrawing heavy water from the sump 164.

The amount of heavy water required to effect the critical size of the active portion of the system having been estimated, the flow of the heavy water into the tank 20 is continued slowly until this calculated amount is approached. The pumping is then stopped and the control rods 168, together with any other safety rods, are slowly withdrawn one at a time from the active portion of the reactor. Neutron density readings are then made in a well known manner, for example, by the use of a previously calibrated ionization chamber. By repeating this process a number of times while the overall size is still below the critical size, the critical size can be predicted from readings thus made by appropriate mathematical calculations. The heavy water is then pumped slowly into the reactor tank 20 until this critical size is reached. The critical size is approached with caution with the control and safety rods in place in the active portion of the reactor and many check readings are made to make certain that the size does not exceed the critical size by more than the desired amount, which, as previously mentioned, is just sufficient to provide a reproduction ratio, with the control and emergency rods removed, of slightly greater than unity, and never greater than about 1.005.

Due to the comparatively small size of the reactor, the expansion and contraction of the various members does not create a serious problem. The presence of the light water surrounding the reactor provides cooling of the metal and serves to control the expansion. The flexibility of the tube sheets forming the end walls and chamber separating walls at the ends of the reactor compensates for expansion of the tubes.

Referring to the modification shown in Figs. 20–23 and first to Fig. 20 of the drawings, numeral 501 denotes a neutronic reactor tank or container of cylindrical shape and of a material that is relatively non-corrosive at low temperatures and that is relatively non-absorbent with respect to neutrons, such as, for example, aluminum or stainless steel. A suitable size of such tank for a self-sustaining chain reaction when D_2O is used as a moderator is one that is 6 feet in diameter and 7 feet 4 inches high, although other sizes may be used as well. Suspended in tank 501 are one hundred and thirty-six rods 502 of uranium metal which are 1.1 inches in diameter and are sheathed by aluminum about .035 inch thick to prevent extreme radioactivity and contamination of the D_2O moderator by fission products emanating from the uranium. Such coating also prevents corrosion of the uranium by the heavy water. Rods 502, described in detail hereinafter, extend to within about ¼ inch of the tank bottom. Sufficient deuterium oxide is introduced into tank 501 to

2,736,696

29

obtain a volume of uranium and heavy water slightly over the critical size. Such critical size may be predicted well in advance of attainment thereof. This is done by taking measurements of the neutron density, preferably adjacent to the perimeter of the tank, for example, in the reflector 504, described hereinafter, as the tank is being filled, and by plotting, for instance, the reciprocals of such neutron densities as ordinates against some measure of the overall size of the filled portion of the tank (e. g., the volume of D₂O) as abscissas, as the size is being increased by raising the level of heavy water in the tank. This will give a curve that, when extrapolated, will indicate or forecast the critical size at the point where the curve crosses the axis of abscissae. When the critical size has been exceeded a self-sustaining nuclear reaction is initiated. In the reactor described, the critical size was obtained when the tank 501 was filled to a D₂O level of 122.4 centimeters from the bottom, and an operating size giving a neutron density doubling time of 37.6 seconds was obtained at a D₂O level of 123.1 centimeters. At a level of 124.7 the doubling time was 6.52 seconds.

Immediately surrounding tank 501 is a neutron reflector 504 of graphite, for example, having substantially a cup shape. Reflector 504 and tank 501, together with its contents, are referred to as the neutronic reactor. The moderator and the uranium immersed in the moderator constitute what may be termed the active portion of the reactor. The thickness of the reflector 504 may be of the order of two or three feet, or more, depending on the size of the active portion and the degree of neutron scattering required. By using this peripheral layer of scattering or reflecting material, the overall size of the active portion of the reactor may be made somewhat smaller than in a case where no scattering layer is employed, since neutron losses to the exterior are effectively reduced. Tank 501 and its contents may be built slightly below the diameter that would be required without a reflector so that the addition of reflector 504 with its neutron reflecting action will convert the reactor from one that is not self-sustaining to one that is self-sustaining.

A concrete shield 505 surrounds the graphite reflector 504 and serves to prevent neutrons and gamma radiations from escaping to the outside of the structure. The carbon in reflector 504, being a relatively light element, absorbs gamma rays only to a relatively small extent. The concrete shield may be of the order of five or ten feet in thickness. The water of crystallization in the concrete absorbs escaping neutrons.

Generally speaking, the higher the atomic weight of an element, the better it serves as a shield to prevent escape of penetrating radiations, such as gamma radiations. Lead, therefore, is an excellent material for a shield for certain purposes while water of the same thickness is only fair. However, thick water shields are sometimes convenient and satisfactory. Interposed between the concrete shield 505 and graphite reflector 504 is a cooled metal shield 507, preferably of a lead-cadmium alloy, having cooling tubes 507a passing therethrough, through which light water, that is, ordinary water, or other coolant, may be circulated. Shield 507 is especially adapted not only to cool the outside of the neutronic reactor but to minimize the escape of gamma and other penetrating radiations. A suitable thickness may be 4 inches or more.

A cover plate 508, for example, of stainless steel, is provided at the top of tank 501 and serves as a thermal shield as well as a support for rods 502 and for control and safety rods pivotally mounted thereunder, described infra. Above the cover plate 508 is a space 509 into which the upper ends of rods 502 project, and through which piping can be conducted. Above space 509 is a shield 509a having alternate layers of iron and "Masonite" brand of steam-exploded wood fiber for preventing the escape of neutrons, gamma rays, and other penetrating radiations from the top of tank 501 to the outside.

30

A tube or well 509b of any suitable diameter, for example 4 inches, extends through shields 509a, space 509, and plate 508 into the tank 501, preferably axially of the tank. Objects may be introduced from the exterior through the tube 509b and into the center of the tank 501 so that they may be bombarded by high intensity neutron radiations for the production of radioactive isotopes, or for other purposes desired. Aluminum is suitable for tube 509b.

A pan 515 of stainless steel or other suitable material is located at the bottom of reflector 504 for collecting any heavy water that may leak from tank 501, inasmuch as heavy water, at present, is relatively expensive. The collected heavy water is drained as indicated by the arrow A.

Helium at substantially atmospheric pressure is introduced through pipes 516 into the top of reactor tank 501, that is, above the level of the heavy water, and thence is circulated to the exterior of the tank 501 through pipes 517. As the result of high neutron densities and heat developed during the operation of the chain reaction in tank 501 some of the deuterium oxide will decompose into D₂ and O₂. These uncombined gases will collect at the top of tank 501. In order to dilute and to remove such uncombined gases, recombine and recondense them without explosive effect, a gas circulating system of any well-known type (not shown) may be used for circulating the helium together with the uncombined gases into a recombiner of any well-known type (not shown) such as a hot grid or platinum-charcoal catalyst, or both, for effecting recombination of the D₂ and O₂ into heavy water, and for returning the condensed D₂O into tank 501. Helium may be circulated, for example, at the rate of 3 cubic feet per minute. A suitable ratio of D₂ to helium may be 1 to 150, although other dilutions may be used instead.

Space 509 also carries D₂O inlet pipe 501a by which D₂O can be supplied to the top of reactor tank 501. Outlet pipe 501b is positioned at the bottom of tank 501. The D₂O is circulated through external heat exchangers (as heat exchangers 27a of Fig. 2) through pipes 501a and 501b to cool the moderator and thereby cool rods 502. Suitable valves and additional piping are provided to direct the heavy water to a storage tank (not shown).

Immediately below coverplate 508 there is provided a cadmium sheet 518 to act as a shield for minimizing the escape of slow neutrons. Immediately below cadmium sheet 518 there is provided an aluminum sheet 519 which is useful primarily to prevent electrolytic action between otherwise dissimilar metals (cadmium and aluminum) in the interior of tank 501.

Shield 509a is supported by two pairs of crossed I beams 520 and 521. Both pairs of I beams have their extremities supported by the concrete shield 505.

Referring to Fig. 21, numeral 502 denotes one of the composite uranium containing rods that is suspended in the deuterium oxide moderator as shown in Fig. 20. Rod 502 comprises a cylindrical rod of uranium or uranium containing material 522 six feet long that is screw-threaded into a supporting rod 523 of non-fissionable material such as, for example, aluminum. A thin tubing of aluminum 524 is drawn on to the outer surface of uranium rod 522 and supporting rod 523 by any well-known drawing process by screwing on an attachment 537 (see Fig. 22) that is subsequently removed. Thereafter, the joints formed at the top of supporting member 523 and at the bottom of rod 522 are welded so as to form an air-tight seal in tubing 524, thus protecting the uranium rod 522 from the effects of the D₂O. A narrow longitudinal groove 522a is provided between the connected elements 522—523 and tubing 524. An axial groove 525 and communicating radial groove 525a are provided at the top of supporting rod 523 that communicate with space 522a thereby making it possible to evacuate the air space and thereby test for possible leaks in the top and bottom welded joints in tubing 524 by noting the pressure change

2,786,696

31

interiorly of the air space after a predetermined evacuation.

A ball valve 526 is provided to seal the space 522a from the atmosphere. A set screw 527 is screwed downwardly of supporting rod 523 so as to firmly seat the ball valve 526. A shank 531 is provided and supporting member 523 is screw-threaded thereto. An upwardly extending aluminum sleeve 534 is screwed to shank 531 and closed by an upper flanged member 533. A lead rod 532 is positioned between the lower shank 531 and upper flanged member 533 for the purpose of reducing gamma ray escape axially of rods 502. The flanges of member 533 are supported on the top of a tube 535 screwed to cover plate 508. Flanged member 533 is provided with an internal thread 533a by which the rod assembly can be attached to a lifting crane, for example, for installation and removal.

The neutron chain reaction may be prevented merely by immersing into the heavy water one or more rods, such as hollow rods 511 and 512 (Figs. 20 and 23), four inches in diameter, containing a 1/16 inch layer of cadmium sandwiched between aluminum walls, and pivotally supported by the cover plate 508. Rod 511, for example, may be operated as a control or regulating rod, being immersed to greater or less extent in the body of heavy water, as desired. Rod 512 may be considered as a safety rod which normally is held out of contact with the body of heavy water in tank 501 and is immersed into the heavy water only for emergency purposes, that is, when control rod 511 per se is insufficient for immediate stoppage of the chain reaction. Rod 511 may be pivotally mounted and rigidly secured to a shaft 513, extending through the tank 501 through sealed bearings 540. A method of automatic control is to make rotation of shaft 513 responsive to the neutron density at a peripheral portion of reflector 504, for example, as indicated by ionization chambers such as chamber 541 having suitable amplifiers (not shown) so as to be effective to control the neutron density of the reactor and keep it substantially constant. When manual control is used the response of ionization chamber 541 is used to monitor reactor power when operating.

In utilizing the output of the reactor, well 509b plays an important role. It extends through the center of the reactor where the highest neutron density exists, and intense neutron bombardment of materials inserted into this well will take place, even at reltaively low reactor powers.

The reactor just described has been operated continuously at 250 kilowatts when filled to higher levels and properly shimmed by a shim rod 550 to compensate for operational poisoning.

Miscellaneous

In the embodiment of the invention disclosed herein as the preferred form, the cooling water is circulated over the outside surface of the uranium rods 22. Thus, there is a film of water between each uranium rod 22 and the walls of the aluminum tube 40. It is therefore necessary for neutrons passing from the heavy water moderator to the uranium rods 22 to pass through this film of light water. As previously mentioned, light water has a high absorption for thermal neutrons and therefore some of the thermal neutrons leaving the moderator are parasitically absorbed in the light water before they have an opportunity to enter the uranium rods 22 to produce fissions. This is not true if the coolant passes through the axis of the uranium only, so as to cool the uranium internally.

Another consideration in an external cooling system of the type disclosed herein where the coolant is passed externally over the uranium bodies 22 is that neutrons must pass through two layers of aluminum or other protective coating before entering the uranium. Some thermal neutrons are parasitically absorbed in the aluminum and therefore are lost to the chain reaction. This is an-

32

other consideration effecting the reproduction ratio of a chain reacting system.

In this description, reference has been made primarily to light water as the coolant for the system. As previously mentioned, diphenyl, also known as biphenyl or phenylbenzine, is also a satisfactory liquid coolant. This substance has the chemical formula $C_6H_5C_6H_5$ and is in the form of a solid at atmospheric temperature, melts at 70° C. and has a boiling point of 225° C. Thus, a system employing diphenyl as the coolant must be operated at a temperature such that the coolant at all times maintains a temperature above 70° C. The upper temperature at which this coolant can be maintained is, of course, limited primarily by the boiling point of the heavy water moderator.

Diphenyl has a lower absorbing characteristic for thermal neutrons than has light water; for example, a loss of neutrons by absorption due to a diphenyl cooling layer 4 millimeters corresponds to the loss due to a 2.2 millimeter layer of light water. Thus, for the same reproduction ratio almost twice as much diphenyl as light water can be circulated through the reactor. The cooling passages then for a system employing diphenyl can be almost twice as wide as those for light water in systems having the same reproduction ratio, and greater power can therefor be removed. About 10 per cent to 15 per cent more pumping power is required to circulate the diphenyl due to its greater viscosity and since diphenyl solidifies at 70° C., special measures are taken to prevent solidification of the coolant while it is disposed in the portion of the cooling circuit outside of the reactor. This, of course, can be accomplished by merely maintaining the temperature of diphenyl at all times safely above 70° C. by regulating the amount of heat extracted from the diphenyl. Since the permissible temperature rise of diphenyl is necessarily limited to the comparatively narrow range between the solidification temperature of diphenyl and the boiling point of heavy water, it may be desirable to place the heavy water under a relatively high pressure so as to raise the boiling point of heavy water and hence the upper limit of the permissible operating temperature range.

During the operation of the neutronic reactor particularly at high neutron densities radioactive elements of exceedingly high capture cross section may be formed in the uranium as an intermediate element in the decay chains of fission fragments and this formation will lower the value of the reproduction factor for the system. Radioactive xenon 135 is an example of such an intermediate element, this product having a half life of about 9 hours and being formed mostly from radioactive iodine which has a half life of about 6.6 hours and decays to barium. There should be sufficient excess in the reproduction ratio of the reactor so that in the event the reproduction factor is reduced as a result of the formation of an intermediate decay element having a high capture cross section for neutrons the control rods may be withdrawn sufficiently to maintain the reproduction ratio at a value of unity while maintaining the power output at the desired level. It might be desirable to initially construct the reactor sufficiently over-sized to supply this excess reproduction ratio when needed and in this event removable impurities for example in the form of additional shim or control rods may be initially placed in the reactor and kept there at all times until the reproduction ratio commences to fall as the result of the formation of these intermediate decay elements.

Although heavy water has been mentioned throughout this description as the deuterium compound employed for the moderator, it is recognized that other compounds of deuterium will also be satisfactory, and for example, as deutero carbons such as deutero diphenyl and deutero benzene.

Throughout this description natural uranium with its U^{235} content has been mentioned as a source of the fis-

2,736,696

33   34

sionable isotope but it is recognized that there are other fissionable isotopes that will serve equally well, such as U^{233} and 94^{239}.

What is claimed is:

1. In a neutronic system, a reactor tank in the form of a horizontally disposed tank, an active portion including a liquid neutron moderator in the tank, a moderator leakage chamber adjacent to an end of the tank, a second chamber adjacent to the leakage chamber, tubes in the tank passing through the moderator and said tank end wall and through both chambers, a gas in the second chamber under pressure greater than the moderator in the leakage chamber, and means for withdrawing moderator from the leakage chamber.

2. In a neutronic system, a reactor including an active portion, a plurality of tubes in the active portion and projecting through one side of said portion, a fluid neutron moderator in the active portion said tubes passing through said moderator, a leakage chamber adjacent to the active portion on the tube projecting side and adapted to receive liquid moderator leaking from the active portion past the tubes, means for withdrawing moderator from the leakage chamber, and means for preventing the loss of moderator from the leakage chamber including a supply of gas on the outside of the leakage chamber under pressure greater than that of the leakage chamber.

3. In a neutronic system, a reactor including an active portion, a plurality of tubes in the active portion projecting through one side of said portion, a fluid neutron moderator in the active portion said tubes passing through said moderator, a leakage chamber adjacent to the active portion on the tube projecting side and adapted to receive liquid moderator leaking from the active portion past the tubes, means for withdrawing moderator from the leakage chamber, and means for preventing the loss of moderator from the leakage chamber including a high pressure gas chamber, and a quantity of gas in said gas chamber under a pressure greater than that of the moderator in the leakage chamber.

4. A neutronic system submerged in light water and including a reactor comprising a reactor tank, a fluid neutron moderator in the tank, a plurality of tubes projecting through a tank wall and through the moderator, a moderator leakage chamber adjacent to said tank wall and adapted to receive moderator leaking past the tubes, a high pressure chamber adjacent to the leakage chamber, a light water leakage chamber adjacent to the high pressure chamber, and a quantity of fluid in the high pressure chamber under a pressure greater than that in either of the other chambers.

5. A neutronic system submerged in light water and including a reactor comprising a reactor tank, a moderator leakage chamber, a high pressure chamber, and a water leakage chamber, the tank and all chambers being disposed in tandem, a quantity of fluid neutron moderator in the reactor tank, a quantity of fluid in the high pressure chamber under a pressure greater than that in either of the other chambers, and a plurality of tubes in the tank projecting through the moderator the tank and the chambers and opening into the water surrounding the system, the moderator leakage chamber being adapted to receive moderator leaking from the reactor tank past the tubes, and the water leakage chamber being adapted to receive light water leaking past the tubes projecting into the water.

6. A neutronic system including a reactor comprising a reactor tank, a moderator leakage chamber, a high pressure chamber, and a water leakage chamber, the tank and all chambers being disposed in tandem in the order named, a quantity of fluid neutron moderator in the reactor tank, a plurality of tubes in the reactor tank projecting through the moderator the tank and the chambers, the moderator leakage chamber being adapted to receive moderator leaking from the reactor tank past the tubes, a body of water disposed outside the water leakage chamber and submerging the tubes projecting from water leakage chamber, the water leakage chamber being adapted to receive water leaking past the tubes, and a quantity of fluid in the high pressure chamber under a pressure greater than that in either of the other chambers.

7. An apparatus for use in a neutronic system comprising a reactor tank having therein a reactor active portion, a moderator leakage chamber, a high pressure chamber, a water leakage chamber, the tank and all chambers being disposed in tandem, and a plurality of tubes projecting through the tank and the chambers, said high pressure chamber having fluid inlet means, the moderator leakage chamber being adapted to receive liquid neutron moderator leaking from the tank past the tubes, and the water leakage chamber being adapted to receive fluid leaking past the tubes projecting through the wall of the chamber opposite the reactor tank.

References Cited in the file of this patent

UNITED STATES PATENTS

2,206,634   Fermi et al. _____ July 2, 1940

FOREIGN PATENTS

114,150   Australia _____ May 2, 1940
114,151   Australia _____ May 3, 1940
861,390   France _____ Oct. 28, 1940
233,011   Switzerland _____ Oct. 2, 1944

OTHER REFERENCES

Nature (May 13, 1939), page 793.
Naturwissenschaften, vol. 27 (1939), pages 402–410.
Business Week (Sept. 1, 1945), pages 57–64.
Goodman: "The Science & Engineering of Nuclear Power," vol. 1, pages 275, 387–501, Addison-Wesley Press, Inc. (1947).
Kelly et al.: Phy. Rev. 73, 1135–9 (1948).

Sept. 4, 1962   E. P. WIGNER ETAL   **3,052,613**

METHOD AND APPARATUS FOR CONDUCTING A NUCLEAR CHAIN REACTION

Filed Aug. 29, 1945   12 Sheets—Sheet 1

FIG.1.

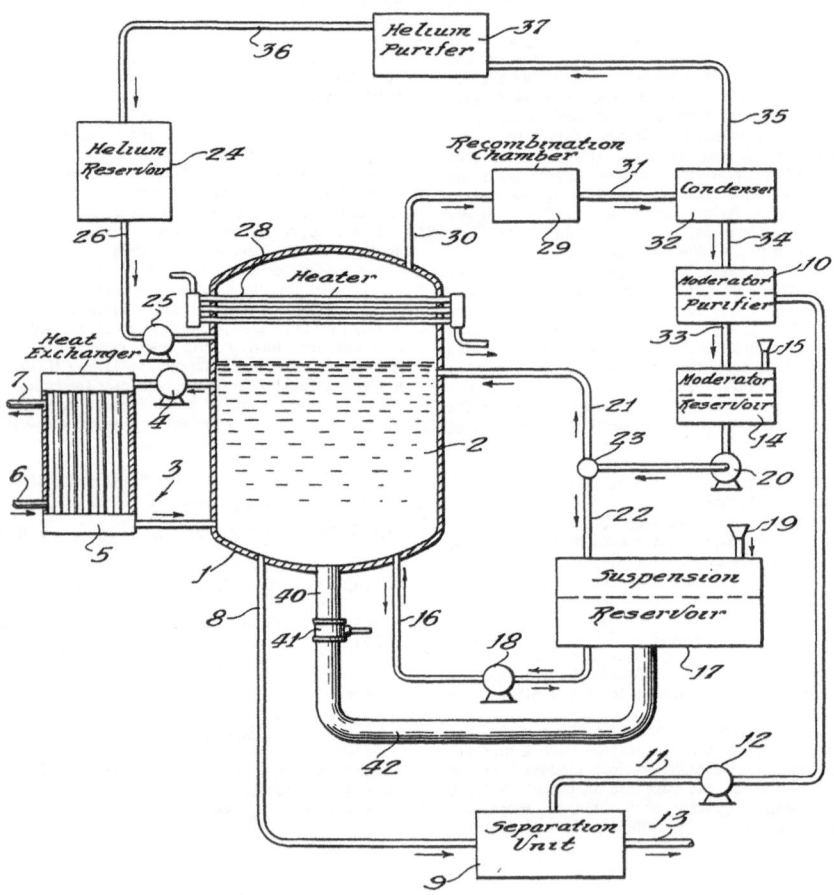

Witnesses:
Hubert E. Metcalf
Henry W. Johnson

Inventors:
Eugene P Wigner
Leo A. Ohlinger
Gale Young
Alvin M. Weinberg
By: Robert H. Lavender
Attorney.

Sept. 4, 1962   E. P. WIGNER ETAL   3,052,613

METHOD AND APPARATUS FOR CONDUCTING A NUCLEAR CHAIN REACTION

Filed Aug. 29, 1945   12 Sheets-Sheet 2

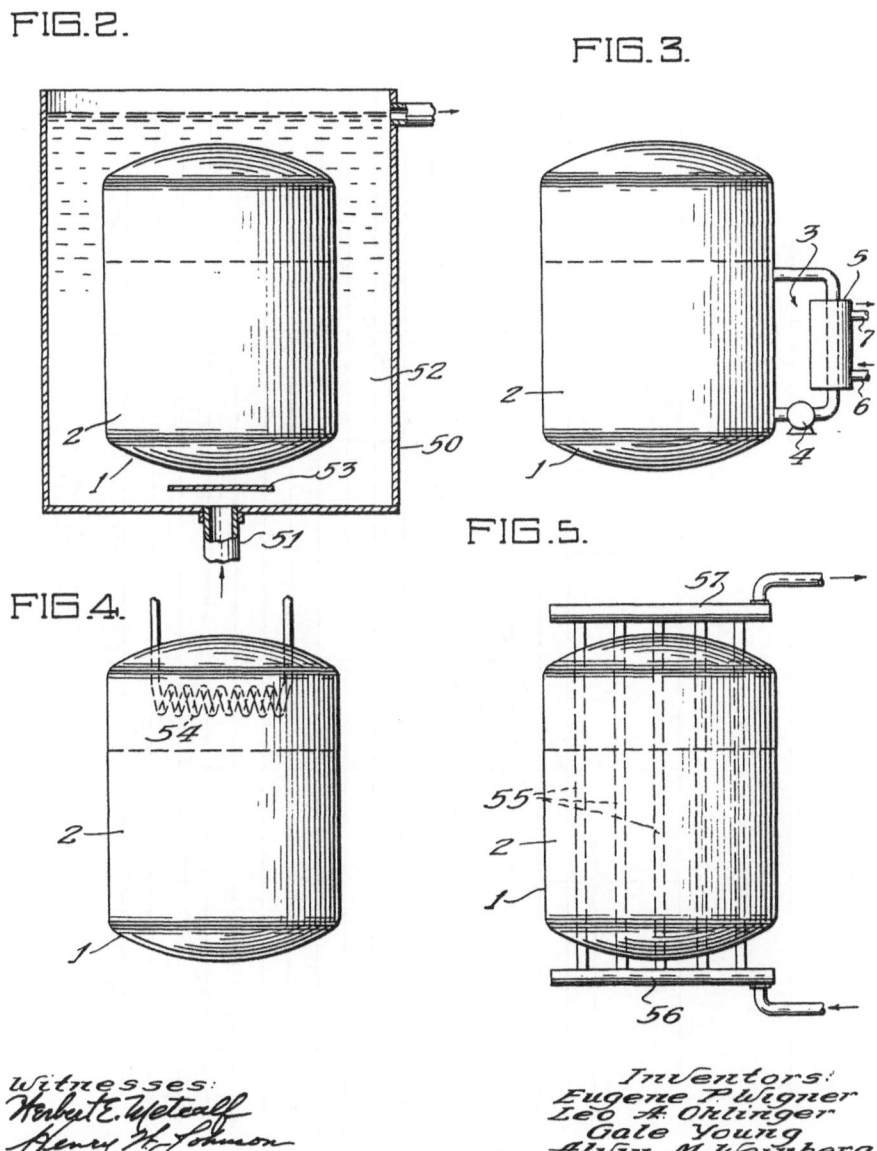

FIG.2.

FIG.3.

FIG.4.

FIG.5.

Inventors:
Eugene P. Wigner
Leo A. Ohlinger
Gale Young
Alvin M. Weinberg
By Robert A. Legende
Attorney.

Sept. 4, 1962   E. P. WIGNER ETAL   3,052,613

METHOD AND APPARATUS FOR CONDUCTING A NUCLEAR CHAIN REACTION

Filed Aug. 29, 1945   12 Sheets—Sheet 3

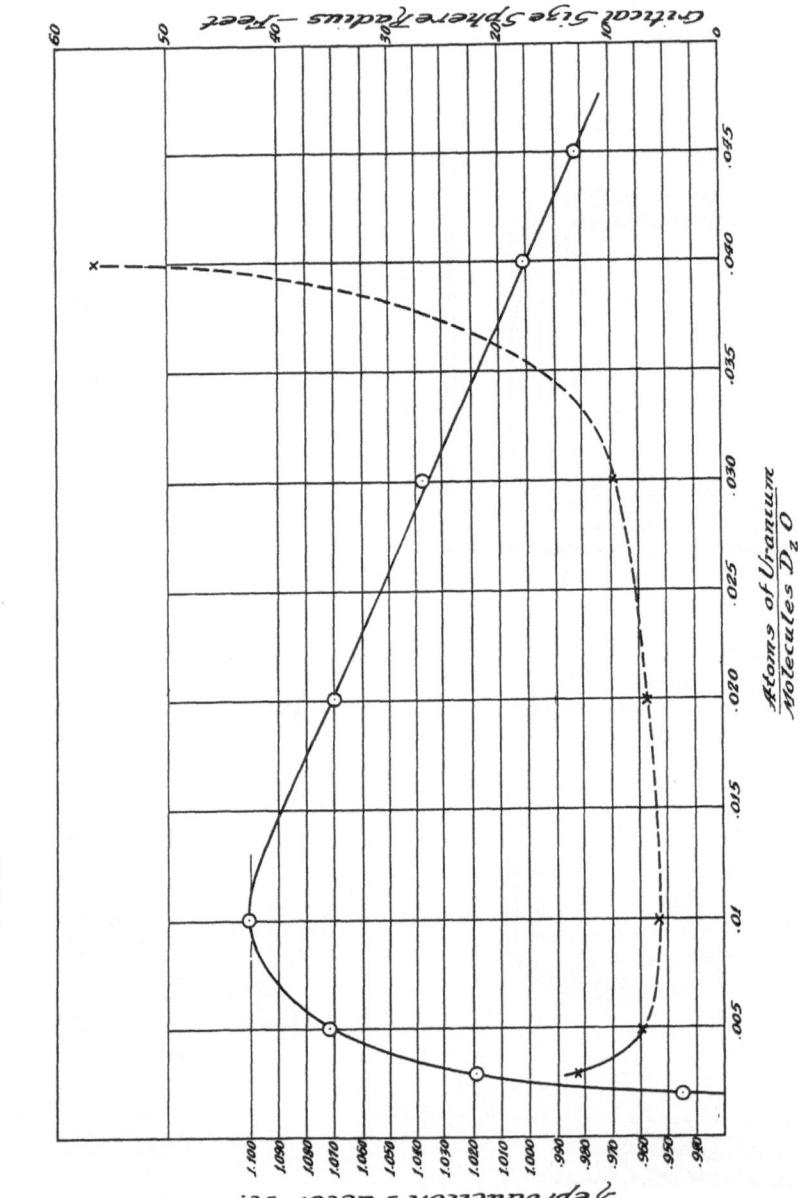

FIG.6.

Witnesses:
Herbert E. Metcalf
Henry W. Johnson

Inventors:
Eugene P. Wigner
Leo A. Ohlinger
Gale Young
Alvin M. Weinberg
By: Robert A. Fernander
Attorney.

Sept. 4, 1962   E. P. WIGNER ETAL   3,052,613

METHOD AND APPARATUS FOR CONDUCTING A NUCLEAR CHAIN REACTION

Filed Aug. 29, 1945   12 Sheets–Sheet 4

FIG.7.

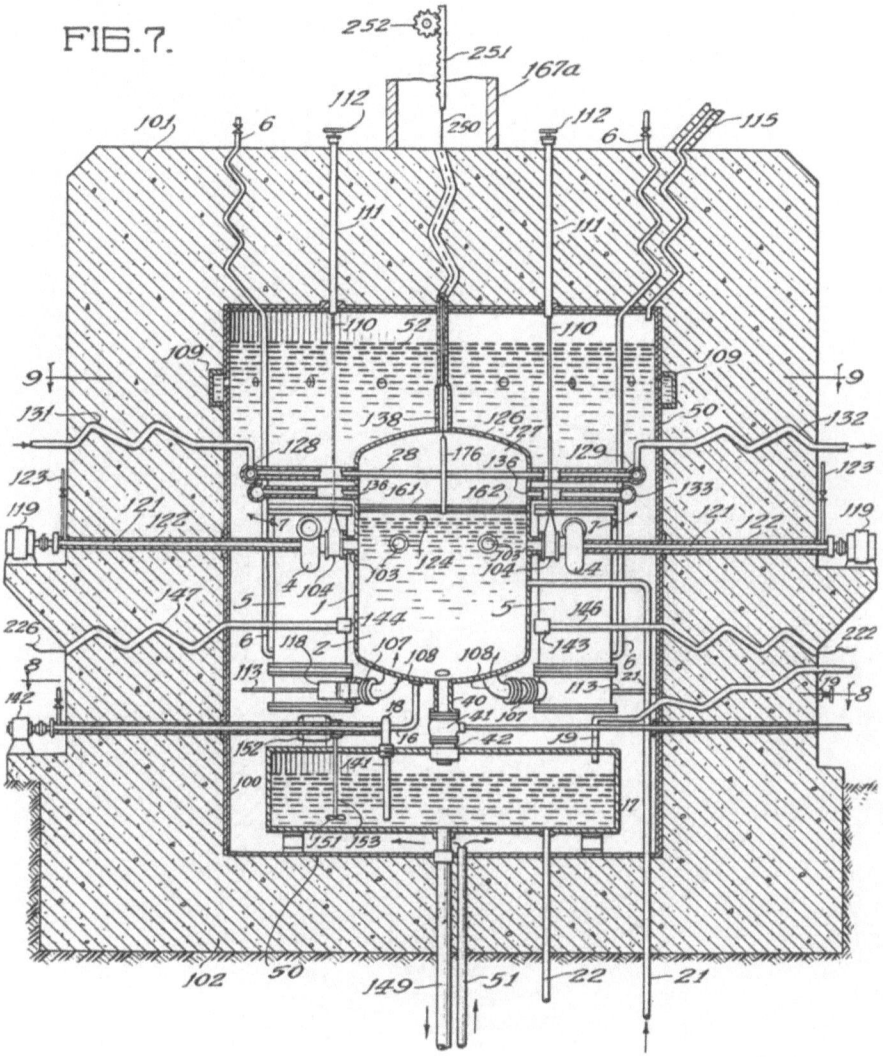

Witnesses:
Hubert E. Metcalf
Henry H. Johnson

Inventors:
Eugene P. Wigner
Leo A. Ohlinger
Gale Young
Alvin M. Weinberg
By: Robert A. Lavender
Attorney:

Sept. 4, 1962   E. P. WIGNER ETAL   3,052,613
METHOD AND APPARATUS FOR CONDUCTING A NUCLEAR CHAIN REACTION
Filed Aug. 29, 1945   12 Sheets—Sheet 5

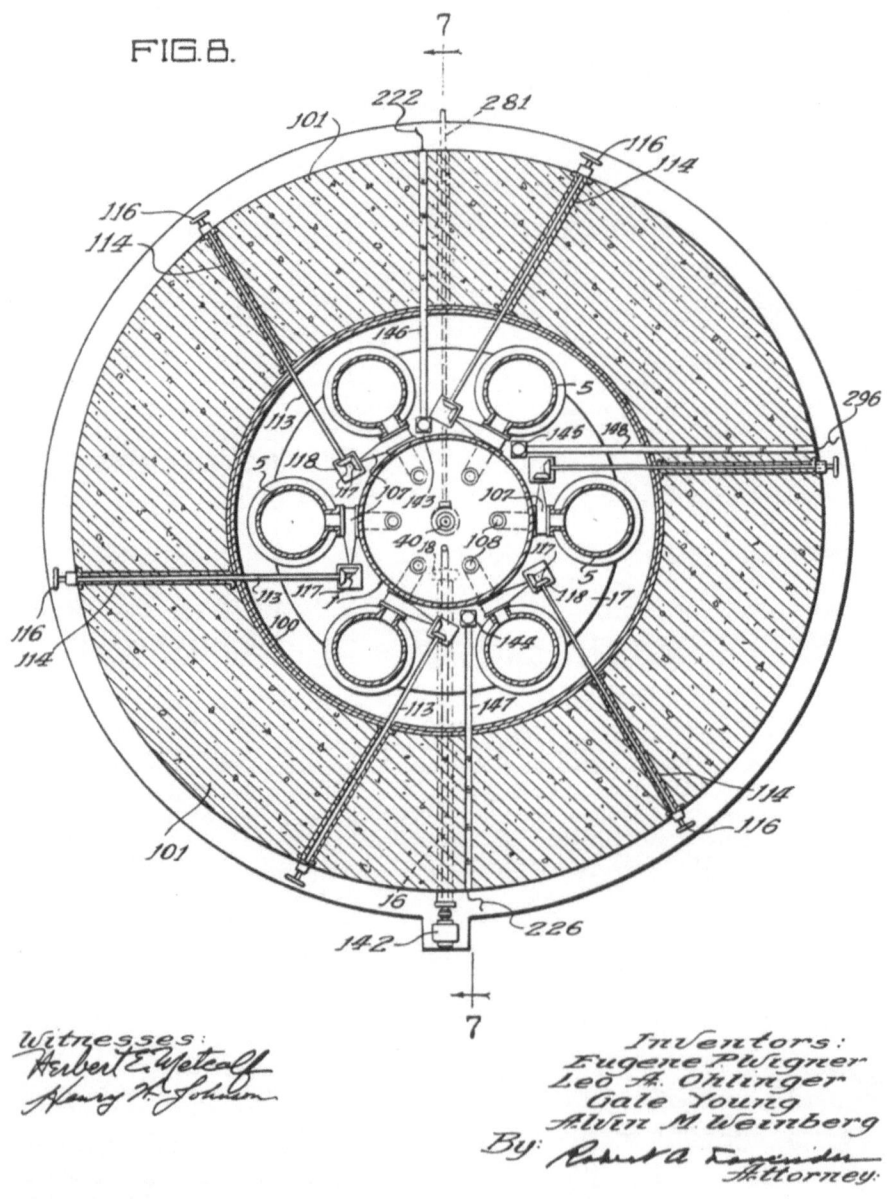

FIG.8.

Witnesses:
Herbert E. Metcalf
Henry H. Johnson

Inventors:
Eugene P. Wigner
Leo A. Ohlinger
Gale Young
Alvin M. Weinberg
By: Robert A. Ramser
Attorney.

Sept. 4, 1962   E. P. WIGNER ETAL   3,052,613

METHOD AND APPARATUS FOR CONDUCTING A NUCLEAR CHAIN REACTION

Filed Aug. 29, 1945   12 Sheets-Sheet 6

FIG.9.

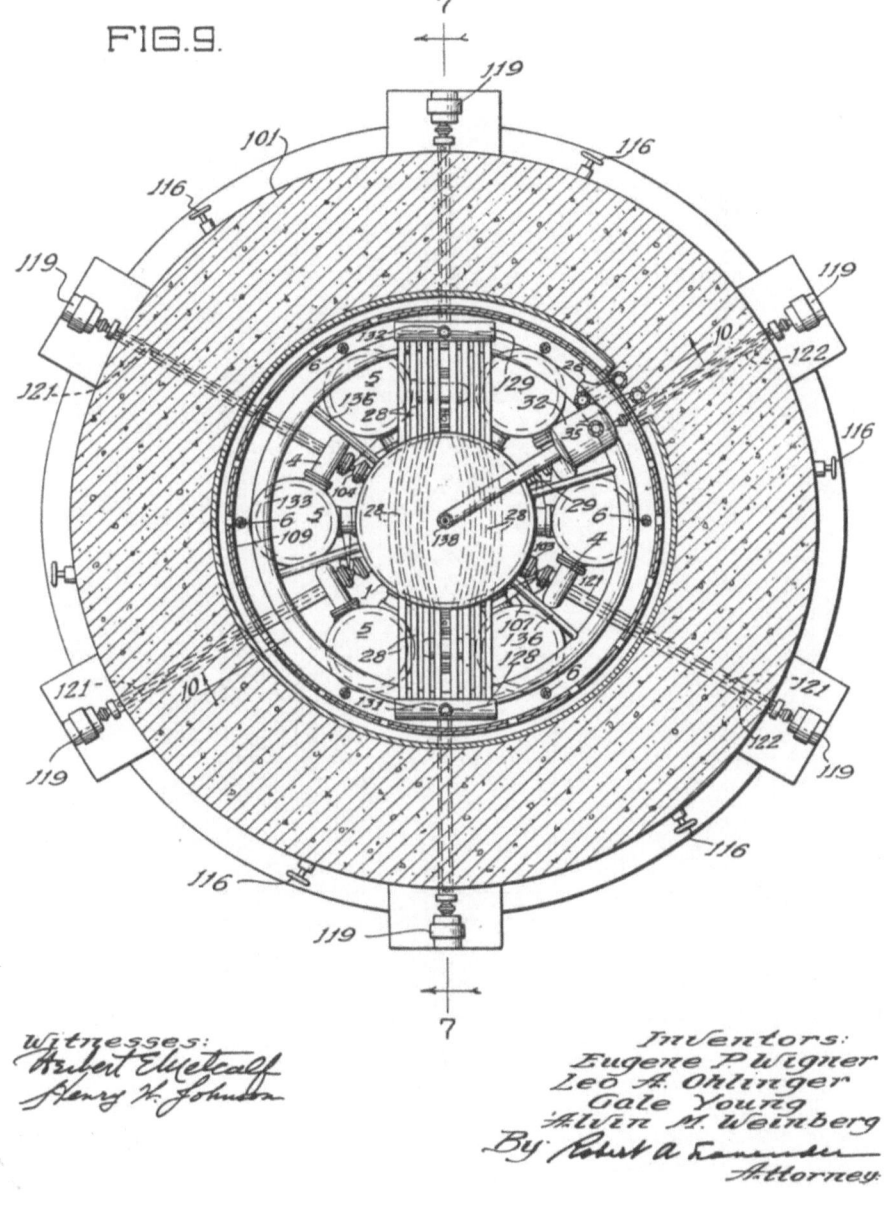

Witnesses:

Inventors:
Eugene P Wigner
Leo A. Ohlinger
Gale Young
Alvin M. Weinberg
By Robert A. —
Attorney.

Sept. 4, 1962   E. P. WIGNER ETAL   3,052,613
            METHOD AND APPARATUS FOR CONDUCTING A NUCLEAR CHAIN REACTION
Filed Aug. 29, 1945   12 Sheets–Sheet 7

FIG.10.

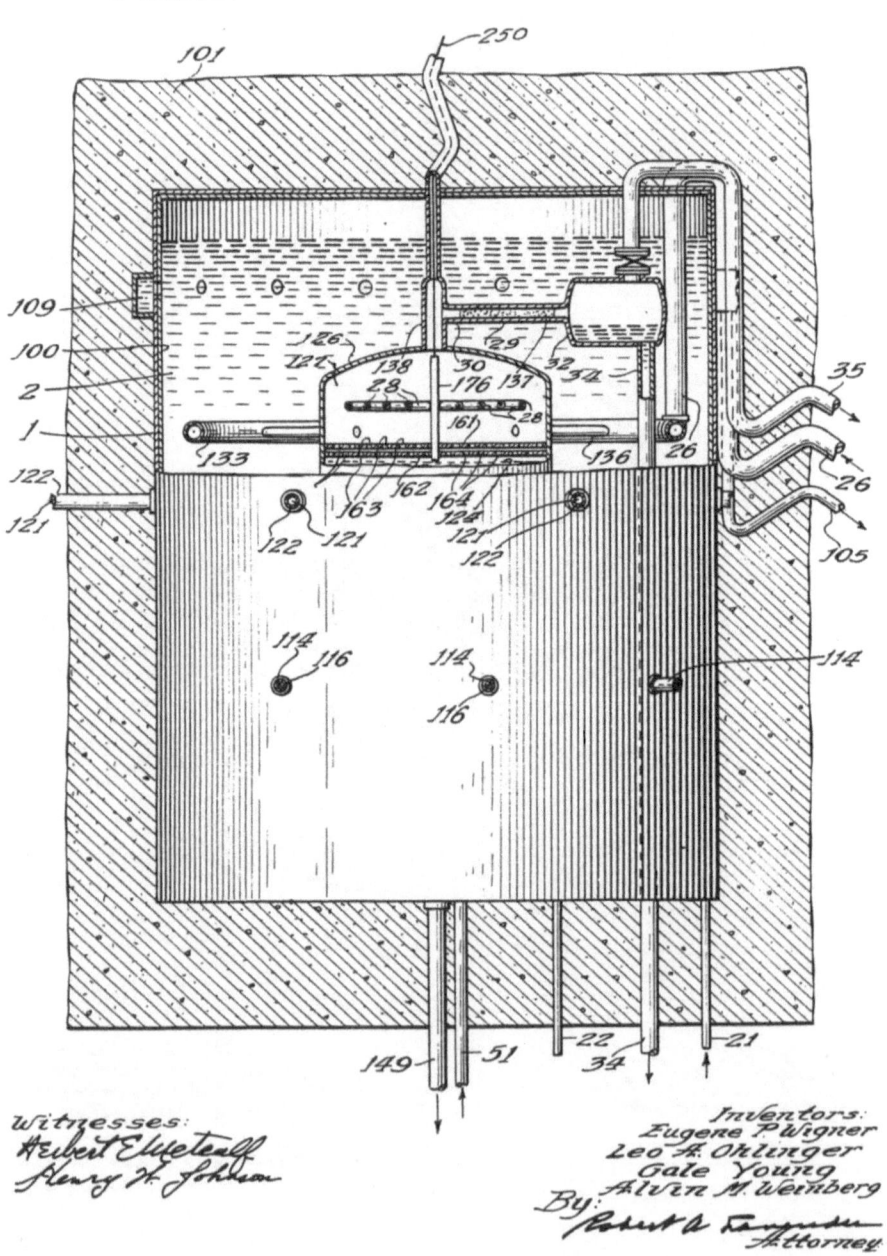

Sept. 4, 1962   E. P. WIGNER ETAL   3,052,613

METHOD AND APPARATUS FOR CONDUCTING A NUCLEAR CHAIN REACTION

Filed Aug. 29, 1945   12 Sheets–Sheet 8

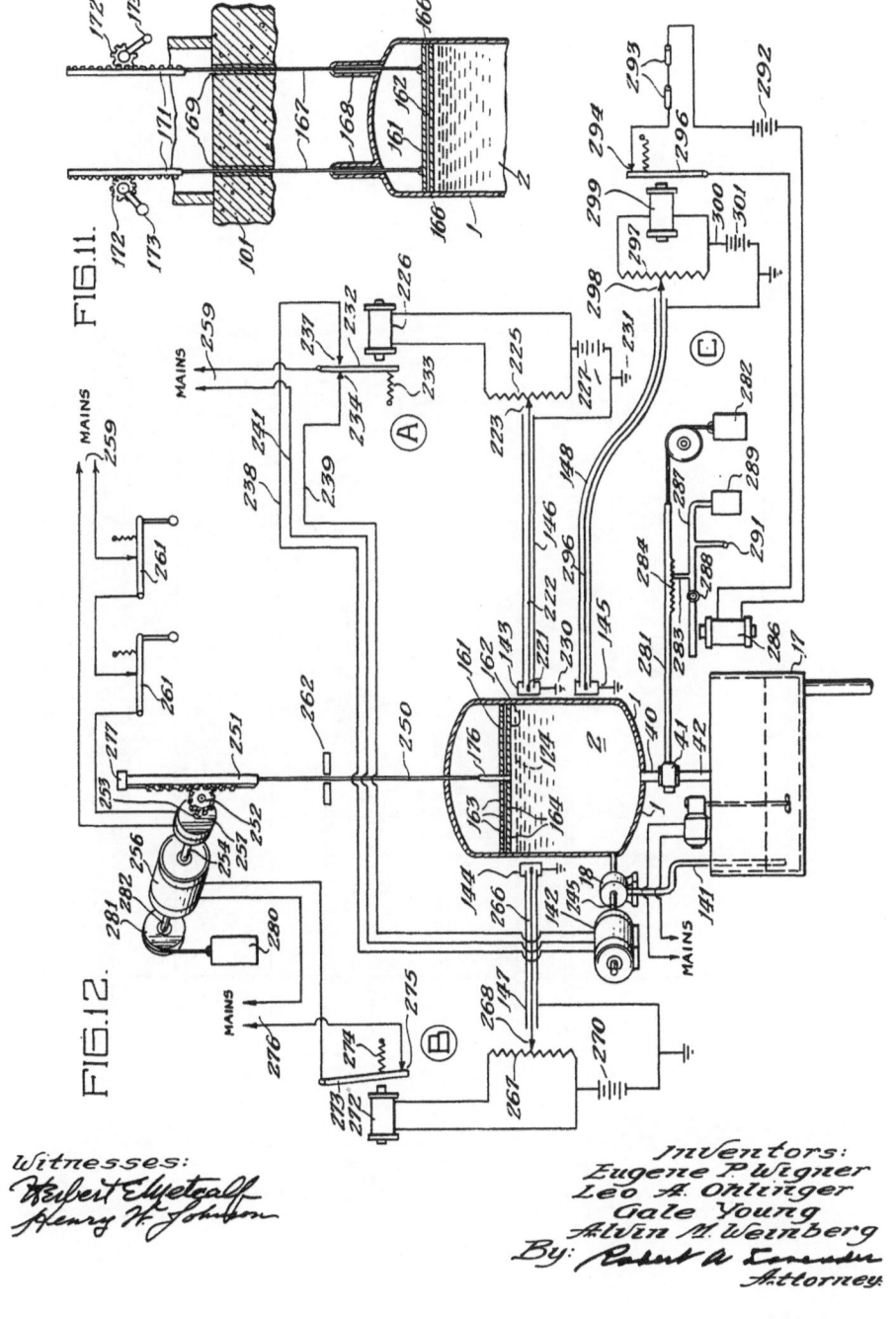

FIG.11.

FIG.12.

MAINS

Witnesses:
Hubert E. Metcalf
Henry W. Johnson

Inventors:
Eugene P. Wigner
Leo A. Ohlinger
Gale Young
Alvin M. Weinberg
By: Robert A. Lorenzen
Attorney.

Sept. 4, 1962   E. P. WIGNER ETAL   3,052,613

METHOD AND APPARATUS FOR CONDUCTING A NUCLEAR CHAIN REACTION

Filed Aug. 29, 1945   12 Sheets—Sheet 9

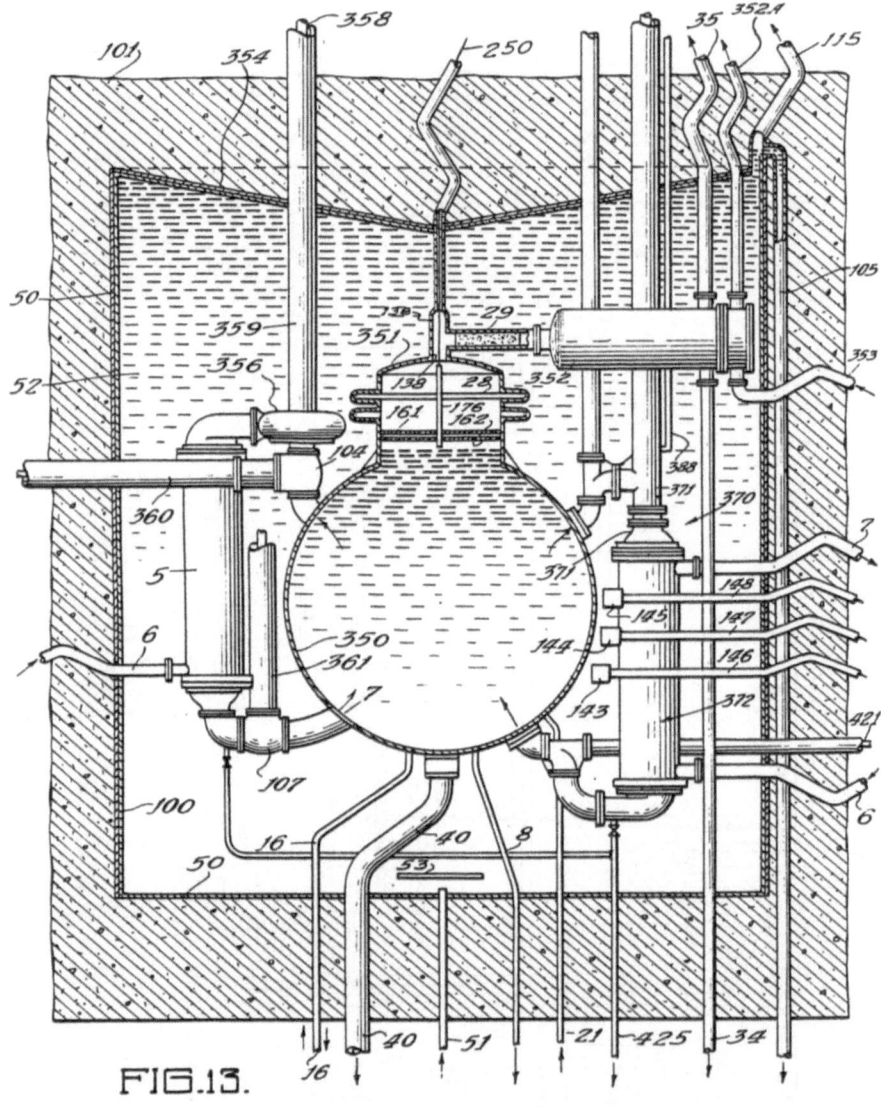

FIG.13.

Witnesses:
Herbert Elliott Metcalf
Henry H. Johnston

Inventors:
Eugene P. Wigner
Leo A. Ohlinger
Gale Young
Alvin M. Weinberg
By: Robert A. Lanudu
Attorney

Sept. 4, 1962   E. P. WIGNER ETAL   3,052,613
METHOD AND APPARATUS FOR CONDUCTING A NUCLEAR CHAIN REACTION
Filed Aug. 29, 1945   12 Sheets—Sheet 10

FIG.14.

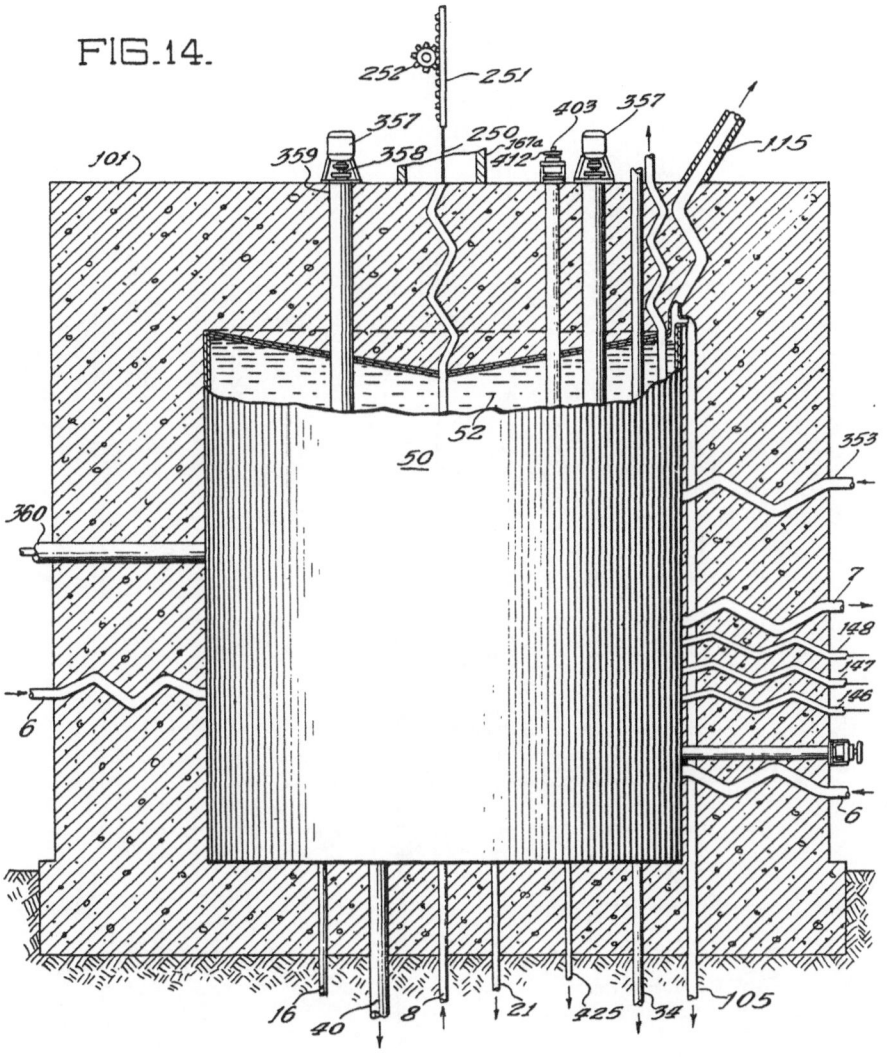

Witnesses:

Inventors:
Eugene P. Wigner
Leo A. Ohlinger
Gale Young
Alvin M. Weinberg
By:
Attorney

Sept. 4, 1962   E. P. WIGNER ETAL   3,052,613

METHOD AND APPARATUS FOR CONDUCTING A NUCLEAR CHAIN REACTION

Filed Aug. 29, 1945   12 Sheets—Sheet 11

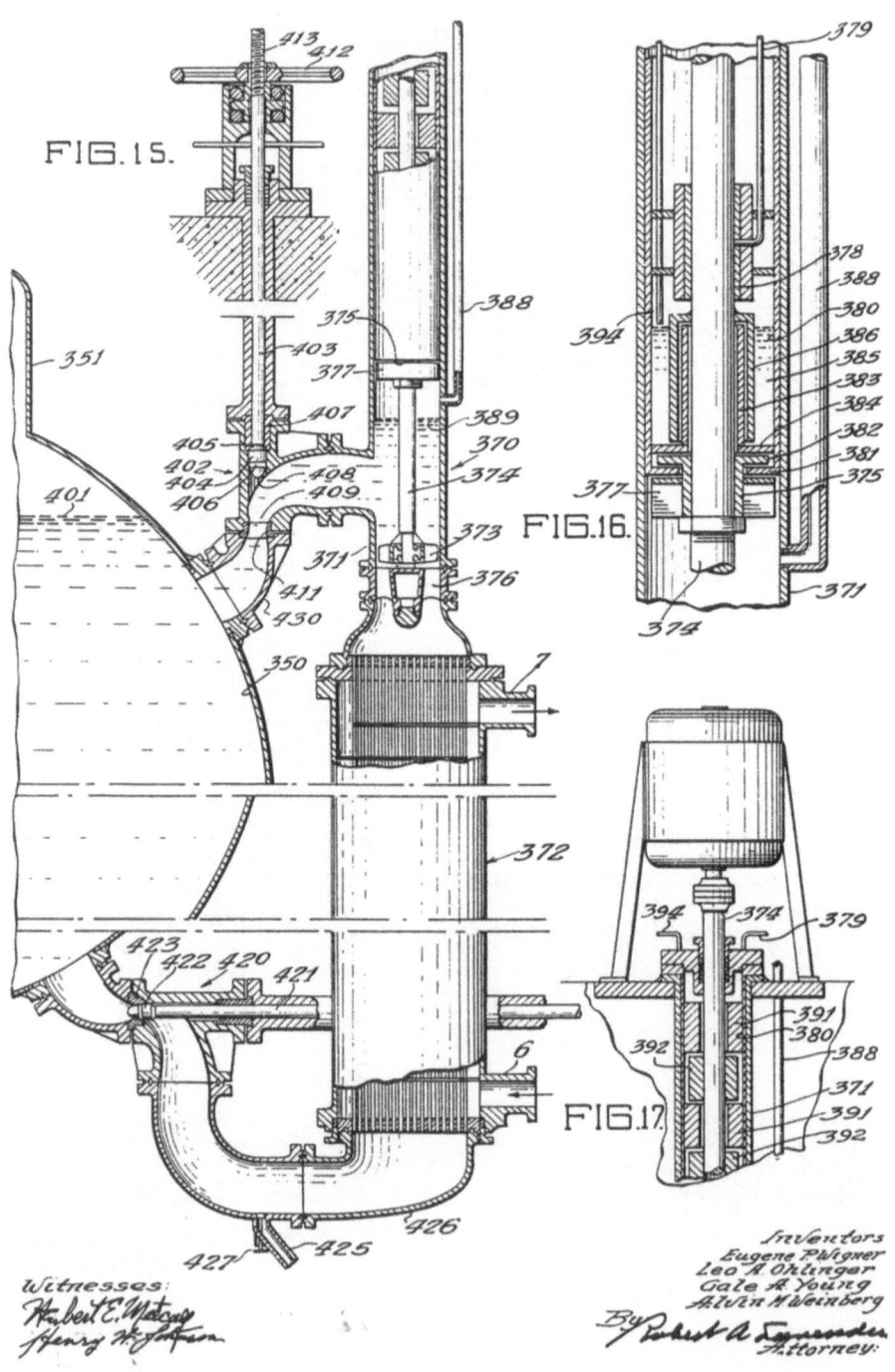

FIG.15.

FIG.16.

FIG.17.

Witnesses:

Inventors
Eugene P.Wigner
Leo A.Ohlinger
Gale A.Young
Alvin M.Weinberg
By
Attorney

Sept. 4, 1962   E. P. WIGNER ETAL   3,052,613

METHOD AND APPARATUS FOR CONDUCTING A NUCLEAR CHAIN REACTION

Filed Aug. 29, 1945   12 Sheets—Sheet 12

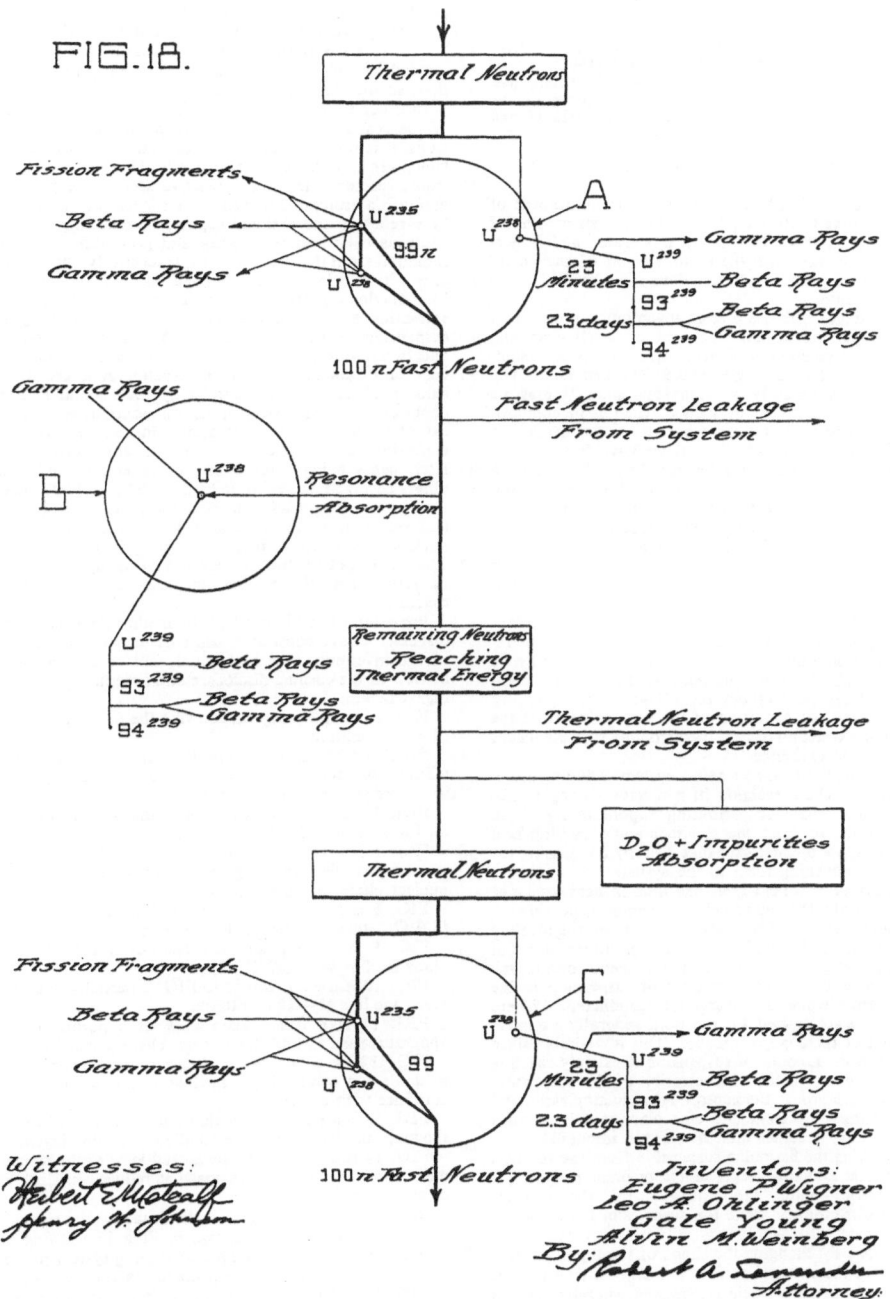

FIG.18.

United States Patent Office

3,052,613
Patented Sept. 4, 1962

1

3,052,613
METHOD AND APPARATUS FOR CONDUCTING
A NUCLEAR CHAIN REACTION
Eugene P. Wigner, Leo A. Ohlinger, Gale J. Young, and
Alvin M. Weinberg, Chicago, Ill., assignors to the
United States of America as represented by the United
States Atomic Energy Commission
Filed Aug. 29, 1945, Ser. No. 613,356
15 Claims. (Cl. 204—154.2)

This invention is concerned with the establishment of a self-sustaining neutron chain reaction in a suspension of a fissionable material in a liquid moderator and is particularly concerned with the establishment of such a reaction in a suspension wherein deuterium oxide is used as the moderator.

In accordance with the present invention, a novel process and apparatus for establishment of a self-sustaining neutron chain reaction of neutrons with a neutron fissionable isotope such as U^{233}, U^{235} and 94^{239} is provided. The invention is particularly advantageous since it may be applied to establishment of such a reaction in compositions such as natural uranium where the concentration of fissionable material is low. Thus, we have found that a self-sustaining reaction may be established by use of a suspension of natural uranium in a liquid moderator containing about 0.0025 to 0.04 atom of uranium per molecule of a moderator such as deuterium oxide or about 0.0013 to 0.02 atom of uranium per atom of deuterium. Where the liquid moderator is less efficient, and absorbs more neutrons than deuterium oxide, this range of uranium concentration is somewhat narrower.

The neutron reaction results in the release of substantial energy in the form of heat and consequently, the reacting suspension must be cooled or at least heat must be extracted. This heat may be removed by one of a combination of several methods including:

(1) Removal of a portion of the suspension from the reactor and extraction of heat therefrom by suitable means such as a heat exchanger.

(2) Removal of heat by a reflux system including establishment of a chain reaction in a reactor having a substantial vapor chamber permitting vapors to rise from the suspension to enter the chamber and removing heat from the vapors by heat exchange thereby condensing the vapors and returning them to the system.

(3) Passage of a heat exchange fluid in heat exchange relationship with the neutron chain reacting suspension.

Whatever method of heat exchange is used, the reaction preferably should be conducted under conditions such that no substantial change occurs in the concentration of the suspension or the effective amount of suspension in the reaction zone while the reaction is conducted. Where such changes are desired, the reactions generally are interrupted before the change is made. This is desirable since variation in the amount of suspension effectively reacting in the reactor or in the concentration of the suspension will cause variation in the neutron reproduction ratio, and if such changes are large and take place rapidly, control of the reaction becomes difficult or even impossible.

In removing the fissioning suspension from the reaction zone for heat exchange, replacement or other purposes, it is preferred to conduct the removal so that the chain reaction is discontinued while the suspension is out of the reaction zone. In accordance with this invention, this may be done by changing the shape of the liquid suspension so that the external surface per unit volume thereof is increased when the liquid is removed whereby neutron leakage from the exterior thereof is increased. Alternatively, the liquid suspension may be withdrawn from a reactor provided with a neutron reflector into a container which has no reflector or which is capable of losing a

2

greater percentage of neutrons. Moreover, the suspension may be prevented from attaining critical size by withdrawal of but a portion of the suspension from the reactor and/or collecting the suspension in a plurality of portions smaller than critical size.

In order to prevent excessive holdup of suspension out of the reaction zone, and also to minimize loss of neutrons, it is found advantageous to utilize heat exchangers which are close to the reactor itself. Thus, the usual neutronic reactor comprises a reactive section comprising a fissionable isotope disposed in a moderator surrounded by a reflector and a shield capable of reflecting escaping neutrons back into the reactor and preventing escape of radiation from the reaction zone, respectively. In accordance with this invention, improved results may be secured by removing a portion of the suspension from the reactor, extracting heat from the suspension and returning the suspension to the reaction zone. This minimizes holdup and in addition permits a saving in neutrons. Thus, circulation of coolant through the reaction zone usually results in absorption of neutrons by the coolant and the cooling tubes. Moreover, in fission reactions some quantity of neutrons are evolved almost instantaneously while a portion, usually about one percent, are evolved much later, some being evolved one or several seconds, and others several minutes after fission. The present process may be conducted so as to cool the removed suspension and return it to the reactor before all of the delayed neutrons are evolved, thereby effecting a neutron saving. The amount of neutrons so saved depends upon the speed of return and the volume of suspension outside the reactor.

Provision of some or all of the methods and apparatus discussed above constitutes some of the principal objects of the invention, others of which will become apparent in view of the ensuing disclosure and accompanying drawings wherein;

FIG. 1 is a flow diagram showing the principal features of our invention;

FIGS. 2 to 5, inclusive, are diagrams representing four methods of cooling a nuclear fission chain reacting system in accordance with our invention;

FIG. 6 is a set of curves representing reproduction factor for various natural uranium to moderator ratios;

FIG. 7 is a vertical sectional view of one embodiment of our invention showing the principal features of our nuclear chain reacting system;

FIG. 8 is a cross-section view of the apparatus shown in FIG. 7 taken on the line 8—8 thereof;

FIG. 9 is a sectional view similar to FIG. 8 taken along the line 9—9 of FIG. 7;

FIG. 10 is a view similar to FIG. 7 partially in section along the line 10—10 of FIG. 9;

FIG. 11 is a fragmentary view of a portion of our apparatus used to limit the nuclear chain reaction therein;

FIG. 12 is a schematic diagram representing three methods for controlling a nuclear chain reaction in accordance with our invention;

FIG. 13 is a fragmentary view similar to that of FIG. 7 showing another physical embodiment of our invention;

FIG. 14 is a partially sectionalized view of the embodiment shown in FIG. 13 (somewhat reduced in size) enclosed in a concrete shield;

FIG. 15 is an enlarged vertical sectional view of a portion of the apparatus shown in FIG. 13 showing one type of circulating, pumping and valving system particularly adapted for use with our chain reacting system;

FIG. 16 is a further enlarged sectional view of a circulating pump casing used in combination with the pump shown in FIG. 15;

FIG. 17 is a partially sectionalized view of the motor drive used in combination with the pump of FIG. 15; and

3,052,613

FIG. 18 is a diagram representing distribution of neutron losses in our self-sustaining chain reacting system.

The invention as described is particularly applicable to solid suspensions of the slurry type in which a solid, fissionable component is suspended in amount substantially in excess of its solubility in the moderator. Various fissionable solids may be used. However, the oxides, for example UO_2, U_3O_8 or UO_3, are particularly suitable where an aqueous suspension is used since such compounds are comparatively stable.

In order that a self-sustaining neutron chain reaction can be established and maintained, the losses of neutrons must be held to a value so low that at least one neutron is available for fission after all losses have been deducted, per neutron consumed in fission. For example, about 2.3 neutrons are evolved per neutron consumed in fission of U^{235} and about 2.8 neutrons are evolved per neutron consumed in fission of 94^{239}. These evolved neutrons are used up in fission of a further quantity of fissionable material or are lost. If the losses do not reduce the ratio of neutrons evolved to neutrons consumed or lost below one, the chain reaction will continue.

Losses may be external or internal. External losses take place when neutrons escape from the reaction zone of a neutronic reactor and are not returned to the reaction zone. For a reactor of infinite size, it is apparent that these exterior losses or neutron leakage would be zero. However, reactors of finite size have a finite leakage and, generally speaking, the magnitude of neutron leakage increases as the ratio of external surface area to volume of a reaction zone increases. Thus, spherical reactors which have a minimum ratio of surface to volume will have a minimum leakage, whereas a body of liquid suspension which is much greater in width and breadth than in depth will have a relatively high leakage. Similarly, even in spherical reactors, the amount of leakage increases as the size of the reactor decreases. For a system which is capable of establishing a chain reaction at infinite size, there is a finite critical size at which leakage becomes sufficiently low to permit maintenance of such a reaction once established.

In the case of a spherical structure employing uranium bodies of any shape or size imbedded in our moderator, the following formula gives approximately the critical overall radius:

$$\text{Critical sphere } R \text{ cm., } R = \frac{56.5}{\sqrt{K-1}}$$

where K is the neutron reproduction factor of the system for infinite size.

For a parallelepiped structure rather than spherical, the critical size can be computed from the formula

$$K - 1 = 323\pi^2\left(\frac{1}{a^2} + \frac{1}{b^2} + \frac{1}{c^2}\right)$$

where a, b, and c are the lengths of the sides in centimeters. The critical size for a cylindrical structure is given, irrespective of the shape of the uranium bodies, by the formula,

$$\text{Cylinder } h \text{ cm. } \quad K - 1 = \frac{323\pi^2}{h^2} + \frac{777}{R^2}$$
$$\text{Radius } R \text{ cm.}$$

Interior losses occur by absorption of neutrons by atoms which do not fission when the neutron has been absorbed. For example, U^{238} which is present in natural uranium absorbs substantial quantities of neutrons to produce 94^{239}. Within reason this absorption offers certain advantages since the 94^{239} formed tends to replace the U^{235} consumed and thus to increase the life of the reaction. At the same time, however, this absorption may in some cases become so great as to prevent establishment of a chain reaction.

Neutron moderators also absorb neutrons. Generally speaking, the moderator used in accordance with this invention should be a fluid of low atomic weight and low

neutron capture cross-section. Bismuth, deuterium and helium are good moderating elements. The ability to slow down neutrons may be expressed by what is known as the scatter cross-section, whereas the ability to absorb or capture neutrons may be expressed as the capture cross-section. The ratios of absorption cross-section to scattering cross-section of various materials are approximately as follows:

Light water (H_2O)	0.00478
Diphenyl	0.00453
Heavy water (D_2O)	0.00017

For natural uranium it is preferred to use materials wherein the above ratio is below about 0.004. However, with enriched uranium compositions containing more than natural amounts of U^{235}, a greater latitude is permissible. Using carbon or deuterium oxide as moderators and natural uranium as the fissionable composition, only about 1.1 or 1.3 neutrons, repectively, are obtained per neutron consumed due to neutron losses in the U^{238} and the moderator. Since the external neutron losses may be substantial, internal neutron losses should be held sufficiently low to prevent these losses from rising so high as to prevent the reaction.

Other components of the reactor including the coolant, impurities in the uranium or moderator, control or limiting rods, fission fragments, etc. absorb neutrons in varying amounts depending upon their neutron capture cross-section. The effect of these impurities or absorbers in a reactor containing natural uranium as the fissionable component has been approximately evaluated for various elements as a danger coefficient. This coefficient is computed according to the formula,

$$\frac{T_i}{T_u} \frac{A_i}{A_u}$$

where

T_i represents the cross-section for absorption of thermal neutrons of the impurity;

T_u represents the cross-section for absorption of thermal neutrons of the uranium;

A_i represents the atomic weight of the impurity or neutron absorber; and

A_u represents the atomic weight of uranium.

The following table gives presently known values for various elements having their natural isotopic content.

Element:	Danger coefficient
He	0
Li	310
B	2150
N	4.0
F	0.02
Na	0.65
Mg	0.48
Al	0.30
Si	0.26
P	0.3
S	0.46
Cl	31
K	2.1
Ca	0.37
Ti	3.8
V	4
Cr	2
Mn	7.5
Fe	1.5
Co	17
Ni	3
Cu	1.8
Zn	0.61
Ga	~1
As	2
Se	6.3

3,052,613

5

Element:	Danger coefficient
Br	2.5
Kr	<6
Rh	50
Ag	18
Cd	870
In	54.2
Sn	0.18
Sb	1.6
I	1.6
Ba	0.30
Sm	~1430
Eu	435
Gd	~6320
Pb	0.03
Bi	0.0025
Th	1.1

From the above it will be apparent that certain elements are objectionable if present in substantial amounts in neutronic reactors. For example, cadmium, boron and gadolinium absorb neutrons to a high degree, and may be used effectively to control the reaction by variation of the amount present in the system. On the other hand, aluminum and beryllium are typical of the elements which can be used in the reactor for cooling pipes or other structure, or may be present as an impurity although the amount thereof must be limited. For example, if a neutronic reactor is capable of supplying 1.06 neutrons per neutron consumed in fission when all losses exclusive of that caused by aluminum have been taken into account, then the loss due to the presence of aluminum can not exceed

$$\frac{0.06}{0.30}$$

part by weight per part of uranium or 20 percent of the weight of the uranium. These principles generally apply to other metals or materials.

From the above it will be apparent that for a neutron chain reaction to remain self-sustaining the equation

$$n - x - y - z - L \geqq 1$$

where

$n=$ number of neutrons evolved by a fission of a fissionable isotope per neutron consumed in fission of such isotope.

$x=$ number of neutrons absorbed by a nonfissioning isotope such as U^{238} in formation of a fissionable isotope per neutron consumed in fission.

$y=$ number of neutrons absorbed by the moderator per neutron consumed in fission.

$z=$ number of neutrons absorbed by other neutron absorbers per neutron consumed in fission, and

$L=$ the number of neutrons lost from the system by leakage per neutron consumed in fission.

Thus, with U^{235} the sum of $x+y+z+L$ cannot exceed about 1.3 and with 94^{239} cannot exceed about 1.8.

The ratio of the fast neutrons produced in one generation by the fissions to the original number of neutrons producing the fission in a system of infinite size from which there can be no loss is called the reproduction factor and is denoted by the symbol K. The K constant of a system of finite size is the reproduction factor which the system would have if expanded to infinite size. Usually this constant is expressed without regard to localized neutron absorbers such as control or limiting rods, which are not uniformly dispersed throughout the entire system. The neutron reproduction ratio (r) is an actual value for a finite system, and differs from K by a factor due to loss of neutrons through leakage and through absorption by localized neutron absorbers. To maintain a chain reaction, r must be at least equal to one but it is preferably maintained below about 1.01 during operation of the reactor.

Computation of K for any system may be determined

6

experimentally in accordance with methods described in copending application of E. Fermi, Serial No. 534,129, filed May 4, 1944, entitled "Nuclear Chain Reacting System," now Patent No. 2,780,595, dated February 8, 1957

The reproduction ratio (r) may be ascertained by observation of the rate of increase of neutron density. It may also be predicted by computation of losses due to local absorbers or leakage which may be deducted from K to secure this value. In such a case allowance for leakage is made depending upon the size of the reactors. For reactors of practical size, leakage ratio usually amounts to about 0.01 to 0.3 neutron per neutron consumed in fission depending upon the amount by which the K of the system exceeds one. Loss due to other absorbers may be computed by computation of the danger sum as heretofore described.

While we will refer to natural uranium or uranium alone as the active material of the suspension, it will be appreciated that any fissionable material such as U^{233}, U^{235}, or 94^{239}, or natural uranium or thorium which is enriched with any of such isotopes, and various compounds thereof that are compatible with the heavy water may be used in following the teachings of our invention.

Referring first to FIG. 1 which shows a flow diagram of a self-sustaining nuclear fission chain reacting system, the chain reaction is caused to occur in a container or reaction tank 1 partially filled with the solid suspension which may be a slurry 2 which is pumped through a circulating system generally designated 3 by a pump or other circulating means 4 for the purpose of cooling the slurry. The circulating system 3 is provided with a heat exchanger 5 through which cooling water may be made to flow as shown by the inlet 6 and outlet 7. Inasmuch as it is an object of our invention to provide means to continuously remove portions of the suspended uranium for recovery of the reaction products, we prefer to provide at the base of the reaction tank 1 an outlet 8 through which the slurry may be withdrawn into a uranium-liquid moderator separating chamber 9. As appears hereinafter, the slurry is withdrawn through the outlet 8 continuously but the quantity thereof is relatively small, being for the purpose of withdrawing a portion of the uranium for removal of elements 93 and 94 as well as fission elements and products produced by the chain reaction. Chamber 9 is provided with separating means for separating the solid matter from the slurry such as by evaporation of the heavy water which is delivered to a purifying tank 10 through the line 11 and pump 12 where the water may be purified by distillation or other methods. The uranium bearing material separated from the slurry is withdrawn from the chamber 9 in any desired manner, as through the piping 13, whereupon the elements 93, 94 and fission products may be separated from one another. Inasmuch as the separation of these products from the uranium of the slurry forms no specific part of our invention, the process of separating these products is not discussed in detail herein.

The purified heavy water is stored in a tank 14, the supply therein being maintained from an outside source such as through the pipe line 15. In addition to the outlet 8, through which a portion of the slurry is removed from the system, we have shown another pipe line 16 for passing the slurry in the tank 1 to and from a slurry reservoir 17 by a reversible pump 18 in the line 16. As appears hereinafter in further detail, the reversible pump 18 is used to control the slurry volume in the reaction tank in accordance with one teaching of our invention. The desired concentration of uranium material may be maintained in the slurry by introducing additional material into the slurry reservoir 17 through the line 19 while additional heavy water may be delivered by pump 20 from the heavy water reservoir 14, either to the reaction tank 1 through the line 21 or to the slurry reservoir 17 through the line 22 controlled by a three-way valve 23. In this manner, the concentration of the uranium may

3,052,813

7

be maintained at any desired value both in the reaction tank and in the slurry reservoir.

As indicated above, the chain reaction is initiated by nuclear fissions produced by neutrons slowed to thermal energy by collision of faster neutrons with the heavy water. Some of these neutrons are effective in decomposing a portion of the heavy water into deuterium and oxygen, and consequently, in accordance with a further teaching of our invention, we provide means causing recombination of the products of decomposition thereby conserving the heavy water for reuse in the system. Referring again to FIG. 1, we provide an atmosphere of helium or other inert gas of low absoprtion over the slurry 2 in tank 1 to dilute the evolved gases to a degree such that formation of an explosive mixture is prevented. The helium is supplied from a reservoir 24 through a pump 25 and pipe line 26 entering the upper portion of the tank 1, thereby directing the gases over a hot grid shown diagrammatically at 28 supported in the upper portion of the tank 1 over the normal level of the slurry 2. The hot grid 28 or equivalent igniter heats the gases to their recombining temperature, but to positively assure recombination, we prefer to provide, in addition, a catalyst chamber 29 connected with the upper portion of the tank 1 by pipe line 30 through which the remaining uncombined gases are driven by the helium. In the presence of the catalyst in the chamber 29 the gases heated but not recombined by the hot grid 28, recombine to form heavy water vapor which is directed by the helium flow through pipe line 31 to a condenser 32 where the vapor is condensed and directed to the purifying tank 10 through the pipe line 34 and thence to the heavy water reservoir 14 through the line 33. The helium is returned to the reservoir 24 by pipe lines 35 and 36 following purification in helium purifier 37.

As indicated above, the nuclear chain reaction in the reaction tank is controlled by varying the level of the slurry therein to change the volume above or below that corresponding to critical size. As an emergency safety precaution we provide means to dump the slurry from the reaction tank into the slurry reservoir. We have shown a pipe line 40 leading to a safety dump valve 41 and thence through line 42 to the slurry reservoir 17. Upon opening the valve 41 such as in response to a safety control circuit described later, the slurry level in the reaction tank 1 is decreased very rapidly, thereby terminating the reaction therein in case of failure or any improper action of the system.

It will be fully appreciated that the self-sustaining chain reaction is initiated merely by exceeding the critical size of the suspension at which the reproduction ratio slightly exceeds unity. Consequently, any volume of this slurry of similar configuration to that contained in the reaction tank, will likewise support a self-sustaining nuclear chain reaction provided the neutron reflectory properties of the container are as good as those of the reacting tanks. It is, therefore, exceedingly important that the slurry reservoir should be of such configuration or construction that even with a maximum quantity of slurry therein a self-sustaining chain reaction cannot be either initiated or maintained in this reservoir. The reservoir 17 should, therefore, be made to have a large surface-to-volume ratio and/or should have less neutron reflecting properties than the reactor so that the neutron losses from the surface thereof exceed the minimum surface losses necessary to allow a self-sustaining reaction. This may be accomplished by making the slurry reserve tank of large cross-sectional area and with minimum depth, or with small cross-section and maximum depth or as a plurality of small mutually separated tanks. For example, if the slurry reservoir is of rectangular form, the maximum allowable dimensions are easily determined from the formula given above for critical size at the maximum attainable reproduction factor whereupon the tank is made somewhat smaller as a precaution against developing a

8

chain reaction therein. In addition, the tank may be constructed of or contain cadmium plates or baffles which effectively absorb neutrons thereby eliminating all possibility of developing a self-sustaining reaction in the slurry outside of the reaction tank. Moreover, if a neutron reflector is used around the reactor, it may be omitted in the reservoir.

In addition to the cooling system shown in FIG. 1, the entire apparatus including the tank 1, heat exchanger 5, associated pump 4, slurry reservoir 17, its associated pump 18, catalyst chamber 29 and condenser 32 is immersed in water for cooling and shielding purposes within a concrete or other good neutron and gamma ray shield for protection of operating personnel. This circulation system and shield is not shown in FIG. 1 for the sake of clarity.

In the operation of the system shown and so far described in FIG. 1, the chain reaction within the reaction tank 1 develops considerable energy in the form of beta and gamma rays, as well as kinetic energy from the fission products. A great portion of this energy is released inside the system in the form of heat and is absorbed by the heavy water of the slurry and withdrawn from the system by circulating the slurry through the heat exchanger 5, the heat being transferred to the cooling water flowing between the inlet and outlet piping 6 and 7. The heat developed by the chain reaction may be removed in a number of different ways or combinations thereof.

We have shown in FIGS. 2–5 wherein corresponding structure shown in FIG. 1 is similarly referenced, various cooling means whereby this heat may be dissipated. For example, the heat may be dissipated by flowing a coolant over the external surface of the chain reaction chamber, by withdrawing a portion of the slurry from the reaction chamber, and cooling the slurry exteriorly, by circulating a coolant through conduits in the slurry chamber and above the slurry level to condense heavy water vapor in the region over the slurry, by flowing the coolant though conduits immersed in the slurry, or by any combination of these methods.

Referring to FIG. 2, we have shown very schematically a reaction tank 1, without any of the auxiliary supply piping partially filled with slurry 2 surrounded by a cooling water tank 50 and with a coolant pipe line conduit 51 entering the tank 50 immediately below the chamber carrying a coolant such as water 52 which flows around the baffle 53 and around the tank 1, absorbing heat therefrom. In this mode of cooling the tank 1 may be wholly immersed in the coolant, the flow being directed in any direction over the external surface thereof. FIG. 3 shows a method wherein the slurry 2 itself may be cooled by withdrawing a portion thereof through a circulating system designated 3 and being driven through a heat exchanger 5 by a pump 4, the heat in the slurry being withdrawn by cooling water or other medium circulated through the heat exchanger between the inlet 6 and outlet 7. The slurry may be circulated by thermo-syphon action without the use of the pump although some sacrifice in cooling capacity ensues. The heat exchanger may, however, be formed integrally with the tank 1 as shown in FIG. 4, wherein a conduit 54 is provided within the tank 1 in the upper portion thereof and supplied with a coolant such as water flowing therethrough to condense heavy water vapor developed within the slurry and collecting in the upper portion of the tank 1. Alternatively, as shown in FIG. 5, the slurry 2 may be cooled by providing a series of tubes 55 within the tank 1 extending between headers 56 and 57 through which the coolant may be made to flow.

During circulation through the heat exchanger, the neutronic reaction is discontinued. This is accomplished by proportioning the size of the reactor and heat exchanger and/or apportioning the volume of suspension in the reactor and exchanger such that the volume of suspension in the reactor is above critical size while the vol-

3,052,613

9   10

ume of suspension in the heat exchanger is below critical size.

Obviously, any of the four cooling systems shown in FIGURES 2 through 5 or combinations thereof may be used for the purpose of cooling the slurry continuously with the chain reaction in progress. For example, and as it appears more fully hereinafter, the cooling systems shown in FIGS. 2 and 3 may be combined to cool the slurry not only from the outside of the reaction chamber but by withdrawing the slurry for cooling in an external heat exchanger.

The nuclear chain reaction within the reaction tank 1 is dependent upon the nuclear fission of the U^{235} constituent of the uranous material of the slurry when subject to thermal neutrons, and also on fission of 94^{239} as U^{235} is used up. During the fission process fast neutrons are emitted by the uranium and these fast neutrons are slowed to thermal energy, this being the function of the moderator. However, there must be sufficient uranium in the slurry to intercept the neutrons once they have reached thermal energy. Consequently, the ratio of uranium atoms to the atoms of the moderator producing the slowing effect must be such that the slowing is sufficient, the availability of uranium in the paths of the slow neutrons is adequate, and the neutron loss occasioned by resonance capture is insufficient to overcome the neutron gain occasioned by the fission process, so that a self-sustaining chain reaction is possible.

The full line curve shown in FIG. 6 has been drawn for a moderator of heavy water of availability purity from calculations based on an absorption cross-section, per molecule of pure heavy water (D_2O), of 0.004×10^{-24} cm.^{3}, the ordinates representing values of reproduction factor K, the abscissae being various ratios of uranium atoms to heavy water molecules of the slurry.

From the full line curve of FIG. 6 it will be appreciated that as the concentration of uranium in the heavy water is increased, the value of K increases from values below unity and reaches a peak of 1.10 at a concentration of about 0.01 atom of uranium per molecule of heavy water. Thus the mass of the suspension and the arrangement of the system must be such that the leakage cannot exceed 1.1–1 or 0.1 K unit, and in any event the leakage cannot be so great that 1 plus the leakage factor in terms of K exceeds the ordinates defined by the curve of FIG. 6. Since the system will operate equally as well, with respect to slurry concentration, with a reproduction factor slightly over unity for a high and low concentration of uranium, and since uranium is at present cheaper to produce than heavy water, the concentration which is used to provide the slurry is preferably in the higher range such as about 0.023 atom of uranium per molecule of heavy water which, neglecting the very low danger coefficients of oxygen content of the uranium oxide preferably used, provides a K factor of 1.060. When using UO_2 as the uranium source this corresponds to approximately one part oxide to four parts heavy water by weight. On a volume basis, the oxide solids represent about 4 percent of the slurry volume.

From the full line curve FIG. 6, it will be appreciated that a low ratio of uranium to moderator such as 0.0025 atom of uranium per molecule of moderator may be used. Such use would reduce erosion of the slurry circulating system and pumps, but small variations in concentration cause greater variations in the reproduction factor over this portion of the curve than to the right of the maximum K value, rendering the system more critical to control. Consequently, we prefer to utilize the higher uranium-to-moderator ratio indicated above. At all events the uranium concentration of the slurry or solution should be maintained substantially constant during the reaction since substantial changes make the reaction difficult or impossible to control.

The particle size of the uranium oxide is preferably below 2 microns (μ). This size of the individual particles is dictated principally by the abrading action of the particles on the pumps, valves and heat exchanger tubes. For larger particle sizes erosion may be excessive. While erosion does not materially affect the mechanical operation of the system, it tends to poison the system by the inclusion of iron and other metals worn from the slurry circulating system. However, continued circulation of the slurry reduces the particle size by abration of the oxide particles upon each other. For example, a slurry originally of 50μ to 70μ size was reduced to a point where 85 percent of the material was below 2μ by pumping through a circulating system for 60 hours at 20 feet per second. Consequently, in the initial stages of operation a somewhat larger slurry particle size may be tolerated although sizes below 2 microns are preferred. As operation continues, the particle size will decrease, and for particle sizes below 0.001μ the erosion may be no more than that produced by a solution of the same densty: Alternatively, the slurry may be pumped through an auxiliary system prior to use in the chain reacting system to reduce the uranium oxide particle size. Such a system may be constructed of materials having very low danger coefficients so that the slurry, while contaminated to a small degree, has less neutron absorbing impurities than it would have if it had been initially run in the chain reacting system. Thus, an auxiliary system may be made of materials such as aluminum having a low danger coefficient whereas such materials used in the reaction circulating system would be subject to early wear and premature failure.

With larger bodies of uranium, the size and distribution of the uranium-bearing particles disposed in a moderator affect the value of the reproduction factor K because the resonance loss (absorption by U^{238}) increases with decrease in particle size. This effect has no practical importance, however, in the present system, because the size range in which this effect occurs is outside the limits imposed by circulation of the slurry. For example, to provide an optimum reproduction factor the particles would have to be one centimeter or larger in diameter. Conversely, while decrease in particle size lowers the reproduction factor somewhat, the loss in neutrons by reason of the larger surface resonance capture provides a net gain in the production of elements 93 and 94. This will be set forth more fully below with reference to a diagram showing distribution of neutron losses in our system.

The structural form of the reaction tank may be of any desired shape such as spherical, cylindrical, parallelepiped, or combination thereof, as long as the volume thereof is sufficient and is concentrated (i.e., so that the surface-to-volume ratio for the reacting material is sufficiently low) to reduce surface losses within the confines as dictated by the maximum reproduction factor. We have shown in FIG. 6 a curve in dashed line detail wherein the relation of critical size of a spherical structure of radius R to the slurry concentration is approximately given. From the value of K corresponding to any slurry concentration, the minimum radius at which the reaction becomes self-sustaining represents the critical size for that condition. A spherical type of reaction tank represents the most economical utilization of the slurry although for ease in construction and control the cylindrical form may be preferred.

Referring generally to FIGS. 7 to 10, inclusive, we have shown a system wherein the reaction tank 1 is cylindrical and adapted to contain a slurry of height approximating the diameter. For a minimum size and consequent savings in heavy water, it may be desirable to provide a slurry concentration providing the highest practical reproduction factor. Consequently, using the formula given above for a cylindrical structure, the minimum cylindrical reaction tank volume would be approximately 11 feet in height and in diameter. However, the reaction tank may be somewhat larger to allow for variations in slurry con-

3,052,613

11

centration, possible poisoning of the reaction by the formation of neutron absorbing fission products, reduction in the reproduction factor by impurities in the uranium oxide including material removed by erosion of the circulating system, and other variables. These variables may be allowed for by calculating the total effect of these variables on the reproduction factor and a reasonable reduction in the factor may be considered to be approximately 4 percent.

Using the 4 percent design safety factor to insure a reaction notwithstanding this reduction in K, the reaction tank would be made approximately 14 feet in diameter, the depth of the slurry in the tank being somewhat less than 14 feet depending on the actual neutron absorbing impurities in the slurry. For absolutely pure materials and for a slurry concentration of about .01 uranium atom per molecule of heavy water, the depth to obtain a self-sustaining neutron reaction in a 14 ft. tank is approximately 6 feet. Consequently, to initiate the reaction in the tank, the level is increased until the critical size for the particular concentration of slurry and impurity content, is exceeded slightly, whereupon the reaction becomes self-sustaining and may be stabilized at any intensity by reducing the slurry to exactly critical size.

The reaction tank, for better utilization of the heavy water, may be hemi-spherical in form with a cylindrical upper portion joined thereto and in which the level of the slurry is varied between upper and lower limits on either side of the level corresponding to critical size. The preferred semi-sphere radius is approximately 7.5 feet, the cylindrical portion being 6 feet in diameter at the point of junction with the semi-sphere for the desired slurry concentration given above. We will describe in the following pages two preferred designs of chain reacting systems, one having a cylindrical reaction tank and the other a semi-spherical tank as above defined referring to various equivalent or preferred auxiliary apparatus for the operation thereof.

While the apparatus made in accordance with our invention may be cooled in any one or more of the ways shown in connection with FIGURES 2 to 5, the particular embodiments of our invention will be described with particular reference to a combination of cooling systems, such as shown in FIGS. 2 and 3. Referring to FIGS. 7 to 10, inclusive, and particularly to FIG. 7, the reaction tank 1 containing the slurry 2 is of stainless steel and is supported within an auxiliary enclosing tank 50 containing water 52 introduced therein by pipe line 51 for cooling the external surface of the tank 1 and other auxiliary apparatus to be described. The internal surface of the tank 50 is lined with lead sheathing 100 to absorb gamma rays liberated by the neutron chain reaction developed in the tank 1 and is surrounded by a massive concrete wall or shield 101 supported on a concrete base 102. This shield is for the purpose of absorbing and limiting the escape of gama rays not absorbed by the water 52 and the lead sheathing 100. Water 52 serves as a cooling medium, as a neutron reflector, and also as a neutron shield. Neutrons striking water body 52 are reflected back into the reactor or are absorbed. In this connection water is found to be an especially effective shield due to the low migration path of neutrons therein since the heat exchangers may be placed close to the reactor without substantial entry of neutrons into the circulating suspension in the exchanger and consequent fission therein. However, other neutron reflectors, such as deuterium oxide or carbon, may be used if desired. The concrete shield also serves as an absorber for neutrons slowed by collision in passing through the water 52 and is preferably composed of materials holding a maximum of water. The entire structure is supported on the earth which serves as an auxiliary shield at the base of the apparatus so that the concrete base 102 need not be as thick as the side walls of the shield 101. Likewise, the lead shield 100 over the side

12

walls and top of the enclosing tank 50 may be omitted on the bottom thereof.

Surrounding the reaction tank 1 we provide means to withdraw heat therefrom, such as by a heat exchanger system, all components of which are of materials such as steel having relatively low neutron absorption danger coefficients, so that particles abraided therefrom will not materially affect the chain reaction. We have shown six heat exchangers 5 in the drawings although it will be appreciated that any other number may be utilized depending entirely upon the rate at which the reaction is carried forth. Each of the heat exchangers 5 is connected with the reaction tank 1 through inlet piping 103, valve 104, and pump 4 near the top thereof and near the base thereof through the outlet valve 107 and piping 108 so that the slurry 2 may be circulated up through the reaction tank 1 and down through the heat exchangers 5. The heat exchangers 5 are cooled by water introduced near the base of the exchangers through the piping 6 and vented near the top thereof through the outlet 7 into the enclosing tank 50 where it merges with the water 52. The cooling water flowing through the heat exchangers as well as the water introduced through the pipe line 51 may be vented from the system through a channel 109 which completely surrounds the enclosing tank, being finally drawn off in the line 105, gases that may result from decomposition of the water 52 being vented through a shielded line 115 to a waste stack not shown. The piping between the reaction tank 1 and the heat exchangers, as well as the internal construction of the heat exchangers, is so designed as to minimize the amount of slurry held over during the heat exchange cycle inasmuch as the heavy water moderator comprising one of the ingredients of the slurry is at present relatively expensive. While the heat exchangers could be located outside of the concrete shield 101, such positioning would increase the slurry holdover while still necessitating additional radiation shielding enclosing these portions of the system.

As best shown in FIGURE 7 the heat exchanger inlet valves 104 are controlled by rod type valve stems 110 extending in an upward direction through close fitting iron guides 111 extending through the concrete shield 101 terminating in hand wheels 112 to actuate the valves. The valve stems 110 are made close fitting with the guides 111 to minimize radiation leakage through the concrete shield. The outlet valves 107 in the pipe lines 108 between the heat exchangers 5 and the reaction tank 1 are similarly controlled through valve stems 113 extending in a horizontal direction through guides 114 imbedded in the concrete shield 101 and hand wheels 116. The connection between the valve stems 113 and the valves 107 is facilitated by bevel gearing 117 housed within the gear boxes 118 closely adjacent the valves 107 as best shown in FIG. 8. The slurry circulating pumps 4 are shown of the centrifugal type and are driven by motors 119 through shafts 121 extending from the pumps 4 to the outside of the shield 101 through closely fitting conduits 122. The motor bearings are lubricated and leakage is prevented by filling the conduits 122 with heavy water introduced therein through the piping 123. Such lubrication assures satisfactory bearing life, prevents loss of slurry into the external cooling water and prevents contamination of this water with fission products produced by neutron bombardment of the uranium in the slurry.

The reaction tank 1 is so designed that under normal operating conditions the slurry level 124 is considerably below the top portion 126 of the tank 1 leaving a gas chamber 127 therein. Under the high neutron intensity present in the slurry volume, some of the heavy water moderator of the slurry becomes decomposed, the dissociated, i.e., uncombined, gases rising through the slurry 2 into the chamber 127 above the normal slurry level. In accordance with our invention, we provide means within the reaction tank 1 to recombine the gases formed by decomposition of the heavy water. We have shown a grid

3,052,613

13

28 of tubular members connected between an inlet header 128 and an outlet header 129. Hot gas is admitted through the line 131 to the inlet header 128, passed through the grid 28 and removed from the header 129 through an outlet line 132. In this manner the grid 28 is maintained above the recombining temperature of the uncombined gases of the heavy water. However, to prevent high concentration of uncombined gases within the chamber 127 we circulate an inert gas such as helium, which has a low neutron capture cross-section, through the chamber to dilute these gases and thereby prevent the formation of an explosive mixture. More particularly, as shown in the drawings, we provide a circular header 133 surrounding the reaction tank 1 about the chamber portion 127, this header being fed with helium which flows into the reaction tank through a plurality of ducts 136 which enter the tank 1 through the periphery thereof, thereby diluting the dissociated gases and driving them over the hot grid 28. To assure a complete recombination of the heavy water gases we provide, in addition to the hot grid 28 which burns the majority of these gases to heavy water, a charcoal platinum catalyst 137 within a chamber 29 connected to the top of the reaction tank 1 through a bell-shaped housing 138. The catalyst 137 may be of platinized charcoal supported within the chamber 29 in the presence of which the hot gases recombine with the evolution of additional heat to form heavy water vapor. This vapor is then directed by the flow of helium into a heavy water condenser 32, which in the modification shown in FIGS. 7 to 10 is cooled by the water 52 in the tank 50. The helium from which the water has been separated by condensation is withdrawn from the system through the line 35 and passed to the purifying system previously described in connection with FIG. 1.

The normal level 124 of the slurry is attained by provision of the slurry in reservoir 17 which may be pumped into the tank 1 by the pump 18 having a suction line 141 extending substantially to the bottom of the reservoir 17, the pump 18 being connected to the bottom of the reaction tank 1 by a line 16. The pump 18 is driven by a reversible motor 142 controlled in accordance with the neutron density in the reaction tank as determined by an ionization chamber 143 positioned adjacent the tank 1 so as to be in a region of relatively high neutron density. Additional ionization chambers 144 and 145 are provided as safety control units, their operation being described in greater detail hereinafter. Each ionization chamber has shielded leads 146, 147 and 148, respectively, extending through the concrete shield 101 for connection to the external control circuit shown in FIG. 12.

It is one of the principal objects of our invention to provide means for continuously removing portions of the slurry for recovery of the newly created elements 93 and 94 as well as the fission products. The concentration of these products of the nuclear reaction may be determined easily after operation for a predetermined time and the concentration of these products maintained constant in the slurry by bleeding off small quantities of the slurry while the chain reaction is continuing. No loss of operation time ensues by this method of removing the products. In prior systems using massive uranium rods or other shapes, it was necessary either to stop the nuclear reaction to remove the uranium or to provide complicated equipment to replace the processed uranium with new uranium for processing. However, in our system portions of the slurry containing the new elements 93 and 94 may be withdrawn from the tank 1 directly through the line 8 as shown in FIG. 1 or from the slurry reservoir 17 through the line 149, FIG. 7.

Inasmuch as solid uranous material in the heavy water moderator in the form of a slurry is more dense than the moderator and tends to settle through the moderator, we provide an agitator 151 immersed in the slurry in the reservoir 17 driven by the motor 152 through the shaft 153. Such agitation of the slurry within the reaction tank 1 may

14

be unnecessary, however, where the circulation through the heat exchangers provides sufficient turbulence to maintain the material in suspension, or where the particle size of the uranium composition is so small that very little settling occurs. At all events the slurry or solution concentration within the reactor should not change substantially during reaction, and, if possible, localized over or under concentration should be avoided to avoid localized hot spots.

We have shown several safety features incorporated within the reaction tank 1 to control the chain reaction to within safe limits as determined either from original design of the apparatus or in accordance with measured neutron intensities during operation. As indicated above, increase in the slurry volume within the reaction tank 1 beyond a predetermined amount may result in a reaction which increases beyond safe limits. Consequently, in accordance with our invention, we provide a safety feature within the reaction tank comprising a plurality of parallel cadmium plates 161–162 which may be perforated, if desired, in a position slightly above and parallel to the maximum desired slurry level so that upon any increase in volume either by failure of control or by expansion of the slurry volume the plates become immersed in the slurry whereby any uranium-heavy water mixture above the plates is removed from the chain reaction volume because cadmium absorbs neutrons originating in the main mass of the slurry. The cadmium plate 161 contains perforations 163 to allow escape of gases liberated within the slurry into the region 127 above the normal slurry level as well as to admit the slurry into this region for abnormally high slurry levels. The cadmium plate 162 is likewise apertured at 164, these apertures being offset from the apertures 163 in the plate 161. Such construction minimizes the number of slow neutrons escaping from the slurry into the upper gas chamber 127 of the tank 1, and, while fast neutrons may pass through the cadmium plates, slow neutrons capable of contributing directly to fission will be substantially prevented from entering the chamber 127. Any increase in volume of the slurry level above the normal level 124 into the chamber 127 will consequently remove this volume from that in the lower portion of the reaction tank with respect to its action in sustaining the chain reaction. Thus, while the reaction will not be terminated, the rate of rise in the reaction will not be as great and the reaction may be controlled more easily.

The plates 161–162 may be supported in such a manner that their position may be varied in a vertical direction within the reaction tank. Referring to FIG. 11, we have shown the plates 161–162 as being connected together as by welding, bolting, or riveting at 166, and supported by rods 167 which extend through the top of the reaction tank 1. Other details of the reaction tank have been omitted for the sake of clarity. The support rods 167 are slidably mounted in thimbles 168 affixed to the top of the reaction tank 1 and extend through closely fitting conduits 169 in the concrete shield 101 to the exterior thereof, terminating in racks 171 engaging pinions 172 which may be rotated by cranks 173 to lower or raise the plates 161–162 and thereby adjust the plates with respect to the normal slurry level 124. As illustrated in FIGURE 7, the operating mechanism outside the shield is enclosed in a gas tight space defined by walls 167a.

In addition to the provision of the cadmium plates 161–162, we provide a safety rod 176 also within walls 167a supported above the slurry 2 and immersible therein should the neutron reaction suddenly increase in density to a point approaching a dangerous condition. The safety rod 176 preferably protrudes through the cadmium plates 161–162 into close proximity with the normal slurry level 124 so that any sudden increase of slurry volume would cause a partial immersion of the safety rod 176 into the slurry. The safety rod is of a material having a high neutron capture cross-section such as boron, cadmium, gadolinium, or other material having high neutron

3,052,613

15

absorption characteristics and when immersed in the slurry reduces the reproduction ratio below unity, thereby terminating the reaction. Alternatively, or as an emergency, the slurry may be dumped from the reaction tank 1 as previously described through one or more dump lines 40—42 joined by the dump valve 41. The preferred control will be described below.

We have shown the safety rod 176 as being suspended over the upper maximum level of the slurry in the reaction tank 1 and within the space between the cadmium plates 161—162 and the top of the reaction tank. In this region the safety rod 176 is subjected to only a low density of slow neutrons to which the safety rod is very absorbent. Neutron absorbers inserted into the slurry are subjected to high neutron densities and they cannot continue to absorb neutrons indefinitely. The continued absorption of neutrons by the absorbing material causes transmutation of the absorbing materials and an element or isotope may be built up within the material which has a smaller neutron capture cross-section than the original material. However, by maintaining the safety rod within the space above the cadmium plates 161—162, this reduction in efficiency of neutron absorption is reduced to a minimum. Consequently, unless the safety rod is left immersed in the slurry, such as may occur if it is used to control the reaction rather than obtaining control by variation of the slurry level, the rod will retain a long effective life and may be depended upon as a positive safety feature. The ultimate safety feature, however, namely the dumping of the slurry through the valve in the base of the reaction tank is independent of this neutron absorption and consequently, may be depended upon as a positive safety feature.

All possible precautions must be taken to prevent an abnormal rise in the slurry volume in the reaction tank and a consequent exponential rise in neutron density either in the case of failure of the cadmium plates 161—162 to provide an adequate control with increase of slurry volume or of the safety rod 176 to become effective in reducing the neutron density to a point where the reproduction factor is equal to or less than unity.

Reference is made to FIG. 12 which shows diagrammatically one form of control and safety circuit which we may use for regulating the output of the system. Referring first to control circuit A, the control ionization chamber 143 referred to above as being placed adjacent the reaction tank 1, is provided with a filling of boron fluoride. A central electrode 221 is provided within the chamber 143 and connected to the wire 222 leading outside of the system enclosed by the concrete shield 101, shown in FIG. 7, to a movable contact 223 on the resistor 225. Resistor 225 is connected across a relay coil 226, one side of which is connected to the battery 227, the other of which is connected to the shield 146 around the wire 222. The shield 146 is grounded preferably at 230 adjacent the end of chamber 143 as well as externally of the system as shown at 231. The tank 1 is permeable to neutrons developed within the slurry 2 and alpha ray ionization due to neutron reaction with the boron within the chamber 143 is proportional to the neutron density. Thus, the current in resistor 225 is varied in accordance with neutron intensity reaching the ionization chamber. Relay coil 226 operates a relay armature 232 which is spring biased by spring 233 to contact one motor-control contact 234, and is urged by current in the relay coil 236 to contact a second motor control contact 237. Contacts 234 and 237 connect to the outside of a split winding of the reversible motor 142 through lines 238 and 239, the center connection 241 of which is connected through power mains 243 to the armature 232. The motor 142 operates shaft 245 directly connected to the reversible pump 18. The pump 18 is connected between the reaction tank 1 and the reservoir 17 through the suction line 141 as previously described. In operation of the system the pump 18 varies the level 124 and consequently the volume of the slurry 2 in the tank 1

16

between predetermined upper and lower limits on either side of the critical size at which the reproduction ratio is unity.

Having described a circuit for controlling the volume of the slurry, we will now describe its operation considering the condition obtaining when the volume of the slurry 2 in the reaction tank 1 is insufficient to support a self-sustaining nuclear chain reaction. The slider contact 223 on resistor 225 is calibrated in accordance with the neutron activity of the slurry. The slider contact 223 is then set in advance corresponding to the desired neutron density at which the system is to operate. While the ionization chamber does not indicate directly the maximum neutron density (i.e., at the center of the body of slurry) within the reaction tank, the ratio of maximum to measured density (the measured density being that at a point, say, just outside the tank) is a known ratio for all operating neutron densities within the reaction tank 1. For a low volume of slurry within the tank 1, the neutron density is much lower than the desired maximum neutron density and the relay coil 226 will not receive enough current to operate the armature 232 since very little ionization takes place within the ionization chamber 143. Consequently, the armature 232 will rest against the contact 234 driving the motor 142 in a direction pumping slurry from the slurry reservoir 17 into the reaction tank 1. However, as the volume of slurry within the tank 1 increases and exceeds the critical volume at which the neutron reproduction ratio is greater than unity the neutron density will rise until the ionization in the chamber 143 becomes so great that at the maximum desired neutron density the armature 232 is drawn into rest with the motor contact 237. Motor 142 is thus energized to reverse the pump 18 and withdraw slurry from the reaction tank 1 into the reservoir 17. The motor 142 will continue to operate until the volume of the slurry 2 in the tank 1 falls to a point at which the reproduction ratio of the chain reaction is less than unity whereupon the neutron density will commence to fall and eventually due to the lower neutron density and lower ionization in the chamber 143, the armature 232 again rests against the contact 234 and the motor 142 is reversed to pump again the slurry into the reaction tank 1. The volume of the slurry 2 in the tank 1 will thereafter hunt between upper and lower limits on either side of a volume corresponding to the critical size of the reaction system. Thus, the volume will vary between a point above the balance position at which the neutron density rises exponentially and a point below the balance position where the neutron density decays, providing an average neutron density within the reaction tank as determined by the setting of the sliding contact 223 on the resistor 225. As the mass of the slurry in the reaction tank causes any temperature change to lag behind any neutron density change, the temperature of the slurry is maintained substantially constant. If desired, any of the well-known anti-hunting circuits may be utilized as will be apparent to those skilled in the art.

It should be distinctly understood that the control circuit A cannot be likened to a throttle control. The rate at which the reaction occurs is not dependent upon the volume of the slurry but rather upon the neutron density attained after exceeding critical size and before decrease to critical size. For example, upon increase beyond critical size the neutron density would continue to increase exponentially with time irrespective of the cooling capabilities of the circulating and heat exchange systems. Control is, therefore, effected by controlling the volume above and below the critical size to maintain a desired neutron density and by decreasing the volume below critical size to decrease the neutron density.

Due to the fact that it might be possible for the control circuit as described to fail, and thereby leave the volume of the slurry at such a high level that the neutron density would continue to rise indefinitely notwithstanding the presence of the cadmium plates 161—162 and the

17

safety rod 176, we prefer to provide a safety circuit which may be brought into action either manually in response to abnormal operating conditions or automatically in response to neutron densities within the pile exceeding the predetermined setting of the sliding contacts 223 on the resistor 225.

One such illustrative safety circuit is shown in FIG. 12 at B. In the circuit B the safety rod 176 is supported over the normal slurry level 124 by a connection 250 to a toothed rack 251 engaging a pinion 252 attached to shafting 253—254 of a motor 256 through a magnetic clutch 257. The two shafts 253—254 are locked together by the magnetic clutch 257 when energized from the main power mains 259 in series with trip switches 261 which may be located at various safety control positions exteriorly of the concrete shielding. Upon failure of the power supply 259 which would deenergize the motor 142 possibly leaving the slurry level at a point at which the neutron density would continue to rise exponentially the magnetic clutch 257 is deenergized allowing the rack 251 to fall to a limiting position at the stop 262 and the safety rod 176 to be immersed in the slurry 2 in the reaction tank 1. The safety rod being of a material such as boron, cadmium, or other material having a very high neutron capture cross-section adds a controllable neutron loss to the system causing the reproduction ratio to fall below unity and the reaction within the tank 1 to decay in neutron density. The abnormal condition causing the high neutron density such as the power failure or the manual tripping of one of the switches 261 having been overcome, the safety rod may be withdraw to allow normal control to become effective by normal variation of the slurry volume within the reaction tank.

We prefer to combine automatic control of the safety rod with the manual or power failure control as also shown in the circuit in FIG. 12 at B. Thus we provide an additional ionization chamber 144 corresponding in construction and positioning to the chamber 143. As previously described in connection with circuit A the ionization chamber 144 is provided with wire line 266 connected to a resistor 267 by sliding contact 268 and with the battery 270 through the shield 147. The resistor 267 is connected across the relay coil 272 and to the other side of the battery 270. The circuit is completed by an armature 273 normally biased by spring 274 into rest with a contact 275 connected to the power mains 276 and thence to the motor 256 returning to the armature 273. The rotation of motor 256 is such that under normal operation, and when energized from the mains 276, the pinion 252 is rotated through the magnetic clutch 257 and shafting 253—254 to raise the rack 251 into a position of rest in contact with stop 277 thereby raising the safety rod without the slurry volume in the reaction tank 1.

The motor 256 is so designed as to slip electrically when the rack 251 is in contact with the stop 277 and the weight of the rack 251, line 250, and safety rod 176 are counterbalanced through the magnetic clutch 257 by weight 280 attached to the periphery of a pulley 281 mounted on the extension 282 of the shaft 254 to reduce the load on the motor.

However, under abnormally high conditions of neutron density within the tank 1, the current in the relay coil 272 increases to such a value that the armature 273 is withdrawn from its position of rest on the contact 275 thereby deenergizing the motor 256 allowing the pinion 252 to rotate dropping rack 251 into contact with stop 262 and safety rod 176 into the slurry 2 in the reaction tank 1. Consequently, by adjusting the sliding contact 268 on the resistor 267 to a position corresponding to a neutron density somewhat higher than the desired maximum, at which the slider contact 223 is set on the resistor 225, the safety system may be brought into control at high neutron densities which might be detrimental to the system.

However, as an abundance of caution and in view of the destructive explosion which might occur in the case

18

of failure of the control and safety systems just described to stop the nuclear chain reaction a portion or all of the slurry may be dumped from the reaction tank in case of emergency. Thus, one or more dump valves such as the valve 41, located in the dump line 40—42 to the slurry reservoir 17 may be opened in emergency to reduce rapidly the volume of slurry in the reaction tank 1 to a safe level.

A circuit for accomplishing dumping is shown in FIG. 12 at C. The dump valve 41 can be used for this purpose by connecting the dump valve rod 281 to a dump counter weight 282 urging the dump valve to an open position. The dump valve is maintained closed by a dump latch 283 held in engagement with rack 284 on the dump valve rod 281 by relay coil 286 which, when energized, holds the lever arm 287 fulcrumed at 288 notwithstanding the weight 289. Latch 283 may be withdrawn manually by a handle 291. To maintain the latch 283 in engagement with rack 284 against the weight 289, the relay coil 286 is energized by battery 292 having a circuit completed through series emergency break switches 293 and contact 294 to a movable armature 296. A third ionization chamber 145 similar to those previously described, is provided having a wire line 296 running through shield 148 connected to the resistor 297 and to a sliding contact 298. The resistor is connected across the relay coil 299 and returned at 300 to the battery 301 and to the ionization chamber 145 through the shield 148.

In operation, the slider contact 298 is adjusted to a position on the resistor 297 corresponding to a maximum safe neutron density slightly above the density setting of resistor 267, circuit B, at which emergency dumping should occur. Under such conditions the magnetic mechanism operates to disengage latch 283 allowing weights 289 and 282 to open the valve 41. The slurry dumped from the reaction tank 1 is not lost but led to the reservoir 17 from which the tank 1 may be replenished by the pump 18 upon manually closing the dump valve 41 for reinitiation of the nuclear chain reaction in the reaction tank.

Other measures may be used to assure complete safety in the operation of the system. Remote recording thermometers should be provided in the inlet 6 and outlet 7 cooling water lines of the heat exchangers 5 as well as remote recording flow meters to determine cooling water flow both in the heat exchanger systems as well as in the water supply line 51 filling the tank 50 in which the reaction tank 1 is immersed. These as well as other controls are dictated by conventional engineering practice and need not be considered in detail except to point out that safety controls operated, for example, by excessive temperature rises or insufficient cooling water flow may be utilized to stop the nuclear chain reaction either by insertion of the neutron absorbing safety rod or by dumping the slurry into the slurry reservoir tank.

To place the system in operation, the slurry reservoir 17 is supplied with sufficient slurry to fill the reaction tank to critical size volume whereupon the slurry is pumped into the reaction tank, by starting the slurry circulation pumps followed by adjusting the flow of the cooling water through the heat exchangers surrounding the reaction chamber. The helium supply to flush the gases produced by decomposition of the heavy water moderator is then adjusted to maintain helium pressure in the reaction tank such that for a temperature rise to 120° C. the moderator will not boil. The control circuit A is then set for a neutron density corresponding to about a 70° F. rise in temperature within the reaction tank and to keep the temperature of the slurry at the intake to the slurry circulation pumps at a safe level. The safety circuit B and emergency circuit C are then set to shut down the system in case of any substantial rise in neutron density through failure of the circulating system or power supply.

While the pumps and heat exchangers could be located on the outside of the concrete shielding walls, such position would increase the hold over while still necessitating

3,052,613

19

additional shielding enclosing these portions of the system, because after the initiation of the action the uranium in the slurry becomes very radioactive. For example, the fission products created by fission of the U²³⁵ produce considerable alpha, beta, and gamma rays which must be absorbed for the safety of operating personnel. Consequently, and inasmuch as the entire apparatus within the shield comprising the concrete walls and the lead lining of the tank becomes radioactive, it is impractical following initiation of the reaction and maintenance of high neutron density within the reaction tank to service or otherwise adjust or rebuild the moving parts within the shield. For example, in the structure shown in FIGS. 7–10, it would be extremely difficult to repair or replace the pump 4 although servicing of the driving mechanism for these pumps is facilitated by placing the driving motors exteriorly of the concrete shield.

In view of the present cost of heavy water moderator, it is desirable to reduce the amount used to a minimum. This may be accomplished by reducing the amount outside the active portion of the system and by utilizing an optimum concentration of uranium-containing material. At the same time, however, a substantial quantity of the suspension or solution must be circulated through the heat exchangers in order to secure rapid and efficient removal of heat from the reactor. We have found that most efficient removal of heat or energy per unit weight of slurry or solution may be secured by using a circulation system of a size such that the amount of suspension or dispersion in the circulation system is about 2 to 6 times the amount of such suspension undergoing neutron chain reaction in the reactor. Maximum utilization may be secured by circulating the suspension through cooling tubes having an internal diameter of ⅜ to ⅝ inch. We have shown the heat exchangers as being as close to the reaction tank as possible to thereby minimize the length of piping carrying the heavy water slurry and also to enable a saving of delayed neutrons as described hereinafter. For a given number of heat exchangers more heat may be removed from the system by using a higher temperature of the circulating slurry, although increase in temperature decreases the reproduction ratio somewhat. The gain in increased heat transfer, however, more than compensates the slight increase in slurry volume then necessary to obtain a reproduction ratio of unity, until boiling of the heavy water occurs. In general, in the preferred type of reactor herein described, boiling of the moderator should not be permitted since the resulting formation of bubbles in the moderator makes this type of reactor very difficult to control. It is thus economically desirable with respect to heavy water utilization, to operate at a temperature below boiling. The utilization may be increased still further by operating the system at a maximum slurry temperature above the normal boiling point, preventing boiling by an increase in pressure. Thus, a pressure greater than that corresponding to a maximum slurry temperature of 120° C. is preferred notwithstanding the necessity for providing a system capable of withstanding the increased pressure. Consequently, we prefer to provide the reaction tank 1 of such construction to withstand an internal pressure of from 100 to 150 pounds per square inch, the slurry reservoir, connecting piping and heat exchangers being designed to withstand a similar pressure. Control of operating pressure is preferably accomplished by pressure control of the helium atmosphere provided to sweep the dissociated heavy water gases over the hot grid, across the catalyst chamber and the vapor into the heavy water condenser.

While we have shown a system incorporating six heat exchangers surrounding the reaction tank, a different number may be used depending upon the maximum or minimum rate of heat dissipation desired. Preferably, a greater number of heat exchangers and associated pumps than the desired minimum capacity is installed inasmuch as pump or exchanger failure necessitates their removal

20

from the system by closing their associated valves. Thus, following initiation of the nuclear reaction radioactive fission fragments will become lodged in the pumps rendering them exceedingly radioactive. Repair or replacement is impractical and the number initially installed must be sufficient to allow for standby in case of failure or for operation at reduced operating levels with failure of one or more of the circulating systems. Furthermore, as previously indicated, savings in the quantity of slurry required to initiate a self-sustaining chain reaction may be obtained by use of a spherical or semi-spherical reaction tank.

We have shown in FIGS. 13–15 a further structural embodiment of our invention wherein the reaction tank is semi-spherical in combination with two types of circulating systems one of which is particularly adapted to the replacement of the working parts.

Referring to FIGS. 13–14, we have shown in schematic outline the principal features of a self-sustaining nuclear reaction system utilizing a semi-spherical reaction tank 350, surrounded by a cylindrical tank portion 351 in place of the cylindrical type of structure previously described. With this type of reaction tank, the slurry volume required for a self-sustaining reaction is considerably less, as previously noted, in comparison with the slurry requirements for a cylindrical structure. The cylindrical tank portion 351 encloses the cadmium plates 161—162, the hot grid 28, and the safety rod 176 previously described and is surmounted by a bell-shaped enclosure 138 leading off from which is the catalyst chamber 29, leading in turn to the condenser 352. The condenser 352 is additionally cooled, not only by the water 52 in the enclosing tank 50, but also by additional cooling medium circulated within the condenser 352 such as through an inlet line 353 and an outlet line 352A. We have likewise shown in FIGS. 13 and 14 tank 50 as being provided with an inwardly extending conical top 354 with a water outlet 105, as well as an outlet 355 for escape of dissociated gases from the water 52 within the tank 50. Such construction assures cooling of the top portion of the tank 50 as well as equalization of temperatures in the top and side portions of the concrete shield 101.

We have shown only two slurry circulating and heat exchange systems in FIG. 13, these being shown as of different types merely for the purpose of indicating flexibility of the system. Obviously, additional circulating and heat exchange systems may be utilized to increase the heat dissipating capacity of the system. Referring in particular to FIG. 13, the heat exchanger 5 is similar to those previously described, although the slurry circulating pump 356 is positioned with its axis vertically so that the drive may be from the top of the concrete shield 101 by a motor 357 having a shaft 358 extending through the casing 359 to the pump 356. The valves 104 and 107 are of the type previously described except provided with housings 360—361, which extend outside of the concrete shield 101 to corresponding hand driven or motor driven controls. By use of the housings 360—361, leakage around the packing glands of these valves may be minimized or prevented by maintaining a helium pressure equal to or slightly exceeding the helium pressure maintained in the cylindrical tank portion 351, as later described. Inasmuch as it is impractical to repair or otherwise adjust the types of circulating pumps previously described, we have shown a construction in FIG. 13 wherein the pump shaft and impeller assembly may be withdrawn from the system for repair or replacement. Reference is made to FIGS. 13 and 15 which show an axial circulating pump generally designated at 370. The pump housing 371 is connected directly to the upper end of the heat exchanger generally designated as 372. The housing 371 encloses an impeller 373 mounted on a shaft 374, the housing likewise enclosing directing vanes 376 immediately below the impeller 373. A second impeller 377 is mounted on a flanged hub 375₂ splined to the shaft 374

21

above the impeller 373 to maintain the slurry in the pump housing 371 below the bearing assembly. Inasmuch as the slurry, especially in the initial stages of operation of the system, may be abrasive in character, if not initially run-in, we take special precautions to prevent its coming in contact with the bearing surfaces by providing a structure such as shown in FIG. 16. Referring to FIG. 16, the shaft 374 extends axially of the pump housing 371, being supported by the bearing 378 which is lubricated through the lubrication line 379 carrying an inorganic lubricant since organic lubricants tend to be unstable to high neutron densities. The bearing is supported within the housing 371 by a sleeve liner 380 extending axially of the housing 371 to the outside of the concrete shield 101 as shown in FIG. 17. This construction allows the complete removal of the impellers and bearing assembly, either for replacement or repair. Between the bearing 378 and the top impeller 377, we apply labyrinth splash and vapor seals between an integral flange 381 extending in a direction toward the shaft 374 on the lower end of the housing liner 380, and the bearing 378. The flange 381 is positioned parallel with and slightly below a flange 382 integral with the impeller hub 375, shaft 374 being surrounded by a cylindrical member 383 having an integral flange 384, the periphery of the integral flange 384 being affixed to the housing liner 380. Rotation of the flanged hub 375 between the flanges 382—384 serves as an effective splash shield to prevent any upward creeping of the slurry in the pump. Creepage of the slurry along the shaft 374 may be further limited and vapor sealed from the bearing by use of a liquid seal. Mercury 385 may be introduced in the chamber formed by the liner 380 and the cylindrical flange member 383, with a telescoping cup-shaped member 386 affixed to the shaft 374, inverted in the mercury, and surrounding the cylindrical member 383. In this manner the assembly is effectively sealed both with respect to the slurry in the casing and with respect to any vapors or gases which may be released from the slurry. To further protect the bearing 378, helium may be admitted to the pump casing 371 through a line 388 and maintained at a predetermined pressure with respect to the helium in the cylindrical portion 351 of the reaction tank 350 to maintain the slurry level well below the top impeller 377.

In the structure just described, it is necessary to provide a casing surrounding the circulating pump shaft which is of relatively large diameter to allow the shaft and impeller to be withdrawn. Consequently, considerable opportunity exists for neutron and gamma ray leakage through the concrete shield interiorly of the pump casing liner. As best shown in FIG. 17, we provide a labyrinth of gamma ray absorbing material to prevent the escape of direct radiation from the system. This labyrinth comprises a plurality of axially separated rings 391, snugly fitted to and stationary with the liner 380, and having therebetween additional rings 392, snugly affixed to the shaft 374. The rings 391—392 are of iron or other dense metal and overlap one another forming a labyrinth or tortuous passage preventing the escape of gamma rays except directly through the metal. Since gamma rays are radiated to best advantage along straight line paths, the members 391 and 392 are interposed in the path of gamma radiation causing the major portion thereof to be absorbed in these members. These members, however, are not completely effective shielding for neutrons and consequently, we prefer to fill the casing with lubricant through the line 379 which extends to the bearing 378, this lubricant being withdrawn through the line 394. The lubrication lines 379 and 394 are positioned between the rings 391 and the liner 380 adjacent the inner surface of the liner. The lubricant used is preferably inorganic such as water although a hydrocarbon oil may be used since the mercury seal and positive helium pressure within the pump casing prevent contamination of the slurry with the lubricant, and the lubricant may be replaced through the lines 379 and 394 to compensate for deterioration

22

under neutron bombardment. If water or hydrocarbon oil is used as a lubricant, the neutrons otherwise escaping through the pump drive casing are slowed by the hydrogen content and substantially absorbed thereby limiting their escape from the system.

The pump casing 371 is joined to the reaction tank 350 somewhat below the normal slurry level 401, through a valve generally designated 402 and inlet line 430. Following initiation of the reaction within the tank 350 substantially the only occasion for closing the valve 402 is in the event of failure of the circulating system, such as by leakage occurring in the heat exchanger 372. Inasmuch as it is impractical to repair any portion of the apparatus within the concrete shield following the operation at high neutron density only a single operation of the valve 402 is required. The valve must, therefore, be positive in its section, must be of such a design as not to become closed by the slurry and must be substantially leakproof.

The valve 402 in FIG. 15 particularly meets these requirements in that it is of the plunger type having a valve stem 403 terminating in a cylindrical valve head 404 having two tapers 405 and 406 on opposite ends thereof. The taper 405 as shown is seated against the packing 407 in the withdrawn or valve-opened position, thereby effectively preventing leaks while open. In this position, the entire valve head 404 is withdrawn into a cylindrical channel 408 so as to be out of the path of the slurry passing through the valve. The valve throat 409 is of constricted cross-sectional area to increase the velocity of the slurry passing through this portion of the valve to maintain the valve seat 411 clear of any slurry particles which might be entrapped in the valve seat portion. The valve seat 411 is likewise tapered to conform with the taper 406 on the cylindrical valve head 404 so that in the closed position not shown, the valve head positively engages the valve seat 411 forming a liquid tight seal therewith. The valve 402 may be driven either electrically or by hand, although we have shown a hand wheel 412 engaging the valve stem 403 at the upper end thereof over a threaded portion 413.

A similar valve generally designated as 420 is of similar construction, although we have shown this valve disposed with its valve stem 421 horizontally and in a closed position showing the method of seating the valve head 422 upon the valve seat 423.

Upon failure of the heat exchanger or the development of leaks in the circulating system the two valves 420 and 402 may be closed and the slurry withdrawn from the circulating system through a drain at the lowermost portion of the system. For example, a drain 425 may be provided in the casing 426 joining the valve 420 with the condenser 372. The drain may be placed in operation by a valve shown schematically at 427 operated, for example, through the concrete shield 101 from the side thereof.

In FIG. 18, A, B, and C represent individual particles of material containing natural uranium, but it is to be distinctly understood that the entire chain may take place in one body, two bodies, or three, as shown, because of the fact that the neutrons during the slowing down process are diffusing over random paths throughout the entire composite mass of the slurry, and are not necessarily passing directly from one uranous body to the next adjacent body. The factor n represents any fixed number of neutrons.

The fission reaction may be represented as follows:

$$_{92}U^{235} + \text{neutron} = A + B + 2.2 \text{ neutrons (average)}$$

A = "Light" fission fragment, e.g., Br, Kr, Rb, Sr, Y, Zr, Cb, Mo, 43, Ru, Rh. Atomic mass, 83–99, inclusive. Atomic number, 35–45, inclusive.

B = "Heavy" fission fragment, e.g., Sb, Te, I, Xe, Cs, Ba, La, Ce, Pr, Nd. Atomic mass, 127–141, inclusive. Atomic number, 51–60, inclusive.

3,052,613

23

The 100n fast neutrons leaving the uranium body, enter the mass of heavy water moderator to diffuse therethrough and through the bodies over paths long in comparison with the spacing of the bodies within the slurry, to undergo successive collisions that slow them down; and a large proportion of the 100n fast neutrons are thus destined to be reduced to thermal energy. During this diffusion, before the neutrons arrive at thermal energies, a portion of the neutrons diffuse out of the system because of its finite size and is lost to the chain reaction. Furthermore, during the course of the extremely irregular diffusion path of the neutrons while they are being slowed down by elastic collisions predominantly in the heavy water, and to a minor extent in other uranium bodies, further neutrons will reach a uranium resonance absorption energy or energies and will be absorbed by U^{238} to form U^{239}. This is known as resonance absorption and includes both surface and volume resonance absorption referred to above.

While resonance absorption is referred to here as a loss, it is only a loss of neutrons with respect to maintenance of the chain reaction, inasmuch as these neutrons provide a gain for the production of new elements. Thus, irrespective of whether the neutron resonance absorption in U^{238} is on the surface, or in the volume of the uranium body, 94^{239} is produced by the resonance absorption according to the following process:

$$_{92}U^{238} + n \rightarrow {}_{92}U^{239}$$
$$_{92}U^{239} \rightarrow 23 \text{ min. } {}_{93}Np^{239}$$
$$_{93}Np^{239} \rightarrow 2.3 \text{ day } {}_{94}Pu^{239}$$

Capture of thermal neutrons by U^{238}, as indicated in bodies A and C of FIG. 18, also results in production of 94^{239} by the same process.

The 94^{239} is a long lived radioactive product (an alphaemitter) with a half life of about 23,000 years. It remains in the uranium body and is fissionable by thermal neutrons in the same general manner as U^{235}. In our system the number of neutrons available for production of 94^{239} during each cycle of the chain reaction is relatively large being 43 percent of the neutrons developed by fission. Consequently, our system is of particular use in producing 94^{239} which may be used for enriching other chain reacting systems to provide a large K factor and a small critical size. The production of 94^{239} in the uranium body greatly prolongs the time during which the reaction can be maintained using the initial uranium in the slurry. The 94^{239} can be removed from the uranium by chemical separation if desired.

As shown in FIG. 18, a substantial portion of the 100n original fast neutrons escape resonance capture and fast neutron leakage, and are reduced to thermal energy within the system. Of these thermal neutrons, some neutrons on the average leak by diffusion out of the system and are lost from the chain reaction.

In any practical system, impurities will be present in both the heavy water and the uranium which capture and/or absorb a quantity of the thermal neutrons. This absorption is due to the fact that a few of the thermal neutrons diffusing through the moderator mass are not in a position to promptly enter a uranium mass when they reach thermal energy, and these thermal neutrons must continue to diffuse through the moderator until they do reach a uranium body or are absorbed.

An analysis of the chain reaction just described shows where the gain in production of 94^{239} is obtained by following our invention. In previous chain reacting systems using uranium in graphite, wherein the graphite was the neutron moderator, the neutron losses in the graphite were so great that it was necessary to aggregate the uranium into large masses distributed through the graphite in regular array to reduce the surface resonance losses to an amount compensating the losses in

24

the graphite. However, by utilizing, in accordance with our invention, a more efficient neutron moderator and one having a very low neutron capture cross-section, large resonance losses are permissible thereby increasing the neutrons available for 94^{239} production. Consequently, by following the teachings of our invention, the 94^{239} production is increased by increasing the surface to volume ratio of the uranium through subdivision of the uranium into fine particles such as by utilizing uranium oxide in the heavy water moderator as a fine particle slurry.

In the chain reaction shown in FIG. 18 for a natural uranium bearing slurry, it is assumed that the reproduction ratio is exactly unity, that is, the neutrons developed by fission are exactly equal to the neutron losses. However, the loss of the neutrons by leakage may be reduced by increasing the slurry volume and effectively raising the reproduction ratio to a value greater than unity.

Consider, for example, the condition which would obtain in the chain reaction shown in FIG. 18 if the volume of slurry were suddenly increased to such an extent that a portion of the neutrons normally lost by fast and thermal neutron leakage were saved by such increase in size. In this case, more neutrons would be available to enter the uranium body C. Instead of the production of 100n fast neutrons by fission at the end of the first cycle, as in the balanced condition, more than 100 fast neutrons would then be produced, and the neutron intensity of the system would then rise exponentially with time as each new cycle starts out with 100 plus an increasing number of fast neutrons. The effective reproduction ratio (r) of the particular system in this event would be correspondingly increased. In practice, however, the slurry volume would never be increased to this extent.

In order to stabilize the reaction by volume control at any desired neutron density within the system, the neutron density is measured as it increases. When a predetermined neutron density within the system is reached, the level and consequently the volume of the slurry is decreased to a point where more neutrons are lost by leakage through an effective increase in total surface to total volume ratio of the reaction volume. The chain reaction will again be in balance at the new neutron density. To reduce the neutron density the level of the slurry may be decreased still more to increase the leakage of neutrons from the system to a volume such that the neutron density will decay, because less than 100n neutrons will be produced for each 100n original neutrons. The system may be again stabilized at the newly desired lower density level by increasing the volume of the slurry effective in maintaining the chain reaction to a point where r is again unity. It should be distinctly understood that the critical size remains essentially constant for any given conditions of temperature, pressure, and density of the slurry and that critical size is a constant irrespective of the power output in the form of heat. Consequently, the level of the slurry for critical size conditions remains quite constant irrespective of output of the pile. Inherent changes in temperature, pressure, and density may, however, be overcome by varying the level of the slurry up and down on either side of the critical size conditions.

An important element in the control of the pile is the fact that not all of the 100n fast neutrons originating in A or C in FIG. 18 are emitted immediately. About one percent or 1n fast neutrons are "delayed neutrons." These delayed fast neutrons may appear at any time up to several minutes after the fission has occurred. Half of these neutrons are emitted within six seconds and 90 percent thereof within about 45 seconds. The mean time of delayed emission is about 5 seconds. The cycle shown in FIG. 18 is completed by 99 percent of the neutrons in about .0015 second, but if the system is near the balanced condition the extra 1 percent may make all the

3,052,613

25

difference between an increase or a decrease in the activity. The fact that the last neutron in the cycle is held back, as it were, imparts a slowness of response to the system that would not be present if the $100n$ neutrons were all emitted instantaneously.

For cases in which the reproduction ratio (r) differs from unity by less than 1 percent, the ratio of rise is given by the formula

$$n = n_0 e^{wt}$$

where

$$w = \frac{r-1}{\alpha - (r-1)} \cdot \frac{1}{T}$$

In this formula α is the fraction of the neutrons that are delayed, $\alpha = .01$, T is the mean time of delayed emission of the delayed neutrons $= 5$ seconds.

As an example, suppose as a result of increasing the critical size by increasing the mass of slurry r becomes 1.001. Then

$$w = \frac{.001}{.01 - .001} \cdot \frac{1}{5} = \frac{1}{45}$$

that is, $n/n_0 = 2.72$ in 45 seconds. Hence, doubling of the neutron density occurs about every 30 seconds and continues indefinitely.

If r were made exactly 1.01, a more detailed theory shows that the neutron density would be tripled each second. However, if the reproduction ratio r is suddenly increased several percent, so that the one percent delayed neutrons are unimportant compared with $r-1$, the neutron density increases at a much more rapid rate as given approximately by $r^{t/e}$ where e is .0015 second, the normal time to complete a cycle. Thus if r were to be made 1.04, the neutron density would increase in 1.5 seconds by a factor of approximately 10^{17} over its original level. However, if r were 1.02 or 1.03, the factor by which the neutron density would be multiplied each second would be 1100 and 700,000 respectively. It is thus apparent that too high a reproduction ratio in a practical system leads to the necessity of providing safety measures, which positively limit all danger of exceeding a permissible rate of neutron density increase. An exceedingly dangerous condition could exist if by accident the volume of slurry was suddenly increased considerably beyond the critical size, as the time required for lowering the level might be too long to prevent destruction of the system. As the same eventual neutron density can be obtained with a reproduction ratio only slightly over unity, as with a higher ratio, only at a slower rate, the lower reproduction ratios are preferred in practice in the interest of safety. Consequently, the neutron reproduction ratio of the system should not exceed about 1.01 during operation of the reactor and control of the reaction should be directed to this end.

As the suspension is withdrawn from the reactor and circulated through the heat exchanger, the neutron reaction substantially ceases due to the fact that the volume suspension being circulated is below a critical size for establishment of a neutron chain reaction. However, delayed neutrons are evolved after withdrawal of the suspension. In order to effect a saving of neutrons, circulation may be conducted at a rate such that the suspension is returned to the reaction zone before all of the delayed neutrons have been evolved. In such a case, a portion of the delayed neutrons will be evolved in the reaction zone and thereby permit use of a smaller volume of slurry or suspension.

In addition to the usual industrial hazards during the operation of the system, operating personnel must be protected from injury by gamma rays and neutrons generated in the reaction tank as well as from radiation from the circulating water surrounding the tank, from beta rays, from close contact with radioactive materials, and from radioactive poisoning due to inhalation of radioactive gases. The major portion of the radiation

26

emitted from the reaction tank is intercepted by the water surrounding the tank and by the concrete shield. The depth or rather, thickness of the water surrounding the reaction tank is determined by the slowing action and capture of the neutrons by the water. Since the heat exchangers contain the radioactive fission products and are nearer the water shield tank, beta rays therefrom must be considered in determining the shielding thickness. The minimum depth of water between the heat exchangers and the lead shield 100 of the tank 50 is 3 feet and the minimum depth from the reaction tank thereto is 8 feet while the thickness of the concrete shield is preferably no less than 10 feet. Inasmuch as the concrete contains water or crystallization and may contain water retaining material, and is relatively dense, it serves as an effective shield for neutrons as well as gamma rays. In addition to these precautions we prefer to provide the lead shield 100 of a thickness of 6 inches, lead being very effective as a gamma ray shield. The slurry removal line through which the slurry is withdrawn for separation of solid matter therefrom, should likewise be shielded after passing through the concrete shield. Such shielding may be effected by passing this line through the concrete base, embedding the line in concrete and covering the concrete with sufficient depth of earth to reduce the radiation at the surface to less than that equivalent to an exposure (for adjacent personnel) of 0.10 roentgen per 8 hour day. Any leakage of the slurry into the circulating water in which the apparatus is immersed will cause contamination of the water and consequent radioactivity therein. In addition, neutrons escaping from the reaction tank may produce radioactive isotopes of any impurities present in the circulating water. Consequently, it is preferred to cool the water, following circulation within the concrete shield, in a cooling tower from which operating personnel may be excluded although this cooling water may be once circulated and discarded to a deep well, for example. In addition to the above precautions a portion of the water within the concrete shield is decomposed into hydrogen and oxygen under high neutron bombardment, and any absorbed gases in the water, such as atmospheric gases, may become radioactive. These gases may be vented with great dilution through a chimney or stack without harm to operating personnel or to those in the surrounding territory. Such precautions will reduce the radiation from the structure to 0.10 roentgen per 8 hour day per person at the point of closest approach, this exposure being the maximum safe radiation permissible to which an individual may be subjected over the whole body.

The system and processes herein contemplated have been particularly described with reference to a slurry. However, it is to be understood that the invention is generally applicable to suspensions of a fissionable isotope or material such as 94^{239}, U^{235} or U^{233} suspended in a fluid, preferably liquid moderator, and thus the invention is applicable to true solutions as well as slurries. Moreover, moderators other than deuterium oxide may be used. Where natural uranium is used as the fissionable component, pure deuterium oxide or at least a mixture of light and heavy water containing 75 to 90 percent or more of deuterium oxide is an effective moderator. Moreover, deuterocarbons, such as deuterodiphenyl, are suitable. In addition, molten suspensions, such as molten thorium or uranium-beryllium alloys, could be used. On the other hand, if the fissionable component is a pure fissionable isotope or comprises a thorium or uranium mixture containing more than 0.7 percent of the fissionable isotope, a wider latitude in choice of moderators may be used, and if the concentration of the fissionable isotope is sufficiently high, ordinary water is a suitable moderator.

The invention has many uses. Radioactive fission products and new elements, both radioactive and stable, are produced by absorption of neutrons in the uranium.

3,052,613

27

Large quantities of neutrons and penetrating gamma rays are produced and can be utilized. With proper heat exchangers, power in the form of heat can be obtained.

If desired, the fission products, radioactive and stable, and the 94^{239} produced by normal operation of the pile can be recovered after a predetermined exposure of the uranium in the slurry to high neutron densities by removing the slurry, separating the uranium therefrom, and extracting the desired elements as referred to above. The radioactive fission products are valuable as gamma radiation sources and for use as biological tracers in medicine, while 94^{239} is useful as a fissionable material to increase K factors when added to natural uranium, or when used alone, in chain reacting systems.

While the theory of the nuclear chain fission mechanism in uranium set forth herein is based on the best presently known experimental evidence, we do not wish to be bound thereby, as additional experimental data later discovered may modify the theory disclosed. Any such modification of theory, however, will in no way affect the results to be obtained in the practice of the invention herein described and claimed.

What is claimed is:

1. A nuclear fission chain reacting system comprising a reaction tank, a liquid slurry consisting of uranous material in heavy water containing 0.04 to 0.0025 atom of U per molecule of D₂O disposed in said tank, the slurry being the only element of the system containing material fissionable by neutrons of thermal energy, slurry circulating means in series flow relation with the slurry in said tank, means for flowing a coolant into heat exchange relation with said slurry circulating means, means for flowing said coolant into heat exchange relation with said tank, and discharge means withdrawing said coolant following heating thereof by said slurry and by said tank, whereby heat is withdrawn from said slurry and said tank.

2. In a nuclear fission chain reacting system a reaction tank having a semispherical portion surmounted by a cylindrical portion, a quantity of slurry including uranium oxide and heavy water in a concentration of 0.0025 to 0.04 atom of uranium per molecule of D₂O in the semispherical portion of said tank to support a chain reaction, and means adjacent to the junction of said tank portions to limit the chain reaction volume to that within the semispherical tank portion.

3. The method of producing element 94 comprising the steps of suspending an oxide of uranium in heavy water to form a slurry in a concentration of 0.0025 to 0.04 atom of uranium per molecule of D₂O, developing a nuclear chain reaction in a volume of said slurry to develop the element 94 therein, withdrawing a portion of said slurry from the volume of said slurry during the maintenance of said chain reaction, and separating the element 94 from other constituents of said slurry.

4. The method of obtaining the products of a nuclear fission chain reacting system wherein the reacting mass comprises a heavy water suspension of active material that includes thermal neutron fissionable material, comprising the steps of circulating the slurry through a chamber, bleeding off a portion of said slurry, removing the active material from said slurry, and returning the heavy water of the bled-off slurry to said system.

5. The method of obtaining fission products from a circulating heavy water thermal neutron fissionable material slurry of a self-sustaining nuclear fission system comprising the steps of initiating a self-sustaining nuclear fission reaction, sustaining said reaction until a predetermined concentration of fission products are formed in said slurry, bleeding off a portion of said slurry while maintaining said chain reaction, and removing the fission products from the bled-off slurry.

6. The method of recovering the products of a nuclear fission chain reaction wherein the active material comprises a suspension of thermal neutron fissionable ma-

28

terial in a fluid moderator, comprising the steps of circulating the suspension through a chamber, bleeding off a portion of said suspension, and separating the active material from the moderator of said suspension.

7. The method of obtaining nuclear fission products which comprises assembling a sufficient volume of a slurry comprising a thermal neutron fissionable material in heavy water to initiate a self-sustaining nuclear chain reaction, maintaining said reaction for a period of time to obtain a quantity of fission products in said material, progressively removing the said material containing said fission products, and replenishing said volume with additional thermal neutron fissionable material substantially free of fission products.

8. A device for effecting a nuclear chain reaction comprising a reaction tank, a quantity of slurry consisting of uranium and heavy water in the proportion of between 0.0025 and 0.04 atom of uranium per molecule of D₂O in said tank, a closed circulating system for circulating said slurry whereby heat developed by nuclear fission of the uranium and by fission products therein may be removed from said slurry, means for continuously removing a portion of said slurry from said tank, means for separating the uranium-containing material from the heavy water of said removed slurry, and means for returning the heavy water separated from said slurry to said tank.

9. In a continuously operating nuclear fission chain reacting system comprising a slurry of uranium and heavy water, said uranium being fissionable with the production of elements 93 and 94 and fission products, means for continuously circulating the slurry in said system for removal of heat therefrom, and continuously operative means for removing elements 93 and 94 and fission products from said system during the continuation of a nuclear fission chain reaction therein.

10. A device for effecting a controlled nuclear chain reaction comprising a container, a closed liquid circulating system leading from and returning to said container, a heat exchanger in said closed system, a slurry consisting of uranium containing particles in heavy water in the proportion of between 0.0025 to 0.04 atom of uranium to molecules of D₂O in said container and said system to support a nuclear fission chain reaction and form elements and fission products therein, means for circulating said slurry through the said circulating system, means for bleeding off some of the slurry for extraction of elements and fission products formed by fission of said uranium, and means for varying the critical size of the slurry in said container to control the nuclear chain reaction therein.

11. A method which comprises establishing a self-sustaining neutron chain reaction in a suspension of a thermal neutron fissionable material suspended in a fluid moderator, withdrawing and cooling a portion of the suspension, and maintaining the volume of the suspension undergoing cooling below the critical size necessary for a self-sustaining chain reaction during the cooling operation.

12. A device for effecting a nuclear chain reaction comprising a slurry containing all the thermal neutron fissionable material in the device consisting of granulated uranium oxide particles of less than 2 microns diameter suspended in deuterium oxide in the proportion of 0.0025 to 0.04 atom of uranium per molecule of deuterium oxide, means to retain said slurry in spherical form with a radius of approximately

$$\frac{56.5}{\sqrt{K-1}}$$

centimeters, where K is the neutron reproduction factor of the slurry and is greater than unity, a closed liquid circulating system including the slurry retaining means and a heat exchanger in said system.

13. A self-sustaining nuclear fission chain reacting system comprising a mass of a suspension consisting of

3,052,613

29

uranium in heavy water in the proportion of 0.04 to 0.0025 atom of uranium per molecule of heavy water, the heavy water having a neutron absorption cross-section of approximately 4 millibarns, said mass being spherical in shape and having a radius not less than that defined 5 relative to the concentration of uranium and heavy water by the dashed line of FIGURE 6, the abscissa setting forth concentrations and the right ordinate setting forth radii.

14. The method of controlling a self-sustaining chain 10 reacting system which comprises disposing within a spherical chamber a suspension consisting of heavy water and uranium wherein the ratio of the uranium atoms to deuterium oxide molecules lies between approximately 0.0025 and 0.040, and varying the volume of the suspension 15 within the chamber to control the neutron-reproduction ratio of the system, the chamber having a radius greater than that set forth by the right ordinates of FIGURE 6 for the concentration of the slurry set forth by the abscissa, the heavy water in said slurry having an absorption 20 cross-section of approximately four millibarns.

15. A neutronic reactor comprising a cylindrical tank 14 feet in diameter and greater than six feet in height, a slurry consisting of heavy water and uranium disposed within the tank, said slurry containing all of the fission- 25 able material in the reactor and having a concentration of about 0.01 uranium atom per molecule of heavy water, the uranium being in the form of particles smaller than two microns, the height of said slurry being approximately six feet, and a closed slurry circulating system including 30 the tank and heat exchanger.

30

References Cited in the file of this patent

UNITED STATES PATENTS

1,953,076	Davies	Apr. 3, 1934
2,097,769	Mitscherling	Nov. 2, 1937
2,206,634	Fermi et al.	July 2, 1940
2,252,740	Teter	Aug. 19, 1941
2,592,112	Bradshaw	Apr. 8, 1952
2,708,656	Fermi et al.	May 17, 1955

FOREIGN PATENTS

114,150	Australia	May 2, 1940
861,390	France	Oct. 28, 1940
233,011	Switzerland	Oct. 2, 1944
114,151	Australia	May 3, 1940

OTHER REFERENCES

Compt. Rend., vol. 208, 1394–6, 1573–5 (1939); vol. 209, 301–3 (1939).

Goodman: "The Science and Engineering of Nuclear Power," vol. 1, page 275, Addison-Wesley (1947).

Kelly et al., Phy. Rev. 73, 1135–9 (1948).

Pollard and Davison: "Applied Nuclear Physics," page 256, John Wiley & Son, New York (1951), 2nd edition.

Physical Review, 56 (1939), pages 284–286.

Nature, 143 (1939), pages 470, 471, 680.

Naturwissenchaften, vol. 27 (1939), pages 402–410.

Nuclear Reactor Development, Atomic Industrial Forum, New York 16, New York, proceedings held at Washington, D.C., May 24, 1954, page 18.

Sept. 27, 1960   E. P. WIGNER   2,954,335
                                  NEUTRONIC REACTOR

Filed Feb. 4, 1946   3 Sheets—Sheet 1

FIG.1.

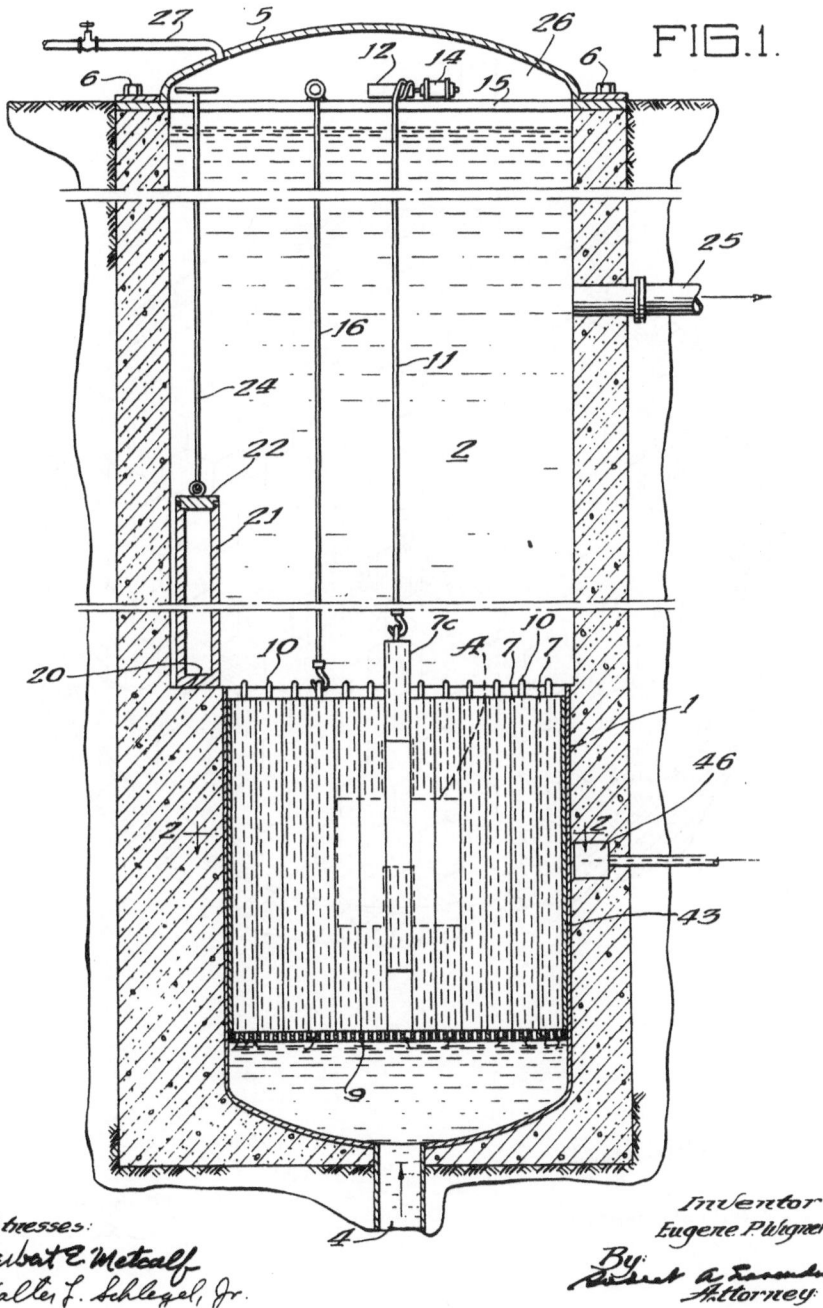

Witnesses:
Herbert E. Metcalf
Walter F. Schlegel, Jr.

Inventor
Eugene P. Wigner
By:
Robert A. Sanders
Attorney

Sept. 27, 1960   E. P. WIGNER   2,954,335
                           NEUTRONIC REACTOR

Filed Feb. 4, 1946   3 Sheets—Sheet 2

FIG.2.

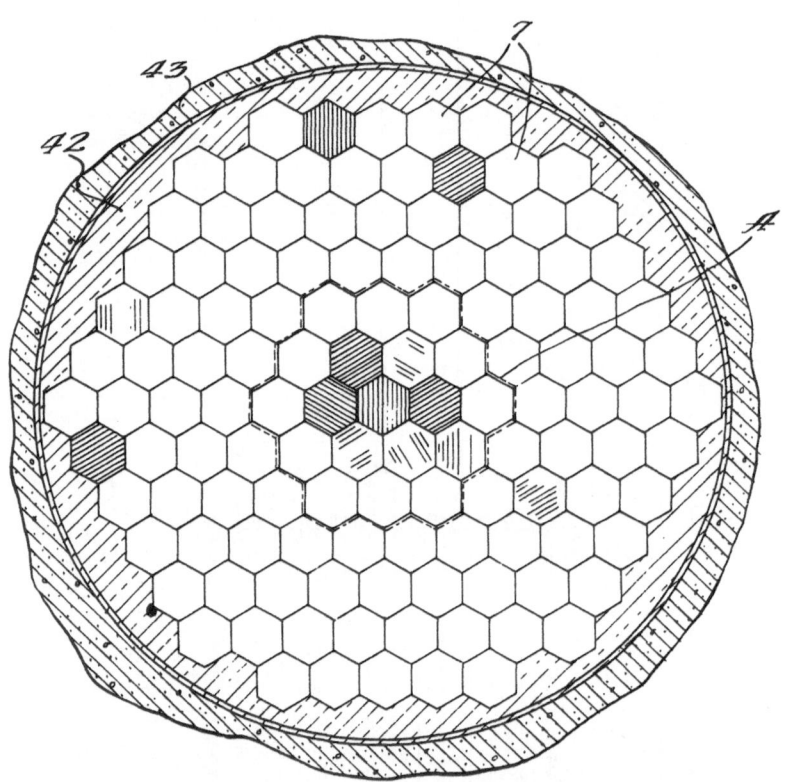

Sept. 27, 1960   E. P. WIGNER   2,954,335

NEUTRONIC REACTOR

Filed Feb. 4, 1946   3 Sheets–Sheet 3

FIG.3.

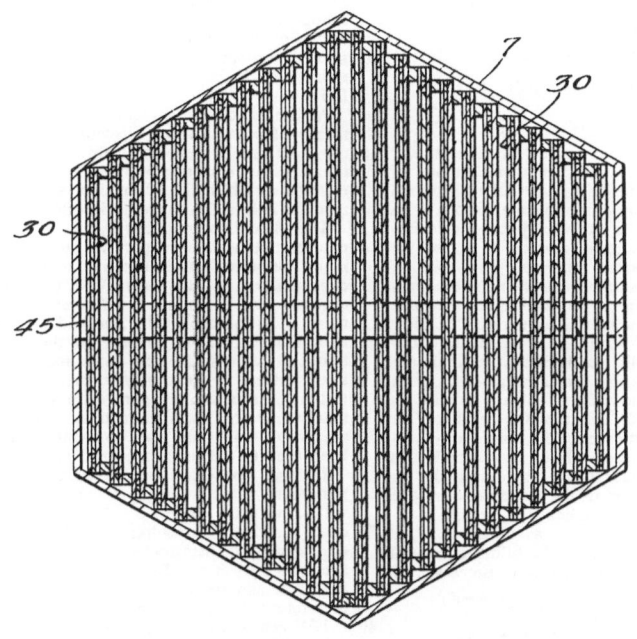

FIG.4.

FIG.5.

FIG.6.

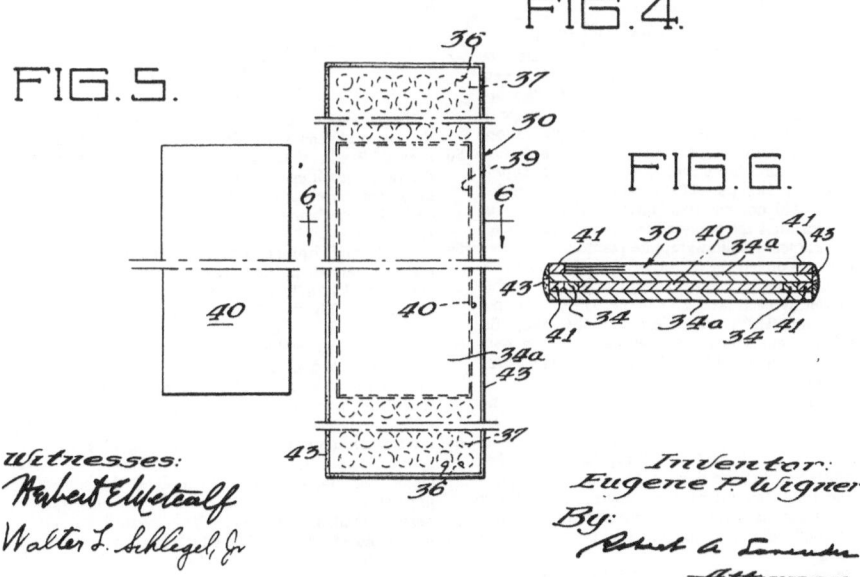

Witnesses:
Herbert E. Metcalf
Walter L. Schlegel, Jr

Inventor:
Eugene P. Wigner
By:
Robert A. Lorenzen
Attorney.

United States Patent Office

2,954,335
Patented Sept. 27, 1960

1

2

2,954,335

NEUTRONIC REACTOR

Eugene P. Wigner, Princeton, N.J., assignor to the United States of America as represented by the United States Atomic Energy Commission

Filed Feb. 4, 1946, Ser. No. 645,459

5 Claims. (Cl. 204—193.2)

The present invention relates to nuclear physics, and more particularly to an improved means and method of converting an isotope fissionable by thermal neutrons to another or the same thermally fissionable isotope in a neutronic reactor. The term "thermally fissionable isotope" as herein used refers, as is common, to an isotope which is fissionable by thermal neutrons.

It is known that a self-sustaining nuclear fission chain reaction can be obtained in devices known as neutronic reactors utilizing natural uranium, as a result of slow neutron fission of the U^{235} content of the natural uranium. In such reactors, discrete bodies of natural uranium of high neutronic purity are disposed, usually in the form of a lattice arrangement of spheres or rods, in a neutron moderator such as graphite, beryllium or heavy water of high neutronic purity, preferably surrounded by a neutron reflector. Neutron absorption in the U^{238} content of the natural uranium during the reaction leads to the production of the transuranic isotope 94^{239}, known as plutonium (symbol Pu), which is fissionable in much the same manner as U^{235}. Pu^{239} or 94^{239} is formed in neutronic reactors utilizing natural uranium in accordance with the following process:

$$92^{238} + n \longrightarrow 92^{239} + 6 \text{ mev. of } \gamma \text{ rays, not necessarily all of one frequency.}$$
$$92^{239} \xrightarrow{23 \text{ min.}} 93^{239} + \beta \text{ 1 mev. } \beta, \text{ no } \gamma \text{ rays}$$
$$93^{239} \xrightarrow{2.3 \text{ day}} 94^{239} + \beta \text{ 600 kv. u pper } \beta \text{ energy limit. Also } 2\gamma \text{ rays, 400 kv. and 270 kv., about } \frac{1}{2} \text{ converted to electrons}$$

A small portion of the 94^{239} produced may also be changed to 94^{240} by absorption of neutrons. The neutronic reactors referred to above may be called "isotope converters" in that one thermally fissionable isotope is formed (94^{239}) as another thermally fissionable isotope (U^{235}) is used up. However, this conversion is not complete, and the natural uranium, which acts to supply both the reaction isotope (U^{235}) and the absorption isotope (U^{238}), will contain two different thermally fissionable isotopes after the reactor has been started. Certain presently known uranium-graphite reactors have been found to have a conversion factor of .78, U^{235} to 94^{239}. However, it may be desirable to form other fissionable isotopes in quantity such, as for example, U^{233}. Isotopes such as U^{238} and Th^{232}, which are not thermally fissionable isotopes, but which, upon absorption of a neutron, produce a thermally fissionable istotope, are called "fertile isotopes."

The term "neutronic purity" as used herein has no necessary relation to chemical purity and merely refers to the absence of foreign material having the characteristic of relatively great neutron absorption. In other words, a substance may be said to have high neutronic purity, if the total amount of other material therein is incapable of absorbing a substantial number of neutrons and is thus ineffective to "poison" the nuclear fission chain reaction.

To obtain conversion of one thermally fissionable iso-

tope to another in the most efficient manner, it is preferred to utilize a substantially neutronically pure thermally fissionable isotope, such as 94^{239}, for the neutronic reaction, and then form the new thermally fissionable isotope U^{233} separately, from a substantially pure non-fissionable isotope, such as thorium^{232}, which can be substantially completely converted to the new fissionable isotope U^{233}, fully recoverable in high purity and concentration from the thorium. Converters using the fissionable isotope in a liquid medium are disclosed and claimed in Patent No. 2,815,321, issued December 3, 1957.

An object of the present invention is to provide a means of converting an element into a fissionable isotope that can be recovered substantially completely and in high purity and concentration, by means of a neutronic reaction in which the fissionable isotope supporting the reaction is in solid form and in high concentration.

The plutonium produced by neutronic reactors using natural uranium to support the reaction is useful for many purposes, but it has one outstanding advantage over, for example, the use of U^{235}, as it exists in natural uranium. As plutonium is a different element from uranium it can be chemically removed from the irradiated natural uranium, and because of that fact can be obtained in substantially pure form and in high concentrations whereas U^{235} can only be obtained in high concentration or substantially pure form (as far as presently known) by the much more difficult process of isotope separation. U^{235} of high concentration, however, has been used to sustain a neutronic reaction.

In high concentrations of substantially pure form, plutonium can also be used, when properly combined with a neutron moderator, to sustain a slow neutron chain reaction in a neutronic reactor of relatively small size wherein the neutron leakage is high. In other words, it can be used as an efficient source of large quantities of neutrons, and the neutrons thus produced can be used to produce another fissionable isotope such as U^{233}.

U^{233} can be formed by irradiating non-fissionable thorium (90^{232}) with slow neutrons according to the following process:

$$_{90}Th^{232} + n \longrightarrow {}_{90}Th^{233} + \gamma \text{ gamma rays}$$
$$_{90}Th^{233} \xrightarrow{23.5 \text{ min.}} {}_{91}Pa^{233} + \beta$$
$$_{91}Pa^{233} \xrightarrow{27.4 \text{ days}} {}_{92}U^{233} + \beta$$

The chemical separation of U^{233} from thorium is readily accomplished with high purity. The fissionable isotope U^{233} will support a chain reaction, and has many desirable qualities. In particular, U^{233} gives a relatively high average neutron yield per fission, the value as presently known being about 2.37–2.4 neutrons per fission (average).

While high isotope conversion factors are desirable in many instances, in other cases simplicity of construction and care of handling the materials involved can be more important. The present application is therefore primarily directed toward the provision of a simply constructed isotope converter wherein the active materials are in solid form, can be easily cooled in place, and easily removed for processing, and replaced with fresh material. It will be understood that the selection of fissionable materials, moderator materials and fertile materials, relative amounts thereof and the critical size of the reactive composition required to produce a self-sustaining chain reaction, are not in themselves the subject of the present invention. These criteria are now familiar to persons skilled in the art. The invention is concerned with a novel construction which may be employed with any of the various combinations of materials which are already well known.

2,954,335

3

Specifically, the present invention utilizes the active fissionable isotope of plutonium, preferably in the form of a sheet of a plutonium-aluminum alloy, or a plating of plutonium on an aluminum sheet. This active aluminum sheet, together with an adjacent quantity of thorium oxide, is sealed between two protecting plates, also preferably of aluminum, to form what may conveniently be termed a sandwich. A plurality of sandwiches are assembled in a tube, preferably of hexagonal shape, and spaced in the tube to permit coolant passage therebetween. Other tubes are filled entirely with sandwichs containing ThO₂ only, and selected tubes are then assembled to form a reactor in which the active plates carrying the plutonium are grouped to form a reaction zone and in which the reaction zone is completely surrounded by an absorption zone containing ThO₂. Water, light or heavy, is then forced through the tubes between the sandwiches to cool the reaction and at least one of the tubes containing plutonium is movable to control the reaction. All of the tubes are removable for processing of their contained materials. A simple, easily assembled converter results. This type of reactor is sometimes referred to as a "sandwich-pile."

Anoher object of the invention is to provide a simple and sensitive means for controlling the neutronic reaction while at the same time producing a useful material by subjecting neutron absorbent material to intense neutron bombardment. This object is accomplished by dividing the reactor into a plurality of sections or cells, the central group of which comprises reactive composition between spaced portions of fertile material, and the outer group of cells comprises fertile material only. Thus a reactive zone is provided which is completely surrounded by an absorption zone of the particular fertile material to be bombarded. The reaction is controlled by moving at least one of the central cells relative to the others thereby at least partially removing the reactive material in said cell from the reactive zone and at the same time moving into said zone a portion of the fertile material in said cell, said material being thus moved to the central part of the reactor at which the neutron density is normally greatest.

Still another object of the invention is to provide a relatively compact sectional reactor structure, the various sections of which are independently removable to accommodate control, treatment, and replacement of the materials therein.

A further object of the invention is to design a sectional reactor structure such as above described wherein the sections or cells are polygonal in transverse cross section with the sides thereof snugly engaged with complementary sides of the adjacent cells to afford a relatively compact structure having a maximum neutron reproduction factor for the structure.

The invention comprehends a novel reactive unit in the form of a sandwich or laminated structure including a plurality of layers with fissionable and/or fertile material carried by the inner of said layers, the outer layers affording fluid-tight seal protecting said material from direct contact with a coolant fluid which is circulated through the reactor.

Reference is made to Fermi et al. Patent 2,708,656, dated May 17, 1955, for a more complete disclosure of reactors and their operations.

The foregoing and other objects and advantages of the present invention will be more fully understood by the following description read with reference to the drawings, wherein:

Fig. 1 is a diagrammatic vertical sectional view, partly in elevation, of a neutronic reactor embodying the present invention;

Fig. 2 is an enlarged fragmentary cross sectional view taken as indicated by line 2—2 in Fig. 1;

Fig. 3 is an enlarged transverse cross sectional view of one fuel element of the reactor shown in Fig. 1;

4

Fig. 4 is a side elevation of an individual sandwich;

Fig. 5 is a side elevation of the active element of the sandwich of Fig. 4; and

Fig. 6 is an enlarged cross sectional view taken as indicated by the line 6—6 in Fig. 4.

Referring first to Figs. 1 and 2 which show the gross structure of the device, concrete walls 1 are provided defining a pit 2 preferably positioned below ground level. Pit 2 is provided at the bottom thereof with a coolant inlet 4 and is normally closed at the top by a pressure head 5 held down by bolts 6. The cross section of the bottom portion of pit 2 is substantially circular to accommodate a plurality of adjacent loosely fitting upright hexagonal tubes 7, each resting on a perforated support 9 adjacent the bottom of the pit and each tube is provided at the top with a removal ring 10.

The central tube 7c is vertically movable by a cable 11 connected at one end to ring 10, the other end being wound on drum 12 rotatable by control motor 14. Motor 14 and drum 12 are mounted on a platform 15 extending across pit 2 just below pressure head 5. Sundry tube removal rods 16 are also provided, extending from platform 15 to engage rings 10 on other tubes as desired so that all tubes can be removed as desired.

As the tubes 7 will be radioactive after use, the pit 2 is widened just above tubes 7 to provide a shelf 20 supporting a coffin 21 of shielding material such as lead and provided with a bayonet locking cap 22 operated by coffin rod 24. The coffin 21 can be opened, a tube 7 deposited therein and the cap 22 replaced, all underneath the coolant.

A coolant exit 25 is provided adjacent the top of the pit 2. An air space 26 is maintained in the pressure head 5 and below platform 15 by air introduced through air pipe 27 attached to pressure head 5. For reactors operating at 10,000 kw. to 100,000 kw. output a coolant head of about 50 ft. provides satisfactory top shielding for radiation from the reaction. Water, light or heavy, is a satisfactory coolant, as both have desirable moderating properties as will be later discussed.

Tubes 7 are filled with vertical, parallel spaced metal sandwiches 30 as shown in Fig. 3. The sandwiches are of two types, one type being illustrated in Figs. 4 to 6 inclusive.

For the center of the reactor, sandwiches are used which contain both plutonium and ThO₂. For the outer portion of the reactor, the sandwiches contain only ThO₂.

The central portion of sandwich 30 is in each case a central aluminum plate 34. When only ThO₂ is to be used in the sandwich the plate 34 may be entirely perforated with holes 36 in which are placed ThO₂ pellets 37 of the same thickness as the plate 34 in abundance to absorb the majority of neutrons. However, certain of the sandwiches 30 contain plutonium, a central space 39 being blanked out of plate 34 in which is disposed a plutonium carrying plate 40. This plate 40 may be an alloy of 0.8 atomic percent plutonium and aluminum or an aluminum base with a plating or paint of plutonium on the exterior, or and aluminum base with holes therein containing pellets of plutonium metal or oxide. Beyond the central space 39 ThO₂ pellets are used in holes 36 up to the ends of the plate 34.

The central plate 34 is disposed between two aluminum cover sheets 34a which are welded around the entire perimeter to spacers 41 at 43, thereby providing a fluid-tight seal around the plate 34, or alternatively, the plate 34 may be aluminum-cladded by hot rolling with the aluminum cover sheets 34a.

The sandwiches 30 are loaded into tubes 7, with, in the particular embodiment shown, eighteen centrally disposed tubes containing the combined thorium and plutonium sandwiches and the remaining tubes containing thorium sandwiches. The sandwiches 30 are held in place by aluminum strips 45 secured respectively to the

2,954,335

5

top and bottom of each tube 7, as by welding or any other convenient manner.

The tubes 7 are assembled as shown in Figs. 1 and 2, with the plutonium containing tubes 7 adjacent, and grouped to form a central reaction zone, as indicated by dotted line A (Fig. 2), and with the central plutonium carrying tube 7c movable as before described.

The tubes 7 containing only thorium sandwiches 30 are grouped around the reaction zone to form an absorption zone extending outwardly to a reflecting zone 42 (Fig. 2) formed of graphite and contoured to fit a reactor tank 43 on the outside and to fit the adjacent surfaces of the peripheral tubes 7 on the inside.

As all the sandwiches 30 are spaced within the tubes 7, the coolant water can pass freely upwardly through the entire assembly and in doing so will be in close heat exchange relationship with all parts of the reactor.

Having described the mechanical aspects of the reactor the nuclear physics of the assembly will next be discussed.

In a reactor utilizing a fissionable isotope in high concentration, the form in which the isotope is used in the moderator is relatively unimportant from a nuclear physics standpoint when the size of the reactor and the amount of the fissionable isotope is above the minimum required. A proper moderator is necessary in order that the fast neutrons of fission be slowed to energies at which new fissions can take place. In the present instance the moderator is the coolant, and either light or heavy water can be used. D_2O will give the least neutron absorption, but if conversion factors are not required to be maximum, H_2O is suitable, as its moderating ability is high, although its neutron absorption is also high. Choice of the type of water will therefore be predicated on the conversion factor required and on economy, as H_2O is, of course, far less expensive than D_2O. Furthermore H_2O does not have to be recirculated.

The reaction zone in the present instance, therefore, consists of plutonium protected by aluminum immersed in water, which also serves as a coolant to remove the heat of reaction.

In one embodiment of the invention the reactor is formed of 2.5 mm. sandwiches spaced 2.5 mm. in the tubes 7. The reaction zone is roughly cylindrical, with a diameter and height of 85 centimeters. The concentration of plutonium in the reaction zone is about 28 mg./cm.², and the total amount of plutonium in the reactor is about 4 kilograms. The absorption zone is about 70 cm. thick radially, and can contain up to equal parts by weight of thorium and water. With a light water (H_2O) flow at a velocity of 15–20 feet per minute, 20,000 kw. will be removed from the reactor during operation. Pressure head 5 permits the entire system to be operated under pressure to prevent boiling of the coolant, if desired. The conversion efficiency of the converter here described is not over 50–60 percent because of the presence of aluminum and H_2O in the reaction and absorption zones, an efficiency which can be increased only a few percent by the use of heavy water as a coolant, and then only at a high cost. However, the simplicity of construction and ease of cooling make the sandwich type converter as herein described highly desirable even when maximum conversion factors are not required.

Only the coolant has to be circulated, and the same circulation cools the reaction and absorption zones. No gas is evolved from the reaction zone; and poisoning factors, due to retention of fission product neutron absorbers in the reaction zone, do not become critical because all or part of the active portion of the reactor can be replaced at daily or lesser intervals as the plutonium is used up and the poisoners produced.

The absorption zone material is also removed from the reactor and the produced U²³³ extracted. U²³³ itself fissions with slow neutrons, and the absorption zone sandwiches are preferably removed when the U²³³ accumu-

6

lates to the extent of 0.05 percent in the thorium. The thorium is then dissolved, the U²³³ and fission products extracted, and the purified thorium inserted in new sandwiches for replacement in the absorption zone.

A larger scale reactor from which about 100,000 kw. can be removed is provided by using sandwiches 2.5 mm. thick spaced 2.5 mm., with a plutonium concentration of 28 mg./cm.², thereby providing in the reactor 1 molecule of fissionable Pu for each 125 molecules of water and 225 molecules of aluminum. The large scale plant has a reaction zone 2 feet wide and 8 feet long, with a 2 foot thick absorption zone. The reaction zone should be cooled with light water moving at about 30 feet per second to abstract 100,000 kw., but if desired, the flow through the absorption zone can be reduced to half that amount.

Such a reactor will contain about 30 kilograms of 94²³⁹ and will use up 150 grams per day. When the total amount of Pu has been 30 percent depleted by use, it must be removed and purified, and fresh amounts replaced to bring the reactor up to 30 kilograms again. 10 to 15 tons of thorium can be used in the absorption zone.

Several distinctive features should be noted in reactors of the type above described. The reactor as a whole is composed of relatively few removable sections, greatly facilitating removal of the materials. The removal can readily be made under water after the pressure head has been removed, the water acting as a biological shield for protection of the operating personnel.

In addition the neutronic reaction may be simply controlled by moving the center section 7c vertically in the manner above described. Thus, the neutron reproduction ratio of the reactive zone A may be regulated by withdrawing or inserting the active material of section 7c into the remaining active material, and also the ThO₂ within the lower portion of this section is subjected to intense neutron bombardment by being drawn into the central part of the zone A where the neutron density is greatest. The method of control described above not only affords a simple and sensitive means of regulating the reaction but also provides a highly efficient means of producing U²³³ by intense neutron bombardment of the ThO₂ in the lower portion of section 7c. It may be noted that, if desired, a plurality of tubes 7 may be connected to suitable actuating means such as motor 14 in order to afford additional control of the reaction. The neutron density is monitored by a conventional ionization chamber 46.

Normally, the reactor will first operate with fresh plutonium, with the active material of the central tube 7c more than half out of the reaction zone. Critical conditions, i.e., where the reaction takes place with a neutron reproduction ratio of unity, can then be maintained at the desired operating power. Then, as the plutonium is used up, and the reproduction ratio drops correspondingly, the center tube 7c can be inserted further into the reactor to maintain a reproduction ratio of unity. Finally, when the center tube is exactly centered, and the reproduction ratio drops below unity the reaction stops, and one or more of the tubes containing the active material must be removed and replaced with tubes having the original amount of fissionable isotope free from fission fragments. A schedule of replacement can readily be worked out to keep the reactor operating with minimum shutdown times, in accordance with the power output and resultant use rate of the plutonium.

It should also be noted that the reaction zone is completely surrounded by the absorption zone, so that substantially all neutron leakage, fast or slow, will be intercepted by the ThO₂ even when the center tube is partially out of the reactor for control purposes.

The absorbing material converts by absorption of slow neutrons, so that the presence of the water in the absorp-

2,954,335

7

tion zone is useful in that the fast neutrons escaping from the reaction zone are slowed to thermal energies by the water in the absorption zone and are thus in condition to be absorbed by the thorium. The minimum thickness of the absorption zone, if the absorbing composition were to be a ThO_2-water slurry of equal parts by weight, is about 38 centimeters to slow all but 1 percent of the escaping fast neutrons to thermal energy. The plate construction used in the present device aggregates the thorium in the water and thus for minimum final escape the absorption zone should be from 50-85 centimeters thick. As some fissions in the U^{233} formed in the absorption zone are bound to take place near the periphery of the absorption zone some of the neutrons produced may escape outwardly. The carbon reflector is used to return as many as possible of such neutrons to the absorption zone.

While the chemical procedures involved in the purification of the plutonium or in the recovery of the U^{233} from the thorium are no part of the present invention, such separations are readily made by liquid-solvent extraction methods, among others, after the sandwiches have been dissolved.

From the above discussion it can be seen that the use of the fissionable isotope in solid form as disclosed and claimed herein has many advantages over the use of the fissionable isotope in a liquid medium such as a solution, for example, even though conversion values are lower. Dissociation of the water and production of gases in the coolant is negligible. The active isotope does not have to be circulated outside of the reactor for cooling, thereby greatly simplifying shielding of the system and completely preventing delayed neutrons being emitted outside of the reaction zone. The reaction zone can be completely surrounded by the absorption zone, and the reactor can be made in easily removable sections.

While the theory of nuclear reaction set forth herein is based on the best presently known experimental evidence, the invention is not limited thereto, as additional experimental data later discovered may modify the theory disclosed.

Obviously, many modifications may be made in the specific embodiments disclosed without departing from the intended scope of the invention.

What is claimed is:

1. In an isotope converter having means for sustaining a nuclear fission chain reaction the improved structure comprising a plurality of plate-like sandwiches, water therebetween, each of the central group of said sandwiches comprising a neutron permeable plate with a central opening and a plurality of other openings at opposite ends thereof, material containing plutonium in the central opening, fertile material in the other openings, and neutron permeable plates joined to opposite sides of the first-mentioned plate and sealing said openings, the outer group of said sandwiches each comprising a neutron permeable plate with a plurality of openings, fertile material in said openings, and neutron permeable plates sealing said last-mentioned openings, whereby the material containing plutonium within said central openings defines a reactive zone entirely surrounded by an absorption zone of said fertile material.

2. In an isotope converter, a plurality of members each comprising longitudinally spaced portions of neu-

8

tron absorbent material containing Th^{232} and an intermediate portion of material containing plutonium, a zone of absorption material containing Th^{232} around said members, water in moderating and heat exchange relation with said members, and means for passing said water, through said members, said intermediate portions defining a reactive zone having a neutron reproduction ratio greater than unity.

3. In an isotope converter, a plurality of hollow members each containing longitudinally spaced portions of neutron absorbent material containing Th^{232} and an intermediate portion of material containing plutonium, a zone of absorption material containing Th^{232} around said members, a neutron moderator and water in moderating and heat exchange relation with said members, said intermediate portions defining a reactive zone having a neutron reproduction ratio greater than unity, and means for regulating the neutron reproduction ratio of said reactive zone comprising means for moving at least one of said members longitudinally thereof.

4. In combination with a neutronic reactor having a chain-reactive composition containing plutonium, the improved control device comprising, in combination, an elongated body having a material containing plutonium in one longitudinal portion thereof and having fertile material and no material containing plutonium in an adjoining longitudinal portion thereof, said body extending through said reactive composition, and a drive mechanism coupled to said body for adjusting the longitudinal position thereof to vary the neutron reproduction factor of the reactive composition.

5. A structural assembly for a neutronic reactor comprising a tube and a plurality of sandwiches positioned in the tube in spaced and generally parallel relation to one another and to the axis of the tube, each sandwich having its opposed side edges in engagement with spaced regions of the tube, each sandwich comprising a middle neutron-permeable plate having a central opening and a plurality of other openings at opposite ends of the plate, material containing plutonium in the central opening, fertile material in said other openings, and outer neutron-permeable plates joined to opposite sides of the middle plate and sealing the openings therein.

References Cited in the file of this patent

FOREIGN PATENTS

114,150	Australia	May 2, 1940
114,151	Australia	May 3, 1940
861,390	France	Oct. 28, 1940
233,011	Switzerland	Oct. 2, 1944
233,278	Switzerland	Oct. 16, 1944

OTHER REFERENCES

Fermi et al.: "Artificial Radioactivity Produced by Neutron Bombardment," Proc. Royal Soc. (London), Series A, No. 868, vol. 149, pages 554–557, April 1935.

Smyth: "Atomic Energy for Military Purposes," August 1945. (Copy may be purchased from Supt. of Documents, Washington 25, D.C.)

"How Atom Splitting Releases Energy," Business Week, pages 57–64, Sept. 1, 1945.

Kelly et al.: Physical Review, 73, 1135–9 (1948).

Dec. 3, 1957   E. P. WIGNER ET AL   2,815,321
                        ISOTOPE CONVERSION DEVICE

Filed Nov. 13, 1945   2 Sheets—Sheet 1

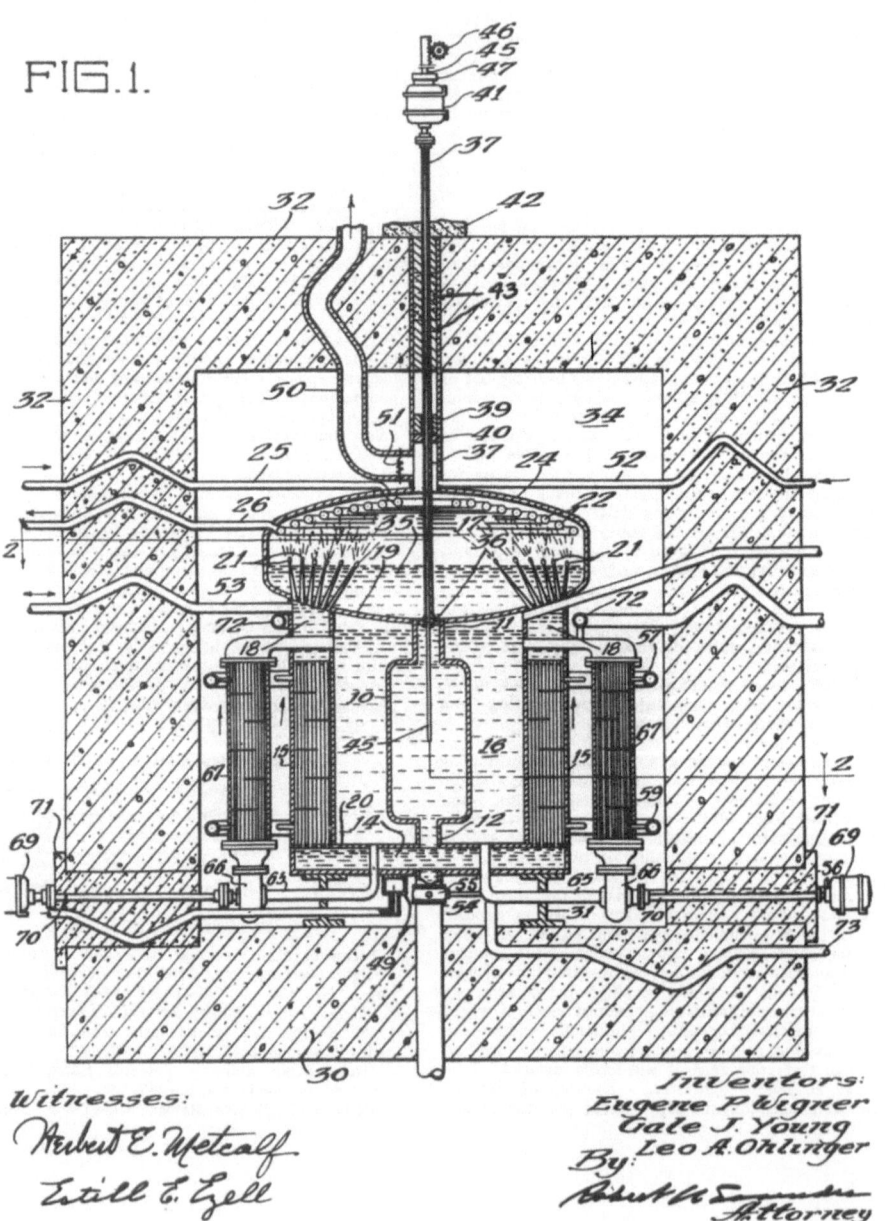

FIG.1.

Dec. 3, 1957   E. P. WIGNER ET AL   2,815,321

ISOTOPE CONVERSION DEVICE

Filed Nov. 13, 1945   2 Sheets–Sheet 2

FIG.2.

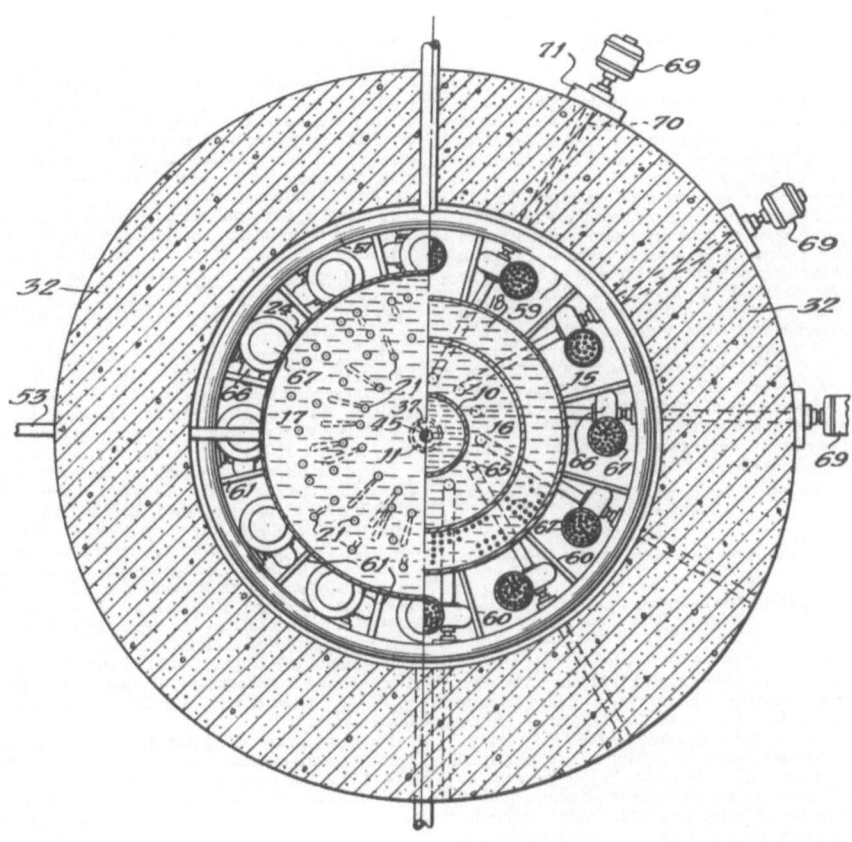

Witnesses:
Herbert E. Metcalf
Estill E. Ezell

Inventors:
Eugene P. Wigner
Gale J. Young
Leo A. Ohlinger
By:
Robert A. Sarandu
Attorney.

United States Patent Office

2,815,321

Patented Dec. 3, 1957

1

2

2,815,321

ISOTOPE CONVERSION DEVICE

Eugene P. Wigner, Leo A. Ohlinger, and Gale J. Young, Chicago, Ill., assignors to the United States of America as represented by the United States Atomic Energy Commission

Application November 13, 1945, Serial No. 628,322

6 Claims.   (Cl. 204—193.2)

The present invention relates to nuclear physics and more particularly to an improved means and method of converting an isotope fissionable by slow neutrons to another or the same fissionable isotope in a neutronic reactor, with a high conversion factor.

It is known that a self-sustaining chain reaction can be obtained in devices known as neutronic reactors utilizing natural uranium, as a result of slow neutron fission of the U^{235} content of the natural uranium. In such reactors, discrete bodies of natural uranium of high purity are disposed, usually in the form of a lattice arrangement of spheres or rods, in a neutron moderator such as graphite, beryllium or heavy water of high purity, surrounded by a neutron reflector. Neutron absorption in the U^{238} content of the natural uranium during the reaction leads to the production of the transuranic isotope 94^{239}, known as plutonium (symbol Pu), which is fissionable in much the same manner as U^{235}. 94^{239} is formed in neutronic reactors utilizing natural uranium in accordance with the following process:

$$92^{238} + n \longrightarrow 92^{239} \quad \text{6 m. e. v. of } \gamma \text{ rays, not necessarily all of one frequency}$$

$$92^{239} \xrightarrow{\text{23 min.}} 93^{239} + \beta \text{ 1 m. e. v. } \beta, \text{ no } \gamma \text{ rays}$$

$$93^{239} \xrightarrow{\text{2.3 day}} 94^{239} + \beta \text{ 600 kv. upper } \beta \text{ energy limit. Also } 2 \gamma \text{ rays, 400 kv. and 270 kv., about } \tfrac{1}{2} \text{ converted to electrons}$$

A small portion of the 94^{239} produced may also be changed to 94^{240} by absorption of neutrons. The neutronic reactors referred to above may be called "isotope converters" in that one fissionable isotope is formed (94^{239}) as another fissionable isotope (U^{235}) is used up. However, this conversion is not complete, and the natural uranium, which in this case acts to supply both the reaction isotope (U^{235}) and the absorption isotope (94^{239}), will contain two different fissionable isotopes after the reaction has been started. Certain presently known uranium graphite reactors have been found to have a conversion factor of .78, U^{235} to 94^{239}.

To obtain a more complete conversion of one fissionable isotope to another it is preferred to utilize a substantially pure fissionable isotope for the neutronic reaction, and then form the new fissionable isotope separately, from a substantially pure non-fissionable isotope which in turn can be substantially completely converted to a fissionable isotope fully recoverable in high purity and concentration.

An object of the present invention is to provide a means and method of converting an element into a fissionable isotope that can be recovered substantially completely and in high purity and concentration by means of a neutronic reaction in which the fissionable isotope supporting the reaction can be substantially completely converted.

The plutonium produced by neutronic reactors using natural uranium to support the reaction is useful for many purposes, but it has one outstanding advantage over, for example, the use of U^{235}, as it exists in natural uranium. As plutonium is a different element from uranium, it can be chemically removed from the irradiated natural uranium, and because of that fact can be obtained in substantially pure form and in high concentrations, whereas U^{235} can only be obtained in high concentration or substantially pure form (as far as presently known), by the much more difficult process of isotope separation. U^{235} of high concentration, however, has been used to sustain a neutronic reaction.

In high concentrations or substantially pure form, plutonium can also be used, when properly combined with a neutron moderator, to sustain a slow neutron chain reaction in a neutronic reactor of small size wherein the neutron leakage is high. In other words, it can be used as an efficient source of large quantities of neutrons, and the neutrons thus produced can be used to produce another fissionable isotope with higher conversion efficiencies than heretofore attained.

In addition, U^{233}, still another fissionable isotope, can be formed by irradiating non-fissionable thorium (90^{232}) with slow neutrons according to the following process:

$$90\text{Th}^{232} + n \longrightarrow 90\text{Th}^{233} + \text{gamma rays}$$

$$90\text{Th}^{233} \xrightarrow{23.5 \text{ min.}} 91\text{Pa}^{233} + \beta; \ 91\text{Pa}^{233} \xrightarrow{27.4 \text{ days}} 92\text{U}^{233} + \beta$$

The chemical separation of U^{233} from thorium is readily accomplished with high purity. The fissionable isotope U^{233} will support a chain reaction, and has many desirable qualities. In particular, U^{233} gives a relatively high average neutron yield per fission, the value as presently known being about 2.37–2.4 neutrons per fission (average).

All of the above isotopes can be used as reaction isotopes in isotope converters. However, as U^{235} is difficult to obtain by isotope separation, U^{233} and 94^{239} are the most readily available reaction isotopes for use in substantially pure form, and are preferred materials.

As to non-fissionable isotopes which can be converted substantially entirely to fissionable isotopes by neutron absorption, U^{238} and 90^{232} can be used. Here again, U^{238} is difficult to obtain free from U^{235}, whereas 90^{232} existing in nature as thorium, converts by slow neutron absorption to U^{233}, with chemical separation being readily accomplished. Consequently, thorium-90^{232} is preferred as the absorption isotope, with the produced fissionable isotope being U^{233}.

From the above discussion, it will be seen that the following uncomplicated conversions can be made, with Nos. 1 and 2 as the preferred types.

	Reaction Isotope	Absorbing Isotope	Produced Isotope
(1)	94^{239}	90^{232}	92^{233}
(2)	92^{233}	90^{232}	92^{233}
(3)	92^{233}	90^{232}	92^{233}
(4)	92^{233}	92^{238}	94^{239}
(5)	92^{233}	92^{238}	94^{239}

Other but more complicated conversions can be made, as will later be brought out.

With proper precautions during the conversion to prevent excessive neutron losses, conversion factors close to unity can be obtained.

When neutron losses can be reduced to the minimum, and when a fissionable isotope giving an average neutron yield per fission of substantially over two, such as U^{233} is used to support the reaction, conversion factors above unity may be obtained in a converter wherein U^{233}, for example, is fissioned to produce U^{233}. This type of improved converter is known as a breeder reactor and is specifically no part of the present invention except insofar

2,815,321

3

as the generic method and structure as claimed herein can be utilized therein, the present application being generic to both types.

The presently described method for obtaining high conversion factors is based on several principles, as follows:

(1) The use in a reaction zone of a substantially pure fissionable isotope to provide a high K factor thereby leading to a reactor of small size and high neutron leakage.

(2) The use of a neutron moderator having a small neutron age to obtain a reactor of small size and high neutron leakage.

(3) The use of an absorber isotope positioned in an absorption zone to intercept the maximum possible number of escaping neutrons in a relatively small amount of absorber.

(4) The use of a structure and materials providing low parasitic neutron losses.

As pointed out above, one principle on which the present type of converter if founded is that the reaction isotope shall be in substantially pure form. In any neutronic reactor, the ratio of fast neutrons produced by fission in one generation to the original number of fast neutrons in a theoretical system of infinite size where there can be no exterior loss of neutrons is called the neutron reproduction factor of the system, and is denoted by the symbol K. In any finite system, some neutrons will escape from the periphery of the system. Consequently, a system of finite size may be said to have a K factor, even though the value thereof would only exist if the system, as built, were extended to infinity without change in composition. Thus, when K is referred to herein as a constant of a system of practical size, it always refers to what would exist in the same type of system of infinite size. In this way, it can be seen that the K factor for systems in which a pure fissionable isotope is used can be in theory equal to the average number of neutrons per fission, by neglecting neutron absorption in impurities and neutron absorption in the moderator and in the fissionable isotope. When a substantially pure reaction isotope is used, it can be replenished as it is destroyed by fission and thus be completely converted. Thus, when a high K factor is used, the size of the reactor can be greatly reduced before the neutron reproduction ratio reaches unity.

However, to obtain the smallest reactors, the moderator used should have a small neutron age, which is defined as being proportional to the root mean square distance a neutron travels in a reacting composition from its birth by fission until it is slowed to thermal energies. The value of the neutron age roughly determines the minimum size of the reactor when the maximum amount of fissionable isotope is used within the structure of minimum size, as determined by moderator characteristics. For this reason, it is customary to evaluate the reactivity of reactors using fissionable isotopes of high concentration in terms of the amount of fissionable isotope used in the reactor. Size alone is not definitive of proper operating conditions as the amount of fissionable material in a structure over minimum size determines the reactivity of the structure. Consequently, critical mass and operating mass is given in terms of grams of fissionable isotope in the reactor or in terms of density of the isotope in the moderator. Light water and heavy water both have small neutron ages and lead to small reactors when used as moderators. A chain reaction in such small reactors can have neutron leakages out of the reaction zone on the order of 50 percent of the neutrons generated during the chain reaction.

If a breeder reactor is to be constructed then D_2O should be used as a moderator, as neutron absorption in D_2O is negligible. However, D_2O is expensive and light water (H_2O) can be used in converters when no absolute necessity for a quantitative isotope "profit" appears,

4

whereas in breeder reactors, by definition, a quantitative isotopic profit must be made. The term water is used herein as a generic term to include both H_2O and D_2O.

U^{233} and 94^{239} may be dispersed in water as soluble uranyl and plutonyl (PuO_2^{++}) salts, such as uranyl and plutonyl nitrate, sulphate or fluoride, or as an alkali metal uranyl or plutonyl carbonate. Such a solution with the proper isotope concentration in water of either type and in a container of the proper size will sustain a chain reaction with a low neutron loss except for exterior neutron leakage, which will be very high.

A third preferred principle to be followed in constructing converters of high conversion factors is to insure the result that the greatest possible number of the leakage neutrons, i. e., those not required for the chain reaction, are usefully absorbed, such as in thorium leading to the production of U^{233}. This is accomplished, for example, by surrounding the reaction zone as completely as possible with an absorbing zone, comprising, for example, a thick layer containing the absorbing isotope, with the layer preferably also having reflecting and moderating properties due to the use of water mixed with the absorber.

While the absorbing zone should have a minimum thickness in order that neutrons of high energy be slowed down in the zone and be absorbed before escape, it is desirable to keep the total amount of absorber as low as possible with respect to the number of neutrons being absorbed. When a reactor of large size is used as a neutron source, then the minimum thickness of the absorbing zone should be the same as when a small reactor is used. However, as the absorbing zone must substantially completely enclose the reaction zone, much more of the absorber must be used when reactors of large radius are used, and this factor complicates removal of the produced isotope. Comparing a large reactor and a small reactor operating with substantially the same total neutron leakage, it can be seen that the amount of the new isotope produced in each case will be about the same, but that in the case of the larger reactor the newly produced isotope will be distributed throughout a much larger amount of absorber isotope. All of the absorber isotope will have to be processed to recover the produced isotope and in the case of the smaller reactor only the smaller amount of the absorber will have to be processed to recover the same amount of produced isotope. This factor is doubly important where the size of the reactor is such that the percentage of produced isotope to total absorber isotope remaining is so low that resort must be had to carrier chemistry to recover the produced isotope. Therefore, a small size due to use of substantially pure fissionable isotope is important not only in providing a high leakage factor, but also in enabling the volume of the absorber isotope to be reduced. A high relative density of produced isotope is formed therein, leading to relatively easier chemical removal processes.

For example, in the reactor herein described about 6½ metric tons of slurry are used, in an absorbing zone 38 to 50 cms. thick. In natural uranium-graphite reactors an absorbing zone of 50 cms. thickness would use 177 metric tons of slurry with a thorium content of 1.0 gm./cc.

While many of the parasitic neutron losses can be reduced by known methods, several of the parasitic neutron losses in a converter of the type described herein can be materially reduced by particular disposition of the absorbing isotope with respect to the reaction isotope.

In neutronic reactors of the solution type operated at elevated powers, it is desirable to circulate the reaction solution outside of the reactor in order to obtain release of gases from the solution, to cool the solution and then return it to the reactor. As about 1 percent of the neutrons emitted in fission are delayed, such circulation indicates that many of these delayed neutrons will be released outside of the reaction zone. When the solution is outside of the reaction zone, it can be conveniently thought of as being in a circulation zone.

2,815,321

5

Another object of the present invention is to provide a means and method of usefully utilizing at least a part of the delayed neutrons in a converter emitted when the reacting component is in a circulation zone.

Furthermore, in conversion reactors where the absorption isotope substantially surrounds the reaction zone, fissions may take place in the already formed new fissionable isotope at or near the periphery of the absorbing zone. When this happens fission neutrons can travel outwardly and be lost for either reaction or absorption. This loss can be reduced by surrounding the absorbing zone with a neutron reflecting material, which in this instance can be the reaction solution itself as it is being circulated outside of the reaction zone.

It is therefore another object of the present invention to provide an isotope converter wherein a maximum number of leakage neutrons are directed to an absorption zone.

Broadly speaking, therefore, the objects of the present invention are accomplished as to method by first almost completely and closely surrounding the reaction zone with an absorption zone, and then circulating the reaction zone composition outside the reaction zone over a path substantially completely and closely surrounding the absorption zone.

One preferred apparatus embodiment is a neutronic reactor having a reaction tank of small size to increase neutron leakage, this small size being obtained by using as a reaction composition a substantially fissionable isotope of high concentration or purity in water, light or heavy, with the reaction zone surrounded as completely as possible and closely with an absorber isotope such as thorium. The reaction composition is circulated outside of the reaction zone, cooled and purified in a circulation zone completely and closely surrounding the absorbing zone, and then returned to the reaction zone.

The present invention may be more fully understood by reference to the drawings in which:

Fig. 1 is a diagrammatic vertical section view partially in elevation of one preferred embodiment of the present invention; and

Fig. 2 is a cross-section taken through two levels of the device of Fig. 1 as indicated by the line 2—2 in Fig. 1.

In order to accomplish this process, apparatus is used whereby parasitic neutron losses are greatly reduced.

Referring to the drawings, which illustrate a solution type reactor surrounded by an absorbing slurry, a vertical, elongated cylindrical reaction tank 10, 120 cm. in diameter and 260 cm. high, is provided with an upper solution inlet pipe 11 and a lower outlet pipe 12. Outlet pipe 12 enters a stocky cylindrical solution distributor or outlet tank 14 extending outwardly beyond the extent of the reaction tank 10 to support and connect with the tubes of a heat exchanger 15 concentrically surrounding the reaction tank 10. The heat exchanger is contained in a conduit means 18 connecting the inlet tank 17 and outlet tank 14. The heat exchanger 15 is spaced from the reaction tank 10 to provide a slurry space or container 16 between the reaction tank 10 and the heat exchanger 15 that completely surrounds the reaction tank 10 except for inlet and outlet pipes 11 and 12, respectively.

Just above the reaction tank 10, the inlet pipe 11 enters a cooling solution inlet tank 17, the bottom 19 of which extends laterally to rest on the top edges of the solution heat exchanger 15. Cooling tank 17 receives solution from the top of the heat exchanger 15 through a plurality of upwardly directed nozzles 21, the solution from the nozzles 21 being projected against a cooling coil 22 positioned just below top 24 of cooling tank 17. Coolant is supplied by coil inlet 25 and coil outlet 26. As shown by Fig. 1, the cooling solution inlet tank 17 has a wall 19 in common with a portion of the wall of the container 16 and the outlet tank 14 has a wall 20 in common with a portion of the wall of the container 16.

The combination of solution distributor 14, inner reaction tank 10, outer heat exchanger 15, and upper cooling

6

tank 17 is supported on a foundation 30 such as concrete, by beams 31, and shield walls 32 of neutron and gamma ray absorbing material are projected upwardly and across the top of the combination to define a reactor space 34. Thick concrete can be used for walls 32.

The reactive solution 35, to be used in tank 10, is circulated by an axial flow impeller 36 positioned in inlet pipe 11 near the bottom 19 of cooling tank 17. Impeller 36 is driven by a hollow shaft 37 extending upwardly in an axial shaft pipe 39 entering cooling tank 17. Gas passage along shaft pipe 39 is blocked a short distance above cooling tank 17 by packing 40. Hollow shaft 37 continues upwardly through shield 32 to an impeller motor 41 positioned outside of the shield 32. Radiation through shaft pipe 39 is blocked by a lead cover 42 and steel and hydrogenous shielding 43.

Impeller motor 41 has a hollow central shaft, thus providing a bore through which a control rod 45 extends through the motor 41. The rod 45 also extends through the hollow shaft 37 to enter the reaction tank 10. The rod 45 is movable vertically by use of rack and pinion 46 to insert more or less of the rod 45 into the reaction zone, as desired. Gas leakage around control rod 45 is prevented by the use of a stuffing box 47 between the motor and the rack and pinion. The portion of the control rod 45 insertable into the reactor incorporates a neutron absorber of high neutron capture cross-section such as cadmium, boron carbide, or boron steel. The neutron density during operation can conveniently be monitored, as by ionization chamber 49.

A gas outlet 50 leaves axial shaft pipe 39 between the cooling tank 17 and packing 40 and extends upwardly through the shield 32 for release of non-condensible gases. Dissociation gases such as H_2 and O_2 or D_2 and O_2 are recombined, as for example, by a burner indicated diagrammatically as a filament 51. The water so recombined runs back into cooling tank 17. Gas dilution to prevent explosions is supplied by dilution gas pipe 52 entering shaft pipe 39 between the cooling tank 17 and gas outlet 50.

Solution is normally supplied to or removed from the solution circulation zone through solution pipe 53 attached to the top of heat exchanger 15. Solution can be more quickly removed from the system by an axial dump pipe 54 entering the bottom of solution distributor 14 under the control of dump valve 55 operated by extended valve shaft 56. The use of dump valve 55 will be later explained.

Coolant is supplied to heat exchanger 15 from an inlet manifold 57 and is discharged into an outlet manifold 59, being connected thereto by solution coolant inlets 61 and solution coolant outlets 60, respectively (Fig. 2).

Thus, when solution is placed in the system, any heat of reaction developed in reaction tank 10 will be removed first by heat exchanger 15 and second by vaporization, condensation and cooling in cooling tank 17 due to the action of nozzles 21 and cooling coil 22.

The slurry to be placed in slurry space 16 is also circulated and cooled. A plurality of slurry outlet pipes 65 each connect the bottom of slurry space 16 with a slurry pump 66, discharging into a slurry heat exchanger 67 which discharges into slurry space 16 at the top thereof. These exchangers 67 are positioned outside of and around exchanger 15. Pumps 66 are driven by pump motors 69 positioned outside of shield 32, each motor and pump being connected by a pump motor shaft 70 extending through shield 32 in a removable plug 71. Slurry can be taken out of the system by means of slurry removal manifold 72 connected to the tops of the heat exchangers 67, and introduced into the system by slurry inlet pipe 73 connected to one of the slurry outlet pipes 65. Coolant is supplied to the slurry heat exchangers 67 from the water manifolds 57 and discharged by manifold 59.

In operation, the control rod 45 is inserted to maxi-

2,815,321

7

mum neutron absorbing position in the reaction tank 10. A reactive composition of fissionable isotope and liquid moderator is inserted in the system through solution pipe 53 and the impeller 36 is rotated. Circulation of the reactive composition is then established through the reaction tank 10, and the heat exchanger 15. Water is circulated through the heat exchanger 15 as a coolant.

Slurry space 16 may be filled with a thorium-D_2O slurry, for example, circulation started, and the device is ready for the start of the nuclear chain reaction. The control rod 45 is then slowly removed from the reaction tank 10 until a point is reached where the neutron reproduction ratio in the reactor tank 10 is greater than unity. The fissionable isotope density in the liquid has been previously determined to be such that, for the size of the tank used, operating conditions are obtained where the reproduction ratio is below unity and above unity for different positions of the control rod. The chain reaction can then be allowed to proceed until a predetermined operating power has been reached with the reproduction ratio above unity. At this power where the heat of the reaction is removed by the heat exchangers to stabilize the operating temperature, the control rod 45 is moved to the position where the reproduction ratio is unity, thereby stabilizing the reaction at the power attained. Slight movements of the control rod then serve to maintain operating power.

During operation, the fissionable isotope is used up in fission as it passes through the reaction tank. Fission products are released, these products being radioactive and neutron absorbing. Consequently, either at intervals or continuously some of the reaction composition is drawn off through solution pipe 53, purified, fortified to proper isotope density, and returned to the reactor. Likewise, the slurry carrying the absorber is periodically withdrawn for chemical extraction of the produced fissionable isotope and purification of the slurry. A slurry having a ThO_2 particle size of .01 to 1 micron is stable and has good viscosity for circulation.

Having described the apparatus used, the specific details of the method will now be set forth more completely. It is important to note here that the present invention is primarily directed to the overall conversion of a fissionable isotope to a fissionable isotope with a high conversion factor and that the heat developed in the reaction is not necessarily utilized. A converter can also use the heat of reaction to produce power by modification of structure. In the present invention, use of the heat of the reaction will be ignored.

In the specific example described, the reaction tank 10 has a volume of about 3 cubic meters and contains 1 gm. Pu per liter D_2O. This concentration of plutonium in heavy water in the reaction tank 10 having the dimensions previously mentioned, will support a sustained chain reaction therein independently of any neutrons supplied from outside the reaction tank and regardless of the amount of thoria contained by the slurry. This 94^{239} can be dissolved in the moderator in the form of a salt such as the nitrate, sulphate or fluoride, and is circulated through the reactor at a high rate. Under these circumstances, the reactor can operate at high power continuously.

The amount of fissionable isotope which should be present in reaction tank 10 in order to establish a self-sustaining neutron chain reaction depends to a substantial degree upon the concentration of the fissionable isotope in the moderator, upon the shape of the tank, and also upon the neutron absorption characteristic of the moderator used. In general, it can be said that the amount of 94^{239} present should be at least about 200 grams with optimum concentration and using pure 94^{239}. The exact amount required will also depend upon the fissionable isotope which is used.

The following table tabulates the critical mass which is

8

required for various concentrations of plutonyl sulphate solution in D_2O using an infinite D_2O reflector. In the table, Z denotes the number of atoms of Pu^{239} present per molecule of D_2O and G denotes the critical quantity of Pu^{239} required in grams. V_X is the critical volume in liters.

$Z \times 10^{-3}$	0.1	0.2	0.25	0.3	0.5	1.0	2.0
V_X Liter	180	80	62	51	32	18	11
G gm	290	210	205	205	210	240	290

From the above data it is shown that a minimum critical mass can be as low as about 200 grams of 94^{239} and be capable of sustaining a reaction in a D_2O moderator with an infinite D_2O reflector. Not less than about 300 grams of U^{235} would be required were U^{235} (12.5 percent U^{235} in U^{238}) to be substituted for the 94 in the above solution. U^{233} values are approximately the same as 94^{239} values.

The variation in critical mass which is required to sustain a neutron chain reaction depends to a very substantial degree upon the nature and thickness of the neutron reflector.

Amounts of the fissionable isotope required will be larger than given above when surrounded by an absorbing-reflecting layer as described herein.

Critical size and mass with ThO_2—D_2O absorbing zone (spherical reactor):

Cm.³ D_2O per gram 94^{239}	Radius in cm.	Kg. 94^{239}
100	33.8	1.62
200	38.6	1.20
300	42.9	1.11
400	46.1	1.07

From the above data, it will be apparent that no hard and fast figure for critical mass may be given since the mass will vary with the nature of the moderator, nature of the fissionable isotope, concentration in moderator, nature and depth of reflector as well as concentration of impurities including U^{238} or Th^{232}.

Generally speaking, however, not less than about 200 grams of the fissionable isotope is required using the best of moderators and securing maximum neutron reflection at optimum concentration. Where ordinary water is used not less than about 300 grams of fissionable isotope will be required and where the fissionable isotope is U^{233}, the minimum critical mass for the best available moderator will be at least about 300 grams and for water it will be at least about 500 grams. These masses must be increased with increasing concentration of impurities and also with variation in the concentration and neutron reflection of the system. Moreover, the actual amounts used in a reactor are somewhat higher since the reactor is desired to be larger than critical size, and in the present instance is cylindrical.

The reactors herein contemplated are operative when using as preferred materials pure fissionable isotopes such as pure U^{233}, pure 94^{239}, etc. However, such purity is not necessary in conversion and frequently it may be desirable to conduct the reaction in the presence of an isotope capable of absorbing neutrons to yield a further quantity of fissionable isotope as the reaction proceeds. Thus, uranium containing U^{238} in concentrations, for example, about 5 to 99 percent, the balance being U^{235}, offers certain advantages since U^{238} is converted to 94^{239} which aids U^{235} to support the reaction. The same is true when Th^{232} is used in lieu of U^{238}, as U^{233} is formed during the reaction.

From the above description it can be seen that fissionable isotopes when used in concentrations in a moderator higher than concentrations obtainable naturally can be used to create a self-sustaining chain reaction in a very small reactor, with amounts only on the order of a kilo-

2,815,321

9

gram of the fissionable isotope. This reactor can then be used to supply the desired neutrons to the absorbing zone of a converter, as herein shown, described and claimed.

In a reactor of the type above described, the main neutron losses may be enumerated as follows:

(1) Delayed neutrons that may be released outside the reaction zone during circulation of the solution.

(2) Neutron absorption by materials present in the reaction zone, such as the moderator and anion in case of solution.

(3) Neutron absorption by fission and corrosion products in the reaction zone.

(4) Neutron absorption in the reactor tank.

(5) Neutron escape from the reaction zone in wrong direction.

(6) Neutron absorption by non-fissionable isotopes formed, including intermediate isotopes, such as protoactinium when thorium is the absorbing isotope.

(7) Neutron absorption by fission products and corrosion products in the absorbing zone.

(8) Neutron escape from the absorbing zone particularly of neutrons formed by fission in the reflectors.

The losses enumerated above and their relationship to the structure herein described will be separately discussed below.

(1) In any reactor where the reaction composition is circulated outside of the reaction zone, delayed neutrons are lost to the reaction if they are released while outside of the reactor. About 1 percent of the neutrons emitted as a result of fission are delayed. Since all of the delayed emission periods, except the first one, are long compared with the uninterrupted sojourn of the U^{233} within the reactor, the percentage of delayed neutrons emitted outside the reactor will be equal to the neutron isotope holdup outside the reactor divided by the total amount of reaction isotope.

The holdup has been somewhat reduced by using an axial flow impeller in the liquid inlet to the reaction tank eliminating a pump holdup. It is preferred to have a substantial holdup, however, in the cooling tank 17 in order that gases formed in the reaction can be readily separated from the solution before it is returned to the reaction zone. In the structure shown and described herein, delayed neutrons emitted while outside the reactor tank 10 are not wholly lost, because the absorbing zone is completely and closely surrounded by the solution while it is outside of the reaction tank 10. In this manner, a significant portion of the delayed neutrons emitted while the solution is outside of the reaction zone can enter the absorption zone through the walls separating the zones. Thus, in the structure shown herein, the absorption zone receives fission neutrons both internally from the reaction zone and externally from the circulation zone.

(2) Absorption in the moderator itself will be low when D_2O is used, somewhat higher if H_2O is used. However, the absorption by the anion also has to be taken into consideration. If the fissionable isotope is used in the form of the nitrate, the nitrogen of the nitrate alone will give a small neutron loss. The nitrate is preferred as it does not give products, under irradiation, which could gum the circulation system. The sulphate, which has negligible neutron absorption, may under certain conditions, precipitate sulphur in the system, although some of this precipitation can be prevented by addition of peroxide to the solution. While the fluoride can be used, it is difficult to avoid precipitation of the peroxide at concentrations of hydrofluoric acid which are tolerable from the point of view of corrosion. Altogether, however, the parasitic losses in the solution are low when a substantially pure isotope is used.

(3) The use of the fissionable isotope in solution greatly aids elimination of xenon 135, a gaseous neutron absorbing "poison" developed during the operation of a

10

neutronic reactor. This poison has an important effect on operation of neutronic reactors.

In neutronic reactors operating at high neutron densities, radioactive elements of exceedingly high capture cross-section are formed relatively quickly in the composition as an intermediate element in the decay chains of the fission fragments. One of the most important of these decay chains is believed to be the 135 fission chain starting with tellurium, as follows:

$$\text{Te(short)} \xrightarrow[\text{half life}]{} \text{I(6.6 hr.)} \xrightarrow[\text{half life}]{} \text{Xe(9.4 hr.)} \xrightarrow[\text{half life}]{}$$
$$\text{Cs(20--30 yr.)} \xrightarrow[\text{half life}]{} \text{barium}$$

The neutron absorptions of tellurium, iodine, caesium and barium are relatively unimportant, but the neutron capture cross-section of radioactive xenon 135 has been measured to be about $2,500,000 \times 10^{-24}$ cm.^{2}, many times larger than that of stable gadolinium, for example, the cross-section of which is about $30,000 \times 10^{-24}$ cm.^{2}. Upon absorption of a neutron, xenon 135 shifts to xenon 136, an element of relatively small capture cross-section.

The rate of production of the originating fragment of the 135 chain is a function of the neutron density in which the fissionable isotope is immersed, and therefore dependent upon the power at which reactors of given type are operated. The radioactive xenon 135 is produced with a noticeable effect on the reaction a few hours after the reaction is started and the effect is, of course, greater as the neutron density is increased and maintained. The xenon 135 effect on the operation of high power reactors when the xenon remains in the reactor can be summarized as follows.

The reaction is started by withdrawing the control rod. The neutron density rises at a rate determined by the reproduction ratio and the effect of the delayed neutrons, until some predetermined neutron density is attained. The control rod is then placed in the unity reproduction ratio position and the reaction is stabilized at the power desired. During this time the radioactive iodine is formed decaying to xenon 135. As more and more iodine decays, more and more xenon 135 is formed, this xenon 135 absorbing sufficient neutrons to reduce the reproduction ratio. This absorption also converts the xenon 135 to xenon 136 which has no excessive capture cross-section. The neutron density drops. If no compensation were made for this drop the neutron density might drop until background conditions prevailed, and then the reaction might automatically start up as the xenon 135 decayed. Normally, the neutron density drop is compensated for by removal of the control or equivalent rod to a new position where the reproduction ratio is again above unity. A neutron density rise occurs, bringing the density back to its former level. Again, more xenon 135 is formed and the process is repeated until an equilibrium condition reached where the xenon 135 formed is transmuted by neutron absorption and by decay into isotopes of lower capture cross-section as fast as it is being formed. In the meantime, the control rod (or equivalent) has to be withdrawn by an amount to remove from the reactor neutron absorbers at least equal in effect to the absorption caused by the equilibrium amount of xenon 135. This requires a large initial reactor size to make the reaction possible with the xenon in the reactor. In the converter herein described, however, a large portion of this gaseous neutron absorber can be removed and the reactor kept small in size. For example, in the solution reactor above described, and operating for example at 100,000 kw., about 100 grams of 94^{239} will undergo fission each day, and the amount of xenon 135 formed is about 3.3 gm. If this were permitted to remain in the system as long as one day, its total cross-section would be five times greater than that of all the 94^{239} present. In order to reduce the total xenon 135 cross-section to one percent of the total cross-section of the fissionable isotope, the

11

xenon 135 formed should be eliminated about every five minutes, or as done here, continuously. The xenon 135, being gaseous, can be flushed out of the solution. This flushing is accomplished continuously by the structure described in that the solution is broken up into fine particles by nozzles 21 in cooling tank 17 and can thereby deliver the gaseous content of the solution to the space above the liquid surface, from which the non-condensible gases can be completely flushed out of the system by the use, for example, of helium circulation through pipe 52.

However, xenon 135 is not the only non-condensible gas produced in the reaction zone during operation. Many other fission products have decay, and daughter decay, chains including gaseous components later decaying to solids. By removal in gaseous form a substantial elimination of these other fission products is obtained that might otherwise eventually poison the reaction.

In addition, nuclear fission of isotopes dissolved in water causes a substantial amount of disintegration or dissociation of the water. The presently described reactor operating for example at 100,000 kw. can release several hundred liters of STP hydrogen (deuterium)-oxygen mixture per second. These gases have a desirable effect in that they carry the xenon and other gaseous fission products with them. Explosions are prevented by dilution of the O-D gases by helium as they are swept out of the tank space in which they are released. The dissociation gases are recombined by ignition and returned to tank 17. Condensible gases are condensed by coil 22 and drip back into tank 17.

However, other radioactive fission products (in addition to those having gaseous stages of decay) with high cross-sections, such as samarium, are produced, and will cause a small neutron loss even if the solution is purified once a day. However, this loss can be tolerated and daily purification will also remove corrosion products before they can build up to any substantial loss factor and permit the replacement of the amount of fissionable isotope destroyed by fission.

With respect to purification of the solution, i. e., decontamination of the solution by removal of the remaining fission and corrosion products, a solvent extraction process has been found satisfactory when water is the U^{233} solvent.

A H_2O solution containing the fissionable isotope which has been reacted and therefore contains solid fission products, is pumped from the converter and made 1 N in HNO_3 and 10 N in NH_4NO_3. It is then charged to the center of a continuous counter-current extraction column. A suitable water-immiscible solvent, such as hexone or dibutyl carbitol, is charged to the column at the bottom and passes up through the water phase.

The fissionable isotope passes from the water solution into the organic solvent, which is separated at the top of the column. The water solution containing the fission products is passed out of the bottom of the column into waste. The organic solvent containing the fissionable isotope is next charged to a similar continuous counter-current extraction column at the bottom and re-extracted into a water solution of the composition used in the breeder pile. The columns are identical in construction, and the water layer from the bottom of the second column can be pumped directly back into the reactor or can be passed through further purification cycles as above, if needed.

When the fissionable isotope is dissolved in deuterium oxide, the extraction column is not used directly as the deuterium will exchange with the hydrogen of the solvent. In this case, the uranium salt is separated first from the heavy water, and then purified as above.

(4) Neutron absorption in the reactor tank itself is very important, as this tank is interposed between the reaction zone and the absorption zone. The tank absorption is, of course, principally determined by the tank

12

material used. The neutron loss for an aluminum tank 1.5 mm. thick however is only about .01 neutron per fission and correspondingly lower for tanks of lead or beryllium. With a thin aluminum tank, the thorium slurry outside of the reactor tank can be placed in immediate contact with the tank, as shown in Figs. 1 and 2.

(5) The use of the reactor surrounded by the absorbing zone, except for the solution inlet and outlet, reduces the escape of neutrons in the directions in which there is absorption to a minimum. The reactor in Figs. 1 and 2 may be said to be, in one sense, totally surrounded by reflecting material in that water is present in both inlet and outlet pipes 11 and 12, respectively. Parasitic neutron escape completely out of the system therefore is reduced.

Furthermore, the reactor shape in the present instance is not spherical and therefore not of minimum size. The reactor is cylindrical and elongated. As the neutron density in a neutronic reactor is highest in the center, the effect of elongation is to reduce the neutron density at the ends of the long dimension of the reactor where the inlet and outlet are located. In this manner, the neutron escape is greatest in line along the short dimension of the reactor midway between the ends thereof and least at the inlet and outlet points. The elongation of the reactor is therefore deliberate, in order to reduce parasitic neutron escape at the inlet and outlet ends of the reactor.

(6) During a chain reaction using a fissionable isotope in high concentration, other uranium isotopes may be formed. For example, 94^{240} may be produced from 94^{239} by absorption of a neutron (without fission) and radioactive decay. U^{234} may be produced from U^{233} in the same fashion. While the extent of formation of such isotopes and their absorption cross-sections and fission capabilities are not at present fully known, experimental evidence indicates a small loss of neutrons by absorption in non-contributing isotopes or uranium formed during reaction operation. This effect, however, only enters into the efficiency of the converter, not its overall operation as these losses can be compensated for by adding new supplies of the fissionable isotope.

Non-productive neutron absorption in the absorption zone, however, has a more serious effect. For example, protactinium is an intermediate stage between the thorium and the uranium isotope desired, and as far as is presently known is not fissionable. This element can cause a neutron loss in two ways. First, a neutron loss by the neutrons which Pa absorbs, and second, by formation of an element decaying into U^{234} instead of into U^{233}, a known fissionable isotope. This effect however can be kept to a minimum by extracting the Pa from the slurry at sufficiently frequent intervals to reduce the absorption by the Pa to about .5 percent of the absorption by the thorium.

(7) Purification of the absorbing slurry is also indicated by the accumulation of fission and corrosion products therein. For example, the total neutron absorption cross-section of thorium is about 100 times smaller than the total neutron absorption cross-section of the reaction isotope. If the optimum amount of thorium in the reflector is divided by 100 times the amount of fissionable isotope in the reaction zone, the number is obtained by which the total cross-section of the thorium in the reflector is larger than the total cross-section of the reaction isotope. This number is about 4. On the other hand, the amount of fission in the reflector is less than 1 percent of the amount of fission in the reaction zone, when all the thorium is distributed uniformly in the reflector. The relative poisoning effect in the reflector by the fission products is, in consequence, about 400 times smaller than the poisoning by the fission products in the reactor, and a correspondingly less frequent purification will be sufficient for the elimination of the fission

2,815,321

13

products. In consequence, the need for purifying the thorium in the reflector will arise primarily from the absorption by the Pa, and to reduce the holdup of U^{233} that has been produced within the reflector. This latter factor will require a purification of the slurry at monthly intervals. This monthly purification will at the same time satisfy the requirement for removal of the intermediate Pa which can then be held for decay into U^{233}.

(8) There are several reasons for loss of neutrons from the absorbing zone. The escape of neutrons perpendicular to the outer surface of the reactor tank is governed by the tail of the slowing down density of neutrons. This means that the thickness of the absorbing zone should be about 38 cm. in all directions to reduce the total number of neutrons escaping outside the system to less than .5 percent of all neutrons entering the absorbing zone when the slurry is approximately equal parts by weight of thorium oxide and water.

Some fission of the U^{233} formed in the absorption zone will take place in the absorption zone. When this happens in the inner portions of the slurry, no harm is done as the neutrons emitted are used either in the reactor or are slowed in the slurry and absorbed to form more U^{233}. However, when the fission of the U^{233} takes place in the outer portions of the slurry some of the neutrons emitted may proceed outwardly and would escape, if other precautions were not taken. A neutron reflecting substance around the absorption zone is, in consequence, indicated. The solution itself, having a large percentage of moderator therein, is for that reason a good reflector, and, in addition is a regenerative reflector in that some fission can take place therein. The device of the present invention, therefore, by positioning the solution while in the circulation zone around and close to the absorbing zone, permits the use of that solution as a peripheral reflector for the absorbing zone while it is being circulated. Even the heat exchanger 15, when filled with solution and when water is used as a coolant will act as a reflector, although not to the same extent as the solution alone with only a single wall between the solution and the absorbing zone, as when in distributor 14 and cooling tank 17. Thus, the solution in the circulation zone as constructed as described above, not only permits delayed neutrons to reach the absorbing isotope, but also acts as a reflector to return escaping neutrons to the absorbing zone.

The converter as above described makes no use of the heat of the reaction. In order to maintain a high overall efficiency in the chain reacting system at least a portion of the heat developed should be used. While the heat exchanger coolant in the herein described system is available for space heating purposes, it is advantageous to be able to extract the heat in more usable form.

Several methods are available to produce power and still maintain a high conversion factor in a converter. For example, the dissociated gases can be burned to produce high temperature combustion gases, which in turn can be used to produce power, or the solution can be reacted under pressure, and flashed to steam in a separate lower pressure boiler, this action producing steam and simultaneously cooling the remaining solution, such as is more fully described and claimed in the co-pending application of H. C. Vernon, Serial No. 628,320, filed November 13, 1945. When power is produced in useful form the use of the U^{233}—U^{233} conversion even when the conversion is less than unity becomes more attractive in that the non-fissionable isotope (thorium) will furnish the bulb energy with only a small loss of U^{233} in the process.

While the herein described reactors have utilized substantially pure isotopes as preferred materials for the reactions, a conversion reaction can also be obtained in the same type of apparatus by using natural uranium,

14

either as the reaction composition (U^{235} content) or the absorbing composition (U^{238} content).

In case absorber is the U^{238} content of the natural uranium, it can be present both inside the reactor in the reacting solution of uranium salt in D_2O, and also can be present in the reflector, which in this case can be a UO_2—D_2O slurry. During the course of the reaction, the 94^{239} will be formed both in the reaction tank and in the reflector slurry. When the 94^{239} is removed periodically from the reaction solution, some U^{235} content of the uranium will remain, of course, but will be depleted, and at some point, dependent on 94^{239} production, will not support a chain reaction in D_2O. Under these circumstances, fresh natural uranium with full U^{235} content must then be supplied to the reaction tank. The uranium depleted in U^{235} is still perfectly suitable for use in the absorbing zone where the 92^{238} is the absorber. Fissions will take place in the U^{235} content of the slurry, although, as pointed out above for U^{233} in the reflector, at a much slower rate than in the reactor. Thus, the U^{235} continues to be converted into 94^{239} via U^{238} when it is in the reflector. The 94^{239} formed in the reflector is removed periodically, as in the thorium reflector and the residue replaced.

The same general conditions could, if desired, be accomplished in the U^{233} thorium conversion reactors as previously described by placing some of the thorium within the reactor along with the U^{233} as a slurry. There are several advantages to be gained, in that the number of neutrons which have to leave the reactor tank is reduced by a factor of about 3, and in consequence the losses in the tank wall and outside will be diminished by the same factor. With thorium, however, the difficulties of rapidly circulating the reactor slurry, and chemical separation difficulties make this type of reactor less desirable than when the pure isotopes are used in the reaction tank. While the converters using mixed isotopes are less desirable in certain cases, the structure used will be substantially the same. Consequently, the present application is intended to cover such use within the scope of the appended claims.

What is claimed is:

1. An isotope conversion device comprising a reaction tank, a thermally fissionable isotope dispersed in a liquid moderator in said reaction tank and in an amount sufficient to support a chain reaction, a container surrounding said reaction tank, an inlet and outlet for said reaction tank passing through said container, a non-thermally fissionable isotope convertible to a thermally fissionable isotope by neutron absorption positioned in said container, an outlet tank connected to said outlet, an inlet tank connected to said inlet, each of said tanks having a wall in common with a portion of the wall of said container, conduit means connecting said inlet and outlet tanks outside of and close to the remaining portion of the wall of said container, said tanks and said conduit means completely enclosing said container, means for circulating said thermally fissionable isotope over the path thus provided, and means for circulating a coolant around said conduit means.

2. An isotope conversion device comprising a reaction tank, a thermally fissionable isotope dispersed in a liquid moderator in said reaction tank and in an amount sufficient to support a chain reaction, a container surrounding said reaction tank, an inlet and outlet for said reaction tank passing through said container, a non-thermally fissionable isotope convertible to a thermally fissionable isotope by neutron absorption positioned in said container, an outlet tank connected to said outlet, an inlet tank connected to said inlet, each of said tanks having a wall in common with a portion of the wall of said container, conduit means connecting said inlet and outlet tanks outside of and close to the remaining wall portion of said container, said conduit means forming with said tanks a complete enclosure for said container, said con-

2,815,321

15

duit means terminating in said inlet tank in a plurality of spray nozzles, a cooling coil in said inlet tank in the path of liquid issuing from said spray nozzles, and means for circulating said thermally fissionable isotope over the path thus provided.

3. An isotope conversion device comprising a reaction tank, a thermally fissionable isotope dispersed in a liquid moderator in said reaction tank and in an amount sufficient to support a chain reaction, a container surrounding said reaction tank, an inlet and outlet for said reaction tank passing through said container, a non-thermally fissionable isotope convertible to a thermally fissionable isotope by neutron absorption positioned in said container, means for circulating said thermally fissionable isotope through said chamber and said outlet over a path closely adjacent the outer surfaces of said container and through said inlet back into said chamber, and means for cooling said thermally fissionable isotope while in said path.

4. A nuclear reactor apparatus comprising a first means for confining in an annular space a first fluid containing atoms capable of being converted by neutrons into atoms fissionable by thermal neutrons, said annular space comprising an absorption zone, a second means for confining within said annular space a second fluid adapted to emit neutrons, said second means defining a reaction zone, a heat exchanger directly surrounding the first confining means and extending the length thereof, means for abstracting the second fluid from the second confining means and passing it radially outwardly in all directions across one end of said space to one end of the heat exchanger, and means for returning the second fluid from the other end of the heat exchanger radially inwardly in all directions across the other end of said annular space to the second confining means.

5. The apparatus of claim 4 wherein the annular space is connected by inlet and outlet means with a plurality of second heat exchangers positioned outside of and around the first heat exchanger, and means for circulating the first fluid through said second heat exchangers.

6. In a nuclear reactor comprising a reaction zone

16

container for a first fluid composed of a thermally fissionable isotope dispersed in a liquid moderator in an amount sufficient to sustain a chain reaction, a heat exchanger capable of cooling said first fluid, and inlet and outlet means connecting said reaction zone container and said heat exchanger and means for circulating the first fluid through said heat exchanger; the combination therewith, of a second container for a second fluid composed of an isotope convertible by neutron absorption into a second isotope fissionable with thermal neutrons dispersed in a liquid moderator, the second container being positioned so as to substantially surround the reaction zone container and to be directly adjacent to said heat exchanger and said inlet and outlet means.

References Cited in the file of this patent

UNITED STATES PATENTS

1,215,140	Giller	Feb. 6, 1917
1,600,106	Fothergill	Sept. 14, 1926
1,927,555	Oetkin	Sept. 19, 1933
2,206,634	Fermi et al.	July 2, 1940
2,708,656	Fermi et al.	May 17, 1955
2,736,696	Wigner et al.	Feb. 28, 1956

FOREIGN PATENTS

114,150	Australia	May 2, 1940
861,390	France	Oct. 28, 1940
648,293	Great Britain	Jan. 3, 1951
233,011	Switzerland	Oct. 2, 1944

OTHER REFERENCES

Naturwissenschaften, vol. 27, 1939, pp. 402–410.

Power, July 1940, pp. 56–59.

A General Account of the Development of Methods of Using Atomic Energy for Military Purposes Under the Auspices of the U. S. Gov't, by H. D. Smyth, pp. 22–26, August 1945.

Goodman: "The Science and Engineering of Nuclear Power," vol. 1, p. 275, Addison-Wesley (1947).

Kelly et al.; Phy. Rev., 73, 1135–9 (1948).

Nov. 11, 1958   E. P. WIGNER ET AL   2,860,093
                    ISOTOPE CONVERSION DEVICE AND METHOD
                           Filed Nov. 13, 1945

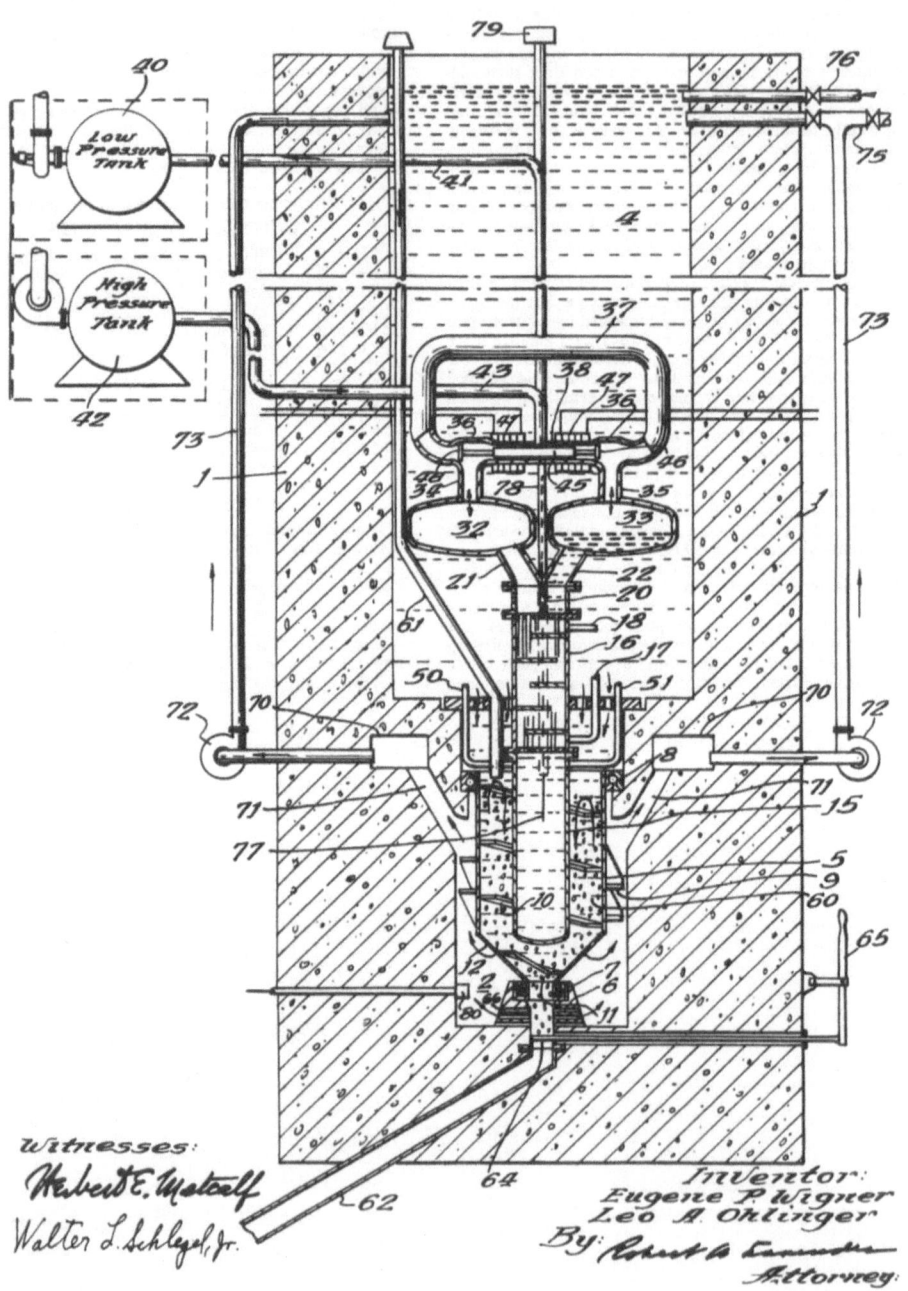

UNITED STATES PATENT OFFICE
CERTIFICATE OF CORRECTION

Patent No. 2,860,093   November 11, 1958

Eugene P. Wigner et al.

It is hereby certified that error appears in the printed specification of the above numbered patent requiring correction and that the said Letters Patent should read as corrected below.

Column 7, line 55, in the table, second column thereof, third line, for "¼ kg" read -- 1¼ kg --.

Signed and sealed this 4th day of April 1961.

(SEAL)

Attest: ERNEST W. SWIDER

XXXXXXXXXX
Attesting Officer

ARTHUR W. CROCKER
Acting Commissioner of Patents

United States Patent Office

2,860,093
Patented Nov. 11, 1958

1

2

2,860,093

ISOTOPE CONVERSION DEVICE AND METHOD

Eugene P. Wigner, Princeton, N. J., and Leo A. Ohlinger, Chicago, Ill., assignors to the United States of America as represented by the United States Atomic Energy Commission

Application November 13, 1945, Serial No. 628,323

17 Claims. (Cl. 204—154.2)

The present invention relates to nuclear physics and more particularly to an improved means and method of cooling a reactive composition used to sustain a chain fission reaction.

It is known that a self-sustaining chain reaction can be obtained in devices known as neutronic reactors utilizing natural uranium, as a result of slow neutron fission of the U^{235} content of the natural uranium. In such reactors, discrete bodies of natural uranium of high purity are disposed, usually in the form of a lattice arrangement of spheres or rods, in a neutron moderator such as graphite, baryllium or heavy water of high purity, surrounded by a neutron reflector. Neutron absorption in the U^{238} content of the natural uranium during the reaction leads to the production of the transuranic isotope 94^{239}, known as plutonium (symbol Pu), which is fissionable in much the same manner as U^{235}. 94^{239} or Pu^{239} is formed in neutronic reactors utilizing natural uranium in accordance with the following process:

$$92^{238} + n \longrightarrow 92^{239} \quad \text{6 m. e. v. of } \gamma \text{ rays, not necessarily all of one frequency}$$

$$92^{239} \xrightarrow{\text{23 min.}} 93^{239} + \beta \text{ 1 m. e. v. } \beta, \text{ no } \gamma \text{ rays}$$

$$93^{239} \xrightarrow{\text{2.3 day}} 94^{239} + \beta \text{ 600 kv. upper } \beta \text{ energy limit. Also 2 } \gamma \text{ rays, 400 kv. and 270 kv., about } \frac{1}{2} \text{ converted to electrons}$$

A small portion of the 94^{239} produced may also be changed to 94^{240} by absorption of neutrons. The neutronic reactors referred to above may be called "isotope converters" in that one fissionable isotope is formed (94^{239}) as another fissionable isotope (U^{235}) is used up. However, this conversion is not complete, and the natural uranium, which in this case acts to supply both the reaction isotope (U^{235}) and the absorption isotope (U^{238}), will contain at least two different fissionable isotopes (U^{235} and Pu^{239}) after the reaction has been started. Certain presently known uranium-graphite reactors have been found to have a conversion factor of .78, U^{235} to 94^{239}.

To obtain a more complete conversion of one fissionable isotope to another, it has been proposed to utilize a substantially pure fissionable isotope in solution or dispersion in a liquid moderator for the neutronic reaction, and then form the new fissionable isotope separately by absorption of leakage neutrons produced as a by-product of the reaction.

As liquid type reactors are small when incorporating a fissionable isotope in high concentration, exterior neutron leakage is high and in consequence such reactors are ideal neutron sources for use in isotope converters. Such converters are more fully described and claimed in the copending application of Wigner et al., filed November 13, 1945, Serial No. 628,322, now Patent No. 2,815,321.

The plutonium produced by neutronic reactors using natural uranium to support the reaction, is useful for many purposes, but it has one outstanding advantage over, for example, the use of U^{235}, as it exists in natural uranium. As plutonium is a different element from uranium, it can

be chemically removed from the irradiated natural uranium, and because of that fact can be obtained in substantially pure form and in high concentrations whereas U^{235} can only be obtained in high concentration or substantially pure form (as far as presently known) by the much more difficult process of isotope separation. U^{235} of high concentration however, has been used to sustain a neutronic reaction.

In high concentrations or substantially pure form, plutonium can also be used, when properly combined with a neutron moderator, to sustain a slow neutron chain reaction in a neutronic reactor of exceptionally small size as compared to the size of reactors using natural uranium. The neutron leakage is high in small reactors. In other words, such reactors can be used as efficient sources of large quantities of neutrons, and the neutrons thus produced can be used to produce another fissionable isotope.

U^{233}, another fissionable isotope, can be formed by irradiating thorium (90^{232}) with slow neutrons according to the following process:

$$90\text{Th}^{232} + n \longrightarrow 90\text{Th}^{233} + \gamma \text{ gamma rays}$$

$$90\text{Th}^{233} \xrightarrow{\text{23.5 min.}} 91\text{Pa}^{233} + \beta; \quad 91\text{Pa}^{233} \xrightarrow{\text{27.4 days}} 92\text{U}^{233} + \beta$$

The chemical separation of U^{233} from thorium is readily accomplished with high purity. The fissionable isotope U^{233} will support a chain reaction, and has many desirable qualities. In particular, U^{233} gives a relatively high average neutron yield per fission, the value as presently known being about 2.37–2.4 neutrons per fission (average).

While operating to produce a self-sustaining chain reaction, a reactive composition generates heat, and this heat must be removed from the reaction zone. One way by which the composition can be cooled is to circulate the composition through the reaction zone by rotating pumps, then through heat exchangers for cooling, and then return the cooled composition to the reactor in a continuous cycle. As the composition becomes extremely radioactive due to the formation of fission products therein, all exposed pump parts become contaminated and are, in consequence, difficult to replace without a shutdown for sufficient time for the radioactivity to decay to safe levels. Exceptional care is required to prevent pump leaks. Furthermore, in any liquid composition type reactor, exterior holdup of the composition is important as high holdup increases the amount of fissionable isotope in the system at one time, only that amount in the reaction zone itself being usefully employed.

It is therefore an object of the present invention to provide a novel means and method of circulating and cooling the reactive composition of a liquid type neutronic reactor without the use of rotating pumps.

In broad terms, the present invention utilizes a liquid of high fissionable isotope concentration in a reaction zone of sufficient size and isotope density to support a chain reaction. While operating, the liquid is oscillated back and forth through the reaction zone by gas pressure, and cooled while outside of the reactor.

Another object of the invention is to provide a novel method and means for withdrawing the gaseous products of nuclear reaction formed within the reactive composition. This object is accomplished by alternately subjecting portions of the fluid composition to high and low gas pressures thus causing the pulsating or oscillating action above described, and the gaseous reaction products are withdrawn from the composition during the low pressure stages of the oscillation cycle.

Still another object of the invention is to design an isotope converter such as above described wherein means are provided for stirring the neutron absorbent material which is disposed adjacent the chain reaction and is

2,860,093

3

adapted to be bombarded by neutrons generated thereby, said means comprising novel means for rotating the tank or container in which said material is disposed.

The present invention will be described for use in an isotope converter where 94^{239} obtained from neutron irradiated natural uranium is used in solution in water, light or heavy, to support a chain reaction and to supply neutrons for absorption in 90^{232} to form the fissionable isotope U^{233}.

The drawing shows diagrammatically a vertical sectional view, partly in elevation, of one embodiment of the present invention.

Referring to the figure, concrete walls 1 are provided defining a pit having a lower cylindrical portion 2, and an upper portion 4 which also can be cylindrical and larger in diameter than the lower portion. In the lower portion is positioned a rotatable cylindrical absorber tank 5 mounted on a lower block 6 by means of lower bearings 7 and maintained upright by upper bearings 8. Absorber tank 5 is smaller than the lower pit portion 2 and is provided on its outer surface with spirally applied fins 9 and on its inner surface with stirring paddles 10. The bottom of the absorber tank is coned and provided with an axial outlet 11 and cone perforations 12.

Axially located inside absorber tank 5 and clearing paddles 10 is an elongated cylindrical reaction tank 15 attached at its upper end to a tube type heat exchanger 16 having vertical tubes and supplied with coolant through inlet 17 and outlet 18.

At its upper end, the tubes of the heat exchanger 16 connect with a divided fitting 20 dividing the flow path into two diverging tubes 21 and 22, respectively, connecting with flattened surge tanks or chambers 32 and 33 positioned side by side in upper pit portion 4. Chambers 32 and 33 are provided with upper gas outlet tubes 34 and 35, respectively.

Each gas outlet tube extends through a two position valve cage 36. Each cage 36 is connected on one side to a low pressure manifold 37 and on the other side to a high pressure manifold 38. Low pressure manifold 37 is connected to a low pressure tank 40 by low pressure pipe 41 and high pressure manifold 38 is connected to a high pressure tank 42 by a high pressure pipe 43, both tanks being positioned outside of the pit. The low pressure tank 40, however, must be shielded, as will later be pointed out.

High pressure manifold 38 connecting the valve cages 36 is straight and contains a horizontal armature 45 having a valve 46 on each end thereof constructed to fit two valve seats in each valve cage 36. The armature is moved by electric current in field coils 47 wound outside of the high pressure manifold 38, which is of nonmagnetic material such as brass, for example.

By proper application of current to the field coils 47, the armature 45 and its attached valves can be reciprocated to connect first one, and then the other surge chamber, to the high pressure tank 42 while the other chamber is under a lower pressure. As shown in the drawing, the valve is positioned to place the left hand chamber under high pressure, while the right hand chamber is under lower pressure. The gas used for the pressure application is preferably helium.

A reactive solution, such as a solution of 94^{239} in water, light or heavy, is placed in the reaction tank 15 and sufficient additional solution is added to completely fill one half of the heat exchanger and to about half fill one surge chamber when the other surge chamber is empty, and the connected other half of the heat exchanger is practically empty.

By cyclically operating the solenoid valves 46 as described, a pulsating differential pressure is placed on the solution, and the solution is forced out of one surge chamber, through the heat exchanger 16 and into the other surge chamber and back again. The movement of the so-

4

lution may be termed oscillative or reciprocative, depending on the particular arrangement of surge tanks, heat exchangers, and reactor tank. The volumes are so coordinated that at each pulsation the entire amount of solution in the reactor is cooled.

Solution is withdrawn and supplied to the system by solution outlet pipe 50 and solution inlet pipe 51, respectively, positioned at the top of the reaction tank 15.

The reaction tank 15, when filled with solution, defines a reaction zone. The space between the reaction tank 15 and the absorber tank 5 defines an absorption zone. In this latter space is placed a quantity of thorium oxide pellets 60 introduced through absorber supply pipe 61. Pellets 60 can be removed from the absorption zone through axial outlet 11 connected with an absorber outlet pipe 62 through bearing block 6. The pellets 60 are held in the absorbing zone by absorber valve 64 operated by lever 65 outside of walls 1.

The thorium in the absorbing zone also needs to be cooled, and the entire device needs to be shielded at the top of the pit. Consequently, the pit is filled with water (H_2O) to provide a shield layer 50 to 75 ft. high over the surge chambers 32 and 33. This water shield is circulated both to cool the thorium pellets, and to rotate the absorber tank 5 to stir up the pellets 60. This water circulation will next be described.

The head of water in the pit is applied to the top of the absorber tank 5 which is open. The water passes through the voids between the pellets 60 and enters lower pit portion 2 through apertures 12, and also through bores 66 in bearing block 6 so that the pellets 60 immediately above the valve 64 can be cooled. The water then passes upwardly between the absorber tank 5 and the pit walls, exerting pressure on spiral fins 9 to rotate the absorber tank 5. The water then passes into a circumferential collector 70 through a plurality of passages 71 near the top of the absorber tank 5. Water from the collector 70 is then pumped up to the top of the pit by a plurality of pumps 72 and risers 73. Thus, as the water circulates the absorber tank 5 is rotated so that paddles 10 will stir up pellets 60.

As will be pointed out later, the circulated shield water will become slightly contaminated in time by radioactive products formed in the absorption zone. Consequently, a certain amount of water is continuously withdrawn from the pit through drawoff pipe 75 leading from one of the pumps, and fresh water is admitted through makeup pipe 76. The radioactivity is thus prevented from building up in the shield water.

The reaction is controlled by the customary control rod 77 extending into the reaction tank at sufficient distance laterally from the central axis so that it can slide in a vertical tube 78 extending upwardly through heat exchanger 16, past the pressure and vacuum manifolds at one side thereof, to emerge above the water level at the top of the pit. The rod passes through a stuffing box 79 at the top of the tube so that it can be vertically moved by any convenient means to position more or less thereof in the reaction tank 15. The lower end of the rod 77 is formed of a high neutron absorbing material, such as cadmium, and absorbs sufficient neutrons when fully inserted to prevent a chain reaction from being sustained. When partially removed from the reaction zone, a chain reaction with a reproduction ratio of over unity will occur. The neutron density will rise to a desired power output and the rod 77 is then inserted to a point where the reproduction ratio is unity. Small movements of the rod 77 thereafter serve to maintain the chain reaction at the desired power level. The neutron density is monitored by a conventional ionization chamber 80.

Having discussed the mechanical structure of the device, the nuclear physics thereof will be next taken up.

The reactive solution is, for example, plutonium sulphate dissolved in water. When heavy water is used for a moderator there is less neutron absorption in the mod-

2,860,093

5

erator than when H_2O is used, but both are suitable. Plutonium sulphate can be used in a thin stainless steel tank, and such a tank absorbs few neutrons. Thus, practically all neutrons leaving the reaction zone will enter the absorption zone on three sides. In a reactor of small size, about 50 percent of the neutrons generated escape outwardly. By the use of an elongated reaction zone, as shown, end leakage is greatly reduced, and the leakage around the middle of the reactor is increased. Thus even though there is no absorber present at the top of the reaction zone, neutron loss at that point can be reduced to about 1 percent.

The excess fast and slow neutrons not used in the chain reaction or parasitically absorbed in the reaction zone escape into the absorption zone. Here the slow neutrons are absorbed in the thorium to form U^{233}. The thickness of the absorption zone should be about 50 centimeters to insure slowing to thermal energy most of the fast neutrons escaping the reaction zone.

Due to the fact that the thorium is continually stirred, the pellets 60 are uniformly irradiated. Otherwise the pellets closest to the reaction zone would have the highest U^{233} content and in turn, would have most fissions of the U^{233} formed therein by absorption of escaping neutrons. If the U^{233} content rises much above .01 percent, the loss by fission requires it to be removed. By continually stirring the pellets, the thorium can stay in the absorption zone until all the pellets 60 reach a U^{233} concentration requiring removal. When that time comes, valve 64 is operated to dump all or part of the pellet 60 charge into discharge pipe 62 and into a shielded coffin (not shown) for chemical removal of the U^{233}.

However, the fact that some fissions do occur in the thorium leads to a small contamination of the shield water flowing over the thorium pellets 60. If the shield water were not continuously drawn off and made up with fresh water, the contamination of the shield water could accumulate and rise to unsafe biological levels. If a body of water is near, water can be passed through once only with discharge into a retention basin for aging, to eliminate all but the longer half-lived products.

It will be noted that when the solution enters a surge chamber, 32 or 33, it is under reduced pressure. This reduced pressure aids in the release of any gases formed during operation of the reactor. The released gases are then passed to the low pressure tank and then are exhausted. Several types of gas are formed during operation and should be removed.

In neutronic reactors operating at high neutron densities, radioactive elements of exceedingly high capture cross-section are formed relatively quickly in the reactive composition as an intermediate element in the decay chains of the fission fragments. One of the most important of these decay chains is believed to be the 135 fission chain starting with tellurium, as follows:

Te(short) $\longrightarrow$ I.(6.6 hr.) $\longrightarrow$
Half life   Half life

Xe(9.4 hr.) $\longrightarrow$ Cs.(20–30 yrs.) $\longrightarrow$ barium
Half life   Half life

The neutron absorptions of tellurium, iodine, caesium and barium are relatively unimportant, but the neutron capture cross-section of radioactive xenon 135 has been measured to be about $2,500,000 \times 10^{-24}$ cm.^{2}, many times larger than that of stable gadolinium, for example, the cross-section of which is about $30,000 \times 10^{-24}$ cm.^{2}. Upon absorption of a neutron, xenon 135 shifts to xenon 136 an element of relatively small capture cross-section.

The rate of production of the originating fragment of the 135 chain is a function of the neutron density in which the fissionable isotope is immersed, and therefore dependent upon the power at which reactors of given type are operated. The radioactive xenon 135 is produced with a noticeable effect on the reaction a few hours after the reaction is started and the effect is, of course, greater

6

as the neutron density is increased and maintained. The xenon 135 effect on the operation of high power reactors, when the xenon 135 remains in the reactor, as before stated, can be summarized as follows:

The reaction is started by withdrawing the control rod 77. The neutron density rises at a rate determined by the reproduction ratio and the effect of the delayed neutrons, until some predetermined neutron density is attained. The control rod 77 is then placed in the unity reproduction ratio position and the reaction is stabilized at the power desired. During this time, the radioactive iodine is formed, decaying to xenon 135. As more and more iodine decays, more and more xenon 135 is formed, this xenon 135 absorbing neutrons, thereby reducing the reproduction ratio. This absorption also converts the xenon 135 to xenon 136 which has no excessive capture cross-section. The neutron density drops. If no compensation were made for this drop, the neutron density might drop until background conditions prevailed, and then the reaction might automatically start up as xenon 135 decayed. Normally, the neutron density drop is compensated for by removal of the control or equivalent rod 77 to a new position where the reproduction ratio is again above unity. As neutron density rise occurs, bringing the density back to its former level, more xenon 135 is formed and the process is repeated until an equilibrium condition is reached where the xenon 135 formed is transmuted by neutron absorption and by decay into isotopes of lower capture cross-section as fast as it is being formed. In the meantime, the control rod 77 (or equivalent) has to be withdrawn by an amount thereby removing from the reactor, neutron absorbers at least equal in effect to the absorption caused by the equilibrium amount of xenon 135. This requires a larger initial reactor size to make the reaction possible with the xenon in the reactor. In the reactor herein described, however, a large portion of this gaseous neutron absorber can be removed and the reactor kept small in size. For example, in the solution reactor above described, and operating, for example, at 100,000 kw., about 100 grams of 94 will undergo fission each day, and the amount of xenon 135 formed is about 3.3 gm. If this were permitted to remain in the system as long as one day, its total cross-section would be five times greater than that of all the 94^{239} present. In order to reduce the total xenon 135 cross-section to one percent of the total cross-section of the fissionable isotope, the xenon 135 formed should be eliminated about every five minutes, or as done here, continuously by release in the chamber.

However, xenon 135 is not the only non-condensible gas produced in the reaction zone during operation. Many other fission products have decay, and daughter decay, chains including gaseous components later decaying to solids. By removal in gaseous form a substantial elimination of these other fission products is obtained that might otherwise eventually poison the reaction.

In addition, nuclear fission of isotopes dissolved in water causes a substantial amount of disintegration or dissociation of the water. The presently described reactor operating, for example, at high power can release many liters of NTP hydrogen (deuterium)-oxygen mixture per second. These gases have a desirable effect in that they carry the xenon and other gaseous fission products with them. Explosions are prevented by dilution of the O–H and O–D gases by the helium gas in the surge chambers as they are swept out of the chambers in which they are released. When D_2O is used, the dissociated gases are recombined by ignition and the D_2O reused. Condensible gases are in most cases condensed on the walls of the chambers as these walls are cooled by the shield water.

Other radioactive fission products (in addition to those having gaseous stages of decay) with high cross sections, such as samarium, are produced, and will cause a small neutron loss even if the solution is purified once a day

2,860,093

7

by continuous removal and resupply through pipes 50 and 51. However, this loss can be tolerated and purification every few days will also remove corrosion products before they can build up to any substantial loss factor, and will permit the replacement of the amount of fissionable isotope destroyed by fission.

With respect to purification of the reactive solution. i. e., decontamination of the solution by removal of the remaining fission and corrosion products, a solvent extraction process has been found satisfactory.

A water solution containing the fissionable isotope which has been reacted and therefore contains solid fission products, is pumped from the converter and made 1 N in HNO$_3$ and 10 N in NH$_4$NO$_3$. It is then charged to the center of a continuous countercurrent extraction column. A suitable water immiscible solvent, such as hexone or dibutyl Carbitol, is charged to the column at the bottom and passes up through the water phase.

The fissionable isotope passes from the water solution into the organic solvent, which is separated at the top of the column. The water solution containing the fission products is passed out of the bottom of the column into waste. The organic solvent containing the fissionable isotope is next charged to a similar continuous countercurrent extraction column at the bottom and re-extracted into a water solution of the composition used in the pile. The columns are identical in construction, and the water layer from the bottom of the second column can be pumped directly back into the reactor or can be passed through further purification cycles as above, if needed.

When the fissionable isotope is dissolved in deuterium oxide, the extraction column is not directly connected as the deuterium will exchange with the hydrogen of the solvent. In this case, the plutonium salt is separated first from the heavy water, and then purified as above.

When the above described reactor is operated with the plutonium dispersed as plutonium sulphate in solution in a light water (H$_2$O) moderator, a typical reactor of this type is as follows:

Volume of reactor _____ 3 cu. meters
                                    (3000 liters)
Plutonium per liter (as sulphate)___ 1 gm.
Total plutonium in reactor_____ 3 kg.
Holdup_____ 1.5 cu. meters.
Pulsation cycle_____ 2 sec.
Velocity (1 atmosphere differential
pressure) _____ 8 m./sec.
Power_____ Up to 100,000 kw.

The following typical data is given for pulsating reactors using a 94^{239} solution or slurry, in a D$_2$O moderator.

Concentration 94^{239}	Large scale .0003 gms./cc.	Small scale .001/gms./cc.
Amount 94	¾ kg.	1 kg.
Amount in reaction zone	4,200 liters	1,000 liters.
Holdup	800 liters	300 liters.
Pulsation cycle	2 sec.	2 sec.
Velocity (1 atmosphere differential pressure).	8 m./sec.	8 m./sec.
Temperature drop (1 pulsation)	40° C.	40° C.
Power	up to 140,000 kw.	up to 54,000 kw.
Length of pipes in Exchanger	100 cm.	100 cm.
Diameter of pipes in Exchanger	1 cm.	1 cm.
Temperature of exit water	70–80° C.	70°–80° C.

The pulsating method of cooling a reactor having a liquid reactive composition can also be applied to a reactor using natural uranium. When a uranium oxide slurry is used as the reactive composition, typical data is as follows:

Volume of reactor_____ 30 cu. meters.
Length of exchanger tubes_____ 200 cm.
Diameter of exchanger tubes_____ 2 cm.
Velocity_____ 14 m./sec.
Pulsation time_____ 2 sec.
Slurry holdup_____ 22 tons
Power_____ Up to 200,000 kw.

8

As uranium slurries tend to erode rotating pump parts, the pulsating method of circulation is particularly useful when slurries are used for the reacting composition.

It can be seen from the above that a principal feature of the present invention is the oscillating circulation of the solution by means of alternating differential gas pressures. This eliminates the need for rotating pumps. It can readily be understood that the reactive solution, even when out of the reaction zone is highly radioactive due to fission fragment content. Rotating pumps used to circulate the solution through heat exchangers must therefore be shielded. They become contaminated and must be left to age before removal from their shielded position if they need replacement. They also cause solution holdup in themselves and in the return piping. In the system above described however, the circulation system can be completely sealed and operated in sealed condition. No stuffing boxes are required at any point.

While the theory of nuclear reactions set forth herein is based on the best presently known experimental evidence, the invention is not limited thereto, as additional experimental data later discovered may modify the theory disclosed.

Obviously, many modifications may be made in the specific embodiments disclosed without departing from the intended scope of the invention.

What is claimed is:

1. The method of cooling the fluid reactive composition of a neutronic reactor which comprises cyclically oscillating the reactive composition through a chain reaction zone while maintaining a sufficient amount of the reactive material in the reaction zone to maintain the chain reaction.

2. The method of cooling a fluid neutronically reactive composition which comprises establishing a chain reaction in said fluid within a reaction zone, cyclically moving said fluid out of said zone by application of gas pressure thereto, and cooling said fluid while outside of said zone while maintaining a sufficient amount of the reactive material in the reaction zone to maintain the chain reaction.

3. The method of cooling a fluid neutronically reactive composition which comprises establishing a chain reaction in said fluid within a reaction zone, cyclically moving said fluid in a reciprocative manner out of and back into said zone by application of gas pressure thereto while maintaining a chain reaction within said zone, and cooling said fluid while moving it out of and back into said zone.

4. A method of cooling a fluid reactive composition in a neutronic reaction system comprising oscillating a portion of said composition between a reaction tank and surge tanks and through a heat exchange device while maintaining a sufficient amount of the composition in the reaction tank to maintain a chain reaction.

5. A method of cooling a body of fluid reactive composition in a neutronic reactor comprising oscillating a portion of said composition between a reaction tank and a heat exchange device connected thereto by substantially simultaneously alternately developing and releasing a charge of compressed gas against said body at a plurality of points while maintaining a chain reaction in the reaction tank.

6. A method of withdrawing the gaseous products of nuclear reaction within a neutronically reactive fluid composition of thermally fissionable material and liquid neutron moderator comprising forcing some of said composition from a reaction chamber alternately into first one and then another of a pair of pressure chambers by alternately applying high and low pressure gas to the respective chambers, and carrying off the gaseous products from the low pressure sides.

7. A method of withdrawing the gaseous products of thermal nuclear reactions within a neutronically reactive fluid composition comprising oscillating said composi-

2,860,093

9   **10**

tion between a reaction zone and a plurality of pressure zones by alternately subjecting respective pressure zones to high and low gas pressures while maintaining the reaction in the reaction zone.

8. A method of sustaining a nuclear chain reaction comprising the steps of loading a reaction chamber with a body of reactive fluid composition having a neutron reproduction ratio greater than unity, and then oscillating a portion of said composition between said chamber and heat exchange means externally thereof without decreasing the size of said body to a point at which neutron losses from the periphery thereof are effective to reduce said ratio to a value below unity.

9. In a system of the class described, a chamber, means therein for sustaining a thermal nuclear chain reaction comprising a fluid composition including fissionable material and liquid neutron moderator, heat exchange means, additional amount of said fluid composition in said heat exchange means, and means for oscillating said composition between said chamber and said heat exchange means.

10. A system of the class described comprising a reaction chamber, means therein for sustaining a thermal nuclear chain reaction comprising a fluid composition including thermally fissionable material and liquid neutron moderator, and a fluid-circulating system associated with said chamber for circulating a portion of said composition in an oscillating manner, said system comprising heat exchange means connected to said chamber, a plurality of pressure tanks independently connected to said chamber through said heat exchange means, and means for alternately subjecting respective tanks to high and low gas pressures.

11. In a system of the class described, a reaction chamber, a pair of tanks having independent connections to said chamber, heat exchange means associated with said connections for cooling fluid passing therethrough, means in said system for sustaining a thermal nuclear chain reaction comprising a body of liquid neutron moderator filling said chamber and partially filling one of said tanks, and thermally fissionable material carried by said liquid, and means for oscillating a portion of said liquid and fissionable material through said heat exchange means comprising means for alternately subjecting respective tanks to high and low gas pressures.

12. In a system of the class described, a reaction chamber, a pair of pressure chambers having independent connections thereto, means in said system for sustaining a thermal nuclear chain reaction comprising reactive composition in said reaction chamber including thermally fissionable material and liquid neutron moderator, and means for withdrawing from said composition the gaseous products of nuclear reactions therein, said last-mentioned means comprising means for alternately subjecting respective chambers to high and low gas pressures.

13. In a neutronic reactor, a chamber, a plurality of heat exchangers externally thereof and independently connected thereto, a body of fluid neutronically reactive composition capable of sustaining a thermal nuclear chain reaction, and means for alternately pulsating said composition between said chamber and respective heat exchangers without reducing the mass of said body below the critical size at which the same is capable of sustaining said reaction.

14. In an isotope converter, a rotatable tank, a fixed reaction tank therein, means in said reaction tank for sustaining a thermal nuclear chain reaction, and means carried by said rotatable tank for stirring material therein.

15. In an isotope converter, a rotatable tank, a neutron-permeable container therein, means in said container for sustaining a thermal nuclear chain reaction, means for circulating coolant fluid through said tank, means on said tank for cooperation with said coolant fluid to impart a rotary motion to said tank, and means carried by said tank for stirring material therein.

16. In an isotope converter, inner and outer tanks, said inner tank being formed of neutron-permeable material, means in one of said tanks for sustaining a nuclear chain reaction, means for passing coolant fluid through the other tank in heat exchange relationship with material therein, and means for stirring material in said other tank comprising means responsive to the motion of said fluid for rotating said other tank.

17. In an isotope converter, a tank, a neutronic reactor in co-axial relationship with said tank and separated therefrom in spaced relationship thereto by a neutron-permeable wall, means for moving said tank, and means operable by the motion of said tank for stirring material therein.

References Cited in the file of this patent
UNITED STATES PATENTS

1,248,851	Heisler	Dec. 4, 1917
1,432,170	Fenton	Oct. 17, 1922
2,042,428	Krekeler	May 26, 1936
2,127,193	Toulmin	Aug. 16, 1938
2,206,634	Fermi et al.	July 2, 1940

FOREIGN PATENTS

114,150	Australia	May 2, 1940
114,151	Australia	May 3, 1940
861,390	France	Oct. 28, 1940
233,011	Switzerland	Oct. 2, 1944

OTHER REFERENCES

Kelly et al.: Phy. Rev. 73, 1135–9 (1948).

Sept. 3, 1963   E. P. WIGNER   3,102,851

NEUTRONIC REACTION SYSTEM

Filed March 19, 1947   4 Sheets–Sheet 1

FIG.1.

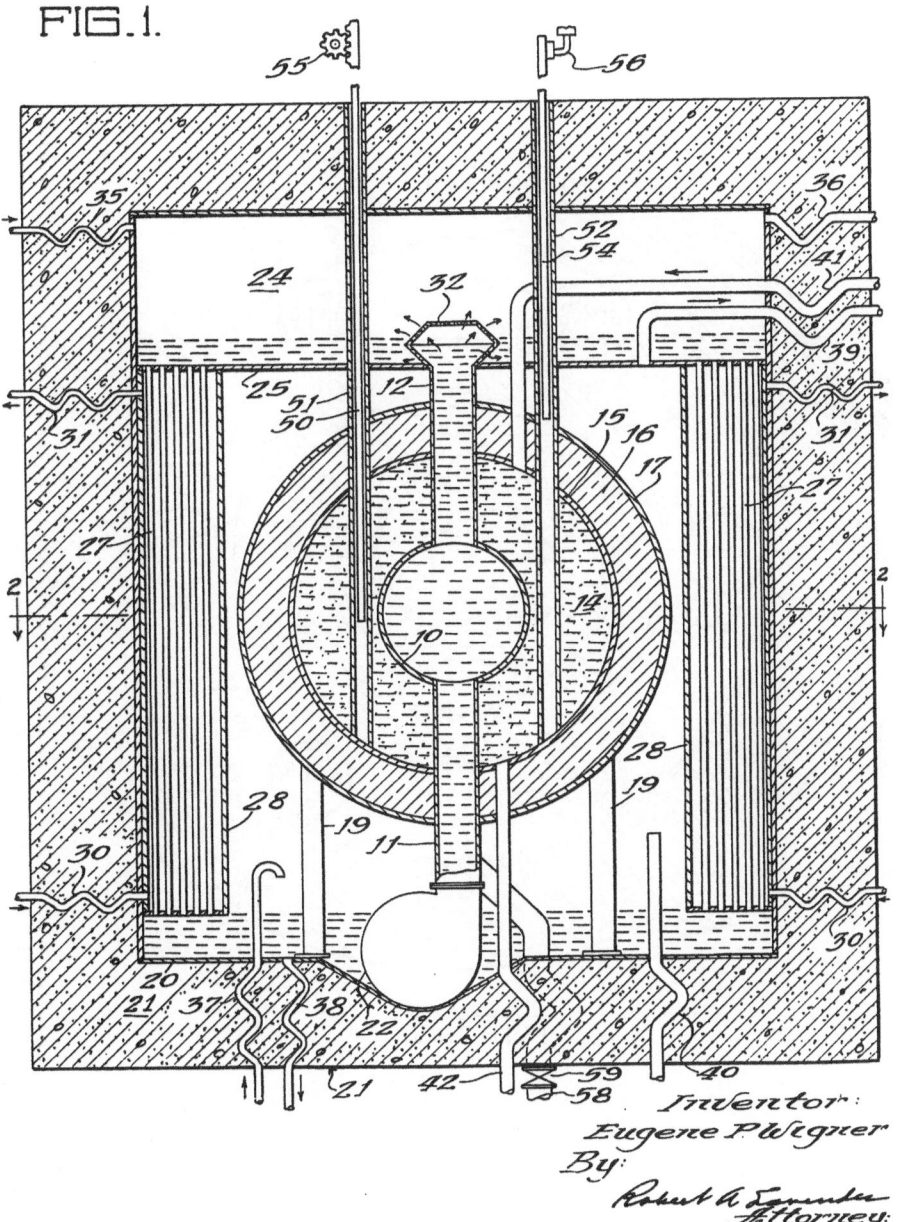

Inventor:
Eugene P. Wigner
By:
Robert A. [signature]
Attorney.

Sept. 3, 1963   E. P. WIGNER   3,102,851
NEUTRONIC REACTION SYSTEM

Filed March 19, 1947   4 Sheets—Sheet 2

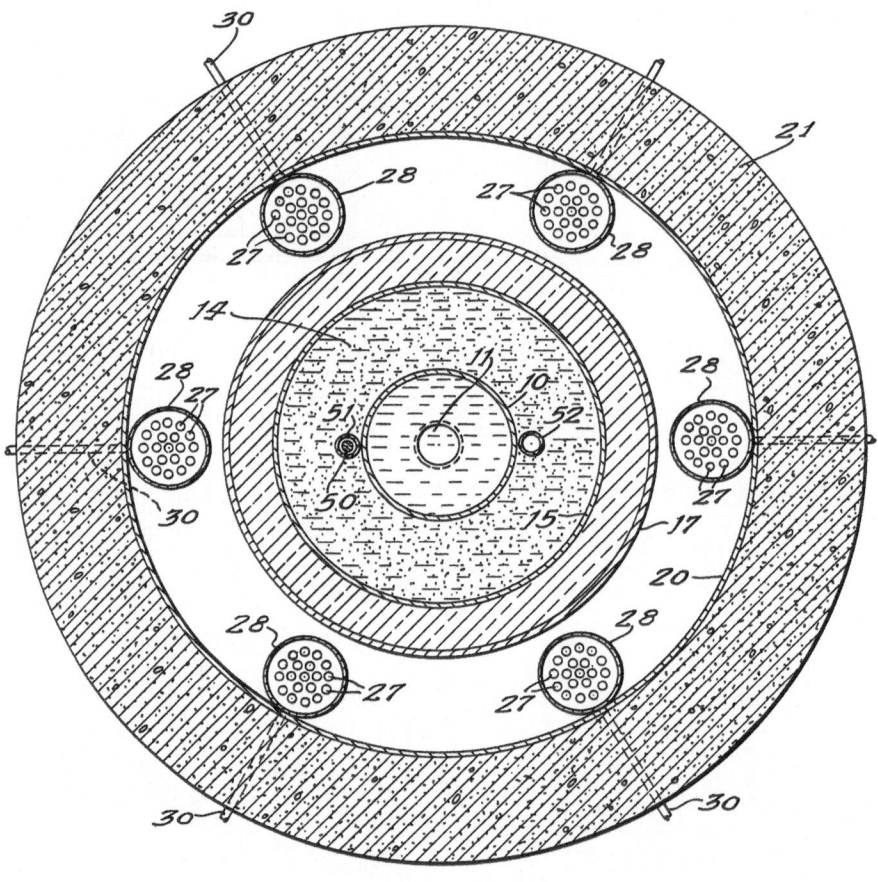

FIG.2.

Inventor
Eugene P. Wigner
By: Robert A. Lavender
Attorney.

Sept. 3, 1963   E. P. WIGNER   3,102,851
NEUTRONIC REACTION SYSTEM

Filed March 19, 1947   4 Sheets—Sheet 3

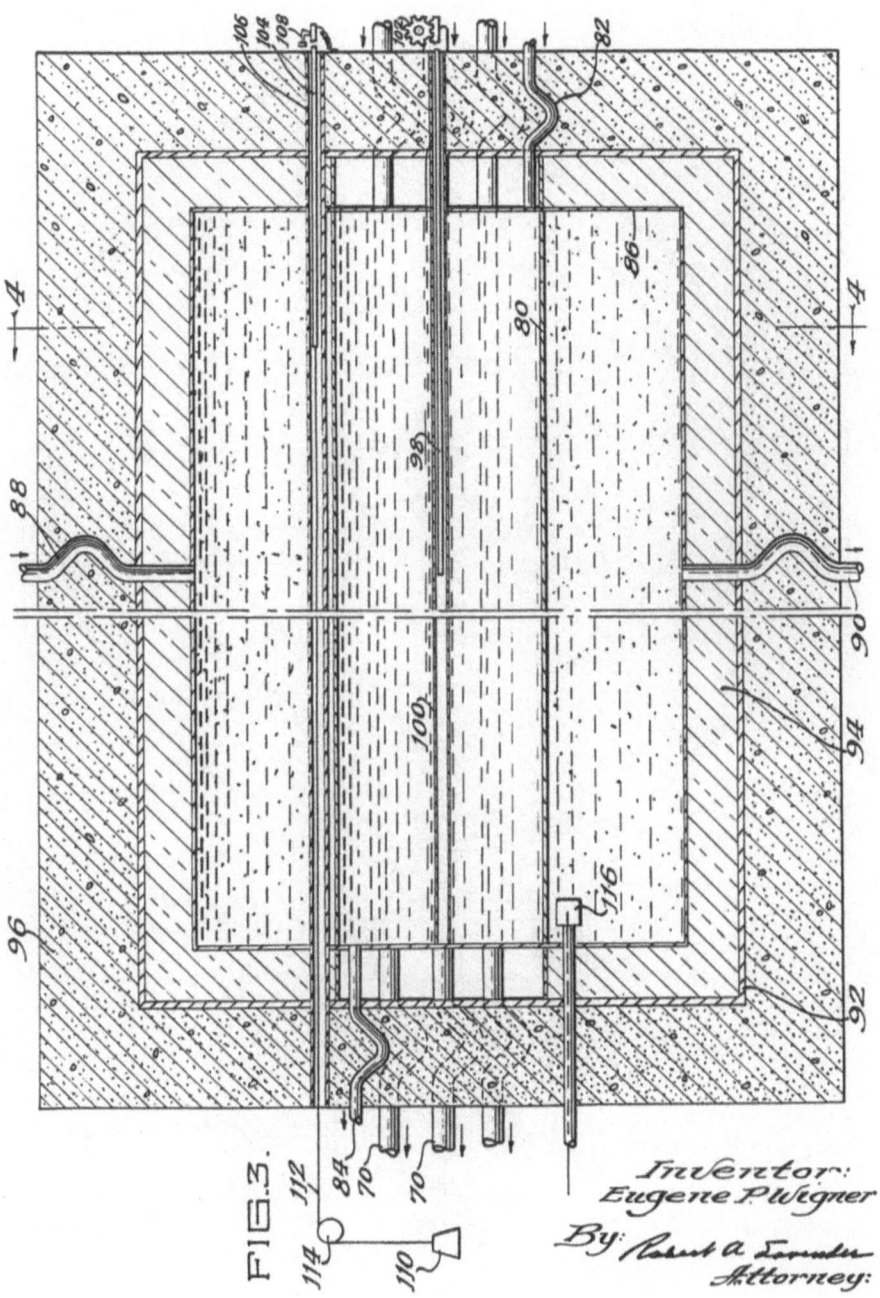

FIG.3.

Inventor:
Eugene P. Wigner
By: Robert A. Lorensen
Attorney:

Sept. 3, 1963   E. P. WIGNER   3,102,851
                         NEUTRONIC REACTION SYSTEM

Filed March 19, 1947   4 Sheets–Sheet 4

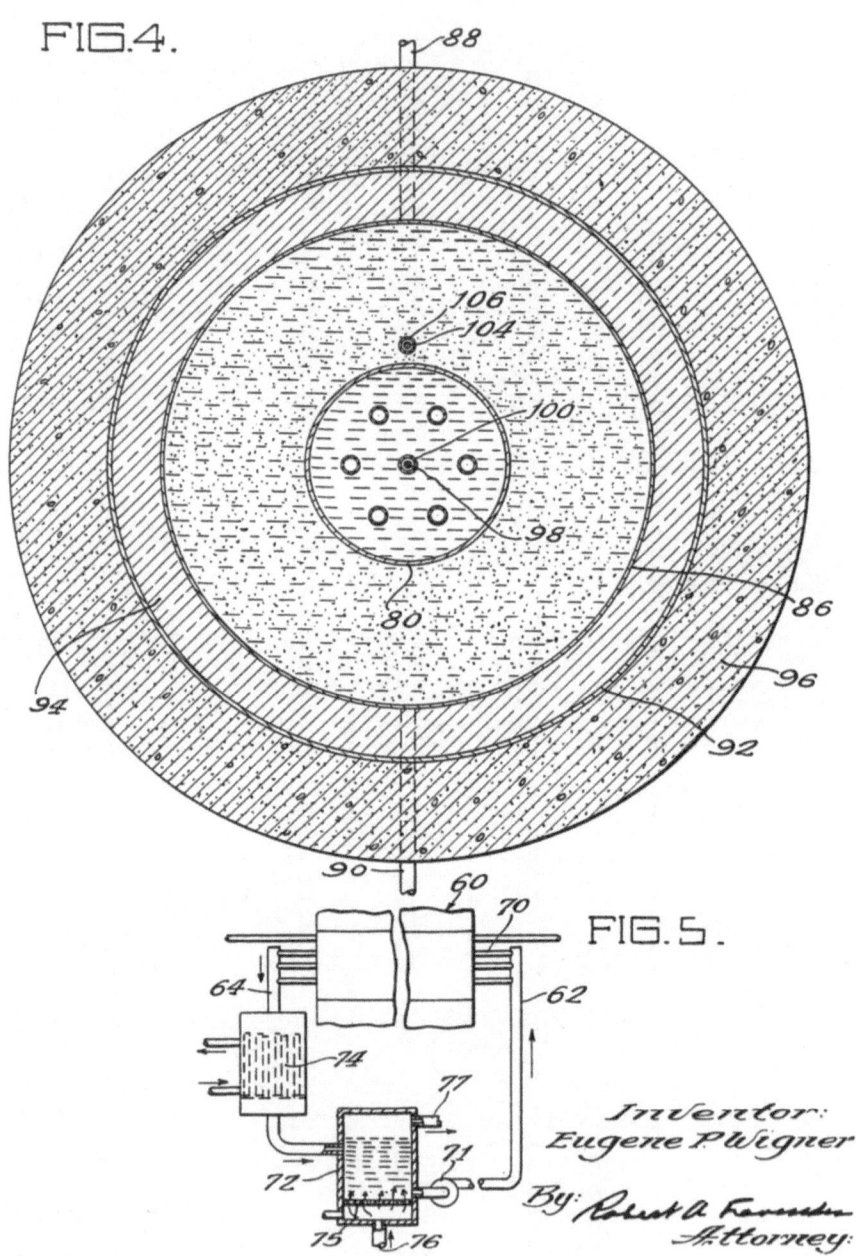

FIG.4.

FIG.5.

Inventor:
Eugene P. Wigner

By: Robert a. Faraday
    Attorney:

United States Patent Office

3,102,851
Patented Sept. 3, 1963

1

3,102,851
NEUTRONIC REACTION SYSTEM
Eugene P. Wigner, Oak Ridge, Tenn., assignor to the
United States of America as represented by the United
States Atomic Energy Commission
Filed Mar. 19, 1947, Ser. No. 735,683
3 Claims. (Cl. 204—193.2)

This invention relates to nuclear physics, and more
particularly to a novel method and means for producing 10
power and fissionable material which may be utilized to
sustain a nuclear fission chain reaction.

It is known in the art that a controlled slow neutron
chain reaction may be established by combining fissionable material and neutron moderator to form a reactive 15
mass, the moderator being effective to reduce the average neutron energy to a value at which the fission cross
section of said material is relatively great.

In slow neutron reactors a thermally fissionable isotope
such as U^{233}, U^{235}, or Pu^{239} or mixtures thereof is sub- 20
jected to fission by absorption of neutrons and a self-
sustaining chain reaction is established by the neutrons
evolved by the fission. In general, such reactors comprise bodies of compositions containing fissionable material such, for example, as natural uranium, disposed in 25
a neutron slowing material which slows the neutrons to
thermal energies (about 0.03 e.v.). Such a slowing material is termed a neutron moderator. Carbon and heavy
water (D_2O) are typical moderators suitable for such
use. Heat is evolved during the reaction which is re- 30
moved by passage of a coolant through the reactor or
in heat exchange relationship therewith. Specific details of the theory and essential characteristics of such
reactors are set forth in a copending application of Enrico Fermi and Leo Szilard, Serial No. 568,904, filed 35
December 19, 1944, now Patent No. 2,708,656.

In neutronic reactors, the ratio of the fast neutrons
produced in one generation by the fissions to the original
number of fast neutrons in a theoretical system of infinite
size where there can be no external loss of neutrons is 40
called the reproduction or multiplication factor or constant of the system, and is denoted by the symbol K.
For any finite system, some neutrons will escape from
the periphery of the system. Consequently, a system of
finite size may be said to have a K constant, even though 45
the value thereof would only exist if the system as built
were extended to infinity without change of geometry or
materials. Thus, when K is referred to herein as a constant of a system of practical size, it always refers to
what would exist in the same type of system of infinite 50
size. If K can be made sufficiently greater than unity to
indicate a net gain in neutrons in the theoretical system
of infinite size, and then an actual system is built to be
sufficiently large so that this gain is not entirely lost by
leakage from the exterior surface of the system, then a 55
self-sustaining chain reacting system of finite and practical size can be built to produce power and related by-
products by nuclear fission of natural uranium. The
actual neutron reproduction ratio in a system of finite
size, therefore, differs from K by the external leakage 60
factor, and by a factor due to the neutron absorption
by localized neutron absorber, and the actual reproduction ratio of the finite-sized system, R, must still be sufficiently greater than unity to permit the neutron density
to rise exponentially with time in the system as built. 65

During the interchange of neutrons in a neutron moderated system of finite size, comprising bodies of any
size disposed in a neutron moderator, neutrons may be
lost to the chain reaction in four ways: by absorption or
capture in the uranium content of the bodies without pro- 70
ducing fission; by absorption or capture in the moderator

2

material; by absorption or capture by the impurities present in both the uranium bodies and the moderator; and
by leakage out of the system through the periphery
thereof. In a fast reactor all of these losses, except the
last, are either substantially eliminated or are greatly
reduced.

A primary object of the present invention is to provide
a "breeder" system wherein a nuclear fission chain reaction is utilized to produce fissionable material at a rate
greater than the rate of consumption of fissionable material within the chain reacting composition. This is
accomplished by neutron bombardment of "fertile" material adapted to undergo nuclear reaction productive of
fissionable material as hereinafter described. Fertile isotopes as herein defined are isotopes such as Th^{232} and
U^{238}, which are converted to thermally fissionable isotopes, U^{233} and Pu^{239}, respectively, by nuclear reaction
under neutron bombardment. These fertile isotopes are
fissionable by fast neutrons and substantially non-fissionable by slow neutrons (below about 1000 e.v.) and absorb neutrons fast or slow to undergo the above-mentioned nuclear reactions.

According to the present invention the novel breeder
system comprises a neutronic reactor wherein U^{233} and
heavy water (D_2O) neutron moderator are combined
in a chain reacting composition surrounded by a neutron
reflector of heavy water containing a fertile material or
isotopes in solution or in suspension. The fertile material absorbs neutrons emanating from the chain reacting
composition and is thus converted to thermally fissionable material.

It has been found in this connection that the η of U^{233}
or, in other words, the number of neutrons emitted per
fission of a U^{233} atom is about 2.37, thus providing a net
increase of 1.37 neutrons per fission. If all of these
neutrons are absorbed in fertile material such as, for
example, thorium atoms, 1.37 U^{233} atoms are produced
for each U^{233} atom consumed or destroyed by the chain
reaction. However, it will be noted that the loss of as
much as 15 percent of the 2.37 neutrons by escape from
the system or by parasitic absorption of other materials
than the thorium results in a net increase of zero thus
preventing the breeding of thermally fissionable material.
An enumeration of the losses follows:

(1) Delayed neutrons which may be released outside
the pile.

(2) Absorption by materials present in the chain reacting part, such as the moderator and anion in case of
solution, bismuth and tubing in case of a second system
to be hereinafter described.

(3) Absorption by fission and corrosion products in
pile.

(4) Absorption by tank.

(5) Escape in wrong direction.

(6) Absorption by U^{234} formed.

(7) Absorption by Pa.

(8) Absorption by fission and corrosion products in
reflector.

(9) Absorption by moderator and coolant in reflector.

(10) Escape from reflectors, particularly of neutrons
formed by fission in reflector.

(11) Losses in chemical separations.

Accordingly it is another object of the present invention to reduce the above-mentioned losses to a minimum
as hereinafter described.

Still another object of the invention is to design a novel
breeder system in which energy released by the reaction
is utilized for the production of power. Such an arrangement requires high temperature operation of the system.

The foregoing and other objects and advantages of the
invention will become apparent from a consideration

3,102,851

3

of the following specification and the accompanying drawings, wherein:

FIG. 1 is a diagrammatic view partly in central vertical cross section and partly in elevation of a neutronic reaction system embodying the invention;

FIG. 2 is a cross sectional view taken on the line 2—2 of FIG. 1;

FIG. 3 is a diagrammatic view partly in central vertical cross section and partly in elevation of a modified form of a reaction system embodying the invention;

FIG. 4 is a cross sectional view taken on the line 4—4 of FIG. 3; and

FIG. 5 is a flow diagram illustrating the reaction system of FIGS. 3 and 4 in combination with a heat exchange system for the production of power and a purifier system for removing gaseous fission products from the chain reacting composition.

Describing the invention in detail and referring first to the embodiment thereof illustrated in FIGS. 1 and 2, a reaction tank 10 is provided, said tank being formed of a material such as beryllium, lead or bismuth having a relatively small neutron capture cross section. The tank contains a reactive composition in the form of uranium in heavy water, the uranium being preferably in a compound such as the nitrate, sulfate or fluoride thereof dissolved in the heavy water. The reactive composition is circulated through the tank 10 by means of inlet and outlet conduits 11 and 12, the inlet conduit being connected to the discharge side of a conventional pump 22 and the outlet conduit 12 being connected to an outlet or discharge header 32 from which the reactive solution flows into a high-pressure chamber 24 within a heat exchanger tank or container 20.

The reactive solution flows downwardly from the chamber 24 through heat exchanger tubes 27 disposed within a plurality of tanks 28 spaced around the reaction tank 10 as best seen in FIG. 2, a heat exchange fluid being circulated through each tank 28 by inlet and outlet conduits 30 and 31. The reactive solution flows from the lower ends of tubes 27 into the bottom or low pressure chamber of the tank 20 to the suction side of the beforementioned pump 22.

The tank 10 is surrounded by a tank 15 formed of a material having a relatively small neutron capture cross section such as those mentioned above. The tank 15 contains a neutron reflector of heavy water and fertile material such as, for example, a slurry of approximately 1 gram of ThO$_2$ per cubic centimeter of heavy water. The slurry within the tank 15 is indicated at 14 (FIG. 1) and is circulated by means of inlet and outlet conduits 41 and 42.

The tank 15 is surrounded by another tank 17 formed of a material having a relatively small neutron capture cross section, said tank 17 containing an outer neutron reflector of a material such as graphite or beryllium adapted to reflect neutrons escaping from the inner reflector 14 thus minimizing escape of neutrons from the system. The tank 17 is supported by legs or pedestals 19 seated on the bottom of the tank 20, which is disposed within and is supported by a concrete vault 21 adapted to afford a biological shield around the system.

The reactive composition within the tank 10 is replenished and renewed as desired by means of a make-up line 37 and a drain line 38 communicating with the bottom of the tank 20.

The nuclear fission chain reaction is regulated by a neutron absorbing control rod 50 reciprocated within a tube 51 formed of a material such as beryllium having a relatively small neutron capture cross section. The rod is actuated by a rack and pinion mechanism 55. Emergency control of the reaction is afforded by a safety rod 54 disposed within a beryllium tube 52, said rod being normally maintained in an elevated position by an electrically operated latch 56 which may be released under emergency conditions to permit the rod 54 to move downwardly

4

by force of gravity, thus terminating the chain reaction. Additional emergency control is afforded by an emergency dump line 58 with a dump valve 59 therein, said valve being adapted to be opened under emergency conditions to drain the reactive solution from the tank 10.

As above noted, in a breeder system the neutron losses must be reduced to a minimum and it has been found that a substantial portion of these losses is due to neutron absorption in gaseous fission products such as xenon[135] formed within the reactive solution 10. These fission products escape within the high pressure chamber 24 and are swept therefrom by an inert gas such as helium pumped through inlet and outlet conduits 35 and 36. Additional gaseous fission products are released within the low pressure chamber at the bottom of the tank 20 and are swept therefrom by an inert gas circulated through the tank 20 by outlet and inlet conduits 39 and 40.

Thus it will be understood that by means of the above-described arrangement, a novel breeder system has been designed wherein neutron losses are reduced to a minimum by utilizing a heavy water neutron moderator within the reactive solution and by surrounding the reactive solution with a heavy water neutron reflector containing fertile material to be converted under neutron bombardment to thermally fissionable material. It will be understood that heavy water has a neutron capture cross section of approximately zero and is an excellent neutron moderator, thus affording a system from which very few neutrons escape. Furthermore, neutron losses due to absorption in gaseous fission products are reduced to a minimum value as above described.

An operative system of the above-described type may be constructed with a reactive solution according to the following table.

D$_2$O (cc.) per gram U^{233}	Critical radius in centimeters	Critical vol. in liters	Vol. of U^{233} in kilograms
100	41.8	162	1.62
200	46.6	241	1.20
300	50.9	332	1.11
400	54.8	430	1.07

It may be noted that the separation of the reactive solution and the fertile reflector 14 by the tank 10 may be eliminated so that the thorium slurry and uranium solution constitutes a single reactive mass, thus eliminating the neutron absorption by the tank 10. However, the critical diameter of the pile in such an arrangement is increased to about 3 meters with a volume of about 14 cubic meters. The amount of U^{233} required for such a pile would be of the order of 70 kilograms and, for this reason, it is preferred to separate the slurry and the solution by the tank 10 to prevent reduction of the neutron multiplication constant within the reactive solution.

Referring now to the modification of the invention shown in FIGS. 3 to 5 inclusive, the neutronic reactor generally designated 60 (FIG. 5), and hereinafter described in detail comprises a reactive solution circulated by means of inlet and outlet lines 62 and 64, the inlet lines or conduits being connected to the discharge side of a pump 71, and the outlet conduit 64 being connected to a heat exchanger 74 adapted to absorb the heat developed within the reactive solution for the purpose of producing power and cooling the solution. The heat exchanger 74 is connected to a purification tank 72 which is connected to the suction side of the before-mentioned pump 71. A perforated wall or screen 75 is disposed within the tank 71 and an inert gas such as helium is pumped through an inlet line 76 upwardly through the screen 75 and through the reactive solution to sweep the gaseous fission products therefrom. The sweeping gas and the gaseous fission products entrained therein are conveyed from the tank 72 by an outlet conduit 77. It may be noted that the reactor 60 comprises a plurality

3,102,851

5

of tubes 70 hereinafter described in detail and adapted to convey the reactive solution through a mass of neutron moderator as hereinafter discussed. Thus, it will be seen that the system diagrammatically illustrated in FIG. 5 comprises a neutronic reactor in the form of a breeder generally designated 60, a heat exchange system 74 for absorbing the heat of the chain reaction to produce power, and a purification tank 72 within which the gaseous fission products of the reaction are removed from the system.

Referring now to FIGS. 3 and 4, the novel breeder system of FIG. 5 is illustrated in detail and comprises a reaction tank 80 through which the before-mentioned tubes 70 extend, said tubes being formed of a material such as beryllium or lead having a relatively low neutron capture cross section. The tank contains neutron moderator in the form of a mass of heavy water surrounding the tube 70 and circulated by means of inlet and outlet conduits 82 and 84. The tank 80 is disposed within a reflector tank 86 containing a slurry of fertile material and heavy water, as in the previously described embodiment, said slurry being circulated by inlet and outlet conduits 88 and 90.

The tank 86 is, in turn, surrounded by another tank 92 which contains an outer neutron reflector 94 in the form of beryllium or graphite. The tank 92 is enclosed within and supported by a concrete vault 96 affording a biological shield around the system.

The nuclear fission chain reaction within the tank 80 is controlled by a neutron absorbent control rod 98 reciprocable within a permeable tube 100 formed of a material such as beryllium having a small neutron capture cross section, and the rod 98 is actuated by a rack and pinion mechanism 102. Emergency control of the reaction is provided by a neutron absorbent control rod 104 disposed within a neutron permeable tube 106 and held in the retracted position shown in FIG. 3 by an electrically operated latch 108. Under emergency conditions, the latch is released and the rod 104 is drawn into the central portion of the system by means of a weight 110 connected to the rod 104 by a line 112 supported by a pulley 114. The neutron density within the system is monitored by a conventional ionization chamber 116 connected in a conventional manner to an electrical circuit (not shown) for measuring the neutron density.

In the arrangement of FIGS. 3 and 4, the uranium compound is dissolved in bismuth in a ratio of about 1 gram to 25 cubic centimeters. The solution is circulated through the tubes 70 which passes as above described through the heavy water neutron moderator in the tank 80 to define therewith a neutronic reactor in which neutron losses are reduced to a minimum as above described. It may be noted that the heavy water within the tank 80 is preferably circulated, degassed, and cooled externally of the tank 80 by a system (not shown) such as that utilized in FIG. 5 for cooling and degassing the uranium bismuth solution. If desired, a mixture of UF_6 and a fluorocarbon of low neutron cross-section may be substituted for the uranium bismuth solution where the

6

reactor is to be operated at a relatively low temperature value.

It will be understood that the above-described embodiments of the invention are merely by way of illustration and not limitation inasmuch as various modifications of the system will be readily apparent to those skilled in the art without departing from the spirit of the invention or the scope of the appended claims.

What is claimed is:

1. A neutronic reactor system comprising a heat exchange tank, high and low pressure chambers therein, a plurality of heat exchangers in said tank comprising tubes interconnecting said chambers, a neutronic reactor within said tank comprising a substantially spherical container, a reactive composition within said container consisting essentially of a solution of a U^{233} compound in heavy water in a concentration sufficient to sustain a chain reaction, a second container surrounding said first container and containing a slurry of thorium dioxide in heavy water, inlet and outlet conduits connected to said reaction tank, the outlet conduit being connected to said high pressure chamber, a pump having its suction side connected to the low pressure chamber and its discharge side connected to the inlet conduit, and means for removing gaseous fission products released in both of said chambers.

2. A neutronic reactor system according to claim 1 wherein the reactive composition completely fills the spherical container.

3. A neutronic reactor system according to claim 2 wherein the spherical container has a radius between approximately 41.8 and 54.8 centimeters, the solution of a U^{233} compound in heavy water has a concentration of from 1 gram of U^{233} per 100 cc. of heavy water to 1 gram of U^{233} per 400 cc. of heavy water, and the thorium dioxide slurry contains about 1 gram of thorium dioxide per cc. of heavy water.

References Cited in the file of this patent

UNITED STATES PATENTS

2,708,656	Fermi et al.	May 17, 1955
2,743,225	Ohlinger	Apr. 24, 1956

FOREIGN PATENTS

233,011	Switzerland	Oct. 2, 1944
861,390	France	Oct. 28, 1940

OTHER REFERENCES

H. D. Smyth: A General Account of the Development of Methods of Using Atomic Energy for Military Purposes, August 1945; pp. 22 and 177.

Harwell: The British Atomic Energy Research Establishment 1946–1951, London, 1952, pp. 34–42.

Sourcebook on Atomic Energy, by Samuel Glasstone, D. Van Nostrand Co., N.Y., 1950, pp. 400–406.

A Forum Report: Nuclear Reactor Development, July 1954, Atomic Industrial Forum, 260 Madison Ave., New York 16, N.Y., page 18.

July 25, 1961   H. SOODAK ET AL   2,993,850

                                        FAST NEUTRON REACTOR

Filed June 14, 1948   6 Sheets—Sheet 1

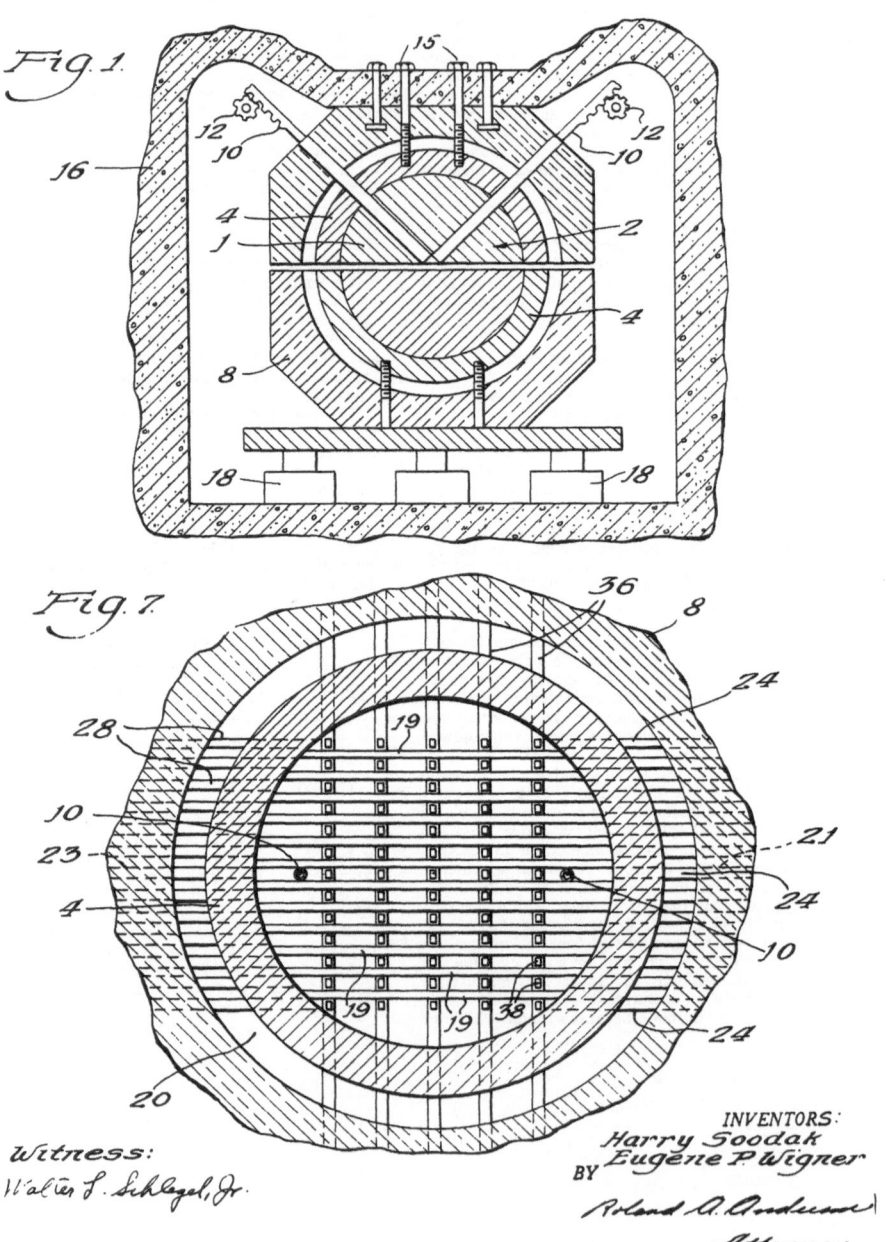

Fig. 1.

Fig. 7.

INVENTORS:
Harry Soodak
Eugene P. Wigner
BY

Witness:
Walter L. Schlegel, Jr.

Roland A. Anderson
Attorney

July 25, 1961   H. SOODAK ET AL   2,993,850

                              FAST NEUTRON REACTOR

Filed June 14, 1948   6 Sheets—Sheet 2

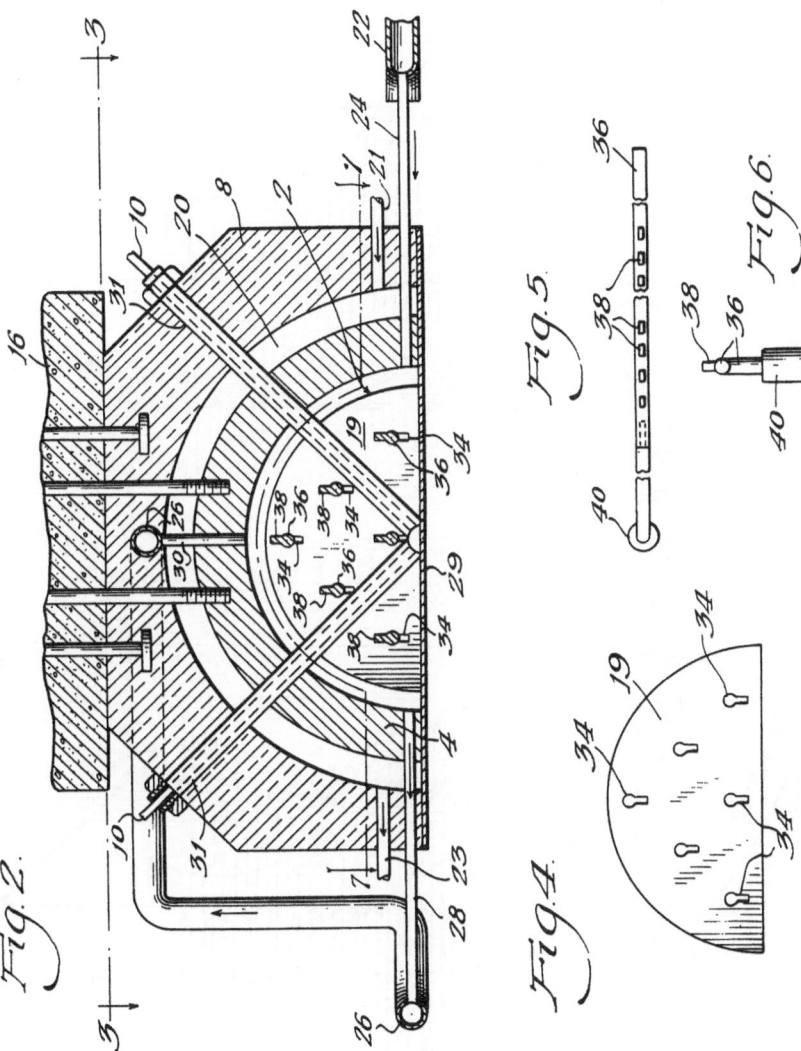

INVENTORS:
Harry Soodak
BY Eugene P. Wigner

Roland A. Anderson
Attorney:

July 25, 1961

H. SOODAK ET AL

FAST NEUTRON REACTOR

2,993,850

Filed June 14, 1948

6 Sheets—Sheet 3

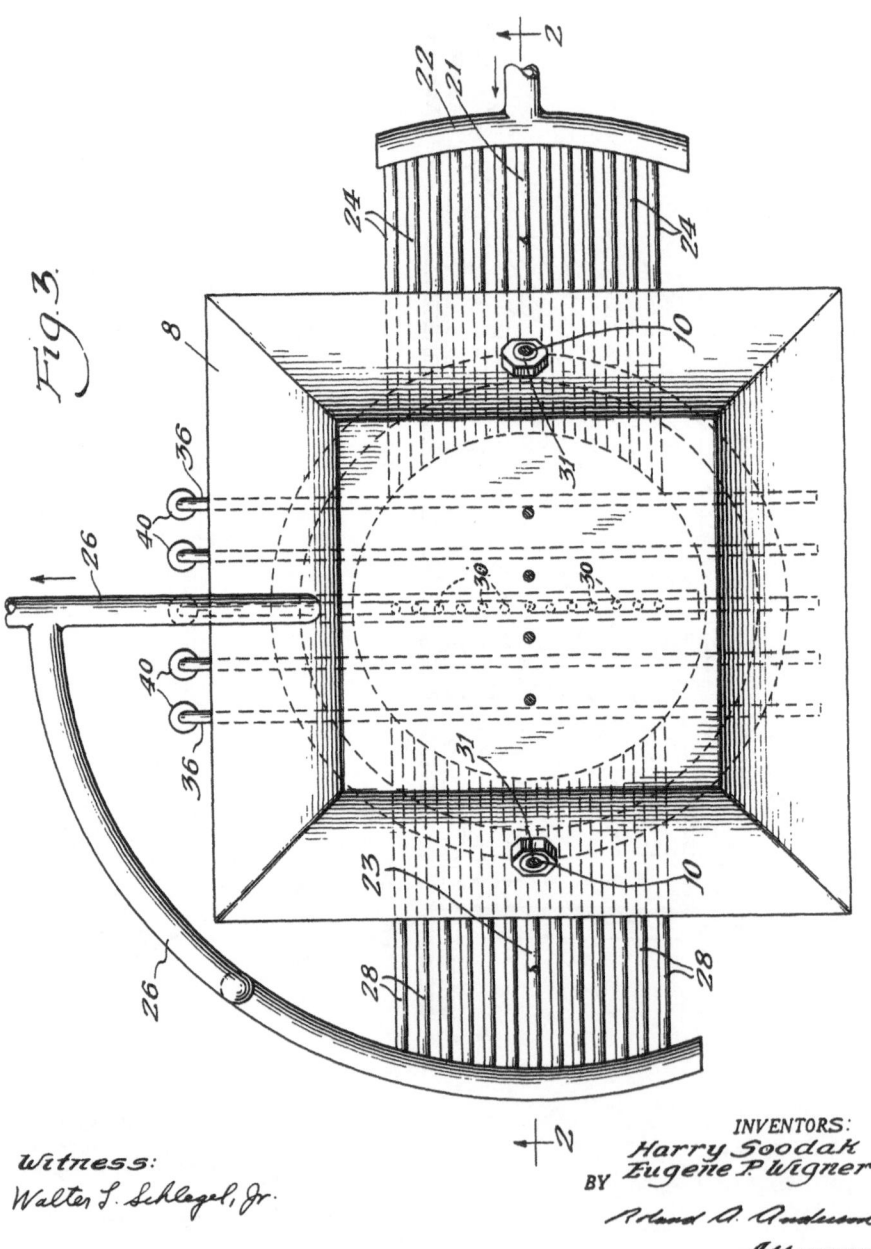

Fig. 3.

INVENTORS:
Harry Soodak
Eugene P. Wigner
BY
Roland A. Anderson
Attorney:

July 25, 1961   H. SOODAK ET AL   2,993,850

FAST NEUTRON REACTOR

Filed June 14, 1948   6 Sheets—Sheet 4

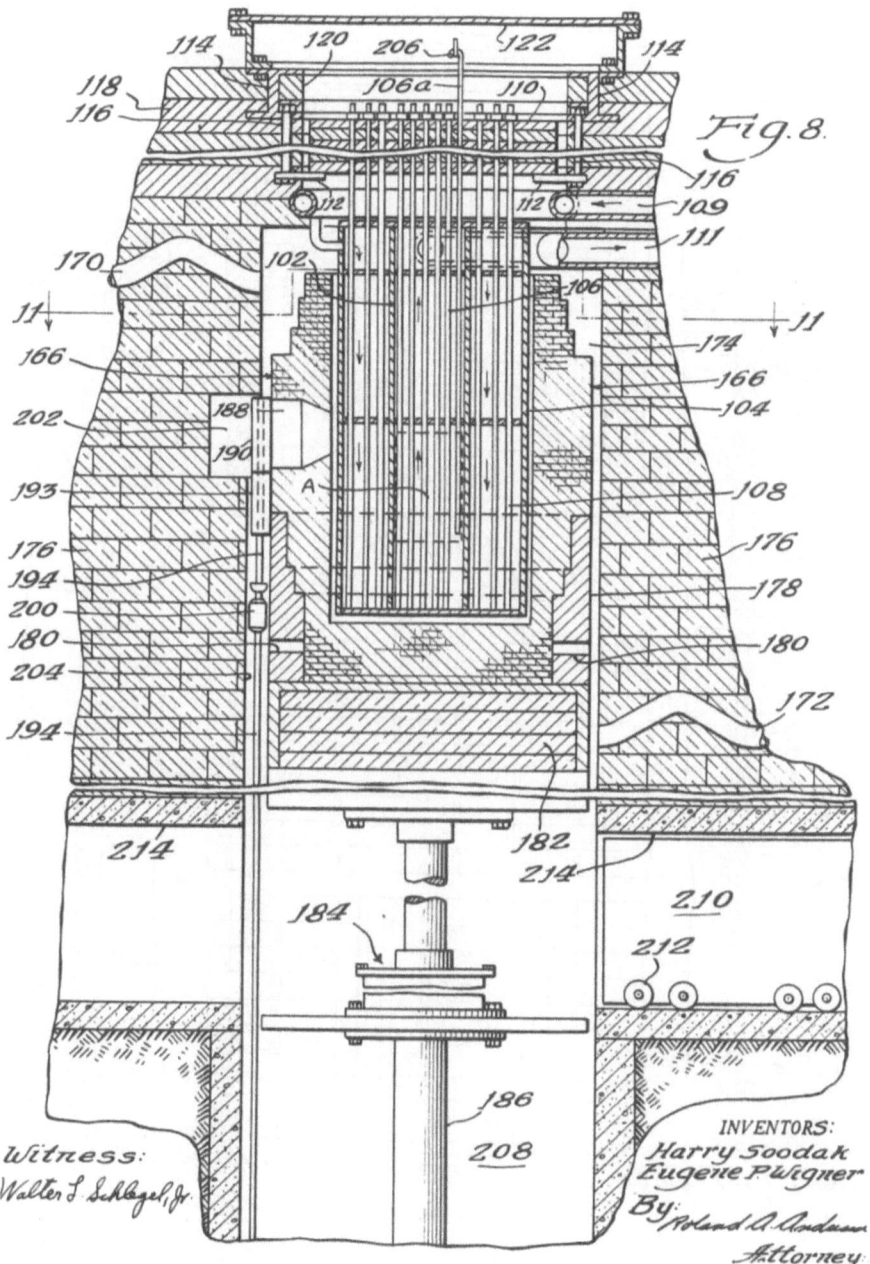

Fig. 8.

INVENTORS:
Harry Soodak
Eugene P. Wigner
By Roland A. Anderson
Attorney

July 25, 1961   H. SOODAK ET AL   2,993,850
                       FAST NEUTRON REACTOR

Filed June 14, 1948   6 Sheets-Sheet 5

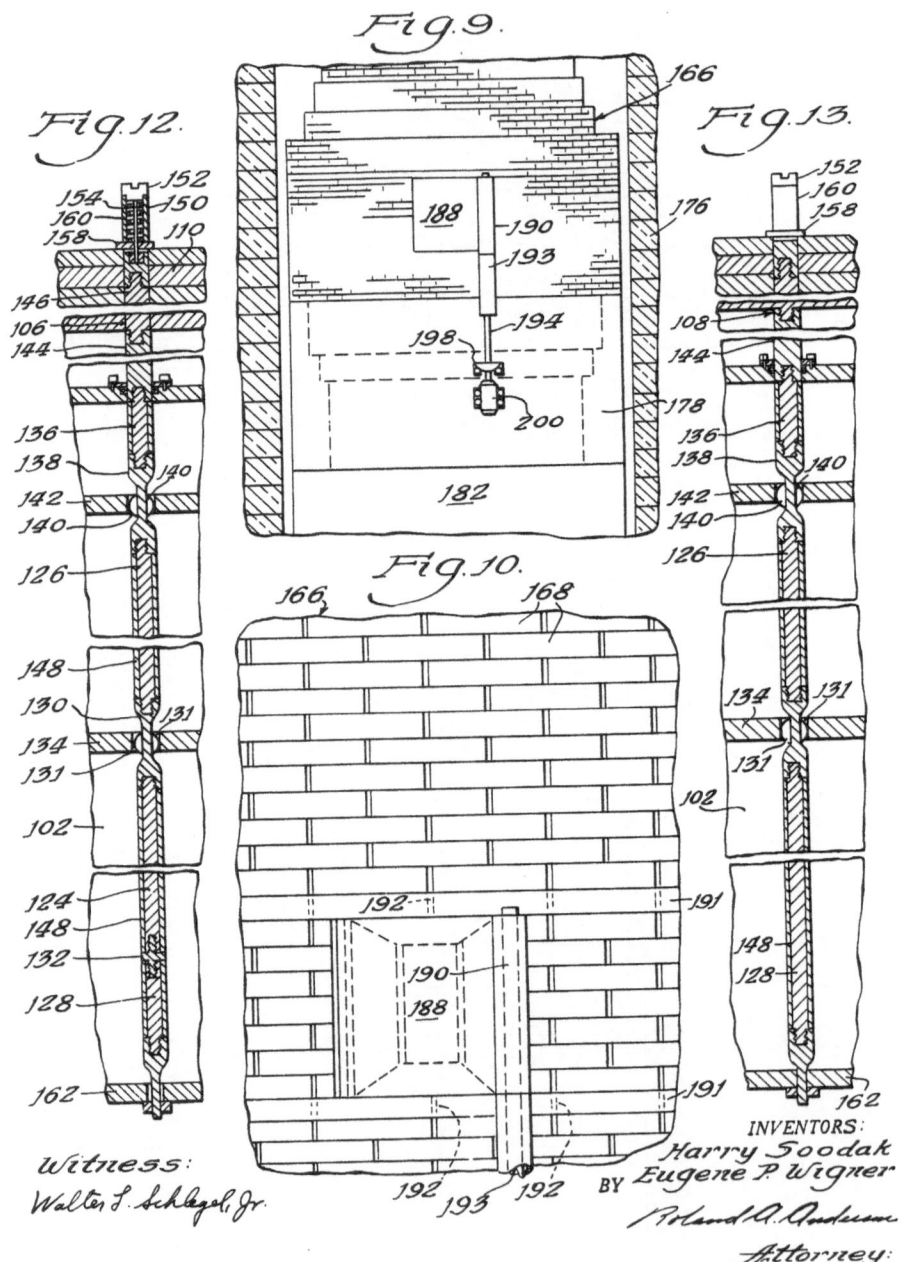

Fig. 9.

Fig. 12.

Fig. 13.

Fig. 10.

Witness:
Walter L. Schlegel, Jr.

INVENTORS:
Harry Soodak
Eugene P. Wigner
BY
Roland A. Anderson
Attorney:

July 25, 1961   H. SOODAK ET AL   2,993,850
                    FAST NEUTRON REACTOR

Filed June 14, 1948   6 Sheets—Sheet 6

Fig.11.

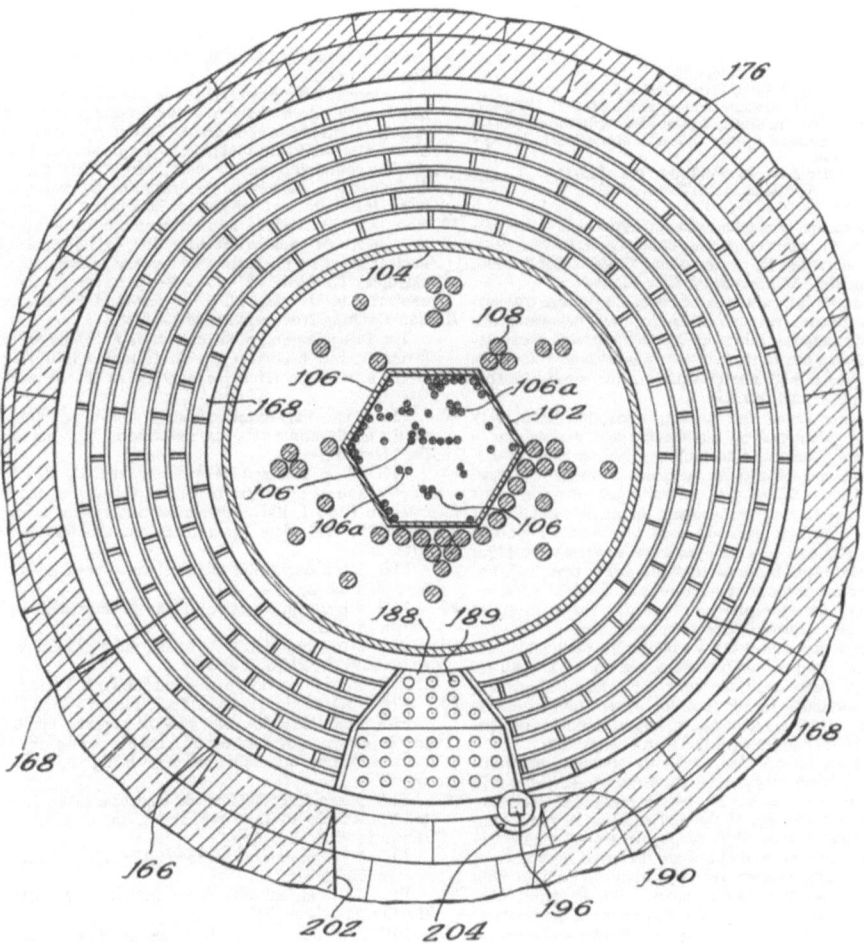

INVENTORS:
Harry Soodak
BY Eugene P. Wigner

Roland A. Anderson
Attorney:

Witness:
Walter S. Schlegel, Jr.

United States Patent Office

2,993,850
Patented July 25, 1961

1

2,993,850
FAST NEUTRON REACTOR
Harry Soodak, Cambridge, Mass., and Eugene P. Wigner,
Princeton, N.J., assignors to the United States of America as represented by the United States Atomic Energy
Commission
Filed June 14, 1948, Ser. No. 32,824
6 Claims.   (Cl. 204—193.2)

This invention relates to nuclear physics and more particularly to fast neutron nuclear fission chain reactors such as those described in a copending Szilard application, Serial No. 698,334, filed September 20, 1946.

As is more fully discussed in said copending application, fast neutron reactors are particularly advantageous for certain purposes due to their small size and compactness, and also due to the fact that relatively few neutrons are absorbed at high energy values in the non-fissionable components of such reactors.

It has been found that neutron absorption losses may be greatly minimized by establishing and maintaining a self-sustaining nuclear fission chain reaction while avoiding the slowing of evolved neutrons below an average energy of about 25,000 ev. At such high energies, it has been discovered that the elements of atomic number of 11 to 83, which are generally used as structural, cooling, or other elements in a reactor, have neutron absorption cross sections, which are substantially lower than their absorption cross sections for neutrons at thermal energies. In contrast, the absorption cross section for neutrons at high energies of certain heavy isotopes, such as Th^{232} or U^{238}, which are converted to thermally fissionable isotopes upon neutron bombardment, are not so greatly reduced; and hence their tendency for neutron absorption is correspondingly higher than for lower atomic number elements. Thus, a substantial saving of neutrons may be effected by maintenance of the high energy level.

Similar advantages may accrue by operating neutronic reactors at lower energies, as for example, as low as 10 ev. or even as low as 0.3 ev., which energies are substantially above the energy of thermal neutrons at room temperature, that is about 0.03 ev. However, the higher energies of 25,000 ev. and above are preferred, inasmuch as non-moderating neutron reflectors may be utilized with reactors operating at these values. For example, U^{238} and Th^{232} may be utilized for this purpose, particularly in view of the fact that neutron absorption occurring in these materials results in the production of thermally fissionable materials, as for example, 94^{239} from U^{238} and U^{233} from Th^{232}. Such materials are called fertile materials.

Materials like U^{238} when principally composed of a single chemical element, substantially all of the atoms of which having the same mass number, are generally known as isotopes. Thus, chemically pure natural uranium comprises mostly two isotopes, U^{238} and U^{235}, in the atomic ratio of about 139 to 1. As a further example, natural occurring thorium consists exclusively of the single isotopes Th^{232}.

A general object of the present invention is to design an effective fast neutron reflector disposed around a fast neutron reactor to diminish neutron losses from the periphery thereof, thereby making possible a reactor of

2

relatively small size particularly suitable for the production of power.

A more specific object of the invention is to provide a novel composite neutron reflector around a fast neutron reactor, said reflector comprising an outer portion of neutron moderator such as graphite or beryllium and an inner portion containing one or more "fertile" isotopes which are capable of scattering fast neutrons by inelastic collision therewith and which are fissionable by fast neutrons, are substantially non-fissionable by slow neutrons, and absorb or capture neutrons slow or fast to undergo nuclear reaction productive of fissionable material. For example, Th^{232} and U^{238} are fertile isotopes which are converted to U^{233} and Pu^{239} respectively by nuclear reaction resulting from neutron absorption.

The foregoing and other objects and advantages of the invention will become apparent from the following description wherein reference is made to the drawings in which:

FIG. 1 is a diagrammatic central vertical sectional view partly in elevation of a fast neutron reactor embodying the invention;

FIG. 2 is an enlarged fragmentary sectional view partly in elevation of the upper half of a pile such as that indicated in FIG. 1, illustrating a coolant system associated therewith, said view being taken on the line 2—2 of FIG. 3;

FIG. 3 is a horizontal sectional view taken on the line 3—3 of FIG. 2;

FIG. 4 is a side elevation of one of the reactive plates shown in FIG. 2;

FIG. 5 is a top plan view of one of the plate retainer rods shown in FIGS. 2 and 3;

FIG. 6 is an end view of the rod shown in FIG. 5;

FIG. 7 is a sectional view taken on line 7—7 of FIG. 2;

FIG. 8 is a side view of a modified reactor embodying the invention, portions of this reactor being shown in vertical section and other portions being shown in side elevation;

FIG. 9 is a fragmentary side elevation taken from the left as seen in FIG. 8, portions of the structure being shown in vertical section;

FIG. 10 is an enlarged fragmentary side elevation corresponding to FIG. 9;

FIG. 11 is an enlarged cross sectional view taken on the line 11—11 of FIG. 8; and

FIGS. 12 and 13 are fragmentary vertical sectional views illustrating respectively a reactive rod and a reflector rod shown in FIGS. 8 and 11.

Describing the invention in detail and referring first to the embodiment thereof illustrated in FIG. 1, the fast neutron reactor is generally designated 1 and comprises a reactive portion 2 containing plutonium or plutonium diluted with natural uranium. The reactive composition 2 is surrounded by a generally spherical reflector 4 formed of natural uranium of high neutronic purity which is principally composed of U^{238}, or other compositions comprising a fertile isotope having the characteristics heretofore set forth.

As illustrated in FIG. 1 the reactor is in the form of two hemispheres, each comprising a hemispherical core of the reactive composition 2 and an outer hemispherical reflector or blanket 4 of uranium contained within re-

2,998,850

3

flector 8 of neutron moderator such as beryllium or graphite preferably of high neutronic purity. Control of the nuclear fission chain reaction may be effected by the movement of the two hemispheres toward and away from each other, the maximum neutron reproduction ratio being obtained when the diametral faces of the hemispheres are disposed in juxtaposition. For adjustment of minor variations in the neutron reproduction ratio, suitable control rods 10 of neutron moderating material such as graphite or beryllium may be provided, the rods 10 being actuated by any conventional means such as rack and pinion mechanisms 12. It will be understood, as hereinafter discussed in detail, that rods 10 are inserted to increase the neutron reproduction ratio and are withdrawn to decrease the ratio or to terminate the reaction.

It may be noted that in the illustrated example of FIG. 1, the reactive composition 2 and the reflector 4 are generally spherical in form. However, if desired, other suitable shapes may be utilized. The spherical structure illustrated requires the smallest mass of plutonium to sustain the chain reaction. No moderator as such is used in the reactive core, although the neutrons may be moderated to some extent by the coolant and by other materials in the reactor as hereinafter discussed. The reactive structure is preferably enclosed within a concrete vault or shield 16 or any suitable material such as concrete, adapted to absorb biologically harmful emanations such as neutrons and alpha, beta and gamma rays, said vault affording support for the upper hemisphere of the reactor 1 by means of the bolts 15, and for a plurality of hydraulic jacks 18 which, in turn, support the lower hemisphere of the reactor 1 and are adapted to actuate the same upwardly and downwardly to afford the before-mentioned control of the chain reaction.

Referring next to FIGS. 2 and 3, the reactor diagrammatically illustrated in FIG. 1 is shown in some detail together with an associated coolant system adapted to absorb heat developed by the nuclear fission chain reaction, and to convey said heat from the reactor for the production of power.

FIGS. 2 and 3 illustrate the upper half of the generally spherical reactor 2, and it will be readily understood from a consideration of FIG. 1 that the lower half of the reactor is generally similar to the upper half illustrated in detail. The upper half of the reactor comprises a plurality of flat semi-circular plates 19 which are preferably of a reactive composition such as natural uranium enriched with Pu^{239} or U^{235}, and are about $1\frac{1}{2}$ to 2 millimeters in thickness. The plates 19 are preferably enclosed by a metal jacket one-quarter millimeter thick in thermal contact therewith, to protect them from the action of the associated coolant. For example, if the coolant utilized is liquid sodium or liquid bismuth, a steel jacket or coating is satisfactory for the plates 19. Instead of plates, rods and other forms of solids with coolant tubes therethrough could be used as the shape makes little difference, but a high ratio of surface area to volume is effective and desirable from the cooling standpoint inasmuch as a large surface area for given volume provides not only better contact for the coolant and better heat dissipation but also reduces internal temperature and thermal stress of the plates. Since the heat conductivity of plutonium is around 0.015 calorie per second per centimeter per degrees centigrade, it is desirable to use thin plates so as to obtain a large surface area for exposure to the coolant. The plates 19 are supported as hereinafter described on suitable rods 36 which may be of uranium or steel and are arranged in upright position and spaced apart from each other horizontally from about 1.5 to 2 millimeters so as to afford passage of coolant therebetween. The plates 19 are semicircular discs in form with progressively smaller diameters so that when assembled, they provide a hemispherical core of plutonium or uranium-plutonium reactive com-

4

position. As mentioned, the pile does not require a moderator as such. Instead each hemispherical assemblage of plates is mounted in the uranium reflector 4 of hemispherical shape, concentric with the associated assembled plates. In the form illustrated, the uranium reflector is preferably about 30 centimeters thick and provides the container for the liquid metal coolant which is circulated among the plates, the reflector itself being cooled thereby from the inside for about half its thickness.

The uranium reflector 4, in turn, is surrounded by a chamber 20 through which a coolant such as liquid sodium or, with suitable sealing in the defining surfaces of the chamber 20, such sealing means being well-known in the art, water is circulated by inlet and outlet conduits 21 and 23 for cooling the outer portion of the reflector. The liquid metal coolant flowing between plates 19 is primarily to distribute the heat and to prevent damage to the plates and to remove the heat from the reactor. The main consideration is that it must not be moderating to any appreciable degree, hence such materials as heavy water or fluorocarbon compositions may not be used. However, non-neutron-absorbent liquid metals and gases are satisfactory. In the form illustrated, allowance is made for about $1\frac{1}{2}$ to 2 millimeters of space between adjacent plates thus providing for the passage of a stream of liquid metal coolant 1 and $\frac{1}{2}$ millimeters thick between each pair of adjacent plates. The coolant should have a heat capacity of around 0.3 calorie per cubic centimeter per degree centigrade and should be capable of reaching a temperature of about 400° C. higher than the coolant inlet temperature. Under such conditions it should be circulated between the plates 19 at a rate of about 10 meters per second and its absorption cross section per cubic centimeter should not exceed about $\frac{1}{4}$ that of natural uranium or about $\frac{1}{80}$ that of plutonium.

It has been found that the liquid form of bismuth, or sodium, or a bismuth-lead eutectic composition circulated at the rate of 10 meters per second provides satisfactory cooling without too great an amount of neutron moderation or absorption. In order to circulate the coolant, a suitable inlet header 22 (FIGS. 2 and 3) is provided from which a plurality of pipes 24 extend through the graphite reflector 8, the chamber 20, and the uranium reflector 4. Liquid metal coolant is supplied through the reactor 2 by the pipes 24 at the velocity desired. In order to effect better distribution of the coolant, the pipes 24 preferably discharge respectively between adjacent plates 19. A discharge header 26 is provided and connected by suitable horizontal pipes 28 and vertical pipes 30 with the interior of the reflector 4 containing the plates 19. A similar cooling system is provided for each of the hemispherical halves of the pile. Because the lower portion of the reactor is movable, it is necessary to introduce flexible connections (not shown), for example steel bellows, in the pipes communicative with headers 22 and 26 and conducting the metal coolant from the reactor to the heat exchanger system and back. Alternatively flexible couplings may be incorporated in tubes 24 and 28 adjacent the reactor and between it and the headers, such flexible connections being required, of course, only for the lower hemisphere of the reactor to accommodate the up and down displacement thereof.

The coolant is circulated by a suitable pump (not shown) connected in a closed system with the headers, a boiler (not shown) being interposed to use the generated heat for useful power. Since the pile is made in the form of two hemispheres instead of a single sphere, for purposes of control later to be described, it is necessary to seal the diametral face of each hemisphere. For this purpose a cover plate 29 (FIG. 2) is provided for each hemisphere. The plate 29 is preferably formed of steel.

The before-mentioned control rods 10 extend through the reflector 4 into the reactor 1 through suitable wells

2,993,850

5

br tubes 31 preferably formed of steel. It may be noted in this connection that the reactor 2 is preferably constructed slightly smaller than the critical size thereof, whereat a chain reaction is produced. Under these conditions the neutron reproduction ratio of the reactor is less than unity. When the control rods 10 are moved inwardly into the reactor, some of the neutrons within the reactive composition 2 are slowed or moderated, thus increasing the effectiveness thereof to fission the plutonium or U^{235}, inasmuch as the fission cross section thereof increases with a decrease in the energy of neutrons. This causes the neutron reproduction ratio of the reactor to rise to a value greater than unity where it may be held until the desired neutron density is attained, at which time the rods may be withdrawn until the ratio is approximately unity. The neutron density may thus be regulated and may be monitored by any conventional means not shown.

If desired, the rods 10 may be formed of fissionable material rather than the neutron moderator. Thus, by inserting the rods into the reactor, the neutron reproduction ratio thereof may be increased to a value greater than unity inasmuch as the fissionable material in the rods is effective to increase the mass of the reactor 2 within the reflector 4 to a size greater than the critical size at which the reactor is capable of sustaining a chain reaction.

For coarse control, the lower portion of the reactor 2 may be displaced relative to the upper hemisphere. When the two parts are separated, neutrons emanating from one part, which would normally enter the other, fail to do so because of escape through the gap between the hemispheres. When the two parts are placed in juxtaposition, this excess leakage or escape of neutrons is reduced so that a larger proportion of those neutrons emanating from one part will enter into the chain reaction in the other hemisphere. In this manner the reproduction ratio may be changed to a value below unity when the hemispheres are separated a maximum amount, to above unity when they are placed close together. As an example, in FIG. 1, this is accomplished by lowering and raising, respectively, the lower portion of the retractor by means of the hydraulic jacks shown schematically at numeral 18. Any other of a plurality of equivalent means well-known in the art could likewise be used, the selection of hydraulic jacks having been made for purposes of illustration only.

As will be clearly seen in FIG. 4, each plate 19 is provided with a plurality of key shaped holes or openings 34, the openings in the respective plates being aligned to accommodate reception of support rods 36, one of which is shown in FIGS. 5 and 6. The rod 36 is formed with a plurality of lugs 38 which are disposed downwardly as the rod is inserted through the openings 34 in the plates 19. The rod 36 is then rotated so that the lugs 38 are disposed upwardly as best seen in FIG. 2, so that the lugs afford spacers for the purpose of maintaining the plates 19 in proper spaced relationship. It will be understood that each rod 36 is supported by complementary openings in the graphite reflector 8, the outer end of each rod being provided with an enlarged portion 40 functioning as a counterweight to maintain the rod 36 in the position illutsrated in FIG. 2 whereat the lugs 38 are disposed upwardly to space the plates 19.

In the above described type of fast reactor or pile it is desirable to dilute the plutonium of the reactive core with natural uranium. This dilution makes it possible to operate the reactor at a higher rate of power production in kilowatts per kilogram of plutonium because the heat conductivity of the plates 19 is increased and the heat production per unit volume of metal in the core may be maintained at a high value consistent with the ability of the coolant to extract the heat.

Table I illustrates how the reactor volume, critical

6

weight of plutonium power obtainable, breeding ratio, and gain in plutonium vary for a cylindrical type of pile for different plutonium-natural uranium compositions, hereinafter referred to as fission metal. The values in Table I are based on reactor plates of 1.5 mm. thickness, a 0.25 mm. thick protective metal jacket enclosing said plates, a space of 1.5 mm. between plates for circulating a liquid metal coolant and a reflecting-absorbing layer of natural uranium of about 30 cm. thickness surrounding the reactor core.

TABLE I

Ratio of Uranium Atoms to Plutonium Atoms in Reactor Plates	Reactor Volume in Liters¹	Critical Weight of Plutonium in Kilograms	Power Obtainable in Kilowatts Per Kilogram of Plutonium	Breeding Ratio. Plutonium Atoms Gained Per Plutonium Atom Consumed	Gain of Plutonium in Grams Per Day
0	2.4	22	400	0.8	6
1	9	40	700	0.7	18
4	70	125	1,600	0.6	100
10	>1,200	>1,000	1,700	0.4	7,600

¹ This is the sum of the volumes of fission metal, protective jacket thereon, and coolant, within the space enclosed by the natural uranium reflector-absorber.

In calculating the values of reactor volume and critical weight of plutonium in the reactor shown in Table I a "two group" method was used. The "two group" method takes into account fast neutrons, i.e. neutrons of above about 1 mev. energy, and slow neutrons, i.e., neutrons of below about 1 mev. energy. In addition the average neutron energy within the reactor is taken to be about 100,000 ev. and account is taken of the neutrons contributed by fissioning of the uranium 235 and uranium 238 isotopes of the natural uranium used as the diluent and in the reflector-absorber layer.

The values for power obtainable given in the fourth column of Table I, are based on using a liquid metal coolant having a heat capacity of about 0.3 calorie per cubic centimeter per degree centigrade. The coolant is assumed to flow in the coolant channels at a linear velocity of 10 meters per second. The maximum temperature of the fission metal is assumed to be about 400° C. higher than the inlet temperature of the coolant.

The breeding ratio values given in the fifth column of Table I are based on the nuclear properties of plutonium and uranium and take into account neutrons lost by leakage from the reactor as well as those neutrons lost by capture in the coolant and in the metallic protective jacket enclosing the fission metal. It is assumed that the coolant has an absorption cross section per cubic centimeter which is about ¼ that of natural uranium, and that the absorption cross section per cubic centimeter of the jacket metal is about ½ that of natural uranium.

Metallic sodium with a melting point of about 100° C. and having a heat capacity of about 0.27 calorie per cubic centimeter per degree centigrade may be used as a liquid metal coolant. Other suitable coolants are sodium-potassium alloys, metallic bismuth and lower melting lead-bismuth alloys.

Table II sets forth critical values for a fast reactor of spherical shape, wherein the volume of fission metal is equal to the combined volume of coolant and jacket metal employed to protect the fission metal. In this case the reactor space which is the sum of the volumes of fission metal, the coolant and the protective jacket is surrounded by a 10 centimeter thick reflector-absorber layer of natural uranium. The coolant and jacket metal are assumed to be pure neutron scatterers, having no neutron absorption and causing no neutron moderation.

2,993,850

7

TABLE II

Ratio of Uranium Atoms to Plutonium Atoms in Reactor Plates	Reactor Volume in Liters	Critical Weight of Plutonium in Kilograms	Power Obtainable in Kilowatts Per Kilogram of Plutonium [1]
0	2.2	20	200
1	7.6	35	500
4	49	91	1,200
10	458	390	1,700
20	(2)	(2)	(2)

[1] Using liquid sodium metal as the coolant, circulated at an average linear rate of 10 meters per second in a channel 2 mm. wide between reactor plates, which are 2 mm. thick and covered by a protective metal jacket ¼ mm. thick. In calculating the values in this column the effect of the protective jacket has been neglected and it is assumed that there is no neutron absorption in the sodium. The temperature of the liquid sodium is taken as 100° C. at the inlet and the maximum temperature allowable in the reactor plates is taken as 579° C.

[2] Not chain reacting.

The values given for the critical volume and critical weight as set forth in Table II have been calculated by the "two group" method as previously described in connection with Table I.

The manner in which the power obtainable varies with the average linear velocity of the coolant for a spherical type of pile (FIGS. 2 and 3, and Table II) is shown in Table III for the case of a 2 mm. thick fission metal plate protected by a ¼ mm. thick metal jacket and a 2 mm. width cooling stream employing a liquid metal coolant such as sodium.

TABLE III

Ratio of Uranium Atoms to Plutonium Atoms in Reactor Plates	Power Produced in Kilowatts Per Kilogram of Plutonium for Varying Velocities of Coolant in Meters Per Second				
	2.5 m./sec.	5 m./sec.	10 m./sec.	20 m./sec.	∞ m./sec.
0	160	190	210	220	230
1	320	430	520	580	660
4	550	860	1,200	1,500	2,000
10	590	1,000	1,700	2,400	4,500

Tables IV and V give the same type of information as Table III except that in Table IV the fission metal plates are 1 mm. thick and the coolant stream is 1 mm. wide, while in Table V the fission metal plates are 3 mm. thick and the coolant stream is 3 mm. wide.

TABLE IV

Ratio of Uranium Atoms to Plutonium Atoms in Reactor Plates	Power Produced in Kilowatts Per Kilogram of Plutonium for Varying Velocities of Coolant in Meters Per Second				
	2.5 m./sec.	5 m./sec.	10 m./sec.	20 m./sec.	∞ m./sec.
0	320	480	640	750	930
1	500	850	1,300	1,700	2,700
4	680	1,300	2,200	3,500	8,200
10	650	1,300	2,300	4,100	18,000

TABLE V

Ratio of Uranium Atoms to Plutonium Atoms in Reactor Plates	Power Produced in Kilowatts Per Kilogram of Plutonium for Varying Velocities of Coolant in Meters Per Second				
	2.5 m./sec.	5 m./sec.	10 m./sec.	20 m./sec.	∞ m./sec.
0	80	90	95	98	100
1	200	240	260	280	290
4	410	570	700	790	910
10	500	800	1,100	1,400	2,000

It may be seen that for fast neutron reactors of the type above described, the specific power obtainable in kilowatts per kilogram of plutonium is higher when the

8

plutonium is diluted with natural uranium in the range of between about 4 to 10 atoms of uranium to 1 atom of plutonium, when the thickness of the fission metal plate is between about 1 mm. to 2 mm., and when the velocity of the stream of liquid metal coolant is between about 10 to 20 meters per second.

The specific power obtainable for a given rate of flow of a given coolant using plates of a given thickness of the fission metal may be varied by utilizing cooling channels that are not the same in width as the thickness of the plates. In general for coolant channels that are less in width than the thickness of the fission metal plates, the ratio of coolant volume plus protective metal jacket volume to fission metal volume will be less than 1, and the specific power obtainable will be higher than when this ratio is 1. Conversely, in the case of using coolant channels that are wider than the thickness of the fission metal plate so that the aforementioned volume ratio is greater than 1, the specific power obtainable will be less than when this ratio is 1.

In general for this type of fast neutron reactor the breeding ratio varies from about 0.9 for reactors employing undiluted plutonium in plates of about 1 mm. thickness to a value of about 0.4 for reactors employing a fission metal composition comprising plutonium and natural uranium of an atomic ratio of about 1 to 10 in the form of plates of about 3 mm. thickness. This type of fast neutron reactor produces useful power and in addition makes it possible to breed plutonium at a useful rate. For example, examination of the data set forth in the third line of Table I shows that the reactor has a power rating of 187,500 kilowatts and that the time required to double the investment of plutonium in the reactor is 1250 days at full operating power.

While the above invention is described and illustrated as employing uranium and plutonium, it is apparent that fissionable isotopes other than plutonium or a plutonium-uranium mixture may be used as the reactive composition of the core of the reactor and other fertile isotopes than uranium U^{238} may be used as the reflector-absorber. As an additional example, but not in limitation of the various isotopes which may be used, the central core could be comprised of U^{233} as a reactive composition and thorium could be used in the absorber reflector. In such a case there would be a production of U^{233} from the thorium at a rate greater than the rate of consumption of U^{233}.

Referring now to FIG. 8, the system shown therein comprises inner and outer steel tanks 102 and 104, the inner tank containing a plurality of composite rods 106 and the outer tank containing a plurality of composite rods 108, all of said rods being supported, as hereinafter described in detail, from a biological shield 110 composed of any suitable material adapted to absorb biologically harmful emanations, such as neutrons and alpha, beta, and gamma rays.

The shield 110 is supported by fingers 112 connected to I beams 114 as by bolts 116. The beams being mounted within a biological shield 118 with a central opening 129 accommodating the before-mentioned shield 110. The top of the opening 120 is closed by a cover plate 122, which may be removed to accommodate assembly and disassembly of the rods 106 and 108.

One of the rods 106 is shown in detail in FIG. 12 and comprises a cylindrical segment 124 composed of thermally fissionable material. It is disposed between cylindrical segments 126 and 128 composed principally of fertile material. The segment 124 is connected to the segments 126 and 128 by steel couplings 130 and 132, respectively, the coupling 130 being provided with spaced fins 131 adapted to center the rod 106 in an opening through a wall or partition 134 within the tank 102. The segment 126 is connected to a cylindrical beryllium segment 136 by a coupling 138 formed with fins 140 adapted to center the rod 106 in an opening within a

9

wall 142 of the tank 102. The beryllium segment 136 is connected to an iron segment 144, which is, in turn, connected to another beryllium segment 146. The beryllium segments 136 and 146 are disposed within a biological shield 110 and form a part thereof. All of the segments are closed within thin walled tubes or sheaths 148 adapted to space the segments from a coolant circulated through the system, as hereinafter described, for the purpose of absorbing the heat of nuclear fission chain reaction.

The rod 106 is supported from the shield 110 by a spring 150 bearing against a cap 152 connected by a pin or spindle 154 at the top beryllium segment 146, said spring being seated against a spring plate 158 bearing against the top of the shield 110. The cap carries a sleeve 160 enclosing the spring 150. The spring 150 functions to place the rod 106 under tension inasmuch as the bottom thereof is interlocked to the bottom wall 162 of the tank 102. Thus, any tendency of the rod to warp due to the heat of the reaction is resisted by the tension on the rod.

The rods 106 are spaced apart .046" and are disposed about .494" apart, center to center. The diameter of each segment is about .394" and the outside diameter of the jacketed segment is about .448". About 170 active rods 106 are utilized in an arrangement wherein the thermally fissionable material is substantially pure U^{235}. The inner tank 102, which is hexagonal is shown in FIG. 11, and is about 6.688" in width between parallel sides thereof.

One of the rods 108 is shown in detail in FIG. 13 and is substantially identical with the rod 106 except for the larger diameter of the rod 108 and the elimination of the thermally fissionable segment 124. The portions of the rod 108 (FIG. 13) corresponding to those of the rod 106 (FIG. 12) are identified by corresponding numerals, and it will be seen that the bottom segment of the rod 108 containing fertile material is longer than the corresponding segment 128 of the rod 106 because of the elimination of the thermally fissionable segment 124 from the rod 108. The rods 108, as above-noted, are contained within the tank 104, and are preferably about .875" in diameter. They are spaced apart about .916" center to center, thus providing a space of about .041" between the closest portions of adjacent rods. The jackets 148 of the rods 108 as well as the rods 106 have a wall thickness of .027". In the instant embodiment of the invention one-hundred and ninety-two rods 108 are utilized.

The rods 106 and 108 are cooled by any suitable coolant such as molten bismuth or a sodium potassium alloy circulated as indicated by the arrows in FIG. 8 by inlet and outlet conduits 109 and 111. The inlet conduit is connected to the outer tank 104 and the coolant flows downwardly therethrough into the tank 102 through ports at the bottom thereof and upwardly therethrough into the outlet conduit 111. Thus, the coolant absorbs the heat of the nuclear fission chain reaction and the energy in the form of heat carried from the reactor by the coolant may be utilized for power or other purposes, if desired.

Referring again to FIG. 8, it will be seen that the thermally fissionable segments 124 of the rods 106 constitute a reactive portion indicated at A, wherein the nuclear chain reaction is sustained, said reactive portion being entirely surrounded by fertile material within the rods 106 and 108. This blanket of fertile material surrounding the reactor A not only absorbs neutrons to produce thermally fissionable material, as above discussed, but also functions as a fast neutron reflector, which reflects or scatters fast neutrons escaping from the reactor A by inelastic collision with the atoms of fertile material, thus reflecting the escaping neutrons back into the reactor to diminish neutron losses from the periphery thereof; thereby decreasing the critical size at which said reactor is capable of sustaining a chain reaction.

The reflector defined by the fertile material in the rods

10

106 and 108 is surrounded, except for the top thereof, by an outer reflector 166 (FIG. 8) composed principally of fertile material, as for example, bricks or blocks 168 (FIGS. 10 and 11) of neutronically pure natural uranium, the principal component of which is U^{238}. The blocks 168 are arranged in annular rows or series spaced radially from the longitudinal axis or center line of the tanks 102 and 104, and the blocks of each row or series are spaced apart, as best seen in FIG. 11. Each row of blocks, as best seen in FIGS. 8 and 10, comprises a plurality of horizontal layers, the spaces between the blocks of the respective layers being vertically staggered or misaligned, as seen in FIG. 10. Thus, the reflector 166 is provided with a multiplicity of passages defined by the space between the blocks to afford passage of a coolant such as air or helium under pressure circulated by inlet and outlet conduits 170 and 172. This coolant passes through a reaction chamber 174 containing the tanks 102 and 104 and the reflector 166, said chamber being defined by a neutron moderating reflector or blanket 176 formed, for example, of blocks of graphite or beryllium.

The reflector 166 is formed with a smaller number of the radially spaced rows of blocks 168 at the top and bottom of the reflector than at the center thereof, inasmuch as the neutron density is normally greatest in the reflector at that point. Thus, the blocks 168 are supported by a steel ring 178, the inner surface of which is stepped to accommodate the above-described arrangement of the blocks. The ring 178 is provided with one or more coolant passages 180 to accommodate flow of the before-mentioned reflector coolant between the bricks 168 constituting the bottom of the reflector 166. The ring 178 is supported by a biological shield 182 (FIG. 8) mounted on a pneumatic or hydraulic hoist 184, which is, in turn, supported by a pedestal or column 186 adapted to be raised and lowered by any suitable device, such as an hydraulic or screw hoist (not shown) for a purpose hereinafter discussed.

The nuclear fission chain reaction within the reactor A is normally controlled by a door 188 in the outer reflector 166, said door preferably being principally composed of fertile material, as for example, a block of neutronically pure natural uranium. The door is bevelled on the inner surface thereof, as clearly seen in FIGS. 8 and 11, to define angular paths with the margins of the complementary opening through the reflector 166, thus substantially preventing the escape of neutrons through said opening when the door is in its closed position as is illustrated in the drawings. The door is preferably provided with a plurality of vertical openings 189 therethrough to accommodate flow of the before-mentioned coolant, thus preventing the development of excessive heat within the door due to neutron bombardment thereof. Referring to FIG. 10, it will be seen that the blocks 168 comprising the layers immediately above and immediately below the door 188 are replaced by a pair of annular uranium plates 191, provided with openings or perforations 192 adapted to accommodate flow of the reflector coolant therethrough, said plates 191 affording a good structural support for the blocks above and at the sides of the opening through the reflector 166.

The door is provided with a hub 190 seated on a sleeve or bearing 193 mounted in any convenient manner, as for example, by welding on the ring 178 adjacent the upper edge thereof. A shaft 194 (FIGS. 8 and 9) extends through the bearing 193 and is provided with a square end 196 (FIG. 11) fitted within a complementary opening in the hub 190. The shaft extends downwardly from the bearing 193 through a conventional roller bearing unit 198 mounted on the ring 178, and the shaft is connected at its lower extremity to a motor 200 adapted to rotate the shaft, thus actuating the door to the open and closed positions thereof. The reflector 176 is provided with a recess 202 and a vertical groove 204. The recess 202 is adapted to receive the door as the latter is

2,993,860

11

actuated to its open position, and the groove 204 is adapted to accommodate vertical movement of the hub 190 and bearing 193 during the elevation and lowering of the reflector 166, as hereinafter described.

It has been found that when the door 188 is open, fast neutrons escaping therethrough are scattered through the reaction chamber 174 and are slowed by elastic collision with the atoms of neutron moderator constituting the reflector 176. Thus, the slowed neutrons are unable to penetrate the fertile material constituting the inner and outer reflectors around the reactor A, and are, for the most part, absorbed by this fertile material or by the wall of the outer tank 104. This loss of neutrons reduces the neutron reproduction ratio of the reactor A and thus, by adjusting the door 108, this ratio may be brought to values above and below unity, thereby accommodating regulation of the neutron density within the system. It will be understood that while one relatively large door 188 is illustrated in the drawings, if desired, any number of such doors of any suitable size may be utilized for the purpose of regulating the reaction.

Emergency control may be effected by dropping the reflector 166 by means of the hoist 184 or, if desired, emergency control may be achieved by elevating one or more of the reactive rods within the tank 102, said rods being designated at 106a in FIGS. 8 and 11; and being identical with the rods 106, except that they are not secured to the bottom wall of the tank 102. Each rod 106a is provided with a coupling 206 (FIG. 8) at its upper extremity for convenient connection to an actuating mechanism (not shown) supported, for example, by the cover plate 122.

It will be understood that from time to time it is desirable to remove the blocks 168 constituting the outer reflector 166, and this is accomplished by actuating the hoist 184 and the column 186 to lower the reflector 166 into a handling chamber 208, whereupon a biological shield 210 is moved, as for example, by means of a wheel 212 through a chamber 214 to a position beneath the reaction tank 104 and above the handling chamber 208. Thereafter the reflector 166 may be disassembled for the purpose of recovering the thermally fissionable material formed within the blocks 168 and a new reflector may be assembled on the ring 178, and may be elevated therewith into position around the reaction tank 104, as shown in FIGS. 8 and 9.

Assuming that the thermally fissionable segments 124 are composed of U^{235}, the amount required for the above-described reactor is determined by the concentration of this isotope in the segments. If the segments 124 are formed of enriched uranium containing U^{235} in concentration of the order of 90 percent, the total amount of enriched uranium would be about 45 kilograms containing about 40 kilograms of U^{235}. Under certain circumstances wherein it is desirable to raise the number of kilowatts per kilogram of U^{235}, it may be desirable to use this material in more dilute form. For example, enriched uranium containing U^{235} in concentration of the order of 35 percent would sustain a reaction in a device such as above-described if 100 kilograms of U^{235} were provided in the reactor. However, the breeding gain and neutron loss in the coolant would not be materially changed by utilizing U^{235} in greater dilution. It may be noted that a smaller quantity of Pu^{239} in the reactor A would be necessary to sustain a reaction, all other factors being equal, inasmuch as a greater number of neutrons are emitted on the average by plutonium atoms during fissioning thereof than are emitted by U^{235} atoms.

Reactors of the above-described type, as more fully discussed in said copending Szilard application, operate at an average neutron energy at least above about 0.3 ev. and preferably above about 25,000 ev., the average energy being largely dependent upon the moderating properties of the coolant and of the structural nonfissionable components of the reactor, as will be readily understood

12

by those skilled in the art. The specific embodiments disclosed operate at an average neutron energy considerably in excess of 25,000 ev.

For an estimation of the breeding gain in thermally fissionable material, the geometry of the reactor and the pertinent nuclear constants must be considered. The following estimates have been made for the above-geometry using reasonable values of the nuclear constants based on actual measurements:

INTERNAL PORTION OF REACTOR

Material	Relative Volume of Components	Relative Neutron Absorption
U^{235}	1	1
Fe	0.4	0.01
Na-K alloy (50% by wgt. of each)	0.1	0.0

REACTOR BLANKET

Natural Uranium	1	1
Fe	0.4	0.08
Na-K alloy (50% by wgt. of each)	0.4	0.01

In the above tables, are listed in the first column the materials of the reactor and in the second column the relative volumes of these materials in terms of the volume of U^{235} and natural uranium in the first and second tables, respectively. The third column indicates the relative number of neutrons absorbed by the various components. It is to be noted that in the reactor proper the absorption of neutrons due to the iron and due to the sodium potassium in the coolant alloy is negligible compared to the neutrons absorbed by the U^{235}. This is not true, however, for the reflector, and it is believed that the numbers given in the table for the fraction of neutrons absorbed in the iron and in the alloy coolant in the reflector are high since these calculations assume that all of the blanket is iron jacketed and cooled by the alloy. As above-described, however, the outer reflector 166 is so designed that neither iron nor cooling metal will be present. The other serious loss of neutrons from the reactor is in the upper direction where channels are provided for removing the rods from the device. It may be noted that by lengthening the segments 126 of these rods sufficiently that the loss in this direction can be held down to that estimated for the other parts of the reflector. For the conditions outlined here the breeding gain is estimated to be between 0.15 and 0.2. More explicitly, this means that for each U^{235} atom consumed between 1.15 and 1.2 new Pu^{239} atoms are generated. It should be pointed out that for most purposes concerning the chain reaction Pu^{239} is considerably more effective or more valuable than U^{235}, so that more is gained by this conversion than is indicated by these numbers.

It will also be apparent that the reactor shown in the above figures comprises a central active region surrounded by a blanket of thorium or uranium which blanket itself is surrounded by a second blanket comprising a neutron moderator capable of slowing neutrons to substantial neutron energies such as carbon, water, deuterium oxide, beryllium or other elements of low thermal neutron cross-section and atomic numbers below about 10. The thorium or uranium blanket not only serves as a fast neutron reflector as above described, but also functions in the nature of a filter or screen for most neutrons of relatively low energies, for example, 0.01 to 10 electron volts. Thus, the fast neutrons leaving the central active region are either reflected or captured by the blanket or passed through the blanket into the neutron moderator where they are slowed to low energies, usually thermal energies. These low energy neutrons are then available for breeding purposes.

Since the critical size of the reactor operating by virtue of thermal neutrons is substantially smaller than the

2,993,850

13

critical size of a reactor operating by virtue of fast neutrons, that is, for example, of the order 100,000 ev. or above, it is usually desirable to prevent or minimize the return of thermal neutrons into the active area. This is also desirable to minimize the problems of radioactivity [5] hazards in connection with the coolant system and the like. The uranium or thorium blanket serves this function and absorbs most of the thermal neutrons which are reflected inwardly from the neutron moderator blanket.

It will be understood that other means may be utilized [10] to prevent or minimize the return of slow or thermal neutrons into the central fast neutron active area, for example, a blanket of cadmium, barium, gadolinium or other isotopes having a high neutron capture cross-section for slow neutrons may be provided between the active [15] area and the neutron moderating blanket for this purpose. Likewise, numerous other high neutron absorbers usually having an atomic number below about 80 may be used for this purpose.

It will be understood that the present invention is not [20] restricted to the specific embodiments disclosed or to the theories of nuclear physics set forth, inasmuch as many modifications of these embodiments may be conceived by those skilled in the art without departing from the spirit of the invention or the scope of the appended claims; [25] and experimental data later discovered may modify the theories set forth without affecting the operativeness of the invention disclosed.

What is claimed is:

1. A neutronic reactor which comprises fissionable [30] material in amount sufficient to be capable of sustaining a nuclear fission chain reaction and in concentration sufficiently high so that the average neutron energy within the reactor is at least about 25,000 ev., a natural uranium blanket substantially surrounding the reactor, [35] and a reflector formed of a neutron moderator around the blanket capable of moderating fast neutrons to thermal energies of about 0.03 ev., said reflector being substantially free from neutron poison, said blanket being sufficiently thick to substantially eliminate flow of neu- [40] trons therethrough from said moderator.

2. A neutronic reactor according to claim 1 wherein the fissionable material is undiluted plutonium, the reactor is cylindrical, the reactor critical volume is 2.4 liters,

14

the critical weight of plutonium is 22 kg., and the thickness of the blanket is 30 cm.

3. A neutronic reactor according to claim 1 wherein the fissionable material is plutonium diluted in a 1:1 proportion with natural uranium, the reactor is cylindrical, the reactor critical volume is 9 liters, the critical weight is 40 kg. and the thickness of the blanket is 30 cm.

4. A neutronic reactor according to claim 1 wherein the fissionable material is plutonium diluted in a 4:1 proportion with natural uranium, the reactor is cylindrical, the reactor critical volume is 70 liters, the critical weight is 125 kg. and the thickness of the blanket is 30 cm.

5. A neutronic reactor according to claim 1 wherein the fissionable material is undiluted plutonium, the reactor is spherical, the reactor critical volume is 2.2 liters, the critical weight of plutonium is 20 kg. and the thickness of the blanket is 10 cm.

6. A neutronic reactor according to claim 1 wherein the fissionable material is plutonium diluted in a 1:1 proportion with natural uranium, the reactor is spherical, the reactor critical volume is 7.6 liters, the critical weight is 35 kg. and the thickness of the blanket is 10 cm.

References Cited in the file of this patent

UNITED STATES PATENTS

1,019,394	Weintraub	Mar. 5, 1912
2,708,656	Fermi et al.	May 17, 1955
2,815,319	Snell	Dec. 3, 1957

FOREIGN PATENTS

861,390	France	Feb. 7, 1941
114,150	Australia	May 2, 1940
114,151	Australia	May 3, 1940
233,011	Switzerland	Oct. 2, 1944

OTHER REFERENCES

Goodman: "The Science and Engineering of Nuclear Power," pp. 274, 278, 302, and 303, Addison-Wesley Press, Inc. (1947).

Smyth: "Atomic Energy for Military Purposes," page 50, August 1945.

Kelly et al.: Physical Review, 73, 1135–9 (1948).

April 22, 1958   E. P. WIGNER   2,831,806
                          NEUTRONIC REACTOR

Filed Oct. 14, 1952   8 Sheets—Sheet 1

FIG.1

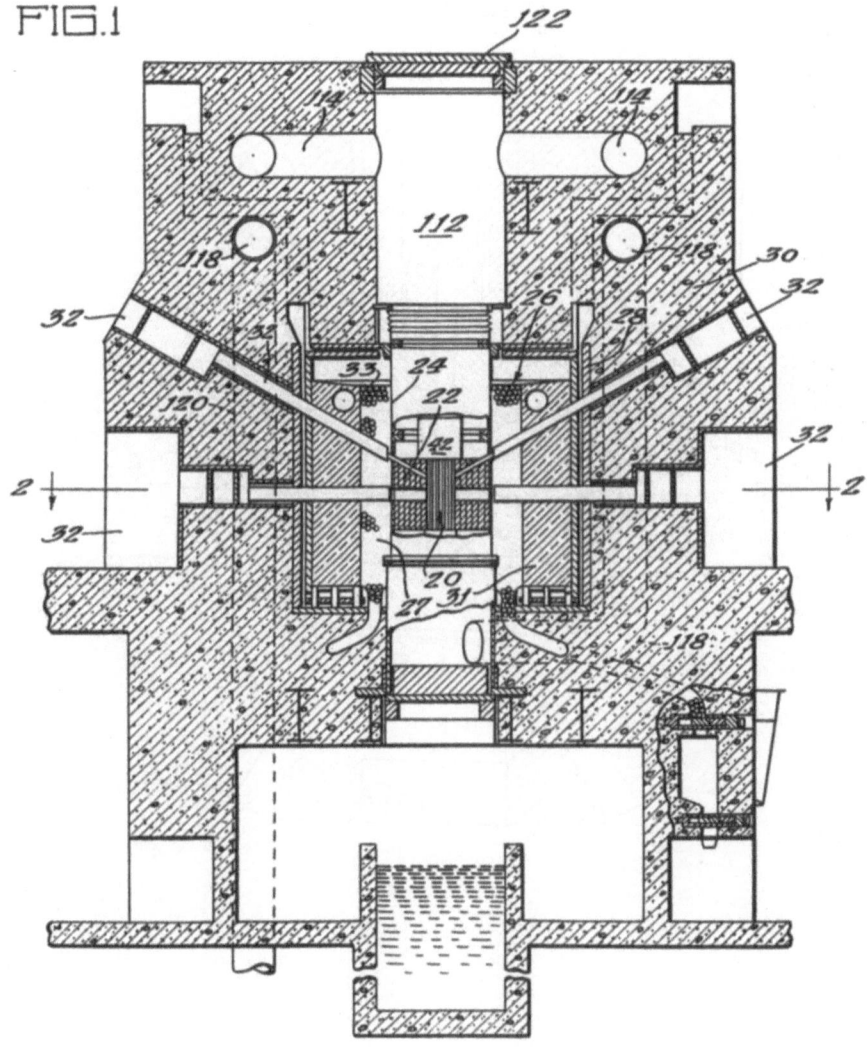

INVENTOR.
Eugene P. Wigner
BY
Roland A. Anderson
Attorney.

April 22, 1958   E. P. WIGNER   2,831,806
                          NEUTRONIC REACTOR

Filed Oct. 14, 1952   8 Sheets—Sheet 2

FIG.2

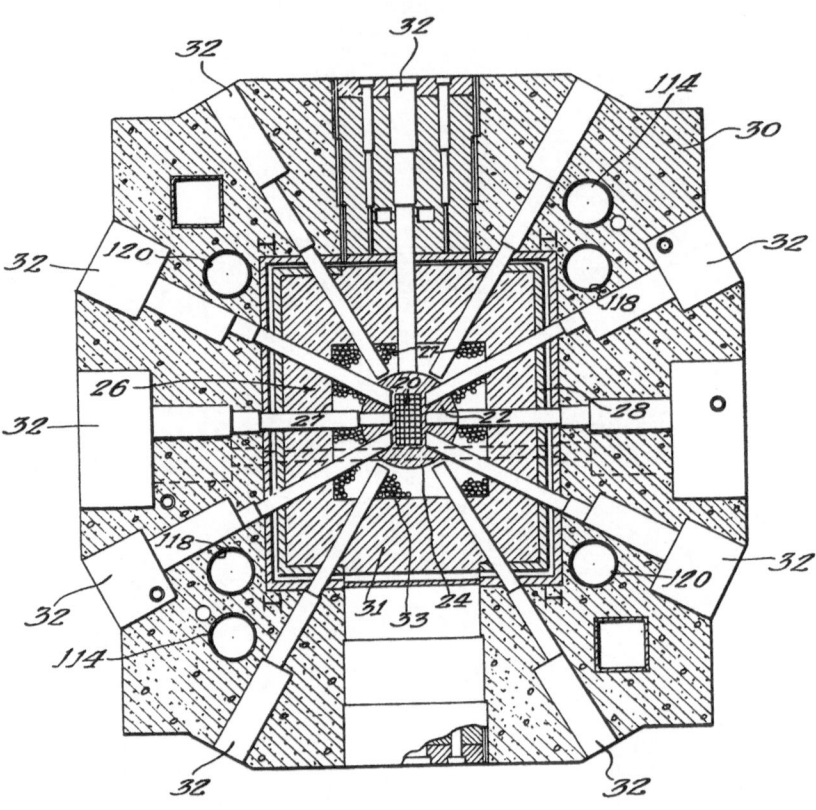

INVENTOR.
Eugene P. Wigner
BY
Roland A. Anderson
Attorney

April 22, 1958

E. P. WIGNER

2,831,806

NEUTRONIC REACTOR

Filed Oct. 14, 1952

8 Sheets—Sheet 3

FIG.3

FIG.5

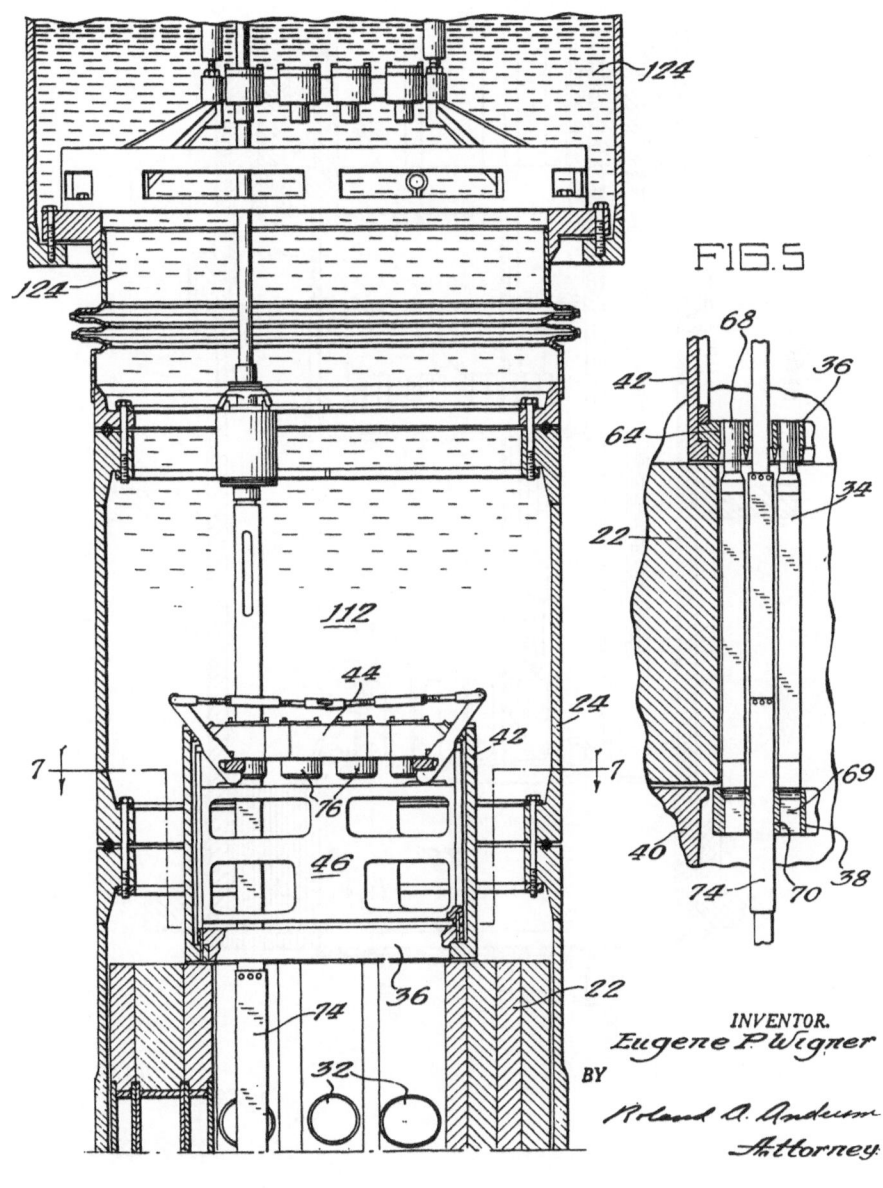

INVENTOR.
Eugene P. Wigner
BY
Roland A. Anderson
Attorney.

April 22, 1958   E. P. WIGNER   **2,831,806**

NEUTRONIC REACTOR

Filed Oct. 14, 1952   8 Sheets—Sheet 4

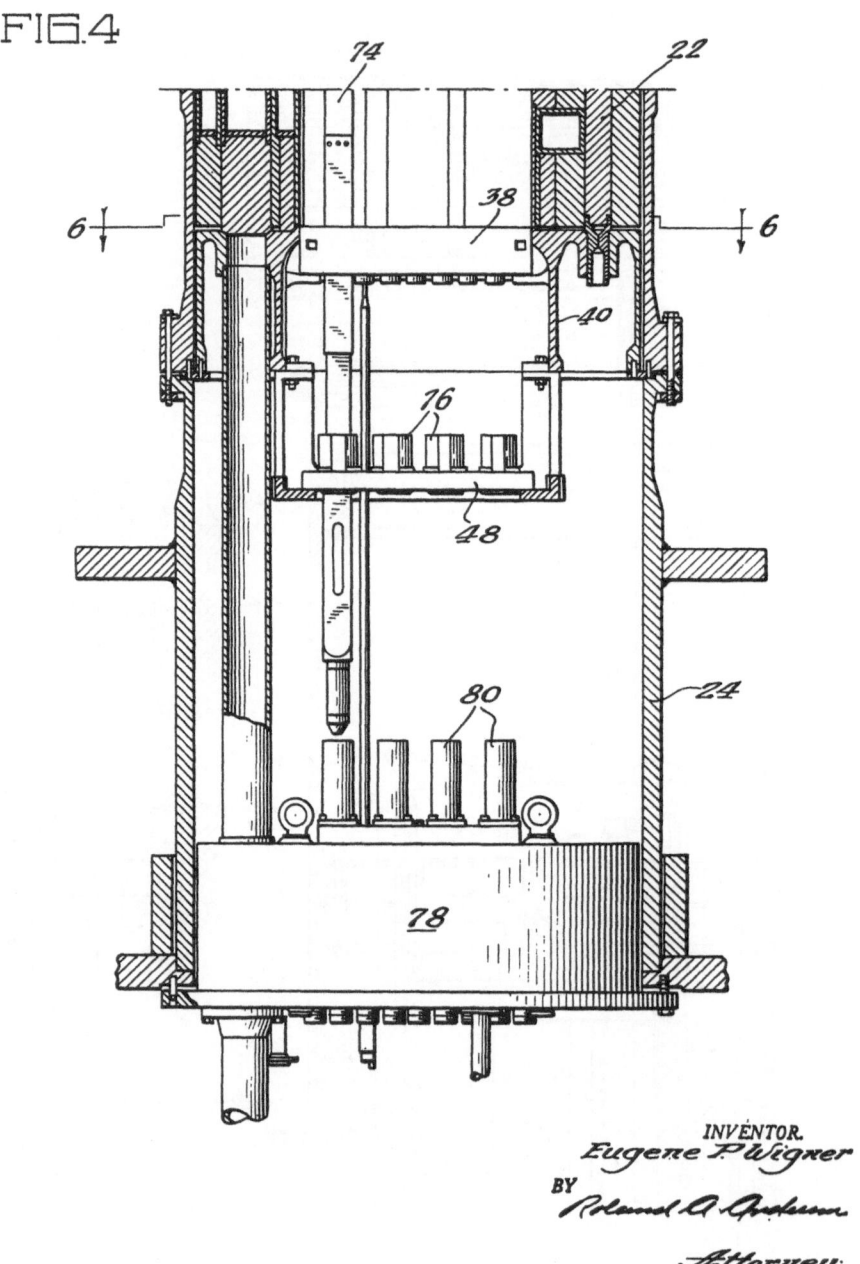

FIG.4

April 22, 1958   E. P. WIGNER   2,831,806
                          NEUTRONIC REACTOR

Filed Oct. 14, 1952   8 Sheets—Sheet 5

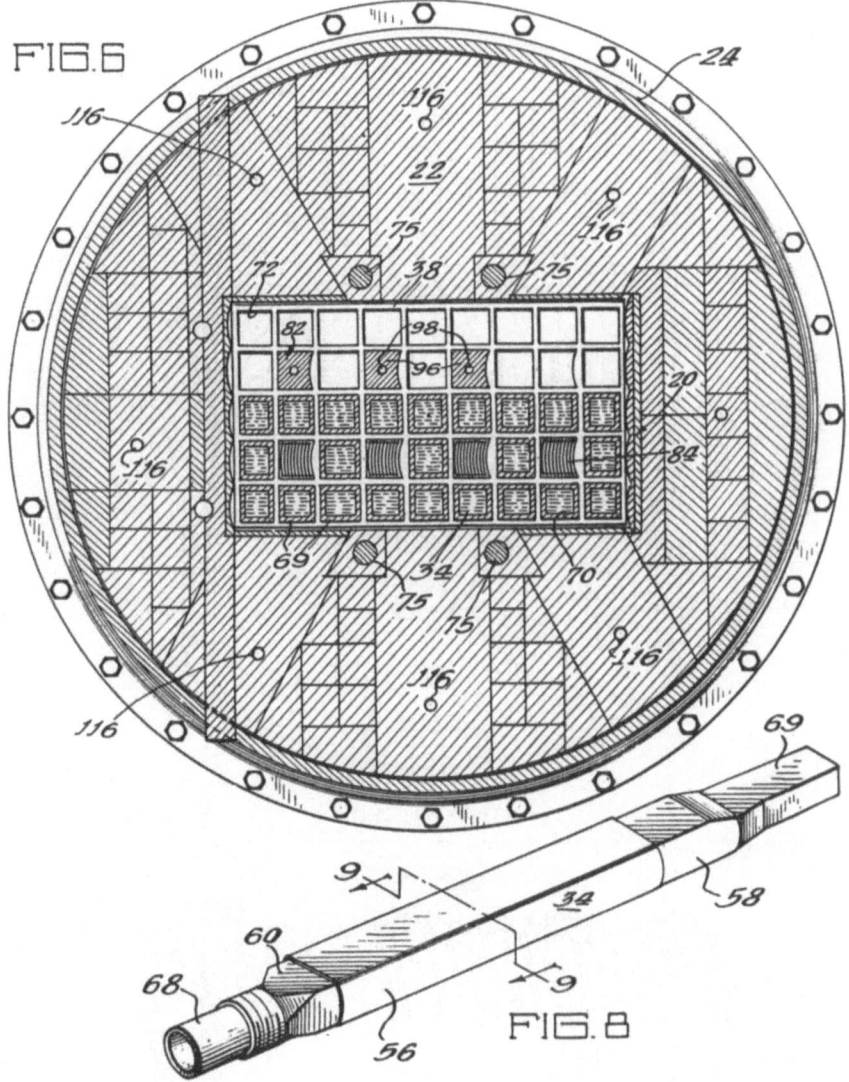

FIG.6

FIG.8

INVENTOR.
Eugene P. Wigner
BY
Roland A. Anderson
Attorney.

April 22, 1958   E. P. WIGNER   2,831,806
                        NEUTRONIC REACTOR

Filed Oct. 14, 1952   8 Sheets—Sheet 6

FIG.7

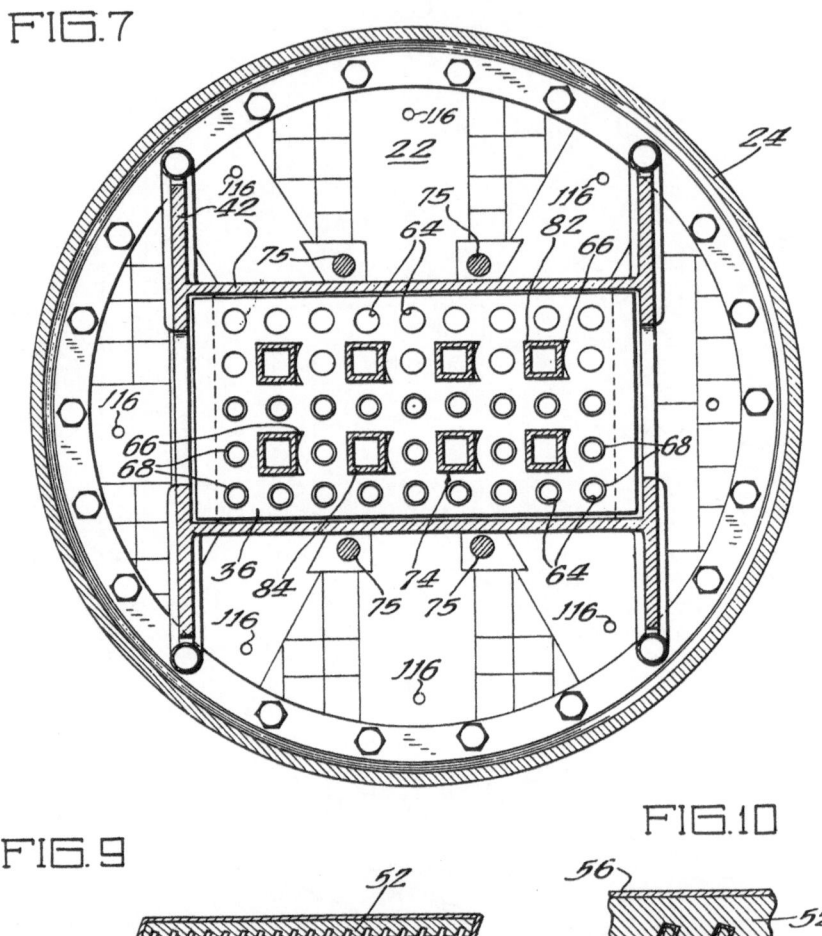

FIG.9

FIG.10

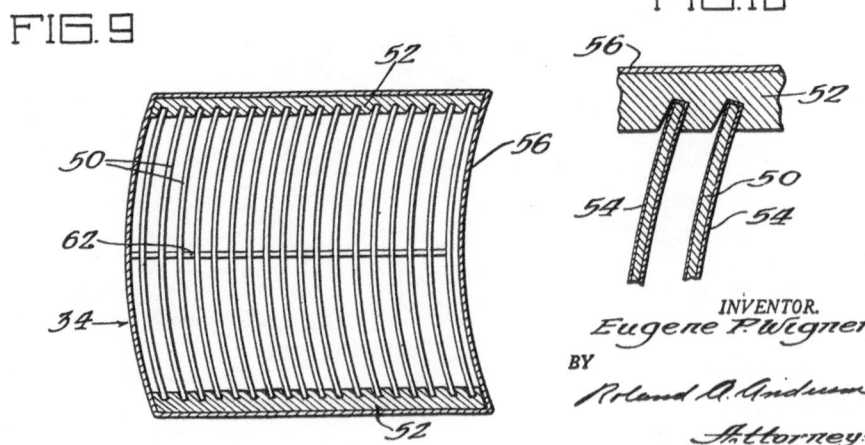

INVENTOR.
Eugene P. Wigner
BY
Roland A. Anderson
Attorney.

April 22, 1958   E. P. WIGNER   2,831,806

NEUTRONIC REACTOR

Filed Oct. 14, 1952   8 Sheets—Sheet 7

FIG.11

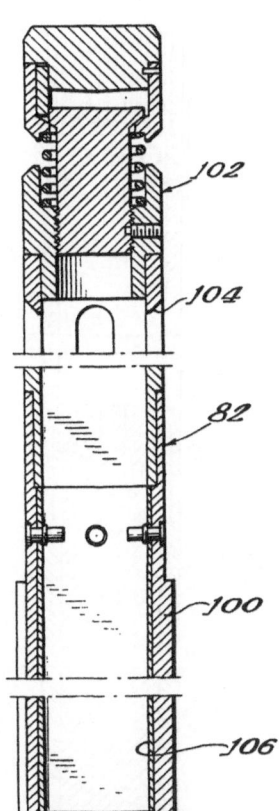

FIG.12

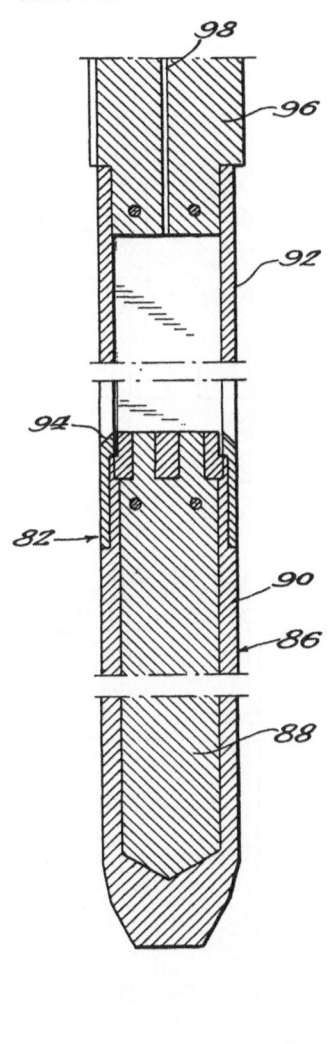

INVENTOR.
Eugene P. Wigner
BY
Roland A. Anderson
Attorney

April 22, 1958

E. P. WIGNER

2,831,806

NEUTRONIC REACTOR

Filed Oct. 14, 1952

8 Sheets—Sheet 8

FIG.13

FIG.14

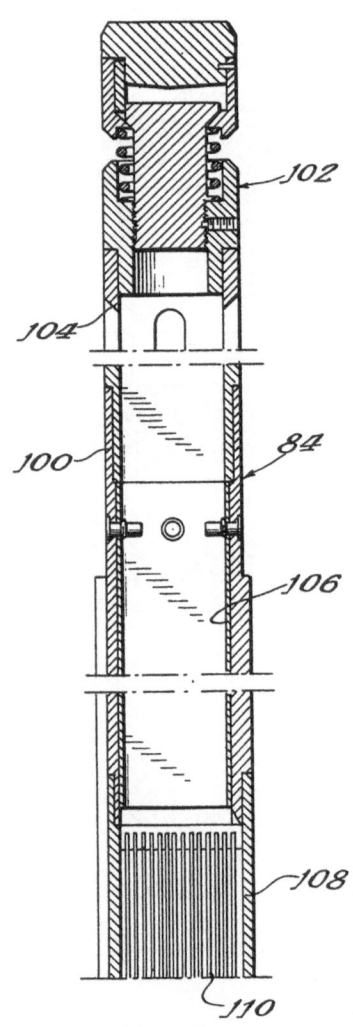

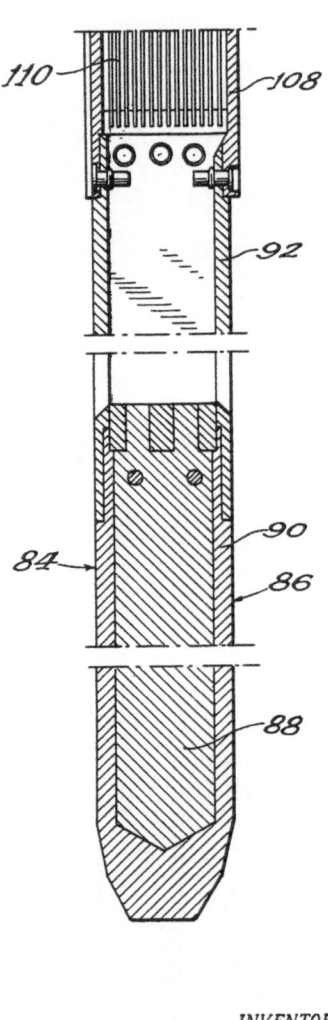

INVENTOR.
Eugene P. Wigner
BY
Roland A. Anderson
Attorney

United States Patent Office

2,831,806
Patented Apr. 22, 1958

1

2,831,806

NEUTRONIC REACTOR

Eugene P. Wigner, Princeton, N. J., assignor to the United States of America as represented by the United States Atomic Energy Commission

Application October 14, 1952, Serial No. 314,595

2 Claims. (Cl. 204—193.2)

The present invention relates generally to neutronic reactors, and more particularly to neutronic reactors designed for the irradiation of materials.

One of the uses of neutronic reactors is to irradiate materials with neutrons and other particles and radiations. In this manner, radioactive isotopes may be produced for all chemical elements with the exception of helium. The physical transformation of elements as a result of irradiation in a neutronic reactor may be accomplished through any one of a number of reactions which are fully described in the published literature. Radioactive isotopes produced by neutronic reactors are receiving large commercial interest, particularly such isotopes as H^3, C^{14}, P^{32}, S^{35}, and I^{131}. As a result, there is a great demand for neutronic reactors which are particularly adapted to produce radioactive isotopes.

The neutrons produced by neutronic reactors may be generally classified in three categories. First, there are virgin neutrons which are those neutrons emitted from the nucleus of an atom of fissionable material which have as yet suffered no collisions with other atoms, and hence possess the energy imparted to them by the fission process. Second, there are the slow, or thermal, neutrons which are neutrons possessing energies approximating the energy imparted by the thermal conditions in which the neutrons exist. Finally, there are epithermal neutrons, which are those neutrons which have suffered collisions with atoms of material but which possess greater than thermal energy.

Radioactive isotopes which are produced by the absorption of neutrons are produced primarily as a result of the absorption of thermal neutrons, rather than neutrons having higher energies. For this reason, a neutronic reactor which is to be used for the production of radioactive isotopes should produce large numbers of thermal neutrons relative to neutrons of higher energy.

For many purposes, it is preferable to subject the materials being irradiated to a thermal neutron flux only, since neutrons of higher energy may produce other effects in the material being irradiated. For example, high energy neutron bombardment is known to displace atoms from the lattice structure of crystalline materials. Hence, it is an object of the present invention to provide a reactor for the production of isotopes which has a region producing substantially only thermal neutron flux.

Also, the inventor has found that it is preferable to construct a neutronic reactor for irradiating materials with a region of maximum thermal neutron flux adjacent to the periphery of the reactor, rather than in the center of the reactor. Materials may be more readily inserted into a region adjacent to the periphery of the reactor than into a region in the center of the reactor. Also, much more space is provided adjacent to the periphery of the reactor for accommodating such materials. Hence, it is a further object of the present invention to provide a neutronic reactor with a maximum thermal neutron flux in a region adjacent to the periphery of the reactor, rather than in a central region of the reactor.

2

The inventor has found that these objects may be accomplished with a neutronic reactor comprising a core and a reflector. Hence, it is also an object of the present invention to provide a neutronic reactor having a core and a reflector in which the maximum thermal neutron flux in the reflector exceeds the maximum thermal neutron flux in the core of the reactor.

The neutronic reactor described hereafter exhibits the foregoing properties, and hence is particularly desirable for the irradiation of materials. Other properties and advantages of neutronic reactors constructed according to the teachings of the present invention will become readily apparent from a study of the following description of the invention, together with the illustrative embodiment shown in the drawings, in which:

Figure 1 is a vertical sectional view of a neutronic reactor constructed according to the teachings of the present invention;

Figure 2 is a horizontal sectional view taken along line 2—2 of Figure 1;

Figure 3 is a vertical sectional view of a portion of the reactor tank shown generally in Figures 1 and 2;

Figure 4 is a vertical sectional view of the remaining portion of the reactor tank shown in Figure 3;

Figure 5 is a fragmentary sectional view of a portion of the core of the reactor shown in Figures 3 and 4;

Figure 6 is a horizontal sectional view taken along line 6—6 of Figure 4;

Figure 7 is a horizontal sectional view taken along line 7—7 of Figure 3;

Figure 8 is an isometric view of one of the fuel assemblies of the neutronic reactor;

Figure 9 is a transverse sectional view of the fuel assembly taken along line 9—9 of Figure 8;

Figure 10 is a fragmentary sectional view of the fuel assembly shown in Figure 9;

Figure 11 is a sectional view of a portion of one type of control element shown generally in Figures 3 through 5;

Figure 12 is a sectional view of the remaining portion of the control element shown in Figure 11;

Figure 13 is a vertical sectional view of a portion of a second type of control element shown generally in Figures 3 through 5; and

Figure 14 is a vertical sectional view of the remaining portion of the control element shown in Figure 13.

As shown in Figure 1, the reactor has a core 20 surrounded by a first reflector 22, both the reflector 22 and the core 20 being disposed within a tank 24. A second reflector 26 is disposed about the outer periphery of the tank 24, and a thermal shield 28 surrounds the second reflector 26. A massive additional shield 30 surrounds the thermal shield 28. Figure 1 also shows a number of passages 32 extending through the shield 30, the thermal shield 28, the second reflector 26, and the first reflector 22 to the core 20 of the reactor. These passages are provided to enable operating personnel to position materials which are to be irradiated at a desired position relative to the core 20 of the reactor.

The core 20 of the reactor comprises a plurality of fuel assemblies 34 immersed in a water moderator. The fuel assemblies 34 are supported between an upper assembly grid 36 and a lower assembly grid 38. A lower support member 40 is secured to the tank 24, and secures the lower assembly grid 38. The upper assembly grid 36 is also secured to the tank 24 by an upper support member 42. An upper guide grid 44 rests upon the upper assembly grid 36 by means of a grid spacer 46. A lower guide grid 48 is positioned beneath the lower assembly grid 38 and attached to the tank 24 by the support member 40.

The fuel assemblies 34 are specifically illustrated in

3

Figures 8 through 10. Plates 50 of material fissionable by neutrons of thermal energy are provided with a corrosion resistant cladding 54 and secured to side plates 52. The plates 50 of fissionable material are curved in shape and wrapped within a covering 56. A comb shaped support 62 maintains the plates 50 in rigid spaced relationship. The ends of the covering 56 are secured to end boxes 58 and 60 which permit the fuel assemblies 34 to be secured between the upper and lower assembly grids 36 and 38.

In one particular construction of the neutronic reactor which will be used to illustrate the present invention throughout this description, the plates 50 are constructed of uranium containing 93.5% U^{235} alloyed with aluminum, and the cladding 54 is aluminum. Each fuel assembly 34 contains eighteen plates 50, each plate being approximately 2.8 inches wide and 24⅝ inches long. Each of the plates 50 is 0.06 inch thick and spaced from adjacent plates by 0.118 inch. The end boxes 58 and 60 are constructed of aluminum. Each of the fuel assemblies 34 contains approximately 140 grams of U^{235}.

Figure 7 indicates that the upper assembly grid 36 is provided with rows of circular apertures 64, some of the circular apertures 64 being separated by apertures 66 which are essentially rectangular but have a curved side. The end boxes 60 of the fuel assemblies 34 are provided with circular connectors 68 which are adapted to fit into the circular apertures 64 of the upper assembly grid 36. The lower assembly grid 38, illustrated in Figure 6, is provided with rows of rectangular orifices 70, some of the orifices 70 being separated by orifices 72, which are essentially the same shape as the apertures 66 in the upper assembly grid 36. The end boxes 58 of the fuel assemblies 34 have rectangular connectors 69 which snugly fit the rectangular orifices 70 of the lower assembly grid 38, thus securing the fuel assemblies 34 between the upper and lower assembly grids 36 and 38.

The apertures 66 in the upper assembly grid 36 and the orifices 72 in the lower assembly grid 38 are provided for control elements 74 which are slidably disposed within the core 20 of the reactor. The control elements 74 are journaled within bearings 76 in the upper and lower guide grids 44 and 48. A bottom plug 78 at the lower end of the tank 24 supports shock absorbers 80 aligned with the bearings 76 in the upper and lower guide grids 44 and 48 for the purpose of absorbing a portion of the shock caused by the falling control elements 74 when the control elements 74 are rapidly inserted into the reactor.

There are three types of control elements 74 used to control the reactor. The one type, illustrated at 82, is constructed of the same material as the first reflector 22 in a portion of the control element 82 which may be disposed in the core 20 of the reactor. The second type of control element 84 contains a material fissionable by neutrons of thermal energy in a portion which may be disposed in the core 20 of the reactor. Finally, regulating rods 75 containing material having a neutron cross section of at least 100 barns, such as cadmium, are slidably disposed in the first reflector 22.

The first type of control element 82, which may be referred to as a reflecting control element, is specifically illustrated in Figures 11 and 12. It is provided with a tip 86 which is constructed of materials which will absorb the shock caused by rapidly inserting the control element 82 into the reactor. A plug 88 of relatively light materials, such as aluminum sheathed in a jacket 90 of durable material, such as iron, provides a tip 86 which will withstand considerable shock. A sleeve 92 is provided with a water outlet aperture 94, and is attached to an element 96 which contains material having neutron reflecting properties approximately the same as those of the first reflector 22. A channel 98 extends centrally through the element 96. A second sleeve 100 is attached to the opposite end of the element 96 and connects the element 96 with a head 102 of the control element 82.

4

A water inlet aperture 104 is disposed in the sleeve 100, so that water may enter into the sleeve 100, flow through the channel 98, through the sleeve 92 and out of the aperture 94, thus cooling the control element 82. A neutron absorbing liner 106 is also disposed within the sleeve 100. The liner 106 should be constructed of a material having a neutron capture cross section of at least 100 barns.

A control element 84 having a portion containing fissionable material is shown in Figures 13 and 14, and is identical with the control elements 82 containing reflecting material, except for the region between the sleeves 92 and 100, and similar numerical designations have been used on the drawings for the identical elements of the two types of control elements 74. The region between the sleeves 92 and 100 of the fuel containing control elements 84 comprises a tube 108 containing fuel plates 110 similar in design and construction to those fuel plates 50 of the fuel assemblies 34.

As illustrated in Figure 6, the core 20 of the reactor comprises three rows of fuel assemblies 34 and control elements 84, there being a total of 27 fuel assemblies 34 and control elements 84. The fuel plates 110 in the control elements 84 are disposed beneath the neutron absorbing liner 106 in these elements 84, so that the control element 84 may be partially withdrawn from the core 20 of the reactor to place the fuel plates 110 in the core 20 of the reactor. In this manner, a maximum of 27 fuel sections may be disposed in the active portion of the reactor. However, when a control element 84 is lowered, the neutron absorbing liner 106 is at least partially disposed within the core 20 of the reactor, thereby at least partially removing the fuel plates 110 and reducing the reactivity of the core 20 of the reactor.

The control elements 82 which contain neutron reflecting material are disposed adjacent to the rows of fuel assemblies 34, and when partially raised, dispose reflecting elements 96 adjacent to the core 20 of the reactor, as illustrated in Figures 6 and 7. In this position, the reflecting elements 96 contribute to the total neutron reflection of the first reflector 22, and add to the reactivity of the core 20 of the reactor. However, if the control elements 82 are lowered into contact with the shock absorbers 80, then the neutron absorbing liners 106 are disposed adjacent to the fuel assemblies 34, and the reactivity of the core of the reactor is reduced.

The neutronic reactor is both cooled and moderated by the same flow of water. The tank 24 is provided with a reservoir portion 112 above the core 20 of the reactor. Coolant pipes 114 are connected to the reservoir portion 112 of the tank 24 and introduce the water coolant into the reservoir portion 112 of the tank 24. The water coolant then flows downward through the fuel assemblies 34 and the control elements 74 of the core 20 of the reactor, and also through channels 116 in the first reflector 22. Return pipes 118 adjacent to the bottom of the tank 24 return the coolant upwardly through the shield 30. The water is then recirculated downwardly through another portion of the shield 30 through coolant exit pipes 120 disposed in the shield 30. In this manner, the core 20 of the reactor, the first reflector 22 and the shield 30 are cooled.

The top of the tank 24 is provided with a cover 122 which is removed when fuel assemblies 34 are to be removed or mounted between the upper and lower assembly grids 36 and 38. Upper guide grid 44 and the upper assembly grid 36 are both removed, and the end boxes 58 of fuel assemblies 34 are inserted into the rectangular orifices 70 of the lower assembly grid 38. The upper assembly grid 36 is then replaced in the upper grid support member 42, the end boxes 60 of the fuel assemblies 34 being disposed within the circular apertures 64 in the upper assembly grid 36. A number of control elements 74 are also disposed within the core 20 of the reactor in order to prevent the core from becoming critical. However, there is less danger of the core becoming critical during charging

2,831,806

5

operations than might be anticipated due to the fact that the moderator has been removed from the core 20 during this step, the water coolant having been removed from the reactor. With the control elements 74 and the fuel assemblies 34 positioned in the core 20, the upper guide grid 44 is replaced in the upper grid support member 42, and the cover 122 replaced upon the reactor. The water coolant may then be circulated through the reactor, forming a pool 124 above the core 20 of the reactor extending to the cover 122. This pool 124 of water forms a shield for the upper portion of the reactor.

As stated above, this reactor has the unusual property of having a higher thermal neutron flux in the reflector 22 than in the core 20 of the reactor. For purposes of this description, the core 20 of the reactor has been defined as that portion of the neutronic reactor in which fission neutrons originate, and the reflector may be defined as a body of neutron moderating materials positioned adjacent to the core of the reactor. In order to have an accumulation of thermal neutrons in the reflector of a reactor which exceeds the maximum thermal neutron flux in the core of the reactor, certain limitations must be placed upon the construction of the reactor. In the first place, it is clear that the absorption of neutrons in the reflector of the reactor must be relatively small, and also that the absorption of neutrons in the core of the reactor must be relatively high. Hence, not all neutron reflecting, or moderating, materials may be used in the reflector of the reactor, nor can all types of core constructions be utilized. It has been found that the maximum thermal neutron flux in the reflector of the reactor will exceed the maximum thermal neutron flux in the core when the ratio of the neutron scattering cross section to neutron absorption cross section averaged over all of the materials in the reflector is approximately twice the ratio of the neutron scattering cross section to neutron absorption cross section averaged over all of the materials in the core of the reactor, and that the former ratio is preferably in the neighborhood of 10 times the latter ratio for such a reactor.

It is also necessary that the reflector be of sufficient thickness to reflect a large proportion of the neutrons back into the core of the reactor. In the particular embodiment of the reactor shown in the figures, a first reflector 22 constructed of beryllium, is disposed immediately adjacent to the core 20 of the reactor, and a second reflector 26 constructed of graphite is disposed about the beryllium reflector 22. The graphite reflector 26 is constructed with a region 27 nearest to the core 20 of the reactor consisting of small balls 33 of graphite. This region 27 is disposed in a relatively high flux portion of the reactor, and considerable distortion of the graphite in this region would normally be anticipated due to the Wigner effect. The second reflector 26 also includes a region 31 of solid graphite, but since the region 31 is more remote from the core 20 of the reactor, distortion of this region due to the Wigner effect will be less pronounced. In the reactor disclosed, the first reflector 22 is constructed of beryllium and is in the form of a cylinder approximately 54 inches in diameter and 40 inches in height. A channel for the core 20 of approximately 18 inches by 30 inches extends longitudinally through the cylindrical first reflector 22. Also, the coolant water flowing through the reflector 22 contributes to the effectiveness of the reflector 22, and constitutes approximately 2% of the volume of the beryllium. The second reflector 26 has overall dimensions of 12 feet by 14 feet by 9 feet 4 inches, a minimum of 40 centimeters immediately adjacent to the tank 24 being the region 27 containing the balls 33 of graphite.

However, the reflector for a reactor having a greater maximum thermal neutron flux in the reflector than in the core need only be adjacent to the side of the core in which the thermal neutron peak is desired, and it has been found that a reflector of 30 centimeters is adequate when constructed of heavy water or beryllium. Also, combinations of heavy water and graphite or beryllium and graphite

6

may be used, where the beryllium or heavy water is adjacent to the core of the reactor. Other materials reflecting neutrons as effectively as these materials may obviously be used.

In order to have an accumulation of thermal neutrons in the reflector, it is also necessary that the core of the reactor have one overall dimension which is not more than one-half of the other overall dimensions of the core, and preferably not more than one-fifth of the other dimensions. The term "overall" dimension is used here to refer to the total region in which virgin neutrons are produced, or in other words the boundaries of the region containing fissionable material. In the particular reactor described to illustrate the present invention, the overall dimensions of the core of the reactor are approximately 22.5 centimeters × 71 centimeters × 66 centimeters, thus clearly meeting this requirement. It is also to be noted that the core of this reactor is in the form of a thin slab, which is believed to be the optimum geometry for a reactor achieving these desired results.

It is clear from the foregoing discussion, that the construction of a neutronic reactor having a maximum thermal neutron flux in the reflector greater than the maximum thermal neutron flux in the core of the reactor requires the reactor to have a higher neutron reproduction factor, K, than a spherical reactor. It has been found, that neutronic reactors with a neutron reproduction factor less than 1.3 cannot be made to attain criticality while exhibiting a higher maximum thermal neutron flux in the reflector of the reactor than in the core thereof.

It has been found that a reactor with 1.35 kilograms of U^{235} disposed in a rectangular core 51 centimeters × 11 centimeters × 66 centimeters with a thirty-centimeter beryllium reflector constructed with aluminum structural material and water moderated, the volume ratio of aluminum to water being 0.65, is critical and exhibits the desired property of having a higher maximum thermal neutron flux in the reflector than in the core. Also, a cylindrical reactor constructed in this manner with a core 23 centimeters in diameter and 53.4 centimeters high with a K of 1.606 has been found to be satisfactory. In such reactors, the maximum thermal neutron flux in the reflector may be approximately 1.5 times the maximum thermal neutron flux in the core of the reactor.

It is possible to determine whether a particular neutronic reactor will exhibit the property of having a higher maximum thermal neutron flux in the reflector than in the core by means of the equation

$$N = \frac{M^2_{ref}}{M^2_c} \times K$$

where N is a number which will be referred to as the "neutron distribution factor" of the reactor, K is the neutron reproduction factor for the reactor, M^2_{ref} is the migration area for neutrons in the reflector of the reactor, and M^2_c is the migration area for neutrons in the core of the reactor. It has been found that for values of N in excess of 9, the maximum thermal neutron flux in the reflector of the reactor will exceed the maximum thermal neutron flux in the core of the reactor.

In the particular embodiment of the present invention described in the drawings, the core had overall dimensions of 22.5 centimeters × 66 centimeters × 71 centimeters and contained approximately 3.95 kilograms of U^{235} disposed in 27 fuel assemblies 34 and control elements 84. The water coolant served both as a moderator and as a coolant. Also, the ratio of aluminum to water by volume was 0.65, aluminum being the only important structural material in the core of the reactor and acting as a part of the moderator.

From the foregoing description, it will be readily apparent to the man skilled in the art that the reactor disclosed satisfies the objects of this invention. Further, the foregoing disclosure, especially in the light of the disclosure of nuclear reactors and theory in the patent

2,831,806

7

of Fermi and Szilard, No. 2,708,656, filed December 19, 1944, will enable the man skilled in the art to construct many neutronic reactors other than the one specifically illustrated, in which the maximum thermal neutron flux in the reflector of the reactor exceeds the maximum thermal neutron flux in the core of the reactor. For this reason, it is intended that the scope of the present invention be not limited by the foregoing description, but only by the appended claims.

What is claimed is:

1. A neutronic reactor comprising, in combination, a core having overall dimensions of at least 11 centimeters × 51 centimeters × 66 centimeters, one of said dimensions being less than one-half of any one of the other dimensions, said core consisting of at least 1.35 kilograms of U^{235}, aluminum and water, the ratio of aluminum to water being not more than 0.65 by volume, and a reflector constructed of beryllium adjacent to said core and surrounding said core for at least 30 centimeters, whereby a neutronic chain reaction within the reactor results in a higher thermal neutron flux in the reflector of the reactor than in the core thereof.

2. A neutronic reactor comprising, in combination, a core comprising an upper grid having at least 27 apertures arranged in 3 rows, said apertures being spaced from each other by approximately 3 inches between centers, a lower grid spaced parallel to the upper grid, said grid having an orifice confronting each of the apertures of the upper grid, 23 elongated fuel assemblies inserted in the orifices and apertures in the grids, each of said assemblies containing approximately 140 grams of U^{235} distributed over a length of approximately 24½ inches, said fuel assemblies comprising a plurality of plates including the U^{235}, said plates being provided with a corrosion resistant cladding and being spaced from each other, a pair of side plates sealed to opposite sides of the U^{235} plates forming closed channels between the U^{235} plates, and a pair of end boxes sealed to the plates of U^{235} and the side plates, one end box in each assembly being disposed in the upper grid aperture and the other end box being disposed in the confronting orifice in the lower grid, 4

8

control elements slidably disposed through the middle row of orifices and apertures in the upper and lower grids and spaced between fuel assemblies in said row, said elements having portions containing approximately 140 grams of U^{235} distributed over a length of approximately 24½ inches and adjacent portions containing material having a neutron capture cross section of at least 100 barns, means to flow water between the fuel assemblies and control elements and through the channels within the fuel assemblies; a reflector disposed about the core, said reflector including a portion containing beryllium and not more than 2% water extending for approximately 30 centimeters from the core of the reactor, and a portion containing graphite extending for at least an additional 25 centimeters from the core of the reactor; and a shield for absorbing neutrons and radiations disposed about the reflector, whereby a neutronic chain reaction in the reactor results in a higher thermal neutron flux in the reflector of the reactor than in the core of the reactor.

References Cited in the file of this patent

Mon P–357, U. S. Atomic Energy Comm., by M. M. Mann et al., Aug. 18, 1947, Clinton Labs. (21 pages).

ANL–4551, U. S. Atomic Energy Comm., by R. K. Winkleback, Dec. 29, 1949, Argonne National Laboratory (39 pages) (rely particularly on pp. 5, 6, 8–10, 27, 28, 30–33).

Atomics, vol. 6, No. 6 November-December 1950, pages 10–15.

U. S. Atomic Energy Comm. AECD 3435, A Low Cost Experimental Neutron Chain Reactor, Part I, by William M. Breazeale, Oak Ridge National Laboratory, Apr. 12, 1952 (date declassified Aug. 13, 1952), pages 1–29.

Atomics, May 1952, pp. 126–129 and 131 (an article entitled "Norwegian Heavy-Water Pile").

Nucleonics, November 1952, Breazeale, pages 56–60.

The Elements of Nuclear Reactor Theory, by Samuel Glasstone and M. C. Edlund; D. Van Nostrand Co., New York, 1952, pages 225, 226, 238, 239, 247.

Feb. 12, 1957   E. P. WIGNER   **2,781,307**

APPARATUS FOR THE MEASUREMENT OF NEUTRON ABSORPTION

Filed March 4, 1947

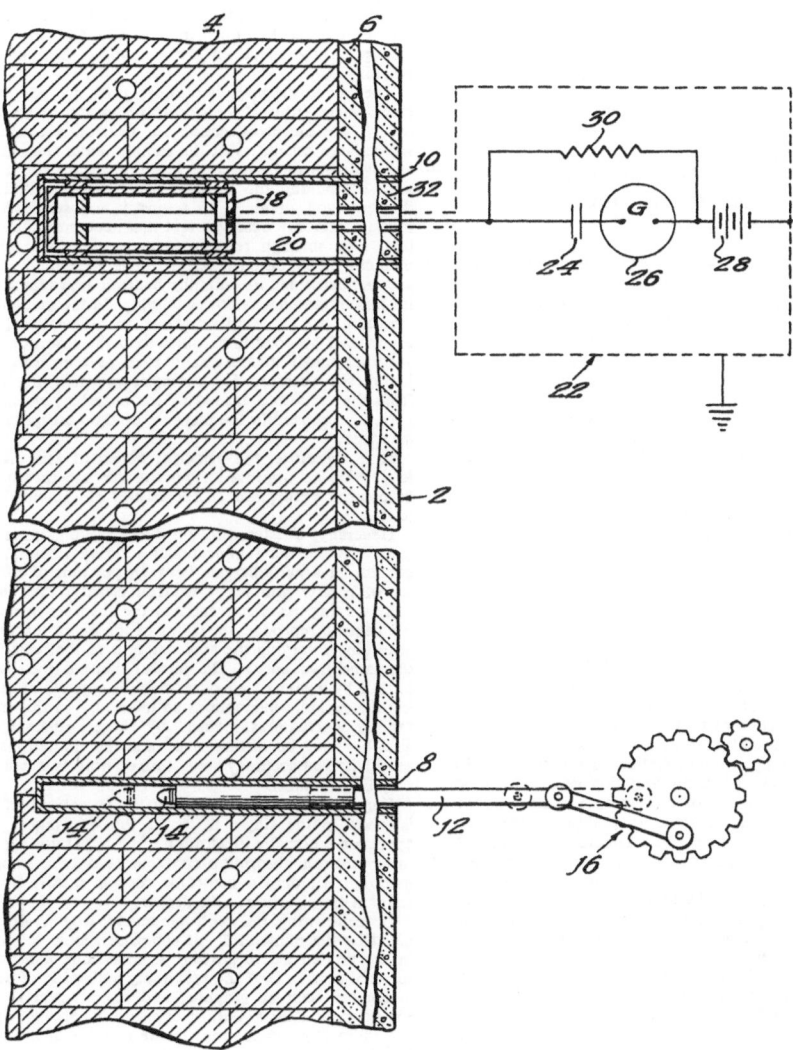

INVENTOR.

Eugene P. Wigner

BY

Robert A. Lavender

Attorney

United States Patent Office

2,781,307

Patented Feb. 12, 1957

1

2,781,307

APPARATUS FOR THE MEASUREMENT OF NEUTRON ABSORPTION

Eugene P. Wigner, Oak Ridge, Tenn., assignor to the United States of America as represented by the United States Atomic Energy Commission

Application March 4, 1947, Serial No. 732,324

1 Claim. (Cl. 204—193)

This invention relates to an improved apparatus for the measurement of neutron absorption characteristics of materials. More specifically the invention relates to an improved apparatus for measuring the effect of the presence of the sample under measurement on the neutron reproduction factor, and thus the power output, of a neutronic reactor capable of sustaining a nuclear fission chain reaction.

An important consideration in selection of non-reacting materials for use in a neutronic reactor is the neutron absorption of the material to be used. As is now well-known, the use or accidental presence of materials having a high absorption for neutrons in a neutronic reactor requires that the size of the reactor be increased in order to have a neutron reproduction factor great enough so that the chain reaction may be self-sustaining. If non-fissionable materials having high absorption for neutrons are present in sufficient quantities production of a divergent chain reaction may be rendered impossible altogether. It is, therefore, necessary in analyzing materials to be used in the construction of a neutronic reactor to determine the neutron absorption of such materials.

The most sensitive and effective general method of measuring the absorption of neutrons of a material now known is the insertion of a sample of the material under scrutiny into an operating neutronic reactor and observation of its effect upon the operation of the reactor. In the past, two ways of employing this general method have been used. In one of these two a sample of the material is plunged into a standard position in the reactor which reactor was theretofore running at a constant power level. As is well known in the art, the insertion of the additional absorber reduces the neutron reproduction factor of the reactor by an amount dependent upon the neutron absorption of the absorber. The neutron reproduction factor is thus reduced below unity and the power output of the reactor falls at an approximately exponential rate. The speed of falling of the power output is a function of the amount of absorber thus inserted. If the speed of falling be calibrated in terms of standard quantities of a material of known neutron absorption characteristics, for example pure boron, then the measurement of the rate of fall with the insertion of a known quantity of the material of unknown neutron absorption characteristics constitutes a measure of the absorption characteristics of such material in terms of the ratio to the absorption characteristics of the known material.

A second way of applying the general method likewise employs the step of plunging the sample under measurement into a reactor theretofore run at a constant power output; but in this method, the insertion of the absorber is compensated by withdrawing another absorber of known characteristics as for example a boron control rod, such withdrawal being adjusted until the effect of the sample is nullified and the reactor is again running at a constant level of power output. Under these circumstances the amount of known absorber withdrawn in order to reinstate the condition of constant power output

2

constitutes a measure of the absorption characteristics of the unknown absorber thus introduced. This latter method may be termed the "null" method of measuring neutron absorption by effect on the operation of a neutronic reactor.

The above methods heretofore used are adequate in cases where either the absorber under measurement has a high nuclear cross-section for neutron absorption or where large quantities of the sample under measurement are available to be inserted into the reactor for the purpose of making the measurement. The limitation on these methods, both as to accuracy and sensitivity, lies in the commonly observed fact that neutronic chain reactors, even through reasonable precautions are taken to maintain all conditions constant, undergo random changes and perturbations in both the neutron reproduction factor and instantaneous power output. In the methods previously in use as described above, such variations, which are caused by conditions other than the insertion of the absorber under measurement, such as temperature and barometric pressure for example, are indistinguishable from the variations caused by the absorber, which latter variations constitute the measure of the absorption characteristics.

It is, therefore, the principal object of this invention to provide an improved apparatus for the measurement of neutron absorption characteristics of samples of materials.

Generally, the teaching of this invention is an apparatus wherein the sample under scrutiny is oscillated in position at a periodic rate between portions of the reactor of greater and lesser neutron density, thus periodically varying the effectiveness of the absorber in reducing the neutron reproduction factor of the chain reactor, and accordingly varying the power output of the reactor with the same frequency. The measurement is then accomplished by means of a device which is capable of measuring the periodic fluctuations so induced. In this manner random fluctuations induced by causes other than the absorber may be distinguished and the limitations upon sensitivity and accuracy heretofore existing as stated above may be minimized.

For a better understanding of the invention, reference is made to the single figure of the drawing in which appears a schematic illustration of an apparatus for measuring neutron absorption characteristics of a sample of material including a neutronic reactor shown fragmentarily in cross-section in the drawing.

In the drawing, the neutronic reactor is generally designated by the numeral 2. It comprises, as is well-known in the art, an active portion 4, containing a fissionable material such as U^{235} and preferably a quantity of neutron moderator such as graphite, and a biological shield 6, for example a thick wall of concrete, to prevent dangerous radioactive emanations to the exterior of the reactor 2. It will be understood that the present invention is not in any way limited to the particular type of neutronic reactor 2 illustrated in the drawing, which is shown for illustrative purposes only.

A shell or thimble 8 extends from the exterior surface of the reactor 2 through the shield 6 and into the active portion 4 of the reactor. A second shell or thimble 10 likewise extends from the face of the reactor 2 into the active portion 4. Both of these thimbles 8 and 10 are preferably of a material of low cross-section for neutrons, for example stainless steel. Into the thimble 8 extends a rod or piston 12 likewise of a material of low neutron absorption. Fastened to the inner end of the piston 12 is a capsule containing the sample 14 under measurement. The piston 12 has imparted to it periodic reciprocating motion by a mechanism 16 illustrated in the drawing merely as a disc-and-crank drive driven by motor means

2,781,307

3

not shown in the drawing. It will be understood that the mechanism 16 for imparting reciprocating motion to the piston 12 is merely illustrative and constitutes in itself no part of the present invention. The teachings of the present invention may be applied with any mechanism whatever for imparting motion to the sample 14, such motion being adapted to change periodically the position of the sample 14 as regards the neutron intensity distribution within the reactor 2. In the illustration of the drawing the thimble 8 acts as a bearing block for the piston 12.

Inserted within the thimble 10 is a neutron responsive ionization chamber 18. The thimble 10 has a shielding plug 32 at the outer end thereof to prevent the existence of a beam of intense radioactivity from the reactor 2. The ionization chamber 18 is connected electrically to the exterior by a shielded coaxial cable 20. The shield of the coaxial cable 20 is connected to one electrode of the ionization chamber 18 and is externally connected to an electromagnetic shield 22, at ground potential, containing the remainder of the measuring equipment. The other electrode of the ionization chamber 18 is connected to a condenser 24, a galvanometer 26, and a power supply 28, all in series. One terminal of the power supply 28 is connected to the shield 22 and thus to ground. A resistor 30 is connected in parallel with the condenser 24 and the galvanometer 26.

Having thus described the elements appearing in the drawing, operation of the device illustrated may now be explained. As is well-known in the art, at any level of power operation the neutron flux density within the reactor 2 is not uniform therein. The maximum flux density occurs at the center of the active portion 4; thence the flux density diminishes. It is much smaller at the periphery of the active portion 4 than at the center. The neutron flux density continues to diminish throughout the shield 6. It will readily be seen that a neutron absorber which is placed at the center of the active portion 4 has a much greater effect upon the neutron reproduction factor of the reactor than it has when placed at the periphery of the active portion 4 or in the shield 6. If the neutron reproduction factor be adjusted by control means well known in the art and therefore not shown in the drawing so that the reactor 2 operates at a constant level of power output with a neutron absorbing sample 14 at some position intermediate between the center of the active portion 4 and the periphery of the active portion 4, and if such sample 14 then be oscillated back and forth, the neutron reproduction factor, and thus the power output, periodically fluctuates above and below the preset value. For any given period and amplitude of mechanical oscillation the amplitude of the periodic variations so induced is a function of the neutron absorption characteristics of the sample 14 so caused to oscillate. Persons skilled in the art will readily observe that since the position of the sample 14 affects the neutron reproduction factor, and thus the rate of change of the power output, rather than the power output itself, the phase of the variation of power output with respect to the position of the sample 14, is such that the latter leads the former.

The ionization chamber 18 is responsive to the instantaneous power output of the reactor 2, and allows a flow of current proportional to the instantaneous power output. When the instantaneous power output is caused to fluctuate periodically around a mean value, the current through the ionization chamber 18 correspondingly fluctuates around the mean value. The current through the ionization chamber flows through the parallel combination of the resistor 30 and the condenser 24 and galvanometer 26, which latter two elements are in series. As is well-known in the art, once the condenser 24 has attained a charge corresponding to the mean level of operation, the current corresponding to this mean level no longer flows through the galvanometer 26. However,

4

fluctuations in the value of power output of the reactor 2, and thus of current through the ionization chamber 18, cause corresponding flow of current through the galvanometer 26, the condenser 24 charging and discharging in accordance with such fluctuations. If the fluctuations have a periodic component the galvanometer 26 thus has a periodically varying indication. The amplitude of the variation is a measure of the amplitude of the variation of the current in the ionization chamber 18 and thus of the variation of the power output of the chain reactor 2. The latter is in turn a measure of the neutron absorption of the sample 14 so caused to oscillate in position.

In order to maximize the discrimination of the system against random variations, it is desirable that the period of the galvanometer 26 be the same as the period of the oscillations of the piston 12. As is well-known in the art, under such conditions the response of the galvanometer 26 to the component of the signal from the ionization chamber 18 of the frequency of the oscillation of the piston 12 is maximized and the response of the galvanometer 26 to signal components of other frequencies is minimized. Thus the neutron absorption of the sample 14 may be measured to an accuracy which is not limited by stray variations of reactor operating conditions except to the extent that such stray variations occur at the same frequency as the variations induced by oscillating the sample 14. It will be understood that in random variations there may be present a component of the same frequency as that of the signal created by the oscillation of the absorbing sample 14. However, such effects, if present, are obviously much smaller and less prejudicial to accuracy than in a system in which such frequency discrimination does not occur.

In the embodiment of the invention illustrated in the drawing, the ionization chamber 18 may be, for example, filled with boron trifluoride gas to render it neutron sensitive, such ionization chambers being well-known in the art. The galvanometer 26 may have a period, for example, of 20 seconds; the condenser 24 may have a value, for example, of 32 microfarads; and the resistor 30 may be, for example, one-half megohm. The voltage value of the power supply 28 will, of course, as is well-known in the art, depend upon the specific design of the ionization chamber 18. The period of the oscillatory motion of the piston 12 may be, for example, 20 seconds, to correspond with the natural period of the galvanometer 26.

It has been stated above that the current through the ionization chamber 18 is proportional to the power output of the reactor 2. However, it should be understood that the term "proportional" is herein used in only an approximate sense because the effect which the oscillating absorbing sample 14 has on the neutron flux through the ionization chamber 18 is to some extent a function of the distance by which these two elements and the thimbles 8 and 10 containing them are separated in the reactor 2. It will be understood that strictly direct proportionality exists only if the ionization chamber 18 is at such a great distance from the absorbing sample 14 that the effect of the absorbing sample 14 on the neutron flux incident upon the ionization chamber 18 is confined to that which arises because of overall change in the reproduction factor of the reactor 2 rather than upon a shielding effect which the sample 14 may assert upon the ionization chamber 18 by reason of being in proximity thereto and therefore absorbing neutrons which would otherwise impinge upon the ionization chamber 18 or would induce fissions which would produce neutrons which would in turn impinge upon the ionization chamber 18. The variation of neutron flux in the vicinity of the absorber will clearly be greater than the variation of the over-all power output of the reactor. Thus if the ionization chamber or other detector is near the oscillating absorber, the amplitude of the signal produced by the oscillation of any given absorber is maximized. However, in such a case errors may be introduced by reason of differences as to neutron

2,781,307

5

energy dependence between the absorption characteristics of the sample and the response characteristics of the detector.

The teachings of the present invention as disclosed in the drawing and in the above description will be readily adapted by persons skilled in the art to many variants of the device and method illustrated and described. For example, many equivalent methods and devices wherein a sample is caused to move periodically between regions of high and low density in a neutronic reactor may readily be devised. Likewise, the ionization chamber and the associated galvanometer circuit are merely illustrative of the many ways in which the variations of the neutron flux or the power output of a neutronic reactor caused by the motion of an absorbing sample may be measured.

What is claimed is:

An apparatus for obtaining an indication of an electrical impulse directly proportional in magnitude to the neutron absorption of a sample of a substance, comprising a neutronic reactor having regions of different neutron flux density, a sample, means for moving the sample at a fixed frequency from a region of one neutron flux density to another region having a different neutron flux density in order to produce a variation at said frequency in the power output of the reactor, the sample size being smaller than the required mass of neutron absorber necessary to

6

depress the neutron reproduction coefficient of the reactor to a value less than unity, an ionization chamber positioned within the neutron atmosphere and responsive to the reactor power output variation, and a meter coupled to the ionization chamber for measuring the amplitude of the reactor power output variation and possessing a natural period equal to the frequency of the periodic sample movement.

References Cited in the file of this patent

UNITED STATES PATENTS

2,275,747	Fearon	Mar. 10, 1942
2,624,847	Jesse et al.	Jan. 6, 1953
2,651,726	Froman et al.	Sept. 8, 1953

FOREIGN PATENTS

114,151	Australia	May 3, 1940
233,011	Switzerland	Oct. 2, 1944
233,278	Switzerland	Oct. 16, 1944

OTHER REFERENCES

Korff et al.: Phy. Rev. 55, 980, May 15, 1939.

A General Account of the Development of Methods of Using Atomic Energy for Military Purposes, H. D. Smyth (August 1945), pp. 85, 179.